Intermediate Algebra

Books in the Tussy and Gustafson Series

In paperback:

Basic Mathematics for College Students
Student edition: ISBN 0-534-36493-4
Instructor's edition: ISBN 0-534-36494-2

Prealgebra
Student edition: ISBN 0-534-34326-0
Instructor's edition: ISBN 0-534-34327-9

Elementary Algebra
Student edition: ISBN 0-534-35528-5
Instructor's edition: ISBN 0-534-35598-6

Intermediate Algebra
Student edition: ISBN 0-534-35581-1
Instructor's edition: ISBN 0-534-36255-9

In hardcover:

Elementary Algebra
Student edition: ISBN 0-534-36882-4
Instructor's edition: ISBN 0-534-37461-1

Intermediate Algebra
Student edition: ISBN 0-534-36881-6
Instructor's edition: ISBN 0-534-37453-0

Elementary and Intermediate Algebra
Student edition: ISBN 0-534-36883-2
Instructor's edition: ISBN 0-534-37471-9

Intermediate Algebra

Alan S. Tussy
Citrus College

R. David Gustafson
Rock Valley College

Brooks/Cole
Thomson Learning™

Australia • Canada • Mexico • Singapore • Spain • United Kingdom • United States

Sponsoring Editor: Robert W. Pirtle

Project Development Editor: Michelle Paolucci

Marketing Manager: Leah Thomson

Marketing Communications: Samantha Cabaluna

Editorial Assistant: Erin Wickersham

Production Editor: Ellen Brownstein

Production Service: Hoyt Publishing Services

Manuscript Editor: David Hoyt

Permissions Editor: Mary Kay Hancharick

Interior Design: Carolyn Deacy

Cover Design: Vernon T. Boes

Cover Photo: H. Sakuramoto/Photonica

Interior Illustration: Lori Heckelman

Print Buyer: Vena Dyer

Typesetting: The Clarinda Company

Cover Printing: Phoenix Color Corp.

Printing and Binding: R.R. Donnelley & Sons

For more information, contact:
BROOKS/COLE
511 Forest Lodge Road
Pacific Grove, CA 93950 USA
www.brookscole.com

For permission to use material from this work, contact us by
Web: www.thomsonrights.com
fax: 1-800-730-2215
phone: 1-800-730-2214

Printed in the United States of America

10 9 8 7 6 5 4 3 2 1

Library of Congress Cataloging-in-Publication Data

Tussy, Alan S., [date]
   Intermediate algebra/Alan S. Tussy, R. David Gustafson.
      p. cm.
   Includes index.
   ISBN 0-534-36881-6 (alk. paper)
      1. Algebra.   I. Gustafson, R. David (Roy David), [date].   II. Title.
QA154.2.T872 1999
512.9—dc21   99-047738
                                                  CIP

Contents

Preface

▶

For the Instructor

Algebra is used to describe numerical relationships. It is a language in its own right. The purpose of this textbook is to teach students how to read, write, speak, and think mathematically using the language of algebra. It presents all the topics associated with a second course in algebra. We have used a blend of the traditional and the reform instructional approaches to do this. In this book, you will find the vocabulary, practice, and well-defined pedagogy of a traditional approach. You will also find that we emphasize the reasoning, modeling, communicating, and technological skills that are such a big part of today's reform movement.

This textbook expands the students' mathematical reasoning abilities and gives them a set of mathematical survival skills that will help them succeed in a world that increasingly requires that every person become a better analytical thinker. We believe it will hold student attrition to a minimum while preparing students to succeed in the next course—whether that is college algebra, trigonometry, statistics, or finite mathematics.

Features of the Text

A Review of Basic Algebra

Chapter 1 begins with a review of the fundamental algebraic concepts of variable, expression, and equation. Students translate English phrases to mathematical symbols and use a five-step problem-solving strategy to solve a wide variety of application problems.

An Innovative Table of Contents

We have made several changes to the usual table of contents found in most intermediate algebra books. An early discussion of graphing on the rectangular coordinate system and of functions appears in Chapter 2. That is followed by Chapter 3, Solving Systems of Equations. Chapter 4 deals exclusively with inequalities.

Interactivity

Most worked examples in the text are accompanied by Self Checks. This feature allows students to practice skills discussed in the example by working a similar problem. Because the Self Check problems immediately follow the worked examples, students can easily refer to the solution and author's notes of the example as they solve the Self Check. Author's notes are used to explain the steps in the solutions of examples. The notes are extensive so as to increase the students' ability to read and write mathematics.

Example titles highlight the ▶
concept being discussed.

Author's notes explain the ▶
steps in the solution process.

Most examples have Self Checks. ▶
The answers are provided.

EXAMPLE 5   **Variables on both sides of the equation.** Solve
$4b - 7 + 2b = 1 + 2b + 8$.

Solution   First, we combine like terms on each side of the equation.

$$4b - 7 + 2b = 1 + 2b + 8$$
$$6b - 7 = 2b + 9 \qquad \text{Combine like terms: } 4b + 2b = 6b \text{ and } 1 + 8 = 9.$$

We note that terms involving b appear on both sides of the equation. To isolate b on the left-hand side, we need to eliminate $2b$ on the right-hand side.

$$6b - 7 = 2b + 9$$
$$6b - 7 - 2b = 2b + 9 - 2b \qquad \text{Subtract } 2b \text{ from both sides.}$$
$$4b - 7 = 9 \qquad \text{Combine like terms on each side: } 6b - 2b = 4b \text{ and } 2b - 2b = 0.$$
$$4b - 7 + 7 = 9 + 7 \qquad \text{To undo the subtraction of 7, add 7 to both sides.}$$
$$4b = 16 \qquad \text{Simplify each side of the equation.}$$
$$b = 4 \qquad \text{To isolate } b, \text{ divide both sides by 4.}$$

Check: $4b - 7 + 2b = 1 + 2b + 8$ The original equation.
$$4(4) - 7 + 2(4) \stackrel{?}{=} 1 + 2(4) + 8 \qquad \text{Substitute 4 for } b.$$
$$16 - 7 + 8 \stackrel{?}{=} 1 + 8 + 8 \qquad \text{Do each multiplication.}$$
$$17 = 17 \qquad \text{Simplify each side.}$$

Since 4 satisfies the equation, the solution set is $\{4\}$.

SELF CHECK   Solve $-6t - 12 - 6t = 1 + 2t - 5$.   *Answer:* $-\frac{4}{7}$ ∎

In-Depth Coverage of Geometry

Perimeter, area, and volume, as well as many other geometric concepts, are used in a variety of contexts throughout the book. We have included many drawings to help the students improve their ability to spot visual patterns in their everyday lives.

Geometric topics appear throughout ▶
the text and are presented in a way
that reinforces important algebra
skills.

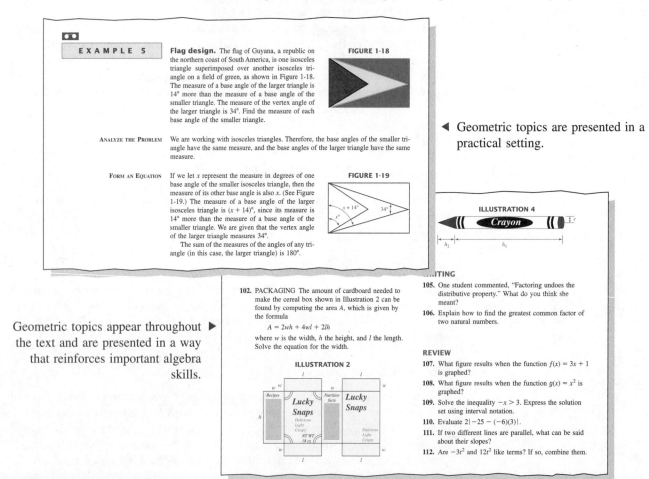

EXAMPLE 5   **Flag design.** The flag of Guyana, a republic on the northern coast of South America, is one isosceles triangle superimposed over another isosceles triangle on a field of green, as shown in Figure 1-18. The measure of a base angle of the larger triangle is 14° more than the measure of a base angle of the smaller triangle. The measure of the vertex angle of the larger triangle is 34°. Find the measure of each base angle of the smaller triangle.

FIGURE 1-18

ANALYZE THE PROBLEM   We are working with isosceles triangles. Therefore, the base angles of the smaller triangle have the same measure, and the base angles of the larger triangle have the same measure.

FORM AN EQUATION   If we let x represent the measure in degrees of one base angle of the smaller isosceles triangle, then the measure of its other base angle is also x. (See Figure 1-19.) The measure of a base angle of the larger isosceles triangle is $(x + 14)°$, since its measure is 14° more than the measure of a base angle of the smaller triangle. We are given that the vertex angle of the larger triangle measures 34°.

The sum of the measures of the angles of any triangle (in this case, the larger triangle) is 180°.

FIGURE 1-19

◀ Geometric topics are presented in a
practical setting.

ILLUSTRATION 4
Crayon

102. **PACKAGING** The amount of cardboard needed to make the cereal box shown in Illustration 2 can be found by computing the area A, which is given by the formula
$$A = 2wh + 4wl + 2lh$$
where w is the width, h the height, and l the length. Solve the equation for the width.

ILLUSTRATION 2

Lucky Snaps

WRITING

105. One student commented, "Factoring undoes the distributive property." What do you think she meant?

106. Explain how to find the greatest common factor of two natural numbers.

REVIEW

107. What figure results when the function $f(x) = 3x + 1$ is graphed?

108. What figure results when the function $g(x) = x^2$ is graphed?

109. Solve the inequality $-x > 3$. Express the solution set using interval notation.

110. Evaluate $2|-25 - (-6)(3)|$.

111. If two different lines are parallel, what can be said about their slopes?

112. Are $-3t^2$ and $12t^2$ like terms? If so, combine them.

Problem-Solving Strategy

One of the major objectives of this textbook is to make students better problem solvers. To this end, we use a five-step problem-solving strategy throughout the book. The five steps are: *Analyze the problem, Form an equation, Solve the equation, State the conclusion,* and *Check the result.*

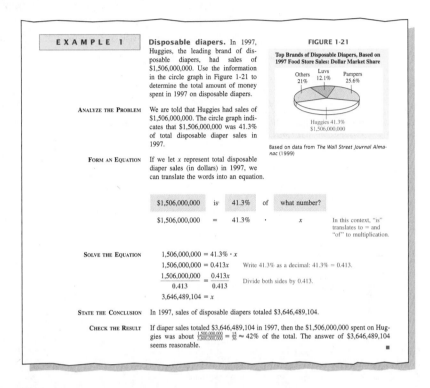

EXAMPLE 1

Disposable diapers. In 1997, Huggies, the leading brand of disposable diapers, had sales of $1,506,000,000. Use the information in the circle graph in Figure 1-21 to determine the total amount of money spent in 1997 on disposable diapers.

FIGURE 1-21

Top Brands of Disposable Diapers, Based on 1997 Food Store Sales: Dollar Market Share

Others 21% — Luvs 12.1% — Pampers 25.6% — Huggies 41.3% $1,506,000,000

Based on data from *The Wall Street Journal Almanac* (1999)

ANALYZE THE PROBLEM We are told that Huggies had sales of $1,506,000,000. The circle graph indicates that $1,506,000,000 was 41.3% of total disposable diaper sales in 1997.

FORM AN EQUATION If we let x represent total disposable diaper sales (in dollars) in 1997, we can translate the words into an equation.

| $1,506,000,000 | is | 41.3% | of | what number? |

$$1,506,000,000 = 41.3\% \cdot x$$

In this context, "is" translates to = and "of" to multiplication.

SOLVE THE EQUATION

$$1,506,000,000 = 41.3\% \cdot x$$
$$1,506,000,000 = 0.413x \quad \text{Write 41.3\% as a decimal: } 41.3\% = 0.413.$$
$$\frac{1,506,000,000}{0.413} = \frac{0.413x}{0.413} \quad \text{Divide both sides by 0.413.}$$
$$3,646,489,104 = x$$

STATE THE CONCLUSION In 1997, sales of disposable diapers totaled $3,646,489,104.

CHECK THE RESULT If diaper sales totaled $3,646,489,104 in 1997, then the $1,506,000,000 spent on Huggies was about $\frac{1,500,000,000}{3,600,000,000} = \frac{15}{36} \approx 42\%$ of the total. The answer of $3,646,489,104 seems reasonable.

Applications and Connections to Other Disciplines

A distinguishing feature of this book is its wealth of application problems. We have included numerous applications from disciplines such as science, economics, business, manufacturing, history, and entertainment, as well as mathematics.

31. NO-FLY ZONES After the Gulf War, U.S. and Allied forces enforced northern and southern "no-fly" zones over Iraq. Iraqi aircraft was prohibited from flying in this air space. If y represents the north latitude parallel measurement, the no-fly zones can be described by

$$y \geq 36 \text{ or } y \leq 33$$

On the map in Illustration 5, shade the regions of Iraq over which there was a no-fly zone.

ILLUSTRATION 5

Turkey — 35th parallel — Iran — ★ Baghdad — Iraq — Saudi Arabia — 29th parallel — Persian Gulf — Kuwait

In Exercises 33–36, graph each system of inequalities and give two possible solutions.

33. COMPACT DISCS Melodic Music has compact discs on sale for either $10 or $15. If a customer wants to spend at least $30 but no more than $60 on CDs, use Illustration 7 to graph a system of inequalities that will show the possible ways a customer can buy $10 CDs ($x$) and $15 CDs ($y$).

ILLUSTRATION 7

34. BOAT SALES Dry Boat Works wholesales aluminum boats for $800 and fiberglass boats for $600. Northland Marina wants to order at least $2,400 worth but no more than $4,800 worth of boats. Use Illustration 8 to graph a system of inequalities that

37. INVESTING MONEY Based on the information in Illustration 12, a woman invested $12,000, some in a money market account and the rest in a 5-year CD. How much was invested in each account if the income from both investments is $1,060 per year?

◄ Every application problem has a title.

ILLUSTRATION 13

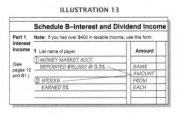

Schedule B—Interest and Dividend Income

Part 1 Interest Income — Note: If you had over $400 in taxable income, use this form.
1 List name of payer.
(See pages 12 and B1.)
① MONEY MARKET ACCT. DEPOSITED $15,000 @ 3.3%
② STOCKS EARNED 5%
Amount — SAME AMOUNT FROM EACH

41. MONEY-LAUNDERING SCHEME Use the evidence compiled by investigators to determine how much money a suspect deposited in the Cayman Islands bank.

■ On 6/1/98, the suspect electronically transferred $300,000 to a Swiss bank account paying an 8% annual yield.

■ That same day, the suspect opened another ac-

Study Sets—More Than Just Exercises

The problems at the end of each section are called Study Sets. Each Study Set includes Vocabulary, Notation, and Writing problems designed to help students improve their ability to read, write, and communicate mathematical ideas. The problems in the Concepts section of the Study Sets encourage students to engage in independent thinking and reinforce major ideas through exploration. In the Practice section of the Study Sets, students get the drill necessary to master the material. In the Applications section, students deal with real-life situations that involve the topics being studied. Each Study Set concludes with a Review section consisting of problems selected from previous sections.

◄ Each Study Set contains Vocabulary, Concepts, Notation, Practice, Applications, Writing, and Review sections.

STUDY SET

Section 3.3

VOCABULARY

In Exercises 1–6, fill in the blanks to make the statements true.

1. $\begin{cases} 2x + y - 3z = 0 \\ 3x - y + 4z = 5 \\ 4x + 2y - 6z = 0 \end{cases}$ is called a _____ of three linear equations.

2. If the first two equations of the system in Exercise 1 are added, the variable y is _____.

3. The equation $2x + 3y + 4z = 5$ is a linear equation with _____ variables.

4. The graph of the equation $2x + 3y + 4z = 5$ is a flat surface called a _____.

5. When three planes coincide, the equations of the system are _____, and there are infinitely many solutions.

6. When three planes intersect in a line, have _____ many solutions.

CONCEPTS

7. For each graph of a system of three whether the solution set contains on nitely many solutions, or no solution

a.   b.

8. Consider the system $\begin{cases} -2x + y + 4z \\ x - y + 2z = \\ x + y - 3z = \end{cases}$
   a. What is the result if Equation 1 added?
   b. What is the result if Equation 2 added?
   c. What variable was eliminated in formed in parts a and b?

NOTATION

9. Write the equation $3z - 2y = x + 6$

10. Fill in the blank to make a true statement: Solutions of a system of three equations in three variables, x, y, and z, are written in the form (x, y, z) and are called ordered _____.

PRACTICE

In Exercises 11–12, tell whether the given ordered triple is a solution of given system.

11. $(2, 1, 1)$, $\begin{cases} x - y + z = 2 \\ 2x + y - z = 4 \\ 2x - 3y + z = 2 \end{cases}$

12. $(-3, 2, -1)$, $\begin{cases} 2x + 2y + 3z = -1 \\ 3x + y - z = -6 \\ x + y + 2z = 1 \end{cases}$

In Exercises 13–28, solve each system. If a system is in-

APPLICATIONS

93. CONVERTING TEMPERATURES In preparing an American almanac for release in Europe, editors need to convert temperature ranges for the planets from degrees Fahrenheit to degrees Celsius. Solve the formula $F = \frac{9}{5}C + 32$ for C. Then use your result to make the conversions for the data shown in Illustration 1. Round to the nearest degree.

ILLUSTRATION 1

Planet	High °F	Low °F	High °C	Low °C
Mercury	810	−290		
Earth	136	−129		
Mars	63	−87		

94. THERMODYNAMICS In thermodynamics, the Gibbs free-energy function is given by the formula $G = U - TS + pV$. Solve for S.

95. WIPER DESIGN The area cleaned by the windshield wiper assembly shown in Illustration 2 is given by the formula
$$A = \frac{d\pi(r_1{}^2 - r_2{}^2)}{360}$$
Engineers have determined the amount of windshield area that needs to be cleaned by the wiper for two different vehicles. Solve the equati your result to find the number of d arm must swing in each case. Rou degree.

Vehicle	Area clean
Luxury car	513 in.²
Sport utility vehicle	586 in.²

96. ELECTRONICS Illustration 3 is a schematic diagram of a resistor connected to a voltage source of 60 volts. As a result, the resistor dissipates power in the form of heat. The power P lost when a voltage E is placed across a resistance R (in ohms) is given by the formula
$$P = \frac{E^2}{R}$$
Solve for R. If P is 4.8 watts and E is 60 volts, find R.

ILLUSTRATION 3

Battery $E = 60$ v   Resistor

97. CHEMISTRY LAB In chemistry, the ideal gas law equation is $PV = nR(T + 273)$, where P is the pressure, V the volume, T the temperature, and n is the number of moles of a gas. R is a constant, 0.082. Solve the equation for n. Then use your result and the data from the student lab notebook in Illustration 4 to find the value of n to the nearest thousandth for trial 1 and trial 2.

---

compute the bill for quantities of 500, 1,200, and 2,500 gallons.

101. SURFACE AREA To find the amount of tin needed to make a coffee can such as that shown in Illustration 6, we use the formula for the surface area of a right circular cylinder,
$$A = 2\pi r^2 + 2\pi rh$$
Solve the formula for h.

ILLUSTRATION 6

102. CARPENTRY A regular polygon has n equal sides and n equal angles. The measure a of an interior angle in degrees is given by $a = 180\left(1 - \frac{2}{n}\right)$. Solve for n. How many sides does the outdoor bandstand

WRITING

103. What does it mean to *solve an equation*?

104. Why doesn't the equation $x = x + 1$ have a real-number solution?

105. If you are asked to solve the equation $2s = 7t - s$ for s, why is $s = \frac{7t - s}{2}$ an incorrect answer?

106. When solving a linear equation in one variable, the objective is to isolate the variable on one side of the equation. What does that mean?

REVIEW

In Exercises 107–114, simplify each expression.

107. $-(4 + t) + 2t$

108. $12(2r + 11) - 11 - 3$

109. $4(b + 8) - 8b$

110. $-2(m - 3) + 8(2m + 7)$

111. $3.8b - 0.9b$ 112. $-5.7p + 5.1p$

113. $\frac{3}{5}t + \frac{2}{5}t$ 114. $-\frac{3}{16}x - \frac{5}{16}x$

Group Work

A one-page feature called Accent on Teamwork appears near the end of each chapter. It gives the instructor a set of problems that can be assigned as group work or to individual students as outside-of-class projects.

Key Concepts

Eleven important algebraic concepts are highlighted in one-page Key Concept features, appearing near the end of each chapter. Each Key Concept page summarizes a concept and gives students the opportunity to review the role it plays in the overall picture.

GROUP ACTIVITIES FOR CHAPTER 2

Accent on Teamwork

Section 2.1

City map Get a map of the city in which you live and use a black marker to draw a rectangular coordinate system on the map. The size of the grid you use will depend on the size of the map. The origin of the coordinate system should be your place of residence. Determine the coordinates of ten important locations in your city, such as the library, fire stations, parks, schools, and city hall.

Section 2.2

Graphing lines Draw rectangular coordinate systems with 1-inch grids on two pieces of white poster board. On the first poster board, graph $y = mx + 2$ for $m = 1, 2, 3, 4$, and 5. Label each line with its respective equation. At the bottom of the poster board, write your observations about how the graph changes for different values of m. On the second poster board, graph $y = 2x + b$ for $b = 1, 2, 3, 4$, and 5. At the bottom of the poster board, write your observations about how the graph changes for different values of b.

Section 2.3

Measuring slope Use a ruler and a level to find the slopes of five ramps or inclines by measuring $\frac{rise}{run}$, as shown in Illustration 1. Record your results in a chart of the form shown below. List the examples in order, from that with the smallest slope to that with the largest slope.

ILLUSTRATION 1

Object/location	Slope
Ramp outside cafeteria	$\dfrac{\text{Rise}}{\text{Run}} = \dfrac{4 \text{ in.}}{24 \text{ in.}} = \dfrac{1}{6}$

Section 2.4

A linear model For this project, you will need a five-gallon pail, a yardstick, a watch that shows seconds, and access to a garden hose.

Tape the yardstick to the pail as shown in Illustration 2. Turn on the water, leaving it running at a constant rate, and begin to fill the pail. Keep track of the time in seconds it takes for the water in the pail to reach heights of 2, 4, 6, 8, 10, and 12 inches.

Draw a coordinate system, with the horizontal axis representing time in seconds and the vertical axis the height of the water in inc[...]
the form (time, [...]
and write the eq[...]
dict the height [...]
run for 24 hour[...]

Section 2.[...]

Linear functio[...]
given in Exerc[...]
function that d[...]
the importance[...]

Section 2.[...]

Graphing func[...]
using a 1-inch g[...]
$y = k|x|$ for k [...]
spective equatio[...]
your observatio[...]
ent values of k.[...]

Parabolas Bor[...]
ematics instruc[...]
school library. I[...]
find some real-[...]
these books. W[...]
clude a sketch [...]

KEY CONCEPT

Functions

In Chapter 2, we introduced one of the most important concepts in mathematics, that of a **function**. Functions are used to describe relationships where one quantity depends on another.

1. Fill in the blanks to make the statements true.
   a. A function is a _____ between a set of _____ values x (called the domain) and a set of output values y (called the _____), where exactly ____ y value in the range corresponds to each number x in the _____.

   b. Since the value of y usually depends on the number x, we call y the _____ variable and x the _____ variable.

2. We can think of a function as a machine. Using the words *input, output, domain,* and *range,* explain how the function machine shown below works.

Four Ways to Represent a Function

Functions can be described in words, with an equation, with a table, or with a graph.

3. The equation $y = 2x + 3$ determines a correspondence between the values of x and y. Find the value of y that corresponds to $x = -10$.

4. The area of a circle is the product of π and the radius squared. Use the variables A and r to describe this relationship with an equation.

5. Use the table below to determine the height of a projectile 1.5 seconds after it was shot vertically into the air.

Time t (seconds)	Height h (feet)
0	0
0.5	28
1.0	48
1.5	60
2.0	64

6. Use the graph below to determine what y-value the function f assigns to $x = 1$.

Function Notation

There is a special notation used with functions. The notation $y = f(x)$ denotes that the variable y is a function of x.

7. Write the equation in Problem 3 using function notation. Find $f(0)$.

8. Use function notation to represent the relationship described in Problem 4.

9. If the function $h(t) = -16t^2 + 64t$ gives the height of the projectile described in Problem 5, find $h(4)$ and interpret the result.

10. Use the graph in Problem 6 to determine $f(-3)$ and $f(3)$.

Increased Emphasis on Graphing

In response to recent trends in mathematics education, we have increased the emphasis on learning mathematics through graphing. Graphing on the number line is discussed in Section 1.3. Chapter 2 introduces graphing using the rectangular coordinate system. Throughout the text, students are asked to graph functions and interpret graphs.

Functions and Modeling

The concept of function is introduced in Chapter 2 and is stressed throughout the text. Students learn to use function notation, graph functions, and write functions that mathematically model many interesting real-life situations. By the end of the course, students will recognize families of functions, their graphs, and areas of application.

Functions are ▶
introduced in Chapter 2.

The letter f used in the notation $y = f(x)$ represents the word *function*. However, other letters can be used to represent functions. For example, the notations $y = g(x)$ and $y = h(x)$ are often used to denote functions involving the independent variable x.

In Example 4, the equation $g(x) = x^2 - 2x$ determines a function, because every possible value of x gives a single value of $g(x)$.

EXAMPLE 4   **Evaluating a function.** Let $g(x) = x^2 - 2x$. Find   **a.** $g\left(\frac{2}{5}\right)$ and   **b.** $g(-2.4)$.

Solution   **a.** To find $g\left(\frac{2}{5}\right)$, we replace x with $\frac{2}{5}$:   **b.** To find $g(-2.4)$, we replace x with -2.4:

$$g(x) = x^2 - 2x$$
$$g\left(\frac{2}{5}\right) = \left(\frac{2}{5}\right)^2 - 2\left(\frac{2}{5}\right)$$
$$= \frac{4}{25} - \frac{4}{5}$$
$$= -\frac{16}{25}$$

$$g(x) = x^2 - 2x$$
$$g(-2.4) = (-2.4)^2 - 2(-2.4)$$
$$= 5.76 + 4.8$$
$$= 10.56$$

SELF CHECK   Let $h(x) = -\dfrac{x^2 + 2}{2}$. Find   **a.** $h(4)$ and   **b.** $h(-0.6)$.   *Answers:*   **a.** -9,   **b.** -1.18 ■

example, the letter A is chosen to name a function that finds the area of tter d is chosen as the independent variable, to help stress the fact that rcle is a function of its diameter.

e area of a circle with a diameter of length d is given by the function ind $A(48)$ to the nearest tenth. What information does it give about the hown in Figure 2-45?

eter of the circular target is 48 ives the area of the target. To replace d with 48.

$$\left(\frac{d}{2}\right)^2$$
$$\left(\frac{48}{2}\right)^2 \qquad \text{Substitute 48 for } d.$$
$$24)^2$$
$$6\pi$$
$$309.557368 \quad \text{Use a calculator to do the multiplication.}$$

tenth, the area of the target is $1{,}809.6$ in.^{2}.

FIGURE 2-45

9.6 in.   48 in.

52. $2(x + 2) = 2(1 - x) + 10$

APPLICATIONS

53. FREEWAY DESIGN A Grand Avenue exit off the 210 Freeway is to be constructed. (See Illustration 5.) Sketch the off-ramp design on the graph if the sides of the pavement are defined by the functions $f(x) = (x - 1)^3$ and $f(x) = (x - 3)^3$.

ILLUSTRATION 5

Grand Ave.

210

54. OPTICS See Illustration 6. The law of reflection states that the angle of reflection is equal to the angle of incidence. What function studied in this section mathematically models the path of the reflected light beam with an angle of incidence measuring 45°?

ILLUSTRATION 6

Incident beam   Angle of incidence   Angle of reflection   Reflected beam

Mirror

55. CENTER OF GRAVITY See Illustration 7. As a diver performs a $1\frac{1}{2}$-somersault in the tuck position,

ILLUSTRATION 7

56. FLASHLIGHTS Light beams coming from a flash-light bulb are reflected outward by a parabolic mirror as parallel rays.
   a. The cross-sectional view of a parabolic mirror is given by the function $f(x) = x^2$ for the following values of x: -0.7, -0.6, -0.5, -0.4, -0.3, -0.2, -0.1, 0, 0.1, 0.2, 0.3, 0.4, 0.5, 0.6, 0.7. Sketch the parabolic mirror using the graph in Illustration 8.
   b. From the light bulb filament at $(0, 0.25)$, draw a line segment representing a beam of light that strikes the mirror at $(-0.4, 0.16)$ and then reflects outward, parallel to the y-axis.

ILLUSTRATION 8

WRITING

57. Explain how to graph an equation by plotting points.
58. Explain why the correct choice of window settings is important when using a graphing calculator.
59. Explain how to solve the equation $2x + 6 = 0$ graphically.
60. What does it mean when we say that the domain of a function is the set of all real numbers?

◀ Exercises demonstrate that
functions model real-life situations.

Systematic Review

Each Study Set ends with a Review section that contains problems similar to those in previous sections. Each chapter ends with a Chapter Review and a Chapter Test. The chapter reviews have been designed to be "user friendly." In a unique format, the reviews list the important concepts of each section of the chapter in one column, with appropriate review problems running parallel in a second column. In addition, Cumulative Review Exercises appear after Chapters 2, 4, 6, 8, and 11.

Calculators

For instructors who wish to use calculators as part of the instruction in this course, the text includes an Accent on Technology feature that introduces keystrokes and shows how scientific calculators and graphing calculators can be used to solve problems.

Logos indicate which study set ▶ problems are to be solved with a calculator.

Student Support

We have included many features that make *Intermediate Algebra* very accessible to students. (See the examples on pages x–xvi.)

Worked Examples

The text contains hundreds of worked examples, many with several parts. Explanatory notes make the examples easy to follow.

Author's Notes

Author's notes, printed in red, are used to explain the steps in the solutions of examples. The notes are extensive; complete sentences are used so as to increase the students' ability to read and write mathematics.

Each step is explained using ▶
detailed author's notes.

Self Checks

There are hundreds of Self Check problems that allow students to practice the skills demonstrated in the worked examples.

Warnings

Throughout the text, students are warned about common mistakes and how to avoid them.

Videotapes

The videotape series that accompanies this book uses eye-catching computer graphics to show students the steps in solving many examples in the text. A video logo ▭ placed next to an example indicates that the example is taught on tape. In addition, the tapes present the solutions of two Study Set problems from each section.

Functional Use of Color

For easy reference, definition boxes (light blue with a green title), strategy boxes (light yellow with a red title), and rule or property boxes (bright yellow with royal blue title), are color-coded. In addition, the book uses color to highlight terms and expressions that you would point to in a classroom discussion.

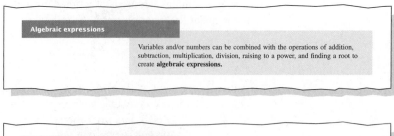

Algebraic expressions

Variables and/or numbers can be combined with the operations of addition, subtraction, multiplication, division, raising to a power, and finding a root to create **algebraic expressions.**

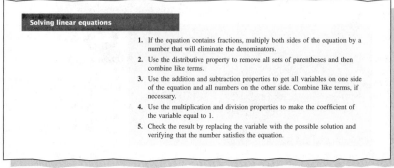

Solving linear equations

1. If the equation contains fractions, multiply both sides of the equation by a number that will eliminate the denominators.
2. Use the distributive property to remove all sets of parentheses and then combine like terms.
3. Use the addition and subtraction properties to get all variables on one side of the equation and all numbers on the other side. Combine like terms, if necessary.
4. Use the multiplication and division properties to make the coefficient of the variable equal to 1.
5. Check the result by replacing the variable with the possible solution and verifying that the number satisfies the equation.

Dividing real numbers

With like signs: Divide their absolute values. The answer is positive.

With unlike signs: Divide their absolute values. Make the answer negative.

Problems and Answers

The book includes thousands of carefully graded exercises. Appendix IV provides the answers to the odd-numbered exercises in the Study Sets, as well as all the answers to the Key Concept, Chapter Review, Chapter Test, and Cumulative Review problems.

Reading and Writing Mathematics

Also included (on pages xxiii–xxiv) are two features to help students improve their ability to read and write mathematics. "Reading Mathematics" helps students get the most out of the examples in this book by showing them how to read the solutions properly. "Writing Mathematics" highlights the characteristics of a well-written solution.

Study Skills and Math Anxiety

These two topics are discussed in detail in the section entitled "For the Student" at the end of this preface. In "Success in Intermediate Algebra," students are asked to design a personal strategy for studying and learning the material. "Taking a Math Test" helps students prepare for a test and then gives them suggestions for improving their performance.

Ancillaries for the Instructor

Complete Solutions Manual

Complete, worked-out solutions for all exercises in the text are provided.

Test Items

These printed test forms include multiple-choice and free-response questions, keyed to the text.

Thomson Learning Testing Tools CD

This integrated package features algorithmic test generation and class management capabilities. Using the CD, a professor can easily post sample tests and examinations to the Internet or the local area network.

Interactive Algebra 3.0

This text-specific tutorial software program is offered in an instructor and a student version. It offers a mathematical guide to students of any technological proficiency, as well as a tracking program enabling professors to follow their students' progress.

Text-Specific Video Series

The tapes work through examples from each section of the text, with additional solutions to two Study Set problems from each section.

Demo Video

This demonstration video, approximately 30 minutes long, features highlights from the text-specific video series.

Ancillaries for the Student

Student Solutions Manual

This manual contains complete, worked-out solutions for the odd-numbered problems.

Greatest Hits Tutorial Video

The video features the concepts and skills students traditionally have the most difficulty comprehending. The tape includes examples, chosen by the authors, from all chapters.

Acknowledgments

We are grateful to the instructors who have reviewed the text at various stages of its development. Their comments and suggestions have proven invaluable in making this a better book. We sincerely thank all of them for lending their time and talent to this project.

Sally Copeland
Johnson County Community College

June Strohm
Pennsylvania State Community College–DuBois

Ben Cornelius
Oregon Institute of Technology

Jo Anne Temple
Texas Technical University

Mary Lou Hammond
Spokane Community College

Sharon Testone
Onondaga Community College

Judith Jones
Valencia Community College

Marilyn Treder
Rochester Community College

Theresa Jones
Amarillo College

Betty Weissbecker
J. Sargeant Reynolds Community College

Janice McFatter
Gulf Coast Community College

Cathleen Zucco
SUNY-New Paltz

We want to express our gratitude to Bob Billups, George Carlson, Robin Carter, Jim Cope, Terry Damron, Marion Hammond, Karl Hunsicker, Doug Keebaugh, Arnold Kondo, John McKeown, Kent Miller, Donna Neff, Steve Odrich, Eric Rabitoy, Dave Ryba, Chris Scott, Liz Tussy, and the Citrus College Library Staff for their help with some of the application problems in the textbook.

We would also like to thank Bob Pirtle, Jennifer Huber, Ellen Brownstein, David Hoyt, Lori Heckelman, Vernon Boes, Erin Wickersham, Melissa Henderson, Michelle Paolucci, and The Clarinda Company for their help in creating this book.

Alan S. Tussy
R. David Gustafson

▶ For the Student

Success in Intermediate Algebra

To be successful in mathematics, you need to know how to study it. The following checklist will help you develop your own personal strategy to study and learn the material. The suggestions listed below require some time and self-discipline on your part, but it will be worth the effort. This will help you get the most out of this course.

As you read each of the following statements, place a check mark in the box if you can truthfully answer Yes. If you can't answer Yes, think of what you might do to make the suggestion part of your personal study plan. You should go over this checklist several times during the term to be sure you are following it.

Preparing for the Class

☐ I have made a commitment to myself to give this course my best effort.

☐ I have the proper materials: a pencil with an eraser, paper, a notebook, a ruler, a calculator, and a calendar or day planner.

☐ I am willing to spend a minimum of two hours doing homework for every hour of class.

☐ I will try to work on this subject every day.

☐ I have a copy of the class syllabus. I understand the requirements of the course and how I will be graded.

☐ I have scheduled a free hour after the class to give me time to review my notes and begin the homework assignment.

Class Participation

☐ I will regularly attend the class sessions and be on time.

☐ When I am absent, I will find out what the class studied, get a copy of any notes or handouts, and make up the work that was assigned when I was gone.

☐ I will sit where I can hear the instructor and see the chalkboard.

☐ I will pay attention in class and take careful notes.

☐ I will ask the instructor questions when I don't understand the material.

☐ When tests, quizzes, or homework papers are passed back and discussed in class, I will write down the correct solutions for the problems I missed so that I can learn from my mistakes.

Study Sessions

☐ I will find a comfortable and quiet place to study.

☐ I realize that reading a math book is different from reading a newspaper or a novel. Quite often, it will take more than one reading to understand the material.

☐ After studying an example in the textbook, I will work the accompanying Self Check.

☐ I will begin the homework assignment only after reading the assigned section.

☐ I will try to use the mathematical vocabulary mentioned in the book and used by my instructor when I am writing or talking about the topics studied in the course.

☐ I will look for opportunities to explain the material to others.

☐ I will check all of my answers to the problems with those provided in the back of the book (or with the *Student Solutions Manual*) and reconcile any differences.

☐ My homework will be organized and neat. My solutions will show all the necessary steps.

☐ I will work some review problems every day.

☐ After completing the homework assignment, I will read the next section to prepare for the coming class session.

☐ I will keep a notebook containing my class notes, homework papers, quizzes, tests, and any handouts—all in order by date.

Special Help

☐ I know my instructor's office hours and am willing to go in to ask for help.

☐ I have formed a study group with classmates that meets regularly to discuss the material and work on problems.

☐ When I need additional explanation of a topic, I view the video and check the web site.

☐ I take advantage of extra tutorial assistance that my school offers for mathematics courses.

☐ I have purchased the *Student Solutions Manual* that accompanies this text, and I use it.

To follow each of these suggestions will take time. It takes a lot of practice to learn mathematics, just as with any other skill.

No doubt, you will sometimes become frustrated along the way. This is natural. When it occurs, take a break and come back to the material after you have had time to clear your thoughts. Keep in mind that the skills and discipline you learn in this course will help make for a brighter future. Good luck!

Taking a Math Test

The best way to relieve anxiety about taking a mathematics test is to know that you are well-prepared for it and that you have a plan. Before any test, ask yourself three questions. When? What? How?

When Will I Study?

1. When is the test?

2. When will I begin to review for the test?

3. What are the dates and times that I will reserve for studying for the test?

What Will I Study?

1. What sections will the test cover?
2. Has the instructor indicated any types of problems that are guaranteed to be on the test?

How Will I Prepare for the Test?

Put a check mark by each method you will use to prepare for the test.

- [ ] Review the class notes.
- [ ] Outline the chapter(s) to see the big picture and to see how the topics relate to one another.
- [ ] Recite the important formulas, definitions, vocabulary, and rules into a tape recorder.
- [ ] Make flash cards for the important formulas, definitions, vocabulary, and rules.
- [ ] Rework problems from the homework assignments.
- [ ] Rework each of the Self Check problems in the text.
- [ ] Form a study group to discuss and practice the topics to be tested.
- [ ] Complete the appropriate Chapter Review(s) and the Chapter Test(s).
- [ ] Review the Warnings given in the text.
- [ ] Work on improving my speed in answering questions.
- [ ] Review the methods that can be used to check my answers.
- [ ] Write a sample test, trying to think of the questions the instructor will ask.
- [ ] Complete the appropriate Cumulative Review Exercises.
- [ ] Get organized the night before the test. Have materials ready to go so that the trip to school will not be hurried.
- [ ] Take some time to relax immediately before the test. Don't study right up to the last minute.

Taking the Test

Here are some tips that can help improve your performance on a mathematics test.

- When you receive the test, scan it, looking for the types of problems you had expected to see. Do them first.
- Write down any formulas or rules as soon as you receive the test.
- Read the instructions carefully.
- Don't spend too much time on any one problem until you have attempted all the problems.
- If your instructor gives partial credit, at least try to begin a solution.
- Save the most difficult problems for last.

- Don't be afraid to skip a problem and come back to it later.
- If you finish early, go back over your work and look for mistakes.

Reading Mathematics

To get the most out of this book, you need to learn how to read it correctly. A mathematics textbook must be read differently than a novel or a newspaper. For one thing, you need to read it slowly and carefully. At times, you will have to reread a section to understand its content. You should also have pencil and paper with you when reading a mathematics book, so that you can work along with the text to understand the concepts presented.

Perhaps the most informative parts of a mathematics book are its examples. Each example in this textbook consists of a problem and its corresponding solution. One form of solution that is used many times in this book is shown in the diagram below. It is important that you follow the "flow" of its steps if you are to understand the mathematics involved. For this solution form, the basic idea is this:

- A property, rule, or procedure is applied to the original expression to obtain an equivalent expression. We show that the two expressions are equivalent by writing an equals sign between them. The property, rule, or procedure that was used is then listed next to the equivalent expression in the form of an author's note, printed in red.
- The process of writing equivalent expressions and explaining the reasons behind them continues, step by step, until the final result is obtained.

The solution in the following diagram consists of three steps, but solutions have varying lengths.

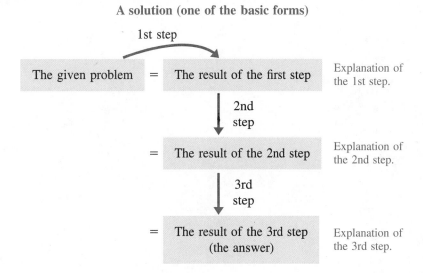

A solution (one of the basic forms)

1st step		
The given problem	= The result of the first step	Explanation of the 1st step.
	2nd step	
	= The result of the 2nd step	Explanation of the 2nd step.
	3rd step	
	= The result of the 3rd step (the answer)	Explanation of the 3rd step.

Writing Mathematics

One of the major objectives of this course is for you to learn how to write solutions to problems properly. A written solution to a problem should explain your thinking in a series of neat and organized mathematical steps. Think of a solution as a mathematical essay—one that your instructor and other students should be able to read and understand. Some solutions will be longer than others, but they must all be in the proper format and use the correct notation. To learn how to do this will take time and practice.

To give you an idea of what will be expected, let's look at two samples of student work. In the first, we have highlighted some important characteristics of a well-written solution. The second sample is poorly done and would not be acceptable.

$$\text{Evaluate: } 35 - 2^2 \cdot 3.$$

A well-written solution:

The problem was copied ▶	$35 - 2^2 \cdot 3 = 35 - 4 \cdot 3$	◀ The first step of the
from the textbook.		solution is written here.
	$= 35 - 12$	◀ The steps are written
		under each other in
	$= 23$	a neat, organized
	▲	manner.
	The equals signs are	
	lined up vertically.	

A poorly written solution:

The problem wasn't ▶	$2^2 = 4 = 35 - 4 \cdot 3$	◀ An equals sign is
copied from the text.	$\underbrace{\qquad}_{12}$	improperly used.
	SUB: 35	
	$\underline{-12}$	◀ The work is disorganized
	$23 \rightarrow (23)$	and difficult to follow.

1

A Review of Basic Algebra

CAMPUS CONNECTION

The Chemistry Department

In a chemistry class, students study many formulas. For example, in one chemistry lab experiment, students measure the temperature, pressure, and volume of a gas sample. Then they find the number of molecules in the sample by solving the formula $PV = nR(T + 273)$ for n and substituting data values for P, V, R, and T. In this chapter, we will work with formulas from a wide variety of disciplines, including some that you would see in a chemistry class.

> *Algebra is a mathematical language that can be used to solve many types of problems.*

▶ 1.1 Describing Numerical Relationships

In this section, you will learn about

> Variables, algebraic expressions, and equations ■ Verbal models
> ■ Constructing tables ■ Graphical models ■ Formulas

Introduction   To solve problems using algebra, you will need to learn how to read it, write it, and speak it. In this section, we discuss several ways in which numerical relationships are described.

Variables, Algebraic Expressions, and Equations

The rental agreement for a banquet hall is shown in Figure 1-1.

FIGURE 1-1

Rental Agreement
ROYAL VISTA BANQUET ROOM
Wedding Receptions•Dances•Reunions•Fashion Shows

Rented To_____ Date_____
Lessee's Address_____

Rental Charges
• $100 per hour
• Nonrefundable $200 cleanup fee

Terms and conditions
Lessor leases the undersigned lessee the above described property upon the terms and conditions set forth on this page and on the back of this page. Lessee promises to pay rental cost stated herein.

In examining the rental agreement, we see that two operations need to be performed to calculate the cost of renting the banquet hall.

■ First, we must *multiply* the hourly rental cost of $100 by the number of hours the hall is to be rented.

■ To that result, we must then *add* the cleanup fee of $200.

In words, we can describe the process as follows.

| The cost of renting the hall | is | 100 | times | the number of hours it is rented | plus | 200. |

We can describe the procedure used to calculate this cost in a more concise way by using *variables* and mathematical symbols. A **variable** is a letter that is used to stand for a number. If we let C stand for the cost of renting the hall and h stand for the number of hours it is rented, the words can be translated to form a *mathematical model*.

The cost of renting the hall	is	100	times	the number of hours it is rented	plus	200.
C	$=$	100	$\cdot$	h	$+$	200

The statement $C = 100h + 200$ is called an **equation.** The $=$ sign indicates that two quantities, C and $100h + 200$, are equal.

Equations

An **equation** is a mathematical sentence that contains an $=$ sign.

Here are some examples of equations.

$$2 + 3 = 5 \qquad 3x - 2 = 4x + 10 \qquad P = a + b + c$$

On the right-hand side of the equation $C = 100h + 200$, the notation $100h + 200$ is called an **algebraic expression**, or more simply, an **expression.** Algebraic expressions are the building blocks of equations.

Algebraic expressions

Variables and/or numbers can be combined with the operations of addition, subtraction, multiplication, division, raising to a power, and finding a root to create **algebraic expressions.**

Here are some examples of algebraic expressions.

$5a - 12$ — This algebraic expression is a combination of the numbers 5 and 12, the variable a, and the operations of multiplication and subtraction.

$\dfrac{50 - y}{3y^3}$ — This algebraic expression is a combination of the numbers 50 and 3, the variable y, and the operations of subtraction, multiplication, division, and raising to a power.

$c = \sqrt{a^2 + b^2}$ — The right-hand side of this equation is the algebraic expression $\sqrt{a^2 + b^2}$.

Verbal Models

Table 1-1 lists some words and phrases that are often used in mathematics to denote the operations of addition, subtraction, multiplication, and division.

TABLE 1-1

Addition +	Subtraction −	Multiplication ·	Division ÷
added to	subtracted from	multiplied by	divided by
plus	difference	product	quotient
the sum of	less than	times	ratio
more than	decreased by	percent (or fraction) of	half
increased by	reduced by	twice	into
greater than	minus	triple	per

In the banquet hall example, the equation $C = 100h + 200$ was used to describe a procedure to calculate the cost of renting the hall. Using vocabulary from Table 1-1, we can write a **verbal model** that also describes this procedure. One such model is:

The cost (in dollars) of renting the hall is the *product* of 100 and the number of hours it is rented, *increased by* 200.

E X A M P L E   1

Describing numerical relationships. The cost to have a dinner catered is $6 per person. For groups of more than 200, a $100 discount is given. Write a mathematical model and a verbal model that describe the relationship between the catering cost and the number of people being served, for groups larger than 200.

Solution   To find the catering cost C (in dollars) for groups larger than 200, we need to *multiply* the number n of people served by $6 and then *subtract* the $100 discount.

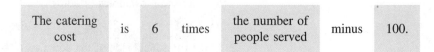

| The catering cost | is | 6 | times | the number of people served | minus | 100. |

In symbols, the mathematical model is:

$C = 6n - 100$

A verbal model is:

The catering cost (in dollars) is the *product* of 6 and the number of people served, *decreased by 100*.

SELF CHECK   After winning a lottery, three friends split the prize equally. Each person then had to pay $2,000 in taxes on his or her share. Write a mathematical model and a verbal model that relate the amount of each person's share, after taxes, to the amount of the lottery prize.

Answers: $S = \frac{p}{3} - 2,000$; each person's share (in dollars), after taxes, is the quotient of the lottery prize and 3, decreased by 2,000   ∎

Constructing Tables

In the banquet hall example, the equation $C = 100h + 200$ can be used to determine the cost of renting the banquet hall for *any* number of hours.

E X A M P L E   2

Using an equation. Find the cost of renting the banquet hall for 3 hours and for 4 hours. Write the results in a table.

Solution   We begin by constructing a table with the appropriate column headings: h for the number of hours the hall is rented and C for the cost (in dollars) to rent the hall. Then we enter the number of hours of each rental time in the left column.

h	C
3	
4	

Next, we use the equation $C = 100h + 200$ to find the total rental cost for 3 hours and for 4 hours.

$C = 100h + 200$
$C = 100(3) + 200$ Replace h with 3.
$\quad = 300 + 200$ Do the multiplication.
$\quad = 500$

$C = 100h + 200$
$C = 100(4) + 200$ Replace h with 4.
$\quad = 400 + 200$ Do the multiplication.
$\quad = 600$

h	C
3	500
4	600

Finally, we enter these results in the right-hand column of the table: $500 for a 3-hour rental and $600 for a 4-hour rental.

SELF CHECK   Find the cost of renting the hall for 6 hours and *Answer:* for 7 hours. Write the results in the table.

h	C
6	800
7	900

Graphical Models

The cost of renting the banquet hall for various lengths of time can also be presented graphically. The **bar graph** in Figure 1-2(a) has a **horizontal axis** labeled "Number of hours the hall is rented." The **vertical axis** of the graph, labeled "Cost to rent the hall ($)," is scaled in units of 50 dollars. The bars directly over each of the times (1, 2, 3, 4, 5, 6, and 7 hours) extend to a height that gives the corresponding cost to rent the hall. For example, if the hall is rented for 5 hours, the height of the bar indicates that the cost is $700.

FIGURE 1-2

Cost of Renting a Banquet Hall

(a)

Cost of Renting a Banquet Hall

(b)

A **line graph** can also be used to show the rental costs. This type of graph consists of a series of dots drawn at the correct "height," connected with line segments. (See Figure 1-2(b).) Not only does the line graph present all the information contained in the bar graph, but it also provides additional information. We can use the line graph to find the cost of renting the banquet hall for a length of time not shown in the bar graph.

EXAMPLE 3   **Reading a line graph.** Use the line graph in Figure 1-2(b) to determine the cost of renting the hall for $4\frac{1}{2}$ hours.

Solution   In Figure 1-3, we locate $4\frac{1}{2}$ on the horizontal axis of the line graph and draw a line straight up to intersect the graph. From the point of intersection with the graph, we draw a horizontal line to the left that intersects the vertical axis. On the vertical axis, we can read that the rental cost is $650 for $4\frac{1}{2}$ hours.

FIGURE 1-3

Number of hours the hall is rented

SELF CHECK Use Figure 1-3 to find the cost of renting the banquet hall for $6\frac{1}{2}$ hours. *Answer:* $850 ∎

Formulas

Equations that express a known relationship between two or more quantities, represented by variables, are called **formulas.** Formulas are used in many fields. Automotive technology, economics, medicine, retail sales, and banking are just a few examples.

EXAMPLE 4

Translating from words to symbols. Use variables to express each relationship as a formula.

a. The distance in miles traveled by a vehicle is the product of its average rate of speed in mph and the time in hours it travels at that rate.

b. The sale price of an item is the difference between the regular price and the discount.

Solution

a. The word *product* indicates multiplication. If we let d stand for the distance traveled in miles, r for the vehicle's average rate of speed in mph, and t for the length of time traveled in hours, we can write the formula as

$$d = rt$$

b. The word *difference* indicates subtraction. If we let s stand for the sale price of the item, p for the regular price, and d for the discount, we have

$$s = p - d$$

SELF CHECK Express each relationship as a formula: **a.** the retail price of an item is the sum of its wholesale cost and the markup, and **b.** the simple interest earned by a deposit is the product of the principal, the annual rate of interest, and the time. *Answers:* **a.** $r = c + m$, **b.** $I = Prt$ ∎

The *perimeter* of a geometric figure is a measure of the distance around it. Table 1-2 shows the formulas for the perimeter (P) of several geometric figures. The perimeter of a circle is called its **circumference** (C).

TABLE 1-2

Square
$P = 4s$

Trapezoid
$P = a + b + c + d$

Rectangle
$P = 2l + 2w$

Circle
$C = \pi D$
(π is approximately 3.1416.)

Triangle
$P = a + b + c$

| EXAMPLE 5 | **Translating from symbols to words.** The formula for the perimeter of a rectangle is $P = 2l + 2w$. Express the formula in words. |

Solution   In the formula, the length and width are each *multiplied by two*. Then we *add* those results.

The perimeter of a rectangle is the sum of twice the length and twice the width.

SELF CHECK   Express the formula for the circumference of a circle, $C = \pi D$, in words.

Answer: The circumference of a circle is the product of pi and the length of the diameter. ■

| EXAMPLE 6 | **Using perimeter formulas.** Find the number of feet of redwood edging needed to outline a square flower bed having sides that are 6.5 feet long. |

Solution   To find the amount of redwood edging needed, we need to find the perimeter of the square flower bed.

$P = 4s$ The formula for the perimeter of a square.

$P = 4(\mathbf{6.5})$ Substitute 6.5 for s, the length of one side of the square.

$= 26$ Do the multiplication.

26 feet of redwood edging is needed to outline the flower bed.

SELF CHECK   Find the amount of fencing needed to enclose a triangular lot with sides that are 150 ft, 205.5 ft, and 165 ft long.

Answer: 520.5 ft ■

STUDY SET

Section 1.1

VOCABULARY

In Exercises 1–8, fill in the blanks to make the statements true.

1. An _____ is a mathematical sentence that contains an = sign.

2. A _____ is a letter that is used to stand for a number.

3. Variables and/or numbers can be combined with mathematical operations to create algebraic _____.

4. Phrases such as *the sum, increased by,* and *more than* are used to indicate the operation of _____.

5. A _____ is an equation that expresses a known relationship between two or more quantities represented by variables.

6. Phrases such as *the difference, decreased by,* and *less than* are used to indicate the operation of _____.

7. The distance around a geometric figure is its _____.

8. The _____ of a circle is the distance around it.

CONCEPTS

In Exercises 9–16, classify each expression as an algebraic expression or an equation.

9. $x - 5$

10. $x - 5 = 5$

11. $P = a + b + c$

12. $\dfrac{x + y}{8}$

13. $d = rt$

14. Prt

15. $2x^2$

16. $T = 6n - 400$

17. a. What type of graph is shown in Illustration 1 on page 8?
   b. What units are used to scale the horizontal axis? The vertical axis?
   c. Estimate the height of the candle after it has burned for $3\frac{1}{2}$ hours. For 8 hours.

18. a. What type of graph is shown in Illustration 2 on page 8?
   b. What units are used to scale the horizontal axis? The vertical axis?
   c. In what year was the sale of hunting and firearms equipment the greatest? The least?

ILLUSTRATION 1

Hours burning

ILLUSTRATION 2

Based on data from the National Sporting Goods Assn.

In Exercises 19–26, translate each verbal model to a mathematical model.

19. The cost each semester is $13 times the number of units taken plus a student services fee of $24.

20. The yearly salary is $25,000 plus $75 times the number of years of experience.

21. The quotient of the number of clients and seventy-five gives the number of social workers needed.

22. The difference between 500 and the number of people in a theater gives the number of unsold tickets.

23. Each test score was increased by 15 points to give a new adjusted test score.

24. The weight of a super-size order of French fries is twice that of a regular-size order.

25. The product of the number of boxes of crayons in a case and 12 gives the number of crayons in a case.

26. The perimeter of an equilateral triangle can be found by tripling the length of one of its sides.

In Exercises 27–28, use the data in each table to find a formula that mathematically describes the relationship between the two quantities. Then state the relationship in words. (Answers may vary.)

27.

Tower height (ft)	Height of base (ft)
15.5	5.5
22	12
25.25	15.25
45.125	35.125

28.

Seasonal employees	Employees
25	75
50	100
60	110
80	130

NOTATION

29. Algebraic expressions are combinations of variables, numbers, and the operations of mathematics. For the algebraic expression $2D + 4$,
  a. What variable is used?
  b. What numbers are used?
  c. What operations are used?

30. Consider the equation $T = 3r - 15$.
  a. How many variables does the equation contain?
  b. What operations are indicated on the right-hand side of the equation?
  c. Is the right-hand side of the equation an algebraic expression?

PRACTICE

In Exercises 31–34, use the given equation (formula) to complete each table.

31. $c = \dfrac{p}{12}$

Number of packages	Cartons
24	
72	
180	

32. $y = 100c$

Number of centuries	Years
1	
6	
21	

33. $n = 22.44 - K$

K	n
0	
1.01	
22.44	

34. $y = x + 15$

x	y
0	
15	
30	

35. The lengths of the two parallel sides of a trapezoid are 10 inches and 15 inches. The other two sides are each 6 inches long. Find the perimeter of the trapezoid.

36. Find the distance the wheel of a bicycle rolls in one revolution if the diameter of the wheel is 20 inches. Round to the nearest tenth.

37. Find the perimeter of a square quilt that has sides 2 yards long.

38. Find the perimeter of a triangular postage stamp with sides 1.8, 1.8, and 1.5 centimeters long.

APPLICATIONS

39. CARPET CLEANING See the advertisement shown in Illustration 3.

ILLUSTRATION 3

Rent the in–home
Carpet Cleaning System
Do it yourself and save!
Safe, effective
Costs only $10 an hour
plus $20 for supplies

a. Write a verbal model that states the relationship between the cost C of renting the carpet-cleaning system and the number of hours h it is rented.

b. Translate the verbal model written in part a to a mathematical model.

c. See Illustration 4. Use your result from part b to complete the table, and then draw a line graph.

ILLUSTRATION 4

h	C
1	
2	
3	
4	
8	

40. UTILITY VEHICLES What geometric concept applies when finding the length of the plastic trim around the cargo area floor mat shown in Illustration 5? Estimate the amount of trim used.

ILLUSTRATION 5

41. FINISH CARPENTRY The miter saw in Illustration 6 can pivot a total of 180° to make an angled cut on a piece of cabinet molding.

ILLUSTRATION 6

a. Write a verbal model that states the relationship between the angle measure s on the scrap piece of molding and the angle measure f on the finish piece of molding after a saw cut is made.

b. Translate the verbal model written in part a to a mathematical model.

c. Use your results from part b to complete the table and draw a line graph in Illustration 7.

ILLUSTRATION 7

f	s
30	
45	
90	
135	
150	

Measure of angle on finish piece (deg)

42. PRODUCTION PLANNING Use the diagram in Illustration 8 to write four formulas involving r that could be used by planners to order the necessary number of oak mounting plates p, bar holders b,

chrome bars c, and mounting screws s for a production run that will produce r towel racks.

ILLUSTRATION 8

Oak mounting plate
Bar holder   Chrome bar
Mounting screws

WRITING

43. Explain the difference between an expression and an equation. Give examples.

44. Name four ways in which numerical relationships were described in this section. Give examples.

45. Use each word below in a sentence so that a mathematical operation is indicated. If you are unsure of the meaning of a word, look it up in a dictionary.

quadrupled   deleted   bisected
confiscated   annexed   docked

46. Consider the formula $h = 7d$, where d is the age of a dog and h is the equivalent "human age" of the dog. Explain what information is given by the formula.

▶ 1.2 The Real Number System

In this section, you will learn about

Natural numbers, whole numbers, and integers ■ Rational numbers ■ Irrational numbers ■ Real numbers ■ The real number line ■ Inequality symbols ■ Opposites ■ Absolute value

---

Introduction In this course, we will work with many different types of numbers. For example, we will use positive numbers to express distances. We will encounter some temperatures that will be negative. We will work with fractional parts of an hour, and we will use decimals to write numbers in scientific notation. The solutions of some geometry problems will contain π or a square root. In this section, we will define many types of numbers. We will show that together, they make up a collection of numbers called the *real numbers*.

Natural Numbers, Whole Numbers, and Integers

The graph in Figure 1-4 shows the daily low temperatures for Anchorage, Alaska for the first seven days of January, 1998. On the horizontal axis, the numbers 1, 2, 3, 4, 5,

6, and 7 have been used to denote the calendar days of the month. This collection of numbers is called a **set,** and the members (or **elements**) of the set can be listed within **braces** { }.

{1, 2, 3, 4, 5, 6, 7}   Read as "the set containing the elements 1, 2, 3, 4, 5, 6, and 7."

Each of the numbers 1, 2, 3, 4, 5, 6, and 7 is a member of a basic set called the **natural numbers.** The natural numbers are the numbers we count with.

FIGURE 1-4

Natural numbers

The set of **natural numbers** is {1, 2, 3, 4, 5, 6, 7, 8, 9, 10, . . .}.

The three dots used in this definition indicate that the list of natural numbers continues on forever.

Two other important sets of numbers are the prime numbers and the composite numbers.

Prime numbers

The **prime numbers** are the natural numbers greater than 1 that are divisible only by themselves and 1.

The first ten prime numbers are 2, 3, 5, 7, 11, 13, 17, 19, 23, and 29. Since every prime number is also a natural number, we say that the set of prime numbers is a **subset** of the set of natural numbers.

Composite numbers

The **composite numbers** are the natural numbers greater than 1 that are not prime numbers.

The first ten composite numbers are 4, 6, 8, 9, 10, 12, 14, 15, 16, and 18.

The natural numbers together with 0 make up a set of numbers called the **whole numbers.**

Whole numbers

The set of **whole numbers** is $\{0, 1, 2, 3, 4, 5, 6, 7, 8, 9, 10, \ldots\}$.

Since every natural number is also a whole number, the set of natural numbers is a subset of the set of whole numbers.

The graph in Figure 1-4 contains both **positive numbers,** numbers greater than 0, and **negative numbers,** numbers less than 0. For example, on January 7, the low temperature in Anchorage was 3°F (3 degrees above zero). On January 3, the low temperature was −9°F (9 degrees below zero). On January 2, the low temperature was 0°F. Zero is neither positive nor negative. The numbers used to represent the temperatures in the graph are members of a set of numbers called the **integers.**

Integers

The set of **integers** is $\{\ldots, -4, -3, -2, -1, 0, 1, 2, 3, 4, \ldots\}$.

Integers that are divisible by 2 are called **even integers,** and integers that are not divisible by 2 are called **odd integers.**

Even integers: $\ldots, -6, -4, -2, 0, 2, 4, 6, \ldots$

Odd integers: $\ldots, -5, -3, -1, 1, 3, 5, \ldots$

Since every whole number is also an integer, the set of whole numbers is a subset of the set of integers.

Rational Numbers

In this course, we will work with positive and negative fractions. For example, if money is deposited in a savings account for 7 months, we say it is deposited for $\frac{7}{12}$ of a year. If a tank loses 40 gallons of water over a 3-minute period, we can describe the change per minute in the number of gallons of water in the tank using the fraction $-\frac{40}{3}$.

We will also work with mixed numbers. For instance, we might speak of $5\frac{7}{8}$ cups of flour or of a river that is $3\frac{1}{2}$ feet below flood stage $\left(-3\frac{1}{2} \text{ ft}\right)$. Fractions and mixed numbers are part of the set of numbers called the **rational numbers.** A **rational number** is any number that can be written as a fraction with an integer numerator and a nonzero integer denominator.

Rational numbers

A **rational number** is any number that can be written as $\frac{a}{b}$, where a and b are integers and $b \neq 0$.

Some examples of rational numbers are $\frac{7}{12}$, $-\frac{40}{3}$ (which could be written $\frac{-40}{3}$), and $\frac{25}{99}$.

The mixed numbers $5\frac{7}{8}$ and $-3\frac{1}{2}$ are also rational numbers, because they can be written as fractions with integer numerators and nonzero integer denominators:

$$5\frac{7}{8} = \frac{47}{8} \quad \text{and} \quad -3\frac{1}{2} = -\frac{7}{2} = \frac{-7}{2}$$

The next example shows that every integer is a rational number.

EXAMPLE 1

Rational numbers. Show that each of the following numbers is a rational number by writing it as a fraction with an integer numerator and a nonzero integer denominator: **a.** -4 and   **b.** 0.

Solution   **a.** -4 is a rational number, because -4 can be written as the fraction $\frac{-4}{1}$.

b.   0 is a rational number, because $0 = \frac{0}{1}$.

We see that all integers are rational numbers, because they can be written as fractions with a denominator of 1. Therefore, the set of integers is a subset of the set of rational numbers.

SELF CHECK   Show that each number is a rational number: *Answers:*   **a.** $10\frac{1}{2} = \frac{21}{2}$,
a. $10\frac{1}{2}$,   **b.** 99,   and   **c.** -3.   **b.** $99 = \frac{99}{1}$,   **c.** $-3 = \frac{-3}{1}$ ∎

We will also work with decimals. Some examples of uses of decimals are as follows.

- The interest rate of a loan was $11\% = 0.11$.
- In baseball, the distance from home plate to second base is 127.279 feet.
- The third-quarter loss for a business was -2.7 million dollars.

Terminating decimals such as 0.11, 127.279, and -2.7 are rational numbers, since they can be written as fractions with integer numerators and nonzero integer denominators.

$$0.11 = \frac{11}{100} \qquad 127.279 = 127\frac{279}{1,000} = \frac{127,279}{1,000} \qquad -2.7 = -2\frac{7}{10} = \frac{-27}{10}$$

Examples of **repeating decimals** are 0.333. . . and 4.252525. . . . The three dots indicate that the digits continue in the pattern shown. Any repeating decimal can be expressed as a fraction with an integer numerator and a nonzero integer denominator. For example, 0.333. . . $= \frac{1}{3}$ and 4.252525. . . $= 4\frac{25}{99} = \frac{421}{99}$. Since every repeating decimal can be written as a fraction, repeating decimals are also rational numbers.

Rational numbers

> The set of **rational numbers** is the set of all terminating and all repeating decimals.

EXAMPLE 2

Terminating and repeating decimals. Change each fraction to a decimal and tell whether the decimal terminates or repeats: **a.** $\frac{4}{5}$ and   **b.** $\frac{17}{6}$.

Solution   **a.**   To change $\frac{4}{5}$ to a decimal, we divide the numerator by the denominator.

$$\begin{array}{r} .8 \\ 5\overline{)4.0} \\ \underline{4\,0} \\ 0 \end{array}$$ Write a decimal point and a 0 to the right of 4.

In decimal form, $\frac{4}{5}$ is 0.8. This is a terminating decimal.

b. To change $\frac{17}{6}$ to a decimal, we use a calculator to do the division and obtain 2.8333. . . . This is a repeating decimal, because the digit 3 repeats forever. This decimal can be written as $2.8\overline{3}$, where the overbar indicates that the 3 repeats.

SELF CHECK   Change each fraction to a decimal and tell whether it terminates or repeats: **a.** $\frac{25}{990}$ and   **b.** $\frac{47}{50}$.

Answers: **a.** $\frac{25}{990} = 0.0\overline{25}$, repeating decimal,   **b.** $\frac{47}{50} = 0.94$, terminating decimal   ∎

FIGURE 1-5

1 ft

⊢—— 2 ft ——⊣

Irrational Numbers

Some decimals cannot be written as fractions. As an example, let's consider the amount of concrete needed to anchor a basketball pole in the ground, as diagrammed in Figure 1-5. Using a formula from geometry, we will soon be able to show that the exact amount of concrete needed is π cubic feet. Expressed in decimal form,

$$\pi = 3.141592654. \ . \ .$$

This **nonterminating, nonrepeating decimal** cannot be written as a fraction with an integer for its numerator and its denominator. Therefore, π is not a rational number. It is an example of an **irrational number.**

Irrational numbers

> The set of **irrational numbers** is the set of all real numbers that are not rational. Irrational numbers are nonterminating, nonrepeating decimals.

ACCENT ON TECHNOLOGY *Approximating Irrational Numbers*

We can approximate the value of irrational numbers by using a scientific calculator. To find the approximate value of π, we simply press the $\boxed{\pi}$ key.

Keystrokes $\boxed{\pi}$ (you may have to use a $\boxed{\text{2nd}}$ or $\boxed{\text{Shift}}$ key first) $\boxed{\text{3.141592654}}$

We see that $\pi \approx 3.141592654$. (Read $\approx$ as "is approximately equal to.") If a problem calls for an approximate answer, we can round this decimal. To the nearest thousandth, $\pi = 3.142$.

Another irrational number is $\sqrt{2}$. To approximate $\sqrt{2}$, we enter 2 and press the $\boxed{\sqrt{\ }}$ key (the square root key).

Keystrokes   2 $\boxed{\sqrt{\ }}$ $\boxed{\text{1.414213562}}$

We see that $\sqrt{2} \approx 1.414213562$. To the nearest hundredth, $\sqrt{2} = 1.41$.

To find π and $\sqrt{2}$ with a graphing calculator, we enter these numbers and press these keys.

Keystrokes $\boxed{\text{2nd}}$ π $\boxed{\text{ENTER}}$

π

3.141592654

Keystrokes $\boxed{\text{2nd}}$ $\boxed{\sqrt{\ }}$ 2 $\boxed{)}$ $\boxed{\text{ENTER}}$

$\sqrt{\ }$ (2 $\boxed{\ }$)

1.414213562

Some other examples of irrational numbers are

$-\sqrt{7} = -2.64575131...$ This is a negative, nonterminating, nonrepeating decimal.

$2\pi = 6.283185307...$ We evaluate 2π using a calculator: $2\pi = 2 \cdot \pi$.

 WARNING! Students often *incorrectly* classify a number such as 4.12122122212222. . . as a repeating decimal. Although the decimal part of this number does exhibit a "pattern," no single block of digits repeats forever. Therefore, it is a nonterminating, nonrepeating decimal—an irrational number.

Real Numbers

Together, the set of rational numbers and the set of irrational numbers form the set of **real numbers.** This means that every real number can be written as either a terminating, a repeating, or a nonterminating, nonrepeating decimal. Thus, the set of real numbers is the set of all decimals.

The real numbers

> A **real number** is any number that is either a rational or an irrational number. All the points on a number line represent the set of real numbers.

Figure 1-6 shows how the sets of numbers discussed in this section are related. It also gives some examples of each type of number.

FIGURE 1-6

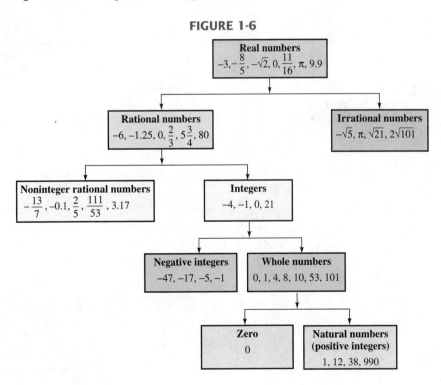

EXAMPLE 3

Classifying real numbers. Which numbers in the following set are natural numbers, whole numbers, integers, rational numbers, irrational numbers, and real numbers?

$$\left\{ \tfrac{5}{8}, -0.03, 45, -9, \sqrt{7}, 5\tfrac{2}{3}, 0, 1.727227222...\right\}$$

Solution

Natural numbers:	45
Whole numbers:	45, 0
Integers:	45, −9, 0
Rational numbers:	$\frac{5}{8}$, −0.03, 45, −9, $5\frac{2}{3}$, 0

Since $-0.03 = \frac{-3}{100}$, $45 = \frac{45}{1}$, $-9 = \frac{-9}{1}$, $5\frac{2}{3} = \frac{17}{3}$, and $0 = \frac{0}{1}$.

Irrational numbers:	$\sqrt{7}$, 1.727227222. . .
Real numbers:	$\frac{5}{8}$, −0.03, 45, −9, $\sqrt{7}$, $5\frac{2}{3}$, 0, 1.727227222. . .

SELF CHECK Use the instructions for Example 3 with the following set: $\left\{-\pi, -\frac{22}{7}, -5, 3.4, 1, \frac{16}{5}, 9.\overline{7}\right\}$.

Answers: natural numbers: 1; whole numbers: 1; integers: −5, 1; rational numbers: $-\frac{22}{7}$, −5, 3.4, 1, $\frac{16}{5}$, $9.\overline{7}$; irrational numbers: $-\pi$; real numbers: all ∎

The Real Number Line

We can illustrate real numbers using a **number line.** To each real number, there corresponds a point on the line. Furthermore, to each point on the line, there corresponds a number, called its **coordinate.** The point labeled 0 is called the **origin.**

Figure 1-7 shows the **graph** of −3 and 4. As we move from left to right on the number line, the coordinates of the points get larger. Since the graph of 4 is to the right of the graph of −3, we know that 4 is greater than −3 and that −3 is less than 4.

FIGURE 1-7

EXAMPLE 4 **Graphing on the number line.** Graph $-\frac{8}{3}$, −1.1, 0.565656. . . , $\frac{\pi}{2}$, $-\sqrt{15}$, and $2\sqrt{2}$.

Solution To help locate the graph of each number, we make some observations.

- Expressed as a mixed number, $-\frac{8}{3} = -2\frac{2}{3}$.
- Since −1.1 is less than −1, its graph is to the *left* of −1.
- 0.565656. . . ≈ 0.6
- From a calculator, $\frac{\pi}{2} \approx 1.6$.
- From a calculator, $-\sqrt{15} \approx -3.9$.
- $2\sqrt{2}$ means $2 \cdot \sqrt{2}$. From a calculator, $2\sqrt{2} \approx 2.8$.

SELF CHECK Graph π, $-2.\overline{1}$, $\sqrt{3}$, $\frac{11}{4}$, and −0.9. *Answers:*

∎

Inequality Symbols

To show that two quantities are not equal, we use one of the **inequality symbols** shown in Table 1-3.

TABLE 1-3

Symbol	Read as	Examples
$\neq$	"is not equal to"	$6 \neq 9$ and $0.33 \neq \frac{3}{5}$
$<$	"is less than"	$\frac{22}{3} < \frac{23}{3}$ and $-7 < -6$
$>$	"is greater than"	$19 > 5$ and $\frac{1}{2} > 0.3$
$\leq$	"is less than or equal to"	$3.5 \leq 3.\overline{5}$ and $1\frac{4}{5} \leq 1.8$
$\geq$	"is greater than or equal to"	$29 \geq 29$ and $-15.2 \geq -16.7$

It is always possible to write an inequality with the inequality symbol pointing in the opposite direction. For example,

$$-3 < 4 \qquad \text{is equivalent to} \qquad 4 > -3$$
$$5.3 \geq 2.9 \qquad \text{is equivalent to} \qquad 2.9 \leq 5.3$$

EXAMPLE 5

Inequality symbols. Use one of the symbols $>$ or $<$ to make each statement true: **a.** $-24 \quad \boxed{} \quad -25$ and **b.** $\frac{3}{4} \quad \boxed{} \quad 0.76$.

Solution

a. Since -24 is to the right of -25 on the number line, $-24 > -25$.

b. If we express the fraction $\frac{3}{4}$ as a decimal, we can easily compare it to 0.76. Since $\frac{3}{4} = 0.75$, $\frac{3}{4} < 0.76$.

SELF CHECK

Use one of the symbols $\geq$ or $\leq$ to make each statement true: **a.** $\frac{2}{3} \quad \boxed{} \quad \frac{4}{3}$ and **b.** $8\frac{1}{2} \quad \boxed{} \quad 8.4$.   *Answers:* **a.** $\leq$, **b.** $\geq$ ■

Opposites

In Figure 1-8, we can see that -3 and 3 are both a distance of 3 units away from zero on the number line. Because of this, we say that -3 and 3 are **opposites** or **additive inverses.**

Parentheses are used to express the opposite of a negative number. For example, the opposite of -3 is written as $-(-3)$. Since -3 and 3 are the same distance from zero, the opposite of -3 is 3. Symbolically, this can be written $-(-3) = 3$. This leads us to the following conclusion.

FIGURE 1-8

The double negative rule

If x is any real number, then $-(-x) = x$.

Absolute Value

The **absolute value** of any real number is the distance between the number and zero on a number line. To indicate absolute value, the number is inserted between two vertical bars. For example, the points shown in Figure 1-8 with coordinates of 3 and -3 both lie 3 units from zero. Thus, $|3| = |-3| = 3$.

The absolute value of a number can be defined more formally.

Absolute value

For any real number x, $\begin{cases} \text{If } x \geq 0, \text{ then } |x| = x. \\ \text{If } x < 0, \text{ then } |x| = -x. \end{cases}$

 WARNING! Since absolute value expresses distance, the absolute value of a number is always positive or zero, but never negative.

EXAMPLE 6

Absolute value. Find the value of each expression: **a.** $|34|$, **b.** $\left|-\frac{4}{5}\right|$, **c.** $|0|$, and **d.** $-|-1.8|$.

Solution **a.** Since 34 is a distance of 34 from 0 on a number line, $|34| = 34$.

b. $-\frac{4}{5}$ is a distance of $\frac{4}{5}$ from 0 on a number line. Therefore, $\left|-\frac{4}{5}\right| = \frac{4}{5}$.

c. $|0| = 0$

d. $-|-1.8| = -(1.8)$ Find $|-1.8|$ first: $|-1.8| = 1.8$.

$\qquad\qquad = -1.8$ Rewrite without parentheses.

SELF CHECK Find the value of each expression: **a.** $|-9.6|$, *Answers:* **a.** 9.6, **b.** -12, **b.** $-|-12|$, and **c.** $\left|\frac{3}{2}\right|$. **c.** $\frac{3}{2}$ ■

STUDY SET

Section 1.2

VOCABULARY

In Exercises 1–8, fill in the blanks to make the statements true.

1. A _____ number is any number that can be written as a fraction with an integer numerator and a nonzero integer denominator.

2. The _____ numbers are the natural numbers greater than 1 that are divisible only by themselves and 1.

3. The _____ of any real number is the distance between the number and zero on a number line.

4. Together, the set of rational numbers and the set of irrational numbers form the set of _____ numbers.

5. _____ numbers are nonterminating, nonrepeating decimals.

6. _____ numbers are greater than 0, and _____ numbers are less than 0.

7. A _____ number is a natural number greater than 1 that is not prime.

8. _____ is neither positive nor negative.

CONCEPTS

In Exercises 9–20, list the elements in the set $\left\{-3, -\frac{8}{5}, 0, \frac{2}{3}, 1, 2, \sqrt{3}, \pi, 4.75, 9, 16.\overline{6}\right\}$ that satisfy the given condition.

9. Natural number

10. Whole number

11. Integer

12. Rational number

13. Irrational number

14. Real number

15. Even natural number

16. Odd integer

17. Prime number

18. Composite number

19. Odd composite number

20. Odd prime number

In Exercises 21–24, tell whether each number is a repeating or a nonrepeating decimal, and whether it is a rational or an irrational number.

21. 0.090090009. . . .

22. $0.\overline{09}$

23. 5.41414141. . .

24. 1.414213562. . .

25. Show that each of the following numbers is a rational number by expressing it as a fraction with an integer numerator and a nonzero integer denominator.

$7, \; -7\frac{3}{5}, \; 0.007, \; 700.1$

26. Tell whether each statement is true or false.
   a. All prime numbers are odd numbers.
   b. $6 \geq 6$
   c. 0 is neither even nor odd.
   d. If x is negative, $|x| = -x$.

27. Find x if $|x| = 3.5$.

28. Name two numbers that are 6 units away from -2 on the number line.

29. The diagram in Illustration 1 can be used to show how the natural numbers, whole numbers, integers, rational numbers, and irrational numbers make up the set of real numbers. If the natural numbers can be represented as shown, label each of the other sets.

ILLUSTRATION 1

Real numbers

Natural numbers

30. ▦ The formula $C = \pi D$ gives the circumference C of a circle, where D is the length of its diameter. Find the circumference of the man's wedding ring shown in Illustration 2. Give an *exact* answer and then an *approximate* answer, rounded to the nearest hundredth of an inch.

ILLUSTRATION 2

1 in.

NOTATION

In Exercises 31–34, fill in the blanks to make the statements true.

31. The symbol $<$ means "_____."

32. $|-2|$ is read as "the _____ value ___ -2."

33. The symbols { } are called _____.

34. The symbol $\geq$ means "_____."

▦ PRACTICE

In Exercises 35–38, change each fraction into a decimal and classify the result as a terminating or a repeating decimal.

35. $\dfrac{7}{8}$

36. $\dfrac{8}{3}$

37. $-\dfrac{11}{15}$

38. $-\dfrac{19}{16}$

In Exercises 39–48, graph each set on a number line.

39. $\left\{ -\dfrac{5}{2}, -0.1, 2.142765. \ldots, \dfrac{\pi}{3}, -\sqrt{11}, 2\sqrt{3} \right\}$

40. $\left\{ 2\dfrac{1}{9}, -3.821134. \ldots, -\dfrac{\pi}{2}, \sqrt{15}, -0.9, \dfrac{\sqrt{2}}{2} \right\}$

41. $\left\{ 3.\overline{15}, \dfrac{22}{7}, 3\dfrac{1}{8}, \pi, \sqrt{10}, 3.1 \right\}$

42. $\left\{ -0.\overline{331}, -0.331, -\dfrac{1}{3}, -\sqrt{0.11} \right\}$

43. The set of prime numbers less than 8

44. The set of integers between -7 and 0

-7 -6 -5 -4 -3 -2 -1   0

45. The set of odd integers between 10 and 18

10 11 12 13 14 15 16 17 18

46. The set of composite numbers less than 10

0   1   2   3   4   5   6   7   8   9   10

47. The set of positive odd integers less than 12

0   1   2   3   4   5   6   7   8   9   10 11 12

48. The set of negative even integers greater than -7

-7 -6 -5 -4 -3 -2 -1   0

In Exercises 49–56, insert either a $<$ or a $>$ symbol to make a true statement.

49. 8 ▨ 9   **50.** 9 ▨ 0

51. $-(-5)$ ▨ -10 **52.** $|-3|$ ▨ $-(-6)$

53. -7.999 ▨ -7.1 **54.** $4\frac{1}{2}$ ▨ $\frac{7}{2}$

55. $6.\overline{1}$ ▨ 6 **56.** -6.07 ▨ $-\frac{17}{6}$

In Exercises 57–60, write each statement with the inequality symbol pointing in the opposite direction.

57. $19 > 12$ **58.** $-3 \geq -5$

59. $-6 \leq -5$ **60.** $-10 < 0$

In Exercises 61–68, write each expression without using absolute value symbols.

61. $|20|$ **62.** $|-20|$

63. $-|-6|$ **64.** $-|-8|$

65. $|-5.9|$ **66.** $-|1.\overline{27}|$

67. $\left|\dfrac{5}{4}\right|$ **68.** $\left|-\dfrac{5}{16}\right|$

APPLICATIONS

69. 🖩 DRAFTING Express each dimension in the drawing of a bracket in Illustration 3 as a four-place decimal.

70. pH SCALE The pH scale, shown in Illustration 4, is used to measure the strength of acids and bases (alkalines) in chemistry. It can be thought of as a number line. On the scale, graph and label each pH measurement given in the table.

ILLUSTRATION 3

ILLUSTRATION 4

Solution	pH
Seawater	8.5
Cola	2.9
Battery acid	1.0
Milk	6.6
Blood	7.4
Ammonia	11.9
Saliva	6.1
Oven cleaner	13.2
Black coffee	5.0
Toothpaste	9.9
Tomato juice	4.1

WRITING

71. Explain why the whole numbers are a subset of the integers.

72. What is a real number? Give examples.

73. Explain why there are no even prime numbers greater than 2.

74. Explain why every integer is a rational number, but not every rational number is an integer.

REVIEW

75. Is $\frac{3x-4}{2}$ an equation or an expression?

76. Translate into mathematical symbols: The weight of an object in ounces is 16 times its weight in pounds.

In Exercises 77–78, complete the table.

77. $T = x - 1.5$

x	T
3.7	
10	
30.6	

78. $j = 3m$

m	j
0	
15	
300	

▶ 1.3   Operations with Real Numbers

In this section, you will learn about

Rules for adding real numbers ■ A rule for subtracting real numbers ■ Rules for multiplying real numbers ■ Rules for dividing real numbers ■ Raising a real number to a power ■ Finding a square root of a real number ■ Order of operations ■ Evaluating algebraic expressions

Introduction   Six operations can be performed with real numbers: addition, subtraction, multiplication, division, raising to a power, and finding a root. In this section, we will review the rules for performing each of these operations. We will also discuss the procedure used to evaluate numerical expressions involving several operations.

Rules for Adding Real Numbers

When two numbers are added, we call the result their **sum.** To find the sum of 2 and 3, we can represent the numbers with arrows, as shown in Figure 1-9(a). Since the endpoint of the second arrow is at 5, we have $2 + 3 = 5$.

To add -2 and -3, we can draw arrows as shown in Figure 1-9(b). Since the endpoint of the second arrow is at -5, we have $-2 + (-3) = -5$.

FIGURE 1-9

(a)   (b)

To add -6 and 2, we can draw arrows as shown in Figure 1-10(a). Since the endpoint of the second arrow is at -4, we have $-6 + 2 = -4$.

To add 7 and -4, we can draw arrows as shown in Figure 1-10(b). Since the endpoint of the second arrow is at 3, we have $7 + (-4) = 3$.

FIGURE 1-10

(a)   (b)

These examples suggest the following rules.

Adding real numbers

With like signs: Add the absolute values of the numbers and keep the common sign.

With unlike signs: Subtract the absolute values of the numbers (the smaller from the larger) and keep the sign of the number with the larger absolute value.

E X A M P L E   1

Adding real numbers. Find each sum: **a.** $-5 + (-3)$,   **b.** $8.9 + (-5.1)$, and   **c.** $-\frac{13}{15} + \frac{3}{5}$.

Solution   **a.** $-5 + (-3) = -8$ Because the numbers have like signs, add their absolute values, 5 and 3, to get 8. Keep the common negative sign.

b. $8.9 + (-5.1) = 3.8$ Since the numbers have unlike signs, subtract their absolute values, 5.1 from 8.9, to get 3.8. Because 8.9 has the larger absolute value, the answer is positive.

c. $-\dfrac{13}{15} + \dfrac{3}{5} = -\dfrac{13}{15} + \dfrac{9}{15}$ Express $\frac{3}{5}$ in terms of the lowest common denominator, 15: $\frac{3}{5} = \frac{3 \cdot 3}{5 \cdot 3} = \frac{9}{15}$.

$= -\dfrac{4}{15}$ Subtract the absolute values, $\frac{9}{15}$ from $\frac{13}{15}$, to get $\frac{4}{15}$. Keep the $-$ sign, because $-\frac{13}{15}$ has the larger absolute value.

SELF CHECK   Add: **a.** $-34 + 25$,   **b.** $-70.4 + (-21.2)$, and   **c.** $\frac{7}{4} + \left(-\frac{3}{2}\right)$.   *Answers:*   **a.** -9,   **b.** -91.6,   **c.** $\frac{1}{4}$ ∎

A Rule for Subtracting Real Numbers

When two numbers are subtracted, we call the result their **difference.** To find a difference, we can change the subtraction into an equivalent addition. For example, the subtraction $7 - 4$ is equivalent to the addition $7 + (-4)$, because they have the same answer:

$$7 - 4 = 3 \quad \text{and} \quad 7 + (-4) = 3$$

This suggests that to subtract two numbers, we can change the sign of the number being subtracted and add. In other words, subtraction is the same as adding the *opposite* of the number being subtracted.

Subtracting real numbers

If x and y are real numbers, then $x - y = x + (-y)$.

| EXAMPLE 2 | **Subtracting real numbers.** Find each difference: **a.** $12 - 4$, **b.** $-1.3 - 5.5$, and **c.** $-\frac{14}{3} - \left(-\frac{7}{3}\right)$. |

Solution In each case, we add the opposite of the number that is being subtracted.

a. $12 - 4 = 12 + (-4)$ Here, 4 is being subtracted, so we change the sign of 4 and add.

$= 8$ Use the rule for adding two numbers with unlike signs.

b. $-1.3 - 5.5 = -1.3 + (-5.5)$ Change the sign of 5.5 and add.

$= -6.8$ Use the rule for adding two numbers with like signs.

c. $-\dfrac{14}{3} - \left(-\dfrac{7}{3}\right) = -\dfrac{14}{3} + \dfrac{7}{3}$ Change the sign of $-\frac{7}{3}$ and add.

$= -\dfrac{7}{3}$ Use the rule for adding two numbers with unlike signs.

| SELF CHECK | Subtract: **a.** $-15 - 4$, **b.** $-12.1 - (-7.6)$, and **c.** $\frac{5}{9} - \frac{7}{9}$. | *Answers:* **a.** -19, **b.** -4.5, **c.** $-\frac{2}{9}$ ■ |

Rules for Multiplying Real Numbers

When two numbers are multiplied, we call the numbers **factors** and the result their **product.** We can find the product of 5 and 4 by using 4 in an addition five times:

$$5(4) = 4 + 4 + 4 + 4 + 4 = 20$$

We can find the product of 5 and -4 by using -4 in an addition five times.

$$5(-4) = (-4) + (-4) + (-4) + (-4) + (-4) = -20$$

Since multiplication by a negative number can be defined as repeated subtraction, we can find the product of -5 and 4 by using 4 in a subtraction five times:

$$-5(4) = -4 - 4 - 4 - 4 - 4$$
$$= -4 + (-4) + (-4) + (-4) + (-4) \quad \text{Add the opposite of each 4.}$$
$$= -20 \quad \text{Do the additions.}$$

We can find the product of -5 and -4 by using -4 in a subtraction five times:

$$-5(-4) = -(-4) - (-4) - (-4) - (-4) - (-4)$$
$$= 4 + 4 + 4 + 4 + 4 \quad \text{Add the opposite of each } -4.$$
$$= 20 \quad \text{Do the additions.}$$

The products $5(4)$ and $-5(-4)$ both equal 20, and the products $5(-4)$ and $-5(4)$ both equal -20. These results suggest the following rules for multiplying real numbers.

Multiplying real numbers

With like signs: Multiply their absolute values. The answer is positive.

With unlike signs: Multiply their absolute values. Make the answer negative.

EXAMPLE 3

Multiplying real numbers. Find each product: **a.** $4(-7)$, **b.** $-5.2(-3)$, and **c.** $-\frac{7}{9}\left(\frac{3}{16}\right)$.

Solution   **a.** $4(-7) = -28$

Multiply the absolute values of 4 and -7: $4 \cdot 7 = 28$. Since the signs are unlike, make the answer negative.

b. $-5.2(-3) = 15.6$

Multiply the absolute values: $5.2 \cdot 3 = 15.6$. Since the signs are like, the answer is positive.

c. $-\frac{7}{9}\left(\frac{3}{16}\right) = -\frac{7 \cdot 3}{9 \cdot 16}$

Multiply the numerators and multiply the denominators. Since the signs of the factors are unlike, the result is negative.

$$= -\frac{7 \cdot \overset{1}{\cancel{3}}}{\cancel{3} \cdot 3 \cdot 16}$$

Factor 9 as $3 \cdot 3$ and divide out the common factor of 3.

$$= -\frac{7}{48}$$

Multiply in the numerator and denominator.

SELF CHECK   Multiply   **a.** $(-6)(5)$,   **b.** $(-4.1)(-8)$,   and   **c.** $\left(\frac{4}{3}\right)\left(-\frac{1}{8}\right)$.

Answers: **a.** -30,   **b.** 32.8, **c.** $-\frac{1}{6}$ ∎

Rules for Dividing Real Numbers

When two numbers are divided, we call the result their **quotient.** In the division $\frac{x}{y} = q$ ($y \neq 0$), the quotient q is a number such that $y \cdot q = x$. We can use this relationship to find rules for dividing real numbers. We consider four divisions:

$$\frac{10}{2} = 5, \text{ because } 2(5) = 10 \qquad \frac{-10}{-2} = 5, \text{ because } -2(5) = -10$$

$$\frac{-10}{2} = -5, \text{ because } 2(-5) = -10 \qquad \frac{10}{-2} = -5, \text{ because } -2(-5) = 10$$

These results suggest the following rules for dividing real numbers.

Dividing real numbers

With like signs: Divide their absolute values. The answer is positive.

With unlike signs: Divide their absolute values. Make the answer negative.

EXAMPLE 4

Dividing real numbers. Find each quotient: **a.** $\frac{-44}{11}$ and   **b.** $\frac{-2.7}{-9}$.

Solution   **a.** $\frac{-44}{11} = -4$

Divide the absolute values of -44 and 11: $\frac{44}{11} = 4$. Since the signs are unlike, make the answer negative.

b. $\frac{-2.7}{-9} = 0.3$

Divide the absolute values: $\frac{2.7}{9} = 0.3$. Since the signs are like, the result is positive.

SELF CHECK   Divide   **a.** $\frac{55}{-5}$ and   **b.** $\frac{-7.2}{-6}$.

Answers: **a.** -11,   **b.** 1.2 ∎

To divide two fractions, we multiply the first fraction by the **reciprocal** of the second fraction. In symbols, if a, b, c, and d are real numbers, then

$$\frac{a}{b} \div \frac{c}{d} = \frac{a}{b} \cdot \frac{d}{c} \quad (b \neq 0, c \neq 0, \text{ and } d \neq 0) \quad \text{$\frac{d}{c}$ is the reciprocal of $\frac{c}{d}$.}$$

EXAMPLE 5

Dividing fractions. Find each quotient: **a.** $\frac{2}{3} \div \left(-\frac{3}{5}\right)$ and **b.** $-\frac{1}{2} \div (-6)$.

Solution **a.** $\dfrac{2}{3} \div \left(-\dfrac{3}{5}\right) = \dfrac{2}{3} \cdot \left(-\dfrac{5}{3}\right)$ Multiply by the reciprocal of $-\frac{3}{5}$, which is $-\frac{5}{3}$.

$= -\dfrac{10}{9}$ The factors have unlike signs, so the answer is negative.

b. $-\dfrac{1}{2} \div (-6) = -\dfrac{1}{2} \cdot \left(-\dfrac{1}{6}\right)$ Multiply by the reciprocal of -6, which is $-\frac{1}{6}$.

$= \dfrac{1}{12}$ The factors have like signs, so the answer is positive.

SELF CHECK Divide **a.** $-\frac{7}{8} \div \frac{2}{3}$ and **b.** $-\frac{1}{10} \div (-5)$. *Answers:* **a.** $-\frac{21}{16}$, **b.** $\frac{1}{50}$ ∎

WARNING! Students often confuse two similar-looking division problems. For example, consider $\frac{0}{4}$ and $\frac{4}{0}$. We know that $\frac{0}{4} = 0$, because $4 \cdot 0 = 0$. On the other hand, $\frac{4}{0}$ is undefined, because there is no real number q such that $0 \cdot q = 4$. In general, if $x \neq 0$, $\frac{0}{x} = 0$ and $\frac{x}{0}$ is undefined.

Raising a Real Number to a Power

Exponents indicate repeated multiplication. For example,

$3^2 = 3 \cdot 3$ Read 3^2 as "3 to the second power" or "3 squared."

$(-9.1)^3 = (-9.1)(-9.1)(-9.1)$ Read $(-9.1)^3$ as "-9.1 to the third power" or "-9.1 cubed."

$\left(\dfrac{2}{3}\right)^4 = \left(\dfrac{2}{3}\right)\left(\dfrac{2}{3}\right)\left(\dfrac{2}{3}\right)\left(\dfrac{2}{3}\right)$ Read $\left(\frac{2}{3}\right)^4$ as "$\frac{2}{3}$ to the fourth power."

These examples suggest the following definition.

Natural-number exponents

If n is a natural number, then

$$x^n = \overbrace{x \cdot x \cdot x \cdot \cdots \cdot x}^{n \text{ factors of } x}$$

The exponential expression x^n is called a **power of x**, and we read it as "x to the nth power." In this expression, x is called the **base**, and n is called the **exponent**. A natural-number exponent tells how many times the base of an exponential expression is to be used as a factor in a product.

Base $\longrightarrow x^n \longleftarrow$ Exponent

EXAMPLE 6

Evaluating exponential expressions. Find each power: **a.** $(-2)^5$, **b.** $\left(\frac{3}{4}\right)^2$, and **c.** -0.1 cubed.

Solution   In each case, we use the fact that an exponent tells how many times the base is to be used as a factor in a product.

a. $(-2)^5 = (-2)(-2)(-2)(-2)(-2) = -32$ The base is -2. The exponent is 5.

b. $\left(\frac{3}{4}\right)^2 = \frac{3}{4}\left(\frac{3}{4}\right) = \frac{9}{16}$ The base is $\frac{3}{4}$. The exponent is 2.

c. -0.1 cubed means $(-0.1)^3$.   The base is -0.1. The exponent is 3.

$$(-0.1)^3 = (-0.1)(-0.1)(-0.1) = -0.001$$

SELF CHECK   Find each power: **a.** $(-3)^3$, **b.** $(0.8)^2$, **c.** 2^4, and **d.** $\frac{7}{5}$ squared.   *Answers:* **a.** -27, **b.** 0.64, **c.** 16, **d.** $\frac{49}{25}$ ∎

ACCENT ON TECHNOLOGY *The Squaring and Exponential Keys*

A homeowner plans to install the cooking island shown in Figure 1-11. To find the number of square feet of kitchen floor space that will be lost, we substitute 3.25 for s in the formula for the area of a square, $A = s^2$. Using the squaring key $\boxed{x^2}$ on a scientific calculator, we can evaluate $(3.25)^2$ by entering these numbers and pressing these keys.

FIGURE 1-11

3.25 ft

3.25 ft

3.25 ft

Keystrokes   3.25 $\boxed{x^2}$ | 10.5625 |

On a graphing calculator, we enter 3.25 and press these keys.

Keystrokes   3.25 $\boxed{x^2}$ $\boxed{\text{ENTER}}$ | 3.25² |
| 10.5625 |

About 10.6 square feet of floor space will be lost.

The number of cubic feet of storage space that the cooking island will add can be found by substituting 3.25 for s in the formula for the volume of a cube, $V = s^3$. Using the exponential key $\boxed{y^x}$ ($\boxed{x^y}$ on some calculators), we can evaluate $(3.25)^3$ on a scientific calculator by entering these numbers and pressing these keys.

Keystrokes   3.25 $\boxed{y^x}$ 3 $\boxed{=}$ | 34.328125 |

On a graphing calculator, we enter 3.25 and press these keys.

Keystrokes   3.25 $\boxed{\wedge}$ 3 $\boxed{\text{ENTER}}$ | 3.25^3 |
| 34.328125 |

The island will add about 34.3 cubic feet of storage space.

WARNING! Although the expressions $(-3)^2$ and -3^2 look alike, they are not. In $(-3)^2$, the base is -3. In -3^2, the base is 3. The $-$ sign in front of 3^2 means the opposite of 3^2. When we evaluate them, we see that the results are different:

$$(-3)^2 = (-3)(-3) \qquad\qquad -3^2 = -(3 \cdot 3)$$
$$= 9 \qquad\qquad\qquad\qquad = -9$$

ACCENT ON TECHNOLOGY *The Parentheses and Negative Keys*

To compute $(-3)^2$ using a scientific calculator, use the left and right *parentheses* keys $(\ )$ and the *negative* key $+/-$. Notice that the negative key is different from the subtraction key $-$. To enter -3, press $+/-$ *after* entering 3.

Keystrokes $(\ )$ 3 $+/-$ $)$ x^2 $=$ | $\quad\quad\quad\quad\quad\quad\quad\quad\quad\quad$ 9 |

If a graphing calculator is used to find $(-3)^2$, press the negative key $(-)$ *before* entering 3.

Keystrokes $(\ )$ $(-)$ 3 $)$ x^2 ENTER | $(-3)^2$ $\quad\quad\quad\quad\quad$ 9 |

To compute -3^2 using a scientific calculator, think of the expression as $-1 \cdot 3^2$. First, find 3^2. Then press $(-)$, which is equivalent to multiplying 3^2 by -1.

Keystrokes 3 x^2 $+/-$ | $\quad\quad\quad\quad\quad\quad\quad\quad\quad$ -9 |

A graphing calculator recognizes -3^2 as $-1 \cdot 3^2$, so we can find -3^2 by entering the following sequence:

Keystrokes $(-)$ 3 x^2 ENTER | -3^2 $\quad\quad\quad\quad\quad$ -9 |

Finding a Square Root of a Real Number

Since the product $3 \cdot 3$ can be denoted by the exponential expression 3^2, we say that 3 is squared. The opposite of squaring a number is called finding its **square root.**

All positive numbers have two square roots: one that is positive and one that is negative. For example, the two square roots of 9 are 3 and -3. The number 3 is a square root of 9, because $3^2 = 9$, and -3 is a square root of 9, because $(-3)^2 = 9$.

The symbol $\sqrt{}$, called a **radical sign,** is used to represent the positive (or *principal*) square root of a number.

Principal square root

A number b is a square root of a if $b^2 = a$.

If $a > 0$, the expression $\sqrt{a}$ represents the **principal** (or positive) **square root** of a. The principal square root of 0 is 0: $\sqrt{0} = 0$.

The principal square root of a positive number is always positive. Although 3 and -3 are both square roots of 9, only 3 is the principal square root. The symbol $\sqrt{9}$ represents 3. To represent -3, we place a $-$ sign in front of the radical:

$$\sqrt{9} = 3 \quad \text{and} \quad -\sqrt{9} = -3$$

EXAMPLE 7

Finding square roots. Evaluate **a.** $\sqrt{121}$, **b.** $-\sqrt{49}$, **c.** $\sqrt{\frac{1}{4}}$, and **d.** $\sqrt{0.09}$.

Solution

a. $\sqrt{121} = 11$, because $11^2 = 121$.

b. Since $\sqrt{49} = 7$, $-\sqrt{49} = -7$.

c. $\sqrt{\frac{1}{4}} = \frac{1}{2}$, because $\left(\frac{1}{2}\right)^2 = \frac{1}{4}$.

d. $\sqrt{0.09} = 0.3$, because $(0.3)^2 = 0.09$.

SELF CHECK

Find each square root: **a.** $\sqrt{64}$, **b.** $-\sqrt{100}$, **c.** $\sqrt{\frac{4}{25}}$, **d.** $\sqrt{1}$, **e.** $\sqrt{0.81}$, and **f.** $-\sqrt{400}$.

Answers: **a.** 8, **b.** -10, **c.** $\frac{2}{5}$, **d.** 1, **e.** 0.9, **f.** -20

■

Order of Operations

Quite often, we will have to evaluate expressions involving more than one operation. For example, consider the expression $3 + 2 \cdot 5$. We can evaluate it in two different ways. We can do the addition first and then do the multiplication. Or we can do the multiplication first and then do the addition. However, we get different results.

Method 1: Do the addition first

$3 + \mathbf{2 \cdot 5} = \mathbf{5 \cdot 5}$ Add 3 and 2 first.

$\qquad\quad = 25$ Do the multiplication.

Method 2: Do the multiplication first

$3 + \mathbf{2 \cdot 5} = 3 + \mathbf{10}$ Multiply 2 and 5 first.

$\qquad\quad = 13$ Do the addition.

└─────── Different results ───────┘

This example shows that we need to establish an order of operations. Otherwise, the same expression can have two different values. To guarantee that calculations will have one correct result, we will use the following set of priority rules.

Rules for the order of operations

Do all calculations within each pair of grouping symbols (parentheses, brackets, or braces), working from the innermost pair to the outermost pair.

1. Evaluate all powers and roots.
2. Do all multiplications and divisions, working from left to right.
3. Do all additions and subtractions, working from left to right.

When all grouping symbols have been removed, repeat the steps above to finish the calculation.

In a fraction, simplify the numerator and the denominator separately. Then simplify the fraction, whenever possible.

To evaluate $3 + 2 \cdot 5$ correctly, we must apply the rules for the order of operations. Since the expression does not contain grouping symbols, we follow steps 1, 2, and 3 of the previous list. The expression does not contain any powers or roots, so we do the multiplication first, followed by the addition.

$3 + \mathbf{2 \cdot 5} = 3 + \mathbf{10}$ Ignore the addition for now and do the multiplication of 2 and 5.

$\qquad\quad = 13$ Next, do the addition.

Using the rules for the order of operations, we see that the correct answer is 13.

EXAMPLE 8

Using the rules for the order of operations. Evaluate each expression:
a. $-5 + 4(-3)^2$ and **b.** $-\frac{10}{5} - 5(3) + 6$.

Solution **a.** Although the expression contains parentheses, there are no operations to perform within the parentheses. So we proceed with steps 1, 2, and 3 of the rules for the order of operations.

$$
\begin{aligned}
-5 + 4(-3)^2 &= -5 + 4(9) && \text{First, evaluate the power: } (-3)^2 = 9. \\
&= -5 + 36 && \text{Do the multiplication: } 4(9) = 36. \\
&= 31 && \text{Do the addition.}
\end{aligned}
$$

b. Since the expression does not contain any powers, we do the multiplications and divisions, working from left to right.

$$
\begin{aligned}
\frac{-10}{5} - 5(3) + 6 &= -2 - 5(3) + 6 && \text{Do the division: } \frac{-10}{5} = -2. \\
&= -2 - 15 + 6 && \text{Do the multiplication: } 5(3) = 15. \\
&= -17 + 6 && \text{Working from left to right, do the subtraction: } -2 - 15 = -17. \\
&= -11 && \text{Do the addition.}
\end{aligned}
$$

SELF CHECK Evaluate each expression: **a.** $-9 + 2(-4)^2$
and **b.** $\frac{20}{-5} - (-6)(-5) + (-12)$. *Answers:* **a.** 23, **b.** -46 ■

EXAMPLE 9

Order of operations within grouping symbols. Evaluate each expression: **a.** $3 - (4 - 8)^2$ and **b.** $2 + 3[-2 - 8(4 - 3^2)]$.

Solution **a.** We begin by doing the operation within the parentheses.

$$
\begin{aligned}
3 - (4 - 8)^2 &= 3 - (-4)^2 && \text{Do the subtraction: } 4 - 8 = -4. \\
&= 3 - 16 && \text{Evaluate the power: } (-4)^2 = 16. \\
&= -13 && \text{Do the subtraction.}
\end{aligned}
$$

b. First, we do the work within the innermost grouping symbols (the parentheses), using the rules for the order of operations.

$$
\begin{aligned}
2 + 3[-2 - 8(4 - 3^2)] &= 2 + 3[-2 - 8(4 - 9)] && \text{Find the power: } 3^2 = 9. \\
&= 2 + 3[-2 - 8(-5)] && \text{Do the subtraction: } 4 - 9 = -5.
\end{aligned}
$$

Next, we work within the brackets.

$$
\begin{aligned}
&= 2 + 3[-2 - (-40)] && \text{Do the multiplication: } 8(-5) = -40. \\
&= 2 + 3[-2 + 40] && \text{Write the subtraction as addition of the opposite.} \\
&= 2 + 3[38] && \text{Do the addition: } -2 + 40 = 38. \\
&= 2 + 114 && \text{Do the multiplication: } 3[38] = 114. \\
&= 116 && \text{Do the addition.}
\end{aligned}
$$

SELF CHECK Evaluate each expression: **a.** $(5 - 3)^3 - 40$
and **b.** $-3[-2(5^3 - 3) + 4] - 1$. *Answers:* **a.** -32, **b.** 719 ■

| EXAMPLE 10 | **Working with absolute value.** Evaluate $|-45 + 30|(2 - 7)$. |

Solution Since the absolute value bars are grouping symbols, we do the operations within the absolute value bars and the parentheses first.

$$|-45 + 30|(2 - 7) = |-15|(-5) \quad \text{Do the addition within the absolute value bars and the subtraction within the parentheses.}$$

$$= 15(-5) \quad \text{Find the absolute value: } |-15| = 15.$$

$$= -75 \quad \text{Do the multiplication.}$$

SELF CHECK Evaluate $2|-25 - (-6)(3)|$. *Answer:* 14 ∎

ACCENT ON TECHNOLOGY *Order of Operations*

Scientific and graphing calculators are programmed to follow the rules for the order of operations. For example, when computing $3 + 2 \cdot 5$, both types of calculators give the correct answer, 13.

Keystrokes 3 $+$ 2 $\times$ 5 $=$

    | 13 |

Keystrokes 3 $+$ 2 $\times$ 5 ENTER

    | 3+2*5 |
    | 13 |

Both types of calculators use left and right parentheses keys $($ $)$ when grouping symbols are needed. To evaluate $3 - (4 - 8)^2$, we enter these numbers and press these keys.

Keystrokes 3 $-$ $($ 4 $-$ 8 $)$ x^2 $=$

    | -13 |

Keystrokes 3 $-$ $($ 4 $-$ 8 $)$ x^2 ENTER

    | 3-(4-8)² |
    | -13 |

Both types of calculators require that we group the terms in the numerator together and the terms in the denominator together when calculating the value of an expression such as $\frac{200 + 120}{20 - 16}$.

Keystrokes $($ 200 $+$ 120 $)$ $\div$ $($ 20 $-$ 16 $)$ $=$

    | 80 |

Keystrokes $($ 200 $+$ 120 $)$ $\div$ $($ 20 $-$ 16 $)$ ENTER

    | (200 + 120)/(20-16 |
    |) |
    | 80 |

If parentheses aren't used when finding $\frac{200 + 120}{20 - 16}$, you will obtain an incorrect result of 190. That is because the calculator will interpret the entry as $200 + \frac{120}{20} - 16$.

Evaluating Algebraic Expressions

Recall that an algebraic expression is a combination of variables and numbers with the operations of arithmetic. To evaluate an algebraic expression, we substitute specific numbers for the variables and then apply the rules for the order of operations.

| EXAMPLE 11 | **Evaluating algebraic expressions.** If $a = -2$, $b = 9$, and $c = -1$, evaluate
a. $-\frac{1}{2}a^2$ and **b.** $\dfrac{-a\sqrt{b} + 3c^3}{c(c - b)}$. |

Solution **a.** We substitute -2 for a and use the rules for the order of operations.

$$-\frac{1}{2}a^2 = -\frac{1}{2}(-2)^2 \qquad \text{Substitute } -2 \text{ for } a. \text{ Write parentheses around } -2 \text{ so that it is squared.}$$

$$= -\frac{1}{2}(4) \qquad \text{Evaluate the power: } (-2)^2 = 4.$$

$$= -2 \qquad \text{Do the multiplication.}$$

b. $\dfrac{-a\sqrt{b} + 3c^3}{c(c-b)} = \dfrac{-(-2)\sqrt{9} + 3(-1)^3}{-1(-1-9)}$ Substitute -2 for a, 9 for b, and -1 for c.

$$= \frac{-(-2)(3) + 3(-1)}{-1(-10)} \qquad \begin{array}{l}\text{In the numerator, evaluate the square root} \\ \text{and the power: } \sqrt{9} = 3 \text{ and } (-1)^3 = -1. \text{ In} \\ \text{the denominator, do the subtraction.}\end{array}$$

$$= \frac{2(3) + 3(-1)}{-1(-10)} \qquad \text{In the numerator, simplify: } -(-2) = 2.$$

$$= \frac{6 + (-3)}{10} \qquad \begin{array}{l}\text{In the numerator, do the multiplications. In} \\ \text{the denominator, do the multiplication.}\end{array}$$

$$= \frac{3}{10} \qquad \text{In the numerator, do the addition.}$$

SELF CHECK   If $r = 2$, $s = -5$, and $t = 3$, evaluate

    a. $-\dfrac{1}{3}s^3t$ and **b.** $\dfrac{\sqrt{-5s}}{(s+t)r^2}$.

                                        Answers: **a.** 125, **b.** $-\frac{5}{8}$ ∎

ACCENT ON TECHNOLOGY *Evaluating Algebraic Expressions*

Graphing calculators can evaluate algebraic expressions. To evaluate

$$\frac{-a\sqrt{b} + 3c^3}{c(c-b)}$$

(Example 11, part b) using a TI-83, we first enter the values of $a = -2$, $b = 9$, and $c = -1$, using the store key $\boxed{\text{STO}}$ and the $\boxed{\text{ALPHA}}$ key. See Figure 1-12(a).

Keystrokes $\boxed{(-)}$ 2 $\boxed{\text{STO}}$ $\boxed{\text{ALPHA}}$ $\boxed{\text{A}}$ $\boxed{\text{ALPHA}}$ $\boxed{:}$ This enters $a = -2$.

              9 $\boxed{\text{STO}}$ $\boxed{\text{ALPHA}}$ $\boxed{\text{B}}$ $\boxed{\text{ALPHA}}$ $\boxed{:}$ This enters $b = 9$.

              $\boxed{(-)}$ 1 $\boxed{\text{STO}}$ $\boxed{\text{ALPHA}}$ $\boxed{\text{C}}$ $\boxed{\text{ALPHA}}$ $\boxed{:}$ This enters $c = -1$.

Next, enter the expression as shown in Figure 1-12(b) and press $\boxed{\text{ENTER}}$ to find that the value of the expression is 0.3. To express the result as a fraction, press $\boxed{\text{MATH}}$, highlight Frac, and then press $\boxed{\text{ENTER}}$ $\boxed{\text{ENTER}}$. See Figure 1-12(c).

FIGURE 1-12

          (a)   **(b)**   **(c)**

For other models of graphing calculators, consult your owner's manual for instructions on storing the value of a variable.

EXAMPLE 12

Surface area. Find the amount of material needed to make the lampshade shown in Figure 1-13 by finding the surface area of the shade using the formula

$$S = \pi(R_1 + R_2)\sqrt{(R_1 - R_2)^2 + h^2}$$

FIGURE 1-13

Solution To find the surface area of the shade, we evaluate the algebraic expression on the right-hand side of the formula using the values given in Figure 1-13.

In the formula, the 1 and 2 in the notation R_1 (read as "R sub-one") and R_2 (read as "R sub-two") are called **subscripts.** They distinguish between the lower radius and the upper radius of the shade.

$$S = \pi(R_1 + R_2)\sqrt{(R_1 - R_2)^2 + h^2}$$

$$S = \pi(9 + 4)\sqrt{(9 - 4)^2 + (12)^2} \qquad \text{Substitute 9 for } R_1, \text{ 4 for } R_2, \text{ and 12 for } h.$$

$$= \pi(13)\sqrt{(5)^2 + (12)^2} \qquad \text{Do the operations inside the parentheses.}$$

$$= \pi(13)\sqrt{25 + 144} \qquad \text{Evaluate the powers within the radical.}$$

$$= \pi(13)\sqrt{169} \qquad \text{Do the addition within the radical.}$$

$$= \pi(13)(13) \qquad \text{Evaluate the square root: } \sqrt{169} = 13.$$

$$= 169\pi \qquad \text{Do the multiplication.}$$

$$\approx 530.9291585 \qquad \text{Use a calculator to approximate } 169\pi.$$

The surface area of the shade is 169π square inches. Rounded to the nearest tenth, 530.9 in.^{2} of material are needed to make the shade. ■

Recall that the area of a two-dimensional geometric figure is a measure of the surface it encloses. Table 1-4 shows the formulas for the area (A) of several geometric figures.

TABLE 1-4

Figure	Name	Area	Figure	Name	Area
	Square	$A = s^2$		Triangle	$A = \frac{1}{2}bh$
	Rectangle	$A = lw$		Trapezoid	$A = \frac{1}{2}h(b_1 + b_2)$
	Circle	$A = \pi r^2$			

The volume of a three-dimensional geometric figure is a measure of its capacity. Table 1-5 shows the formulas for the volume (V) of several geometric figures.

TABLE 1-5

Figure	Name	Volume	Figure	Name	Volume
	Cube	$V = s^3$		Cylinder	$V = Bh*$
	Rectangular solid	$V = lwh$		Cone	$V = \frac{1}{3}Bh*$
	Sphere	$V = \frac{4}{3}\pi r^3$		Pyramid	$V = \frac{1}{3}Bh*$

*B represents the area of the base.

EXAMPLE 13

Finding area and volume. Find the amount of skin covered by a rectangular band-aid $\frac{5}{8}$ inches wide and 3 inches long.

Solution

To find the amount of skin covered by the band-aid, we need to find the area of the band-aid.

$A = lw$ The formula for the area of a rectangle.

$A = 3 \cdot \dfrac{5}{8}$ Substitute 3 for l and $\frac{5}{8}$ for w.

$A = \dfrac{15}{8}$

The band-aid covers $\frac{15}{8}$ or $1\frac{7}{8}$ in.^{2} (square inches) of skin.

SELF CHECK

Find the volume of a medicine cabinet that is 23 inches tall, $13\frac{1}{2}$ inches wide, and 4 inches deep.   *Answer:* 1,242 in.^{3} ■

STUDY SET

Section 1.3

VOCABULARY

In Exercises 1–10, fill in the blanks to make the statements true.

1. When we add two numbers, the result is called the _____. When we subtract two numbers, the result is called the _____.

2. When we multiply two numbers, the result is called the _____. When we divide two numbers, the result is called the _____.

3. To _____ an algebraic expression, we substitute values for the variables and then apply the rules for the order of operations.

4. In the expression $9 + 6[22 - (6 - 1)]$, the _____ are the innermost grouping symbols, and the brackets are the _____ grouping symbols.

5. 6^2 can be read as "six _____," and 6^3 can be read as "six _____."

6. 4^5 is the fifth _____ of four.

7. In the exponential expression x^2, x is the _____, and 2 is the _____.

8. An _____ is used to represent repeated multiplication.

9. Subtraction is the same as adding the _____ of the number being subtracted.

10. The principal _____ of 16 is 4.

CONCEPTS

11. Consider the expression $6 + 3 \cdot 2$.
   a. In what two different ways might we evaluate the given expression?

   b. Which result from part a is correct and why?

12. **a.** What operations does the expression $60 - (-9)^2 + 5(-1)$ contain?

   b. In what order should they be performed?

13. What are we finding when we calculate
   a. the amount of surface a circle encloses?
   b. the capacity of a cylinder?

14. **a.** What is the related multiplication statement for the division statement $\frac{0}{6} = 0$?
   b. Why isn't there a related multiplication statement for $\frac{6}{0}$?

NOTATION

In Exercises 15–16, complete each solution.

15. $60 - 20 \cdot 2^3 = 60 - 20\left(\ \ \right)$
$$= 60 - \ \ \ \ $$
$$= -100$$

16. $9 + 8[(1 + 4) \cdot 5] = 9 + 8\left[\left(\ \ \right) \cdot 5\right]$
$$= 9 + 8\left(\ \ \right)$$
$$= 9 + \ \ \ \ $$
$$= 209$$

17. What is the name of the symbol $\sqrt{\ \ }$?

18. What is the one number that a fraction cannot have as its denominator?

PRACTICE

In Exercises 19–20, complete the table. Then draw a bar graph on the axes provided.

19.

Length of edge of cube	1	2	3	4
Volume of cube				

20.

Length of side of square	1	2	3	4	5	6
Area of square						

In Exercises 21–40, do the operations.

21. $-3 + (-5)$ 22. $-2 + (-8)$

23. $-7.1 + 2.8$ 24. $3.1 + (-5.2)$

25. $-3 - 4$ 26. $-11 - (-17)$

27. $-3.3 - (-3.3)$ 28. $0.14 - (-0.13)$

29. $-2(6)$ 30. $-3(-7)$

31. $\dfrac{-8}{4}$ 32. $\dfrac{-16}{-4}$

33. $\dfrac{1}{2} + \left(-\dfrac{1}{3}\right)$ 34. $-\dfrac{3}{4} + \left(-\dfrac{1}{5}\right)$

35. $\dfrac{1}{2} - \left(-\dfrac{3}{5}\right)$ 36. $\dfrac{1}{26} - \dfrac{11}{13}$

37. $\left(-\dfrac{3}{5}\right)\left(\dfrac{10}{7}\right)$ 38. $\left(-\dfrac{6}{7}\right)\left(-\dfrac{5}{12}\right)$

39. $-\dfrac{16}{5} \div \left(-\dfrac{10}{3}\right)$ 40. $-\dfrac{5}{24} \div \dfrac{10}{3}$

In Exercises 41–52, evaluate each expression.

41. 12^2

42. 9^2

43. -5^2

44. $(-5)^2$

45. $4 \cdot 2^3$

46. $(4 \cdot 2)^3$

47. $(1.3)^2$

48. $\left(\dfrac{3}{5}\right)^2$

49. $\sqrt{64}$

50. $\sqrt{121}$

51. $-\sqrt{\dfrac{9}{16}}$

52. $-\sqrt{0.16}$

In Exercises 53–64, evaluate each expression.

53. $3 - 5 \cdot 4$

54. $12 - 2 \cdot 3$

55. $\left(-3 - \sqrt{25}\right)^2$

56. $4^2 - (-2)^2$

57. $2 + 3\left(\dfrac{25}{5}\right) + (-4)$

58. $(-2)^3\left(\dfrac{-6}{-2}\right)(-1)$

59. $\dfrac{-\sqrt{49} - 3^2}{2 \cdot 4}$

60. $\dfrac{1}{2}\left(\dfrac{1}{8}\right) + \left(-\dfrac{1}{4}\right)^2$

61. $-2|4 - 8|$

62. $\left|\sqrt{49} - 8(4 - 7)\right|$

63. $(4 + 2 \cdot 3)^4$

64. $|9 - 5(1 - 8)|$

In Exercises 65–76, evaluate each expression.

65. $3 + 2[-1 - 4(5)]$

66. $-3[5^2 - (7 - 3)^2]$

67. $3 - [3^3 + (3 - 1)^3]$

68. $3|-(3 \cdot 5 - 2 \cdot 6)^2|$

69. $\dfrac{|-25| - 2(-5)}{2^4 - 9}$

70. $\dfrac{2[-4 - 2(3 - 1)]}{3(3)(2)}$

71. $\dfrac{3[-9 + 2(7 - 3)]}{(8 - 5)(9 - 7)}$

72. $\dfrac{5 \cdot 4 \cdot 3 \cdot 2 \cdot 1}{1 \cdot 2 \cdot 3 \cdot 4}$

73. $\dfrac{(6 - 5)^4 + 21}{27 - \left(\sqrt{16}\right)^2}$

74. $\dfrac{3(3,246 - 1,111)}{561 - 546}$

75. $54^3 - 16^4 + 19(3)$

76. $\dfrac{36^2 - 2(48)}{(25)^2 - \sqrt{105,625}}$

In Exercises 77–86, evaluate each expression for the given values of the variables.

77. $-\dfrac{2}{3}a^2$ for $a = -6$

78. $\left(-\dfrac{2}{3}a\right)^2$ for $a = -6$

79. $\dfrac{y_2 - y_1}{x_2 - x_1}$ for $x_1 = -3$, $x_2 = 5$, $y_1 = 12$, $y_2 = -4$

80. $P_0\left(1 + \dfrac{r}{k}\right)^{kt}$ for $P_0 = 500$, $r = 4$, $k = 2$, $t = 3$

81. $(x + y)(x^2 - xy + y^2)$ for $x = -4$, $y = 5$

82. $\dfrac{-b + \sqrt{b^2 - 4ac}}{2a}$ for $a = 1$, $b = 2$, $c = -3$

83. $\dfrac{x^2}{a^2} + \dfrac{y^2}{b^2}$ for $x = -3$, $y = -4$, $a = 5$, $b = -5$

84. $\dfrac{n}{2}[2a_1 + (n - 1)d]$ for $n = 50$, $a_1 = -4$, $d = 5$

85. $\sqrt{(x_2 - x_1)^2 + (y_2 - y_1)^2}$ for $x_1 = -2$, $x_2 = 4$, $y_1 = 4$, $y_2 = -4$

86. $\dfrac{|Ax_0 + By_0 + C|}{\sqrt{A^2 + B^2}}$ for $A = 3$, $B = 4$, $C = -5$, $x_0 = 2$, and $y_0 = -1$

In Exercises 87–88, find each area to the nearest tenth.

87. The area of a triangle with a base of 2.75 centimeters (cm) and a height of 8.25 cm

88. The area of a circle with a radius of 5.7 meters

In Exercises 89–92, find each volume to the nearest hundredth.

89. The volume of a rectangular solid with dimensions of 2.5 cm, 3.7 cm, and 10.2 cm

90. The volume of a pyramid whose base is a square with each side measuring 2.57 cm and with a height of 12.32 cm

91. The volume of a sphere with a radius of 5.7 meters

92. The volume of a cone whose base has a radius of 5.5 in. and whose height is 8.52 in.

APPLICATIONS

93. ALUMINUM FOIL Find the number of *square feet* of aluminum foil on a roll if the dimensions printed on the box are $8\frac{1}{3}$ yards $\times$ 12 inches.

94. HOCKEY A goal is scored in hockey when the puck, a vulcanized rubber disk 2.5 cm (1 in.) thick and 7.6 cm (3 in.) in diameter, is driven into the opponent's goal. Find the volume of a puck in cubic centimeters and cubic inches. Round to the nearest tenth.

95. PAPER PRODUCTS When folded, the paper sheet shown in Illustration 1 on the next page forms a rectangular-shaped envelope. The formula
$$A = \tfrac{1}{2}h_1(b_1 + b_2) + b_3h_3 + \tfrac{1}{2}b_1h_2 + b_1b_3$$
gives the amount of paper (in square units) used in the design. Explain what each of the four terms in the formula finds. Then evaluate the formula for $b_1 = 6$, $b_2 = 2$, $b_3 = 3$, $h_1 = 2$, $h_2 = 2.5$, and $h_3 = 3$. All dimensions are in inches.

ILLUSTRATION 1

96. BOND MARKET The graph in Illustration 2 shows the quarterly investment in U.S. bond funds from 1995 through 1997.
   a. In the graph, what does a negative net cash inflow indicate?

   b. Find the net cash inflow for each of the years 1995, 1996, and 1997.

ILLUSTRATION 2

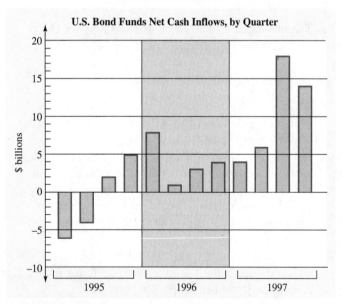

Based on data from the *Los Angeles Times* (Jan. 6, 1998)

   c. What was the difference in the net cash inflow for the first quarter of 1995 compared to the first quarter of 1996 and the first quarter of 1996 compared to the first quarter of 1997?

97. ACCOUNTING On a financial balance sheet, debts (negative numbers) are denoted within parentheses.

Assets (positive numbers) are written without parentheses. What is the 1999 fund balance for the preschool whose financial records are shown in Illustration 3?

ILLUSTRATION 3

Community Care Preschool Balance Sheet, June, 1999	
Fund balances	
Classroom supplies	$ 5,889
Emergency needs	927
Holiday program	(2,928)
Insurance	1,645
Janitorial	(894)
Licensing	715
Maintenance	(6,321)
BALANCE	?

98. TEMPERATURE EXTREMES The highest and lowest temperatures ever recorded in several cities are shown in Illustration 4. List the cities in order, from the smallest to the largest range in temperature extremes.

ILLUSTRATION 4

City	Extreme temperatures	
	Highest	**Lowest**
Atlanta, Georgia	105	−8
Boise, Idaho	111	−25
Helena, Montana	105	−42
New York, New York	107	−3
Omaha, Nebraska	114	−23

99. ICE CREAM See Illustration 5. If the two equal-sized scoops of ice cream melt completely into the cone, will they overflow the cone?

ILLUSTRATION 5

100. PHYSICS Waves are motions that carry energy from one place to another. Illustration 6 shows an example of a wave called a *standing wave*. What is the difference in the height of the crest of the wave and the depth of the trough of the wave?

ILLUSTRATION 6

101. **PEDIATRICS** Young's rule, shown below, is used by some doctors to calculate dosage for infants and children.

$$\frac{\text{Age of child}}{\text{Age of child} + 12} \left(\begin{array}{c} \text{average} \\ \text{adult dose} \end{array} \right) = \text{child's dose}$$

The syringe in Illustration 7 shows the adult dose of a certain medication. Use Young's rule to determine the dosage for a 6-year-old child. Then use an arrow to locate the dosage on the calibration.

ILLUSTRATION 7

Adult dose

102. **DOSAGES** The adult dosage of procaine penicillin is 300,000 units daily. Calculate the dosage for a 12-year-old child using Young's rule. (See Exercise 101.)

WRITING

103. Explain what the statement $x - y = x + (-y)$ means.

104. Explain why rules for the order of operations are necessary.

REVIEW

105. What two numbers are a distance of 5 away from -2 on the number line?

106. Place the proper symbol ($>$ or $<$) in the blank: -4.6 ▇ -4.5.

107. List the set of integers.

108. Translate into mathematical symbols: ten less than twice x.

109. True or false: The real numbers is the set of all decimals.

110. True or false: Irrational numbers are nonterminating, nonrepeating decimals.

▶ **1.4**

Simplifying Algebraic Expressions

In this section, you will learn about

> Properties of real numbers ■ Properties of 0 and 1 ■ More properties of real numbers ■ Simplifying algebraic expressions ■ The distributive property ■ Combining like terms

Introduction Suppose we are asked to find the total dollar amount of the checks that have been recorded in the check register shown in Figure 1-14. As the notes in the figure point out, we can simplify the computation by adding the numbers in a different order than the way they are entered. The *commutative* and *associative* properties of addition guarantee that we will obtain the same result whether we add them in their original order or in this more convenient way. In this section, we will introduce some important properties of real numbers. Then we will see how they can be used to simplify expressions containing variables.

FIGURE 1-14

Number	Date	Description of Transaction	Payment/Debit	
101	3/6	DR. OKAMOTO, DDS	$64	00
102	3/6	UNION OIL CO.	$25	00
103	3/8	STATER BROS.	$16	00
104	3/9	LITTLE LEAGUE	$75	00

Add $64.00 and $16.00 to get $80.00.

Add $25.00 and $75.00 to get $100.00.

Now add the two subtotals to get the total dollar amount of the checks: $80.00 + $100.00 = $180.00.

Properties of Real Numbers

When working with real numbers, we will use the following properties.

Properties of real numbers

If a, b, and c are real numbers, then we have

The associative properties for addition and multiplication

$$(a + b) + c = a + (b + c) \qquad (ab)c = a(bc)$$

The commutative properties for addition and multiplication

$$a + b = b + a \qquad\qquad ab = ba$$

The distributive property of multiplication over addition

$$a(b + c) = ab + ac$$

The *associative properties* enable us to group the numbers in a sum or a product any way that we wish and get the same result.

E X A M P L E   1

Using the associative property. Evaluate $(14 + 94) + 6$ in two ways.

Solution

$$(14 + 94) + 6 = \mathbf{108} + 6 \quad \text{Work within the parentheses first.}$$
$$= 114$$

To evaluate the expression in a second way, we apply the associative property of addition.

$$(14 + 94) + 6 = 14 + (\mathbf{94 + 6}) \quad \text{Use parentheses to group 94 with 6.}$$
$$= 14 + \mathbf{100} \qquad \text{Do the addition within the parentheses.}$$
$$= 114$$

Notice that the results are the same.

SELF CHECK   Evaluate $2 \cdot (50 \cdot 37)$ in two ways. *Answer:* 3,700   ■

 WARNING! Subtraction and division are not associative, because different groupings give different results. For example,

$$(8 - 4) - 2 = 4 - 2 = 2 \quad \text{but} \quad 8 - (4 - 2) = 8 - 2 = 6$$
$$(8 \div 4) \div 2 = 2 \div 2 = 1 \quad \text{but} \quad 8 \div (4 \div 2) = 8 \div 2 = 4$$

The *commutative properties* enable us to add or multiply two numbers in either order and obtain the same result. Here are two examples.

$$3 + (-5) = -2 \qquad \text{and} \qquad -5 + 3 = -2$$
$$-2.6(-8) = 20.8 \qquad \text{and} \qquad -8(-2.6) = 20.8$$

 WARNING! Subtraction and division are not commutative, because doing these operations in different orders will give different results. For example,

$$8 - 4 = 4 \quad \text{but} \quad 4 - 8 = -4$$
$$8 \div 4 = 2 \quad \text{but} \quad 4 \div 8 = \frac{1}{2}$$

The *distributive property* enables us to evaluate certain types of expressions involving a multiplication and an addition. As an example, let's consider the expression $4(5 + 3)$, which can be evaluated in two ways.

Method 1: Rules for the Order of Operations

In this method, we compute the sum inside the parentheses first.

$4(\mathbf{5 + 3}) = 4(\mathbf{8})$ Do the addition inside the parentheses first.

$\qquad\quad\; = 32$ Do the multiplication.

Method 2: The Distributive Property

In this method, we distribute the 4 across the 5 and the 3, find each product separately, and add the results.

$$\underset{\text{First product}}{} \quad \underset{\text{Second product}}{}$$

$4(5 + 3) = 4 \cdot 5 \;+\; 4 \cdot 3$ To apply the distributive property, we multiply each term inside the parentheses by the factor outside the parentheses.

$\qquad\qquad = 20 \;+\; 12$ Do the multiplications first: $4 \cdot 5 = 20$ and $4 \cdot 3 = 12$.

$\qquad\qquad = 32$ Do the addition.

Notice that each method gives a result of 32.

Properties of 0 and 1

The real numbers 0 and 1 have important special properties.

Properties of 0 and 1

Additive identity: The sum of 0 and any number is the number itself.

$0 + a = a + 0 = a$

Multiplicative identity: The product of 1 and any number is the number itself.

$1 \cdot a = a \cdot 1 = a$

Multiplication property of 0: The product of any number and 0 is 0.

$a \cdot 0 = 0 \cdot a = 0$

For example,

$$7 + 0 = 7, \qquad 1(5.4) = 5.4, \qquad \left(-\frac{7}{3}\right)1 = -\frac{7}{3}, \qquad \text{and} \qquad -19(0) = 0$$

More Properties of Real Numbers

If the sum of two numbers is 0, they are called **additive inverses, negatives,** or **opposites** of each other. For example, 6 and -6 are additive inverses, because $6 + (-6) = 0$.

The additive inverse property

For every real number a, there is a real number $-a$ such that

$a + (-a) = -a + a = 0$

If the product of two numbers is 1, the numbers are called **multiplicative inverses** or **reciprocals** of each other.

The multiplicative inverse property

For every nonzero real number a, there exists a real number $\dfrac{1}{a}$ such that

$$a \cdot \frac{1}{a} = \frac{1}{a} \cdot a = 1$$

Some example of reciprocals (multiplicative inverses) are

- 5 and $\dfrac{1}{5}$ are reciprocals, because $5\left(\dfrac{1}{5}\right) = 1$.

- $\dfrac{3}{2}$ and $\dfrac{2}{3}$ are reciprocals, because $\dfrac{3}{2}\left(\dfrac{2}{3}\right) = 1$.

- -0.25 and -4 are reciprocals, because $-0.25(-4) = 1$.

The reciprocal of 0 does not exist, because $\frac{1}{0}$ is undefined.

Recall that when a number is divided by 1, the result is the number itself, and when a number is divided by itself, the result is 1.

Division properties

Division by 1: If x represents any real number, then $\dfrac{x}{1} = x$.

Division of a number by itself: If x represents any real number, then $\dfrac{x}{x} = 1$ ($x \neq 0$).

Examples of three types of division involving zero that we encounter in mathematics are

$$\frac{0}{5}, \qquad \frac{5}{0}, \qquad \text{and} \qquad \frac{0}{0}$$

To help explain the concept of division of zero, we consider the division $\frac{0}{5} = ?$ and its equivalent multiplication fact $5(?) = 0$.

Multiplication fact	**Division fact**
$5(?) = 0$	$\dfrac{0}{5} = 0$

$\uparrow$ $\uparrow$

This must be 0 if   So the quotient is 0.
the product is to be 0.

This example suggests that the quotient of zero and any nonzero number is zero.

To illustrate that division by zero is not possible, we consider the division $\frac{5}{0} = ?$ and its equivalent multiplication fact $0(?) = 5$.

Multiplication fact	*Division fact*
$0(?) = 5$	$\dfrac{5}{0} = $ undefined
$\uparrow$	$\uparrow$
There is no number that gives 5 when multiplied by 0.	There is no quotient.

This example illustrates that the quotient of any number divided by zero is *undefined*.

To show that division of zero by zero doesn't have a precisely determined result, we consider $\frac{0}{0} = ?$ and its equivalent multiplication fact $0(?) = 0$.

Multiplication fact	*Division fact*
$0(?) = 0$	$\dfrac{0}{0} = $ indeterminate
$\uparrow$	$\uparrow$
Any number multiplied by 0 gives 0.	We can not determine this— it could be any number.

This example illustrates that zero divided by zero is *indeterminate*.

Division with 0

Division of 0: For any nonzero real number x, $\frac{0}{x} = 0$.

Division by 0: For any nonzero real number x, $\frac{x}{0}$ is undefined.

Division of 0 by 0: $\frac{0}{0}$ is indeterminate.

Simplifying Algebraic Expressions

To **simplify algebraic expressions,** we use properties of the real numbers to write the expressions in a less complicated form. As an example, let's consider the expression $6(5x)$ and simplify it.

$$6(5x) = 6 \cdot (5 \cdot x) \quad \text{$6(5x) = 6 \cdot 5x$ and $5x = 5 \cdot x$.}$$
$$= (6 \cdot 5) \cdot x \quad \text{Apply the associative property of multiplication to group the 5 with the 6.}$$
$$= 30x \quad \text{Do the multiplication inside the parentheses: $6 \cdot 5 = 30$.}$$

Since $6(5x) = 30x$, we say that $6(5x)$ simplifies to $30x$.

EXAMPLE 2

Simplifying algebraic expressions. Simplify **a.** $9(10t)$, **b.** $-5.3r(-2s)$, and **c.** $-\frac{21}{2}a\left(\frac{1}{3}\right)$.

Solution

a. $9(10t) = (9 \cdot 10)t$ — Use the associative property of multiplication to regroup the factors.

$= 90t$ — Do the multiplication inside the parentheses: $9 \cdot 10 = 90$.

b. $-5.3r(-2s) = [-5.3(-2)](r \cdot s)$ — Use the commutative and associative properties to group the numbers and group the variables.

$= 10.6rs$ — Do the multiplications: $-5.3(-2) = 10.6$ and $r \cdot s = rs$.

c. $-\dfrac{21}{2}a\left(\dfrac{1}{3}\right) = -\dfrac{21}{2}\left(\dfrac{1}{3}\right)a$

Use the commutative property of multiplication to change the order of the factors a and $\frac{1}{3}$.

$= -\dfrac{7}{2}a$

Do the multiplication:

$-\dfrac{21}{2} \cdot \dfrac{1}{3} = -\dfrac{21 \cdot 1}{2 \cdot 3} = -\dfrac{7 \cdot \cancel{3}^{1} \cdot 1}{2 \cdot \cancel{3}_{1}} = -\dfrac{7}{2}.$

SELF CHECK   Simplify   **a.** $14 \cdot 3s$,   **b.** $-1.6b(3t)$,   and   *Answers:*   **a.** $42s$,   **b.** $-4.8bt$,
c. $-\frac{2}{3}x(-9)$.   **c.** $6x$ ∎

The Distributive Property

We can use the distributive property to *remove the parentheses* when an algebraic expression is multiplied by a quantity. For example, to remove the parentheses in the expression $5(x + 2)$, we proceed as follows:

$\overset{\frown}{5(x + 2)} = 5 \cdot x + 5 \cdot 2$ Distribute the 5. The arrows help us visualize the distribution.

$= 5x + 10$ Do the multiplications.

Since subtraction is the same as adding the opposite, the distributive property also holds for subtraction.

$\overset{\frown}{5(x - 2)} = 5 \cdot x - 5 \cdot 2$ Distribute the 5.

$= 5x - 10$ Do the multiplications.

E X A M P L E   3   **Applying the distributive property.** Use the distributive property to remove parentheses: **a.** $6(a + 9)$ and   **b.** $-15(4b - 1)$.

Solution   **a.** $6(a + 9) = 6 \cdot a + 6 \cdot 9$ Distribute the 6.

$= 6a + 54$ Do the multiplications.

b. $-15(4b - 1) = -15(4b) - (-15)(1)$ Distribute the -15.

$= -60b - (-15)$ Do the multiplications.

$= -60b + 15$ Add the opposite of -15, which is 15.

SELF CHECK   Remove parentheses: **a.** $9(r + 4)$ and   *Answers:*   **a.** $9r + 36$,
b. $-11(-3x - 5)$.   **b.** $33x + 55$ ∎

WARNING! If an expression contains parentheses, it does not necessarily mean that the distributive property can be applied. For example, the distributive property does not apply to

$5(12x)$ or $5(-4 \cdot y)$ Here, a product is multiplied by 5.

However, the distributive property does apply to

$5(12 + x)$ and $5(-4 - y)$ Here, a sum and a difference are multiplied by 5.

A more general form of the distributive property is the **extended distributive property.**

$a(b + c + d + e + \cdots) = ab + ac + ad + ae + \cdots$

EXAMPLE 4

Applying the extended distributive property. Remove parentheses: $-0.5(7 - 5y + 6z)$.

Solution

$-0.5(7 - 5y + 6z)$

$= -0.5(7) - (-0.5)(5y) + (-0.5)(6z)$ Distribute the -0.5.

$= -3.5 - (-2.5y) + (-3z)$ Do the multiplications.

$= -3.5 + 2.5y - 3z$ Subtracting $-2.5y$ is the same as adding $2.5y$. Adding $-3z$ is the same as subtracting $3z$.

SELF CHECK   Remove parentheses: $\frac{1}{3}(-6t + 3s - 9)$.   *Answer:* $-2t + s - 3$ ■

Since multiplication is commutative, we can write the distributive property in the following forms.

$$(b + c)a = ba + ca, \qquad (b - c)a = ba - ca, \qquad (b + c + d)a = ba + ca + da$$

EXAMPLE 5

Applying the distributive property. Remove parentheses: $(-8 - 3y)(-30)$.

Solution

$(-8 - 3y)(-30) = -8(-30) - 3y(-30)$ Distribute the -30.

$= 240 - (-90y)$ Do the multiplications.

$= 240 + 90y$ Add the opposite of $-90y$.

SELF CHECK   Remove parentheses: $(-5s + 4t)(-10)$.   *Answer:* $50s - 40t$ ■

To use the distributive property to simplify $-(x + 3)$, we note that the negative sign in front of the parentheses represents the number -1.

The $-$ sign represents -1.

$-(x + 3) = -1(x + 3)$

$= -1(x) + (-1)(3)$ Distribute the -1.

$= -x + (-3)$ Do the multiplications.

$= -x - 3$

EXAMPLE 6

Distributing a negative sign. Simplify $-(-21 - 20m)$.

Solution

$-(-21 - 20m) = -1(-21 - 20m)$ Write the $-$ sign in front of the parentheses as -1.

$= -1(-21) - (-1)(20m)$ Distribute the -1.

$= 21 - (-20m)$ Do the multiplications.

$= 21 + 20m$

SELF CHECK   Simplify $-(-27k + 15)$.   *Answer:* $27k - 15$ ■

Combining Like Terms

Addition signs separate algebraic expressions into parts called **terms.** For example, the expression $3x^2 + 2x + 4$ has three terms: $3x^2$, $2x$, and 4. A term may be

■ a number; examples are -6, 45.7, 35, and $\frac{2}{3}$.

- a variable or a product of variables (which may be raised to powers); examples are x, bh, s^2, Prt, and a^3bc^4.

- a product of a number and one or more variables (which may be raised to powers); examples are $3x$, $-7y$, $2.5y^2$, and $\pi r^2 h$.

Since subtraction can be written as addition of the opposite, the expression $6a - 5b$ can be written in the equivalent form $6a + (-5b)$. We can then see that the algebraic expression $6a - 5b$ contains two terms, $6a$ and $-5b$.

The **numerical coefficients** (or simply the **coefficients**) of the terms of the expression $x^3 - 5x^2 - x + 28$ are 1, -5, -1, and 28, respectively.

Terms with exactly the same variables raised to exactly the same powers are called **like terms** or **similar terms**:

$5x$ and $6x$ are like terms	$27x^2y^3$ and $-326x^2y^3$ are like terms
$4x$ and $-17y$ are unlike terms, because they have different variables	$15x^2y$ and $6xy^2$ are unlike terms, because the variables have different exponents

If we are to add (or subtract) objects, they must have the same units. For example, we can add dollars to dollars and inches to inches, but we cannot add dollars to inches. The same is true when working with terms of an algebraic expression. They can be added or subtracted only when they are like terms.

This expression can be simplified, because it contains like terms.	This expression cannot be simplified, because its terms are not like terms.
$5x + 6x$	$5x + 6y$
↑ ↑	↑ ↑
Like terms.	Unlike terms.
The variable parts are the same.	The variable parts are not the same.

To simplify expressions containing like terms, we use the distributive property. For example,

$$5x + 6x = (5 + 6)x \qquad \text{and} \qquad 32y - 16y = (32 - 16)y$$
$$= 11x \qquad\qquad\qquad\qquad = 16y$$

These examples suggest the following rule.

Combining like terms

To add or subtract like terms, combine their coefficients and keep the same variables with the same exponents.

EXAMPLE 7

Combining like terms. Simplify each expression: **a.** $-8f + (-12f)$, **b.** $0.56s^3 - 0.2s^3$, and **c.** $-\frac{1}{2}ab + \frac{1}{3}ab$.

Solution

a. $-8f + (-12f) = -20f$ — Add the coefficients of the like terms: $-8 + (-12) = -20$. Keep the variable f.

b. $0.56s^3 - 0.2s^3 = 0.36s^3$ — Subtract: $0.56 - 0.2 = 0.36$. Keep s^3.

c. $-\dfrac{1}{2}ab + \dfrac{1}{3}ab = -\dfrac{1 \cdot 3}{2 \cdot 3}ab + \dfrac{1 \cdot 2}{3 \cdot 2}ab$ — Express each fraction in terms of the LCD, 6.

$$= -\frac{3}{6}ab + \frac{2}{6}ab$$ — Do the multiplications.

$$= -\frac{1}{6}ab$$ — Add the coefficients: $-\frac{3}{6} + \frac{2}{6} = -\frac{1}{6}$. Keep ab.

SELF CHECK Simplify by combining like terms: **a.** $5k + 8k$,   *Answers:* **a.** $13k$,   **b.** $200a^2$,
b. $-600a^2 - (-800a^2)$,   and   **c.** $\frac{2}{3}xy - \frac{3}{4}xy$.   **c.** $-\frac{1}{12}xy$ ■

EXAMPLE 8   **Combining like terms.** Simplify $9b - B - 14b + 34B$.

Solution   This expression has four terms. The uppercase B and the lowercase b are different variables. So the first and third terms are like terms, and the second and fourth terms are like terms.

$$9b - B - 14b + 34B = -5b + 33B \quad \text{Combine like terms: } 9b - 14b = -5b \text{ and}$$
$$-B + 34B = 33B.$$

SELF CHECK Simplify $8R + 7r - 14R - 21r$.   *Answer:* $-6R - 14r$ ■

EXAMPLE 9   **Simplifying algebraic expressions.** Simplify $9(x + 1) - 3(7x - 1)$.

Solution   We apply the distributive property and then combine like terms.

$$9(x + 1) - 3(7x - 1) = 9x + 9 - 21x + 3 \quad \text{Use the distributive property twice.}$$
$$= -12x + 12 \quad \text{Combine like terms: } 9x - 21x = -12x$$
$$\text{and } 9 + 3 = 12.$$

SELF CHECK Simplify $-5(y - 4) + 2(4y + 8)$.   *Answer:* $3y + 36$ ■

STUDY SET

Section 1.4

VOCABULARY

In Exercises 1–8, fill in the blanks to make the statements true.

1. _____ terms are terms with exactly the same variables raised to exactly the same powers.

2. To add or subtract like terms, combine their _____ and keep the same variables and exponents.

3. A number or the product of numbers and variables is called a _____.

4. $\frac{1}{3}$ and 3 are _____, because $\frac{1}{3} \cdot 3 = 1$.

5. To _____ algebraic expressions, we use properties of real numbers to write the expressions in a less complicated form.

6. The distributive property is used to _____ parentheses when a sum or difference is multiplied by a quantity.

7. Division by 0 is _____.

8. The _____ of the term $-8c$ is -8.

CONCEPTS

9. Using the variables x, y, and z, write the associative property of addition.

10. Using the variables x and y, write the commutative property of multiplication.

11. Using the variables r, s, and t, write the distributive property of multiplication over addition.

12. **a.** What is the additive identity?
    b. What is the multiplicative identity?
    c. Simplify $-(-10)$.

13. What number should be
    a. subtracted from 5 to obtain 0?
    b. added to 5 to obtain 0?

14. By what number should
    a. 5 be divided to obtain 1?
    b. 5 be multiplied to obtain 1?

15. Give the reciprocal.
    a. $\frac{15}{16}$ **b.** -20
    c. 0.5 **d.** x

16. Does the distributive property apply?
    a. $2(3)(5)$ **b.** $2(3 \cdot 5)$
    c. $2(3x)$ **d.** $2(x - 3)$

17. Consider the expression $2x^2 - x + 6$.
    a. What are the terms of the expression?
    b. Give the coefficient of each term.

18. Which properties of real numbers involve changing *order* and which involve changing *grouping*?

In Exercises 19–26, tell whether the terms are like terms. If they are, combine them.

19. $2x$, $6x$ **20.** $-3x$, $5y$

21. $-5xy$, $-7yz$ **22.** $-3t^2$, $12t^2$

23. $3x^2$, $-5x^2$ **24.** $5y^2$, $7xy$

25. xy, $3xt$ **26.** $-4x$, $-5x$

NOTATION

27. In $-(x - 7)$, what does the negative sign in front of the parentheses represent?

28. Do the division, if possible.
    a. $\dfrac{0}{8}$ **b.** $\dfrac{8}{0}$

PRACTICE

In Exercises 29–40, fill in the blanks to make the statements true by applying the given property of the real numbers.

29. $3 + 7 =$ _____
    Commutative property of addition

30. $2(5 \cdot 97) =$ _____
    Associative property of multiplication

31. $3(2 + d) =$ _____
    Distributive property

32. $1 \cdot y =$ _____
    Commutative property of multiplication

33. $c + 0 =$ ___
    Additive identity property

34. $-4(x - 2) =$ _____
    Distributive property and simplifying

35. $25 \cdot \dfrac{1}{25} =$ ___
    Multiplicative inverse property

36. $z + (9 - 27) =$ _____
    Commutative property of addition

37. $8 + (7 + a) =$ _____
    Associative property of addition

38. ___ $\cdot 3 = 3$
    Multiplicative identity property

39. $(x + y)2 =$ _____
    Commutative property of multiplication

40. $h + (-h) =$ ___
    Additive inverse property

In Exercises 41–44, evaluate each side of the equation separately to show that the same result is obtained. Identify the property of real numbers that is being illustrated.

41. $(37.9 + 25.2) + 14.3 = 37.9 + (25.2 + 14.3)$

42. $7.1(3.9 + 8.8) = 7.1 \cdot 3.9 + 7.1 \cdot 8.8$

43. $2.73(4.534 + 57.12) = 2.73 \cdot 4.534 + 2.73 \cdot 57.12$

44. $(6.789 + 345.1) + 27.347 = (345.1 + 6.789) + 27.347$

In Exercises 45–54, remove the parentheses.

45. $-4(t - 3)$ **46.** $-4(-t + 3)$

47. $-(t - 3)$ **48.** $-(-t + 3)$

49. $\dfrac{2}{3}(3s - 9)$ **50.** $\dfrac{1}{5}(5s - 15)$

51. $0.7(s + 2)$ **52.** $2.5(6s - 8)$

53. $3\left(\dfrac{4}{3}x - \dfrac{5}{3}y + \dfrac{1}{3}\right)$

54. $6\left(-\dfrac{4}{3} + \dfrac{7}{6}s + \dfrac{16}{3}t\right)$

In Exercises 55–62, simplify each expression.

55. $9(8m)$ **56.** $12n(4)$

57. $5(-9q)$ **58.** $-7(2t)$

59. $(-5p)(-6b)$ **60.** $(-7d)(-7e)$

61. $-5(8r)(-2y)$ **62.** $-7s(-4t)(-1)$

In Exercises 63–92, simplify each expression by combining like terms.

63. $3x + 15x$ **64.** $12y - 17y$

65. $18x^2 - 5x^2$ **66.** $37x^2 + 3x^2$

67. $-9x + 9x$ **68.** $-26y + 26y$

69. $-b^2 + b^2$ **70.** $-3c^3 + 3c^3$

71. $8x + 5x - 7x$

72. $-y + 3y + 6y$

73. $3x^2 + 2x^2 - 5x^2$

74. $8x^3 - x^3 + 2x^3$

75. $3.8h - 0.7h$

76. $-5.7m + 5.3m$

77. $\frac{2}{5}ab - \left(-\frac{1}{2}ab\right)$

78. $-\frac{3}{4}st - \frac{1}{3}st$

79. $\frac{3}{5}t + \frac{1}{3}t$

80. $\frac{3}{16}x - \frac{5}{4}x$

81. $4(y + 9) - 8y$

82. $-(4 + z) + 2z$

83. $2z + 5(z - 4)$

84. $12(2m + 11) - 11$

85. $8(2c + 7) - 2(c - 3)$

86. $9(z + 3) - 5(3 - z)$

87. $2x^2 + 4(3x - x^2) + 3x$

88. $3p^2 - 6(5p^2 + p) + p^2$

89. $-(a + 2) - (a - b)$

90. $3z - 2(y - z) + y$

91. $-3(p - 2) + 2(p + 3) - 5(p - 1)$

92. $5(q + 7) - 3(q - 1) - (q + 2)$

APPLICATIONS

93. PARKING AREA A restaurant parking lot is shown in Illustration 1.
   a. Express the area of the entire parking lot as the product of its length and width.
   b. Express the area of the entire lot as the sum of the areas of the self-parking space and the valet parking space.
   c. Write an equation that shows that your answers to parts a and b are equal. What property of real numbers is illustrated by this example?

ILLUSTRATION 1

Length 20 meters

6 meters

x meters

94. CROSS SECTION When the steel casting shown in Illustration 2 is cut down the middle, we see that it has a uniform cross section consisting of two identical trapezoids. What is the area of the cross section? (The measurements are in inches.)

ILLUSTRATION 2

WRITING

95. Explain why the distributive property does not apply when simplifying $6(2 \cdot x)$.

96. In each case, explain what you can conclude about one or both of the numbers.
   a. When the two numbers are added, the result is 0.
   b. When the two numbers are subtracted, the result is 0.
   c. When the two numbers are multiplied, the result is 0.
   d. When the two numbers are divided, the result is 0.

97. What are like terms?

98. Use each of the words *commute, associate,* and *distribute* in a sentence in which the context is nonmathematical.

REVIEW

In Exercises 99–106, evaluate each expression.

99. $-5.6 - (-5.6)$

100. $-(-4 \cdot 2)^3$

101. $\left(-\frac{3}{2}\right)\left(\frac{7}{12}\right)$

102. $-\sqrt{\frac{9}{16}}$

103. $(4 + 2 \cdot 3)^3$

104. $-3|4 - 8|$

105. $\frac{-\sqrt{64} - 5^2}{2 \cdot 4 + 3}$

106. $\frac{1}{2} - \left(-\frac{4}{5}\right)$

▶ 1.5 Solving Linear Equations and Formulas

In this section, you will learn about

Solutions of equations ■ Linear equations ■ Properties of equality
■ Solving linear equations ■ Simplifying expressions to solve
equations ■ Identities and impossible equations ■ Solving formulas

Introduction   In one method that we will often use to solve problems, we begin by letting a variable stand for an unknown quantity. Then we write an equation involving the variable to describe the situation mathematically. Finally, we perform a series of steps on the equation to find the value represented by the variable. The process of determining the value (or values) represented by a variable is called *solving the equation.* In this section, we will discuss an equation-solving strategy to solve one basic type of equation called a *linear equation* in one variable.

Solutions of Equations

An **equation** is a statement indicating that two quantities are equal. The equation $2 + 4 = 6$ is true, and the equation $2 + 5 = 6$ is false. If an equation contains a variable (say, x), it can be either true or false, depending on the value of x. For example, if $x = 1$, the equation $7x - 3 = 4$ is true.

$$7x - 3 = 4$$

$7(1) - 3 \overset{?}{=} 4$ Substitute 1 for x. At this stage, we don't know whether the left- and right-hand sides of the equation are equal, so we use an "is possibly equal to" sign $\overset{?}{=}$.

$7 - 3 \overset{?}{=} 4$ Do the multiplication.

$4 = 4$ We obtain a true statement.

Since 1 makes the equation true, we say that 1 *satisfies* the equation. However, the equation is false for all other values of x.

The set of numbers that satisfy an equation is called its **solution set.** The elements of the solution set are called **solutions** or **roots** of the equation. Finding the solution set of an equation is called **solving the equation.**

EXAMPLE 1   **Checking a solution.** Determine whether 2 is a solution of $3x + 2 = 2x + 5$.

Solution   We substitute 2 for x wherever it appears in the equation and see whether it satisfies the equation.

$3x + 2 = 2x + 5$ The original equation.

$3(2) + 2 \overset{?}{=} 2(2) + 5$ Substitute 2 for x.

$6 + 2 \overset{?}{=} 4 + 5$ Do the multiplication.

$8 = 9$ Do the addition.

Since $8 = 9$ is a false statement, the number 2 does not satisfy the equation. It is not a solution of $3x + 2 = 2x + 5$.

SELF CHECK   Is -5 a solution of $2x - 5 = 3x$?   *Answer:* yes   ■

Linear Equations

In practice, we are not told the solutions of an equation—we need to find the solutions (that is, solve the equation) ourselves. In this text, you will learn how to solve many different types of equations. The easiest to solve are **linear equations.**

Linear equations

A **linear equation in one variable** (say, x) is any equation that can be written in the form

$$ax - c = 0 \quad (a \text{ and } c \text{ are real numbers and } a \neq 0)$$

Some examples of linear equations are

$$-2x - 8 = 0 \qquad 3(a + 2) = 20 \qquad 4b - 7 + 2b = 1 + 2b + 8$$

Linear equations are also called **first-degree equations,** because the highest power on the variable is 1.

Properties of Equality

When solving linear equations, the objective is to *isolate* the variable on one side of the equation. This is achieved by "undoing" the operations performed on the variable. As we undo the operations, we produce a series of simpler equations, all having the same solutions. Such equations are called **equivalent equations.**

Equivalent equations

Equations with the same solutions are called **equivalent equations.**

The solution of the equation $x = 2$ is obviously 2, because replacing x with 2 yields a true statement, $2 = 2$. The equation $x + 4 = 6$ also has a solution of 2. Since $x = 2$ and $x + 4 = 6$ have the same solution, they are equivalent equations.

The following properties are used to isolate a variable on one side of an equation.

Properties of equality

If a, b, and c are real numbers and $a = b$, then

$$a + c = b + c \qquad \text{(Addition property of equality)}$$
$$a - c = b - c \qquad \text{(Subtraction property of equality)}$$

If $c \neq 0$, then

$$ca = cb \qquad \text{(Multiplication property of equality)}$$
$$\frac{a}{c} = \frac{b}{c} \qquad \text{(Division property of equality)}$$

In words, we have

If any quantity is added to (or subtracted from) both sides of an equation, a new equation is formed that is equivalent to the original equation.

If both sides of an equation are multiplied (or divided) by the same nonzero quantity, a new equation is formed that is equivalent to the original equation.

Solving Linear Equations

| EXAMPLE 2 | **Isolating the variable by "undoing" operations.** Solve $-2x - 8 = 0$. |

Solution   We note that x is multiplied by -2 and then 8 is subtracted from that product. To isolate x on the left-hand side of the equation, we use the rules for the order of operations in reverse.

- To undo the subtraction of 8, we add 8 to both sides.
- To undo the multiplication by -2, we divide both sides by -2.

$$-2x - 8 + 8 = 0 + 8 \quad \text{Add 8 to both sides.}$$
$$-2x = 8 \quad \text{Simplify both sides of the equation.}$$
$$\frac{-2x}{-2} = \frac{8}{-2} \quad \text{Divide both sides by } -2.$$
$$x = -4 \quad \text{Simplify both sides of the equation. } x \text{ is isolated.}$$

Check: We substitute -4 for x to verify that it satisfies the original equation.

$$-2x - 8 = 0$$
$$-2(-4) - 8 \overset{?}{=} 0 \quad \text{Substitute } -4 \text{ for } x.$$
$$8 - 8 \overset{?}{=} 0 \quad \text{Do the multiplication.}$$
$$0 = 0 \quad \text{Do the subtraction.}$$

Since -4 satisfies the original equation, the solution set is $\{-4\}$.

| SELF CHECK | Solve $-3a + 15 = 0$. | *Answer:* 5 | ■ |

| EXAMPLE 3 | **Multiplying by the reciprocal.** Solve $\dfrac{3}{4}y = -7$. |

Solution   On the left-hand side, y is multiplied by $\frac{3}{4}$. We can undo the multiplication by dividing both sides by $\frac{3}{4}$. Since division by $\frac{3}{4}$ is equivalent to multiplication by its **reciprocal,** we can isolate y by multiplying both sides by $\frac{4}{3}$.

$$\frac{3}{4}y = -7$$

$$\frac{4}{3}\left(\frac{3}{4}y\right) = \frac{4}{3}(-7) \quad \text{Use the multiplication property of equality: Multiply both sides by the reciprocal of } \frac{3}{4}, \text{ which is } \frac{4}{3}.$$

$$\left(\frac{4}{3} \cdot \frac{3}{4}\right)y = \frac{4}{3}(-7) \quad \text{Use the associative property of multiplication to regroup.}$$

$$1y = \frac{4}{3}(-7) \quad \text{The product of a number and its reciprocal is 1: } \frac{4}{3} \cdot \frac{3}{4} = 1.$$

$$y = -\frac{28}{3} \quad \text{On the right-hand side, do the multiplication.}$$

Check:
$$\frac{3}{4}y = -7 \quad \text{The original equation.}$$

$$\frac{3}{4}\left(-\frac{28}{3}\right) \overset{?}{=} -7 \quad \text{Substitute } -\frac{28}{3} \text{ for } y.$$

$$-\frac{\overset{1}{\cancel{3}} \cdot \overset{1}{\cancel{4}} \cdot 7}{\cancel{4} \cdot \cancel{3}} \overset{?}{=} -7 \quad \text{Multiply the numerators and the denominators. Factor 28 and divide out the common factors.}$$

$$-7 = -7 \quad \text{Simplify the left-hand side.}$$

Since $-\frac{28}{3}$ satisfies the original equation, the solution set is $\left\{-\frac{28}{3}\right\}$.

SELF CHECK   Solve $\frac{2}{3}b - 3 = -15$.   *Answer:* -18 ■

The equation in Example 3 can be solved using an alternate two-step approach.

$$\frac{3}{4}y = -7$$

$$4\left(\frac{3}{4}y\right) = 4(-7) \quad \text{Multiply both sides by 4 to undo the division by 4.}$$

$$3y = -28 \quad \text{Simplify: } 4\left(\frac{3}{4}y\right) = \frac{4}{1}\left(\frac{3}{4}y\right) = \frac{\overset{1}{\cancel{4}} \cdot 3}{1 \cdot \underset{1}{\cancel{4}}}y = 3y.$$

$$\frac{3y}{3} = -\frac{28}{3} \quad \text{Divide both sides by 3 to undo the multiplication by 3.}$$

$$y = -\frac{28}{3}$$

Simplifying Expressions to Solve Equations

To solve more complicated equations, we often need to apply the distributive property and combine like terms.

EXAMPLE 4   **Using the distributive property.** Solve $3(a + 2) = 20$.

Solution   We begin by applying the distributive property to remove parentheses on the left-hand side.

$$3(a + 2) = 20$$
$$3a + 6 = 20 \qquad \text{Distribute the 3.}$$
$$3a + 6 - 6 = 20 - 6 \qquad \text{To undo the addition of 6, subtract 6 from both sides.}$$
$$3a = 14 \qquad \text{Simplify each side of the equation.}$$
$$\frac{3a}{3} = \frac{14}{3} \qquad \text{To undo the multiplication by 3, divide both sides by 3.}$$
$$a = \frac{14}{3} \qquad \text{Simplify the left-hand side.}$$

Check: $3(a + 2) = 20 \qquad$ The original equation.

$$3\left(\frac{14}{3} + 2\right) \overset{?}{=} 20 \qquad \text{Substitute } \tfrac{14}{3} \text{ for } a.$$

$$3\left(\frac{14}{3} + \frac{6}{3}\right) \overset{?}{=} 20 \qquad \text{Get a common denominator: } 2 = \tfrac{6}{3}.$$

$$3\left(\frac{20}{3}\right) \overset{?}{=} 20 \qquad \text{Add the fractions: } \tfrac{14}{3} + \tfrac{6}{3} = \tfrac{20}{3}.$$

$$20 = 20 \qquad \text{Simplify the left-hand side.}$$

Since $\frac{14}{3}$ satisfies the equation, the solution set is $\left\{\frac{14}{3}\right\}$.

SELF CHECK   Solve $-2(x + 3) = 18$.   *Answer:* -12 ■

E X A M P L E   5

Variables on both sides of the equation. Solve
$4b - 7 + 2b = 1 + 2b + 8$.

Solution   First, we combine like terms on each side of the equation.

$$4b - 7 + 2b = 1 + 2b + 8$$
$$6b - 7 = 2b + 9 \qquad \text{Combine like terms: } 4b + 2b = 6b \text{ and } 1 + 8 = 9.$$

We note that terms involving b appear on both sides of the equation. To isolate b on the left-hand side, we need to eliminate $2b$ on the right-hand side.

$6b - 7 = 2b + 9$	
$6b - 7 - \mathbf{2b} = 2b + 9 - \mathbf{2b}$	Subtract $2b$ from both sides.
$4b - 7 = 9$	Combine like terms on each side: $6b - 2b = 4b$ and $2b - 2b = 0$.
$4b - 7 + \mathbf{7} = 9 + \mathbf{7}$	To undo the subtraction of 7, add 7 to both sides.
$4b = 16$	Simplify each side of the equation.
$b = 4$	To isolate b, divide both sides by 4.

Check: $4b - 7 + 2b = 1 + 2b + 8$	The original equation.
$4(4) - 7 + 2(4) \stackrel{?}{=} 1 + 2(4) + 8$	Substitute 4 for b.
$16 - 7 + 8 \stackrel{?}{=} 1 + 8 + 8$	Do each multiplication.
$17 = 17$	Simplify each side.

Since 4 satisfies the equation, the solution set is {4}.

SELF CHECK   Solve $-6t - 12 - 6t = 1 + 2t - 5$.   *Answer:* $-\frac{4}{7}$ ∎

In general, to solve linear equations in one variable, we will follow these steps.

Solving linear equations

1. If the equation contains fractions, multiply both sides of the equation by a number that will eliminate the denominators.

2. Use the distributive property to remove all sets of parentheses and then combine like terms.

3. Use the addition and subtraction properties to get all variables on one side of the equation and all numbers on the other side. Combine like terms, if necessary.

4. Use the multiplication and division properties to make the coefficient of the variable equal to 1.

5. Check the result by replacing the variable with the possible solution and verifying that the number satisfies the equation.

EXAMPLE 6   **Using the equation-solving strategy.** Solve $\frac{1}{3}(6x - 15) = \frac{3}{2}(x - 2) + 2$.

Solution   **Step 1:** One way to clear the equation of fractions is to multiply both sides by the least common denominator (LCD) of $\frac{1}{3}$ and $\frac{3}{2}$. The LCD of these fractions is the smallest number that can be divided by both 2 and 3 exactly. That number is 6.

$$\frac{1}{3}(6x - 15) = \frac{3}{2}(x - 2) + 2$$

$$6\left[\frac{1}{3}(6x \div 15)\right] = 6\left[\frac{3}{2}(x - 2) + 2\right] \qquad \text{To eliminate the fractions, multiply both sides by the LCD, 6.}$$

$$2(6x - 15) = 6 \cdot \frac{3}{2}(x - 2) + 6 \cdot 2 \qquad \begin{array}{l}\text{On the left-hand side, do the multiplication: } 6 \cdot \frac{1}{3} = 2. \text{ On the right-hand side, distribute the 6.}\end{array}$$

$$2(6x - 15) = 9(x - 2) + 12 \qquad \begin{array}{l}\text{Do the multiplications on the right-hand side.}\end{array}$$

Step 2: We use the distributive property to remove parentheses and then combine like terms.

$$12x - 30 = 9x - 18 + 12 \qquad \begin{array}{l}\text{On the left-hand side, distribute the 2. On the right-hand side, distribute the 9.}\end{array}$$

$$12x - 30 = 9x - 6 \qquad \text{Combine like terms: } -18 + 12 = -6.$$

Step 3: We use the addition and subtraction properties of equality by subtracting $9x$ from both sides and by adding 30 to both sides.

$$12x - 30 - 9x + 30 = 9x - 6 - 9x + 30$$

$$3x = 24 \qquad \text{On each side, combine like terms.}$$

Step 4: The coefficient of the variable x is 3. To undo the multiplication by 3, we divide both sides by 3.

$$\frac{3x}{3} = \frac{24}{3} \qquad \text{Divide both sides by 3.}$$

$$x = 8 \qquad \text{Do the divisions.}$$

Step 5: We check by substituting 8 for x in the original equation and simplifying:

$$\frac{1}{3}(6x - 15) = \frac{3}{2}(x - 2) + 2$$

$$\frac{1}{3}[6(8) - 15] \stackrel{?}{=} \frac{3}{2}(8 - 2) + 2$$

$$\frac{1}{3}(48 - 15) \stackrel{?}{=} \frac{3}{2}(6) + 2$$

$$\frac{1}{3}(33) \stackrel{?}{=} 9 + 2$$

$$11 = 11$$

Since 8 satisfies the equation, the solution set is $\{8\}$.   ∎

EXAMPLE 7

Using the equation-solving strategy. Solve $\dfrac{x+2}{5} - 4x = \dfrac{8}{5} - \dfrac{x+9}{2}$.

Solution   Some of the steps used to solve an equation can be done in your head, as you will see in this example.

$$\frac{x+2}{5} - 4x = \frac{8}{5} - \frac{x+9}{2}$$

$$10\left(\frac{x+2}{5} - 4x\right) = 10\left(\frac{8}{5} - \frac{x+9}{2}\right)$$ To eliminate the fractions, multiply both sides by the LCD, 10.

$$10 \cdot \frac{x+2}{5} - 10 \cdot 4x = 10 \cdot \frac{8}{5} - 10 \cdot \frac{x+9}{2}$$ On each side, distribute the 10.

$$2(x+2) - 40x = 2(8) - 5(x+9)$$ Do each multiplication by 10.

$$2x + 4 - 40x = 16 - 5x - 45$$ On each side, remove parentheses.

$$-38x + 4 = -5x - 29$$ On each side, combine like terms.

$$-33x = -33$$ Add $5x$ to both sides. Subtract 4 from both sides. These steps are done in your head—we don't show them.

$$x = 1$$ Divide both sides by -33. This step is also done in your head.

Check: $\dfrac{x+2}{5} - 4x = \dfrac{8}{5} - \dfrac{x+9}{2}$

$$\frac{1+2}{5} - 4(1) \overset{?}{=} \frac{8}{5} - \frac{1+9}{2}$$ Substitute 1 for x.

$$\frac{3}{5} - 4 \overset{?}{=} \frac{8}{5} - 5$$

$$\frac{3}{5} - \frac{20}{5} \overset{?}{=} \frac{8}{5} - \frac{25}{5}$$ Write 4 as $\frac{20}{5}$ and 5 as $\frac{25}{5}$.

$$-\frac{17}{5} = -\frac{17}{5}$$ Do each subtraction.

Since 1 satisfies the equation, it is the solution.

SELF CHECK   Solve $\dfrac{a+3}{2} + 2a = \dfrac{3}{2} - \dfrac{a+27}{5}$.

Answer: -2 ∎

EXAMPLE 8

Solving equations involving decimals. Solve $-35.6 = 77.89 - x$.

Solution

$$-35.6 = 77.89 - x$$

$$-35.6 - 77.89 = 77.89 - x - 77.89$$ Subtract 77.89 from both sides.

$$-113.49 = -x$$ Simplify each side of the equation.

$$-113.49 = -1x$$ $-x = -1x$.

$$\frac{-113.49}{-1} = \frac{-1x}{-1}$$ To isolate x, divide both sides by -1.

$$113.49 = x$$ Simplify each side of the equation.

$$x = 113.49$$

Check to see that $x = 113.49$ satisfies the equation.

SELF CHECK   Solve $-1.3 = -2.6 - x$.

Answer: -1.3 ∎

For more complicated equations involving decimals, we can multiply both sides of the equation by a power of 10 to clear the equation of decimals.

EXAMPLE 9

Clearing an equation of decimals. Solve $0.04(12) + 0.01x = 0.02(12 + x)$.

Solution   The equation contains the decimals 0.04, 0.01, and 0.02. Multiplying both sides by $10^2 = 100$ changes the decimals in the equation to integers, which are easier to work with.

$$0.04(12) + 0.01x = 0.02(12 + x)$$

$$\mathbf{100}[0.04(12) + 0.01x] = \mathbf{100}[0.02(12 + x)] \qquad \text{To make 0.04, 0.01, and 0.02 integers, multiply both sides by 100.}$$

$$100 \cdot 0.04(12) + 100 \cdot 0.01x = 100 \cdot 0.02(12 + x) \qquad \text{On the left-hand side, distribute the 100.}$$

$$4(12) + 1x = 2(12 + x) \qquad \text{Do the three multiplications by 100.}$$

$$48 + x = 24 + 2x \qquad \text{Remove parentheses.}$$

$$48 + x - 24 - x = 24 + 2x - 24 - x \qquad \text{Subtract 24 and } x \text{ from both sides.}$$

$$24 = x \qquad \text{Simplify each side.}$$

$$x = 24$$

Check by substituting 24 for x in the original equation.   ■

Identities and Impossible Equations

The equations discussed so far are called **conditional equations.** For these equations, some numbers satisfy the equation and others do not. An **identity** is an equation that is satisfied by every number for which both sides of the equation are defined.

EXAMPLE 10

An equation that is an identity. Solve $-2(x - 1) - 4 = -4(1 + x) + 2x + 2$.

Solution
$$-2(x - 1) - 4 = -4(1 + x) + 2x + 2$$

$$-2x + 2 - 4 = -4 - 4x + 2x + 2 \qquad \text{Use the distributive property.}$$

$$-2x - 2 = -2x - 2 \qquad \text{On each side, combine like terms.}$$

$$-2 = -2 \qquad \text{Add } 2x \text{ to both sides.}$$

The result $-2 = -2$ indicates that the equation will be true for every value of x. Since the equation is an identity, the solution set is the set of real numbers.

SELF CHECK   Solve $3(a + 4) + 5 = 2(a - 1) + a + 19$ and give the solution set.   *Answer:* all real numbers   ■

An **impossible equation** or a **contradiction** is an equation that has no solution.

EXAMPLE 11

An impossible equation. Solve $-6.2(-x - 1) - 4 = 4.2x - (-2x)$.

Solution
$$-6.2(-x - 1) - 4 = 4.2x - (-2x)$$

$$6.2x + 6.2 - 4 = 4.2x + 2x \qquad \text{On the left-hand side, remove parentheses. On the right-hand side, write the subtraction as addition of the opposite.}$$

$$6.2x + 2.2 = 6.2x \qquad \text{On each side, combine like terms.}$$

$$6.2x + 2.2 - \mathbf{6.2x} = 6.2x - \mathbf{6.2x} \qquad \text{Subtract } 6.2x \text{ from both sides.}$$

$$2.2 \neq 0 \qquad \text{Simplify each side.}$$

The result $2.2 \neq 0$ indicates that there is no value for x that will make the equation true. The solution set is the empty set, denoted as $\varnothing$.

SELF CHECK   Solve $3(a + 4) + 2 = 2(a - 1) + a + 19$ and give the solution set.

Answer: $\varnothing$ ■

Solving Formulas

To solve a formula for a variable means to isolate that variable on one side of the equation and have all other quantities on the other side. We can use the skills discussed in this section to solve many types of formulas for a specified variable.

EXAMPLE 12   **Solving a formula for a specified variable.** Solve $A = \frac{1}{2}bh$ for h.

Solution

$$A = \frac{1}{2}bh$$

$2A = bh$ To eliminate the fraction, multiply both sides by 2.

$\dfrac{2A}{b} = h$ To isolate h, divide both sides by b.

$h = \dfrac{2A}{b}$ Write the equation with h on the left-hand side.

SELF CHECK   Solve $A = \frac{1}{2}bh$ for b.

Answer: $b = \frac{2A}{h}$ ■

 WARNING! When asked to solve a formula for a specific variable, we must isolate that variable on one side of the equation. For example, it would be incorrect to say that $t = \frac{4A - S}{t}$ is solved for t, because t appears on *both* sides of the equation.

EXAMPLE 13   **Solving a formula for a specified variable.** For simple interest, the formula $A = P + Prt$ gives the amount of money in an account at the end of a specific time. A represents the amount, P the principal, r the rate of interest, and t the time. We can solve the formula for t as follows:

Solution

$$A = P + Prt$$

$A - P = Prt$ To isolate the term involving t, subtract P from both sides.

$\dfrac{A - P}{Pr} = t$ To isolate t, divide both sides by Pr.

$t = \dfrac{A - P}{Pr}$ Write the equation with t on the left-hand side.

SELF CHECK   Solve $A = P + Prt$ for r.

Answer: $r = \frac{A - P}{Pt}$ ■

EXAMPLE 14   **Solving a formula for a specified variable.** The formula $F = \frac{9}{5}C + 32$ converts degrees Celsius to degrees Fahrenheit. Solve the formula for C.

Solution

$$F = \frac{9}{5}C + 32$$

$F - 32 = \dfrac{9}{5}C$ To isolate the term involving C, subtract 32 from both sides.

$\dfrac{5}{9}(F - 32) = \dfrac{5}{9}\left(\dfrac{9}{5}C\right)$ To isolate C, multiply both sides by $\frac{5}{9}$.

$\dfrac{5}{9}(F - 32) = C$ $\frac{5}{9} \cdot \frac{9}{5} = 1$.

$C = \dfrac{5}{9}(F - 32)$

To convert degrees Fahrenheit to degrees Celsius, we can use the formula $C = \frac{5}{9}(F - 32)$.

SELF CHECK   Solve $S = \frac{180(t - 2)}{7}$ for t.   *Answer:* $t = \frac{7S + 360}{180}$ or $t = \frac{7S}{180} + 2$ ■

STUDY SET
Section 1.5

VOCABULARY

In Exercises 1–6, fill in the blanks to make the statements true.

1. An _____ is a statement that two quantities are equal.

2. If two equations have the same solution set, they are called _____ equations.

3. If a number is substituted for a variable in an equation and the equation is true, we say that the number _____ the equation.

4. An _____ is an equation that is true for all values of its variable.

5. An impossible equation is true for no _____ of its variable.

6. $3x + 1 = 10$ is a _____ equation in one variable.

CONCEPTS

In Exercises 7–8, fill in the blanks to make the statements true.

7. If a, b, and c are real numbers, and $a = b$, then $a + c = b +$ ▢ and $a - c = b -$ ▢.

8. If a, b, and c are real numbers, and $a = b$, then $c \cdot a =$ ▢ $\cdot\, b$ and $\frac{a}{c} = $ ▢ $(c \neq 0)$.

9. **a.** Simplify $5y + 2 - 3y$.
   b. Solve $5y + 2 - 3y = 8$.
   c. Evaluate $5y + 2 - 3y$ for $y = 8$.

10. **a.** In $-5b + 3 = -18$, what can we do to both sides to undo the addition of 3?
    b. In $-5b = -18$, what can we do to both sides to undo the multiplication by -5?

11. Is $a = \frac{3x + 2}{a}$ solved for a? Why or why not?

12. Is $2x = 3m + 2$ solved for x? Why or why not?

13. When solving $\frac{x+1}{3} - \frac{2}{15} = \frac{x-1}{5}$, why would we multiply both sides by 15?

14. When solving $1.45x - 0.5(1 - x) = 0.7x$, why would we multiply both sides by 100?

NOTATION

In Exercises 15–16, complete each solution.

15. Solve the equation $-2(x + 7) = 20$.

$$-2(x + 7) = 20$$
$$\boxed{} - 14 = 20$$
$$-2x - 14 + \boxed{} = 20 + \boxed{}$$
$$-2x = \boxed{}$$
$$\frac{-2x}{\boxed{}} = \frac{34}{\boxed{}}$$
$$x = -17$$

16. Solve the equation $\frac{d}{3} + 4 = 1$.

$$\frac{d}{3} + 4 = 1$$
$$\frac{d}{3} + 4 - \boxed{} = 1 - \boxed{}$$
$$\frac{d}{3} = \boxed{}$$
$$\boxed{}\left(\frac{d}{3}\right) = \boxed{}(-3)$$
$$d = -9$$

17. Fill in the blanks to make the statements true.
    a. $-x = \boxed{}\, x$
    b. $\frac{2t}{3} = \boxed{}\, t$

18. When checking a solution of an equation, the symbol $\overset{?}{=}$ is used. What does it mean?

PRACTICE

In Exercises 19–22, tell whether 5 is a solution of each equation.

19. $3x + 2 = 17$

20. $7x - 2 = 53 - 5x$

21. $3(2m - 3) = 15$

22. $\frac{3}{5}p - 5 = -2$

In Exercises 23–70, solve each equation.

23. $\frac{x}{4} = 7$

24. $-\frac{x}{6} = 8$

25. $-\frac{4}{5}s = 16$

26. $-3 = -\frac{9}{8}s$

27. $2x - 1 = 0$

28. $3x - 4 = 0$

29. $5y + 6 = 0$

30. $7y + 3 = 0$

31. $8x = x$

32. $-z = 5z$

33. $0 = -x$

34. $-\frac{2}{3}x = 0$

35. $2x + 2(1) = 6$

36. $3x - 4(1) = 8$

37. $4(3) + 2y = -6$

38. $5(2) + 10y = -10$

39. $200 = 34 - t$

40. $8 - x = -12$

41. $1.6a = 4.032$

42. $0.52 = 0.05y$

43. $3x + 1 = 3$

44. $8k - 2 = 13$

45. $3(k - 4) = -36$

46. $4(x + 6) = 84$

47. $4j + 12.54 = 18.12$

48. $9.8 - 15r = -15.7$

49. $4a - 22 - a = -2a - 7$

50. $a + 18 = 5a - 3 + a$

51. $2(2x + 1) = x + 15 + 2x$

52. $-2(x + 5) = x + 30 - 2x$

53. $2(a - 5) - (3a + 1) = 0$

54. $8(3a - 5) - 4(2a + 3) = 12$

55. $9(x + 2) = -6(4 - x) + 18$

56. $3(x + 2) - 2 = -(5 + x) + x$

57. $\frac{1}{2}x - 4 = -1 + 2x$

58. $2x + 3 = \frac{2}{3}x - 1$

59. $\frac{b}{2} - \frac{b}{3} = 4$

60. $\frac{w}{2} + \frac{w}{3} = 10$

61. $\frac{a + 1}{3} + \frac{a - 1}{5} = \frac{2}{15}$

62. $\frac{2z + 3}{3} + \frac{3z - 4}{6} = \frac{z - 2}{2}$

63. $\frac{5a}{2} - 12 = \frac{a}{3} + 1$

64. $5 - \frac{x + 2}{3} = 7 - x$

65. $\frac{3 + p}{3} - 4p = 1 - \frac{p + 7}{2}$

66. $\frac{4 - t}{2} - \frac{3t}{5} = 2 + \frac{t + 1}{3}$

67. $0.45 = 16.95 - 0.25(75 - 3x)$

68. $0.02x + 0.0175(15,000 - x) = 277.5$

69. $0.04(12) + 0.01t - 0.02(12 + t) = 0$

70. $0.25(t + 32) = 3.2 + x$

In Exercises 71–76, solve each equation. If the equation is an identity or an impossible equation, so indicate.

71. $4(2 - 3t) + 6t = -6t + 8$

72. $2x - 6 = -2x + 4(x - 2)$

73. $3(x - 4) + 6 = -2(x + 4) + 5x$

74. $2(x - 3) = \frac{3}{2}(x - 4) + \frac{x}{2}$

75. $2y + 1 = 5(0.2y + 1) - (4 - y)$

76. $-3x = -2x + 1 - (5 + x)$

In Exercises 77–92, solve each formula for the indicated variable.

77. $V = \frac{1}{3}Bh$ for B

78. $A = \frac{1}{2}bh$ for b

79. $I = Prt$ for t

80. $E = mc^2$ for m

81. $P = 2l + 2w$ for w

82. $T - W = ma$ for W

83. $A = \frac{1}{2}h(B + b)$ for B

84. $l = a + (n - 1)d$ for n

85. $y = mx + b$ for x

86. $\lambda = Ax + AB$ for B

87. $\bar{v} = \frac{1}{2}(v + v_0)$ for v_0

88. $l = a + (n - 1)d$ for d

89. $S = \frac{a - lr}{1 - r}$ for l

90. $s = \frac{1}{2}gt^2 + vt$ for g

91. $S = \frac{n(a + l)}{2}$ for l

92. $K = \frac{Mv_0^2}{2} + \frac{Iw^2}{2}$ for I

APPLICATIONS

93. CONVERTING TEMPERATURES In preparing an American almanac for release in Europe, editors need to convert temperature ranges for the planets from degrees Fahrenheit to degrees Celsius. Solve the formula $F = \frac{9}{5}C + 32$ for C. Then use your result to make the conversions for the data shown in Illustration 1. Round to the nearest degree.

ILLUSTRATION 1

Planet	High °F	Low °F	High °C	Low °C
Mercury	810	−290		
Earth	136	−129		
Mars	63	−87		

94. THERMODYNAMICS In thermodynamics, the Gibbs free-energy function is given by the formula $G = U − TS + pV$. Solve for S.

95. WIPER DESIGN The area cleaned by the windshield wiper assembly shown in Illustration 2 is given by the formula

$$A = \frac{d\pi(r_1^{\,2} - r_2^{\,2})}{360}$$

Engineers have determined the amount of windshield area that needs to be cleaned by the wiper for two different vehicles. Solve the equation for d and use your result to find the number of degrees d the wiper arm must swing in each case. Round to the nearest degree.

Vehicle	Area cleaned	d (deg)
Luxury car	513 in.^{2}	
Sport utility vehicle	586 in.^{2}	

ILLUSTRATION 2

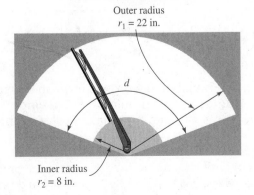

Outer radius
$r_1 = 22$ in.

d

Inner radius
$r_2 = 8$ in.

96. ELECTRONICS Illustration 3 is a schematic diagram of a resistor connected to a voltage source of 60 volts. As a result, the resistor dissipates power in the form of heat. The power P lost when a voltage E is placed across a resistance R (in ohms) is given by the formula

$$P = \frac{E^2}{R}$$

Solve for R. If P is 4.8 watts and E is 60 volts, find R.

ILLUSTRATION 3

Battery $E = 60$ v   Resistor

97. CHEMISTRY LAB In chemistry, the ideal gas law equation is $PV = nR(T + 273)$, where P is the pressure, V the volume, T the temperature, and n is the number of moles of a gas. R is a constant, 0.082. Solve the equation for n. Then use your result and the data from the student lab notebook in Illustration 4 to find the value of n to the nearest thousandth for trial 1 and trial 2.

ILLUSTRATION 4

Ideal gas law Lab #1		Betsy Kinsell Chem 1 Section A	
Data:	Pressure (Atmosph.)	Volume (Liters)	Temp (°C)
Trial 1	0.900	0.250	90
Trial 2	1.250	1.560	−10
$R = 0.082$ (Constant)			

98. INVESTMENT An amount P, invested at a simple interest rate r, will grow to an amount A in t years according to the formula $A = P(1 + rt)$. Solve for P. Suppose a man invested some money at 5.5%. If after 5 years, he had $6,693.75 on deposit, what amount did he originally invest?

99. COST OF ELECTRICITY The cost of electricity in a city is given by the formula $C = 0.07n + 6.50$, where C is the cost and n is the number of kilowatt hours used. Solve for n. Then find the number of kilowatt hours used each month by the homeowner whose checks to pay the monthly electric bills are shown in Illustration 5 on page 60.

ILLUSTRATION 5

design shown in Illustration 7 have if the performance platform is a regular polygon with interior angles measuring 135°?

ILLUSTRATION 7

100. COST OF WATER A monthly water bill in a certain city is calculated by using the formula $n = \frac{5,000C - 17,500}{6}$, where n is the number of gallons used and C is the monthly cost. Solve for C and compute the bill for quantities of 500, 1,200, and 2,500 gallons.

101. SURFACE AREA To find the amount of tin needed to make a coffee can such as that shown in Illustration 6, we use the formula for the surface area of a right circular cylinder,

$$A = 2\pi r^2 + 2\pi rh$$

Solve the formula for h.

ILLUSTRATION 6

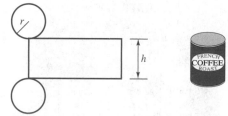

102. CARPENTRY A regular polygon has n equal sides and n equal angles. The measure a of an interior angle in degrees is given by $a = 180\left(1 - \frac{2}{n}\right)$. Solve for n. How many sides does the outdoor bandstand

WRITING

103. What does it mean to *solve an equation?*

104. Why doesn't the equation $x = x + 1$ have a real-number solution?

105. If you are asked to solve the equation $2s = 7t - s$ for s, why is $s = \frac{7t - s}{2}$ an incorrect answer?

106. When solving a linear equation in one variable, the objective is to isolate the variable on one side of the equation. What does that mean?

REVIEW

In Exercises 107–114, simplify each expression.

107. $-(4 + t) + 2t$

108. $12(2r + 11) - 11 - 3$

109. $4(b + 8) - 8b$

110. $-2(m - 3) + 8(2m + 7)$

111. $3.8b - 0.9b$ **112.** $-5.7p + 5.1p$

113. $\frac{3}{5}t + \frac{2}{5}t$ **114.** $-\frac{3}{16}x - \frac{5}{16}x$

▶ 1.6   Using Equations to Solve Problems

In this section, you will learn about

     A problem-solving strategy ■ Translating words to form an equation ■ Analyzing a problem ■ Number–value problems ■ Drawing diagrams ■ Geometry problems ■ Using formulas to solve problems

---

Introduction One of the objectives of this course is to improve your problem-solving abilities. In the next two sections, you will have the opportunity to do that as we discuss how to use equations to solve many different types of problems.

A Problem-Solving Strategy

The key to problem solving is understanding the problem and devising a plan for solving it. The following list provides a strategy for solving problems.

Problem solving

1. *Analyze the problem* by reading it carefully to understand the given facts. What information is given? What vocabulary is given? What are you asked to find? Often a diagram or table will help you visualize the facts of the problem.
2. *Form an equation* by picking a variable to represent the quantity to be found. Then express all other quantities mentioned as expressions involving the variable. Finally, write an equation expressing a quantity in two different ways.
3. *Solve the equation.*
4. *State the conclusion.*
5. *Check the result* in the words of the problem.

Translating Words to Form an Equation

In order to solve problems, which are almost always given in words, we must translate those words into mathematical symbols. In the next example, we use translation to write an equation that mathematically models the situation.

EXAMPLE 1

Top U.S. employers. As of early 1998, the top two employers in the United States were the United States Postal Service and Wal-Mart Stores, employing a combined total of 1,375,000 people. If Wal-Mart employed 25,000 fewer people than the Postal Service, how many people worked for the Postal Service?

ANALYZE THE PROBLEM We are to find the number of people working for the Postal Service. Translating two key phrases in the problem will help us write an equation.

- The phrase "employing a combined total of 1,375,000 people" indicates that if we add the number of Postal Service employees and the number of Wal-Mart employees, the result will be 1,375,000.

- The phrase "Wal-Mart employed 25,000 fewer people than the Postal Service" indicates that if the Postal Service employed x people, Wal-Mart employed $x - 25,000$ people.

FORM AN EQUATION To form an equation, we translate the words into mathematical symbols. We let x represent the number of people employed by the Postal Service. Then $x - 25,000$ represents the number of people employed by Wal-Mart.

The number of people employed by the Postal Service	plus	the number of people employed by Wal-Mart	is	1,375,000.
x	$+$	$x - 25,000$	$=$	1,375,000

SOLVE THE EQUATION

$$x + x - 25,000 = 1,375,000$$
$$2x - 25,000 = 1,375,000 \quad \text{Combine like terms.}$$
$$2x = 1,400,000 \quad \text{Add 25,000 to both sides.}$$
$$x = 700,000 \quad \text{Divide both sides by 2.}$$

STATE THE CONCLUSION   As of 1998, the Postal Service employed 700,000 people.

CHECK THE RESULT   If the Postal Service employed 700,000, Wal-Mart employed 25,000 less, or 675,000 people. Adding 700,000 and 675,000, we get 1,375,000. The answer checks. ∎

Analyzing a Problem

Sometimes the wording of a problem doesn't contain key phrases that translate directly to an equation. In such cases, an analysis of the problem often gives clues that help us write an equation.

EXAMPLE 2

Travel promotion. The price of a 7-day Alaskan cruise, normally $2,752 per person, is reduced by $1.75 per person for large groups traveling together. How large a group is needed for the price to be $2,500 per person?

ANALYZE THE PROBLEM   For each member of the group, the cost is reduced by $1.75. For a group of 20 people, the $2,752 price is reduced by 20($1.75) = $35.

The per-person price of the cruise = $2,752 − 20($1.75)

For a group of 30 people, the $2,752 cost is reduced by 30($1.75) = $52.50.

The per-person price of the cruise = $2,752 − 30($1.75)

FORM AN EQUATION   We can describe this situation in words. If we let x represent the group size necessary for the price of the cruise to be $2,500 per person, we can translate the words into an equation.

The price of the cruise	is	$2,752	minus	the number of people in the group	times	$1.75.
2,500	=	2,752	−	x	·	1.75

SOLVE THE EQUATION

$$2,500 = 2,752 - 1.75x$$
$$2,500 - \mathbf{2,752} = 2,752 - 1.75x - \mathbf{2,752} \qquad \text{Subtract 2,752 from both sides.}$$
$$-252 = -1.75x \qquad \text{Simplify each side.}$$
$$144 = x \qquad \text{Divide both sides by } -1.75.$$

STATE THE CONCLUSION   If 144 people travel together, the price will be $2,500 per person.

CHECK THE RESULT   For 144 people, the cruise cost of $2,752 will be reduced by 144($1.75) = $252. If we subtract, $2,752 − $252 = $2,500. The answer checks. ∎

Number–Value Problems

Some problems deal with quantities that have a value. In these problems, we must distinguish between the *number of* and the *value of* the unknown quantity. For problems such as these, we will use the relationship

Number · value = total value

EXAMPLE 3

Portfolio analysis. A college foundation owns stock in Kodak (selling at $58 per share), Coca-Cola (selling at $65 per share), and Microsoft (selling at $125 per share). The foundation owns an equal number of shares of Kodak and Coca-Cola stock, but five times as many shares of Microsoft stock. If this portfolio is worth $523,600, how many shares of each stock does the foundation own?

ANALYZE THE PROBLEM The value of the Kodak stock plus the value of the Coca-Cola stock plus the value of the Microsoft stock must equal $523,600. We need to find the number of shares of each of these stocks that the foundation has in its portfolio.

FORM AN EQUATION If we let x represent the number of shares of Kodak stock, then x also represents the number of shares of Coca-Cola stock. Since the foundation owns five times as many shares of Microsoft stock as Kodak or Coca-Cola stock, $5x$ represents the number of shares of Microsoft. The value of the shares of each company's stock would be the *product* of the number of shares of that stock and its per-share value. See Figure 1-15.

FIGURE 1-15

Stock	Number of shares	·	Value per share	=	Total value of the stock
Kodak	x		$58		$58x$
Coca-Cola	x		$65		$65x$
Microsoft	$5x$		$125		$125(5x)$

We can now form the equation.

The value of Kodak stock	plus	the value of Coca-Cola stock	plus	the value of Microsoft stock	is	the total value of all of the stock.
$58x$	$+$	$65x$	$+$	$125(5x)$	$=$	$523,600$

SOLVE THE EQUATION

$$58x + 65x + 125(5x) = 523,600$$
$$58x + 65x + 625x = 523,600 \quad \text{125(5x) = 625x.}$$
$$748x = 523,600 \quad \text{Combine like terms on the left-hand side.}$$
$$x = 700 \quad \text{Divide both sides by 748.}$$

STATE THE CONCLUSION The foundation owns 700 shares of Kodak, 700 shares of Coca-Cola, and $5(700) = 3,500$ shares of Microsoft.

CHECK THE RESULT The value of 700 shares of Kodak stock is $700(\$58) = \$40,600$. The value of 700 shares of Coca-Cola stock is $700(\$65) = \$45,500$. The value of 3,500 shares of Microsoft is $3,500(\$125) = \$437,500$. The sum is $\$40,600 + \$45,500 + \$437,500 = \$523,600$. The answers check. ■

Drawing Diagrams

When solving problems, diagrams are often helpful, because they allow us to visualize the facts of the problem.

EXAMPLE 4

Triathlon course. A triathlon in Hawaii includes swimming, long-distance running, and cycling. The long-distance run is 11 times longer than the distance the competitors swim. The distance they cycle is 85.8 miles longer than the run. Overall, the competition covers 140.6 miles. Find the length of each part of the triathlon and round each length to the nearest tenth of a mile.

ANALYZE THE PROBLEM   The entire triathlon course covers a distance of 140.6 miles. We note that the distance the competitors run is related to the distance they swim, and the distance they cycle is related to the distance they run.

FORM AN EQUATION   If x represents the distance the competitors swim, then $11x$ represents the length of the long-distance run, and $11x + 85.8$ represents the distance they cycle. From the diagram in Figure 1-16, we see that the sum of the individual parts of the triathlon must equal the total distance covered.

FIGURE 1-16

The distance they swim	plus	the distance they run	plus	the distance they cycle	is	the total length of the course.
x	$+$	$11x$	$+$	$11x + 85.8$	$=$	140.6

SOLVE THE EQUATION
$$x + 11x + 11x + 85.8 = 140.6$$
$$23x + 85.8 = 140.6 \qquad \text{Combine like terms.}$$
$$23x = 54.8 \qquad \text{Subtract 85.8 from both sides.}$$
$$x \approx 2.382608696 \qquad \text{Divide both sides by 23.}$$

STATE THE CONCLUSION   To the nearest tenth, the distance the competitors swim is 2.4 miles. The distance they run is $11x$, or approximately $11(2.382608696) = 26.20869565$ miles. To the nearest tenth, that is 26.2 miles. The distance they cycle is $11x + 85.8$, or approximately $26.20869565 + 85.8 = 112.0086957$ miles. To the nearest tenth, that is 112.0 miles.

CHECK THE RESULT   If we add the lengths of the three parts of the triathlon and round to the nearest tenth, we get 140.6 miles. The answers check.   ∎

Geometry Problems

Sometimes a geometric fact or formula is helpful in solving a problem. Figure 1-17 shows several geometric figures. A **right angle** is an angle whose measure is 90°. A **straight angle** is an angle whose measure is 180°. An **acute angle** is an angle whose measure is greater than 0° and less than 90°. An angle whose measure is greater than 90° and less than 180° is called an **obtuse angle.**

If the sum of two angles equals 90°, the angles are called **complementary,** and each angle is called the **complement** of the other. If the sum of two angles equals 180°, the angles are called **supplementary,** and each angle is the **supplement** of the other.

A **right triangle** is a triangle with one right angle. An **isosceles triangle** is a triangle with two sides of equal measure that meet to form the **vertex angle.** The angles opposite the equal sides, called the **base angles,** are also equal. An **equilateral triangle** is a triangle with three equal sides and three equal angles.

FIGURE 1-17

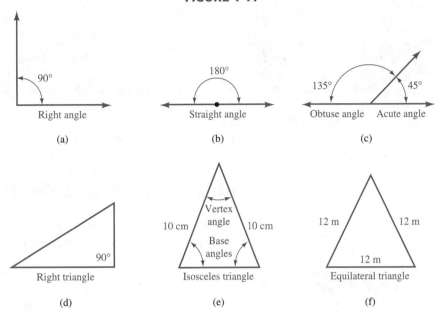

Right angle (a)

Straight angle (b)

Obtuse angle Acute angle (c)

Right triangle (d)

Isosceles triangle (e)

Equilateral triangle (f)

EXAMPLE 5

Flag design. The flag of Guyana, a republic on the northern coast of South America, is one isosceles triangle superimposed over another isosceles triangle on a field of green, as shown in Figure 1-18. The measure of a base angle of the larger triangle is 14° more than the measure of a base angle of the smaller triangle. The measure of the vertex angle of the larger triangle is 34°. Find the measure of each base angle of the smaller triangle.

FIGURE 1-18

ANALYZE THE PROBLEM We are working with isosceles triangles. Therefore, the base angles of the smaller triangle have the same measure, and the base angles of the larger triangle have the same measure.

FORM AN EQUATION If we let x represent the measure in degrees of one base angle of the smaller isosceles triangle, then the measure of its other base angle is also x. (See Figure 1-19.) The measure of a base angle of the larger isosceles triangle is $(x + 14)°$, since its measure is 14° more than the measure of a base angle of the smaller triangle. We are given that the vertex angle of the larger triangle measures 34°.

The sum of the measures of the angles of any triangle (in this case, the larger triangle) is 180°.

FIGURE 1-19

The measure of one base angle	plus	the measure of the other base angle	plus	the measure of the vertex angle	is	180°.
$x + 14$	$+$	$x + 14$	$+$	34	$=$	180

SOLVE THE EQUATION

$$x + 14 + x + 14 + 34 = 180$$

$$2x + 62 = 180 \quad \text{Combine like terms.}$$

$$2x = 118 \quad \text{Subtract 62 from both sides.}$$

$$x = 59 \quad \text{Divide both sides by 2.}$$

STATE THE CONCLUSION

The measure of each base angle of the smaller triangle is 59°.

CHECK THE RESULT

If $x = 59$, then $x + 14 = 73$. The sum of the measures of each base angle and the vertex angle of the *larger* triangle is $73° + 73° + 34° = 180°$. The answer checks. ∎

Using Formulas to Solve Problems

When preparing to write an equation to solve a problem, the given facts of the problem often suggest a formula that can be used to model the situation mathematically.

EXAMPLE 6

Designing a kennel. A man has a 50-foot roll of fencing to make a rectangular kennel. If he wants the kennel to be 6 feet longer than it is wide, find its dimensions.

ANALYZE THE PROBLEM

The perimeter P of the rectangular kennel is 50 feet. Recall that the formula for the perimeter of a rectangle is $P = 2l + 2w$. We need to find its length and width.

FORM AN EQUATION

We let w represent the width of the kennel. Then the length, which is 6 feet more than the width, is represented by the expression $w + 6$. See Figure 1-20.

FIGURE 1-20

We can now form the equation by substituting 50 for P, $w + 6$ for the length, and w for the width in the formula for the perimeter of a rectangle.

$$P = 2l + 2w$$
$$50 = 2(w + 6) + 2w$$

SOLVE THE EQUATION

$$50 = 2(w + 6) + 2w$$

$$50 = 2w + 12 + 2w \quad \text{Use the distributive property to remove parentheses.}$$

$$50 = 4w + 12 \quad \text{Combine like terms.}$$

$$38 = 4w \quad \text{Subtract 12 from both sides.}$$

$$9.5 = w \quad \text{Divide both sides by 4.}$$

STATE THE CONCLUSION

The width of the kennel is 9.5 feet. The length would be 6 feet more than this, or 15.5 feet.

CHECK THE RESULT

If a rectangle has a width of 9.5 feet and a length of 15.5 feet, its length is 6 feet more than its width, and the perimeter is $2(9.5)$ feet $+ 2(15.5)$ feet $= 50$ feet. ∎

VOCABULARY

In Exercises 1–8, fill in the blanks to make the statements true.

1. An _____ angle has a measure of more than 0° and less than 90°.

2. A _____ angle is an angle whose measure is 90°.

3. If the sum of the measures of two angles equals 90°, the angles are called _____ angles.

4. If the sum of the measures of two angles equals 180°, the angles are called _____ angles.

5. If a triangle has a right angle, it is called a _____ triangle.

6. If a triangle has two sides with equal measures, it is called an _____ triangle.

7. The sum of the measures of the _____ of a triangle is 180°.

8. An _____ triangle has three sides of equal length and three angles of equal measure.

CONCEPTS

9. The unit used to measure the intensity of sound is called the *decibel*. In Illustration 1, translate the comments in the right-hand column into mathematical symbols to complete the decibel column.

ILLUSTRATION 1

	Decibels	Compared to conversation
Conversation	d	—
Vacuum cleaner		15 decibels more
Circular saw		10 decibels less than twice
Jet takeoff		20 decibels more than twice
Whispering		10 decibels less than half
Rock band		Twice the decibel level

10. Illustration 2 shows the four types of problems an instructor put on a history test.
    a. Complete the table.
    b. Which type of question appears the most on the test?

c. Write an algebraic expression that represents the total number of points on the test.

ILLUSTRATION 2

Type of question	Number	Value	Total value
Multiple choice	x	5	
True/false	$3x$	2	
Essay	$x - 2$	10	
Fill-in	x	5	

APPLICATIONS

11. Illustration 3 shows the three pieces that fit together to make a flute.
    a. What is the length of the shortest piece?
    b. What is the length of the longest piece?
    c. Write an algebraic expression that represents the length of the flute.

ILLUSTRATION 3

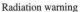

12. For each picture in Illustration 4, what geometric concept studied in this section is illustrated?

ILLUSTRATION 4

Radiation warning

(a)   (b)

13. Write an algebraic expression that represents the length in inches of the head of the tennis racquet shown in Illustration 5.

ILLUSTRATION 5

x in.

26.5 in.

14. The surface area of the earth is 510,066,000 square kilometers (km^2). If we let x represent the number of km^2 covered by water, what would the algebraic expression 510,066,000 − x represent?

15. CEREAL SALES In 1997, the two top-selling cereals were General Mills' Cheerios and Kellogg's Frosted Flakes, with combined sales of $549,000,000. Frosted Flakes sales were $39,000,000 less than sales of Cheerios. What were the 1997 sales for each brand?

16. MARKET SHARE In 1997, the top three brands of toothpaste captured 57.5% of the market share. Crest, the top seller, had 3.3% more of the market than the number two seller, Colgate. Aquafresh, the number three seller, had a market share that was 9.4% less than that of Colgate. What was the market share in 1997 for each of these brands?

17. SPRING TOUR A group of junior high students will be touring Washington, DC. Their chaperons will have the $1,810 cost of the tour reduced by $15.50 for each student they personally supervise. How many students will a chaperon have to supervise so that his or her cost to take the tour will be $1,500?

18. MACHINING Each pass through a lumber plane shaves off 0.015 inch of thickness from a board. How many times must a board, originally 0.875 inch thick, be run through the planer if a board of thickness 0.74 inch is desired?

19. MOVING EXPENSES To help move his furniture, a man rents a truck for $29.25 per day plus 15¢ per mile. If he has budgeted $75 for transportation expenses, how many miles will he be able to drive the truck if the move takes 1 day?

20. COMPUTING SALARIES A student working for a delivery company earns $47.50 per day plus $1.75 for each package she delivers. How many deliveries must she make each day to earn $100 a day?

21. VALUE OF AN IRA In an Individual Retirement Account (IRA) valued at $53,900, a couple has 500 shares of stock, some in Big Bank Corporation and some in Safe Savings and Loan. If Big Bank sells for $115 per share and Safe Savings sells for $97 per share, how many shares of each does the couple own?

22. ASSETS OF A PENSION FUND A pension fund owns 2,000 fewer shares in mutual stock funds than mutual bond funds. Currently, the stock funds sell for $12 per share, and the bond funds sell for $15 per share. How many shares of each does the pension fund own if the value of the securities is $165,000?

23. SELLING CALCULATORS Last month, a bookstore ran the ad shown in Illustration 6. Sales of $4,980 were generated, with 15 more graphing calculators sold than scientific calculators. How many of each type of calculator did the bookstore sell?

ILLUSTRATION 6

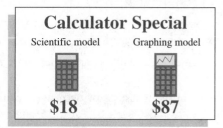

24. SELLING GRASS SEED A seed company sells two grades of grass seed. A 100-pound bag of a mixture of rye and Kentucky bluegrass sells for $245, and a 100-pound bag of bluegrass sells for $347. How many bags of each are sold in a week when the receipts for 19 bags are $5,369?

25. WOODWORKING The carpenter in Illustration 7 saws a board into two pieces. He wants one piece to be 1 foot longer than twice the length of the shorter piece. Find the length of each piece.

ILLUSTRATION 7

26. STATUE OF LIBERTY From the foundation of the large pedestal on which it sits, to the top of the torch, the Statue of Liberty National Monument measures 305 feet. The pedestal is 3 feet taller than the statue. Find the height of the pedestal and the height of the statue.

27. NURSING Illustration 8 shows the angle a needle should make with the skin when administering a certain type of intradermal injection. Find the measure of both angles labeled in the illustration.

ILLUSTRATION 8

28. SUPPLEMENTARY ANGLES Refer to Illustration 9 and find x.

ILLUSTRATION 9

29. ARCHITECTURE Because of soft soil and a shallow foundation, the Leaning Tower of Pisa in Italy is not vertical, as shown in Illustration 10. How many degrees from vertical is the tower?

ILLUSTRATION 10

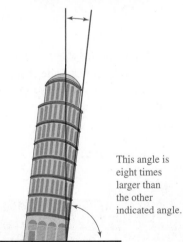

This angle is eight times larger than the other indicated angle.

30. KITCHEN STEPSTOOL The sum of the measures of the three angles of any triangle is 180°. In Illustration 11, the measure of angle 2 is 10° larger than the measure of angle 1. The measure of angle 3 is 10° larger than the measure of angle 2. Find each angle measure.

ILLUSTRATION 11

31. SUPPLEMENTARY ANGLES AND PARALLEL LINES In Illustration 12, lines r and s are cut by a third line l to form angles 1 and 2. When lines r and s are parallel, angles 1 and 2 are supplementary. If angle $1 = (x + 50)°$, angle $2 = (2x - 20)°$, and lines r and s are parallel, find x.

ILLUSTRATION 12

32. ANGLES AND PARALLEL LINES In Illustration 13, $r \parallel s$ (read as "line r is parallel to line s"), and $a = 103$. Find b, c, and d. (*Hint:* See Exercise 31.)

ILLUSTRATION 13

33. VERTICAL ANGLES When two lines intersect as in Illustration 14, four angles are formed. Angles that are side-by-side, such as ∠1 (angle 1) and ∠2, are called **adjacent angles.** Angles that are nonadjacent, such as ∠1 and ∠3 or ∠2 and ∠4, are called **vertical angles.** From geometry, we know that if two lines intersect, vertical angles have the same measure. If $m(\angle 1) = (3x + 10)°$ and $m(\angle 3) = (5x - 10)°$, find x. (Read $m(\angle 1)$ as "the measure of ∠1.")

ILLUSTRATION 14

34. ANGLES AND PARALLEL LINES In Illustration 15, $r \parallel s$ (read as "line r is parallel to line s"), and $b = 137$. Find a and c.

ILLUSTRATION 15

35. ANGLES OF A QUADRILATERAL The sum of the angles of any four-sided figure (called a *quadrilateral*) is 360°. The quadrilateral shown in Illustration 16 has two equal base angles. Find x.

ILLUSTRATION 16

36. HEIGHT OF A TRIANGLE If the height of a triangle with a base of 8 inches is tripled, its area is increased by 96 square inches. Find the height of the triangle.

37. GOLDEN RECTANGLES Throughout history, most artists and designers have felt that rectangles having a length 1.618 times as long as their width have the most visually attractive shape. Such rectangles are known as *golden rectangles*. Measure the length and width of the rectangles in Illustration 17. Which of the rectangles is closest to being a golden rectangle? Do you agree that it is more visually attractive?

ILLUSTRATION 17

i.

ii.

38. QUILTING A woman is planning to make a quilt in the shape of a golden rectangle. (See Exercise 37.) She has exactly 22 feet of a special lace that she plans to sew around the edge of the quilt. What should the length and width of the quilt be? Round both answers up to the nearest hundredth.

39. FENCING A PASTURE A farmer has 624 feet of fencing to enclose the pasture shown in Illustration 18. Because a river runs along one side, fencing will be needed on only three sides. Find the dimensions of the pasture if its length is double its width.

ILLUSTRATION 18

40. FENCING A PEN A man has 150 feet of fencing to build the pen shown in Illustration 19. If one end is a square, find the outside dimensions.

ILLUSTRATION 19

41. ENCLOSING A SWIMMING POOL A woman wants to enclose the pool shown in Illustration 20 and have a walkway of uniform width all the way around. How wide will the walkway be if the woman uses 180 feet of fencing?

ILLUSTRATION 20

42. INSTALLING SOLAR HEATING One solar panel in Illustration 21 is to be 3 feet wider than the other. To be equally efficient, they must have the same area. Find the width of each.

ILLUSTRATION 21

43. MAKING FURNITURE A woodworker wants to put two partitions crosswise in a drawer that is 28 inches deep, as shown in Illustration 22. He wants to place the partitions so that the spaces created increase by 3 inches from front to back. If the thickness of each partition is $\frac{1}{2}$ inch, how far from the front end should he place the first partition?

ILLUSTRATION 22

ILLUSTRATION 23

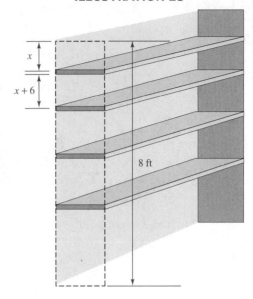

44. BUILDING SHELVES A carpenter wants to put four shelves on an 8-foot wall so that the five spaces created decrease by 6 inches as we move up the wall. (See Illustration 23.) If the thickness of each shelf is $\frac{3}{4}$ inch, how far will the bottom shelf be from the floor?

WRITING

45. Briefly explain what should be accomplished in each of the steps (*analyze, form, solve, state,* and *check*) of the problem-solving strategy used in this section.

46. Write a problem that can be represented by the following verbal model and equation.

Measure of 1st angle	plus	measure of 2nd angle	plus	measure of 3rd angle	is 180°.
x	$+$	$2x$	$+$	$x + 10$	$= 180$

REVIEW

47. When expressed as a decimal, is $\frac{7}{9}$ a terminating or repeating decimal?

48. Solve $x + 20 = 4x - 1 + 2x$.

49. List the integers.

50. Solve $2x + 2 = \frac{2}{3}x - 2$.

51. Evaluate $2x^2 + 5x - 3$ for $x = -3$.

52. Solve $T - R = ma$ for R.

▶ **1.7**

More Applications of Equations

In this section, you will learn about

■ Percent problems ■ Statistics problems ■ Investment problems
■ Uniform motion problems ■ Mixture problems

Introduction In this section, we will again use equations as we solve a variety of problems.

Percent Problems

Numeric information is often expressed using percents. Percent means per one hundred. We can use the **percent formula** (Amount = percent · base) to solve many types of percent problems.

EXAMPLE 1

Disposable diapers. In 1997, Huggies, the leading brand of disposable diapers, had sales of $1,506,000,000. Use the information in the circle graph in Figure 1-21 to determine the total amount of money spent in 1997 on disposable diapers.

FIGURE 1-21

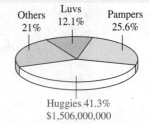

Top Brands of Disposable Diapers, Based on 1997 Food Store Sales: Dollar Market Share

Others 21% Luvs 12.1% Pampers 25.6%

Huggies 41.3%
$1,506,000,000

Based on data from *The Wall Street Journal Almanac* (1999)

ANALYZE THE PROBLEM

We are told that Huggies had sales of $1,506,000,000. The circle graph indicates that $1,506,000,000 was 41.3% of total disposable diaper sales in 1997.

FORM AN EQUATION

If we let x represent total disposable diaper sales (in dollars) in 1997, we can translate the words into an equation.

| $1,506,000,000 | is | 41.3% | of | what number? |

$1,506,000,000 = 41.3% · x

In this context, "is" translates to = and "of" to multiplication.

SOLVE THE EQUATION

$1,506,000,000 = 41.3\% \cdot x$

$1,506,000,000 = 0.413x$ Write 41.3% as a decimal: 41.3% = 0.413.

$\dfrac{1,506,000,000}{\mathbf{0.413}} = \dfrac{0.413x}{\mathbf{0.413}}$ Divide both sides by 0.413.

$3,646,489,104 = x$

STATE THE CONCLUSION

In 1997, sales of disposable diapers totaled $3,646,489,104.

CHECK THE RESULT

If diaper sales totaled $3,646,489,104 in 1997, then the $1,506,000,000 spent on Huggies was about $\frac{1,500,000,000}{3,600,000,000} = \frac{15}{36} \approx 42\%$ of the total. The answer of $3,646,489,104 seems reasonable. ■

When the regular price of merchandise is reduced, the amount of reduction is called **markdown** (or discount).

| Sale price | = | regular price | − | markdown |

Usually, the markdown is expressed as a percent of the regular price.

| Markdown | = | percent of markdown | · | regular price |

EXAMPLE 2

Wedding gowns. At a bridal shop, a wedding gown that normally sells for $397.98 is on sale for $265.32. Find the percent of markdown.

ANALYZE THE PROBLEM

In this case, $265.32 is the sale price, $397.98 is the regular price, and the markdown is the *product* of $397.98 and the percent of markdown.

FORM AN EQUATION   We let r represent the percent of markdown, expressed as a decimal. We then substitute $265.32 for the sale price and $397.98 for the regular price in the formula.

Sale price	is	regular price	minus	markdown.	
265.32	=	397.98	−	$r \cdot 397.98$	Markdown = percent of markdown · regular price.

SOLVE THE EQUATION

$$265.32 = 397.98 - r \cdot 397.98$$

$$265.32 = 397.98 - 397.98r \qquad \text{Rewrite } r \cdot 397.98 \text{ as } 397.98r.$$

$$-132.66 = -397.98r \qquad \text{Subtract } 397.98 \text{ from both sides.}$$

$$\frac{-132.66}{-397.98} = r \qquad \text{Divide both sides by } -397.98.$$

$$0.333333\ldots = r \qquad \text{Do the division using a calculator.}$$

$$33.3333\ldots \% = r \qquad \text{To write the decimal as a percent, multiply } 0.333333\ldots \text{ by 100 and insert a \% sign.}$$

STATE THE CONCLUSION   The percent of markdown on the wedding gown is $33.3333\ldots\%$ or $33\frac{1}{3}\%$.

CHECK THE RESULT   The markdown is $33\frac{1}{3}\%$ of $397.98, or $132.66. The sale price is $397.98 − $132.66, or $265.32. The answer checks.   ∎

Percents are often used to describe how a quantity has changed. To describe such changes, we use **percent increase** or **percent decrease.**

EXAMPLE 3

Home video rentals. The home video rental industry experienced its most dramatic growth from 1985 through 1990. Annual rental revenue increased from $2.55 billion in 1985 to $6.63 billion in 1990. What was the percent increase in rental revenue?

ANALYZE THE PROBLEM   To find the percent increase, we first find the *amount of increase* by subtracting the smaller number from the larger. Note that we are working in billions of dollars.

$$\$6.63 - \$2.55 = \$4.08 \qquad \text{Subtract the 1985 revenue from the 1990 revenue.}$$

FORM AN EQUATION   Next, we find what percent of the original $2.55 billion revenue the $4.08 billion increase was. We will let x = the unknown percent and translate the words into an equation.

What percent	of	2.55	is	4.08?	
x	·	2.55	=	4.08	x is the percent, 2.55 is the base, and 4.08 is amount.

SOLVE THE EQUATION

$$x \cdot 2.55 = 4.08$$

$$2.55x = 4.08$$

$$x = \frac{4.08}{2.55} \qquad \text{Divide both sides by 2.55.}$$

$$x = 1.6 \qquad \text{Use a calculator to do the division.}$$

$$x = 160\% \qquad \text{Write 1.6 as a percent: Multiply 1.6 by 100 and insert a \% sign.}$$

STATE THE CONCLUSION   There was a 160% increase in annual video rental revenue between 1985 and 1990.

CHECK THE RESULT   A 100% increase in the 1985 revenue would be $2.55 billion. A 200% increase would be 2($2.55 billion) = $5.1 billion. Our result of a 160% increase, which corresponds to a dollar increase of $4.08 billion, seems reasonable.   ∎

Statistics Problems

Statistics is a branch of mathematics that deals with the analysis of numerical data. Three types of averages are commonly used in statistics as measures of central tendency of a distribution of numbers: the **mean,** the **median,** and the **mode.**

Mean, median, and mode

The **mean** of a collection of values is the sum of those values divided by the number of values.

$$\text{Mean} = \frac{\text{sum of the values}}{\text{number of values}}$$

The **median** of a collection of values is the middle value. To find the median,

1. Arrange the values in increasing order.

2. If there is an odd number of values, choose the middle value.

3. If there is an even number of values, add the middle two values and divide by 2.

The **mode** of a collection of values is the value that occurs most often.

EXAMPLE 4

Physiology experiment. As part of a project for a physiology class, a student measured ten people's reaction time to a visual stimulus. Their reaction times, in hundredths of a second, are listed below. Find **a.** the mean, **b.** the median, and **c.** the mode of the distribution.

0.29, 0.22, 0.19, 0.36, 0.28, 0.23, 0.16, 0.28, 0.33, 0.26

Solution **a.** To find the mean, we add the values and divide by the number of values, which is 10.

$$\text{Mean} = \frac{0.29 + 0.22 + 0.19 + 0.36 + 0.28 + 0.23 + 0.16 + 0.28 + 0.33 + 0.26}{10} = 0.26 \text{ second}$$

b. To find the median, we first arrange the values in increasing order:

0.16, 0.19, 0.22, 0.23, 0.26, 0.28 , 0.28, 0.29, 0.33, 0.36

Because there is an even number of measurements, the median will be the sum of the middle two values, 0.26 and 0.28, divided by 2. Thus, the median is

$$\text{Median} = \frac{0.26 + 0.28}{2} = 0.27 \text{ second}$$

c. Since the time 0.28 second occurs most often, it is the mode. ∎

EXAMPLE 5

Bank service charges. When the average (mean) daily balance of a customer's checking account falls below $500 in any week, the bank assesses a $15 service charge. What minimum balance will the account shown in Figure 1-22 need to have on Friday to avoid the service charge?

FIGURE 1-22

Security Savings

☐ Weekly Statement ☐

Acct: 201-234-002   Type: checking

Day	Date	Daily balance	Comments
Mon	3/11	$730.70	
Tue	3/12	$350.19	
Wed	3/13	–$50.19	overdrawn
Thu	3/14	$275.55	
Fri	3/15		

ANALYZE THE PROBLEM   We can find the average (mean) daily balance for the week by adding the daily balances and dividing by 5. We want the mean to be $500 so that there is no service charge.

FORM AN EQUATION   We will let x = the minimum balance needed on Friday. Then we translate the words into mathematical symbols.

The sum of the five daily balances	divided by	5	is	$500.

$$\frac{730.70 + 350.19 + (-50.19) + 275.55 + x}{5} = 500$$

SOLVE THE EQUATION

$$\frac{730.70 + 350.19 + (-50.19) + 275.55 + x}{5} = 500$$

$$\frac{1{,}306.25 + x}{5} = 500 \qquad \text{Combine like terms in the numerator.}$$

$$5\left(\frac{1{,}306.25 + x}{5}\right) = 5(500) \qquad \text{Multiply both sides by 5.}$$

$$1{,}306.25 + x = 2{,}500$$

$$x = 1{,}193.75 \qquad \text{Subtract 1,306.25 from both sides.}$$

STATE THE CONCLUSION   On Friday, the account balance needs to be $1,193.75 to avoid a service charge.

CHECK THE RESULT   Check the result by adding the five daily balances and dividing by 5. ■

Investment Problems

The money an investment earns is called *interest*. **Simple interest** is computed by the formula $I = Prt$, where I is the interest earned, P is the principal (amount invested), r is the annual interest rate, and t is the length of time the principal is invested.

EXAMPLE 6

Interest income. To protect against a major loss, a financial analyst suggests a diversified plan for a client who has $50,000 to invest for 1 year.

1. Alco Development, Inc. Builds mini-malls. High yield: 12% per year. Risky!
2. Certificate of deposit (CD). Insured, safe. Low yield: 4.5% annual interest.

If the client puts some money in each investment and wants to earn $3,600 in interest, how much should be invested at each rate?

ANALYZE THE PROBLEM   In this case, we are working with two investments made at two different rates for 1 year. If we add the interest from the two investments, the sum should equal $3,600.

FORM AN EQUATION   If we let x represent the number of dollars invested at 12%, the interest earned is $I = Prt = \$x(12\%)(1) = \$0.12x$. If x is invested at 12%, there is $\$(50{,}000 - x)$ to invest at 4.5%, which will earn $\$0.045(50{,}000 - x)$ in interest. These facts are listed in Figure 1-23.

FIGURE 1-23

	P $\cdot$	r $\cdot$	t $=$	I
Alco Development, Inc.	x	0.12	1	$0.12x$
Certificate of deposit	$50{,}000 - x$	0.045	1	$0.045(50{,}000 - x)$

The sum of the two amounts of interest should equal $3,600. We now translate the verbal model into an equation.

The interest earned at 12%	plus	the interest earned at 4.5%	is	the total interest earned.
$0.12x$	$+$	$0.045(50,000 - x)$	$=$	$3,600$

SOLVE THE EQUATION

$$0.12x + 0.045(50,000 - x) = 3,600$$

$$\mathbf{1,000}[0.12x + 0.045(50,000 - x)] = \mathbf{1,000}(3,600)$$ To eliminate the decimals, multiply both sides by 1,000.

$$120x + 45(50,000 - x) = 3,600,000$$ Distribute the 1,000 and simplify both sides.

$$120x + 2,250,000 - 45x = 3,600,000$$ Remove parentheses.

$$75x + 2,250,000 = 3,600,000$$ Combine like terms.

$$75x = 1,350,000$$ Subtract 2,250,000 from both sides.

$$x = 18,000$$ Divide both sides by 75.

STATE THE CONCLUSION $18,000 should be invested at 12% and $(50,000 - 18,000) = $32,000 should be invested at 4.5%.

CHECK THE RESULT   The annual interest on $18,000 is $0.12($18,000) = $2,160$. The interest earned on $32,000 is $0.045($32,000) = $1,440$. The total interest is $2,160 + $1,440 = $3,600$. The answers check.   ■

Uniform Motion Problems

Problems that involve an object traveling at a constant rate for a specified period of time over a certain distance are called **uniform motion** problems. To solve these problems, we use the formula $d = rt$, where d is distance, r is rate, and t is time.

EXAMPLE 7

Travel time. After a weekend stay on her grandparents' farm, a girl is to return home, 385 miles away. To split up the drive, the parents and grandparents start at the same time and drive toward each other, planning to meet somewhere along the way. If the parents travel at an average rate of 60 mph and the grandparents at 50 mph, how long will it take them to reach each other so that the child can be reunited with her parents?

ANALYZE THE PROBLEM   The vehicles are traveling toward each other as shown in Figure 1-24(a). We know the rates the cars are traveling (60 mph and 50 mph). We also know that they will travel for the same amount of time.

FORM AN EQUATION   We can let t represent the time that each vehicle travels. Then the distance traveled by the parents is $60t$ miles, and the distance traveled by the grandparents is $50t$ miles. This information is organized in the table in Figure 1-24(b). The sum of the distances traveled by the parents and grandparents is 385 miles.

The distance the parents travel	plus	the distance the grandparents travel	is	the distance between the child's home and the grandparents' farm.
$60t$	$+$	$50t$	$=$	385

FIGURE 1-24

Home   Farm

|———— 385 mi ————|

	r	$\cdot$	t	$=$	d
Parents	60		t		$60t$
Grandparents	50		t		$50t$

(a)   (b)

SOLVE THE EQUATION

$$60t + 50t = 385$$
$$110t = 385 \quad \text{Combine like terms.}$$
$$t = 3.5 \quad \text{Divide both sides by 110.}$$

STATE THE CONCLUSION   The parents and grandparents will meet in $3\frac{1}{2}$ hours.

CHECK THE RESULT   The parents travel $3.5(60) = 210$ miles. The grandparents travel $3.5(50) = 175$ miles. The total distance traveled is $210 + 175 = 385$ miles. The answer checks. ■

WARNING! When using the formula $d = rt$, make sure that the units match. For example, if the rate is given in miles per *hour*, the time must be expressed in *hours* and the distance in miles.

Mixture Problems

We now discuss two types of mixture problems. In the first example, a *dry mixture* of a specified value is created from two differently priced components.

EXAMPLE 8

Mixing nuts. The owner of a candy store notices that 20 pounds of gourmet cashews are getting stale. They did not sell because of their high price of $12 per pound. The owner decides to mix peanuts with the cashews to lower the price per pound. If peanuts sell for $3 per pound, how many pounds of peanuts must be mixed with the cashews to make a mixture that could be sold for $6 per pound?

ANALYZE THE PROBLEM   To solve this problem, we will use the formula $v = pn$, where v represents value, p represents the price per pound, and n represents the number of pounds.

FORM AN EQUATION   We can let x represent the number of pounds of peanuts to be used. Then $20 + x$ represents the number of pounds in the mixture. We enter the known information in the table shown in Figure 1-25. The value of the cashews plus the value of the peanuts will be equal to the value of the mixture.

FIGURE 1-25

	p	$\cdot$	n	$=$	v
Cashews	12		20		240
Peanuts	3		x		$3x$
Mixture	6		$20 + x$		$6(20 + x)$

Next we translate the verbal model into mathematical symbols.

The value of the cashews	plus	the value of the peanuts	is	the value of the mixture.
240	$+$	$3x$	$=$	$6(20 + x)$

SOLVE THE EQUATION

$240 + 3x = 6(20 + x)$

$240 + 3x = 120 + 6x$ Use the distributive property to remove parentheses.

$120 = 3x$ Subtract $3x$ and 120 from both sides.

$40 = x$ Divide both sides by 3.

STATE THE CONCLUSION

The owner should mix 40 pounds of peanuts with the 20 pounds of cashews.

CHECK THE RESULT

The cashews are valued at $12(20) = 240, and the peanuts are valued at $3(40) = 120. The mixture is valued at $6(60) = 360. Since the value of the cashews plus the value of the peanuts equals the value of the mixture, the answer checks. ∎

In the next example, a *liquid mixture* of a desired strength is to be made from two solutions with different concentrations.

EXAMPLE 9

Milk production. Owners of a dairy find that milk with a 2% butterfat content is their best seller. Suppose the dairy has large quantities of whole milk having a 4% butterfat content and milk having a 1% butterfat content. How much of each type of milk should be mixed to obtain 120 gallons of milk that is 2% butterfat?

ANALYZE THE PROBLEM

We are to find the amount of 4% milk to mix with 1% milk to get 120 gallons of a milk that has a 2% butterfat content. In Figure 1-26(a), if we let g represent the number of gallons of the 4% milk used in the mixture, then $(120 - g)$ represents the number of gallons of the 1% milk needed to obtain the desired concentration.

FIGURE 1-26

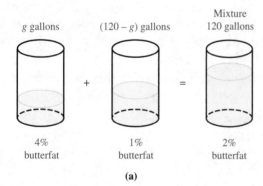

g gallons   (120 − g) gallons   Mixture 120 gallons

4% butterfat   1% butterfat   2% butterfat

(a)

	Percent butterfat ·	Gallons of milk =	Gallons of butterfat
4% milk	0.04	g	$0.04g$
1% milk	0.01	$(120 - g)$	$0.01(120 - g)$
2% milk	0.02	120	$0.02(120)$

(b)

FORM AN EQUATION

The amount of butterfat in a tank is the *product* of the percent butterfat and the number of gallons of milk in the tank. In the first tank shown in Figure 1-26(a), 4% of the g gallons, or $0.04g$ gallons, is butterfat. In the second tank, 1% of the $(120 - g)$ gallons, or $0.01(120 - g)$ gallons, is butterfat. Upon mixing, the third tank will have $0.02(120)$ gallons of butterfat in it. These results are recorded in the last column of the table in Figure 1-26(b).

The sum of the amounts of butterfat in the first two tanks should equal the amount of butterfat in the third tank.

The amount of butterfat in g gallons of 4% milk	plus	the amount of butterfat in (120 − g) gallons of 1% milk	is	the amount of butterfat in 120 gallons of the mixture.
$0.04(g)$	$+$	$0.01(120 - g)$	$=$	$0.02(120)$

SOLVE THE EQUATION $0.04(g) + 0.01(120 - g) = 0.02(120)$

$4(g) + 1(120 - g) = 2(120)$	Multiply both sides by 100 to clear the equation of decimals.
$4g + 120 - g = 240$	Use the distributive property to remove parentheses.
$3g + 120 = 240$	Combine like terms.
$3g = 120$	Subtract 120 from both sides.
$g = 40$	Divide both sides by 3.

STATE THE CONCLUSION   40 gallons of 4% milk and $(120 - 40) = 80$ gallons of 1% milk should be mixed to get 120 gallons of milk with a 2% butterfat content.

CHECK THE RESULT   The 40 gallons of 4% milk contains $0.04(40) = 1.6$ gallons of butterfat, and 80 gallons of 1% milk contains $0.01(80) = 0.8$ gallons of butterfat—a total of $1.6 + 0.8 = 2.4$ gallons of butterfat. The 120 gallons of the 2% mixture contains 2.4 gallons of butterfat. The answers check. ■

STUDY SET

Section 1.7

VOCABULARY

In Exercises 1–6, fill in the blanks to make the statements true.

1. When an investment is made, the amount of money invested is called the _____.

2. The value that occurs the most in a distribution of numbers is called the _____.

3. The middle value of a distribution of numbers is called the _____.

4. The _____ of several values is the sum of those values divided by the number of values.

5. In the statement, "10 is 20% of 50," 10 is the _____, and 50 is the _____.

6. When the regular price of an item is reduced, the amount of reduction is called the _____.

CONCEPTS

7. The three top-grossing films for the weekend of November 7–9, 1997, are shown in Illustration 1. See the "% chg. from last" column in the table.
   a. Why is there a blank for *Starship Troopers*?

   b. Explain what $+1,118\%$ means for the movie *Bean*.

   c. What does the negative entry for *I Know What You Did Last Summer* mean?

ILLUSTRATION 1

Film	Wks. in release	Wknd. total millions	% chg. from last
1 Starship Troopers	1	$22.1	—
2 Bean	3	12.7	+1,118
3 I Know What You Did Last Summer	4	6.5	−31

Based on data from Entertainment Data Inc.

8. Complete the table in Illustration 2 for each 1-year investment.

ILLUSTRATION 2

Account	Principal	Rate	Interest earned
CD	$1,500	6%	
Bonds	$x	5.65%	
Stocks	$(850 − x)	7%	

9. a. If $5,000 of $30,000 is invested at 5%, how much is left to be invested at another rate?
   b. If $x of $30,000 is invested at 5%, how much is left to be invested at another rate?

10. Complete the table in Illustration 3, given the speed of light and sound through air and water.

ILLUSTRATION 3

	Rate	Time	Distance
Light (air)	180,693 mi/sec	60 sec	
Light (water)	136,711 mi/sec	60 sec	
Sound (air)	1,088 ft/sec	x sec	
Sound (water)	5,440 ft/sec	$(x - 3)$ sec	

11. Complete the table in Illustration 4.

ILLUSTRATION 4

	Price	Pounds	Value
M & M plain	$2.49	30	
M & M peanut	$2.79	p	
Mixture	$2.59	$p + 30$	

12. The two coolers shown in Illustration 5 contain a mixture of punch concentrate and water.

   a. What percent water is the punch in cooler 1?

   b. Find the amount of punch concentrate in cooler 1 and cooler 2.

   c. If the contents of each cooler are mixed together, how much punch will there be?

   d. How much punch concentrate will the mixture contain?

ILLUSTRATION 5

20 pints of punch, x pints of punch,
20% concentrate   10% concentrate

NOTATION

In Exercises 13–16, translate each statement into mathematical symbols.

13. What number is 5% of 10.56?

14. 16 is what percent of 55?

15. 32.5 is 74% of what number?

16. What is 83.5% of 245?

In Exercises 17–18, complete each solution.

17. Solve $0.09x + 0.08(2,000 - x) = 400$.

$$\boxed{}[0.09x + 0.08(2,000 - x)] = \boxed{}(400)$$
$$\boxed{} + 8(2,000 - x) =$$
$$9x + \boxed{} - 8x = 40,000$$
$$x = 24,000$$

18. Solve $0.2(5) + 0.6l = 0.4(5 + l)$.

$$\boxed{}[0.2(5) + 0.6l] = \boxed{}[0.4(5 + l)]$$
$$\boxed{}(5) + 6l = \boxed{}(5 + l)$$
$$10 + 6l = 20 + \boxed{}$$
$$\boxed{} = 10$$
$$l = 5$$

19. What formula is used to solve simple interest problems?

20. What formula is used to solve uniform motion problems?

21. What formula is used to solve dry mixture problems?

22. Give the percent formula.

APPLICATIONS

23. ENERGY In 1996, the United States alone accounted for 25% of the world's total energy consumption, using 93.9 quadrillion British thermal units (Btu). A quadrillion is 1,000,000,000,000,000. What was the world's energy consumption in 1996?

24. PERSONAL COMPUTERS In 1998, 47.8 million, or 47.3%, of U.S. households had a personal computer. How many U.S. households were there in 1998? Round to the nearest tenth of a million.

25. BUYING A WASHER AND A DRYER Find the percent of markdown of the sale in Illustration 6.

ILLUSTRATION 6

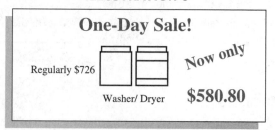

One-Day Sale!

Regularly $726

Now only

Washer/ Dryer   **$580.80**

26. BUYING FURNITURE A bedroom set regularly sells for $983. If it is on sale for $737.25, what is the percent of markdown?

27. FLEA MARKET A vendor sells tool chests at a flea market for $65. If she makes a profit of 30% on each unit sold, what does she pay the manufacturer for each tool chest? (*Hint:* The retail price = the wholesale price + the markup.)

28. MANAGING A BOOKSTORE A bookstore sells a textbook for $39.20. If the bookstore makes a profit of 40% on each sale, what does the bookstore pay the publisher for each book? (*Hint:* The retail price = the wholesale price + the markup.)

29. IMPROVING PERFORMANCE The graph in Illustration 7 shows how the installation of a special computer chip increases the horsepower of a truck. What is the percent increase in horsepower for the engine running at 4,000 revolutions per minute (rpm)? Round to the nearest percent.

ILLUSTRATION 7

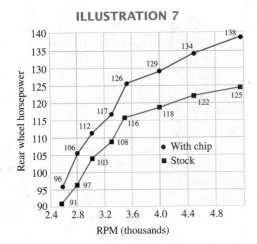

30. GREENHOUSE GASES Over the next 15 years, the United States is striving to cut its carbon dioxide emissions from a high of 4,970.0 million tons per year to 4,671.8 million tons per year. This will help to lessen air pollution. What percent of decrease is this?

31. BROADWAY SHOWS Complete the table in Illustration 8 to find the percent increase in attendance at Broadway shows for each season compared to the previous season. Round to the nearest tenth.

ILLUSTRATION 8

Season	Broadway attendance	% increase
1995–96	9.5 million	—
1996–97	10.6 million	
1997–98	11.5 million	

Source: League of American Theater Producers Inc.

32. THE ART MARKET The line graph in Illustration 9 shows the annual auction sales at Christie's, a major art auction house. Between what two years was there the greatest drop in sales? Approximate the percent of decrease for this period.

ILLUSTRATION 9

Christie's Annual Auction Sales ($ millions)

Based on data from *The Wall Street Journal Almanac* (1999)

33. FUEL EFFICIENCY The ten most fuel-efficient cars in 1999, based on manufacturer's estimated city and highway average miles per gallon (mpg), are shown in Illustration 10. Find the mean, median, and mode of both sets of data.

ILLUSTRATION 10

Model	mpg city/hwy
Chevy Metro	41/44
Suzuki Swift	39/43
Honda Civic	33/40
Toyota Tercel	31/39
Nissan Sentra	30/40
Saturn Coupe	29/40
Saturn Sedan	29/40
Mazda Protege	29/34
Ford Escort	28/38
Mercury Tracer	28/38

Based on data from *Autoland Buying and Leasing Guide, 1999*

34. SPORT FISHING The report shown below lists the fishing conditions at Pyramid Lake for a Saturday in January. Find the median and the mode of the weights of the stripped bass caught at the lake.

Pyramid Lake—Some striped bass are biting but are on the small side. Striking jigs and plastic worms. Water is cold: 38°. Weights of fish caught (lb): 6, 9, 4, 7, 4, 3, 3, 5, 6, 9, 4, 5, 8, 13, 4, 5, 4, 6, 9

35. JOB TESTING To be accepted into a police training program, a recruit must have an average score of 85 on a battery of four tests. If a candidate scored 78 on the oral test, 91 on the physical fitness test, and 87 on the psychological test, what is the lowest score she can obtain on the written test and still be accepted into the training program?

36. GAS MILEAGE The city mileage estimates of Ford models selling for $20,000–$25,000 are listed in Illustration 11. Suppose management wants to add a new model in this price range. What mileage should the new model get if the city mileage average for this group must then be 18.5 mpg?

ILLUSTRATION 11

Model	City mileage (mpg)
Explorer	14.5
F-Series (P/U)	15.0
Mustang	20.5
Taurus	19.0

37. INVESTING MONEY Based on the information in Illustration 12, a woman invested $12,000, some in a money market account and the rest in a 5-year CD. How much was invested in each account if the income from both investments is $1,060 per year?

ILLUSTRATION 12

First Republic Savings and Loan	
Account	**Rate**
NOW	5.5%
Savings	7.5%
Money Market	8.0%
Checking	4.0%
5-year CD	9.0%

38. ENTREPRENEURS Last year, a women's professional organization made two small-business loans totaling $28,000 to young women beginning their own businesses. The money was lent at 7% and 10% simple interest rates. If the annual income the organization received from these loans was $2,560, what was each loan amount?

39. INHERITANCE Paula split an inheritance between two investments, one a certificate of deposit paying 7% annual interest, and the other a promising biotech company offering an annual return of 10%. She invested twice as much in the 10% investment as she did in the 7% investment. If her combined annual income from the two investments was $4,050, how much did she inherit?

40. INCOME TAX RETURN On the federal income tax form Schedule B in Illustration 13, a taxpayer forgot to write in the amount of interest income he earned for the year. From what is written on the form, determine the amount of interest earned from each investment and the amount he invested in stocks.

ILLUSTRATION 13

Schedule B–Interest and Dividend Income

Part 1 Interest Income (See pages 12 and B1.)	**Note:** If you had over $400 in taxable income, use this form.	
	1 List name of payer.	**Amount**
	① MONEY MARKET ACCT. DEPOSITED $15,000 @ 3.3%	SAME AMOUNT FROM EACH
	② STOCKS EARNED 5%	

41. MONEY-LAUNDERING SCHEME Use the evidence compiled by investigators to determine how much money a suspect deposited in the Cayman Islands bank.

- On 6/1/98, the suspect electronically transferred $300,000 to a Swiss bank account paying an 8% annual yield.

- That same day, the suspect opened another account in a Cayman Islands bank that offered a 5% annual yield.

- A document, dated 6/3/99, was seized during a raid of the suspect's home. It stated, "The total interest earned in one year from the two overseas accounts was 7.25% of the total amount deposited."

42. FINANCIAL PRESENTATION A financial planner showed her client the investment plan in Illustration 14. Find the total amount the client will have to invest to earn $2,700 in interest.

ILLUSTRATION 14

3-Part Investment Plan

| 50% of investment in tax-free account: 4% annual yield | 30% of investment in CD: 6% annual yield | 20% of investment in bonds: 8% annual yield |

Total interest earned in year 1: $2,700

43. TRAVEL TIME At lunchtime, a man called his wife from his office to tell her that they needed to switch vehicles so that he could use the family van to pick up some building materials after work. The wife then left their home, traveling towards his office in their van at 35 mph. At the same time, the husband left his office in his car, traveling towards their home at 45 mph. If his office is 20 miles from their home, how long will it take them to meet so that they can switch vehicles?

44. AIR TRAFFIC CONTROL An airliner leaves Los Angeles bound for Caracas, Venezuela, flying at an average rate of 500 mph. At the same time, another airliner leaves Caracas bound for Los Angeles, averaging 550 mph. If the airports are 3,675 miles apart, when will the air traffic controllers have to make the pilots aware that the planes are passing each other?

45. CYCLING A cyclist leaves his training base for a morning workout, riding at the rate of 18 mph. One hour later, her support staff leaves the base in a car going 45 mph in the same direction. How long will it take the support staff to catch up with the cyclist?

46. RUNNING A MARATHON Two marathon runners leave the starting gate, one running 12 mph and the other 10 mph. If they maintain the pace, how long will it take for them to be one-quarter of a mile apart?

47. RADIO COMMUNICATION At 2 P.M., two military convoys leave Eagle River, WI, one headed north and one headed south. The convoy headed north averages 50 mph, and the convoy headed south averages 40 mph. They will lose radio contact when the distance between them is more than 135 miles. When will this occur?

48. SEARCH AND RESCUE Two search-and-rescue teams leave base camp at the same time, looking for a lost child. The first team, on horseback, heads north at 3 mph, and the other team, on foot, heads south at 1.5 mph. How long will it take them to search a distance of 18 miles between them?

49. JET SKIING A jet ski can go 12 mph in still water. If a rider goes upstream for 3 hours against a current of 4 mph, how long will it take the rider to return? (*Hint:* Upstream speed is $(12 - 4)$ mph; how far can the rider go in 3 hours?)

50. PHYSICAL FITNESS For her workout, Sarah walks north at the rate of 3 mph and returns at the rate of 4 mph. How many miles does she walk if the round trip takes 3.5 hours?

51. SWEET AND SOUR The owner of a candy store wants to make a 30-pound mixture of two candies to sell for $1 per pound. If red licorice bits sell for 95¢ per pound and lemon gumdrops sell for $1.10 per pound, how many pounds of each should be used?

52. HEALTH FOOD A pound of dried pineapple bits sells for $6.19, a pound of dried banana chips sells for $4.19, and a pound of raisins sells for $2.39 a pound. Two pounds of raisins are to be mixed with equal amounts of pineapple and banana to create a trail mix that will sell for $4.19 a pound. How many pounds of pineapple and banana chips should be used?

53. GARDENING A wholesaler of premium organic planting mix notices that the retail garden centers are not buying her product because of its high price of $1.57 per cubic foot. She decides to mix sawdust with the planting mix to lower the price per cubic foot. If the wholesaler can buy the sawdust for $0.10 per cubic foot, how many cubic feet of each must be mixed to have 6,000 cubic feet of planting mix that could be sold to retailers for $1.08 per cubic foot?

54. METALLURGY A one-ounce commemorative coin is to be made of a combination of pure gold, costing $380 dollars an ounce, and a gold alloy that costs $140 dollars an ounce. If the cost of the coin is to be $200, and 500 are to be minted, how many ounces of gold and gold alloy are needed to make the coins?

55. DILUTING SOLUTIONS In Illustration 15, how much water should be added to 20 ounces of a 15% solution of alcohol to dilute it to a 10% solution?

ILLUSTRATION 15

56. INCREASING CONCENTRATION The beaker shown in Illustration 16 contains a 2% saltwater solution.
  a. How much water must be boiled away to increase the concentration of the salt solution from 2% to 3%?
  b. Where on the beaker would the new water level be?

ILLUSTRATION 16

57. DAIRY FOODS Cream is approximately 22% butterfat. How many gallons of cream must be mixed with milk testing at 2% butterfat to get 20 gallons of milk containing 4% butterfat?

58. LOWERING FAT How many pounds of extra-lean hamburger that is 7% fat must be mixed with 30 pounds of lean hamburger that is 15% fat to obtain a mixture that is 10% fat?

WRITING

59. If a car travels at 60 mph for 30 minutes, explain why the distance traveled is *not* $60 \cdot 30 = 1,800$ miles.

60. If a mixture is to be made from solutions with concentrations of 12% and 30%, can the mixture have a concentration less than 12% or greater than 30%? Explain.

61. Write a mixture problem that can be represented by the following verbal model and equation.

The value of the regular coffee	plus	the value of the gourmet coffee	equals	the value of the blend.
$4x$	$+$	$7(40 - x)$	$=$	$5(40)$

62. Write a uniform motion problem that can be represented by the following verbal model and equation.

The distance traveled by the 1st train	plus	the distance traveled by the 2nd train	equals	330 miles.
$45t$	$+$	$55t$	$=$	330

REVIEW

In Exercises 63–66, solve each equation.

63. $9x = 6x$

64. $7a + 2 = 12 - 4(a - 3)$

65. $\dfrac{8(y - 5)}{3} = 2(y - 4)$

66. $\dfrac{t - 1}{3} = \dfrac{t + 2}{6} + 2$

Let x =

In Chapter 1, we have discussed one of the most important problem-solving techniques used in algebra. In this method, the first step is to let a variable represent the unknown quantity. Then we use the variable in writing an equation that mathematically describes the situation in question. Finally, we solve the equation to find the value represented by the variable and state the solution called for in the problem. Let's review some of the key parts of this problem-solving method.

Analyzing the Problem

Always begin problem solving by reading the problem carefully. What facts are given? What are we asked to find? Can we infer anything? In Exercises 1 and 2, what do we know and what are we asked to find?

1. GEOGRAPHY Of the 48 contiguous states, 4 more lie east of the Mississippi River than lie west of the Mississippi. How many states are west of the Mississippi River?

2. GEOMETRY In a right triangle, the measure of one acute angle is 5° more than twice that of the other angle. Find the measure of the smallest angle.

Letting a Variable Represent an Unknown Quantity

We usually let the variable represent what we are asked to find. In Exercises 3 and 4, what quantity should the variable represent? State your response in the form "Let $x = \dots$".

3. CAMPING To make anchor lines for a tent, a 60-foot rope is cut into four pieces, each successive piece twice as long as the previous one. Find the length of each anchor line.

4. INSURANCE COVERAGE While waiting for his van to be repaired, a man rents a car for $12 per day and 10 cents per mile. His insurance company will pay up to $100 of the rental fee. If he needs the car for two days, how many miles of driving will his policy cover?

Forming an Equation

As Exercises 5–8 show, several methods can be used to help form an equation.

5. TRANSLATION For each phrase, what operation is indicated?
  a. less than   **b.** of
  c. increased by   **d.** ratio

6. FORMULAS What formula is suggested by each type of problem?
  a. Uniform motion   **b.** Simple interest

  c. Dry mixture   **d.** Perimeter of a rectangle

7. TABLES Complete the table. What equation is suggested?

	% alcohol	Amount	Amount alcohol
Spray 1	0.15	15 oz	
Spray 2	0.50	x oz	
Mixture	0.40	$(15 + x)$ oz	

8. DIAGRAMS What equation is suggested by the diagram below?

2,850 mi

450x mi   500x mi

Accent on Teamwork

Section 1.1

Production planning In the Study Set for Section 1.1, Exercise 42 asks for a series of equations to be written to help a production planner order the correct number of parts for a production run of *r* towel racks. See Illustration 1.

ILLUSTRATION 1

Oak mounting plate

Bar holder   Chrome bar

Mounting screws

a. Decide on a product for which you will be the production planner. Make a detailed drawing of it, like the one in Illustration 1.
b. For one component of your product, create a table that gives the number that should be ordered for production runs of 50, 100, 150, and 200 units. Do the same for a second component of your product, using a bar graph, and for a third component, using a line graph.
c. For each of the three components in part b, write an equation that will give the number of components to be ordered for a production run of *u* units.

Section 1.2

Number lines Exercise 70 in Study Set 1.2 shows how a number line is used in chemistry to illustrate pH. Go to the library and look in some textbooks and encyclopedias to find examples of number lines that are used in other disciplines, such as history, medicine, psychology, and cosmetology.

Section 1.3

Order of operations In making a cake from a mix, the instructions must be followed carefully. Otherwise, the results can be disastrous. Think of two other multistep processes. Explain why the steps must be performed in the proper order or the outcome is greatly affected. Think of two processes for which the outcome is not affected by the order in which the steps are performed.

Section 1.4

Properties of real numbers Use the numbers 100 and 10 to show that there is no commutative property of subtraction or of division. Use 100, 10, and 2 to show that there is no associative property of subtraction or of division.

Section 1.5

Subtraction property of equality Borrow a scale and some weights from your school's science department and use them as part of a class presentation to model how the subtraction property of equality is used to solve the equation $x + 2 = 5$.

Section 1.6

Golden rectangle See Exercise 37 in Study Set 1.6. Cut out a set of six rectangles, each with the same width but with different lengths. One of the rectangles should have a length 1.618 times its width; this is the golden rectangle. Paste the rectangles on a piece of poster board randomly. (See Illustration 2.) Survey 50 people. Ask them to pick the rectangle that is "the most visually appealing." See if the golden rectangle is picked most often. Make a table showing the results.

ILLUSTRATION 2

Section 1.7

Mixture problem Make a 10% solution of lemonade by mixing 1 paper cup of lemonade concentrate with 9 paper cups of water in a pitcher. In a similar manner, using a 4-to-6 ratio, make a 40% solution in another pitcher. Then use algebra to determine how many ounces of the 40% lemonade must be mixed with 8 ounces of the 10% lemonade to get a 20% solution. Measure out the appropriate amounts and make the 20% mixture. Taste each of the three solutions. Can you tell the difference in the concentration?

Section 1.1

CONCEPTS

A *variable* is a letter that stands for a number.

An *equation* is a mathematical sentence that contains an = sign.

Formulas are equations that express a known relationship between two or more quantities.

Bar graphs and *line graphs* are used to describe numerical relationships.

The *perimeter* of a geometric figure is the distance around it.

Describing Numerical Relationships

REVIEW EXERCISES

1. Translate each verbal model into a mathematical model.
   a. The cost C (in dollars) to rent t tables is $15 more than the product of $2 and t.
   b. A rectangle has an area of 25 in.^{2}. The length of the rectangle is the quotient of its area and its width.
   c. The waiting period for a business license is now 3 weeks less than it used to be.

2. To determine the proper cooking time for prime rib, a cookbook suggests using the formula $T = 30p$, where T is the cooking time in minutes and p is the weight of the prime rib in pounds. Use this formula to complete the table.

p	T
6.0	
6.5	
7.0	
7.5	
8.0	

3. Use the data from the table in Exercise 2 to draw each type of graph.
   a. Bar graph

   b. Line graph

4. The owner of a new business wants to frame the first dollar bill her business ever received. How long a piece of molding will she need if a dollar is 2.625 inches wide and 6.125 inches long?

Section 1.2

Natural numbers:
{1, 2, 3, . . .}

Whole numbers:
{ 0, 1, 2, 3, . . .}

Integers:
{. . . , −3, −2, −1, 0, 1, 2, 3, . . .}

Prime numbers:
2, 3, 5, 7, 11, 13, . . .

Composite numbers:
4, 6, 8, 9, 10, 12, . . .

Integers divisible by 2 are *even integers*. Integers not divisible by 2 are *odd integers*.

The Real Number System

5. List the numbers in $\left\{-5, 0, -\sqrt{3}, 2.4, 7, -\frac{2}{3}, -3.\overline{6}, \pi, \frac{15}{4}, 0.13242368. . .\right\}$ that satisfy the given condition.
   a. Natural numbers
   b. Whole numbers
   c. Integers
   d. Rational numbers
   e. Irrational numbers
   f. Real numbers
   g. Negative numbers
   h. Positive numbers
   i. Prime numbers
   j. Composite numbers
   k. Even integers
   l. Odd integers

Rational numbers are numbers that can be written as $\frac{a}{b}$, where a and b are integers and $b \neq 0$. They form the set of all terminating and all repeating decimals.

Irrational numbers are numbers that can be written as nonterminating, nonrepeating decimals.

A *real number* is any number that is either a rational or an irrational number. All points on the number line represent the set of real numbers.

For any real number x:

$$\begin{cases} \text{If } x \geq 0, \text{ then } |x| = x. \\ \text{If } x < 0, \text{ then } |x| = -x. \end{cases}$$

6. Use one of the symbols $>$ or $<$ to make each statement true.
  a. $-16 \ \boxed{} \ -17$ **b.** $-(-1.8) \ \boxed{} \ 2\frac{1}{2}$

7. Tell whether each statement is true or false.
  a. $23.000001 \geq 23.1$ **b.** $-11 \leq -11$

8. Graph the prime numbers between 20 and 30 on the number line.

9. Graph the set $\left\{2.75, 2.\overline{3}, \sqrt{7}, \frac{8}{3}, \frac{3\pi}{4}\right\}$ on the number line.

10. Write each expression without using absolute value symbols.
  a. $|-18|$ **b.** $-|-6.26|$

Section 1.3

Operations with Real Numbers

Adding real numbers: With like signs, add the absolute values and keep the common sign. With unlike signs, subtract the absolute values and keep the sign of the number that has the greatest absolute value.

Subtracting real numbers:

$$x - y = x + (-y)$$

Multiplying and dividing real numbers: With like signs, multiply (or divide) their absolute values. The answer is positive. With unlike signs, multiply (or divide) their absolute values. The answer is negative.

x^n is a *power of x. x* is the base, and *n* is the *exponent.*

$$x^n = \overbrace{x \cdot x \cdot x \cdot \cdots \cdot x}^{n \text{ factors of } x}$$

A number b is a *square root* of a if $b^2 = a$. $\sqrt{a}$ represents the *principal* (positive) *square root* of a.

11. Do the operations.
  a. $-3 + (-4)$ **b.** $-70.5 + 80.6$

  c. $-\dfrac{1}{2} - \dfrac{1}{4}$ **d.** $-6 - (-8)$

  e. $(-4.2)(-3.0)$ **f.** $-\dfrac{1}{10} \cdot \dfrac{5}{16}$

  g. $\dfrac{-2.2}{-11}$ **h.** $-\dfrac{9}{8} \div 21$

  i. $15 - 25 - 23$ **j.** $-3.5 + (-7.1) + 4.9$

  k. $-3(-5)(-8)$ **l.** $-1(-1)(-1)(-1)$

12. Evaluate each expression.
  a. $(-3)^5$ **b.** $\left(-\dfrac{2}{3}\right)^2$

  c. 0.3 cubed   **d.** -5^2

13. Evaluate each expression.
  a. $\sqrt{4}$ **b.** $-\sqrt{100}$

  c. $\sqrt{\dfrac{9}{25}}$ **d.** $\sqrt{0.64}$

Order of operations: Work from the innermost pair to the outermost pair of grouping symbols in the following order.

1. Evaluate all powers and roots.
2. Do all multiplications and divisions, working from left to right.
3. Do all additions and subtractions, working from left to right.

When the grouping symbols have been removed, repeat the steps above to finish the calculation. In a fraction, simplify the numerator and denominator separately, and then simplify the fraction.

To *evaluate an algebraic expression,* substitute the values for the variables and then apply the rules for the order of operations.

The *area* of a figure is the amount of surface it encloses. The *volume* of a figure is its capacity.

14. Evaluate each expression.

a. $-6 + 2(-5)^2$

b. $\dfrac{-20}{4} - (-3)(-2)\left(-\sqrt{1}\right)$

c. $4 - (5 - 9)^2$

d. $4 + 6[-1 - 5(25 - 3^3)]$

e. $2|-1.3 + (-2.7)|$

f. $\dfrac{(7 - 6)^4 + 32}{36 - \left(\sqrt{16} + 1\right)^2}$

g. $-2(4)^2$

h. $-(-2 \cdot 4)^2$

15. Evaluate the algebraic expression for the given values of the variables.

a. $(x + y)(x^2 - xy + y^2)$ for $x = -2$ and $y = 4$

b. $\dfrac{-b - \sqrt{b^2 - 4ac}}{2a}$ for $a = 2$, $b = -3$, and $c = -2$

16. SAFETY CONES See Illustration 1.
   a. Find the area covered by the square rubber base if its sides are 10 inches long.
   b. Find the volume of the cone that is centered atop the base. Round to the nearest tenth.

ILLUSTRATION 1

1 in.

15 in.

Section 1.4

Simplifying Algebraic Expressions

Properties of real numbers:
1. *Associative properties:*
   $(a + b) + c = a + (b + c)$
   $(ab)c = a(bc)$
2. *Commutative properties:*
   $a + b = b + a$
   $ab = ba$
3. *Distributive property:*
   $a(b + c) = ab + ac$
4. 0 is the *additive identity:*
   $a + 0 = 0 + a = a$
5. 1 is the *multiplicative identity:*
   $1 \cdot a = a \cdot 1 = a$

17. Fill in the blanks to make the statements true by applying the indicated property of the real numbers.

a. $3(x + 7) = $ _____
Distributive property (and simplify)

b. $t \cdot 5 = $ ___
Commutative property of multiplication

c. $-x + x = $ ___
Additive inverse property

d. $(27 + 1) + 99 = $ _____
Associative property of addition

e. $\frac{1}{8} \cdot 8 = $ ___
Multiplicative inverse property

f. $0 + m = $ ___
Additive identity property

g. ___ $\cdot 9.87 = 9.87$
Multiplicative identity property

h. $5(-9)(0)(2,345) = $ ___
Multiplication property of 0

i. $(-3 \cdot 5)2 = $ _____
Associative property of multiplication

j. $(t + z) \cdot t = $ _____
Commutative property of addition

6. *Multiplication property of* 0:
$$a \cdot 0 = 0 \cdot a = 0$$

7. $-a$ is the *negative* (or *additive inverse*) of a:
$$a + (-a) = 0$$

8. If $a \neq 0$, then $\frac{1}{a}$ is the reciprocal (or multiplicative inverse of a):
$$a \cdot \frac{1}{a} = \frac{1}{a} \cdot a = 1$$

To *simplify algebraic expressions*, we use properties of real numbers to write them in a less complicated form.

Terms with exactly the same variables raised to exactly the same powers are called *like* (similar) *terms*.

To add or subtract *like terms*, combine their coefficients and keep the same variables with the same exponents.

18. Do each division, if possible.

  a. $\dfrac{102}{102}$

  b. $\dfrac{0}{6}$

  c. $\dfrac{-25}{1}$

  d. $\dfrac{5.88}{0}$

19. Remove the parentheses and simplify.

  a. $8(x + 6)$ **b.** $-6(x - 2)$
  c. $-(-4 + 3y)$ **d.** $(3x - 2y)1.2$
  e. $10(2c^2 - c + 1)$ **f.** $\frac{2}{3}(3t + 9)$

20. Simplify each expression.

  a. $8(6k)$ **b.** $(-7x)(-10y)$
  c. $-9(-3p)(-7)$ **d.** $15a + 30a + 7$
  e. $3g^2 - 3g^2$ **f.** $-m + 4(m - 12)$
  g. $\frac{7}{8}x + \frac{1}{8}x$ **h.** $21.45l - 45.99l$
  i. $5t^3 + 6t^2 - 2t^2 + t^3$ **j.** $8(2h + 9) - 5(h - 1)$

Section 1.5

Solving Linear Equations and Formulas

The set of numbers that satisfy an equation is called its *solution set*.

Properties of equality:
If $a = b$, then
$$a + c = b + c$$
$$a - c = b - c$$
$$ca = cb \quad (c \neq 0)$$
$$\frac{a}{c} = \frac{b}{c} \quad (c \neq 0)$$

To solve a linear equation:
1. Clear the equation of any fractions.
2. Remove all parentheses and combine like terms.
3. Get all variables on one side and all numbers on the other. Combine like terms.
4. Make the coefficient of the variable 1.
5. Check the result.

21. Determine whether -6 is a solution of each equation.

  a. $6 - x = 2x + 24$ **b.** $\dfrac{5}{3}(x - 3) = -12$

22. Solve each equation and give the solution set.

  a. $\dfrac{x}{5} = -45$ **b.** $t - 3.67 = 4.23$
  c. $0.0035 = 0.25g$ **d.** $0 = x + 4$

23. Solve each equation.

  a. $5x + 12 = 0$ **b.** $-3x - 7 + x = 6x + 20 - 5x$
  c. $4(y - 1) = 28$ **d.** $2 - 13(x - 1) = 4 - 6x$
  e. $\dfrac{8(x - 5)}{3} = 2(x - 4)$ **f.** $\dfrac{3y}{4} - 14 = -\dfrac{y}{3} - 1$
  g. $-k = -0.06$ **h.** $\dfrac{5}{4}p = -10$
  i. $\dfrac{4t + 1}{3} - \dfrac{t + 5}{6} = \dfrac{t - 3}{6}$ **j.** $33.9 - 0.5(75 - 3x) = 0.9$

An *identity* is an equation that is satisfied by every number for which both sides are defined. An *impossible equation* has no solution.

To *solve a formula* for a variable, isolate that variable on one side and isolate all other quantities on the other side.

24. Solve each equation. If the equation is an identity or an impossible equation, so state.

a. $2(x - 6) = 10 + 2x$ **b.** $-5x + 2x - 1 = -(3x + 1)$

25. Solve each formula for the indicated variable.

a. $V = \pi r^2 h$ for h **b.** $Y + 2g = m$ for g

c. $\dfrac{T}{6} = \dfrac{1}{6} ab(x + y)$ for x **d.** $V = \dfrac{4}{3} \pi r^3$ for r^3

Section 1.6   Using Equations to Solve Problems

Problem-solving strategy:
1. Analyze the problem.

2. Form an equation.

3. Solve the equation.

4. State the conclusion.

5. Check the result.

Number · value = total value

26. AIRPORTS The world's two busiest airports are Chicago O'Hare International and Hartsfield Atlanta International. Together they served 132 million passengers in 1996, with O'Hare handling 6 million more than Atlanta. How many passengers did each airport serve?

27. TUITION A private school reduces the monthly tuition cost of $245 by $5 per child if a family has more than one child attending the school. Write an algebraic expression that gives the monthly tuition cost per child for a family having c children.

28. TREASURY BILLS What is the value of five $1,000 T-bills? What is the value of x $1,000 T-bills?

29. CABLE TV A 186-foot television cable is to be cut into four pieces. Find the length of each piece if each successive piece is 3 feet longer than the previous one.

30. TOOLING Illustration 2 shows the angle at which a drill is to be held when drilling a hole into a piece of aluminum. Find the measures of both angles labeled in the illustration.

ILLUSTRATION 2

The measure of this angle is 15° less than half of the other angle.

Section 1.7   More Applications of Equations

Interest = Principal · rate · time

31. INVESTMENTS Sally has $25,000 to invest. She invests some money at 10% interest and the rest at 9%. If her total annual income from these two investments is $2,430, how much does she invest at each rate?

Percent means per one hundred.

Amount = percent · base

32. APPAREL SALES Use the information in the graph in Illustration 3 to determine whether the largest percent increase in apparel sales from 1996 to 1997 occurred in men's, women's, or children's clothing.

ILLUSTRATION 3

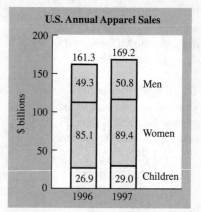

Based on data from *The Wall Street Journal Almanac* (1999)

$$Mean = \frac{\text{sum of the values}}{\text{number of values}}.$$

The *mode* is the value that occurs most often.

The *median* is the middle value.

33. SCHOLASTIC APTITUDE TEST The mean SAT verbal test scores of college-bound seniors for the years 1993–1998 are listed below. Find the mean, median, and mode.

1993	1994	1995	1996	1997	1998
500	499	504	505	505	505

Distance = rate · time

34. PAPARAZZI A celebrity leaves a nightclub in his car and travels at 1 mile per minute (60 mph) trying to avoid a tabloid photographer. One minute later, the photographer leaves the nightclub on his motorcycle, traveling at 1.5 miles per minute (90 mph) in pursuit of the celebrity. How long will it take the photographer to catch up with the celebrity?

35. PEST CONTROL How much water must be added to 20 gallons of a 12% pesticide/water solution to dilute it to an 8% solution?

$v = pn$, where v is the value, p is the price per pound, and n is the number of pounds.

36. CANDY SALES Write an algebraic expression that gives the value of a mixture of 3 pounds of Tootsie Rolls with x pounds of Bit O'Honey, if the mix is to sell for $1.95 per pound.

CHAPTER 1

Test

In Problems 1–2, translate each verbal model into a mathematical model.

1. Each test score T was increased by 10 points to give a new adjusted test score s.

2. The area A of a triangle is the product of one-half the length of the base b and the height h.

In Problems 3–6, consider the set $\left\{-2, \pi, 0, -3\frac{3}{4}, 9.2, \frac{14}{5}, 5, -\sqrt{7}\right\}$.

3. Which numbers are integers?

4. Which numbers are rational numbers?

5. Which numbers are irrational numbers?

6. Which numbers are real numbers?

In Problems 7 and 8, graph each set on the number line.

7. $\left\{ \dfrac{7}{6}, \dfrac{\pi}{2}, 1.8234503.\ldots, \sqrt{3}, 1.\overline{91} \right\}$

8. The set of prime numbers less than 12

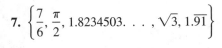

In Problems 9–10, write each expression without using absolute value symbols.

9. $-|8|$ **10.** $|-5.5|$

In Problems 11–16, evaluate each expression.

11. $7 - (-5.3)$

12. $-\dfrac{5}{3}\left(-\dfrac{4}{25}\right)$

13. $\dfrac{1}{2} - \left(-\dfrac{3}{5}\right)$

14. $(-4)^3$

15. $\dfrac{2[-4 - 2(3 - 1)]}{3(\sqrt{9})(2)}$

16. $7 + 2[-1 - 4(5)]$

17. Evaluate the expression for $a = 2$, $b = -3$, and $c = 4$.

$$\dfrac{-3b + a}{ac - b}$$

18. PEDIATRICS Some doctors use Young's rule in calculating dosage for infants and children.

$$\dfrac{\text{Age of child}}{\text{Age of child} + 12}\left(\dfrac{\text{average}}{\text{adult dose}}\right) = \text{child's dose}$$

The adult dose of Achromycin is 250 mg. What is the dose for an 8-year old child?

In Problems 19–20, tell which property of real numbers justifies each statement.

19. $3 + 5 = 5 + 3$

20. $a(b + c) = ab + ac$

In Problems 21–24, simplify each expression.

21. $-y + 3y + 9y$

22. $5(-9q)$

23. $-(4 + t) + t$

24. $12\left(-\dfrac{4}{3} + \dfrac{1}{6}a + \dfrac{3}{2}b\right)$

In Problems 25–26, solve each equation.

25. $9(x + 4) + 4 = 4(x - 5)$

26. $\dfrac{y - 1}{5} + 2 = \dfrac{2y - 3}{3}$

In Problems 27–29, refer to the data in the table.

U.S. Unemployment Rate (1990–1997)							
1990	1991	1992	1993	1994	1995	1996	1997
5.6	6.8	7.5	6.9	6.1	5.6	5.4	4.9

27. Find the mean.   **28.** Find the median.

29. Find the mode.

30. Solve $P = L + \dfrac{s}{f}i$ for i.

31. CALCULATORS The viewing window of a calculator has a perimeter of 26 centimeters and is 5 centimeters longer than it is wide. Find the dimensions of the window.

32. THE U.S. FEDERAL DEBT The graph in Illustration 1 shows the federal debt, in billions of dollars. Find the percent increase in the debt from 1960 through 1997. Round to the nearest percent.

ILLUSTRATION 1

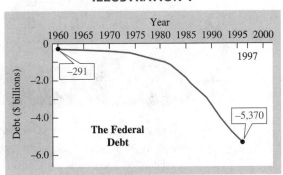

33. INVESTING An investment club invested part of $10,000 at 9% annual interest and the rest at 8%. If the annual income from these investments was $860, how much was invested at 8%?

34. GOLDSMITH How many ounces of a 40% gold alloy must be mixed with 10 ounces of a 10% gold alloy to obtain an alloy that is 25% gold?

35. Explain why the formula is not solved for A.

$$A = \dfrac{1 + A}{2b - 10}$$

36. What does it mean when we say that $x = 3$ is a solution of the equation $2x - 3 = x$?

2

Graphs, Equations of Lines, and Functions

CAMPUS CONNECTION

The Earth Science Department

In a physical geography class, students study the form and structure of the earth. To designate locations on the surface of the earth, geologists use a *coordinate system* consisting of a series of lines representing longitude and latitude. In Chapter 2, we will discuss the rectangular coordinate system. It allows us to make visual representations of mathematical relationships. Learning how to use the rectangular coordinate system will give you the insight you'll need when you encounter coordinate systems in other disciplines such as Earth Science.

Many relationships between two quantities can be described by using a table, a graph, or an equation.

▶ 2.1   The Rectangular Coordinate System

In this section, you will learn about

The rectangular coordinate system ■ Graphing mathematical relationships ■ Reading graphs ■ Step graphs ■ The midpoint formula

Introduction   It is often said that a picture is worth a thousand words. In this chapter, we will show how numerical relationships can be described by mathematical pictures called **graphs.** We will also show how graphs are constructed and how we can obtain important information from them.

The Rectangular Coordinate System

Many cities are laid out on a rectangular grid. For example, on the east side of Rockford, IL, all streets run north and south, and all avenues run east and west. (See Figure 2-1.) If we agree to list the street numbers first, every address can be identified by using an ordered pair of numbers. If Jose Quevedo lives on the corner of Third Street and Sixth Avenue, his address is given by the ordered pair (3, 6).

This is the street.┐
         ↓
      (3, 6)
         ↑
         └── This is the avenue.

If Lisa has an address of (6, 3), we know that she lives on the corner of Sixth Street and Third Avenue. From the figure, we can see that

■ Bob Anderson's address is (4, 1).

■ Rosa Vang's address is (7, 5).

■ The address of the store is (8, 2).

FIGURE 2-1

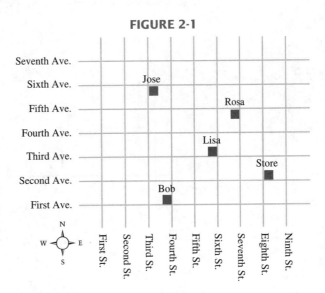

The idea of associating an ordered pair of numbers with points on a grid is attributed to the 17th-century French mathematician René Descartes. The grid is often called a **rectangular coordinate system,** or **Cartesian coordinate system** after its inventor.

In general, a rectangular coordinate system is formed by two intersecting perpendicular number lines. (See Figure 2-2.)

- The horizontal number line is usually called the **x-axis.**
- The vertical number line is usually called the **y-axis.**

The positive direction on the x-axis is to the right, and the positive direction on the y-axis is upward. If no scale is indicated on the axes, we assume that the axes are scaled in units of 1.

The point where the axes cross is called the **origin.** This is the 0 point on each axis. The two axes form a **coordinate plane** and divide it into four regions called **quadrants,** which are numbered using Roman numerals, as shown in Figure 2-2.

Every point on a coordinate plane can be identified by a pair of real numbers x and y, written as (x, y). The first number in the pair is the **x-coordinate,** and the second number is the **y-coordinate.** The numbers are called the **coordinates** of the point. Some examples of ordered pairs are $(-4, 6)$, $(2, 3)$, and $(6, -4)$.

FIGURE 2-2

$(-4, 6)$

In an ordered pair, the ⌐ ⌐ The y-coordinate
x-coordinate is listed first.   is listed second.

The process of locating a point in the coordinate plane is called **graphing** or **plotting** the point. In Figure 2-3(a), we show how to graph the point Q with coordinates of $(-4, 6)$. Since the x-coordinate is negative, we start at the origin and move 4 units to the left along the x-axis. Since the y-coordinate is positive, we then move up 6 units to locate point Q. Point Q is the **graph** of the ordered pair $(-4, 6)$ and lies in quadrant II.

To plot the point $P(2, 3)$, we start at the origin, move 2 units to the right along the x-axis, and then move up 3 units to locate point P. Point P lies in quadrant I. To plot point $R(6, -4)$, we start at the origin and move 6 units to the right and then 4 units down. Point R lies in quadrant IV.

FIGURE 2-3

(a)   (b)

WARNING! Note that point Q with coordinates of $(-4, 6)$ is not the same as point R with coordinates $(6, -4)$. This illustrates that the order of the coordinates of a point is important. This is why we call the pairs **ordered pairs.**

In Figure 2-3(b) on the previous page, no scale is indicated on the axes, so we assume that they are scaled in units of 1. Here, we see that the points $(-5, 0)$, $(0, 0)$, and $(4.5, 0)$ all lie on the x-axis. In fact, every point with a y-coordinate of 0 will lie on the x-axis. Note that the coordinates of the origin are $(0, 0)$.

In Figure 2-3(b), we also see that the points $\left(0, -\frac{8}{3}\right)$, $(0, 0)$, and $(0, 3)$ all lie on the y-axis. In fact, every point with an x-coordinate of 0 will lie on the y-axis.

A point may lie in one of the four quadrants or it may lie on one of the axes, in which case the point is not considered to be in any quadrant. For points in quadrant I, the x- and y-coordinates are positive. Points in quadrant II have a negative x-coordinate and a positive y-coordinate. In quadrant III, both coordinates are negative. In quadrant IV, the x-coordinate is positive and the y-coordinate is negative.

EXAMPLE 1

Halley's comet. Halley's comet passes the earth every 76 years as it travels in an elliptical orbit about the sun. Figure 2-4 shows its approximate path. Use the graph to determine the comet's position for the years 1912, 1930, 1948, 1966, 1978, and the most recent time it passed by the earth, 1986.

FIGURE 2-4

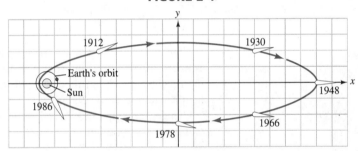

Scale: 1 unit = 161,400,000 mi

Solution To find the coordinates of each position, we start at the origin and move left or right along the x-axis to find the x-coordinate and then up or down to find the y-coordinate.

Year	Position of comet on graph	Coordinates
1912	5 units to the *left*, then 2 units *up*	$(-5, 2)$
1930	5 units to the *right*, then 2 units *up*	$(5, 2)$
1948	9 units to the *right*, no units *up* or *down*	$(9, 0)$
1966	5 units to the *right*, then 2 units *down*	$(5, -2)$
1978	No units *left* or *right*, then 2.5 units *down*	$(0, -2.5)$
1986	8 units to the *left*, then 1 unit *down*	$(-8, -1)$

■

Graphing Mathematical Relationships

Every day, we deal with quantities that are related.

■ The distance we travel depends on how fast we are going.

■ Your test score depends on the amount of time you study.

■ The height of a toy rocket depends on the time since it was shot into the air.

We can use graphs to visualize relationships between two quantities. For example, suppose that we know the height of a toy rocket at 1-second intervals from 0 second to 6 seconds. We can list this information in the **table of values** shown in Figure 2-5.

FIGURE 2-5

Time (in seconds)	Height of rocket (in feet)		
0	0	⟶	(0, 0)
1	80	⟶	(1, 80)
2	128	⟶	(2, 128)
3	144	⟶	(3, 144)
4	128	⟶	(4, 128)
5	80	⟶	(5, 80)
6	0	⟶	(6, 0)
↑ *x*-coordinate	↑ *y*-coordinate		↑ The data in the table can be expressed as ordered pairs.

This information can also be used to construct a graph that shows the relationship between the height of the rocket and the time since it was shot into the air. Because the height depends on time, we will associate time with the horizontal axis and the height with the vertical axis.

To get the graph in Figure 2-6, we plot the seven ordered pairs shown in the table and draw a smooth curve through the data points.

FIGURE 2-6

This graph shows the height of the rocket in relation to the time since it was shot into the air. It does not show the path of the rocket.

From the graph, we can see that the height of the rocket increases as the time increases from 0 second to 3 seconds. Then the height decreases until the rocket hits the ground in 6 seconds. We can also use the graph to make observations about the height of the rocket at other times. For example, the blue lines on the graph show that in 1.5 seconds, the height of the rocket will be approximately 108 feet.

Reading Graphs

As we have seen, graphs can be used to describe how quantities change with time. From such a graph, we can determine when a quantity is increasing and when it is decreasing.

EXAMPLE 2

Reading a graph. The graph in Figure 2-7 shows the water level of a reservoir before, during, and after a storm. On the x-axis, zero represents the day the storm began. On the y-axis, zero represents the normal water level that operators try to maintain.

a. In anticipation of the storm, operators released water to lower the level of the reservoir. By how many feet was the water lowered prior to the storm?

b. After the storm ended, on what day did the water level begin to fall?

c. When was the water level exactly 2 feet above normal?

FIGURE 2-7

Solution
a. On the left, the graph passes through the point $(-4, 0)$. This means that four days before the storm began, the water was at the normal level. If we look directly below zero on the x-axis (the day the storm began), we see that the point $(0, -2)$ is on the graph. So that day, the water level was two feet below normal. Thus, over the four-day period before the storm, the water level was lowered two feet.

b. If we look at the x-axis, we see that the storm lasted 3 days. From the third to the fifth day, the water level remained constant, 4 feet above normal. The graph does not begin to decrease until day 5.

c. We can draw a horizontal line passing through 2 on the y-axis. This line intersects the graph in two places—at the points $(2, 2)$ and $(7, 2)$. This means that 2 days and 7 days after the storm began, the water level was exactly 2 feet above normal.

SELF CHECK

Refer to the graph in Figure 2-7 to answer the following questions. **a.** When was the water at the normal level? **b.** By how many feet did the water level rise during the storm? **c.** After the storm ended and the water level began to fall, how long did it take for the water level to return to normal?

Answers: **a.** 4 days before the storm began and 1 day and 9 days after the storm began, **b.** 6 ft, **c.** 4 days ∎

EXAMPLE 3

Automobile manufacturers. The graph in Figure 2-8 shows the annual earnings for General Motors and Ford for the years 1990–1998. Use the graph to answer the following questions.

a. How much money did Ford make in 1990?

b. Which company lost the most money in 1992?

c. In what years did the earnings of General Motors surpass those of Ford?

d. In 1996, which company saw an improvement in its earnings compared to the previous year?

FIGURE 2-8

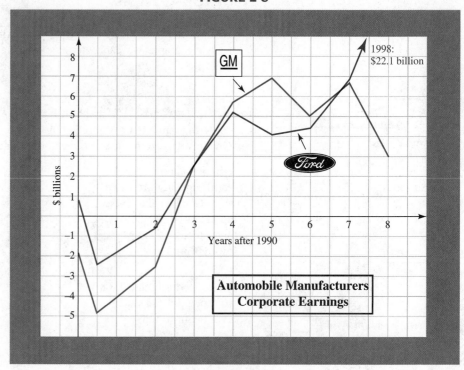

Based on data from *Standard & Poor's Stock Report* (1999)

Solution **a.** 1990 is 0 years after 1990, so Ford's earnings can be found by locating the point where the blue line intersects the vertical axis. This occurs at approximately (0, 0.8). Since 0.8 is positive, it indicates a profit; and since the axis is scaled in billions of dollars, Ford made about $0.8 billion in 1990.

b. 1992 is 2 years after 1990. From 2 on the horizontal axis, we move downward until we reach the point on the two graphs that is lowest. That point, (2, −2.5), is on the red line. Since −2.5 is negative, it represents a loss of $2.5 billion by GM. So GM lost the most money in 1992, about $2.5 billion.

c. We need to see where the graph of GM's earnings (in red) is above the graph of Ford's earnings (in blue). This is true in 1994, 1995, and 1996. In those years, GM's earnings were larger than Ford's.

d. We find the positions on the horizontal axis representing the years 1995 and 1996 and see that the graph of GM's earnings falls. On the other hand, the graph of Ford's earnings rises. So Ford saw an improvement in earnings from 1995 to 1996.

SELF CHECK See Figure 2-8. **a.** In what year did the two companies earn the same amount? **b.** How do Ford's 1996 earnings compare to its 1997 earnings? **c.** In what years were GM's earnings greater than the previous year?

Answers: **a.** 1993, **b.** they are less, **c.** 1992, 1993, 1994, 1995, 1997   ∎

EXAMPLE 4 **Sunburn.** The ultraviolet index shown in Figure 2-9 is printed in the weather section of a newspaper each day. For a given UV level and skin sensitivity, the graph gives the number of minutes it takes for a person exposed to the sun to get sunburned. How long will it take a sunbather with most sensitive skin to get sunburned on a Los Angeles beach?

FIGURE 2-9

Solution   The UV level for Los Angeles is 7. We locate 7 on the vertical axis of the graph and extend a horizontal line that intersects the graph for the most sensitive skin type. The horizontal line intersects the graph at the point (10, 7). This means that the sunbather will burn in 10 minutes of exposure to the sun unless he or she applies some sunscreen.

SELF CHECK   See Figure 2-9. How long will it take a person with least sensitive skin to get sunburned in Las Vegas?   *Answer:* 70 min   ■

Step Graphs

The graph in Figure 2-10 shows the cost of renting a rototiller for different periods of time. The horizontal axis is labeled with the variable d. This reinforces the fact that it is associated with the number of *days* the rototiller is rented. The vertical axis is labeled with the variable c, for *cost*. In this case, ordered pairs on the graph will be of the form (d, c). For example, the point (3, 50) on the graph tells us that the cost of renting a rototiller for 3 days is $50. The cost of renting the rototiller for more than 4 days up to 5 days is $70. Since the jumps in cost form steps in the graph, we call the graph a **step graph.**

The solid dot at the end of each step indicates the rental cost for 1, 2, 3, 4, and 5 days. Each open circle indicates that that point is not on the graph.

FIGURE 2-10

E X A M P L E   5   **Rental costs.** Use the information in Figure 2-10 to answer the following questions. Write the results in a table of values.   **a.** Find the cost of renting the rototiller for 2 days.   **b.** Find the cost of renting the rototiller for $3\frac{1}{2}$ days.   **c.** How long can you rent the rototiller if you have budgeted $40 for the rental?   **d.** Is the cost of renting the rototiller the same each day?

Solution   **a.** We locate 2 on the d-axis and move up to locate the point on the graph directly above the 2. Since that point has coordinates (2, 40), a two-day rental costs $40. We enter this ordered pair in the table on the next page.

b. We locate $3\frac{1}{2}$ on the d-axis and move straight up to locate the point on the graph with coordinates $\left(3\frac{1}{2}, 60\right)$, which indicates that a $3\frac{1}{2}$-day rental would cost $60. We enter this ordered pair in the table on the next page.

c. We draw a horizontal line through the point labeled 40 on the c-axis. Since this line intersects one of the steps of the graph, we can look down to the d-axis to find the d-values that correspond to a c-value of 40. From the graph, we see that the rototiller can be rented for more than 1 and up to 2 days for $40. We write (2, 40) in the table.

d Length of rental (days)	c Cost (dollars)
2	40
$3\frac{1}{2}$	60
2	40

d. No, the cost each day is not the same. If we look at how the c-coordinates change, we see that the first-day rental fee is $20. The second day, the cost jumps another $20. The third day, and all subsequent days, the cost jumps $10.

SELF CHECK   Referring to Figure 2-10, find the cost of renting the rototiller for $2\frac{1}{2}$ days.   *Answer:* $50   ■

The Midpoint Formula

If point M in Figure 2-11 lies midway between point P with coordinates $(-2, 5)$ and point Q with coordinates $(4, -2)$, it is called the **midpoint** of line segment PQ. To find the coordinates of point M, we find the mean of the x-coordinates and the mean of the y-coordinates of points P and Q.

$$x = \frac{-2 + 4}{2} \quad \text{and} \quad y = \frac{5 + (-2)}{2}$$

$$= \frac{2}{2} \qquad\qquad\qquad = \frac{3}{2}$$

$$= 1$$

Thus, point M has coordinates of $\left(1, \frac{3}{2}\right)$.

To distinguish between the coordinates of two general points on a line segment, we often use subscript notation. In Figure 2-12, point $P(x_1, y_1)$ is read as "point P with coordinates of x sub 1 and y sub 1," and point $Q(x_2, y_2)$ is read as "point Q with coordinates of x sub 2 and y sub 2." Using this notation, we can write the midpoint formula in the following way.

FIGURE 2-11

FIGURE 2-12

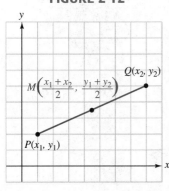

The midpoint formula

The **midpoint** of a line segment with endpoints at $P(x_1, y_1)$ and $Q(x_2, y_2)$ is the point M with coordinates

$$\left(\frac{x_1 + x_2}{2}, \frac{y_1 + y_2}{2}\right)$$

EXAMPLE 6

Finding the endpoint of a line segment. The midpoint of the line segment joining $P(-5, -3)$ and $Q(x_2, y_2)$ is the point $M(-1, 2)$. Find the coordinates of point Q.

Solution   We can let $P(x_1, y_1) = P(-5, -3)$ and $M(x_M, y_M) = M(-1, 2)$, where x_M represents the x-coordinate of point M, and y_M represents the y-coordinate of point M. We can then find the coordinates of point Q using the midpoint formula.

$$x_M = \frac{x_1 + x_2}{2} \quad \text{and} \quad y_M = \frac{y_1 + y_2}{2}$$

$$-1 = \frac{-5 + x_2}{2} \qquad\qquad 2 = \frac{-3 + y_2}{2}$$

$$-2 = -5 + x_2 \qquad\qquad 4 = -3 + y_2 \quad \text{Multiply both sides by 2.}$$

$$3 = x_2 \qquad\qquad\qquad 7 = y_2$$

Since $x_2 = 3$ and $y_2 = 7$, the coordinates of point Q are $(3, 7)$.

SELF CHECK   If the midpoint of a segment PQ is $M(-2, 5)$ and one endpoint is $Q(6, -2)$, find the coordinates of point P.   *Answer:* $(-10, 12)$ ∎

STUDY SET

Section 2.1

VOCABULARY

In Exercises 1–8, fill in the blanks to make the statements true.

1. The pair of numbers $(6, -2)$ is called an _____ pair.

2. In the ordered pair $(-2, -9)$, -9 is called the ___ coordinate.

3. The point with coordinates $(0, 0)$ is the _____.

4. The x- and y-axes divide the coordinate plane into four regions called _____.

5. Ordered pairs of numbers can be graphed on a _____ coordinate system.

6. The process of locating a point on a coordinate plane is called _____ the point.

7. If a point is midway between two points P and Q, it is called the _____ of segment PQ.

8. If a line segment joins points P and Q, points P and Q are called _____ of the segment.

CONCEPTS

In Exercises 9–16, fill in the blanks to make the statements true.

9. To plot the point with coordinates $(6, -3.5)$, we start at the _____ and move 6 units to the _____ and then 3.5 units _____.

10. To plot the point with coordinates $\left(-6, \frac{3}{2}\right)$, we start at the _____ and move 6 units to the _____ and then $\frac{3}{2}$ units ___.

11. In which quadrant do points with a negative x-coordinate and a positive y-coordinate lie?

12. In which quadrant do points with a positive x-coordinate and a negative y-coordinate lie?

13. Use the graph to complete the table.

x	y
0	
1	
2	
	3

14. Use the graph to complete the table.

x	y
-2	
-1	
	-2
1	
2	

15. The x-coordinate of the midpoint of the line segment

joining $P(x_1, y_1)$ and $Q(x_2, y_2)$ is _____.

16. The y-coordinate of the midpoint of the line segment

joining $P(x_1, y_1)$ and $Q(x_2, y_2)$ is _____.

NOTATION

17. For the ordered pair (t, d), which axis is each variable associated with?

18. Do these ordered pairs name the same point? $\left(5.25, -\frac{3}{2}\right)$, $\left(5\frac{1}{4}, -1.5\right)$, $\left(\frac{21}{4}, -1\frac{1}{2}\right)$

19. Fill in the blank: The expression x_1 is read as

_____.

20. Fill in the blank: If $M(x_M, y_M)$ is the midpoint of a

line segment, _____ is the y-coordinate of the midpoint.

PRACTICE

In Exercises 21–28, plot each point on the rectangular coordinate system shown in Illustration 1.

ILLUSTRATION 1

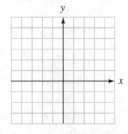

21. $A(4, 3)$

22. $B(-2, 1)$

23. $C(3.5, -2)$

24. $D(-2.5, -3)$

25. $E(5, 0)$

26. $F(-4, 0)$

27. $G\left(\frac{8}{3}, 0\right)$

28. $H\left(0, \frac{10}{3}\right)$

In Exercises 29–36, give the coordinates of each point shown in Illustration 2.

ILLUSTRATION 2

29. A

30. B

31. C

32. D

33. E

34. F

35. G

36. H

37. The graph in Illustration 3 shows the depths of a submarine at certain times.
  a. Where is the sub when $t = 2$?
  b. What is the sub doing as t increases from $t = 2$ to $t = 3$?
  c. How deep is the sub when $t = 4$?
  d. How large an ascent does the sub begin to make when $t = 6$?

ILLUSTRATION 3

38. The graph in Illustration 4 shows the altitudes of a plane at certain times.
  a. Where is the plane when $t = 0$?
  b. What is the plane doing as t increases from $t = 1$ to $t = 2$?
  c. What is the altitude of the plane when $t = 2$?
  d. How much of a descent does the plane begin to make when $t = 4$?

ILLUSTRATION 4

ILLUSTRATION 6

39. Refer to the graph in Illustration 5.
  a. When did U.S. petroleum imports decline?

  b. When did U.S. petroleum production increase?

  c. When did imports surpass production?
  d. Estimate the difference in U.S. petroleum imports and production for 1997.

ILLUSTRATION 5

Based on data from the United States Department of Energy

40. Refer to the graph in Illustration 6.
  a. Which runner ran faster at the start of the race?
  b. Which runner stopped to rest first?
  c. Which runner dropped the baton and had to go back and get it?
  d. At what times was runner 1 stopped and runner 2 running?
  e. Describe what was happening at time D.

  f. Which runner won the race?

In Exercises 41–50, find the midpoint of line segment PQ.

41. $P(0, 0)$, $Q(6, 8)$
42. $P(10, 12)$, $Q(0, 0)$
43. $P(6, 8)$, $Q(12, 16)$
44. $P(10, 4)$, $Q(2, -2)$
45. $P(2, 4)$, $Q(5, 8)$

46. $P(5, 9)$, $Q(8, 13)$
47. $P(-2, -8)$, $Q(3, 4)$
48. $P(-5, -2)$, $Q(7, 3)$
49. $Q(-3, 5)$, $P(-5, -5)$
50. $Q(2, -3)$, $P(4, -8)$

51. If $M(-2, 3)$ is the midpoint of segment PQ and the coordinates of P are $(-8, 5)$, find the coordinates of Q.
52. If $M(6, -5)$ is the midpoint of segment PQ and the coordinates of Q are $(-5, -8)$, find the coordinates of P.
53. If $M(-7, -3)$ is the midpoint of segment PQ and the coordinates of Q are $(6, -3)$, find the coordinates of P.
54. If $M\left(\frac{1}{2}, -2\right)$ is the midpoint of segment PQ and the coordinates of P are $\left(-\frac{5}{2}, 5\right)$, find the coordinates of Q.

APPLICATIONS

55. ROAD MAPS Road maps have a built-in coordinate system to help locate cities. Use the map in Illustration 7 to find the coordinates of these cities in South Carolina: Jonesville, Easley, Hodges, and Union. Express each answer in the form (number, letter).

ILLUSTRATION 7

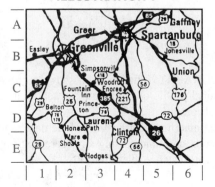

56. TRAIL OF DESTRUCTION A coordinate system that designates the location of places on the surface of the earth uses a series of latitude and longitude lines, as shown in Illustration 8.

a. If we agree to list longitude first, what are the coordinates of New Orleans, expressed as an ordered pair?

ILLUSTRATION 8

b. In August of 1992, Hurricane Andrew destroyed Homestead, Florida. Estimate the coordinates of Homestead.

c. Estimate the coordinates of where the hurricane hit Louisiana.

57. EARTHQUAKE DAMAGE The map in Illustration 9 shows the area where damage was caused by an earthquake.

a. Find the coordinates of the epicenter (the source of the quake).

b. Was damage done at the point (4, 5)?

c. Was damage done at the point (−1, −4)?

ILLUSTRATION 9

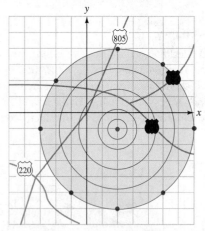

58. GEOGRAPHY Illustration 10 shows a cross-sectional profile of the Sierra Nevada mountain range in California.

a. Estimate the coordinates of blue oak, sagebrush scrub, and tundra using an ordered pair of the form (distance, elevation).

b. The *treeline* is the highest elevation at which trees grow. Estimate the treeline for this mountain range.

ILLUSTRATION 10

59. VIDEO RENTAL The charges for renting a video are shown in the graph in Illustration 11.

a. Find the charge for a 1-day rental.

b. Find the charge for a 2-day rental.

ILLUSTRATION 11

c. Find the charge if the video is kept for 5 days.

d. Find the charge if the video is kept for a week.

60. POSTAGE RATES The graph shown in Illustration 12 gives the first-class postage rates in 1999 for mailing parcels weighing up to 5 ounces.

a. Find the cost of postage to mail a 3-oz letter.

b. Find the difference in cost for a 2.75-oz letter and a 3.75-oz letter.

c. What is the heaviest letter that can be mailed first class for 99¢?

ILLUSTRATION 12

61. BOEING C-17 Engineers use a coordinate system with three axes when designing airplanes. As shown in Illustration 13, the x-axis is used to describe left/right on the airplane, the y-axis forward/backward, and the z-axis up/down. Any point on the airplane can be described by an *ordered triple* of the form (x, y, z). The coordinates of three points on the plane are $(0, 181, 56)$, $(-46, 48, 19)$, and $(84, 94, 24)$. Which highlighted part of the plane corresponds with which ordered triple?

ILLUSTRATION 13

62. ROAST TURKEY The thawing guidelines that appear on the label of a frozen turkey are listed in the table. In Illustration 14, draw a step graph that illustrates these instructions.

Size	Refrigerator thawing
10 lb to just under 18 lb	3 days
18 lb to just under 22 lb	4 days
22 lb to just under 24 lb	5 days
24 lb to just under 30 lb	6 days

ILLUSTRATION 14

63. MULTICULTURAL STUDIES Social scientists use the diagram in Illustration 15 to characterize cultures. The amount of group/family loyalty in a culture is measured on the horizontal *group* axis. The amount of social mobility is measured on the vertical *social grid* axis. In the diagram, four cultures are classified.

In which culture, R, S, T, or U, would you expect that. . .
a. anyone can grow up to be president, and parents expect their children to get out on their own as soon as possible?
b. only the upper class attends college, and people must marry within their own social class?

ILLUSTRATION 15

64. INDUSTRIAL PSYCHOLOGY The results of a personal profile test taken by an employee are plotted as an ordered pair on the grid in Illustration 16. The test shows whether the employee is more task oriented or people oriented. From the results, would you expect the employee to agree or disagree with each of the following statements?
a. Completing a project is almost an obsession with me, and I cannot be content until I am finished.

b. Even if I'm in a hurry while running errands, I will stop to talk with a friend.

ILLUSTRATION 16

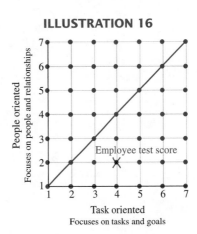

WRITING

65. Explain how to plot the point with coordinates of $(-2, 5)$.

66. Explain why the coordinates of the origin are $(0, 0)$.

REVIEW

In Exercises 67–70, evaluate each expression.

67. $-5 - 5(-5)$

68. $(-5)^2 + (-5)$

71. Solve $-4x + 0.7 = -2.1$.

69. $\dfrac{-3 + 5(2)}{9 + 5}$

70. $|-1 - 9|$

72. Solve $P = 2l + 2w$ for w.

▶ **2.2**

Graphing Linear Equations

In this section, you will learn about

Graphing linear equations ■ The intercept method ■ Horizontal and vertical lines ■ Linear models

Introduction   In this section, we will discuss equations that contain two variables. Such equations are used to describe relationships between two quantities. To see a picture of the mathematical relationships, we will construct graphs of the equations.

Graphing Linear Equations

The equation $y = -\frac{1}{2}x + 4$ contains the variables x and y. The solutions of this equation can be written as ordered pairs of real numbers. For example, the ordered pair $(-4, 6)$ is a solution, because the equation is satisfied when $x = -4$ and $y = 6$.

$$y = -\frac{1}{2}x + 4$$

$$6 = -\frac{1}{2}(-4) + 4 \quad \text{Substitute } -4 \text{ for } x \text{ and } 6 \text{ for } y.$$

$$6 = 2 + 4 \qquad\qquad \text{Do the multiplication: } -\tfrac{1}{2}(-4) = 2.$$

$$6 = 6$$

This pair and others that satisfy the equation are listed in the table of values shown in Figure 2-13.

FIGURE 2-13

$$y = -\frac{1}{2}x + 4$$

x	y	(x, y)
-4	6	$(-4, 6)$
-2	5	$(-2, 5)$
0	4	$(0, 4)$
2	3	$(2, 3)$
4	2	$(4, 2)$

Note that we choose x-values that are multiples of the denominator 2. This makes the computations easier when multiplying the x value by $-\frac{1}{2}$ to find the corresponding y-value.

Pick values for x.   Compute each y-value.   Write each solution as an ordered pair.

The **graph of the equation** $y = -\frac{1}{2}x + 4$ is the graph of all points (x, y) on the rectangular coordinate system whose coordinates satisfy the equation.

EXAMPLE 1

Graphing equations. Graph $y = -\frac{1}{2}x + 4$.

Solution To graph the equation, we plot the five or-dered pairs listed in the table in Figure 2-13. These points lie on the line shown in Figure 2-14. In fact, if we were to plot more pairs that satisfied the equation, it would become obvious that the resulting points will all lie on the line.

When we say that the graph of an equa-tion is a line, we imply two things:

1. Every point with coordinates that sat-isfy the equation will lie on the line.

2. Any point on the line will have coordi-nates that satisfy the equation.

FIGURE 2-14

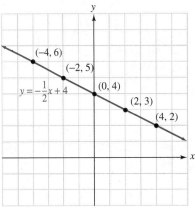

SELF CHECK

Complete the table of values for $y = 2x - 3$. Then graph the equation.

x	y
-1	
0	
1	
2	
3	

Answer:

x	y
-1	-5
0	-3
1	-1
2	1
3	3

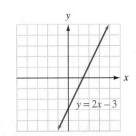

When the graph of an equation is a line, we call the equation a **linear equation.** Linear equations are often written in the form $Ax + By = C$, called **general form,** where A, B, and C represent numbers (called **constants**) and x and y are variables.

EXAMPLE 2

Graphing linear equations. Graph $3x + 2y = 12$.

Solution We can pick values for either x or y, substitute them into the equation, and solve for the other variable. For example, if $x = 2$,

$$3x + 2y = 12$$
$$3(\mathbf{2}) + 2y = 12 \quad \text{Substitute 2 for } x.$$
$$6 + 2y = 12 \quad \text{Do the multiplication: } 3(2) = 6.$$
$$2y = 6 \quad \text{Subtract 6 from both sides.}$$
$$y = 3 \quad \text{Divide both sides by 2.}$$

The ordered pair $(2, 3)$ satisfies the equation. If $y = 6$,

$$3x + 2y = 12$$
$$3x + 2(\mathbf{6}) = 12 \quad \text{Substitute 6 for } y.$$
$$3x + 12 = 12 \quad \text{Do the multiplication: } 2(6) = 12.$$
$$3x = 0 \quad \text{Subtract 12 from both sides.}$$
$$x = 0 \quad \text{Divide both sides by 3.}$$

A second ordered pair that satisfies the equation is $(0, 6)$.

These pairs and others that satisfy the equation are shown in the table in Figure 2-15. After we plot each pair, we see that they all lie on a line. The graph of the equa-tion is the line shown in the figure.

FIGURE 2-15

$$3x + 2y = 12$$

x	y	(x, y)
-2	9	$(-2, 9)$
0	6	$(0, 6)$
2	3	$(2, 3)$
4	0	$(4, 0)$
6	-3	$(6, -3)$

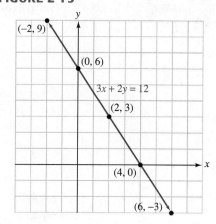

SELF CHECK   Graph $2x - 3y = -6$.

Answer:

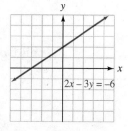

ACCENT ON TECHNOLOGY *Generating Tables of Values Using a Graphing Calculator*

If an equation in x and y is solved for y, we can use a graphing calculator to generate tables of values like the one shown in Figure 2-15. Several brands of graphing calculators are available, and each one has its own sequence of keystrokes to make tables. The instructions in this discussion are for a TI-83 graphing calculator. For specific details about other brands, please consult the owner's manual.

To construct a table of values for $3x + 2y = 12$, we first solve the equation for y.

$$3x + 2y = 12$$
$$2y = -3x + 12 \quad \text{Subtract } 3x \text{ from both sides.}$$
$$y = -\frac{3}{2}x + 6 \quad \text{Divide both sides by 2 and simplify.}$$

To construct a table of values for $y = -\frac{3}{2}x + 6$, we must enter the x-values that are to appear in the table. To do this, we press $\boxed{\text{2nd}}$ $\boxed{\text{TBLSET}}$ and enter the first value for x on the line labeled TblStart =. In Figure 2-16(a), -2 has been entered on this line. Other values for x that are to appear in the table are determined by setting an **increment value,** on the line labeled $\triangle$Tbl =. Figure 2-16(a) shows that an increment of 2 was entered. This means that each x-value in the table will be 2 units larger than the previous x-value.

To enter the equation $y = -\frac{3}{2}x + 6$, we press $\boxed{y =}$ and enter $-(3/2)x + 6$, as shown in Figure 2-16(b). (Ignore the subscript 1 on y; it is not relevant at this time.)

The final step is to press the keys $\boxed{\text{2nd}}$ $\boxed{\text{TABLE}}$. This will display a table of values, as shown in Figure 2-16(c). The table contains all of the entries that we obtained in Figure 2-15, as well as two additional solutions: $(8, -6)$ and $(10, -9)$.

FIGURE 2-16

(a)   (b)   (c)

The Intercept Method

In Example 2, the graph intersected the y-axis at the point with coordinates $(0, 6)$ (called the **y-intercept**) and intersected the x-axis at the point with coordinates $(4, 0)$ (called the **x-intercept**). In general, we have the following definitions.

Intercepts of a line

> The **y-intercept** of a line is the point $(0, b)$, where the line intersects the y-axis. To find b, substitute 0 for x in the equation of the line and solve for y.
>
> The **x-intercept** of a line is the point $(a, 0)$, where the line intersects the x-axis. To find a, substitute 0 for y in the equation of the line and solve for x.

EXAMPLE 3

Graphing using the intercept method. Use the x- and y-intercepts to graph $2x + 5y = 10$.

Solution   To find the y-intercept, we substitute 0 for x and solve for y:

$$2x + 5y = 10$$
$$2(0) + 5y = 10 \quad \text{Substitute 0 for } x.$$
$$5y = 10 \quad \text{Do the multiplication: } 2(0) = 0.$$
$$y = 2 \quad \text{Divide both sides by 5.}$$

The y-intercept is the point $(0, 2)$. To find the x-intercept, we substitute 0 for y and solve for x:

$$2x + 5y = 10$$
$$2x + 5(0) = 10 \quad \text{Substitute 0 for } y.$$
$$2x = 10 \quad \text{Do the multiplication: } 5(0) = 0.$$
$$x = 5 \quad \text{Divide both sides by 2.}$$

The x-intercept is the point $(5, 0)$.

Although two points are enough to draw the line, it is a good idea to find and plot a third point as a check. To find the coordinates of a third point, we can substitute any convenient number (such as -5) for x and solve for y:

$$2x + 5y = 10$$
$$2(-5) + 5y = 10 \quad \text{Substitute } -5 \text{ for } x.$$
$$-10 + 5y = 10 \quad \text{Do the multiplication: } 2(-5) = -10.$$
$$5y = 20 \quad \text{Add 10 to both sides.}$$
$$y = 4 \quad \text{Divide both sides by 5.}$$

The line will also pass through the point $(-5, 4)$.

A table of ordered pairs and the graph of $2x + 5y = 10$ are shown in Figure 2-17.

FIGURE 2-17

$2x + 5y = 10$

x	y	(x, y)
0	2	$(0, 2)$
5	0	$(5, 0)$
-5	4	$(-5, 4)$

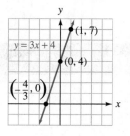

SELF CHECK Find the x- and y-intercepts and graph $y = 3x + 4$.

Answer:

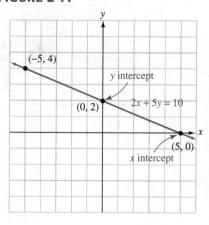

Horizontal and Vertical Lines

EXAMPLE 4 **Graphing lines.** Graph **a.** $y = 3$ and **b.** $x = -2$.

Solution **a.** Since the equation $y = 3$ does not contain x, the numbers chosen for x have no effect on y. The value of y is always 3. (See the table in Figure 2-18.)

FIGURE 2-18

x	y	(x, y)		x	y	(x, y)
-3	3	$(-3, 3)$		-2	-2	$(-2, -2)$
0	3	$(0, 3)$		-2	0	$(-2, 0)$
2	3	$(2, 3)$		-2	2	$(-2, 2)$
4	3	$(4, 3)$		-2	6	$(-2, 6)$

$y = 3$ $x = -2$

The value of x can be any number.

The value of y can be any number.

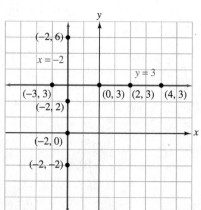

After plotting the pairs (x, y) shown in the table in Figure 2-18, we see that the graph is a horizontal line, parallel to the x-axis, with a y-intercept of $(0, 3)$. The line has no x-intercept.

b. Since the equation $x = -2$ does not contain y, the value of y can be any number.

After plotting the pairs (x, y) shown in the table in Figure 2-18, we see that the graph is a vertical line, parallel to the y-axis, with an x-intercept of $(-2, 0)$. The line has no y-intercept.

SELF CHECK

Graph $x = 4$ and $y = -3$ on one set of coordinate axes.

Answer:

The results of Example 4 suggest the following facts.

Horizontal and vertical lines

If a and b are real numbers, then

The graph of the equation $x = a$ is a vertical line with x-intercept at $(a, 0)$. If $a = 0$, the line is the y-axis.

The graph of the equation $y = b$ is a horizontal line with y-intercept at $(0, b)$. If $b = 0$, the line is the x-axis.

Linear Models

In the next two examples, we will see how linear equations can mathematically model real-life situations. In each case, the equations describe a *linear relationship* between two quantities; when they are graphed, the result is a line. We can make observations about what has taken place in the past and what might take place in the future by carefully inspecting the graph.

EXAMPLE 5

U.S. workforce demographics. The linear equation $p = \frac{3}{5}t + 38$ models the percent of females 16 years or older who were part of the civilian workforce for each of the years 1960–1996. In the equation, t represents the number of years after 1960, and p represents the percent. Graph this equation. What has been the trend over this period?

Solution

The variables t and p are used in the equation. We will associate t with the horizontal axis and p with the vertical axis. Ordered pairs will be of the form (t, p).

To graph the equation, we pick three values for t, substitute them into the equation, and find each corresponding value of p. The results are listed in the table in Figure 2-19.

For t = 0	*For t = 10*	*For t = 20*
$p = \dfrac{3}{5}t + 38$	$p = \dfrac{3}{5}t + 38$	$p = \dfrac{3}{5}t + 38$
$p = \dfrac{3}{5}(0) + 38$	$p = \dfrac{3}{5}(10) + 38$	$p = \dfrac{3}{5}(20) + 38$
$p = 38$	$p = 6 + 38$	$p = 12 + 38$
	$p = 44$	$p = 50$

The pairs (0, 38), (10, 44), and (20, 50) satisfy the equation. Next, we plot these points and draw a line through them, as shown in Figure 2-19. From the graph, we see that there has been a steady increase in the percent of the female population 16 years or older that is part of the workforce.

FIGURE 2-19

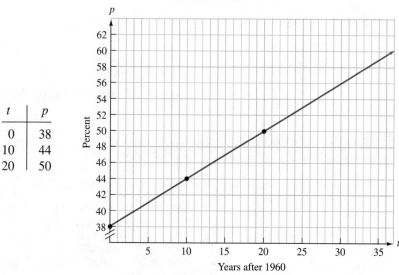

t	p
0	38
10	44
20	50

SELF CHECK

a. Referring to Example 5, use the equation $p = \frac{3}{5}t + 38$ to determine the percent of females 16 years or older who were part of the workforce in 1975.   **b.** Use the graph in Figure 2-18 to find the percent of females who were part of the workforce in 1990.

Answers: **a.** 47%,   **b.** 56%

EXAMPLE 6

Linear depreciation. A copy machine that has been purchased for $6,750 is expected to depreciate according to the formula $y = -950x + 6,750$, where y is the value of the copier after x years. When will the copier have no value?

Solution   The copier will have no value when y is 0. To find x when $y = 0$, we substitute 0 for y and solve for x.

$$y = -950x + 6,750$$
$$0 = -950x + 6,750$$
$$-6,750 = -950x \qquad \text{Subtract 6,750 from both sides.}$$
$$7.105263158 = x \qquad \text{Divide both sides by } -950.$$

The copier will have no value in about 7.1 years.

The equation $y = -950x + 6,750$ is graphed in Figure 2-20. Important information can be obtained from the intercepts of the graph.

FIGURE 2-20

The y-intercept of the graph
is (0, 6,750). This indicates
that the purchase price of the
copier was $6,750.

The x-intercept of the graph
is approximately (7.1, 0).
This indicates that the value
of the copier will be $0 in
about 7.1 years.

$y = -950 + 6,750$

Value of copier
($ thousands)

Age of copier

SELF CHECK   **a.** Referring to Example 6, use the equation $y = -950x + 6,750$ to determine when the copier will be worth $3,900.   **b.** Use the graph in Figure 2-20 to determine when the copier will be worth $2,000.

Answers: **a.** 3 years,
b. 5 years   ∎

ACCENT ON TECHNOLOGY *Graphing Lines*

We have graphed linear equations by finding ordered pairs, plotting points, and drawing lines through those points. Graphing is much easier if we use a graphing calculator.

Graphing calculators have a window to display graphs. To see the proper picture of a graph, we must decide on the minimum and maximum values for the x- and y-coordinates. A window with standard settings of

Xmin = −10   Xmax = 10   Ymin = −10   Ymax = 10

will produce a graph where the values of x and the values of y are between −10 and 10, inclusive. We can use the notation [−10, 10] to describe each of these intervals.

To graph $3x + 2y = 12$, we must first solve the equation for y.

$3x + 2y = 12$

$2y = -3x + 12$ Subtract $3x$ from both sides.

$y = -\dfrac{3}{2}x + 6$ Divide both sides by 2.

To graph the equation, we enter the right-hand side of the equation after a symbol such as $Y_1 =$ or $f(x) = $. After entering the right-hand side, the display should look like

$Y_1 = -(3/2)X + 6$ or $f(x) = -(3/2)X + 6$

We then press the $\boxed{\text{GRAPH}}$ key to get the graph shown in Figure 2-21(a). To show more detail, we can draw the graph in a different window. A window with settings of [−1, 5] for x and [−2, 7] for y will give the graph shown in Figure 2-21(b).

(continued)

FIGURE 2-21

(a)   (b)

Finding the coordinates of a point on the graph. We can use the TRACE command to find the approximate coordinates of any point on a graph. For example, to find the x-intercept of the graph of $2y = -5x - 7$, we first solve for y to get $y = -\frac{5}{2}x - \frac{7}{2}$. Then we graph the equation using the standard window settings of $[-10, 10]$ for x and $[-10, 10]$ for y. We then press the $\boxed{\text{TRACE}}$ key to get Figure 2-22(a). We can then use the $\boxed{\blacktriangleleft}$ and $\boxed{\blacktriangleright}$ keys to move the cursor along the line toward the x-intercept until we arrive at a point with the coordinates shown in Figure 2-22(b).

To get better results, we can zoom in to get a magnified picture, trace again and move the cursor to the point with coordinates shown in Figure 2-22(c). Since the y-coordinate is nearly 0, this point is nearly the x-intercept. We can achieve better results with more zooms.

FIGURE 2-22

| Y1=⁻(5/2)X-(7/2) | Y1=⁻(5/2)X-(7/2) | Y1=⁻(5/2)X-(7/2) |
| X=0   Y=⁻3.5 | X=⁻1.489362   Y=.22340426 | X=⁻1.382979   Y=⁻.0425532 |

(a)   (b)   (c)

Determining the intercepts of a graph. To determine the x-intercept of the graph of $2y = -5x - 7$ $\left(y = -\frac{5}{2}x - \frac{7}{2}\right)$, we can use the ZERO feature, found under the CALC menu. After we guess left and right bounds, the cursor automatically moves to the x-intercept of the graph, and the coordinates of that point are displayed at the bottom of the screen. See Figure 2-23(a).

To determine the y-intercept of the graph, we can use the VALUE feature, which is also found under the CALC menu. With this option, we first enter an x-value of 0, as shown in Figure 2-23(b). Then the cursor highlights the y-intercept, and the coordinates of the y-intercept are given. See Figure 2-23(c).

FIGURE 2-23

(a)   (b)   (c)

STUDY SET

Section 2.2

VOCABULARY

In Exercises 1–6, fill in the blanks to make the statements true.

1. The graph of an equation is the graph of all points (x, y) on the rectangular coordinate system whose coordinates _____ the equation.

2. Any equation whose graph is a line is called a _____ equation.

3. The point where a graph intersects the y-axis is called the _____.

4. The point where a graph intersects the x-axis is called the _____.

5. The graph of any equation of the form $x = a$ is a _____ line.

6. The graph of any equation of the form $y = b$ is a _____ line.

CONCEPTS

7. **a.** Consider the equation $2x + 4 = 8$, studied in Chapter 1. How many variables does it contain? How many solutions does it have?
   b. Consider $2x + 4y = 8$. How many variables does it contain? How many solutions does it have?

8. Consider the linear equation $6x - 4y = -12$.
   a. Find the x-intercept of its graph.
   b. Find the y-intercept of its graph.
   c. Does its graph pass through $(2, 6)$?

9. Linear equations can be written in the form $Ax + By = C$. Write a linear equation with $A = 3$, $B = 5$, and $C = -10$.

10. On which axis does each point lie?
    a. $(0, b)$ **b.** $(a, 0)$

11. See Illustration 1.
    a. What is the x-intercept and what is the y-intercept of the line?

    b. If the coordinates of point M are substituted into the equation of the line that is graphed here, will a true or a false statement result?

ILLUSTRATION 1

12. See Illustration 2. Does the graph indicate a linear relationship between the two quantities? Explain why or why not.

ILLUSTRATION 2

13. The graph of $8x + 4y = -5$ is shown in Illustration 3.
    a. Explain why the display shows $Y_1 = -2x - (5/4)$ not $8x + 4y = -5$ as being graphed.

    b. What important feature of the line does the cursor highlight?

ILLUSTRATION 3

14. The graph of $8x + 4y = -5$ is shown in Illustration 4. What important feature of the line does the cursor highlight?

ILLUSTRATION 4

NOTATION

In Exercises 15–16, complete each solution.

15. Verify that $(-3, -1)$ is a solution of $2x + 2y = -8$.

$$2x + 2y = -8$$
$$2(\quad) + 2(-1) \stackrel{?}{=} -8$$
$$-6 + (\quad) \stackrel{?}{=} -8$$
$$-8 = -8$$

16. To find the coordinates of a point on the graph of $5x + 2y = 10$, choose $x = 1$ and find y.

$$5x + 2y = 10$$
$$5(\quad) + 2y = 10$$
$$5 + \quad = 10$$
$$2y = \quad$$
$$y = \tfrac{5}{2}$$

The point $\quad$ is on the graph of $5x + 2y = 10$.

17. The graph of the equation $x = 0$ is which axis?

18. Solve the equation $3x + 2y = 9$ for y.

PRACTICE

In Exercises 19–22, complete the table of solutions for each equation.

19. $y = -x + 4$

x	y
-1	
0	
2	

20. $y = x - 2$

x	y
-2	
0	
4	

21. $y = -\dfrac{1}{3}x - 1$

x	y
-3	
0	
3	

22. $y = -\dfrac{1}{2}x + \dfrac{5}{2}$

x	y
-1	
3	
5	

In Exercises 23–26, use the results from Exercises 19–22 to graph each equation.

23. $y = -x + 4$

24. $y = x - 2$

25. $y = -\dfrac{1}{3}x - 1$

26. $y = -\dfrac{1}{2}x + \dfrac{5}{2}$

In Exercises 27–34, construct a table of solutions and then graph the equation.

27. $y = x$

28. $y = -2x$

29. $y = -3x + 2$

30. $y = 2x - 3$

31. $x = 3$

32. $y = -4$

33. $-3y + 2 = 5$

34. $-2x + 3 = 11$

In Exercises 35–42, graph each equation using the intercept method.

35. $3x + 4y = 12$

36. $4x - 3y = 12$

37. $3y = 6x - 9$

38. $2x = 4y - 10$

39. $2y + x = -2$

40. $4y + 2x = -8$

41. $3x + 4y - 8 = 0$

42. $-2y - 3x + 9 = 0$

In Exercises 43–46, use a graphing calculator to graph each equation, and then find the x-coordinate of the x-intercept to the nearest hundredth.

43. $y = 3.7x - 4.5$

44. $y = \dfrac{3}{5}x + \dfrac{5}{4}$

45. $1.5x - 3y = 7$

46. $0.3x + y = 7.5$

APPLICATIONS

47. HOURLY WAGES The following table gives amount y (in dollars) that a student can earn for working x hours. Plot the ordered pairs and estimate how much the student will earn for working 8 hours.

x	2	4	5	6
y	12	24	30	36

48. VALUE OF A CAR The following table shows the value y (in dollars) of a car that is x years old. Plot the ordered pairs and estimate the value of the car when it is 4 years old.

x	0	1	3
y	15,000	12,000	6,000

49. FARMING The equation $f = -\frac{1}{50}t + 2.4$ models the number of farms in the United States (t is the number of years after 1980, and f is the number of farms in millions). Graph the equation in Illustration 5. Use the graph to estimate the number of farms in 1984 and 1993. Describe the trend.

ILLUSTRATION 5

50. TV COVERAGE See Illustration 6. A blimp, hovering directly over a sports stadium, sends a television signal to the ground along the path described by $y = -\frac{5}{4}x + 5$.
   a. Sketch the path of the signal.
   b. How far from the entrance of the stadium should a transmitter truck be parked to receive the signal from the blimp?

ILLUSTRATION 6

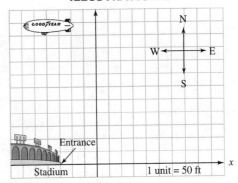

51. HOUSE APPRECIATION A house purchased for $125,000 is expected to appreciate according to the formula $y = 7,500x + 125,000$, where y is the value of the house after x years. Find the value of the house 5 years later.

52. CAR DEPRECIATION A car purchased for $17,000 is expected to depreciate according to the formula $y = -1,360x + 17,000$. When will the car have no value?

53. DEMAND EQUATION The number of television sets that consumers buy depends on price. The higher the price, the fewer TVs people will buy. The equation that relates price to the number of TVs sold at that price is called a **demand equation.** If the demand equation for a 13-inch TV is $p = -\frac{1}{10}q + 170$, where p is the price and q is the number of TVs sold at that price, how many TVs will be sold at a price of $150?

54. SUPPLY EQUATION The number of television sets that manufacturers produce depends on price. The higher the price, the more TVs manufacturers will produce. The equation that relates price to the number of TVs produced at that price is called a **supply equation.** If the supply equation for a 13-inch TV is $p = \frac{1}{10}q + 130$, where p is the price and q is the number of TVs produced for sale at that price, how many TVs will be produced if the price is $150?

55. BUYING TICKETS Tickets to a circus cost $5 each from Ticketron plus a $2 service fee for each block of tickets.
   a. Write a linear equation that gives the cost c when t tickets are purchased.
   b. Complete the table in Illustration 7 and graph the equation.
   c. Use the graph to estimate the cost of buying 5 tickets.

ILLUSTRATION 7

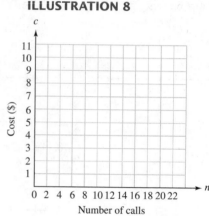

56. TELEPHONE COSTS In a community, the monthly cost of local telephone service is $5 per month, plus 25¢ per call.
   a. Write a linear equation that gives the cost c for a person making n calls. Then graph the equation.
   b. Complete the table in Illustration 8.
   c. Use the graph to estimate the cost of service in a month when 20 calls were made.

ILLUSTRATION 8

WRITING

57. Explain how to graph a line using the intercept method.

58. When graphing a line by plotting points, why is it a good practice to find three solutions instead of two?

REVIEW

59. List the prime numbers between 10 and 30.

60. Write the first ten composite numbers.

61. In what quadrant does the point $(-2, -3)$ lie?

62. What is the formula that gives the area of a circle?

63. Simplify $-4(-20s)$.

64. Approximate π to the nearest thousandth.

65. Remove parentheses: $-(-3x - 8)$.

66. Simplify $\frac{1}{3}b + \frac{1}{3}b + \frac{1}{3}b$.

▶ 2.3   Rate of Change and the Slope of a Line

In this section, you will learn about

> Average rate of change ■ Slope of a line ■ Interpretation of slope ■ Horizontal and vertical lines ■ Slopes of parallel lines ■ Slopes of perpendicular lines

Introduction   Our world is one of constant change. In this section, we will show how to describe the amount of change in one quantity in relation to the amount of change in another by finding an *average rate of change.*

Average Rate of Change

The line graphs in Figure 2-24 are models that approximate the number of daily morning newspapers and the number of evening newspapers published in the United States for the years 1990–1997. From the graph, we can see that the number of morning newspapers increased and the number of evening newspapers decreased from 1990 to 1997.

FIGURE 2-24

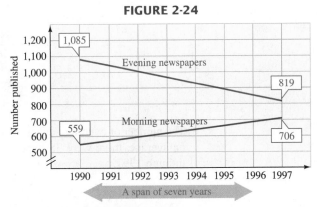

Based on data from *Editor & Publisher*

If we want to know the rate at which the number of morning newspapers increased or the rate at which the number of evening newspapers decreased over this period, we can do so by finding an **average rate of change.** To find an average rate of change, we find the **ratio** of the change in the number of newspapers to the length of time in which that change took place.

Ratios and rates

A **ratio** is a comparison of two numbers by their indicated quotient. In symbols, if a and b are two numbers, the ratio of a to b is $\frac{a}{b}$. Ratios that are used to compare quantities with different units are called **rates.**

In Figure 2-24, we see that in 1990, the number of morning newspapers published was 559. In 1997, the number grew to 706. This is a change of $706 - 559$ or 147 over a 7-year time span. So we have

$$\text{Average rate of change} = \frac{\text{change in number of morning newspapers}}{\text{change in time}}$$

The rate of change is a ratio.

$$= \frac{147 \text{ newspapers}}{7 \text{ years}}$$

$$= \frac{21 \cdot \overset{1}{\cancel{7}} \cdot \text{newspapers}}{\underset{1}{\cancel{7}} \text{ years}}$$

Factor 147 as $21 \cdot 7$ and divide out the common factor of 7.

$$= \frac{21 \text{ newspapers}}{1 \text{ year}}$$

The number of morning newspapers published in the United States increased, on average, at a rate of 21 newspapers per year (written 21 newspapers/year) from 1990 through 1997.

In Figure 2-24, we see that in 1990 the number of evening newspapers published was 1,085. In 1997, the number fell to 819. To find the change, we subtract the earlier number from the later number: $819 - 1,085 = -266$. The negative result indicates a decline in the number of evening newspapers over the 7-year time span. So we have

$$\text{Average rate of change} = \frac{819 - 1,085}{7}$$

$$= \frac{-266}{7}$$

$$= \frac{-38 \cdot \overset{1}{\cancel{7}}}{\underset{1}{\cancel{7}}}$$

Factor -266 as $-38 \cdot 7$.

$$= \frac{-38}{1}$$

Divide out the common factor of 7.

The number of evening newspapers changed at a rate of -38 newspapers/year. That is, on average, there were 38 fewer evening newspapers per year, every year, from 1990 through 1997.

Slope of a Line

In the newspaper example, we measured the steepness of the two lines in the graph in Figure 2-24 to determine the average rates of change. In doing so, we found the **slope** of each line. The slope of a nonvertical line is a number that measures the line's steepness.

To calculate the slope of a line (usually denoted by the letter m), we must first pick two points on the line. To distinguish between the coordinates of two points, we use **subscript notation.** The first point can be denoted as (x_1, y_1), and the second point can be denoted as (x_2, y_2). After picking two points on the line, we write the ratio of the vertical change to the corresponding horizontal change as we move from one point to the other.

Slope of a line

The **slope** of a line passing through points (x_1, y_1) and (x_2, y_2) is

$$m = \frac{\text{change in } y}{\text{change in } x} = \frac{y_2 - y_1}{x_2 - x_1} \quad (x_2 \neq x_1)$$

EXAMPLE 1

Using the slope formula. Find the slope of the line shown in Figure 2-25, passing through $(-2, 4)$ and $(3, -4)$.

Solution We can let $(x_1, y_1) = (-2, 4)$ and $(x_2, y_2) = (3, -4)$. Then

$$x_1 = -2 \qquad x_2 = 3$$
$$\text{and}$$
$$y_1 = 4 \qquad y_2 = -4$$

$$m = \frac{\text{change in } y}{\text{change in } x}$$

$$= \frac{y_2 - y_1}{x_2 - x_1} \qquad \text{The slope formula.}$$

$$= \frac{-4 - 4}{3 - (-2)} \qquad \begin{array}{l}\text{Now we substitute} \\ -4 \text{ for } y_2, 4 \text{ for } y_1, \\ 3 \text{ for } x_2, \text{ and } -2 \text{ for } x_1.\end{array}$$

$$= \frac{-8}{5}$$

$$= -\frac{8}{5}$$

The slope of the line is $-\frac{8}{5}$.

FIGURE 2-25

SELF CHECK

Find the slope of the line passing through the points $(-3, 6)$ and $(4, -8)$.

Answer: -2 ∎

When calculating slope, it does not matter which point that we call (x_1, y_1) and which point we call (x_2, y_2). To illustrate this, we would have obtained the same result in Example 1 if we had let $(x_1, y_1) = (3, -4)$ and $(x_2, y_2) = (-2, 4)$.

$$m = \frac{y_2 - y_1}{x_2 - x_1} = \frac{4 - (-4)}{-2 - 3} = \frac{8}{-5} = -\frac{8}{5}$$

WARNING! When using the slope formula, we must be careful to subtract the y-coordinates and the x-coordinates in the same order. For instance, in Example 1 with $(x_1, y_1) = (-2, 4)$ and $(x_2, y_2) = (3, -4)$, it would be incorrect to write

This is $y_2 - y_1$.
↓
$$m = \frac{-4 - 4}{-2 - 3}$$
↑
This is $x_1 - x_2$. The subtraction is not in the same order.

This is $y_1 - y_2$. The subtraction is not in the same order.
↓
$$m = \frac{4 - (-4)}{3 - (-2)}$$
↑
This is $x_2 - x_1$.

The change in y (often denoted as Δy and read as "delta y") is the **rise** of the line between two points on the line. The change in x (often denoted as Δx and read as "delta x") is the **run.** Using this terminology, we can define slope as the ratio of the rise to the run:

$$m = \frac{\Delta y}{\Delta x} = \frac{\text{rise}}{\text{run}} \qquad (\Delta x \neq 0)$$

EXAMPLE 2

Finding the slope of a line. Find the slope of the line determined by $3x - 4y = 12$.

Solution We first find the coordinates of two points on the line.

- If we choose $x = 0$, then $y = -3$. The point $(0, -3)$ is on the line.
- If we choose $y = 0$, then $x = 4$. The point $(4, 0)$ is on the line.

We then refer to Figure 2-26 and find the slope of the line passing through $(0, -3)$ and $(4, 0)$ by substituting 0 for y_2, -3 for y_1, 4 for x_2, and 0 for x_1 in the formula for slope.

FIGURE 2-26

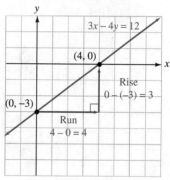

$$m = \frac{\text{rise}}{\text{run}}$$

$$= \frac{y_2 - y_1}{x_2 - x_1}$$

$$= \frac{0 - (-3)}{4 - (0)}$$

$$= \frac{3}{4}$$

The slope of the line is $\frac{3}{4}$.

SELF CHECK

Find the coordinates of two other points on the line $3x - 4y = 12$. Then calculate the slope of the line.   *Answer:* $\frac{3}{4}$ ∎

The answers to Example 2 and the Self Check illustrate an important fact. When finding the slope of a line, *any* pair of points on the line can be used—the result will be the same.

Interpretation of Slope

For applied problems, slope can be thought of as the average rate of change in one quantity per unit change in another quantity.

EXAMPLE 3

Carpeting cost. A store sells a high-quality carpet for $25 per square yard, plus a $20 delivery charge. The total cost c of n square yards is given by the following formula.

Total cost	=	cost per square yard	·	the number of square yards	+	the delivery charge.
c	=	25	·	n	+	20

Graph this equation and interpret the slope of the line.

Solution The total cost c depends on the number n of square yards of carpet purchased. This relationship is described by $c = 25n + 20$. We can graph the equation on a coordinate system with a horizontal n-axis and a vertical c-axis. Figure 2-27 shows a table of ordered pairs and the graph.

FIGURE 2-27

$$c = 25n + 20$$

n	c	(n, c)
10	270	(10, 270)
30	770	(30, 770)
40	1,020	(40, 1,020)

If we pick the points $(30, 770)$ and $(40, 1,020)$ to find the slope, we have

$$m = \frac{\Delta c}{\Delta n}$$

$$= \frac{c_2 - c_1}{n_2 - n_1}$$ This is the form the slope formula takes when working with ordered pairs of the form (n, c).

$$= \frac{1,020 - 770}{40 - 30}$$ Substitute 1,020 for c_2, 770 for c_1, 40 for n_2, and 30 for n_1.

$$= \frac{250}{10}$$

$$= 25$$

The slope of 25 (in dollars/square yard) is the price per square yard of the carpet.

SELF CHECK   What information is given by the   *Answer:* The c-coordinate of the
c-intercept of the graph in Figure 2-27? c-intercept is the delivery charge, $20. ■

EXAMPLE 4   **Rate of descent.** It takes a skier 25 minutes to complete the course shown in Figure 2-28. Find his average rate of descent in feet per minute.

FIGURE 2-28

Solution   To find the average rate of descent, we must find the ratio of the change in altitude (ΔA) to the change in time (Δt). To find this ratio, we calculate the slope of the line passing through the points (0, 12,000) and (25, 8,500).

$$\text{Average rate of descent} = \frac{\Delta A}{\Delta t}$$

$$= \frac{8,500 - 12,000}{25 - 0}$$ In the numerator, write the change in altitude; in the denominator, the change in time.

$$= \frac{-3,500}{25}$$ Do the subtractions.

$$= -140$$

The average rate of descent is 140 feet per minute.

SELF CHECK   In Example 4, find the average rate of descent if the skier completes the course in 20 minutes.   *Answer:* 175 feet per minute   ■

Horizontal and Vertical Lines

If (x_1, y_1) and (x_2, y_2) are points on a horizontal line (see Figure 2-29(a)), then $y_1 = y_2$, and the numerator of the fraction

$$\frac{y_2 - y_1}{x_2 - x_1}$$ On a horizontal line, $x_2 \neq x_1$.

is 0. Thus, the value of the fraction is 0, and the slope of the horizontal line is 0.

If (x_1, y_1) and (x_2, y_2) are two points on a vertical line (see Figure 2-29(b)), then $x_1 = x_2$, and the denominator of the fraction

$$\frac{y_2 - y_1}{x_2 - x_1}$$ On a vertical line, $y_2 \neq y_1$.

is 0. Since the denominator of a fraction cannot be 0, a vertical line has no defined slope.

FIGURE 2-29

(a)   (b)

Slopes of horizontal and vertical lines

All horizontal lines (lines with equations of the form $y = b$) have a slope of 0.

All vertical lines (lines with equations of the form $x = a$) have no defined slope.

If a line rises as we follow it from left to right, as in Figure 2-30(a), its slope is positive. If a line drops as we follow it from left to right, as in Figure 2-30(b), its slope is negative. If a line is horizontal, as in Figure 2-30(c), its slope is 0. If a line is vertical, as in Figure 2-30(d), it has no defined slope.

FIGURE 2-30

Positive slope	Negative slope	Zero slope	Undefined slope
(a)	(b)	(c)	(d)

Slopes of Parallel Lines

To see a relationship between parallel lines and their slopes, we refer to the parallel lines l_1 and l_2 shown in Figure 2-31, with slopes of m_1 and m_2, respectively. Because right triangles ABC and DEF are similar, it follows that

$$
\begin{aligned}
m_1 &= \frac{\Delta y \text{ of } l_1}{\Delta x \text{ of } l_1} \\
&= \frac{\Delta y \text{ of } l_2}{\Delta x \text{ of } l_2} \qquad \text{Since the triangles are similar, corresponding sides of } \triangle ABC \text{ and} \\
& \qquad\qquad\qquad \triangle DEF \text{ are proportional: } \frac{CB}{BA} = \frac{FE}{ED}. \\
&= m_2
\end{aligned}
$$

FIGURE 2-31

Thus, if two nonvertical lines are parallel, they have the same slope. It is also true that when two lines have the same slope, they are parallel.

Slopes of parallel lines

Nonvertical parallel lines have the same slope, and different lines having the same slope are parallel. Since vertical lines are parallel, lines with no defined slope are parallel.

EXAMPLE 5

Slopes of parallel lines. The lines in Figure 2-32 are parallel. Find y.

Solution Since the lines are parallel, they have equal slopes. To find y, we find the slope of each line, set them equal, and solve the resulting equation.

FIGURE 2-32

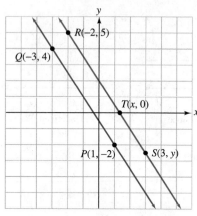

$$\textbf{\textit{Slope of PQ = Slope of RS}}$$

$$\frac{-2 - 4}{1 - (-3)} = \frac{y - 5}{3 - (-2)}$$

$$\frac{-6}{4} = \frac{y - 5}{5} \qquad \text{Do the subtractions.}$$

$$-30 = 4(y - 5) \qquad \text{Multiply both sides by 20.}$$

$$-30 = 4y - 20 \qquad \text{Use the distributive property.}$$

$$-10 = 4y \qquad \text{Add 20 to both sides.}$$

$$-\frac{5}{2} = y \qquad \text{Divide both sides by 4 and simplify.}$$

Thus, $y = -\dfrac{5}{2}$.

SELF CHECK   In Example 5, find x.   *Answer:* $\frac{4}{3}$ ■

Slopes of Perpendicular Lines

Two real numbers a and b are called **negative reciprocals** if $ab = -1$. For example,

$$-\frac{4}{3} \qquad \text{and} \qquad \frac{3}{4}$$

are negative reciprocals, because $-\frac{4}{3}\left(\frac{3}{4}\right) = -1$.

The following fact relates perpendicular lines and their slopes.

Slopes of perpendicular lines

If two nonvertical lines are perpendicular, their slopes are negative reciprocals.

If the slopes of two lines are negative reciprocals, the lines are perpendicular.

We can also state the fact given above symbolically: If the slopes of two nonvertical lines are m_1 and m_2, then the lines are perpendicular if

$$m_1 \cdot m_2 = -1 \qquad \text{or} \qquad m_2 = -\frac{1}{m_1}$$

Because a horizontal line is perpendicular to a vertical line, a line with a slope of 0 is perpendicular to a line with no defined slope.

EXAMPLE 6

Slopes of perpendicular lines. Are the lines OP and PQ shown in Figure 2-33 perpendicular?

Solution We find the slopes of the lines and see whether they are negative reciprocals.

$$\text{Slope of } OP = \frac{\Delta y}{\Delta x} \qquad \text{and} \qquad \text{Slope of } PQ = \frac{\Delta y}{\Delta x}$$

FIGURE 2-33

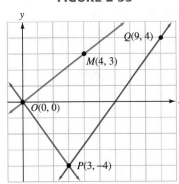

$$= \frac{y_2 - y_1}{x_2 - x_1} \qquad\qquad\qquad = \frac{y_2 - y_1}{x_2 - x_1}$$

$$= \frac{-4 - 0}{3 - 0} \qquad\qquad\qquad = \frac{4 - (-4)}{9 - 3}$$

$$= -\frac{4}{3} \qquad\qquad\qquad\quad = \frac{8}{6}$$

$$= \frac{4}{3}$$

Since their slopes are not negative reciprocals $\left(-\dfrac{4}{3} \cdot \dfrac{4}{3} \neq -1\right)$, the lines are not perpendicular.

SELF CHECK In Figure 2-33, is line *MO* perpendicular to line *OP*? *Answer:* yes ■

STUDY SET

Section 2.3

VOCABULARY

In Exercises 1–6, fill in the blanks to make the statements true.

1. _____ is defined as the change in *y* divided by the change in *x*.

2. A slope is an average _____ of change.

3. The _____ in *x* (denoted as Δx) is the horizontal run of the line between two points on the line.

4. The change in *y* (denoted as Δy) is the vertical _____ of the line between two points on the line.

5. $\frac{7}{8}$ and $-\frac{8}{7}$ are negative _____.

6. _____ lines have the same slope.

CONCEPTS

7. See Illustration 1.

    a. Which line is horizontal? What is its slope?

    b. Which line is vertical? What is its slope?

    c. Which line has a positive slope? What is it?

    d. Which line has a negative slope? What is it?

ILLUSTRATION 1

8. See Illustration 2.

    a. Find the slopes of lines l_1 and l_2. Are they parallel?

    b. Find the slopes of lines l_2 and l_3. Are they perpendicular?

ILLUSTRATION 2

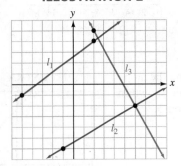

9. EDUCATION The graphs in Illustration 3 are models that approximate the growth in weekly earnings of Americans 25 years and older, by educational attainment, for 1980–1997. Which educational level's weekly earnings are

    a. increasing at the fastest rate? Estimate the average rate of change.

    b. increasing at the slowest rate? Estimate the average rate of change.

ILLUSTRATION 3

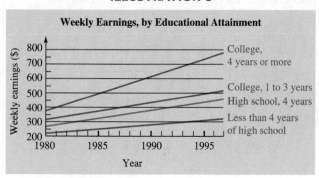

Weekly Earnings, by Educational Attainment

Based on data from *The Wall Street Journal Almanac* (1999)

10. HALLOWEEN A couple kept records of the number of trick-or-treaters who came to their door on Halloween night. (See Illustration 4.) Find the slope of the line. What information does the slope give?

ILLUSTRATION 4

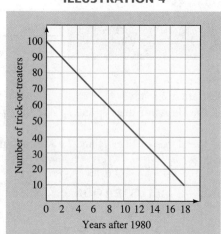

11. Determine the slope of the line shown in Illustration 5.

ILLUSTRATION 5

12. Determine the slope of the line shown in Illustration 6.

ILLUSTRATION 6

NOTATION

13. What formula is used to find the slope of a line?

14. Tell the difference between x^2 and x_2.

15. See Illustration 7.
   a. What is Δy?
   b. What is Δx?
   c. What is $\dfrac{\Delta y}{\Delta x}$?

ILLUSTRATION 7

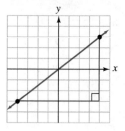

16. See Illustration 8.
   a. What is Δy?
   b. What is Δx?
   c. What is $\dfrac{\Delta y}{\Delta x}$?

ILLUSTRATION 8

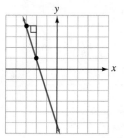

PRACTICE

In Exercises 17–28, find the slope of the line that passes through the given points, if possible.

17. $(0, 0)$, $(3, 9)$ **18.** $(9, 6)$, $(0, 0)$

19. $(-1, 8)$, $(6, 1)$ **20.** $(-5, -8)$, $(3, 8)$

21. $(3, -1)$, $(-6, 2)$ **22.** $(0, -8)$, $(-5, 0)$

23. $(7, 5)$, $(-9, 5)$ **24.** $(2, -8)$, $(3, -8)$

25. $(-7, -5)$, $(-7, -2)$ **26.** $(3, -5)$, $(3, 14)$

27. (a, b), (b, a) **28.** (a, b), $(-b, -a)$

In Exercises 29–36, find the slope of the line determined by each equation.

29. $3x + 2y = 12$ **30.** $2x - y = 6$

31. $3x = 4y - 2$ **32.** $x = y$

33. $y = \dfrac{x - 4}{2}$ **34.** $x = \dfrac{3 - y}{4}$

35. $4y = 3(y + 2)$ **36.** $x + y = \dfrac{2 - 3y}{3}$

In Exercises 37–42, tell whether the lines with the given slopes are parallel, perpendicular, or neither.

37. $m_1 = 3$, $m_2 = -\dfrac{1}{3}$

38. $m_1 = \dfrac{1}{4}$, $m_2 = 4$

39. $m_1 = 4$, $m_2 = 0.25$

40. $m_1 = -5$, $m_2 = \dfrac{1}{-0.2}$

41. $m_1 = \dfrac{1}{a}$, $m_2 = a$

42. $m_1 = a$, $m_2 = -\dfrac{1}{a}$

In Exercises 43–48, tell whether the line that passes through the two given points is parallel or perpendicular (or neither) to a line with a slope of −2.

43. $(3, 4)$, $(4, 2)$

44. $(6, 4)$, $(8, 5)$

45. $(-2, 1)$, $(6, 5)$

46. $(3, 4)$, $(-3, -5)$

47. $(5, 4)$, $(6, 6)$

48. $(-2, 3)$, $(4, -9)$

APPLICATIONS

49. LANDING A PLANE A jet descends in a stairstep pattern, as shown in Illustration 9. The required elevations of the plane's path are given. Find the slope of the descent in each of the three parts of its landing that are labeled. Which part is the steepest?

50. DROP IN PRICE The price of computers has been dropping for the past ten years. If a desktop PC cost $5,700 ten years ago, and the same computing power cost $400 two years ago, find the rate of decrease per year. (Assume a straight-line model.)

51. MAPS Topographic maps have contour lines that connect points of equal elevation on a mountain. The vertical distance between contour lines in Illustration 10 is 50 feet. Find the slope of the west face and the east face of the mountain peak.

ILLUSTRATION 10

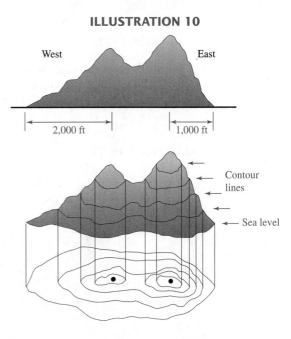

52. SKIING The men's giant slalom course shown in Illustration 11 (on the next page) is longer than the women's course. Does this mean that the men's course is steeper? Use the concept of the slope of a line to explain.

ILLUSTRATION 9

Based on data from *Los Angeles Times* (August 7, 1997), p. A8

ILLUSTRATION 11

53. ROAD SIGNS Find the slope of the road shown in Illustration 12. Use this information to complete the road warning sign for truckers. (*Hint:* 1 mi = 5,280 ft.)

ILLUSTRATION 12

54. GREENHOUSE EFFECT The graphs in Illustration 13 are estimates of future average global temperature rise due to the greenhouse effect. Assume that the models are straight lines. Estimate the average rate of change of each model. Express your answers as fractions.

ILLUSTRATION 13

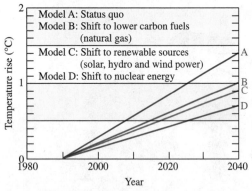

Based on data from *The Blue Planet* (Wiley, 1995)

55. DECK DESIGN See Illustration 14. Find the slopes of the cross-brace and the supports. Is the cross-brace perpendicular to either support?

ILLUSTRATION 14

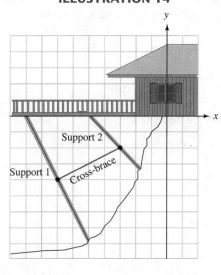

56. AIR PRESSURE Air pressure, measured in units called pascals (Pa), decreases with altitude. Find the rate of change in Pascals for the fastest and the slowest decreasing steps of the graph in Illustration 15.

ILLUSTRATION 15

Based on data from *The Blue Planet* (Wiley, 1995)

WRITING

57. POLITICS Illustration 16 shows how federal Medicare spending would have continued if the Republican-sponsored Balanced Budget Act hadn't become law in 1997. Write a brief statement explaining why Democrats could argue that the budget act "cut spending." Then write a brief statement explaining why Republicans could respond by saying, "There was no cut in spending—only a reduction in the rate of growth of spending."

58. NUCLEAR ENERGY Since 1992, the number of nuclear reactors licensed for operation in the United States has remained about the same. Knowing this, what can be said about the rate of change in the number of reactors since 1992? Explain your answer.

59. Explain why a vertical line has no defined slope.

60. Explain how to determine from their slopes whether two lines are parallel, perpendicular, or neither.

ILLUSTRATION 16

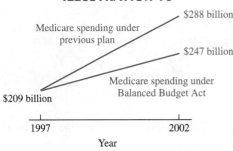

Based on information supplied by Congressman David Dreier's office

REVIEW

61. HALLOWEEN CANDY A candy maker wants to make a 60-pound mixture of two candies to sell for $2 per pound. If black licorice bits sell for $1.90 per pound and orange gumdrops sell for $2.20 per pound, how many pounds of each should be used?

62. MEDICATIONS A doctor prescribes an ointment that is 2% hydrocortisone. A pharmacist has 1% and 5% concentrations in stock. How many ounces of each should the pharmacist use to make a 1-ounce tube?

63. READING GRAPHS In Illustration 17, d is the distance a person has walked after t hours. How many hours did it take for the person to walk 10 miles?

ILLUSTRATION 17

64. CIRCLE GRAPHS In Illustration 18, each part of the circle represents the amount of money spent in each of five categories. Approximately what percent was spent on rent?

ILLUSTRATION 18
Monthly Expenses of Joe Sigueri

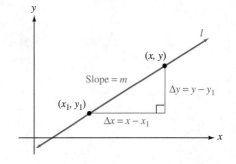

▶ 2.4   Writing Equations of Lines

In this section, you will learn about

> Point–slope form of the equation of a line ■ Slope–intercept form of the equation of a line ■ Using slope as an aid in graphing ■ Parallel and perpendicular lines ■ Straight-line depreciation ■ Curve fitting

Introduction   We have seen that linear relationships are often presented in graphs. In this section, we begin a discussion of how to write an equation to model a linear relationship.

Point–Slope Form of the Equation of a Line

Suppose that line l in Figure 2-34 has a slope of m and passes through (x_1, y_1). If (x, y) is a second point on line l, we have

$$m = \frac{y - y_1}{x - x_1}$$

or if we multiply both sides by $x - x_1$, we have

1. $y - y_1 = m(x - x_1)$

FIGURE 2-34

Because Equation 1 displays the coordinates of the point (x_1, y_1) on the line and the slope m of the line, it is called the **point–slope form** of the equation of a line.

Point–slope form

The equation of the line passing through (x_1, y_1) and with slope m is

$$y - y_1 = m(x - x_1)$$

EXAMPLE 1

Using the point–slope form. Write the equation of the line that has a slope of $-\frac{2}{3}$ and passes through $(-4, 5)$.

Solution   We substitute $-\frac{2}{3}$ for m, -4 for x_1, and 5 for y_1 into the point–slope form and simplify.

$$y - y_1 = m(x - x_1) \qquad \text{Point–slope form.}$$

$$y - 5 = -\frac{2}{3}[x - (-4)] \quad \text{Substitute } -\frac{2}{3} \text{ for } m, -4 \text{ for } x_1, \text{ and 5 for } y_1.$$

$$y - 5 = -\frac{2}{3}(x + 4) \qquad \text{Simplify the expression inside the brackets.}$$

$$y - 5 = -\frac{2}{3}x - \frac{8}{3} \qquad \text{Use the distributive property to remove parentheses.}$$

$$y = -\frac{2}{3}x + \frac{7}{3} \qquad \text{To solve for } y, \text{ add } 5 = \frac{15}{3} \text{ to both sides and simplify.}$$

The equation of the line is $y = -\dfrac{2}{3}x + \dfrac{7}{3}$.

SELF CHECK

Write the equation of the line that has a slope of $\frac{5}{4}$ and passes through $(0, 5)$.   *Answer:* $y = \frac{5}{4}x + 5$ ∎

EXAMPLE 2

Given two points on a line. Write the equation of the line passing through $P(-5, 4)$ and $Q(8, -6)$.

Solution   First we find the slope of the line.

$$m = \frac{y_2 - y_1}{x_2 - x_1} \qquad \text{The slope formula.}$$

$$= \frac{-6 - 4}{8 - (-5)} \qquad \text{Substitute } -6 \text{ for } y_2, 4 \text{ for } y_1, 8 \text{ for } x_2, \text{ and } -5 \text{ for } x_1.$$

$$= -\frac{10}{13}$$

Since the line passes through P and Q, we can choose either point and substitute its coordinates into the point–slope form. If we choose $P(-5, 4)$, we substitute -5 for x_1, 4 for y_1, and $-\frac{10}{13}$ for m and proceed as follows.

$$y - y_1 = m(x - x_1) \qquad \text{Point–slope form.}$$

$$y - 4 = -\frac{10}{13}[x - (-5)] \quad \text{Substitute } -\frac{10}{13} \text{ for } m, -5 \text{ for } x_1, \text{ and 4 for } y_1.$$

$$y - 4 = -\frac{10}{13}(x + 5) \qquad \text{Simplify the expression inside the brackets.}$$

$$y - 4 = -\frac{10}{13}x - \frac{50}{13} \qquad \text{Remove parentheses: distribute } -\frac{10}{13}.$$

$$y = -\frac{10}{13}x + \frac{2}{13} \qquad \text{To solve for } y, \text{ add } 4 = \frac{52}{13} \text{ to both sides and simplify.}$$

The equation of the line is $y = -\dfrac{10}{13}x + \dfrac{2}{13}$.

SELF CHECK

SELF CHECK Write the equation of the line passing through
$(-2, 5)$ and $(4, -3)$.   *Answer:* $y = -\frac{4}{3}x + \frac{7}{3}$ ∎

Slope–Intercept Form of the Equation of a Line

FIGURE 2-35

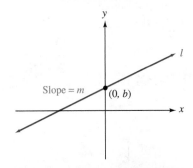

Since the y-intercept of the line l shown in Figure 2-35 is the point $(0, b)$, we can
write its equation by substituting 0 for x_1 and b for y_1 in the point–slope form and
simplifying.

$$y - y_1 = m(x - x_1)$$
$$y - b = m(x - 0)$$
$$y - b = mx$$
2. $y = mx + b$ To solve for y, add b to both sides.

Because Equation 2 displays the slope m and the y-coordinate b of the y-intercept,
it is called the **slope–intercept form** of the equation of a line.

Slope–intercept form

> The equation of the line with slope m and y-intercept $(0, b)$ is
>
> $$y = mx + b$$

E X A M P L E   3

Using the slope–intercept form. Write the equation of the line that has slope
4 and passes through $(5, 9)$.

Solution   Since we are given that $m = 4$ and that $(5, 9)$ satisfies the equation, we can substitute
5 for x, 9 for y, and 4 for m in the equation $y = mx + b$ and solve for b.

$$y = mx + b$$ Slope–intercept form.
$$9 = 4(5) + b$$ Substitute 9 for y, 4 for m, and 5 for x.
$$9 = 20 + b$$ Do the multiplication.
$$-11 = b$$ To solve for b, subtract 20 from both sides.

Because $m = 4$ and $b = -11$, the equation is $y = 4x - 11$.

SELF CHECK Write the equation of the line that has slope -2
and passes through $(-2, 8)$.   *Answer:* $y = -2x + 4$ ∎

When an equation describing a linear relationship between two quantities is written
in slope–intercept form, two pieces of information about the relationship are easily seen.
As an example, let's consider the equation $L = -0.05t + 7.25$. If we begin with a pencil
7.25 inches long, this linear model gives the new length L in inches of the pencil after
it has been inserted into a sharpener and the handle turned t times. (See Figure 2-36.)

FIGURE 2-36

The new length L of the pencil is related to the number
of times t the handle of the sharpener is turned.

The original length of the pencil
7.25 in.

The new length of the pencil
L in.

t turns of
the handle

The value of m (in this case, -0.05) gives the change in the length of the pencil for one turn of the handle. Because the slope is negative, we know that the length of the pencil *decreases* by 0.05 inch for each turn of the handle. The value of b (in this case, 7.25) tells us that before any turns were made (when $t = 0$), the length of the pencil was 7.25 inches.

$$L = -0.05t + 7.25$$

The slope gives the rate of change of the length of the pencil.

The intercept gives the original length of the pencil.

Using Slope As an Aid in Graphing

FIGURE 2-37

It is easy to graph a linear equation when it is written in slope–intercept form. For example, to graph $y = \frac{4}{3}x - 2$, we note that $b = -2$ and that the y-intercept is $(0, b) = (0, -2)$. (See Figure 2-37.)

Because the slope of the line is $\frac{\Delta y}{\Delta x} = \frac{4}{3}$, we can locate another point Q on the line by starting at point $P(0, -2)$ and counting 3 units to the right and 4 units up. The change in x from point P to point Q is $\Delta x = 3$, and the corresponding change in y is $\Delta y = 4$. The line joining points P and Q is the graph of the equation.

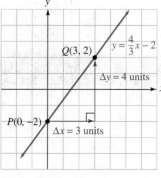

EXAMPLE 4

Find the slope and the y-intercept of the line with the equation $2x + 3y = -9$. Then graph the line.

Solution

We write the equation in the form $y = mx + b$ to find the slope m and the y-intercept $(0, b)$.

$2x + 3y = -9$	The given equation in general form.
$3y = -2x - 9$	Subtract $2x$ from both sides.
$\dfrac{3y}{3} = \dfrac{-2x}{3} - \dfrac{9}{3}$	Divide both sides by 3.
$y = -\dfrac{2}{3}x - 3$	Simplify both sides. We see that $m = -\frac{2}{3}$ and $b = -3$.

The slope is $-\frac{2}{3}$, and the y-intercept is $(0, -3)$. To draw the graph, we plot the y-intercept $(0, -3)$ and locate a second point on the line by moving 3 units to the right and 2 units down. We draw a line through the two points to obtain the graph shown in Figure 2-38.

FIGURE 2-38

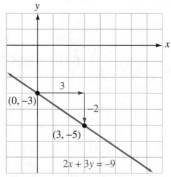

SELF CHECK

Find the slope and the y-intercept of the line with the equation $3x - 2y = -4$. Then graph the line.

Answer: $m = \frac{3}{2}$, $(0, 2)$

Parallel and Perpendicular Lines

| EXAMPLE 5 | **Parallel lines.** Show that the lines represented by $4x + 8y = 10$ and $2x = 12 - 4y$ are parallel. |

Solution We solve each equation for y to see that the lines are distinct and that their slopes are equal.

$$4x + 8y = 10 \qquad\qquad 2x = 12 - 4y$$
$$8y = -4x + 10 \qquad\qquad 4y = -2x + 12$$
$$y = -\frac{1}{2}x + \frac{5}{4} \qquad\qquad y = -\frac{1}{2}x + 3$$

Since the values of b in these equations are different $\left(\frac{5}{4} \text{ and } 3\right)$, the lines are distinct. Since the slope of each line is $-\frac{1}{2}$, they are parallel.

SELF CHECK Are the lines represented by $3x - 2y = 4$ and $2x = 5(y + 1)$ parallel?   *Answer:* no   ■

| EXAMPLE 6 | **Perpendicular lines.** Show that the lines represented by $4x + 8y = 10$ and $4x - 2y = 21$ are perpendicular. |

Solution We solve each equation for y to see that the slopes of their straight-line graphs are negative reciprocals.

$$4x + 8y = 10 \qquad\qquad 4x - 2y = 21$$
$$8y = -4x + 10 \qquad\qquad -2y = -4x + 21$$
$$y = -\frac{1}{2}x + \frac{5}{4} \qquad\qquad y = 2x - \frac{21}{2}$$

Since the slopes are $-\frac{1}{2}$ and 2 (which are negative reciprocals), the lines are perpendicular.

SELF CHECK Are the lines represented by $3x + 2y = 6$ and $2x - 3y = 6$ perpendicular?   *Answer:* yes   ■

| EXAMPLE 7 | **Parallel lines.** Write the equation of the line that passes through $(-2, 5)$ and is parallel to the line $y = 8x - 3$. |

Solution Since the slope of the line given by $y = 8x - 3$ is the coefficient of x, the slope is 8. Since the desired equation is to have a graph that is parallel to the graph of $y = 8x - 3$, its slope must also be 8.

We substitute -2 for x_1, 5 for y_1, and 8 for m in the point–slope form and simplify.

$$y - y_1 = m(x - x_1)$$
$$y - 5 = 8[x - (-2)] \quad \text{Substitute 5 for } y_1, \text{ 8 for } m, \text{ and } -2 \text{ for } x_1.$$
$$y - 5 = 8(x + 2) \quad\quad \text{Simplify the expression inside the brackets.}$$
$$y - 5 = 8x + 16 \quad\quad \text{Use the distributive property to remove parentheses.}$$
$$y = 8x + 21 \quad\quad\quad \text{Add 5 to both sides.}$$

The equation is $y = 8x + 21$.

SELF CHECK Write the equation of the line that is parallel to the line $y = 8x - 3$ and passes through the origin.   *Answer:* $y = 8x$ ■

We now summarize the various forms for the equation of a line.

Forms for the equation of a line

General form of a linear equation	$Ax + By = C$ A and B cannot both be 0.
Slope–intercept form of a linear equation	$y = mx + b$ The slope is m, and the y-intercept is $(0, b)$.
Point–slope form of a linear equation	$y - y_1 = m(x - x_1)$ The slope is m, and the line passes through (x_1, y_1).
A **horizontal line**	$y = b$ The slope is 0, and the y-intercept is $(0, b)$.
A **vertical line**	$x = a$ There is no defined slope, and the x-intercept is $(a, 0)$.

Straight-Line Depreciation

For tax purposes, many businesses use *straight-line depreciation* to find the declining value of aging equipment.

EXAMPLE 8

Value of a lathe. The owner of a machine shop buys a lathe for $1,970 and expects it to last 10 years. The lathe can then be sold as scrap for an estimated *salvage value* of $270. If y represents the value of the lathe after x years of use, and y and x are related by the equation of a line,

a. Find the equation of the line.

b. Find the value of the lathe after $2\frac{1}{2}$ years.

c. Find the economic meaning of the y-intercept of the line.

d. Find the economic meaning of the slope of the line.

Solution **a.** To find the equation of the line, we first find its slope and then use point–slope form to find its equation.

When the lathe is new, its age x is 0, and its value y is $1,970. When the lathe is 10 years old, $x = 10$, and its value is $y = \$270$. Since the line passes through the points $(0, 1,970)$ and $(10, 270)$, as shown in Figure 2-39, the slope of the line is

FIGURE 2-39

$$m = \frac{y_2 - y_1}{x_2 - x_1}$$
$$= \frac{270 - 1,970}{10 - 0}$$
$$= \frac{-1,700}{10}$$
$$= -170$$

To find the equation of the line, we substitute -170 for m, 0 for x_1, and 1,970 for y_1 in the point–slope form and simplify.

$$y - y_1 = m(x - x_1)$$
$$y - 1,970 = -170(x - 0)$$
3. $$y = -170x + 1,970$$

The current value y of the lathe is related to its age x by the linear model $y = -170x + 1,970$.

b. To find the value of the lathe after $2\frac{1}{2}$ years, we substitute 2.5 for x in Equation 3 and solve for y.

$$y = -170x + 1{,}970$$
$$= -170(2.5) + 1{,}970$$
$$= -425 + 1{,}970$$
$$= 1{,}545$$

In $2\frac{1}{2}$ years, the lathe will be worth \$1,545.

c. The y-intercept of the graph is $(0, b)$, where b is the value of y when $x = 0$.

$$y = -170x + 1{,}970$$
$$y = -170(0) + 1{,}970$$
$$y = 1{,}970$$

Thus, b is the value of a 0-year-old lathe, which is the lathe's original cost, \$1,970.

d. Each year, the value of the lathe decreases by \$170, because the slope of the line is -170. The slope of the depreciation line is the *annual depreciation rate*. ■

Curve Fitting

In statistics, the process of using one variable to predict another is called **regression.** For example, if we know a man's height, we can usually make a good prediction about his weight, because taller men usually weigh more than shorter men.

Figure 2-40 shows the results of sampling ten men at random and finding their heights and weights. The graph of the ordered pairs (h, w) is called a **scattergram.**

FIGURE 2-40

Man	Height (h) in inches	Weight (w) in pounds
1	66	140
2	68	150
3	68	165
4	70	180
5	70	165
6	71	175
7	72	200
8	74	190
9	75	210
10	75	215

(a) (b)

To write a *prediction equation* (sometimes called a *regression equation*), we must find the equation of the line that comes closer to all of the points in the scattergram than any other possible line. In statistics, there are exact methods to find this equation. However, in this course we can only approximate this equation.

To write an approximation of the regression equation, we place a straightedge on the scattergram shown in Figure 2-40 and draw the line joining two points that seems to best fit all of the points. In the figure, line PQ is drawn, where point P has coordinates of (66, 140) and point Q has coordinates of (75, 210).

Our approximation of the regression equation will be the equation of the line passing through points P and Q. To find the equation of this line, we first find its slope.

$$m = \frac{y_2 - y_1}{x_2 - x_1}$$

$$= \frac{210 - 140}{75 - 66}$$

$$= \frac{70}{9}$$

We can then use point–slope form to find the equation of the line. Since the line passes through (66, 140) and (75, 210), we can use either point when substituting values for x_1 and y_1.

$$y - y_1 = m(x - x_1)$$

$$y - 140 = \frac{70}{9}(x - 66) \qquad \text{Choose (66, 140) for } (x_1, y_1).$$

$$y = \frac{70}{9}x - \frac{4{,}620}{9} + 140 \qquad \text{Remove parentheses and add 140 to both sides.}$$

4. $\qquad y = \frac{70}{9}x - \frac{1{,}120}{3} \qquad \text{Add: } -\frac{4{,}620}{9} + 140 = -\frac{4{,}620}{9} + \frac{1{,}260}{9} = -\frac{3{,}360}{9} = -\frac{1{,}120}{3}.$

Our approximation of the regression equation is $y = \frac{70}{9}x - \frac{1{,}120}{3}$.

To predict the weight of a man who is 73 inches tall, for example, we substitute 73 for x in Equation 4 and simplify.

$$y = \frac{70}{9}x - \frac{1{,}120}{3}$$

$$y = \frac{70}{9}(73) - \frac{1{,}120}{3}$$

$$y = \frac{5{,}110}{9} - \frac{1{,}120}{3}$$

$$y \approx 194.4444444$$

We would predict that a 73-inch-tall man chosen at random will weigh about 194 pounds.

STUDY SET

Section 2.4

VOCABULARY

In Exercises 1–4, fill in the blanks to make the statements true.

1. The point–slope form of the equation of a line is

_____.

2. The _____ form of the equation of a line is $y = mx + b$.

3. Two lines are _____ when their slopes are negative reciprocals.

4. Two lines are _____ when they have the same slope.

CONCEPTS

5. If you know the slope of a line, is that enough information about the line to write its equation?

6. If you know a point that a line passes through, is that enough information about the line to write its equation?

7. The line graphed in Illustration 1 passes through the point $(-2, -3)$. Find its slope. Then write its equation. Express your answer in point–slope form.

ILLUSTRATION 1

8. For the line graphed in Illustration 2, find the slope and the y-intercept. Then write the equation of the line. Express your answer in slope–intercept form.

ILLUSTRATION 2

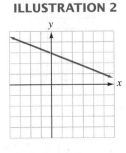

9. When the graph of the line $y = -\frac{2}{3}x + 1$ is drawn, what slope and y-intercept will the line have?

10. When the graph of the line $y - 3 = -\frac{2}{3}(x + 1)$ is drawn, what slope will it have? What point does the equation indicate it will pass through?

11. Is the following statement true? The equations $y - 2 = 3(x - 2)$, $y = 3x - 4$, and $3x - y = 4$ all describe the same line.

12. See the linear model graphed in Illustration 3.
    a. What information does the y-intercept give?

    b. What information does the slope give?

ILLUSTRATION 3

13. When each equation is graphed, what will the y-intercept be?
    a. $y = 2x$ **b.** $x = -3$

14. When each equation is graphed, what will be the slope of the line?
    a. $y = -x$ **b.** $x = -3$

15. The two lines graphed in Illustration 4(a) appear to be perpendicular. Their equations are displayed in Illustration 4(b). Are the lines actually perpendicular? Explain.

ILLUSTRATION 4

(a)   (b)

16. The two lines graphed in Illustration 5(a) appear to be parallel. Their equations are displayed in Illustration 5(b). Are the lines actually parallel? Explain.

ILLUSTRATION 5

(a)   (b)

NOTATION

In Exercises 17–18, complete each solution.

17. Write $y + 2 = \frac{1}{3}(x + 3)$ in slope–intercept form.

$$y + 2 = \frac{1}{3}(x + 3)$$

$$y + 2 = \boxed{} + 1$$

$$y + 2 - \boxed{} = \frac{1}{3}x + 1 - \boxed{}$$

$$y = \frac{1}{3}x - \boxed{}$$

$$m = \boxed{}, \quad b = \boxed{}$$

18. Write the equation of the line that has slope -2 and passes through the point $(3, 1)$.

$$y - y_1 = m(x - x_1)$$
$$y - \boxed{} = -2(x - \boxed{})$$
$$y - 1 = \boxed{} + 6$$
$$y = -2x + 7$$

PRACTICE

In Exercises 19–22, use point–slope form to write the equation of the line with the given properties. Then write each equation in slope–intercept form.

19. $m = 5$, passing through $(0, 7)$

20. $m = -8$, passing through $(0, -2)$

21. $m = -3$, passing through $(2, 0)$

22. $m = 4$, passing through $(-5, 0)$

In Exercises 23–28, use point–slope form to write the equation of the line passing through the two given points. Then write each equation in slope–intercept form.

23. $(0, 0)$, $(4, 4)$ **24.** $(-5, 5)$, $(0, 0)$

25.

x	y
3	4
0	-3

26.

x	y
4	0
6	-8

27.

28.

In Exercises 29–36, use the slope–intercept form to write the equation of the line with the given properties.

29. $m = 3$, $b = 17$

30. $m = -2$, $b = 11$

31. $m = -7$, passing through $(7, 5)$

32. $m = 3$, passing through $(-2, -5)$

33. $m = 0$, passing through $(2, -4)$

34. $m = -7$, passing through the origin

35. passing through $(6, 8)$ and $(2, 10)$

36. passing through $(-4, 5)$ and $(2, -6)$

In Exercises 37–40, write each equation in slope–intercept form. Then find the slope and the y-intercept of the line determined by the equation.

37. $3x - 2y = 8$ **38.** $-2x + 4y = 12$

39. $-2(x + 3y) = 5$ **40.** $5(2x - 3y) = 4$

In Exercises 41–46, find the slope and y-intercept and use them to draw the line.

41. $y = x - 1$

42. $y = -x + 2$

43. $y = \dfrac{2}{3}x + 2$

44. $y = -\dfrac{5}{4}x + \dfrac{5}{2}$

45. $4y - 3 = -3x - 11$

46. $-2x + 4y = 12$

In Exercises 47–54, tell whether the graphs of each pair of equations are parallel, perpendicular, or neither.

47. $y = 3x + 4$, $y = 3x - 7$

48. $y = 4x - 13$, $y = \dfrac{1}{4}x + 13$

49. $x + y = 2$, $y = x + 5$

50. $x = y + 2$, $y = x + 3$

51. $3x + 6y = 1$, $y = \dfrac{1}{2}x$

52. $2x + 3y = 9$, $3x - 2y = 5$

53. $y = 3$, $x = 4$

54. $y = -3$, $y = -7$

In Exercises 55–60, write the equation of the line that passes through the given point and is parallel to the given line. Write the answer in slope–intercept form.

55. $(0, 0)$, $y = 4x - 7$

56. $(0, 0)$, $x = -3y - 12$

57. $(2, 5)$, $4x - y = 7$

58. $(-6, 3)$, $y + 3x = -12$

59. $(4, -2)$, $x = \dfrac{5}{4}y - 2$

60. $(1, -5)$, $x = -\dfrac{3}{4}y + 5$

In Exercises 61–66, write the equation of the line that passes through the given point and is perpendicular to the given line. Write the answer in slope–intercept form.

61. $(0, 0)$, $y = 4x - 7$

62. $(0, 0)$, $x = -3y - 12$

63. $(2, 5)$, $4x - y = 7$

64. $(-6, 3)$, $y + 3x = -12$

65. $(4, -2)$, $x = \dfrac{5}{4}y - 2$

66. $(1, -5)$, $x = -\dfrac{3}{4}y + 5$

APPLICATIONS

In Exercises 67–70, assume straight-line depreciation or straight-line appreciation.

67. BIG-SCREEN TV Find the linear depreciation equation for the TV in the want ad shown in Illustration 6.

ILLUSTRATION 6

For Sale: 3-year-old 45-inch TV, with matrix surround sound & picture within picture, remote. $1,750 new. Asking $800. Call 875-5555. Ask for Mike.

68. SALVAGE VALUE A truck was purchased for $19,984. Its salvage value at the end of 8 years is expected to be $1,600. Find the depreciation equation.

69. ART In 1987, the painting *Rising Sunflowers* by Vincent van Gogh (Illustration 7) sold for $36,225,000. Suppose that an art appraiser expected the painting to double in value in 20 years. Let *x* represent the time in years after 1987. Find the appreciation equation.

ILLUSTRATION 7

70. REAL ESTATE LISTING See Illustration 8. Use the information given in the description of the property to write an appreciation equation for the house.

ILLUSTRATION 8

Vacation Home
$122,000
Only 2 years old

- Great investment property!
- Expected to appreciate $4,000/yr

Sq ft: 1,635	Fam rm: yes	Den: no
Bdrm: 3	Ba: 1.5	Gar: enclosed
A/C: yes	Firepl: yes	Kit: built-ins

71. CRIMINOLOGY City growth and the number of burglaries are related by a linear equation. Records show that 575 burglaries were reported in a year when the local population was 77,000 and that the rate of increase in the number of burglaries was 1 for every 100 new residents.

a. Using the variables *p* for population and *B* for burglaries, write an equation (in slope–intercept form) that police can use to predict future burglary statistics.

b. How many burglaries can be expected when the population reaches 110,000?

72. CABLE TV Since 1990, when the average monthly basic cable TV rate in the United States was $16.75, the cost has risen by about $1.32 a year.

a. Write an equation in slope–intercept form to predict cable TV costs in the future. Use *t* to represent time in years after 1990 and *C* to represent the average basic monthly cost.

b. If the equation in part a were graphed, what would be the meaning of the *C*-intercept and the slope of the line?

73. PSYCHOLOGY EXPERIMENT The scattergram in Illustration 9 (on the next page) shows the performance of a rat in a maze.

a. Draw a line through $(1, 10)$ and $(19, 1)$. Write its equation using the variables *t* and *E*. In psychology, this equation is called the *learning curve* for the rat.

b. What does the slope of the line tell us?

c. What information does the *x*-intercept of the graph give?

ILLUSTRATION 9

Food

Entrance

35°F and 15°F feel like when a 15-mph wind is blowing. The relationship between the actual temperature and the wind-chill temperature can be modeled with a linear equation.
   a. Write the equation that models this relationship. Answer in slope–intercept form.

   b. What information is given by the *y*-intercept of the graph of the equation found in part a?

ILLUSTRATION 11

Actual temperature	Wind-chill temperature
35°F	16°F
15°F	−11°F

76. COMPUTER-AIDED DRAFTING Illustration 12 shows a computer-generated drawing of an airplane part. When the designer clicks the mouse on a line on the drawing, the computer finds the equation of the line. Use a calculator to determine whether the angle where the weld is to be made is a right angle.

74. UNDERSEA DIVING Illustration 10 shows that the pressure *p* that divers experience is related to the depth *d* of the dive. A linear model can be used to describe this relationship.
   a. Write the linear model in slope–intercept form.

   b. Pearl and sponge divers often reach depths of 100 feet. What pressure do they experience? Round to the nearest tenth.

   c. Scuba divers can safely dive to depths of 250 feet. What pressure do they experience? Round to the nearest tenth.

ILLUSTRATION 12

$y = 0.351x - 0.652$

weld

$y = -2.799x + 2.000$

ILLUSTRATION 10

Sea level (*d* = 0)
Pressure = 14.7 pounds per square inch (psi)

d = 33 ft
Pressure = 29.4 psi

d = 66 ft
Pressure = 44.1 psi

WRITING

77. Explain how to find the equation of a line passing through two given points.

78. Explain what *m*, x_1, and y_1 represent in the point–slope form of the equation of a line.

75. WIND-CHILL FACTOR A combination of cold and wind makes a person feel colder than the actual temperature. Illustration 11 shows what temperatures of

79. Explain what m and b represent in the slope–intercept form of the equation of a line.

80. Linear relationships between two quantities can be described by an equation or a graph. Which do you think is the more informative? Why?

REVIEW

81. INVESTMENTS Equal amounts are invested at 6%, 7%, and 8% annual interest. The three investments yield a total of $2,037 annual interest. Find the total amount of money invested.

82. MIXING COFFEE To make a mixture of 80 pounds of coffee worth $272, a grocer mixes coffee worth $3.25 a pound with coffee worth $3.85 a pound. How many pounds of cheaper coffee should the grocer use?

83. Find b if the point $(-2, 3)$ is on the line $y = 3x + b$.

84. Find the area of a rectangle that has a perimeter of 32 centimeters and a length of 12 centimeters.

▶ 2.5   Introduction to Functions

In this section, you will learn about

> Representing functions ■ Function notation ■ The graph of a function ■ Finding the domain and range of a function ■ The vertical line test ■ Linear functions

Introduction In Chapter 1, we examined the rental hall agreement shown in Figure 2-41. We determined that the cost C of renting the hall depended on the number of hours h the hall was to be rented. We described this dependence in four ways: using words, using an equation, using a table, and using a graph.

In this example, a specific rule assigns to each number of hours h the hall is rented a unique number C, the cost to rent the hall. Correspondences of this type are called **functions**, and they are of great importance in mathematics and its applications. In this section, we introduce the concept of function, and we discuss a special notation that is used when working with functions.

FIGURE 2-41

Rental Agreement
ROYAL VISTA BANQUET ROOM
Wedding Receptions•Dances•Reunions•Fashion Shows

Rented To_____Date_____
Lessee's Address_____

Rental Charges
• $100 per hour
• Nonrefundable $200 cleanup fee

Terms and conditions
Lessor leases the undersigned lessee the above described property upon the terms and conditions set forth on this page and on the back of this page. Lessee promises to pay rental cost stated herein.

Words						*Equation*
The cost to rent the hall	is	100	times	the number of hours it is rented	plus	200.

$$C = 100h + 200$$

(continued)

FIGURE 2-41 (continued)

Table   *Graph*

h	C
1	300
2	400
3	500
4	600
5	700
6	800
7	900

Representing Functions

If x and y are real numbers, an equation in x and y determines a correspondence between the values of x and y. To see how, we consider the equation $y = \frac{1}{2}x + 3$. To find the value of y (called an **output value**) that corresponds to $x = 4$ (called an **input value**), we multiply 4 by $\frac{1}{2}$ and then add 3.

$$y = \frac{1}{2}x + 3$$

$$y = \frac{1}{2}(4) + 3 \qquad \text{Substitute the input value of 4 for } x.$$

$$= 2 + 3$$

$$= 5 \qquad\qquad \text{The output value is 5.}$$

The ordered pair (4, 5) satisfies the equation and shows that a y-value of 5 corresponds to an x-value of 4. This ordered pair and others that satisfy the equation appear in the table shown in Figure 2-42. The graph of the equation also appears in the figure.

To see how the table determines the correspondence, we simply find an input in the x column and then read across to find the corresponding output in the y column. For example, if we select 2 as an input value, we get 4 as an output value. Thus, a y-value of 4 corresponds to an x-value of 2.

To see how the graph determines the correspondence, we draw a vertical and a horizontal line through any point (say, point P) on the graph, as shown in Figure 2-42. Because these lines intersect the x-axis at 4 and the y-axis at 5, the point $P(4, 5)$ associates 5 on the y-axis with 4 on the x-axis. This shows that a y value of 5 is assigned to an x-value of 4.

FIGURE 2-42

$y = \frac{1}{2}x + 3$

x	y
−2	2	A y-value of 2 corresponds to an x-value of −2.
0	3	A y-value of 3 corresponds to an x-value of 0.
2	4	A y-value of 4 corresponds to an x-value of 2.
4	5	A y-value of 5 corresponds to an x-value of 4.
6	6	A y-value of 6 corresponds to an x-value of 6.

↑ ↑
We choose   We compute
the inputs.   each output.

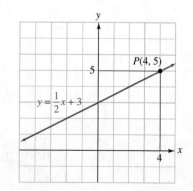

When a correspondence is described by an equation, by a table, using words, or with a graph, and only one *y*-value corresponds to each *x*-value, we call the correspondence a **function**. The set of input values *x* is called the **domain** of the function, and the set of output values *y* is called the **range**. Since the value of *y* usually depends on the number *x*, we call *y* the **dependent variable** and *x* the **independent variable**.

Functions

> A **function** is a correspondence between a set of input values *x* (called the **domain**) and a set of output values *y* (called the **range**), where exactly one *y*-value in the range corresponds to each number *x* in the domain.

EXAMPLE 1

Recognizing functions. Does $y = 2x - 3$ define *y* to be a function of *x*? If so, illustrate the function with a table and a graph.

Solution   For $y = 2x - 3$ to define a function, every input number *x* must determine one output value of *y*. To find *y* in the equation $y = 2x - 3$, we multiply *x* by 2 and then subtract 3. Since this arithmetic gives one result, each choice of *x* determines one value of *y*. Thus, the equation defines *y* to be a function of *x*.

A table of values and the graph appear in Figure 2-43.

FIGURE 2-43

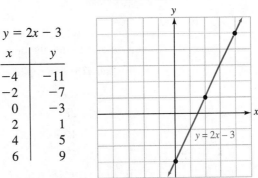

$y = 2x - 3$

x	*y*
−4	−11
−2	−7
0	−3
2	1
4	5
6	9

SELF CHECK   Does $y = -2x + 3$ define *y* to be a function of *x*?   *Answer:* yes   ■

As you will see in the next example, not every equation in two variables defines a function.

EXAMPLE 2

Recognizing functions. Does $y^2 = x$ define *y* to be a function of *x*?

Solution   For a function to exist, each input *x* must determine one output *y*. If we let $x = 16$, for example, *y* could be either 4 or −4, because $4^2 = 16$ and $(-4)^2 = 16$. Since more than one value of *y* is determined when $x = 16$, the equation does not represent a function.

SELF CHECK   Does $|y| = x$ define *y* to be a function of *x*? Explain your answer.   *Answer:* No; if $x = 2$, *y* could be either 2 or −2.   ■

Function Notation

There is a special notation that we will use to denote functions.

Function notation

> The notation $y = f(x)$ denotes that the variable y is a function of x.

The notation $y = f(x)$ is read as "y equals f of x." Note that y and $f(x)$ are two notations for the same quantity. Thus, the equations $y = 4x + 3$ and $f(x) = 4x + 3$ are equivalent. We read $f(x) = 4x + 3$ as "f of x is equal to $4x + 3$."

This is the variable used to represent input values.
$$f(x) = 4x + 3$$
This is the name of the function.   This expression shows how to obtain an output value from a given input value.

WARNING! The notation $f(x)$ does not mean "f times x."

The notation $y = f(x)$ provides a way of denoting the value of y (the dependent variable) that corresponds to some number x (the independent variable). For example, if $y = f(x)$, the value of y that is determined by $x = 3$ is denoted by $f(3)$.

EXAMPLE 3

Evaluating a function. Let $f(x) = 4x + 3$. Find **a.** $f(3)$, **b.** $f(-1)$, **c.** $f(0)$, and **d.** $f(r + 1)$.

Solution In each case, we will substitute the value for x within the parentheses of the function notation and in the algebraic expression $4x + 3$. Then we evaluate the expression.

a. To find $f(3)$, we replace x with 3:

$$f(x) = 4x + 3$$
$$f(3) = 4(3) + 3$$
$$= 12 + 3$$
$$= 15$$

b. To find $f(-1)$, we replace x with -1:

$$f(x) = 4x + 3$$
$$f(-1) = 4(-1) + 3$$
$$= -4 + 3$$
$$= -1$$

c. To find $f(0)$, we replace x with 0:

$$f(x) = 4x + 3$$
$$f(0) = 4(0) + 3$$
$$= 3$$

d. To find $f(r + 1)$, we replace x with $r + 1$:

$$f(x) = 4x + 3$$
$$f(r + 1) = 4(r + 1) + 3$$
$$= 4r + 4 + 3$$
$$= 4r + 7$$

SELF CHECK If $f(x) = -2x - 1$, find **a.** $f(2)$, **b.** $f(-3)$, and **c.** $f(-t)$.

Answers: **a.** -5, **b.** 5, **c.** $2t - 1$ ∎

To see why function notation is helpful, we consider the following equivalent sentences:

1. In the equation $y = 4x + 3$, find the value of y when x is 3.

2. In the equation $f(x) = 4x + 3$, find $f(3)$.

Statement 2, which uses $f(x)$ notation, is much more concise.

We can think of a function as a machine that takes some input x and turns it into some output $f(x)$, as shown in Figure 2-44(a). The machine shown in Figure 2-44(b) turns the input number 6 into the output value -11 and turns the input number 2 into the output value -3. The set of numbers that we can put into the machine is the domain of the function, and the set of numbers that comes out is the range.

FIGURE 2-44

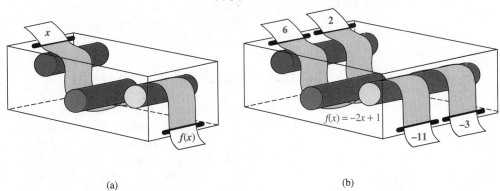

(a)   (b)

The letter f used in the notation $y = f(x)$ represents the word *function*. However, other letters can be used to represent functions. For example, the notations $y = g(x)$ and $y = h(x)$ are often used to denote functions involving the independent variable x.

In Example 4, the equation $g(x) = x^2 - 2x$ determines a function, because every possible value of x gives a single value of $g(x)$.

EXAMPLE 4

Evaluating a function. Let $g(x) = x^2 - 2x$. Find **a.** $g\left(\frac{2}{5}\right)$ and **b.** $g(-2.4)$.

Solution

a. To find $g\left(\frac{2}{5}\right)$, we replace x with $\frac{2}{5}$:

$$g(x) = x^2 - 2x$$

$$g\left(\frac{2}{5}\right) = \left(\frac{2}{5}\right)^2 - 2\left(\frac{2}{5}\right)$$

$$= \frac{4}{25} - \frac{4}{5}$$

$$= -\frac{16}{25}$$

b. To find $g(-2.4)$, we replace x with -2.4:

$$g(x) = x^2 - 2x$$

$$g(-2.4) = (-2.4)^2 - 2(-2.4)$$

$$= 5.76 + 4.8$$

$$= 10.56$$

SELF CHECK

Let $h(x) = -\dfrac{x^2 + 2}{2}$. Find **a.** $h(4)$ and **b.** $h(-0.6)$.   *Answers:* **a.** -9, **b.** -1.18 ∎

In the next example, the letter A is chosen to name a function that finds the area of a circle. The letter d is chosen as the independent variable, to help stress the fact that the area of a circle is a function of its diameter.

EXAMPLE 5

Archery. The area of a circle with a diameter of length d is given by the function $A(d) = \pi\left(\frac{d}{2}\right)^2$. Find $A(48)$ to the nearest tenth. What information does it give about the archery target shown in Figure 2-45?

Solution Since the diameter of the circular target is 48 inches, $A(48)$ gives the area of the target. To find $A(48)$, we replace d with 48.

$$A(d) = \pi\left(\frac{d}{2}\right)^2$$

$$A(48) = \pi\left(\frac{48}{2}\right)^2 \qquad \text{Substitute 48 for } d.$$

$$= \pi(24)^2$$

$$= 576\pi$$

$$\approx 1,809.557368 \qquad \text{Use a calculator to do the multiplication.}$$

To the nearest tenth, the area of the target is 1,809.6 in.^{2}.

FIGURE 2-45

9.6 in.   48 in.

FIGURE 2-46
This axis can be labeled
$f(x)$ or y.

In Example 5, find $A(9.6)$ to the nearest tenth. What information does it give about the archery target in Figure 2-45?

Answers: 72.4; the area of the "bull's eye" is 72.4 in.^{2} ■

The Graph of a Function

The *graph of a function* is a "picture" of the ordered pairs $(x, f(x))$ that define the function. From the graph of a function, we can determine function values. To illustrate this, let's consider the graph of function f shown in Figure 2-46 and find $f(4)$.

To find $f(4)$, we need to find the y value that f assigns to an x value of 4. The blue line in the figure illustrates that the graph is 3 units above the point marked 4 on the x-axis. So the point $(4, 3)$ is on the graph of the function, and we can conclude that f assigns a y value of 3 to an x value of 4. Therefore, $f(4) = 3$.

Finding the Domain and Range of a Function

EXAMPLE 6

Domain and range. Find the domain and range of each function: **a.** the ordered pairs $(-2, 4)$, $(0, 6)$, and $(2, 8)$, **b.** the equation $y = 3x + 1$, and **c.** the equation $y = \frac{1}{x-2}$.

Solution **a.** This function consists of only three ordered pairs. The ordered pairs set up a correspondence between x and y where a single value of y corresponds to each x. The domain is the set of numbers x:

x	y	(x, y)
-2	4	$(-2, 4)$
0	6	$(0, 6)$
2	8	$(2, 8)$

4 corresponds to -2.
6 corresponds to 0.
8 corresponds to 2.

$\{-2, 0, 2\}$. The range is the set of values y: $\{4, 6, 8\}$.

b. We will be able to evaluate $3x + 1$ for any real-number input x. So the domain of the function is the set of real numbers. Since the output y can be any real number, the range is the set of real numbers.

c. To find the domain of $y = \frac{1}{x-2}$, we exclude any real-number x inputs for which we would be unable to compute $\frac{1}{x-2}$. The number 2 cannot be substituted for x, because that would make the denominator equal to zero. Since any real number except 2 can be substituted for x in the equation $y = \frac{1}{x-2}$, the domain is the set of all real numbers except 2.

Since a fraction with a numerator of 1 cannot be 0, the range is the set of all real numbers except 0.

Find the domain and range of each function: **a.** the ordered pairs $(-3, 5)$, $(-2, 7)$, and $(1, 11)$ and **b.** the equation $y = \frac{2}{x+3}$.

Answers: **a.** $\{-3, -2, 1\}$, $\{5, 7, 11\}$, **b.** the set of all real numbers except -3, the set of all real numbers except 0. ■

We can determine the domain and the range of a function from its graph. For the graph in Figure 2-47, the domain is shown on the x-axis, and the range is shown on the y-axis. Note that for any x in the domain, there corresponds a value $y = f(x)$ in the range.

FIGURE 2-47

EXAMPLE 7

Finding the domain and range from a graph. Find the domain and range of the function graphed in Figure 2-48.

FIGURE 2-48

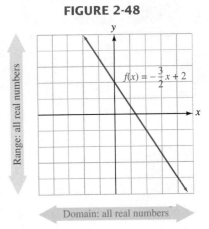

Solution Since every real number x on the x-axis determines a corresponding value of y, the domain is the set of real numbers. Since the values of y can be any real number on the y-axis, the range is the set of real numbers.

The Vertical Line Test

The **vertical line test** can be used to determine whether the graph of an equation represents a function. If any vertical line intersects a graph more than once, the graph cannot represent a function, because to one number x there would correspond more than one value of y.

The graph in Figure 2-49(a) represents a function, because every vertical line that intersects the graph does so exactly once. The graph in Figure 2-49(b) does not represent a function, because some vertical lines intersect the graph more than once.

FIGURE 2-49

Function
(a)

Not a function
(b)

Linear Functions

In Section 2.2, we graphed equations whose graphs were lines. Such equations define a basic type of function called a **linear function.**

Linear functions

A **linear function** is a function defined by an equation that can be written in the form

$$f(x) = mx + b \qquad \text{or} \qquad y = mx + b$$

where m is the slope of the line graph and $(0, b)$ is the y-intercept.

EXAMPLE 8

Manicurist. A recent graduate of a cosmetology school rents a station from the owner of a beauty salon for $18 a day. She expects to make $12 profit from each customer she serves. Write a linear function describing her daily income if she serves c customers per day. Then graph the function.

Solution   The manicurist makes a profit of $12 per customer, so if she serves c customers a day, she will make $12c$. To find her income, we must *subtract* the $18 rental fee she pays from the profit. Therefore, the income function is $I(c) = 12c - 18$.

The graph of this linear function, shown in Figure 2-50, is a line with slope 12 and intercept $(0, -18)$. Since the manicurist cannot have a negative number of customers, we do not extend the line into quadrant III.

FIGURE 2-50

ACCENT ON TECHNOLOGY *Evaluating Functions*

We can use a graphing calculator to find the value of a function at different numbers x. For example, to find the income earned by the manicurist in Example 8 for different numbers of customers, we first graph the income function $I(c) = 12c - 18$ as $y = 12x - 18$, using window settings of $[0, 10]$ for x and $[0, 100]$ for y to obtain Figure 2-51(a). To find her income when she serves seven customers, we trace and move the cursor until the x-coordinate on the screen is nearly 7, as in Figure 2-51(b). From the screen, we can read that her income is about $66.25.

To find her income when she serves nine customers, we trace and move the cursor until the x-coordinate is nearly 9, as in Figure 2-51(c). From the screen, we can read that her income is about $90.51.

FIGURE 2-51

     (a)   (b)   (c)

With some graphing calculator models, we can evaluate a function by entering function notation. To find $I(15)$, the income earned by the manicurist of Example 8 if she serves 15 customers, we use the following steps on a TI-83.

With $I(c) = 12c - 18$ entered as $Y_1 = 12x - 18$, we call up the home screen by pressing [2nd] [QUIT]. Then we enter [VARS] [▶] [1] [ENTER]. The symbolism Y_1 will be displayed. See Figure 2-52(a). Next, we enter the input value 15, as shown in Figure 2-52(b), and press [ENTER]. In Figure 2-52(c) we see that $Y_1(15) = 162$. That is, $I(15) = 162$. The manicurist will earn $162 if she serves 15 customers in one day.

FIGURE 2-52

(a)   (b)   (c)

STUDY SET

Section 2.5

VOCABULARY

In Exercises 1–8, fill in the blanks to make the statements true.

1. A _____ is a correspondence between a set of input values and a set of output values, where each input value determines ___ output value.

2. When graphing a function, input values are associated with the _____ axis and output values with the _____ axis.

3. For a function, the set of all output values is called the _____ of the function.

4. For a function, the set of all inputs is called the _____ of the function.

5. For $y = 2x - 9$, the independent variable is ___.

6. For $y = \frac{x}{3} - 9$, if $x = 3$, the corresponding value of ___ is called the output value.

7. For $y = -x + 3$, the dependent variable is ___.

8. Any substitution for x in $f(x) = 5x - 4$ is called an _____ value.

CONCEPTS

9. Consider the following problems. Fill in the blank so that they ask for the same thing.
   1. In the equation $y = -5x + 1$, find the value of y when $x = -1$.
   2. In the equation $f(x) = -5x + 1$, find _____.

10. For the function $f(x) = \frac{1}{x+4}$, why isn't -4 in the domain of f?

11. Consider the graph of the function in Illustration 1.
    a. Label each arrow in the illustration with the appropriate term: *domain* or *range*.
    b. Give the domain and range.

ILLUSTRATION 1

12. Use the graph of function f shown in Illustration 2 to find each of the following.
    a. $f(-2)$ b. $f(0)$
    c. $f(1)$

ILLUSTRATION 2

NOTATION

In Exercises 13–14, complete each solution.

13. If $f(x) = x^2 - 3x$, find $f(-5)$.
$$f(x) = x^2 - 3x$$
$$f(-5) = (-5)^2 - 3(\quad)$$
$$= \quad + 15$$
$$= 40$$

14. If $g(x) = \dfrac{2 - x}{6}$, find $g(8)$.

$$g(x) = \frac{2 - x}{6}$$

$$g(8) = \frac{2 - \blacksquare}{6}$$

$$= \frac{\blacksquare}{6}$$

$$= -1$$

15. Complete this sentence: $f(5) = 6$ is read "f ___ 5 is 6."

16. Give two forms in which a linear function can be written.

PRACTICE

In Exercises 17–24, tell whether the equation determines y to be a function of x.

17. $y = 2x + 3$ **18.** $y = 4x - 1$

19. $y = 2x^2$ **20.** $y^2 = x + 1$

21. $y = 3 + 7x^2$ **22.** $y^2 = 3 - 2x$

23. $x = |y|$ **24.** $y = |x|$

In Exercises 25–32, find f(3) and f(−1).

25. $f(x) = 3x$ **26.** $f(x) = -4x$

27. $f(x) = 2x - 3$ **28.** $f(x) = 3x - 5$

29. $f(x) = 7 + 5x$ **30.** $f(x) = 3 + 3x$

31. $f(x) = 9 - 2x$ **32.** $f(x) = 12 + 3x$

In Exercises 33–40, find g(2) and g(3).

33. $g(x) = x^2$ **34.** $g(x) = x^2 - 2$

35. $g(x) = x^3 - 1$ **36.** $g(x) = x^3$

37. $g(x) = (x + 1)^2$ **38.** $g(x) = (x - 3)^2$

39. $g(x) = 2x^2 - x$ **40.** $g(x) = 5x^2 + 2x$

In Exercises 41–48, find h(2) and h(−2).

41. $h(x) = |x| + 2$ **42.** $h(x) = |x| - 5$

43. $h(x) = x^2 - 2$ **44.** $h(x) = x^2 + 3$

45. $h(x) = \dfrac{1}{x + 3}$ **46.** $h(x) = \dfrac{3}{x - 4}$

47. $h(x) = \dfrac{x}{x - 3}$ **48.** $h(x) = \dfrac{x}{x^2 + 2}$

In Exercises 49–52, complete each table.

49. $f(t) = |t - 2|$

t	$f(t)$
-1.7	
0.9	
5.4	

50. $f(r) = -2r^2 + 1$

Input	Output
-1.7	
0.9	
5.4	

51. $g(x) = x^3$

Input	Output
$-\frac{3}{4}$	
$\frac{1}{6}$	
$\frac{5}{2}$	

52. $g(x) = 2\left(-x - \dfrac{1}{4}\right)$

x	$g(x)$
$-\frac{3}{4}$	
$\frac{1}{8}$	
$\frac{5}{2}$	

In Exercises 53–56, find g(w) and g(w + 1).

53. $g(x) = 2x$

54. $g(x) = -3x$

55. $g(x) = 3x - 5$

56. $g(x) = 2x - 7$

In Exercises 57–64, find the domain of each function.

57. $g(x) = \dfrac{x}{6 - x}$

58. $H(x) = \dfrac{4}{3 - 2x}$

59. $s(x) = |x - 7|$

60. $t(x) = \left| \dfrac{2x}{3} + 1 \right|$

61. $f(x) = x^2$

62. $g(x) = x^3$

63. $s(x) = 3x + 6$

64. $h(x) = \frac{4}{5}x - 8$

In Exercises 65–68, find the domain and range of each function.

65. $\{(-2, 3), (4, 5), (6, 7)\}$

66. $\{(0, 2), (1, 2), (3, 4)\}$

67. $f(x) = \dfrac{1}{x - 4}$

68. $f(x) = \dfrac{5}{x + 1}$

In Exercises 69–76, use the vertical line test to tell whether the given graph represents a function.

69.

70.

71.

72.

73.

74.

75.

76.

In Exercises 77–80, graph each function. Then find the domain and range.

77. $f(x) = 2x - 1$

78. $f(x) = -x + 2$

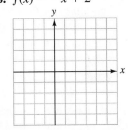

79. $f(x) = \dfrac{2}{3}x - 2$

80. $f(x) = -\dfrac{3}{2}x - 3$

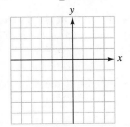

In Exercises 81–84, tell whether each equation defines a linear function.

81. $y = 3x^2 + 2$

82. $y = \dfrac{x - 3}{2}$

83. $y = x$

84. $y = 3x^3 - 4$

APPLICATIONS

85. DECONGESTANT The temperature in degrees Celsius that is equivalent to a temperature in degrees Fahrenheit is given by the linear function $C(F) = \frac{5}{9}(F - 32)$. Use this function to find the temperature range, in degrees Celsius, at which a bottle of Dimetapp should be stored. The label directions are shown in Illustration 3.

ILLUSTRATION 3

DIRECTIONS: Adults and children 12 years of age and over: Two teaspoons every 4 hours. DO NOT EXCEED 6 DOSES IN A 24-HOUR PERIOD. Store at a controlled room temperature between 68°F and 77°F.

86. BODY TEMPERATURES The temperature in degrees Fahrenheit that is equivalent to a temperature in degrees Celsius is given by the linear function $F(C) = \frac{9}{5}C + 32$. Convert each of the temperatures in the following excerpt from *The Good Housekeeping Family Health and Medical Guide* to degrees Fahrenheit. (Round to the nearest degree.)

In disease, the temperature of the human body may vary from about 32.2°C to 43.3°C for a time, but there is grave danger to life should it drop and remain below 35°C or rise and remain at or above 41°C.

87. CONCESSIONAIRE A baseball club pays a peanut vendor $50 per game for selling bags of peanuts for $1.75 each.
a. Write a linear function that describes the income the vendor makes for the baseball club during a game if she sells *b* bags of peanuts.
b. Find the income the baseball club will make if the vendor sells 110 bags of peanuts during a game.

88. NEW HOME CONSTRUCTION In a proposal to some prospective clients, a housing contractor listed the following costs.
- Fees, permits, miscellaneous $12,000
- Construction, per square foot $75

a. Write a linear function that the clients could use to determine the cost of building a home having *f* square feet.
b. Find the cost to build a home having 1,950 square feet.

89. EARTH'S ATMOSPHERE Illustration 4 shows a graph of the temperatures of the atmosphere at various altitudes above the earth's surface. The temperature is expressed in degrees Kelvin, a scale widely used in scientific work.

a. Estimate the coordinates of three points on the graph that have an x-coordinate of 200.

b. Explain why this is not the graph of a function.

ILLUSTRATION 4

(Height (km) vs. Temperature (°K) graph showing atmospheric layers: Ionosphere, Thermosphere, Mesosphere, Ozone layer, Stratosphere, Troposphere, with Meteors labeled.)

90. CHEMICAL REACTIONS When students in a chemistry laboratory mixed solutions of acetone and chloroform, they found that heat was immediately generated. As time went by, the mixture cooled down. Illustration 5 shows a graph of data points of the form (time, temperature) taken by the students during the experiment.

a. The linear function $T(t) = -\frac{t}{240} + 30$ models the relationship between the elapsed time t since the solutions were combined and the temperature $T(t)$ of the mixture. Graph the function in Illustration 5.

b. Predict the temperature of the mixture immediately after the two solutions are combined.

c. Is $T(180)$ more or less than the temperature recorded by the students for $t = 300$?

ILLUSTRATION 5

(Temperature (°C) vs. Elapsed time (sec) graph. Initial temperatures: Acetone, Chloroform } 23.6°C)

91. INCOME TAX The function

$$T(a) = 3,697.50 + 0.28(a - 24,650)$$

(where a is adjusted gross income) is a mathematical model of the instructions given on the second line of the tax rate Schedule X shown in Illustration 6.

a. Find $T(35,000)$ and interpret the result.

b. From the directions in the tax rate schedule, what are the domain and the range of the function?

ILLUSTRATION 6

Schedule X–Use if your filing status is **Single**			
If your adjusted gross income is: Over —	But not over —	Your tax is	of the amount over —
$0	$24,650	.... 15%	$0
24,650	59,750	$3,697.50 + 28%	24,650
59,750	124,650	13,525.50 + 31%	59,750

c. Write a function that would compute the tax according to the instructions given in the third line of the schedule.

92. COST FUNCTION An electronics firm manufactures tape recorders, receiving $120 for each recorder it makes. If x represents the number of recorders produced, the income received is determined by the *revenue function* $R(x) = 120x$. The manufacturer has fixed costs of $12,000 per month and variable costs of $57.50 for each recorder manufactured. Thus, the *cost function* is $C(x) = 57.50x + 12,000$. How many recorders must the company sell for revenue to equal cost? (*Hint:* Set $R(x) = C(x)$.)

93. BALLISTICS The height of a bullet shot from the ground straight upward is given by the function $f(t) = -16t^2 + 256t$.

a. Find the height of the bullet 3 seconds after it is shot.

b. Find $f(16)$. Interpret the result.

94. PLATFORM DIVING The number of feet a diver is above the surface of the water is given by the function $h(t) = -16t^2 + 16t + 32$, where t is the elapsed time in seconds after the diver jumped. Find the height of the diver for the times shown in the table.

t	$h(t)$
0	
0.5	
1.5	
2.0	

WRITING

95. Give four ways in which a function can be described. Which do you think is the most informative? Why?

96. Explain why we can think of a function as a machine.

REVIEW

In Exercises 97–100, show that each number is a rational number by expressing it as a ratio of two integers.

97. $-3\dfrac{3}{4}$ **98.** 4.7

99. $0.333\ldots$ **100.** $-0.\overline{6}$

▶ 2.6   Graphs of Functions

In this section, you will learn about

Graphs of nonlinear functions ■ Translations of graphs
■ Reflections of graphs ■ Solving equations graphically

Introduction   In the previous section, we saw that many real-world situations can be modeled by linear functions. When these functions are graphed, we can learn information by examining the resulting line. In this section, we will discuss three more types of functions. Like linear functions, they can be used as mathematical models. But unlike linear functions, their graphs are not straight lines; therefore, they are called **nonlinear functions.**

Graphs of Nonlinear Functions

The first nonlinear function we will discuss is the function $f(x) = x^2$ (or $y = x^2$), called the **squaring function.**

EXAMPLE 1

The squaring function. Graph the function $f(x) = x^2$ and find the domain and range.

Solution   We substitute values for x in the equation and compute the corresponding values of $f(x)$. For example, if $x = -3$, we have

$$f(x) = x^2$$
$$f(-3) = (-3)^2 \quad \text{Substitute } -3 \text{ for } x.$$
$$= 9$$

The ordered pair $(-3, 9)$ satisfies the equation and will lie on the graph. We list this pair and others that satisfy the equation in the table shown in Figure 2-53. We plot the points and draw a smooth curve through them to get the graph, called a **parabola.**

FIGURE 2-53

$f(x) = x^2$

x	$f(x)$	$(x, f(x))$
-3	9	$(-3, 9)$
-2	4	$(-2, 4)$
-1	1	$(-1, 1)$
0	0	$(0, 0)$
1	1	$(1, 1)$
2	4	$(2, 4)$
3	9	$(3, 9)$

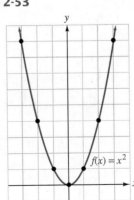

From the graph, we can see that x can be any real number. This indicates that the domain of the squaring function is the set of real numbers. We can also see that y is always positive or zero. This indicates that the range is the set of nonnegative real numbers.

SELF CHECK

Graph $g(x) = x^2 - 2$. Find the domain and the range. Compare the graph to the graph of $f(x) = x^2$.

Answers: D = the set of real numbers, R = the set of all real numbers greater than or equal to -2; the graph has the same shape but is 2 units lower

Another important nonlinear function is $f(x) = x^3$ (or $y = x^3$), called the **cubing function.**

EXAMPLE 2

The cubing function. Graph the function $f(x) = x^3$ and find the domain and range.

Solution

We substitute values for x in the equation and compute the corresponding values of $f(x)$. For example, if $x = -2$, we have

$$f(x) = x^3$$
$$f(-2) = (-2)^3 \quad \text{Substitute } -2 \text{ for } x.$$
$$= -8$$

The ordered pair $(-2, -8)$ satisfies the equation and will lie on the graph. We list this pair and others that satisfy the equation in the table shown in Figure 2-54. We plot the points and draw a smooth curve through them to get the graph.

FIGURE 2-54

$$f(x) = x^3$$

x	$f(x)$	$(x, f(x))$
-2	-8	$(-2, -8)$
-1	-1	$(-1, -1)$
0	0	$(0, 0)$
1	1	$(1, 1)$
2	8	$(2, 8)$

From the graph, we can see that x can be any real number. This indicates that the domain of the cubing function is the set of real numbers. We can also see that y can be any real number. This indicates that the range is the set of real numbers.

SELF CHECK Graph $g(x) = x^3 + 1$. Find the domain and the range. Compare the graph to the graph of $f(x) = x^3$.

Answers: D = the set of real numbers, R = the set of real numbers; the graph has the same shape but is 1 unit higher

A third nonlinear function is $f(x) = |x|$ (or $y = |x|$), called the **absolute value function.**

E X A M P L E   3 **The absolute value function.** Graph the function $f(x) = |x|$ and find the domain and range.

Solution We substitute values for x in the equation and compute the corresponding values of $f(x)$. For example, if $x = -3$, we have

$$f(x) = |x|$$
$$f(-3) = |-3| \quad \text{Substitute } -3 \text{ for } x.$$
$$= 3$$

The ordered pair $(-3, 3)$ satisfies the equation and will lie on the graph. We list this pair and others that satisfy the equation in the table shown in Figure 2-55. We plot the points and connect them to get the graph.

FIGURE 2-55

$$f(x) = |x|$$

x	$f(x)$	$(x, f(x))$
-3	3	$(-3, 3)$
-2	2	$(-2, 2)$
-1	1	$(-1, 1)$
0	0	$(0, 0)$
1	1	$(1, 1)$
2	2	$(2, 2)$
3	3	$(3, 3)$

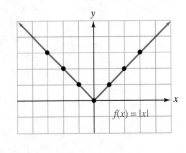

From the graph, we can see that x can be any real number. So the domain of the absolute value function is the set of real numbers. We can also see that y is always positive or zero. This indicates that the range is the set of nonnegative real numbers.

SELF CHECK

Graph $g(x) = |x - 2|$. Find the domain and the range. Compare the graph to the graph of $f(x) = |x|$.

Answers: D = the set of real numbers, R = the set of nonnegative real numbers; the graph has the same shape but is 2 units to the right

---

ACCENT ON TECHNOLOGY *Graphing Functions*

We can graph nonlinear functions with a graphing calculator. For example, to graph $f(x) = x^2$ in a standard window of $[-10, 10]$ for x and $[-10, 10]$ for y, we enter the function by typing x $\boxed{x^2}$ and then press the $\boxed{\text{GRAPH}}$ key. We will obtain the graph shown in Figure 2-56(a).

To graph $f(x) = x^3$, we enter the function by typing x ^ 3 and then press the $\boxed{\text{GRAPH}}$ key to obtain the graph in Figure 2-56(b). To graph $f(x) = |x|$, we enter the function by pressing the $\boxed{\text{ABS}}$ key (or selecting abs from a menu), typing x, and pressing the $\boxed{\text{GRAPH}}$ key to obtain the graph in Figure 2-56(c).

FIGURE 2-56

(a)   (b)   (c)

When using a graphing calculator, we must be sure that the viewing window does not show a misleading graph. For example, if we graph $f(x) = |x|$ in the window $[0, 10]$ for x and $[0, 10]$ for y, we will obtain a misleading graph that looks like a line. See Figure 2-57. This is not correct. The proper graph is the V-shaped graph shown in Figure 2-56(c). One of the challenges of using graphing calculators is finding an appropriate viewing window.

FIGURE 2-57

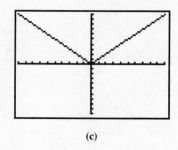

Translations of Graphs

Examples 1, 2, and 3 and their Self Checks suggest that the graphs of different functions may be identical except for their positions in the xy-plane. For example, Figure

2-58 shows the graph of $f(x) = x^2 + k$ for three different values of k. If $k = 0$, we get the graph of $f(x) = x^2$. If $k = 3$, we get the graph of $f(x) = x^2 + 3$, which is identical to the graph of $f(x) = x^2$ except that it is shifted 3 units upward. If $k = -4$, we get the graph of $f(x) = x^2 - 4$, which is identical to the graph of $f(x) = x^2$ except that it is shifted 4 units downward. These shifts are called **vertical translations.**

In general, we can make these observations.

FIGURE 2-58

Vertical translations

If f is a function and k is a positive number, then

- The graph of $y = f(x) + k$ is identical to the graph of $y = f(x)$ except that it is translated k units upward.
- The graph of $y = f(x) - k$ is identical to the graph of $y = f(x)$ except that it is translated k units downward.

EXAMPLE 4

Vertical translations. Graph $f(x) = |x| + 2$.

Solution   The graph of $f(x) = |x| + 2$ will be the same V-shaped graph as $f(x) = |x|$, except shifted 2 units up. The graph appears in Figure 2-59.

FIGURE 2-59

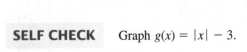

SELF CHECK   Graph $g(x) = |x| - 3$.

Answer:

Figure 2-60 shows the graph of $f(x) = (x + h)^2$ for three different values of h. If $h = 0$, we get the graph of $f(x) = x^2$. The graph of $f(x) = (x - 3)^2$ is identical to the graph of $f(x) = x^2$ except that it is shifted 3 units to the right. The graph of $f(x) = (x + 2)^2$ is identical to the graph of $f(x) = x^2$ except that it is shifted 2 units to the left. These shifts are called **horizontal translations**.

In general, we can make these observations.

FIGURE 2-60

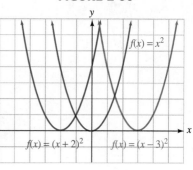

$f(x) = x^2$

$f(x) = (x + 2)^2$ $f(x) = (x - 3)^2$

Horizontal translations

$y = f(x + k)$ $y = f(x)$ $y = f(x - k)$

If f is a function and k is a positive number, then

■ The graph of $y = f(x - k)$ is identical to the graph of $y = f(x)$ except that it is translated k units to the right.

■ The graph of $y = f(x + k)$ is identical to the graph of $y = f(x)$ except that it is translated k units to the left.

E X A M P L E   5

Horizontal translations. Graph $f(x) = (x + 3)^3$.

Solution The graph of $f(x) = (x + 3)^3$ will be the same shape as the graph of $f(x) = x^3$ except that it is shifted 3 units to the left. The graph appears in Figure 2-61.

FIGURE 2-61

$f(x) = (x + 3)^3$

SELF CHECK Graph $g(x) = (x - 2)^2$.

Answer:

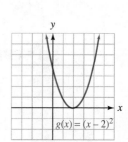

$g(x) = (x - 2)^2$

In the next example, two translations are made to a basic graph.

EXAMPLE 6

Two translations. Graph $f(x) = (x - 3)^2 + 2$.

Solution   We can graph this equation by translating the graph of $f(x) = x^2$ to the right 3 units and then 2 units up, as shown in Figure 2-62.

FIGURE 2-62

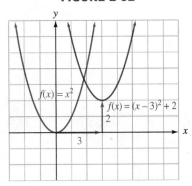

SELF CHECK   Graph $f(x) = |x + 2| - 3$.

Answer:

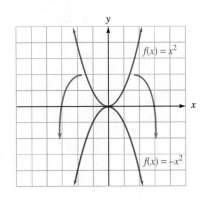

Reflections of Graphs

Figure 2-63 shows a table of solutions for $f(x) = x^2$ and for $f(x) = -x^2$. We note that for a given value of x, the corresponding y-values in the tables are opposites. When graphed, we see that the $-$ in $f(x) = -x^2$ has the effect of "flipping" the graph of $f(x) = x^2$ over the x-axis so that the parabola opens downward. We say that the graph of $f(x) = -x^2$ is a **reflection** of the graph of $f(x) = x^2$ in the x-axis.

FIGURE 2-63

\(f(x) = x^2\)			\(f(x) = -x^2\)		
x	$f(x)$	$(x, f(x))$	x	$f(x)$	$(x, f(x))$
-2	4	$(-2, 4)$	-2	-4	$(-2, -4)$
-1	1	$(-1, 1)$	-1	-1	$(-1, -1)$
0	0	$(0, 0)$	0	0	$(0, 0)$
1	1	$(1, 1)$	1	-1	$(1, -1)$
2	4	$(2, 4)$	2	-4	$(2, -4)$

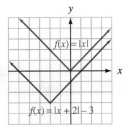

EXAMPLE 7	**Reflection of a graph.** Graph $f(x) = -x^3$.

Solution To graph $f(x) = -x^3$, we use the graph of $f(x) = x^3$ from Example 2. First, we reflect the portion of the graph of $f(x) = x^3$ in quadrant I to quadrant IV, as shown in Figure 2-64. Then we reflect the portion of the graph of $f(x) = x^3$ in quadrant III to quadrant II.

FIGURE 2-64

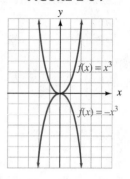

SELF CHECK Graph $f(x) = -|x|$.

Answer:

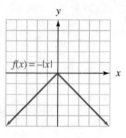

Reflection of a graph

The graph of $y = -f(x)$ is the graph of $y = f(x)$ reflected about the x-axis.

Solving Equations Graphically

Some of the graphing concepts discussed in this chapter can be used to solve equations. For example, the solution of $-2x - 4 = 0$ is the number x that will make y equal to 0 in the equation $y = -2x - 4$. To find this number, we inspect the graph of $y = -2x - 4$ and locate the point on the graph that has a y-coordinate of 0. In Figure 2-65, we see that the point is $(-2, 0)$, which is the x-intercept of the graph. We can conclude that the x-coordinate of the x-intercept, $x = -2$, is the solution of $-2x - 4 = 0$.

FIGURE 2-65

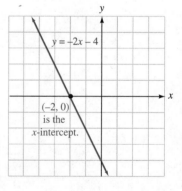

Check:
$$-2x - 4 = 0$$
$$-2(-2) - 4 \stackrel{?}{=} 0$$
$$4 - 4 \stackrel{?}{=} 0$$
$$0 = 0$$

To solve more complex equations in this way, a graphing calculator is helpful.

ACCENT ON TECHNOLOGY *Solving Equations Graphically*

To solve $2(x - 3) - 4 = 7x$ with a graphing calculator, we subtract $7x$ from both sides so that the right-hand side is 0.

$$2(x - 3) - 4 = 7x$$

$$2(x - 3) - 4 - 7x = 0 \quad \text{Subtract } 7x \text{ from both sides.}$$

Next, we enter the left-hand side of the equation into a graphing calculator in the form $y = 2(x - 3) - 4 - 7x$. There is no need to simplify the expression to graph it. See Figure 2-66(a).

If we use window settings of $[-5, 5]$ for x and $[-5, 5]$ for y and press $\boxed{\text{GRAPH}}$, we will obtain the graph in Figure 2-66(b).

Since y will be 0 when the graph intersects the x-axis, the solution of the equation $0 = 2(x - 3) - 4 - 7x$ will be the x-coordinate of the x-intercept of the graph. To find the x-intercept, we trace as shown in Figure 2-66(c). After repeated zooms and traces, we will see that the x-coordinate of the x-intercept is -2. Verify that the solution of $2(x - 3) - 4 = 7x$ is $x = -2$.

FIGURE 2-66

(a)

(b)

(c)

Another way to find the solution of $2(x - 3) - 4 = 7x$ is to use the ZERO feature, found under the CALC menu. After we guess left and right bounds, the cursor automatically locates the x-intercept of the graph and displays its coordinates. See Figure 2-66(d). The x-coordinate of the x-intercept of the graph is called a *zero* of $y = 2(x - 3) - 4 - 7x$. The zero (in this case, -2) is the solution of the given equation. Please see your owner's manual for more specific instructions on how to use this feature.

FIGURE 2-66(d)

Another approach that can be used to solve equations involves a graphical determination of the value of the variable that will make the sides of the equation equal.

EXAMPLE 8

Solving an equation from a graph. Solve $2x + 4 = -2$ graphically.

Solution The graphs of $y = 2x + 4$ and $y = -2$ are shown in Figure 2-67. To solve the equation $2x + 4 = -2$, we need to find the value of x that makes $2x + 4$ equal -2. The point of intersection of the graphs is $(-3, -2)$. This tells us that if $x = -3$, the expression $2x + 4$ equals -2. So the solution of $2x + 4 = -2$ is $x = -3$.

FIGURE 2-67

SELF CHECK   Solve $2x + 4 = 2$ graphically.   *Answer: −1*   ■

ACCENT ON TECHNOLOGY *Solving Equations Graphically*

To solve the equation $2(x − 3) + 3 = 7$ with a graphing calculator, we graph both the left-hand side and the right-hand side of the equation in the same window, as shown in Figure 2-68(a). We then trace to find the coordinates of the point where the two graphs intersect, as shown in Figure 2-68(b).

After another zoom and trace, we would conclude from the display shown in Figure 2-68(c) that the intersection point is (5, 7). When $x = 5$, the expression $2(x − 3) + 3 = 2(5 − 3) + 3 = 2(2) + 3 = 7$. So $x = 5$ is the solution of $2(x − 3) + 3 = 7$.

FIGURE 2-68

(a)   (b)   (c)

An easier way to find the coordinates of the point of intersection is with the INTERSECT feature found on most graphing calculators. With this feature, the cursor automatically highlights the intersection point, and the x- and y-coordinates are displayed. See Figure 2-68(d). Consult your owner's manual for the specific instructions concerning this feature.

FIGURE 2-68(d)

STUDY SET

Section 2.6

VOCABULARY

In Exercises 1–8, fill in the blanks to make the statements true.

1. The function $f(x) = x^2$ is called the _____ function.

2. The function $f(x) = x^3$ is called the _____ function.

3. The function $f(x) = |x|$ is called the _____ function.

4. Functions whose graphs are not straight lines are called _____ functions.

5. Shifting the graph of an equation up or down is called a _____ translation.

6. Shifting the graph of an equation to the left or to the right is called a horizontal _____.

7. The graph of $f(x) = −x^2$ is a _____ of the graph of $f(x) = x^2$ in the x-axis.

8. The set of _____ real numbers is the set of real numbers greater than or equal to 0.

CONCEPTS

In Exercises 9–12, fill in the blanks to make the statements true.

9. The graph of $f(x) = (x + 4)^3$ is the same as the graph of $f(x) = x^3$ except that it is shifted ___ units to the _____.

10. The graph of $f(x) = x^3 - 2$ is the same as the graph of $f(x) = x^3$ except that it is shifted ___ units _____.

11. The graph of $f(x) = x^2 + 5$ is the same as the graph of $f(x) = x^2$ except that it is shifted ___ units ___.

12. The graph of $f(x) = |x - 5|$ is the same as the graph of $f(x) = |x|$ except that it is shifted ___ units to the _____.

13. Use the graphs in Illustration 1 to solve each equation.
    a. $3x - 2 = 4$
    b. $3x - 2 = -2$

ILLUSTRATION 1

14. Use the graph of $y = 2(x - 6) - 4x$, shown below, to solve $2(x - 6) - 4x = 0$. Check your answer.

15. Use the information in the given calculator displays to solve $9x - 5(x - 6) - 18 = 0$. Check your answer.

16. Use the information in the given calculator displays to solve $-4x - 3(2 - x) = -9$. Check your answer.

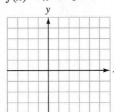

17. Illustration 2 shows the graph of $f(x) = x^2 + k$, for three values of k. What are the three values?

ILLUSTRATION 2

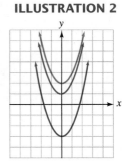

18. Use the graph in Illustration 3 to find each function value.
    a. $f(-3)$
    b. $f(0)$
    c. $f(1)$

ILLUSTRATION 3

19. Use the graph in Illustration 4 to find each function value.
    a. $g(-2)$
    b. $g(1)$
    c. $g(2.5)$

ILLUSTRATION 4

20. Use a graphing calculator to sketch the *reflection* of the graph shown below.

PRACTICE

In Exercises 21–28, graph each function by plotting points. Give the domain and range. Check your work with a graphing calculator.

21. $f(x) = x^2 - 3$

22. $f(x) = x^2 + 2$

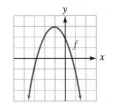

23. $f(x) = (x - 1)^3$

24. $f(x) = (x + 1)^3$

25. $f(x) = |x| - 2$

26. $f(x) = |x| + 1$

27. $f(x) = |x - 1|$

28. $f(x) = |x + 2|$

In Exercises 29–36, graph each function using window settings of $[-4, 4]$ for x and $[-4, 4]$ for y. The graph is not what it appears to be. Pick a better viewing window and find a better representation of the true graph.

29. $f(x) = x^2 + 8$

30. $f(x) = x^3 - 8$

31. $f(x) = |x + 5|$

32. $f(x) = |x - 5|$

33. $f(x) = (x - 6)^2$

34. $f(x) = (x + 9)^2$

35. $f(x) = x^3 + 8$

36. $f(x) = x^3 - 12$

In Exercises 37–48, for each function, first sketch the graph of its associated function, $f(x) = x^2$, $f(x) = x^3$, or $f(x) = |x|$. Then draw each graph using a translation or a reflection.

37. $f(x) = x^2 - 5$

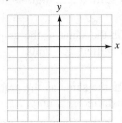

38. $f(x) = x^3 + 4$

39. $f(x) = (x - 1)^3$

40. $f(x) = (x + 4)^2$

41. $f(x) = |x - 2| - 1$

42. $f(x) = (x + 2)^2 - 1$

43. $f(x) = (x + 1)^3 - 2$

44. $f(x) = |x + 4| + 3$

45. $f(x) = -x^3$

46. $f(x) = -|x|$

47. $f(x) = -x^2$

48. $f(x) = -(x + 1)^2$

In Exercises 49–52, use a graphing calculator to solve each equation.

49. $4(x - 1) = 3x$

50. $4(x - 3) - x = x - 6$

51. $11x + 6(3 - x) = 3$

52. $2(x + 2) = 2(1 - x) + 10$

APPLICATIONS

53. FREEWAY DESIGN A Grand Avenue exit off the 210 Freeway is to be constructed. (See Illustration 5.) Sketch the off-ramp design on the graph if the sides of the pavement are defined by the functions $f(x) = (x - 1)^3$ and $f(x) = (x - 3)^3$.

ILLUSTRATION 5

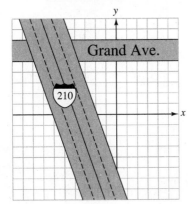

54. OPTICS See Illustration 6. The law of reflection states that the angle of reflection is equal to the angle of incidence. What function studied in this section mathematically models the path of the reflected light beam with an angle of incidence measuring 45°?

ILLUSTRATION 6

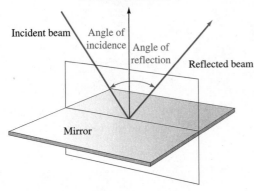

55. CENTER OF GRAVITY See Illustration 7. As a diver performs a $1\frac{1}{2}$-somersault in the tuck position,

her center of gravity follows a path that can be described by a graph shape studied in this section. What graph shape is that?

ILLUSTRATION 7

56. FLASHLIGHTS Light beams coming from a flashlight bulb are reflected outward by a parabolic mirror as parallel rays.
   a. The cross-sectional view of a parabolic mirror is given by the function $f(x) = x^2$ for the following values of x: $-0.7, -0.6, -0.5, -0.4, -0.3, -0.2, -0.1, 0, 0.1, 0.2, 0.3, 0.4, 0.5, 0.6, 0.7$. Sketch the parabolic mirror using the graph in Illustration 8.
   b. From the light bulb filament at $(0, 0.25)$, draw a line segment representing a beam of light that strikes the mirror at $(-0.4, 0.16)$ and then reflects outward, parallel to the y-axis.

ILLUSTRATION 8

WRITING

57. Explain how to graph an equation by plotting points.

58. Explain why the correct choice of window settings is important when using a graphing calculator.

59. Explain how to solve the equation $2x + 6 = 0$ graphically.

60. What does it mean when we say that the domain of a function is the set of all real numbers?

REVIEW

Solve each formula for the indicated variable.

61. $T - W = ma$ for W

62. $a + (n - 1)d = l$ for n

63. $s = \dfrac{1}{2}gt^2 + vt$ for g

64. $e = mc^2$ for m

65. Last year, Rock Valley College had an operating budget of $4.5 million. Due to salary increases and a new robotics program, the budget was increased by 20%. Find the operating budget for this year.

66. In Illustration 9, the line passing through points R, C, and S is parallel to line segment AB. Find the measure of $\angle ACB$. (Read $\angle ACB$ as "angle ACB.")

ILLUSTRATION 9

Functions

In Chapter 2, we introduced one of the most important concepts in mathematics, that of a **function.** Functions are used to describe relationships where one quantity depends on another.

1. Fill in the blanks to make the statements true.
   a. A function is a _____ between a set of _____ values x (called the domain) and a set of output values y (called the _____), where exactly ___ y value in the range corresponds to each number x in the _____.

   b. Since the value of y usually depends on the number x, we call y the _____ variable and x the _____ variable.

2. We can think of a function as a machine. Using the words *input, output, domain,* and *range,* explain how the function machine shown below works.

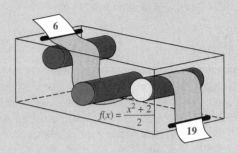

$$f(x) = \frac{x^2 + 2}{2}$$

Four Ways to Represent a Function

Functions can be described in words, with an equation, with a table, or with a graph.

3. The equation $y = 2x + 3$ determines a correspondence between the values of x and y. Find the value of y that corresponds to $x = -10$.

4. The area of a circle is the product of π and the radius squared. Use the variables A and r to describe this relationship with an equation.

5. Use the table below to determine the height of a projectile 1.5 seconds after it was shot vertically into the air.

6. Use the graph below to determine what y-value the function f assigns to $x = 1$.

Time t (seconds)	Height h (feet)
0	0
0.5	28
1.0	48
1.5	60
2.0	64

Function Notation

There is a special notation used with functions. The notation $y = f(x)$ denotes that the variable y is a function of x.

7. Write the equation in Problem 3 using function notation. Find $f(0)$.

9. If the function $h(t) = -16t^2 + 64t$ gives the height of the projectile described in Problem 5, find $h(4)$ and interpret the result.

8. Use function notation to represent the relationship described in Problem 4.

10. Use the graph in Problem 6 to determine $f(-3)$ and $f(3)$.

Accent on Teamwork

Section 2.1

City map Get a map of the city in which you live and use a black marker to draw a rectangular coordinate system on the map. The size of the grid you use will depend on the size of the map. The origin of the coordinate system should be your place of residence. Determine the coordinates of ten important locations in your city, such as the library, fire stations, parks, schools, and city hall.

Section 2.2

Graphing lines Draw rectangular coordinate systems with 1-inch grids on two pieces of white poster board. On the first poster board, graph $y = mx + 2$ for $m = 1, 2, 3, 4,$ and 5. Label each line with its respective equation. At the bottom of the poster board, write your observations about how the graph changes for different values of m. On the second poster board, graph $y = 2x + b$ for $b = 1, 2, 3, 4,$ and 5. At the bottom of the poster board, write your observations about how the graph changes for different values of b.

Section 2.3

Measuring slope Use a ruler and a level to find the slopes of five ramps or inclines by measuring $\frac{\text{rise}}{\text{run}}$, as shown in Illustration 1. Record your results in a chart of the form shown below. List the examples in order, from that with the smallest slope to that with the largest slope.

ILLUSTRATION 1

Object/location	Slope		
Ramp outside cafeteria	$\dfrac{\text{Rise}}{\text{Run}} =$	$\dfrac{4 \text{ in.}}{24 \text{ in.}} =$	$\dfrac{1}{6}$

Section 2.4

A linear model For this project, you will need a five-gallon pail, a yardstick, a watch that shows seconds, and access to a garden hose.

Tape the yardstick to the pail as shown in Illustration 2. Turn on the water, leaving it running at a constant rate, and begin to fill the pail. Keep track of the time in seconds it takes for the water in the pail to reach heights of 2, 4, 6, 8, 10, and 12 inches.

Draw a coordinate system, with the horizontal axis representing time in seconds and the vertical axis the height of the water in inches. Then plot your data as ordered pairs of the form (time, height). Draw the line that best fits the data and write the equation of the line. Use the equation to predict the height the column of water would be if it were to run for 24 hours, which is 86,400 seconds.

ILLUSTRATION 2

Section 2.5

Linear functions Think of a real-life situation like that given in Exercise 87 of Study Set 2.5 and write a linear function that describes it. Graph the function and explain the importance of the slope of the line and its y-intercept.

Section 2.6

Graphing functions Draw a rectangular coordinate system using a 1-inch grid on a piece of white poster board. Graph $y = k|x|$ for $k = \frac{1}{2}, 1, 2,$ and 3. Label each graph with its respective equation. At the bottom of the poster board, write your observations about how the graph changes for different values of k.

Parabolas Borrow some calculus books from your mathematics instructor, or see if you can find several in your school library. Look up the word *parabola* in the index and find some real-life applications of parabolas mentioned in these books. Write a brief description of each use, and include a sketch of the application.

Section 2.1

The Rectangular Coordinate System

CONCEPTS

A *rectangular coordinate system* is formed by two intersecting perpendicular number lines called the *x-axis* and the *y-axis,* which divide the plane into four *quadrants.*

The process of locating a point in the coordinate plane is called *plotting* or *graphing* that point.

Data in a *table of values* can be expressed as ordered pairs.

Graphs can be used to visualize relationships between two quantities.

REVIEW EXERCISES

1. Plot each point on the rectangular coordinate system in Illustration 1.
  a. $(0, 3)$
  b. $(-2, -4)$
  c. $\left(\frac{5}{2}, -1.75\right)$
  d. the origin
  e. $(2.5, 0)$

ILLUSTRATION 1

2. Use the graph in Illustration 2 to complete the table.

x	y	(x, y)
-3		$(-3, \quad)$
-2		$(-2, \quad)$
	3	$(\quad, 3)$
0		$(0, \quad)$
1		$(1, \quad)$
	0	$(\quad, 0)$
	-5	$(\quad, -5)$

ILLUSTRATION 2

3. The graph in Illustration 3 shows how the height of the water in a flood control channel changed over a seven-day period.
  a. Describe the height of the water at the beginning of day 2.

  b. By how much did the water level increase or decrease from day 4 to day 5?

  c. During what time period did the water level stay the same?

ILLUSTRATION 3

4. AUCTIONS The dollar increments used by an auctioneer during the bidding process depend on what initial price the auctioneer began with for the item. See the step graph in Illustration 4.
  a. What increments are used by the auctioneer if the bidding on an item began at $150?

  b. If the first bid on an item being auctioned is $750, what will be the next price asked for by the auctioneer?

ILLUSTRATION 4

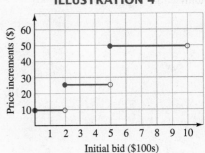

Midpoint formula: The midpoint of a line segment with endpoints at $P(x_1, y_1)$ and $Q(x_2, y_2)$ is the point M with coordinates

$$\left(\frac{x_1 + x_2}{2}, \frac{y_1 + y_2}{2}\right)$$

5. Find the midpoint of the line segment joining $P(8, -2)$ and $Q(-4, 6)$.

6. The midpoint of a line segment has coordinates of $\left(-\frac{1}{2}, \frac{1}{2}\right)$. If one endpoint has coordinates of $(-3, 7)$, what are the coordinates of the other endpoint?

Section 2.2

Graphing Linear Equations

The *graph of an equation* is the graph of all points on the rectangular coordinate system whose coordinates satisfy the equation.

7. Graph each equation.

a. $y = 3x + 4$

b. $y = -\frac{1}{3}x - 1$

To find the *y-intercept* of a line, substitute 0 for x in the equation and solve for y. To find the *x-intercept* of a line, substitute 0 for y in the equation and solve for x.

8. Graph each equation using the intercept method.

a. $2x + y = 4$

b. $3x - 4y - 8 = 0$

The graph of the equation $x = a$ is a *vertical line* with x-intercept at $(a, 0)$.

The graph of the equation $y = b$ is a *horizontal line* with y-intercept at $(0, b)$.

9. Graph each equation.

a. $y = 4$

b. $x = -2$

10. Complete the table of solutions for each equation.

a. $y = -3x$

x	y
-3	
0	
3	

b. $y = \frac{1}{2}x - \frac{5}{2}$

x	y
-3	
0	
3	

Section 2.3

The *slope* of a nonvertical line is defined to be

$$m = \frac{\text{rise}}{\text{run}} = \frac{\Delta y}{\Delta x}$$

The *slope formula:*
If $(x_2 \neq x_1)$,

$$m = \frac{y_2 - y_1}{x_2 - x_1}$$

The slope of a line gives the *average rate of change.*

Horizontal lines have a slope of 0. Vertical lines have no defined slope.

Parallel lines have the same slope. The slopes of two nonvertical *perpendicular lines* are negative reciprocals.

Rate of Change and the Slope of a Line

11. Find the slope of lines l_1 and l_2 in Illustration 5.

ILLUSTRATION 5

12. U.S. VEHICLE SALES On the graph in Illustration 6, draw a line through the points $(0, 21.2)$ and $(16, 43.2)$. Use this linear model to estimate the rate of increase in the market share of minivans, sport utility vehicles, and light trucks over the years 1980–1996.

ILLUSTRATION 6

Trucks, minivans, sport utility vehicles, and pickup trucks: Market share of U.S. vehicle sales

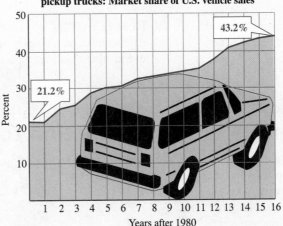

Years after 1980

Based on data from the American Automotive Association

13. Find the slope of the line passing through points P and Q.
  a. $P(2, 5)$ and $Q(5, 8)$
  b. $P(3, -2)$ and $Q(-6, 12)$
  c. $P(-2, 4)$ and $Q(8, 4)$
  d. $P(-5, -4)$ and $Q(-5, 8)$

14. Find the slope of the graph of each equation, if one exists.
  a. $y = \frac{2}{3}x + 18$
  b. $4x + 2y = 8$
  c. $x = 10$
  d. $y = 7$

15. Tell whether the lines with the given slopes are parallel, perpendicular, or neither.
  a. $m_1 = 4, m_2 = -\frac{1}{4}$
  b. $m_1 = 0.5, m_2 = \frac{1}{2}$

Section 2.4

Equations of a line:

Point–slope form:

$$y - y_1 = m(x - x_1)$$

Slope–intercept form:

$$y = mx + b$$

General form:

$$Ax + By = C$$

Many real-life situations can be modeled by linear equations.

Writing Equations of Lines

16. Write the equation of the line with the given properties. Express the result in slope–intercept form.
    a. Slope of 3; passing through $P(-8, 5)$
    b. Passing through $(-2, 4)$ and $(6, -9)$

17. Write the equation of the line with the given properties. Write the equation in general form.
    a. Passing through $(-3, -5)$; parallel to the graph of $3x - 2y = 7$

    b. Passing through $(-3, -5)$; perpendicular to the graph of $3x - 2y = 7$

18. Write $3x + 4y = -12$ in slope–intercept form. Give the slope and y-intercept of the graph of the equation. Then use this information to graph the line.

19. DEPRECIATION A business purchased a copy machine for \$8,700 and will depreciate it on a straight-line basis over the next 5 years. At the end of its useful life, it will be sold as scrap for \$100. Find its depreciation equation.

Section 2.5

A *function* is a correspondence between a set of input values x and a set of output values y, where exactly one value of y in the *range* corresponds to each number x in the *domain*.

The notation $y = f(x)$ denotes that the variable y (the *dependent* variable) is a function of x (the *independent* variable).

The *domain* of a function is the set of input values. The *range* is the set of output values.

Introduction to Functions

20. Tell whether each equation determines y to be a function of x.
    a. $y = 6x - 4$
    b. $y = 4 - x^2$
    c. $y^2 = x$
    d. $|y| = x$

21. Assume that $f(x) = 3x + 2$ and $g(x) = \dfrac{x^2 - 4x + 4}{2}$ and find each value.
    a. $f(-3)$
    b. $g(8)$
    c. $g(-2)$
    d. $f(t)$

22. Find the domain and range of each function.
    a. $f(x) = 4x - 1$
    b. $f(x) = x^2 + 1$

    c. $f(x) = \dfrac{4}{2 - x}$
    d. $y = -|4x|$

The *vertical line test* can be used to determine whether a graph represents a function.

23. Use the vertical line test to determine whether each graph represents a function.

a.

b.

24. MARKET SHARE In Exercise 12, a line was drawn on the graph to estimate the rate of increase in the market share of minivans, sport utility vehicles, and light trucks. Use this information to write an equation of the line. Express your result using function notation. Then use the function to predict the market share for the year 2000 if the trend continues.

25. Tell which are linear functions.
  a. $f(x) = 3x + 2$ **b.** $y = x^2 - 25$

Section 2.6

Graphs of Functions

Graphs of *nonlinear functions* are not lines.

The *squaring function:*

$$f(x) = x^2$$

The *cubing function:*

$$f(x) = x^3$$

The *absolute value function:*

$$f(x) = |x|$$

A *horizontal translation* shifts a graph left or right. A *vertical translation* shifts a graph upward or downward. A *reflection* "flips" a graph in the *x*-axis.

26. Graph each function.
  a. $f(x) = x^2 - 3$

  b. $f(x) = (x - 2)^3 + 1$

27. Graph $f(x) = |x + 2|$, $g(x) = |x - 2|$, and $h(x) = -|x|$ on the same coordinate system.

28. Use the graph in Illustration 7 to find each function value.
  a. $f(-2)$ **b.** $f(3)$

ILLUSTRATION 7

29. Use the graphs in Illustration 8 to solve each equation. Check each answer.

a. $2(2 - x) + x = 0$ **b.** $2(2 - x) + x = 5$

c. $2(2 - x) + x = x$

ILLUSTRATION 8

CHAPTER 2

Test

In Problems 1–4, refer to the graph shown in Illustration 1, which shows the height of an object at different times after it was shot straight up into the air.

ILLUSTRATION 1

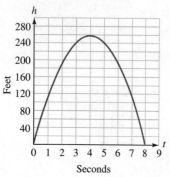

1. How high was the object 3 seconds into the flight?

2. At what times was the object about 110 feet above the ground?

3. What was the maximum height reached by the object?

4. How long did the flight take?

5. Find the coordinates of the midpoint of the line segment joining $(-2, -5)$ and $(7, 8)$.

6. The coordinates of one endpoint of a line segment are $(-2, 3)$. If the coordinates of the segment's midpoint are $(2, -3)$, find the coordinates of the other endpoint.

7. Find the x- and y-intercepts of the graph of $2x - 5y = 10$. Then graph the equation.

8. Graph the equation $y = -2$.

9. Find the slope of the line shown in Illustration 2.

ILLUSTRATION 2

10. Find the rate of change of the temperature for the period of time shown in the graph in Illustration 3.

ILLUSTRATION 3

In Problems 11–14, find the slope of each line, if possible.

11. The line through $(-2, 4)$ and $(6, 8)$

12. The graph of $2x - 3y = 8$

13. The graph of $x = 12$

14. The graph of $y = 12$

15. Write the equation of the line that has slope of $\frac{2}{3}$ and passes through $(4, -5)$. Give the answer in slope–intercept form.

16. Write the equation of the line that passes through $(-2, 6)$ and $(-4, -10)$. Give the answer in general form.

17. Find the slope and the y-intercept of the graph of $-2x + 6 = 6y + 15$.

18. Determine whether the graphs of $4x - y = 12$ and $y = \frac{1}{4}x + 3$ are parallel, perpendicular, or neither.

19. Write the equation of the line that passes through the origin and is parallel to the graph of $y = \frac{3}{2}x - 7$.

20. Does $|y| = x$ define y to be a function of x?

21. Find the domain and range of the function $f(x) = |x|$.

22. Find the domain and range of the function $f(x) = x^3$.

In Problems 23–26, $f(x) = 3x + 1$ and $g(x) = x^2 - 2x - 1$. Find each value.

23. $f(3)$

24. $g(0)$

25. $f\left(\dfrac{2}{3}\right)$

26. $g(r)$

In Problems 27–28, tell whether each graph represents a function.

27.

28.

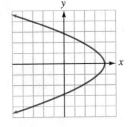

29. Graph $f(x) = x^2 + 3$.

30. Graph $g(x) = -|x + 2|$.

31. See Illustration 4. Explain how to solve $3(x - 2) - 2(-2 + x) = 0$ graphically. What is the solution?

ILLUSTRATION 4

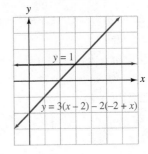

32. See Illustration 4. Explain how to solve $3(x - 2) - 2(-2 + x) = 1$ graphically.

33. Describe four different ways to represent a function.

34. Explain why the graph of a circle does not represent a function.

CHAPTERS 1–2

Cumulative Review Exercises

In Exercises 1–10, tell which numbers in the set
$\left\{-2, 0, 1, 2, \frac{13}{12}, 6, 7, \sqrt{5}, \pi\right\}$ *are in each category.*

1. Natural numbers

2. Whole numbers

3. Rational numbers

4. Irrational numbers

5. Negative numbers

6. Real numbers

7. Prime numbers

8. Composite numbers

9. Even numbers

10. Odd numbers

In Exercises 11–12, simplify each expression.

11. $-|5| + |-3|$

12. $\dfrac{|-5| + |-3|}{-|4|}$

In Exercises 13–16, evaluate each expression.

13. $2 + 4 \cdot 5$

14. $\dfrac{8 - 4}{2 - 4}$

15. $-\dfrac{16}{5} \div \left(-\dfrac{10}{3}\right)$

16. $\dfrac{(9 - 8)^4 + 21}{3^3 - \left(\sqrt{16}\right)^2}$

In Exercises 17–18, evaluate each expression for $x = 2$ and $y = -3$.

17. $-x - 2y$

18. $\dfrac{x^2 - y^2}{2x + y}$

In Exercises 19–22, tell which property of real numbers justifies each statement.

19. $(a + b) + c = a + (b + c)$

20. $3(x + y) = 3x + 3y$

21. $(a + b) + c = c + (a + b)$

22. $(ab)c = a(bc)$

In Exercises 23–26, simplify each expression.

23. $12y - 17y$

24. $-7s(-4t)(-1)$

25. $3x^2 + 2x^2 - 5x^2$

26. $-(4 + z) + 2z$

In Exercises 27–32, solve each equation.

27. $2x - 5 = 11$

28. $\dfrac{2x - 6}{3} = x + 7$

29. $4(y - 3) + 4 = -3(y + 5)$

30. $2x - \dfrac{3(x - 2)}{2} = 7 - \dfrac{x - 3}{3}$

31. $-3 = -\dfrac{9}{8}s$

32. $0.04(24) + 0.02x = 0.04(12 + x)$

In Exercises 33–34, solve each formula.

33. $S = \dfrac{n(a + l)}{2}$ for a

34. $A = \dfrac{1}{2}h(b_1 + b_2)$ for h

35. INVESTMENTS A woman invested part of $20,000 at 6% and the rest at 7%. If her annual interest is $1,260, how much did she invest at 6%?

36. DRIVING RATES John drove to a distant city in 5 hours. When he returned, there was less traffic, and the trip took only 3 hours. If he drove 26 mph faster on the return trip, how fast did he drive each way?

37. Graph $2x - 3y = 6$.

38. Find the slope of the line shown in Illustration 1.

ILLUSTRATION 1

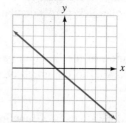

39. Write the equation of the line passing through $(-2, 5)$ and $(8, -9)$.

40. Write the equation of the line passing through $(-2, 3)$ and parallel to the graph of $3x + y = 8$.

In Exercises 41–44, $f(x) = 3x^2 + 2$ and $g(x) = -2x - 1$. Find each function value.

41. $f(-1)$

42. $g(0)$

43. $g(-2)$

44. $f(-r)$

In Exercises 45–46, graph each equation and tell whether it is a function. If it is a function, give the domain and range.

45. $y = -x^2 + 1$

46. $y = |x - 3|$

47. See Illustration 2. Explain why there is not a linear relationship between the height of the antenna and its maximum range of reception.

ILLUSTRATION 2

48. Explain how to use Illustration 3 to solve each equation. Give each solution.
   a. $-7 - 5(x - 1) + 4x = 0$

   b. $-7 - 5(x - 1) + 4x = -4$

ILLUSTRATION 3

3

Systems of Equations

CAMPUS CONNECTION

The Cosmetology Department

Cosmetology students learn the techniques of cutting hair, facials, manicuring, and permanents. They also study business practices used in the operation of a beauty shop. In Chapter 3, we will discuss a mathematical procedure that a shop owner can use to determine how much business he or she must have to make a profit. To answer such a question, we must write two equations, each containing two variables. We call the pair of equations a *system of linear equations*. After learning how to solve systems of linear equations, you will be able to solve many problems that could not be solved using just one variable.

> *To solve many problems, we must use two and sometimes three variables. This requires that we solve a system of equations.*

▶ 3.1 Solving Systems by Graphing

In this section, you will learn about

The graphing method ■ Consistent systems ■ Inconsistent systems ■ Dependent equations

Introduction   The red line graphed in Figure 3-1 shows the cost for a company to produce a given number of skateboards. The blue line shows the revenue the company will earn for selling a given number of those skateboards. The graph offers the company important financial information.

■ The production costs exceed the revenue earned if less than 400 skateboards are sold. In this case, the company loses money.

■ The revenue earned exceeds the production costs if more than 400 skateboards are sold. In this case, the company makes a profit.

■ Production costs equal revenue earned if exactly 400 skateboards are sold. This fact is indicated by the point of intersection of the two lines, (400, 20,000), which is called the **break-even point**.

FIGURE 3-1

This example shows that important information can be learned by finding the point of intersection of two lines. In this section, we will discuss how to use the *graphing method* to do this.

The Graphing Method

In Chapter 2, we discussed linear equations containing the variables *x* and *y*. We found that such equations had infinitely many solutions (x, y), and that we could graph them on a rectangular coordinate system. In this chapter, we will discuss **systems of linear equations** involving two or three equations. A left brace { is used when writing a system of equations.

In the pair of equations

$$\begin{cases} x + 2y = 4 \\ 2x - y = 3 \end{cases}$$ (called a system of two linear equations)

there are infinitely many pairs (x, y) that satisfy the first equation and infinitely many pairs (x, y) that satisfy the second equation. However, there is only one pair (x, y) that satisfies both equations. The process of finding this pair is called *solving the system*.

To solve a system of two equations in two variables by graphing, we use the following steps.

The graphing method

1. On a single set of coordinate axes, graph each equation.
2. Find the coordinates of the point (or points) where the graphs intersect. These coordinates give the solution of the system.
3. If the graphs have no point in common, the system has no solution.
4. Check the solution in both of the original equations.

Consistent Systems

When a system of equations (as in Example 1) has a solution, the system is called a **consistent system**.

EXAMPLE 1

The graphing method. Solve the system $\begin{cases} x + 2y = 4 \\ 2x - y = 3 \end{cases}$.

Solution
We graph both equations on one set of coordinate axes, as shown in Figure 3-2.

Although infinitely many ordered pairs (x, y) satisfy $x + 2y = 4$, and infinitely many ordered pairs (x, y) satisfy $2x - y = 3$, only the coordinates of the point where the graphs intersect satisfy both equations. Since the intersection point has coordinates of $(2, 1)$, the solution is the ordered pair $(2, 1)$, or $x = 2$ and $y = 1$.

FIGURE 3-2

$x + 2y = 4$

x	y	(x, y)
4	0	(4, 0)
0	2	(0, 2)
−2	3	(−2, 3)

$2x - y = 3$

x	y	(x, y)
$\frac{3}{2}$	0	$\left(\frac{3}{2}, 0\right)$
0	−3	(0, −3)
−1	−5	(−1, −5)

Use the intercept method to graph each line.

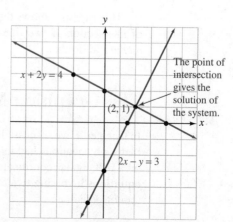

The point of intersection gives the solution of the system.

To check the solution, we substitute 2 for x and 1 for y in both equations and verify that (2, 1) satisfies each one.

Check: *The first equation*   ***The second equation***

The first equation	*The second equation*
$x + 2y = 4$	$2x - y = 3$
$2 + 2(1) \overset{?}{=} 4$	$2(2) - 1 \overset{?}{=} 3$
$2 + 2 \overset{?}{=} 4$	$4 - 1 \overset{?}{=} 3$
$4 = 4$	$3 = 3$

SELF CHECK   Solve $\begin{cases} x - 3y = -5 \\ 2x + y = 4 \end{cases}$.

Answer: (1, 2)

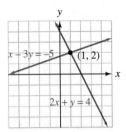

Inconsistent Systems

When a system has no solution (as in Example 2), it is called an **inconsistent system.**

EXAMPLE 2   **The graphing method.** Solve the system $\begin{cases} 2x + 3y = 6 \\ 4x + 6y = 24 \end{cases}$, if possible.

Solution   We graph both equations on one set of coordinate axes, as shown in Figure 3-3. In this example, the graphs are parallel, because the slopes of the two lines are equal and they have different y-intercepts. We can see that the slope of each line is $-\frac{2}{3}$ by writing each equation in slope–intercept form.

$$2x + 3y = 6 \qquad\qquad 4x + 6y = 24$$
$$3y = -2x + 6 \qquad\qquad 6y = -4x + 24$$
$$y = -\frac{2}{3}x + 2 \qquad\qquad y = -\frac{2}{3}x + 4$$

Since the graphs are parallel lines, the lines do not intersect, and the system does not have a solution. It is an inconsistent system.

FIGURE 3-3

$2x + 3y = 6$				$4x + 6y = 24$			
x	y	(x, y)		x	y	(x, y)	Use the intercept method to graph each line.
3	0	(3, 0)		6	0	(6, 0)	
0	2	(0, 2)		0	4	(0, 4)	
−3	4	(−3, 4)		−3	6	(−3, 6)	

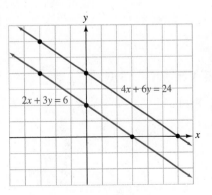

SELF CHECK   Solve $\begin{cases} 3y - 2x = 6 \\ 2x - 3y = 6 \end{cases}$.   *Answer:* no solutions

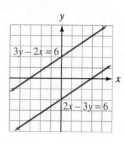

Dependent Equations

When the equations of a system have different graphs (as in Examples 1 and 2), the equations are called **independent equations**. Two equations with the same graph are called **dependent equations**.

E X A M P L E   3   **The graphing method.** Solve the system $\begin{cases} y = \frac{1}{2}x + 2 \\ 2x + 8 = 4y \end{cases}$.

Solution   We graph each equation on one set of coordinate axes, as shown in Figure 3-4. Since the graphs coincide (are the same), the system has infinitely many solutions. Any ordered pair (x, y) that satisfies one equation also satisfies the other.

From the graph shown in Figure 3-4, we see that $(-4, 0)$, $(0, 2)$, and $(2, 3)$ are solutions. We can find infinitely many more solutions by finding additional ordered pairs (x, y) that satisfy both equations.

Because the two equations have the same graph, they are dependent equations.

FIGURE 3-4

Graph by using the slope
and y-intercept.

$$y = \tfrac{1}{2}x + 2$$

$$m = \tfrac{1}{2} \qquad b = 2$$

Slope $= \tfrac{1}{2}$ y-intercept: $(0, 2)$

Graph by using the
intercept method.

$$2x + 8 = 4y$$

x	y	(x, y)
-4	0	$(-4, 0)$
0	2	$(0, 2)$
2	3	$(2, 3)$

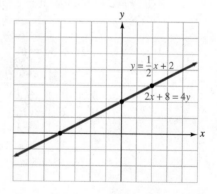

SELF CHECK   Solve $\begin{cases} 2x - y = 4 \\ y = 2x - 4 \end{cases}$.   *Answer:* There are infinitely many solutions; three of them are $(0, -4)$, $(2, 0)$, and $(4, 4)$.

We now summarize the possibilities that can occur when two linear equations, each with two variables, are graphed.

Solving a system of equations by the graphing method

If the lines are different and intersect, the equations are independent, and the system is consistent. **One solution exists.**

If the lines are different and parallel, the equations are independent, and the system is inconsistent. **No solution exists.**

If the lines coincide, the equations are dependent, and the system is consistent. **Infinitely many solutions exist.**

If each equation in one system is equivalent to a corresponding equation in another system, the systems are called **equivalent**.

EXAMPLE 4

Equivalent systems. Solve the system $\begin{cases} \frac{3}{2}x - y = \frac{5}{2} \\ x + \frac{1}{2}y = 4 \end{cases}$.

Solution   We multiply both sides of $\frac{3}{2}x - y = \frac{5}{2}$ by 2 to eliminate the fractions and obtain the equation $3x - 2y = 5$. We multiply both sides of $x + \frac{1}{2}y = 4$ by 2 to eliminate the fractions and obtain the equation $2x + y = 8$.

The new system

$$\begin{cases} 3x - 2y = 5 \\ 2x + y = 8 \end{cases}$$

is equivalent to the original system and is easier to solve, since it has no fractions. If we graph each equation in the new system, as in Figure 3-5, we see that the coordinates of the point where the two lines intersect are $(3, 2)$. Verify that $x = 3$ and $y = 2$ satisfy each equation in the original system.

FIGURE 3-5

$3x - 2y = 5$		
x	y	(x, y)
$\frac{5}{3}$	0	$\left(\frac{5}{3}, 0\right)$
0	$-\frac{5}{2}$	$\left(0, -\frac{5}{2}\right)$
1	-1	$(1, -1)$

$2x + y = 8$		
x	y	(x, y)
4	0	$(4, 0)$
0	8	$(0, 8)$
1	6	$(1, 6)$

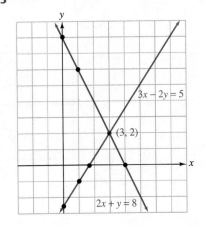

SELF CHECK Solve $\begin{cases} \frac{5}{2}x - y = 2 \\ x + \frac{1}{3}y = 3 \end{cases}$ *Answer:* (2, 3)

by the graphing method.

 WARNING! When checking a solution of a system of equations, always substitute the values of the variables into the original equations.

ACCENT ON TECHNOLOGY *Solving Systems by Graphing*

The graphing method has limitations. First, the method is limited to equations with two variables. Systems with three or more variables cannot be solved graphically. Second, it is often difficult to find exact solutions graphically. However, the TRACE and ZOOM capabilities of graphing calculators enable us to get very good approximations of such solutions.

To solve the system $\begin{cases} 3x + 2y = 12 \\ 2x - 3y = 12 \end{cases}$

with a graphing calculator, we must first solve each equation for y so that we can enter the equations into the calculator. After solving for y, we obtain the following equivalent system:

$$\begin{cases} y = -\frac{3}{2}x + 6 \\ y = \frac{2}{3}x - 4 \end{cases}$$

If we use window settings of $[-10, 10]$ for x and $[-10, 10]$ for y, the graphs of the equations will look like those in Figure 3-6(a). If we zoom in on the intersection point of the two lines and trace, we will get an approximate solution like the one shown in Figure 3-6(b). To get better results, we can do more zooms. We would then find that, to the nearest hundredth, the solution is $(4.63, -0.94)$. Verify that this is reasonable, and then verify that the exact solution is $x = \frac{60}{13}$ and $y = -\frac{12}{13}$.

We can also find the intersection of two lines by using the INTERSECT feature found on most graphing calculators. To learn how to use this feature, please consult your owner's manual. After graphing the lines and using INTERSECT, we obtain a graph similar to Figure 3-6(c). The display shows the approximate coordinates of the point of intersection.

FIGURE 3-6

(a)   (b)   (c)

STUDY SET

Section 3.1

VOCABULARY

In Exercises 1–6, fill in the blanks to make the statements true.

1. $\begin{cases} x + 2y = 4 \\ 2x - y = 3 \end{cases}$ is called a _____ of linear equations.

2. When a system of equations has one or more solutions, it is called a _____ system.

3. If a system has no solutions, it is called an _____ system.

4. If two equations have different graphs, they are called _____ equations.

5. Two equations with the same graph are called _____ equations.

6. When solving a system of two linear equations by the graphing method, we look for the point of _____ of the two lines.

CONCEPTS

7. Refer to Illustration 1. Tell whether a true or a false statement would be obtained when the coordinates of
   a. point A are substituted into the equation for line l_1.
   b. point B are substituted into the equation for line l_1.
   c. point C are substituted into the equation for line l_1.
   d. point C are substituted into the equation for line l_2.

ILLUSTRATION 1

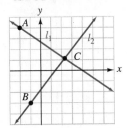

8. Refer to Illustration 2.
   a. How many ordered pairs satisfy the equation $3x + y = 3$? Name three.

   b. How many ordered pairs satisfy the equation $\frac{2}{3}x - y = -3$? Name three.

ILLUSTRATION 2

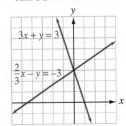

   c. How many ordered pairs satisfy both equations? Name it or them.

9. a. The intercept method can be used to graph the equation $2x - 4y = -8$. Complete the following table.

x	y	(x, y)
	0	
0		
2		

   b. What is the x-intercept of the graph of $2x - 4y = -8$? What is the y-intercept?

10. a. To graph the equation $y = 3x + 1$, we can pick three numbers for x and find the corresponding values of y. Complete the following table.

x	y	(x, y)
-1		
0		
2		

   b. We can also graph the equation $y = 3x + 1$ if we know the slope and the y-intercept of the line. What are they?

11. Write a system of two linear equations that has
   a. only one solution, $(2, 3)$.

   b. an infinite number of solutions.

   c. no solution.

12. Estimate the solution of the system of linear equations shown in the display below. Then check your answer.

NOTATION

In Exercises 13–14, fill in the blanks to make the statement true.

13. A left _____ is often used when writing a system of equations.

14. The solution of a system of linear equations is written as an ordered _____.

PRACTICE

In Exercises 15–18, tell whether the ordered pair is a solution of the system of equations.

15. $(1, 2)$; $\begin{cases} 2x - y = 0 \\ y = \frac{1}{2}x + \frac{3}{2} \end{cases}$

16. $(-1, 2)$; $\begin{cases} y = 3x + 5 \\ y = x + 4 \end{cases}$

17. $(2, -3)$; $\begin{cases} y + 2 = \frac{1}{2}x \\ 3x + 2y = 0 \end{cases}$

18. $(-4, 3)$; $\begin{cases} 4x - y = -19 \\ 3x + 2y = -6 \end{cases}$

In Exercises 19–42, solve each system by graphing, if possible. If a system is inconsistent or if the equations are dependent, so indicate.

19. $\begin{cases} x + y = 6 \\ x - y = 2 \end{cases}$

20. $\begin{cases} x - y = 4 \\ 2x + y = 5 \end{cases}$

21. $\begin{cases} 2x + y = 1 \\ x - 2y = -7 \end{cases}$

22. $\begin{cases} 3x - y = -3 \\ 2x + y = -7 \end{cases}$

23. $\begin{cases} x + y = 0 \\ 2x - y = 6 \end{cases}$

24. $\begin{cases} 4x - 3y = 5 \\ 2x + y = 0 \end{cases}$

25. $\begin{cases} 3x + y = 3 \\ 3x + 2y = 0 \end{cases}$

26. $\begin{cases} 2x + 2y = -1 \\ 3x + 4y = 0 \end{cases}$

27. $\begin{cases} x = 13 - 4y \\ 3x = 4 + 2y \end{cases}$

28. $\begin{cases} 3x = 7 - 2y \\ 2x = 2 + 4y \end{cases}$

29. $\begin{cases} x = 3 - 2y \\ 2x + 4y = 6 \end{cases}$

30. $\begin{cases} 3x = 5 - 2y \\ 3x + 2y = 7 \end{cases}$

31. $\begin{cases} x = 2 \\ y = -\dfrac{1}{2}x + 2 \end{cases}$

32. $\begin{cases} y = -2 \\ y = \dfrac{2}{3}x - \dfrac{4}{3} \end{cases}$

33. $\begin{cases} y = 3 \\ x = 2 \end{cases}$

34. $\begin{cases} 2x + 3y = -15 \\ 2x + y = -9 \end{cases}$

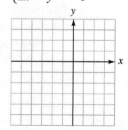

35. $\begin{cases} x = \dfrac{11 - 2y}{3} \\ y = \dfrac{11 - 6x}{4} \end{cases}$

36. $\begin{cases} x = \dfrac{1 - 3y}{4} \\ y = \dfrac{12 + 3x}{2} \end{cases}$

37. $\begin{cases} y = -\dfrac{5}{2}x + \dfrac{1}{2} \\ 2x - \dfrac{3}{2}y = 5 \end{cases}$

38. $\begin{cases} \dfrac{5}{2}x + 3y = 6 \\ y = -\dfrac{5}{6}x + 2 \end{cases}$

39. $\begin{cases} x = \dfrac{5y - 4}{2} \\ x - \dfrac{5}{3}y + \dfrac{1}{3} = 0 \end{cases}$

40. $\begin{cases} 2x = 5y - 11 \\ 3x = 2y \end{cases}$

41. $\begin{cases} x = -\dfrac{3}{2}y \\ x = \dfrac{3}{2}y - 2 \end{cases}$

42. $\begin{cases} 4x = 3y - 1 \\ 3y = 4 - 8x \end{cases}$

In Exercises 43–46, use a graphing calculator to solve each system. Give all answers to the nearest hundredth.

43. $\begin{cases} y = 3.2x - 1.5 \\ y = -2.7x - 3.7 \end{cases}$

44. $\begin{cases} y = -0.45x + 5 \\ y = 5.55x - 13.7 \end{cases}$

45. $\begin{cases} 1.7x + 2.3y = 3.2 \\ y = 0.25x + 8.95 \end{cases}$

46. $\begin{cases} 2.75x = 12.9y - 3.79 \\ 7.1x - y = 35.76 \end{cases}$

APPLICATIONS

47. TAKE-OUT FOOD
   a. Estimate the point of intersection of the two graphs shown in Illustration 3. Express your answer in the form (year, number of meals).
   b. What information about dining out does the point of intersection give?

ILLUSTRATION 3

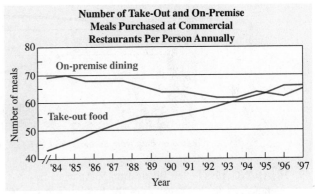

Based on data from *The Wall Street Journal Almanac* (1999)

48. ROAD ATLAS See Illustration 4. Name the cities that lie along Interstate 40. Name the cities that lie along Interstate 25. What city lies on both interstate highways?

ILLUSTRATION 4

ILLUSTRATION 5

ILLUSTRATION 6

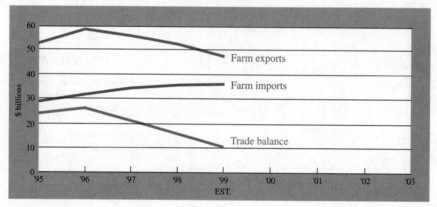

Based on data from *Business Week* (June 28, 1999)

49. HEARING TEST See Illustration 5. At what frequency and decibel level were the hearing test results the same for the left and right ear? Write your answer as an ordered pair.

50. U.S. FARMERS See Illustration 6. Assume that recent trends continue.
   a. Estimate the coordinates of the point where the graphs of farm exports and farm imports will intersect. Interpret your answer.

   b. Where will the graph of trade balance intersect the horizontal axis? (*Hint:* Trade balance is the difference between farm export and farm import totals.)

51. LAW OF SUPPLY AND DEMAND The demand function, graphed in Illustration 7, describes the relationship between the price x of a certain camera and the demand for the camera.
   a. The supply function, $S(x) = \frac{25}{4}x - 525$, describes the relationship between the price x of the camera and the number of cameras the manufacturer is willing to supply. Graph this function in Illustration 7.
   b. For what price will the supply of cameras equal the demand?

c. As the price of the camera is increased, what happens to supply and what happens to demand?

ILLUSTRATION 7

52. COST AND REVENUE The function
   $C(x) = 200x + 400$ gives the cost for a college to offer x sections of an introductory class in CPR (cardiopulmonary resuscitation). The function
   $R(x) = 280x$ gives the amount of revenue the college brings in when offering x sections of CPR.
   a. Find the *break-even point* (where cost = revenue) by graphing each function on the same coordinate system.

b. How many sections does the college need to offer to make a profit on the CPR training course?

53. NAVIGATION The paths of two ships are tracked on the same coordinate system. One ship is following a path described by the equation $2x + 3y = 6$, and the other is following a path described by the equation $y = \frac{2}{3}x - 3$.
    a. Is there a possibility of a collision?
    b. What are the coordinates of the danger point?

    c. Is a collision a certainty?
54. AIR TRAFFIC CONTROL Two airplanes are tracked using the same coordinate system on a radar screen. One plane is following a path described by the equation $y = \frac{2}{5}x - 2$, and the other is following a path described by the equation $2x = 5y + 7$. Is there a possibility of a collision?

WRITING

55. Suppose the solution of a system of two linear equations is $\left(\frac{14}{5}, -\frac{8}{3}\right)$. Knowing this, explain any drawbacks with solving the system by the graphing method.
56. Can a system of two linear equations have exactly two solutions? Why or why not?

REVIEW

REVIEW

In Exercises 57–60, $f(x) = -x^3 + 2x - 2$ and $g(x) = \dfrac{2 - x}{9 + x}$. Find each value.

57. $f(-1)$ 58. $f(10)$
59. $g(2)$ 60. $g(-20)$

61. Determine the domain and range of $f(x) = x^2 - 2$.

62. Find the slope of the line passing through the points $(-4, 8)$ and $(3, 8)$.

63. In Illustration 8, the area of the square is 81 square centimeters. Find the area of the shaded triangle.

ILLUSTRATION 8

64. If the area of the circle in Illustration 9 is 49π square centimeters, find the area of the square.

ILLUSTRATION 9

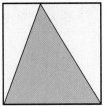

▶ 3.2 Solving Systems Algebraically

In this section, you will learn about

The substitution method ■ The addition method ■ An inconsistent system ■ A system with infinitely many solutions ■ Problem solving

---

Introduction   The graphing method provides a way to visualize the process of solving systems of equations. However, when using the graphing method, it can sometimes be difficult to determine the exact coordinates of the point of intersection. It is also true that we cannot use the graphing method to solve systems of higher order, such as three equations, each with three variables. In this section, we will discuss two other methods, called the *substitution* and the *addition* methods. They can be used to find the exact solutions of systems of equations algebraically.

The Substitution Method

To solve a system of two equations (each containing two variables) by the **substitution method,** we follow these steps.

The substitution method

1. If necessary, solve one equation for one of its variables—preferably a variable with a coefficient of 1 or -1.

2. Substitute the resulting expression for that variable into the other equation and solve it.

3. Find the value of the other variable by substituting the value of the variable found in step 2 into the equation found in step 1.

4. State the solution.

5. Check the solution in both of the original equations.

EXAMPLE 1

The substitution method. Solve the system $\begin{cases} 4x + y = 13 \\ -2x + 3y = -17 \end{cases}$.

Solution **Step 1:** We solve the first equation for y, because y has a coefficient of 1.

$$4x + y = 13$$

1. $\qquad y = -4x + 13$ To isolate y, subtract $4x$ from both sides.

Step 2: We then substitute $-4x + 13$ for y in the second equation of the system. This step will eliminate the variable y from that equation. The result will be an equation containing only one variable, x.

$$-2x + 3y = -17 \quad \text{The second equation.}$$
$$-2x + 3(-4x + 13) = -17 \quad \text{Substitute } -4x + 13 \text{ for } y. \text{ The variable } y \text{ is eliminated from the equation.}$$
$$-2x - 12x + 39 = -17 \quad \text{Use the distributive property to remove parentheses.}$$
$$-14x = -56 \quad \text{To solve for } x, \text{ first combine like terms and then subtract 39 from both sides.}$$
$$x = 4 \quad \text{Divide both sides by } -14.$$

Step 3: To find y, we substitute 4 for x in Equation 1 and simplify:

$$y = -4x + 13$$
$$= -4(4) + 13 \quad \text{Substitute 4 for } x.$$
$$= -3$$

Step 4: The solution is $x = 4$ and $y = -3$, or just $(4, -3)$. The graphs of the equations of the given system would intersect at the point $(4, -3)$.

Step 5: To verify that this solution satisfies both equations, we substitute $x = 4$ and $y = -3$ into each equation in the system and simplify.

Check: *The first equation* 　　　　　*The second equation*

$$4x + y = 13 \qquad\qquad -2x + 3y = -17$$
$$4(4) + (-3) \overset{?}{=} 13 \qquad -2(4) + 3(-3) \overset{?}{=} -17$$
$$16 - 3 \overset{?}{=} 13 \qquad\qquad -8 - 9 \overset{?}{=} -17$$
$$13 = 13 \qquad\qquad\qquad -17 = -17$$

Since the ordered pair $(4, -3)$ satisfies both equations of the system, it checks.

SELF CHECK   Solve $\begin{cases} x + 3y = 9 \\ 2x - y = -10 \end{cases}$.

　　　　　　　　　　　　　　　　　　　Answer: $(-3, 4)$ ■

EXAMPLE 2

The substitution method. Solve the system $\begin{cases} \frac{2}{9}x - \frac{2}{9}y = \frac{2}{3} \\ 0.1x = 0.2 - 0.1y \end{cases}$.

Solution   First we find an equivalent system without fractions or decimals. To do this, we multiply both sides of the first equation by 9, which is the lowest common denominator of the fractions in the equation. Then we multiply both sides of the second equation by 10.

2. $\begin{cases} 2x - 2y = 6 \\ x = 2 - y \end{cases}$
3.

Since the variable x is isolated in Equation 3, we will substitute $2 - y$ for x in Equation 2. This step will eliminate x from Equation 2, leaving an equation containing only one variable, y. We then solve for y.

$2x - 2y = 6$	Equation 2.
$2(2 - y) - 2y = 6$	Substitute $2 - y$ for x.
$4 - 2y - 2y = 6$	Use the distributive property to remove parentheses.
$-4y = 2$	To solve for y, combine like terms and then subtract 4 from both sides.
$y = -\dfrac{1}{2}$	Divide both sides by -4 and then simplify the fraction.

We can find x by substituting $-\frac{1}{2}$ for y in Equation 3 and simplifying:

$x = 2 - y$	Equation 3.
$x = 2 - \left(-\dfrac{1}{2}\right)$	Substitute $-\frac{1}{2}$ for y.
$= 2 + \dfrac{1}{2}$	
$= \dfrac{5}{2}$	$2 + \frac{1}{2} = \frac{4}{2} + \frac{1}{2} = \frac{5}{2}.$

The solution is the ordered pair $\left(\frac{5}{2}, -\frac{1}{2}\right)$. Verify that this solution satisfies both equations in the original system.

SELF CHECK   Solve $\begin{cases} \frac{x}{8} + \frac{y}{4} = \frac{1}{2} \\ 0.01y = -0.02x + 0.04 \end{cases}$.   *Answer:* $\left(\frac{4}{3}, \frac{4}{3}\right)$ ∎

The Addition Method

Another method for solving a system of linear equations is the **addition method**. In this method, we combine the equations in a way that will eliminate the terms involving one of the variables.

The addition method

1. Write both equations of the system in general form: $Ax + By = C$.
2. Multiply the terms of one or both of the equations by constants chosen to make the coefficients of x (or y) differ only in sign.
3. Add the equations and solve the resulting equation, if possible.
4. Substitute the value obtained in step 3 into either of the original equations and solve for the remaining variable.
5. State the solution obtained in steps 3 and 4.
6. Check the solution in both of the original equations.

EXAMPLE 3

The addition method. Solve the system $\begin{cases} 4x + y = 13 \\ -2x + 3y = -17 \end{cases}$.

Solution   **Step 1:** This is the system discussed in Example 1. In this example, we will solve it by the addition method. Since both equations are already written in general form, step 1 is unnecessary.

Step 2: We note that the coefficient of x in the first equation is 4. If we multiply both sides of the second equation by 2, the coefficient of x in that equation will be -4. Then the coefficients of x will differ only in sign.

4. $\begin{cases} 4x + y = 13 \\ -4x + 6y = -34 \end{cases}$
5.

Step 3: When these equations are added, the terms involving x drop out (or are eliminated), and we get an equation that contains only the variable y. We then proceed by solving for y.

$$\begin{array}{r} 4x + y = 13 \\ + \underline{-4x + 6y = -34} \\ 7y = -21 \\ y = -3 \end{array}$$

Add the like terms, column by column:
$4x + (-4x) = 0$, $y + 6y = 7y$, and $13 + (-34) = -21$.

To solve for y, divide both sides by 7.

Step 4: To find x, we substitute -3 for y in either of the original equations and solve for x. If we use Equation 4, we have

$$\begin{array}{rl} 4x + y = 13 \\ 4x + (-3) = 13 & \text{Substitute } -3 \text{ for } y. \\ 4x = 16 & \text{To solve for } x, \text{ add 3 to both sides.} \\ x = 4 & \text{Divide both sides by 4.} \end{array}$$

Step 5: The solution is $x = 4$ and $y = -3$, or just $(4, -3)$.

Step 6: The check was completed in Example 1.

SELF CHECK

Solve $\begin{cases} 3x + 2y = 0 \\ 2x - y = -7 \end{cases}$.   *Answer:* $(-2, 3)$ ∎

EXAMPLE 4

The addition method. Solve the system $\begin{cases} 4x = 3(2 + y) \\ 3(x - 10) = -2y \end{cases}$.

Solution   To use the addition method, we must write each equation in general form. In each case, the first step is to remove the parentheses.

The first equation	*The second equation*
$4x = 3(2 + y)$	$3(x - 10) = -2y$
$4x = 6 + 3y$	$3x - 30 = -2y$
$4x - 3y = 6$	$3x + 2y = 30$

We now solve the equivalent system

6. $\begin{cases} 4x - 3y = 6 \\ 3x + 2y = 30 \end{cases}$
7.

Since the coefficients of y already have opposite signs, we choose to eliminate y. To make the y-terms drop out when we add the equations, we multiply both sides of Equation 6 by 2 and both sides of Equation 7 by 3 to get

$$\begin{cases} 8x - 6y = 12 \\ 9x + 6y = 90 \end{cases}$$

When these equations are added, the y-terms drop out, and we get

$$17x = 102 \quad 8x + 9x = 17x, \; -6y + 6y = 0, \text{ and } 12 + 90 = 102.$$
$$x = 6 \quad \text{To solve for } x, \text{ divide both sides by 17.}$$

To find y, we can substitute 6 for x in either of the two original equations, or in Equation 6 or Equation 7. If we substitute 6 for x in Equation 7, we get

$$3x + 2y = 30$$
$$3(6) + 2y = 30 \quad \text{Substitute 6 for } x.$$
$$18 + 2y = 30 \quad \text{Do the multiplication.}$$
$$2y = 12 \quad \text{Subtract 18 from both sides.}$$
$$y = 6 \quad \text{Divide both sides by 2.}$$

The solution is the ordered pair $(6, 6)$.

SELF CHECK   Solve $\begin{cases} 4(2x - y) = 18 \\ 3(x - 3) = 2y - 1 \end{cases}$.   *Answer:* $\left(1, -\frac{5}{2}\right)$ ■

An Inconsistent System

EXAMPLE 5   **A system with no solution.** Solve the system $\begin{cases} y = 2x + 4 \\ 8x - 4y = 7 \end{cases}$, if possible.

Solution   Because the first equation is already solved for y, we use the substitution method.

$$8x - 4y = 7 \quad \text{The second equation.}$$
$$8x - 4(2x + 4) = 7 \quad \text{Substitute } 2x + 4 \text{ for } y.$$

We then solve this equation for x:

$$8x - 8x - 16 = 7 \quad \text{Use the distributive property to remove parentheses.}$$
$$-16 \neq 7 \quad \text{Combine like terms.}$$

This result indicates that the equations in the system are independent and that the system is inconsistent. Since the system has no solution, the graphs of the equations in the system will be parallel.

SELF CHECK   Solve $\begin{cases} x = -2.5y + 8 \\ y = -0.4x + 2 \end{cases}$.   *Answer:* no solution   ■

A System with Infinitely Many Solutions

EXAMPLE 6   **Dependent equations.** Solve the system $\begin{cases} 4x + 6y = 12 \\ -2x - 3y = -6 \end{cases}$.

Solution   Since the equations are written in general form, we use the addition method. We copy the first equation and multiply both sides of the second equation by 2 to get

$$\begin{array}{r} 4x + 6y = 12 \\ -4x - 6y = -12 \\ \hline \end{array}$$

After adding the left-hand sides and the right-hand sides, we get

$$0x + 0y = 0$$
$$0 = 0$$

Here, both the x- and y-terms drop out. The true statement $0 = 0$ indicates that the equations are dependent and that the system is consistent.

Note that the equations of the system are equivalent, because when the second equation is multiplied by -2, it becomes the first equation. The graphs of these equations would coincide. Any ordered pair that satisfies one of the equations also satisfies the other. Some solutions are $(0, 2)$, $(3, 0)$, and $(-3, 4)$.

SELF CHECK   Solve $\begin{cases} x - \frac{5}{2}y = \frac{19}{2} \\ -\frac{2}{5}x + y = -\frac{19}{5} \end{cases}$

Answer: There are infinitely many solutions; three of them are $(2, -3)$, $(12, 1)$, and $\left(\frac{19}{2}, 0\right)$. ∎

Problem Solving

To solve problems using two variables, we follow the same problem-solving strategy discussed in Chapters 1 and 2, except that we form two equations instead of one.

EXAMPLE 7

Wedding pictures. In Figure 3-7, a professional photographer offers two different packages for wedding pictures. Use the information in the figure to determine the cost of an 8×10-in. and a 5×7-in. photograph.

FIGURE 3-7

WEDDING PICTURES	
Package 1 includes...	Package 2 includes...
8 - 8 x 10's 12 - 5 x 7's	6 - 8 x 10's 22 - 5 x 7's
Only **$133.00**	Only **$168.00**

ANALYZE THE PROBLEM   From the figure, we see that eight 8×10 and twelve 5×7 pictures cost $133, and six 8×10 and twenty-two 5×7 pictures cost $168. We need to find the cost of an 8×10 and a 5×7 photograph.

FORM TWO EQUATIONS   We can let x represent the cost of an 8×10 photograph and let y represent the cost of a 5×7 photograph. For the first package, the cost of eight 8×10 pictures is $8 \cdot \$x = \$8x$, and the cost of twelve 5×7 pictures is $12 \cdot \$y = \$12y$. For the second package, the cost of six 8×10 pictures is $\$6x$, and the cost of twenty-two 5×7 pictures is $\$22y$. To find x and y, we must write and solve two equations.

The cost of eight 8×10 photographs	plus	the cost of twelve 5×7 photographs	is	the value of the the first package.
$8x$	$+$	$12y$	$=$	133

The cost of six 8×10 photographs	plus	the cost of twenty-two 5×7 photographs	is	the value of the second package.
$6x$	$+$	$22y$	$=$	168

SOLVE THE SYSTEM   To find the cost of the 8×10 and the 5×7 photographs, we must solve the following system:

8. $\begin{cases} 8x + 12y = 133 \\ 6x + 22y = 168 \end{cases}$
9.

We will use the addition method to solve this system. To make the x-terms drop out, we multiply both sides of Equation 8 by 3. Then we multiply both sides of Equation 9 by -4. We then add the resulting equations and solve for y:

$$24x + 36y = 399$$
$$\underline{-24x - 88y = -672}$$
$$-52y = -273 \quad \text{Add like terms, column by column. The } x\text{-terms drop out.}$$
$$y = 5.25 \quad \text{Divide both sides by } -52.$$

To find x, we substitute 5.25 for y in Equation 8 and solve for x:

$$8x + 12y = 133$$
$$8x + 12(\mathbf{5.25}) = 133 \quad \text{Substitute 5.25 for } y.$$
$$8x + 63 = 133 \quad \text{Do the multiplication.}$$
$$8x = 70 \quad \text{Subtract 63 from both sides.}$$
$$x = 8.75 \quad \text{Divide both sides by 8.}$$

State the Conclusion   The cost of an 8×10 photo is $8.75, and the cost of a 5×7 photo is $5.25.

Check the Result   If the first package contains eight 8×10 and twelve 5×7 photographs, the value of the package is $8($8.75) + 12($5.25) = $70 + $63 = 133. If the second package contains six 8×10 and twenty-two 5×7 photographs, the value of the package is $6($8.75) + 22($5.25) = $52.50 + $115.50 = 168. The answers check.   ∎

EXAMPLE 8

Water treatment. A technician determines that 50 fluid ounces of a 15% muriatic acid solution needs to be added to the water in a swimming pool to kill a growth of algae. If the technician has 5% and 20% muriatic solutions on hand, how many ounces of each must be combined to create the 15% solution?

Analyze the Problem   We need to find the number of ounces of a 5% solution and the number of ounces of a 20% solution that must be combined to obtain 50 ounces of a 15% solution.

Form Two Equations   We can let x represent the number of ounces of the 5% solution and let y represent the number of ounces of the 20% solution that are to be mixed. (See Figure 3-8(a).) Then the amount of muriatic acid in the 5% solution is $0.05x$ ounces, and the amount of muriatic acid in the 20% solution is $0.20y$ ounces. The sum of these amounts is also the amount of muriatic acid in the final mixture, which is 15% of 50 ounces. This information is shown in the table in Figure 3-8(b).

FIGURE 3-8

Solution	% acid	Ounces	Amount of acid
5% solution	0.05	x	$0.05x$
20% solution	0.20	y	$0.20y$
15% mixture	0.15	50	$0.15(50)$

One equation comes from the information in this column.

Another equation comes from the information in this column.

(a)

(b)

The facts of the problem give the following two equations:

The number of ounces of 5% solution	plus	the number of ounces of 20% solution	is	the total number of ounces in the 15% mixture.
x	$+$	y	$=$	50

The acid in the 5% solution	plus	the acid in the 20% solution	is	the acid in the 15% mixture.
$0.05x$	$+$	$0.20y$	$=$	$0.15(50)$

SOLVE THE SYSTEM   To find out how many ounces of each are needed, we solve the following system:

10. $\begin{cases} x + y = 50 \\ 0.05x + 0.20y = 7.5 \quad 0.15(50) = 7.5. \end{cases}$
11.

To solve this system by substitution, we can solve the first equation for y:

$$x + y = 50$$
12. $\qquad y = 50 - x$ Subtract x from both sides.

Then we substitute $50 - x$ for y in Equation 11 and solve for x.

$0.05x + 0.20y = 7.5$	Equation 11.
$0.05x + 0.20(50 - x) = 7.5$	Substitute $50 - x$ for y.
$5x + 20(50 - x) = 750$	Multiply both sides by 100.
$5x + 1{,}000 - 20x = 750$	Use the distributive property to remove parentheses.
$-15x = -250$	Combine like terms and subtract 1,000 from both sides.
$x = \dfrac{-250}{-15}$	Divide both sides by -15.
$x = \dfrac{50}{3}$	Simplify: $\dfrac{250}{15} = \dfrac{\overset{1}{\cancel{5}} \cdot 50}{\underset{1}{\cancel{5}} \cdot 3}$.

To find y, we can substitute $\frac{50}{3}$ for x in Equation 12:

$y = 50 - x$	Equation 12.
$= 50 - \dfrac{50}{3}$	Substitute $\frac{50}{3}$ for x.
$= \dfrac{100}{3}$	$50 = \frac{150}{3}$.

STATE THE CONCLUSION   To obtain 50 ounces of a 15% solution, the technician must mix $\frac{50}{3}$ or $16\frac{2}{3}$ ounces of the 5% solution with $\frac{100}{3}$ or $33\frac{1}{3}$ ounces of the 20% solution.

CHECK THE RESULT   We note that $16\frac{2}{3}$ ounces of solution plus $33\frac{1}{3}$ ounces of solution equals the required 50 ounces of solution. We also note that 5% of $16\frac{2}{3} \approx 0.83$ and 20% of $33\frac{1}{3} \approx 6.67$, giving a total of 7.5, which is 15% of 50. The answers check. ■

EXAMPLE 9

Parallelograms. Refer to the parallelogram shown in Figure 3-9 and find the values of x and y.

Solution   To solve this problem, we will use two important facts about parallelograms.

FIGURE 3-9

- When a diagonal intersects two parallel sides of a parallelogram, pairs of *alternate interior angles* have the same measure. In Figure 3-9, $\angle BAC$ and $\angle DCA$ are alternate interior angles and therefore have the same measure. Thus, $(x - y)° = 30°$.

- *Opposite angles* of a parallelogram have the same measure. Since $\angle B$ and $\angle D$ in Figure 3-9 are opposite angles of the parallelogram, $(x + y)° = 110°$.

We can form the following system of equations and solve it by addition.

13. $\quad x - y = 30$
14. $\underline{\quad x + y = 110}$
$\qquad 2x = 140$ Add Equations 13 and 14. The y-terms drop out.
$\qquad\quad x = 70$ Divide both sides by 2.

We can substitute 70 for x in Equation 14 and solve for y.

$$x + y = 110$$

$$70 + y = 110 \quad \text{Substitute 70 for } x.$$

$$y = 40 \quad \text{Subtract 70 from both sides.}$$

Thus, $x = 70$ and $y = 40$. ∎

Running a machine involves both *setup costs* and *unit costs*. Setup costs include the cost of preparing a machine to do a certain job. The costs to make one item are unit costs. They depend on the number of items to be manufactured, including costs of raw materials and labor.

E X A M P L E   1 0

Break point. The setup cost of a machine that makes wooden coathangers is $400. After setup, it costs $1.50 to make each hanger (the unit cost). Management is considering the purchase of a new machine that can manufacture the same type of coathanger at a cost of $1.25 per hanger. If the setup cost of the new machine is $500, find the number of coathangers that the company would need to manufacture to make the cost the same using either machine. This is called the **break point**.

ANALYZE THE PROBLEM   We are to find the number of coathangers that will cost equal amounts to produce on either machine. The machines have different setup costs and different unit costs.

FORM TWO EQUATIONS   The cost C_1 of manufacturing x coathangers on the machine currently in use is $1.50x + $400 (the number of coathangers manufactured times $1.50, plus the setup cost of $400). The cost C_2 of manufacturing the same number of coathangers on the new machine is $1.25x + $500. The break point occurs when the costs to make the same number of hangers using either machine are equal ($C_1 = C_2$).

If x represents the number of coathangers to be manufactured, the cost C_1 using the machine currently in use is

The cost of using the current machine	is	the cost of manufacturing x coathangers	plus	the setup cost.
C_1	$=$	$1.5x$	$+$	400

The cost C_2 using the new machine to make x coathangers is

The cost of using the new machine	is	the cost of manufacturing x coathangers	plus	the setup cost.
C_2	$=$	$1.25x$	$+$	500

SOLVE THE SYSTEM   To find the break point, we must solve the system $\begin{cases} C_1 = 1.5x + 400 \\ C_2 = 1.25x + 500 \end{cases}$.

Since the break point occurs when $C_1 = C_2$, we can substitute $1.5x + 400$ for C_2 in the second equation to get

$$1.5x + 400 = 1.25x + 500$$

$$1.5x = 1.25x + 100 \quad \text{Subtract 400 from both sides.}$$

$$0.25x = 100 \quad \text{Subtract } 1.25x \text{ from both sides.}$$

$$x = 400 \quad \text{Divide both sides by 0.25.}$$

STATE THE CONCLUSION   The break point is 400 coathangers.

CHECK THE RESULT To make 400 coathangers, the cost on the current machine would be $400 + $1.50(400) = $400 + $600 = $1,000$. The cost using the new machine would be $500 + $1.25(400) = $500 + $500 = $1,000$. Since the costs are equal, the break point is 400. ∎

STUDY SET

Section 3.2

VOCABULARY

In Exercises 1–4, fill in the blanks to make the statements true.

1. $Ax + By = C$ is the _____ form of a linear equation.

2. In the equation $x + 3y = -1$, the x-term has an understood _____ of 1.

3. When we add the two equations of the system $\begin{cases} x + y = 5 \\ x - y = -3 \end{cases}$, the y-terms are _____.

4. To solve $\begin{cases} y = 3x \\ x + y = 4 \end{cases}$, we can _____ $3x$ for y in the second equation.

CONCEPTS

5. If the system $\begin{cases} 4x - 3y = 7 \\ 3x - 2y = 6 \end{cases}$ is to be solved using the addition method, by what constant should each equation be multiplied if
   a. the x-terms are to drop out?

   b. the y-terms are to drop out?

6. If the system $\begin{cases} 4x - 3y = 7 \\ 3x + y = 6 \end{cases}$ is to be solved using the substitution method, what variable in what equation would it be easier to solve for?

7. Can the system $\begin{cases} 2x + 5y = 7 \\ 4x - 3y = 16 \end{cases}$ be solved more easily by the substitution or the addition method?

8. Given: the equation $3x + y = -4$.
   a. Solve for x.
   b. Solve for y.
   c. Which variable was easier to solve for? Explain why.

9. The substitution method was used to solve three systems of linear equations. The results after y was eliminated and the remaining equation was solved for x are listed below. Match each result with a possible graph of the system from Illustration 1.
   a. $-2 = 3$ b. $x = 3$ c. $3 = 3$

ILLUSTRATION 1

Possible graphs

10. Consider the system $\begin{cases} \frac{2}{3}x - \frac{y}{6} = \frac{16}{9} \\ 0.03x + 0.02y = 0.03 \end{cases}$.
    a. What algebraic step should be performed to clear the first equation of fractions?

    b. What algebraic step should be performed to clear the second equation of decimals?

PRACTICE

In Exercises 11–18, solve each system by substitution, if possible. If a system is inconsistent or if the equations are dependent, so indicate.

11. $\begin{cases} y = x \\ x + y = 4 \end{cases}$ 12. $\begin{cases} y = x + 2 \\ x + 2y = 16 \end{cases}$

13. $\begin{cases} x = 2 + y \\ 2x + y = 13 \end{cases}$ 14. $\begin{cases} x = -4 + y \\ 3x - 2y = -5 \end{cases}$

15. $\begin{cases} x + 2y = 6 \\ 3x - y = -10 \end{cases}$ 16. $\begin{cases} 2x - y = -21 \\ 4x + 5y = 7 \end{cases}$

17. $\begin{cases} \frac{3}{2}x + 2 = y \\ 0.6x - 0.4y = -0.4 \end{cases}$ 18. $\begin{cases} 2x - \frac{5}{2} = y \\ 0.04x - 0.02y = 0.05 \end{cases}$

In Exercises 19–26, solve each system by addition, if possible. If a system is inconsistent or if the equations are dependent, so indicate.

19. $\begin{cases} x - y = 3 \\ x + y = 7 \end{cases}$ 20. $\begin{cases} x + y = 1 \\ x - y = 7 \end{cases}$

21. $\begin{cases} 2x + y = -10 \\ 2x - y = -6 \end{cases}$ 22. $\begin{cases} x + 2y = -9 \\ x - 2y = -1 \end{cases}$

23. $\begin{cases} 2x + 3y = 8 \\ 3x - 2y = -1 \end{cases}$

24. $\begin{cases} 5x - 2y = 19 \\ 3x + 4y = 1 \end{cases}$

25. $\begin{cases} 4(x - 2) = -9y \\ 2(x - 3y) = -3 \end{cases}$

26. $\begin{cases} 2(2x + 3y) = 5 \\ 8x = 3(1 + 3y) \end{cases}$

In Exercises 27–42, solve each system by any method, if possible. If a system is inconsistent or if the equations are dependent, so indicate.

27. $\begin{cases} 3x - 4y = 9 \\ x + 2y = 8 \end{cases}$

28. $\begin{cases} 3x - 2y = -10 \\ 6x + 5y = 25 \end{cases}$

29. $\begin{cases} 2(x + y) + 1 = 0 \\ 3x + 4y = 0 \end{cases}$

30. $\begin{cases} 5x + 3y = -7 \\ 3(x - y) - 7 = 0 \end{cases}$

31. $\begin{cases} 0.16x - 0.08y = 0.32 \\ 2x - 4 = y \end{cases}$

32. $\begin{cases} 0.6y - 0.9x = -3.9 \\ 3x - 17 = 4y \end{cases}$

33. $\begin{cases} x = \frac{3}{2}y + 5 \\ 2x - 3y = 8 \end{cases}$

34. $\begin{cases} x = \frac{2}{3}y \\ y = 4x + 5 \end{cases}$

35. $\begin{cases} 0.5x + 0.5y = 6 \\ \frac{x}{2} - \frac{y}{2} = -2 \end{cases}$

36. $\begin{cases} \frac{x}{2} - \frac{y}{3} = -4 \\ \frac{x}{2} + \frac{y}{9} = 0 \end{cases}$

37. $\begin{cases} \frac{3}{4}x + \frac{2}{3}y = 7 \\ \frac{3}{5}x - \frac{1}{2}y = 18 \end{cases}$

38. $\begin{cases} \frac{2}{3}x - \frac{1}{4}y = -8 \\ 0.5x - 0.375y = -9 \end{cases}$

39. $\begin{cases} \frac{3x}{2} - \frac{2y}{3} = 0 \\ \frac{3x}{4} + \frac{4y}{3} = \frac{5}{2} \end{cases}$

40. $\begin{cases} \frac{3x}{5} + \frac{5y}{3} = 2 \\ \frac{6x}{5} - \frac{5y}{3} = 1 \end{cases}$

41. $\begin{cases} 12x - 5y - 21 = 0 \\ \frac{3}{4}x - \frac{2}{3}y = \frac{19}{8} \end{cases}$

42. $\begin{cases} 4y + 5x - 7 = 0 \\ \frac{10}{7}x - \frac{4}{9}y = \frac{17}{21} \end{cases}$

In Exercises 43–46, the systems involve variables other than x and y. When writing the solution as an ordered pair, write the values for the variables in alphabetical order.

43. $\begin{cases} \frac{3}{2}p + \frac{1}{3}q = 2 \\ \frac{2}{3}p + \frac{1}{9}q = 1 \end{cases}$

44. $\begin{cases} a + \frac{b}{3} = \frac{5}{3} \\ \frac{a + b}{3} = 3 - a \end{cases}$

45. $\begin{cases} \frac{m - n}{5} + \frac{m + n}{2} = 6 \\ \frac{m - n}{2} - \frac{m + n}{4} = 3 \end{cases}$

46. $\begin{cases} \frac{r - 2}{5} + \frac{s + 3}{2} = 5 \\ \frac{r + 3}{2} + \frac{s - 2}{3} = 6 \end{cases}$

In Exercises 47–50, solve each system. To do this, substitute a for $\frac{1}{x}$ and b for $\frac{1}{y}$ and solve for a and b. Then find x and y using the fact that $a = \frac{1}{x}$ and $b = \frac{1}{y}$.

47. $\begin{cases} \frac{1}{x} + \frac{1}{y} = \frac{5}{6} \\ \frac{1}{x} - \frac{1}{y} = \frac{1}{6} \end{cases}$

48. $\begin{cases} \frac{1}{x} + \frac{1}{y} = \frac{9}{20} \\ \frac{1}{x} - \frac{1}{y} = \frac{1}{20} \end{cases}$

49. $\begin{cases} \frac{1}{x} + \frac{2}{y} = -1 \\ \frac{2}{x} - \frac{1}{y} = -7 \end{cases}$

50. $\begin{cases} \frac{3}{x} - \frac{2}{y} = -30 \\ \frac{2}{x} - \frac{3}{y} = -30 \end{cases}$

APPLICATIONS

Use two variables to solve each problem.

51. ADVERTISING Use the information in the fee schedule shown in Illustration 2 to find the cost of a 15-second and a 30-second radio commercial on radio station KLIZ.

ILLUSTRATION 2

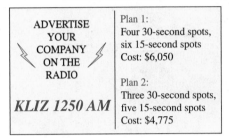

ADVERTISE YOUR COMPANY ON THE RADIO

KLIZ 1250 AM

Plan 1:
Four 30-second spots, six 15-second spots
Cost: $6,050

Plan 2:
Three 30-second spots, five 15-second spots
Cost: $4,775

52. TEMPORARY HELP A law firm had to hire several workers to help finish a large project. From the billing records shown in Illustration 3, determine the daily fee charged by the employment agency for a clerk-typist and for a computer programmer.

ILLUSTRATION 3

TEMPORARY EMPLOYMENT, INC.
We meet your employment needs!

Billed to: __Archer Law Offices__ Attn: __B. Kinsell__

Day	Posistion/Employee Name	Total cost
Mon. 3/22	*Clerk-typists:* K. Amad, B. Tran, S. Smith *Programmers:* T. Lee, C. Knox	$685
Tues. 3/23	*Clerk-typists:* K. Amad, B. Tran, S. Smith, W. Morada *Programmers:* T. Lee, C. Knox, B. Morales	$975

53. PETS According to the Pet Industry Joint Advisory Council, as of 1998, there were an estimated 124 million dogs and cats in the United States. If there were 8 million more cats than dogs, how many of each type of pet were there in 1998?

54. ELECTRONICS In Illustration 4 on the next page, two resistors in the voltage divider circuit have a total resistance of 1,375 ohms. To provide the required voltage, R_1 must be 125 ohms greater than R_2. Find both resistances.

ILLUSTRATION 4

ILLUSTRATION 7

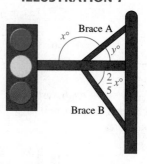

55. FENCING A FIELD The rectangular field in Illustration 5 is surrounded by 72 meters of fencing. If the field is partitioned as shown, a total of 88 meters of fencing is required. Find the dimensions of the field.

ILLUSTRATION 5

56. GEOMETRY In a right triangle, one acute angle is 15° greater than two times the other acute angle. Find the difference between the measures of the angles.

57. BRACING The bracing of a basketball backboard shown in Illustration 6 forms a parallelogram. Find the values of x and y.

ILLUSTRATION 6

58. TRAFFIC SIGNAL In Illustration 7, brace A and brace B are perpendicular. Find the values of x and y.

59. INVESTMENT CLUB Part of $8,000 was invested by an investment club at 10% interest and the rest at 12%. If the annual income from these investments is $900, how much was invested at each rate?

60. RETIREMENT INCOME A retired couple invested part of $12,000 at 6% interest and the rest at 7.5%. If their annual income from these investments is $810, how much was invested at each rate?

61. TV NEWS A news van and a helicopter left a TV station parking lot at the same time, headed in opposite directions to cover breaking news stories that were 145 miles apart. If the helicopter had to travel 55 miles farther than the van, how far did the van have to travel to reach the location of the news story?

62. DELIVERY SERVICE A delivery truck travels 50 miles in the same time that a cargo plane travels 180 miles. The speed of the plane is 143 mph faster than the speed of the truck. Find the speed of the delivery truck.

63. PRODUCTION PLANNING A bicycle manufacturer builds racing bikes and mountain bikes, with the per-unit manufacturing costs shown in Illustration 8. The company has budgeted $15,900 for labor and $13,075 for materials. How many bicycles of each type can be built?

ILLUSTRATION 8

Model	Cost of materials	Cost of labor
Racing	$55	$60
Mountain	$70	$90

64. FARMING A farmer keeps some animals on a strict diet. Each animal is to recieve 15 grams of protein and 7.5 grams of carbohydrates. The farmer uses two food mixes, with nutrients as shown in Illustration 9. How many grams of each mix should be used to provide the correct nutrients for each animal?

ILLUSTRATION 9

Mix	Protein	Carbohydrates
Mix A	12%	9%
Mix B	15%	5%

65. RECORDING COMPANY Three people invest a total of $105,000 to start a recording company that will produce reissues of classic jazz. Each release will be a set of 3 CDs that will retail for $45 per set. If each set can be produced for $18.95, how many sets must be sold for the investors to make a profit?

66. MACHINE SHOP Two machines can mill a brass plate. One machine has a setup cost of $300 and a cost per plate of $2. The other machine has a setup cost of $500 and a cost per plate of $1. Find the break point.

67. PUBLISHING A printer has two presses. One has a setup cost of $210 and can print the pages of a certain book for $5.98. The other press has a setup cost of $350 and can print the pages of the same book for $5.95. Find the break point.

68. MIXING CANDY How many pounds of each candy shown in Illustration 10 must be mixed to obtain 60 pounds of candy that would be worth $3 per pound?

ILLUSTRATION 10

69. COSMETOLOGY A beauty shop specializing in permanents has fixed costs of $2,101.20 per month. The owner estimates that the cost for each permanent is $23.60, which covers labor, chemicals, and electricity. If her shop can give as many permanents as she wants at a price of $44 each, how many must be given each month to break even?

70. PRODUCTION PLANNING A paint manufacturer can choose between two processes for manufacturing house paint, with monthly costs as shown in Illustration 11. Assume that the paint sells for $18 per gallon.

ILLUSTRATION 11

Process	Fixed costs	Unit cost (per gallon)
A	$32,500	$13
B	$80,600	$5

a. Find the break point for process A.

b. Find the break point for process B.

c. If expected sales are 7,000 gallons per month, which process should the company use?

71. DERMATOLOGY Tests of an antibacterial face-wash cream showed that a mixture containing 0.3% Triclosan (active ingredient) gave the best results. How many grams of cream from each tube shown in Illustration 12 should be used to make an equal-size tube of the 0.3% cream?

ILLUSTRATION 12

72. MIXING SOLUTIONS How many ounces of the two alcohol solutions in Illustration 13 must be mixed to obtain 100 ounces of a 12.2% solution?

ILLUSTRATION 13

73. RETOOLING A manufacturer of automobile water pumps is considering retooling for one of two manufacturing processes, with monthly fixed costs and unit costs as indicated in Illustration 14. Each water pump can be sold for $50.

ILLUSTRATION 14

Process	Fixed costs	Unit cost
A	$12,390	$29
B	$20,460	$17

a. Find the break point for process A.

b. Find the break point for process B.

c. If expected sales are 550 per month, which process should be used?

74. SALARY OPTIONS A sales clerk can choose from two salary plans: a straight 7% commision, or $150 + 2% commission. How much would the clerk have to sell for each plan to produce the same monthly paycheck?

WRITING

75. Which method would you use to solve the system
$\begin{cases} 4x + 6y = 5 \\ 8x - 3y = 3 \end{cases}$? Explain why.

76. Which method would you use to solve the system
$\begin{cases} x - 2y = 2 \\ 2x + 3y = 11 \end{cases}$? Explain why.

77. When solving a problem using two variables, why must we write two equations?

78. Write a problem that can be solved by solving the system $\begin{cases} x + y = 36 \\ \$1.29x + \$2.29y = \$72.44 \end{cases}$

79. Write a problem to fit the information given in Illustration 15.

ILLUSTRATION 15

Solution	% insecticide	Ounces	Amount of insecticide
2% solution	0.02	x	0.02x
10% solution	0.10	y	0.10y
7% mixture	0.07	80	0.07(80)

80. Make a table like the one in Illustration 16, listing one advantage and one disadvantage for each of the methods that can be used to solve a system of two linear equations.

ILLUSTRATION 16

Method	Advantage	Disadvantage
Graphing		
Substitution		
Addition		

REVIEW

In Exercises 81–86, find the slope of each line.

81.

82.

83. The line passing through $(0, -8)$ and $(-5, 0)$

84. The line with equation $y = -3x + 4$

85. The line with equation $4x - 3y = -3$

86. The line with equation $y = 3$

▶ **3.3**

Systems with Three Variables

In this section, you will learn about

    Solving three equations with three variables ■ Consistent systems
    ■ An inconsistent system ■ Systems with dependent equations
    ■ Problem solving ■ Curve fitting

Introduction In the two preceding sections, we solved systems of two linear equations with two variables. In this section, we will solve systems of linear equations with three variables by using a combination of the addition method and the substitution method. We will then use that procedure to solve problems involving three variables.

Solving Three Equations with Three Variables

We now extend the definition of a linear equation to include equations of the form $Ax + By + Cz = D$. The solution of a system of three linear equations with three variables is an **ordered triple** of numbers. For example, the solution of the system

$$\begin{cases} 2x + 3y + 4z = 20 \\ 3x + 4y + 2z = 17 \\ 3x + 2y + 3z = 16 \end{cases}$$

is the triple $(1, 2, 3)$, since each equation is satisfied if $x = 1$, $y = 2$, and $z = 3$.

$2x + 3y + 4z = 20$	$3x + 4y + 2z = 17$	$3x + 2y + 3z = 16$
$2(1) + 3(2) + 4(3) = 20$	$3(1) + 4(2) + 2(3) = 17$	$3(1) + 2(2) + 3(3) = 16$
$2 + 6 + 12 = 20$	$3 + 8 + 6 = 17$	$3 + 4 + 9 = 16$
$20 = 20$	$17 = 17$	$16 = 16$

The graph of an equation of the form $Ax + By + Cz = D$ is a flat surface called a *plane*. A system of three linear equations with three variables is consistent or inconsistent, depending on how the three planes corresponding to the three equations intersect. Figure 3-10 illustrates some of the possibilities.

FIGURE 3-10

Consistent system   Consistent system   Inconsistent systems

The three planes intersect at a single point P: One solution

The three planes have a line l in common: Infinitely many solutions

The three planes have no point in common: No solutions

(a)   (b)   (c)

To solve a system of three linear equations with three variables, we follow these steps.

Solving three equations with three variables

1. Pick any two equations and eliminate a variable.
2. Pick a different pair of equations and eliminate the same variable as in step 1.
3. Solve the resulting pair of two equations in two variables.
4. To find the value of the third variable, substitute the values of the two variables found in step 3 into any equation containing all three variables and solve the equation.
5. Check the solution in all three of the original equations.

Consistent Systems

| EXAMPLE 1 |

Solving a system of three equations. Solve the system
$$\begin{cases} 2x + y + 4z = 12 \\ x + 2y + 2z = 9 \\ 3x - 3y - 2z = 1 \end{cases}.$$

Solution   **Step 1:** We are given the system

1. $\begin{cases} 2x + y + 4z = 12 \end{cases}$
2. $\begin{cases} x + 2y + 2z = 9 \end{cases}$ To clarify the solution process, we number each equation.
3. $\begin{cases} 3x - 3y - 2z = 1 \end{cases}$

If we pick Equations 2 and 3 and add them, the variable z is eliminated.

2. $\quad x + 2y + 2z = \ 9$
3. $\underline{\quad 3x - 3y - 2z = \ 1}$
4. $\quad 4x - \ y \qquad = 10$ This equation does not contain z.

Step 2: We now pick a different pair of equations (Equations 1 and 3) and eliminate z again. If each side of Equation 3 is multiplied by 2 and the resulting equation is added to Equation 1, z is eliminated.

1. $\quad 2x + \ y + 4z = 12$
$\underline{\quad 6x - 6y - 4z = \ 2}$ Multiply both sides of Equation 3 by 2.
5. $\quad 8x - 5y \qquad = 14$ This equation does not contain z.

Step 3: Equations 4 and 5 form a system of two equations with two variables, x and y.

4. $\begin{cases} 4x - y = 10 \end{cases}$
5. $\begin{cases} 8x - 5y = 14 \end{cases}$

To solve this system, we multiply Equation 4 by -5 and add the resulting equation to Equation 5 to eliminate y:

$$\quad -20x + 5y = -50 \quad \text{Multiply both sides of Equation 4 by } -5.$$
5. $\underline{\quad 8x - 5y = \ \ 14}$
$$\quad -12x \qquad = -36$$
$$\qquad x = 3 \qquad \text{To find } x, \text{ divide both sides by } -12.$$

To find y, we substitute 3 for x in any equation containing x and y (such as Equation 5) and solve for y:

5. $\quad 8x - 5y = 14$
$\quad 8(3) - 5y = 14$ Substitute 3 for x.
$\quad 24 - 5y = 14$ Simplify.
$\quad -5y = -10$ Subtract 24 from both sides.
$\quad y = 2$ Divide both sides by -5.

Step 4: To find z, we substitute 3 for x and 2 for y in an equation containing x, y, and z (such as Equation 1) and solve for z:

1. $\quad 2x + y + 4z = 12$
$\quad 2(3) + 2 + 4z = 12$ Substitute 3 for x and 2 for y.
$\quad 8 + 4z = 12$ Simplify.
$\quad 4z = 4$ Subtract 8 from both sides
$\quad z = 1$ Divide both sides by 4.

The solution of the system is $(x, y, z) = (3, 2, 1)$. Because this system has a solution, it is a consistent system.

Step 5: Verify that these values satisfy each equation in the original system.

SELF CHECK   Solve $\begin{cases} 2x + y + 4z = 16 \\ x + 2y + 2z = 11 \\ 3x - 3y - 2z = -9 \end{cases}$.

Answer: $(1, 2, 3)$ ∎

When one or more of the equations of a system is missing a term, the elimination of a variable that is normally performed in step 1 of the solution process can be skipped.

E X A M P L E   2

Missing terms. Solve the system $\begin{cases} 3x = 6 - 2y + z \\ -y - 2z = -8 - x \\ x = 1 - 2z \end{cases}$.

Solution   **Step 1:** First, we write each equation in $Ax + By + Cz = D$ form.

1. $\begin{cases} 3x + 2y - z = 6 \\ x - y - 2z = -8 \\ x + 2z = 1 \end{cases}$
2.
3.

Since Equation 3 does not have a y-term, we can proceed to step 2, where we will find another equation that does not contain a y-term.

Step 2: If each side of Equation 2 is multiplied by 2 and the resulting equation is added to Equation 1, y is eliminated.

1. $3x + 2y - \ z = \quad 6$
   $\underline{2x - 2y - 4z = -16}$ Multiply both sides of Equation 2 by 2.
4. $5x \qquad - 5z = -10$

Step 3: Equations 3 and 4 form a system of two equations with two variables, x and z:

3. $\begin{cases} x + 2z = 1 \\ 5x - 5z = -10 \end{cases}$
4.

To solve this system, we multiply Equation 3 by -5 and add the resulting equation to Equation 4 to eliminate x:

$\quad -5x - 10z = \ -5$ Multiply both sides of Equation 3 by -5.
4. $\underline{\quad 5x - \ 5z = -10}$
$\qquad\qquad -15z = -15$
$\qquad\qquad\quad z = 1$ To find z, divide both sides by -15.

To find x, we substitute 1 for z in Equation 3.

3. $\quad x + 2z = 1$
   $\quad x + 2(1) = 1$ Substitute 1 for z.
   $\qquad x + 2 = 1$ Multiply.
   $\qquad\quad x = -1$ Subtract 2 from both sides.

Step 4: To find y, we substitute -1 for x and 1 for z in Equation 1:

1. $\qquad 3x + 2y - z = 6$
   $3(-1) + 2y - 1 = 6$ Substitute -1 for x and 1 for z.
   $\quad -3 + 2y - 1 = 6$ Multiply.
   $\qquad\qquad 2y = 10$ Add 4 to both sides.
   $\qquad\qquad\ y = 5$ Divide both sides by 2.

The solution of the system is $(-1, 5, 1)$.

Step 5: Check the solution in all three of the original equations.

SELF CHECK   Solve $\begin{cases} x + 2y - z = 1 \\ 2x - y + z = 3 \\ x + z = 3 \end{cases}$.

Answer: (1, 1, 2)   ■

An Inconsistent System

EXAMPLE 3   **A system with no solution.** Solve the system $\begin{cases} 2a + b - 3c = -3 \\ 3a - 2b + 4c = 2 \\ 4a + 2b - 6c = -7 \end{cases}$.

Solution   We can multiply the first equation of the system by 2 and add the resulting equation to the second equation to eliminate *b*:

$$4a + 2b - 6c = -6 \quad \text{Multiply both sides of the first equation by 2.}$$
$$\underline{3a - 2b + 4c = 2}$$
1. $7a - 2c = -4$

Now add the second and third equations of the system to eliminate *b* again:

$$3a - 2b + 4c = 2$$
$$\underline{4a + 2b - 6c = -7}$$
2. $7a - 2c = -5$

Equations 1 and 2 form the system

1. $\begin{cases} 7a - 2c = -4 \\ 7a - 2c = -5 \end{cases}$
2.

Since $7a - 2c$ cannot equal both -4 and -5, the system is inconsistent and has no solution.

SELF CHECK   Solve $\begin{cases} 2a + b - 3c = 8 \\ 3a - 2b + 4c = 10 \\ 4a + 2b - 6c = -5 \end{cases}$.

Answer: no solution   ■

Systems with Dependent Equations

When the equations in a system of two equations with two variables are dependent, the system has infinitely many solutions. This is not always true for systems of three equations with three variables. In fact, a system can have dependent equations and still be inconsistent. Figure 3-11 illustrates the different possibilities.

FIGURE 3-11

Consistent system

Consistent system

Inconsistent system

When three planes coincide, the equations are dependent, and there are infinitely many solutions.

When three planes intersect in a common line, the equations are dependent, and there are infinitely many solutions.

When two planes coincide and are parallel to a third plane, the system is inconsistent, and there are no solutions.

(a)

(b)

(c)

EXAMPLE 4

A system with infinitely many solutions. Solve the system

$$\begin{cases} 3x - 2y + z = -1 \\ 2x + y - z = 5 \\ 5x - y = 4 \end{cases}.$$

Solution   We can add the first two equations to get

$$3x - 2y + z = -1$$
$$\underline{2x + \; y - z = \; \; 5}$$
1. $5x - \; y \qquad = \; \; 4$

Since Equation 1 is the same as the third equation of the system, the equations of the system are dependent, and there will be infinitely many solutions. From a graphical perspective, the equations represent three planes that intersect in a common line, as shown in Figure 3-11(b).

To write the general solution of this system, we can solve Equation 1 for y to get

$$5x - y = 4$$
$$-y = -5x + 4 \quad \text{Subtract } 5x \text{ from both sides.}$$
$$y = 5x - 4 \quad \text{Multiply both sides by } -1.$$

We can then substitute $5x - 4$ for y in the first equation of the system and solve for z to get

$$3x - 2y + z = -1$$
$$3x - 2(5x - 4) + z = -1 \quad \text{Substitute } 5x - 4 \text{ for } y.$$
$$3x - 10x + 8 + z = -1 \quad \text{Use the distributive property to remove parentheses.}$$
$$-7x + 8 + z = -1 \quad \text{Combine like terms.}$$
$$z = 7x - 9 \quad \text{Add } 7x \text{ and } -8 \text{ to both sides.}$$

Since we have found the values of y and z in terms of x, every solution of the system has the form $(x, 5x - 4, 7x - 9)$, where x can be any real number. For example,

If $x = 1$, a solution is $(1, 1, -2)$. $5(1) - 4 = 1$, and $7(1) - 9 = -2$.
If $x = 2$, a solution is $(2, 6, 5)$. $5(2) - 4 = 6$, and $7(2) - 9 = 5$.
If $x = 3$, a solution is $(3, 11, 12)$. $5(3) - 4 = 11$, and $7(3) - 9 = 12$.

SELF CHECK   Solve $\begin{cases} 3x + 2y + z = -1 \\ 2x - y - z = 5 \\ 5x + y = 4 \end{cases}$.   *Answer:* There are infinitely many solutions. A general solution is $(x, 4 - 5x, -9 + 7x)$. Three solutions are $(1, -1, -2)$, $(2, -6, 5)$, and $(3, -11, 12)$.   ∎

Problem Solving

EXAMPLE 5

Tool manufacturing. A tool company makes three types of hammers, which are marketed as "good," "better," and "best." The cost of manufacturing each type of hammer is $4, $6, and $7, respectively, and the hammers sell for $6, $9, and $12. Each day, the cost of manufacturing 100 hammers is $520, and the daily revenue from their sale is $810. How many hammers of each type are manufactured?

ANALYZE THE PROBLEM   We need to find how many of each type of hammer are manufactured daily. We must write three equations to find three unknowns.

FORM THREE EQUATIONS   If we let x represent the number of good hammers, y represent the number of better hammers, and z represent the number of best hammers, we know that

The total number of hammers is $x + y + z$.

The cost of manufacturing the good hammers is $4x$ ($4 times x hammers).

The cost of manufacturing the better hammers is $6y ($6 times y hammers).

The cost of manufacturing the best hammers is $7z ($7 times z hammers).

The revenue received by selling the good hammers is $6x ($6 times x hammers).

The revenue received by selling the better hammers is $9y ($9 times y hammers).

The revenue received by selling the best hammers is $12z ($12 times z hammers).

We can assemble the facts of the problem to write three equations.

The number of good hammers	plus	the number of better hammers	plus	the number of best hammers	is	the total number of hammers.
x	$+$	y	$+$	z	$=$	100

The cost of good hammers	plus	the cost of better hammers	plus	the cost of best hammers	is	the total cost.
$4x$	$+$	$6y$	$+$	$7z$	$=$	520

The revenue from good hammers	plus	the revenue from better hammers	plus	the revenue from best hammers	is	the total revenue.
$6x$	$+$	$9y$	$+$	$12z$	$=$	810

SOLVE THE SYSTEM   We must now solve the system

$$\textbf{1.} \quad \begin{cases} x + y + z = 100 \\ \textbf{2.} \quad 4x + 6y + 7z = 520 \\ \textbf{3.} \quad 6x + 9y + 12z = 810 \end{cases}$$

If we multiply Equation 1 by -7 and add the result to Equation 2, we get

$$-7x - 7y - 7z = -700$$
$$\underline{\quad 4x + 6y + 7z = \quad 520}$$
$$\textbf{4.} \quad -3x - y \qquad = -180$$

If we multiply Equation 1 by -12 and add the result to Equation 3, we get

$$-12x - 12y - 12z = -1{,}200$$
$$\underline{\quad 6x + \quad 9y + 12z = \qquad 810}$$
$$\textbf{5.} \quad -6x - \quad 3y \qquad = \quad -390$$

If we multiply Equation 4 by -3 and add it to Equation 5, we get

$$9x + 3y = \quad 540$$
$$\underline{-6x - 3y = -390}$$
$$3x \qquad = \quad 150$$
$$\quad x = 50 \qquad \text{To find } x, \text{ divide both sides by 3.}$$

To find y, we substitute 50 for x in Equation 4:

$$-3x - y = -180$$
$$-3(\textbf{50}) - y = -180 \quad \text{Substitute 50 for } x.$$
$$-150 - y = -180 \quad -3(50) = -150.$$
$$-y = -30 \quad \text{Add 150 to both sides.}$$
$$y = 30 \quad \text{Divide both sides by } -1.$$

To find z, we substitute 50 for x and 30 for y in Equation 1:

$$x + y + z = 100$$
$$50 + 30 + z = 100$$
$$z = 20 \quad \text{Subtract 80 from both sides.}$$

STATE THE CONCLUSION   The company manufactures 50 good hammers, 30 better hammers, and 20 best hammers each day.

CHECK THE RESULT   Check the solution in each equation in the original system.   ■

Curve Fitting

EXAMPLE 6

Finding the equation of a parabola. The equation of a parabola opening upward or downward is of the form $y = ax^2 + bx + c$. Find the equation of the parabola shown in Figure 3-12 by determining the values of a, b, and c.

FIGURE 3-12

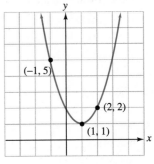

Solution   Since the parabola passes through the points $(-1, 5)$, $(1, 1)$, and $(2, 2)$, each pair of coordinates must satisfy the equation $y = ax^2 + bx + c$. If we substitute the x- and y-coordinates of each point into the equation and simplify, we obtain the following system of three equations wih three variables.

1. $\begin{cases} a - b + c = 5 & \text{Substitute the coordinates of } (-1, 5) \text{ into } y = ax^2 + bx + c \text{ and simplify.} \\ a + b + c = 1 & \text{Substitute the coordinates of } (1, 1) \text{ into } y = ax^2 + bx + c \text{ and simplify.} \\ 4a + 2b + c = 2 & \text{Substitute the coordinates of } (2, 2) \text{ into } y = ax^2 + bx + c \text{ and simplify.} \end{cases}$
2.
3.

If we add Equations 1 and 2, we obtain

$$a - b + c = 5$$
$$\underline{a + b + c = 1}$$
4. $2a \quad\quad + 2c = 6$

If we multiply Equation 1 by 2 and add the result to Equation 3, we get

$$2a - 2b + 2c = 10$$
$$\underline{4a + 2b + c = 2}$$
5. $6a \quad\quad + 3c = 12$

We can then divide both sides of Equation 4 by 2 to get Equation 6 and divide both sides of Equation 5 by 3 to get Equation 7. We now have the system

6. $\begin{cases} a + c = 3 \\ 2a + c = 4 \end{cases}$
7.

To eliminate c, we multiply Equation 6 by -1 and add the result to Equation 7. We get

$$-a - c = -3$$
$$\underline{2a + c = 4}$$
$$a \quad\quad = 1$$

To find c, we can substitute 1 for a in Equation 6 and find that $c = 2$. To find b, we can substitute 1 for a and 2 for c in Equation 2 and find that $b = -2$.

After we substitute these values of a, b, and c into the equation $y = ax^2 + bx + c$, we have the equation of the parabola.

$$y = ax^2 + bx + c$$
$$y = 1x^2 - 2x + 2$$
$$y = x^2 - 2x + 2$$

■

VOCABULARY

In Exercises 1–6, fill in the blanks to make the statements true.

1. $\begin{cases} 2x + y - 3z = 0 \\ 3x - y + 4z = 5 \\ 4x + 2y - 6z = 0 \end{cases}$ is called a _____ of three linear equations.

2. If the first two equations of the system in Exercise 1 are added, the variable y is _____.

3. The equation $2x + 3y + 4z = 5$ is a linear equation with _____ variables.

4. The graph of the equation $2x + 3y + 4z = 5$ is a flat surface called a _____.

5. When three planes coincide, the equations of the system are _____, and there are infinitely many solutions.

6. When three planes intersect in a line, the system will have _____ many solutions.

CONCEPTS

7. For each graph of a system of three equations, tell whether the solution set contains one solution, infinitely many solutions, or no solution.

a.   b.

8. Consider the system $\begin{cases} -2x + y + 4z = 3 \\ x - y + 2z = 1 \\ x + y - 3z = 2 \end{cases}$.

   a. What is the result if Equation 1 and Equation 2 are added?
   b. What is the result if Equation 2 and Equation 3 are added?
   c. What variable was eliminated in the steps performed in parts a and b?

NOTATION

9. Write the equation $3z - 2y = x + 6$ in $Ax + By + Cz = D$ form.

10. Fill in the blank to make a true statement: Solutions of a system of three equations in three variables, x, y, and z, are written in the form (x, y, z) and are called ordered _____.

PRACTICE

In Exercises 11–12, tell whether the given ordered triple is a solution of given system.

11. $(2, 1, 1)$, $\begin{cases} x - y + z = 2 \\ 2x + y - z = 4 \\ 2x - 3y + z = 2 \end{cases}$

12. $(-3, 2, -1)$, $\begin{cases} 2x + 2y + 3z = -1 \\ 3x + y - z = -6 \\ x + y + 2z = 1 \end{cases}$

In Exercises 13–28, solve each system. If a system is inconsistent or if the equations are dependent, so indicate.

13. $\begin{cases} x + y + z = 4 \\ 2x + y - z = 1 \\ 2x - 3y + z = 1 \end{cases}$

14. $\begin{cases} x + y + z = 4 \\ x - y + z = 2 \\ x - y - z = 0 \end{cases}$

15. $\begin{cases} 2x + 2y + 3z = 10 \\ 3x + y - z = 0 \\ x + y + 2z = 6 \end{cases}$

16. $\begin{cases} x - y + z = 4 \\ x + 2y - z = -1 \\ x + y - 3z = -2 \end{cases}$

17. $\begin{cases} b + 2c = 7 - a \\ a + c = 8 - 2b \\ 2a + b + c = 9 \end{cases}$

18. $\begin{cases} 2a = 2 - 3b - c \\ 4a + 6b + 2c - 5 = 0 \\ a + c = 3 + 2b \end{cases}$

19. $\begin{cases} 2x + y - z = 1 \\ x + 2y + 2z = 2 \\ 4x + 5y + 3z = 3 \end{cases}$

20. $\begin{cases} 4x + 3z = 4 \\ 2y - 6z = -1 \\ 8x + 4y + 3z = 9 \end{cases}$

21. $\begin{cases} a + b + c = 180 \\ \frac{a}{4} + \frac{b}{2} + \frac{c}{3} = 60 \\ 2b + 3c - 330 = 0 \end{cases}$

22. $\begin{cases} 2a + 3b - 2c = 18 \\ 5a - 6b + c = 21 \\ 4b - 2c - 6 = 0 \end{cases}$

23. $\begin{cases} 0.5a + 0.3b = 2.2 \\ 1.2c - 8.5b = -24.4 \\ 3.3c + 1.3a = 29 \end{cases}$

24. $\begin{cases} 4a - 3b = 1 \\ 6a - 8c = 1 \\ 2b - 4c = 0 \end{cases}$

25. $\begin{cases} 2x + 3y + 4z = 6 \\ 2x - 3y - 4z = -4 \\ 4x + 6y + 8z = 12 \end{cases}$

26. $\begin{cases} x - 3y + 4z = 2 \\ 2x + y + 2z = 3 \\ 4x - 5y + 10z = 7 \end{cases}$

27. $\begin{cases} x + \frac{1}{3}y + z = 13 \\ \frac{1}{2}x - y + \frac{1}{3}z = -2 \\ x + \frac{1}{2}y - \frac{1}{3}z = 2 \end{cases}$ **28.** $\begin{cases} x - \frac{1}{5}y - z = 9 \\ \frac{1}{4}x + \frac{1}{5}y - \frac{1}{2}z = 5 \\ 2x + y + \frac{1}{6}z = 12 \end{cases}$

APPLICATIONS

29. MAKING STATUES An artist makes three types of ceramic statues at a monthly cost of $650 for 180 statues. The manufacturing costs for the three types are $5, $4, and $3. If the statues sell for $20, $12, and $9, respectively, how many of each type should be made to produce $2,100 in monthly revenue?

30. POTPOURRI The owner of a home decorating shop wants to mix dried rose petals selling for $6 per pound, dried lavender selling for $5 per pound, and buckwheat hulls selling for $4 per pound to get 10 pounds of a mixture that would sell for $5.50 per pound. She wants to use twice as many pounds of rose petals as lavender. How many pounds of each should she use?

31. DIETITIAN A hospital dietitian is to design a meal that will provide a patient with exactly 14 grams (g) of fat, 9 g of carbohydrates, and 9 g of protein. She is to use a combination of the three foods listed in the table. If one ounce of each of the foods has the nutrient content shown in Illustration 1, how many ounces of each should be used?

ILLUSTRATION 1

Food	Fat	Carbohydrates	Protein
A	2 g	1 g	2 g
B	3 g	2 g	1 g
C	1 g	1 g	2 g

32. NUTRITIONAL PLANNING One ounce of each of three foods has the vitamin and mineral content shown in Illustration 2. How many ounces of each must be used to provide exactly 22 milligrams (mg) of niacin, 12 mg of zinc, and 20 mg of vitamin C?

ILLUSTRATION 2

Food	Niacin	Zinc	Vitamin C
A	1 mg	1 mg	2 mg
B	2 mg	1 mg	1 mg
C	2 mg	1 mg	2 mg

33. CHAINSAW SCULPTING A north woods sculptor carves three types of statues with a chainsaw. The number of hours required for carving, sanding, and painting a totem pole, a bear, and a deer are shown in Illustration 3. How many of each should be produced to use all available labor hours?

ILLUSTRATION 3

	Totem pole	Bear	Deer	Time available
Carving	2 hr	2 hr	1 hr	14 hr
Sanding	1 hr	2 hr	2 hr	15 hr
Painting	3 hr	2 hr	2 hr	21 hr

34. CLOTHING MAKER A clothing manufacturer makes coats, shirts, and slacks. The time required for cutting, sewing, and packaging each item is shown in Illustration 4. How many of each should be made to use all available labor hours?

ILLUSTRATION 4

	Coats	Shirts	Slacks	Time available
Cutting	20 min	15 min	10 min	115 hr
Sewing	60 min	30 min	24 min	280 hr
Packaging	5 min	12 min	6 min	65 hr

35. MAIL The circle graph in Illustration 5 shows the types of mail the average American household receives each week. Use the information in the graph to determine what percent of the week's mail is advertising, is bills and statements, and is personal.

ILLUSTRATION 5

This is 10% more than twice the sum of the other two types of mail.

Advertisements

Bills and statements

Personal

This is 4% more than the personal mail.

Based on information from *Time* (July 14, 1997)

36. NFL RECORD As of the end of the 1998 National Football League season, Jerry Rice of the San Francisco 49ers had scored an all-time record 175 touchdowns. Here are some interestings facts about this feat.

- He had caught 29 more TD passes from Steve Young than from Joe Montana.

- He had caught 25 TD passes from quarterbacks other than Montana or Young.

- The number of TD passes he had caught from Joe Montana was five times the number of touchdowns he had scored on running plays.

Determine the number touchdown passes Rice had caught from Joe Montana and from Steve Young as of 1998.

37. GRAPHS OF SYSTEMS Explain how each picture in Illustration 6 could be thought of as an example of the graph of a system of three equations. Then describe the solution, if there is any.

ILLUSTRATION 6

a.

b.

c.

d.

38. ZOOLOGY An X-ray of a laboratory mouse revealed a cancerous tumor located at the intersection of the coronal, sagittal, and transverse planes, as shown in Illustration 7. From this description, would you expect the tumor to be at the base of the tail, on the back, in the stomach, on the tip of the right ear, or in the mouth of the mouse?

ILLUSTRATION 7

39. ASTRONOMY Comets have elliptical orbits, but the orbits of some comets are so vast that they are indistinguishable from parabolas. Find the equation of the parabola that describes the orbit of the comet shown in Illustration 8.

ILLUSTRATION 8

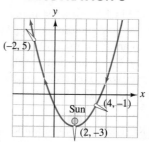

40. CURVE FITTING Find the equation of the parabola shown in Illustration 9.

ILLUSTRATION 9

41. PARK WALKWAY A circular sidewalk is to be constructed in a city park. The walk is to pass by three particular areas of the park, as shown in the graph in Illustration 10. If an equation of a circle is of the form $x^2 + y^2 + Cx + Dy + E = 0$, find the equation that describes the path of the sidewalk by determining C, D, and E.

ILLUSTRATION 10

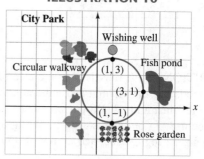

42. CURVE FITTING The equation of a circle is of the form $x^2 + y^2 + Cx + Dy + E = 0$. Find the equation of the circle shown in Illustration 11 by determining C, D, and E.

ILLUSTRATION 11

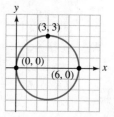

43. TRIANGLES The sum of the angles in any triangle is 180°. In triangle *ABC*, angle *A* is 100° less than the sum of angles *B* and *C*, and angle *C* is 40° less than twice angle *B*. Find each angle.

44. QUADRILATERALS The sum of the angles of any four-sided figure is 360°. In the quadrilateral shown in Illustration 12, the measures of angle *A* and angle *B* are the same, angle *C* is 20° greater than angle *A*, and angle *D* measures 40°. Find the angles.

ILLUSTRATION 12

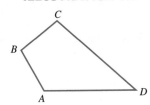

45. INTEGER PROBLEM The sum of three integers is 48. If the first integer is doubled, the sum is 60. If the second integer is doubled, the sum is 63. Find the integers.

46. INTEGER PROBLEM The sum of three integers is 18. The third integer is four times the second, and the second integer is 6 more than the first. Find the integers.

WRITING

47. Explain how a system of three equations with three variables can be reduced to a system of two equations with two variables.

48. What makes a system of three equations with three variables inconsistent?

REVIEW

In Exercises 49–52, graph each function.

49. $f(x) = |x|$

50. $g(x) = x^2$

51. $h(x) = x^3$

52. $S(x) = x$

▶ **3.4**

Solving Systems Using Matrices

In this section, you will learn about

> Matrices ■ Solving a system of two equations ■ Gaussian elimination ■ Solving a system of three equations ■ Inconsistent systems and dependent equations

Introduction   In this section, we will discuss another method for solving systems of linear equations. This technique uses a mathematical tool called a *matrix* in a series of steps that are based on the addition method.

Matrices

Another method of solving systems of equations involves rectangular arrays of numbers called *matrices*.

Matrices

A **matrix** is any rectangular array of numbers arranged in rows and columns, written within brackets.

Some examples of matrices are

$$A = \begin{bmatrix} 1 & -3 & 8 \\ 2 & 5 & -1 \end{bmatrix} \begin{matrix} \leftarrow \text{Row 1} \\ \leftarrow \text{Row 2} \end{matrix} \qquad B = \begin{bmatrix} 1 & 4 & -2 & -4 \\ 6 & -2 & 6 & 1 \\ 3 & 8 & -3 & 12 \end{bmatrix} \begin{matrix} \leftarrow \text{Row 1} \\ \leftarrow \text{Row 2} \\ \leftarrow \text{Row 3} \end{matrix}$$

$$\begin{matrix} \uparrow & \uparrow & \uparrow \\ \text{Column} & \text{Column} & \text{Column} \\ 1 & 2 & 3 \end{matrix} \qquad \begin{matrix} \uparrow & \uparrow & \uparrow & \uparrow \\ \text{Column} & \text{Column} & \text{Column} & \text{Column} \\ 1 & 2 & 3 & 4 \end{matrix}$$

The numbers in each matrix are called **elements.** Because matrix A has two rows and three columns, it is called a 2×3 matrix (read "2 by 3" matrix). Matrix B is a 3×4 matrix (three rows and four columns).

Solving a System of Two Equations

To show how to use matrices to solve systems of linear equations, we consider the system

$$\begin{cases} x - y = 4 \\ 2x + y = 5 \end{cases}$$

which can be represented by the following matrix, called an **augmented matrix:**

$$\begin{bmatrix} 1 & -1 & \vdots & 4 \\ 2 & 1 & \vdots & 5 \end{bmatrix}$$

Each row of the augmented matrix represents one equation of the system. The first two columns of the augmented matrix are determined by the coefficients of x and y in the equations of the system. The last column is determined by the constants in the equations.

$$\begin{bmatrix} 1 & -1 & \vdots & 4 \\ 2 & 1 & \vdots & 5 \end{bmatrix} \begin{matrix} \leftarrow \text{This row represents the equation } x - y = 4. \\ \leftarrow \text{This row represents the equation } 2x + y = 5. \end{matrix}$$

$$\begin{matrix} \uparrow & \uparrow & \uparrow \\ \text{Coefficients} & \text{Coefficients} & \text{Constants} \\ \text{of } x & \text{of } y & \end{matrix}$$

EXAMPLE 1

Augmented matrices. Represent each system of equations using an augmented matrix:

a. $\begin{cases} 3x + y = 11 \\ x - 8y = 0 \end{cases}$ and **b.** $\begin{cases} 2a + b - 3c = -3 \\ 9a + 4c = 2 \\ a - b - 6c = -7 \end{cases}$.

Solution **a.** $\begin{cases} 3x + y = 11 & \longleftrightarrow \\ x - 8y = 0 & \longleftrightarrow \end{cases} \begin{bmatrix} 3 & 1 & \vdots & 11 \\ 1 & -8 & \vdots & 0 \end{bmatrix}$

b. $\begin{cases} 2a + b - 3c = -3 & \longleftrightarrow \\ 9a + 4c = 2 & \longleftrightarrow \\ a - b - 6c = -7 & \longleftrightarrow \end{cases} \begin{bmatrix} 2 & 1 & -3 & \vdots & -3 \\ 9 & 0 & 4 & \vdots & 2 \\ 1 & -1 & -6 & \vdots & -7 \end{bmatrix}$

SELF CHECK

Represent each system using an augmented matrix: **a.** $\begin{cases} 2x - 4y = 9 \\ 5x - y = -2 \end{cases}$ and

b. $\begin{cases} a + b - c = -4 \\ -2b + 7c = 0 \\ 10a + 8b - 4c = 5 \end{cases}$.

Answers:

a. $\begin{bmatrix} 2 & -4 & \vdots & 9 \\ 5 & -1 & \vdots & -2 \end{bmatrix}$,

b. $\begin{bmatrix} 1 & 1 & -1 & \vdots & -4 \\ 0 & -2 & 7 & \vdots & 0 \\ 10 & 8 & -4 & \vdots & 5 \end{bmatrix}$

■

Gaussian Elimination

To solve a 2×2 system of equations by **Gaussian elimination,** we transform the augmented matrix into the following matrix, which has 1's down its main diagonal and a 0 below the 1 in the first column.

$$\begin{bmatrix} 1 & a & \vdots & b \\ 0 & 1 & \vdots & c \end{bmatrix} \qquad (a, b, \text{ and } c \text{ are real numbers})$$

Main diagonal

To write the augmented matrix in this form, we use three operations called **elementary row operations.**

Elementary row operations

Type 1: Any two rows of a matrix can be interchanged.
Type 2: Any row of a matrix can be multiplied by a nonzero constant.
Type 3: Any row of a matrix can be changed by adding a nonzero constant multiple of another row to it.

■ A type 1 row operation corresponds to interchanging two equations of the system.
■ A type 2 row operation corresponds to multiplying both sides of an equation by a nonzero constant.
■ A type 3 row operation corresponds to adding a nonzero multiple of one equation to another.

None of these row operations will change the solution of the given system of equations.

EXAMPLE 2

Row operations. Consider the matrices

$$A = \begin{bmatrix} 2 & 4 & \vdots & -3 \\ 1 & -8 & \vdots & 0 \end{bmatrix} \quad B = \begin{bmatrix} 1 & -1 & \vdots & 2 \\ 4 & -8 & \vdots & 0 \end{bmatrix} \quad C = \begin{bmatrix} 2 & 1 & -8 & \vdots & 4 \\ 0 & 1 & 4 & \vdots & -2 \\ 0 & 0 & -6 & \vdots & 24 \end{bmatrix}$$

a. Interchange rows 1 and 2 of matrix A.
b. Multiply row 3 of matrix C by $-\frac{1}{6}$.
c. To the numbers in row 2 of matrix B, add the results of multiplying each number in row 1 by -4.

Solution
a. Interchanging the rows of matrix A, we obtain $\begin{bmatrix} 1 & -8 & \vdots & 0 \\ 2 & 4 & \vdots & -3 \end{bmatrix}$.

b. We multiply each number in row 3 by $-\frac{1}{6}$. Rows 1 and 2 remain unchanged.

$$\begin{bmatrix} 2 & 1 & -8 & \vdots & 4 \\ 0 & 1 & 4 & \vdots & -2 \\ 0 & 0 & 1 & \vdots & -4 \end{bmatrix}$$
We can represent the instruction to multiply the third row by $-\frac{1}{6}$ with the symbolism $-\frac{1}{6}R_3$.

c. If we multiply each number in row 1 of matrix B by -4, we get

$$-4 \quad 4 \quad -8$$

We then add these numbers to row 2. (Note that row 1 remains unchanged.)

$$\begin{bmatrix} 1 & -1 & \vdots & 2 \\ 4 + (-4) & -8 + 4 & \vdots & 0 + (-8) \end{bmatrix}$$
We can abbreviate this procedure using the notation $-4R_1 + R_2$, which means "Multiply row 1 by -4 and add the result to row 2."

After simplifying, we have the matrix

$$\begin{bmatrix} 1 & -1 & \vdots & 2 \\ 0 & -4 & \vdots & -8 \end{bmatrix}$$

SELF CHECK

Refer to Example 2.
a. Interchange the rows of matrix B.
b. To the numbers in row 1 of matrix A, add the results of multiplying each number in row 2 by -2.
c. Interchange rows 2 and 3 of matrix C.

Answers: **a.** $\begin{bmatrix} 4 & -8 & \vdots & 0 \\ 1 & -1 & \vdots & 2 \end{bmatrix}$,

b. $\begin{bmatrix} 0 & 20 & \vdots & -3 \\ 1 & -8 & \vdots & 0 \end{bmatrix}$,

c. $\begin{bmatrix} 2 & 1 & -8 & \vdots & 4 \\ 0 & 0 & -6 & \vdots & 24 \\ 0 & 1 & 4 & \vdots & -2 \end{bmatrix}$ ∎

We now solve a system of two linear equations using the **Gaussian elimination** process, which involves a series of elementary row operations.

EXAMPLE 3

Solving a system using Gaussian elimination. Solve the system
$$\begin{cases} 2x + y = 5 \\ x - y = 4 \end{cases}.$$

Solution We can represent the system with the following augmented matrix:

$$\begin{bmatrix} 2 & 1 & \vdots & 5 \\ 1 & -1 & \vdots & 4 \end{bmatrix}$$

First, we want to get a 1 in the top row of the first column where the red 2 is. This can be achieved by applying a type 1 row operation: Interchange rows 1 and 2.

$$\begin{bmatrix} 1 & -1 & \vdots & 4 \\ 2 & 1 & \vdots & 5 \end{bmatrix}$$ Interchanging row 1 and row 2 can be abbreviated as $R_1 \leftrightarrow R_2$.

To get a 0 under the 1 in the first column, we use a type 3 row operation. To row 2, we add the results of multiplying each number in row 1 by -2.

$$\begin{bmatrix} 1 & -1 & \vdots & 4 \\ 0 & 3 & \vdots & -3 \end{bmatrix}$$ $-2R_1 + R_2$.

To get a 1 in the bottom row of the second column, we use a type 2 row operation: Multiply row 2 by $\frac{1}{3}$.

$$\begin{bmatrix} 1 & -1 & \vdots & 4 \\ 0 & 1 & \vdots & -1 \end{bmatrix}$$ $\frac{1}{3}R_2$.

This augmented matrix represents the equations

$$1x - 1y = 4$$
$$0x + 1y = -1$$

Writing the equations without the coefficients of 1 and -1, we have

1. $x - y = 4$

2. $y = -1$

From Equation 2, we see that $y = -1$. We can **back substitute** -1 for y in Equation 1 to find x.

$$x - y = 4$$
$$x - (-1) = 4 \quad \text{Substitute } -1 \text{ for } y.$$
$$x + 1 = 4 \quad -(-1) = 1.$$
$$x = 3 \quad \text{Subtract 1 from both sides.}$$

The solution of the system is $(3, -1)$. Verify that this ordered pair satisfies the original system.

SELF CHECK   Solve $\begin{cases} 3x - 2y = -5 \\ x - y = -4 \end{cases}$.   *Answer:* $(3, 7)$ ■

In general, if a system of linear equations has a single solution, we can use the following steps to solve the system using matrices.

Solving systems of linear equations using matrices

1. Write an augmented matrix for the system.
2. Use elementary row operations to transform the augmented matrix into a matrix with 1's down its main diagonal and 0's under the 1's.
3. When step 2 is complete, write the resulting system. Then use back substitution to find the solution.

Solving a System of Three Equations

To show how to use matrices to solve systems of three linear equations containing three variables, we consider the system

$$\begin{cases} x - 2y - z = 6 \\ 2x + 2y - z = 1 \\ -x - y + 2z = 1 \end{cases}$$

which can be represented by the augmented matrix

$$\begin{bmatrix} 1 & -2 & -1 & \vdots & 6 \\ 2 & 2 & -1 & \vdots & 1 \\ -1 & -1 & 2 & \vdots & 1 \end{bmatrix}$$

To solve a 3×3 system of equations by Gaussian elimination, we transform the augmented matrix into a matrix with 1's down its main diagonal and 0's below its main diagonal.

$$\begin{bmatrix} 1 & a & b & \vdots & c \\ 0 & 1 & d & \vdots & e \\ 0 & 0 & 1 & \vdots & f \end{bmatrix} \qquad (a, b, c, \ldots, f \text{ are real numbers})$$

Main diagonal

E X A M P L E   4   **Solving a system using Gaussian elimination.**
Solve the system $\begin{cases} x - 2y - z = 6 \\ 2x + 2y - z = 1 \\ -x - y + 2z = 1 \end{cases}$.

Solution   This system can be represented by the augmented matrix

$$\begin{bmatrix} 1 & -2 & -1 & \vdots & 6 \\ 2 & 2 & -1 & \vdots & 1 \\ -1 & -1 & 2 & \vdots & 1 \end{bmatrix}$$ We need to get a 0 where the red 2 is.

To get a 0 under the 1 in the first column where the red 2 is, we perform a type 3 row operation: Multiply row 1 by -2 and add the results to row 2.

$$\begin{bmatrix} 1 & -2 & -1 & \vdots & 6 \\ 0 & 6 & 1 & \vdots & -11 \\ -1 & -1 & 2 & \vdots & 1 \end{bmatrix} \quad -2R_1 + R_2.$$

To get a 0 under the 0 in the first column where the red -1 is, we perform another type 3 row operation: To row 3 we add row 1.

$$\begin{bmatrix} 1 & -2 & -1 & \vdots & 6 \\ 0 & 6 & 1 & \vdots & -11 \\ 0 & -3 & 1 & \vdots & 7 \end{bmatrix} \quad R_1 + R_3.$$

To get a 0 under the 6 where the red -3 is, we use another type 3 row operation: Multiply row 2 by $\frac{1}{2}$ and add it to row 3.

$$\begin{bmatrix} 1 & -2 & -1 & \vdots & 6 \\ 0 & 6 & 1 & \vdots & -11 \\ 0 & 0 & \frac{3}{2} & \vdots & \frac{3}{2} \end{bmatrix} \quad \frac{1}{2}R_2 + R_3.$$

We can use a type 2 row operation to simplify further: Multiply row 3 by $\frac{2}{3}$.

$$\begin{bmatrix} 1 & -2 & -1 & \vdots & 6 \\ 0 & 6 & 1 & \vdots & -11 \\ 0 & 0 & 1 & \vdots & 1 \end{bmatrix} \quad \frac{2}{3}R_3.$$

The final matrix represents the system

1. $\begin{cases} x - 2y - z = 6 \\ 0x + 6y + z = -11 \\ 0x + 0y + z = 1 \end{cases}$
2.
3.

From Equation 3, we can read that $z = 1$. To find y, we back substitute 1 for z in Equation 2 and solve for y:

$6y + z = -11$ Equation 2.
$6y + 1 = -11$ Substitute 1 for z.
$6y = -12$ Subtract 1 from both sides.
$y = -2$ Divide both sides by 6.

Thus, $y = -2$. To find x, we back substitute 1 for z and -2 for y in Equation 1 and solve for x:

$x - 2y - z = 6$ Equation 1.
$x - 2(-2) - 1 = 6$ Substitute 1 for z and -2 for y.
$x + 3 = 6$ Simplify.
$x = 3$ Subtract 3 from both sides.

Thus, $x = 3$. The solution of the given system is $(3, -2, 1)$. Verify that this triple satisfies each equation of the original system.

SELF CHECK   Solve $\begin{cases} x - 2y - z = 2 \\ 2x + 2y - z = -5. \\ -x - y + 2z = 7 \end{cases}$

Answer: $(1, -2, 3)$ ■

Inconsistent Systems and Dependent Equations

In the next example, we consider a system with no solution.

EXAMPLE 5

An inconsistent system. Using matrices, solve the system $\begin{cases} x + y = -1 \\ -3x - 3y = -5 \end{cases}$.

Solution This system can be represented by the augmented matrix

$$\left[\begin{array}{cc:c} 1 & 1 & -1 \\ -3 & -3 & -5 \end{array} \right]$$

Since the matrix has a 1 in the top row of the first column, we proceed to get a 0 under it by multiplying row 1 by 3 and adding the results to row 2.

$$\left[\begin{array}{cc:c} 1 & 1 & -1 \\ 0 & 0 & -8 \end{array} \right] \quad 3R_1 + R_2.$$

This matrix represents the system

$$\begin{cases} x + y = -1 \\ 0 + 0 = -8 \end{cases}$$

This system has no solution, because the second equation is never true. Therefore, the system is inconsistent. It has no solutions.

SELF CHECK Solve $\begin{cases} 4x - 8y = 9 \\ x - 2y = -5 \end{cases}$. *Answer:* no solution ■

In the next example, we consider a system with infinitely many solutions.

EXAMPLE 6

Dependent equations. Using matrices, solve the system
$$\begin{cases} 2x + 3y - 4z = 6 \\ 4x + 6y - 8z = 12 \\ -6x - 9y + 12z = -18 \end{cases}.$$

Solution This system can be represented by the augmented matrix

$$\left[\begin{array}{ccc:c} 2 & 3 & -4 & 6 \\ 4 & 6 & -8 & 12 \\ -6 & -9 & 12 & -18 \end{array} \right]$$

To get a 1 in the top row of the first column, we multiply row 1 by $\frac{1}{2}$.

$$\left[\begin{array}{ccc:c} 1 & \frac{3}{2} & -2 & 3 \\ 4 & 6 & -8 & 12 \\ -6 & -9 & 12 & -18 \end{array} \right] \quad \frac{1}{2}R_1.$$

Next, we want to get 0's under the 1 in the first column. This can be achieved by multiplying row 1 by -4 and adding the results to row 2, and multiplying row 1 by 6 and adding the results to row 3.

$$\left[\begin{array}{ccc:c} 1 & \frac{3}{2} & -2 & 3 \\ 0 & 0 & 0 & 0 \\ 0 & 0 & 0 & 0 \end{array} \right] \quad \begin{array}{l} -4R_1 + R_2. \\ 6R_1 + R_3. \end{array}$$

The last matrix represents the system

$$\begin{cases} x + \frac{3}{2}y - 2z = 3 \\ 0x + 0y + 0z = 0 \\ 0x + 0y + 0z = 0 \end{cases}$$

If we clear the first equation of fractions, we have the system

$$\begin{cases} 2x + 3y - 4z = 6 \\ 0 = 0 \\ 0 = 0 \end{cases}$$

This system has dependent equations and infinitely many solutions. Solutions of this system would be any triple (x, y, z) that satisfies the equation $2x + 3y - 4z = 6$. Two such solutions would be $(0, 2, 0)$ and $(1, 0, -1)$.

SELF CHECK Solve $\begin{cases} 5x - 10y + 15z = 35 \\ -3x + 6y - 9z = -21. \\ 2x - 4y + 6z = 14 \end{cases}$

Answer: There are infinitely many solutions—any triple satisfying the equation $x - 2y + 3z = 7$. ∎

STUDY SET

Section 3.4

VOCABULARY

In Exercises 1–6, fill in the blanks to make the statements true.

1. A _____ is a rectangular array of numbers.

2. The numbers in a matrix are called its _____.

3. A 3×4 matrix has 3 _____ and 4 _____.

4. Elementary _____ operations are used to produce new matrices that lead to the solution of a system.

5. A matrix that represents the equations of a system is called an _____ matrix.

6. The augmented matrix $\begin{bmatrix} 1 & 3 & \vdots & -2 \\ 0 & 1 & \vdots & 4 \end{bmatrix}$ has 1's down its main _____.

CONCEPTS

7. For each matrix, tell the number of rows and the number of columns.

   a. $\begin{bmatrix} 4 & 6 & \vdots & -1 \\ \frac{1}{2} & 9 & \vdots & -3 \end{bmatrix}$

   b. $\begin{bmatrix} 1 & -2 & 3 & \vdots & 1 \\ 0 & 1 & 6 & \vdots & 4 \\ 0 & 0 & 1 & \vdots & \frac{1}{3} \end{bmatrix}$

8. For each augmented matrix, give the system of equations it represents.

   a. $\begin{bmatrix} 1 & 6 & \vdots & 7 \\ 0 & 1 & \vdots & 4 \end{bmatrix}$

   b. $\begin{bmatrix} 2 & -2 & 9 & \vdots & 1 \\ 3 & 1 & 1 & \vdots & 0 \\ 2 & -6 & 8 & \vdots & -7 \end{bmatrix}$

9. Write the system of equations represented by the augmented matrix. Then use back substitution to find the solution.

   $\begin{bmatrix} 1 & -1 & \vdots & -10 \\ 0 & 1 & \vdots & 6 \end{bmatrix}$

10. Write the system of equations represented by the augmented matrix. Then use back substitution to find the solution.

    $\begin{bmatrix} 1 & -2 & 1 & \vdots & -16 \\ 0 & 1 & 2 & \vdots & 8 \\ 0 & 0 & 1 & \vdots & 4 \end{bmatrix}$

11. Matrices were used to solve a system of two linear equations. The final matrix is shown here. Explain what the result tells about the system.

    $\begin{bmatrix} 1 & 2 & \vdots & -4 \\ 0 & 0 & \vdots & 2 \end{bmatrix}$

12. Matrices were used to solve a system of two linear equations. The final matrix is shown here. What does the result tell about the equations?

    $\begin{bmatrix} 1 & 2 & \vdots & -4 \\ 0 & 0 & \vdots & 0 \end{bmatrix}$

NOTATION

13. Consider the matrix $A = \begin{bmatrix} 3 & 6 & -9 & \vdots & 0 \\ 1 & 5 & -2 & \vdots & 1 \\ -2 & 2 & -2 & \vdots & 5 \end{bmatrix}$.

    a. Explain what is meant by $\frac{1}{3}R_1$. Then perform the operation on matrix A.

b. Explain what is meant by $-R_1 + R_2$. Then perform the operation on the answer to part a.

14. Consider the matrix $B = \begin{bmatrix} -3 & 1 & \vdots & -6 \\ 1 & -4 & \vdots & 4 \end{bmatrix}$.

a. Explain what is meant by $R_1 \leftrightarrow R_2$. Then perform the operation on matrix B.

b. Explain what is meant by $3R_1 + R_2$. Then perform the operation on the answer to part a.

In Exercises 15–16, complete each solution.

15. Solve $\begin{cases} 4x - y = 14 \\ x + y = 6 \end{cases}$.

$$\begin{bmatrix} 4 & & \vdots & 14 \\ 1 & 1 & \vdots & 6 \end{bmatrix}$$

$$\begin{bmatrix} & 1 & \vdots & 6 \\ 4 & -1 & \vdots & 14 \end{bmatrix} \quad R_1 \leftrightarrow R_2.$$

$$\begin{bmatrix} 1 & 1 & \vdots & 6 \\ 0 & & \vdots & -10 \end{bmatrix} \quad -4R_1 + R_2.$$

$$\begin{bmatrix} 1 & 1 & \vdots & 6 \\ 0 & 1 & \vdots & \end{bmatrix} \quad -\tfrac{1}{5}R_2.$$

This matrix represents the system

$$\begin{cases} x + y = 6 \\ = 2 \end{cases}$$

The solution is $(, 2)$.

16. Solve $\begin{cases} 2x + 2y = 18 \\ x - y = 5 \end{cases}$.

$$\begin{bmatrix} 2 & 2 & \vdots & 18 \\ & -1 & \vdots & 5 \end{bmatrix}$$

$$\begin{bmatrix} 1 & 1 & \vdots & 9 \\ & -1 & \vdots & 5 \end{bmatrix} \quad \tfrac{1}{2}R_1.$$

$$\begin{bmatrix} 1 & 1 & \vdots & 9 \\ 0 & -2 & \vdots & -4 \end{bmatrix} \quad -R_1 + R_2.$$

$$\begin{bmatrix} 1 & 1 & \vdots & 9 \\ 0 & 1 & \vdots & \end{bmatrix} \quad -\tfrac{1}{2}R_2.$$

This matrix represents the system

$$\begin{cases} x + y = \\ y = 2 \end{cases}$$

The solution is $(, 2)$.

PRACTICE

In Exercises 17–32, use matrices to solve each system of equations.

17. $\begin{cases} x + y = 2 \\ x - y = 0 \end{cases}$

18. $\begin{cases} x + y = 3 \\ x - y = -1 \end{cases}$

19. $\begin{cases} 2x + y = 1 \\ x + 2y = -4 \end{cases}$

20. $\begin{cases} 5x - 4y = 10 \\ x - 7y = 2 \end{cases}$

21. $\begin{cases} 2x - y = -1 \\ x - 2y = 1 \end{cases}$

22. $\begin{cases} 2x - y = 0 \\ x + y = 3 \end{cases}$

23. $\begin{cases} 3x + 4y = -12 \\ 9x - 2y = 6 \end{cases}$

24. $\begin{cases} 2x - 3y = 16 \\ -4x + y = -22 \end{cases}$

25. $\begin{cases} x + y + z = 6 \\ x + 2y + z = 8 \\ x + y + 2z = 9 \end{cases}$

26. $\begin{cases} x - y + z = 2 \\ x + 2y - z = 6 \\ 2x - y - z = 3 \end{cases}$

27. $\begin{cases} 3x + y - 3z = 5 \\ x - 2y + 4z = 10 \\ x + y + z = 13 \end{cases}$

28. $\begin{cases} 2x + y - 3z = -1 \\ 3x - 2y - z = -5 \\ x - 3y - 2z = -12 \end{cases}$

29. $\begin{cases} 3x - 2y + 4z = 4 \\ x + y + z = 3 \\ 6x - 2y - 3z = 10 \end{cases}$

30. $\begin{cases} 2x + 3y - z = -8 \\ x - y - z = -2 \\ -4x + 3y + z = 6 \end{cases}$

31. $\begin{cases} 2a + b + 3c = 3 \\ -2a - b + c = 5 \\ 4a - 2b + 2c = 2 \end{cases}$

32. $\begin{cases} 3a + 2b + c = 8 \\ 6a - b + 2c = 16 \\ -9a + b - c = -20 \end{cases}$

In Exercises 33–44, use matrices to solve each system of equations. If the equations of a system are dependent or if a system is inconsistent, so indicate.

33. $\begin{cases} x - 3y = 9 \\ -2x + 6y = 18 \end{cases}$

34. $\begin{cases} -6x + 12y = 10 \\ 2x - 4y = 8 \end{cases}$

35. $\begin{cases} 4x + 4y = 12 \\ -x - y = -3 \end{cases}$

36. $\begin{cases} 5x - 15y = 10 \\ 2x - 6y = 4 \end{cases}$

37. $\begin{cases} 6x + y - z = -2 \\ x + 2y + z = 5 \\ 5y - z = 2 \end{cases}$

38. $\begin{cases} 2x + 3y - 2z = 18 \\ 5x - 6y + z = 21 \\ 4y - 2z = 6 \end{cases}$

39. $\begin{cases} 2x + y - z = 1 \\ x + 2y + 2z = 2 \\ 4x + 5y + 3z = 3 \end{cases}$

40. $\begin{cases} x - 3y + 4z = 2 \\ 2x + y + 2z = 3 \\ 4x - 5y + 10z = 7 \end{cases}$

41. $\begin{cases} 5x + 3y = 4 \\ 3y - 4z = 4 \\ x + z = 1 \end{cases}$ **42.** $\begin{cases} y + 2z = -2 \\ x + y = 1 \\ 2x - z = 0 \end{cases}$

43. $\begin{cases} x - y = 1 \\ 2x - z = 0 \\ 2y - z = -2 \end{cases}$ **44.** $\begin{cases} x + y - 3z = 4 \\ 2x + 2y - 6z = 5 \\ -3x + y - z = 2 \end{cases}$

In Exercises 45–48, remember these facts from geometry. Two angles whose measures add up to 90° are complementary. Two angles whose measures add up to 180° are supplementary. The sum of the measures of the interior angles in a triangle is 180°.

45. One angle is 46° larger than its complement. Find the angles.

46. One angle is 28° larger than its supplement. Find the angles.

47. In Illustration 1, angle *B* is 25° more than angle *A*, and angle *C* is 5° less than twice angle *A*. Find each angle in the triangle.

ILLUSTRATION 1

48. In Illustration 2, angle *A* is 10° less than angle *B*, and angle *B* is 10° less than angle *C*. Find each angle in the triangle.

ILLUSTRATION 2

APPLICATIONS

Write a system of equations to solve each problem. Then use matrices to solve the system.

49. PHYSICAL THERAPY After an elbow injury, a volleyball player has restricted movement of her arm. (See Illustration 3.) Her range of motion (angle 1) is 28° less than angle 2. Find the measure of each angle.

ILLUSTRATION 3

50. TRIANGLES In Illustration 4, ∠*B* (read "angle *B*") is 25° more than ∠*A*, and ∠*C* is 5° less than twice ∠*A*. Find each angle in the triangle. (*Hint:* The sum of the angles in a triangle is 180°.)

ILLUSTRATION 4

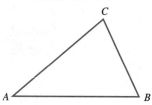

51. THEATER SEATING Illustration 5 shows the cash receipts from two sold-out performances of a play and the ticket prices. Find the number of seats in each of the three sections of the 800-seat theater.

ILLUSTRATION 5

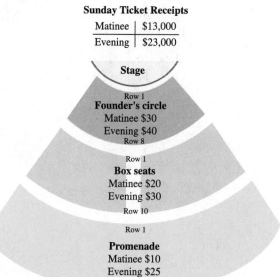

52. ICE SKATING Illustration 6 shows three circles traced out by a figure skater during her performance. If the centers of the circles are the given distances apart, find the radius of each circle.

ILLUSTRATION 6

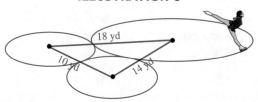

WRITING

53. Explain what is meant by the phrase *back substitution*.

54. Explain how a type 3 row operation is similar to the addition method of solving a system of equations.

REVIEW

55. What is the formula used to find the slope of a line, given two points on the line?

56. What is the form of the equation of a horizontal line? Of a vertical line?

57. What is the point–slope form of the equation of a line?

58. What is the slope–intercept form of the equation of a line?

▶ **3.5**

Solving Systems Using Determinants

In this section, you will learn about

Determinants ▪ Evaluating a determinant ▪ Using Cramer's rule to solve a system of two equations ▪ Using Cramer's rule to solve a system of three equations

Introduction   In this section, we will discuss another method for solving systems of linear equations. With this method, called *Cramer's rule,* we work with combinations of the coefficients and the constants of the equations written as *determinants.*

Determinants

An idea closely related to the concept of matrix is the **determinant.** A determinant is a number that is associated with a **square matrix,** a matrix that has the same number of rows and columns. For any square matrix A, the symbol $|A|$ represents the determinant of A. To write a determinant, we put the elements of a square matrix between two vertical lines.

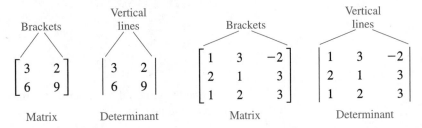

Brackets

Vertical lines

Brackets

Vertical lines

$$\begin{bmatrix} 3 & 2 \\ 6 & 9 \end{bmatrix} \qquad \begin{vmatrix} 3 & 2 \\ 6 & 9 \end{vmatrix} \qquad \begin{bmatrix} 1 & 3 & -2 \\ 2 & 1 & 3 \\ 1 & 2 & 3 \end{bmatrix} \qquad \begin{vmatrix} 1 & 3 & -2 \\ 2 & 1 & 3 \\ 1 & 2 & 3 \end{vmatrix}$$

Matrix   Determinant   Matrix   Determinant

Like matrices, determinants are classified according to the number of rows and columns they contain. The determinant on the left is a 2×2 determinant. The other is a 3×3 determinant.

Evaluating a Determinant

The determinant of a 2×2 matrix is the number that is equal to the product of the numbers on the main diagonal minus the product of the numbers on the other diagonal.

$$\begin{vmatrix} a & b \\ c & d \end{vmatrix} \qquad\qquad \begin{vmatrix} a & b \\ c & d \end{vmatrix}$$

Main diagonal   Other diagonal

Value of a 2 × 2 determinant

If a, b, c, and d are numbers, the **determinant** of the matrix $\begin{bmatrix} a & b \\ c & d \end{bmatrix}$ is

$$\begin{vmatrix} a & b \\ c & d \end{vmatrix} = ad - bc$$

E X A M P L E   1

Evaluating 2 × 2 determinants. Find each value: **a.** $\begin{vmatrix} 3 & 2 \\ 6 & 9 \end{vmatrix}$ and **b.** $\begin{vmatrix} -5 & \frac{1}{2} \\ -1 & 0 \end{vmatrix}$.

Solution   From the product of the numbers along the main diagonal, we subtract the product of the numbers along the other diagonal.

a. $\begin{vmatrix} 3 & 2 \\ 6 & 9 \end{vmatrix} = 3(9) - 2(6)$

$= 27 - 12$

$= 15$

b. $\begin{vmatrix} -5 & \frac{1}{2} \\ -1 & 0 \end{vmatrix} = -5(0) - \frac{1}{2}(-1)$

$= 0 + \frac{1}{2}$

$= \frac{1}{2}$

SELF CHECK   Evaluate $\begin{vmatrix} 4 & -3 \\ 2 & 1 \end{vmatrix}$.

Answer: 10 ∎

A 3 × 3 determinant is evaluated by **expanding by minors.**

Value of a 3 × 3 determinant

$$\begin{vmatrix} a_1 & b_1 & c_1 \\ a_2 & b_2 & c_2 \\ a_3 & b_3 & c_3 \end{vmatrix} = a_1 \overset{\text{Minor of } a_1}{\begin{vmatrix} b_2 & c_2 \\ b_3 & c_3 \end{vmatrix}} - b_1 \overset{\text{Minor of } b_1}{\begin{vmatrix} a_2 & c_2 \\ a_3 & c_3 \end{vmatrix}} + c_1 \overset{\text{Minor of } c_1}{\begin{vmatrix} a_2 & b_2 \\ a_3 & b_3 \end{vmatrix}}$$

To find the minor of a_1, we cross out the elements of the determinant that are in the same row and column as a_1:

$$\begin{vmatrix} a_1 & b_1 & c_1 \\ a_2 & b_2 & c_2 \\ a_3 & b_3 & c_3 \end{vmatrix}$$ The minor of a_1 is $\begin{vmatrix} b_2 & c_2 \\ b_3 & c_3 \end{vmatrix}$.

To find the minor of b_1, we cross out the elements of the determinant that are in the same row and column as b_1:

$$\begin{vmatrix} a_1 & b_1 & c_1 \\ a_2 & b_2 & c_2 \\ a_3 & b_3 & c_3 \end{vmatrix}$$ The minor of b_1 is $\begin{vmatrix} a_2 & c_2 \\ a_3 & c_3 \end{vmatrix}$.

To find the minor of c_1, we cross out the elements of the determinant that are in the same row and column as c_1:

$$\begin{vmatrix} a_1 & b_1 & c_1 \\ a_2 & b_2 & c_2 \\ a_3 & b_3 & c_3 \end{vmatrix}$$ The minor of c_1 is $\begin{vmatrix} a_2 & b_2 \\ a_3 & b_3 \end{vmatrix}$.

EXAMPLE 2

Using minors to evaluate a 3 × 3 determinant. Find the value of

$$\begin{vmatrix} 1 & 3 & -2 \\ 2 & 1 & 3 \\ 1 & 2 & 3 \end{vmatrix}.$$

Solution   We evaluate this determinant by expanding by minors along the first row of the determinant.

$$\begin{vmatrix} 1 & 3 & -2 \\ 2 & 1 & 3 \\ 1 & 2 & 3 \end{vmatrix} = 1 \overset{\text{Minor of 1}}{\begin{vmatrix} 1 & 3 \\ 2 & 3 \end{vmatrix}} - 3 \overset{\text{Minor of 3}}{\begin{vmatrix} 2 & 3 \\ 1 & 3 \end{vmatrix}} + (-2) \overset{\text{Minor of } -2}{\begin{vmatrix} 2 & 1 \\ 1 & 2 \end{vmatrix}}$$

$$= 1(3 - 6) - 3(6 - 3) - 2(4 - 1) \qquad \text{Evaluate each } 2 \times 2 \text{ determinant.}$$

$$= 1(-3) - 3(3) - 2(3)$$

$$= -3 - 9 - 6$$

$$= -18$$

SELF CHECK   Evaluate $\begin{vmatrix} 2 & -1 & 3 \\ 1 & 2 & -2 \\ 3 & 1 & 1 \end{vmatrix}.$ *Answer:* 0   ■

We can evaluate a 3 × 3 determinant by expanding it along any row or column. To determine the signs between the terms of the expansion of a 3 × 3 determinant, we use the following array of signs.

Array of signs for a 3 × 3 determinant

$$\begin{array}{ccc} + & - & + \\ - & + & - \\ + & - & + \end{array}$$

EXAMPLE 3

Expanding along a column. Evaluate the determinant $\begin{vmatrix} 1 & 3 & -2 \\ 2 & 1 & 3 \\ 1 & 2 & 3 \end{vmatrix}$ by expanding on the middle column.

Solution   This is the determinant of Example 2. To expand it along the middle column, we use the signs of the middle column of the array of signs:

$$\begin{vmatrix} 1 & 3 & -2 \\ 2 & 1 & 3 \\ 1 & 2 & 3 \end{vmatrix} = -3 \overset{\text{Minor of 3}}{\begin{vmatrix} 2 & 3 \\ 1 & 3 \end{vmatrix}} + 1 \overset{\text{Minor of 1}}{\begin{vmatrix} 1 & -2 \\ 1 & 3 \end{vmatrix}} - 2 \overset{\text{Minor of 2}}{\begin{vmatrix} 1 & -2 \\ 2 & 3 \end{vmatrix}} \quad \begin{array}{l}\text{Use the} \\ \text{sign pattern} \\ - + -.\end{array}$$

$$= -3(6 - 3) + 1[3 - (-2)] - 2[3 - (-4)] \qquad \begin{array}{l}\text{Evaluate} \\ \text{each} \\ 2 \times 2 \\ \text{determinant.}\end{array}$$

$$= -3(3) + 1(5) - 2(7)$$

$$= -9 + 5 - 14$$

$$= -18$$

As expected, we get the same value as in Example 2.

SELF CHECK
Evaluate
$$\begin{vmatrix} 1 & 3 & -2 \\ 2 & 1 & 3 \\ 1 & 2 & 3 \end{vmatrix}$$ by

expanding along the last column.   *Answer:* -18

ACCENT ON TECHNOLOGY *Evaluating Determinants*

It is possible to use a graphing calculator to evaluate determinants. For example, to evaluate the determinant in Example 3, we first enter the matrix by pressing the MATRIX key, selecting EDIT, and pressing the ENTER key. We then enter the dimensions and the elements of the matrix to get Figure 3-13(a). We then press 2nd QUIT to clear the screen. We then press MATRIX, select MATH, and press 1 to get Figure 3-13(b). We then press MATRIX, select NAMES, and press 1 to get Figure 3-13(c). To get the value of the determinant, we now press ENTER to get Figure 3-13(d), which shows that the value of the matrix is -18.

Figure 3-13

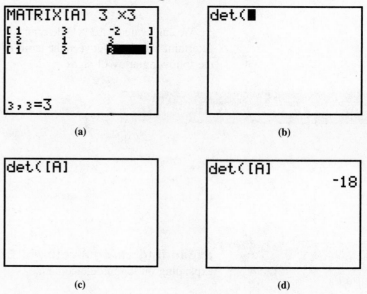

(a)   (b)

(c)   (d)

Using Cramer's Rule to Solve a System of Two Equations

The method of using determinants to solve systems of linear equations is called **Cramer's rule,** named after the 18th-century mathematician Gabriel Cramer. To develop Cramer's rule, we consider the system

$$\begin{cases} ax + by = e \\ cx + dy = f \end{cases}$$

where x and y are variables and a, b, c, d, e, and f are constants.

If we multiply both sides of the first equation by d and multiply both sides of the second equation by $-b$, we can add the equations and eliminate y:

$$
\begin{array}{rl}
adx + bdy = & ed \\
\underline{-bcx - bdy =} & \underline{-bf} \\
adx - bcx = & ed - bf
\end{array}
$$

To solve for x, we use the distributive property to write $adx - bcx$ as $(ad - bc)x$ on the left-hand side and divide each side by $ad - bc$:

$$(ad - bc)x = ed - bf$$

$$x = \frac{ed - bf}{ad - bc} \qquad (ad - bc \neq 0)$$

We can find y in a similar manner. After eliminating the variable x, we get

$$y = \frac{af - ec}{ad - bc} \qquad (ad - bc \neq 0)$$

Determinants provide an easy way of remembering these formulas. Note that the denominator for both x and y is

$$\begin{vmatrix} a & b \\ c & d \end{vmatrix} = ad - bc$$

The numerators can be expressed as determinants also:

$$x = \frac{ed - bf}{ad - bc} = \frac{\begin{vmatrix} e & b \\ f & d \end{vmatrix}}{\begin{vmatrix} a & b \\ c & d \end{vmatrix}} \qquad \text{and} \qquad y = \frac{af - ec}{ad - bc} = \frac{\begin{vmatrix} a & e \\ c & f \end{vmatrix}}{\begin{vmatrix} a & b \\ c & d \end{vmatrix}}$$

If we compare these formulas with the original system

$$\begin{cases} ax + by = e \\ cx + dy = f \end{cases}$$

we note that in the expressions for x and y above, the denominator determinant is formed by using the coefficients a, b, c, and d of the variables in the equations. The numerator determinants are the same as the denominator determinant, except that the column of coefficients of the variable for which we are solving is replaced with the column of constants e and f.

Cramer's rule for two equations in two variables

The solution of the system $\begin{cases} ax + by = e \\ cx + dy = f \end{cases}$ is given by

$$x = \frac{D_x}{D} = \frac{\begin{vmatrix} e & b \\ f & d \end{vmatrix}}{\begin{vmatrix} a & b \\ c & d \end{vmatrix}} \qquad \text{and} \qquad y = \frac{D_y}{D} = \frac{\begin{vmatrix} a & e \\ c & f \end{vmatrix}}{\begin{vmatrix} a & b \\ c & d \end{vmatrix}}$$

If every determinant is 0, the system is consistent, but the equations are dependent.

If $D = 0$ and D_x or D_y is nonzero, the system is inconsistent. If $D \neq 0$, the system is consistent, and the equations are independent.

EXAMPLE 4

Cramer's rule. Use Cramer's rule to solve $\begin{cases} 4x - 3y = 6 \\ -2x + 5y = 4 \end{cases}$.

Solution   The value of x is the quotient of two determinants, D and D_x. The denominator determinant D is made up of the coefficients of x and y:

$$D = \begin{vmatrix} 4 & -3 \\ -2 & 5 \end{vmatrix}$$

To solve for x, we form the numerator determinant D_x from D by replacing its first column (the coefficients of x) with the column of constants (6 and 4).

To solve for y, we form the numerator determinant D_y from D by replacing the second column (the coefficients of y) with the column of constants (6 and 4).

To find the values of x and y, we evaluate each determinant:

$$x = \frac{D_x}{D} = \frac{\begin{vmatrix} 6 & -3 \\ 4 & 5 \end{vmatrix}}{\begin{vmatrix} 4 & -3 \\ -2 & 5 \end{vmatrix}} = \frac{6(5) - (-3)(4)}{4(5) - (-3)(-2)} = \frac{30 + 12}{20 - 6} = \frac{42}{14} = 3$$

$$y = \frac{D_y}{D} = \frac{\begin{vmatrix} 4 & 6 \\ -2 & 4 \end{vmatrix}}{\begin{vmatrix} 4 & -3 \\ -2 & 5 \end{vmatrix}} = \frac{4(4) - 6(-2)}{14} = \frac{16 + 12}{14} = \frac{28}{14} = 2$$

The solution of this system is (3, 2). Verify that $x = 3$ and $y = 2$ satisfy both equations.

SELF CHECK   Solve $\begin{cases} 2x - 3y = -16 \\ 3x + 5y = 14 \end{cases}$ using Cramer's rule.   *Answer:* $(-2, 4)$ ■

EXAMPLE 5

An inconsistent system. Use Cramer's rule to solve $\begin{cases} 7x = 8 - 4y \\ 2y = 3 - \frac{7}{2}x \end{cases}$.

Solution   We multiply both sides of the second equation by 2 to eliminate the fraction and write the system in the form

$$\begin{cases} 7x + 4y = 8 \\ 7x + 4y = 6 \end{cases}$$

When we attempt to use Cramer's rule to solve this system for x, we obtain

$$x = \frac{D_x}{D} = \frac{\begin{vmatrix} 8 & 4 \\ 6 & 4 \end{vmatrix}}{\begin{vmatrix} 7 & 4 \\ 7 & 4 \end{vmatrix}} = \frac{8}{0}$$ which is undefined

Since the denominator determinant D is 0 and the numerator determinant D_x is not 0, the system is inconsistent. It has no solutions.

We can see directly from the system that it is inconsistent. For any values of x and y, it is impossible that 7 times x plus 4 times y could be both 8 and 6.

SELF CHECK   Solve $\begin{cases} 3x = 8 - 4y \\ y = \frac{5}{2} - \frac{3}{4}x \end{cases}$ using Cramer's rule.   *Answer:* no solutions   ■

Using Cramer's Rule to Solve a System of Three Equations

Cramer's rule can be extended to solve systems of three linear equations with three variables.

Cramer's rule for three equations with three variables

The solution of the system $\begin{cases} ax + by + cz = j \\ dx + ey + fz = k \\ gx + hy + iz = l \end{cases}$ is given by

$$x = \frac{D_x}{D}, \qquad y = \frac{D_y}{D}, \qquad \text{and} \qquad z = \frac{D_z}{D}$$

where

$$D = \begin{vmatrix} a & b & c \\ d & e & f \\ g & h & i \end{vmatrix} \qquad D_x = \begin{vmatrix} j & b & c \\ k & e & f \\ l & h & i \end{vmatrix}$$

$$D_y = \begin{vmatrix} a & j & c \\ d & k & f \\ g & l & i \end{vmatrix} \qquad D_z = \begin{vmatrix} a & b & j \\ d & e & k \\ g & h & l \end{vmatrix}$$

If every determinant is 0, the system is consistent, but the equations are dependent.

If $D = 0$ and D_x or D_y or D_z is nonzero, the system is inconsistent. If $D \neq 0$, the system is consistent, and the equations are independent.

EXAMPLE 6

A system of three equations. Use Cramer's rule to solve
$$\begin{cases} 2x + y + 4z = 12 \\ x + 2y + 2z = 9 \\ 3x - 3y - 2z = 1 \end{cases}.$$

Solution
The denominator determinant D is the determinant formed by the coefficients of the variables. The numerator determinants, D_x, D_y, and D_z, are formed by replacing the coefficients of the variable being solved for by the column of constants. We form the quotients for x, y, and z and evaluate each determinant by expanding by minors about the first row:

$$x = \frac{D_x}{D} = \frac{\begin{vmatrix} 12 & 1 & 4 \\ 9 & 2 & 2 \\ 1 & -3 & -2 \end{vmatrix}}{\begin{vmatrix} 2 & 1 & 4 \\ 1 & 2 & 2 \\ 3 & -3 & -2 \end{vmatrix}}$$

$$= \frac{12 \begin{vmatrix} 2 & 2 \\ -3 & -2 \end{vmatrix} - 1 \begin{vmatrix} 9 & 2 \\ 1 & -2 \end{vmatrix} + 4 \begin{vmatrix} 9 & 2 \\ 1 & -3 \end{vmatrix}}{2 \begin{vmatrix} 2 & 2 \\ -3 & -2 \end{vmatrix} - 1 \begin{vmatrix} 1 & 2 \\ 3 & -2 \end{vmatrix} + 4 \begin{vmatrix} 1 & 2 \\ 3 & -3 \end{vmatrix}}$$

$$= \frac{12(2) - 1(-20) + 4(-29)}{2(2) - 1(-8) + 4(-9)}$$

$$= \frac{-72}{-24}$$

$$= 3$$

$$y = \frac{D_y}{D} = \frac{\begin{vmatrix} 2 & 12 & 4 \\ 1 & 9 & 2 \\ 3 & 1 & -2 \end{vmatrix}}{\begin{vmatrix} 2 & 1 & 4 \\ 1 & 2 & 2 \\ 3 & -3 & -2 \end{vmatrix}}$$

$$= \frac{2\begin{vmatrix} 9 & 2 \\ 1 & -2 \end{vmatrix} - 12\begin{vmatrix} 1 & 2 \\ 3 & -2 \end{vmatrix} + 4\begin{vmatrix} 1 & 9 \\ 3 & 1 \end{vmatrix}}{-24}$$

$$= \frac{2(-20) - 12(-8) + 4(-26)}{-24}$$

$$= \frac{-48}{-24}$$

$$= 2$$

$$z = \frac{D_z}{D} = \frac{\begin{vmatrix} 2 & 1 & 12 \\ 1 & 2 & 9 \\ 3 & -3 & 1 \end{vmatrix}}{\begin{vmatrix} 2 & 1 & 4 \\ 1 & 2 & 2 \\ 3 & -3 & -2 \end{vmatrix}}$$

$$= \frac{2\begin{vmatrix} 2 & 9 \\ -3 & 1 \end{vmatrix} - 1\begin{vmatrix} 1 & 9 \\ 3 & 1 \end{vmatrix} + 12\begin{vmatrix} 1 & 2 \\ 3 & -3 \end{vmatrix}}{-24}$$

$$= \frac{2(29) - 1(-26) + 12(-9)}{-24}$$

$$= \frac{-24}{-24}$$

$$= 1$$

The solution of this system is (3, 2, 1).

SELF CHECK Solve $\begin{cases} x + y + 2z = 6 \\ 2x - y + z = 9 \\ x + y - 2z = -6 \end{cases}$ using Cramer's rule.

Answer: (2, −2, 3) ∎

STUDY SET

Section 3.5

VOCABULARY

In Exercises 1–6, fill in the blanks to make the statements true.

1. $\begin{vmatrix} 2 & 1 \\ -6 & 1 \end{vmatrix}$ is a 2 × 2 _____.

2. A _____ matrix has the same number of rows and columns.

3. The _____ of b_1 in $\begin{vmatrix} a_1 & b_1 & c_1 \\ a_2 & b_2 & c_2 \\ a_3 & b_3 & c_3 \end{vmatrix}$ is $\begin{vmatrix} a_2 & c_2 \\ a_3 & c_3 \end{vmatrix}$.

4. In $\begin{vmatrix} 7 & -3 \\ 1 & 2 \end{vmatrix}$, 7 and 2 lie along the main _____.

5. A 3 × 3 determinant has 3 _____ and 3
_____.

6. _____ rule uses determinants to solve systems of linear equations.

CONCEPTS

In Exercises 7–8, fill in the blanks to make the statements true.

7. If the denominator determinant D for a system of equations is zero, the equations of the system are _____ or the system is _____.

8. To find the minor of 5, we _____ the elements of the determinant that are in the same row and column as 5.

$$\begin{vmatrix} 3 & 5 & 1 \\ 6 & -2 & 2 \\ 8 & -1 & 4 \end{vmatrix}$$

9. What is the value of $\begin{vmatrix} a & b \\ c & d \end{vmatrix}$?

10. $\begin{vmatrix} 5 & 1 & -1 \\ 8 & 7 & 4 \\ 9 & 7 & 6 \end{vmatrix} = -1\begin{vmatrix} 8 & 7 \\ 9 & 7 \end{vmatrix} - 4\begin{vmatrix} 5 & 1 \\ 9 & 7 \end{vmatrix} +$

$6\begin{vmatrix} 5 & 1 \\ 8 & 7 \end{vmatrix}$

In evaluating this determinant, about what row or column was it expanded?

11. What is the denominator determinant D for the system $\begin{cases} 3x + 4y = 7 \\ 2x - 3y = 5 \end{cases}$?

12. What is the denominator determinant D for the system $\begin{cases} x + 2y = -8 \\ 3x + y - z = -2 \\ 8x + 4y - z = 6 \end{cases}$?

13. For the system $\begin{cases} 3x + 2y = 1 \\ 4x - y = 3 \end{cases}$, $D_x = -7$, $D_y = 5$, and $D = -11$. What is the solution of the system?

14. For the system $\begin{cases} 2x + 3y - z = -8 \\ x - y - z = -2 \\ -4x + 3y + z = 6 \end{cases}$,

$D_x = -28$, $D_y = -14$, $D_z = 14$, and $D = 14$. What is the solution?

NOTATION

In Exercises 15–16, complete the evaluation of each determinant.

15. $\begin{vmatrix} 5 & -2 \\ -2 & 6 \end{vmatrix}$

$= 5(\quad) - (-2)(-2)$

$= \quad - 4$

$= 26$

16. $\begin{vmatrix} 2 & 1 & 3 \\ 3 & 4 & 2 \\ 1 & 5 & 3 \end{vmatrix}$

$= 2\begin{vmatrix} 4 & \quad \\ 5 & 3 \end{vmatrix} - \quad 1\begin{vmatrix} 3 & 2 \\ \quad & 3 \end{vmatrix} + 3\begin{vmatrix} 3 & 4 \\ 1 & \quad \end{vmatrix}$

$= 2(\quad - 10) - 1(9 - \quad) + 3(15 - \quad)$

$= 2(2) - 1(\quad) + \quad(11)$

$= 4 - 7 + \quad$

$= 30$

PRACTICE

In Exercises 17–32, evaluate each determinant.

17. $\begin{vmatrix} 2 & 3 \\ -2 & 1 \end{vmatrix}$

18. $\begin{vmatrix} 3 & -2 \\ -2 & 4 \end{vmatrix}$

19. $\begin{vmatrix} -1 & 2 \\ 3 & -4 \end{vmatrix}$

20. $\begin{vmatrix} -1 & -2 \\ -3 & -4 \end{vmatrix}$

21. $\begin{vmatrix} 10 & 0 \\ 1 & 20 \end{vmatrix}$

22. $\begin{vmatrix} 1 & 15 \\ 15 & 0 \end{vmatrix}$

23. $\begin{vmatrix} -6 & -2 \\ 15 & 4 \end{vmatrix}$

24. $\begin{vmatrix} 3 & -2 \\ 12 & -8 \end{vmatrix}$

25. $\begin{vmatrix} 1 & 2 & 0 \\ 0 & 1 & 2 \\ 0 & 0 & 1 \end{vmatrix}$

26. $\begin{vmatrix} -1 & 2 & 1 \\ 2 & 1 & -3 \\ 1 & 1 & 1 \end{vmatrix}$

27. $\begin{vmatrix} 1 & -2 & 3 \\ -2 & 1 & 1 \\ -3 & -2 & 1 \end{vmatrix}$

28. $\begin{vmatrix} 1 & 1 & 2 \\ 2 & 1 & -2 \\ 3 & 1 & 3 \end{vmatrix}$

29. $\begin{vmatrix} 1 & 0 & 1 \\ 0 & 1 & 0 \\ 1 & 1 & 1 \end{vmatrix}$

30. $\begin{vmatrix} 3 & 5 & 1 \\ 6 & -2 & 2 \\ 8 & -1 & 4 \end{vmatrix}$

31. $\begin{vmatrix} 1 & 2 & 1 \\ -3 & 7 & 3 \\ -4 & 3 & -5 \end{vmatrix}$

32. $\begin{vmatrix} 1 & 4 & 7 \\ 2 & 5 & 8 \\ 3 & 6 & 9 \end{vmatrix}$

In Exercises 33–54, use Cramer's rule to solve each system of equations, if possible. If a system is inconsistent or if the equations are dependent, so indicate.

33. $\begin{cases} x + y = 6 \\ x - y = 2 \end{cases}$

34. $\begin{cases} x - y = 4 \\ 2x + y = 5 \end{cases}$

35. $\begin{cases} 2x + 3y = 0 \\ 4x - 6y = -4 \end{cases}$

36. $\begin{cases} 4x - 3y = -1 \\ 8x + 3y = 4 \end{cases}$

37. $\begin{cases} 3x + 2y = 11 \\ 6x + 4y = 11 \end{cases}$

38. $\begin{cases} 5x + 6y = 12 \\ 10x + 12y = 24 \end{cases}$

39. $\begin{cases} y = \frac{-2x+1}{3} \\ 3x - 2y = 8 \end{cases}$

40. $\begin{cases} 2x + 3y = -1 \\ x = \frac{y-9}{4} \end{cases}$

41. $\begin{cases} x + y + z = 4 \\ x + y - z = 0 \\ x - y + z = 2 \end{cases}$

42. $\begin{cases} x + y + z = 4 \\ x - y + z = 2 \\ x - y - z = 0 \end{cases}$

43. $\begin{cases} x + y + 2z = 7 \\ x + 2y + z = 8 \\ 2x + y + z = 9 \end{cases}$

44. $\begin{cases} x + 2y + 2z = 10 \\ 2x + y + 2z = 9 \\ 2x + 2y + z = 1 \end{cases}$

45. $\begin{cases} 2x + y + z = 5 \\ x - 2y + 3z = 10 \\ x + y - 4z = -3 \end{cases}$

46. $\begin{cases} 3x + 2y - z = -8 \\ 2x - y + 7z = 10 \\ 2x + 2y - 3z = -10 \end{cases}$

47. $\begin{cases} 4x - 3y = 1 \\ 6x - 8z = 1 \\ 2y - 4z = 0 \end{cases}$

48. $\begin{cases} 4x + 3z = 4 \\ 2y - 6z = -1 \\ 8x + 4y + 3z = 9 \end{cases}$

49. $\begin{cases} 2x + 3y + 4z = 6 \\ 2x - 3y - 4z = -4 \\ 4x + 6y + 8z = 12 \end{cases}$

50. $\begin{cases} x - 3y + 4z - 2 = 0 \\ 2x + y + 2z - 3 = 0 \\ 4x - 5y + 10z - 7 = 0 \end{cases}$

51. $\begin{cases} 2x + y - z - 1 = 0 \\ x + 2y + 2z - 2 = 0 \\ 4x + 5y + 3z - 3 = 0 \end{cases}$

52. $\begin{cases} 2x - y + 4z + 2 = 0 \\ 5x + 8y + 7z = -8 \\ x + 3y + z + 3 = 0 \end{cases}$

53. $\begin{cases} x + y = 1 \\ \frac{1}{2}y + z = \frac{5}{2} \\ x - z = -3 \end{cases}$

54. $\begin{cases} \frac{1}{2}x + y + z + \frac{3}{2} = 0 \\ x + \frac{1}{2}y + z - \frac{1}{2} = 0 \\ x + y + \frac{1}{2}z + \frac{1}{2} = 0 \end{cases}$

APPLICATIONS

Write a system of equations to solve each problem. Then use Cramer's rule to solve the system.

55. INVENTORY Illustration 1 shows an end-of-the-year inventory report for a warehouse that supplies electronics stores. If the warehouse stocks two models of cordless telephones, one valued at $67 and the other at $100, how many of each model of phone did the warehouse have at the time of the inventory?

ILLUSTRATION 1

Item	Number	Merchandise value
Televisions	800	$1,005,450
Radios	200	$15,785
Cordless phones	360	$29,400

56. SIGNALING A system of sending signals uses two flags held in various positions to represent letters of the alphabet. Illustration 2 shows how the letter U is signaled. Find x and y, if y is to be 30° more than x.

ILLUSTRATION 2

57. INVESTING A student wants to average a 6.6% return by investing $20,000 in the three stocks listed in Illustration 3. Because HiTech is a high-risk investment, he wants to invest three times as much in SaveTel and OilCo combined as he invests in HiTech. How much should he invest in each stock?

ILLUSTRATION 3

Stock	Rate of return
HiTech	10%
SaveTel	5%
OilCo	6%

58. INVESTING A woman wants to average a $7\frac{1}{3}\%$ return by investing \$30,000 in three certificates of deposit. (See Illustration 4.) She wants to invest five times as much in the 8% CD as in the 6% CD. How much should she invest in each CD?

ILLUSTRATION 4

Type of CD	Rate of return
12-month	6%
24-month	7%
36-month	8%

In Exercises 59–62, use a calculator with matrix capabilities to evaluate each determinant.

59. $\begin{vmatrix} 2 & -3 & 4 \\ -1 & 2 & 4 \\ 3 & -3 & 1 \end{vmatrix}$

60. $\begin{vmatrix} -3 & 2 & -5 \\ 3 & -2 & 6 \\ 1 & -3 & 4 \end{vmatrix}$

61. $\begin{vmatrix} 2 & 1 & -3 \\ -2 & 2 & 4 \\ 1 & -2 & 2 \end{vmatrix}$

62. $\begin{vmatrix} 4 & 2 & -3 \\ 2 & -5 & 6 \\ 2 & 5 & -2 \end{vmatrix}$

WRITING

63. Tell how to find the minor of an element of a determinant.

64. Tell how to find x when solving a system of three linear equations by Cramer's rule. Use the words *coefficients* and *constants* in your explanation.

REVIEW

65. Are the lines $y = 2x - 7$ and $x - 2y = 7$ perpendicular?

66. Are the lines $y = 2x - 7$ and $2x - y = 10$ parallel?

67. Are the equations $y = 2x - 7$ and $f(x) = 2x - 7$ the same?

68. How are the graphs of $f(x) = x^2$ and $g(x) = x^2 - 2$ related?

69. For the linear function $y = 2x - 7$, what variable is associated with the domain?

70. Is the graph of a circle the graph of a function?

71. The graph of a line passes through $(0, -3)$. Is this the x-intercept or the y-intercept of the line?

72. What is the name of the function $f(x) = |x|$?

73. For the function $y = 2x^2 + 6x + 1$, what is the independent variable and what is the dependent variable?

74. If $f(x) = x^3 - x$, what is $f(-1)$?

Systems of Equations

In Chapter 3, we have solved problems involving two and three variables by writing and solving a **system of equations.**

Solutions of a System of Equations

A solution of a system of equations involving two or three variables is an ordered pair or an ordered triple whose coordinates satisfy each equation of the system. In Exercises 1 and 2, decide whether the given ordered pair or ordered triple is a solution of the system.

1. $\begin{cases} 2x - y = 1 \\ 4x + 2y = 0 \end{cases} \left(\dfrac{1}{4}, -\dfrac{1}{2} \right)$

2. $\begin{cases} 2x - y + z = 9 \\ 3x + y - 4z = 8 \\ 2x - 7z = -1 \end{cases} (4, 0, 1)$

Methods of Solving Systems of Linear Equations

We have studied several methods for solving systems of two and three linear equations.

3. Solve $\begin{cases} 2x + 5y = 8 \\ y = 3x + 5 \end{cases}$ using the *graphing method.*

4. Solve $\begin{cases} 9x - 8y = 1 \\ 6x + 12y = 5 \end{cases}$ using the *addition method.*

5. Solve $\begin{cases} 4x - y - 10 = 0 \\ 3x + 5y = 19 \end{cases}$ using the *substitution method.*

6. Solve $\begin{cases} -x + 3y + 2z = 5 \\ 3x + 2y + z = -1 \\ 2x - y + 3z = 4 \end{cases}$ using the *addition method.*

7. Solve $\begin{cases} x - 6y = 3 \\ x + 3y = 21 \end{cases}$ using matrices.

8. Solve $\begin{cases} x + 2z = 7 \\ 2x - y + 3z = 9 \\ y - z = 1 \end{cases}$ using Cramer's rule.

Dependent Equations and Inconsistent Systems

If the equations in a system of two linear equations are dependent, the system has infinitely many solutions. An inconsistent system has no solutions.

9. Suppose you are solving a system of two equations by the addition method, and you obtain the following.

$$2x - 3y = 4$$
$$\underline{-2x + 3y = -4}$$
$$0 = 0$$

What can you conclude?

10. Suppose you are solving a system of two equations by the substitution method, and you obtain

$$-2(x - 3) + 2x = 7$$
$$-2x + 6 + 2x = 7$$
$$6 = 7$$

What can you conclude?

Accent on Teamwork

Section 3.1

Line graphs Find five examples of line graphs where two lines intersect. (See Illustration 1.) Your school library is a good resource. Ask to look through the library's collection of recent magazines and newspapers. For each example, explain the information given by the point of intersection of the graphs.

ILLUSTRATION 1

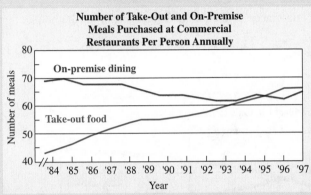

Section 3.2

Comparing solutions Example 8 of Section 3.2, involving water treatment, was solved using two variables. This problem can also be solved using one variable. Solve it using one variable and then comment on which method you like better. What are the advantages and the drawbacks of each method?

Break-point analysis Suppose you are a financial analyst for the coathanger company mentioned in Example 10 of Section 3.2. It is your job to decide whether the company should purchase the new machine.

First, graph the equations

$$C = 1.5x + 400$$
$$C = 1.25x + 500$$

on the same coordinate system. Then write a report that could be given to company managers, explaining their options concerning the purchase of the new machine. Under what conditions should they keep the machine

now in use? Under what conditions should they buy the new machine?

Section 3.3

Graphs of systems of equations From your textbook, find and then graph examples of each of the following types of systems of two linear equations:

- Consistent system
- Inconsistent system
- Dependent equations

Then make cardboard models of each of the graphs of the systems of three linear equations illustrated in Figure 3-10 and Figure 3-11 of Section 3.3. Make a presentation to your class using the graphs and cardboard models as visual aids to help you explain whether each system has a solution, and if so, what form the solution takes.

Section 3.4

Gaussian elimination Use matrices and elementary row operations to show that the solution of this 4×4 system of linear equations is $(1, 1, 0, 1)$.

$$\begin{cases} x + y + z + w = 3 \\ x - y - z - w = -1 \\ x + y - z - w = 1 \\ x + y - z + w = 3 \end{cases}$$

Section 3.5

Methods of solution Have each person in your group solve the system

$$\begin{cases} x - y = 4 \\ 2x + y = 5 \end{cases}$$

in a different way. The methods to use are graphing, addition, substitution, matrices, and Cramer's rule. Have each person briefly explain his or her method of solution. After everyone has presented a solution, discuss the advantages and drawbacks of each method. Can your group come to a consensus? Is there a favorite method?

Section 3.1

CONCEPTS

The graph of a linear equation is the graph of all points (x, y) on the rectangular coordinate system whose coordinates satisfy the equation.

To solve a system of two linear equations by the *graphing method*, find the coordinates of the point where the two graphs intersect.

Solving Systems by Graphing

REVIEW EXERCISES

1. See Illustration 1.
   a. Give three points that satisfy the equation $2x + y = 5$.

   b. Give three points that satisfy the equation $x - y = 4$.

   c. What is the solution of $\begin{cases} 2x + y = 5 \\ x - y = 4 \end{cases}$?

ILLUSTRATION 1

2. POLITICS Explain the importance of the points of intersection of the graphs shown in Illustration 2.

ILLUSTRATION 2
President Clinton's Job Approval Rating*

Based on data from *Los Angeles Times* (September 12, 1997)

If a system of equations has at least one solution, the system is a *consistent system*. When a system has no solution, it is called an *inconsistent system*.

If the graphs of the equations of a system are distinct, the equations are *independent equations*. Otherwise, the equations are *dependent equations*.

3. Solve each system by the graphing method, if possible. If a system is inconsistent or if the equations are dependent, so indicate.

   a. $\begin{cases} 2x + y = 11 \\ -x + 2y = 7 \end{cases}$

   b. $\begin{cases} y = -\frac{3}{2}x \\ 2x - 3y + 13 = 0 \end{cases}$

   c. $\begin{cases} \frac{1}{2}x + \frac{1}{3}y = 2 \\ y = 6 - \frac{3}{2}x \end{cases}$

   d. $\begin{cases} \frac{x}{3} - \frac{y}{2} = 1 \\ 6x - 9y = 3 \end{cases}$

Section 3.2

Solving Systems Algebraically

To solve a system by the *substitution method:*

1. Solve one equation for one of its variables.
2. Substitute the resulting expression for that variable into the other equation and solve that equation.
3. Find the value of the other variable by substituting the value of the variable found in step 2 into the equation from step 1.

To solve a system by the *addition method:*

1. Write both equations in general form.
2. Multiply the terms of one or both equations by constants so that the coefficients of one variable differ only in sign.
3. Add the equations from step 2 and solve the resulting equation.
4. Substitute the value obtained in step 3 into either original equation and solve for the remaining variable.

4. Solve each system using the substitution method, if possible. If a system is inconsistent or if the equations are dependent, so indicate.

a. $\begin{cases} x = y - 4 \\ 2x + 3y = 7 \end{cases}$

b. $\begin{cases} y = 2x + 5 \\ 3x - 5y = -4 \end{cases}$

c. $\begin{cases} 0.1x + 0.2y = 1.1 \\ 2x - y = 2 \end{cases}$

d. $\begin{cases} x = -2 - 3y \\ -2x - 6y = 4 \end{cases}$

5. Solve each system using the addition method, if possible.

a. $\begin{cases} x + y = -2 \\ 2x + 3y = -3 \end{cases}$

b. $\begin{cases} 2x - 3y = 5 \\ 2x - 3y = 8 \end{cases}$

c. $\begin{cases} x + \dfrac{1}{2}y = 7 \\ -2x = 3y - 6 \end{cases}$

d. $\begin{cases} y = \dfrac{x - 3}{2} \\ x = \dfrac{2y + 7}{2} \end{cases}$

6. To solve $\begin{cases} 5x - 2y = 19 \\ 3x + 4y = 1 \end{cases}$, which method, addition or substitution, would you use? Explain why.

7. Estimate the solution of the system

$\begin{cases} y = -\frac{2}{3}x \\ 2x - 3y = -4 \end{cases}$

from the graphs in Illustration 3. Then solve the system algebraically.

ILLUSTRATION 3

In Exercises 8–9, use two equations to solve each problem.

8. MILEAGE MAP See Illustration 4. The distance between Austin and Houston is 4 miles less than twice the distance between Austin and San Antonio. The round trip from Houston to Austin to San Antonio and back to Houston is 442 miles. Determine the mileages between Austin and Houston and between Austin and San Antonio.

ILLUSTRATION 4

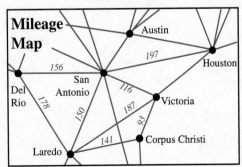

9. RIVERBOAT RIDE A Mississippi riverboat travels 30 miles downstream in three hours and then makes the return trip upstream in five hours. Find the speed of the riverboat in still water and the speed of the current.

Section 3.3

The solution of a system of three linear equations is an *ordered triple.*

To solve a system of linear equations with three variables:
1. Pick any two equations and eliminate a variable.
2. Pick a different pair of equations and eliminate the same variable.
3. Solve the resulting pair of equations.
4. Use substitution to find the value of the third variable.

Systems with Three Variables

10. Tell whether $(2, -1, 1)$ is a solution of the system $\begin{cases} x - y + z = 4 \\ x + 2y - z = -1. \\ x + y - 3z = -1 \end{cases}$

11. Solve each system, if possible.

a. $\begin{cases} x + y + z = 6 \\ x - y - z = -4 \\ -x + y - z = -2 \end{cases}$

b. $\begin{cases} 2x + 3y + z = -5 \\ -x + 2y - z = -6 \\ 3x + y + 2z = 4 \end{cases}$

c. $\begin{cases} x + y - z = -3 \\ x + z = 2 \\ 2x - y + 2z = 3 \end{cases}$

d. $\begin{cases} 3x + 3y + 6z = -6 \\ -x - y - 2z = 2 \\ 2x + 2y + 4z = -4 \end{cases}$

12. MIXING NUTS The owner of a produce store wanted to mix peanuts selling for $3 per pound, cashews selling for $9 per pound, and Brazil nuts selling for $9 per pound to get 50 pounds of a mixture that would sell for $6 per pound. She used 15 fewer pounds of cashews than peanuts. How many pounds of each did she use?

Section 3.4

A *matrix* is a rectangular array of numbers.

A system of linear equations can be represented by an *augmented matrix.*

Systems of linear equations can be solved using *Gaussian elimination* and *elementary row operations:*
1. Any two rows can be interchanged.
2. Any row can be multiplied by a nonzero constant.
3. Any row can be changed by adding a nonzero constant multiple of another row to it.

Solving Systems Using Matrices

13. Represent each system of equations using an augmented matrix.

a. $\begin{cases} 5x + 4y = 3 \\ x - y = -3 \end{cases}$

b. $\begin{cases} x + 2y + 3z = 6 \\ x - 3y - z = 4 \\ 6x + y - 2z = -1 \end{cases}$

14. Solve each system using matrices, if possible.

a. $\begin{cases} x - y = 4 \\ 3x + 7y = -18 \end{cases}$

b. $\begin{cases} x + 2y - 3z = 5 \\ x + y + z = 0 \\ 3x + 4y + 2z = -1 \end{cases}$

c. $\begin{cases} 16x - 8y = 32 \\ -2x + y = -4 \end{cases}$

d. $\begin{cases} x + 2y + 2z = 2 \\ 4x + 5y + 3z = 3 \\ 2x + y - z = 1 \end{cases}$

15. INVESTING One year, a couple invested a total of $10,000 in two projects. The first investment, a mini-mall, made a 6% profit. The other investment, a skateboard park, made a 12% profit. If their investments made $960, how much was invested at each rate? To answer this question, write a system of two equations and solve it using matrices.

Section 3.5

A *determinant* of a *square matrix* is a number.

To evaluate a 2×2 determinant:

$$\begin{vmatrix} a & b \\ c & d \end{vmatrix} = ad - bc$$

To evaluate a 3×3 determinant, we expand it by *minors* along any row or column using the *array of signs*.

Cramer's rule can be used to solve systems of linear equations.

Solving Systems Using Determinants

16. Evaluate each determinant.

a. $\begin{vmatrix} 2 & 3 \\ -4 & 3 \end{vmatrix}$

b. $\begin{vmatrix} -3 & -4 \\ 5 & -6 \end{vmatrix}$

c. $\begin{vmatrix} -1 & 2 & -1 \\ 2 & -1 & 3 \\ 1 & -2 & 2 \end{vmatrix}$

d. $\begin{vmatrix} 3 & -2 & 2 \\ 1 & -2 & -2 \\ 2 & 1 & -1 \end{vmatrix}$

17. Use Cramer's rule to solve each system, if possible.

a. $\begin{cases} 3x + 4y = 10 \\ 2x - 3y = 1 \end{cases}$

b. $\begin{cases} -6x - 4y = -6 \\ 3x + 2y = 5 \end{cases}$

c. $\begin{cases} x + 2y + z = 0 \\ 2x + y + z = 3 \\ x + y + 2z = 5 \end{cases}$

d. $\begin{cases} 2x + 3y + z = 2 \\ x + 3y + 2z = 7 \\ x - y - z = -7 \end{cases}$

18. VETERINARY MEDICINE The daily requirements of a balanced diet for an animal are shown in the nutritional pyramid in Illustration 5. The number of grams per cup of nutrients in three food mixes are shown in the table. How many cups of each mix should be used to meet the daily requirements for protein, carbohydrates, and essential fatty acids in the animal's diet? To answer this problem, write a system of three equations and solve it using Cramer's rule.

	Grams per cup		
	Protein	**Carbohydrates**	**Fatty Acids**
Mix A	5	2	1
Mix B	6	3	2
Mix C	8	3	1

ILLUSTRATION 5

| Vitamins |
| Minerals |
| Essential fatty acids 5 grams |
| Carbohydrates 10 grams |
| Quality protein 24 grams |

CHAPTER 3

Test

1. Solve $\begin{cases} 2x + y = 5 \\ y = 2x - 3 \end{cases}$ by graphing.

2. Use substitution to solve $\begin{cases} 2x - 4y = 14 \\ x + 2y = 7 \end{cases}$.

3. Use addition to solve $\begin{cases} 2x + 3y = -5 \\ 3x - 2y = 12 \end{cases}$.

4. Are the equations of the system
$$\begin{cases} 3(x + y) = x - 3 \\ -y = \dfrac{2x + 3}{3} \end{cases}$$
dependent or independent?

5. Is $\left(-1, -\frac{1}{2}, 5\right)$ a solution of $\begin{cases} x - 2y + z = 5 \\ 2x + 4y = -4 \\ -6y + 4z = 22 \end{cases}$?

6. Solve the system $\begin{cases} x + y + z = 4 \\ x + y - z = 6 \\ 2x - 3y + z = -1 \end{cases}$
using the addition method.

In Problems 7–8, write a system of equations to solve each problem.

7. In Illustration 1, find x and y, if y is 15 more than x.

ILLUSTRATION 1

8. ANTIFREEZE How much of a 40% antifreeze solution must a mechanic mix with an 80% antifreeze solution if 20 gallons of a 50% antifreeze solution are needed?

In Problems 9–10, use matrices to solve each system.

9. $\begin{cases} x + y = 4 \\ 2x - y = 2 \end{cases}$

10. $\begin{cases} x + y + 2z = -1 \\ x + 3y - 6z = 7 \\ 2x - y + 2z = 0 \end{cases}$

In Problems 11–12, evaluate each determinant.

11. $\begin{vmatrix} 2 & -3 \\ 4 & 5 \end{vmatrix}$

12. $\begin{vmatrix} 1 & 2 & 0 \\ 2 & 0 & 3 \\ 1 & -2 & 2 \end{vmatrix}$

In Problems 13–16, consider the system $\begin{cases} x - y = -6 \\ 3x + y = -6 \end{cases}$, which is to be solved using Cramer's rule.

13. When solving for x, what is the numerator determinant D_x? (**Don't evaluate it.**)

14. When solving for y, what is the denominator determinant D? (**Don't evaluate it.**)

15. Solve the system for x.

16. Solve the system for y.

17. Solve the following system for z only, using Cramer's rule.
$$\begin{cases} x + y + z = 4 \\ x + y - z = 6 \\ 2x - 3y + z = -1 \end{cases}$$

18. MOVIE TICKETS The receipts for one showing of a movie were $410 for an audience of 100 people. The ticket prices are given in the table. If twice as many children's tickets as general admission tickets were purchased, how many of each type of ticket were sold?

Ticket prices	
Children	$3.00
General Admission	$6.00
Seniors	$5.00

19. CANDY SALES Summarize the information that can be learned from the graph in Illustration 2. What do the points of intersection of the graphs tell us?

ILLUSTRATION 2

Holiday Candy Sales

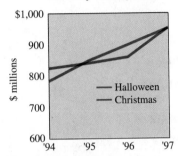

Based on data from *Los Angeles Times* (October 30, 1997)

20. Which method, substitution or addition, would you use to solve the following system?

$$\begin{cases} \dfrac{x}{2} - \dfrac{y}{4} = -4 \\ y = -2 - x \end{cases}$$

Explain your reasoning.

21. What does it mean to say that a system of two linear equations is an *inconsistent* system?

22. Suppose that two variables are used to solve an application problem. Why must two equations be written to solve the problem?

4

Inequalities

CAMPUS CONNECTION

The Physical Education Department

In an aerobics class, students perform exercises that improve their cardiovascular fitness. Cardiologists have determined what heart rate range participants need to maintain to get the most out of the training. For example, a 30-year-old woman needs to raise her heart rate to between 137 to 165 beats per minute and then sustain that rate for 10 to 12 minutes. In this chapter, you will see that these suggested ranges can be described using *inequalities*. Learning how to write and solve inequalities will allow you to solve many new types of application problems, including some from physical education.

When working with unequal quantities, we use inequalities instead of equations to describe the situation mathematically.

▶ 4.1 Solving Linear Inequalities

In this section, you will learn about

Inequalities ■ Graphs and interval notation ■ Properties of inequalities ■ Solving linear inequalities ■ Problem solving

Introduction Traffic signs like the one shown in Figure 4-1 often appear in front of schools. From the sign, a motorist knows that

FIGURE 4-1

- A speed *greater than* 25 miles per hour breaks the law and could possibly result in a ticket for speeding.
- A speed *less than or equal to* 25 miles per hour is within the posted speed limit.

Statements such as these can be expressed mathematically using *inequality symbols*.

Inequalities

Inequalities are statements indicating that two quantities are unequal. Inequalities can be recognized by one or more of the following symbols.

Inequality symbols

$a \neq b$	means	"a is not equal to b."
$a < b$	means	"a is less than b."
$a > b$	means	"a is greater than b."
$a \leq b$	means	"a is less than or equal to b."
$a \geq b$	means	"a is greater than or equal to b."

By definition, $a < b$ means that "a is less than b," but it also means that $b > a$. Furthermore, if a is to the left of b on the number line, then $a < b$. If a is to the right of b on a number line, then $a > b$.

We can use inequality symbols to describe the warning that the traffic sign in Figure 4-1 gives to drivers. If x represents the motorist's speed in miles per hour, he or she is in danger of receiving a speeding ticket if $x > 25$, and he or she is observing the posted speed limit if $x \leq 25$.

Graphs and Interval Notation

The graph of a set of real numbers that is a portion of a number line is called an **interval.** The graph shown in Figure 4-2(a) on the next page represents all real numbers that are greater than -5. This interval contains the numbers that *satisfy* the inequality

$x > -5$, such as -4, -1.8, 0, $2\frac{3}{4}$, π, and $1,050$. The parenthesis at -5 indicates that -5 is not included in the interval. We can also express this interval in **interval notation** as $(-5, \infty)$, where ∞ (read as **positive infinity**) indicates that the interval extends indefinitely to the right. The left parenthesis indicates that the endpoint -5 is not included.

FIGURE 4-2

Three ways to describe the real numbers greater than -5	*Three ways to describe the real numbers less than or equal to 7*
1. Graph: -5	1. Graph: $\overset{}{\underset{7}{\vdash}}$
2. Inequality: $x > -5$	2. Inequality: $x \le 7$
3. Interval notation: $(-5, \infty)$	3. Interval notation: $(-\infty, 7]$
(a)	**(b)**

The interval shown in Figure 4-2(b) is the graph of the real numbers less than or equal to 7. This interval contains the numbers that satisfy the inequality $x \le 7$. The bracket at 7 indicates that 7 is included in the interval. To express this interval in interval notation, we write $(-\infty, 7]$, where $-\infty$ (read as **negative infinity**) indicates that the interval extends indefinitely to the left. The bracket indicates that 7 is in the interval.

WARNING! The symbol ∞ does not represent a number. It indicates that an interval extends indefinitely to the right. We always use a parenthesis after the symbol ∞. For similar reasons, a parenthesis is always used in front of the symbol $-\infty$ when writing interval notation.

E X A M P L E   1

Graphing intervals. Write the inequality $x \ge 8$ in interval notation and then graph it.

Solution   The inequality $x \ge 8$ is satisfied by all real numbers that are greater than or equal to 8. This is the interval $[8, \infty)$. The graph is shown in Figure 4-3.

FIGURE 4-3

$\overset{}{\underset{8}{\vdash}}$

SELF CHECK   Write the inequality $x < 0$ in interval notation and then graph it.

Answer: $(-\infty, 0)$ $\overset{}{\underset{0}{)}}$ ∎

If an interval extends forever in one direction, as in the previous examples, it is called an **unbounded interval.** The following chart illustrates the various types of unbounded intervals and shows how they are described using an inequality and a graph.

Unbounded intervals

The interval (a, ∞) includes all real numbers x such that $x > a$.	$\overset{}{\underset{a}{(}}$
The interval $[a, \infty)$ includes all real numbers x such that $x \ge a$.	$\overset{}{\underset{a}{[}}$
The interval $(-\infty, a)$ includes all real numbers x such that $x < a$.	$\overset{}{\underset{a}{)}}$
The interval $(-\infty, a]$ includes all real numbers x such that $x \le a$.	$\overset{}{\underset{a}{]}}$
The interval $(-\infty, \infty)$ includes all real numbers x. The graph of this interval is the entire number line.	$\overset{}{\underset{0}{\leftrightarrow}}$

When graphing intervals, an open circle can be used to show that a point is not included in a graph, and a solid circle can be used to show that a point *is* included. For example,

We will use parentheses and brackets when graphing intervals, because they are consistent with interval notation.

Properties of Inequalities

To solve inequalities, we will use three properties of inequalities.

Property 1 of inequalities

> Any real number can be added to (or subtracted from) both sides of an inequality to produce another inequality with the same direction.

Property 1 indicates that any number can be added to both sides of a true inequality to get another true inequality with the same direction. To illustrate this property, we consider $3 < 12$ and add 4 to both sides.

$3 < 12$ A true inequality.

$3 + 4 < 12 + 4$ Add 4 to both sides.

$7 < 16$ The result is a true inequality. Note that the $<$ symbol is unchanged in this process.

Subtracting the same number from both sides of $3 < 12$ doesn't change the direction of the inequality either.

$3 < 12$ A true inequality.

$3 - 4 < 12 - 4$ Subtract 4 from both sides.

$-1 < 8$ The result is a true inequality.

Property 2 of inequalities

> If both sides of an inequality are multiplied (or divided) by a positive number, another inequality results, with the same direction as the original one.

Property 2 indicates that both sides of a true inequality can be multiplied by any positive number to get another true inequality with the same direction. To illustrate this property, we consider $-4 < 6$ and multiply both sides by 2.

$-4 < 6$ A true inequality.

$2(-4) < 2(6)$ Multiply both sides by 2.

$-8 < 12$ The result is a true inequality. The $<$ symbol is unchanged in this process.

Dividing both sides by the same positive number does not change the direction of the inequality either.

$-4 < 6$ A true inequality.

$\dfrac{-4}{2} < \dfrac{6}{2}$ Divide both sides by 2.

$-2 < 3$ The result is a true inequality. The $<$ symbol is unchanged in this process.

Property 3 of inequalities

If both sides of an inequality are multiplied (or divided) by a negative number, another inequality results, but with the opposite direction from the original inequality.

Property 3 indicates that if both sides of a true inequality are multiplied by a negative number, another true inequality results, but with the opposite direction. For example, we consider the inequality $-4 < 6$ and multiply both sides by -2.

$-4 < 6$ A true inequality containing a $<$ symbol.

$-2(-4) > -2(6)$ Multiply both sides by -2 and change $<$ to $>$.

$8 > -12$ The result is a true inequality. The inequality would not be true if we did not change the original symbol $<$ to $>$.

Dividing both sides by the same negative number also changes the direction of the inequality.

$-4 < 6$ A true inequality.

$\dfrac{-4}{-2} > \dfrac{6}{-2}$ Divide both sides by -2 and change $<$ to $>$.

$2 > -3$ The result is a true inequality.

WARNING! We must remember to change the direction of the inequality symbol every time we multiply or divide both sides of an inequality by a negative number.

Solving Linear Inequalities

In this section, we will work with **linear inequalities** in one variable.

Linear inequalities

A **linear inequality** in x is any inequality that can be expressed in one of the following forms, where a and c are real numbers and $a \neq 0$.

$$ax + c < 0 \qquad ax + c > 0 \qquad ax + c \leq 0 \qquad \text{or} \qquad ax + c \geq 0$$

Some examples of linear inequalities are $3x < 0$, $3(2x - 9) < 9$, and $-12x - 8 \leq 16$.

To **solve a linear inequality** means to find all the values that, when substituted for the variable, make the inequality true. Most of the inequalities we will solve have infinitely many solutions. The steps we will use to solve an inequality are the same as those used to solve an equation, with one exception. If we multiply or divide both sides of an inequality by a negative number, we must reverse the direction of the inequality symbol.

EXAMPLE 2

Solving linear inequalities. Solve $3(2x - 9) < 9$. Give the result in interval notation and then graph the interval.

Solution Our objective is to isolate x on the left-hand side of the inequality. To do that, we use the same strategy as we used to solve equations.

$$3(2x - 9) < 9$$
$$6x - 27 < 9 \qquad \text{Use the distributive property to remove parentheses.}$$
$$6x < 36 \qquad \text{To undo the subtraction of 27, add 27 to both sides.}$$
$$x < 6 \qquad \text{To undo the multiplication by 6, divide both sides by 6.}$$

The solution set is the interval $(-\infty, 6)$, whose graph is shown in Figure 4-4. The parenthesis at 6 indicates that 6 is not included in the solution set.

The solution set contains infinitely many real numbers. We cannot check to see if all of them satisfy the original inequality. As an informal check, we pick one number in the graph, such as 4, and see whether it satisfies the inequality.

$$3(2x - 9) < 9 \qquad \text{The original inequality.}$$
$$3[2(4) - 9] \overset{?}{<} 9 \qquad \text{Substitute 4 for } x. \text{ Read } \overset{?}{<} \text{ as "is possibly less than."}$$
$$3(8 - 9) \overset{?}{<} 9 \qquad \text{Do the multiplication: } 2(4) = 8.$$
$$3(-1) \overset{?}{<} 9 \qquad \text{Do the subtraction: } 8 - 9 = -1.$$
$$-3 < 9 \qquad \text{This is a true statement.}$$

FIGURE 4-4

Since $-3 < 9$, 4 satisfies the inequality. The solution appears to be correct.

SELF CHECK Solve $2(3x + 2) > -44$ and graph the interval. *Answer:* $(-8, \infty)$

EXAMPLE 3

Reversing the inequality symbol. Solve the inequality $-12x - 8 \leq 16$. Give the result in interval notation and then graph the interval.

Solution To solve this inequality, we need to isolate x.

$$-12x - 8 \leq 16$$
$$-12x \leq 24 \qquad \text{To undo the subtraction of 8, add 8 to both sides.}$$
$$x \geq -2 \qquad \text{To undo the multiplication by } -12, \text{ divide both sides by } -12. \text{ Because we are dividing by a negative number, we reverse the } \leq \text{ symbol.}$$

FIGURE 4-5

The solution set is the interval $[-2, \infty)$, whose graph is shown in Figure 4-5. The bracket at -2 indicates that -2 is included in the solution set.

SELF CHECK Solve $-6x + 6 \leq 0$ and graph it. *Answer:* $[1, \infty)$

EXAMPLE 4

Clearing an inequality of fractions. Solve $\frac{2}{3}(x + 2) > \frac{4}{5}(x - 3)$.

Solution   It will be easier to solve the inequality if we clear it of fractions. We do that by multiplying both sides by the LCD of $\frac{2}{3}$ and $\frac{4}{5}$.

$$\frac{2}{3}(x + 2) > \frac{4}{5}(x - 3)$$

$$15 \cdot \frac{2}{3}(x + 2) > 15 \cdot \frac{4}{5}(x - 3) \qquad \text{Multiply both sides by the LCD of } \frac{2}{3} \text{ and } \frac{4}{5}, \text{ which is 15.}$$

$$10(x + 2) > 12(x - 3) \qquad \text{Simplify: } 15 \cdot \frac{2}{3} = 10 \text{ and } 15 \cdot \frac{4}{5} = 12.$$

$$10x + 20 > 12x - 36 \qquad \text{Distribute the 10 and the 12.}$$

$$-2x + 20 > -36 \qquad \text{To eliminate } 12x \text{ on the right-hand side, subtract } 12x \text{ from both sides.}$$

$$-2x > -56 \qquad \text{Subtract 20 from both sides.}$$

$$x < 28 \qquad \text{Divide both sides by } -2 \text{ and reverse the } > \text{ symbol.}$$

FIGURE 4-6

The solution set is the interval $(-\infty, 28)$, whose graph is shown in Figure 4-6.

SELF CHECK   Solve $\frac{3}{2}(x + 2) < \frac{3}{5}(x - 3)$ and graph it.   *Answer:* $\left(-\infty, -\frac{16}{3}\right)$

$\xleftarrow{\qquad}\xrightarrow{\qquad}$
$-16/3$ ∎

WARNING! When solving an inequality, the variable sometimes ends up on the right-hand side. For instance, suppose we solve an inequality and obtain $-3 < x$. This inequality can be expressed in the equivalent form $x > -3$, which most students find easier to graph and to express in interval notation.

ACCENT ON TECHNOLOGY *Solving Linear Inequalities*

There are several ways to solve linear inequalities graphically. For example, to solve $3(2x - 9) < 9$ we can subtract 9 from both sides and solve instead the equivalent inequality $3(2x - 9) - 9 < 0$. Using window settings of $[-10, 10]$ for x and $[-10, 10]$ for y, we graph $y = 3(2x - 9) - 9$ and then use TRACE. Moving the cursor closer and closer to the x-axis, as shown in Figure 4-7(a), we see that the graph is below the x-axis for x-values in the interval $(-\infty, 6)$. This interval is the solution, because in this interval, $3(2x - 9) - 9 < 0$.

FIGURE 4-7

(a)   (b)   (c)

Anther way to solve $3(2x - 9) < 9$ is to graph $y = 3(2x - 9)$ and $y = 9$. We can then trace to see that the graph of $y = 3(2x - 9)$ is below the graph of $y = 9$ for x-values in the interval $(-\infty, 6)$. See Figure 4-7(b). This interval is the solution, because in this interval, $3(2x - 9) < 9$.

A third approach is to enter and then graph

$Y_1 = 3(2x - 9)$

$Y_2 = 9$

$Y_3 = Y_1 < Y_2$ To do this, use the $\boxed{\text{VARS}}$ key. Consult your owner's manual for the specific directions.

The graphs of $y = 3(2x - 9)$, $y = 9$, and a horizontal line 1 unit above the x-axis will be displayed, as shown in Figure 4-7(c). In the TRACE mode, we then move the cursor to the rightmost endpoint of the horizontal line to determine that the interval $(-\infty, 6)$ is the solution of $3(2x - 9) < 9$.

Problem Solving

In previous chapters, we have used a five-step problem-solving strategy to solve problems. This process involved writing and then solving equations. We will now show how inequalities can be used to solve problems. To decide whether to use an equation or an inequality to solve a problem, you must spot key words and phrases. Here are some common statements that translate to inequalities.

The statement	*translates to*
a does not exceed b.	$a \leq b$
a is at most b.	$a \leq b$

The statement	*translates to*
a is at least b.	$a \geq b$
a is no less than b.	$a \geq b$

EXAMPLE 5

Translating from words to symbols. Translate the sentence to mathematical symbols: *The instructor said that the test would take no more than 50 minutes.*

Solution   Since the test will take no more than 50 minutes, it will take 50 minutes or less to complete. If we let t represent the time it takes to complete the test, then $t \leq 50$.

SELF CHECK   Translate the sentence to mathematical symbols:
A PG-13 movie rating means that you must be at least 13 years old to see the movie.   *Answer:* $a \geq 13$ ■

EXAMPLE 6

Political contributions. Some volunteers are making long-distance telephone calls to solicit contributions for their candidate. The calls are billed at the rate of 42¢ for the first three minutes and 11¢ for each additional minute or part thereof. If the campaign chairperson has ordered that the cost of each call is not to exceed $2.00, for how many minutes can a volunteer talk to a prospective donor on the phone?

ANALYZE THE PROBLEM   We are given the rate at which a call is billed. Since the cost of a call is not to exceed $2.00, the cost must be *less than or equal to* $2.00. This phrase indicates that we should write an *inequality* to find how long a volunteer can talk to a prospective donor.

FORM AN INEQUALITY   We will let x represent the total number of minutes that a call can last. Then the cost of a call will be 42¢ for the first three minutes plus 11¢ times the number of additional minutes, where the number of *additional* minutes is $x - 3$ (the total number of minutes minus the first 3 minutes). With this information, we can form an inequality.

The cost of the first three minutes	plus	the cost of the additional minutes	is not to exceed	$2.
0.42	+	$0.11(x - 3)$	$\leq$	2

SOLVE THE INEQUALITY   To simplify the computations, we first clear the inequality of decimals.

$$0.42 + 0.11(x - 3) \leq 2$$

$42 + 11(x - 3) \leq 200$ To eliminate the decimals, multiply both sides by 100.

$42 + 11x - 33 \leq 200$ Use the distributive property to remove parentheses.

$11x + 9 \leq 200$ Combine like terms.

$11x \leq 191$ Subtract 9 from both sides.

$x \leq 17.\overline{36}$ Divide both sides by 11.

STATE THE CONCLUSION   Since the phone company doesn't bill for part of a minute, the longest time a call can last is 17 minutes. If a call lasts for $17.\overline{36}$ minutes, it will be charged as an 18-minute call, and the cost will be $\$0.42 + \$0.11(15) = \$2.07$.

CHECK THE RESULT   If the call lasts 17 minutes, the cost will be $\$0.42 + \$0.11(14) = \$1.96$. This is less than $2.00. The result checks. ∎

STUDY SET

Section 4.1

VOCABULARY

In Exercises 1–8, fill in the blanks to make the statements true.

1. $<$, $>$, $\leq$, and $\geq$ are _____ symbols.

2. $(-\infty, 5)$ is an example of an unbounded _____.

3. The _____ on the right of the interval notation $(-\infty, 5)$, indicates that 5 is not included in the interval.

4. To _____ an inequality means to find all values of the variable that make the inequality true.

5. $3x + 2 \geq 7$ is an example of a _____ inequality.

6. ∞ is a symbol representing positive _____.

7. The symbol for "_____" is $<$.

8. The symbol for "_____" is $\geq$.

CONCEPTS

9. Describe each set of real numbers using interval notation and then graph it.
   a. All real numbers greater than 4

b. All real numbers less than -4

c. All real numbers less than or equal to 4

10. Match each interval with its graph.
   a. $(-\infty, -1]$ i
   b. $(-\infty, 1)$ ii
   c. $[-1, \infty)$ iii

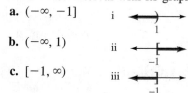

11. Classify each of the following as an equation, an expression, or an inequality.
   a. $-6 - 5x = 8$
   b. $5 - 2x$
   c. $7x - 5x > -4x$
   d. $-(7x - 9)$

12. In each case, tell what is wrong with the interval notation.
   a. $(\infty, -3)$

   b. $[-\infty, -3)$

13. In Illustration 1, which of the following are true?
   i. $b > 0$
   ii. $a - b < 0$
   iii. $ab > 0$

ILLUSTRATION 1

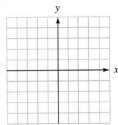

14. In Illustration 1, which of the following are false?
   i. $b - a > 0$
   ii. $ab < 0$
   iii. $b - a < 0$

15. Perform each step listed below on the inequality $4 > -2$. *Do not reverse the inequality symbol.* Is the resulting statement true?
   a. Add 2 to both sides.
   b. Subtract 4 from both sides.
   c. Multiply both sides by 4.
   d. Divide both sides by -2.

16. Consider the linear inequality $3x + 6 \leq 6$. Tell whether each value is a solution of the inequality.
   a. 0   **b.** $\dfrac{2}{3}$
   c. -10 **d.** 1.5

17. What inequality symbol is suggested by each sentence?
   a. As many as 16 people were seriously injured.

   b. There are no fewer than 10 references to carpools in the speech.

18. Write an equivalent inequality with the variable on the left-hand side.
   a. $-10 > x$
   b. $\dfrac{7}{8} < x$
   c. $0 \leq x$

NOTATION

In Exercises 19–20, complete each solution.

19. Solve $-5x - 1 \geq -11$.
$$-5x - 1 \geq -11$$
$$-5x \geq \boxed{}$$
$$\frac{-5x}{-5} \boxed{} \frac{-10}{-5}$$
$$x \leq 2$$
Using interval notation, the result is $\left(\boxed{}, 2\right]$.

20. Solve $3 - 6x < 17 + x$.
$$3 - 6x < 17 + x$$
$$3 - \boxed{} < 17$$
$$-7x < \boxed{}$$
$$\frac{-7x}{-7} \boxed{} \frac{14}{-7}$$
$$x > -2$$
Using interval notation, the result is $\left(-2, \boxed{}\right)$.

21. Explain the meaning of $(2, 3)$ in each context.
   a.

   b.

   $\xleftarrow{} \overset{\displaystyle -3\ -2\ -1\ \ 0\ \ 1\ \ 2\ \ 3}{|\ \ |\ \ |\ \ |\ \ |\ \ |\ \ |} \xrightarrow{}$

22. The solution set of a linear inequality in x is graphed in Illustration 2. Tell whether a true or false statement results when
   a. -4 is substituted for x.   **ILLUSTRATION 2**

   b. -3 is substituted for x.

   c. 0 is substituted for x.

PRACTICE

In Exercises 23–48, solve each inequality. Give the result in interval notation and then graph the solution set.

23. $3x > -9$

24. $4x < -36$

25. $-30y \leq -600$

26. $-6y \geq -600$

27. $0.6x \geq 36$

28. $0.2x < 8$

29. $3 > -\dfrac{9}{10}x$

30. $-\dfrac{2}{5} < -\dfrac{4}{5}x$

31. $x + 4 < 5$

32. $x - 5 > 2$

33. $-5t + 3 \leq 5$

34. $-9t + 6 \geq 16$

35. $-3x - 1 \leq 5$

36. $-2y + 6 < 16$

37. $7 < \dfrac{5}{3}a - 3$

38. $5 > \dfrac{7}{2}a - 9$

39. $0.4x + 0.4 \le 0.1x + 0.85$

40. $0.05 - 0.5x \le -0.7 - 0.8x$

41. $3(z - 2) \le 2(z + 7)$

42. $5(3 + z) > -3(z + 3)$

43. $-11(2 - b) < 4(2b + 2)$

44. $-9(h - 3) + 2h \le 8(4 - h)$

45. $\dfrac{1}{2}y + 2 \ge \dfrac{1}{3}y - 4$

46. $\dfrac{1}{4}x - \dfrac{1}{3} \le x + 2$

47. $\dfrac{2}{3}x + \dfrac{3}{2}(x - 5) \le x$

48. $\dfrac{5}{9}(x + 3) - \dfrac{4}{3}(x - 3) \ge x - 1$

APPLICATIONS

49. REAL ESTATE Refer to the graph in Illustration 3. For which regions of the country was the following inequality true in 1997?

Median sales price < U.S. median price

ILLUSTRATION 3

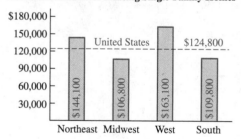

1997 Median Price of Existing Single-Family Homes

Based on data from the National Association of Realtors

50. MUSIC INDUSTRY Refer to the graph in Illustration 4. Determine the first full year in which the following inequality was true.

CDs shipped > cassettes shipped

51. GEOMETRY The **triangle inequality** states an important relationship between the sides of any triangle:

The sum of the lengths of any two sides of a triangle > the length of the third side

Use the triangle inequality to show that the dimensions of the shuffleboard court shown in Illustration 5 must be mislabeled.

ILLUSTRATION 4

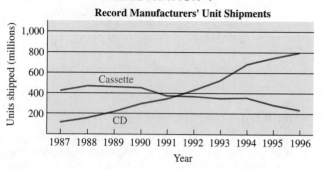

Record Manufacturers' Unit Shipments

Based on information from *The Wall Street Journal Almanac* (1998), p. 915

ILLUSTRATION 5

52. COMPUTER PROGRAMMING Flow charts like the one in Illustration 6 are used by programmers to show the step-by-step instructions of a computer program. For row 1 in the table, work through the steps of the flow chart using the values of a, b, and c, and tell what the computer printout would be. Now do the same for row 2, and then for row 3.

	a	b	c
Row 1	1	1	1
Row 2	9	−12	4
Row 3	11	−25	−24

ILLUSTRATION 6

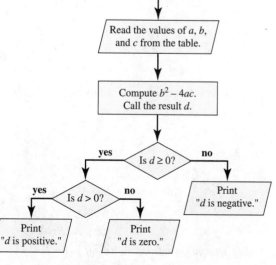

53. FUND RAISING A school PTA wants to rent a dunking tank for its annual school fund raising carnival. The cost is $85.00 for the first three hours and then $19.50 for each additional hour or part thereof. How long can the tank be rented if up to $185 is budgeted for this expense?

54. INVESTMENTS If a woman has invested $10,000 at 8% annual interest, how much more must she invest at 9% so that her annual income will exceed $1,250?

55. BUYING A COMPUTER A student who can afford to spend up to $2,000 sees the ad shown in Illustration 7. If she decides to buy the computer, find the greatest number of CD-ROMs that she can also purchase. (Disregard sales tax.)

ILLUSTRATION 7

Big Sale!!!!

◀ ||| $1,695.95

All CD-ROMs
$19.95

56. AVERAGING GRADES A student has scores of 70, 77, and 85 on three government exams. What score does she need on a fourth exam to give her an average of 80 or better?

57. WORK SCHEDULE A student works two part-time jobs. He earns $7 an hour for working at the college library and $12 an hour for construction work. To save time for study, he limits his work to 20 hours a week. If he enjoys the work at the library more, how many hours can he work at the library and still earn at least $175 a week?

58. SCHEDULING EQUIPMENT An excavating company charges $300 an hour for the use of a backhoe and $500 an hour for the use of a bulldozer. (Part of an hour counts as a full hour.) The company employs one operator for 40 hours per week to operate the machinery. If the company wants to bring in at least $18,500 each week from equipment rental, how many hours per week can it schedule the operator to use a backhoe?

59. MEDICAL PLANS A college provides its employees with a choice of the two medical plans shown in Illustration 8. For what size hospital bills is Plan 2 better for the employee than Plan 1? (*Hint:* The cost to the employee includes both the deductible payment and the employee's coinsurance payment.)

ILLUSTRATION 8

Plan 1	Plan 2
Employee pays $100 Plan pays 70% of the rest	Employee pays $200 Plan pays 80% of the rest

60. MEDICAL PLANS To save costs, the college in Exercise 59 raised the employee deductible, as shown in Illustration 9. For what size hospital bills is Plan 2 better for the employee than Plan 1? (*Hint:* The cost to the employee includes both the deductible payment and the employee's coinsurance payment.)

ILLUSTRATION 9

Plan 1	Plan 2
Employee pays $200 Plan pays 70% of the rest	Employee pays $400 Plan pays 80% of the rest

In Exercises 61–64, use a graphing calculator to solve each inequality.

61. $2x + 3 < 5$ **62.** $3x - 2 > 4$

63. $5x + 2 \geq 4x - 2$ **64.** $3x - 4 \leq 2x + 4$

WRITING

65. The techniques for solving linear equations and linear inequalities are similar, yet different. Explain.

66. Explain how the symbol ∞ is used in this section. Is ∞ a real number?

67. Explain how to use the graph in Illustration 10 to solve $2x + 1 < 3$.

ILLUSTRATION 10

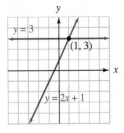

68. Explain each property of inequalities that is symbolically illustrated below.
   a. If $a < b$, and c is any real number, then $a + c < b + c$.
   b. If $a < b$, and c is any negative real number, then $ac > bc$.

In Exercises 69–70, use the graph of the function to find $f(-1)$, $f(0)$, *and* $f(2)$.

69.

70.

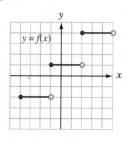

In Exercises 71–72, complete each input/output table.

71. $f(x) = x - x^3$

72. $g(t) = \dfrac{t^2 - 1}{5}$

input	output
-2	
2	

input	output
-6	
4	

▶ 4.2

Solving Compound Inequalities

In this section, you will learn about

Solving compound inequalities containing the word *and* ■ Double linear inequalities ■ Compound inequalities containing the word *or* ■ Solving compound inequalities containing the word *or*

Introduction   The label on the tube of antibiotic ointment shown in Figure 4-8 advises the user about the temperature at which the medication should be stored. A careful reading of the statement reveals that the storage instruction consists of two parts:

The storage temperature should be at least 59°F

and

The storage temperature should be at most 77°F

FIGURE 4-8

DIRECTIONS: Clean the affected area thoroughly. Apply a small amount of this product (an amount equal to the surface area of the tip of a finger) on the area 1 to 3 times daily. Do not use in eyes. **Store at 59° to 77°F.** Do not use longer than 1 week. Keep this and all drugs out of the reach of children.

In mathematics, when the words *and* or *or* are used to connect pairs of inequalities, we call these statements *compound inequalities*. In this section, we will discuss the procedures used to solve three types of compound inequalities, as well as the notation used to express their solution sets.

Solving Compound Inequalities Containing the Word *and*

When two inequalities are joined with the word *and*, we call the statement a **compound inequality.** Some examples are

$$x \geq -3 \text{ and } x \leq 6$$

$$\frac{x}{2} + 1 > 0 \text{ and } 2x - 3 < 5$$

$$x + 3 \leq 2x - 1 \text{ and } 3x - 2 < 5x - 4$$

The solution set of these inequalities contains all the numbers that make *both* of the inequalities true. For example, we can find the solution set of the compound inequality $x \geq -3$ and $x \leq 6$ by first graphing the solution sets of each inequality on the same number line and then looking for the numbers common to both graphs.

In Figure 4-9(a), the graph of the solution set of $x \geq -3$ is shown in red, and the graph of the solution set of $x \leq 6$ is shown in blue. Figure 4-9(b) shows the graph of the solution of $x \geq -3$ and $x \leq 6$. The shaded interval in Figure 4-9(b) is where the red and blue graphs overlap in Figure 4-9(a). It represents the numbers common to the graphs of $x \geq -3$ and $x \leq 6$.

FIGURE 4-9

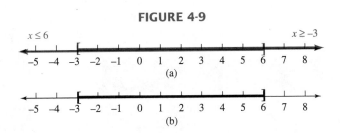

The solution set of $x \geq -3$ and $x \leq 6$ can be denoted by the **bounded interval** $[-3, 6]$, where the brackets indicate that the endpoints, -3 and 6, are included. It represents all real numbers between -3 and 6, including -3 and 6. Intervals such as this, which contain both endpoints, are called **closed intervals.**

When solving a compound inequality containing *and,* the solution set is the **intersection** of the solution sets of the two inequalities. The intersection of two sets is the set of elements that are common to both sets. We can denote the intersection of two sets using the symbol $\cap$, which is read as "intersection." For the compound inequality $x \geq -3$ and $x \leq 6$, we can write

$$[-3, \infty) \cap (-\infty, 6] = [-3, 6]$$

The solution set of the compound inequality $x \geq -3$ and $x \leq 6$ can be expressed in three other ways:

1. *As a graph:* ◄─────┤═══════╟─────►
                              -3   6

2. *In interval notation:* $[-3, 6]$

3. *In words:* all real numbers between -3 and 6, including -3 and 6

EXAMPLE 1

Solving compound inequalities containing the word *and.* Solve $\frac{x}{2} + 1 > 0$ and $2x - 3 < 5$.

Solution   We begin by solving each linear inequality separately.

$$\frac{x}{2} + 1 > 0 \qquad \text{and} \qquad 2x - 3 < 5$$

$$\frac{x}{2} > -1 \qquad\qquad\qquad 2x < 8$$

$$x > -2 \qquad\qquad\qquad x < 4$$

Next, we graph the solutions of each inequality on the same number line and determine their intersection. See Figure 4-10.

FIGURE 4-10

FIGURE 4-11

The intersection of the graphs in Figure 4-10 is the set of all real numbers between -2 and 4. Using interval notation, the solution set is the interval $(-2, 4)$, whose graph is shown in Figure 4-11. This bounded interval, which does not include either endpoint, is called an **open interval.**

SELF CHECK   Solve $3x > -18$ and $\frac{x}{5} - 1 \leq 1$ *Answer:* $(-6, 10]$
and graph the solution set.

The solution of the compound inequality in the Self Check of Example 1 is the interval $(-6, 10]$. A bounded interval such as this, which includes only one endpoint, is called a **half-open interval.** The following chart shows the various types of bounded intervals, along with the inequalities and interval notation that describes them.

Intervals

Open intervals	The interval (a, b) includes all real numbers x such that $a < x < b$.	
Half-open intervals	The interval $[a, b)$ includes all real numbers x such that $a \leq x < b$.	
	The interval $(a, b]$ includes all real numbers x such that $a < x \leq b$.	
Closed intervals	The interval $[a, b]$ includes all real numbers x such that $a \leq x \leq b$.	

WARNING! Remember that in interval notation, the notation $(-3, 4)$ represents the set of real numbers between -3 and 4, not the coordinates of a point on a graph.

EXAMPLE 2   **Solving compound inequalities containing the word *and*.** Solve $x + 3 \leq 2x - 1$ and $3x - 2 < 5x - 4$.

Solution   We solve each inequality separately.

$$x + 3 \leq 2x - 1 \qquad \text{and} \qquad 3x - 2 < 5x - 4$$
$$4 \leq x \qquad\qquad\qquad 2 < 2x$$
$$x \geq 4 \qquad\qquad\qquad 1 < x$$
$$\qquad\qquad\qquad\qquad\qquad x > 1$$

FIGURE 4-12

Only those x where $x \geq 4$ and $x > 1$ are in the solution set. Since all numbers greater than or equal to 4 are also greater than 1, the solutions are the numbers x where $x \geq 4$. The solution set is the interval $[4, \infty)$, whose graph is shown in Figure 4-12.

SELF CHECK   Solve $2x + 3 < 4x + 2$ and $3x + 1 < 5x + 3$ *Answer:* $\left(\frac{1}{2}, \infty\right)$
and graph the solution set.

EXAMPLE 3   **A compound inequality with no solution.** Solve $x - 1 > -3$ and $2x < -8$.

Solution   We solve each inequality separately.

$$x - 1 > -3 \qquad \text{and} \qquad 2x < -8$$
$$x > -2 \qquad\qquad\qquad x < -4$$

We note that the graphs of the solution sets shown in Figure 4-13 do not intersect.

FIGURE 4-13

This means that there are no numbers that make both parts of the original compound inequality true. So the solution set is the empty set, which can be denoted ∅.

SELF CHECK Solve $2x - 3 < x - 2$ and $0 < x - 3.5$. *Answer:* no solution ∎

Double Linear Inequalities

Inequalities that contain two inequality symbols are called **double inequalities.** An example of a double inequality is

$-3 \leq 2x + 5 < 7$ Read as "-3 is less than or equal to $2x + 5$ and $2x + 5$ is less than 7."

Any double linear inequality can be rewritten as a compound inequality containing the word *and.* In general, the following is true.

Double linear inequalities

The compound inequality $c < x < d$ is equivalent to $c < x$ and $x < d$.

EXAMPLE 4 **Solving double linear inequalities.** Solve $-3 \leq 2x + 5 < 7$ and graph the solution set.

Solution This double inequality $-3 \leq 2x + 5 < 7$ means that

$-3 \leq 2x + 5$ and $2x + 5 < 7$

We could solve each linear inequality separately, but we note that each solution would involve the same steps: subtracting 5 from both sides and dividing both sides by 2. We can solve the double inequality more efficiently by leaving it in its original form and applying these steps to each of its three parts to isolate x in the middle.

$-3 \leq 2x + 5 < 7$

$-3 - 5 \leq 2x + 5 - 5 < 7 - 5$ To undo the addition of 5, subtract 5 from all three parts.

$-8 \leq 2x < 2$ Do the subtractions.

$\dfrac{-8}{2} \leq \dfrac{2x}{2} < \dfrac{2}{2}$ To undo the multiplication by 2, divide all three parts by 2.

$-4 \leq x < 1$ Do the divisions.

FIGURE 4-14

The solution set is the half-open interval $[-4, 1)$, whose graph is shown in Figure 4-14.

SELF CHECK Solve $-5 \leq 3x - 8 \leq 7$ and graph the solution set. *Answer:* [1, 5]   ∎

When multiplying or dividing all three parts of a double inequality by a negative number, don't forget to reverse the direction of *both* inequalities. As an example, we will solve $-15 < -5x \leq 25$.

$-15 < -5x \leq 25$

$\dfrac{-15}{-5} > \dfrac{-5x}{-5} \geq \dfrac{25}{-5}$ Divide all three parts by -5 to isolate x in the middle. Reverse both inequality signs.

$3 > x \geq -5$ Do the divisions.

$-5 \leq x < 3$ Write an equivalent compound inequality with the smallest number on the left.

Compound Inequalities Containing the Word *or*

A warning on the water temperature gauge of a commercial dishwasher, shown in Figure 4-15, cautions the operator to shut down the unit if

The water temperature goes below 140°

or

The water temperature goes above 160°

FIGURE 4-15

When two inequalities are joined with the word *or,* we also call the statement a compound inequality. Some examples are

$x < 140$ or $x > 160$

$x \le -3$ or $x \ge 2$

$\dfrac{x}{3} > \dfrac{2}{3}$ or $-(x - 2) > 3$

Solving Compound Inequalities Containing the Word *or*

The solution set of a compound inequality containing the word *or* contains all the numbers that make *one or the other or both* inequalities true. For example, we can find the solution set of the compound inequality $x \le -3$ or $x \ge 2$ by putting the graphs of each inequality on the same number line.

In Figure 4-16(a), the graph of the solution set of $x \le -3$ is shown in red, and the graph of the solution set of $x \ge 2$ is shown in blue. Figure 4-16(b) shows the graph of the solution set of $x \le -3$ or $x \ge 2$. This graph is a combination of the graph of $x \le -3$ with the graph of $x \ge 2$.

FIGURE 4-16

When solving a compound inequality containing *or,* the solution set is the **union** of the solution sets of the two inequalities. The union of two sets is the set of elements that are in either of the sets or both. We can denote the union of two sets using the symbol ∪, which is read as "union." For the compound inequality $x \le -3$ or $x \ge 2$, we can write the solution set using interval notation as

$(-\infty, -3] \cup [2, \infty)$

We can express the solution set of the compound inequality $x \leq -3$ or $x \geq 2$ in three other ways:

1. *As a graph:*

2. *In interval notation:* $(-\infty, -3] \cup [2, \infty)$

3. *In words:* all real numbers less than or equal to -3 or greater than or equal to 2

WARNING! In the statement $x \leq -3$ or $x \geq 2$, it is incorrect to string the inequalities together as $2 \leq x \leq -3$, because that would imply that $2 \leq -3$, which is false.

EXAMPLE 5   **Solving compound inequalities containing the word *or*.** Solve: $\frac{x}{3} > \frac{2}{3}$ or $-(x - 2) > 3$.

Solution   We solve each inequality separately.

$$\frac{x}{3} > \frac{2}{3} \qquad \text{or} \qquad -(x - 2) > 3$$
$$x > 2 \qquad\qquad\qquad -x + 2 > 3$$
$$-x > 1$$
$$x < -1$$

The graph of the solution set is obtained by graphing the solution sets of each inequality on the same number line. See Figure 4-17.

FIGURE 4-17

The union of the two solution sets consists of all real numbers less than -1 or greater than 2. Using interval notation, the solution set is the interval $(-\infty, -1) \cup (2, \infty)$. Its graph appears in Figure 4-18.

FIGURE 4-18

SELF CHECK   Solve: $\frac{x}{2} > 2$ or $-3(x - 2) > 0$ and graph the solution set.   *Answer:* $(-\infty, 2) \cup (4, \infty)$

EXAMPLE 6   **A compound inequality satisfied by all real numbers.** Solve: $x + 3 \geq -3$ or $-x > 0$.

Solution   We solve each inequality separately.

$$x + 3 \geq -3 \qquad \text{or} \qquad -x > 0$$
$$x \geq -6 \qquad\qquad\qquad x < 0$$

We graph the solution set of each inequality on the same number line in Figure 4-19.

FIGURE 4-19

The entire number line is shaded, which indicates that all real numbers satisfy the original compound inequality. Using interval notation the solution set is denoted $(-\infty, \infty)$. Its graph is shown in Figure 4-20.

FIGURE 4-20

SELF CHECK   Solve: $x - 1 < 5$ or $-2x \leq 10$ *Answer:* $(-\infty, \infty)$
and graph the solution set.

STUDY SET

Section 4.2

VOCABULARY

In Exercises 1–4, fill in the blanks to make the statements true.

1. $x \geq 3$ and $x \leq 4$ is a _____ inequality.

2. $-6 \leq x + 1 < 1$ is a _____ linear inequality.

3. The bounded _____ (2, 8] includes all real numbers x such that $2 < x \leq 8$.

4. $x \leq 3$ or $x > 5$ is a compound _____.

CONCEPTS

In Exercises 5–8, fill in the blanks to make the statements true.

5. The word *and* between two inequality statements requires that _____ of the inequalities must be true for the entire statement to be true.

6. The word *or* between two inequality statements requires that only _____ of the inequalities must be true for the entire statement to be true.

7. If the three parts of a double inequality are divided by a negative number, the direction of both inequality symbols must be _____.

8. The double inequality $-2 < 3x + 4 < 10$ can be written as $-2 < 3x + 4$ _____ $3x + 4 < 10$.

9. In each case, tell whether $x = -3$ is a solution of the compound inequality.

a. $\dfrac{x}{3} + 1 \geq 0$ and $2x - 3 < -10$

b. $2x \leq 0$ or $-3x < -5$

10. In each case, tell whether $x = -3$ is a solution of the double linear inequality.

a. $-1 < -3x + 4 < 12$

b. $-1 < -3x + 4 < 14$

11. Give the solution of each inequality in interval notation, if possible.

a. $x < -3$ and $x > 3$

b. $x < 3$ or $x > -3$

12. Give the solution of each inequality in interval notation, if possible.

a. $x < 0$ or $x \geq 0$

b. $x < 0$ and $x > 0$

13. Match each interval with its corresponding graph.

a. [2, 3)   i

b. (2, 3)   ii

c. [2, 3]   iii

14. Give the interval notation that describes each set. Then graph it.

a. The real numbers between -3 and 3

b. The real numbers less than -3 or greater than 3

c. The real numbers between -3 and 3, including 3

NOTATION

15. a. Graph $(-\infty, 2) \cup [3, \infty)$.

b. Graph $(-\infty, 3) \cap [-2, \infty)$.

16. Classify each interval as open, half-open, or closed.
 a. $(-2, 15]$
 b. $[-2, 15]$
 c. $(-2, 15)$
 d. $[-2, 15)$

17. What is incorrect about the double inequality $3 < -3x + 4 < -3$?

18. What set is denoted by the interval notation $(-\infty, \infty)$? Graph it.

PRACTICE

In Exercises 19–56, solve each compound inequality. Give the result (if one exists) in interval notation and graph the solution set.

19. $x > -2$ and $x \le 5$

20. $x \le -4$ and $x \ge -7$

21. $2.2x < -19.8$ and $-4x < 40$

22. $\dfrac{1}{2}x \le 2$ and $0.75x \ge -6$

23. $x + 3 < 3x - 1$ and $4x - 3 \le 3x$

24. $4x \ge -x + 5$ and $6 \ge 4x - 3$

25. $x + 2 < -\dfrac{1}{3}x$ and $-6x < 9x$

26. $5(x - 2) \ge 0$ and $-3x < 9$

27. $x - 1 \le 2(x + 2)$ and $x \le 2x - 5$

28. $5(x + 1) \le 4(x + 3)$ and $x + 12 < -3$

29. $4 \le x + 3 \le 7$

30. $-5.3 < x - 2.3 < -1.3$

31. $15 > 2x - 7 > 9$

32. $25 > 3x - 2 > 7$

33. $-2 < -b + 3 < 5$

34. $2 < -t - 2 < 9$

35. $-6 < -3(x - 4) \le 24$

36. $-4 \le -2(x + 8) < 8$

37. $2x + 1 \ge 5$ and $-3(x + 1) \ge -9$

38. $2(-2) \le 3x - 1$ and $3x - 1 \le -1 - 3$

39. $\dfrac{x}{0.7} + 5 > 4$ and $-4.8 \le \dfrac{3x}{-0.125}$

40. $-x < -2x$ and $3x > 2x$

41. $-4 > \dfrac{2}{3}x - 2 > -6$

42. $-6 \le \dfrac{1}{3}a + 1 < 0$

43. $0 \le \dfrac{4 - x}{3} \le 2$

44. $-2 \le \dfrac{5 - 3x}{2} \le 2$

45. $x \le -2$ or $x > 6$

46. $x \ge -1$ or $x \le -3$

47. $x - 3 < -4$ or $x - 2 > 0$

48. $4x < -12$ or $\dfrac{x}{2} > 4$

49. $3x + 2 < 8$ or $2x - 3 > 11$

50. $3x + 4 < -2$ or $3x + 4 > 10$

51. $x > 3$ or $x < 5$

52. $x < -15$ or $x > -100$

53. $-4(x + 2) \ge 12$ or $3x + 8 < 11$

54. $4.5x - 1 < -10$ or $6 - 2x \ge 12$

55. $4.5x - 2 > 2.5$ or $\dfrac{1}{2}x \le 1$

56. $0 < x$ or $3x - 5 > 4x - 7$

APPLICATIONS

57. BABY FURNITURE See Illustration 1. A company manufactures various sizes of playpens having perimeters between 128 and 192 inches, inclusive.

a. Complete the double inequality that mathematically describes the range of the perimeters of the playpens.

ILLUSTRATION 1

$$? \leq 4s \leq ?$$

b. Solve the double inequality to find the range of the side lengths of the playpens.

58. TRUCKING The distance that a truck can travel in 8 hours, at a constant rate of r mph, is given by $8r$. A trucker wants to travel at least 350 miles, and company regulations don't allow him to exceed 450 miles in one 8-hour shift.

a. Complete the double inequality that describes the mileage range of the truck.

$$? \leq 8r \leq ?$$

b. Solve the double inequality to find the range of the average rate (speed) of the truck for the 8-hour trip.

59. TREATING A FEVER Use the flow chart shown in Illustration 2 to determine what action should be taken for a 13-month-old child who has had a 99.8° temperature for 3 days and is not suffering any other symptoms. T represents the child's temperature, A the child's age in months, and S the number of hours the child has experienced the symptoms.

ILLUSTRATION 2

Is $T \geq 101°$ and $A < 3$? →**yes**→ See doctor now.

↓ **no**

Is there stiffness of the neck, confusion, marked irritability, rapid breathing, lethargy, or a seizure? →**yes**→ See doctor now.

↓ **no**

Is $3 \leq A \leq 12$ and $S > 24$? →**yes**→ Call doctor today.

↓ **no**

Has fever shown no improvement in last 72 hours or is $S > 120$? →**yes**→ See doctor today.

↓ **no**

Is there sore throat, ear pain, cough, abdominal pains, skin rash, diarrhea, urinary frequency, or other symptoms? →**yes**→ See the applicable article about the specific problem.

↓ **no**

Apply home treatment.

Based on information from *Take Care of Yourself* (Addison-Wesley, 1993)

60. THERMOSTAT The *Temp range* control on the thermostat shown in Illustration 3 directs the heater to come on when the room temperature gets 5 degrees below the *Temp setting*; it directs the air conditioner to come on when the room temperature gets 5 degrees above the *Temp setting*. Use interval notation to describe

a. the temperature range for the room when neither the heater nor the air conditioner will be on.

b. the temperature range for the room when either the heater or air conditioner will be on. (*Note:* The lowest temperature theoretically possible is −460° F, called *absolute zero.*)

ILLUSTRATION 3

61. HEALTH CARE Refer to the graph in Illustration 4. Let P represent the percent of people covered by private insurance, M the percent covered by Medicare/Medicaid, and N the percent not covered. For what years are the following true?

a. $P \geq 70$ and $N \geq 15$

b. $P \geq 70$ or $N \geq 15$

c. $P \geq 72$ and $M \leq 10$

d. $P \geq 72$ or $M \leq 10$

ILLUSTRATION 4

U.S. Health Insurance Coverage Status (in percent), 1987-1997		
Private insurance ☐ Medicaid/Medicare ☐ Not covered ☐		
1987	75.5	8.4	12.9
1989	74.6	8.6	13.6
1991	72.1	10.7	14.1
1993	70.2	12.2	15.3
1995	70.3	12.1	15.4
1997	70.1	10.6	16.1

Based on data from U.S. Bureau of the Census

62. POLLS For each response to the poll question shown in Illustration 5, the *margin of error* is +/− (read as "plus or minus") 3.2%. This means that for the statistical methods used to do the polling, the actual response could be as much as 3.2 points more or 3.2 points less than shown. Use interval notation to describe the possible interval (in percent) for each response.

ILLUSTRATION 5

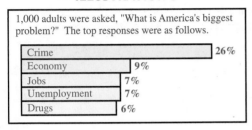

1,000 adults were asked, "What is America's biggest problem?" The top responses were as follows.

Crime	26%
Economy	9%
Jobs	7%
Unemployment	7%
Drugs	6%

WRITING

63. Explain how to find the union and how to find the intersection of $(-\infty, 5)$ and $(-2, \infty)$ graphically.

64. Explain why the double inequality

$$2 < x < 8$$

can be written in the equivalent form

$$2 < x \text{ and } x < 8.$$

65. Each of the shaded regions in the **Venn diagram** in Illustration 6 represents a set. Describe the intersection of set A and set B.

ILLUSTRATION 6

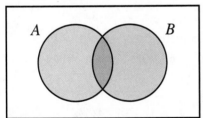

66. See Exercise 65. Describe the union of set A and set B.

REVIEW

In Exercises 67–70, refer to the chart in Illustration 7, which shows the results of each of the games of the eventual champion, the University of Kentucky, in the 1998 NCAA Men's Basketball Tournament. Round to the nearest tenth when necessary.

ILLUSTRATION 7

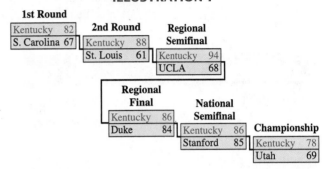

67. What are the mean, median, and mode of the set of Kentucky scores?

68. What are the mean and the median of the set of scores of Kentucky's opponents?

69. Find the margin of victory for Kentucky in each of its games. Then find the average (mean) margin of victory for Kentucky in the tournament.

70. What was the average (mean) combined score for Kentucky and its opponents in the tournament?

▶ 4.3 Solving Absolute Value Equations and Inequalities

In this section, you will learn about

> Absolute value ■ Absolute value functions ■ Equations of the form $|x| = k$ ■ Equations with two absolute values ■ Inequalities of the form $|x| < k$ ■ Inequalities of the form $|x| > k$

Introduction Many quantities studied in mathematics, science, and engineering are expressed as positive numbers. To guarantee that a quantity is positive, we often use the concept of absolute value. In this section, we will work with equations and inequalities containing expressions involving absolute value. Using the definition of absolute value, we will develop procedures to solve absolute value equations and absolute value inequalities.

Absolute Value

FIGURE 4-21

4 units   4 units

−4   0   4

Recall that the absolute value of any real number is the distance between the number and zero on the number line. For example, the points shown in Figure 4-21 with coordinates of 4 and −4 both lie 4 units from 0. Thus, $|4| = |-4| = 4$.

The absolute value of a real number can be defined more formally.

Absolute value

If $x \geq 0$, then $|x| = x$.
If $x < 0$, then $|x| = -x$.

This definition gives a way to associate a nonnegative real number with any real number.

- If $x \geq 0$, then x (which is positive or 0) is its own absolute value.
- If $x < 0$, then $-x$ (which is positive) is the absolute value.

Either way, $|x|$ is positive or 0. That is, $|x| \geq 0$, for all real numbers x.

EXAMPLE 1   **Finding absolute values.** Find   **a.** $|9|$,   **b.** $|-5.68|$,   and   **c.** $|0|$.

Solution   **a.** Since $9 \geq 0$, the number 9 is its own absolute value: $|9| = 9$.

b. Since $-5.68 < 0$, the negative of -5.68 is the absolute value:

$$|-5.68| = -(-5.68) = 5.68$$

c. Since $0 \geq 0$, 0 is its own absolute value: $|0| = 0$.

SELF CHECK   Find   **a.** $|-3|$,   **b.** $|100.99|$,   and   *Answers:*   **a.** 3,   **b.** 100.99,
c. $|-2\pi|$.   **c.** 2π ■

WARNING! The placement of a − sign in an expression containing an absolute value symbol is important. For example, $|-19| = 19$, but $-|19| = -19$.

Absolute Value Functions

In Section 2.6, we graphed the absolute value function $f(x) = |x|$ and considered many translations of its graph. The work in Example 2 reviews these concepts.

EXAMPLE 2

Graph the absolute value function $f(x) = |x - 1| + 3$.

Solution We graph this function by translating the graph of $f(x) = |x|$ one unit to the right and three units up, as shown in Figure 4-22.

FIGURE 4-22

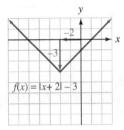

$f(x) = |x - 1| + 3$

SELF CHECK Graph $f(x) = |x + 2| - 3$.

Answer:

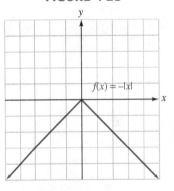

$f(x) = |x + 2| - 3$

We also saw that the graph of $f(x) = -|x|$ is the same as the graph of $f(x) = |x|$ reflected about the x-axis, as shown in Figure 4-23.

FIGURE 4-23

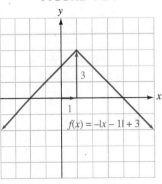

$f(x) = -|x|$

EXAMPLE 3

Graph the absolute value function $f(x) = -|x - 1| + 3$.

Solution We graph this function by translating the graph of $f(x) = -|x|$ one unit to the right and three units up, as shown in Figure 4-24.

FIGURE 4-24

$f(x) = -|x - 1| + 3$

SELF CHECK Graph $f(x) = -|x + 2| - 3$. *Answer:*

FIGURE 4-25

FIGURE 4-26

Equations of the Form $|x| = k$

The absolute value of a real number represents the distance on the number line from a point to the origin. To solve the **absolute value equation** $|x| = 5$, we must find the coordinates of all points on the number line that are exactly 5 units from zero. See Figure 4-25. The only two points that satisfy this condition have coordinates 5 and -5. That is, $x = 5$ or $x = -5$.

In general, the solution set of the absolute value equation $|x| = k$, where $k \geq 0$, includes the coordinates of the points on the number line that are k units from the origin. (See Figure 4-26.)

Absolute value equations

If $k \geq 0$, then

$$|x| = k \qquad \text{is equivalent to} \qquad x = k \text{ or } x = -k$$

EXAMPLE 4

Solving an absolute value equation. Solve **a.** $|x| = 8$, **b.** $|s| = 0.003$, and **c.** $|c| = -15$.

Solution
a. If $|x| = 8$, then $x = 8$ or $x = -8$.

b. If $|s| = 0.003$, then $s = 0.003$ or $s = -0.003$.

c. The absolute value of a number is either positive or zero but never negative. Therefore, there is no value for c for which $|c| = -15$. This equation has no solution.

SELF CHECK Solve **a.** $|y| = 24$, **b.** $|x| = \frac{1}{2}$, and **c.** $|a| = -1.1$.

Answers: **a.** 24, -24
b. $\frac{1}{2}, -\frac{1}{2}$, **c.** no solution

The equation $|x - 3| = 7$ indicates that a point on the number line with a coordinate of $x - 3$ is 7 units from the origin. Thus, $x - 3$ can be either 7 or -7.

$$x - 3 = 7 \qquad \text{or} \qquad x - 3 = -7$$
$$x = 10 \qquad\qquad x = -4$$

FIGURE 4-27

The solutions of the absolute value equation are 10 and -4. We can graph them on a number line, as shown in Figure 4-27. If either of these numbers is substituted for x in $|x - 3| = 7$, the equation is satisfied.

$$|x - 3| = 7 \qquad\qquad |x - 3| = 7$$
$$|10 - 3| \stackrel{?}{=} 7 \qquad\qquad |-4 - 3| \stackrel{?}{=} 7$$
$$|7| \stackrel{?}{=} 7 \qquad\qquad |-7| \stackrel{?}{=} 7$$
$$7 = 7 \qquad\qquad 7 = 7$$

EXAMPLE 5	**Solving an absolute value equation.** Solve **a.** $\lvert 3x - 2 \rvert = 5$ and **b.** $\lvert 5 - x \rvert = -10$.

Solution   **a.** We can write $\lvert 3x - 2 \rvert = 5$ as

$$3x - 2 = 5 \quad \text{or} \quad 3x - 2 = -5$$

and solve each equation for x:

$$
\begin{array}{ll}
3x - 2 = 5 \quad \text{or} & 3x - 2 = -5 \\
3x = 7 & 3x = -3 \\
x = \dfrac{7}{3} & x = -1
\end{array}
$$

Verify that both solutions check.

b. For any real number x, $\lvert 5 - x \rvert$ will be nonnegative. For this reason, $\lvert 5 - x \rvert = -10$ has no solution.

SELF CHECK	Solve **a.** $\lvert 2x - 3 \rvert = 7$ and **b.** $\lvert \frac{x}{4} - 1 \rvert = -3$.	*Answers:* **a.** 5, -2, **b.** no solution ■

When solving an absolute value equation, we want the absolute value isolated on one side. If this is not the case in a given equation, we use the equation-solving procedures studied earlier to isolate the absolute value first.

EXAMPLE 6	**Isolating the absolute value.** Solve $\lvert \frac{2}{3}x + 3 \rvert + 4 = 10$.

Solution   We can isolate $\lvert \frac{2}{3}x + 3 \rvert$ on the left-hand side of the equation by subtracting 4 from both sides.

$$\left\lvert \frac{2}{3}x + 3 \right\rvert + 4 = 10$$

1. $\left\lvert \dfrac{2}{3}x + 3 \right\rvert = 6$ Subtract 4 from both sides.

Now that the absolute value is isolated, we can write Equation 1 as

$$\frac{2}{3}x + 3 = 6 \quad \text{or} \quad \frac{2}{3}x + 3 = -6$$

and solve each equation for x:

$$
\begin{array}{ll}
\dfrac{2}{3}x + 3 = 6 \quad \text{or} & \dfrac{2}{3}x + 3 = -6 \\[2mm]
\dfrac{2}{3}x = 3 & \dfrac{2}{3}x = -9 \\[2mm]
2x = 9 & 2x = -27 \\[2mm]
x = \dfrac{9}{2} & x = -\dfrac{27}{2}
\end{array}
$$

Verify that both solutions check.

SELF CHECK	Solve $\lvert 0.4x - 2 \rvert - 0.6 = 0.4$.	*Answer:* 7.5, 2.5 ■

WARNING! Since the absolute value of a quantity cannot be negative, equations such as $\lvert 7x + \frac{1}{2} \rvert = -4$ have no solution. Their solution sets are empty.

EXAMPLE 7

An absolute value equal to 0. Solve $3\left|\frac{1}{2}x - 5\right| - 4 = -4$.

Solution We first isolate $\left|\frac{1}{2}x - 5\right|$ on the left-hand side.

$$3\left|\frac{1}{2}x - 5\right| - 4 = -4$$

$$3\left|\frac{1}{2}x - 5\right| = 0 \qquad \text{Add 4 to both sides.}$$

$$\left|\frac{1}{2}x - 5\right| = 0 \qquad \text{Divide both sides by 3.}$$

Since 0 is the only number whose absolute value is 0, $\frac{1}{2}x - 5$ must be 0, and we have

$$\frac{1}{2}x - 5 = 0$$

$$\frac{1}{2}x = 5 \qquad \text{Add 5 to both sides.}$$

$$x = 10 \qquad \text{Multiply both sides by 2.}$$

Verify that 10 satisfies the original equation.

SELF CHECK Solve $-5\left|\frac{2x}{3} + 4\right| + 1 = 1$. *Answer:* -6 ∎

Equations with Two Absolute Values

The equation $|a| = |b|$ is true when $a = b$ or when $a = -b$. For example,

$$|3| = |3| \qquad \text{or} \qquad |3| = |-3|$$
$$\uparrow \qquad \uparrow \qquad\qquad\qquad \uparrow \qquad \uparrow$$

The same number.   These numbers are opposites.

In general, the following statement is true.

Equations with two absolute values

If a and b represent algebraic expressions, the equation $|a| = |b|$ is equivalent to the pair of equations

$$a = b \qquad \text{or} \qquad a = -b$$

EXAMPLE 8

Solving equations with two absolute values. Solve $|5x + 3| = |3x + 25|$.

Solution This equation is true when $5x + 3 = 3x + 25$, or when $5x + 3 = -(3x + 25)$. We solve each equation for x.

$$5x + 3 = 3x + 25 \qquad \text{or} \qquad 5x + 3 = -(3x + 25)$$
$$2x = 22 \qquad\qquad\qquad 5x + 3 = -3x - 25$$
$$x = 11 \qquad\qquad\qquad\qquad 8x = -28$$
$$x = -\frac{28}{8}$$
$$x = -\frac{7}{2}$$

Verify that both solutions check.

SELF CHECK Solve $|2x - 3| = |4x + 9|$. *Answer:* $-1, -6$ ∎

Inequalities of the Form $|x| < k$

FIGURE 4-28

To solve the **absolute value inequality** $|x| < 5$, we must find the coordinates of all points on a number line that are less than 5 units from the origin. See Figure 4-28. Thus, x is between -5 and 5 and

$|x| < 5$ is equivalent to $-5 < x < 5$

FIGURE 4-29

In general, the solution set of the absolute value inequality $|x| < k \ (k > 0)$ includes the coordinates of the points on the number line that are less than k units from the origin. See Figure 4-29.

Solving $|x| < k$ and $|x| \le k$

$|x| < k$ is equivalent to $-k < x < k \quad (k > 0)$
$|x| \le k$ is equivalent to $-k \le x \le k \quad (k \ge 0)$

EXAMPLE 9

Solving an absolute value inequality. Solve $|2x - 3| < 9$ and graph the solution set.

Solution We write the absolute value inequality as a double inequality and solve for x.

$|2x - 3| < 9$ is equivalent to $-9 < 2x - 3 < 9$
$-9 < 2x - 3 < 9$
$-6 < 2x < 12$ Add 3 to all three parts.
$-3 < x < 6$ Divide all parts by 2.

FIGURE 4-30

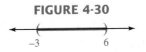

Any number between -3 and 6 is in the solution set. This is the interval $(-3, 6)$; its graph is shown in Figure 4-30.

SELF CHECK Solve $|3x + 2| < 4$ and graph the solution set.   *Answer:* $\left(-2, \frac{2}{3}\right)$

∎

EXAMPLE 10

Tolerances. When manufactured parts are inspected by a quality control engineer, they are classified as acceptable if each dimension falls within a given *tolerance range* of the dimensions listed on the blueprint. For the bracket shown in Figure 4-31, the distance between the two drilled holes is given as 2.900 inches. Because the tolerance is ± 0.015 inch, this distance can be as much as 0.015 inch longer or 0.015 inch shorter, and the part will be considered acceptable. The acceptable distance d between holes can be represented by the absolute value inequality $|d - 2.900| \le 0.015$. Solve the inequality and explain the result.

FIGURE 4-31

Unless otherwise specified, dimensions are in inches.	Bracket Assembly
	Drawing CC14-568
	Date: 8/15
Tolerances ±0.015	Sheet 1 Size A

Solution   We can write the absolute value inequality as a double inequality and solve for *d*:

$$|d - 2.900| \le 0.015 \quad \text{is equivalent to} \quad -0.015 \le d - 2.900 \le 0.015$$
$$-0.015 \le d - 2.900 \le 0.015$$
$$2.885 \le d \le 2.915 \qquad \text{Add 2.900 to all three parts.}$$

The solution set is the interval [2.885, 2.915]. This means that the distance between the two holes should be between 2.885 and 2.915 inches, inclusive. If the distance is less than 2.885 inches or more than 2.915 inches, the part should be rejected.   ■

Inequalities of the Form $|x| > k$

To solve the absolute value inequality $|x| > 5$, we must find the coordinates of all points on a number line that are more than 5 units from the origin. See Figure 4-32.

FIGURE 4-32

FIGURE 4-33

Thus, $x < -5$ or $x > 5$.

In general, the solution set of $|x| > k$ includes the coordinates of the points on the number line that are more than *k* units from the origin. See Figure 4-33. Thus,

$$|x| > k \quad \text{is equivalent to} \quad x < -k \text{ or } x > k$$

The word *or* indicates an either/or situation. It is only necessary that *x* satisfy one of the two conditions to be in the solution set.

Solving $|x| > k$ and $|x| \ge k$

If $k \ge 0$, then

$$|x| > k \quad \text{is equivalent to} \quad x < -k \text{ or } x > k$$
$$|x| \ge k \quad \text{is equivalent to} \quad x \le -k \text{ or } x \ge k$$

EXAMPLE 11   **Solving an absolute value inequality.** Solve $\left|\frac{3-x}{5}\right| \ge 6$ and graph the solution set.

Solution   We write the absolute value inequality as two separate inequalities connected with the word *or*.

$$\left|\frac{3-x}{5}\right| \ge 6 \quad \text{is equivalent to} \quad \frac{3-x}{5} \le -6 \text{ or } \frac{3-x}{5} \ge 6$$

Then we solve each inequality for *x*:

$$\frac{3-x}{5} \le -6 \qquad \text{or} \qquad \frac{3-x}{5} \ge 6$$

$3 - x \le -30$	$3 - x \ge 30$	Multiply both sides by 5.
$-x \le -33$	$-x \ge 27$	Subtract 3 from both sides.
$x \ge 33$	$x \le -27$	Divide both sides by -1 and reverse the direction of the inequality symbol.

FIGURE 4-34

The solution set is the interval $(-\infty, -27] \cup [33, \infty)$. Its graph appears in Figure 4-34.

SELF CHECK Solve $\left|\frac{2-x}{4}\right| \geq 1$ and graph the solution set.

Answer: $(-\infty, -2] \cup [6, \infty)$

$-2 \qquad 6$ ∎

EXAMPLE 12

Solving an absolute value inequality. Solve $\left|\frac{2}{3}x - 2\right| - 3 > 6$ and graph the solution set.

Solution We begin by adding 3 to both sides to isolate the absolute value on the left-hand side.

$$\left|\frac{2}{3}x - 2\right| - 3 > 6$$

$$\left|\frac{2}{3}x - 2\right| > 9 \quad \text{Add 3 to both sides to isolate the absolute value.}$$

We then proceed as follows:

$$\frac{2}{3}x - 2 < -9 \qquad \text{or} \qquad \frac{2}{3}x - 2 > 9$$

$$\frac{2}{3}x < -7 \qquad\qquad \frac{2}{3}x > 11 \quad \text{Add 2 to both sides.}$$

$$2x < -21 \qquad\qquad 2x > 33 \quad \text{Multiply both sides by 3.}$$

$$x < -\frac{21}{2} \qquad\qquad x > \frac{33}{2} \quad \text{Divide both sides by 2.}$$

FIGURE 4-35

$-21/2 \qquad 33/2$

The solution set is $\left(-\infty, -\frac{21}{2}\right) \cup \left(\frac{33}{2}, \infty\right)$. Its graph appears in Figure 4-35.

SELF CHECK Solve $\left|\frac{3}{4}x + 2\right| - 1 > 3$ and graph the solution set.

Answer: $(-\infty, -8) \cup \left(\frac{8}{3}, \infty\right)$

$-8 \qquad 8/3$ ∎

ACCENT ON TECHNOLOGY *Solving Absolute Value Inequalities*

We can also solve absolute value inequalities using a graphing calculator. For example, to solve $|2x - 3| < 9$, we graph the equations $y = |2x - 3|$ and $y = 9$ on the same coordinate system. If we use settings of $[-5, 15]$ for x and $[-5, 15]$ for y, we get the graph shown in Figure 4-36.

The inequality $|2x - 3| < 9$ will be true for all x-coordinates of points that lie on the graph of $y = |2x - 3|$ and below the graph of $y = 9$. Using the TRACE or INTERSECT feature, we can see that these values of x are in the interval $(-3, 6)$.

FIGURE 4-36

STUDY SET

Section 4.3

VOCABULARY

In Exercises 1–4, fill in the blanks to make the statements true.

1. $|2x - 1| = 10$ is an absolute value _____.
2. $|2x - 1| > 10$ is an absolute value _____.
3. To _____ the absolute value in $|3 - x| - 4 = 5$, we add 4 to both sides.
4. $|x| = 2$ is _____ to $x = 2$ or $x = -2$.

CONCEPTS

In Exercises 5–10, fill in the blanks to make the statements true.

5. $|x| \geq$ ___ for all real numbers x.
6. If $x < 0$, $|x| =$ _____.
7. To solve $|x| > 5$, we must find the coordinates of all points on a number line that are _____ 5 units from the origin.
8. To solve $|x| < 5$, we must find the coordinates of all points on a number line that are _____ 5 units from the origin.
9. To solve $|x| = 5$, we must find the coordinates of all points on a number line that are ___ units from the origin.
10. The equation $|a| = |b|$ is true when _____ or when _____.

11. Tell whether $x = -3$ is a solution of the given equation or inequality.
    a. $|x - 1| = 4$
    b. $|x - 1| > 4$
    c. $|x - 1| \leq 4$
    d. $|5 - x| = |x + 12|$

12. Write each equation or inequality in its equivalent form.
    a. $|x| = 8$
    b. $|x| \geq 8$
    c. $|x| \leq 8$
    d. $|5x - 1| = |x + 3|$

NOTATION

13. Match each equation or inequality with its graph.
    a. $|x| = 1$
    b. $|x| > 1$
    c. $|x| < 1$

14. Match each graph with its corresponding equation or inequality.
    a.
    b.
    c.

    i. $|x| \geq 2$
    ii. $|x| \leq 2$
    iii. $|x| = 2$

In Exercises 15–18, write each compound inequality as an inequality containing absolute value symbols.

15. $-4 < x < 4$
16. $x < -4$ or $x > 4$
17. $x + 3 < -6$ or $x + 3 > 6$
18. $-5 \leq x - 3 \leq 5$

PRACTICE

In Exercises 19–26, find the value of each expression.

19. $|8|$
20. $|-18|$
21. $-|0.02|$
22. $-|-3.14|$
23. $-\left| -\dfrac{31}{16} \right|$
24. $-\left| \dfrac{25}{4} \right|$
25. $|\pi|$
26. $\left| -\dfrac{\pi}{2} \right|$

In Exercises 27–30, graph each absolute value function.

27. $f(x) = |x| - 2$

28. $f(x) = -|x| + 1$

29. $f(x) = -|x + 4|$

30. $f(x) = |x - 1| + 2$

In Exercises 31–48, solve each equation, if possible.

31. $|x| = 23$

32. $|x| = 90$

33. $|x - 3.1| = 6$

34. $|x + 4.3| = 8.9$

35. $|3x + 2| = 16$

36. $|5x - 3| = 22$

37. $\left| \dfrac{7}{2}x + 3 \right| = -5$

38. $\left| \dfrac{2x}{3} + 10 \right| = 0$

39. $|3 - 4x| = 5$

40. $|8 - 5x| = 18$

41. $2|3x + 24| = 0$

42. $5|x - 21| = -8$

43. $\left| \dfrac{3x + 48}{3} \right| = 12$

44. $\left| \dfrac{4x - 64}{4} \right| = 32$

45. $|x + 3| + 7 = 10$

46. $|2 - x| + 3 = 5$

47. $8 = -1 + |0.3x - 3|$

48. $-1 = 1 - |0.1x + 8|$

In Exercises 49–56, solve each equation, if possible.

49. $|2x + 1| = |3x + 3|$

50. $|5x - 7| = |4x + 1|$

51. $|2 - x| = |3x + 2|$

52. $|4x + 3| = |9 - 2x|$

53. $\left| \dfrac{x}{2} + 2 \right| = \left| \dfrac{x}{2} - 2 \right|$

54. $|7x + 12| = |x - 6|$

55. $\left| x + \dfrac{1}{3} \right| = |x - 3|$

56. $\left| x - \dfrac{1}{4} \right| = |x + 4|$

In Exercises 57–84, solve each inequality. Write the solution set in interval notation (if possible) and graph it.

57. $|x| < 4$

58. $|x| < 9$

59. $|x + 9| \le 12$

60. $|x - 8| \le 12$

61. $|3x - 2| < 10$

62. $|4 - 3x| \le 13$

63. $|3x + 2| \le -3$

64. $|5x - 12| < -5$

65. $|x| > 3$

66. $|x| > 7$

67. $|x - 12| > 24$

68. $|x + 5| \ge 7$

69. $|3x + 2| > 14$

70. $|2x - 5| > 25$

71. $|4x + 3| > -5$

72. $|7x + 2| > -8$

73. $|2 - 3x| \ge 8$

74. $|-1 - 2x| > 5$

75. $-|2x - 3| < -7$

76. $-|3x + 1| < -8$

77. $\left| \dfrac{x - 2}{3} \right| \le 4$

78. $\left| \dfrac{x - 2}{3} \right| > 4$

79. $|3x + 1| + 2 < 6$

80. $1 + \left| \dfrac{1}{7}x + 1 \right| \le 1$

81. $\left| \dfrac{1}{3}x + 7 \right| + 5 > 6$

82. $-2|3x - 4| < 16$

83. $|0.5x + 1| + 2 \le 0$

84. $15 \ge 7 - |1.4x + 9|$

APPLICATIONS

85. TEMPERATURE RANGES The temperatures on a sunny summer day satisfied the inequality $|t - 78°| \le 8°$, where t is a temperature in degrees Fahrenheit. Solve this inequality and express the range of temperatures as a double inequality.

86. OPERATING TEMPERATURES A car CD player has an operating temperature of $|t - 40°| < 80°$, where t is a temperature in degrees Fahrenheit. Solve the inequality and express this range of temperatures as an interval.

87. AUTO MECHANICS On most cars, the bottoms of the front wheels are closer together than the tops, creating a *camber angle*. This lessens road shock to the steering system. (See Illustration 1.) The specifications for a certain car state that the camber angle c of its wheels should be $0.6° \pm 0.5°$.
a. Express the range with an inequality containing absolute value symbols.
b. Solve the inequality and express this range of camber angles as an interval.

ILLUSTRATION 1

88. STEEL PRODUCTION A sheet of steel is to be 0.250 inch thick with a tolerance of 0.025 inch.
a. Express this specification with an inequality containing absolute value symbols, using x to represent the thickness of a sheet of steel.
b. Solve the inequality and express the range of thickness as an interval.

89. ERROR ANALYSIS In a lab, students measured the percent of copper p in a sample of copper sulfate. The students know that copper sulfate is actually 25.46% copper by mass. They are to compare their results to the actual value and find the amount of *experimental error*.
a. Which measurements shown in Illustration 2 satisfy the absolute value inequality $|p - 25.46| \le 1.00$?
b. What can be said about the amount of error for each of the trials listed in part a?

ILLUSTRATION 2

Lab 4	Section A
Title:	
"Percent copper (Cu) in copper sulfate ($CuSO_4 \cdot 5H_2O$)"	

Results

	% Copper
Trial #1:	22.91%
Trial #2:	26.45%
Trial #3:	26.49%
Trial #4:	24.76%

90. ERROR ANALYSIS See Exercise 89.
a. Which measurements satisfy the absolute value inequality $|p - 25.46| > 1.00$?
b. What can be said about the amount of error for each of the trials listed in part a?

WRITING

91. Explain how to find the absolute value of a given number.

92. Explain why the equation $|x - 4| = -5$ has no solutions.

93. Explain the use of parentheses and brackets when graphing inequalities.

94. Explain the differences between the solution sets of $|x| < 8$ and $|x| > 8$.

95. Explain how to use the graph in Illustration 3 to solve $|x - 2| < 3$.

96. Explain how to use the graph in Illustration 3 to solve $|x - 2| \ge 3$.

ILLUSTRATION 3

REVIEW

97. RAILROAD CROSSING The warning sign in Illustration 4 is to be painted on the street in front of a railroad crossing. If y is 30° more than twice x, find x and y.

98. GEOMETRY Refer to Illustration 4. What is $2x + 2y$?

ILLUSTRATION 4

▶ 4.4   Linear Inequalities in Two Variables

In this section, you will learn about

Graphing linear inequalities ■ Horizontal and vertical boundary lines ■ Problem solving

Introduction   In the first three sections of this chapter, we have worked with linear inequalities in one variable. Some examples are

$$x \geq -7, \quad 5 < \frac{7}{2}a - 9, \quad \text{and} \quad 5(3 + z) > -3(z + 3)$$

These inequalities have infinitely many solutions. When their solutions are graphed on a real number line, we obtain an interval.

In this section, we will discuss linear inequalities in two variables. Some examples are

$$y > 3x + 2, \quad 2x - 3y \leq 6, \quad \text{and} \quad y < 2x$$

The solutions of these inequalities are ordered pairs. We can graph their solutions on a rectangular coordinate system.

Graphing Linear Inequalities

The **graph of a linear inequality** in x and y is the graph of all ordered pairs (x, y) that satisfy the inequality.

Linear inequalities

A **linear inequality** in x and y is any inequality that can be written in the form

$$Ax + By < C \quad \text{or} \quad Ax + By > C \quad \text{or} \quad Ax + By \leq C \quad \text{or} \quad Ax + By \geq C$$

where A, B, and C are real numbers and A and B are not both 0.

Because the inequality $y > 3x + 2$ can be written in the form $-3x + y > 2$, it is an example of a linear inequality in x and y.

To graph the linear inequality $y > 3x + 2$, we begin by graphing the linear *equation* $y = 3x + 2$. The graph of $y = 3x + 2$, shown in Figure 4-37(a), is a boundary line that separates the rectangular coordinate plane into two regions called **half-planes.** It is drawn with a broken line to show that it is not part of the graph of $y > 3x + 2$.

To find which half-plane is the graph of $y > 3x + 2$, we can substitute the coordinates of any point in either half-plane. We will choose the origin as the test point because its coordinates, $(0, 0)$, make the computations easy. We substitute 0 for x and 0 for y into the inequality and simplify.

$y > 3x + 2$ The original inequality.

$0 > 3(0) + 2$ Substitute 0 for y and 0 for x.

$0 > 2$ This statement is false.

Since the coordinates of the origin don't satisfy $y > 3x + 2$, the origin is not part of the graph of the inequality. Thus, the half-plane on the other side of the broken line is its graph. We then shade that region, as shown in Figure 4-37(b).

FIGURE 4-37

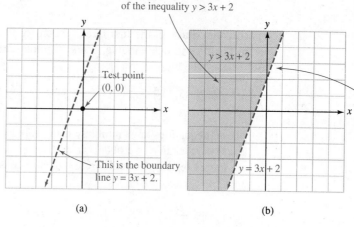

The shaded half-plane represents all the solutions of the inequality $y > 3x + 2$

Test point (0, 0)

This is the boundary line $y = 3x + 2$.

$y > 3x + 2$

$y = 3x + 2$

The boundary line is often called an edge of the half-plane. In this case, the edge is not included in the graph.

(a)   (b)

EXAMPLE 1

Graphing a linear inequality. Graph $2x - 3y \le 6$.

Solution   This inequality is the combination of the inequality $2x - 3y < 6$ and the equation $2x - 3y = 6$.

We begin by graphing $2x - 3y = 6$ to find the boundary line that separates the two half-planes. We do so by noting that the line's x-intercept is $(3, 0)$ and its y-intercept is $(0, -2)$. This time, we draw the solid line shown in Figure 4-38(a), because equality is permitted by the symbol $\le$. To decide which half-plane to shade, we check to see whether the coordinates of the origin satisfy the inequality.

$$2x - 3y \le 6$$
$$2(0) - 3(0) \le 6 \quad \text{Substitute 0 for } x \text{ and 0 for } y.$$
$$0 \le 6 \quad \text{This statement is true.}$$

The coordinates of the origin satisfy the inequality. In fact, the coordinates of every point on the same side of the boundary line as the origin satisfy the inequality. We then shade that half-plane to complete the graph of $2x - 3y \le 6$, shown in Figure 4-38(b).

FIGURE 4-38

Test point (0, 0)

$2x - 3y = 6$

$2x - 3y \le 6$

$2x - 3y = 6$

In this case, the edge is included.

(a)   (b)

SELF CHECK   Graph $3x - 2y \geq 6$.

Answer:

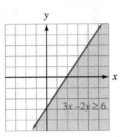

EXAMPLE 2   **Picking a test point other than the origin.** Graph $y < 2x$.

Solution   To graph $y = 2x$, we use the fact that the equation is in slope–intercept form and that $m = 2 = \frac{2}{1}$ and $b = 0$. Since the symbol $<$ does not include an equals sign, the points on the graph of $y = 2x$ are not on the graph of $y < 2x$. We draw the boundary line as a broken line to show this, as in Figure 4-39(a).

To decide which half-plane is the graph of $y < 2x$, we check to see whether the coordinates of some fixed point satisfy the inequality. We cannot use the origin as a test point, because the boundary line passes through the origin. However, we can choose a different point—say, (3, 1).

$y < 2x$
$1 < 2(3)$ Substitute 1 for y and 3 for x.
$1 < 6$ This is a true statement.

Since $1 < 6$ is a true inequality, the point (3, 1) satisfies the inequality and is in the graph of $y < 2x$. We then shade the half-plane containing (3, 1), as shown in Figure 4-39(b).

FIGURE 4-39

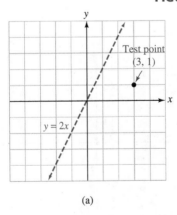

(a)

In this case, the edge is not included. (b)

SELF CHECK   Graph $y > -x$.

Answer:

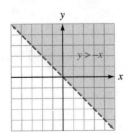

The following is a summary of the procedure for graphing linear inequalities.

Graphing linear inequalities in two variables

1. Graph the boundary line of the region. If the inequality allows the possibility of equality (the symbol is either ≤ or ≥), draw the boundary line as a solid line. If equality is not allowed (< or >), draw the boundary line as a broken line.

2. Pick a test point that is on one side of the boundary line. (Use the origin if possible.) Replace x and y in the original inequality with the coordinates of that point. If the inequality is satisfied, shade the side that contains that point. If the inequality is not satisfied, shade the other side.

Horizontal and Vertical Boundary Lines

Recall from Chapter 2 that the graph of the equation $x = a$ is a vertical line with x-intercept at $(a, 0)$, and the graph of the equation $y = b$ is a horizontal line with y-intercept at $(0, b)$.

EXAMPLE 3

A vertical boundary. Graph $x \geq -1$.

Solution   The graph of the boundary $x = -1$ is a vertical line passing through $(-1, 0)$. We draw the boundary as a solid line to show that it is part of the solution. See Figure 4-40(a).

In this case, we need not pick a test point. The inequality $x \geq -1$ is satisfied by points with an x-coordinate greater than or equal to -1. Points satisfying this condition lie to the right of the boundary. We shade that half-plane, as shown in Figure 4-40(b), to complete the graph of $x \geq -1$.

FIGURE 4-40

(a)

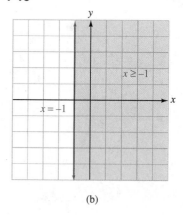

(b)

SELF CHECK   Graph $y < 4$.

Answer:

Problem Solving

In the next example, we will solve a problem by writing a linear inequality in two variables to model a situation mathematically.

EXAMPLE 4

Social Security. Retirees, ages 62–65, can earn as much as $9,120 and still receive their full Social Security benefits. If their annual earnings exceed $9,120, their benefits are reduced. A 64-year-old retired woman receiving Social Security works two part-time jobs: one at the library, paying $380 per week and another at a pet store, paying $285 per week. Write an inequality representing the number of weeks the woman can work at each job during the year without losing any of her Social Security benefits.

ANALYZE THE PROBLEM

We need to find the various combinations of the number of weeks she can work at the library and at the pet store so that the annual income from these jobs is less than or equal to $9,120.

FORM AN INEQUALITY

If we let x represent the number of weeks she works at the library, she will earn $380x$ on the first job annually. If we let y represent the number of weeks she works at the pet store, she will earn $285y$ on the second job annually. Combining the income from the two jobs, the total is not to exceed $9,120.

The weekly rate on the library job	·	the weeks worked on the library job	plus	the weekly rate on the pet store job	·	the weeks worked on the pet store job	should not exceed	$9,120.
$380	·	x	+	$285	·	y	≤	$9,120

SOLVE THE INEQUALITY

The graph of $380x + 285y \leq 9,120$ is shown in Figure 4-41. Any point in the shaded region indicates a way that she can schedule her work weeks and earn $9,120 or less annually. For example, if she works 8 weeks at the library and 20 weeks at the pet store (represented by the ordered pair $(8, 20)$), she will earn

$$\$380(8) + \$285(20) = \$3,040 + \$5,700$$
$$= \$8,740$$

If she works 16 weeks at the library and 8 weeks at the pet store, she will earn

$$\$380(16) + \$285(8) = \$6,080 + \$2,280$$
$$= \$8,360$$

FIGURE 4-41

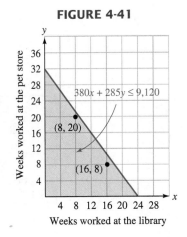

Since she cannot work a negative number of weeks, the graph has no meaning when x or y is negative, so only the first quadrant is shown. ∎

ACCENT ON TECHNOLOGY *Graphing Inequalities*

Some graphing calculators (such as the TI-83) have a graph style icon in the $y =$ editor. Some of the different graph styles are

\	line	A straight line or curved graph is shown.	\Y$_1$ =
◥	above	Shading covers the area above a graph.	◥Y$_1$ =
◣	below	Shading covers the area below a graph.	◣Y$_1$ =

We can change the icon in front of Y_1 by placing the cursor on it and pressing the ENTER key.

To graph $2x - 3y \leq 6$ of Example 1, we first write it in an equivalent form, with y isolated on the left-hand side.

$2x - 3y \leq 6$

$\quad -3y \leq -2x + 6$ Subtract $2x$ from both sides.

$\quad\quad y \geq \frac{2}{3}x - 2$ Divide both sides by -3. Change the direction of the inequality symbol.

We then change the graph style icon to above (◥), because the inequality $y \geq \frac{2}{3}x - 2$ contains a $\geq$ symbol. Using window settings of $[-10, 10]$ for x and $[-10, 10]$ for y, we enter the boundary equation $y = \frac{2}{3}x - 2$. See Figure 4-42(a). Finally, we press the GRAPH key to get Figure 4-42(b).

To graph $y < 2x$ from Example 2, we change the graph style icon to below (◣), because the inequality contains a $<$ symbol. Using window settings of $[-10, 10]$ for x and $[-10, 10]$ for y, we enter the boundary equation $y = 2x$ and press the GRAPH key to get Figure 4-42(c).

FIGURE 4-42

(a)   (b)

(c)

If your calculator does not have a graph style icon, you can graph linear inequalities with a SHADE feature. To do so, consult your owner's manual.

It is important to note that graphing calculators do not distinguish between solid and broken lines to show whether or not the edge of a region is included within the graph.

STUDY SET

Section 4.4

VOCABULARY

In Exercises 1–4, fill in the blanks to make the statements true.

1. $4x - 2y \geq -8$ is an example of a _____ inequality in _____ variables.

2. Graphs of linear inequalities are _____.

3. The boundary line of a half-plane is called an _____.

4. The graph of a linear inequality in x and y contains the points (x, y) whose coordinates _____ the inequality.

CONCEPTS

5. Tell whether each ordered pair is a solution of $3x - 2y \geq 5$.
   a. $(3, 1)$ **b.** $(0, 3)$
   c. $(-1, -4)$ **d.** $\left(1, \frac{1}{2}\right)$

6. A linear inequality has been graphed in Illustration 1. Tell whether each point satisfies the inequality.
   a. $(-1, 4)$
   b. $(3, -2)$
   c. $(0, 0)$
   d. $(-3, -3)$

ILLUSTRATION 1

7. To graph the inequality $y > 3x - 1$, we begin by graphing the boundary line $y = 3x - 1$. What is the slope m of the line? What is its y-intercept?

8. To graph the inequality $2x + 3y \leq -6$, we begin by graphing the boundary line $2x + 3y = -6$. What are its x- and y-intercepts?

9. ZOOS To determine the allowable number of juvenile chimpanzees x and adult chimpanzees y that can live in an enclosure, a zookeeper refers to the graph in Illustration 2. Can 7 juvenile and 4 adult chimps be kept in the enclosure?

ILLUSTRATION 2

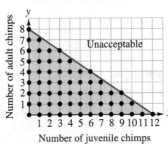

Number of juvenile chimps

10. The boundary for the graph of a linear inequality is shown in Illustration 3. Why can't the origin be used as a test point to decide which side to shade?

ILLUSTRATION 3

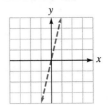

NOTATION

11. **a.** Solve the inequality in one variable and graph its solution set: $2x + 4 \geq 8$.

    b. Graph the inequality in two variables: $2x + 4y \geq 8$.

12. Tell whether the graph of each inequality includes the boundary line. In each case, would the boundary be a solid or a broken line?
    a. $y < 3x - 1$
    b. $2x + 3y \geq -6$
    c. $y \leq -10$
    d. $x > 1$

PRACTICE

In Exercises 13–32, graph each inequality.

13. $y > x + 1$

14. $y < 2x - 1$

15. $y \geq x$

16. $y \leq 2x$

17. $2x + y \le 6$

18. $x - 2y \ge 4$

29. $x < 4$

30. $y \ge -2$

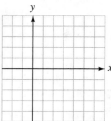

19. $3x \ge -y + 3$

20. $2x \le -3y - 12$

31. $y < 0$

32. $x \ge 0$

21. $y \ge 1 - \dfrac{3}{2}x$

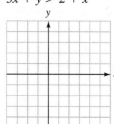

22. $y < \dfrac{x}{3} - 1$

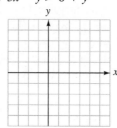

In Exercises 33–36, find the equation of the boundary line. Then give the inequality whose graph is shown.

33.

34.

23. $3x + y > 2 + x$

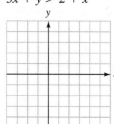

24. $3x - y > 6 + y$

35.

36.

In Exercises 37–40, use a graphing calculator to graph each inequality.

25. $y < -\dfrac{x}{2}$

26. $y > \dfrac{x}{3}$

37. $y < 0.27x - 1$

38. $y > -3.5x + 2.7$

39. $y \ge -2.37x + 1.5$

40. $y \le 3.37x - 1.7$

APPLICATIONS

41. GEOGRAPHY A region of the continental United States is shaded in the map shown in Illustration 4.

a. What is the boundary that separates the shaded and unshaded regions?

b. In words, describe the shaded area with respect to the boundary.

27. $\dfrac{x}{2} + \dfrac{y}{2} \le 2$

28. $\dfrac{x}{3} - \dfrac{y}{2} \ge 1$

ILLUSTRATION 4

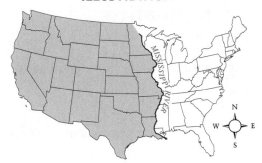

42. KOREAN WAR After World War II, the 38th parallel of north latitude was established as the boundary between North Korea and South Korea. (See Illustration 5.) The Korean War began on June 25, 1950, when the North Korean army crossed this line and invaded South Korea. In the Illustration, shade the region of the Korean Peninsula south of the 38th parallel.

ILLUSTRATION 5

In Exercises 43–46, write a linear inequality that models the situation. Then graph each inequality for nonnegative values of x and y and give three ordered pairs that satisfy the inequality.

43. RESTAURANT SEATING As part of a remodeling project, a restaurant owner will be installing new booths that seat 4 persons, and new tables that seat 6 persons. The overall seating must conform to the sign shown in Illustration 6. Write an inequality that describes the possible combinations of booths (x) and tables (y) that the owner can install.

ILLUSTRATION 6

MAXIMUM OCCUPANCY
NOT TO EXCEED
120
By order of Clake County Fire Marshal

44. GARDENING During an Arbor Day sale, a garden store sold more than $2,000 worth of trees. If a 6-foot maple costs $100 and a 5-foot pine costs $125, write an inequality that shows the possible ways that maple trees (x) and pine trees (y) were sold.

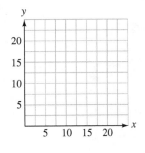

45. SPORTING GOODS A sporting goods manufacturer allocates at least 1,200 units of time per day to make fishing rods and reels. If it takes 10 units of time to make a rod and 15 units of time to make a reel, write an inequality that describes the possible ways to schedule the time to make rods (x) and reels (y).

46. HOUSEKEEPING One housekeeper charges $6 per hour, and another charges $7 per hour. If Sarah can afford no more than $42 per week to clean her house, write an inequality that describes the possible ways that she can hire the first housekeeper (x) and the second housekeeper (y).

WRITING

47. Explain how to decide where to draw the boundary of the graph of a linear inequality, and whether to draw it as a solid or a broken line.

48. Explain how to decide which side of the boundary of the graph of a linear inequality should be shaded.

REVIEW

In Exercises 49–50, tell whether the ordered pair $(-4, 3)$ is a solution of the system of linear equations.

49. $\begin{cases} 4x - y = -19 \\ 3x + 2y = -6 \end{cases}$ **50.** $\begin{cases} y = 2x + 11 \\ \frac{x}{2} + y = 0 \end{cases}$

In Exercises 51–52, solve each system of equations.

51. $\begin{cases} x + y = 4 \\ x - y = 2 \end{cases}$

52. $\begin{cases} x - \frac{y}{2} = -2 \\ 0.01x + 0.02y = 0.03 \end{cases}$

▶ **4.5**

Systems of Linear Inequalities

In this section, you will learn about

> Solving systems of linear inequalities ■ Compound inequalities
> ■ Problem solving

Introduction In Chapter 3, we discussed how to solve systems of linear equations by the graphing method. For example, to solve

$$\begin{cases} y = -x + 1 \\ 2x - y = 2 \end{cases}$$

by this method, we graph both equations on the same set of coordinate axes and then find the coordinates of the point of intersection of the straight lines.

In this section, we will discuss how to graphically solve systems of linear *inequalities,* such as

$$\begin{cases} y \le -x + 1 \\ 2x - y > 2 \end{cases}$$

Solving Systems of Linear Inequalities

When the solution of a linear inequality in x and y is graphed, the result is a half-plane. To solve a system of linear inequalities, we graph each of the inequalities on one set of coordinate axes and look for the intersection, or overlap, of the shaded half-planes.

EXAMPLE 1

Graphing a system of linear inequalities. Graph the solution set of $\begin{cases} y \le -x + 1 \\ 2x - y > 2 \end{cases}$.

Solution We graph each inequality on one set of coordinate axes, as shown in Figure 4-43. To make it easy to see the intersection of the two half-planes, we graph one solution set in red and the other in blue.

To graph $y \le -x + 1$, we first graph the boundary $y = -x + 1$, as shown below. Since the edge is to be included, we draw it as a solid line. To determine which half-plane to shade, we use the origin as a test point. Because the coordinates of the origin satisfy $y \le -x + 1$, we shade the half-plane containing the origin.

To graph $2x - y > 2$, we first graph the boundary $2x - y = 2$, as shown in the figure. Since the edge is not included, we draw it as a broken line. To determine which half-plane to shade, we again use the origin as a test point. Because the coordinates of the origin don't satisfy $2x - y > 2$, we shade the opposite half-plane.

The area where the half-planes intersect represents the solution of the system of inequalities, because any point in that region has coordinates that will satisfy both inequalities.

$y = -x + 1$

We graph this line using the slope and y-intercept.

$$m = -1 = -\frac{1}{1}$$

$b = 1$

y-intercept: $(0, 1)$

$2x - y = 2$

We graph this line using the intercept method.

x	y	(x, y)
0	-2	$(0, -2)$
1	0	$(1, 0)$

FIGURE 4-43

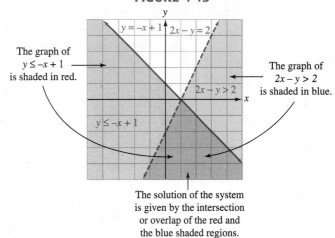

The graph of
$y \le -x + 1$
is shaded in red.

The graph of
$2x - y > 2$
is shaded in blue.

The solution of the system
is given by the intersection
or overlap of the red and
the blue shaded regions.

SELF CHECK   Graph $\begin{cases} x + y \ge 1 \\ 2x - y < 2 \end{cases}$.

Answer:

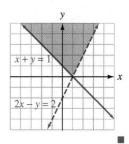

ACCENT ON TECHNOLOGY *Solving Systems of Inequalities*

To solve the system of Example 1 with a graphing calculator, we can use window settings of $x = [-10, 10]$ and $y = [-10, 10]$. To graph $y \le -x + 1$, we enter the boundary equation $y = -x + 1$ and change the graph style icon to below (▲). See Figure 4-44(a). To graph $2x - y > 2$, we first write it in equivalent form as $y < 2x - 2$. Then we enter the boundary equation $y = 2x - 2$ and change the graph style icon to below (▲). See Figure 4-44(a). Finally, we press the [GRAPH] key to obtain Figure 4-44(b).

FIGURE 4-44

(a)   (b)

EXAMPLE 2   **A system of three inequalities.** Graph the solution set of $\begin{cases} x \ge 1 \\ y \ge x \\ 4x + 5y < 20 \end{cases}$.

Solution We will find the graph of the solution set of the system in stages, using several graphs. Normally, a problem such as this is solved using just one set of axes.

The graph of $x \geq 1$ includes the points that lie on the graph of $x = 1$ and to the right, as shown in red in Figure 4-45(a).

FIGURE 4-45

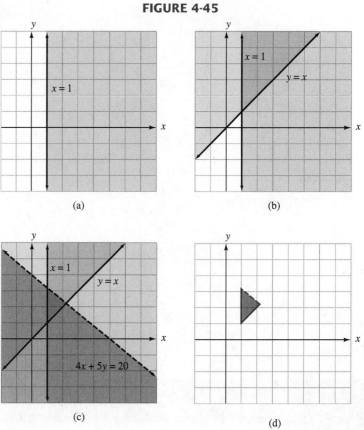

Figure 4-45(b) shows the graph of $x \geq 1$ and the graph of $y \geq x$. The graph of $y \geq x$, in blue, includes the points that lie on the graph of the boundary $y = x$ and above it.

Figure 4-45(c) shows the graphs of $x \geq 1$, $y \geq x$, and $4x + 5y < 20$. The graph of $4x + 5y < 20$ includes the points that lie below the graph of the boundary $4x + 5y = 20$.

The graph of the solution of the system includes the points that lie within the shaded triangle together with the points on the two sides of the triangle that are drawn with solid line segments, as shown in Figure 4-45(d).

SELF CHECK Graph $\begin{cases} x \geq 0 \\ y \leq 0 \\ y \geq -2 \end{cases}$.   *Answer:*

∎

Compound Inequalities

We have graphed the solution set of double linear inequalities, such as $2 < x \le 5$, on a number line. These inequalities contained only one variable. In the next example, we will graph the solution set of $2 < x \le 5$ in the context of two variables. In this case, we use the rectangular coordinate system.

EXAMPLE 3

Graphing a compound inequality. Graph $2 < x \le 5$ in the rectangular coordinate system.

Solution   The compound inequality $2 < x \le 5$ is equivalent to the following system of two linear inequalities:

$$\begin{cases} 2 < x \\ x \le 5 \end{cases}$$

FIGURE 4-46

The graph of $2 < x$, shown in Figure 4-46 in red, is the half-plane to the right of the vertical line $x = 2$. The graph of $x \le 5$, shown in the figure in blue, includes the line $x = 5$ and the half-plane to its left. The graph of $2 < x \le 5$ will contain all points in the plane that satisfy the inequalities $2 < x$ and $x \le 5$ simultaneously. These points are in the purple-shaded region of the figure.

SELF CHECK   Graph $-2 \le y < 3$ in the rectangular coordinate system.

Answer:

To graph a compound inequality containing the word *or* in the rectangular coordinate system, we sketch the *union* of the solution sets of the inequalities involved. For example, Figure 4-47 shows the graph of the compound inequality

$$x \le -2 \text{ or } x > 3$$

in the rectangular coordinate system.

FIGURE 4-47

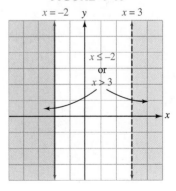

Problem Solving

EXAMPLE 4

Landscaping. A homeowner has a budget of $300 to $600 for trees and bushes to landscape his yard. After some comparison shopping, he finds that good trees cost $150 and mature bushes cost $75. What combinations of trees and bushes can he afford to buy?

ANALYZE THE PROBLEM

We need to find the number of trees and the number of bushes that the homeowner can afford. This suggests we should use two variables. We know that he is willing to spend *at least* $300 and *at most* $600 for trees and bushes. These phrases suggest that we should write two inequalities that model the situation.

FORM TWO INEQUALITIES

If *x* represents the number of trees purchased, then $150*x* will be the cost of the trees. If *y* represents the number of bushes purchased, then $75*y* will be the cost of the bushes. We know that the homeowner wants the sum of these costs to be from $300 to $600. We can then form the following system of linear inequalities.

The cost of a tree	·	the number of trees purchased	plus	the cost of a bush	·	the number of bushes purchased	should be at least	$300.
$150	·	x	+	$75	·	y	$\geq$	$300

The cost of a tree	·	the number of trees purchased	plus	the cost of a bush	·	the number of bushes purchased	should be at most	$600.
$150	·	x	+	$75	·	y	$\leq$	$600

SOLVE THE SYSTEM

We graph the system

$$\begin{cases} 150x + 75y \geq 300 \\ 150x + 75y \leq 600 \end{cases}$$

as in Figure 4-48. The coordinates of each point shown in the graph give a possible combination of trees (*x*) and bushes (*y*) that can be purchased.

STATE THE CONCLUSION

The possible combinations of trees and bushes that can be purchased are given by

(0, 4), (0, 5), (0, 6), (0, 7), (0, 8)
(1, 2), (1, 3), (1, 4), (1, 5), (1, 6)   The ordered pair (1, 6), for example, indicates that
(2, 0), (2, 1), (2, 2), (2, 3), (2, 4)   the homeowner can afford 1 tree and 6 bushes.
(3, 0), (3, 1), (3, 2), (4, 0)

Only these points can be used, because the homeowner cannot buy a portion of a tree or a bush.

CHECK THE RESULT

Check some of the ordered pairs to verify that they satisfy both inequalities.

Because the homeowner cannot buy a negative number of trees or bushes, we graph the system for $x \geq 0$ and $y \geq 0$.

FIGURE 4-48

■

STUDY SET

Section 4.5

VOCABULARY

In Exercises 1–4, fill in the blanks to make the statements true.

1. $\begin{cases} x + y \le 2 \\ x - 3y > 10 \end{cases}$ is a system of linear _____.

2. If an edge is included in the graph of an inequality, we draw it as a _____ line.

3. To solve a system of inequalities by graphing, we graph each inequality. The solution is the region where the graphs overlap or _____.

4. To determine which half-plane to shade when graphing a linear inequality, we see whether the coordinates of a test _____ satisfy the inequality.

CONCEPTS

5. Tell whether each point satisfies the system of linear inequalities $\begin{cases} x + y \le 2 \\ x - 3y > 10 \end{cases}$.

  a. $(2, -3)$
  b. $(12, -1)$
  c. $(0, -3)$
  d. $(-0.5, -5)$

6. a. Tell whether $(-3, 10)$ satisfies the compound inequality $-5 < x \le 8$ in the rectangular coordinate system.

  b. Tell whether $(-3, 3)$ satisfies the compound inequality $y \le 0$ or $y > 4$ in the rectangular coordinate system.

7. In Illustration 1, the solution of one linear inequality is shaded in red, and the solution of a second linear inequality is shaded in blue. Tell whether a true or false statement results if the coordinates of the given point are substituted into the given inequality.

  a. A, inequality 1   **b.** A, inequality 2
  c. B, inequality 1   **d.** B, inequality 2
  e. C, inequality 1   **f.** C, inequality 2

ILLUSTRATION 1

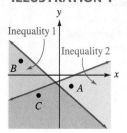

8. Match each equation, inequality, or system with the graph of its solution in Illustration 2.

  a. $2x + y = 2$ **b.** $2x + y \ge 2$

  c. $\begin{cases} 2x + y = 2 \\ 2x - y = 2 \end{cases}$ **d.** $\begin{cases} 2x + y \ge 2 \\ 2x - y \le 2 \end{cases}$

ILLUSTRATION 2

PRACTICE

In Exercises 9–20, graph the solution set of each system of inequalities.

9. $\begin{cases} y < 3x + 2 \\ y < -2x + 3 \end{cases}$ **10.** $\begin{cases} y \le x - 2 \\ y \ge 2x + 1 \end{cases}$

11. $\begin{cases} 3x + 2y > 6 \\ x + 3y \le 2 \end{cases}$ **12.** $\begin{cases} x + y < 2 \\ x + y \le 1 \end{cases}$

13. $\begin{cases} 3x + y \le 1 \\ -x + 2y \ge 6 \end{cases}$

14. $\begin{cases} x + 2y < 3 \\ 2x + 4y < 8 \end{cases}$

23. $y < -2$ or $y > 3$

24. $-x \le 1$ or $x \ge 2$

In Exercises 25–28, use a graphing calculator to solve each system.

25. $\begin{cases} y < 3x + 2 \\ y < -2x + 3 \end{cases}$

26. $\begin{cases} y > -x + 2 \\ y < -x + 4 \end{cases}$

27. $\begin{cases} 2x + y \ge 6 \\ y \le 2(2x - 3) \end{cases}$

28. $\begin{cases} 3x + y < -2 \\ y > 3(1 - x) \end{cases}$

15. $\begin{cases} x > 0 \\ y > 0 \end{cases}$

16. $\begin{cases} x \le 0 \\ y < 0 \end{cases}$

17. $\begin{cases} 2x + 3y \le 6 \\ 3x + y \le 1 \\ x \le 0 \end{cases}$

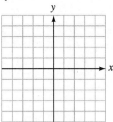

18. $\begin{cases} 2x + y \le 2 \\ y \ge x \\ x \ge 0 \end{cases}$

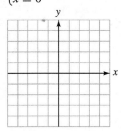

APPLICATIONS

29. PROFESSIONAL FOOTBALL The 1997 Denver Broncos football team scored either a field goal or a touchdown 89.3% of the time when they had possession of the football in their "red zone." If x represents the yard line the ball is on, the red zone is an area on their half of the field that can be described by the system

$$\begin{cases} x > 0 \\ x \le 20 \end{cases}$$

Shade the red zone on the field shown in Illustration 3.

ILLUSTRATION 3

← Broncos moving this direction

19. $\begin{cases} x - y < 4 \\ y \le 0 \\ x \ge 0 \end{cases}$

20. $\begin{cases} x \ge 0 \\ y \ge 0 \\ 9x + 3y \le 18 \\ 3x + 6y \le 18 \end{cases}$

30. TRACK AND FIELD In the shot put, the solid metal ball must land in a marked sector for it to be a fair throw. In Illustration 4, graph the system of inequalities that describes the region in which a shot must land.

ILLUSTRATION 4

$$\begin{cases} y \le \dfrac{3}{8}x \\ y \ge -\dfrac{3}{8}x \\ x \ge 1 \end{cases}$$

In Exercises 21–24, graph each inequality in the rectangular coordinate system.

21. $-2 \le x < 0$

22. $-3 < y \le -1$

31. NO-FLY ZONES After the Gulf War, U.S. and Allied forces enforced northern and southern "no-fly" zones over Iraq. Iraqi aircraft was prohibited from flying in this air space. If y represents the north latitude parallel measurement, the no-fly zones can be described by

$$y \geq 36 \text{ or } y \leq 33$$

On the map in Illustration 5, shade the regions of Iraq over which there was a no-fly zone.

ILLUSTRATION 5

32. CARDIOVASCULAR FITNESS The graph in Illustration 6 shows the range of pulse rates that persons ages 20–90 should maintain during aerobic exercise to get the most benefit from the training. The shaded region "Effective Training Heart Rate Zone" can be described by a system of linear inequalities. Determine what inequality symbol should be inserted in each blank.

$$\begin{cases} x \rule{1cm}{0.4pt} 20 \\ x \rule{1cm}{0.4pt} 90 \\ y \rule{1cm}{0.4pt} -0.87x + 191 \\ y \rule{1cm}{0.4pt} -0.72x + 158 \end{cases}$$

ILLUSTRATION 6

In Exercises 33–36, graph each system of inequalities and give two possible solutions.

33. COMPACT DISCS Melodic Music has compact discs on sale for either $10 or $15. If a customer wants to spend at least $30 but no more than $60 on CDs, use Illustration 7 to graph a system of inequalities that will show the possible ways a customer can buy $10 CDs ($x$) and $15 CDs ($y$).

ILLUSTRATION 7

34. BOAT SALES Dry Boat Works wholesales aluminum boats for $800 and fiberglass boats for $600. Northland Marina wants to order at least $2,400 worth but no more than $4,800 worth of boats. Use Illustration 8 to graph a system of inequalities that will show the possible combination of aluminum boats (x) and fiberglass boats (y) that can be ordered.

ILLUSTRATION 8

35. FURNITURE SALES A distributor wholesales desk chairs for $150 and side chairs for $100. Best Furniture wants to order no more than $900 worth of chairs, including more side chairs than desk chairs. Use Illustration 9 to graph a system of inequalities that will show the possible combinations of desk chairs (x) and side chairs (y) that can be ordered.

ILLUSTRATION 9

36. FURNACE EQUIPMENT J. Bolden Heating Company wants to order no more than $2,000 worth of electronic air cleaners and humidifiers from a wholesaler that charges $500 for air cleaners and $200 for humidifiers. If Bolden wants more humidifiers than air cleaners, use Illustration 10 to graph a system of inequalities that will show the possible combinations of air cleaners (x) and humidifiers (y) that can be ordered.

ILLUSTRATION 10

WRITING

37. Explain how to solve a system of two linear inequalities graphically.

38. Explain how a system of two linear inequalities might have no solution.

REVIEW

In Exercises 39–42, use the given conditions to determine in which quadrant of a rectangular coordinate system each point (x, y) is located.

39. $x > 0$ and $y < 0$

40. $x < 0$ and $y < 0$

41. $x < 0$ and $y > 0$

42. $x > 0$ and $y > 0$

Inequalities

Types of Inequalities

An **inequality** is a statement indicating that quantities are unequal. In Chapter 4, we have worked with several different types of inequalities and combinations of inequalities.

1. Classify each statement as one of the following: linear inequality in one variable, compound inequality, double linear inequality, absolute value inequality, linear inequality in two variables, system of linear inequalities.

   a. $x - 3 < -4$ or $x - 2 > 0$

   b. $\begin{cases} y < 3x + 2 \\ y < -2x + 3 \end{cases}$

   c. $|x - 8| \leq 12$

   d. $y < \dfrac{x}{3} - 1$

   e. $\dfrac{1}{2}x + 2 \geq \dfrac{1}{3}x - 4$

   f. $-6 < -3(x - 4) \leq 24$

   g. $5(x - 2) \geq 0$ and $-3x < 9$

   h. $|-1 - 2x| > 5$

   i. $y > -x$

Solutions of Inequalities

A solution of a linear inequality in one variable is a value that, when substituted for the variable, makes the inequality true. A solution of a linear inequality in two variables (or a system of linear inequalities) is an ordered pair whose coordinates satisfy the inequality (or inequalities).

2. Tell whether $x = -2$ is a solution of the inequalities in one variable. Tell whether $(-1, 3)$ is a solution of the inequalities (or system of inequalities) in two variables.

   a. $x - 3 < -4$ or $x - 2 > 0$

   b. $\begin{cases} y < 3x + 2 \\ y < -2x + 3 \end{cases}$

   c. $|x - 8| \leq 12$

   d. $y < \dfrac{x}{3} - 1$

   e. $\dfrac{1}{2}x + 2 \geq \dfrac{1}{3}x - 4$

   f. $-6 < -3(x - 4) \leq 24$

   g. $5(x - 2) \geq 0$ and $-3x < 9$

   h. $|-1 - 2x| > 5$

   i. $y > -x$

Graphs of Inequalities

To graph the solution set of an inequality in one variable, we use a number line. To graph the solution set of an inequality in two variables, we use a rectangular coordinate system.

3. Graph the solution set of the linear inequality in one variable: $2x + 1 > 4$.

4. Graph the solution set of the linear inequality in two variables: $2x + y \geq 4$.

Accent on Teamwork

Section 4.1

Inequality statements The warning given on the street sign in Illustration 1 can be described using an inequality. If we let w stand for the weight of a truck, then its weight must be such than $w \leq 3,000$ pounds if the truck is to use that street. Find six more real-life situations that can be described using an inequality.

ILLUSTRATION 1

NO TRUCKS
OVER
3,000 LB

Triangle inequality The triangle inequality is discussed in Exercise 51 of Study Set 4.1. To demonstrate it, cut two pipe cleaners to lengths of 2 inches and 4 inches, to serve as two sides of a triangle. (See Illustration 2.) Cut a third pipe cleaner to a length of 5 inches. Show that there is no triangle with sides of 2, 4, and 5 inches. Cut another pipe cleaner to a length of 6 inches. Show that there is no triangle with sides of 2, 4, and 6 inches. How long must the third pipe cleaner be for you to be able to form a triangle?

ILLUSTRATION 2

4 in.

2 in.

Section 4.2

Polls In Exercise 62 of Study Set 4.2, the margin of error for a poll of 1,000 adults was discussed. Find a published poll in a magazine or a newspaper in which the margin of error is given. Use interval notation to describe the possible intervals for each response in the poll. Taking the possible error into account, could the rankings of the responses possibly change?

Room temperature What temperature range do you like to keep inside your home? If t is the temperature, write a double linear inequality that describes the acceptable temperature range. Graph the solution set and express it as an interval. Then write a compound inequality containing the word *or* to describe the unacceptable temperature range inside your home. Graph the solution set and express it using interval notation.

Vocabulary
a. How are the connecting words *and* and *or* used with inequalities? Give examples.
b. Explain the concepts of union and intersection of sets. Give some examples using a number line.
c. How are the symbols ∞ and $-\infty$ used in the context of a number line?

Section 4.3

Tolerances Visit a machine shop or automobile repair shop and ask the employees to show you some examples of how they work with tolerances in their profession. Videotape their explanations and then play the video for your class. Show how the tolerances can be described using an absolute value inequality and interval notation. See Example 10 in Section 4.3 for an example.

Section 4.4

Inequalities in two variables
a. Can an inequality in two variables be an identity, one that is satisfied by all (x, y) pairs? Illustrate.
b. Can an inequality in two variables have no solutions? Illustrate.

Section 4.5

Intersection Sketch six examples of objects that intersect (overlap). Shade the region where they intersect. For instance, you could draw the intersection of two major streets in your city or the intersection of a chair and the floor.

Section 4.1

CONCEPTS

To solve an inequality, apply the *properties of inequalities*.

If both sides of an inequality are multiplied (or divided) by a negative number, another inequality results, but with the opposite direction from the original inequality.

Solving Linear Inequalities

REVIEW EXERCISES

1. Solve each inequality. Give each solution set in interval notation and graph it.

    a. $5(x - 2) \le 5$ **b.** $0.3x - 0.4 \ge 1.2 - 0.1x$

    c. $-16 < -\dfrac{4}{5}x$ **d.** $\dfrac{7}{4}(x + 3) < \dfrac{3}{8}(x - 3)$

2. INVESTMENTS A woman has invested \$10,000 at 6% annual interest. How much more must she invest at 7% so that her annual income is at least \$2,000?

Section 4.2

A solution of a compound inequality containing *and* makes both of the inequalities true.

The solution set of a compound inequality containing *and* is the *intersection* of the two solution sets.

Solving Compound Inequalities

3. Tell whether $x = -4$ is a solution of the compound inequality.

    a. $x < 0$ and $x > -5$ **b.** $x + 3 < -3x - 1$ and $4x - 3 > 3x$

4. Solve each compound inequality. Give the result in interval notation and graph the solution set.

    a. $-2x > 8$ and $x + 4 \ge -6$ **b.** $5(x + 2) \le 4(x + 1)$ and $11 + x < 0$

Double linear inequalities:

  $c < x < d$

is equivalent to

  $c < x$ and $x < d$

5. Solve each compound inequality. Give the result in interval notation and graph the solution set.

    a. $3 < 3x + 4 < 10$ **b.** $-2 \le \dfrac{5 - x}{2} \le 2$

A solution of a compound inequality containing the word *or* makes one, or the other, or both inequalities true.

6. Tell whether $x = -4$ is a solution of the compound inequality.

    a. $x < 1.6$ or $x > -3.9$ **b.** $x + 1 < 2x - 1$ or $4x - 3 > 3x$

The solution set of a compound inequality containing *or* is the *union* of the two solution sets.

7. Solve each compound inequality. Give the result in interval notation and graph the solution set.

    a. $x + 1 < -4$ or $x - 4 > 0$ **b.** $\dfrac{x}{2} + 3 > -2$ or $4 - x > 4$

8. INTERIOR DECORATING
A manufacturer makes a line of
decorator rugs that are 4 feet
wide and of varying lengths l
(in feet). See Illustration 1. The
floor area covered by the rugs
ranges from 17 ft^2 to 25 ft^2.
Write and then solve a double
linear inequality to find the
range of the lengths of the rugs.

ILLUSTRATION 1

Section 4.3

Solving Absolute Value Equations and Inequalities

Definition of *absolute value:*

$$\begin{cases} \text{If } x \geq 0, |x| = x. \\ \text{If } x < 0, |x| = -x. \end{cases}$$

9. Find each absolute value.

   a. $|-7|$ **b.** $\left|\dfrac{5}{16}\right|$

   c. $-|71.05|$ **d.** $-|-12|$

10. Graph each function.

   a. $f(x) = |x + 1| - 3$ **b.** $f(x) = |x - 2| + 1$

Absolute value equations:

If $k \geq 0$, $|x| = k$

is equivalent to

$x = k$ or $x = -k$

$|a| = |b|$ is equivalent to
$a = b$ or $a = -b$

11. Solve each absolute value equation. Check the results.

   a. $|4x| = 8$ **b.** $2|3x + 1| = 20$

   c. $\left|\dfrac{3}{2}x - 4\right| - 10 = -1$ **d.** $\left|\dfrac{2 - x}{3}\right| = 4$

   e. $|3x + 2| = |2x - 3|$ **f.** $\left|\dfrac{3 - 2x}{2}\right| = \left|\dfrac{3x - 2}{3}\right|$

Absolute value inequalities:

If $k > 0$, $|x| < k$

is equivalent to

$-k < x < k$

12. Solve each absolute value inequality. Give the solution in interval notation and graph it.

   a. $|x| \leq 3$ **b.** $|2x + 7| < 3$

   c. $|5 - 3x| \leq 14$ **d.** $\left|\dfrac{2}{3}x + 14\right| < 0$

Absolute value inequalities:

$$|x| > k$$

is equivalent to

$$x < -k \text{ or } x > k$$

e. $|x| > 1$

f. $\left| \dfrac{1 - 5x}{3} \right| \geq 7$

g. $|3x - 8| > 4$

h. $\left| \dfrac{3}{2}x - 14 \right| \geq 0$

13. PRODUCE Before packing, freshly picked tomatoes are weighed on the scale shown in Illustration 2. Tomatoes having a weight w (in ounces) that falls within the high-lighted range are sold to grocery stores.
   a. Express this acceptable weight range using an absolute value inequality.

   b. Solve the inequality and express this range as an interval.

ILLUSTRATION 2

Section 4.4

Linear Inequalities in Two Variables

To graph a *linear inequality* in x and y, graph the *boundary line,* and then use a *test point* to decide which side of the boundary should be shaded.

14. Graph each inequality in the rectangular coordinate system.
   a. $2x + 3y > 6$

   b. $y \leq 4 - x$

   c. $y < \dfrac{1}{2}x$

   d. $x \geq -\dfrac{3}{2}$

15. CONCERT TICKETS Tickets to a concert cost \$6 for reserved seats and \$4 for general admission. If receipts must be at least \$10,200 to meet expenses, find an inequality that shows the possible ways that the box office can sell reserved seats (x) and general admission tickets (y). Then graph each inequality for nonnegative values of x and y and give three ordered pairs that satisfy the inequality.

Section 4.5

To solve a *system of linear inequalities,* graph each of the inequalities on the same set of coordinate axes and look for the intersection of the shaded *half-planes.*

Systems of Linear Inequalities

16. Graph the solution set of each system of inequalities.

a. $\begin{cases} y \geq x + 1 \\ 3x + 2y < 6 \end{cases}$

b. $\begin{cases} x - y < 3 \\ y \leq 0 \\ x \geq 0 \end{cases}$

Compound inequalities can be graphed in the rectangular coordinate system.

17. Graph each compound inequality in the rectangular coordinate system.

a. $-2 < x < 4$

b. $y \leq -2$ or $y > 1$

18. PETROLEUM EXPLORATION Organic matter converts to oil and gas within a specific range of temperature and depth called the *petroleum window.* The petroleum window in Illustration 3 can be described by a system of linear inequalities, where x is the temperature in °C of the soil at a depth of y meters. Determine what inequality symbol should be inserted in each blank.

$$\begin{cases} x \,\boxed{\phantom{<}}\, 35 \\ x \,\boxed{\phantom{<}}\, 130 \\ y \,\boxed{\phantom{<}}\, -56x + 280 \\ y \,\boxed{\phantom{<}}\, -18x + 90 \end{cases}$$

ILLUSTRATION 3

Based on data from *The Blue Planet* (Wiley, 1995)

CHAPTER 4

Test

1. Tell whether the statement is true or false.

$-5.67 \geq -5$

2. Tell whether $x = -2$ is a solution of the inequality.

$3(x - 2) \leq 2(x + 7)$

In Problems 3–4, graph the solution set of each inequality. Also give the solution in interval notation.

3. $7 < \dfrac{2}{3}t - 1$

4. $-2(2x + 3) \geq 14$

5. AVERAGING GRADES Use the information from the gradebook in Illustration 1 to determine what score Karen Nelson-Sims needs on the fifth exam to keep her exam average above 80 in the class.

ILLUSTRATION 1

Sociology 101 8:00-10:00 pm MW	Exam 1	Exam 2	Exam 3	Exam 4	Exam 5
Nelson-Sims, Karen	70	79	85	88	

In Problems 6–8, solve each compound inequality. Give the result in interval notation and graph the solution set.

6. $3x \geq -2x + 5$ and $7 \geq 4x - 2$

7. $3x < -9$ or $-\dfrac{x}{4} < -2$

8. $-2 < \dfrac{x - 4}{3} < 4$

In Problems 9–10, find the value of each expression.

9. $|8|$

10. $-|-4.75|$

In Problems 11–12, solve each equation.

11. $|4 - 3x| = 19$

12. $|3x + 4| = |x + 12|$

In Problems 13–16, graph the solution set of each inequality. Also give the solution in interval notation.

13. $|x + 3| \leq 4$

14. $|2x - 4| > 22$

15. $|4 - 2x| + 1 > 3$

16. $2|2x - 4| \leq 4$

In Problems 17–20, graph the solution set in the rectangular coordinate system.

17. $3x + 2y \geq 6$

18. $y < x$

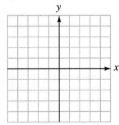

19. $\begin{cases} 2x - 3y \geq 6 \\ y \leq -x + 1 \end{cases}$

20. $-2 \leq y < 5$

21. ACCOUNTING On average, it takes an accountant 1 hour to complete a simple tax return and 3 hours to complete a complicated return. If the accountant wants to work less than 9 hours per day, find an inequality that shows the number of possible ways that simple returns (x) and complicated returns (y) can be completed each day. Then graph the inequality and give three ordered pairs that satisfy it.

22. Two linear inequalities are graphed on the same coordinate axes in Illustration 2. The solution set of the first inequality is shaded in red, and the solution set of the second in blue. Can we determine from the graph whether the point $(3, -4)$ is a solution of either inequality? Is $(3, -4)$ a solution of the system of two linear inequalities? Explain your answer.

ILLUSTRATION 2

23. Graph $f(x) = |x + 1| - 2$.

24. INDOOR CLIMATE The general zone of comfort acceptable to most people when working in an office can be described by a system of linear inequalities where x is the dry bulb temperature and y is the percent relative humidity. See Illustration 3. Determine what inequality symbol should be inserted in each blank.

$$\begin{cases} y \;\rule{1cm}{0.4pt}\; 60 \\ y \;\rule{1cm}{0.4pt}\; 27 \\ y \;\rule{1cm}{0.4pt}\; -11x + 852 \\ y \;\rule{1cm}{0.4pt}\; -5x + 445 \end{cases}$$

ILLUSTRATION 3

CHAPTERS 1–4

Cumulative Review Exercises

1. The diagram in Illustration 1 shows the sets that compose the set of real numbers. Which of the indicated sets make up the *rational numbers* and the *irrational numbers*?

ILLUSTRATION 1

2. SCREWS The thread profile of a screw is determined by the distance between threads. This distance, indicated by the letter p, is known as the *pitch*. If $p = 0.125$, find each of the dimensions labeled in Illustration 2.

ILLUSTRATION 2

In Exercises 3–4, evaluate each expression when x = 2 and y = −4.

3. $|x| − xy$

4. $\dfrac{x^2 − y^2}{3x + y}$

In Exercises 5–6, simplify each expression.

5. $3p^2 − 6(5p^2 + p) + p^2$

6. $−(a + 2) − (a − b)$

7. PLASTIC WRAP Estimate the number of *square feet* of plastic wrap on a roll if the dimensions printed on the box describe the roll as 205 feet long by $11\frac{3}{4}$ inches wide.

8. INVESTMENTS Find the amount of money that was invested at $8\frac{7}{8}\%$ if it earned $1,775 in simple interest in one year.

In Exercises 9–12, solve each equation, if possible.

9. $3x − 6 = 20$

10. $6(x − 1) = 2(x + 3)$

11. $\dfrac{5b}{2} − 10 = \dfrac{b}{3} + 3$

12. $2a − 5 = −2a + 4(a − 2) + 1$

In Exercises 13–14, tell whether the lines represented by the equations are parallel or perpendicular.

13. $3x + 2y = 12$, $2x − 3y = 5$

14. $3x = y + 4$, $y = 3(x − 4) − 1$

15. Write the equation of the line that passes through $(−2, 3)$ and is perpendicular to the graph of $3x + y = 8$. Answer in slope–intercept form.

16. Find the slope of the line that passes through $(0, −8)$ and $(−5, 0)$.

17. PRISONS The graph in Illustration 3 shows the growth of the U.S. prison population from 1970 to 1995. Find the rate of change in the prison population from 1970 to 1975.

18. PRISONS Refer to the graph in Illustration 3. During what five-year period was the rate of change in the U.S. prison population the greatest? Find the rate of change.

ILLUSTRATION 3

Based on data from *U.S. Statistical Abstract*

In Exercises 19–20, $f(x) = 3x^2 − x$. Find each value.

19. $f(2)$

20. $f(−2)$

21. Use graphing to solve $\begin{cases} 2x + y = 5 \\ x − 2y = 0 \end{cases}$.

22. Use addition to solve $\begin{cases} \dfrac{x}{10} + \dfrac{y}{5} = \dfrac{1}{2} \\ \dfrac{x}{2} − \dfrac{y}{5} = \dfrac{13}{10} \end{cases}$.

23. Use substitution to solve $\begin{cases} y = 4 − 3x \\ 2x − 3y = −1 \end{cases}$.

24. Solve $\begin{cases} x + y + z = 1 \\ 2x − y − z = −4 \\ x − 2y + z = 4 \end{cases}$.

In Exercises 25–26, use Cramer's rule to solve each system.

25. $\begin{cases} 4x − 3y = −1 \\ 3x + 4y = −7 \end{cases}$

26. $\begin{cases} x − 2y − z = −2 \\ 3x + y − z = 6 \\ 2x − y + z = −1 \end{cases}$

27. U.S. WORKERS The graph in Illustration 4 on the next page shows how the makeup of the U.S. workforce changed over the years 1900–1990. Estimate the coordinates of the points of intersection in the graph. Explain their significance.

28. WHITE-COLLAR JOBS Refer to the graph in Illustration 4. To model the growth in the percent of white-collar workers, draw a line through $(0, 20)$ and $(90, 70)$ and write an equation for this line. Use the equation to predict the percent of white-collar workers in the year 2010.

ILLUSTRATION 4

Based on data from *U.S. Statistical Abstract*

29. ENTREPRENEURS A person invests $18,375 to set up a small business producing a piece of computer software that will sell for $29.95. If each piece can be produced for $5.45, how many pieces must be sold to break even?

30. CONCERT TICKETS Tickets for a concert cost $5, $3, and $2. Twice as many $5 tickets were sold as $2 tickets. The receipts for 750 tickets were $2,625. How many tickets were sold at each price?

In Exercises 31–32, solve each equation.

31. $|4x - 3| = 9$

32. $|2x - 1| = |3x + 4|$

In Exercises 33–36, solve each inequality. Give the solution in interval notation and graph it.

33. $-3(x - 4) \geq x - 32$

34. $-8 < -3x + 1 < 10$

35. $|3x - 2| \leq 4$

36. $|2x + 3| - 1 > 4$

In Exercises 37–38, use graphing to solve each inequality or system of inequalities.

37. $2x - 3y \leq 12$

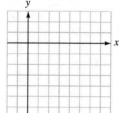

38. $\begin{cases} y < x + 2 \\ 3x + y \leq 6 \end{cases}$

5

Exponents, Polynomials, and Polynomial Functions

CAMPUS CONNECTION

The Astronomy Department

In science classes, we often encounter very large and very small numbers. For example, in astronomy, students learn that one of the stars in the constellation known as the Big Dipper is 465,000,000,000,000 miles from earth. Such numbers are awkward to work with when making calculations. In this chapter, we will use *exponents* to write them in a compact form called scientific notation.

> *Polynomials are algebraic expressions that can be used to model many real-world situations. They often contain terms in which the variables have exponents.*

▶ 5.1

Exponents

In this section, you will learn about

Exponents ■ Rules for exponents ■ Zero exponents ■ Negative exponents ■ More rules for exponents

Introduction   In Chapter 1, we evaluated exponential expressions having natural-number exponents. In this section, we will extend the definition of exponent to include negative-integer exponents (as in 3^{-2}) and zero exponents (as in 3^0). We will also apply the definition of exponent to develop several rules that can be used to simplify exponential expressions.

Exponents

Exponents provide a way to write products of *repeated factors* in a concise form. For example,

$y \cdot y = y^2$ Read y^2 as "y to the second power" or "y squared."

$z \cdot z \cdot z = z^3$ Read z^3 as "z to the third power" or "z cubed."

$x \cdot x \cdot x \cdot x = x^4$ Read x^4 as "x to the fourth power."

These examples suggest the following definition.

Natural-number exponents

If n is a natural number, then

$$x^n = \overbrace{x \cdot x \cdot x \cdot \cdots \cdot x}^{n \text{ factors of } x}$$

The **exponential expression** x^n is called a **power of x,** and we read it as "x to the nth power." In this expression, x is called the **base,** and n is called the **exponent.**

Base $\longrightarrow x^n \longleftarrow$ Exponent

A natural-number exponent tells how many times the base of an exponential expression is to be used as a factor in a product. When $n = 1$, the exponent is usually omitted. For example, $x^1 = x$.

EXAMPLE 1

Identifying the base and the exponent. Identify the base and the exponent in each expression: **a.** $(-a)^2$, **b.** $-a^2$, **c.** $5x^3$, and **d.** $(5x)^3$.

Solution   **a.** For $(-a)^2$, $-a$ is the base and the exponent is 2: $(-a)^2 = (-a)(-a)$.

   b. For $-a^2$, the base is a and the exponent is 2: $-a^2 = -(a \cdot a)$.

   c. For $5x^3$, x is the base and the exponent is 3: $5x^3 = 5 \cdot x \cdot x \cdot x$.

   d. For $(5x)^3$, the base is $5x$ and the exponent is 3: $(5x)^3 = (5x)(5x)(5x)$.

SELF CHECK   Identify the base and the exponent in each expression: **a.** $(-kt)^4$,   **b.** πr^2,   and   **c.** $-h^8$.

Answers: **a.** $-k^2 t$, 4,   **b.** r, 2,   **c.** h, 8   ∎

Rules for Exponents

Several important rules for exponents come directly from the definition of exponent. The first rule we will discuss, called the **product rule for exponents,** gives a quick way to find the result when multiplying exponential expressions that have the same base.

Since x^5 means that x is to be used as a factor five times, and since x^3 means that x is to be used as a factor three times, $x^5 \cdot x^3$ means that x will be used as a factor eight times.

$$x^5 x^3 = \overbrace{x \cdot x \cdot x \cdot x \cdot x}^{\text{5 factors of } x} \cdot \overbrace{x \cdot x \cdot x}^{\text{3 factors of } x} = \overbrace{x \cdot x \cdot x \cdot x \cdot x \cdot x \cdot x \cdot x}^{\text{8 factors of } x} = x^8$$

In general,

$$x^m x^n = \overbrace{x \cdot x \cdot x \cdot \cdots \cdot x}^{m \text{ factors of } x} \cdot \overbrace{x \cdot x \cdot x \cdot \cdots \cdot x}^{n \text{ factors of } x} = \overbrace{x \cdot x \cdot x \cdot x \cdot \cdots \cdot x}^{m+n \text{ factors of } x} = x^{m+n}$$

Thus, *to multiply exponential expressions with the same base, keep the base and add the exponents.*

The product rule for exponents

If m and n are natural numbers, then

$$x^m x^n = x^{m+n}$$

EXAMPLE 2   **Applying the product rule for exponents.** Simplify each expression.

   a. $x^{11} x^5 = x^{11+5}$ Keep the base.
   Add the exponents.

   $\qquad = x^{16}$

   b. $y^5 y^4 y^3 = (y^5 y^4) y^3$
   $\qquad = y^9 y^3$
   $\qquad = y^{12}$

   c. $a^2 b^3 a^3 b^2 = a^2 a^3 b^3 b^2$
   $\qquad = a^5 b^5$

   d. $-8x^4(x^3) = -8(x^4 x^3)$
   $\qquad = -8x^7$

SELF CHECK   Simplify each expression: **a.** $2^3 2^5$,   **b.** $k \cdot k^4$,   **c.** $a^2 b^3 a^3 b^4$,   and   **d.** $-8a^4(a^2 b)$.

Answers: **a.** $2^8 = 256$,   **b.** k^5,   **c.** $a^5 b^7$,   **d.** $8a^6 b$ ∎

WARNING! The product rule for exponents applies only to exponential expressions with the same base. The expression $x^5 y^3$, for example, cannot be simplified, because the bases of the exponential expressions are different.

To find another property of exponents, we simplify $(x^4)^3$, which means x^4 cubed or $x^4 \cdot x^4 \cdot x^4$.

$$(x^4)^3 = x^4 \cdot x^4 \cdot x^4 = \overbrace{x \cdot x \cdot x \cdot x}^{x^4} \cdot \overbrace{x \cdot x \cdot x \cdot x}^{x^4} \cdot \overbrace{x \cdot x \cdot x \cdot x}^{x^4} = x^{12}$$

In general, we have

$$(x^m)^n = \overbrace{x^m \cdot x^m \cdot x^m \cdot \cdots \cdot x^m}^{n \text{ factors of } x^m} = \overbrace{x \cdot x \cdot x \cdot x \cdot x \cdot \cdots \cdot x}^{mn \text{ factors of } x} = x^{mn}$$

Thus, *to raise an exponential expression to a power, keep the base and multiply the exponents.*

To find a third property of exponents, we square $3x$ to get

$$(3x)^2 = (3x)(3x) = 3 \cdot 3 \cdot x \cdot x = 3^2 x^2 = 9x^2$$

In general, we have

$$(xy)^n = \overbrace{(xy)(xy)(xy) \cdot \cdots \cdot (xy)}^{n \text{ factors of } xy} = \overbrace{xxx \cdot \cdots \cdot x}^{n \text{ factors of } x} \cdot \overbrace{yyy \cdot \cdots \cdot y}^{n \text{ factors of } y} = x^n y^n$$

To find a fourth property of exponents, we cube $\dfrac{x}{3}$ to get

$$\left(\frac{x}{3}\right)^3 = \frac{x}{3} \cdot \frac{x}{3} \cdot \frac{x}{3} = \frac{x \cdot x \cdot x}{3 \cdot 3 \cdot 3} = \frac{x^3}{3^3} = \frac{x^3}{27}$$

In general, we have

$$\left(\frac{x}{y}\right)^n = \overbrace{\left(\frac{x}{y}\right)\left(\frac{x}{y}\right)\left(\frac{x}{y}\right) \cdot \cdots \cdot \left(\frac{x}{y}\right)}^{n \text{ factors of } \frac{x}{y}} \quad (y \neq 0)$$

$$= \frac{\overbrace{xxx \cdot \cdots \cdot x}^{n \text{ factors of } x}}{\underbrace{yyy \cdot \cdots \cdot y}_{n \text{ factors of } y}} \quad \text{Multiply the numerators and multiply the denominators.}$$

$$= \frac{x^n}{y^n}$$

The previous results are called the **power rules for exponents.**

The power rules for exponents

If m and n are natural numbers, then

$$(x^m)^n = x^{mn} \qquad (xy)^n = x^n y^n \qquad \left(\frac{x}{y}\right)^n = \frac{x^n}{y^n} \quad (y \neq 0)$$

EXAMPLE 3

Applying the power rules for exponents. Simplify each expression.

a. $(3^2)^3 = 3^{2 \cdot 3}$ Keep the base. Multiply
the exponents.

$= 3^6$

$= 729$

b. $(x^{11})^5 = x^{11 \cdot 5}$

$= x^{55}$

c. $(x^2 x^3)^6 = (x^5)^6$

$= x^{30}$

d. $(-2x^2)^4(x^3)^2 = (-2)^4 x^8 x^6$

$= 16x^{14}$

SELF CHECK Simplify each expression: **a.** $(a^5)^8$, **b.** $(6^3)^5$, *Answers:* **a.** a^{40}, **b.** 6^{15},
c. $(a^4a^3)^3$, and **d.** $(-3a^3)^3(a^2)^3$.   **c.** a^{21}, **d.** $-27a^{15}$ ∎

E X A M P L E   4   **Applying the power rules for exponents.** Simplify each expression. Assume that no denominators are zero.

a. $(x^2y)^3 = (x^2)^3y^3$ **b.** $(x^3y^4)^4 = (x^3)^4(y^4)^4$
$\qquad\quad = x^6y^3$ $\qquad\qquad\quad = x^{12}y^{16}$

c. $\left(\dfrac{x}{y^2}\right)^4 = \dfrac{x^4}{(y^2)^4}$ **d.** $\left(\dfrac{6x^3}{y^4}\right)^2 = \dfrac{6^2(x^3)^2}{(y^4)^2}$

$\qquad\qquad = \dfrac{x^4}{y^8}$ $\qquad\qquad\quad = \dfrac{36x^6}{y^8}$

SELF CHECK Simplify each expression: **a.** $(a^4b^5)^2$
and **b.** $\left(\dfrac{-6a^5}{b^7}\right)^3$.   *Answers:* **a.** a^8b^{10}, **b.** $-\dfrac{216a^{15}}{b^{21}}$ ∎

Zero Exponents

Since the rules for exponents hold for exponents of 0, we have

$$x^0x^n = x^{0+n} = x^n = 1x^n$$

Because $x^0x^n = 1x^n$, it follows that $x^0 = 1$ $(x \neq 0)$.

Zero exponents

If $x \neq 0$, then $x^0 = 1$.

 WARNING! 0^0 is undefined.

Because of the previous definition, any nonzero base raised to the 0th power is 1. For example, if no variables are zero, then

$$3^0 = 1, \qquad (-7)^0 = 1, \qquad (3ax^3)^0 = 1, \qquad \left(\dfrac{1}{2}x^5y^7z^9\right)^0 = 1$$

E X A M P L E   5   **Zero exponents.** Simplify each expression: **a.** $(5x)^0$, **b.** $5x^0$, and
c. $-5x^0y$.

Solution **a.** $(5x)^0 = 1$ The base is $5x$ and the exponent is 0.
b. $5x^0 = 5 \cdot x^0 = 5 \cdot 1 = 5$ The base is x and the exponent is 0.
c. $-5x^0y = -5 \cdot x^0 \cdot y = -5 \cdot 1 \cdot y = -5y$

SELF CHECK Simplify each expression: **a.** $2xy^0$ and **b.** $-(xy)^0$. *Answers:* **a.** $2x$, **b.** -1
∎

Negative Exponents

Since the rules for exponents are true for negative-integer exponents, we have

$$x^{-n}x^n = x^{-n+n} = x^0 = 1 \quad (x \neq 0)$$

Because $x^{-n} \cdot x^n = 1$ and $\frac{1}{x^n} \cdot x^n = 1$, we define x^{-n} to be the reciprocal of x^n.

Negative exponents

If n is an integer and $x \neq 0$, then

$$x^{-n} = \frac{1}{x^n} \qquad \text{and} \qquad \frac{1}{x^{-n}} = x^n$$

WARNING! By the definition of negative exponents, a base cannot be 0. Thus, an expression such as 0^{-5} is undefined.

Because of the definition, we can write expressions containing negative exponents as expressions without negative exponents. For example,

$$3^{-2} = \frac{1}{3^2} = \frac{1}{9} \qquad 10^{-3} = \frac{1}{10^3} = \frac{1}{1,000} \qquad \frac{1}{4^{-4}} = 4^4 = 256$$

and if b, c, and x are not 0, we have

$$(2c)^{-3} = \frac{1}{(2c)^3} = \frac{1}{8c^3} \qquad 3x^{-1} = 3 \cdot \frac{1}{x} = \frac{3}{x} \qquad \frac{7}{b^{-2}} = 7 \cdot \frac{1}{b^{-2}} = 7b^2$$

E X A M P L E   6

Negative exponents. Write each expression without negative exponents.

a. $-2m^{-8} = -2 \cdot \dfrac{1}{m^8} = -\dfrac{2}{m^8}$

b. $\dfrac{6a}{y^{-5}} = 6a \cdot \dfrac{1}{y^{-5}} = 6a \cdot y^5 = 6ay^5$

SELF CHECK

Write each expression without negative exponents: **a.** $-3.14t^{-7}$ and **b.** $\dfrac{32j}{y^{-9}}$.   *Answers:* **a.** $-\dfrac{3.14}{t^7}$, **b.** $32jy^9$ ∎

E X A M P L E   7

Simplifying expressions involving negative exponents. Simplify each exponential expression.

a. $x^{-5}x^3 = x^{-5+3}$ Keep the common base and add the exponents.

$$= x^{-2}$$

$$= \frac{1}{x^2}$$

b. $(x^{-3})^{-2} = x^{(-3)(-2)}$ Keep the base and multiply the exponents.

$$= x^6$$

SELF CHECK

Simplify each expression: **a.** $a^{-7}a^3$ and **b.** $(a^{-5})^{-3}$.   *Answers:* **a.** $\dfrac{1}{a^4}$, **b.** a^{15} ∎

More Rules for Exponents

To develop a rule for dividing exponential expressions, we proceed as follows:

$$\frac{x^m}{x^n} = x^m\left(\frac{1}{x^n}\right) = x^m x^{-n} = x^{m+(-n)} = x^{m-n}$$

Thus, *to divide exponential expressions with the same nonzero base, keep the common base and subtract the exponent in the denominator from the exponent in the numerator.*

The quotient rule for exponents

If m and n are integers, then

$$\frac{x^m}{x^n} = x^{m-n} \quad (x \neq 0)$$

EXAMPLE 8

Applying the quotient rule for exponents. Simplify each expression. Write each answer without using negative exponents.

a. $\dfrac{a^5}{a^3} = a^{5-3}$ Keep the base. Subtract the exponents.

$= a^2$

b. $\dfrac{2x^{-5}}{x^{11}} = 2x^{-5-11}$

$= 2x^{-16}$

$= \dfrac{2}{x^{16}}$

SELF CHECK

Simplify each expression: **a.** $\dfrac{b^7}{b^5}$ and
b. $\dfrac{3b^{-3}}{b^3}$.

Answers: **a.** b^2,   **b.** $\dfrac{3}{b^6}$ ■

EXAMPLE 9

Applying the quotient rule for exponents. Simplify each expression. Write each answer without using negative exponents.

a. $\dfrac{x^4 x^3}{x^{-5}} = \dfrac{x^7}{x^{-5}}$

$= x^{7-(-5)}$

$= x^{12}$

b. $\dfrac{(x^2)^3}{(x^3)^2} = \dfrac{x^6}{x^6}$

$= x^{6-6}$

$= x^0$

$= 1$

c. $\dfrac{x^2 y^3}{xy^4} = x^{2-1} y^{3-4}$

$= xy^{-1}$

$= x \cdot \dfrac{1}{y}$

$= \dfrac{x}{y}$

d. $\left(\dfrac{a^{-2}b^3}{a^5 b^4}\right)^3 = (a^{-2-5}b^{3-4})^3$

$= (a^{-7}b^{-1})^3$

$= \left(\dfrac{1}{a^7 b}\right)^3$

$= \dfrac{1}{a^{21}b^3}$

SELF CHECK

Simplify each expression: **a.** $\dfrac{(a^{-2})^3}{(a^2)^{-3}}$ and
b. $\left(\dfrac{a^{-2}b^5}{b^8}\right)^{-3}$.

Answers: **a.** 1,   **b.** $a^6 b^9$ ■

To illustrate another property of exponents, we consider the following simplification of $\left(\frac{2}{3}\right)^{-4}$.

$$\left(\frac{2}{3}\right)^{-4} = \frac{1}{\left(\frac{2}{3}\right)^4} = \frac{1}{\frac{2^4}{3^4}} = 1 \div \frac{2^4}{3^4} = 1 \cdot \frac{3^4}{2^4} = \frac{3^4}{2^4} = \left(\frac{3}{2}\right)^4$$

The example suggests that to raise a fraction to a negative power, we can invert the fraction and then raise it to a positive power.

Fractions to negative powers

If n is an integer, then

$$\left(\frac{x}{y}\right)^{-n} = \left(\frac{y}{x}\right)^n \qquad (x \neq 0, \, y \neq 0)$$

EXAMPLE 10

Fractions to negative powers. Write each expression without using parentheses or negative exponents.

a. $\left(\frac{2}{3}\right)^{-4} = \left(\frac{3}{2}\right)^4$

$= \frac{3^4}{2^4}$

$= \frac{81}{16}$

b. $\left(\frac{y^2}{x^3}\right)^{-3} = \left(\frac{x^3}{y^2}\right)^3$

$= \frac{x^9}{y^6}$

c. $\left(\frac{2x^2}{3y^{-3}}\right)^{-4} = \left(\frac{3y^{-3}}{2x^2}\right)^4$

$= \frac{3^4 y^{-12}}{2^4 x^8}$

$= \frac{81}{16x^8} \cdot y^{-12}$

$= \frac{81}{16x^8} \cdot \frac{1}{y^{12}}$

$= \frac{81}{16x^8 y^{12}}$

d. $\left(\frac{a^{-2}b^3}{a^2a^3b^4}\right)^{-3} = \left(\frac{a^2a^3b^4}{a^{-2}b^3}\right)^3$

$= \left(\frac{a^5b^4}{a^{-2}b^3}\right)^3$

$= (a^{5-(-2)}b^{4-3})^3$

$= (a^7b)^3$

$= a^{21}b^3$

SELF CHECK Write $\left(\frac{3a^3}{2b^{-2}}\right)^{-5}$ without using parentheses or negative exponents.

Answer: $\dfrac{32}{243a^{15}b^{10}}$ ∎

We summarize the rules for exponents as follows.

Rules for exponents

If there are no divisions by 0, then for all integers m and n,

$$x^m x^n = x^{m+n} \qquad (x^m)^n = x^{mn} \qquad (xy)^n = x^n y^n \qquad \left(\frac{x}{y}\right)^n = \frac{x^n}{y^n}$$

$$x^0 = 1 \quad (x \neq 0) \qquad x^{-n} = \frac{1}{x^n} \qquad \frac{x^m}{x^n} = x^{m-n} \qquad \left(\frac{x}{y}\right)^{-n} = \left(\frac{y}{x}\right)^n$$

STUDY SET

Section 5.1

VOCABULARY

In Exercises 1–4, fill in the blanks to make the statements true.

1. In the exponential expression x^n, x is called the _____, and n is called the _____.

2. The expression $x \cdot x \cdot x \cdot x \cdot x$ contains five _____ of x.

3. $\{1, 2, 3, 4, 5, 6, \ldots\}$ is the set of _____ numbers.

4. $\{\ldots, -3, -2, -1, 0, 1, 2, 3, \ldots\}$ is the set of _____.

CONCEPTS

In Exercises 5–12, complete the rules for exponents. Assume that $x \neq 0$ and $y \neq 0$.

5. $x^m x^n = $

6. $(x^m)^n = $

7. $(xy)^n = $

8. $\left(\dfrac{x}{y}\right)^n = $

9. $x^0 = $

10. $x^{-n} = $

11. $\dfrac{x^m}{x^n} = $

12. $\left(\dfrac{x}{y}\right)^{-n} = $

13. An exponential expression with a negative exponent can be written as an equivalent expression with a positive exponent. Explain.

14. Explain the difference between the two expressions.
    a. $2x$ and x^2
    b. $-2x$ and x^{-2}

15. A cube is shown in Illustration 1.
    a. Find the area of its base.
    b. Find its volume.

ILLUSTRATION 1

x^3 ft
x^3 ft
x^3 ft

16. A rectangular solid is shown in Illustration 2.
    a. Find the area of its base.
    b. Find its volume.

ILLUSTRATION 2

y^3 ft
y^2 ft
y^4 ft

NOTATION

In Exercises 17–18, complete each simplification.

17. $\dfrac{x^5 x^4}{x^{-2}} = \dfrac{x^{}}{x^{-2}}$

$= x^{9-}$

$= x^{11}$

18. $\left(\dfrac{a^{-4}}{a^3}\right)^2 = (a^{-4-3})^2$

$= (a^{})^2$

$= a^{}$

$= \dfrac{1}{a^{14}}$

PRACTICE

In Exercises 19–26, identify the base and the exponent.

19. 5^3

20. -7^2

21. $-x^5$

22. $(-t)^4$

23. $2b^6$

24. $(3xy)^5$

25. $\left(\dfrac{n}{4}\right)^3$

26. $(-pq)^2$

In Exercises 27–94, simplify each expression. Assume that no denominators are zero. Write each answer without using negative exponents.

27. 3^2

28. 3^4

29. -3^2

30. -3^4

31. $(-3)^2$

32. $(-3)^3$

33. 5^{-2}

34. 5^{-4}

35. -5^{-2}

36. -5^{-4}

37. $(-5)^{-2}$

38. $(-5)^{-4}$

39. 8^0

40. -9^0

41. $(-8)^0$

42. $(-9)^0$

43. $(-2x)^5$

44. $(-3a)^3$

45. $x^2 x^3$

46. $y^3 y^4$

47. $x^2 x^3 x^5$

48. $y^3 y^7 y^2$

49. $k^0 k^7$

50. $x^8 x^{11}$

51. $2aba^3b^4$

52. $2x^2y^3x^3y^22$

53. p^9pp^0

54. z^7z^0z

55. $(-x)^2y^4x^3$

56. $-x^2y^7y^3x^{-2}$

57. $(b^{-8})^9$

58. $(z^{12})^2$

59. $(x^4)^7$

60. $(y^7)^5$

61. $(r^{-3}s)^3$

62. $(m^5n^2)^{-3}$

63. $(a^2a^3)^4$

64. $(bb^2b^3)^4$

65. $(-d^2)^3(d^{-3})^3$

66. $(c^3)^2(c^4)^{-2}$

67. $(3x^3y^4)^3$

68. $\left(\dfrac{1}{2}a^2b^5\right)^4$

69. $\left(-\dfrac{1}{3}mn^2\right)^6$

70. $(-3p^2q^3)^5$

71. $\left(\dfrac{a^3}{b^2}\right)^5$

72. $\left(\dfrac{a^2}{b^3}\right)^4$

73. $\left(\dfrac{a^{-3}}{b^{-2}}\right)^{-2}$

74. $\left(\dfrac{k^{-3}}{k^{-4}}\right)^{-1}$

75. $\dfrac{a^8}{a^3}$

76. $\dfrac{c^7}{c^2}$

77. $\dfrac{c^{12}c^5}{c^{10}}$

78. $\dfrac{a^{33}}{a^2a^3}$

79. $\left(\dfrac{2}{3}\right)^{-2}$

80. $\left(\dfrac{4}{5}\right)^{-3}$

81. $\dfrac{1}{a^{-4}}$

82. $\dfrac{3}{b^{-5}}$

83. $\dfrac{(3x^2)^{-2}}{x^3x^{-4}x^0}$

84. $\dfrac{y^{-3}y^{-4}y^0}{(2y^{-2})^3}$

85. $\left(\dfrac{4a^{-2}b}{3ab^{-3}}\right)^3$

86. $\left(\dfrac{2ab^{-3}}{3a^{-2}b^2}\right)^2$

87. $\left(\dfrac{3a^{-2}b^2}{17a^2b^3}\right)^0$

88. $\dfrac{a^0 + b^0}{2(a + b)^0}$

89. $\left(\dfrac{-2a^4b}{a^{-3}b^2}\right)^3$

90. $\left(\dfrac{-3x^4y^2}{-9x^5y^{-2}}\right)^2$

91. $\left(-\dfrac{2a^3b^2}{3a^{-3}b^2}\right)^{-3}$

92. $\left(\dfrac{3x^5y^2}{6x^5y^{-2}}\right)^{-4}$

93. $\dfrac{(3x^2)^{-2}}{x^3x^{-4}x^0}$

94. $\dfrac{y^{-3}y^{-4}y^0}{(-2y^{-2})^3}$

 In Exercises 95–98, use a calculator to find each value.

95. 1.23^6

96. 0.0537^4

97. -6.25^3

98. $(-25.1)^5$

 In Exercises 99–104, use a calculator to verify that each statement is true by showing that the values on either side of the equation are equal.

99. $(3.68)^0 = 1$

100. $(2.1)^4(2.1)^3 = (2.1)^7$

101. $(7.2)^2(2.7)^2 = [(7.2)(2.7)]^2$

102. $\left(\dfrac{5.4}{2.7}\right)^{-4} = \left(\dfrac{2.7}{5.4}\right)^4$

103. $(3.2)^2(3.2)^{-2} = 1$

104. $(7.23)^{-3} = \dfrac{1}{(7.23)^3}$

APPLICATIONS

105. MICROSCOPES Illustration 3 shows the relative sizes of some chemical and biological structures, expressed as fractions of a meter (m). Express each fraction shown in the illustration as a power of 10, from the largest to the smallest.

ILLUSTRATION 3

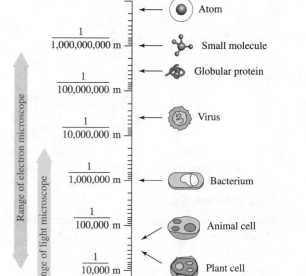

106. ASTRONOMY See Illustration 4. The distance d, in miles, of the nth planet from the sun is given by the formula

$$d = 9{,}275{,}200[3(2^{n-2}) + 4]$$

Find the distance of earth and the distance of Mars from the sun.

ILLUSTRATION 4

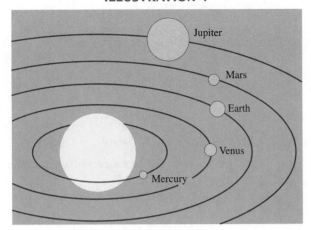

107. LICENSE PLATES The number of different license plates of the form three digits followed by three letters, as in Illustration 5, is $10 \cdot 10 \cdot 10 \cdot 26 \cdot 26 \cdot 26$. Write this expression using exponents. Then evaluate it.

ILLUSTRATION 5

108. PHYSICS Albert Einstein's work in the area of special relativity resulted in the observation that the total energy E of a body is equal to its total mass m times the square of the speed of light c. This relationship is given by the famous equation $E = mc^2$. Identify the base and exponent on the right-hand side.

WRITING

109. Explain how an exponential expression with a negative exponent can be expressed as an equivalent expression with a positive exponent. Give an example.

110. In the definition of x^{-n}, x cannot be 0. Why not?

111. Explain the **error** in the following solution. Write $-8ab^{-3}$ without using negative exponents.

$$-8ab^{-3} = \frac{a}{8b^3}$$

112. Is a positive number (greater than 1) raised to a negative power greater than or less than 1? Explain your answer.

REVIEW

In Exercises 113–116, solve each inequality. Give the result in interval notation and then graph the solution set.

113. $a + 5 < 6$

114. $-9x + 5 \geq 15$

115. $6(t - 2) \leq 4(t + 7)$

116. $\frac{1}{4}p - \frac{1}{3} \leq p + 2$

▶ 5.2   Scientific Notation

In this section, you will learn about

> Writing numbers in scientific notation ■ Converting from scientific notation ■ Using scientific notation to simplify computations

---

Introduction   Very large and very small numbers occur often in science and other disciplines. For example, the star nearest to the earth (excluding the sun) is Proxima Centauri, about 24,793,000,000,000 miles away, and the mass of a hydrogen atom is approximately 0.00000000000000000000001673 gram.

Because these numbers contain many zeros, they are difficult to read and cumbersome to work with in computations. In this section, we will discuss a notation that allows us to express such numbers in a more manageable form, using exponents.

Writing Numbers in Scientific Notation

Scientific notation provides a compact way of writing large and small numbers.

Scientific notation

> A number is written in **scientific notation** when it is written in the form $N \times 10^n$, where $1 \leq |N| < 10$ and n is an integer.

Each of the following numbers is written in scientific notation.

$$3.67 \times 10^6 \qquad 2.24 \times 10^{-4} \qquad 9.875 \times 10^{22}$$

Every positive number written in scientific notation is the product of a number between 1 (including 1) and 10 and an integer power of 10.

An integer exponent

A decimal between 1
(including 1) and 10

EXAMPLE 1

Writing numbers in scientific notation. Write each number in scientific notation: **a.** 24,793,000,000,000 and **b.** 0.000000000000000000000001673.

Solution
a. The number 2.4793 is between 1 and 10. To get 24,793,000,000,000, the decimal point in 2.4793 must be moved thirteen places to the *right*. This can be done by multiplying 2.4793 by 10^{13}.

$$24{,}793{,}000{,}000{,}000 = 2.4793 \times 10^{13}$$

b. The number 1.673 is between 1 and 10. To get 0.000000000000000000000001673, the decimal point in 1.673 must be moved twenty-four places to the *left*. This can be done by multiplying 1.673 by 10^{-24}.

$$0.000000000000000000000001673 = 1.673 \times 10^{-24}$$

SELF CHECK

Write each italicized number in scientific notation.
a. In 1996, the country earning the most money in tourism was the United States, *$64,400,000,000*.
b. DNA molecules contain and transmit the information that allows cells to reproduce. They are only *0.000000002* meter wide.

Answers: **a.** 6.44×10^{10}, **b.** 2×10^{-9} ∎

Numbers such as 47.2×10^3 and 0.063×10^{-2} appear to be written in scientific notation, because they are the product of a number and a power of 10. However, they are not. Their first factors (47.2 and 0.063) are not between 1 and 10.

EXAMPLE 2

Writing numbers in scientific notation. Write **a.** 47.2×10^3 and **b.** 0.063×10^{-2} in scientific notation.

Solution
Since the first factors are not between 1 and 10, neither number is in scientific notation. However, we can change them to scientific notation as follows:

a. $\mathbf{47.2 \times 10^3} = (\mathbf{4.72 \times 10^1}) \times 10^3$ Write 47.2 in scientific notation.

$\qquad\qquad = 4.72 \times (10^1 \times 10^3)$ Group the powers of 10 together.

$\qquad\qquad = 4.72 \times 10^4$ Apply the product rule for exponents: $10^1 \times 10^3 = 10^{1+3} = 10^4$.

b. $0.063 \times 10^{-2} = (6.3 \times 10^{-2}) \times 10^{-2}$ Write 0.063 in scientific notation.

$= 6.3 \times (10^{-2} \times 10^{-2})$

$= 6.3 \times 10^{-4}$

SELF CHECK   Write   **a.** 27.3×10^2 and   *Answers:*   **a.** 2.73×10^3,
b. 0.0025×10^{-3} in scientific notation.   **b.** 2.5×10^{-6} ∎

Converting from Scientific Notation

We can change a number written in scientific notation to **standard notation.** For example, to write 9.3×10^7 in standard notation, we multiply 9.3 by 10^7.

$9.3 \times 10^7 = 9.3 \times 10,000,000 = 93,000,000$ 10^7 is 1 followed by 7 zeros.

EXAMPLE 3   **Converting to standard notation.** Change   **a.** 8.706×10^5 and
b. 1.1×10^{-3} to standard notation.

Solution   **a.** Since multiplication by 10^5 (100,000) moves the decimal point 5 places to the right,

$8.706 \times 10^5 = 8.70600. = 870,600$

b. Since multiplication by 10^{-3} (0.001) moves the decimal point 3 places to the left,

$1.1 \times 10^{-3} = 0.001.1 = 0.0011$

SELF CHECK   Change each number in scientific notation to
standard notation.   **a.** The country with the
largest area of forest is Russia, with 1.9×10^9
acres.   **b.** The average distance between mol-   *Answers:*   **a.** 1,900,000,000,
ecules of air in a room is 3.937×10^{-7} inch.   **b.** 0.0000003937   ∎

Each of the following numbers is written in both scientific and standard notation. In each case, the exponent gives the number of places that the decimal point moves, and the sign of the exponent indicates the direction that it moves:

$5.32 \times 10^4 = 5.3200.$ $6.45 \times 10^7 = 6.4500000.$
4 places to the right   7 places to the right

$2.37 \times 10^{-4} = 0.0002.37$ $9.234 \times 10^{-2} = 0.09.234$
4 places to the left   2 places to the left

$4.89 \times 10^0 = 4.89$

No movement of the decimal point

Using Scientific Notation to Simplify Computations

Scientific notation is useful when multiplying and dividing very large or very small numbers.

EXAMPLE 4   **Astronomy.** The wheel-shaped galaxy in which we live is called the Milky Way. See Figure 5-1 on the next page. This system of some 10^{11} stars, one of which is the sun, has a diameter of approximately 100,000 light years. (A light year is the distance light travels in a vacuum in one year: 9.46×10^{15} meters.) What is the diameter of the Milky Way in meters?

FIGURE 5-1

100,000 light years

A cross-sectional representation of the Milky Way Galaxy

Solution To find the diameter of the Milky Way in meters, we will multiply the diameter of the Milky Way, expressed in light years, by the number of meters in a light year. To perform the calculation, we write 100,000 in scientific notation as 1.0×10^5.

$1.0 \times 10^5 \cdot 9.46 \times 10^{15}$

$= (1.0 \cdot 9.46) \times (10^5 \cdot 10^{15})$ Apply the commutative and associative properties of multiplication to group the first factors together and the powers of 10 together.

$= 9.46 \times 10^{5+15}$ Do the multiplication: $1.0 \cdot 9.46 = 9.46$. For the powers of 10, keep the base and add the exponents.

$= 9.46 \times 10^{20}$ Do the addition.

The Milky Way Galaxy is about 9.46×10^{20} meters in diameter.

SELF CHECK A light year is 5.88×10^{12} miles. What is the diameter of the Milky Way in miles?   *Answer:* 5.88×10^{17} mi   ■

EXAMPLE 5

In an article in *Scientific American,* oil industry experts estimated that there were 1.02×10^{12} barrels of oil reserves in the ground at the start of 1998. At that time, world production was 2.36×10^{10} barrels per year. If annual production remains the same, and if no new oil discoveries are made, when do these experts predict that the world's oil supply will run out?

Solution If we divide the estimated number of barrels of oil in reserve, 1.02×10^{12}, by the number of barrels produced each year, 2.36×10^{10}, we can find the number of years of oil supply left.

$\dfrac{1.02 \times 10^{12}}{2.36 \times 10^{10}} = \dfrac{1.02}{2.36} \times \dfrac{10^{12}}{10^{10}}$ Divide the first factors and the second factors in the numerator and denominator separately.

$\approx 0.43 \times 10^{12-10}$ Do the division: $\frac{1.02}{2.36} \approx 0.43$. For the powers of 10, keep the base and subtract the exponents.

$\approx 0.43 \times 10^2$ Do the subtraction.

≈ 43 Write 0.43×10^2 in standard notation.

According to industry estimates, as of 1998, there were 43 years of oil reserves left. Under these conditions, the world's oil supply will run out in the year 2041.   ■

EXAMPLE 6   **Computing using scientific notation.** Use scientific notation to evaluate

$$\frac{(0.00000064)(24,000,000,000)}{(400,000,000)(0.0000000012)}$$

Solution After writing each number in scientific notation, we can do the arithmetic on the numbers and the exponential expressions separately.

$$\frac{(0.00000064)(24,000,000,000)}{(400,000,000)(0.0000000012)} = \frac{(6.4 \times 10^{-7})(2.4 \times 10^{10})}{(4 \times 10^{8})(1.2 \times 10^{-9})}$$

$$= \frac{(6.4)(2.4)}{(4)(1.2)} \times \frac{10^{-7}10^{10}}{10^{8}10^{-9}}$$

$$= \frac{15.36}{4.8} \times 10^{-7+10-8-(-9)}$$

$$= 3.2 \times 10^{4}$$

The result is 3.2×10^{4}. In standard notation, this is 32,000.

SELF CHECK   Use scientific notation to evaluate

$$\frac{(320)(25,000)}{0.00004}$$

Answer: $2 \times 10^{11} = 200,000,000,000$ ∎

ACCENT ON TECHNOLOGY *Using Scientific Notation*

Scientific calculators and graphing calculators often give answers in scientific notation. For example, if we use a calculator to find 301.2^{8}, the display will read

6.77391496 ¹⁹

On a scientific calculator.

301.2 ∧ 8
6.77391496E19

On a graphing calculator.

In either case, the answer is given in scientific notation and is to be interpreted as

$6.77391496 \times 10^{19}$

Numbers can also be entered into a calculator in scientific notation. For example, to enter 24,000,000,000 (which is 2.4×10^{10} in scientific notation), we enter these numbers and press these keys:

Keystrokes   2.4 [EXP] 10   On most scientific calculators.

2.4 [EE] 10   On a graphing calculator and on some scientific calculators.

To use a scientific calculator to evaluate

$$\frac{(24,000,000,000)(0.00000006495)}{0.00000004824}$$

we must enter each number in scientific notation, because each number has too many digits to be entered directly. In scientific notation, the three numbers are

2.4×10^{10} 6.495×10^{-8} 4.824×10^{-8}

Using a scientific calculator, we enter these numbers and press these keys:

Keystrokes   2.4 [EXP] 10 [×] 6.495 [EXP] 8 [+/−] [÷] 4.824 [EXP] 8 [+/−] [=]

The display will read | 3.231343284 ¹⁰ |. In standard notation, the answer is 32,313,432,840.

The steps are similar on a graphing calculator.

STUDY SET

Section 5.2

VOCABULARY

In Exercises 1–2, fill in the blanks to make the statements true.

1. 7.4×10^{10} is written in _____ notation.
2. 10^{-3}, 10^0, 10^1, and 10^4 are _____ of 10.

CONCEPTS

In Exercises 3–6, fill in the blanks to make the statements true.

3. A number is written in scientific notation when it is written in the form $N \times$ ___, where $1 \leq |N| < 10$ and n is an integer.
4. The number 5.3×10^2 ___ (> or <) the number 5.3×10^{-2}.
5. To change 6.31×10^{-4} to standard notation, we move the decimal point four places to the
   _____.
6. To change 9.7×10^3 to standard notation, we move the decimal point three places to the
   _____.

NOTATION

7. Explain why the number 60.22×10^{22} is not written in scientific notation.
8. Explain why the number 0.6022×10^{24} is not written in scientific notation.

PRACTICE

In Exercises 9–24, write each number in scientific notation.

9. 3,900
10. 1,700
11. 0.0078
12. 0.068
13. 173,000,000,000,000
14. 89,800,000,000
15. 0.0000096
16. 0.000000046
17. 323×10^5
18. 689×10^9
19. $6,000 \times 10^{-7}$
20. 765×10^{-5}
21. 0.0527×10^5
22. 0.0298×10^3
23. 0.0317×10^{-2}
24. 0.0012×10^{-3}

In Exercises 25–36, write each number in standard notation.

25. 2.7×10^2
26. 7.2×10^3
27. 3.23×10^{-3}
28. 6.48×10^{-2}
29. 7.96×10^5
30. 9.67×10^6
31. 3.7×10^{-4}
32. 4.12×10^{-5}
33. 5.23×10^0
34. 8.67×10^0
35. 23.65×10^6
36. 75.6×10^{-5}

In Exercises 37–42, give all answers in scientific notation. Use a calculator to check your results.

37. $(7.9 \times 10^5)(2.3 \times 10^6)$
38. $(6.1 \times 10^8)(3.9 \times 10^5)$
39. $(9.1 \times 10^{-5})(5.5 \times 10^{12})$
40. $(8.4 \times 10^{-13})(4.8 \times 10^9)$
41. $\dfrac{4.2 \times 10^{-12}}{8.4 \times 10^{-5}}$
42. $\dfrac{1.21 \times 10^{-15}}{1.1 \times 10^2}$

In Exercises 43–50, write each numeral in scientific notation and do the operations. Give all answers in scientific notation and in standard form. Use a calculator to check your results.

43. $(89,000,000,000)(4,500,000,000)$
44. $(0.000000061)(3,500,000,000)$
45. $\dfrac{0.00000129}{0.0003}$
46. $\dfrac{4,400,000,000,000}{0.0002}$
47. $\dfrac{(220,000)(0.000009)}{0.00033}$
48. $\dfrac{(640,000)(2,700,000)}{120,000}$
49. $\dfrac{(0.00024)(96,000,000)}{640,000,000}$
50. $\dfrac{(0.0000013)(0.00009)}{0.00039}$

APPLICATIONS

51. FIVE-CARD POKER The odds against being dealt the hand shown in Illustration 1 are about 2.6×10^6 to 1. Express the odds using standard notation.

ILLUSTRATION 1

52. ENERGY See Illustration 2. Express each of the following using scientific notation. (1 quadrillion is 10^{15}.)
   a. 1996 U.S. energy consumption
   b. 1996 U.S. energy production
   c. The difference in 1996 consumption and production

ILLUSTRATION 2

1996 U.S. Energy Consumption and Production
(petroleum, natural gas, coal, hydroelectric, nuclear, geothermal, solar, wind)

Based on data from the Energy Information Administration, United States Department of Energy

53. THE YEAR 2000 Express in scientific notation each of the dollar amounts that appeared in the following excerpt from the *Federal Computer Week* web page (February 16, 1998).

President Clinton's fiscal 1999 budget proposal of $1.7 trillion includes expenditures of about $3.9 billion to ensure that federal computers can accept dates after Dec. 31, 1999. Clinton has proposed spending $275 million at the Defense Department and $312 million at the Treasury Department to fix the year 2000 problem.

54. STAR TREK In the science fiction series *Star Trek*, crew members talk of their spacecraft, the U.S.S. Enterprise, traveling at various "warp speeds." To convert a warp speed, W, to an equivalent velocity in miles per second, v, we can use the equation

$$v = W^3 c$$

where c is the speed of light, 1.86×10^5 miles per second. Find the velocity of a spacecraft traveling at warp 2.

55. ATOMS A simple model of a helium atom is shown in Illustration 3. If a proton has a mass of 1.7×10^{-24}

grams, and if the mass of an electron is only about $\frac{1}{2,000}$ that of a proton, find the mass of an electron.

ILLUSTRATION 3

56. OCEANS The mass of the earth's oceans is only about $\frac{1}{4,400}$ that of the earth. If the mass of the earth is 6.578×10^{21} tons, find the mass of the oceans.

57. LIGHT YEAR Light travels about 300,000,000 meters per second. A **light year** is the distance that light can travel in one year. Estimate the number of meters in one light year.

58. AQUARIUM Express the volume of the fish tank shown in Illustration 4 in scientific notation.

ILLUSTRATION 4

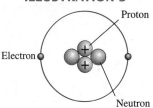

4,000 mm

7,000 mm

3,000 mm

59. THE BIG DIPPER One of the stars in the Big Dipper is named Merak. (See Illustration 5.) It is approximately 4.65×10^{14} miles from the earth.
   a. If light travels about 1.86×10^5 miles/sec, how many seconds does it take light emitted from Merak to reach the earth? (*Hint:* Use the formula $t = \frac{d}{r}$.)

ILLUSTRATION 5

Merak

   b. Convert your result from part a to years.

60. BIOLOGY A paramecium is a single-celled organism that propels itself with hair-like projections called *cilia*. Use the scale in Illustration 6 (on the next page) to estimate the length of the paramecium. Express the result in scientific and in standard notation.

ILLUSTRATION 6

5×10^{-5} m

61. COMET HALE-BOPP On March 23, 1997, Comet Hale-Bopp made its closest approach to earth, coming within 1.3 **astronomical units.** One astronomical unit (AU) is the distance from the earth to the sun—about 9.3×10^7 miles. Express this distance in miles, using scientific notation.

62. DIAMONDS The approximate number of atoms of carbon in a $\frac{1}{2}$-carat diamond can be found by computing

$$\frac{6.0 \times 10^{23}}{1.2 \times 10^2}$$

Express the number of carbon atoms in scientific and in standard notation.

63. ATOMS A hydrogen atom is so small that a single drop of water contains more than a million million billion hydrogen atoms. Express this number in scientific notation.

64. ASTRONOMY The American Physical Society recently honored first-year graduate student Gwen Bell for coming up with what it considers the most accurate estimate of the mass of the Milky Way. In pounds, her estimate is a 3 with 42 zeros after it. Express this number in scientific notation.

WRITING

65. Explain how to change a number from standard notation to scientific notation.

66. Explain how to change a number from scientific notation to standard notation.

67. Explain why 9.99×10^n represents a number less than 1 but greater than 0 if n is a negative integer.

68. Explain the advantages of writing very large and very small numbers in scientific notation.

REVIEW

In Exercises 69–72, solve each compound inequality. Give the result in interval notation and graph the solution set.

69. $4x \geq -x + 5$ and $6 \geq 4x - 3$

70. $15 > 2x - 7 > 9$

71. $3x + 2 < 8$ or $2x - 3 > 11$

72. $-4(x + 2) \geq 12$ or $3x + 8 < 11$

▶ **5.3**

Polynomials and Polynomial Functions

In this section, you will learn about

Polynomials ■ Degree of a polynomial ■ Polynomial functions
■ Evaluating polynomial functions ■ Graphing polynomial functions
■ Combining like terms ■ Adding and subtracting polynomials

Introduction In arithmetic, we learned how to add, subtract, multiply, divide, and find powers of numbers. In the next several sections, we will learn how to perform these operations on algebraic expressions called *polynomials*. To begin the discussion of polynomials, we introduce some vocabulary that is used to classify them. Then we will see how *polynomial functions* can be used to model many real-life situations.

Polynomials

An **algebraic term** is a number or a product of a number and a variable (or variables) raised to a power. Some examples are

$$17, \qquad 9x, \qquad \frac{15}{16}y^2, \qquad \text{and} \qquad -2.4x^4y^5$$

The **numerical coefficient**, or simply the **coefficient**, of 17 is 17. The numerical coefficients of the remaining terms are the numbers 9, $\frac{15}{16}$, and -2.4, respectively.

Polynomials

A **polynomial** is an algebraic term or the sum of two or more algebraic terms whose variables have whole-number exponents. No variable appears in a denominator.

The following expressions are polynomials in x:

$$-6x, \qquad 3x^2 + 2x, \qquad \frac{3}{2}x^5 - \frac{7}{3}x^4 - \frac{8}{3}x^3, \qquad \text{and} \qquad 19x^{20} + \sqrt{3}x^{14} + 4.5x^{11} - x^2$$

 WARNING! The following expressions are not polynomials:

$$\frac{2x}{x^2 + 1}, \qquad x^{1/2} - 8, \qquad \text{and} \qquad x^{-3} + 2x + 24$$

The first expression is a quotient and has a variable in the denominator. The last two have exponents that are not whole numbers.

If any terms of a polynomial contain more than one variable, we say that the polynomial is in more than one variable. Some examples are

$$3xy, \qquad 5x^2y^2 + 2xy - 3y, \qquad \text{and} \qquad u^2v^2w^2 + uv + 1$$

Polynomials can be classified according to the number of terms they have. A polynomial with one term is called a **monomial,** a polynomial with two terms is called a **binomial,** and a polynomial with three terms is called a **trinomial.**

Monomials	*Binomials*	*Trinomials*
$2x^3$	$2x + 5$	$2x^2 + 4x + 3$
a^2b	$-17x^4 - \dfrac{3}{5}x$	$3mn^3 - m^2n^3 + 7n$
$3x^3y^5z^2$	$32x^{13}y^5z^3 + 47x^3yz$	$-12x^5y^2 + 13x^4y^3 - 7x^3y^3$

Degree of a Polynomial

Because the variable x occurs three times as a factor in the monomial $2x^3$, the monomial is called a *third-degree monomial* or a *monomial of degree 3.* The monomial $3x^3y^5z^2$ is called a *monomial of degree 10,* because the variables x, y, and z occur as factors a total of ten times $(3 + 5 + 2)$. These examples illustrate the following definition.

Degree of a monomial

The **degree** of a monomial with one variable is the exponent on the variable. The degree of a monomial in several variables is the sum of the exponents on those variables. If the monomial is a nonzero constant, its degree is 0. The constant 0 has no defined degree.

| EXAMPLE 1 | **Determining the degree of a monomial.** Find the degree of **a.** $3x^4$, **b.** $-4x^2y^3$, and **c.** 3. |

Solution

a. $3x^4$ is a monomial of degree 4, because the exponent on the variable is 4.

b. $-4x^2y^3$ is a monomial of degree 5, because the sum of the exponents on the variables is 5.

c. 3 is a monomial of degree 0, because $3 = 3x^0$.

SELF CHECK

Find the degree of **a.** $-12a^2$, **b.** $8a^3b^2$, and **c.** $\frac{1}{2}x^3y^2z^{12}$. *Answers:* **a.** 2, **b.** 5, **c.** 17 ∎

We determine the degree of a polynomial by considering the degrees of each of its terms.

Degree of a polynomial

The **degree of a polynomial** is the same as the degree of the term in the polynomial with largest degree.

| EXAMPLE 2 | **Determining the degree of a polynomial.** Find the degree of each polynomial: **a.** $3x^5 + 4x^2 + 7$, **b.** $7x^2y^8 - 3x^2y^2$, and **c.** $3x + 2y - xy$. |

Solution

a. The terms of $3x^5 + 4x^2 + 7$ have degree 5, 2, and 0, respectively. This trinomial is of degree 5, because the largest degree of the three terms is 5.

b. $7x^2y^8 - 3x^2y^2$ is a binomial of degree 10.

c. $3x + 2y - xy$ is a trinomial of degree 2. (Recall that $xy = x^1y^1$.)

SELF CHECK

Find the degree of **a.** $x^2 - x + 1$ and **b.** $x^2y^3 - 12x^7y^2 + 3x^9y^3 - 3$. *Answers:* **a.** 2, **b.** 12 ∎

If the terms of a polynomial in one variable are written so that the exponents decrease as we move from left to right, we say that the terms are written with their exponents in *descending order.* If the terms are written so that the exponents increase as we move from left to right, we say that the terms are written with their exponents in *ascending order.*

$-5x^4 + 2x^3 + 7x^2 + 3x - 1$ This polynomial is written in descending powers of x.

$-1 + 3x + 7x^2 + 2x^3 - 5x^4$ The same polynomial is now written in ascending powers of x.

Polynomial Functions

In Chapter 2, we saw that linear functions are defined by equations of the form $f(x) = mx + b$. Some examples of linear functions are

$$f(x) = 3x + 1 \qquad g(x) = -\frac{1}{2}x - 1 \qquad h(x) = 5x$$

In each case, the right-hand side of the equation is a polynomial. For this reason, linear functions are members of a larger class of functions known as **polynomial functions.**

> A **polynomial function** is a function whose equation is defined by a polynomial in one variable.

Another example of a polynomial function is $f(x) = -x^2 + 6x - 8$. This is a second-degree polynomial function, called a **quadratic function.** Quadratic functions are of the form $f(x) = ax^2 + bx + c$, where $a \neq 0$.

An example of a third-degree polynomial function is $f(x) = x^3 - 3x^2 - 9x + 2$. Third-degree polynomial functions, also called **cubic functions,** are of the form $f(x) = ax^3 + bx^2 + cx + d$, where $a \neq 0$. Polynomial functions of degree 4 or higher, such as $f(x) = x^4 + 2x^3 - 3x + 1$, do not have special names.

Polynomial functions can be used to model many real-life situations. For example, the polynomial function $h(t) = -16t^2 + 128t$ gives the height h of a toy rocket above the ground t seconds after it is shot upward. To find the height h for specific times t, we evaluate the function at those values of t.

Evaluating Polynomial Functions

To *evaluate a polynomial function* at a specific value, we replace the variable in the defining equation with the value, called the **input.** Then we simplify the resulting expression to find the **output.**

EXAMPLE 3

Rocketry. If a toy rocket is shot straight up with an initial velocity of 128 feet per second, its height $h(t)$ (in feet) above the ground t seconds after liftoff is given by the function

$$h(t) = -16t^2 + 128t$$

Find the height of the rocket at   **a.** 0 second,   **b.** 3 seconds,   and   **c.** 7.9 seconds.

Solution   **a.** To find the height of the rocket at 0 second, we substitute 0 for t and evaluate the right-hand side.

$h(t) = -16t^2 + 128t$	The given function.
$h(0) = -16(0)^2 + 128(0)$	The input is 0.
$\quad = 0$	The output is 0.

At 0 second, the rocket's height is 0. It is on the ground waiting to be launched.

b. To find the height at 3 seconds, we substitute 3 for t and simplify.

$h(t) = -16t^2 + 128t$	The given function.
$h(3) = -16(3)^2 + 128(3)$	The input is 3.
$\quad = -16(9) + 384$	
$\quad = -144 + 384$	
$\quad = 240$	The output is 240.

At 3 seconds after liftoff, the height of the rocket is 240 feet.

c. To find the height at 7.9 seconds, we substitute 7.9 for t and simplify.

$h(t) = -16t^2 + 128t$	The given function.
$h(7.9) = -16(7.9)^2 + 128(7.9)$	The input is 7.9.
$\quad = -16(62.41) + 1,011.2$	
$\quad = -998.56 + 1,011.2$	
$\quad = 12.64$	The output is 12.64.

At 7.9 seconds, the height is 12.64 feet. The rocket has almost fallen back to earth.

SELF CHECK

In Example 3, find the height of the rocket 4 seconds after it is shot upward.

Answer: 256 ft ∎

EXAMPLE 4

Packaging. To make boxes, a manufacturer cuts equal-sized squares from each corner of a 10 in. × 12 in. piece of cardboard and then folds up the sides. (See Figure 5-2.) The polynomial function $f(x) = 4x^3 - 44x^2 + 120x$ gives the volume (in cubic inches) of the resulting box when a square with sides x inches long is cut from each corner. Find the volume of a box if 3-inch squares are cut out.

FIGURE 5-2

Fold on dotted lines.

Solution   To find the volume of the box, we evaluate the function for $x = 3$.

$$f(x) = 4x^3 - 44x^2 + 120x$$
$$f(3) = 4(3)^3 - 44(3)^2 + 120(3) \quad \text{Substitute 3 for } x.$$
$$= 4(27) - 44(9) + 120(3)$$
$$= 108 - 396 + 360$$
$$= 72$$

If 3-inch squares are cut out, the box will have a volume of 72 in.^{3}.

SELF CHECK

In Example 4, find the volume of the resulting box if 2-inch squares are cut from each corner of the cardboard.   *Answer:* 96 in.^{3} ∎

Graphing Polynomial Functions

We have previously graphed three basic polynomial functions. Figure 5-3 shows the graph of a linear function $f(x) = 3x - 1$, the graph of the squaring function $f(x) = x^2$, and the graph of the cubing function $f(x) = x^3$.

FIGURE 5-3

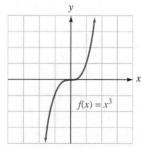

A linear function
The domain is $(-\infty, \infty)$.
The range is $(-\infty, \infty)$.

(a)

The squaring function
The domain is $(-\infty, \infty)$.
The range is $[0, \infty)$.

(b)

The cubing function
The domain is $(-\infty, \infty)$.
The range is $(-\infty, \infty)$.

(c)

When graphing a linear function, we need to plot only two points, because the graph is a straight line. The graphs of polynomial functions of degree greater than 1 are smooth, continuous curves. To graph them, we must plot more points.

In Example 3, we saw that the polynomial function $h(t) = -16t^2 + 128t$ gives the height of the rocket t seconds after it is shot upward. Since the height of the rocket depends on time, we say that the height is a function of time. To graph this function, we can make a table of values, plot the points, and join them with a smooth curve.

EXAMPLE 5

Graphing a polynomial function. Graph $h(t) = -16t^2 + 128t$.

Solution   From Example 3, we have seen that

When $t = 0$, $h(t) = 0$.

When $t = 3$, $h(t) = 240$.

To graph the function, we select other values of t, evaluate the function at those values, and write the ordered pairs in the table shown in Figure 5-4. Then we plot the pairs and join the resulting points to get the parabola shown in the figure. From the graph, we can see that 4 seconds into the flight, the rocket attains a maximum height of 256 feet.

FIGURE 5-4

$h(t) = -16t^2 + 128t$

t	$h(t)$	$(t, h(t))$
0	0	$(0, 0)$
1	112	$(1, 112)$
2	192	$(2, 192)$
3	240	$(3, 240)$
4	256	$(4, 256)$
5	240	$(5, 240)$
6	192	$(6, 192)$
7	112	$(7, 112)$
8	0	$(8, 0)$

From the graph, we see that the domain of the function is $[0, 8]$ and the range is $[0, 256]$.

WARNING! The parabola shown in the figure describes the height of the rocket in relation to time. It does not show the path of the rocket. The rocket goes straight up and then falls straight down.

SELF CHECK   Use the graph in Figure 5-4 to estimate the height of the rocket at 6.5 seconds.   *Answer:* 150 ft   ∎

EXAMPLE 6

Graphing a polynomial function. Graph $f(x) = x^3 - 3x^2 - 9x + 2$.

Solution   To graph this cubic function, we begin by evaluating it for $x = -3$.

$$f(x) = x^3 - 3x^2 - 9x + 2$$
$$f(-3) = (-3)^3 - 3(-3)^2 - 9(-3) + 2$$
$$= -27 - 3(9) - 9(-3) + 2$$
$$= -27 - 27 + 27 + 2$$
$$= -25$$

In the table of values in Figure 5-5 on the next page, we enter the ordered pair $(-3, -25)$. We continue the evaluation process for $x = -2, -1, 0, 1, 2, 3, 4$, and 5, and list the results in the table. After plotting the ordered pairs, we draw a smooth curve through the points to get the graph of function f.

FIGURE 5-5

$f(x) = x^3 - 3x^2 - 9x + 2$

x	$f(x)$	$(x, f(x))$
-3	-25	$(-3, -25)$
-2	0	$(-2, 0)$
-1	7	$(-1, 7)$
0	2	$(0, 2)$
1	-9	$(1, -9)$
2	-20	$(2, -20)$
3	-25	$(3, -25)$
4	-18	$(4, -18)$
5	7	$(5, 7)$

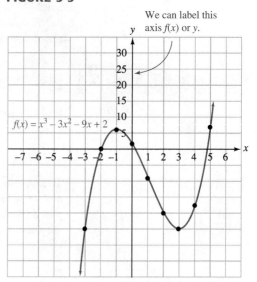

We can label this axis $f(x)$ or y.

$f(x) = x^3 - 3x^2 - 9x + 2$

SELF CHECK

What are the domain and the range of the function graphed in Figure 5-5?

Answers: domain: $(-\infty, \infty)$, range: $(-\infty, \infty)$

ACCENT ON TECHNOLOGY *Graphing Polynomial Functions*

We can graph polynomial functions with a graphing calculator. For example, to graph the function from Example 5, $h(t) = -16t^2 + 128t$, we can rewrite it as $f(x) = -16x^2 + 128x$ and use window settings of $[0, 8]$ for x and $[0, 260]$ for y to get the parabola shown in Figure 5-6(a).

We can trace to estimate the height of the rocket for any number of seconds into the flight. Figure 5-6(b) shows that the height of the rocket 1.6 seconds into the flight is approximately 165 feet.

FIGURE 5-6

(a)

(b)

Combining Like Terms

Recall that when terms have the same variables with the same exponents, they are called **like** or **similar terms.**

- $3x^2$ and $7x^2$ are like terms, because they have the same variables with the same exponents.
- $4x^4y^2$ and $98x^7y^9$ are unlike terms. They have the same variables, but with different exponents.
- $3xy$ and $5xz$ are unlike terms. They have different variables.

The distributive property enables us to combine like terms. For example,

$$3x^2 + 7x^2 = (3 + 7)x^2 \qquad \text{Use the distributive property.}$$
$$= 10x^2 \qquad \text{Do the addition.}$$
$$5x^2y^3 + 22x^2y^3 = (5 + 22)x^2y^3 \qquad \text{Use the distributive property.}$$
$$= 27x^2y^3 \qquad \text{Do the addition.}$$

The results of the previous examples suggest that to add like terms, *we add their numerical coefficients and keep the same variables with the same exponents.*

 WARNING! The terms in the following binomials cannot be combined, because they are not like terms. (Tell why.) In each case, no further simplification is possible.

$$3x^2 - 5y^2, \qquad -2a^2 + 3a^3, \qquad \text{and} \qquad 5y^2 + 17xy$$

Adding and Subtracting Polynomials

Adding polynomials

To add polynomials, remove parentheses and combine like terms.

EXAMPLE 7

Adding polynomials. Add $3x^2 - 2x + 4$ and $2x^2 + 4x - 3$.

Solution
$$(3x^2 - 2x + 4) + (2x^2 + 4x - 3) \qquad \text{We are to add two trinomials.}$$
$$= 3x^2 - 2x + 4 + 2x^2 + 4x - 3 \qquad \text{Remove the parentheses.}$$
$$= 3x^2 + 2x^2 - 2x + 4x + 4 - 3 \qquad \text{Use the commutative and associative properties of addition to rearrange terms so that like terms are together.}$$
$$= 5x^2 + 2x + 1 \qquad \text{Combine like terms.}$$

SELF CHECK Add $2a^2 - 3a + 5$ and $5a^2 + 4a - 2$. *Answer:* $7a^2 + a + 3$ ∎

EXAMPLE 8

Adding polynomials in two variables. Add $-5x^3y^2 - 4x^2y^3$ and $2x^3y^2 + 5x^2y^3$.

Solution
$$(-5x^3y^2 - 4x^2y^3) + (2x^3y^2 + 5x^2y^3)$$
$$= -5x^3y^2 + 2x^3y^2 - 4x^2y^3 + 5x^2y^3 \qquad \text{Remove parentheses and rearrange terms.}$$
$$= -3x^3y^2 + x^2y^3 \qquad \text{Combine like terms.}$$

SELF CHECK Add $-6a^2b^3 - 5a^3b^2$ and $3a^2b^3 + 2a^3b^2$. *Answer:* $-3a^2b^3 - 3a^3b^2$ ∎

The additions in Examples 7 and 8 can be done by aligning the terms vertically.

$$+\ \begin{array}{r} 3x^2 - 2x + 4 \\ 2x^2 + 4x - 3 \\ \hline 5x^2 + 2x + 1 \end{array}$$
$$+\ \begin{array}{r} -5x^3y^2 - 4x^2y^3 \\ 2x^3y^2 + 5x^2y^3 \\ \hline -3x^3y^2 + \ x^2y^3 \end{array}$$

To subtract one monomial from another, we add the negative (or opposite) of the monomial that is to be subtracted.

EXAMPLE 9	**Subtracting monomials.** Do each subtraction.

a. $8x^2 - 3x^2 = 8x^2 + (-3x^2)$ Add the opposite of $3x^2$, which is $-3x^2$.

$\qquad\qquad\ = 5x^2$ Combine like terms.

b. $3x^2y - 9x^2y = 3x^2y + (-9x^2y)$

$\qquad\qquad\quad = -6x^2y$

c. $-5x^5y^3z^2 - 3x^5y^3z^2 = -5x^5y^3z^2 + (-3x^5y^3z^2)$

$\qquad\qquad\qquad\quad = -8x^5y^3z^2$

SELF CHECK   Subtract: **a.** $-2a^2b^3 - 5a^2b^3$ and   *Answers:* **a.** $-7a^2b^3$,
b. $-2a^2b^3 - (-5a^2b^3)$.   **b.** $3a^2b^3$

Subtracting polynomials

To subtract two polynomials, we add the first polynomial and the negative (or opposite) of the second polynomial.

EXAMPLE 10	**Subtracting two polynomials.** Do each subtraction.

a. $(8x^3y + 2x^2y) - (2x^3y - 3x^2y)$

$\quad = 8x^3y + 2x^2y - 2x^3y + 3x^2y$ Remove the parentheses. Change every sign in the second polynomial.

$\quad = 8x^3y - 2x^3y + 2x^2y + 3x^2y$ Rearrange terms.

$\quad = 6x^3y + 5x^2y$ Combine like terms.

b. $(3rt^2 + 4r^2t^2) - (8rt^2 - 4r^2t^2 + r^3t^2)$

$\quad = 3rt^2 + 4r^2t^2 - 8rt^2 + 4r^2t^2 - r^3t^2$

$\quad = 3rt^2 - 8rt^2 + 4r^2t^2 + 4r^2t^2 - r^3t^2$ Rearrange terms.

$\quad = -5rt^2 + 8r^2t^2 - r^3t^2$ Combine like terms.

SELF CHECK   Subtract: $(6a^2b^3 - 2a^2b^2) - (-2a^2b^3 + a^2b^2)$.   *Answer:* $8a^2b^3 - 3a^2b^2$

To subtract polynomials in vertical form, we add the negative (or opposite) of the polynomial that is being subtracted.

$$-\ \begin{array}{r} 8x^3y + 2x^2y \\ 2x^3y - 3x^2y \\ \hline \end{array} \quad \Rightarrow \quad +\ \begin{array}{r} 8x^3y + 2x^2y \\ -2x^3y + 3x^2y \\ \hline 6x^3y + 5x^2y \end{array}$$

STUDY SET

Section 5.3

VOCABULARY

In Exercises 1–8, fill in the blanks to make the statements true.

1. A _____ is the sum of one or more algebraic terms whose variables have whole-number exponents.

2. A _____ is a polynomial with one term. A _____ is a polynomial with two terms. A _____ is a polynomial with three terms.

3. The _____ of a monomial with one variable is the exponent on the variable.

4. A second-degree polynomial function is also called a _____ function.

5. A third-degree polynomial function is also called a _____ function.

6. The _____ of the term $-15x^2y^3$ is -15. The _____ of the term is 5.

7. Terms having the same variables with the same exponents are called _____ terms.

8. The _____ of $x^2 + x - 3$ is $-x^2 - x + 3$.

CONCEPTS

In Exercises 9–20, classify each polynomial as a monomial, binomial, trinomial, or none of these. Then determine the degree of the polynomial.

9. $3x^2$

10. $2y^3 + 4y^2$

11. $3x^2y - 2x + 3y$

12. $a^2 + ab + b^2$

13. $x^2 - y^2$

14. $\dfrac{17}{2}x^3 + 3x^2 - x - 4$

15. 5

16. $8x^3y^5$

17. $9x^2y^4 - x - y^{10} + 1$

18. x^{17}

19. $4x^9 + 3x^2y^4$

20. -12

21. Write each polynomial with the exponents on x in descending order.
    a. $3x - 2x^4 + 7 - 5x^2$
    b. $a^2x - ax^3 + 7a^3x^5 - 5a^3x^2$

22. Write each polynomial with the exponents on y in ascending order.
    a. $4y^2 - 2y^5 + 7y - 5y^3$
    b. $x^3y^2 + x^2y^3 - 2x^3y + x^7y^6 - 3x^6$

In Exercises 23–30, tell whether the terms are like or unlike terms. If they are like terms, combine them.

23. $3x, 7x$

24. $-8x, 3y$

25. $7x, 7y$

26. $3mn, 5mn$

27. $3r^2t^3, -8r^2t^3$

28. $9u^2v, 10u^2v$

29. $9x^2y^3, 3x^2y^2$

30. $27x^6y^4z, 8x^6y^4z^2$

31. Write a polynomial that represents the perimeter of the triangle shown in Illustration 1.

ILLUSTRATION 1

$2x^2 + 3x + 1$ $3x^2 + x - 1$

$4x^2 - x - 2$

32. Write a polynomial that represents the perimeter of the rectangle shown in Illustration 2.

ILLUSTRATION 2

$2x^3 - x$

$x^3 + 3x$

NOTATION

In Exercises 33–34, complete each solution.

33. If $f(x) = 2x^2 + x + 2$, find $f(-1)$.
$$f(x) = 2x^2 + x + 2$$
$$f(-1) = 2\big(\big)^2 + \big(\big) + 2$$
$$= 2\big(\big) + (-1) + 2$$
$$= 3$$

34. If $h(t) = -t^3 - t^2 + 2t + 1$, find $h(3)$.
$$h(t) = -t^3 - t^2 + 2t + 1$$
$$h(3) = -\big(\big)^3 - \big(\big)^2 + 2(3) + 1$$
$$= - 9 + 6 + 1$$
$$= -29$$

PRACTICE

In Exercises 35–38, complete each table of values. Then graph each polynomial function. Check your work with a graphing calculator.

35. $f(x) = 2x^2 - 4x + 2$

x	f(x)
−1	
0	
1	
2	
3	

36. $f(x) = -x^2 + 2x + 6$

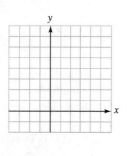

x	f(x)
−2	
−1	
0	
1	
2	
3	
4	

37. $f(x) = 2x^3 - 3x^2 - 11x + 6$

x	f(x)
−3	
−2	
−1	
0	
1	
2	
3	
4	

38. $f(x) = -x^3 - x^2 + 6x$

x	f(x)
−4	
−3	
−2	
−1	
0	
1	
2	
3	

In Exercises 39–40, use a graphing calculator to graph each polynomial function. Use window settings of $[-4, 6]$ for x and $[-5, 5]$ for y.

39. $f(x) = 2.75x^2 - 4.7x + 1.5$

40. $f(x) = 0.37x^3 - 1.4x + 1.5$

In Exercises 41–62, do each operation.

41. $(3x^2 + 2x + 1) + (-2x^2 - 7x + 5)$

42. $(-2a^2 - 5a - 7) + (-3a^2 + 7a + 1)$

43. $(-a^2 + 2a + 3) - (4a^2 - 2a - 1)$

44. $(x^2 - 3x + 8) - (3x^2 + x + 3)$

45. $(2a^2 + 4a - 7) + (3a^2 - a - 2)$

46. $(6x^3 + 3x - 2) - (2x^3 + 3x^2 + 5)$

47. $(7y^3 + 4y^2 + y + 3) + (-8y^3 - y + 3)$

48. $(-8p^3 - 2p - 4) - (2p^3 + p^2 - p)$

49. $(3pq + p - q) + (-pq - p - q)$

50. $(-2mn + 2m - n) - (-2mn - 2m + n)$

51. $(-2x^2y^3 + 6xy + 5y^2) - (-4x^2y^3 - 7xy + 2y^2)$

52. $(3ax^3 - 2ax^2 + 3a^3) + (4ax^3 + 3ax^2 - 2a^3)$

53. $(3x^2 + 4x - 3) + (2x^2 - 3x - 1) - (x^2 + x + 7)$

54. $(-2x^2 + 6x + 5) - (-4x^2 - 7x + 2) - (4x^2 + 10x + 5)$

55.
$$\begin{array}{r} 3x^3 - 2x^2 + 4x - 3 \\ + \quad -2x^3 + 3x^2 + 3x - 2 \\ \hline 5x^3 - 7x^2 + 7x - 12 \end{array}$$

56.
$$\begin{array}{r} 7a^3 \qquad + 3a + 7 \\ + \quad -2a^3 + 4a^2 \qquad - 13 \\ \hline 3a^3 - 3a^2 + 4a + 5 \end{array}$$

57.
$$\begin{array}{r} 3x^2 - 4x + 17 \\ - \quad 2x^2 + 4x - 5 \\ \hline \end{array}$$

58.
$$\begin{array}{r} -2y^2 - 4y + 3 \\ - \quad 3y^2 + 10y - 5 \\ \hline \end{array}$$

59.
$$\begin{array}{r} -5y^3 + 4y^2 - 11y + 3 \\ - \quad -2y^3 - 14y^2 + 17y - 32 \\ \hline \end{array}$$

60.
$$\begin{array}{r} 17x^4 - 3x^2 - 65x - 12 \\ - \quad 23x^4 + 14x^2 + 3x - 23 \\ \hline \end{array}$$

61. $-$ $\dfrac{4x^2 + 0x}{4x^2 - 2x}$

62. $-$ $\dfrac{-2a^3 + 0a^2}{-2a^3 + 3a^2}$

APPLICATIONS

63. JUGGLING During a performance, a juggler tosses one
ball straight upward while
continuing to juggle three
others. (See Illustration 3.)
The height $f(t)$, in feet, of
the ball is given by the
polynomial function
$f(t) = -16t^2 + 32t + 4$, where t is the time in seconds since the ball was thrown. Find the height of the
ball 1 second after it is tossed upward.

ILLUSTRATION 3

64. STOPPING DISTANCE The number of feet that a car
travels before stopping depends on the driver's reaction
time and the braking distance. (See Illustration 4.) For
one driver, the stopping distance $d(v)$, in feet, is given
by the polynomial function $d(v) = 0.04v^2 + 0.9v$,
where v is the velocity of the car. Find the stopping
distance at 60 mph.

ILLUSTRATION 4

65. STORAGE TANK
See Illustration 5.
The volume $V(r)$ of
the gasoline storage
tank, in cubic feet, is
given by the polyno-
mial function
$V(r) = 4.2r^3 + 37.7r^2$, where r is the radius in feet of
the cylindrical part of the tank. What is the capacity
of the tank if its radius is 4 feet?

ILLUSTRATION 5

66. ROLLER COASTER The polynomial function
$f(x) = 0.001x^3 - 0.12x^2 + 3.6x + 10$ models the path
of a portion of the track of a roller coaster, as shown
in Illustration 6. Find the height of the track for
$x = 0, 20, 40,$ and 60.

ILLUSTRATION 6

$f(x) = 0.001x^3 - 0.12x^2 + 3.6x + 10$

67. RAIN GUTTERS A rectangular sheet of metal will
be used to make a rain gutter by bending up its sides,
as shown in Illustration 7. If the ends are covered,
the capacity $f(x)$ of the gutter is a polynomial function of x: $f(x) = -240x^2 + 1,440x$. Find the capacity
of the gutter if x is 3 inches.

ILLUSTRATION 7

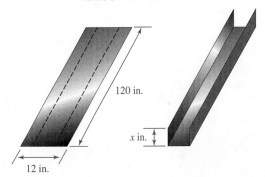

68. CUSTOMER SERVICE A software service hotline
has found that on Mondays, the polynomial function
$C(t) = -0.0625t^4 + t^3 - 6t^2 + 16t$ approximates the
number of callers to the hotline at any one time.
Here, t represents the time, in hours, since the hotline
opened at 8:00 A.M. How many service technicians
should be on duty on Mondays at noon if the company doesn't want any callers to the hotline waiting
to be helped by a technician?

69. REAL ESTATE A computer analysis of two properties on the market generated functions to predict the
value, in dollars, of each property after x years.

Rental home: $R(x) = 1,100x + 125,000$

Duplex: $D(x) = 1,400x + 150,000$

a. Find one polynomial function V that will give the
combined value of the two properties after x
years.

b. Use your answer to part a to find the combined
value of the two properties after 20 years.

70. BUSINESS EXPENSES A company purchased two
cars for its sales force to use. The following functions
give the respective values of the vehicles after x years.

Toyota Camry LE: $T(x) = -2,100x + 16,600$

Ford Explorer Sport: $F(x) = -2,700x + 19,200$

a. Find one polynomial function V that will give the value of both cars after x years.

b. Use your answer in part a to find the combined value of the two cars after 3 years.

WRITING

71. Explain why the terms x^2y and xy^2 are not like terms.

72. The family of polynomial functions contains linear, quadratic, and cubic functions. Explain.

73. Explain what is **wrong** in the following solution. Subtract $2x - 3$ from $3x + 4$.

$$(2x - 3) - (3x + 4) = 2x - 3 - 3x - 4$$
$$= -x - 7$$

74. Explain why $f(x) = \frac{1}{x+1}$ is not a polynomial function.

75. Use the word *descending* in a sentence in which the context is not mathematical. Do the same for the word *ascending*.

76. Look up the meaning of the prefix *poly* in a dictionary. Why do you think the name *polynomial* was given to expressions such as $x^3 - x^2 + 2x + 15$?

REVIEW

In Exercises 77–80, solve each inequality. Give the result in interval notation.

77. $|x| \leq 5$ **78.** $|x| > 7$

79. $|x - 4| < 5$ **80.** $|2x + 1| \geq 7$

▶ 5.4   Multiplying Polynomials

In this section, you will learn about

> Multiplying monomials ■ Multiplying a polynomial by a monomial
> ■ Multiplying a polynomial by a polynomial ■ The FOIL method
> ■ Special products ■ Applications of multiplying polynomials
> ■ Multiplying expressions that are not polynomials

Introduction In this section, we discuss the procedures used to multiply polynomials. These procedures involve the application of several algebraic concepts introduced in earlier chapters, such as the commutative and associative properties of multiplication, the rules for exponents, and the distributive property.

Multiplying Monomials

In Section 1.4, we saw that to multiply one monomial by another, *we multiply the numerical factors and then multiply the variable factors.*

EXAMPLE 1

Multiplying monomials. Do each multiplication. We can use the commutative and associative properties of multiplication to rearrange the terms and regroup the factors.

a. $\quad (3x^2)(6x^3) = 3 \cdot x^2 \cdot 6 \cdot x^3$
$$= (3 \cdot 6)(x^2 \cdot x^3)$$
$$= 18x^5 \qquad \text{To simplify } x^2 \cdot x^3, \text{ keep the base and add the exponents.}$$

b. $\quad (-8x)(2y)(xy) = -8 \cdot x \cdot 2 \cdot y \cdot x \cdot y$
$$= (-8 \cdot 2) \cdot x \cdot x \cdot y \cdot y$$
$$= -16x^2y^2$$

c. $\quad (2a^3b)(-7b^2c)(-12ac^4) = 2 \cdot a^3 \cdot b \cdot (-7) \cdot b^2 \cdot c \cdot (-12) \cdot a \cdot c^4$
$$= 2(-7)(-12) \cdot a^3 \cdot a \cdot b \cdot b^2 \cdot c \cdot c^4$$
$$= 168a^4b^3c^5$$

SELF CHECK
Multiply **a.** $(-2a^3)(4a^2)$ and
b. $(-5b^3)(-3a)(a^2b)$.

Answers: **a.** $-8a^5$, **b.** $15a^3b^4$ ∎

Multiplying a Polynomial by a Monomial

To multiply a polynomial by a monomial, *we multiply each term of the polynomial by the monomial.*

EXAMPLE 2

Multiplying by a monomial. Do each multiplication. We can use the distributive property to remove parentheses.

a. $3x^2(6xy + 3y^2) = 3x^2 \cdot 6xy + 3x^2 \cdot 3y^2$
$\qquad\qquad\qquad = 18x^3y + 9x^2y^2$ Do the multiplications.

Since $18x^3y$ and $9x^2y^2$ are not like terms, we cannot add them.

b. $5x^3y^2(xy^3 - 2x^2y) = 5x^3y^2 \cdot xy^3 - 5x^3y^2 \cdot 2x^2y$
$\qquad\qquad\qquad\qquad = 5x^4y^5 - 10x^5y^3$

c. $-2ab^2(3bz - 2az + 4z^3)$
$\qquad = -2ab^2 \cdot 3bz - (-2ab^2) \cdot 2az + (-2ab^2) \cdot 4z^3$
$\qquad = -6ab^3z + 4a^2b^2z - 8ab^2z^3$

SELF CHECK
Multiply $-2a^2(a^2 - a + 3)$.

Answer: $-2a^4 + 2a^3 - 6a^2$ ∎

Multiplying a Polynomial by a Polynomial

To multiply a polynomial by a polynomial, we use the distributive property repeatedly.

EXAMPLE 3

Multiplying polynomials. Do each multiplication. We can use the distributive property to remove parentheses.

a. $(3x + 2)(4x + 9) = (3x + 2) \cdot 4x + (3x + 2) \cdot 9$ Distribute $3x + 2$.
$\qquad\qquad\qquad\quad = 12x^2 + 8x + 27x + 18$ Distribute $4x$ and distribute 9.
$\qquad\qquad\qquad\quad = 12x^2 + 35x + 18$ Combine like terms.

b. $(2a - b)(3a^2 - 4ab + b^2)$
$\qquad = (2a - b)3a^2 - (2a - b)4ab + (2a - b)b^2$
$\qquad = 6a^3 - 3a^2b - 8a^2b + 4ab^2 + 2ab^2 - b^3$
$\qquad = 6a^3 - 11a^2b + 6ab^2 - b^3$

SELF CHECK
Multiply $(2a + b)(3a - 2b)$.

Answer: $6a^2 - ab - 2b^2$ ∎

The results of Example 3 suggest that to multiply one polynomial by another, *we multiply each term of one polynomial by each term of the other polynomial.*

In the next example, we organize the work done in Example 3 vertically.

EXAMPLE 4

Multiplying vertically. Do each multiplication.

a. $\quad 3x + 2$
$\quad\underline{\quad 4x + 9\quad}$
$\quad 12x^2 + \;\; 8x$ ⟵ This is the result of $4x(3x + 2)$.
$\quad\underline{\qquad\quad + 27x + 18}$ ⟵ This is the result of $9(3x + 2)$.
$\quad 12x^2 + 35x + 18$ Combine like terms, column by column.

b. $3a^2 - 4ab + b^2$

$\underline{\qquad 2a \quad - b \qquad}$

$6a^3 - 8a^2b + 2ab^2 \qquad \longleftarrow 2a(3a^2 - 4ab + b^2)$

$\underline{\qquad - 3a^2b + 4ab^2 - b^3 \quad} \longleftarrow -b(3a^2 - 4ab + b^2)$

$6a^3 - 11a^2b + 6ab^2 - b^3$

SELF CHECK Multiply:

$3x^2 + 2x - 5$

$\underline{\qquad 2x + 1 \qquad}$

Answer: $6x^3 + 7x^2 - 8x - 5$ ∎

The FOIL Method

When we multiply two binomials, the distributive property requires that each term of one binomial be multiplied by each term of the other binomial. This fact can be emphasized by drawing arrows to show the indicated products. For example, to multiply $3x + 2$ and $x + 4$, we can write

First terms   Last terms

$(3x + 2)(x + 4) = \mathbf{3}x(x) + \mathbf{3}x(4) + \mathbf{2}(x) + \mathbf{2}(4)$

Inner terms $= 3x^2 + 12x + 2x + 8$

$= 3x^2 + 14x + 8$ Combine like terms:

Outer terms $12x + 2x = 14x$.

We note that

- the product of the **First** terms is $3x \cdot x = 3x^2$,
- the product of the **Outer** terms is $3x \cdot 4 = 12x$,
- the product of the **Inner** terms is $2 \cdot x = 2x$, and
- the product of the **Last** terms is $2 \cdot 4 = 8$.

The procedure is called the **FOIL** method of multiplying two binomials. Foil is an acronym for **First** terms, **Outer** terms, **Inner** terms, and **Last** terms. Of course, the resulting terms of the product must be combined, if possible.

It is easy to multiply binomials by sight using the FOIL method. We find the product of the first terms, then find the products of the outer terms and the inner terms and add them (when possible), and then find the product of the last terms.

EXAMPLE 5

The FOIL method. Find each product.

a. $(2x - 3)(3x + 2) = 6x^2 - 5x - 6$

The product of the first terms is $2x \cdot 3x = 6x^2$. The middle term in the result comes from combining the outer and inner products of $4x$ and $-9x$:

$4x + (-9x) = -5x$

The product of the last terms is $-3 \cdot 2 = -6$.

b. $(3x + 1)(3x + 4) = 9x^2 + 15x + 4$

The product of the first terms is $3x \cdot 3x = 9x^2$. The middle term in the result comes from combining the products $12x$ and $3x$:

$12x + 3x = 15x$

The product of the last terms is $1 \cdot 4 = 4$.

c. $(4x - y)(2x + 3y) = 8x^2 + 10xy - 3y^2$

The product of the first terms is $4x \cdot 2x = 8x^2$. The middle term in the result comes from combining the products $12xy$ and $-2xy$:

$$12xy - 2xy = 10xy$$

The product of the last terms is $-y \cdot 3y = -3y^2$.

SELF CHECK Multiply $(3a + 4b)(2a - b)$. *Answer:* $6a^2 + 5ab - 4b^2$ ∎

Special Products

It is easy to square a binomial using the FOIL method.

EXAMPLE 6 **Squaring a binomial.** Find each square: **a.** $(x + y)^2$ and **b.** $(x - y)^2$.

Solution We use the FOIL method to multiply each term of one binomial by each term of the other binomial, and then we combine like terms.

a. $(x + y)^2 = (x + y)(x + y)$ Write $(x + y)$ as a factor twice.

$\qquad\qquad = x^2 + xy + xy + y^2$ Use the FOIL method.

$\qquad\qquad = x^2 + 2xy + y^2$ Combine like terms: $xy + xy = 2xy$.

We see that the square of the binomial is the square of the first term, plus twice the product of the terms, plus the square of the last term. This product is illustrated graphically in Figure 5-7.

FIGURE 5-7

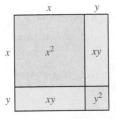

The area of the largest square is the product of its length and width: $(x + y)(x + y) = (x + y)^2$.

The area of the largest square is also the sum of its four pieces: $x^2 + xy + xy + y^2 = x^2 + 2xy + y^2$.

Thus, $(x + y)^2 = x^2 + 2xy + y^2$.

b. $(x - y)^2 = (x - y)(x - y)$

$\qquad\qquad = x^2 - xy - xy + y^2$ Use the FOIL method.

$\qquad\qquad = x^2 - 2xy + y^2$ Combine like terms: $-xy - xy = -2xy$.

We see that the square of the binomial is the square of the first term, minus twice the product of the terms, plus the square of the last term.

SELF CHECK Find the squares: **a.** $(a + 2)^2$ and **b.** $(a - 4)^2$.

Answers: **a.** $a^2 + 4a + 4$, **b.** $a^2 - 8a + 16$ ∎

EXAMPLE 7 **Multiplying the sum and difference of two terms.** Multiply $(x + y)(x - y)$.

Solution $(x + y)(x - y) = x^2 - xy + xy - y^2$ Use the FOIL method.

$\qquad\qquad\qquad = x^2 - y^2$ Combine like terms: $-xy + xy = 0$.

From this example, we see that the product of the sum of two quantities and the difference of two quantities is the square of the first quantity minus the square of the second quantity.

SELF CHECK   Multiply $(a + 3)(a - 3)$.   *Answer:* $a^2 - 9$ ∎

The products discussed in Examples 6 and 7 are called **special products.** Because they occur so often, it is useful to learn their forms.

Special product formulas

$$(x + y)^2 = (x + y)(x + y) = x^2 + 2xy + y^2$$
$$(x - y)^2 = (x - y)(x - y) = x^2 - 2xy + y^2$$
$$(x + y)(x - y) = x^2 - y^2$$

Because $x^2 + 2xy + y^2 = (x + y)^2$ and $x^2 - 2xy + y^2 = (x - y)^2$, the two trinomials are called **perfect square trinomials.**

 WARNING! The squares $(x + y)^2$ and $(x - y)^2$ have trinomials for their products. Don't forget to write the middle terms in these products. Remember that

$$(x + y)^2 \neq x^2 + y^2 \qquad \text{and} \qquad (x - y)^2 \neq x^2 - y^2$$

Also remember that the product $(x + y)(x - y)$ is the binomial $x^2 - y^2$.

At first, the expression $3[x^2 - 2(x + 3)]$ doesn't look like a polynomial. However, if we remove the parentheses and the brackets and simplify, it takes on the form of a polynomial.

$$3[x^2 - 2(x + 3)] = 3[x^2 - 2x - 6]$$
$$= 3x^2 - 6x - 18$$

If an expression has one set of grouping symbols that is enclosed within another set, we always eliminate the inner set first.

EXAMPLE 8   **Expressions containing grouping symbols.** Find the product of $-2[y^3 + 3(y^2 - 2)]$ and $5[y^2 - 2(y + 1)]$.

Solution   We change each expression into polynomial form

$$-2[y^3 + 3(y^2 - 2)] \qquad\qquad 5[y^2 - 2(y + 1)]$$
$$= -2(y^3 + 3y^2 - 6) \qquad\qquad = 5(y^2 - 2y - 2)$$
$$= -2y^3 - 6y^2 + 12 \qquad\qquad = 5y^2 - 10y - 10$$

and then do the multiplication vertically:

$$
\begin{array}{r}
-2y^3 - 6y^2 + 12 \\
5y^2 - 10y - 10 \\
\hline
-10y^5 - 30y^4 \qquad\qquad + 60y^2 \\
+ 20y^4 + 60y^3 \qquad\qquad - 120y \\
+ 20y^3 + 60y^2 \qquad\qquad - 120 \\
\hline
-10y^5 - 10y^4 + 80y^3 + 120y^2 - 120y - 120
\end{array}
$$

SELF CHECK   Find the product of $2[a^2 + 3(a - 2)]$ and $3[a^2 + 3(a - 1)]$.   *Answer:* $6a^4 + 36a^3 - 162a + 108$ ∎

Applications of Multiplying Polynomials

Profit, revenue, and cost are terms used in the business world. The profit p earned on the sale of one or more items is given by the formula

$$p = r - c$$

where r is the revenue taken in and c is the wholesale cost. If a salesperson has 12 vacuum cleaners and sells them for \$225 each, the revenue will be $r = \$(12 \cdot 225) = \$2,700$. This illustrates the following formula for finding the revenue, r:

$$r = \boxed{\begin{array}{c} \text{number of} \\ \text{items sold } (x) \end{array}} \cdot \boxed{\begin{array}{c} \text{selling price} \\ \text{of each item } (p) \end{array}} = xp = px$$

E X A M P L E   9

Selling vacuum cleaners. Over the years, a saleswoman has found that the number of vacuum cleaners she can sell depends on price. The lower the price, the more she can sell. She has determined that the number of vacuums (x) that she can sell at a price (p) is related by the equation $x = -\frac{2}{25}p + 28$.

a. Find a formula for the revenue r.

b. How much revenue will be taken in if the vacuums are priced at \$250?

Solution   **a.** To find a formula for revenue, we substitute $-\frac{2}{25}p + 28$ for x in the formula $r = px$ and solve for r.

$$r = px \qquad \qquad \text{The formula for revenue.}$$

$$r = p\left(-\frac{2}{25}p + 28\right) \quad \text{Substitute } -\frac{2}{25}p + 28 \text{ for } x.$$

$$= -\frac{2}{25}p^2 + 28p \quad \text{Multiply the polynomials.}$$

b. To find how much revenue will be taken in if the vacuums are priced at \$250, we substitute 250 for p in the formula for revenue.

$$r = -\frac{2}{25}p^2 + 28p \qquad \text{The formula for revenue.}$$

$$r = -\frac{2}{25}(\mathbf{250})^2 + 28(\mathbf{250}) \quad \text{Substitute 250 for } p.$$

$$= -5{,}000 + 7{,}000$$

$$= 2{,}000$$

The revenue will be \$2,000.   ∎

Multiplying Expressions That Are Not Polynomials

The following example shows how to use the FOIL method to multiply expressions that are not polynomials.

E X A M P L E   1 0

Using the FOIL method. Find each product: **a.** $(x^{-2} + y)(x^2 - y^{-2})$ and **b.** $(x^n + 2x)(x^n + 3x^{-n})$.

Solution We multiply each term of the second expression by each term of the first expression and then simplify.

a. $(x^{-2} + y)(x^2 - y^{-2}) = x^{-2}x^2 - x^{-2}y^{-2} + yx^2 - yy^{-2}$
$$= x^{-2+2} - x^{-2}y^{-2} + yx^2 - y^{1+(-2)}$$
$$= x^0 - \frac{1}{x^2y^2} + x^2y - y^{-1}$$
$$= 1 - \frac{1}{x^2y^2} + x^2y - \frac{1}{y}$$

b. $(x^n + 2x)(x^n + 3x^{-n}) = x^nx^n + x^n(3x^{-n}) + 2x(x^n) + 2x(3x^{-n})$
$$= x^{n+n} + 3x^{n+(-n)} + 2x^{1+n} + 6xx^{-n}$$
$$= x^{2n} + 3x^0 + 2x^{n+1} + 6x^{1+(-n)}$$
$$= x^{2n} + 3 + 2x^{n+1} + 6x^{1-n}$$

SELF CHECK Find each product: **a.** $(a^{-3} + b)(a^2 - b^{-1})$ *Answers:* **a.** $\frac{1}{a} - \frac{1}{a^3b} + a^2b - 1,$
and **b.** $(a^n + b)(a^n + b^n)$. **b.** $a^{2n} + a^nb^n + a^nb + b^{n+1}$ ∎

STUDY SET

Section 5.4

VOCABULARY

In Exercises 1–4, fill in the blanks to make the statements true.

1. The expression $(x + 4)(x - 5)$ is the _____ of two binomials.

2. The expression $(x + 4)^2$ is the _____ of a binomial.

3. The polynomial $3x^2 - 3x + 2$ contains three _____.

4. To find $-2x(3x^2 - 2)$, we use the _____ property to remove parentheses.

CONCEPTS

In Exercises 5–12, fill in the blanks to make the statements true.

5. To multiply a monomial by a monomial, we multiply the numerical _____ and then multiply the variable factors.

6. To multiply a polynomial by a monomial, we multiply each _____ of the polynomial by the monomial.

7. To multiply a polynomial by a polynomial, we multiply each _____ of one polynomial by each term of the other polynomial.

8. FOIL is an acronym for _____ terms, _____ terms, _____ terms, and _____ terms.

9. $(x + y)^2 = (x + y)(x + y) =$ _____

10. $(x - y)^2 = (x - y)(x - y) =$ _____

11. $(x + y)(x - y) =$ _____

12. $x^2 + 2xy + y^2$ and $x^2 - 2xy + y^2$ are called _____ trinomials.

13. Write a polynomial that represents the area of the rectangle shown in Illustration 1.

ILLUSTRATION 1

$x + 4$

$x - 2$

14. Write a polynomial that represents the area of the triangle shown in Illustration 2.

ILLUSTRATION 2

$b - 2$

$b + 5$

15. Write a polynomial that represents the area of the square shown in Illustration 3.

ILLUSTRATION 3

$4a + 3$

16. Write a polynomial that represents the area of the rectangle shown in Illustration 4.

ILLUSTRATION 4

$2a + 3$

$2a - 3$

17. Consider $(2x + 4)(4x - 3)$. Give the
    a. First terms   **b.** Outer terms
    c. Inner terms   **d.** Last terms

18. Find
    a. $(4b - 1) + (2b - 1)$
    b. $(4b - 1)(2b - 1)$
    c. $(4b - 1) - (2b - 1)$

PRACTICE

In Exercises 19–36, find each product.

19. $(2a^2)(-3ab)$

20. $(-3x^2y)(3xy)$

21. $(-3ab^2c)(5ac^2)$

22. $(-2m^2n)(-4mn^3)$

23. $(4a^2b)(-5a^3b^2)(6a^4)$

24. $(2x^2y^3)(4xy^5)(-5y^6)$

25. $(-5xx^2)(-3xy)^4$

26. $(-2a^2ab^2)^3(-3ab^2b^2)$

27. $3(x + 2)$

28. $-5(a + b)$

29. $3x(x^2 + 3x)$

30. $-2x(3x^2 - 2)$

31. $-2x(3x^2 - 3x + 2)$

32. $3a(4a^2 + 3a - 4)$

33. $7rst(r^2 + s^2 - t^2)$

34. $3x^2yz(x^2 - 2y + 3z^2)$

35. $4m^2n(-3mn)(m + n)$

36. $-3a^2b^3(2b)(3a + b)$

In Exercises 37–52, find each product. If possible, find the product by sight.

37. $(x + 2)(x + 3)$

38. $(y - 3)(y + 4)$

39. $(3t - 2)(2t + 3)$

40. $(p + 3)(3p - 4)$

41. $(3y - z)(2y - z)$

42. $(2m - n)(3m - n)$

43. $(x + 2)^2$

44. $(x - 3)^2$

45. $(a - 4)^2$

46. $(y + 5)^2$

47. $(2a + b)^2$

48. $(a - 2b)^2$

49. $(x + 2)(x - 2)$

50. $(z + 3)(z - 3)$

51. $(x^3 - 1)(x + 1)$

52. $(y^3 + 1)(1 - y)$

In Exercises 53–60, find each product.

53. $(x - y)(x^2 + xy + y^2)$

54. $(x + y)(x^2 - xy + y^2)$

55. $(3y + 1)(2y^2 + 3y + 2)$

56. $(a + 2)(3a^2 + 4a - 2)$

57. $(2a - b)(4a^2 + 2ab + b^2)$

58. $(x - 3y)(x^2 + 3xy + 9y^2)$

59. $(a + b)(a - b)(a - 3b)$

60. $(x - y)(x + 2y)(x - 2y)$

In Exercises 61–68, find each product. Write all answers without negative exponents.

61. $x^3(2x^2 + x^{-2})$

62. $x^{-4}(2x^{-3} - 5x^2)$

63. $x^3y^{-6}z^{-2}(3x^{-2}y^2z - x^3y^{-4})$

64. $ab^{-2}c^{-3}(a^{-4}bc^3 + a^{-3}b^4c^3)$

65. $(x^{-1} + y)(x^{-1} - y)$

66. $(x^{-1} - y)(x^{-1} - y)$

67. $(2x^{-3} + y^3)(2x^3 - y^{-3})$

68. $(5x^{-4} - 4y^2)(5x^2 - 4y^{-4})$

In Exercises 69–78, find each product. Consider n to be a whole number.

69. $x^n(x^{2n} - x^n)$

70. $a^{2n}(a^n + a^{2n})$

71. $(x^n + 1)(x^n - 1)$

72. $(x^n - a^n)(x^n + a^n)$

73. $(x^n - y^n)(x^n - y^{-n})$

74. $(x^n + y^n)(x^n + y^{-n})$

75. $(x^{2n} + y^{2n})(x^{2n} - y^{2n})$

76. $(a^{3n} - b^{3n})(a^{3n} + b^{3n})$

77. $(x^n + y^n)(x^n + 1)$

78. $(1 - x^n)(x^{-n} - 1)$

In Exercises 79–86, simplify each expression. (Hint: Do any multiplications first and then combine like terms.)

79. $3x(2x + 4) - 3x^2$

80. $2y - 3y(y^2 + 4)$

81. $3pq - p(p - q)$

82. $-4rs(r - 2) + 4rs$

83. $(x + 3)(x - 3) + (2x - 1)(x + 2)$

84. $(2b + 3)(b - 1) - (b + 2)(3b - 1)$

85. $(3x - 4)^2 - (2x + 3)^2$

86. $(3y + 1)^2 + (2y - 4)^2$

In Exercises 87–90, use a calculator to help find each product.

87. $(3.21x - 7.85)(2.87x + 4.59)$

88. $(7.44y + 56.7)(-2.1y - 67.3)$

89. $(-17.3y + 4.35)^2$

90. $(-0.31x + 29.3)(-81x - 0.2)$

APPLICATIONS

91. THE YELLOW PAGES Refer to Illustration 5.
  a. Describe the area occupied by the ads for movers by using a product of two binomials.
  b. Describe the area occupied by the ad for Budget Moving Co. by using a product. Then do the multiplication.
  c. Describe the area occupied by the ad for Snyder Movers by using a product. Then do the multiplication.

ILLUSTRATION 5

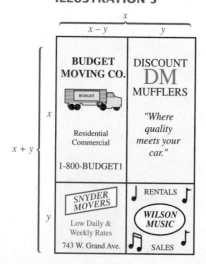

  d. Explain why your answer to part a is equal to the sum of your answers to parts b and c. What special product does this exercise illustrate?

92. HELICOPTER PAD To determine the amount of fluorescent paint needed to paint the circular ring on the landing pad design shown in Illustration 6, painters must find its area. The area of the ring is given by the expression $\pi(R + r)(R - r)$.
  a. Find the product $\pi(R + r)(R - r)$.
  b. If $R = 25$ feet and $r = 20$ feet, find the area to be painted. Round to the nearest tenth.
  c. If a quart of flourescent paint covers 65 ft^2, how many quarts will be needed to paint the ring?

ILLUSTRATION 6

93. GIFT BOXES The corners of a 12-in.-by-12-in. piece of cardboard are creased, folded inward, and glued to make a gift box. (See Illustration 7.) Write a polynomial that gives the volume of the resulting box.

ILLUSTRATION 7

94. CALCULATING REVENUE A salesperson has found that the number (x) of televisions she can sell at a certain price (p) is related by the equation $x = -\frac{1}{5}p + 90$.
  a. Find the number of TVs she will sell if the price is $375.
  b. Write a formula for the revenue when x TVs are sold.
  c. Find the revenue generated by TV sales if they are priced at $400 each.

WRITING

95. Explain how to use the FOIL method.

96. Explain how you would multiply two trinomials.

97. On a test, when asked to find $(x - y)^2$, a student answered $x^2 - y^2$. What error did the student make?

98. Explain how the distributive property is used to find the product: $2x^3(x^2 - 5x + 1)$.

REVIEW

In Exercises 99–102, graph each inequality or system of inequalities.

99. $2x + y \leq 2$

100. $x \geq 2$

101. $\begin{cases} y - 2 < 3x \\ y + 2x < 3 \end{cases}$

102. $\begin{cases} y < 0 \\ x < 0 \end{cases}$

 5.5

The Greatest Common Factor and Factoring by Grouping

In this section, you will learn about

> Prime-factored form of a natural number ■ Factoring out the greatest common factor ■ Factoring by grouping ■ Formulas

Introduction   In Section 5.4, we discussed how to multiply polynomials. In the next four sections, we will reverse the operation of multiplication and show how to find the factors of a known product. The process of finding the individual factors of a known product is called **factoring.**

Prime-Factored Form of a Natural Number

If one number a divides a second number b, then a is called a **factor** of b. For example, because 3 divides 24, it is a factor of 24. Each number in the following list is a factor of 24, because each number divides 24.

> 1, 2, 3, 4, 6, 8, 12, and 24

To factor a natural number means to write it as a product of other natural numbers. If each factor is a prime number, the natural number is said to be written in **prime-factored form.** Example 1 shows how to find the prime-factored forms of 60, 84, and 180, respectively.

EXAMPLE 1

Prime factors. Find the prime factorization of each number.

a. $60 = 6 \cdot 10$
$= 2 \cdot 3 \cdot 2 \cdot 5$
$= 2^2 \cdot 3 \cdot 5$

b. $84 = 4 \cdot 21$
$= 2 \cdot 2 \cdot 3 \cdot 7$
$= 2^2 \cdot 3 \cdot 7$

c. $180 = 10 \cdot 18$
$= 2 \cdot 5 \cdot 3 \cdot 6$
$= 2 \cdot 5 \cdot 3 \cdot 3 \cdot 2$
$= 2^2 \cdot 3^2 \cdot 5$

SELF CHECK   Find the prime factorization of 120.   *Answer:* $2^3 \cdot 3 \cdot 5$ ■

The largest natural number that divides 60, 84, and 180 is called the **greatest common factor (GCF)** of the numbers. Because 60, 84, and 180 all have two factors of 2 and one factor of 3, the GCF of these three numbers is $2^2 \cdot 3 = 12$. We note that

$$\frac{60}{12} = 5, \qquad \frac{84}{12} = 7, \qquad \text{and} \qquad \frac{180}{12} = 15$$

There is no natural number greater than 12 that divides 60, 84, and 180.

Algebraic monomials can also have greatest common factors.

EXAMPLE 2

The greatest common factor. Find the GCF of $6a^2b^3c$, $9a^3b^2c$, and $18a^4c^3$.

Solution

We begin by factoring each monomial.

$$6a^2b^3c = 3 \cdot 2 \cdot a \cdot a \cdot b \cdot b \cdot b \cdot c$$
$$9a^3b^2c = 3 \cdot 3 \cdot a \cdot a \cdot a \cdot b \cdot b \cdot c$$
$$18a^4c^3 = 2 \cdot 3 \cdot 3 \cdot a \cdot a \cdot a \cdot a \cdot c \cdot c \cdot c$$

Since each monomial has one factor of 3, two factors of a, and one factor of c in common, the GCF is

$$3^1 \cdot a^2 \cdot c^1 = 3a^2c$$

SELF CHECK

Find the GCF of $24x^2y^3$, $3x^3y$, and $18x^2y^2$.   *Answer:* $3x^2y$ ∎

To find the GCF of several monomials, we follow these steps.

Steps for finding the GCF

1. Find the prime-factored form of each monomial.
2. Identify the prime factors that are common to each monomial.
3. Find the product of the factors found in step 2, with each factor raised to the smallest power that occurs in any one monomial.

Factoring Out the Greatest Common Factor

We have seen that the distributive property provides a method for multiplying a polynomial by a monomial. For example,

$$2x^3y^3(3x^2 - 4y^3) = 2x^3y^3 \cdot 3x^2 - 2x^3y^3 \cdot 4y^3$$
$$= 6x^5y^3 - 8x^3y^6$$

If the product of a multiplication is $6x^5y^3 - 8x^3y^6$, we can use the distributive property to find the individual factors.

$$6x^5y^3 - 8x^3y^6 = 2x^3y^3 \cdot 3x^2 - 2x^3y^3 \cdot 4y^3$$
$$= 2x^3y^3(3x^2 - 4y^3)$$

Since $2x^3y^3$ is the GCF of the terms of $6x^5y^3 - 8x^3y^6$, this process is called **factoring out the greatest common factor.**

EXAMPLE 3

Factoring out the greatest common factor. Factor $25a^3b + 15ab^3$.

Solution

We begin by factoring each monomial:

$$25a^3b = 5 \cdot 5 \cdot a \cdot a \cdot a \cdot b$$
$$15ab^3 = 5 \cdot 3 \cdot a \cdot b \cdot b \cdot b$$

Since each term has one factor of 5, one factor of a, and one factor of b in common, and there are no other common factors, $5ab$ is the GCF of the two terms. We can use the distributive property to factor it out.

$$25a^3b + 15ab^3 = 5ab \cdot 5a^2 + 5ab \cdot 3b^2$$
$$= 5ab(5a^2 + 3b^2)$$

SELF CHECK   Factor $9x^4y^2 - 12x^3y^3$.   *Answer:* $3x^3y^2(3x - 4y)$ ■

E X A M P L E   4   **Factoring out the greatest common factor.** Factor $3xy^2z^3 + 6xyz^3 - 3xz^2$.

Solution   We begin by factoring each monomial:

$$3xy^2z^3 = 3 \cdot x \cdot y \cdot y \cdot z \cdot z \cdot z$$
$$6xyz^3 = 3 \cdot 2 \cdot x \cdot y \cdot z \cdot z \cdot z$$
$$-3xz^2 = -3 \cdot x \cdot z \cdot z$$

Since each term has one factor of 3, one factor of x, and two factors of z in common, and because there are no other common factors, $3xz^2$ is the GCF of the three terms. We can use the distributive property to factor it out.

$$3xy^2z^3 + 6xyz^3 - 3xz^2 = 3xz^2 \cdot y^2z + 3xz^2 \cdot 2yz - 3xz^2 \cdot 1$$
$$= 3xz^2(y^2z + 2yz - 1)$$

WARNING! When the $3xz^2$ is factored out, remember to write the -1.

SELF CHECK   Factor $2a^4b^2 + 6a^3b^2 - 4a^2b$.   *Answer:* $2a^2b(a^2b + 3ab - 2)$ ■

A polynomial that cannot be factored is called a **prime polynomial** or an **irreducible polynomial.**

E X A M P L E   5   **Factoring out the greatest common factor.** Factor $3x^2 + 4y + 7$, if possible.

Solution   We factor each monomial:

$$3x^2 = 3 \cdot x \cdot x \qquad 4y = 2 \cdot 2 \cdot y \qquad 7 = 7$$

Since there are no common factors other than 1, this polynomial cannot be factored. It is a prime polynomial.

SELF CHECK   Factor $6a^3 + 7b^2 + 5$.   *Answer:* a prime polynomial ■

E X A M P L E   6   **The negative of the greatest common factor.** Factor the negative of the GCF from $-6u^2v^3 + 8u^3v^2$.

Solution   Because the GCF of the two terms is $2u^2v^2$, the negative of the GCF is $-2u^2v^2$. To factor out $-2u^2v^2$, we proceed as follows:

$$-6u^2v^3 + 8u^3v^2 = -2u^2v^2 \cdot 3v + 2u^2v^2 \cdot 4u$$
$$= -2u^2v^2 \cdot 3v - (-2u^2v^2)4u$$
$$= -2u^2v^2(3v - 4u)$$

SELF CHECK Factor out the negative of the GCF from
$-8a^2b^2 - 12ab^3$. *Answer:* $-4ab^2(2a + 3b)$ ∎

We can also factor out common factors that have variable or negative exponents.

EXAMPLE 7 **Variable exponents.** Factor x^{2n} from $x^{4n} + x^{3n} + x^{2n}$.

Solution We write the trinomial as $x^{2n} \cdot x^{2n} + x^{2n} \cdot x^n + x^{2n} \cdot 1$ and factor out x^{2n}.

$$x^{4n} + x^{3n} + x^{2n} = x^{2n} \cdot x^{2n} + x^{2n} \cdot x^n + x^{2n} \cdot 1$$
$$= x^{2n}(x^{2n} + x^n + 1)$$

SELF CHECK Factor a^n from $a^{6n} + a^{3n} - a^n$. *Answer:* $a^n(a^{5n} + a^{2n} - 1)$ ∎

EXAMPLE 8 **Negative exponents.** Factor $a^{-2}b^{-2}$ from $a^{-2}b - a^3b^{-2}$.

Solution We write $a^{-2}b - a^3b^{-2}$ as $a^{-2}b^{-2} \cdot b^3 - a^{-2}b^{-2} \cdot a^5$ and factor out $a^{-2}b^{-2}$.

$$a^{-2}b - a^3b^{-2} = a^{-2}b^{-2} \cdot b^3 - a^{-2}b^{-2} \cdot a^5$$
$$= a^{-2}b^{-2}(b^3 - a^5)$$

SELF CHECK Factor $x^{-3}y^{-2}$ from $x^{-3}y^{-2} - x^6y^{-2}$. *Answer:* $x^{-3}y^{-2}(1 - x^9)$ ∎

A common factor can have more than one term. For example, in the expression

$$x(a + b) + y(a + b)$$

the binomial $a + b$ is a factor of both terms. We can factor it out to get

$$x(a + b) + y(a + b) = (a + b)x + (a + b)y \quad \text{Use the commutative property of multiplication.}$$

$$= (a + b)(x + y)$$

EXAMPLE 9 **A common factor having more than one term.** Factor
$a(x - y + z) - b(x - y + z) + 3(x - y + z)$.

Solution We can factor out the GCF of the three terms, which is $(x - y + z)$.

$$a(x - y + z) - b(x - y + z) + 3(x - y + z)$$
$$= (x - y + z)a - (x - y + z)b + (x - y + z)3$$
$$= (x - y + z)(a - b + 3)$$

SELF CHECK Factor $x(a + b - c) - y(a + b - c)$. *Answer:* $(a + b - c)(x - y)$ ∎

Factoring by Grouping

Suppose that we wish to factor

$$ac + ad + bc + bd$$

Although there is no factor common to all four terms, there is a common factor of a in the first two terms and a common factor of b in the last two terms. We can factor out these common factors to get

$$ac + ad + bc + bd = a(c + d) + b(c + d)$$

We can now factor out the common factor of $c + d$ on the right-hand side:

$$ac + ad + bc + bd = (c + d)(a + b)$$

The grouping in this type of problem is not always unique. For example, if we write the expression $ac + ad + bc + bd$ in the form

$$ac + bc + ad + bd$$

and factor c from the first two terms and d from the last two terms, we obtain

$$ac + bc + ad + bd = c(a + b) + d(a + b)$$
$$= (a + b)(c + d) \quad \text{This is equivalent to } (c+d)(a+b).$$

The method used in the previous examples is called **factoring by grouping.**

EXAMPLE 10

Factoring by grouping. Factor $3ax^2 + 3bx^2 + a + 5bx + 5ax + b$.

Solution   Although there is no factor common to all six terms, $3x^2$ can be factored out of the first two terms, and $5x$ can be factored out of the fourth and fifth terms to get

$$3ax^2 + 3bx^2 + a + 5bx + 5ax + b = 3x^2(a + b) + a + 5x(b + a) + b$$

This result can be written in the form

$$3ax^2 + 3bx^2 + a + 5bx + 5ax + b = 3x^2(a + b) + 5x(a + b) + 1(a + b)$$

Since $a + b$ is common to all three terms, it can be factored out to get

$$3ax^2 + 3bx^2 + a + 5bx + 5ax + b = (a + b)(3x^2 + 5x + 1)$$

SELF CHECK   Factor $2x^3 + 3x^2 + x + 2x^2y + 3xy + y$.   *Answer:* $(x + y)(2x^2 + 3x + 1)$ ■

To factor an expression completely, it is often necessary to factor more than once, as the following example illustrates.

EXAMPLE 11

Factoring completely. Factor $3x^3y - 4x^2y^2 - 6x^2y + 8xy^2$.

Solution   We begin by factoring out the common factor of xy.

$$3x^3y - 4x^2y^2 - 6x^2y + 8xy^2 = xy(3x^2 - 4xy - 6x + 8y)$$

We can now factor $3x^2 - 4xy - 6x + 8y$ by grouping:

$$3x^3y - 4x^2y^2 - 6x^2y + 8xy^2$$
$$= xy(3x^2 - 4xy - 6x + 8y)$$
$$= xy[x(3x - 4y) - 2(3x - 4y)] \quad \text{Factor } x \text{ from } 3x^2 - 4xy \text{ and } -2 \text{ from } -6x + 8y.$$
$$= xy(3x - 4y)(x - 2) \quad \text{Factor out } 3x - 4y.$$

Because no more factoring can be done, the factorization is complete.

SELF CHECK   Factor $3a^3b + 3a^2b - 2a^2b^2 - 2ab^2$.   *Answer:* $ab(3a - 2b)(a + 1)$ ■

WARNING! Whenever you factor an expression, always factor it completely. Each factor of a completely factored expression will be prime.

Formulas

Factoring is often required to solve a literal equation for one of its variables.

EXAMPLE 12

Electronics The formula $r_1r_2 = rr_2 + rr_1$ is used in electronics to relate the combined resistance, r, of two resistors wired in parallel. The variable r_1 represents the resistance of the first resistor, and the variable r_2 represents the resistance of the second. Solve for r_2.

Solution To isolate r_2 on one side of the equation, we get all terms involving r_2 on the left-hand side and all terms not involving r_2 on the right-hand side. We proceed as follows:

$$r_1r_2 = rr_2 + rr_1$$
$$r_1r_2 - rr_2 = rr_1 \qquad \text{Subtract } rr_2 \text{ from both sides.}$$
$$r_2(r_1 - r) = rr_1 \qquad \text{Factor out } r_2 \text{ on the left-hand side.}$$
$$r_2 = \frac{rr_1}{r_1 - r} \qquad \text{Divide both sides by } r_1 - r.$$

SELF CHECK Solve $A = p + prt$ for p.

Answer: $p = \dfrac{A}{1 + rt}$ ∎

STUDY SET

Section 5.5

VOCABULARY

In Exercises 1–6, fill in the blanks to make the statements true.

1. When we write $2x + 4$ as $2(x + 2)$, we say that we have _____ $2x + 4$.

2. When we write 100 as $2^2 \cdot 5^2$, we say that we have written 100 in _____ form.

3. Because 5 divides 20, we say that 5 is a _____ of 20.

4. The polynomial $2x^2y^3 - 4xy^2 + 6xy$ has three _____.

5. The abbreviation GCF stands for _____.

6. If a polynomial cannot be factored, it is called a _____ polynomial or an irreducible polynomial.

CONCEPTS

7. The prime factorizations of three monomials are shown here. Find their GCF.

$$2 \cdot 2 \cdot 3 \cdot x \cdot x \cdot y \cdot y \cdot y$$
$$2 \cdot 3 \cdot 3 \cdot x \cdot y \cdot y \cdot y \cdot y$$
$$2 \cdot 3 \cdot 3 \cdot 7 \cdot x \cdot x \cdot x \cdot y \cdot y$$

8. **a.** What property is shown here?
$$4a^2b(2ab^3 - 3a^2b^4) = 4a^2b \cdot 2ab^3 - 4a^2b \cdot 3a^2b^4$$

   b. Explain how we use the distributive property in reverse to factor $8a^3b^4 - 12a^4b^5$.

9. Explain why each factorization of $30t^2 - 20t^3$ is not complete.
   a. $5t^2(6 - 4t)$

   b. $10t(3t - 2t^2)$

10. **a.** Factor $-5y^3 - 10y^2 + 15y$ by factoring out the positive GCF.
    b. Factor $-5y^3 - 10y^2 + 15y$ by factoring out the negative GCF.

NOTATION

In Exercises 11–14, complete each factorization.

11. $3a - 12 = 3(a - \boxed{})$

12. $8z^3 + 4z^2 + 2z = 2z(4z^2 + 2z + \boxed{})$

13. $x^3 - x^2 + 2x - 2 = \boxed{}(x - 1) + \boxed{}(x - 1)$
    $$= (x - 1)(x^2 + 2)$$

14. $-24a^3b^2 + 12ab^2 = -12ab^2(2a^2 \boxed{} 1)$

PRACTICE

In Exercises 15–22, find the prime-factored form of each number.

15. 6
16. 10
17. 135
18. 98
19. 128
20. 357
21. 325
22. 288

In Exercises 23–30, find the GCF of each set of monomials.

23. 36, 48

24. 45, 75

25. 42, 36, 98

26. 16, 40, 60

27. $4a^2b$, $8a^3c$

28. $6x^3y^2z$, $9xyz^2$

29. $18x^4y^3z^2$, $-12xy^2z^3$

30. $6x^2y^3$, $24xy^3$, $40x^2y^2z^3$

In Exercises 31–44, factor each polynomial, if possible.

31. $2x + 8$

32. $3y - 9$

33. $2x^2 - 6x$

34. $3y^3 + 3y^2$

35. $5xy + 12ab^2$

36. $7x^2 + 14x$

37. $15x^2y - 10x^2y^2$

38. $11m^3n^2 - 12x^2y$

39. $14r^2s^3 + 15t^6$

40. $13ab^2c^3 - 26a^3b^2c$

41. $27z^3 + 12z^2 + 3z$

42. $25t^6 - 10t^3 + 5t^2$

43. $45x^{10}y^3 - 63x^7y^7 + 81x^{10}y^{10}$

44. $48u^6v^6 - 16u^4v^4 - 3u^6v^3$

In Exercises 45–54, factor out the negative of the greatest common factor.

45. $-3a - 6$

46. $-6b + 12$

47. $-3x^2 - x$

48. $-4a^3 + a^2$

49. $-6x^2 - 3xy$

50. $-15y^3 + 25y^2$

51. $-18a^2b - 12ab^2$

52. $-21t^5 + 28t^3$

53. $-63u^3v^6z^9 + 28u^2v^7z^2 - 21u^3v^3z^4$

54. $-56x^4y^3z^2 - 72x^3y^4z^5 + 80xy^2z^3$

In Exercises 55–64, factor out the designated factor.

55. x^2 from $x^{n+2} + x^{n+3}$

56. y^3 from $y^{n+3} + y^{n+5}$

57. y^n from $2y^{n+2} - 3y^{n+3}$

58. x^n from $4x^{n+3} - 5x^{n+5}$

59. x^{-2} from $x^4 - 5x^6$

60. y^{-4} from $7y^4 + y$

61. t^{-3} from $t^5 + 4t^{-6}$

62. p^{-5} from $6p^3 - p^{-2}$

63. $4y^{-2n}$ from $8y^{2n} + 12 + 16y^{-2n}$

64. $7x^{-3n}$ from $21x^{6n} + 7x^{3n} + 14$

In Exercises 65–74, factor each expression.

65. $4(x + y) + t(x + y)$

66. $5(a - b) - t(a - b)$

67. $(a - b)r - (a - b)s$

68. $(x + y)u + (x + y)v$

69. $3(m + n + p) + x(m + n + p)$

70. $x(x - y - z) + y(x - y - z)$

71. $(u + v)^2 - (u + v)$

72. $a(x - y) - (x - y)^2$

73. $-a(x + y) + b(x + y)$

74. $-bx(a - b) - cx(a - b)$

In Exercises 75–86, factor by grouping.

75. $ax + bx + ay + by$

76. $ar - br + as - bs$

77. $x^2 + yx + 2x + 2y$

78. $2c + 2d - cd - d^2$

79. $3c - cd + 3d - c^2$

80. $x^2 + 4y - xy - 4x$

81. $a^2 - 4b + ab - 4a$

82. $7u + v^2 - 7v - uv$

83. $ax + bx - a - b$

84. $x^2y - ax - xy + a$

85. $x^2 + xy + xz + xy + y^2 + zy$

86. $ab - b^2 - bc + ac - bc - c^2$

In Exercises 87–92, factor by grouping. Factor out all common monomials first.

87. $mpx + mqx + npx + nqx$

88. $abd - abe + acd - ace$

89. $x^2y + xy^2 + 2xyz + xy^2 + y^3 + 2y^2z$

90. $a^3 - 2a^2b + a^2c - a^2b + 2ab^2 - abc$

91. $2n^4p - 2n^2 - n^3p^2 + np + 2mn^3p - 2mn$

92. $a^2c^3 + ac^2 + a^3c^2 - 2a^2bc^2 - 2bc^2 + c^3$

In Exercises 93–100, solve for the indicated variable.

93. $r_1r_2 = rr_2 + rr_1$ for r_1

94. $r_1r_2 = rr_2 + rr_1$ for r

95. $d_1d_2 = fd_2 + fd_1$ for f

96. $d_1 d_2 = f d_2 + f d_1$ for d_1

97. $b^2 x^2 + a^2 y^2 = a^2 b^2$ for a^2

98. $b^2 x^2 + a^2 y^2 = a^2 b^2$ for b^2

99. $S(1 - r) = a - lr$ for r

100. $Tx = (x - 2)180°$ for x.

APPLICATIONS

101. GEOMETRIC FORMULAS
   a. Write an expression that gives the area of the portion of the figure in Illustration 1 that is shaded red.
   b. Do the same for the portion of the figure that is shaded blue.
   c. Add the results from parts a and b and then factor that expression. What important formula from geometry do you obtain?

ILLUSTRATION 1

102. PACKAGING The amount of cardboard needed to make the cereal box shown in Illustration 2 can be found by computing the area A, which is given by the formula

$$A = 2wh + 4wl + 2lh$$

where w is the width, h the height, and l the length. Solve the equation for the width.

ILLUSTRATION 2

103. LANDSCAPING See Illustration 3. The combined area of the portions of the square lot that the sprinkler doesn't reach is given by $4r^2 - \pi r^2$, where r is the radius of the circular spray. Factor this expression.

ILLUSTRATION 3

104. CRAYONS The amount of colored wax used to make the crayon shown in Illustration 4 can be found by computing its volume using the formula

$$V = \pi r^2 h_1 + \frac{1}{3}\pi r^2 h_2$$

Factor the expression on the right-hand side of this equation.

ILLUSTRATION 4

WRITING

105. One student commented, "Factoring undoes the distributive property." What do you think she meant?

106. Explain how to find the greatest common factor of two natural numbers.

REVIEW

107. What figure results when the function $f(x) = 3x + 1$ is graphed?

108. What figure results when the function $g(x) = x^2$ is graphed?

109. Solve the inequality $-x > 3$. Express the solution set using interval notation.

110. Evaluate $2|-25 - (-6)(3)|$.

111. If two different lines are parallel, what can be said about their slopes?

112. Are $-3t^2$ and $12t^2$ like terms? If so, combine them.

▶ **5.6**

The Difference of Two Squares; the Sum and Difference of Two Cubes

In this section, you will learn about

  Perfect squares ■ The difference of two squares ■ Perfect cubes
  ■ The sum and difference of two cubes

Introduction In this section, we will discuss some special rules of factoring. These rules are applied to polynomials that can be written as the difference of two squares or as the sum or difference of two cubes. To apply these special factoring methods, we must first be able to recognize such polynomials. We begin with a discussion that will help you recognize polynomials with terms that are *perfect squares.*

Perfect Squares

To factor the difference of two squares, it is helpful to know the first 20 integers that are **perfect squares.**

1, 4, 9, 16, 25, 36, 49, 64, 81, 100, 121, 144, 169, 196, 225, 256, 289, 324, 361, 400

Expressions such as $x^6y^4z^2$ are also perfect squares, because they can be written as the square of another quantity:

$$x^6y^4z^2 = (x^3y^2z)^2$$

The Difference of Two Squares

In Section 5.4, we developed the special product formula

1. $(x + y)(x - y) = x^2 - y^2$

The binomial $x^2 - y^2$ is called the **difference of two squares,** because x^2 represents the square of x, y^2 represents the square of y, and $x^2 - y^2$ represents the difference of these squares.

Equation 1 can be written in reverse order to give a formula for factoring the difference of two squares.

Factoring the difference of two squares

$$x^2 - y^2 = (x + y)(x - y)$$

If we think of the difference of two squares as the square of a **First** quantity minus the square of a **Last** quantity, we have the formula

$$F^2 - L^2 = (F + L)(F - L)$$

and we say: *To factor the square of a First quantity minus the square of a Last quantity, we multiply the First plus the Last by the First minus the Last.*

EXAMPLE 1

Factoring a difference of two squares. Factor $49x^2 - 16$.

Solution We begin by rewriting the binomial $49x^2 - 16$ as a difference of two squares: $(7x)^2 - (4)^2$. Then we use the formula for factoring the difference of two squares:

$$F^2 - L^2 = (F + L)(F - L)$$
$$(7x)^2 - 4^2 = (7x + 4)(7x - 4)$$

We can verify this result using the FOIL method to do the multiplication.

$$(7x + 4)(7x - 4) = 49x^2 - 28x + 28x - 16$$
$$= 49x^2 - 16$$

SELF CHECK Factor $81p^2 - 25$. *Answer:* $(9p + 5)(9p - 5)$ ∎

 WARNING! Many expressions that represent the sum of two squares, such as $(7x)^2 + 4^2$, cannot be factored in the real number system. The binomial $49x^2 + 16$ is a prime binomial.

EXAMPLE 2

Factoring a difference of two squares. Factor $64a^4 - 25b^2$.

Solution We can write $64a^4 - 25b^2$ in the form $(8a^2)^2 - (5b)^2$ and use the formula for factoring the difference of two squares.

$$F^2 - L^2 = (F + L)(F - L)$$
$$(8a^2)^2 - (5b)^2 = (8a^2 + 5b)(8a^2 - 5b)$$

Verify by multiplication.

SELF CHECK Factor $36r^4 - s^2$. *Answer:* $(6r^2 + s)(6r^2 - s)$ ∎

EXAMPLE 3

Using the formula for the difference of two squares twice. Factor $x^4 - 1$.

Solution Because the binomial is the difference of the squares of x^2 and 1, it factors into the sum of x^2 and 1 and the difference of x^2 and 1.

$$x^4 - 1 = (x^2)^2 - (1)^2$$
$$= (x^2 + 1)(x^2 - 1)$$

The factor $x^2 + 1$ is the sum of two quantities and is prime. However, the factor $x^2 - 1$ is the difference of two squares and can be factored as $(x + 1)(x - 1)$. Thus,

$$x^4 - 1 = (x^2 + 1)(x^2 - 1)$$
$$= (x^2 + 1)(x + 1)(x - 1)$$

SELF CHECK Factor $a^4 - 81$. *Answer:* $(a^2 + 9)(a + 3)(a - 3)$ ∎

 WARNING! When asked to factor a polynomial, we must be sure to factor it completely. After factoring a polynomial, always check to see whether any of the factors in the result can be factored further.

EXAMPLE 4	**Using the formula for the difference of two squares twice.** Factor $(x + y)^4 - z^4$.

Solution This expression is the difference of two squares and can be factored:

$$(x + y)^4 - z^4 = [(x + y)^2]^2 - (z^2)^2$$
$$= [(x + y)^2 + z^2][(x + y)^2 - z^2]$$

The factor $(x + y)^2 + z^2$ is the sum of two squares and is prime. However, the factor $(x + y)^2 - z^2$ is the difference of two squares and can be factored as $(x + y + z)(x + y - z)$. Thus,

$$(x + y)^4 - z^4 = [(x + y)^2 + z^2][(x + y)^2 - z^2]$$
$$= [(x + y)^2 + z^2](x + y + z)(x + y - z)$$

SELF CHECK   Factor $(a - b)^4 - c^4$.   *Answer:* $[(a - b)^2 + c^2](a - b + c)(a - b - c)$ ∎

When possible, we always factor out a common factor before factoring the difference of two squares. The factoring process is easier when all common factors are factored out first.

EXAMPLE 5	**Factoring out the GCF first.** Factor $2x^4y - 32y$.

Solution

$2x^4y - 32y = 2y(x^4 - 16)$	Factor out the GCF, which is $2y$.
$= 2y(x^2 + 4)(x^2 - 4)$	Factor $x^4 - 16$.
$= 2y(x^2 + 4)(x + 2)(x - 2)$	Factor $x^2 - 4$.

SELF CHECK   Factor $3a^4 - 3$.   *Answer:* $3(a^2 + 1)(a + 1)(a - 1)$ ∎

EXAMPLE 6	**Factoring by grouping.** Factor $x^2 - y^2 + x - y$.

Solution If we group the first two terms and factor the difference of two squares, we have

$x^2 - y^2 + x - y = (x + y)(x - y) + (x - y)$	Factor $x^2 - y^2$.
$= (x - y)(x + y + 1)$	Factor out $x - y$.

SELF CHECK   Factor $a^2 - b^2 + a + b$.   *Answer:* $(a + b)(a - b + 1)$ ∎

EXAMPLE 7	**Variable exponents.** Factor $x^{2n} - y^{2n}$.

Solution We note that

$$x^{2n} = x^n \cdot x^n = (x^n)^2 \qquad \text{and} \qquad y^{2n} = y^n \cdot y^n = (y^n)^2$$

We can rewrite $x^{2n} - y^{2n}$ as a difference of two squares: $(x^n)^2 - (y^n)^2$. Using the formula for factoring the difference of two squares, we have

$$
\begin{array}{ccccccc}
F^2 & - & L^2 & = & (F & + & L)(F & - & L) \\
\downarrow & & \downarrow & & \downarrow & & \downarrow & & \downarrow \\
(x^n)^2 & - & (y^n)^2 & = & (x^n & + & y^n)(x^n & - & y^n)
\end{array}
$$

Verify this result using multiplication.

SELF CHECK   Factor $a^{2n} - 4$.   *Answer:* $(a^n + 2)(a^n - 2)$ ∎

Perfect Cubes

The number 64 is called a perfect cube, because $4^3 = 64$. To factor the sum or difference of two cubes, it is helpful to know the first ten perfect cubes:

1, 8, 27, 64, 125, 216, 343, 512, 729, 1,000

Expressions such as $x^9y^6z^3$ are also perfect cubes, because they can be written as the cube of another quantity:

$$x^9y^6z^3 = (x^3y^2z)^3$$

The Sum and Difference of Two Cubes

To find formulas for factoring the sum or difference of two cubes, we use the following product formulas:

2. $(x + y)(x^2 - xy + y^2) = x^3 + y^3$

3. $(x - y)(x^2 + xy + y^2) = x^3 - y^3$

To verify Equation 2, we multiply $x^2 - xy + y^2$ by $x + y$.

$$
\begin{aligned}
(x + y)(x^2 - xy + y^2) &= (x + y)x^2 - (x + y)xy + (x + y)y^2 \\
&= x \cdot x^2 + y \cdot x^2 - x \cdot xy - y \cdot xy + x \cdot y^2 + y \cdot y^2 \\
&= x^3 + x^2y - x^2y - xy^2 + xy^2 + y^3 \\
&= x^3 + y^3
\end{aligned}
$$

Equation 3 can also be verified by multiplication.

If we write Equations 2 and 3 in reverse order, we have the formulas for factoring the sum and difference of two cubes.

Sum and difference of two cubes

$$x^3 + y^3 = (x + y)(x^2 - xy + y^2)$$
$$x^3 - y^3 = (x - y)(x^2 + xy + y^2)$$

If we think of the sum of two cubes as the sum of the cube of a **First** quantity plus the cube of a **Last** quantity, we have the formula

$$F^3 + L^3 = (F + L)(F^2 - FL + L^2)$$

*To factor the cube of a **First** quantity plus the cube of a **Last** quantity, we multiply the sum of the **First** and **Last** by*

- *the **First** squared*
- *minus the **First** times the **Last***
- *plus the **Last** squared.*

The formula for the difference of two cubes is

$$F^3 - L^3 = (F - L)(F^2 + FL + L^2)$$

*To factor the cube of a **First** quantity minus the cube of a **Last** quantity, we multiply the difference of the **First** and **Last** by*

- *the **First** squared*
- *plus the **First** times the **Last***
- *plus the **Last** squared.*

EXAMPLE 8

Factoring a sum of two cubes. Factor $a^3 + 8$.

Solution Since $a^3 + 8$ can be written as $a^3 + 2^3$, we have the sum of two cubes, which factors as follows:

$$\mathbf{F}^3 + \mathbf{L}^3 = (\mathbf{F} + \mathbf{L})(\mathbf{F}^2 - \mathbf{F}\,\mathbf{L} + \mathbf{L}^2)$$
$$a^3 + 2^3 = (a + 2)(a^2 - a2 + 2^2)$$
$$= (a + 2)(a^2 - 2a + 4)$$

Thus, $a^3 + 8 = (a + 2)(a^2 - 2a + 4)$. Check by multiplication.

SELF CHECK Factor $p^3 + 27$.   *Answer:* $(p + 3)(p^2 - 3p + 9)$ ∎

EXAMPLE 9

Factoring a difference of two cubes. Factor $27a^3 - 64b^3$.

Solution Since $27a^3 - 64b^3$ can be written as $(3a)^3 - (4b)^3$, we have the difference of two cubes, which factors as follows:

$$\mathbf{F}^3 - \mathbf{L}^3 = (\mathbf{F} - \mathbf{L})(\mathbf{F}^2 + \mathbf{F}\,\mathbf{L} + \mathbf{L}^2)$$
$$(3a)^3 - (4b)^3 = (3a - 4b)[(3a)^2 + (3a)(4b) + (4b)^2]$$
$$= (3a - 4b)(9a^2 + 12ab + 16b^2)$$

Thus, $27a^3 - 64b^3 = (3a - 4b)(9a^2 + 12ab + 16b^2)$. Check by multiplication.

SELF CHECK Factor $8p^3 - 27q^3$.   *Answer:* $(2p - 3q)(4p^2 + 6pq + 9q^2)$ ∎

EXAMPLE 10

Factoring a difference of two cubes. Factor $a^3 - (c + d)^3$.

Solution $a^3 - (c + d)^3 = [a - (c + d)][a^2 + a(c + d) + (c + d)^2]$

Now we simplify the expressions inside both sets of brackets.

$$a^3 - (c + d)^3 = (a - c - d)(a^2 + ac + ad + c^2 + 2cd + d^2)$$

SELF CHECK Factor $(p + q)^3 - r^3$.   *Answer:* $(p + q - r)(p^2 + 2pq + q^2 + pr + qr + r^2)$ ∎

EXAMPLE 11

Factoring completely. Factor $x^6 - 64$.

Solution This expression is both the difference of two squares and the difference of two cubes. It is easier to factor it as the difference of two squares first. This expression factors into the product of a sum and a difference.

$$x^6 - 64 = (x^3)^2 - 8^2$$
$$= (x^3 + 8)(x^3 - 8)$$

Each of these factors further, however, for one is the sum of two cubes and the other is the difference of two cubes:

$$x^6 - 64 = (x + 2)(x^2 - 2x + 4)(x - 2)(x^2 + 2x + 4)$$

SELF CHECK Factor $x^6 - 1$.   *Answer:* $(x + 1)(x^2 - x + 1)(x - 1)(x^2 + x + 1)$ ∎

| EXAMPLE 12 | **Factoring out the GCF.** Factor $2a^5 + 128a^2$. |

Solution   We first factor out the common monomial factor of $2a^2$ to obtain

$$2a^5 + 128a^2 = 2a^2(a^3 + 64)$$

Then we factor $a^3 + 64$ as the sum of two cubes to obtain

$$2a^5 + 128a^2 = 2a^2(a + 4)(a^2 - 4a + 16)$$

| SELF CHECK | Factor $3x^5 + 24x^2$.   *Answer:* $3x^2(x + 2)(x^2 - 2x + 4)$ ∎ |

| EXAMPLE 13 | **Variable exponents.** Factor $x^{3n} + y^{3n}$. |

Solution   We note that

$$x^{3n} = x^n \cdot x^n \cdot x^n = (x^n)^3 \qquad \text{and} \qquad y^{3n} = y^n \cdot y^n \cdot y^n = (y^n)^3$$

Since $x^{3n} + y^{3n}$ can be written as $(x^n)^3 + (y^n)^3$, we have the sum of two cubes, which factors as follows:

$$F^3 + L^3 = (F + L)(F^2 - F L + L^2)$$

$$(x^n)^3 + (y^n)^3 = (x^n + y^n)[(x^n)^2 - x^n y^n + (y^n)^2]$$
$$= (x^n + y^n)(x^{2n} - x^n y^n + y^{2n})$$

Verify this result using multiplication.

| SELF CHECK | Factor $27 - b^{3n}$.   *Answer:* $(3 - b^n)(9 + 3b^n + b^{2n})$ ∎ |

STUDY SET

Section 5.6

VOCABULARY

In Exercises 1–2, fill in the blanks to make the statements true.

1. When the polynomial $4x^2 - 25$ is rewritten as $(2x)^2 - (5)^2$, we see that it is the difference of two _____.

2. When the polynomial $8x^3 + 125$ is rewritten as $(2x)^3 + (5)^3$, we see that it is the sum of two _____.

CONCEPTS

3. Write the first ten perfect square natural numbers.

4. Write the first ten perfect cube natural numbers.

5. Use multiplication to verify that the sum of two squares $x^2 + 25$ does not factor as $(x + 5)(x + 5)$.

6. Use multiplication to verify that the difference of two squares $x^2 - 25$ factors as $(x + 5)(x - 5)$.

7. Explain why each factorization is not complete.
   a. $4g^2 - 16 = (2g + 4)(2g - 4)$

   b. $1 - t^8 = (1 + t^4)(1 - t^4)$

8. When asked to factor $81t^2 - 16$, one student answered $(9t - 4)(9t + 4)$, and another answered $(9t + 4)(9t - 4)$. Explain why both students are correct.

9. Factor each polynomial.
   a. $5p^2 + 20$
   b. $5p^2 - 20$

10. Factor each polynomial.
   a. $5p^3 + 20$
   b. $5p^3 + 40$

NOTATION

In Exercises 11–16, complete each factorization.

11. $p^3 + q^3 = (p + q)$ ▢

12. $p^3 - q^3 = (p - q)$ ▢

13. $p^2 - q^2 = (p + q)$ ▢

14. $p^2q + pq^2 =$ ▢ $(p + q)$

15. $36y^2 - 49m^2 = ($ ▢ $)^2 - (7m)^2$

$= (6y$ ▢ $7m)(6y -$ ▢ $)$

16. $h^3 - 27k^3 = (h)^3 - ($ ▢ $)^3$

$= (h$ ▢ $3k)(h^2 +$ ▢ $+ 9k^2)$

PRACTICE

In Exercises 17–40, factor each polynomial, if possible.

17. $x^2 - 4$

18. $y^2 - 9$

19. $9y^2 - 64$

20. $16x^4 - 81y^2$

21. $x^2 + 25$

22. $144a^2 - b^4$

23. $625a^2 - 169b^4$

24. $4y^2 + 9z^4$

25. $81a^4 - 49b^2$

26. $64r^6 - 121s^2$

27. $36x^4y^2 - 49z^4$

28. $4a^2b^4c^6 - 9d^8$

29. $(x + y)^2 - z^2$

30. $a^2 - (b - c)^2$

31. $(a - b)^2 - c^2$

32. $(m + n)^2 - p^4$

33. $x^4 - y^4$

34. $16a^4 - 81b^4$

35. $256x^4y^4 - z^8$

36. $225a^4 - 16b^8c^{12}$

37. $a^{2x} - b^{2x}$

38. $16 - x^{2n}$

39. $4x^{2n} - 9y^{2n}$

40. $25 - x^{6n}$

In Exercises 41–48, factor each polynomial.

41. $2x^2 - 288$

42. $8x^2 - 72$

43. $2x^3 - 32x$

44. $3x^3 - 243x$

45. $5x^3 - 125x$

46. $6x^4 - 216x^2$

47. $r^2s^2t^2 - t^2x^4y^2$

48. $16a^4b^3c^4 - 64a^2bc^6$

In Exercises 49–54, factor each polynomial by grouping.

49. $a^2 - b^2 + a + b$

50. $x^2 - y^2 - x - y$

51. $a^2 - b^2 + 2a - 2b$

52. $m^2 - n^2 + 3m + 3n$

53. $2x + y + 4x^2 - y^2$

54. $m - 2n + m^2 - 4n^2$

In Exercises 55–68, factor each polynomial.

55. $r^3 + s^3$

56. $t^3 - v^3$

57. $x^3 - 8y^3$

58. $27a^3 + b^3$

59. $64a^3 - 125b^6$

60. $8x^6 + 125y^3$

61. $125x^3y^6 + 216z^9$

62. $1{,}000a^6 - 343b^3c^6$

63. $x^6 + y^6$

64. $x^9 + y^9$

65. $a^{3b} - c^{3b}$

66. $8 - x^{3n}$

67. $27x^{3n} + y^{3n}$

68. $a^{3b} + b^{3c}$

In Exercises 69–76, factor each expression.

69. $5x^3 + 625$

70. $2x^3 - 128$

71. $4x^5 - 256x^2$

72. $2x^6 + 54x^3$

73. $128u^2v^3 - 2t^3u^2$

74. $56rs^2t^3 + 7rs^2v^6$

75. $(a + b)x^3 + 27(a + b)$

76. $(c - d)r^3 - (c - d)s^3$

APPLICATIONS

77. CANDY To find the amount of chocolate used in the outer coating of the malted-milk ball shown in Illustration 1 (on the next page), we can find the volume V of the chocolate shell using the formula

$$V = \frac{4}{3}\pi r_1{}^3 - \frac{4}{3}\pi r_2{}^3$$

Factor the expression on the right-hand side of the formula.

ILLUSTRATION 1

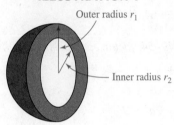

Outer radius r_1

Inner radius r_2

78. MOVIE STUNTS See Illustration 2. The function that gives the distance a stuntwoman is above the ground t seconds after she falls over the side of a 144-foot tall building is

$$h(t) = 144 - 16t^2$$

Factor the right-hand side of the equation.

ILLUSTRATION 2

144 ft

WRITING

79. Describe the pattern used to factor the difference of two squares.

80. Describe the patterns used to factor the sum and the difference of two cubes.

REVIEW

In Exercises 81–84, graph the line with the given characteristics.

81. Passing through $(-2, -1)$; slope $= -\dfrac{2}{3}$

82. y-intercept $(0, -4)$; slope $= 3$

83. Horizontal; y-intercept $(0, -2)$

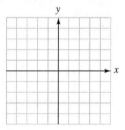

84. Parallel to the y-axis, passing through $(1, 4)$

85. Write the equation of line l shown in Illustration 3.

86. Write the equation of line r shown in Illustration 3.

ILLUSTRATION 3

▶ **5.7**

Factoring Trinomials

In this section, you will learn about

Perfect square trinomials ■ Factoring trinomials with lead coefficients of 1 ■ Factoring trinomials with lead coefficients other than 1 ■ Test for factorability ■ Using substitution to factor trinomials ■ Factoring by grouping ■ Using grouping to factor trinomials

Introduction   In this section, we will discuss several techniques for factoring trinomials. These techniques are based on the fact that the product of two binomials is often a trinomial. With that observation in mind, we begin the study of trinomial factoring by considering two special products.

Perfect Square Trinomials

Many trinomials can be factored by using the following special product formulas.

1. $(x + y)(x + y) = x^2 + 2xy + y^2$

2. $(x - y)(x - y) = x^2 - 2xy + y^2$

To factor $x^2 + 6x + 9$, we note that it can be written in the form $x^2 + 2(3)x + 3^2$. If $y = 3$, this form matches the right-hand side of Equation 1. Thus, $x^2 + 6x + 9$ factors as

$$x^2 + 6x + 9 = x^2 + 2(3)x + 3^2$$
$$= (x + 3)(x + 3)$$
$$= (x + 3)^2$$

Since $x^2 + 6x + 9$ is the square of $x + 3$, $x^2 + 6x + 9$ is called a **perfect square trinomial.** This result can be verified by multiplication:

$$(x + 3)(x + 3) = x^2 + 3x + 3x + 9$$
$$= x^2 + 6x + 9$$

EXAMPLE 1

Factoring a perfect square trinomial. Factor $x^2 - 4xz + 4z^2$.

Solution   To factor the perfect square trinomial $x^2 - 4xz + 4z^2$, we note that it can be written in the form $x^2 - 2x(2z) + (2z)^2$. If $y = 2z$, this form matches the right-hand side of Equation 2.

$$x^2 - 4xz + 4z^2 = x^2 - 2x(2z) + (2z)^2$$
$$= (x - 2z)(x - 2z)$$
$$= (x - 2z)^2$$

This result can be verified by multiplication.

SELF CHECK   Factor $b^2 - 10b + 25$.   *Answer:* $(b - 5)^2$ ■

We begin our discussion of *general trinomials* by considering trinomials with lead coefficients (the coefficient of the squared term) of 1.

Factoring Trinomials with Lead Coefficients of 1

Since the product of two binomials is often a trinomial, we expect that many trinomials will factor as two binomials. For example, to factor $x^2 + 7x + 12$, we must find two binomials $x + a$ and $x + b$ such that

$$x^2 + 7x + 12 = (x + a)(x + b)$$

where $ab = 12$ and $ax + bx = 7x$.

To find the numbers a and b, we list the possible factorizations of 12 and find the one where the sum of the factors is 7.

The one to choose
↓

$$12(1) \quad 6(2) \quad 4(3) \quad -12(-1) \quad -6(-2) \quad -4(-3)$$

Thus, $a = 4$, $b = 3$, and

$$x^2 + 7x + 12 = (x + a)(x + b)$$

3. $x^2 + 7x + 12 = (x + 4)(x + 3)$

This factorization can be verified by multiplying $x + 4$ and $x + 3$ and observing that the product is $x^2 + 7x + 12$.

Because of the commutative property of multiplication, the order of the factors in Equation 3 is not important.

To factor trinomials with lead coefficients of 1, we follow these steps.

Factoring trinomials with lead coefficients of 1

1. Write the trinomial in descending powers of one variable.
2. List the factorizations of the third term of the trinomial.
3. Pick the factorization where the sum of the factors is the coefficient of the middle term.

EXAMPLE 2

Factoring a trinomial with lead coefficient of 1. Factor $x^2 - 6x + 8$.

Solution   Since the trinomial is written in descending powers of x, we can move to step 2 and list the possible factorizations of the third term, which is 8.

The one to choose
$$8(1) \qquad 4(2) \qquad -8(-1) \qquad -4(-2)$$

In the trinomial, the coefficient of the middle term is -6. The only factorization where the sum of the factors is -6 is $-4(-2)$. Thus, $a = -4$, $b = -2$, and

$$x^2 - 6x + 8 = (x + a)(x + b)$$
$$= (x - 4)(x - 2)$$

We can verify this result by multiplication:

$$(x - 4)(x - 2) = x^2 - 2x - 4x + 8 \quad \text{Use the FOIL method.}$$
$$= x^2 - 6x + 8$$

SELF CHECK   Factor $a^2 - 7a + 12$.   *Answer:* $(a - 4)(a - 3)$ ∎

EXAMPLE 3

Factoring a trinomial. Factor $-x + x^2 - 12$.

Solution   We begin by writing the trinomial in descending powers of x:

$$-x + x^2 - 12 = x^2 - x - 12$$

The possible factorizations of the third term are

The one to choose
$$12(-1) \qquad 6(-2) \qquad 4(-3) \qquad 1(-12) \qquad 2(-6) \qquad 3(-4)$$

In the trinomial, the coefficient of the middle term is -1. The only factorization where the sum of the factors is -1 is $3(-4)$. Thus, $a = 3$, $b = -4$, and

$$-x + x^2 - 12 = (x + a)(x + b)$$
$$= (x + 3)(x - 4)$$

SELF CHECK   Factor $-3a + a^2 - 10$.   *Answer:* $(a + 2)(a - 5)$ ∎

EXAMPLE 4

Factoring completely. Factor $30x - 4xy - 2xy^2$.

Solution   We begin by writing the trinomial in descending powers of y:

$$30x - 4xy - 2xy^2 = -2xy^2 - 4xy + 30x$$

Each term in this trinomial has a common factor of $-2x$, which we will factor out.

$$30x - 4xy - 2xy^2 = -2x(y^2 + 2y - 15)$$

To factor $y^2 + 2y - 15$, we list the factors of -15 and find the pair whose sum is 2.

The one to choose

$$15(-1) \qquad 5(-3) \qquad 1(-15) \qquad 3(-5)$$

The only factorization where the sum of the factors is 2 (the coefficient of the middle term of $y^2 + 2y - 15$) is $5(-3)$. Thus, $a = 5$, $b = -3$, and

$$30x - 4xy - 2xy^2 = -2x(y^2 + 2y - 15)$$
$$= -2x(y + 5)(y - 3)$$

SELF CHECK   Factor $18a + 3ab - 3ab^2$.   *Answer:* $-3a(b + 2)(b - 3)$ ∎

WARNING! In Example 4, be sure to include all factors in the final result. It is a common error to forget to write the $-2x$.

Factoring Trinomials with Lead Coefficients Other Than 1

There are more combinations of factors to consider when factoring trinomials with lead coefficients other than 1. To factor $5x^2 + 7x + 2$, for example, we must find two binomials of the form $ax + b$ and $cx + d$ such that

$$5x^2 + 7x + 2 = (ax + b)(cx + d)$$

Since the first term of the trinomial $5x^2 + 7x + 2$ is $5x^2$, the first terms of the binomial factors must be $5x$ and x.

$$5x^2$$
$$5x^2 + 7x + 2 = (5x + b)(x + d)$$

Since the product of the last terms must be 2, and the sum of the products of the outer and inner terms must be $7x$, we must find two numbers whose product is 2 that will give a middle term of $7x$.

$$2$$
$$5x^2 + 7x + 2 = (5x + b)(x + d)$$
$$O + I = 7x$$

Since $2(1)$ and $(-2)(-1)$ give a product of 2, there are four possible combinations to consider:

$$\textbf{(5x + 2)(x + 1)} \qquad (5x - 2)(x - 1)$$
$$(5x + 1)(x + 2) \qquad (5x - 1)(x - 2)$$

Of these possibilities, only the first one gives the correct middle term of $7x$. Thus,

4. $5x^2 + 7x + 2 = (5x + 2)(x + 1)$

We can verify this result by multiplication:

$$(5x + 2)(x + 1) = 5x^2 + 5x + 2x + 2$$
$$= 5x^2 + 7x + 2$$

Test for Factorability

If a trinomial has the form $ax^2 + bx + c$, with integer coefficients and $a \neq 0$, we can test to see whether it is factorable.

- If the value of $b^2 - 4ac$ is a perfect square, the trinomial can be factored using only integers.
- If the value of $b^2 - 4ac$ is not a perfect square, the trinomial cannot be factored using only integers.

For example, $5x^2 + 7x + 2$ is a trinomial in the form $ax^2 + bx + c$ with

$$a = 5, \qquad b = 7, \qquad \text{and} \qquad c = 2$$

For this trinomial, the value of $b^2 - 4ac$ is

$$b^2 - 4ac = 7^2 - 4(5)(2)$$
$$= 49 - 40$$
$$= 9$$

Since 9 is a perfect square, the trinomial is factorable. Its factorization is shown in Equation 4.

Test for factorability

A trinomial of the form $ax^2 + bx + c$, with integer coefficients and $a \neq 0$, will factor into two binomials with integer coefficients if the value of $b^2 - 4ac$ is a perfect square. If $b^2 - 4ac = 0$, the factors will be the same.

EXAMPLE 5

A lead coefficient that is not 1. Factor $3p^2 - 4p - 4$.

Solution   In the trinomial, $a = 3$, $b = -4$, and $c = -4$. To see whether it factors, we evaluate $b^2 - 4ac$.

$$b^2 - 4ac = (-4)^2 - 4(3)(-4)$$
$$= 16 + 48$$
$$= 64$$

Since 64 is a perfect square, the trinomial is factorable.

To factor the trinomial, we note that the first terms of the binomial factors must be $3p$ and p to give the first term of $3p^2$.

$$3p^2 - 4p - 4 = (3p + ?)(p + ?)$$

The product of the last terms must be -4, and the sum of the products of the outer terms and the inner terms must be $-4p$.

$$3p^2 - 4p - 4 = (3p + ?)(p + ?)$$

$$O + I = -4p$$

Because $1(-4)$, $-1(4)$, and $-2(2)$ all give a product of -4, there are six possible combinations to consider:

$$(3p + 1)(p - 4) \qquad (3p - 4)(p + 1)$$
$$(3p - 1)(p + 4) \qquad (3p + 4)(p - 1)$$
$$(3p - 2)(p + 2) \qquad \mathbf{(3p + 2)(p - 2)}$$

Of these possibilities, only the last gives the required middle term of $-4p$. Thus,

$$3p^2 - 4p - 4 = (3p + 2)(p - 2)$$

SELF CHECK Factor $4q^2 - 9q - 9$. *Answer:* $(4q + 3)(q - 3)$ ∎

EXAMPLE 6

Testing for factorability. Factor $4t^2 - 3t - 5$, if possible.

Solution In the trinomial, $a = 4$, $b = -3$, and $c = -5$. To see whether the trinomial is factorable, we evaluate $b^2 - 4ac$ by substituting the values of a, b, and c.

$$b^2 - 4ac = (-3)^2 - 4(4)(-5)$$
$$= 9 + 80$$
$$= 89$$

Since 89 is not a perfect square, the trinomial is not factorable using only integer coefficients.

SELF CHECK Factor $5a^2 - 8a + 2$, if possible. *Answer:* a prime polynomial ∎

It is not easy to give specific rules for factoring general trinomials. However, the following hints are helpful.

Factoring a general trinomial

1. Write the trinomial in descending powers of one variable.
2. Factor out any greatest common factor (including -1, if that is necessary to make the coefficient of the first term positive).
3. Test the trinomial for factorability.
4. When the sign of the third term is $+$, the signs between the terms of each binomial factor are the same as the sign of the middle term of the trinomial.
    When the sign of the third term is $-$, the signs between the terms of the binomials are opposite.
5. Try various combinations of the factors of the first terms and the last terms until you find the one that works.
6. Check the factorization by multiplication.

EXAMPLE 7

Factoring a trinomial containing two variables. Factor $24y^2 + 10xy - 6x^2$.

Solution We write the trinomial in descending powers of x and factor out -2.

$$24y^2 + 10xy - 6x^2 = -6x^2 + 10xy + 24y^2$$
$$= -2(3x^2 - 5xy - 12y^2)$$

In the trinomial $3x^2 - 5xy - 12y^2$, $a = 3$, $b = -5$, and $c = -12$.

$$b^2 - 4ac = (-5)^2 - 4(3)(-12)$$
$$= 25 + 144$$
$$= 169$$

Since 169 is a perfect square, the trinomial will factor.

Since the sign of the first term of $3x^2 - 5xy - 12y^2$ is positive and the sign of the third term is negative, the signs between the binomial factors will be opposite. Because the first term is $3x^2$, the first terms of the binomial factors must be $3x$ and x.

$$-2(3x^2 - 5xy - 12y^2) = -2(3x \qquad)(x \qquad)$$

The product of the last terms must be $-12y^2$, and the sum of the product of the outer terms and the product of the inner terms must be $-5xy$.

$$24y^2 + 10xy - 6x^2 = -2(3x \qquad ?y)(x \qquad ?y)$$

$$O + I = -5xy$$

Since $1(-12)$, $2(-6)$, $3(-4)$, $12(-1)$, $6(-2)$, and $4(-3)$ all give a product of -12, there are 12 possible combinations to consider.

$(3x + 1y)(x - 12y)$	$(3x - 12y)(x + 1y)$
$(3x + 2y)(x - 6y)$	$(3x - 6y)(x + 2y)$
$(3x + 3y)(x - 4y)$	$(3x - 4y)(x + 3y)$
$(3x + 12y)(x - 1y)$	$(3x - 1y)(x + 12y)$
$(3x + 6y)(x - 2y)$	$(3x - 2y)(x + 6y)$
The one to choose $\longrightarrow$ $(3x + 4y)(x - 3y)$	$(3x - 3y)(x + 4y)$

The combinations marked in color cannot work, because one of the factors has a common factor. This implies that $3x^2 - 5xy - 12y^2$ would have a common factor, which it doesn't.

After mentally trying the remaining combinations, we find that only $(3x + 4y)(x - 3y)$ gives the proper middle term of $-5xy$.

$$24y^2 + 10xy - 6x^2 = -2(3x^2 - 5xy - 12y^2)$$
$$= -2(3x + 4y)(x - 3y)$$

Verify this result by multiplication.

SELF CHECK

Factor $-6x^2 - 15xy - 6y^2$. *Answer:* $-3(x + 2y)(2x + y)$ ∎

EXAMPLE 8

A lead coefficient that is not 1. Factor $6y + 13x^2y + 6x^4y$.

Solution

We write the trinomial in descending powers of x and factor out the common factor of y to obtain

$$6y + 13x^2y + 6x^4y = 6x^4y + 13x^2y + 6y$$
$$= y(6x^4 + 13x^2 + 6)$$

A test for factorability will show that $6x^4 + 13x^2 + 6$ will factor.

Since the coefficients of the first and last terms of $6x^4 + 13x^2 + 6$ are positive, the signs between the terms in each binomial will be $+$.

Since the first term of the trinomial is $6x^4$, the first terms of the binomial factors must be either $2x^2$ and $3x^2$ or x^2 and $6x^2$.

Since the product of the last terms of the binomial factors must be 6, we must find two numbers whose product is 6 that will lead to a middle term of $7x^2$. After trying some combinations, we find the one that works.

$$6y + 13x^2y + 6x^4y = y(6x^4 + 13x^2 + 6)$$
$$= y(2x^2 + 3)(3x^2 + 2)$$

Verify this result by multiplication.

SELF CHECK   Factor $4b + 11a^2b + 6a^4b$.   *Answer:* $b(2a^2 + 1)(3a^2 + 4)$ ■

EXAMPLE 9   **Variable exponents.** Factor $x^{2n} + x^n - 2$.

Solution   Since the first term is x^{2n}, the first terms of the binomial must be x^n and x^n.

$$x^{2n} + x^n - 2 = (x^n \quad\ )(x^n \quad\ )$$

Since the third term of the trinomial is -2, the last terms of the binomial factors must have opposite signs, have a product of -2, and lead to a middle term of x^n. The only combination that works is

$$x^{2n} + x^n - 2 = (x^n + 2)(x^n - 1)$$

Verify this result by multiplication.

SELF CHECK   Factor $a^{2n} + a^n - 6$.   *Answer:* $(a^n + 3)(a^n - 2)$ ■

Using Substitution to Factor Trinomials

For more complicated expressions, a substitution sometimes helps to simplify the factoring process.

EXAMPLE 10   **Using substitution.** Factor $(x + y)^2 + 7(x + y) + 12$.

Solution   We rewrite the trinomial $(x + y)^2 + 7(x + y) + 12$ as $z^2 + 7z + 12$, where $z = x + y$. The trinomial $z^2 + 7z + 12$ factors as $(z + 4)(z + 3)$.
To find the factorization of $(x + y)^2 + 7(x + y) + 12$, we substitute $x + y$ for z in the expression $(z + 4)(z + 3)$ to obtain

$$z^2 + 7z + 12 = (z + 4)(z + 3)$$
$$(x + y)^2 + 7(x + y) + 12 = (x + y + 4)(x + y + 3)$$

SELF CHECK   Factor $(a + b)^2 - 3(a + b) - 10$.   *Answer:* $(a + b + 2)(a + b - 5)$ ■

Factoring by Grouping

EXAMPLE 11   **Factoring by grouping.** Factor $x^2 + 6x + 9 - z^2$.

Solution   We group the first three terms together and factor the trinomial to get

$$x^2 + 6x + 9 - z^2 = (x + 3)(x + 3) - z^2$$
$$= (x + 3)^2 - z^2$$

We can now factor the difference of two squares to get

$$x^2 + 6x + 9 - z^2 = (x + 3 + z)(x + 3 - z)$$

SELF CHECK   Factor $a^2 + 4a + 4 - b^2$.   *Answer:* $(a + 2 + b)(a + 2 - b)$ ■

Using Grouping to Factor Trinomials

The method of factoring by grouping can be used to help factor trinomials of the form $ax^2 + bx + c$. For example, to factor the trinomial $6x^2 + 7x - 3$, we proceed as follows:

1. First find the product ac: $6(-3) = -18$. This number is called the **key number.**
2. Find two factors of the key number -18 whose sum is $b = 7$:

   $$9(-2) = -18 \quad \text{and} \quad 9 + (-2) = 7$$

3. Use the factors 9 and -2 as coefficients of two terms to be placed between $6x^2$ and -3:

   $$6x^2 + 7x - 3 = 6x^2 + 9x - 2x - 3$$

4. Factor by grouping:

   $$6x^2 + 9x - 2x - 3 = 3x(2x + 3) - 1(2x + 3)$$
   $$= (2x + 3)(3x - 1) \qquad \text{Factor out } 2x + 3.$$

We can verify this factorization by multiplication.

EXAMPLE 12

The grouping method. Factor $10x^2 + 13x - 3$.

Solution   Since $a = 10$ and $c = -3$ in the trinomial, $ac = -30$. We now find two factors of -30 whose sum is 13. Two such factors are 15 and -2. We use these factors as coefficients of two terms to be placed between $10x^2$ and -3:

$$10x^2 + 15x - 2x - 3$$

Finally, we factor by grouping.

$$5x(2x + 3) - 1(2x + 3) = (2x + 3)(5x - 1)$$

SELF CHECK   Factor $15a^2 + 17a - 4$.   *Answer:* $(3a + 4)(5a - 1)$ ∎

STUDY SET

Section 5.7

VOCABULARY

In Exercises 1–4, fill in the blanks to make the statements true.

1. A polynomial with three terms, such as $3x^2 - 2x + 4$, is called a _____.
2. Since $y^2 + 2y + 1$ is the square of $y + 1$, we call $y^2 + 2y + 1$ a _____ square trinomial.
3. For $a^2 - a - 6$, the _____ coefficient (the coefficient of the a^2 term) is 1.
4. The trinomial $4a^2 - 5a - 6$ is written in _____ powers of a.

CONCEPTS

5. Consider $3x^2 - x + 16$. What is the sign of the
   a. First term?
   b. Middle term?
   c. Last term?
6. Explain what is meant when we say that the trinomial $h^2 - 12h + 27$ can be written as the product of two binomials.
7. If $b^2 - 4ac$ is a perfect square, the trinomial $ax^2 + bx + c$ can be factored using what type of coefficients?
8. Use the substitution $x = a + b$ to rewrite the trinomial $6(a + b)^2 - 17(a + b) - 3$.

NOTATION

In Exercises 9–12, find each product.

9. $(x + y)(x + y) = x^2 + $ ▊

10. $(x - y)(x - y) = x^2 - $ ▊

11. $(x + y)(x - y) = $ ▊

12. $(a + b)(a + b) = $ ▊ $+ b^2$

13. The trinomial $4m^2 - 4m + 1$ is written in $ax^2 + bx + c$ form. Identify a, b, and c.

14. Consider the trinomial $15s^2 + 4s - 4$. Is $b^2 - 4ac$ a perfect square?

PRACTICE

In Exercises 15–20, complete each factorization.

15. $x^2 + 5x + 6 = (x + 3)$ ▊

16. $x^2 - 6x + 8 = (x - 4)$ ▊

17. $x^2 + 2x - 15 = (x + 5)$ ▊

18. $x^2 - 3x - 18 = (x - 6)$ ▊

19. $2a^2 + 9a + 4 = $ ▊ $(a + 4)$

20. $6p^2 - 5p - 4 = $ ▊ $(2p + 1)$

In Exercises 21–28, use a special product formula to factor each perfect square trinomial.

21. $x^2 + 2x + 1$

22. $y^2 - 2y + 1$

23. $a^2 - 18a + 81$

24. $b^2 + 12b + 36$

25. $4y^2 + 4y + 1$

26. $9x^2 + 6x + 1$

27. $9b^2 - 12b + 4$

28. $4a^2 - 12a + 9$

In Exercises 29–40, test each trinomial for factorability and factor it, if possible.

29. $x^2 - 5x + 6$

30. $y^2 + 7y + 6$

31. $x^2 - 7x + 10$

32. $c^2 - 7c + 12$

33. $b^2 + 8b + 18$

34. $x^2 + 4x - 28$

35. $x^2 - x - 30$

36. $a^2 + 4a - 45$

37. $a^2 + 5a - 50$

38. $b^2 + 9b - 36$

39. $x^2 - 4xy - 21y^2$

40. $a^2 + 4ab - 5b^2$

In Exercises 41–50, factor each trinomial. If the lead coefficient is negative, begin by factoring out -1.

41. $3x^2 + 12x - 63$

42. $2y^2 + 4y - 48$

43. $b^2x^2 - 12bx^2 + 35x^2$

44. $c^3x^2 + 11c^3x - 42c^3$

45. $-a^2 + 4a + 32$

46. $-x^2 - 2x + 15$

47. $-3x^2 + 15x - 18$

48. $-2y^2 - 16y + 40$

49. $-2p^2 - 2pq + 4q^2$

50. $-6m^2 + 3mn + 3n^2$

In Exercises 51–70, factor each trinomial. Factor out all common monomials first (including -1 if the lead coefficient is negative). If a trinomial is prime, so indicate.

51. $6y^2 + 7y + 2$

52. $6x^2 - 11x + 3$

53. $8a^2 + 6a - 9$

54. $15b^2 + 4b - 4$

55. $6x^2 - 5xy - 4y^2$

56. $18y^2 - 3yz - 10z^2$

57. $5x^2 + 4x + 1$

58. $6z^2 + 17z + 12$

59. $8x^2 - 10x + 3$

60. $4a^2 + 20a + 3$

61. $a^2 - 3ab - 4b^2$

62. $b^2 + 2bc - 80c^2$

63. $3x^3 - 10x^2 + 3x$

64. $3t^3 - 3t^2 + t$

65. $-3a^2 + ab + 2b^2$

66. $-2x^2 + 3xy + 5y^2$

67. $5a^2 + 45b^2 - 30ab$

68. $-4x^2 - 9 + 12x$

69. $21x^4 - 10x^3 - 16x^2$

70. $16x^3 - 50x^2 + 36x$

In Exercises 71–76, factor each trinomial.

71. $x^4 + 8x^2 + 15$

72. $x^4 + 11x^2 + 24$

73. $y^4 - 13y^2 + 30$

74. $y^4 - 13y^2 + 42$

75. $a^4 - 13a^2 + 36$

76. $b^4 - 17b^2 + 16$

In Exercises 77–84, factor each expression. Assume that n is a natural number.

77. $x^{2n} + 2x^n + 1$

78. $2x^{2n} + x^n - 1$

79. $2a^{6n} - 3a^{3n} - 2$

80. $b^{2n} - b^n - 6$

81. $x^{4n} + 2x^{2n}y^{2n} + y^{4n}$

82. $y^{6n} + 2y^{3n}z + z^2$

83. $6x^{2n} + 7x^n - 3$

84. $6y^{4n} + 5y^{2n} + 1$

In Exercises 85–88, use a substitution to help factor each expression.

85. $(x + a)^2 + 2(x + a) + 1$

86. $(a + b)^2 - 2(a + b) + 1$

87. $(a + b)^2 - 2(a + b) - 24$

88. $(x - y)^2 + 3(x - y) - 10$

In Exercises 89–94, factor each expression by using grouping.

89. $x^2 + 4x + 4 - y^2$

90. $x^2 - 6x + 9 - 4y^2$

91. $x^2 + 2x + 1 - 9z^2$

92. $x^2 + 10x + 25 - 16z^2$

93. $c^2 - 4a^2 + 4ab - b^2$

94. $4c^2 - a^2 - 6ab - 9b^2$

In Exercises 95–100, use grouping to help factor each trinomial.

95. $a^2 - 17a + 16$

96. $b^2 - 4b - 21$

97. $2u^2 + 5u + 3$

98. $6y^2 + 5y - 6$

99. $20r^2 - 7rs - 6s^2$

100. $6s^2 + st - 12t^2$

APPLICATIONS

101. ICE The surface area of the cubical block of ice shown in Illustration 1 is $6x^2 + 36x + 54$. Find the length of an edge of the block.

ILLUSTRATION 1

102. CHECKERS The area of the square checkerboard in Illustration 2 is $25x^2 - 40x + 16$. Find the length of a side.

ILLUSTRATION 2

WRITING

103. Explain how you would factor -1 from a trinomial.

104. Explain how you would test the polynomial $ax^2 + bx + c$ for factorability.

REVIEW

105. If $f(x) = |2x - 1|$, find $f(-2)$.

106. If $g(x) = 2x^2 - 1$, find $g(-2)$.

107. Solve $-3 = -\dfrac{9}{8}s$.

108. Solve $2x + 3 = \dfrac{2}{3}x - 1$.

109. Simplify $3p^2 - 6(5p^2 + p) + p^2$.

110. Solve the system $\begin{cases} 2(2x + 3y) = 5 \\ 8x = 3(1 + 3y) \end{cases}$.

▶ **5.8**

Summary of Factoring Techniques

In this section, you will learn about

A general factoring strategy

Introduction Factoring some polynomials involves several steps in which two or more factoring techniques must be used. At times, it can be difficult to decide how to proceed with the factoring process for such polynomials. In this section, we will discuss a general factoring strategy—a step-by-step plan to follow when factoring any polynomial.

A General Factoring Strategy

In this section we will discuss ways to approach a randomly chosen factoring problem. For example, suppose we wish to factor the trinomial

$$x^2y^2z^3 + 7xy^2z^3 + 6y^2z^3$$

We begin by attempting to identify the problem type. The first possibility to look for is **factoring out a common factor**. Because the trinomial has a common factor of y^2z^3, we factor it out:

$$x^2y^2z^3 + 7xy^2z^3 + 6y^2z^3 = y^2z^3(x^2 + 7x + 6)$$

We note that $x^2 + 7x + 6$ is a trinomial that can be factored as $(x + 6)(x + 1)$. Thus,

$$x^2y^2z^3 + 7xy^2z^3 + 6y^2z^3 = y^2z^3(x^2 + 7x + 6)$$
$$= y^2z^3(x + 6)(x + 1)$$

To identify the type of factoring problem, we follow these steps.

A general factoring strategy

1. Factor out all common factors.
2. If an expression has two terms, check for the following problem types:
   a. **The difference of two squares:** $x^2 - y^2 = (x + y)(x - y)$
   b. **The sum of two cubes:** $x^3 + y^3 = (x + y)(x^2 - xy + y^2)$
   c. **The difference of two cubes:** $x^3 - y^3 = (x - y)(x^2 + xy + y^2)$
3. If an expression has three terms, attempt to factor the trinomial as a **trinomial.**
4. If an expression has four or more terms, try factoring by **grouping.**
5. Continue until each individual factor is prime.
6. Check the results by multiplying.

EXAMPLE 1

Applying the factoring strategy. Factor $48a^4c^3 - 3b^4c^3$.

Solution We begin by factoring out the common factor of $3c^3$:

$$48a^4c^3 - 3b^4c^3 = 3c^3(16a^4 - b^4)$$

Since the expression $16a^4 - b^4$ has two terms, we check to see whether it is the difference of two squares, which it is. As the difference of two squares, it factors as $(4a^2 + b^2)(4a^2 - b^2)$.

$$48a^4c^3 - 3b^4c^3 = 3c^3(16a^4 - b^4)$$
$$= 3c^3(4a^2 + b^2)(4a^2 - b^2)$$

The binomial $4a^2 + b^2$ is the sum of two squares and is prime. However, $4a^2 - b^2$ is the difference of two squares and factors as $(2a + b)(2a - b)$.

$$48a^4c^3 - 3b^4c^3 = 3c^3(16a^4 - b^4)$$
$$= 3c^3(4a^2 + b^2)(4a^2 - b^2)$$
$$= 3c^3(4a^2 + b^2)(2a + b)(2a - b)$$

Since each of the individual factors is prime, the factorization is complete.

SELF CHECK Factor $3p^4r^3 - 3q^4r^3$. *Answer:* $3r^3(p^2 + q^2)(p + q)(p - q)$ ∎

EXAMPLE 2

Applying the factoring strategy. Factor $x^5y + x^2y^4 - x^3y^3 - y^6$.

Solution   We begin by factoring out the common factor of y:

$$x^5y + x^2y^4 - x^3y^3 - y^6 = y(x^5 + x^2y^3 - x^3y^2 - y^5)$$

Because the expression $x^5 + x^2y^3 - x^3y^2 - y^5$ has four terms, we try factoring by grouping to obtain

$$x^5y + x^2y^4 - x^3y^3 - y^6$$
$$= y(x^5 + x^2y^3 - x^3y^2 - y^5) \qquad \text{Factor out } y.$$
$$= y[x^2(x^3 + y^3) - y^2(x^3 + y^3)] \quad \text{Factor by grouping.}$$
$$= y(x^3 + y^3)(x^2 - y^2) \qquad \text{Factor out } x^3 + y^3.$$

Finally, we factor $x^3 + y^3$ (the sum of two cubes) and $x^2 - y^2$ (the difference of two squares) to obtain

$$x^5y + x^2y^4 - x^3y^3 - y^6 = y(x + y)(x^2 - xy + y^2)(x + y)(x - y)$$

Because each of the individual factors is prime, the factorization is complete.

SELF CHECK   Factor $a^5p - a^3b^2p + a^2b^3p - b^5p$.   *Answer:* $p(a + b)(a^2 - ab + b^2)(a + b)(a - b)$ ∎

EXAMPLE 3

Applying the factoring strategy. Factor $x^3 + 5x^2 + 6x + x^2y + 5xy + 6y$.

Solution   There are no common factors. Since there are more than three terms, we try factoring by grouping. We can factor x from the first three terms and y from the last three terms.

$$x^3 + 5x^2 + 6x + x^2y + 5xy + 6y$$
$$= x(x^2 + 5x + 6) + y(x^2 + 5x + 6)$$
$$= (x^2 + 5x + 6)(x + y) \qquad \text{Factor out } x^2 + 5x + 6.$$
$$= (x + 3)(x + 2)(x + y) \qquad \text{Factor } x^2 + 5x + 6.$$

SELF CHECK   Factor $a^3 - 5a^2 + 6a + a^2b - 5ab + 6b$.   *Answer:* $(a - 2)(a - 3)(a + b)$ ∎

EXAMPLE 4

Applying the factoring strategy. Factor $x^4 + 2x^3 + x^2 + x + 1$.

Solution   There are no common factors. Since there are more than three terms, we try factoring by grouping. We can factor x^2 from the first three terms.

$$x^4 + 2x^3 + x^2 + x + 1 = x^2(x^2 + 2x + 1) + (x + 1)$$
$$= x^2(x + 1)(x + 1) + (x + 1) \quad \text{Factor } x^2 + 2x + 1.$$
$$= (x + 1)[x^2(x + 1) + 1] \qquad \text{Factor out } x + 1.$$
$$= (x + 1)(x^3 + x^2 + 1)$$

SELF CHECK   Factor $a^4 - a^3 - 2a^2 + a - 2$.   *Answer:* $(a - 2)(a^3 + a^2 + 1)$ ∎

EXAMPLE 5

Variable exponents. Factor $-nx^{2n} - 6nx^n - 9n$.

Solution   We begin by factoring out a common factor of $-n$ from each term.

$$-nx^{2n} - 6nx^n - 9n = -n(x^{2n} + 6x^n + 9)$$

Since the expression $x^{2n} + 6x^n + 9$ can be written in the form

$$(x^n)^2 + 2(x^n)(3) + (3)^2$$

it is a perfect square trinomial, and it factors in the following way:

$$-nx^{2n} - 6nx^n - 9n = -n(x^{2n} + 6x^n + 9)$$

$$= -n(x^n + 3)(x^n + 3) \quad \text{Factor the perfect square trinomial.}$$

$$= -n(x^n + 3)^2 \quad \begin{array}{l}\text{Write the result in simpler form using an}\\\text{exponent.}\end{array}$$

SELF CHECK   Factor $6a^{2n} - 12a^n + 6$.   *Answer:* $6(a^n - 1)^2$ ∎

STUDY SET

Section 5.8

VOCABULARY

In Exercises 1–4, fill in the blanks to make the statements true.

1. The process of finding the individual factors of a known product is called _____.

2. $x^3 + y^3$ is called a sum of two _____.

3. $x^3 - y^3$ is called a difference of two _____.

4. $x^2 - y^2$ is called a _____ of two squares.

CONCEPTS

In Exercises 5–8, fill in the blanks to make the statements true.

5. In any factoring problem, always factor out any _____ factors first.

6. If an expression has two terms, check to see whether the problem type is the _____ of two squares, the sum of two _____, or the _____ of two cubes.

7. If an expression has three terms, try to factor it as a _____.

8. If an expression has four or more terms, try factoring it by _____.

9. Explain how to verify that $y^2z^3(x + 6)(x + 1)$ is the factored form of $x^2y^2z^3 + 7xy^2z^3 + 6y^2z^3$.

10. Why is the polynomial $x + 6$ classified as prime?

NOTATION

In Exercises 11–12, complete each factorization.

11. $18a^3b + 3a^2b^2 - 6ab^3 = \boxed{}(6a^2 + ab - 2b^2)$

$$= 3ab(3a + \boxed{})(\boxed{} - b)$$

12. $2x^4 - 1{,}250 = 2(\boxed{})$

$$= 2(\boxed{})(x^2 - 25)$$

$$= 2(x^2 + 25)(x + 5)(\boxed{})$$

PRACTICE

In Exercises 13–52, factor each polynomial, if possible.

13. $x^2 + 16 + 8x$

14. $20 + 11x - 3x^2$

15. $8x^3y^3 - 27$

16. $3x^2y + 6xy^2 - 12xy$

17. $xy - ty + xs - ts$

18. $bc + b + cd + d$

19. $25x^2 - 16y^2$

20. $27x^9 - y^3$

21. $12x^2 + 52x + 35$

22. $12x^2 + 14x - 6$

23. $6x^2 - 14x + 8$

24. $12x^2 - 12$

25. $4x^2y^2 + 4xy^2 + y^2$

26. $100z^2 - 81t^2$

27. $x^3 + (a^2y)^3$

28. $4x^2y^2z^2 - 26x^2y^2z^3$

29. $2x^3 - 54$

30. $4(xy)^3 + 256$

31. $ae + bf + af + be$

32. $a^2x^2 + b^2y^2 + b^2x^2 + a^2y^2$

33. $2(x + y)^2 + (x + y) - 3$

34. $(x - y)^3 + 125$

35. $625x^4 - 256y^4$

36. $2(a - b)^2 + 5(a - b) + 3$

37. $36x^4 - 36$

38. $6x^2 - 63 - 13x$

39. $a^4 - 13a^2 + 36$

40. $x^4 - 17x^2 + 16$

41. $x^2 + 6x + 9 - y^2$

42. $x^2 + 10x + 25 - y^8$

43. $4x^2 + 4x + 1 - 4y^2$

44. $9x^2 - 6x + 1 - 25y^2$

45. $x^2 - y^2 - 2y - 1$

46. $a^2 - b^2 + 4b - 4$

47. $2a^{2n} + 2a^n - 24$

48. $ma^{2n} - mb^{2n}$

49. $54a^{3n} + 16b^{3n}$

50. $-a^{2n} - 2a^n b^n - b^{2n}$

51. $12m^{4x} + 10m^{2x} + 2$

52. $nx^{4n} + 2nx^{2n}y^{2n} + ny^{4n}$

WRITING

53. What is your strategy for factoring a polynomial?

54. For the factorization below, explain why the polynomial is not factored completely.

$$48a^4 c^3 - 3b^4 c^3 = 3c^3(16a^4 - b^4)$$

REVIEW

55. Tell whether the graphs of $x + y = 2$ and $y = x + 5$ are parallel or perpendicular.

56. When expressed as a decimal, is $\frac{7}{8}$ a terminating or a repeating decimal?

57. Evaluate $\begin{vmatrix} 1 & 15 \\ 15 & 0 \end{vmatrix}$.

58. If a triangle has exactly two sides with equal measures, what type of triangle is it?

▶ **5.9**

Solving Equations by Factoring

In this section, you will learn about

Solving quadratic equations ■ Solving higher-degree polynomial equations ■ Problem solving

---

Introduction Equations that involve *first-degree* polynomials, such as $3x + 6 = 0$, are called linear equations. Equations such as $3x^2 + 6x = 0$ that involve *second-degree* polynomials are called quadratic equations. The techniques that we have used to solve linear equations cannot be used to solve quadratic equations, because those techniques cannot isolate the variable on one side of the equation. However, we can solve many quadratic equations using factoring.

Solving Quadratic Equations

An equation such as $3x^2 + 4x - 7 = 0$ or $-5y^2 + 3y + 8 = 0$ is called a **quadratic** or **second-degree** equation.

Quadratic equations

A **quadratic equation** is any equation that can be written in the form

$$ax^2 + bx + c = 0$$

where a, b, and c are real numbers and $a \neq 0$.

Many quadratic equations can be solved by factoring and then by using the **zero-factor property.**

Zero-factor property

If a and b are real numbers, then

If $ab = 0$, then $a = 0$ or $b = 0$.

The zero-factor property states that *if the product of two or more numbers is 0, then at least one of the numbers must be 0.*

To solve the quadratic equation $x^2 + 5x + 6 = 0$, we factor its left-hand side to obtain

$$(x + 3)(x + 2) = 0$$

Since the product of $x + 3$ and $x + 2$ is 0, at least one of the factors must be 0. Thus, we can set each factor equal to 0 and solve each resulting linear equation for x:

$$x + 3 = 0 \quad \text{or} \quad x + 2 = 0$$
$$x = -3 \qquad\qquad x = -2$$

To check these solutions, we substitute -3 and -2 for x in the equation and verify that each number satisfies the equation.

$$x^2 + 5x + 6 = 0 \quad \text{or} \quad x^2 + 5x + 6 = 0$$
$$(-3)^2 + 5(-3) + 6 \overset{?}{=} 0 \qquad (-2)^2 + 5(-2) + 6 \overset{?}{=} 0$$
$$9 - 15 + 6 \overset{?}{=} 0 \qquad\qquad 4 - 10 + 6 \overset{?}{=} 0$$
$$0 = 0 \qquad\qquad\qquad 0 = 0$$

Both -3 and -2 are solutions, because both satisfy the equation.

EXAMPLE 1

Solving quadratic equations by factoring. Solve $3x^2 + 6x = 0$.

Solution

To solve the equation, we factor the left-hand side, set each factor equal to 0, and solve each resulting equation for x.

$$3x^2 + 6x = 0$$
$$3x(x + 2) = 0 \qquad \text{Factor out the common factor of } 3x.$$
$$3x = 0 \quad \text{or} \quad x + 2 = 0 \qquad \text{By the zero-factor property, at least one of the factors must be equal to zero.}$$
$$x = 0 \qquad\qquad x = -2 \qquad \text{Solve each linear equation.}$$

Verify that both solutions check.

SELF CHECK

Solve $4p^2 - 12p = 0$. *Answer:* 0, 3 ∎

WARNING! In Example 1, do not attempt to solve the equation by dividing both sides by $3x$, or you will lose the solution $x = 0$.

EXAMPLE 2

Solving quadratic equations by factoring. Solve $x^2 - 16 = 0$.

Solution

To solve the equation, we factor the difference of two squares on the left-hand side, set each factor equal to 0, and solve each resulting equation.

$$x^2 - 16 = 0$$
$$(x + 4)(x - 4) = 0$$
$$x + 4 = 0 \quad \text{or} \quad x - 4 = 0$$
$$x = -4 \qquad\qquad x = 4$$

Verify that both solutions check.

SELF CHECK

Solve $a^2 - 81 = 0$. *Answer:* 9, -9 ∎

The following steps can be used to solve a quadratic equation by factoring.

> **Solving a quadratic equation by the factoring method**
>
> 1. Write the equation in $ax^2 + bx + c = 0$ form (called *quadratic* form).
> 2. Factor the polynomial.
> 3. Use the zero-factor property to set each factor equal to zero.
> 4. Solve each resulting equation.
> 5. Check each solution in the original equation.

Many equations that do not appear to be quadratic can be put into quadratic form and then solved by factoring.

EXAMPLE 3

Writing the equation in quadratic form. Solve $x = \frac{6}{5} - \frac{6}{5}x^2$.

Solution   We must write the equation in quadratic form. To clear the equation of fractions, we multiply both sides by 5.

$$x = \frac{6}{5} - \frac{6}{5}x^2$$

$$5x = 6 - 6x^2 \qquad \text{Multiply both sides by 5.}$$

To use factoring to solve this quadratic equation, one side of the equation must be 0. Since it is easier to factor a second-degree polynomial if the coefficient of the squared term is positive, we add $6x^2$ to both sides and subtract 6 from both sides to obtain

$$6x^2 + 5x - 6 = 0$$

$$(3x - 2)(2x + 3) = 0 \qquad \text{Factor the trinomial.}$$

$$3x - 2 = 0 \quad \text{or} \quad 2x + 3 = 0 \qquad \text{Set each factor equal to 0 and solve for } x.$$

$$3x = 2 \qquad\qquad 2x = -3$$

$$x = \frac{2}{3} \qquad\qquad x = -\frac{3}{2}$$

Verify that both solutions check.

SELF CHECK   Solve $x = \frac{6}{7}x^2 - \frac{3}{7}$.   *Answer:* $\frac{3}{2}, -\frac{1}{3}$ ∎

WARNING! To solve a quadratic equation by factoring, be sure to set the quadratic polynomial equal to 0 before factoring and applying the zero-factor property. Do not make the following error:

$$6x^2 + 5x = 6$$

$$x(6x + 5) = 6 \qquad\qquad \text{If the product of two numbers is 6, neither number need be}$$
$$\qquad\qquad\qquad\qquad 6. \text{ For example, } 2 \cdot 3 = 6.$$

$$x = 6 \quad \text{or} \quad 6x + 5 = 6$$

$$x = \frac{1}{6}$$

Neither solution checks.

ACCENT ON TECHNOLOGY *Solving Quadratic Equations*

To solve a quadratic equation such as $x^2 + 4x - 5 = 0$ with a graphing calculator, we can use standard window settings of $[-10, 10]$ for x and $[-10, 10]$ for y and graph the quadratic function $y = x^2 + 4x - 5$, as shown in Figure 5-8(a). We can then trace to find the x-coordinates of the x-intercepts of the parabola. See Figures 5-8(b) and 5-8(c). For better results, we can zoom in. Since these are the numbers x that make $y = 0$, they are the solutions of the equation.

FIGURE 5-8

(a)   (b)   (c)

We can also find the x-intercepts of the graph of $y = x^2 + 4x - 5$ by using the ZERO feature found on most graphing calculators. Figures 5-9(a) and 5-9(b) show how this feature locates the x-intercept and displays its coordinates. (Consult your owner's manual for the specific instructions about how to use this feature.) From the displays, we can conclude that $x = -5$ and $x = 1$ are solutions of $x^2 + 4x - 5 = 0$. Verify this by checking them.

FIGURE 5-9

(a)   (b)

Solving Higher-Degree Polynomial Equations

We can solve many polynomial equations with degree greater than 2 by factoring and applying an extension of the zero-factor property.

EXAMPLE 4   **Solving third-degree equations by factoring.** Solve $6x^3 - x^2 = 2x$.

Solution   First, we subtract $2x$ from both sides so that the right-hand side of the equation is 0.

$$6x^3 - x^2 - 2x = 0$$

Then we factor x from the third-degree polynomial on the left-hand side and proceed as follows:

$$6x^3 - x^2 - 2x = 0$$
$$x(6x^2 - x - 2) = 0 \qquad \text{Factor out } x.$$
$$x(3x - 2)(2x + 1) = 0 \qquad \text{Factor } 6x^2 - x - 2.$$
$$x = 0 \quad \text{or} \quad 3x - 2 = 0 \quad \text{or} \quad 2x + 1 = 0 \qquad \text{Set each of the three factors equal to 0.}$$
$$x = \frac{2}{3} \qquad\qquad x = -\frac{1}{2} \quad \text{Solve each equation.}$$

Verify that the three solutions check.

SELF CHECK Solve $5x^3 + 13x^2 = 6x$. _Answer:_ $0, \frac{2}{5}, -3$ ∎

EXAMPLE 5

Solving higher-degree equations by factoring. Solve $x^4 + 4 - 5x^2 = 0$.

Solution First, we write the powers of x in descending order. Then we factor the trinomial on the left-hand side and proceed as follows:

$$x^4 - 5x^2 + 4 = 0$$
$$(x^2 - 1)(x^2 - 4) = 0$$
$$(x + 1)(x - 1)(x + 2)(x - 2) = 0 \quad \text{Factor } x^2 - 1 \text{ and } x^2 - 4.$$
$$x + 1 = 0 \quad \text{or} \quad x - 1 = 0 \quad \text{or} \quad x + 2 = 0 \quad \text{or} \quad x - 2 = 0$$
$$x = -1 \qquad\quad x = 1 \qquad\quad x = -2 \qquad\quad x = 2$$

Verify that each solution checks.

SELF CHECK Solve $a^4 + 36 - 13a^2 = 0$. _Answer:_ $2, -2, 3, -3$ ∎

ACCENT ON TECHNOLOGY _Solving Equations_

To solve the equation $x^4 - 5x^2 + 4 = 0$ with a graphing calculator, we can use window settings of $[-6, 6]$ for x and $[-5, 10]$ for y and graph the polynomial function $y = x^4 - 5x^2 + 4$ as shown in Figure 5-10. We can then read the values of x that make $y = 0$. They are $x = -2, -1, 1,$ and 2. If the x-coordinates of the x-intercepts were not obvious, we could approximate their values by using TRACE and ZOOM or by using the ZERO feature.

FIGURE 5-10

Problem Solving

EXAMPLE 6

Stained glass. The triangular stained glass window shown in Figure 5-11 is to be installed in a chapel. The length of the base of the window is 3 times its height. The area of the window is 96 square feet. Find its base and height.

FIGURE 5-11

ANALYZE THE PROBLEM
We are to find the length of the base and the height of the window. The formula that gives the area of a triangle is $A = \frac{1}{2}bh$, where b is the length of the base and h the height.

FORM AN EQUATION
We can let h be the positive number that represents the height of the window. Then $3h$ represents the length of the base. To form an equation in terms of h, we can substitute $3h$ for b and 96 for A in the formula for the area of a triangle.

$$A = \frac{1}{2}bh$$

$$96 = \frac{1}{2}(3h)h$$

SOLVE THE EQUATION
To solve this equation, we must write it in quadratic form.

$$96 = \frac{1}{2}(3h)h$$

$192 = 3h^2$	$(3h)h = 3h^2$. Multiply both sides by 2.
$64 = h^2$	Divide both sides by 3.
$0 = h^2 - 64$	To obtain 0 on the left-hand side, subtract 64 from both sides.
$0 = (h + 8)(h - 8)$	Factor the difference of two squares.

$$h + 8 = 0 \quad \text{or} \quad h - 8 = 0$$
$$h = -8 \qquad\qquad h = 8$$

STATE THE CONCLUSION
Since the height of a triangle cannot be negative, we must discard the negative solution. Thus, the height of the window is 8 feet, and the length of its base is 3(8), or 24 feet.

CHECK THE RESULT
The area of a triangle with a base of 24 feet and a height of 8 feet is 96 square feet:

$$A = \frac{1}{2}bh = \frac{1}{2}(24)(8) = 12(8) = 96$$

The solution checks. ∎

EXAMPLE 7

Ballistics. If the initial velocity of an object thrown straight up into the air is 176 feet per second, when will the object strike the ground?

ANALYZE THE PROBLEM
The height of an object thrown straight up into the air with an initial velocity of v feet per second is given by the formula

$$h = vt - 16t^2$$

The height h is in feet, and t represents the number of seconds since the object was released. When the object hits the ground, its height will be 0.

FORM AN EQUATION   In the formula, we set h equal to 0 and set v equal to 176.

$$h = vt - 16t^2$$
$$0 = 176t - 16t^2$$

SOLVE THE EQUATION   To solve this equation, we will use the factoring method.

$$0 = 176t - 16t^2$$
$$0 = 16t(11 - t) \qquad \text{Factor out } 16t.$$
$$16t = 0 \quad \text{or} \quad 11 - t = 0 \qquad \text{Set each factor equal to 0.}$$
$$t = 0 \qquad\qquad t = 11$$

STATE THE CONCLUSION   When $t = 0$, the object's height above the ground is 0 feet, because it has not been released. When $t = 11$, the height is again 0 feet, and the object has returned to the ground. The solution is 11 seconds.

CHECK THE RESULT   Verify that $h = 0$ when $t = 11$.   ∎

STUDY SET

Section 5.9

VOCABULARY

In Exercises 1–2, fill in the blanks to make the statements true.

1. A _____ equation is any equation that can be written in the form $ax^2 + bx + c = 0$ $(a \neq 0)$.

2. To _____ an equation means to find all the values of the variable that make the equation true.

CONCEPTS

3. If the product of two numbers is 0, what must be true about at least one of the numbers?

4. Use a check to determine whether -5 and 4 are solutions of $a^2 - 9a + 20 = 0$.

5. Tell whether each equation is a quadratic equation.
   a. $w^2 + 7w + 12 = 0$
   b. $6t + 11 = 0$
   c. $x(x + 3) = -2$
   d. $k^3 - 4k^2 + k - 15 = 0$

6. What is **wrong** with the work shown here?

Solve for x: $x^2 + 2x = 8$.

$x(x + 2) = 8$

$x = 8 \quad$ or $\quad x + 2 = 8$

$x = 8 \quad$ or $\quad x = 6$

7. Use the graph in Illustration 1 to solve the quadratic equation $x^2 - 2x - 3 = 0$.

ILLUSTRATION 1

8. Use the graph in Illustration 2 to solve the polynomial equation $x^3 - 4x^2 + 4x = 0$.

ILLUSTRATION 2

NOTATION

In Exercises 9–10, complete each solution.

9. Solve $y^2 - 3y - 54 = 0$

$$(y - 9)(\rule{1.5cm}{0.4pt}) = 0$$

$$\rule{1.5cm}{0.4pt} = 0 \quad \text{or} \quad y + 6 = 0$$

$$y = 9 \qquad\qquad y = \rule{1cm}{0.4pt}$$

10. Solve $x^2 - x = 12$.

$$x^2 - x - 12 = \boxed{}$$
$$(x - 4)\left(\boxed{}\right) = 0$$
$$x - 4 = 0 \quad \text{or} \quad \boxed{} = 0$$
$$x = 4 \qquad\qquad x = \boxed{}$$

PRACTICE

In Exercises 11–28, solve each equation.

11. $4x^2 + 8x = 0$ **12.** $x^2 - 9 = 0$

13. $y^2 - 16 = 0$ **14.** $y^2 - 25 = 0$

15. $x^2 + x = 0$ **16.** $x^2 - 3x = 0$

17. $5y^2 - 25y = 0$ **18.** $y^2 - 36 = 0$

19. $z^2 + 8z + 15 = 0$ **20.** $w^2 + 7w + 12 = 0$

21. $x^2 + 6x + 8 = 0$ **22.** $x^2 + 9x + 20 = 0$

23. $3m^2 + 10m + 3 = 0$ **24.** $2r^2 + 5r + 3 = 0$

25. $2y^2 - 5y + 2 = 0$ **26.** $2x^2 - 3x + 1 = 0$

27. $2x^2 - x - 1 = 0$ **28.** $2x^2 - 3x - 5 = 0$

In Exercises 29–40, write each equation in quadratic form and then solve it by factoring.

29. $x(x - 6) + 9 = 0$ **30.** $x^2 + 8(x + 2) = 0$

31. $8a^2 = 3 - 10a$ **32.** $5z^2 = 6 - 13z$

33. $b(6b - 7) = 10$ **34.** $2y(4y + 3) = 9$

35. $\dfrac{3a^2}{2} = \dfrac{1}{2} - a$ **36.** $x^2 = \dfrac{1}{2}(x + 1)$

37. $x^2 + 1 = \dfrac{5}{2}x$ **38.** $\dfrac{3}{5}(x^2 - 4) = -\dfrac{9}{5}x$

39. $x\left(3x + \dfrac{22}{5}\right) = 1$ **40.** $x\left(\dfrac{x}{11} - \dfrac{1}{7}\right) = \dfrac{6}{77}$

In Exercises 41–52, solve each equation by using factoring.

41. $x^3 + x^2 = 0$ **42.** $2x^4 + 8x^3 = 0$

43. $y^3 - 49y = 0$ **44.** $2z^3 - 200z = 0$

45. $x^3 - 4x^2 - 21x = 0$ **46.** $x^3 + 8x^2 - 9x = 0$

47. $z^4 - 13z^2 + 36 = 0$ **48.** $y^4 - 10y^2 + 9 = 0$

49. $3a(a^2 + 5a) = -18a$ **50.** $7t^3 = 2t\left(t + \dfrac{5}{2}\right)$

51. $\dfrac{x^2(6x + 37)}{35} = x$ **52.** $x^2 = -\dfrac{4x^3(3x + 5)}{3}$

53. INTEGER PROBLEM The product of two consecutive even integers is 288. Find the integers. (*Hint:* Let $x =$ the first even integer. Then represent the second even integer in terms of x.)

54. INTEGER PROBLEM The product of two consecutive odd integers is 143. Find the integers. (*Hint:* Let $x =$ the first odd integer. Then represent the second odd integer in terms of x.)

In Exercises 55–58, use a graphing calculator to find the solutions of each equation, if one exists. If an answer is not exact, give the answer to the nearest hundredth.

55. $2x^2 - 7x + 4 = 0$

56. $x^2 - 4x + 7 = 0$

57. $-3x^3 - 2x^2 + 5 = 0$

58. $-2x^3 - 3x - 5 = 0$

APPLICATIONS

59. COOKING The electric griddle shown in Illustration 3 has a cooking surface of 160 square inches. Find the length and the width of the griddle.

ILLUSTRATION 3

60. STRUCTURAL ENGINEERING The formula for the area of a trapezoid is $A = \dfrac{h(B + b)}{2}$. The area of the trapezoidal truss in Illustration 4 is 44 square feet. Find the height of the truss if the shorter base is the same as the height.

ILLUSTRATION 4

61. SWIMMING POOL DESIGN Building codes require that the rectangular swimming pool in Illustration 5 be surrounded by a uniform-width walkway of at least 516 square feet. The length of the pool is 10 feet less than twice the width. How wide should the border be?

ILLUSTRATION 5

62. FINE ARTS An artist intends to paint a 60-square-foot mural on the large wall shown in Illustration 6. Find the dimensions of the mural if the artist leaves a border of uniform width around it.

ILLUSTRATION 6

63. ARCHITECTURE The rectangular room shown in Illustration 7 is twice as long as it is wide. It is divided into two rectangular parts by a partition, positioned as shown. If the larger part of the room contains 560 square feet, find the dimensions of the entire room.

ILLUSTRATION 7

64. WINTER RECREATION The length of the rectangular ice-skating rink in Illustration 8 is 20 meters greater than twice its width. Find the width.

ILLUSTRATION 8

65. BALLISTICS The muzzle velocity of a cannon is 480 feet per second. If a cannonball is fired vertically, at what times will it be at a height of 3,344 feet?

66. SLINGSHOT A slingshot can provide an initial velocity of 128 feet per second. At what times will a stone, shot vertically upward, be 192 feet above the ground?

67. BUNGEE JUMPING See Illustration 9. The formula $h = -16t^2 + 212$ gives the distance a bungee jumper is from the ground for the free-fall portion of the jump, t seconds after leaping off a bridge. We can find the number of seconds it takes the jumper to reach the point in the fall where the 64-foot bungee cord starts to stretch by substituting 148 for h and solving for t. Find t.

ILLUSTRATION 9

68. MAJOR LEAGUE BASEBALL In 1998, pitcher Roger Clemens, of the Toronto Blue Jays, threw a fastball that was clocked at 97 miles per hour. This is a velocity of approximately 144 feet per second. If he could throw the baseball vertically into the air with this velocity, how long would it take for the ball to fall to the ground?

69. FORENSIC MEDICINE The kinetic energy E of a moving object is given by $E = \frac{1}{2}mv^2$, where m is the mass of the object (in kilograms) and v is the object's velocity (in meters per second). Kinetic energy is measured in joules. By measuring the damage done to a victim who has been struck by a 3-kilogram club, a police pathologist finds that the energy at impact was 54 joules. Find the velocity of the club at impact.

70. TRAFFIC ACCIDENTS Investigators at a traffic accident used the function $d(v) = 0.04v^2 + 0.8v$, where v is the velocity of the car (in mph) and $d(v)$ is the stopping distance of the car (in feet), to reconstruct the events leading up to a collision. From physical evidence, it was concluded that it took one car 32 feet to stop. At what velocity was the car traveling prior to the accident?

71. BREAK-EVEN POINT The cost for a guitar maker to hand-craft x guitars is given by the function $C(x) = \frac{1}{8}x^2 - x + 6$. The revenue taken in with the sale of x guitars is given by the function $R(x) = \frac{1}{4}x^2$. Find the number of guitars that must be sold so that the cost equals the revenue.

72. REVENUE Over the years, the manager of a crafts store has found that the number of scented candles x she can sell in a month depends on the price p according to the formula $x = 200 - 10p$. At what price should she sell the candles if she needs to bring in $750 in revenue a month from their sale? (*Hint:* Revenue = price · number sold = px.)

WRITING

73. Explain the zero-factor property.

74. In the work shown below, explain why the student has not solved for x.

Solve $x^2 + x - 6 = 0$.

$$x^2 + x = 6$$
$$\boxed{x = 6 - x^2}$$

75. Explain what is **wrong** with the following solution.

Solve $x^2 - x = 0$.

$$\frac{x^2}{x} - \frac{x}{x} = \frac{0}{x}$$
$$x - 1 = 0$$
$$x = 1$$

76. Explain what is **wrong** with the following solution.

Solve $x^2 - x = 6$.

$$x(x - 1) = 6$$
$$\boxed{x = 6} \quad \text{or} \quad x - 1 = 6$$
$$\boxed{x = 7}$$

77. The graphs of the two polynomial functions $[f(x) = 2x^3 - 8x$ and $f(x) = 2x(x + 2)(x - 2)]$ in Illustration 10 appear to be the same. After examining the defining equations, explain why we know that they indeed are identical graphs.

ILLUSTRATION 10

78. Explain why the x-coordinate of the x-intercept of the graph of $y = 8x^2 + 10x - 3$ (indicated as a *zero* in Illustration 11) is a solution of $8x^2 + 10x - 3$.

ILLUSTRATION 11

REVIEW

79. ALUMINUM FOIL Find the number of square feet of aluminum foil on a roll if it has dimensions of $8\frac{1}{3}$ yards × 12 inches.

80. HOCKEY A hockey puck is a vulcanized rubber disk 2.5 cm (1 in.) thick and 7.6 cm (3 in.) in diameter. Find the volume of a puck in cubic centimeters and cubic inches. Round to the nearest tenth.

Polynomials

A **polynomial** is an algebraic term or the sum of two or more algebraic terms whose variables have whole-number exponents. No variable appears in a denominator.

Operations with Polynomials

In arithmetic, we learned how to add, subtract, multiply, divide, and find powers of numbers. In algebra, we need to be able to perform these operations on polynomials.

In Exercises 1–8, do each operation.

1. $(-2x^2 - 5x - 7) + (-3x^2 + 7x + 1)$

2. $(6s^3 + 3s - 2) - (2s^3 + 3s^2 + 5)$

3. $(3m - 4)(m + 3)$

4. $3r^2st(r^2 - 2s + 3t^2)$

5. $(a - 2d)^2$

6. $(x - 3y)(x^2 + 3xy + 9y^2)$

7. $(3b + 1)(2b^2 + 3b + 2)$

8. $(2y + 3)(y - 1) - (y + 2)(3y - 1)$

Polynomial Functions

Polynomial functions can be used to model many real-world situations.

9. WINDOW WASHERS A man on a scaffold, washing the outside windows of a skyscraper, drops a squeegee. As it falls, its distance in feet from the ground $d(t)$, t seconds after being dropped, is given by the polynomial function $d(t) = -16t^2 + 576$. Find $d(6)$ and explain the result.

10. Write a polynomial function $V(x)$ that gives the volume of the ice chest shown in Illustration 1. Then find $V(3)$.

ILLUSTRATION 1

$2x + 2$

$4x - 1$

$2x$

Solving Equations by Factoring

Quadratic equations can be written in the form $ax^2 + bx + c = 0$. Many quadratic equations can be solved by factoring the polynomial $ax^2 + bx + c$ and applying the zero-factor property. Some higher-degree equations can also be solved by using an extension of this procedure.

In Exercises 11–16, solve each equation by factoring.

11. $x^2 - 81 = 0$

12. $5x^2 - 25x = 0$

13. $z^2 + 8z + 15 = 0$

14. $2r^2 + 5r + 3 = 0$

15. $\dfrac{3t^2}{2} + t = \dfrac{1}{2}$

16. $m^3 = 9m - 8m^2$

Accent on Teamwork

Section 5.1

Exponents Have a student in your group write each of the eight rules for exponents listed on page 314 on separate 3×5 cards. On another set of cards, write an explanation of each rule using words. On a third set of cards, write an example of the use of each rule for exponents. Shuffle the cards and work together to match the symbolic description, the word description, and the example for each of the eight rules for exponents.

Section 5.2

Scientific notation Go to the library and find five examples of extremely large and five examples of extremely small numbers. Encyclopedias, government statistics books, and science books are good places to look. Write each number in scientific notation on a separate piece of paper. Include a brief explanation of what the number represents. Present the 10 examples in numerical order, beginning with the smallest number first.

Section 5.3

Adding polynomials According to an old adage, "You can't add apples and oranges." Give some examples of how this concept applies when adding two polynomials.

Section 5.4

Multiplying binomials
Use colored construction paper to make a model like that shown in Illustration 1. Use the model as part of a presentation to demonstrate why $(x + y)(x + y) = x^2 + 2xy + y^2$.

ILLUSTRATION 1

Section 5.5

Solving formulas Examine the student's work shown below. Write some comments to the student about what it means to *solve for the indicated variable*. Then write out the correct solution.

Solve for r_1: $r_1 r_2 = rr_2 + rr_1$.

$$\frac{r_1 r_2}{r_2} = \frac{rr_2 + rr_1}{r_2}$$

$$r_1 = \frac{rr_2 + rr_1}{r_2}$$

Section 5.6

Factoring formulas Use multiplication to prove that in each statement given below, the expression on the left is not equal to the expression on the right.

$$x^2 - y^2 \neq (x - y)^2$$
$$x^3 - y^3 \neq (x - y)^3$$
$$x^3 + y^3 \neq (x + y)^3$$

For each statement, give the correct factoring formula.

Section 5.7

Authoring a textbook Assign each of the 12 examples in Section 5.7 to members of your group. Have them write a new but similar problem for each example and then write a solution, complete with an explanation and author notes using the same format as in this book. They should also create an accompanying Self Check problem and include the answer. Compile all 12 examples into a booklet. Make copies of your booklet for the other members of the class.

Section 5.8

Factoring flowchart Create a factoring flowchart that would lead a student through the correct steps to identify the type(s) of factoring necessary to factor any given polynomial. The factoring strategy shown on page 371 may be helpful as guideline. To get you started, the top of the flowchart should begin as shown in Illustration 2.

ILLUSTRATION 2

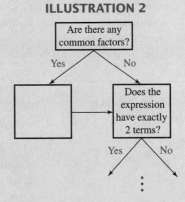

Section 5.9

Quadratic equations Suppose you know that the two solutions of a quadratic equation are 4 and -3. Consider the method used to solve quadratic equations studied in Section 5.9. Work backward to find the original equation.

Section 5.1

Exponents

CONCEPTS

If n is a natural number,

$$\overbrace{x^n = x \cdot x \cdot x \cdot \cdots \cdot x}^{n \text{ factors of } x}$$

where x is the *base* and n the *exponent*.

Rules for exponents: If there are no divisions by 0, then for all integers m and n,

$$x^m x^n = x^{m+n} \quad (x^m)^n = x^{mn}$$

$$(xy)^n = x^n y^n \quad \left(\frac{x}{y}\right)^n = \frac{x^n}{y^n}$$

$$x^0 = 1 \qquad x^{-n} = \frac{1}{x^n}$$

$$\frac{x^m}{x^n} = x^{m-n}$$

$$\left(\frac{x}{y}\right)^{-n} = \left(\frac{y}{x}\right)^n$$

REVIEW EXERCISES

1. Evaluate each expression.
   a. 3^6
   b. -2^5
   c. $(-4)^3$
   d. 15^1

2. Simplify each expression and write all answers without negative exponents.
   a. $x^4 \cdot x^2$
   b. $a^3 b^5 a^2 b$
   c. $(m^6)^3$
   d. $(-t^2)^2(t^3)^3$
   e. $(3x^2 y^3)^2$
   f. $\left(\frac{x^4}{b}\right)^4$
   g. $-3x^0$
   h. $(x^2)^{-5}$
   i. -5^{-4}
   j. $\frac{70}{x^{-4}}$
   k. $(3x^{-3})^{-2}$
   l. $2x^{-4} x^3$
   m. $-\left(\frac{c^{-3}}{c^{-5}}\right)^5$
   n. $\left(\frac{4}{5}\right)^{-2}$
   o. $\frac{y^{-3}}{y^4 y}$
   p. $\left(\frac{-2a^4 b}{a^{-3} b^2}\right)^{-3}$

Section 5.2

Scientific Notation

Scientific notation is a compact way of writing large and small numbers. Numbers are written in the form

$$N \times 10^n$$

where $1 \le |N| < 10$ and n is an integer.

3. Write each number in scientific notation.
   a. 19,300,000,000
   b. 0.00000002735

4. Write each number in standard notation.
   a. 7.277×10^7
   b. 8.3×10^{-9}

In Exercises 5–7, write each number in scientific notation and do the operations. Give answers in scientific notation.

5. THE SPEED OF LIGHT Light travels at about 300,000 kilometers per second. If the average distance from the sun to the planet Mars is approximately 228,000,000 kilometers, how long does it take light from the sun to reach Mars?

6. PROTONS If the mass of 1 proton is 0.000000000000000000000000167248 gram, find the mass of 1 million protons.

7. Evaluate $\dfrac{(616,000,000)(0.000009)}{0.00066}$.

Section 5.3

A *polynomial* is an algebraic term or the sum of two or more algebraic terms whose variables have whole-number exponents. No variable appears in a denominator.

The *degree of a polynomial* is the degree of the term with the highest degree contained within the polynomial.

To *evaluate a polynomial* function, we replace the variable in the defining equation with its value, called the *input*. Then we simplify to find the *output*.

Polynomials and Polynomial Functions

8. Tell whether each expression is a polynomial.

a. $\dfrac{2x^2}{x+1}$

b. $-5x^3 + x^2 - 5x - 4$

c. $2.8y^{15} - y^{10} + y^8 - \dfrac{3}{2}y^6$

d. $x^{-3} + x^{-2} - x^{-1} - 1$

9. Classify each polynomial as a monomial, binomial, trinomial, or none of these. Then determine the degree of the polynomial.

a. $x^2 - 8$

b. $-15a^3b$

c. $x^4 + x^3 - x^2 + x - 4$

d. $9x^2y + 13x^3y^2 + 8x^4y^4$

10. SQUIRT GUN See Illustration 1. The volume, in cubic inches, of the reservoir on top of the squirt gun is given by the polynomial function $V(r) = 4.19r^3 + 25.12r^2$, where r is the radius in inches. Find $V(2)$ to the nearest cubic inch.

ILLUSTRATION 1

11. Graph each polynomial function.

a. $f(x) = x^2 - 2x$

b. $f(x) = x^3 - 3x^2 + 4$

To add polynomials, remove parentheses and *combine like terms* (terms having the same variables with the same exponents).

To subtract polynomials, add the first polynomial and the negative (opposite) of the second polynomial.

12. Do each operation.

a. $(3x^2 + 4x + 9) + (2x^2 - 2x + 7)$

b. $(2x^2y^3 - 5x^2y + 9y) + (x^2y^3 - 3x^2y - y)$

c. $(4x^3 + 4x^2 + 7) - (-2x^3 - x - 2)$

d. $\quad \begin{array}{r} -10k^4 - 4k^3 + 5k^2 - k + 1 \\ -\ \underline{-16k^4 + 2k^3 - 4k^2 - k + 3} \end{array}$

Section 5.4

To *multiply monomials,* multiply their numerical factors and multiply their variable factors.

To *multiply a polynomial by a monomial,* multiply each term of the polynomial by the monomial.

Multiplying Polynomials

13. Find each product.

a. $(8a^2)\left(-\dfrac{1}{2}a\right)$

b. $(-3xy^2z)(-2xz^3)(xz)$

14. Find each product.

a. $2xy^2(x^3y - 4xy^5)$

b. $-a^2b(-a^2 - 2ab + b^2)$

The *FOIL method* is used to multiply two binomials.

To *multiply polynomials,* multiply each term of one polynomial by each term of the other polynomial.

15. Find each product.
  a. $(8x - 5)(2x + 3)$ **b.** $(3x^2 + 2)(2x - 4)$

  c. $(5a - 6)^2$ **d.** $(7c^2 - d)(7c^2 + d)$
  e. $(5x^2 - 4x)(3x^2 - 2x + 10)$ **f.** $(r + s)(r - s)(r - 3s)$

  g. $(a^{-2} + b)(a - b^{-2})$ **h.** $(p^n + 2q^{-n})(p^n - q^n)$

16. SHAVING A razor blade is made from a thin piece of platinum steel. Before its center is punched out, the blade has the shape shown in Illustration 2. Write a polynomial that gives the area.

ILLUSTRATION 2

Section 5.5

The Greatest Common Factor and Factoring by Grouping

To *prime factor* a natural number means to write it as a product of prime numbers.

17. Find the prime factorization of 350.

The largest natural number that divides each number in a set of numbers is called their *greatest common factor (GCF).*

18. Find the GCF of each set of monomials.
  a. 42, 36, 54   **b.** $6x^2y^5, 15xy^3$

The process of finding the individual factors of a known product is called *factoring.*

19. Factor each polynomial, if possible.
  a. $4x + 8$ **b.** $3x^3 - 6x^2 + 9x$

Always factor out *common factors* as the first step in a factoring problem. Use the distributive property to do this.

  c. $5x^2y^3 - 11mn^2$ **d.** $7a^4b^2 + 49a^3b$
  e. $5x^2(x + y) - 15x^3(x + y)$ **f.** $27x^3y^3z^3 + 81x^4y^5z^2 - 90x^2y^3z^7$

A polynomial that cannot be factored is a *prime polynomial.*

20. Factor out the negative of the greatest common factor.
  a. $-7b + 14$ **b.** $-49a^3b^2(a - b)^4 + 63a^2b^4(a - b)^3$

21. Factor a^n from $a^{2n} - a^nb^n$.

22. Factor x^{-1} from $x^4 + x^2 - x^{-1}$.

If an expression has four or more terms, try to factor the expression by *grouping.*

23. Factor each polynomial by grouping.
  a. $xy + 2y + 4x + 8$ **b.** $r^2y - ar - ry + a$

24. Solve $m_1m_2 = mm_2 + mm_1$ for m_1.

Section 5.6

The Difference of Two Squares; the Sum and Difference of Two Cubes

Factoring the *difference of two squares:*

$$x^2 - y^2 = (x + y)(x - y)$$

25. Factor each expression, if possible.

a. $z^2 - 16$ **b.** $y^2 - 121$

c. $x^2y^4 - 64z^6$ **d.** $a^2b^2 + c^2$

e. $c^2 - (a + b)^2$ **f.** $3x^6 - 300x^2$

g. $4 - x^{2n}$ **h.** $x^{4m} - y^{2m}$

Factoring the *sum of two cubes:*

$$x^3 + y^3 = (x + y)(x^2 - xy + y^2)$$

the *difference of two cubes:*

$$x^3 - y^3 = (x - y)(x^2 + xy + y^2)$$

26. Factor each polynomial, if possible.

a. $t^3 + 64$ **b.** $2x^3y - 54yz^3$

c. $x^{3m} - y^{3m}$ **d.** $27 + a^{6n}$

Section 5.7

Factoring Trinomials

Perfect square trinomials are the squares of binomials:

$$x^2 + 2xy + y^2 = (x + y)^2$$
$$x^2 - 2xy + y^2 = (x - y)^2$$

27. Factor each trinomial, if possible.

a. $x^2 + 10x + 25$ **b.** $a^2 - 14a + 49$

Test for factorability: A trinomial of the form $ax^2 + bx + c$ will factor with integer coefficients if $b^2 - 4ac$ is a perfect square.

To factor trinomials with a *lead coefficient of 1,* list the factorizations of the third term.

To factor trinomials with a *lead coefficient other than 1,* use the procedure for factoring a general trinomial.

28. Factor each trinomial, if possible.

a. $y^2 + 21y + 20$ **b.** $z^2 - 11z + 30$

c. $-x^2 - 3x + 28$ **d.** $a^2 - 24b^2 - 5ab$

e. $4a^2 - 5a + 1$ **f.** $3b^2 + 2b + 1$

g. $y^3 + y^2 - 2y$ **h.** $15x^2 - 57xy - 12y^2$

i. $2a^4 + 4a^3 - 6a^2$ **j.** $v^4 - 13v^2 + 42$

k. $y^{2n} - y^n - 6$

29. Use a substitution to factor $(s + t)^2 - 2(s + t) + 1$.

30. Use grouping to factor $k^2 + 2k + 1 - 9m^2$.

Section 5.8

Use these steps to factor a random expression:
1. Factor out all common factors.
2. If an expression has two terms, check to see if it is
   a. The difference of two squares:
   $(x^2 - y^2) = (x + y)(x - y)$
   b. The sum of two cubes:
   $(x^3 + y^3)$
   $= (x + y)(x^2 - xy + y^2)$
   c. The difference of two cubes:
   $(x^3 - y^3)$
   $= (x - y)(x^2 + xy + y^2)$
3. If an expression has three terms, attempt to factor it as a *general trinomial*.
4. If an expression has four or more terms, try factoring by *grouping*.
5. Continue until each individual factor is prime.
6. Check the results by multiplying.

Summary of Factoring Techniques

31. Factor each expression, if possible.

   a. $x^3 + 5x^2 - 6x$
   b. $3x^2y - 12xy - 63y$

   c. $z^2 - 4 + zx - 2x$
   d. $x^2 + 2x + 1 - p^2$

   e. $x^2 + 4x + 4 - 4p^4$
   f. $y^2 + 3y + 2 + 2x + xy$

   g. $4a^3b^3 + 256$
   h. $36z^4 - 36$

   i. $6x^{2n} + 7x^n - 3$
   j. $-3x^{4n} + 48$

32. SPANISH ROOF TILE The amount of clay used to make the roof tile shown in Illustration 3 is given by

$$V = \frac{\pi}{2} r_1{}^2 h - \frac{\pi}{2} r_2{}^2 h$$

Factor the right-hand side of the formula completely.

ILLUSTRATION 3

Section 5.9

To solve a *quadratic equation* by factoring:
1. Write the equation in the form $ax^2 + bx + c = 0$.
2. Factor the polynomial.
3. Use the zero-factor property to set each factor equal to zero.
4. Solve each resulting equation.
5. Check each solution.

The *zero-factor property:*
If a and b are real numbers, then

   If $ab = 0$, then $a = 0$ or $b = 0$.

Solving Equations by Factoring

33. Solve each equation by factoring.

   a. $4x^2 - 3x = 0$
   b. $x^2 - 36 = 0$
   c. $12x^2 = 5 - 4x$
   d. $7y^2 - 37y + 10 = 0$

   e. $t^2(15t - 2) = 8t$
   f. $u^3 = \dfrac{u}{3}(19u + 14)$

34. PYRAMIDS The volume of the pyramid in Illustration 4 is given by the formula $V = \frac{Bh}{3}$, where B is the area of its base and h is its height. The volume of the pyramid is 1,020 cubic meters. Find the dimensions of its rectangular base if one edge of the base is 3 meters longer than the other, and the height of the pyramid is 9 meters.

ILLUSTRATION 4

35. Use the graph of $y = 2x^2 - x - 1$, shown in Illustration 5, to estimate the solutions of $2x^2 - x - 1$. Check your answers.

ILLUSTRATION 5

Test

In Problems 1–4, simplify each expression. Write all answers without using negative exponents. Assume that no denominators are zero.

1. $x^3 x^5 x$

2. $(-2x^2 y^3)^3$

3. $m^3 (m^{-4})^2$

4. $\left(\dfrac{3m^2 n^3}{m^4 n^{-2}} \right)^{-2}$

5. Write 4,706,000,000,000 in scientific notation.

6. Write 2.45×10^{-4} in standard notation.

7. Evaluate $\dfrac{3.19 \times 10^{15}}{2.2 \times 10^{-4}}$. Express the answer in scientific notation.

8. SPEED OF LIGHT Light travels 1.86×10^5 miles per second. How far does it travel in a minute? Express the answer in scientific notation.

In Problems 9–10, find the degree of each polynomial.

9. $3x^3 - 4x^5 - 3x^2 - 5$

10. $3x^5 y^3 - x^8 y^2 + 2x^9 y^4 - 3x^2 y^5 + 4$

11. BOATING The height (in feet) of a warning flare from the surface of the ocean t seconds after being shot into the air is given by the polynomial function $h(t) = -16t^2 + 80t + 10$. What is the height of the flare 2.5 seconds after being fired?

12. STRUCTURAL ENGINEERING Write a polynomial that gives the cross-sectional area of the wooden beam shown in Illustration 1.

ILLUSTRATION 1

$3x - 2$

$x + 2$

13. Graph the function $f(x) = x^2 + 2x$.

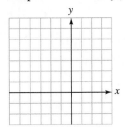

14. a. Graph the function $f(x) = x^3 + 4x^2 + 4x$.

 b. From the graph, determine the solutions of $x^3 + 4x^2 + 4x = 0$.

In Problems 15–22, do the operations.

15. $(2y^2 + 4y + 3) + (3y^2 - 3y - 4)$

16. $(-3u^2 + 2u - 7) - (u^2 + 7)$

17. $(3x^3 y^2 z)(-2xy^{-1} z^3)$

18. $-5a^2 b(3ab^3 - 2ab^4)$

19. $(z + 4)(z - 4)$

20. $(3x^n - 2)(4x^n - 3)$

21. $(4t - 9)^2$

22. $2s(s - t)(s + t)$

In Problems 23–34, factor each expression, if possible.

23. $3x + 6x^2$

24. $12a^3 b^2 c - 3a^2 b^2 c^2 + 6abc^3$

25. $(u - v)r + (u - v)s$

26. $ax - xy + ay - y^2$

27. $x^2 - 49$

28. $4y^4 - 64$

29. $b^2 + 25$

30. $b^3 + 125$

31. $3u^3 - 24$

32. $a^2 - 5a - 6$

33. $6b^2 + bc - 2c^2$

34. $x^2 + 6x + 9 - y^2$

35. Factor $a^n b$ from $a^{2n}b^2 - a^{3n}b$.

36. Factor $b^{2n} + b^n - 2$.

37. Solve for x: $x^2 - 5x - 6 = 0$

38. Solve for x: $2x(4x + 3) = 9$

39. Solve for x: $x^2 + 4x = 0$

40. Solve for v: $v_1 v_3 - v_3 v = v_1 v$

41. PREFORMED CONCRETE The slab of concrete in Illustration 2 is twice as long as it is wide. The area in which it is placed includes a 1-foot-wide border of 70 square feet. Find the dimensions of the slab.

ILLUSTRATION 2

42. Explain how a factorization of a polynomial can be verified using a check. Give an example.

6

Rational Expressions

CAMPUS CONNECTION

The Computer Science Department

In a computer science class, students learn how to access the Internet. Using online services such as CompuServe and America Online, they can browse the World Wide Web to explore vast information resources, including products and services, communication, support forums, and entertainment. In this chapter, we will use a *rational expression* to determine the average cost per hour to use an online service. From a graph, we will see that the average cost decreases as the usage increases. Using rational expressions, we will describe many other interesting applications from engineering, photography, drafting, electronics, cooking, and transportation.

> *In this chapter, we extend the concept of fractions to include quotients of two polynomials, which are called rational expressions.*

▶6.1 Rational Functions and Simplifying Rational Expressions

In this section, you will learn about

> Rational expressions ■ Rational functions ■ Graphing rational functions ■ Finding the domain of a rational function ■ Simplifying rational expressions ■ Simplifying rational expressions by factoring out −1

Introduction We have seen that linear and polynomial functions can be used to model many real-world situations. In this section, we introduce another family of functions known as *rational functions*. Rational functions get their name from the fact that their defining equation contains a *ratio* (fraction) of two polynomials.

Rational Expressions

Rational expressions are fractions that indicate the quotient (or ratio) of two polynomials, such as

$$\frac{-8y^3z^5}{6y^4z^3}, \qquad \frac{3x}{x-7}, \qquad \frac{5m+n}{8m+16}, \qquad \text{and} \qquad \frac{6a^2-13a+6}{3a^2+a-2}$$

 WARNING! Since division by 0 is undefined, the value of a polynomial in the denominator of a rational expression cannot be 0. For example, x cannot be 7 in the rational expression $\frac{3x}{x-7}$, because the value of the denominator would be 0. In the rational expression $\frac{5m+n}{8m+16}$, m cannot be -2, because the value of the denominator would be 0.

Rational Functions

Rational expressions often define functions. For example, if the cost of subscribing to an online information network is $6 per month plus $1.50 per hour of access time, the average (mean) hourly cost of the service is the total monthly cost, divided by the number of hours of access time used that month:

$$\frac{C}{n} = \frac{1.50n+6}{n} \qquad \text{\small C is the total monthly cost, and n is the number of hours the service is used that month.}$$

The right-hand side of this equation is a rational expression: the quotient of the binomial $1.50n + 6$ and the monomial n.

The rational function that gives the average hourly cost of using the online information network for n hours per month can be written

$$f(n) = \frac{1.50n+6}{n}$$

We are assuming that at least one access call will be made each month, so the function is defined for $n > 0$.

Rational functions

A **rational function** is a function whose equation is defined by a rational expression in one variable, where the value of the polynomial in the denominator is never zero.

EXAMPLE 1

Evaluating rational expressions. Use the function $f(n) = \dfrac{1.50n + 6}{n}$ to find the average hourly cost when the network described earlier is used for **a.** 1 hour and **b.** 9 hours.

Solution **a.** To find the average hourly cost for 1 hour of access time, we find $f(1)$:

$$f(1) = \frac{1.50(1) + 6}{1} = 7.5 \qquad \text{Input 1 for } n \text{ and simplify.}$$

The average hourly cost for 1 hour of access time is \$7.50.

b. To find the average hourly cost for 9 hours of access time, we find $f(9)$:

$$f(9) = \frac{1.50(9) + 6}{9} = 2.166666666 \ldots \quad \text{Input 9 for } n \text{ and simplify.}$$

The average hourly cost for 9 hours of access time is approximately \$2.17.

SELF CHECK Find the average hourly cost when the network is used for **a.** 3 hours and **b.** 100 hours. *Answers:* **a.** \$3.50, **b.** \$1.56 ∎

Graphing Rational Functions

To graph the rational function $f(n) = \frac{1.50n + 6}{n}$, we substitute values for n (the inputs) in the equation, compute the corresponding values of $f(n)$ (the outputs), and express the results as ordered pairs. From the evaluations in Example 1 and its Self Check, we know four ordered pairs that satisfy the equation: $(1, 7.50)$, $(3, 3.50)$, $(9, 2.17)$, and $(100, 1.56)$. Those pairs and others are listed in the table in Figure 6-1. We then plot the points and draw a smooth curve through them to get the graph.

FIGURE 6-1

$$f(n) = \frac{1.50n + 6}{n}$$

n	$f(n)$	$(n, f(n))$
1	7.50	(1, 7.50)
2	4.50	(2, 4.50)
3	3.50	(3, 3.50)
4	3.00	(4, 3.00)
5	2.70	(5, 2.70)
6	2.50	(6, 2.50)
7	2.36	(7, 2.36)
8	2.25	(8, 2.25)
9	2.17	(9, 2.17)
10	2.10	(10, 2.10)
100	1.56	(100, 1.56)

As the access time increases, the graph approaches the line $y = 1.5$, which indicates that the average hourly cost approaches \$1.50 as the hours of use increase.

From the graph, we can see that the average hourly cost decreases as the number of hours of access time increases. Since the cost of each extra hour of access time is

$1.50, the average hourly cost can approach $1.50 but never drop below it. Thus, the graph of the function approaches the line $y = 1.5$ as n increases. When a graph approaches a line, we call the line an **asymptote.** The line $y = 1.5$ is a **horizontal asymptote** of the graph.

As n gets smaller and approaches 0, the graph approaches the y-axis. The y-axis is a **vertical asymptote** of the graph.

Finding the Domain of a Rational Function

Since division by 0 is undefined, any values that make the denominator 0 in a rational function must be excluded from the domain of the function.

EXAMPLE 2

Finding the domain of a rational function. Find the domain of
$f(x) = \dfrac{3x + 2}{x^2 + x - 6}.$

Solution
From the set of real numbers, we must exclude any values of x that make the denominator 0. To find these values, we set $x^2 + x - 6$ equal to 0 and solve for x.

$$x^2 + x - 6 = 0$$

$$(x + 3)(x - 2) = 0 \qquad \text{Factor the trinomial.}$$

$$x + 3 = 0 \quad \text{or} \quad x - 2 = 0 \qquad \text{Set each factor equal to 0.}$$

$$x = -3 \qquad\qquad x = 2 \qquad \text{Solve each linear equation.}$$

Thus, the domain of the function is the set of all real numbers except -3 and 2. In interval notation, the domain is $(-\infty, -3) \cup (-3, 2) \cup (2, \infty)$.

SELF CHECK
Find the domain of $f(x) = \dfrac{x^2 + 1}{x - 2}.$

Answer: $(-\infty, 2) \cup (2, \infty)$ ∎

ACCENT ON TECHNOLOGY *Finding the Domain and Range of a Rational Function*

We can find the domain and range of the function in Example 2 by looking at its graph. If we use window settings of $[-10, 10]$ for x and $[-10, 10]$ for y and graph the function

$$f(x) = \frac{3x + 2}{x^2 + x - 6}$$

we will obtain the graph in Figure 6-2(a).

From the figure, we can see that

- As x approaches -3 from the left, the values of y decrease, and the graph approaches the vertical line $x = -3$.

- As x approaches -3 from the right, the values of y increase, and the graph approaches the vertical line $x = -3$.

From the figure, we can also see that

- As x approaches 2 from the left, the values of y decrease, and the graph approaches the vertical line $x = 2$.

- As x approaches 2 from the right, the values of y increase, and the graph approaches the vertical line $x = 2$.

The lines $x = -3$ and $x = 2$ are vertical asymptotes. Although the vertical lines in the graph appear to be the graphs of $x = -3$ and $x = 2$, they are not. Graphing calculators draw graphs by connecting dots whose x-coordinates are close together. Often when two such points straddle a vertical asymptote and their y-coordinates are far apart, the calculator draws a line between them anyway, producing what appears to be a vertical asymptote. If you set your calculator to dot mode instead of connected mode, the vertical lines will not appear.

From Figure 6-2(a), we can also see that

■ As x increases to the right of 2, the values of y decrease and approach the line $y = 0$.

■ As x decreases to the left of -3, the values of y increase and approach the line $y = 0$.

The line $y = 0$ (the x-axis) is a horizontal asymptote. Graphing calculators do not draw lines that appear to be horizontal asymptotes.

From the graph, we can see that every real number x, except -3 and 2, gives a value of y. This observation confirms that the domain of the function is $(-\infty, -3) \cup (-3, 2) \cup (2, \infty)$. We can also see that y can be any value. Thus, the range is $(-\infty, \infty)$.

To find the domain and range of the function $f(x) = \frac{2x+1}{x-1}$, we use a calculator to draw the graph shown in Figure 6-2(b). From this graph, we can see that the line $x = 1$ is a vertical asymptote and that the line $y = 2$ is a horizontal asymptote. Since x can be any real number except 1, the domain is the interval $(-\infty, 1) \cup (1, \infty)$. Since y can be any value except 2, the range is $(-\infty, 2) \cup (2, \infty)$.

FIGURE 6-2

(a)   (b)

Simplifying Rational Expressions

When working with rational expressions, we will use the familiar rules from arithmetic.

Properties of fractions

If a, b, c, d, and k are real numbers, and if there are no divisions by 0, then

1. $\dfrac{a}{b} = \dfrac{c}{d}$ if and only if $ad = bc$ **2.** $\dfrac{a}{1} = a$ and $\dfrac{a}{a} = 1$

3. $\dfrac{ak}{bk} = \dfrac{a}{b} \cdot \dfrac{k}{k} = \dfrac{a}{b}$ **4.** $-\dfrac{a}{b} = \dfrac{-a}{b} = \dfrac{a}{-b}$

Property 3 of fractions is true because

$$\frac{ak}{bk} = \frac{a}{b} \cdot \frac{k}{k} = \frac{a}{b} \cdot 1 = \frac{a}{b} \quad (b \neq 0, k \neq 0) \quad \text{Any number times 1 is the number.}$$

Property 3, which is known as the **fundamental property of fractions,** is used to simplify rational expressions. It enables us to divide out factors that are common to the numerator and the denominator of a fraction.

Simplifying a rational expression

To simplify a rational expression,

1. Factor the numerator and the denominator.
2. Divide out the common factors of the numerator and the denominator.

EXAMPLE 3

Dividing out common factors Simplify **a.** $\dfrac{10k}{25k^2}$ and **b.** $\dfrac{-8y^3z^5}{6y^4z^3}$.

Solution To simplify these rational expressions, we factor each numerator and denominator and divide out all common factors.

a. $\dfrac{10k}{25k^2} = \dfrac{5 \cdot 2 \cdot k}{5 \cdot 5 \cdot k \cdot k}$

$$= \dfrac{\overset{1}{\cancel{5}} \cdot 2 \cdot \overset{1}{\cancel{k}}}{\underset{1}{\cancel{5}} \cdot 5 \cdot \underset{1}{\cancel{k}} \cdot k} \qquad \text{Divide out the common factors of 5 and } k. \text{ Show this using slashes and 1's.}$$

$$= \dfrac{2}{5k} \qquad \text{Do the multiplications in the numerator and the denominator.}$$

b. $\dfrac{-8y^3z^5}{6y^4z^3} = \dfrac{-2 \cdot 4 \cdot y \cdot y \cdot y \cdot z \cdot z \cdot z \cdot z \cdot z}{2 \cdot 3 \cdot y \cdot y \cdot y \cdot y \cdot z \cdot z \cdot z}$

$$= \dfrac{\overset{1}{\cancel{-2}} \cdot 4 \cdot \overset{1}{\cancel{y}} \cdot \overset{1}{\cancel{y}} \cdot \overset{1}{\cancel{y}} \cdot \overset{1}{\cancel{z}} \cdot \overset{1}{\cancel{z}} \cdot \overset{1}{\cancel{z}} \cdot z \cdot z}{\underset{1}{\cancel{2}} \cdot 3 \cdot \underset{1}{\cancel{y}} \cdot \underset{1}{\cancel{y}} \cdot \underset{1}{\cancel{y}} \cdot y \cdot \underset{1}{\cancel{z}} \cdot \underset{1}{\cancel{z}} \cdot \underset{1}{\cancel{z}}}$$

$$= -\dfrac{4z^2}{3y}$$

SELF CHECK Simplify $\dfrac{-12a^4b^2}{-3ab^4}$. *Answer:* $\dfrac{4a^3}{b^2}$ ∎

The fractions in Example 3 can also be simplified using the rules of exponents:

$$\dfrac{10k}{25k^2} = \dfrac{5 \cdot 2}{5 \cdot 5} k^{1-2} \qquad\qquad \dfrac{-8y^3z^5}{6y^4z^3} = \dfrac{-2 \cdot 4}{2 \cdot 3} y^{3-4}z^{5-3}$$

$$= \dfrac{2}{5} \cdot k^{-1} \qquad\qquad\qquad = \dfrac{-4}{3} \cdot y^{-1}z^2$$

$$= \dfrac{2}{5} \cdot \dfrac{1}{k} \qquad\qquad\qquad = -\dfrac{4}{3} \cdot \dfrac{1}{y} \cdot \dfrac{z^2}{1}$$

$$= \dfrac{2}{5k} \qquad\qquad\qquad\qquad = -\dfrac{4z^2}{3y}$$

EXAMPLE 4

Simplifying rational expressions. Simplify $\dfrac{x^2 - 16}{x + 4}$.

Solution We factor $x^2 - 16$ and use the fact that $\frac{x+4}{x+4} = 1$.

$$\dfrac{x^2 - 16}{x + 4} = \dfrac{\overset{1}{\cancel{(x + 4)}}(x - 4)}{\underset{1}{\cancel{(x + 4)}}} \qquad \text{Factor the difference of two squares.}$$

$$= \dfrac{x - 4}{1} \qquad \text{Divide out the common factor of } x + 4.$$

$$= x - 4$$

SELF CHECK Simplify $\dfrac{x^2 - 9}{x - 3}$.

Answer: $x + 3$ ∎

ACCENT ON TECHNOLOGY *Checking an Algebraic Simplification*

After simplifying an expression, we can use a scientific calculator to check the answer. One way to check whether $\frac{x^2 - 16}{x + 4} = x - 4$ is correct in Example 4 is to evaluate $\frac{x^2 - 16}{x + 4}$ and $x - 4$ for a value of x (say, 3). The expressions should give identical results.

Keystrokes $(\boxed{3} \boxed{x^2} \boxed{-} 16 \boxed{)} \boxed{\div} (\boxed{3} \boxed{+} 4 \boxed{)} \boxed{=}$ $\boxed{\qquad\qquad -1}$

Since $x - 4$ is -1 when $x = 3$, the results of the evaluations are the same. Evaluate the expressions for several other values of x. If the results differ for any given value, the original expression was not simplified correctly.

A graphing calculator can also be used to show that the simplification in Example 4 is correct. To do this, we enter the functions $f(x) = \frac{x^2 - 16}{x + 4}$ and $g(x) = x - 4$ as Y_1 and Y_2, respectively. See Figure 6-3(a). Then select the TABLE feature. Reading across the table, the values of Y_1 and Y_2 should be the same for each value of x as shown in Figure 6-3(b).

· **FIGURE 6-3**

Plot1 Plot2 Plot3
\Y₁◻(X²−16)/(X+4)
\Y₂◻X−4
\Y₃=
\Y₄=
\Y₅=
\Y₆=

X	Y₁	Y₂
-2	-6	-6
-1	-5	-5
0	-4	-4
1	-3	-3
2	-2	-2
3	-1	-1
4	0	0
X=-2		

(a)   (b)

To use a third method, we can compare the graphs of $f(x) = \frac{x^2 - 16}{x + 4}$, shown in Figure 6-4(a), and $g(x) = x - 4$, shown in Figure 6-4(b). Except for the point where $x = -4$, the graphs are the same. The point where $x = -4$ is excluded from the graph of $f(x) = \frac{x^2 - 16}{x + 4}$, because -4 is not in the domain of f. However, graphing calculators do not show that this point is excluded. The point where $x = -4$ is included in the graph of $g(x) = x - 4$, because -4 is in the domain of g.

(continued)

FIGURE 6-4

(a)   (b)

| EXAMPLE 5 | **Simplifying rational expressions.** Simplify $\dfrac{6a^2 - 13a + 6}{3a^2 + a - 2}$. |

Solution   We factor the trinomials in the numerator and the denominator and then divide out the common factor.

$$\frac{6a^2 - 13a + 6}{3a^2 + a - 2} = \frac{\overset{1}{\cancel{(3a - 2)}}(2a - 3)}{\underset{1}{\cancel{(3a - 2)}}(a + 1)}$$

$$= \frac{2a - 3}{a + 1}$$

 WARNING! Do not divide out the a's in $\frac{2a-3}{a+1}$. The a in the numerator is a factor of the first term only, not a factor of the entire numerator. Likewise, the a in the denominator is a factor of the first term only, not a factor of the entire denominator.

SELF CHECK   Simplify $\dfrac{2b^2 + 7b - 15}{2b^2 + 13b + 15}$.   *Answer:* $\dfrac{2b - 3}{2b + 3}$ ■

We will encounter many fractions that are already in simplified form. For example, to attempt to simplify

$$\frac{x^2 + xa + 2x + 2a}{x^2 + x - 6}$$

we factor the numerator and denominator and divide out any common factors:

$$\frac{x^2 + xa + 2x + 2a}{x^2 + x - 6} = \frac{x(x + a) + 2(x + a)}{(x - 2)(x + 3)} = \frac{(x + a)(x + 2)}{(x - 2)(x + 3)}$$

Since there are no common factors in the numerator and denominator, the fraction is in *lowest terms*. It cannot be simplified.

 WARNING! Only factors that are common to the entire numerator and the entire denominator can be divided out. *Terms* common to both the numerator and denominator cannot be divided out. It is incorrect to divide out the common term of 3 in the following simplification, because it gives a wrong answer.

$$\frac{3 + 7}{3} = \frac{\overset{1}{\cancel{3}} + 7}{\underset{1}{\cancel{3}}} = \frac{1 + 7}{1} = 8 \quad \text{The correct simplification is } \frac{3 + 7}{3} = \frac{10}{3}.$$

The 3's in the fraction $\frac{5 + 3(2)}{3(4)}$ cannot be divided out, because the 3 in the numerator is a factor of the second term only. To be divided out, the 3 must be a factor of the entire numerator.

It is not correct to divide out the y in the fraction $\frac{x^2y + 6x}{y}$, because y is not a factor of the entire numerator.

Simplifying Rational Expressions by Factoring Out −1

To simplify $\frac{b-a}{a-b}$ $(a \neq b)$, we factor −1 from the numerator and divide out any factors common to both the numerator and the denominator:

$$\frac{b-a}{a-b} = \frac{-a+b}{a-b} \qquad \text{Rewrite the numerator.}$$

$$= \frac{-1\overset{1}{\cancel{(a-b)}}}{\underset{1}{\cancel{(a-b)}}} \qquad \begin{array}{l} \text{Factor out } -1 \text{ from each term in the numerator.} \\ \text{Divide out the common factor.} \end{array}$$

$$= \frac{-1}{1}$$

$$= -1$$

In general, we have the following principle.

Quotient of a quantity and its opposite

The quotient of any nonzero quantity and its negative (or opposite) is −1.

EXAMPLE 6

Factoring out −1. Simplify $\dfrac{3x^2 - 10xy - 8y^2}{4y^2 - xy}$.

Solution We factor the numerator and denominator. Because $x - 4y$ and $4y - x$ are negatives, their quotient is −1.

$$\frac{3x^2 - 10xy - 8y^2}{4y^2 - xy} = \frac{(3x + 2y)\overset{-1}{\cancel{(x-4y)}}}{y\underset{1}{\cancel{(4y-x)}}}$$

$$= \frac{-(3x + 2y)}{y}$$

$$= \frac{-3x - 2y}{y}$$

SELF CHECK Simplify $\dfrac{2a^2 - 3ab - 9b^2}{3b^2 - ab}$.

Answer: $-\dfrac{2a + 3b}{b}$ or $\dfrac{-2a - 3b}{b}$ ∎

STUDY SET

Section 6.1

VOCABULARY

In Exercises 1–4, fill in the blanks to make the statements true.

1. A fraction that is the quotient of two polynomials, such as

$$\frac{x^2 - x - 6}{x^3 - 8}$$

is called a _____ expression.

2. In the rational expression

$$\frac{(x + 2)(3x - 1)}{(x + 2)(4x + 2)}$$

$x + 2$ is a common _____ of the numerator and the denominator.

3. If a graph approaches a line, the line is called an
_____.

4. The _____ of a fraction can never be 0.

CONCEPTS

5. The graph of rational function f (for $x > 0$) is shown in Illustration 1. Find each of the following.
  a. $f(1)$
  b. $f(2)$
  c. $f(4)$

ILLUSTRATION 1

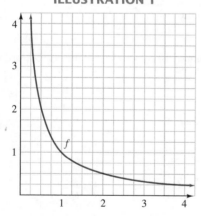

6. Show that $\dfrac{x - y}{y - x} = -1$ by factoring out -1 from each term in the numerator.

7. Simplify each rational expression.
  a. $\dfrac{x + 8}{x + 8}$
  b. $\dfrac{x + 8}{8 + x}$
  c. $\dfrac{x - 8}{x - 8}$
  d. $\dfrac{x - 8}{8 - x}$

8. Simplify each rational expression, if possible.
  a. $\dfrac{x + 8}{x}$
  b. $\dfrac{x + 8}{8}$
  c. $\dfrac{a^3 + 8}{2}$
  d. $\dfrac{x^2 + 5x + 6}{x^2 + x - 12}$

In Exercises 9–10, refer to the graphs below. Each graph shows the average cost to manufacture a certain item for a given number of units produced.

Item 1

Item 2

Item 3

Item 4

9. MANUFACTURING For each graph, briefly describe how the average cost per unit changes as the number of units produced increases.

10. Which graph is best described as the graph of a
  a. linear function? **b.** quadratic function?

  c. rational function? **d.** polynomial function?

NOTATION

11. Tell whether each statement is true or false.

a. $-\dfrac{x-4}{x+4} = \dfrac{4-x}{x+4}$ b. $\dfrac{a-3b}{2b-a} = \dfrac{3b-a}{a-2b}$

12. Explain what the slashes and the 1's show.

$$\dfrac{t^2-4}{t^2+2t} = \dfrac{\overset{1}{\cancel{(t+2)}}(t-2)}{t\underset{1}{\cancel{(t+2)}}} = \dfrac{t-2}{t}$$

PRACTICE

In Exercises 13–16, the time t it takes to travel 600 miles is a function of the average rate of speed, $t = \frac{600}{r}$. Find t for each value of r.

13. 30 mph

14. 40 mph

15. 50 mph

16. 60 mph

In Exercises 17–20, complete the table of values for each rational function (round to the nearest hundredth when applicable). Then graph it. Each function is defined for $x > 0$. Label the horizontal asymptote.

17. $f(x) = \dfrac{6}{x}$

 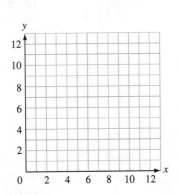

x	f(x)
1	
2	
4	
6	
8	
10	
12	

18. $f(x) = \dfrac{12}{x}$

x	f(x)
1	
4	
8	
12	
16	
20	
24	

19. $f(x) = \dfrac{x+2}{x}$

x	f(x)
1	
2	
4	
6	
8	
10	
12	

20. $f(x) = \dfrac{2x+4}{x}$

x	f(x)
1	
4	
8	
12	
16	
20	
24	

In Exercises 21–28, find the domain of each rational function. Use interval notation.

21. $f(x) = \dfrac{2}{x}$

22. $f(x) = \dfrac{8}{x-1}$

23. $f(x) = \dfrac{2}{x+2}$

24. $f(x) = \dfrac{2}{x^2-2x}$

25. $f(x) = \dfrac{2}{x-x^2}$

26. $f(x) = \dfrac{2}{x^2-36}$

27. $f(x) = \dfrac{2}{x^2-x-56}$

28. $f(x) = \dfrac{2}{x^2+2x-24}$

In Exercises 29–76, simplify each rational expression when possible.

29. $\dfrac{12}{18}$ **30.** $\dfrac{25}{55}$

31. $-\dfrac{112}{36}$

32. $-\dfrac{49}{21}$

33. $\dfrac{12x^3}{3x}$

34. $-\dfrac{15a^2}{25a^3}$

35. $\dfrac{-24x^3y^4}{18x^4y^3}$

36. $\dfrac{15a^5b^4}{21b^3c^2}$

37. $-\dfrac{11x(x-y)}{22(x-y)}$

38. $\dfrac{x(x-2)^2}{(x-2)^3}$

39. $\dfrac{(a-b)(d-c)}{(c-d)(a-b)}$

40. $\dfrac{(p+q)(p-r)}{(r-p)(p+q)}$

41. $\dfrac{y+x}{x^2-y^2}$

42. $\dfrac{x-y}{x^2-y^2}$

43. $\dfrac{5x-10}{x^2-4x+4}$

44. $\dfrac{y-xy}{xy-x}$

45. $\dfrac{12-3x^2}{x^2-x-2}$

46. $\dfrac{x^2+2x-15}{25-x^2}$

47. $\dfrac{x^2+y^2}{x+y}$

48. $\dfrac{3x+6y}{2y+x}$

49. $\dfrac{x^3+8}{x^2-2x+4}$

50. $\dfrac{x^2+3x+9}{x^3-27}$

51. $\dfrac{x^2+2x+1}{x^2+4x+3}$

52. $\dfrac{6x^2+x-2}{8x^2+2x-3}$

53. $\dfrac{3m-6n}{3n-6m}$

54. $\dfrac{ax+by+ay+bx}{a^2-b^2}$

55. $\dfrac{4x^2+24x+32}{16x^2+8x-48}$

56. $\dfrac{a^2-4}{a^3-8}$

57. $\dfrac{3x^2-3y^2}{x^2+2y+2x+yx}$

58. $\dfrac{x^2+x-30}{x^2-x-20}$

59. $\dfrac{4x^2+8x+3}{6+x-2x^2}$

60. $\dfrac{6x^2+13x+6}{6-5x-6x^2}$

61. $\dfrac{a^3+27}{4a^2-36}$

62. $\dfrac{a-b}{b^2-a^2}$

63. $\dfrac{2x^2-3x-9}{2x^2+3x-9}$

64. $\dfrac{6x^2-7x-5}{2x^2+5x+2}$

65. $\dfrac{(m+n)^3}{m^2+2mn+n^2}$

66. $\dfrac{x^3-27}{3x^2-8x-3}$

67. $\dfrac{m^3-mn^2}{mn^2+m^2n-2m^3}$

68. $\dfrac{p^3+p^2q-2pq^2}{pq^2+p^2q-2p^3}$

69. $\dfrac{x^4-y^4}{(x^2+2xy+y^2)(x^2+y^2)}$

70. $\dfrac{(x^2-1)(x+1)}{(x^2-2x+1)^2}$

71. $\dfrac{6xy-4x-9y+6}{6y^2-13y+6}$

72. $\dfrac{x^2+2xy}{x+2y+x^2-4y^2}$

73. $\dfrac{(2x^2+3xy+y^2)(3a+b)}{(x+y)(2xy+2bx+y^2+by)}$

74. $\dfrac{(x-1)(6ax+9x+4a+6)}{(3x+2)(2ax-2a+3x-3)}$

75. $\dfrac{(x^2+2x+1)(x^2-2x+1)}{(x^2-1)^2}$

76. $\dfrac{2x^2+2x-12}{x^3+3x^2-4x-12}$

In Exercises 77–80, use a graphing calculator to graph each rational function. From the graph, determine its domain and range.

77. $f(x)=\dfrac{x}{x-2}$

78. $f(x)=\dfrac{x+2}{x}$

79. $f(x)=\dfrac{x+1}{x^2-4}$

80. $f(x)=\dfrac{x-2}{x^2-3x-4}$

APPLICATIONS

81. ENVIRONMENTAL CLEANUP Suppose the cost (in dollars) of removing $p\%$ of the pollution in a river is given by the rational function

$$f(p)=\frac{50{,}000p}{100-p}\quad(0\le p<100)$$

Find the cost of removing each percent of pollution.
a. 50%   **b.** 80%

82. DIRECTORY COSTS The average (mean) cost for a service club to publish a directory of its members is given by the rational function

$$f(x)=\frac{1.25x+700}{x}$$

where x is the number of directories printed. Find the average cost per directory if
a. 500 directories are printed.
b. 2,000 directories are printed.

83. UTILITY COSTS An electric company charges $7.50 per month plus 9¢ for each kilowatt hour (kwh) of electricity used.

a. Find a linear function that gives the total cost of n kwh of electricity.

b. Find a rational function that gives the average cost per kwh when using n kwh.

c. Find the average cost per kwh when 775 kwh are used.

84. SCHEDULING WORK CREWS The rational function

$$f(t) = \frac{t^2 + 2t}{2t + 2}$$

gives the number of days it would take two construction crews, working together, to frame a house that crew 1 (working alone) could complete in t days and crew 2 (working alone) could complete in $t + 2$ days.

a. If crew 1 could frame a certain house in 15 days, how long would it take both crews working together?

b. If crew 2 could frame a certain house in 20 days, how long would it take both crews working together?

85. FILLING A POOL The rational function

$$f(t) = \frac{t^2 + 3t}{2t + 3}$$

gives the number of hours it would take two pipes, working together, to fill a pool that the larger pipe (working alone) could fill in t hours and the smaller pipe (working alone) could fill in $t + 3$ hours.

a. If the smaller pipe could fill a pool in 7 hours, how long would it take both pipes to fill the pool?

b. If the larger pipe could fill a pool in 8 hours, how long would it take both pipes to fill the pool?

86. RETENTION STUDY After learning a list of words, two subjects were tested over a 28-day period to see what percent of the list they remembered. In both cases, their percent recall could be modeled by rational functions, as shown in Illustration 2.

ILLUSTRATION 2

a. Use the graphs to complete the table.

Days since learning	0	1	2	4	7	14	28
% recall—subject 1							
% recall—subject 2							

b. After 28 days, which subject had the better recall?

WRITING

87. A student simplified $\frac{6x^2 - 7x - 5}{2x^2 + 5x + 2}$ and obtained $\frac{3x - 5}{x - 2}$. As a check, she graphed $Y_1 = \frac{6x^2 - 7x - 5}{2x^2 + 5x + 2}$ and $Y_2 = \frac{3x - 5}{x - 2}$. (See Illustration 3.) What conclusion can be drawn from the graphs? Explain your answer.

ILLUSTRATION 3

88. Simplify $\frac{6x^2 + x - 2}{8x^2 + 2x - 3}$. Then explain how the table of values for $Y_1 = \frac{6x^2 + x - 2}{8x^2 + 2x - 3}$ and $Y_2 = \frac{3x + 2}{4x + 3}$ shown in Illustration 4 can be used to check your result.

ILLUSTRATION 4

X	Y1	Y2
-2	.8	.8
-1	1	1
0	.66667	.66667
1	.71429	.71429
2	.72727	.72727
3	.73333	.73333
4	.73684	.73684

X= -2

REVIEW

In Exercises 89–92, factor each expression.

89. $3x^2 - 9x$

90. $-6t^2 + 5t + 6$

91. $27x^6 + 64y^3$

92. $x^2 + ax + 2x + 2a$

▶ 6.2 Proportion and Variation

In this section, you will learn about

Ratios ■ Proportions ■ Solving proportions ■ Similar triangles ■ Direct variation ■ Inverse variation ■ Joint variation ■ Combined variation

Introduction In this section, we discuss five mathematical models that have a variety of applications. First, we show how a *ratio-proportion model* can be used to solve shopping problems and to determine the height of a tree given the length of its shadow. Then we introduce four types of *variation models,* each of which expresses a special relationship between two or more quantities. We use these models to solve problems involving travel, lighting, geometry, and highway construction.

Ratios

The quotient of two numbers or two quantities is often referred to as a **ratio.** For example, the fraction $\frac{2}{3}$ can be read as "the ratio of 2 to 3." The notation $2:3$ (read as "2 is to 3") is another common way to denote a ratio. Some more examples of ratios are

$$\frac{4x}{7y} \text{ (the ratio of } 4x \text{ to } 7y) \qquad \text{and} \qquad \frac{x-2}{3x} \text{ (the ratio of } x-2 \text{ to } 3x)$$

Ratios are often used to express **unit costs,** such as the cost per pound of ground beef.

The cost of a package of ground beef $\longrightarrow$ $\dfrac{\$7.47}{5 \text{ lb}} \approx \$1.49$ per lb $\longleftarrow$ The cost per pound
The weight of the package $\longrightarrow$

Ratios are also used to express **rates,** such as an average rate of speed.

A distance traveled $\longrightarrow$ $\dfrac{372 \text{ miles}}{6 \text{ hours}} = 62$ mph $\longleftarrow$ The average rate of speed
in a period of time $\longrightarrow$

Proportions

An equation indicating that two ratios are equal is called a **proportion.** Two examples of proportions are

$$\frac{1}{4} = \frac{2}{8} \qquad \text{and} \qquad \frac{4}{7} = \frac{12}{21}$$

In the proportion $\frac{a}{b} = \frac{c}{d}$, a and d are called the **extremes** of the proportion, and b and c are called the **means.**

To develop a fundamental property of proportions, we suppose that

$$\frac{a}{b} = \frac{c}{d}$$

is a proportion and multiply both sides by bd to obtain

$$bd\left(\frac{a}{b}\right) = bd\left(\frac{c}{d}\right)$$

$$\frac{\cancel{b}da}{\cancel{b}} = \frac{b\cancel{d}c}{\cancel{d}} \qquad \text{Divide out common factors.}$$

$$ad = bc$$

Thus, if $\frac{a}{b} = \frac{c}{d}$, then $ad = bc$. This illustrates the following property.

Fundamental property of proportions

In a proportion, the product of the extremes is equal to the product of the means.

Solving Proportions

We can solve many problems by writing and then solving a proportion. To solve a proportion, we apply the fundamental property of proportions.

EXAMPLE 1

Solving proportions. Solve for x: $\dfrac{x+3}{x} = \dfrac{x}{x+6}$.

Solution

$$\frac{x+3}{x} = \frac{x}{x+6}$$

$(x+3)(x+6) = x \cdot x$ In a proportion, the product of the extremes equals the product of the means.

$x^2 + 9x + 18 = x^2$ Do the multiplications.

$9x + 18 = 0$ Subtract x^2 from both sides.

$x = -2$ Subtract 18 from both sides and then divide by 9.

Thus, $x = -2$. To check, we substitute -2 for x in the proportion and simplify:

$$\frac{x+3}{x} = \frac{x}{x+6}$$

$$\frac{-2+3}{-2} \overset{?}{=} \frac{-2}{-2+6}$$

$$\frac{-1}{2} \overset{?}{=} \frac{-2}{4}$$

$$-\frac{1}{2} = -\frac{1}{2}$$

SELF CHECK   Solve for x: $\frac{x-1}{x} = \frac{x}{x+3}$.   *Answer:* $\frac{3}{2}$ ■

WARNING! In Example 1, the expression $\frac{x+3}{x}$ is undefined if $x = 0$, because division by 0 would be indicated. Similarly, $\frac{x}{x+6}$ is undefined if $x = -6$. For these reasons, before beginning the solution process, we can rule out $x = 0$ and $x = -6$ as possible solutions of $\frac{x+3}{x} = \frac{x}{x+6}$.

| EXAMPLE 2 | **Solving proportions.** Solve $\dfrac{5a + 2}{2a} = \dfrac{18}{a + 4}$. |

Solution   Because $\frac{5a+2}{2a}$ is undefined if $a = 0$, we state the restriction that $a \neq 0$. Because $\frac{18}{a+4}$ is undefined if $a = -4$, we also note that $a \neq -4$.

$$\frac{5a + 2}{2a} = \frac{18}{a + 4}$$

$(5a + 2)(a + 4) = 2a(18)$	In a proportion, the product of the extremes equals the product of the means.
$5a^2 + 22a + 8 = 36a$	Multiply.
$5a^2 - 14a + 8 = 0$	Subtract $36a$ from both sides.
$(5a - 4)(a - 2) = 0$	Factor to solve the quadratic equation.
$5a - 4 = 0$ or $a - 2 = 0$	Set each factor equal to 0.
$5a = 4$ $a = 2$	Solve each linear equation.
$a = \dfrac{4}{5}$	

Thus, $a = \frac{4}{5}$ or $a = 2$. Check each solution.

SELF CHECK   Solve $\frac{3x+1}{12} = \frac{x}{x+2}$.   *Answer:* $\frac{2}{3}$, 1   ■

| EXAMPLE 3 | **Gourmet cooking.** To make a dessert of Pears Hélène, a chef needs to purchase 14 pears. If they are on sale at 6 for \$2.34, what will 14 cost? |

Solution   A proportion can be used to model this situation. First, we let c represent the cost of 14 pears. The price per pear when purchasing 6 pears is $\frac{\$2.34}{6}$, and the price per pear when purchasing 14 pears is $\frac{\$c}{14}$. Since these ratios are equal, we have the following proportion.

$\dfrac{2.34}{6} = \dfrac{c}{14}$	\$2.34 is to 6 as \$$c$ is to 14.
$14(2.34) = 6c$	In a proportion, the product of the extremes is equal to the product of the means.
$32.76 = 6c$	Multiply.
$\dfrac{32.76}{6} = c$	Divide both sides by 6.
$c = 5.46$	Simplify.

Fourteen pears will cost \$5.46.

SELF CHECK   A model railroad engine is 9 inches long. The scale is 87 feet to 1 foot. How long is a real engine? (*Hint:* Note the difference in units.)   *Answer:* $65\frac{1}{4}$ ft   ■

Similar Triangles

If two angles of one triangle have the same measure as two angles of a second triangle, the triangles will have the same shape. In this case, we call the triangles **similar triangles.** Here are some facts about similar triangles.

Similar triangles

If two triangles are similar, then

1. the three angles of the first triangle have the same measure, respectively, as the three angles of the second triangle.
2. the lengths of all corresponding sides are in proportion.

The triangles shown in Figure 6-5 are similar triangles.

FIGURE 6-5

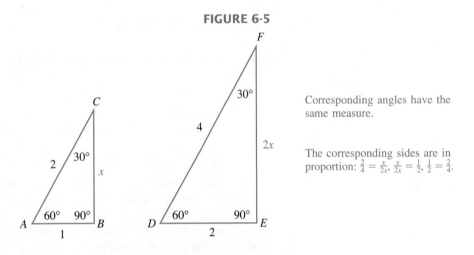

Corresponding angles have the same measure.

The corresponding sides are in proportion: $\frac{2}{4} = \frac{x}{2x}$, $\frac{x}{2x} = \frac{1}{2}$, $\frac{1}{2} = \frac{2}{4}$.

The properties of similar triangles often enable us to determine the lengths of the sides of triangles indirectly. For example, on a sunny day, we can find the height of a tree and stay safely on the ground.

EXAMPLE 4

Height of a tree. A tree casts a shadow of 29 feet at the same time as a vertical yardstick casts a shadow of 2.5 feet. Find the height of the tree.

Solution Refer to Figure 6-6 on the next page, which shows the triangles determined by the tree and its shadow and the yardstick and its shadow. Because the triangles have the same shape, they are similar, and the measures of their corresponding sides are in proportion. If we let h represent the height of the tree, we can find h by setting up and solving the following proportion.

$$\frac{h}{3} = \frac{29}{2.5} \qquad h \text{ is to 3 as 29 is to 2.5.}$$

$2.5h = 3(29)$ In a proportion, the product of the extremes is equal to the product of the means.

$2.5h = 87$ Multiply.

$h = 34.8$ Divide both sides by 2.5.

The tree is about 35 feet tall.

FIGURE 6-6

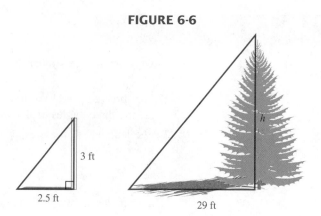

3 ft

2.5 ft

29 ft

h

■

Direct Variation

To introduce direct variation, we consider the formula for the circumference of a circle

$$C = \pi D$$

where C is the circumference, D is the diameter, and $\pi \approx 3.14159$. If we double the diameter of a circle, we determine another circle with a larger circumference C_1 such that

$$C_1 = \pi(2D) = 2\pi D = 2C$$

Thus, doubling the diameter results in doubling the circumference. Likewise, if we triple the diameter, we will triple the circumference.

In this formula, we say that the variables C and D *vary directly,* or that they are *directly proportional.* This is because as one variable gets larger, so does the other, in a predictable way. In this example, the constant π is called the *constant of variation* or the *constant of proportionality.*

Direct variation

> The words "y varies directly with x" or "y is directly proportional to x" mean that $y = kx$ for some nonzero constant k. The constant k is called the **constant of variation** or the **constant of proportionality.**

Since the formula for direct variation ($y = kx$) defines a linear function, its graph is always a line with a y-intercept at the origin. The graph of $y = kx$ appears in Figure 6-7 for three positive values of k.

One example of direct variation is Hooke's law from physics. Hooke's law states that the distance a spring will stretch varies directly with the force that is applied to it.

If d represents a distance and f represents a force, this verbal model of Hooke's law can be expressed mathematically as

$$d = kf \qquad \text{This is a model for direct variation.}$$

where k is the constant of variation. If the spring stretches 10 inches when a weight of 6 pounds is attached, k can be found as follows:

$$d = kf$$

$$\mathbf{10} = k(\mathbf{6}) \qquad \text{Substitute 10 for } d \text{ and 6 for } f.$$

$$\frac{5}{3} = k$$

FIGURE 6-7

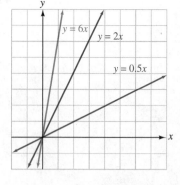

y

$y = 6x$

$y = 2x$

$y = 0.5x$

x

To find the force required to stretch the spring a distance of 35 inches, we can solve the equation $d = kf$ for f, with $d = 35$ and $k = \frac{5}{3}$.

$$d = kf$$

$$35 = \frac{5}{3}f \quad \text{Substitute 35 for } d \text{ and } \tfrac{5}{3} \text{ for } k.$$

$$105 = 5f \quad \text{Multiply both sides by 3.}$$

$$21 = f \quad \text{Divide both sides by 5.}$$

Thus, the force required to stretch the spring a distance of 35 inches is 21 pounds.

EXAMPLE 5

Direct variation. The distance traveled in a given time is directly proportional to the speed. If a car travels 70 miles at 30 mph, how far will it travel in the same time at 45 mph?

Solution

The verbal model *distance is directly proportional to speed* can be expressed by the equation

1. $d = ks$ This is the direct variation model.

where d is distance, k is the constant of variation, and s is the speed. To find k, we substitute 70 for d and 30 for s and solve for k.

$$d = ks$$

$$70 = k(30)$$

$$k = \frac{7}{3}$$

To find the distance traveled at 45 mph, we substitute $\frac{7}{3}$ for k and 45 for s in Equation 1 and simplify.

$$d = ks$$

$$d = \frac{7}{3}(45)$$

$$= 105$$

In the time it takes to go 70 miles at 30 mph, the car could travel 105 miles at 45 mph.

SELF CHECK

In Example 5, how far will the car travel in the same time at 60 mph?

Answer: 140 mi ∎

Solving variation problems

To solve a variation problem:

1. Translate the verbal model into an equation.
2. Substitute the first set of values into the equation from step 1 to determine the value of k.
3. Substitute the value of k into the equation from step 1.
4. Substitute the remaining set of values into the equation from step 3 and solve for the unknown.

Inverse Variation

In the formula $w = \frac{12}{l}$, w gets smaller as l gets larger, and w gets larger as l gets smaller. Since these variables vary in opposite directions in a predictable way, we say that the variables *vary inversely,* or that they are *inversely proportional.* The constant 12 is the constant of variation.

Inverse variation

The words "y varies inversely with x" or "y is inversely proportional to x" mean that $y = \frac{k}{x}$ for some nonzero constant k. The constant k is called the **constant of variation.**

FIGURE 6-8

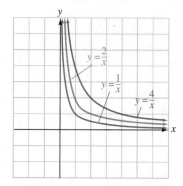

The formula for inverse variation $\left(y = \frac{k}{x}\right)$ defines a rational function whose graph will have the x- and y-axes as asymptotes. The graph of $y = \frac{k}{x}$ appears in Figure 6-8 for three positive values of k.

Because of gravity, an object in space is attracted to the earth. The force of this attraction varies inversely with the square of the distance from the object to the center of the earth. If f represents the force and d represents the distance, the relationship between f and d can be expressed by the equation

$$f = \frac{k}{d^2}$$ This is an inverse variation model.

If we know that an object 4,000 miles from the center of the earth is attracted to the earth with a force of 90 pounds, we can find k.

$$f = \frac{k}{d^2}$$

$$90 = \frac{k}{4,000^2}$$ Substitute 90 for f and 4,000 for d.

$$k = 90(4,000)^2$$ Multiply both sides by $4,000^2$ to solve for k.

$$= 1,440,000,000$$

$$= 1.44 \times 10^9$$ Write the value of k using scientific notation.

To find the force of attraction when the object is 5,000 miles from the center of the earth, we proceed as follows:

$$f = \frac{k}{d^2}$$

$$f = \frac{1.44 \times 10^9}{5,000^2}$$ Substitute 1.44×10^9 for k and 5,000 for d.

$$= \frac{1.44 \times 10^9}{2.5 \times 10^7}$$ Write $5,000^2 = 25,000,000$ using scientific notation.

$$= 57.6$$

The object will be attracted to earth with a force of 57.6 pounds when it is 5,000 miles from the earth's center.

E X A M P L E   6

Photography. The intensity I of light received from a light source varies inversely with the square of the distance from the light source. If a photographer, 16 feet away from his subject, has a light meter reading of 4 foot-candles of illuminance, what will the meter read if the photographer moves in for a close-up, 4 feet away from the subject?

Solution   The words *intensity varies inversely with the square of the distance d* can be expressed by the equation

$$I = \frac{k}{d^2} \quad \text{This is inverse variation.}$$

To find k, we substitute 4 for I and 16 for d and solve for k.

$$I = \frac{k}{d^2}$$

$$4 = \frac{k}{16^2}$$

$$4 = \frac{k}{256}$$

$$1,024 = k$$

To find the intensity when the photographer is 4 feet away from the subject, we substitute 4 for d and 1,024 for k and simplify.

$$I = \frac{k}{d^2}$$

$$I = \frac{1,024}{4^2}$$

$$= 64$$

The intensity at 4 feet is 64 foot-candles.

SELF CHECK   In Example 6, find the intensity when the photographer is 8 feet away from the subject.   *Answer:* 16 foot-candles   ■

Joint Variation

There are times when one variable varies with the product of several variables. For example, the area of a triangle varies directly with the product of its base and height:

$$A = \frac{1}{2} bh$$

Such variation is called *joint variation.*

Joint variation

> If one variable varies directly with the product of two or more variables, the relationship is called **joint variation.** If y varies jointly with x and z, then $y = kxz$. The nonzero constant k is called the **constant of variation.**

EXAMPLE 7   **Joint variation.** The volume V of a cone varies jointly with its height h and the area of its base B. If $V = 6$ cm^3 when $h = 3$ cm and $B = 6$ cm^2, find V when $h = 2$ cm and $B = 8$ cm^2.

Solution   The words *V varies jointly with h and B* mean that *V* varies directly as the product of *h* and *B*. Thus,

$$V = khB$$ The joint variation model can also be read as *V is directly proportional to the product of h and B.*

We can find *k* by substituting 6 for *V*, 3 for *h* and 6 for *B*.

$$V = khB$$
$$6 = k(3)(6)$$
$$6 = k(18)$$
$$\tfrac{1}{3} = k$$ Divide both sides by 18.

To find *V* when *h* = 2 and *B* = 8, we substitute these values into the formula $V = \tfrac{1}{3}hB$.

$$V = \tfrac{1}{3}hB$$
$$V = \left(\tfrac{1}{3}\right)(2)(8)$$
$$= \tfrac{16}{3}$$

The volume will be $\tfrac{16}{3}$ cm³ or $5\tfrac{1}{3}$ cm³.   ■

Combined Variation

Many applied problems involve a combination of direct and inverse variation. Such variation is called **combined variation.**

EXAMPLE 8

Highway construction. The time it takes to build a highway varies directly with the length of the road, but inversely with the number of workers. If it takes 100 workers 4 weeks to build 2 miles of highway, how long will it take 80 workers to build 10 miles of highway?

Solution   We can let *t* represent the time in weeks, *l* represent the length in miles, and *w* represent the number of workers. The relationship between these variables can be expressed by the equation

$$t = \frac{kl}{w}$$ This is a combined variation model.

We substitute 4 for *t*, 100 for *w*, and 2 for *l* to find *k*:

$$4 = \frac{k(2)}{100}$$
$$400 = 2k$$ Multiply both sides by 100.
$$200 = k$$ Divide both sides by 2.

We now substitute 80 for *w*, 10 for *l*, and 200 for *k* in the equation $t = \frac{kl}{w}$ and simplify:

$$t = \frac{kl}{w}$$
$$t = \frac{200(10)}{80}$$
$$= 25$$

It will take 25 weeks for 80 workers to build 10 miles of highway.

SELF CHECK   In Example 8, how long will it take 60 workers to build 6 miles of highway?   *Answer:* 20 weeks   ■

STUDY SET

Section 6.2

VOCABULARY

In Exercises 1–8, fill in the blanks to make the statements true.

1. _____ are used to express unit costs and rates.

2. An equation that states that two ratios are equal, such as $\frac{1}{2} = \frac{4}{8}$, is called a _____.

3. In a proportion, the product of the _____ is equal to the product of the _____.

4. If two angles of one triangle have the same measure as two angles of a second triangle, the triangles are _____.

5. The equation $y = kx$ defines _____ variation, and the equation $y = \frac{k}{x}$ defines _____ variation.

6. The equation $y = kxz$ defines _____ variation, and the equation $y = \frac{kx}{z}$ defines _____ variation.

7. _____ variation is represented by a rational function.

8. _____ variation is represented by a linear function.

CONCEPTS

In Exercises 9–12, decide whether direct or inverse variation applies, and then sketch a reasonable graph for the situation.

9.

10.

11.

12.

NOTATION

In Exercises 13–14, complete each solution.

13. Solve $\dfrac{-7}{6} = \dfrac{x + 3}{12}$.

$$\frac{-7}{6} = \frac{x + 3}{12}$$

$$\boxed{}(12) = \boxed{}(x + 3)$$

$$-84 = 6x + \boxed{}$$

$$\boxed{} = 6x$$

$$-17 = x$$

14. Solve $\dfrac{18}{2x + 1} = \dfrac{3}{14}$.

$$\frac{18}{2x + 1} = \frac{3}{14}$$

$$\boxed{}(14) = \left(\boxed{}\right)3$$

$$252 = \boxed{} + 3$$

$$249 = \boxed{}$$

$$41.5 = x$$

PRACTICE

In Exercises 15–30, solve each proportion for the variable, if possible.

15. $\dfrac{x}{5} = \dfrac{15}{25}$

16. $\dfrac{4}{y} = \dfrac{6}{27}$

17. $\dfrac{r - 2}{3} = \dfrac{r}{5}$

18. $\dfrac{x + 1}{x - 1} = \dfrac{6}{4}$

19. $\dfrac{5}{5z + 3} = \dfrac{2z}{2z^2 + 6}$

20. $\dfrac{9t + 6}{t} = \dfrac{7}{3}$

21. $\dfrac{2(y + 3)}{3} = \dfrac{4(y - 4)}{5}$

22. $\dfrac{b + 4}{5} = \dfrac{3(b - 2)}{3}$

23. $\dfrac{2}{3x} = \dfrac{6x}{36}$

24. $\dfrac{y}{4} = \dfrac{4}{y}$

25. $\dfrac{2}{c} = \dfrac{c - 3}{2}$

26. $\dfrac{2}{x + 6} = \dfrac{-2x}{5}$

27. $\dfrac{1}{x + 3} = \dfrac{-2x}{x + 5}$

28. $\dfrac{x - 1}{x + 1} = \dfrac{2}{3x}$

29. $\dfrac{9z + 6}{z(z + 3)} = \dfrac{7}{z + 3}$

30. $\dfrac{3}{n(n + 3)} = \dfrac{2}{(n + 1)(n + 3)}$

In Exercises 31–36, express each verbal model in symbols.

31. A varies directly with the square of p.

32. z varies inversely with the cube of t.

33. v varies inversely with the square of r.

34. C varies jointly with x, y, and z.

35. P varies directly with the square of a and inversely with the cube of j.

36. M varies inversely with the cube of n and jointly with x and the square of z.

In Exercises 37–40, express each variation model in words. In each equation, k is the constant of variation.

37. $L = kmn$

38. $P = \dfrac{km}{n}$

39. $R = \dfrac{kL}{d^2}$

40. $U = krs^2t$

APPLICATIONS

In Exercises 41–48, set up and solve the required proportion.

41. CAFFEINE Many convenience stores sell super-size 44-ounce soft drinks in refillable cups. For each of the products listed in Illustration 1, find the amount of caffeine contained in one of the large cups. Round to the nearest milligram.

ILLUSTRATION 1

Soft drink, 12 oz	Caffeine (mg)
Mountain Dew	55
Coca-Cola Classic	47
Pepsi	37

Based on data from *Los Angeles Times* (November 11, 1997) p. S4

42. TELEPHONES As of 1997, Sweden had 683 telephone lines per 1,000 people—the highest ratio of any country in the world. If Sweden's population is about 8,901,000, how many telephone lines does the country have?

43. WALL PAPERING The instructions on the label of wallpaper adhesive read as shown in Illustration 2. Estimate the amount of adhesive needed to paper 500 square feet of kitchen walls if a heavy wallpaper will be used.

ILLUSTRATION 2

COVERAGE: One-half gallon will hang approximately 4 single rolls (140 sq ft), depending on the weight of the wall covering and the condition of the wall.

44. RECOMMENDED DOSAGE The recommended child's dose of the sedative hydroxine is 0.006 gram per kilogram of body mass. Find the dosage for a 30-kg child in milligrams.

45. ERGONOMICS The science of ergonomics coordinates the design of working conditions with the requirements of the worker. Illustration 3 gives guidelines for the dimensions (in inches) of a computer workstation to be used by a person whose height is 69 inches. Find a set of workstation dimensions for a person 5 feet 11 inches tall. Round to the nearest tenth.

ILLUSTRATION 3

48.5 Eye height

27.0 Working/ elbow height

17.1 Seat height

Based on information from the Anthropometric Survey of the U.S. Army personnel database

46. SHOPPING A recipe for guacamole dip calls for 5 avocados. If they are advertised at 3 for $1.98, what will 5 avocados cost?

47. DRAWING See Illustration 4. To make an enlargement of the sailboat, an artist drew a grid over the smaller picture and transferred the contents of each small box to its corresponding larger box on another sheet of paper. If the smaller picture is 3 in. × 5 in. and if the width of the enlargement is 7.5 in., what are the dimensions of the enlargement?

48. DRAFTING In a scale drawing, a 280-foot antenna tower is drawn $7\frac{1}{2}$ inches high. The building next to it is drawn $2\frac{1}{4}$ inches high. How tall is the actual building?

ILLUSTRATION 4

In Exercises 49–54, use similar triangles to help solve each problem.

49. WASHINGTON, DC The Washington Monument casts a shadow of $166\frac{1}{2}$ feet at the same time as a 5-foot-tall tourist casts a shadow of $1\frac{1}{2}$ feet. (See Illustration 5.) Find the height of the monument.

ILLUSTRATION 5

5 ft

$1\frac{1}{2}$ ft

$166\frac{1}{2}$ ft

50. HEIGHT OF A FLAGPOLE A man places a mirror on the ground and sees the reflection of the top of a flagpole, as in Illustration 6. The two triangles in the illustration are similar. Find the height h of the flagpole.

ILLUSTRATION 6

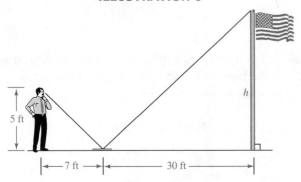

5 ft

7 ft

30 ft

h

51. WIDTH OF A RIVER Use the dimensions in Illustration 7 to find w, the width of the river. The two triangles in the illustration are similar.

ILLUSTRATION 7

20 ft

32 ft

75 ft

w ft

52. FLIGHT PATH An airplane ascends 150 feet as it flies a horizontal distance of 1,000 feet. How much altitude will it gain as it flies a horizontal distance of 1 mile? (See Illustration 8.) (*Hint:* 5,280 feet = 1 mile.)

ILLUSTRATION 8

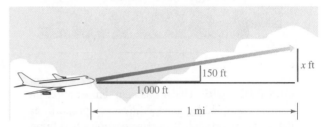

x ft

150 ft

1,000 ft

1 mi

53. SKI RUNS A ski course with $\frac{1}{2}$ mile of horizontal run falls 100 feet in every 300 feet of run. Find the height of the hill.

54. GRAPHIC ARTS The compass shown in Illustration 9 is used to draw circles with different radii (plural for radius). For the setting shown, what radius will the resulting circle have?

ILLUSTRATION 9

2.25 cm

6 cm

1.5 cm

In Exercises 55–68, solve each problem by writing a variation model of the situation.

55. FREE FALL An object in free fall travels a distance s that is directly proportional to the square of the time t. If an object falls 1,024 feet in 8 seconds, how far will it fall in 10 seconds?

56. FINDING DISTANCE The distance that a car can go is directly proportional to the number of gallons of gasoline it consumes. If a car can go 288 miles on 12 gallons of gasoline, how far can it go on a full tank of 18 gallons?

57. FARMING The length of time that a given number of bushels of corn will last when feeding cattle varies inversely with the number of animals. If x bushels will feed 25 cows for 10 days, how long will the feed last for 10 cows?

58. ORGAN PIPES The frequency of vibration of air in an organ pipe is inversely proportional to the length of the pipe. (See Illustration 10.) If a pipe 2 feet long vibrates 256 times per second, how many times per second will a 6-foot pipe vibrate?

ILLUSTRATION 10

59. GAS PRESSURE Under constant temperature, the volume occupied by a gas is inversely proportional to the pressure applied. If the gas occupies a volume of 20 cubic inches under a pressure of 6 pounds per square inch, find the volume when the gas is subjected to a pressure of 10 pounds per square inch.

60. REAL ESTATE Illustration 11 shows the listing price for three homes in the same general locality. Write the variation model (direct or inverse) that describes the relationship between the listing price and the number of square feet of a house in this area.

ILLUSTRATION 11

Number of square feet	Listing price
1,720	$129,000
1,205	$90,375
1,080	$81,000

61. TRUCKING COSTS The costs incurred by a trucking company vary jointly with the number of trucks in service and the number of hours they are used. When 4 trucks are used for 6 hours each, the costs are $1,800. Find the costs of using 10 trucks, each for 12 hours.

62. OIL STORAGE The number of gallons of oil that can be stored in a cylindrical tank varies jointly with the height of the tank and the square of the radius of

its base. The constant of proportionality is 23.5. Find the number of gallons that can be stored in the cylindrical tank in Illustration 12.

ILLUSTRATION 12

20 ft

15 ft

63. ELECTRONICS The voltage (in volts) measured across a resistor is directly proportional to the current (in amperes) flowing through the resistor. The constant of variation is the **resistance** (in ohms). If 6 volts is measured across a resistor carrying a current of 2 amperes, find the resistance.

64. ELECTRONICS The power (in watts) lost in a resistor (in the form of heat) is directly proportional to the square of the current (in amperes) passing through it. The constant of proportionality is the resistance (in ohms). What power is lost in a 5-ohm resistor carrying a 3-ampere current?

65. STRUCTURAL ENGINEERING The deflection of a beam is inversely proportional to its width and the cube of its depth. If the deflection of a 4-inch-by-4-inch beam is 1.1 inches, find the deflection of a 2-inch-by-8-inch beam positioned as in Illustration 13.

ILLUSTRATION 13

Width   Force

Depth

66. STRUCTURAL ENGINEERING Find the deflection of the beam in Exercise 65 when the beam is positioned as in Illustration 14.

ILLUSTRATION 14

Width   Force

Depth

67. TENSION IN A STRING When playing with a Skip It toy, a child swings a weighted ball on the end of a string in a circular motion around one leg while jumping over the revolving string with the other leg. (See Illustration 15.) The tension T in the string is directly proportional to the square of the speed s of the ball and inversely proportional to the radius r of the circle. If the tension in the string is 6 pounds when the speed of the ball is 6 feet per second and the radius is 3 feet, find the tension when the speed is 8 feet per second and the radius is 2.5 feet.

ILLUSTRATION 15

68. GAS PRESSURE The pressure of a certain amount of gas is directly proportional to the temperature (measured in degrees Kelvin) and inversely proportional to the volume. A sample of gas at a pressure of 1 atmosphere occupies a volume of 1 cubic meter at a temperature of 273 Kelvin. When heated, the gas expands to twice its volume, but the pressure remains constant. To what temperature is it heated?

WRITING

69. Distinguish between a *ratio* and a *proportion*.

70. From everyday life, give examples of two quantities that vary directly and two quantities that vary inversely.

REVIEW

Simplify each expression.

71. $(x^2x^3)^2$

72. $\left(\dfrac{a^3a^5}{a^{-2}}\right)^3$

73. $\dfrac{b^0 - 2b^0}{b^0}$

74. $\left(\dfrac{2r^{-2}r^{-3}}{4r^{-5}}\right)^{-3}$

▶ **6.3**

Multiplying and Dividing Rational Expressions

In this section, you will learn about

Multiplying rational expressions ■ Finding powers of rational expressions ■ Dividing rational expressions ■ Mixed operations

Introduction   In this section, we begin with a review of the rules for multiplying and dividing arithmetic fractions—fractions whose numerators and denominators are integers. Then we use these rules, in combination with the simplification skills learned in Section 6.1, to find products and quotients of rational expressions.

Multiplying Rational Expressions

In Section 6.1, we introduced four basic properties of fractions. We now present the rule for multiplying fractions.

Multiplying fractions

If a, b, c, and d are real numbers and if no denominators are 0, then

$$\frac{a}{b} \cdot \frac{c}{d} = \frac{a \cdot c}{b \cdot d} = \frac{ac}{bd}$$

To multiply fractions, we multiply the numerators and multiply the denominators.

$$\frac{3}{5} \cdot \frac{2}{7} = \frac{3 \cdot 2}{5 \cdot 7} \qquad \frac{4}{7} \cdot \frac{5}{8} = \frac{4 \cdot 5}{7 \cdot 8}$$

$$= \frac{6}{35} \qquad\qquad = \frac{\overset{1}{\cancel{2}} \cdot \overset{1}{\cancel{2}} \cdot 5}{7 \cdot \underset{1}{\cancel{2}} \cdot \underset{1}{\cancel{2}} \cdot 2} \quad \frac{2}{2} = 1.$$

$$= \frac{5}{14}$$

The same rule applies to rational expressions. If $t \neq 0$, then

$$\frac{x^2 y}{t} \cdot \frac{xy^3}{t^3} = \frac{x^2 y \cdot xy^3}{tt^3}$$

$$= \frac{x^2 x \cdot yy^3}{t^4}$$

$$= \frac{x^3 y^4}{t^4}$$

E X A M P L E   1

Multiplying rational expressions. Find the product of $\dfrac{x^2 - 6x + 9}{x}$ and $\dfrac{x^2}{x - 3}$.

Solution We multiply the numerators and multiply the denominators and then simplify the resulting fraction.

$$\frac{x^2 - 6x + 9}{x} \cdot \frac{x^2}{x - 3} = \frac{(x^2 - 6x + 9)x^2}{x(x - 3)} \qquad \text{Multiply the numerators and multiply the denominators.}$$

$$= \frac{(x - 3)(x - 3)xx}{x(x - 3)} \qquad \text{Factor in the numerator.}$$

$$= \frac{\overset{1}{\cancel{(x - 3)}}(x - 3)\overset{1}{\cancel{x}}x}{\underset{1}{\cancel{x}}\underset{1}{\cancel{(x - 3)}}} \qquad \text{Divide out common factors: } \frac{x-3}{x-3} = 1 \text{ and } \frac{x}{x} = 1.$$

$$= x(x - 3)$$

SELF CHECK   Multiply $\dfrac{a^2 + 6a + 9}{a} \cdot \dfrac{a^3}{a + 3}$.

Answer: $a^2(a + 3)$ ∎

A C C E N T   O N   T E C H N O L O G Y *Checking an Algebraic Simplification*

We can check the simplification in Example 1 by graphing the rational functions $f(x) = \left(\frac{x^2 - 6x + 9}{x}\right)\left(\frac{x^2}{x-3}\right)$, shown in Figure 6-9(a), and $g(x) = x(x - 3)$, shown in Figure 6-9(b), and observing that the graphs are the same, except that 0 and 3 are not included in the domain of the first function.

FIGURE 6-9

(a)

(b)

The split-screen G-T (graph, table) mode can also be used to check a simplification. If we enter $Y_3 = Y_1 - Y_2$, use the cursor to highlight the $=$ sign as shown in Figure 6-10(a), and then press $\boxed{\text{GRAPH}}$, we get the display shown in Figure 6-10(b). The zeros under the Y_3 column indicate that the value of $\left(\frac{x^2 - 6x + 9}{x}\right)\left(\frac{x^2}{x - 3}\right)$ and the value of $x(x - 3)$ are the same for different values of x. (The error message is given because when $x = 0$, $\left(\frac{x^2 - 6x + 9}{x}\right)\left(\frac{x^2}{x - 3}\right)$ is undefined.)

The graph of $Y_3 = Y_1 - Y_2$ is difficult to see because it lies on the x-axis. The graph indicates that for all x-values (except those that make the fractions undefined), $Y_3 = 0$, or more specifically, $\left(\frac{x^2 - 6x + 9}{x}\right)\left(\frac{x^2}{x - 3}\right) = x(x - 3)$.

FIGURE 6-10

(a)

(b)

EXAMPLE 2   **Multiplying rational expressions.** Multiply $\dfrac{x^2 - x - 6}{x^2 - 4} \cdot \dfrac{x^2 + x - 6}{x^2 - 9}$.

Solution $\dfrac{x^2 - x - 6}{x^2 - 4} \cdot \dfrac{x^2 + x - 6}{x^2 - 9}$

$= \dfrac{(x^2 - x - 6)(x^2 + x - 6)}{(x^2 - 4)(x^2 - 9)}$ Multiply the numerators and multiply the denominators.

$= \dfrac{(x - 3)(x + 2)(x + 3)(x - 2)}{(x + 2)(x - 2)(x + 3)(x - 3)}$ Factor the polynomials.

$= \dfrac{\overset{1}{\cancel{(x - 3)}}\overset{1}{\cancel{(x + 2)}}\overset{1}{\cancel{(x + 3)}}\overset{1}{\cancel{(x - 2)}}}{\underset{1}{\cancel{(x + 2)}}\,\underset{1}{\cancel{(x - 2)}}\,\underset{1}{\cancel{(x + 3)}}\,\underset{1}{\cancel{(x - 3)}}}$ Divide out common factors: $\frac{x - 3}{x - 3} = 1$, $\frac{x + 2}{x + 2} = 1$, $\frac{x + 3}{x + 3} = 1$, and $\frac{x - 2}{x - 2} = 1$.

$= 1$

SELF CHECK   Multiply $\dfrac{a^2 + a - 56}{a^2 - 49} \cdot \dfrac{a^2 - a - 56}{a^2 - 64}$.   *Answer:* 1   ■

WARNING! Note that when all factors divide out, the result is 1 and not 0.

EXAMPLE 3

Multiplying rational expressions. Multiply

$$\frac{6x^2 + 5x - 4}{2x^2 + 5x + 3} \cdot \frac{8x^2 + 6x - 9}{12x^2 + 7x - 12}.$$

Solution
$$\frac{6x^2 + 5x - 4}{2x^2 + 5x + 3} \cdot \frac{8x^2 + 6x - 9}{12x^2 + 7x - 12}$$

$$= \frac{(6x^2 + 5x - 4)(8x^2 + 6x - 9)}{(2x^2 + 5x + 3)(12x^2 + 7x - 12)}$$
Multiply the numerators and multiply the denominators.

$$= \frac{(3x + 4)(2x - 1)(4x - 3)(2x + 3)}{(2x + 3)(x + 1)(3x + 4)(4x - 3)}$$
Factor the polynomials.

$$= \frac{\overset{1}{\cancel{(3x + 4)}}(2x - 1)\overset{1}{\cancel{(4x - 3)}}\overset{1}{\cancel{(2x + 3)}}}{\underset{1}{\cancel{(2x + 3)}}(x + 1)\underset{1}{\cancel{(3x + 4)}}\underset{1}{\cancel{(4x - 3)}}}$$
Divide out common factors.

$$= \frac{2x - 1}{x + 1}$$

SELF CHECK Multiply $\dfrac{2a^2 + 5a - 12}{2a^2 + 11a + 12} \cdot \dfrac{2a^2 - 3a - 9}{2a^2 - a - 3}.$ *Answer:* $\frac{a-3}{a+1}$ ∎

EXAMPLE 4

Writing a fraction with a denominator of 1.

Multiply $(2x - x^2) \cdot \dfrac{x}{x^2 - 5x + 6}.$

Solution
$$(2x - x^2) \cdot \frac{x}{x^2 - 5x + 6}$$

$$= \frac{2x - x^2}{1} \cdot \frac{x}{x^2 - 5x + 6}$$
Write $2x - x^2$ as $\dfrac{2x - x^2}{1}$.

$$= \frac{(2x - x^2)x}{1(x^2 - 5x + 6)}$$
Multiply the fractions.

$$= \frac{x\overset{-1}{\cancel{(2 - x)}}x}{1\underset{1}{\cancel{(x - 2)}}(x - 3)}$$
Factor out x in the numerator and factor the trinomial in the denominator. Recall that the quotient of any nonzero quantity and its negative is -1: $\frac{2-x}{x-2} = -1$.

$$= \frac{-x^2}{x - 3}$$

Since $\dfrac{-a}{b} = -\dfrac{a}{b}$, the $-$ sign can be written in front of the fraction. Thus, the final result can be written as

$$-\frac{x^2}{x - 3}$$

SELF CHECK Multiply $\dfrac{x^2 + 5x + 6}{(x^2 + 4x)(x + 2)}(x^3 + 4x^2).$ *Answer:* $x(x + 3)$ ∎

In Examples 1–4, we would obtain the same answers if we had factored first and divided out the common factors before we multiplied.

Finding Powers of Rational Expressions

EXAMPLE 5

Squaring a rational expression. Find $\left(\dfrac{x^2 + x - 1}{2x + 3}\right)^2$.

Solution To square the rational expression, we write it as a factor twice and do the multiplication.

$$\left(\frac{x^2 + x - 1}{2x + 3}\right)^2 = \left(\frac{x^2 + x - 1}{2x + 3}\right)\left(\frac{x^2 + x - 1}{2x + 3}\right)$$

$$= \frac{(x^2 + x - 1)(x^2 + x - 1)}{(2x + 3)(2x + 3)}$$

$$= \frac{x^4 + 2x^3 - x^2 - 2x + 1}{4x^2 + 12x + 9}$$

SELF CHECK Find $\left(\dfrac{x + 5}{x^2 - 6x}\right)^2$.

Answer: $\dfrac{x^2 + 10x + 25}{x^4 - 12x^3 + 36x^2}$ ∎

Dividing Rational Expressions

Here is the rule for dividing fractions.

Dividing fractions

If a, b, c, and d are real numbers and if no denominators are 0, then

$$\frac{a}{b} \div \frac{c}{d} = \frac{a}{b} \cdot \frac{d}{c} = \frac{ad}{bc}$$

We can prove this rule as follows:

$$\frac{a}{b} \div \frac{c}{d} = \frac{\dfrac{a}{b}}{\dfrac{c}{d}} = \frac{\dfrac{a}{b}}{\dfrac{c}{d}} \cdot 1 = \frac{\dfrac{a}{b}}{\dfrac{c}{d}} \cdot \frac{\dfrac{d}{c}}{\dfrac{d}{c}} = \frac{\dfrac{a}{b} \cdot \dfrac{d}{c}}{\dfrac{c}{d} \cdot \dfrac{d}{c}} = \frac{\dfrac{a}{b} \cdot \dfrac{d}{c}}{\dfrac{cd}{cd}} = \frac{\dfrac{a}{b} \cdot \dfrac{d}{c}}{1} = \frac{a}{b} \cdot \frac{d}{c}$$

Thus, *to divide two fractions, we can invert the divisor (the second fraction) and multiply.*

$$\frac{3}{5} \div \frac{2}{7} = \frac{3}{5} \cdot \frac{7}{2} \qquad\qquad \frac{4}{7} \div \frac{2}{21} = \frac{4}{7} \cdot \frac{21}{2}$$

$$= \frac{3 \cdot 7}{5 \cdot 2} \qquad\qquad\qquad = \frac{4 \cdot 21}{7 \cdot 2}$$

$$= \frac{21}{10} \qquad\qquad\qquad = \frac{\overset{1}{2} \cdot 2 \cdot 3 \cdot \overset{1}{7}}{\underset{1}{7} \cdot \underset{1}{2}}$$

$$= 6$$

We can state the rule for dividing two fractions in another way: *To divide two fractions, we multiply the first fraction and the reciprocal of the second.* The **reciprocal** of a fraction can be found by interchanging the numerator and denominator. For example, the reciprocal of $\frac{2}{7}$ is $\frac{7}{2}$. Similarly, the reciprocal of the rational expression $\frac{x^2 - 2x + 4}{2x^2 - 2}$ is $\frac{2x^2 - 2}{x^2 - 2x + 4}$.

The rule for division of fractions applies to rational expressions.

$$\frac{x^2}{y^3 z^2} \div \frac{x^2}{yz^3} = \frac{x^2}{y^3 z^2} \cdot \frac{yz^3}{x^2}$$

Invert the divisor and multiply (or multiply the first fraction by the reciprocal of the second).

$$= \frac{x^2 yz^3}{x^2 y^3 z^2}$$

Multiply the numerators and the denominators.

$$= x^{2-2} y^{1-3} z^{3-2}$$

To divide exponential expressions with the same base, keep the base and subtract the exponents.

$$= x^0 y^{-2} z^1$$

Simplify the exponents.

$$= 1 \cdot y^{-2} \cdot z \qquad x^0 = 1.$$

$$= \frac{z}{y^2}$$

Write the result without the negative exponent.

EXAMPLE 6

Dividing rational expressions. Divide $\dfrac{x^3 + 8}{x + 1} \div \dfrac{x^2 - 2x + 4}{2x^2 - 2}$.

Solution   We invert the divisor, which is $\dfrac{x^2 - 2x + 4}{2x^2 - 2}$, and multiply.

$$\frac{x^3 + 8}{x + 1} \div \frac{x^2 - 2x + 4}{2x^2 - 2}$$

$$= \frac{x^3 + 8}{x + 1} \cdot \frac{2x^2 - 2}{x^2 - 2x + 4}$$

$$= \frac{(x^3 + 8)(2x^2 - 2)}{(x + 1)(x^2 - 2x + 4)}$$

Multiply the numerators and the denominators.

$$= \frac{(x + 2)\overset{1}{\cancel{(x^2 - 2x + 4)}}2\overset{1}{\cancel{(x + 1)}}(x - 1)}{\underset{1}{\cancel{(x + 1)}}\underset{1}{\cancel{(x^2 - 2x + 4)}}}$$

Factor $x^3 + 8$ and $2x^2 - 2$. The polynomial $x^2 - 2x + 4$ does not factor. Then divide out common factors.

$$= 2(x + 2)(x - 1)$$

SELF CHECK   Divide $\dfrac{x^3 - 8}{x - 1} \div \dfrac{x^2 + 2x + 4}{3x^2 - 3x}$.

Answer: $3x(x - 2)$ ∎

EXAMPLE 7

Division involving two variables. Divide $\dfrac{b^3 - 4b}{x - 1} \div (b - 2)$.

Solution $\dfrac{b^3 - 4b}{x - 1} \div (b - 2) = \dfrac{b^3 - 4b}{x - 1} \div \dfrac{b - 2}{1}$

Write $b - 2$ as a fraction with a denominator of 1.

$$= \frac{b^3 - 4b}{x - 1} \cdot \frac{1}{b - 2}$$

Invert the divisor and multiply.

$$= \frac{b^3 - 4b}{(x - 1)(b - 2)}$$

Multiply the numerators and the denominators.

$$= \frac{b(b + 2)\cancel{(b - 2)}^{1}}{(x - 1)\cancel{(b - 2)}_{1}} \quad \text{Factor } b^3 - 4b \text{ and then divide out common factors.}$$

$$= \frac{b(b + 2)}{x - 1}$$

SELF CHECK   Divide $\dfrac{m^4 - 9m^2}{a^2 - 3a} \div (m^2 + 3m)$.

Answer: $\frac{m(m-3)}{a(a-3)}$ ■

Mixed Operations

E X A M P L E 8   Simplify $\dfrac{x^2 + 2x - 3}{6x^2 + 5x + 1} \div \dfrac{2x^2 - 2}{2x^2 - 5x - 3} \cdot \dfrac{6x^2 + 4x - 2}{x^2 - 2x - 3}$.

Solution   The rules for the order of operations require multiplications and divisions to be done in order from left to right, so we begin by focusing on the division. We introduce grouping symbols to emphasize this. To divide the rational expressions in the parentheses, we invert $\frac{2x^2 - 2}{2x^2 - 5x - 3}$ and multiply.

$$\left(\frac{x^2 + 2x - 3}{6x^2 + 5x + 1} \div \frac{2x^2 - 2}{2x^2 - 5x - 3} \right) \frac{6x^2 + 4x - 2}{x^2 - 2x - 3} = \left(\frac{x^2 + 2x - 3}{6x^2 + 5x + 1} \cdot \frac{2x^2 - 5x - 3}{2x^2 - 2} \right) \frac{6x^2 + 4x - 2}{x^2 - 2x - 3}$$

Next, we multiply the three fractions and simplify the result.

$$= \frac{(x^2 + 2x - 3)(2x^2 - 5x - 3)(6x^2 + 4x - 2)}{(6x^2 + 5x + 1)(2x^2 - 2)(x^2 - 2x - 3)}$$

$$= \frac{(x + 3)\cancel{(x - 1)}^{1}\cancel{(2x + 1)}^{1}\cancel{(x - 3)}^{1}2\cancel{(3x - 1)}^{1}\cancel{(x + 1)}^{1}}{(3x + 1)\cancel{(2x + 1)}_{1}2(x + 1)\cancel{(x - 1)}_{1}\cancel{(x - 3)}_{1}\cancel{(x + 1)}_{1}}$$

$$= \frac{(x + 3)(3x - 1)}{(3x + 1)(x + 1)} \qquad ■$$

STUDY SET

Section 6.3

VOCABULARY

In Exercises 1–4, fill in the blanks to make the statements true.

1. In the division statement

$$\frac{a^2 - 9}{a^2 - 49} \div \frac{a + 3}{a + 7}$$

the second fraction is called the _____.

2. The _____ of $\dfrac{a + 3}{a + 7}$ is $\dfrac{a + 7}{a + 3}$.

3. In a fraction, the part above the fraction bar is called the _____.

4. In a fraction, the part below the fraction bar is called the _____.

CONCEPTS

In Exercises 5–8, fill in the blanks to make the statements true.

5. To multiply two fractions, we _____ their numerators and multiply their denominators. In symbols, $\frac{a}{b} \cdot \frac{c}{d} =$ _____.

6. To divide two fractions, we invert the divisor and _____. In symbols, $\frac{a}{b} \div \frac{c}{d} =$ _____.

7. The denominator of a fraction cannot be ___.

8. $\dfrac{a+1}{a+1} =$ ___, provided $a \neq -1$

NOTATION

In Exercises 9–10, complete each solution.

9. $\dfrac{x^2+3x}{5x-25} \cdot \dfrac{x-5}{x+3} = \dfrac{(x^2+3x)\rule{1.5cm}{0.3mm}}{\rule{1.5cm}{0.3mm}\,(x+3)}$

$\qquad = \dfrac{\rule{1.5cm}{0.3mm}\,(x-5)}{\rule{1.5cm}{0.3mm}\,(x+3)}$

$\qquad = \dfrac{x}{5}$

10. $\dfrac{x^2-x-6}{4x^2+16x} \div \dfrac{x-3}{x+4} = \dfrac{x^2-x-6}{4x^2+16x} \cdot \rule{1.5cm}{0.3mm}$

$\qquad = \dfrac{\rule{1.5cm}{0.3mm}\,(x+4)}{(4x^2+16x)\rule{1.5cm}{0.3mm}}$

$\qquad = \dfrac{\rule{1.5cm}{0.3mm}\,(x+2)(x+4)}{\rule{1.5cm}{0.3mm}\,(x-3)}$

$\qquad = \dfrac{x+2}{4x}$

11. A student checks her answers with those in the back of her textbook. Tell whether they are equivalent.

Student's answer	Book's answer	Equivalent?
$\dfrac{-x^{10}}{y^2}$	$-\dfrac{x^{10}}{y^2}$	
$\dfrac{x-3}{x+3}$	$\dfrac{3-x}{3+x}$	
$\dfrac{b+a}{(2-x)(d+c)}$	$-\dfrac{a+b}{(x-2)(c+d)}$	

12. a. Write $5x^2 + 35x$ as a fraction.

   b. What is the reciprocal of $5x^2 + 35x$?

PRACTICE

In Exercises 13–54, do the operations and simplify.

13. $\dfrac{3}{4} \cdot \dfrac{5}{3}$

14. $-\dfrac{5}{6} \cdot \dfrac{3}{7}$

15. $-\dfrac{6}{11} \div \dfrac{36}{55}$

16. $\dfrac{17}{12} \div \dfrac{34}{3}$

17. $\dfrac{x^2y^2}{cd} \cdot \dfrac{c^{-2}d^2}{x}$

18. $\dfrac{a^{-2}b^2}{x^{-1}y} \cdot \dfrac{a^4b^4}{x^2y^3}$

19. $\dfrac{-x^2y^{-2}}{x^{-1}y^{-3}} \div \dfrac{x^{-3}y^2}{x^4y^{-1}}$

20. $\dfrac{(a^3)^2}{b^{-1}} \div \dfrac{(a^3)^{-2}}{b^{-1}}$

21. $\dfrac{x^2+2x+1}{x} \cdot \dfrac{x^2-x}{x^2-1}$

22. $\dfrac{a+6}{16-a^2} \cdot \dfrac{3a-12}{3a+18}$

23. $\dfrac{2x^2-x-3}{x^2-1} \cdot \dfrac{x^2+x-2}{2x^2+x-6}$

24. $\dfrac{9x^2+3x-20}{3x^2-7x+4} \cdot \dfrac{3x^2-5x+2}{9x^2+18x+5}$

25. $\dfrac{x^2-16}{x^2-25} \div \dfrac{x+4}{x-5}$

26. $\dfrac{a^2-9}{a^2-49} \div \dfrac{a+3}{a+7}$

27. $-\dfrac{a^2+2a-35}{12x} \div \dfrac{ax-3x}{a^2+4a-21}$

28. $\dfrac{x^2-4}{2b-bx} \div \dfrac{x^2+4x+4}{2b+bx}$

29. $\dfrac{3t^2-t-2}{6t^2-5t-6} \cdot \dfrac{4t^2-9}{2t^2+5t+3}$

30. $\dfrac{2p^2-5p-3}{p^2-9} \cdot \dfrac{2p^2+5p-3}{2p^2+5p+2}$

31. $\dfrac{3n^2+5n-2}{12n^2-13n+3} \div \dfrac{n^2+3n+2}{4n^2+5n-6}$

32. $\dfrac{8y^2-14y-15}{6y^2-11y-10} \div \dfrac{4y^2-9y-9}{3y^2-7y-6}$

33. $\dfrac{2x^2+5x+3}{3x^2-5x+2} \div \dfrac{2x^2+x-3}{-3x^2+5x-2}$

34. $\dfrac{2p^2-5p-3}{p^2-9} \div \dfrac{2p^2+5p+2}{2p^2+5p-3}$

35. $(x+1) \cdot \dfrac{1}{x^2+2x+1}$

36. $-\dfrac{x-2}{x} \div (x^2-4)$

37. $(x^2-x-2) \cdot \dfrac{x^2+3x+2}{x^2-4}$

38. $(2x^2-9x-5) \cdot \dfrac{x}{2x^2+x}$

39. $\dfrac{a^2+2a-35}{12x} \div \dfrac{ax-3x}{a^2+4a-21}$

40. $\dfrac{x^2-4}{2b-bx} \div \dfrac{x^2+4x+4}{2b+bx}$

41. $\dfrac{x^3+y^3}{x^3-y^3} \div \dfrac{x^2-xy+y^2}{x^2+xy+y^2}$

42. $\dfrac{x^2-6x+9}{4-x^2} \div \dfrac{x^2-9}{x^2-8x+12}$

43. $\dfrac{ax+ay+bx+by}{x^3-27} \div \dfrac{xc+xd+yc+yd}{x^2+3x+9}$

44. $\dfrac{x^2 + 3x + yx + 3y}{x^2 - 9} \div \dfrac{x + 3}{x - 3}$

45. $\dfrac{x^2 - x - 6}{x^2 - 4} \cdot \dfrac{x^2 - x - 2}{9 - x^2}$

46. $\dfrac{p^3 - q^3}{q^2 - p^2} \cdot \dfrac{q^2 + pq}{p^3 + p^2 q + pq^2}$

47. $(4x + 12) \cdot \dfrac{x^2}{2x - 6} \div \dfrac{2}{x - 3}$

48. $(4x^2 - 9) \div \dfrac{2x^2 + 5x + 3}{x + 2} \div (2x - 3)$

49. $(x^2 - x - 6) \div (x - 3) \div (x - 2)$

50. $(x^2 - x - 6) \div [(x - 3) \div (x - 2)]$

51. $\dfrac{2x^2 - 2x - 4}{x^2 + 2x - 8} \cdot \dfrac{3x^2 + 15x}{x + 1} \div \dfrac{4x^2 - 100}{x^2 - x - 20}$

52. $\dfrac{6a^2 - 7a - 3}{a^2 - 1} \div \dfrac{4a^2 - 12a + 9}{a^2 - 1} \cdot \dfrac{2a^2 - a - 3}{3a^2 - 2a - 1}$

53. $\dfrac{x^2 - x - 12}{x^2 + x - 2} \div \dfrac{x^2 - 6x + 8}{x^2 - 3x - 10} \cdot \dfrac{x^2 - 3x + 2}{x^2 - 2x - 15}$

54. $\dfrac{4a^2 - 10a + 6}{a^4 - 3a^3} \div \dfrac{3 - 2a}{2a^3} \cdot \dfrac{a - 3}{2a - 2}$

In Exercises 55–58, find each power.

55. $\left(\dfrac{x - 3}{x^3 + 4}\right)^2$

56. $\left(\dfrac{2t^2 + t}{t - 1}\right)^2$

57. $\left(\dfrac{2m^2 - m - 3}{x^2 - 1}\right)^2$

58. $\left(\dfrac{-k - 3}{x^2 - x + 1}\right)^2$

APPLICATIONS

59. PHYSICS EXPERIMENT Illustration 1 contains data from a physics experiment. k_1 and k_2 are constants. Complete the table.

ILLUSTRATION 1

Trial	Rate (m/sec)	Time (sec)	Distance (m)
1	$\dfrac{k_1^2 + 3k_1 + 2}{k_1 - 3}$	$\dfrac{k_1^2 - 3k_1}{k_1 + 1}$	
2	$\dfrac{k_2^2 + 6k_2 + 5}{k_2 + 1}$		$k_2^2 + 11k_2 + 30$

60. TRUNK CAPACITY The shape of the storage space in the trunk of the car shown in Illustration 2 is approximately a rectangular solid. Write a simplified rational expression that gives the number of cubic units of storage space in the trunk.

ILLUSTRATION 2

WRITING

61. Explain how to multiply two rational expressions.

62. Write some comments to the student who wrote the following solution, explaining the **error**.

$$\dfrac{x^2 + x - 2}{x^2 - 4} \cdot \dfrac{x - 2}{x - 1} = \dfrac{(x + 2)(x - 1)}{(x + 2)(x - 2)} \cdot \dfrac{x - 2}{x - 1}$$

$$= \dfrac{(x + 2)(x - 1)(x - 2)}{(x + 2)(x - 2)(x - 1)}$$

$$= 0$$

63. The graph of $Y_3 = Y_1 - Y_2$, where

$$Y_1 = \left(\dfrac{2x^2 - 5x - 3}{x^2 - 9}\right)\left(\dfrac{2x^2 + 5x - 3}{2x^2 + 5x + 2}\right)$$

$$Y_2 = \dfrac{2x - 1}{x + 2}$$

is shown in Illustration 3. Explain how the graph and table can be used to verify that

$$\left(\dfrac{2x^2 - 5x - 3}{x^2 - 9}\right) \cdot \left(\dfrac{2x^2 + 5x - 3}{2x^2 + 5x + 2}\right) = \dfrac{2x - 1}{x + 2}$$

ILLUSTRATION 3

64. A student obtained an answer of $\frac{x+3}{x+7}$ after performing $\frac{x^2-9}{x^2-49} \div \frac{x+3}{x+7}$. As a check, he graphed $Y_3 = Y_1 - Y_2$, where

$$Y_1 = \left(\frac{x^2-9}{x^2-49}\right) \div \left(\frac{x+3}{x+7}\right)$$

$$Y_2 = \frac{x+3}{x+7}$$

See Illustration 4. Explain what conclusion can be drawn from the graph and the table.

ILLUSTRATION 4

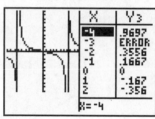

REVIEW

Do each operation.

65. $-2a^2(3a^3 - a^2)$

66. $(2t - 1)^2$

67. $(2g - n)(3g - n)$

68. $(2c - b)(4c^2 + 2cb + b^2)$

▶ **6.4**

Adding and Subtracting Rational Expressions

In this section, you will learn about

Adding and subtracting rational expressions with like denominators ■ Adding and subtracting rational expressions with unlike denominators ■ Finding the least common denominator ■ Mixed operations

---

Introduction The procedures used to add and subtract rational expressions are based on the rules for adding and subtracting arithmetic fractions. In this section, we will add and subtract rational expressions with *like* and *unlike* denominators.

Adding and Subtracting Rational Expressions with Like Denominators

Fractions with like denominators are added and subtracted according to the following rules.

Adding and subtracting fractions

If a, b, and c are real numbers and if there are no divisions by 0, then

$$\frac{a}{b} + \frac{c}{b} = \frac{a+c}{b} \qquad \text{and} \qquad \frac{a}{b} - \frac{c}{b} = \frac{a-c}{b}$$

In words, *we add (or subtract) fractions with like denominators by adding (or subtracting) the numerators and keeping the common denominator.* Whenever possible, we should simplify the result. These two rules apply to addition and subtraction of rational expressions with like denominators.

EXAMPLE 1

Rational expressions with like denominators. Perform the operations:

a. $\dfrac{4}{3x} + \dfrac{7}{3x}$ and **b.** $\dfrac{a^2}{a^2 - 1} - \dfrac{a}{a^2 - 1}$.

Solution

a. $\dfrac{4}{3x} + \dfrac{7}{3x} = \dfrac{4 + 7}{3x}$ Add the numerators and keep the common denominator.

$\qquad\qquad = \dfrac{11}{3x}$ Do the addition in the numerator.

b. $\dfrac{a^2}{a^2 - 1} - \dfrac{a}{a^2 - 1} = \dfrac{a^2 - a}{a^2 - 1}$ Subtract the numerators and keep the common denominator.

We note that the polynomials factor in the numerator and the denominator of the result.

$$\dfrac{a^2}{a^2 - 1} - \dfrac{a}{a^2 - 1} = \dfrac{a\overset{1}{\cancel{(a - 1)}}}{(a + 1)\underset{1}{\cancel{(a - 1)}}} \quad \text{Simplify by dividing out the common factor of } (a - 1).$$

$$= \dfrac{a}{a + 1}$$

SELF CHECK

Perform the operations: **a.** $\frac{17}{22} + \frac{13}{22}$, **b.** $\frac{1}{6a} - \frac{7}{6a}$, and **c.** $\frac{3a}{a - 2} + \frac{2a}{a - 2}$.

Answers: **a.** $\frac{15}{11}$, **b.** $-\frac{1}{a}$, **c.** $\frac{5a}{a - 2}$ ∎

ACCENT ON TECHNOLOGY *Checking Algebra*

We can check the subtraction in part b of Example 1 by graphing the rational functions $f(a) = \frac{a^2}{a^2 - 1} - \frac{a}{a^2 - 1}$, shown in Figure 6-11(a), and $g(a) = \frac{a}{a + 1}$, shown in Figure 6-11(b), and observing that the graphs are the same. Note that -1 and 1 are not in the domain of the first function and that -1 is not in the domain of the second function.

FIGURE 6-11

(a) (b) (c)

Figure 6-11(c) shows the display when the G-T mode is used to check the simplification. Here, $Y_3 = Y_1 - Y_2$, where $Y_1 = \frac{x^2}{x^2 - 1} - \frac{x}{x^2 - 1}$ and $Y_2 = \frac{x}{x + 1}$.

Adding and Subtracting Rational Expressions with Unlike Denominators

To add or subtract fractions with unlike denominators, we change them into fractions with a common denominator. This is done using the **fundamental property of fractions.**

The fundamental property of fractions

If a, b, and k are real numbers, and $b \neq 0$ and $k \neq 0$, then

$$\frac{a}{b} = \frac{a \cdot k}{b \cdot k}$$

In words, *multiplying the numerator and the denominator of a fraction by the same nonzero number does not change the value of the fraction.* This property is true because multiplying the numerator and denominator of a fraction by the same number is equivalent to multiplying it by 1. When a number is multiplied by 1, its value does not change. We use this property to add and subtract rational expressions with unlike denominators.

When adding or subtracting two rational expressions with denominators that are negatives (opposites), we can multiply the numerator and the denominator of one of the expressions by -1 to get a common denominator.

EXAMPLE 2

Denominators that are negatives (opposites). Add $\dfrac{x}{x-y} + \dfrac{y}{y-x}$.

Solution

$$\frac{x}{x-y} + \frac{y}{y-x} = \frac{x}{x-y} + \frac{(-1)y}{(-1)(y-x)}$$ Multiply the numerator and the denominator of the second rational expression by -1.

$$= \frac{x}{x-y} + \frac{-y}{-y+x}$$ Do the multiplication.

$$= \frac{x}{x-y} + \frac{-y}{x-y}$$ Rewrite the second denominator, $-y+x$, as $x-y$. The fractions now have a common denominator.

$$= \frac{x-y}{x-y}$$ Add the numerators and keep the common denominator.

$$= 1$$ Simplify.

SELF CHECK   Add $\dfrac{2a}{a-b} + \dfrac{b}{b-a}$.   *Answer:* $\dfrac{2a-b}{a-b}$ ∎

When adding or subtracting two rational expressions with different denominators, we must often multiply one or both of them by an appropriate expression to get a common denominator. This process is called **building a fraction.**

EXAMPLE 3

Subtracting rational expressions. Subtract $3 - \dfrac{7}{x-2}$.

Solution   If we write 3 as $\frac{3}{1}$ and then multiply its numerator and denominator by $x - 2$, the fractions will have a common denominator of $x - 2$.

$$3 - \frac{7}{x-2} = \frac{3}{1} - \frac{7}{x-2}$$ $3 = \frac{3}{1}$.

$$= \frac{3(x-2)}{1(x-2)} - \frac{7}{x-2}$$ Build $\frac{3}{1}$ to a fraction with denominator $x-2$ by multiplying its numerator and denominator by $x-2$.

$$= \frac{3x-6}{x-2} - \frac{7}{x-2}$$ Distribute the 3.

$$= \frac{3x-6-7}{x-2}$$ Subtract the numerators and keep the common denominator.

$$= \frac{3x-13}{x-2}$$ Combine like terms in the numerator.

SELF CHECK   Subtract $6 - \dfrac{5y}{6-y}$.   *Answer:* $\dfrac{-11y+36}{6-y}$ ∎

EXAMPLE 4

Adding rational expressions. Add $\dfrac{3}{x} + \dfrac{4}{y}$.

Solution   A common denominator for the fractions is xy. We multiply each numerator and denominator by the appropriate factor so that each denominator builds to xy.

$$\frac{3}{x} + \frac{4}{y} = \frac{3 \cdot y}{x \cdot y} + \frac{4 \cdot x}{y \cdot x}$$

$$= \frac{3y}{xy} + \frac{4x}{xy} \qquad \text{Multiply in the numerators and in the denominators.}$$

$$= \frac{3y + 4x}{xy} \qquad \text{Add the numerators and keep the common denominator.}$$

SELF CHECK   Add $\frac{5}{a} + \frac{7}{b}$.   *Answer:* $\frac{5b + 7a}{ab}$ ∎

EXAMPLE 5

Subtracting rational expressions. Subtract $\dfrac{4x}{x + 2} - \dfrac{7x}{x - 2}$.

Solution   By inspection, we see that a common denominator is $(x + 2)(x - 2)$. We multiply the numerator and denominator of each rational expression by the appropriate factor, so that each one has a denominator of $(x + 2)(x - 2)$.

$$\frac{4x}{x + 2} - \frac{7x}{x - 2}$$

$$= \frac{4x(x - 2)}{(x + 2)(x - 2)} - \frac{(x + 2)7x}{(x + 2)(x - 2)}$$

$$= \frac{4x^2 - 8x}{(x + 2)(x - 2)} - \frac{7x^2 + 14x}{(x + 2)(x - 2)} \qquad \begin{array}{l}\text{Use the distributive property to remove} \\ \text{parentheses in each numerator.}\end{array}$$

$$= \frac{(4x^2 - 8x) - (7x^2 + 14x)}{(x + 2)(x - 2)} \qquad \begin{array}{l}\text{Subtract the numerators and keep the} \\ \text{common denominator. Since the } - \text{ sign} \\ \text{between the fractions in the previous step} \\ \text{applies to both terms of } 7x^2 + 14x, \text{ we} \\ \text{must insert parentheses to show this.}\end{array}$$

$$= \frac{4x^2 - 8x - 7x^2 - 14x}{(x + 2)(x - 2)} \qquad \begin{array}{l}\text{In the numerator, to subtract the polyno-} \\ \text{mials, add the first and the opposite of} \\ \text{the second.}\end{array}$$

$$= \frac{-3x^2 - 22x}{(x + 2)(x - 2)} \qquad \text{Combine like terms.}$$

If the common factor of $-x$ is factored out of the terms in the numerator, this result can be written in two other equivalent forms.

$$\frac{-3x^2 - 22x}{(x + 2)(x - 2)} = \frac{-x(3x + 22)}{(x + 2)(x - 2)} = -\frac{x(3x + 22)}{(x + 2)(x - 2)}$$

SELF CHECK   Subtract $\frac{3a}{a + 3} - \frac{5a}{a - 3}$.   *Answer:* $\dfrac{-2a(a + 12)}{(a + 3)(a - 3)}$ ∎

Finding the Least Common Denominator

When adding or subtracting rational expressions with unlike denominators, it is easiest if we write the rational expressions in terms of the smallest common denominator possible, called the **least** (or lowest) **common denominator (LCD).** To find the least common denominator of several rational expressions, we follow these steps.

Finding the LCD

1. Factor each denominator.
2. List the different factors of each denominator.
3. Write each factor found in step 2 to the highest power that occurs in any one factorization.
4. The LCD is the product of the factors to the highest powers found in step 3.

E X A M P L E   6

Finding the LCD of rational expressions. Find the LCD of
a. $\dfrac{5a}{24b}$ and $\dfrac{11a}{18b^2}$ and **b.** $\dfrac{1}{x^2 - 12x + 36}$ and $\dfrac{3 - x}{x^2 - 6x}$.

Solution   **a.** We write each denominator as the product of prime numbers and variables.

$$24b = 2 \cdot 2 \cdot 2 \cdot 3 \cdot b = 2^3 \cdot 3 \cdot b$$
$$18b^2 = 2 \cdot 3 \cdot 3 \cdot b \cdot b = 2 \cdot 3^2 \cdot b^2$$

To find the LCD, we form a product using each of these factors the greatest number of times it appears in any one factorization. We use 2 three times, because it appears three times as a factor of 24. We use 3 twice, because it occurs twice as a factor of 18. We use b twice, because it appears twice as a factor of $18b^2$.

$$\text{LCD} = 2 \cdot 2 \cdot 2 \cdot 3 \cdot 3 \cdot b \cdot b$$
$$= 72b^2$$

b. We factor each denominator completely:

$$x^2 - 12x + 36 = (x - 6)^2 \quad x - 6 \text{ occurs as a factor twice.}$$
$$x^2 - 6x = x(x - 6)$$

To find the LCD, we form a product using the highest power of each of the factors:

$$\text{LCD} = x(x - 6)^2$$

SELF CHECK

Find the LCD of
a. $\dfrac{3y}{28z^3}$ and $\dfrac{5x}{21z}$ and
b. $\dfrac{a - 1}{a^2 - 25}$ and $\dfrac{3 - a^2}{a^2 + 7a + 10}$.

Answers: **a.** $84z^3$,
b. $(a - 5)(a + 5)(a + 2)$ ∎

To add or subtract rational expressions with unlike denominators, we follow these steps.

Adding or subtracting rational expressions with unlike denominators

1. Find the LCD.
2. Express each rational expression with a denominator that is the LCD.
3. Add (or subtract) the resulting rational expressions.
4. Simplify the result if possible.

EXAMPLE 7

Adding rational expressions with unlike denominators.

Add $\dfrac{5a}{24b} + \dfrac{11a}{18b^2}$.

Solution   In Example 6, we saw that the LCD of these rational expressions is $72b^2$. We multiply each numerator and denominator by whatever it takes to build the denominator to $72b^2$.

$$\frac{5a}{24b} + \frac{11a}{18b^2} = \frac{5a \cdot 3b}{24b \cdot 3b} + \frac{11a \cdot 4}{18b^2 \cdot 4}$$

$$= \frac{15ab}{72b^2} + \frac{44a}{72b^2} \qquad \text{Do the multiplications in the numerators and the denominators.}$$

$$= \frac{15ab + 44a}{72b^2} \qquad \text{Add the numerators and keep the common denominator. The result does not simplify.}$$

SELF CHECK   Add $\dfrac{3y}{28z^3} + \dfrac{5x}{21z}$.   *Answer:* $\dfrac{9y + 20xz^2}{84z^3}$ ∎

EXAMPLE 8

Subtracting rational expressions with unlike denominators.

Subtract $\dfrac{x}{x^2 - 2x + 1} - \dfrac{4}{x^2 - 1}$.

Solution   We factor each denominator to find the LCD:

$$x^2 - 2x + 1 = (x - 1)(x - 1) = (x - 1)^2$$
$$x^2 - 1 = (x + 1)(x - 1)$$

The LCD is $(x - 1)^2(x + 1)$ or $(x - 1)(x - 1)(x + 1)$.

We now write each rational expression with its denominator in factored form. Then we multiply each numerator and denominator by the missing factor, so that each has a denominator of $(x - 1)(x - 1)(x + 1)$.

$$\frac{x}{x^2 - 2x + 1} - \frac{4}{x^2 - 1}$$

$$= \frac{x}{(x - 1)(x - 1)} - \frac{4}{(x + 1)(x - 1)}$$

$$= \frac{x(x + 1)}{(x - 1)(x - 1)(x + 1)} - \frac{4(x - 1)}{(x + 1)(x - 1)(x - 1)}$$

$$= \frac{x^2 + x}{(x - 1)(x - 1)(x + 1)} - \frac{4x - 4}{(x + 1)(x - 1)(x - 1)}$$

$$= \frac{(x^2 + x) - (4x - 4)}{(x - 1)(x - 1)(x + 1)} \qquad \text{Subtract the numerators and keep the common denominator.}$$

$$= \frac{x^2 + x - 4x + 4}{(x - 1)(x - 1)(x + 1)} \qquad \text{In the numerator, subtract the polynomials.}$$

$$= \frac{x^2 - 3x + 4}{(x - 1)^2(x + 1)} \qquad \text{Combine like terms. The result does not simplify.}$$

SELF CHECK   Subtract $\dfrac{a}{a^2 - 4a + 4} - \dfrac{2}{a^2 - 4}$.   *Answer:* $\dfrac{a^2 + 4}{(a - 2)^2(a + 2)}$ ∎

Mixed Operations

EXAMPLE 9

Combine: $\dfrac{2x}{x^2-4} - \dfrac{1}{x^2-3x+2} + \dfrac{x+1}{x^2+x-2}$.

Solution   We factor each denominator to find the LCD:

$$\left.\begin{array}{l} x^2-4 = (x-2)(x+2) \\ x^2-3x+2 = (x-2)(x-1) \\ x^2+x-2 = (x-1)(x+2) \end{array}\right\} \text{LCD} = (x-2)(x+2)(x-1)$$

We then write each rational expression as an equivalent rational expression with the LCD as its denominator and do the subtraction and addition.

$$\dfrac{2x}{x^2-4} - \dfrac{1}{x^2-3x+2} + \dfrac{x+1}{x^2+x-2}$$

$$= \dfrac{2x}{(x-2)(x+2)} - \dfrac{1}{(x-2)(x-1)} + \dfrac{x+1}{(x-1)(x+2)}$$

$$= \dfrac{2x(x-1)}{(x-2)(x+2)(x-1)} - \dfrac{1(x+2)}{(x-2)(x-1)(x+2)} + \dfrac{(x+1)(x-2)}{(x-1)(x+2)(x-2)}$$

$$= \dfrac{2x(x-1) - 1(x+2) + (x+1)(x-2)}{(x+2)(x-2)(x-1)}$$

$$= \dfrac{2x^2 - 2x - x - 2 + x^2 - x - 2}{(x+2)(x-2)(x-1)}$$

$$= \dfrac{3x^2 - 4x - 4}{(x+2)(x-2)(x-1)} \qquad \text{This result can be simplified.}$$

$$= \dfrac{\overset{1}{(3x+2)\cancel{(x-2)}}}{(x+2)\underset{1}{\cancel{(x-2)}}(x-1)} \qquad \text{Factor the trinomial and divide out the common factor.}$$

$$= \dfrac{3x+2}{(x+2)(x-1)}$$ ∎

STUDY SET

Section 6.4

VOCABULARY

In Exercises 1–2, fill in the blanks to make the statements true.

1. The least common denominator of a group of rational expressions is _____ by each of their denominators.

2. The rational expressions $\frac{x}{x-5}$ and $\frac{x+1}{x-5}$ have _____ denominators. The rational expressions $\frac{2x}{x+5}$ and $\frac{5}{x-5}$ have _____ denominators.

CONCEPTS

In Exercises 3–8, fill in the blanks to make the statements true.

3. To subtract fractions with like denominators, we _____ the numerators and _____ the common denominator. In symbols,

$$\dfrac{a}{b} - \dfrac{c}{b} = \boxed{}$$

4. To add fractions with like denominators, we
_____ the numerators and keep the
_____ denominator. In symbols,

$$\frac{a}{b} + \frac{c}{b} = \boxed{}$$

5. Multiplying the numerator and the denominator of a fraction by the _____ nonzero number does not change the value of the fraction.

6. To find the LCD of several rational expressions, we _____ each denominator and use each factor to the _____ power that it appears in any one factorization.

7. The abbreviation for "least common denominator" is _____.

8. The denominators of the rational expressions $\frac{a-2}{a-8}$ and $\frac{a}{8-a}$ are _____.

In Exercises 9–10, consider the two procedures shown here.

i. $$\frac{x^2 - 2x}{x^2 + 4x - 12} = \frac{x\overset{1}{\cancel{(x-2)}}}{(x+6)\underset{1}{\cancel{(x-2)}}} = \frac{x}{x+6}$$

ii. $$\frac{x}{x+6} = \frac{x(x-2)}{(x+6)(x-2)} = \frac{x^2 - 2x}{x^2 + 4x - 12}$$

9. **a.** In which of these procedures are we *building* a rational expression?
   b. For what type of problem is this procedure often necessary?
   c. What name is used to describe the other procedure?

10. We can think of these two procedures as having the opposite effect. Explain.

NOTATION

In Exercises 11–12, complete each solution.

11. $$\frac{6x-1}{3x-1} + \frac{3x-2}{3x-1} = \frac{6x - 1 + \boxed{}}{3x - 1}$$
$$= \frac{9x - \boxed{}}{3x - 1}$$
$$= \frac{3\left(\boxed{}\right)}{3x - 1}$$
$$= 3$$

12. $$\frac{8}{3v} - \frac{1}{4v^2} = \frac{8\left(\boxed{}\right)}{3v(4v)} - \frac{1(3)}{4v^2\left(\boxed{}\right)}$$
$$= \frac{\boxed{}}{12v^2} - \frac{3}{\boxed{}}$$
$$= \frac{32v - 3}{12v^2}$$

PRACTICE

In Exercises 13–24, do the operations and simplify the result when possible.

13. $\dfrac{3}{4} + \dfrac{7}{4}$

14. $\dfrac{10}{33} - \dfrac{21}{33}$

15. $\dfrac{3}{4y} + \dfrac{8}{4y}$

16. $\dfrac{5}{3z^2} - \dfrac{6}{3z^2}$

17. $\dfrac{3x}{2x + 2} + \dfrac{x + 4}{2x + 2}$

18. $\dfrac{4y}{y - 4} - \dfrac{16}{y - 4}$

19. $\dfrac{3x}{x - 3} - \dfrac{9}{x - 3}$

20. $\dfrac{9x}{x - y} - \dfrac{9y}{x - y}$

21. $\dfrac{5x}{x + 1} + \dfrac{3}{x + 1} - \dfrac{2x}{x + 1}$

22. $\dfrac{4}{a + 4} - \dfrac{2a}{a + 4} + \dfrac{3a}{a + 4}$

23. $\dfrac{3(x^2 + x)}{x^2 - 5x + 6} + \dfrac{-3(x^2 - x)}{x^2 - 5x + 6}$

24. $\dfrac{2x + 4}{x^2 + 13x + 12} - \dfrac{x + 3}{x^2 + 13x + 12}$

In Exercises 25–32, the denominators of several fractions are given. Find the LCD.

25. $12x,\ 18x^2$

26. $15ab^2,\ 27a^2b$

27. $x^2 + 3x,\ x^2 - 9$

28. $3y^2 - 6y,\ 3y(y - 4)$

29. $x^3 + 27,\ x^2 + 6x + 9$

30. $x^3 - 8,\ x^2 - 4x + 4$

31. $2x^2 + 5x + 3,\ 4x^2 + 12x + 9,\ x^2 + 2x + 1$

32. $2x^2 + 5x + 3,\ 4x^2 + 12x + 9,\ 4x + 6$

In Exercises 33–76, do the operations and simplify the result when possible.

33. $\dfrac{1}{2} + \dfrac{1}{3}$

34. $\dfrac{8}{9} - \dfrac{5}{12}$

35. $\dfrac{3a}{2} - \dfrac{4b}{7}$

36. $\dfrac{a}{2} + \dfrac{2a}{5}$

37. $\dfrac{3}{4x} + \dfrac{2}{3x}$

38. $\dfrac{2}{5a} + \dfrac{3}{2b}$

39. $\dfrac{3a}{2b} - \dfrac{2b}{3a}$

40. $\dfrac{5m}{2n} - \dfrac{3n}{4m}$

41. $\dfrac{3}{ab^2} - \dfrac{5}{a^2b}$

42. $\dfrac{1}{xy^3} - \dfrac{2}{x^2y}$

43. $\dfrac{r}{4b^2} + \dfrac{s}{6b}$

44. $\dfrac{t}{12c^3} + \dfrac{t}{15c^2}$

45. $\dfrac{a+b}{3} + \dfrac{a-b}{7}$

46. $\dfrac{x-y}{2} + \dfrac{x+y}{3}$

47. $\dfrac{3}{x+2} + \dfrac{5}{x-4}$

48. $\dfrac{2}{a+4} - \dfrac{6}{a+3}$

49. $\dfrac{x+2}{x+5} - \dfrac{x-3}{x+7}$

50. $\dfrac{7}{x+3} + \dfrac{4x}{x+6}$

51. $4 + \dfrac{1}{x}$

52. $2 - \dfrac{1}{x+1}$

53. $\dfrac{x+8}{x-3} - \dfrac{x-14}{3-x}$

54. $\dfrac{3-x}{2-x} + \dfrac{x-1}{x-2}$

55. $\dfrac{2a+1}{3a-2} - \dfrac{a-4}{2-3a}$

56. $\dfrac{4}{x-3} + \dfrac{5}{3-x}$

57. $\dfrac{x}{x^2+5x+6} + \dfrac{x}{x^2-4}$

58. $\dfrac{x}{3x^2-2x-1} + \dfrac{4}{3x^2+10x+3}$

59. $\dfrac{4}{x^2-2x-3} - \dfrac{x}{3x^2-7x-6}$

60. $\dfrac{2a}{a^2-2a-8} + \dfrac{3}{a^2-5a+4}$

61. $2x + 3 + \dfrac{1}{x+1}$

62. $x + 1 + \dfrac{1}{x-1}$

63. $1 + x - \dfrac{x}{x-5}$

64. $2 - x + \dfrac{3}{x-9}$

65. $\dfrac{8}{x^2-9} + \dfrac{2}{x-3} - \dfrac{6}{x}$

66. $\dfrac{x}{x^2-4} - \dfrac{x}{x+2} + \dfrac{2}{x}$

67. $\dfrac{3x}{2x-1} + \dfrac{x+1}{3x+2} - \dfrac{2x}{6x^3+x^2-2x}$

68. $\dfrac{x+3}{2x^2-5x+2} - \dfrac{3x-1}{x^2-x-2}$

69. $\dfrac{3}{x+1} - \dfrac{2}{x-1} + \dfrac{x+3}{x^2-1}$

70. $\dfrac{2}{x-2} + \dfrac{3}{x+2} - \dfrac{x-1}{x^2-4}$

71. $\dfrac{a}{a-b} + \dfrac{b}{a+b} + \dfrac{a^2+b^2}{b^2-a^2}$

72. $\dfrac{1}{x+y} - \dfrac{1}{x-y} + \dfrac{2y}{y^2-x^2}$

73. $\dfrac{7n^2}{m-n} + \dfrac{3m}{n-m} - \dfrac{3m^2-n}{m^2-2mn+n^2}$

74. $\dfrac{3b}{2a-b} + \dfrac{2a-1}{b-2a} - \dfrac{3a^2+b}{b^2-4ab+4a^2}$

75. $\dfrac{m+1}{m^2+2m+1} + \dfrac{m-1}{m^2-2m+1} + \dfrac{2}{m^2-1}$

(*Hint:* Simplify first.)

76. $\dfrac{a+2}{a^2+3a+2} + \dfrac{a-1}{a^2-1} + \dfrac{3}{a+1}$

(*Hint:* Simplify first.)

APPLICATIONS

77. DRAFTING Among the tools used in drafting are the 45°–45°–90° and the 30°–60°–90° triangles shown in Illustration 1. Find the perimeter of each triangle. Express each result as a single rational expression.

ILLUSTRATION 1

For a 45°-45°-90° triangle, these two sides are the same length.

For a 30°-60°-90° triangle, this side is half as long as the hypotenuse.

78. THE AMAZON The Amazon River flows in a general eastern direction to the Atlantic Ocean. In Brazil, when the river is at low stage, the rate of flow is about 5 mph. Suppose that a river guide can canoe in still water at a rate of r mph.

a. Complete the table to find rational expressions that represent the time it would take the guide to canoe 3 miles downriver and to canoe 3 miles upriver on the Amazon.

	Rate (mph)	Time (hr)	Distance (mi)
Downriver			3
Upriver			3

b. Find the difference in the times for the trips downriver and upriver. Express the result as a single rational expression.

WRITING

79. Explain how to find the least common denominator of a set of fractions.

80. Explain how to add two rational expressions with unlike denominators.

81. Write some comments to the student who wrote the following solution, explaining his **misunderstanding**.

$$\frac{1}{x} \cdot \frac{3}{2} = \frac{1 \cdot 2}{x \cdot 2} \cdot \frac{3 \cdot x}{2 \cdot x}$$

$$= \frac{2}{2x} \cdot \frac{3x}{2x}$$

$$= \frac{6x}{2x}$$

82. Write some comments to the student who wrote the following solution, pointing out where she made an **error**.

$$\frac{1}{x} - \frac{x+1}{x} = \frac{1-x+1}{x}$$

$$= \frac{2-x}{x}$$

REVIEW

In Exercises 83–86, solve each equation.

83. $a(a - 6) = -9$

84. $x^2 - \frac{1}{2}(x + 1) = 0$

85. $y^3 + y^2 = 0$

86. $5x^2 = 6 - 13x$

▶ **6.5**

Complex Fractions

In this section, you will learn about

Complex fractions ■ Simplifying complex fractions

Introduction We have seen that rational expressions have a polynomial in the numerator and in the denominator. In this section, we will consider *complex rational expressions*, called *complex fractions*, which have numerators and/or denominators that contain rational expressions.

Complex Fractions

A **complex fraction** is a fraction that has a fraction (rational expression) in its numerator or its denominator or both. Examples of complex fractions are

$$\frac{\dfrac{3a}{b}}{\dfrac{6ac}{b^2}}, \qquad \frac{\dfrac{2}{x} + 1}{3 + x}, \qquad \text{and} \qquad \frac{\dfrac{1}{x} + \dfrac{1}{y}}{\dfrac{1}{x} - \dfrac{1}{y}}$$

Simplifying Complex Fractions

To *simplify a complex fraction* means to express it as a simple fraction. We can use two methods to simplify the complex fraction

$$\frac{\dfrac{3a}{b}}{\dfrac{6ac}{b^2}}$$

With the first method, we eliminate the fractions in the numerator and denominator by writing the complex fraction as a division and using the division rule for fractions:

$$\frac{\dfrac{3a}{b}}{\dfrac{6ac}{b^2}} = \frac{3a}{b} \div \frac{6ac}{b^2}$$

$$= \frac{3a}{b} \cdot \frac{b^2}{6ac} \qquad \text{Invert the divisor and multiply.}$$

$$= \frac{b}{2c} \qquad \text{Multiply the fractions and simplify.}$$

In the other method, we eliminate the fractions in the numerator and denominator by multiplying the fraction by 1, written in the form $\frac{b^2}{b^2}$. We use $\frac{b^2}{b^2}$ because b^2 is the LCD of $\frac{3a}{b}$ and $\frac{6ac}{b^2}$.

$$\frac{\dfrac{3a}{b}}{\dfrac{6ac}{b^2}} = \frac{b^2 \cdot \dfrac{3a}{b}}{b^2 \cdot \dfrac{6ac}{b^2}}$$

$$= \frac{\dfrac{3ab^2}{b}}{\dfrac{6acb^2}{b^2}}$$

$$= \frac{3ab}{6ac} \qquad \text{Simplify the fractions in the numerator and denominator.}$$

$$= \frac{b}{2c} \qquad \text{Divide out the common factor of } 3a.$$

With either method, the result is the same.

Methods for simplifying complex fractions

Method 1

Write the numerator and denominator of the complex fraction as single fractions. Then divide the fractions and simplify.

Method 2

Multiply the numerator and denominator of the complex fraction by the LCD of the fractions in its numerator and denominator. Then simplify the results, if possible.

EXAMPLE 1	**Simplifying a complex fraction.** Simplify $\dfrac{\frac{2}{x}+1}{3+x}$.

Solution **Method 1**

We add the fractions in the numerator and proceed as follows:

$$\frac{\frac{2}{x}+1}{3+x} = \frac{\frac{2}{x}+\frac{x}{x}}{\frac{3+x}{1}} \qquad \text{Write 1 as } \tfrac{x}{x} \text{ and } 3+x \text{ as } \tfrac{3+x}{1}.$$

$$= \frac{\frac{2+x}{x}}{\frac{3+x}{1}} \qquad \text{Add } \tfrac{2}{x} \text{ and } \tfrac{x}{x} \text{ to get } \tfrac{2+x}{x}.$$

$$= \frac{2+x}{x} \div \frac{3+x}{1} \qquad \text{Write the complex fraction as a division.}$$

$$= \frac{2+x}{x} \cdot \frac{1}{3+x} \qquad \text{Invert the divisor and multiply.}$$

$$= \frac{2+x}{x^2+3x} \qquad \text{Multiply the numerators and multiply the denominators.}$$

Method 2

To eliminate the denominator of x, we multiply the numerator and the denominator of the complex fraction by x.

$$\frac{\frac{2}{x}+1}{3+x} = \frac{x\left(\frac{2}{x}+1\right)}{x(3+x)}$$

$$= \frac{2+x}{x^2+3x} \qquad \begin{array}{l}\text{Use the distributive property to remove parentheses.}\\ \text{Write } 3x+x^2 \text{ as } x^2+3x.\end{array}$$

SELF CHECK Simplify $\dfrac{\frac{3}{a}+2}{2+a}$. *Answer:* $\dfrac{2a+3}{a^2+2a}$ ∎

ACCENT ON TECHNOLOGY *Checking Algebra*

An informal check of the simplification done in Example 1 can be performed using a scientific calculator. If

$$\frac{\frac{2}{x}+1}{3+x} = \frac{2+x}{x^2+3x}$$

then the expressions on each side will have identical values when evaluated for a given value of x (say, $x = 5$). To evaluate the expression on the left-hand side, we enter these numbers and press these keys.

Keystrokes $\boxed{(}\ 2\ \boxed{\div}\ 5\ \boxed{+}\ 1\ \boxed{)}\ \boxed{\div}\ \boxed{(}\ 3\ \boxed{+}\ 5\ \boxed{)}\ \boxed{=}$ $\boxed{\qquad 0.175}$

To evaluate the expression on the right-hand side, we enter these numbers and press these keys.

(continued)

Keystrokes $(2 + 5) \div (5 x^2 + 3 \times 5) =$ ⬜ 0.175

The results are the same, so it appears that the simplification is correct. We say "appears" because checking for only a single value of x is not definitive. The expressions should yield identical values when evaluated for any value of x for which the fractions are defined.

We can also check the simplification in Example 1 by graphing the functions

$$f(x) = \dfrac{\dfrac{2}{x} + 1}{3 + x} \qquad \text{shown in Figure 6-12(a), and}$$

$$g(x) = \dfrac{2 + x}{x^2 + 3x} \qquad \text{shown in Figure 6-12(b),}$$

and observing that the graphs are the same. Each graph has window settings of $[-10, 10]$ for x and $[-10, 10]$ for y.

FIGURE 6-12

(a)

(b)

EXAMPLE 2

Simplifying a complex fraction. Simplify $\dfrac{\dfrac{1}{x} + \dfrac{1}{y}}{\dfrac{1}{x} - \dfrac{1}{y}}$.

Solution   **Method 1**

To write the numerator and denominator of the complex fraction as single fractions, we add the fractions in the numerator and subtract the fractions in the denominator.

$$\dfrac{\dfrac{1}{x} + \dfrac{1}{y}}{\dfrac{1}{x} - \dfrac{1}{y}} = \dfrac{\dfrac{1 \cdot y}{x \cdot y} + \dfrac{1 \cdot x}{y \cdot x}}{\dfrac{1 \cdot y}{x \cdot y} - \dfrac{1 \cdot x}{y \cdot x}}$$

Rewrite the fractions in the numerator in terms of the LCD, xy. Do the same in the denominator.

$$= \dfrac{\dfrac{y + x}{xy}}{\dfrac{y - x}{xy}}$$

Add the fractions in the numerator, and subtract the fractions in the denominator.

$$= \dfrac{y + x}{xy} \div \dfrac{y - x}{xy}$$

Write the complex fraction as a division.

$$= \dfrac{y + x}{xy} \cdot \dfrac{xy}{y - x}$$

Invert the divisor and multiply.

$$= \dfrac{(y + x)\overset{11}{\cancel{xy}}}{\underset{11}{\cancel{xy}}(y - x)}$$

Multiply the fractions and divide out the common factors.

$$= \dfrac{y + x}{y - x}$$

Method 2

We multiply the numerator and denominator by xy (the LCD of the fractions appearing in the complex fraction) and simplify.

$$\frac{\dfrac{1}{x} + \dfrac{1}{y}}{\dfrac{1}{x} - \dfrac{1}{y}} = \frac{xy\left(\dfrac{1}{x} + \dfrac{1}{y}\right)}{xy\left(\dfrac{1}{x} - \dfrac{1}{y}\right)} \qquad \text{The LCD of } \tfrac{1}{x} \text{ and } \tfrac{1}{y} \text{ is } xy.$$

$$= \frac{\dfrac{xy}{x} + \dfrac{xy}{y}}{\dfrac{xy}{x} - \dfrac{xy}{y}} \qquad \text{Use the distributive property to remove parentheses.}$$

$$= \frac{y + x}{y - x} \qquad \text{Simplify each fraction.}$$

SELF CHECK Simplify $\dfrac{\dfrac{1}{a} - \dfrac{1}{b}}{\dfrac{1}{a} + \dfrac{1}{b}}$.

Answer: $\frac{b - a}{b + a}$ ■

EXAMPLE 3   **A fraction containing negative exponents.** Simplify $\dfrac{x^{-1} + y^{-1}}{x^{-2} - y^{-2}}$.

Solution   **Method 1**

We need to write the numerator and denominator of the complex fraction as single fractions.

$$\frac{x^{-1} + y^{-1}}{x^{-2} - y^{-2}} = \frac{\dfrac{1}{x} + \dfrac{1}{y}}{\dfrac{1}{x^2} - \dfrac{1}{y^2}} \qquad \begin{array}{l}\text{Write the fraction without using negative} \\ \text{exponents.}\end{array}$$

$$= \frac{\dfrac{y}{xy} + \dfrac{x}{xy}}{\dfrac{y^2}{x^2y^2} - \dfrac{x^2}{x^2y^2}} \qquad \begin{array}{l}\text{Get a common denominator } (xy) \text{ in the} \\ \text{numerator and a common denominator } (x^2y^2) \\ \text{in the denominator.}\end{array}$$

$$= \frac{\dfrac{y + x}{xy}}{\dfrac{y^2 - x^2}{x^2y^2}} \qquad \begin{array}{l}\text{Add the fractions in the numerator and sub-} \\ \text{tract the fractions in the denominator.}\end{array}$$

$$= \frac{y + x}{xy} \div \frac{y^2 - x^2}{x^2y^2} \qquad \text{Write the fraction as a division.}$$

$$= \frac{y + x}{xy} \cdot \frac{xxyy}{(y - x)(y + x)} \qquad \text{Invert and multiply; factor } y^2 - x^2 \text{ and } x^2y^2.$$

$$= \frac{\overset{1}{\cancel{(y + x)}}\overset{1}{\cancel{xx}}\overset{1}{\cancel{yy}}}{\underset{1}{\cancel{xy}}(y - x)\underset{1}{\cancel{(y + x)}}} \qquad \begin{array}{l}\text{Multiply the numerators and the denomina-} \\ \text{tors. Divide out the common factors.}\end{array}$$

$$= \frac{xy}{y - x}$$

Method 2

We multiply both numerator and denominator by x^2y^2, the LCD of the fractions, and proceed as follows:

$$\frac{x^{-1} + y^{-1}}{x^{-2} - y^{-2}} = \frac{\dfrac{1}{x} + \dfrac{1}{y}}{\dfrac{1}{x^2} - \dfrac{1}{y^2}}$$ Write the fraction without negative exponents.

$$= \frac{x^2y^2\left(\dfrac{1}{x} + \dfrac{1}{y}\right)}{x^2y^2\left(\dfrac{1}{x^2} - \dfrac{1}{y^2}\right)}$$ Multiply numerator and denominator by x^2y^2.

$$= \frac{xy^2 + yx^2}{y^2 - x^2}$$ Use the distributive property to remove parentheses.

$$= \frac{xy(y + x)}{(y + x)(y - x)}$$ Factor the numerator and denominator.

$$= \frac{xy}{y - x}$$ Divide out the common factor $y + x$.

SELF CHECK   Simplify $\dfrac{a^{-2} + b^{-2}}{a^{-1} - b^{-1}}$.   *Answer:* $\dfrac{b^2 + a^2}{ab(b - a)}$ ∎

 WARNING! $x^{-1} + y^{-1}$ means $\frac{1}{x} + \frac{1}{y}$, and $(x + y)^{-1}$ means $\frac{1}{x+y}$. Thus,

$$x^{-1} + y^{-1} \neq (x + y)^{-1}$$

E X A M P L E   4   **Simplifying a complex fraction.** Simplify $\dfrac{\dfrac{1}{a^2 - 3a + 2}}{\dfrac{3}{a - 2} - \dfrac{2}{a - 1}}$.

Solution   We will use Method 2 to do the simplification. To determine the LCD for all the fractions appearing in the complex fraction, we must factor $a^2 - 3a + 2$.

$$\frac{\dfrac{1}{a^2 - 3a + 2}}{\dfrac{3}{a - 2} - \dfrac{2}{a - 1}} = \frac{\dfrac{1}{(a - 2)(a - 1)}}{\dfrac{3}{a - 2} - \dfrac{2}{a - 1}}$$

The LCD of the fractions in the numerator and denominator of the complex fraction is $(a - 2)(a - 1)$. We multiply the numerator and the denominator by the LCD.

$$= \frac{(a - 2)(a - 1)\left[\dfrac{1}{(a - 2)(a - 1)}\right]}{(a - 2)(a - 1)\left[\dfrac{3}{a - 2} - \dfrac{2}{a - 1}\right]}$$

$$= \frac{\dfrac{(a - 2)(a - 1)}{(a - 2)(a - 1)}}{\dfrac{3(a - 2)(a - 1)}{a - 2} - \dfrac{2(a - 2)(a - 1)}{a - 1}}$$ Do the multiplication in the numerator. In the denominator, distribute the LCD.

$$= \frac{1}{3(a-1) - 2(a-2)}$$ Simplify each of the three rational expressions by dividing out the common factors.

$$= \frac{1}{3a - 3 - 2a + 4}$$ In the denominator, remove parentheses.

$$= \frac{1}{a + 1}$$ Combine like terms.

SELF CHECK Simplify $\dfrac{\dfrac{b}{b+4} + \dfrac{3}{b+3}}{\dfrac{b}{b^2 + 7b + 12}}$.

Answer: $\dfrac{b^2 + 6b + 12}{b}$ ∎

If a fraction has a complex fraction in its numerator or denominator, it is often called a **continued fraction**.

EXAMPLE 5 **Simplifying a continued fraction.** Simplify $\dfrac{\dfrac{2x}{1 - \dfrac{1}{x}} + 3}{3 - \dfrac{2}{x}}$.

Solution We begin by multiplying the numerator and denominator of the fraction

$$\frac{2x}{1 - \dfrac{1}{x}}$$

by x to eliminate the complex fraction in the numerator of the given fraction.

$$\frac{\dfrac{2x}{1 - \dfrac{1}{x}} + 3}{3 - \dfrac{2}{x}} = \frac{\dfrac{x(2x)}{x\left(1 - \dfrac{1}{x}\right)} + 3}{3 - \dfrac{2}{x}}$$

$$= \frac{\dfrac{2x^2}{x - 1} + 3}{3 - \dfrac{2}{x}}$$

We then multiply the numerator and denominator of the previous fraction by $x(x-1)$, the LCD of $\frac{2x^2}{x-1}$, 3, and $\frac{2}{x}$, and simplify.

$$\frac{\dfrac{2x}{1 - \dfrac{1}{x}} + 3}{3 - \dfrac{2}{x}} = \frac{x(x-1)\left(\dfrac{2x^2}{x-1} + 3\right)}{x(x-1)\left(3 - \dfrac{2}{x}\right)}$$

$$= \frac{2x^3 + 3x(x-1)}{3x(x-1) - 2(x-1)}$$

$$= \frac{2x^3 + 3x^2 - 3x}{3x^2 - 5x + 2}$$

This result does not simplify.

SELF CHECK Simplify $\dfrac{3 + \dfrac{2}{a}}{\dfrac{2a}{1 + \dfrac{1}{a}} + 3}$.

Answer: $\dfrac{(3a + 2)(a + 1)}{a(2a^2 + 3a + 3)}$ ∎

STUDY SET
Section 6.5

VOCABULARY

In Exercises 1–2, fill in the blanks to make the statements true.

1. A _____ fraction is a fraction that has fractions (rational expressions) in its numerator and/or its denominator.

2. To _____ a complex fraction means to express it as a simple fraction.

CONCEPTS

3. The first step in simplifying a complex fraction using Method 2 is shown below. With this method, the fractions in the numerator and denominator are to be eliminated by multiplying the complex fraction by 1. How is the 1 written in this case?

$$\frac{\dfrac{4}{t^2}}{\dfrac{3b}{t}} = \frac{t^2 \cdot \dfrac{4}{t^2}}{t^2 \cdot \dfrac{3b}{t}}$$

4. Determine the LCD of the rational expressions appearing in each complex fraction.

a. $\dfrac{1 + \dfrac{4}{c}}{\dfrac{2}{c} + c}$

b. $\dfrac{\dfrac{6}{m^2} + \dfrac{1}{2m}}{\dfrac{m^2 - 1}{4}}$

c. $\dfrac{\dfrac{p}{p + 2} + \dfrac{12}{p + 3}}{\dfrac{p - 1}{p^2 + 5p + 6}}$

d. $\dfrac{2 + \dfrac{3}{x + 1}}{\dfrac{1}{x} + x + x^2}$

NOTATION

In Exercises 5–6, complete each solution.

5. $\dfrac{\dfrac{3}{x} - \dfrac{x}{y}}{6} = \dfrac{\dfrac{3 \cdot \boxed{}}{x \cdot \boxed{}} - \dfrac{x \cdot x}{y \cdot x}}{6}$

$= \dfrac{\dfrac{3y - x^2}{xy}}{\boxed{} \cdot 6}$

$= \dfrac{3y - x^2}{xy} \cdot \boxed{}$

$= \dfrac{3y - x^2}{6xy}$

6. $\dfrac{\dfrac{2}{a^2} - \dfrac{1}{b}}{\dfrac{2}{a} + \dfrac{1}{b^2}} = \dfrac{\boxed{}\left(\dfrac{2}{a^2} - \dfrac{1}{b}\right)}{\boxed{}\left(\dfrac{2}{a} + \dfrac{1}{b^2}\right)}$

$= \dfrac{\dfrac{2a^2b^2}{a^2} - \boxed{}}{\dfrac{2a^2b^2}{a} + \boxed{}}$

$= \dfrac{2b^2 - a^2b}{2ab^2 + a^2}$

7. The fraction $\dfrac{\dfrac{a}{b}}{\dfrac{c}{d}}$ is equivalent to $\dfrac{a}{b} \boxed{} \dfrac{c}{d}$.

8. What is the numerator and what is the denominator of the following complex fraction?

$$\frac{6 - k - \dfrac{5}{k}}{\dfrac{6}{k^2} + \dfrac{4}{k} - 4}$$

PRACTICE

In Exercises 9–54, simplify each complex fraction.

9. $\dfrac{\frac{1}{2}}{\frac{3}{4}}$

10. $-\dfrac{\frac{3}{4}}{\frac{1}{2}}$

11. $\dfrac{\frac{1}{2} - \frac{2}{3}}{\frac{2}{3} + \frac{1}{2}}$

12. $\dfrac{\frac{1}{4} - \frac{1}{5}}{\frac{1}{3}}$

13. $\dfrac{\frac{4x}{y}}{\frac{6xz}{y^2}}$

14. $\dfrac{\frac{5t^4}{9x}}{\frac{2t}{18x}}$

15. $\dfrac{5ab^2}{\frac{ab}{25}}$

16. $\dfrac{\frac{6a^2b}{4t}}{3a^2b^2}$

17. $\dfrac{\frac{x-y}{xy}}{\frac{y-x}{x}}$

18. $\dfrac{\frac{x^2 + 5x + 6}{3xy}}{\frac{x^2 - 9}{6xy}}$

19. $\dfrac{\frac{1}{x} - \frac{1}{y}}{xy}$

20. $\dfrac{xy}{\frac{1}{x} - \frac{1}{y}}$

21. $\dfrac{\frac{1}{a} + \frac{1}{b}}{\frac{1}{a}}$

22. $\dfrac{\frac{1}{b}}{\frac{1}{a} - \frac{1}{b}}$

23. $\dfrac{1 + \frac{x}{y}}{1 - \frac{x}{y}}$

24. $\dfrac{\frac{x}{y} + 1}{1 - \frac{x}{y}}$

25. $\dfrac{\frac{y}{x} - \frac{x}{y}}{\frac{1}{x} + \frac{1}{y}}$

26. $\dfrac{\frac{y}{x} - \frac{x}{y}}{\frac{1}{y} - \frac{1}{x}}$

27. $\dfrac{\frac{1}{a} - \frac{1}{b}}{\frac{a}{b} - \frac{b}{a}}$

28. $\dfrac{\frac{1}{a} + \frac{1}{b}}{\frac{a}{b} - \frac{b}{a}}$

29. $\dfrac{x + 1 - \frac{6}{x}}{\frac{1}{x}}$

30. $\dfrac{x - 1 - \frac{2}{x}}{\frac{x}{3}}$

31. $\dfrac{5xy}{1 + \frac{1}{xy}}$

32. $\dfrac{3a}{a + \frac{1}{a}}$

33. $\dfrac{a - 4 + \frac{1}{a}}{-\frac{1}{a} - a + 4}$

34. $\dfrac{a + 1 + \frac{1}{a^2}}{\frac{1}{a^2} + a - 1}$

35. $\dfrac{1 + \frac{6}{x} + \frac{8}{x^2}}{1 + \frac{1}{x} - \frac{12}{x^2}}$

36. $\dfrac{1 - x - \frac{2}{x}}{\frac{6}{x^2} + \frac{1}{x} - 1}$

37. $\dfrac{\frac{1}{a+1} + 1}{\frac{3}{a-1} + 1}$

38. $\dfrac{2 + \frac{4}{y-7}}{\frac{4}{y-7}}$

39. $\dfrac{2 + \frac{3}{x+1}}{\frac{1}{x} + x}$

40. $\dfrac{\frac{1}{x} - \frac{4}{x-1}}{\frac{3}{x-1} + \frac{2}{x}}$

41. $\dfrac{y}{x^{-1} - y^{-1}}$

42. $\dfrac{x^{-1} + y^{-1}}{(x+y)^{-1}}$

43. $\dfrac{x - y^{-2}}{y - x^{-2}}$

44. $\dfrac{x^{-2} - y^{-2}}{x^{-1} - y^{-1}}$

45. $\dfrac{\frac{t}{x^2 - y^2}}{\frac{t}{x+y}}$

46. $\dfrac{\frac{7}{a-b}}{\frac{b}{a^3 - b^3}}$

47. $\dfrac{\frac{2}{x+3} - \frac{1}{x-3}}{\frac{3}{x^2 - 9}}$

48. $\dfrac{2 + \frac{1}{x^2 - 1}}{1 + \frac{1}{x-1}}$

49. $\dfrac{\frac{h}{h^2 + 3h + 2}}{\frac{4}{h+2} - \frac{4}{h+1}}$

50. $\dfrac{\frac{1}{r^2 + 4r + 4}}{\frac{r}{r+2} + \frac{r}{r+2}}$

51. $a + \dfrac{a}{1 + \dfrac{a}{a+1}}$

52. $b + \dfrac{b}{1 - \dfrac{b+1}{b}}$

53. $\dfrac{x - \dfrac{1}{1 - \dfrac{x}{2}}}{\dfrac{3}{x + \dfrac{2}{3}} - x}$

54. $\dfrac{3x - \dfrac{1}{3 - \dfrac{x}{2}}}{\dfrac{3}{\dfrac{x}{2} - 3} + x}$

APPLICATIONS

55. ENGINEERING The stiffness k of the shaft shown in Illustration 1 is given by the formula

$$k = \dfrac{1}{\dfrac{1}{k_1} + \dfrac{1}{k_2}}$$

ILLUSTRATION 1

Section 1 · · · Section 2

where k_1 and k_2 are the individual stiffnesses of each section. Simplify the complex fraction.

56. TRANSPORTATION If a bus travels a distance d_1 at a speed s_1, and then travels a distance d_2 at a speed s_2, the average (mean) speed $\bar{s}$ is given by the formula

$$\bar{s} = \dfrac{d_1 + d_2}{\dfrac{d_1}{s_1} + \dfrac{d_2}{s_2}}$$

Simplify the complex fraction.

57. KITCHEN UTENSILS See Illustration 2. What is the ratio of the width of the opening of the ice tongs to the width of the opening of the handles? Express the result in simplest form.

ILLUSTRATION 2

$8 - \dfrac{2}{d}$ ⎰ ⎱ $6 - \dfrac{2}{d}$

58. DATA ANALYSIS Use the data in the table to find the average measurement for the three-trial experiment. Express the answer as a rational expression.

	Trial 1	Trial 2	Trial 3
Measurement	$\frac{k}{3}$	$\frac{k}{5}$	$\frac{k}{6}$

WRITING

59. What is a complex fraction?

60. Two methods can be used to simplify a complex fraction. Which method do you think is simpler? Why?

61. Evaluate

$$\dfrac{\dfrac{4}{3} + x}{\dfrac{x}{2} + \dfrac{2}{x}} \quad \text{and} \quad \dfrac{x^2 + 4}{x + 2}$$

for $x = 2$. Tell whether the result verifies that

$$\dfrac{\dfrac{4}{3} + x}{\dfrac{x}{2} + \dfrac{2}{x}} = \dfrac{x^2 + 4}{x + 2}$$

Explain your answer.

62. To check a simplification to verify that

$$\dfrac{\dfrac{4}{3} + x}{\dfrac{x}{2} + \dfrac{2}{x}} = \dfrac{x^2 + 4}{x + 2}$$

a student graphed each expression's associated function. See Illustration 3. What conclusion can be made after examining the graphs?

ILLUSTRATION 3

REVIEW

In Exercises 63–66, solve each equation.

63. $\dfrac{8(a - 5)}{3} = 2(a - 4)$

64. $\dfrac{3t^2}{5} + \dfrac{7t}{10} = \dfrac{3t + 6}{5}$

65. $a^4 - 13a^2 + 36 = 0$

66. $|2x - 1| = 9$

▶ 6.6

Equations Containing Rational Expressions

In this section, you will learn about

Solving rational equations ■ Extraneous solutions ■ Solving formulas for a specified variable ■ Problem solving

---

Introduction In this section, we will use the five-step problem-solving strategy to solve problems from disciplines such as business, photography, aviation, electronics, and publishing. We will encounter a new type of equation when we write mathematical models of each of these situations. The equations will contain rational expressions and are therefore called *rational equations*. To solve them, we use a technique that clears them of fractions.

Solving Rational Equations

If an equation contains one or more rational expressions, it is called a **rational equation.** Some examples of rational equations are

$$\frac{3}{5} + \frac{7}{x+2} = 2, \qquad \frac{x+3}{x-3} = \frac{2}{x^2-4}, \qquad \text{and} \qquad \frac{-x^2+10}{x^2-1} + \frac{3x}{x-1} = \frac{2x}{x+1}$$

To solve a rational equation, we can multiply both sides of the equation by the LCD of the rational expressions in the equation to clear it of fractions.

EXAMPLE 1

Solving rational equations. Solve $\frac{3}{5} + \frac{7}{x+2} = 2$.

Solution We note that x cannot be -2, because this would give a 0 in the denominator of $\frac{7}{x+2}$. If $x \neq -2$, we can multiply both sides of the equation by $5(x+2)$ to get

$$5(x+2)\left(\frac{3}{5} + \frac{7}{x+2}\right) = 5(x+2)(2) \qquad \text{Multiply both sides by the LCD.}$$

$$5(x+2)\left(\frac{3}{5}\right) + 5(x+2)\left(\frac{7}{x+2}\right) = 5(x+2)2 \qquad \text{On the left-hand side, distribute } 5(x+2).$$

$$\overset{1}{\cancel{5}}(x+2)\left(\frac{3}{\underset{1}{\cancel{5}}}\right) + 5\overset{1}{\cancel{(x+2)}}\left(\frac{7}{\underset{1}{\cancel{(x+2)}}}\right) = 5(x+2)2 \qquad \text{On the left-hand side, divide out the common factors: } \frac{5}{5} = 1 \text{ and } \frac{x+2}{x+2} = 1.$$

$$3(x+2) + 5(7) = 10(x+2) \qquad \text{Simplify each side.}$$

The resulting equation does not contain any fractions. We now solve this *linear equation* for x.

$$3x + 6 + 35 = 10x + 20 \qquad \text{Use the distributive property and simplify.}$$

$$3x + 41 = 10x + 20 \qquad \text{Combine like terms.}$$

$$-7x = -21 \qquad \text{Subtract } 10x \text{ and } 41 \text{ from both sides.}$$

$$x = 3 \qquad \text{Divide both sides by } -7.$$

Check: To check, we substitute 3 for x in the original equation and simplify:

$$\frac{3}{5} + \frac{7}{x+2} = 2$$

$$\frac{3}{5} + \frac{7}{3+2} \stackrel{?}{=} 2$$

$$\frac{3}{5} + \frac{7}{5} \stackrel{?}{=} 2$$

$$2 = 2$$

SELF CHECK   Solve $\frac{2}{5} + \frac{5}{x-2} = \frac{29}{10}$.   *Answer:* 4

ACCENT ON TECHNOLOGY *Solving Rational Equations Graphically*

To use a graphing calculator to approximate the solution of $\frac{3}{5} + \frac{7}{x+2} = 2$, we graph the functions $f(x) = \frac{3}{5} + \frac{7}{x+2}$ and $g(x) = 2$ using window settings of $[-10, 10]$ for x and $[-10, 10]$ for y. If we trace and move the cursor closer to the intersection point of the two graphs, we will get the approximate value of x shown in Figure 6-13(a). If we zoom twice and trace again, we get the results shown in Figure 6-13(b). Algebra will show that the exact solution is 3.

An alternate way of finding the point of intersection of the two graphs is to use the INTERSECT feature. In Figure 6-13(c), the display shows that the graphs intersect at the point (3, 2). This implies that the solution of the rational equation is $x = 3$. Check the result.

FIGURE 6-13

(a)   (b)   (c)

EXAMPLE 2   **Solving rational equations.** Solve $\dfrac{-x^2 + 10}{x^2 - 1} + \dfrac{3x}{x - 1} = \dfrac{2x}{x + 1}$.

Solution   We start by noting that x cannot be 1 or -1, because this would give a 0 in the denominator of a fraction. If $x \neq 1$ and $x \neq -1$, we can clear the equation of fractions by multiplying both sides by the LCD of the three rational expressions and proceeding as follows:

$$\frac{-x^2 + 10}{x^2 - 1} + \frac{3x}{x - 1} = \frac{2x}{x + 1}$$

$$\frac{-x^2 + 10}{(x+1)(x-1)} + \frac{3x}{x - 1} = \frac{2x}{x + 1}$$

Factor the denominator $x^2 - 1$. We determine the LCD to be $(x + 1)(x - 1)$.

$$\frac{(x+1)(x-1)(-x^2+10)}{(x+1)(x-1)} + \frac{3x(x+1)(x-1)}{x-1} = \frac{2x(x+1)(x-1)}{x+1}$$ Multiply both sides by the LCD.

$$\frac{\overset{1}{\cancel{(x+1)}}\overset{1}{\cancel{(x-1)}}(-x^2+10)}{\underset{1}{\cancel{(x+1)}}\underset{1}{\cancel{(x-1)}}} + \frac{3x(x+1)\overset{1}{\cancel{(x-1)}}}{\underset{1}{\cancel{(x-1)}}} = \frac{2x\overset{1}{\cancel{(x+1)}}(x-1)}{\underset{1}{\cancel{(x+1)}}}$$ Divide out the common factors.

$$-x^2 + 10 + 3x(x+1) = 2x(x-1)$$ Simplify each side. The resulting equation does not contain any fractions.

$$-x^2 + 10 + 3x^2 + 3x = 2x^2 - 2x$$ Remove parentheses.

$$2x^2 + 10 + 3x = 2x^2 - 2x$$ Combine like terms.

$$10 + 3x = -2x$$ Subtract $2x^2$ from both sides.

$$10 + 5x = 0$$ Add $2x$ to both sides.

$$5x = -10$$ Subtract 10 from both sides.

$$x = -2$$ Divide both sides by 5.

Verify that -2 is a solution to the original equation. ∎

ACCENT ON TECHNOLOGY *Checking Apparent Solutions*

We can use a scientific calculator to check the solution found in Example 2 by evaluating

$$\frac{-x^2+10}{x^2-1} + \frac{3x}{x-1} \quad \text{and} \quad \frac{2x}{x+1}$$

for $x = -2$. In each case, the result is 4. Since the results are the same, -2 is a solution of the equation.

A check can also be done using a graphing calculator. One way of doing this is to enter

$$Y_1 = \frac{-x^2+10}{x^2-1} + \frac{3x}{x-1} \quad \text{and} \quad Y_2 = \frac{2x}{x+1}$$

and compare the values of the expressions when $x = -2$ in the table mode. See Figure 6-14. We know that -2 is a solution of

$$\frac{-x^2+10}{x^2-1} + \frac{3x}{x-1} = \frac{2x}{x+1}$$

because the value of Y_1 and Y_2 are the same (namely, 4) for $x = -2$.

FIGURE 6-14

X	Y₁	Y₂
-2	4	4
-1	ERROR	ERROR
0	-10	0
1	ERROR	1
2	8	1.3333
3	4.625	1.5
4	3.6	1.6

X=-2

E X A M P L E   3

A rational equation that leads to a quadratic equation. Solve $\dfrac{x+1}{5} - 2 = -\dfrac{4}{x}$.

Solution   We start by noting that x cannot be 0, because this would give a 0 in the denominator of a fraction. If $x \neq 0$, we can clear the equation of fractions by multiplying both sides by $5x$.

$$\frac{x+1}{5} - 2 = -\frac{4}{x}$$

$$5x\left(\frac{x+1}{5} - 2\right) = 5x\left(-\frac{4}{x}\right) \qquad \text{Multiply both sides by the LCD, } 5x.$$

$$5x\left(\frac{x+1}{5}\right) - 5x(2) = 5x\left(-\frac{4}{x}\right) \qquad \text{Distribute } 5x.$$

$$\overset{1}{\cancel{5}}x\left(\frac{x+1}{\underset{1}{\cancel{5}}}\right) - 5x(2) = 5\overset{1}{\cancel{x}}\left(-\frac{4}{\cancel{x}}\right) \qquad \text{Divide out the common factors.}$$

$$x(x+1) - 10x = -20 \qquad \text{Simplify.}$$

$$x^2 + x - 10x = -20 \qquad \text{Remove parentheses.}$$

To use factoring to solve this quadratic equation, we must write it in quadratic form $(ax^2 + bx + c = 0)$.

$$x^2 - 9x + 20 = 0 \qquad \text{Combine like terms and add 20 to both sides.}$$

$$(x - 5)(x - 4) = 0 \qquad \text{Factor } x^2 - 9x + 20.$$

$$x - 5 = 0 \quad \text{or} \quad x - 4 = 0 \qquad \text{Set each factor equal to 0.}$$

$$x = 5 \qquad\qquad x = 4$$

Verify that 4 and 5 both satisfy the original equation.

SELF CHECK   Solve $a + \frac{2}{3} = \frac{2a - 12}{3(a - 3)}$.   *Answer:* 1, 2   ∎

We can now summarize the procedure used to solve rational equations.

Solving rational equations

1. Factor all denominators.
2. Multiply both sides of the equation by the LCD of all rational expressions in the equation.
3. Use the distributive property to remove parentheses, divide out any common factors, and write the result in simplified form.
4. Solve the resulting linear or quadratic equation.
5. Check the result(s) in the original equation.

Extraneous Solutions

When we multiply both sides of an equation by a quantity that contains a variable, we can get false solutions, called **extraneous solutions.** This happens when we multiply both sides of an equation by 0 and get a solution that gives a 0 in the denominator of a rational expression. Extraneous solutions must be discarded when solving rational equations.

EXAMPLE 4

Extraneous solutions. Solve $\dfrac{2(x+1)}{x-3} = \dfrac{x+5}{x-3}$.

Solution   We start by noting that x cannot be 3, because this would give a 0 in the denominator of a fraction. If $x \neq 3$, we can clear the equation of fractions by multiplying both sides by $x - 3$.

$$\frac{2(x+1)}{x-3} = \frac{x+5}{x-3}$$

$$(x-3)\frac{2(x+1)}{x-3} = (x-3)\frac{x+5}{x-3} \qquad \text{Multiply both sides by the LCD.}$$

$$2(x+1) = x+5 \qquad \text{Divide out the common factor of } x-3.$$

$$2x+2 = x+5 \qquad \text{Remove parentheses.}$$

$$x+2 = 5 \qquad \text{Subtract } x \text{ from both sides.}$$

$$x = 3 \qquad \text{Subtract 2 from both sides.}$$

Since x cannot be 3, the 3 must be discarded. This equation has no solutions.

SELF CHECK   Solve $\frac{5}{a} - \frac{3}{2} = 2 + \frac{5}{a}$.   *Answer:* no solutions; 0 is extraneous ∎

Solving Formulas for a Specified Variable

Many formulas must be cleared of fractions before we can solve them for a specific variable.

EXAMPLE 5

Physics. The *law of gravitation,* formulated by Sir Isaac Newton in 1684, states that if two masses, m_1 and m_2, are separated by a distance of r, the force F exerted by one mass on the other is

$$F = \frac{Gm_1m_2}{r^2}$$

where G is the gravitational constant. (See Figure 6-15.) Solve for m_2.

FIGURE 6-15

Solution

$$F = \frac{Gm_1m_2}{r^2}$$

$$r^2(F) = r^2\left(\frac{Gm_1m_2}{r^2}\right) \qquad \text{Multiply both sides by the LCD, } r^2. \text{ Divide out the common factors.}$$

$$\frac{r^2F}{Gm_1} = \frac{Gm_1m_2}{Gm_1} \qquad \text{To isolate } m_2, \text{ divide both sides by } Gm_1.$$

$$\frac{r^2F}{Gm_1} = m_2 \qquad \text{Simplify the right-hand side.}$$

$$m_2 = \frac{r^2F}{Gm_1}$$

SELF CHECK   Solve the law of gravitation formula for r^2.   *Answer:* $r^2 = \dfrac{Gm_1m_2}{F}$ ∎

EXAMPLE 6

Electronics. In electronic circuits, resistors oppose the flow of an electric current. The total resistance R of a parallel combination of two resistors as shown in Figure 6-16 is given by

$$\frac{1}{R} = \frac{1}{R_1} + \frac{1}{R_2}$$

where R_1 is the resistance of the first resistor and R_2 is the resistance of the second resistor. Solve the formula for R.

FIGURE 6-16

Current → ☐ Total resistance?

Resistor 1
Resistor 2

Solution

We begin by clearing the equation of fractions by multiplying both sides by the LCD, which is RR_1R_2.

$$\frac{1}{R} = \frac{1}{R_1} + \frac{1}{R_2}$$

$$RR_1R_2\left(\frac{1}{R}\right) = RR_1R_2\left(\frac{1}{R_1} + \frac{1}{R_2}\right) \quad \text{Multiply both sides by the LCD.}$$

$$\frac{\overset{1}{\cancel{R}}R_1R_2}{\underset{1}{\cancel{R}}} = \frac{R\overset{1}{\cancel{R_1}}R_2}{\underset{1}{\cancel{R_1}}} + \frac{RR_1\overset{1}{\cancel{R_2}}}{\underset{1}{\cancel{R_2}}} \quad \text{Distribute } RR_1R_2. \text{ Then divide out the common factors.}$$

$$R_1R_2 = RR_2 + RR_1 \quad \text{Simplify.}$$

$$R_1R_2 = R(R_2 + R_1) \quad \text{Factor out } R \text{ on the right-hand side.}$$

$$\frac{R_1R_2}{R_2 + R_1} = R \quad \text{To isolate } R, \text{ divide both sides by } R_2 + R_1.$$

$$R = \frac{R_1R_2}{R_2 + R_1}$$

SELF CHECK

The *two-intercept form* for the equation of a line is $\frac{x}{a} + \frac{y}{b} = 1$. Solve for b.

Answer: $b = -\frac{ay}{x-a}$ or $b = \frac{ay}{a-x}$ ∎

Problem Solving

We can use rational equations to model *shared-work* problems. In this case, we assume that the work is being performed at a constant rate by all of those involved.

EXAMPLE 7

Drywalling a house. A contractor knows that one crew can drywall a house in 4 days and that another crew can drywall the same house in 5 days. One day must be allowed for the plaster coat to dry. If the contractor uses both crews, can the house be ready for painting in 4 days?

ANALYZE THE PROBLEM

Because 1 day is necessary for drying, the drywallers must complete their work in 3 days. Since the first crew can drywall the house in 4 days, it can do $\frac{1}{4}$ of the job in 1 day. Since the second crew can drywall the house in 5 days, it can do $\frac{1}{5}$ of the job in 1 day. If it takes x days for both crews to finish the house, together they can do $\frac{1}{x}$ of the job in 1 day. The amount of work the first crew can do in 1 day plus the amount of work the second crew can do in 1 day equals the amount of work both crews can do in 1 day working together.

FORM AN EQUATION If x represents the number of days it takes for both crews, working together, to drywall the house, we can form the equation

What crew 1 can do in one day	plus	what crew 2 can do in one day	is	what they can do together in one day.
$\dfrac{1}{4}$	$+$	$\dfrac{1}{5}$	$=$	$\dfrac{1}{x}$

SOLVE THE EQUATION

$$20x\left(\frac{1}{4} + \frac{1}{5}\right) = 20x\left(\frac{1}{x}\right)$$ Multiply both sides by the LCD, 20x.

$$\frac{20x \cdot 1}{4} + \frac{20x \cdot 1}{5} = \frac{20x \cdot 1}{x}$$ Remove parentheses.

$$\frac{\overset{1}{\cancel{4}} \cdot 5x \cdot 1}{\underset{1}{\cancel{4}}} + \frac{\overset{1}{\cancel{5}} \cdot 4x \cdot 1}{\underset{1}{\cancel{5}}} = \frac{20 \cdot \overset{1}{\cancel{x}} \cdot 1}{\underset{1}{\cancel{x}}}$$ Factor and divide out common factors.

$$5x + 4x = 20$$ Simplify.

$$9x = 20$$ Combine like terms.

$$x = \frac{20}{9}$$ Divide both sides by 9.

STATE THE CONCLUSION Since it will take only $\frac{20}{9}$ or $2\frac{2}{9}$ days for both crews to drywall the house and it takes 1 day for drying, it will be ready for painting in $3\frac{2}{9}$ days, which is less than 4 days.

CHECK THE RESULT In $2\frac{2}{9}$ days, the first crew will do $\dfrac{2\frac{2}{9}}{4}$ or about 56% of the job. The second crew will do $\dfrac{2\frac{2}{9}}{5}$ or about 44% of the job. When we combine the two efforts, it appears that 100% of the job would be completed. The result seems reasonable. ∎

In the next two examples, rational equations are used to model situations involving *uniform motion*.

EXAMPLE 8

Driving to a convention. A doctor drove 200 miles to attend a national convention. Because of poor weather, her average speed on the return trip was 10 mph less than her average speed going to the convention. If the return trip took 1 hour longer, how fast did she drive in each direction?

ANALYZE THE PROBLEM We need to find her rates of speed going to and returning from the convention. They can be represented using a variable. The distance traveled was 200 miles each way. To describe the travel times, we note that

$$rt = d \quad r \text{ is the rate of speed, } t \text{ is the time, and } d \text{ is the distance.}$$

or

$$t = \frac{d}{r} \quad \text{Divide both sides by } r.$$

FORM AN EQUATION Let r represent the average rate of speed going to the meeting. Then $r - 10$ represents the average rate of speed on the return trip. We can organize the facts of the problem in the table shown in Figure 6-17.

FIGURE 6-17

	Rate	·	Time	=	Distance
Going	r		$\dfrac{200}{r}$		200
Returning	$r - 10$		$\dfrac{200}{r - 10}$		200

We obtained these two entries by
dividing the distance by the rate.

r mph

$(r - 10)$ mph

200 mi

Because the return trip took 1 hour longer, we can form the following equation:

The time it took to travel to the convention	plus	1	is	the time it took to return.
$\dfrac{200}{r}$	$+$	1	$=$	$\dfrac{200}{r - 10}$

SOLVE THE EQUATION   We can solve the equation as follows:

$$r(r - 10)\left(\frac{200}{r} + 1\right) = r(r - 10)\left(\frac{200}{r - 10}\right)$$ Multiply both sides by $r(r - 10)$.

$$\overset{1}{\cancel{r}}(r - 10)\frac{200}{\underset{1}{\cancel{r}}} + r(r - 10)1 = r\overset{1}{\cancel{(r - 10)}}\left(\frac{200}{\underset{1}{\cancel{r - 10}}}\right)$$ Distribute $r(r - 10)$. Then divide out the common factors.

$$200(r - 10) + r(r - 10) = 200r$$ Simplify.

$$200r - 2{,}000 + r^2 - 10r = 200r$$ Remove parentheses. Note that this is a quadratic equation.

$$r^2 - 10r - 2{,}000 = 0$$ Subtract $200r$ from both sides.

$$(r - 50)(r + 40) = 0$$ Factor $r^2 - 10r - 2{,}000$.

$$r - 50 = 0 \quad \text{or} \quad r + 40 = 0$$ Set each factor equal to 0.

$$r = 50 \qquad\qquad r = -40$$

STATE THE CONCLUSION   We must exclude the solution of -40, because a speed cannot be negative. Thus, the doctor averaged 50 mph going to the convention, and she averaged $50 - 10$ or 40 mph returning.

CHECK THE RESULT   At 50 mph, the 200-mile trip took 4 hours. At 40 mph, the return trip took 5 hours, which is 1 hour longer.   ■

EXAMPLE 9

Riverboat cruise. The Forest City Queen can make a 9-mile trip down the Rock River and return in a total of 1.6 hours. If the riverboat travels 12 mph in still water, find the speed of the current in the Rock River.

ANALYZE THE PROBLEM   We can represent the upstream and downstream rates of speed using a variable. In each case, the distance traveled is 9 miles. To write an expression for the time traveled, divide the distance by the rate of speed.

FORM AN EQUATION   We can let c represent the speed of the current. Since the boat travels 12 mph and a current of c mph pushes the boat while it is going downstream, the speed of the boat going downstream is $(12 + c)$ mph. On the return trip, the current pushes against the boat, and its speed is $(12 - c)$ mph. Since $t = \frac{d}{r}$ $\left(\text{time} = \frac{\text{distance}}{\text{rate}}\right)$, the time required for the downstream leg of the trip is $\frac{9}{12 + c}$ hours, and the time required for the upstream leg of the trip is $\frac{9}{12 - c}$ hours. We can organize this information in the table shown in Figure 6-18.

FIGURE 6-18

	Rate	·	Time	=	Distance
Going downstream	$12 + c$		$\dfrac{9}{12 + c}$		9
Going upstream	$12 - c$		$\dfrac{9}{12 - c}$		9

Furthermore, we know that the total time required for the round trip is 1.6 or $\frac{8}{5}$ hours.

The time it takes to travel downstream	plus	the time it takes to travel upstream	is	the total time for the round trip.
$\dfrac{9}{12 + c}$	$+$	$\dfrac{9}{12 - c}$	$=$	$\dfrac{8}{5}$

SOLVE THE EQUATION   Multiply both sides of this equation by $5(12 + c)(12 - c)$ to clear it of fractions.

$$5(12 + c)(12 - c)\left(\frac{9}{12 + c} + \frac{9}{12 - c}\right) = 5(12 + c)(12 - c)\left(\frac{8}{5}\right)$$

$$\frac{5\overset{1}{\cancel{(12 + c)}}(12 - c)9}{\underset{1}{\cancel{12 + c}}} + \frac{5(12 + c)\overset{1}{\cancel{(12 - c)}}9}{\underset{1}{\cancel{12 - c}}} = \frac{\overset{1}{\cancel{5}}(12 + c)(12 - c)8}{\underset{1}{\cancel{5}}}$$

Distribute, and then divide out the common factors.

$$45(12 - c) + 45(12 + c) = 8(12 + c)(12 - c) \qquad \text{Simplify.}$$

$$540 - 45c + 540 + 45c = 8(144 - c^2)$$

On the left-hand side, distribute. On the right-hand side, use the FOIL method.

$$1{,}080 = 1{,}152 - 8c^2$$

Combine like terms and multiply. This is a quadratic equation.

$$8c^2 - 72 = 0$$

Add $8c^2$ and subtract 1,152 from both sides.

$$c^2 - 9 = 0 \qquad \text{Divide both sides by 8.}$$

$$(c + 3)(c - 3) = 0 \qquad \text{Factor } c^2 - 9.$$

$$c + 3 = 0 \quad \text{or} \quad c - 3 = 0 \qquad \text{Set each factor equal to 0.}$$

$$c = -3 \qquad\qquad c = 3$$

STATE THE CONCLUSION   Since the current cannot be negative, the apparent solution -3 must be discarded. The current in the Rock River is 3 mph.

CHECK THE RESULT   Verify the result.   ■

STUDY SET

Section 6.6

VOCABULARY

In Exercises 1–4, fill in the blanks to make the statements true.

1. An equation that contains rational expressions, such as $\frac{2}{x} + \frac{3}{2} = \frac{3}{4x}$, is called a _____ equation.

2. A proposed solution to an equation that does not satisfy the equation is called an _____ solution.

3. The lowest _____ of a group of rational expressions is divisible by each of their denominators.

4. To solve an equation means to find all the values of the variable that make the equation a _____ statement.

CONCEPTS

5. Is $x = 2$ a solution of the following equations?
   a. $\dfrac{x+2}{x+3} + \dfrac{1}{x^2 + 2x - 3} = 1$
   b. $\dfrac{x+2}{x-2} + \dfrac{1}{x^2 - 4} = 1$

6. To clear the equation
   $$\frac{4}{10} + y = \frac{4y - 50}{5y - 25}$$
   of fractions, by what should both sides be multiplied?

7. Complete the table in Illustration 1.

ILLUSTRATION 1

	r	$\cdot$	t	$=$	d
Running	x				12
Bicycling	$x + 15$				12

8. Illustration 2 shows the length of time it takes each child in a family to wash their mother's minivan. Complete the table.

ILLUSTRATION 2

	Time to wash the van alone (min)	Amount of van washed in 1 minute
Glenn	25	
Brandon	30	
Kevin	x	

9. Solve $\dfrac{x}{x+2} = \dfrac{7}{9}$ by
   a. setting the product of the extremes equal to the product of the means.
   b. multiplying both sides by the LCD.

10. a. What type of equation is $2x + x^2 = 15$?

    b. How should it be rewritten if we want to use factoring to solve it?
    c. Solve the equation.

NOTATION

In Exercises 11–12, complete each solution.

11.
$$\frac{10}{3y} - \frac{7}{30} = \frac{9}{2y}$$
$$\boxed{}\left(\frac{10}{3y} - \frac{7}{30}\right) = 30y\left(\boxed{}\right)$$
$$30y\left(\boxed{}\right) - 30y\left(\frac{7}{30}\right) = 30y\left(\frac{9}{2y}\right)$$
$$100 - \boxed{} = 135$$
$$-7y = \boxed{}$$
$$y = -5$$

12.
$$\frac{2}{u-1} + \frac{1}{u} = \frac{1}{u^2 - u}$$
$$\boxed{} + \frac{1}{u} = \frac{1}{u(u-1)}$$
$$\boxed{}\left(\frac{2}{u-1} + \frac{1}{u}\right) = u(u-1)\left[\frac{1}{u(u-1)}\right]$$
$$u(u-1)\left(\frac{2}{u-1}\right) + u(u-1)\left(\boxed{}\right) = u(u-1)\left[\frac{1}{u(u-1)}\right]$$
$$2u + \left(\boxed{}\right) = \boxed{}$$
$$\boxed{} = 2$$
$$u = \frac{2}{3}$$

PRACTICE

In Exercises 13–42, solve each equation. If a solution is extraneous, so indicate.

13. $\dfrac{1}{4} + \dfrac{9}{x} = 1$

14. $\dfrac{1}{3} - \dfrac{10}{x} = -3$

15. $\dfrac{34}{x} - \dfrac{3}{2} = -\dfrac{13}{20}$

16. $\dfrac{1}{2} + \dfrac{7}{x} = 2 + \dfrac{1}{x}$

17. $\dfrac{3}{y} + \dfrac{7}{2y} = 13$

18. $\dfrac{2}{x} + \dfrac{1}{2} = \dfrac{7}{2x}$

19. $\dfrac{x+1}{x} - \dfrac{x-1}{x} = 0$

20. $\dfrac{2}{x} + \dfrac{1}{2} = \dfrac{9}{4x} - \dfrac{1}{2x}$

21. $\dfrac{7}{5x} - \dfrac{1}{2} = \dfrac{5}{6x} + \dfrac{1}{3}$

22. $\dfrac{x-3}{x-1} - \dfrac{2x-4}{x-1} = 0$

23. $\dfrac{3-5y}{2+y} = \dfrac{3+5y}{2-y}$

24. $\dfrac{x}{x-2} = 1 + \dfrac{1}{x-3}$

25. $\dfrac{a+2}{a+1} - \dfrac{a-4}{a-3} = 0$

26. $\dfrac{z+2}{z+8} - \dfrac{z-3}{z-2} = 0$

27. $\dfrac{x+2}{x+3} - 1 = \dfrac{1}{3 - 2x - x^2}$

28. $\dfrac{x-3}{x-2} - \dfrac{1}{x} = \dfrac{x-3}{x}$

29. $\dfrac{x}{x+2} = 1 - \dfrac{3x+2}{x^2 + 4x + 4}$

30. $\dfrac{3+2a}{a^2 + 6 + 5a} + \dfrac{2-5a}{a^2 - 4} = \dfrac{2-3a}{a^2 - 6 + a}$

31. $\dfrac{2}{x-2} + \dfrac{1}{x+1} = \dfrac{1}{x^2 - x - 2}$

32. $\dfrac{5}{y-1} + \dfrac{3}{y-3} = \dfrac{8}{y-2}$

33. $\dfrac{a-1}{a+3} - \dfrac{1-2a}{3-a} = \dfrac{2-a}{a-3}$

34. $\dfrac{5}{2z^2 + z - 3} - \dfrac{2}{2z+3} = \dfrac{z+1}{z-1} - 1$

35. $\dfrac{5}{x+4} + \dfrac{1}{x+4} = x - 1$

36. $\dfrac{2}{x-1} + \dfrac{x-2}{3} = \dfrac{4}{x-1}$

37. $\dfrac{3}{x+1} - \dfrac{x-2}{2} = \dfrac{x-2}{x+1}$

38. $\dfrac{x-4}{x-3} + \dfrac{x-2}{x-3} = x - 3$

39. $\dfrac{2}{x-3} + \dfrac{3}{4} = \dfrac{17}{2x}$

40. $\dfrac{30}{y-2} + \dfrac{24}{y-5} = 13$

41. $\dfrac{x+4}{x+7} - \dfrac{x}{x+3} = \dfrac{3}{8}$

42. $\dfrac{5}{x+4} - \dfrac{1}{3} = \dfrac{x-1}{x}$

In Exercises 43–50, solve each formula for the indicated variable.

43. $I = \dfrac{E}{R_L + r}$ for r (from physics)

44. $P = \dfrac{R-C}{n}$ for C (from business)

45. $S = \dfrac{a-lr}{1-r}$ for r (from mathematics)

46. $\mu_R = \dfrac{n_1(n_1 + n_2 + 1)}{2}$ for n_2 (from statistics)

47. $P = \dfrac{Q_1}{Q_2 - Q_1}$ for Q_1 (from refrigeration/heating)

48. $\dfrac{P_1 V_1}{T_1} = \dfrac{P_2 V_2}{T_2}$ for T_2 (from chemistry)

49. $\dfrac{1}{R} = \dfrac{1}{R_1} + \dfrac{1}{R_2} + \dfrac{1}{R_3}$ for R (from electronics)

50. $P + \dfrac{a}{V^2} = \dfrac{RT}{V-b}$ for b (from physics)

APPLICATIONS

51. PHOTOGRAPHY Illustration 3 shows the relationship between distances when taking a photograph. The design of a camera lens uses the equation

$$\dfrac{1}{f} = \dfrac{1}{s_1} + \dfrac{1}{s_2}$$

which relates the focal length f of a lens to the image distance s_1 and the object distance s_2. Find the focal length of the lens in the illustration. (*Hint:* Convert feet to inches.)

ILLUSTRATION 3

Object   Lens   Image

52. OPTICS The focal length, f, of a lens is given by the lensmaker's formula,

$$\frac{1}{f} = 0.6\left(\frac{1}{r_1} + \frac{1}{r_2}\right)$$

where f is the focal length of the lens and r_1 and r_2 are the radii of the two circular surfaces. Find the focal length of the lens in Illustration 4.

ILLUSTRATION 4

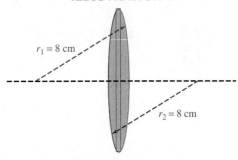

$r_1 = 8$ cm

$r_2 = 8$ cm

53. TAX ACCOUNTING As a piece of equipment gets older, its value usually lessens. One way to calculate *depreciation* is to use the formula

$$V = C - \left(\frac{C - S}{L}\right)N$$

where V denotes the value of the equipment at the end of year N, L is its useful lifetime (in years), C is its cost new, and S is its salvage value at the end of its useful life. Solve for L. Then determine what an accountant considered the useful lifetime of a forklift that cost $25,000 new, was worth $13,000 after 4 years, and has a salvage value of $1,000.

54. MECHANICAL ENGINEERING The equation

$$a = \frac{9.8m_2 - f}{m_2 + m_1}$$

models the system shown in Illustration 5, where a is the acceleration of the suspended block, m_1 and m_2 are the masses of the blocks, and f is the friction force. Solve the equation for m_2.

ILLUSTRATION 5

55. HOUSEPAINTING Illustration 6 shows two bids to paint a house.
   a. To get the job done quicker, the homeowner hired both the painters who submitted bids. How long will it take them to paint the house working together?
   b. What will the homeowner have to pay each painter?

ILLUSTRATION 6

Santos Painting	Mays House Painting
Residential	Bid:
Bid:	$200 per day
3 days	5 days work
@ $220 a day	Total: $1,000
Total: $660	

56. ROOFING A HOUSE A homeowner estimates that it will take him 7 days to roof his house. A professional roofer estimates that he could roof the house in 4 days. How long will it take if the homeowner helps the roofer?

57. OYSTERS According to the *Guinness Book of World Records,* the record for opening oysters is 100 in 140 seconds by Mike Racz in Invercargill, New Zealand on July 16, 1990. If it would take a novice $8\frac{1}{2}$ minutes to perform the same task, how long would it take them working together to open 100 oysters?

58. FARMING In 10 minutes, a conveyor belt can move 1,000 bushels of corn into the storage bin shown in Illustration 7. A smaller belt can move 1,000 bushels to the storage bin in 14 minutes. If both belts are used, how long will it take to move 1,000 bushels to the storage bin?

ILLUSTRATION 7

59. FILLING A POND One pipe can fill a pond in 3 weeks, and a second pipe can fill it in 5 weeks. However, evaporation and seepage can empty the pond in 10 weeks. If both pipes are used, how long will it take to fill the pond?

60. HOUSECLEANING Sally can clean the house in 6 hours, and her father can clean the house in 4 hours. Sally's younger brother, Dennis, can completely mess up the house in 8 hours. If Sally and her father clean and Dennis plays, how long will it take to clean the house?

61. BOXING For his morning workout, a boxer bicycles for 8 miles and then jogs back to camp along the same route. If he bicycles 6 mph faster than he jogs, and the entire workout lasts 2 hours, how fast does he jog?

62. DELIVERIES A FedEx delivery van traveled from Rockford to Chicago in 3 hours less time than it took a second FedEx van to travel from Rockford to St. Louis. If the vans traveled at the same average speed, use the information in the mileage chart in Illustration 8 to help determine how long the first driver was on the road.

ILLUSTRATION 8

63. FINDING RATES OF SPEED Two trains made the same 315-mile run. Since one train traveled 10 mph faster than the other, it arrived 2 hours earlier. Find the speed of each train.

64. TRAIN TRAVEL A train traveled 120 miles from Freeport to Chicago and returned the same distance in a total time of 5 hours. If the train traveled 20 mph slower on the return trip, how fast did the train travel in each direction?

65. CROP DUSTING A helicopter spraying fertilizer over a field can fly 0.5 mile downwind in the same time as it can fly 0.4 mile upwind. Find the speed of the wind if the helicopter travels 45 mph in still air when dusting crops.

66. BOATING A man can drive a motorboat 45 miles down the Rock River in the same amount of time that he can drive 27 miles upstream. Find the speed of the current if the speed of the boat is 12 mph in still water.

67. UNIT COST One month, an appliance store manager bought several microwave ovens for a total of $1,800. The next month, because the unit cost of the same model of microwave increased by $25, she bought one fewer oven for the same total price. How many ovens did she buy the first month? (*Hint:* Write an expression for the unit cost of a microwave for the second month, then use the formula: Unit cost · number = total cost.)

68. EXTENDED VACATION Use the facts in the E-mail message in Illustration 9 to determine how long the student had originally planned to stay in Europe. (*Hint:* Unit cost · number = total cost.)

ILLUSTRATION 9

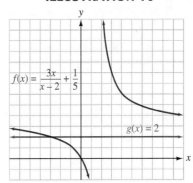

E-Mail

Hi Mom and Dad,
After working so hard to earn $1,200 to take this trip to Europe, it was all worth it! I've been very frugal and been able to cut $20 from my daily expenses. Because of this, I'll be able to stay three extra days. Please pick me up on Friday instead. Love, Liz

WRITING

69. Why is it necessary to check the solutions of a rational equation?

70. Explain what it means to *clear* a rational equation of fractions.

71. See Illustration 10. Explain how to solve the rational equation graphically:

$$\frac{3x}{x-2} + \frac{1}{5} = 2$$

ILLUSTRATION 10

$f(x) = \dfrac{3x}{x-2} + \dfrac{1}{5}$

$g(x) = 2$

72. Would you use the same approach to answer the following problems? Explain why or why not.

Simplify: $\dfrac{x^2 - 10}{x^2 - 1} - \dfrac{3x}{x - 1} - \dfrac{2x}{x + 1}$

Solve: $\dfrac{x^2 - 10}{x^2 - 1} - \dfrac{3x}{x - 1} = -\dfrac{2x}{x + 1}$

REVIEW

In Exercises 73–76, write each italicized number in scientific notation.

73. OIL The total cost of the Alaskan pipeline, running 800 miles from Prudhoe Bay to Valdez, was *$9,000,000,000*.

74. NATURAL GAS The TransCanada Pipeline transported a record *2,352,000,000,000* cubic feet of gas in 1995.

75. RADIOACTIVITY The least stable radioactive isotope is lithium 5, which decays in *0.0000000000000000000000044* second.

76. BALANCES The finest balances in the world are made in Germany. They can weigh objects to an accuracy of *35 × 10⁻¹¹* ounce.

▶ 6.7   Dividing Polynomials

In this section, you will learn about

> Dividing a monomial by a monomial ■ Dividing a polynomial by a monomial ■ Dividing a polynomial by a polynomial ■ Missing terms

---

Introduction In Chapter 5, we discussed addition, subtraction, and multiplication of polynomials. We will now introduce the procedures used to divide polynomials. This topic appears in a chapter about rational expressions because rational expressions indicate division of polynomials. For example,

$$\frac{x^2 - 3x + 7}{x + 1} = (x^2 - 3x + 7) \div (x + 1)$$

We begin the discussion of division of polynomials with the simplest case, a monomial divided by a monomial.

Dividing a Monomial by a Monomial

In Example 1, we review two methods that can be used to divide a monomial by a monomial.

> **E X A M P L E   1**

A monomial divided by a monomial. Simplify $3a^2b^3 \div (2a^3b)$.

Solution   **Method 1**

We write $3a^2b^3 \div (2a^3b)$ as a rational expression and divide out all common factors:

$$\frac{3a^2b^3}{2a^3b} = \frac{3\overset{1}{\cancel{a}}\overset{1}{\cancel{a}}\cancel{b}\,b\,b}{2\cancel{a}\cancel{a}\underset{1}{a}\cancel{b}}$$

$$= \frac{3b^2}{2a}$$

Method 2

We write $3a^2b^3 \div (2a^3b)$ as a rational expression and use the rules for exponents:

$$\frac{3a^2b^3}{2a^3b} = \frac{3}{2}a^{2-3}b^{3-1} \quad \text{When dividing like bases, keep the base and subtract exponents.}$$

$$= \frac{3}{2}a^{-1}b^2 \quad \text{Subtract.}$$

$$= \frac{3}{2}\left(\frac{1}{a}\right)\frac{b^2}{1} \quad \text{We don't want negative exponents in the result.}$$

$$= \frac{3b^2}{2a} \quad \text{Multiply the numerators. Multiply the denominators.}$$

SELF CHECK   Simplify $\dfrac{6x^3y^2}{8x^2y^3}$.   *Answer:* $\frac{3x}{4y}$ ■

Dividing monomials

> **Method 1**   Factor the numerator and denominator completely. Then divide out all common factors.
>
> **Method 2**   Divide the coefficients. Then use the rules for exponents to simplify the divisions of variable factors with like bases.

Dividing a Polynomial by a Monomial

In Example 2, we use the fact that $\frac{a}{b} = \frac{1}{b} \cdot a$ $(b \neq 0)$ to divide a polynomial by a monomial.

EXAMPLE 2

A polynomial divided by a monomial. Divide $4x^3y^2 + 3x^2y^5 - 12xy$ by $3x^2y^3$.

Solution   We rewrite the division as a product and use the distributive property to remove parentheses.

$$\frac{4x^3y^2 + 3x^2y^5 - 12xy}{3x^2y^3} = \frac{1}{3x^2y^3}(4x^3y^2 + 3x^2y^5 - 12xy)$$

$$= \frac{4x^3y^2}{3x^2y^3} + \frac{3x^2y^5}{3x^2y^3} - \frac{12xy}{3x^2y^3}$$

We then simplify each of the rational expressions on the right-hand side of the equals sign to get

$$\frac{4x^3y^2 + 3x^2y^5 - 12xy}{3x^2y^3} = \frac{4x}{3y} + y^2 - \frac{4}{xy^2}$$

SELF CHECK   Divide $\dfrac{8a^3b^4 - 4a^4b^2 + a^2b^2}{4a^2b^2}$.   *Answer:* $2ab^2 - a^2 + \frac{1}{4}$ ■

Dividing a polynomial by a monomial

1. Multiply the polynomial by the reciprocal of the monomial.
2. Use the distributive property to remove parentheses.
3. Simplify each of the resulting rational expressions using the rules for simplifying a monomial divided by a monomial.

Dividing a Polynomial by a Polynomial

There is an **algorithm** (a repeating series of steps) to use when the divisor is not a monomial. To use the division algorithm to divide $x^2 + 7x + 12$ (the **dividend**) by $x + 4$ (the **divisor**), we write the division in long division form and proceed as follows:

$$\begin{array}{r} x \\ x + 4 \overline{)\, x^2 + 7x + 12\,} \end{array}$$
How many times does x divide x^2? $x^2/x = x$. Place the x in the quotient.

$$\begin{array}{r} x \\ x + 4 \overline{)\, x^2 + 7x + 12\,} \\ \underline{x^2 + 4x} \\ 3x + 12 \end{array}$$
Multiply each term in the divisor by x to get $x^2 + 4x$, subtract $x^2 + 4x$ from $x^2 + 7x$, and bring down the 12.

$$\begin{array}{r} x \; + 3 \\ x + 4 \overline{)\, x^2 + 7x + 12\,} \\ \underline{x^2 + 4x} \\ 3x + 12 \end{array}$$
How many times does x divide $3x$? $3x/x = +3$. Place the $+3$ in the quotient.

$$\begin{array}{r} x \; + 3 \\ x + 4 \overline{)\, x^2 + 7x + 12\,} \\ \underline{x^2 + 4x} \\ 3x + 12 \\ \underline{3x + 12} \\ 0 \end{array}$$
Multiply each term in the divisor by 3 to get $3x + 12$, and subtract $3x + 12$ from $3x + 12$ to get 0.

The division process stops when the result of the subtraction is a constant or a polynomial with degree less than the degree of the divisor. Here, the quotient is $x + 3$ and the remainder is 0.

We can check the quotient by multiplying the divisor by the quotient. The product should be the dividend.

$$\overbrace{(x + 4)}^{\text{Divisor}} \cdot \overbrace{(x + 3)}^{\text{quotient}} = \overbrace{x^2 + 7x + 12}^{\text{dividend}} \quad \text{The quotient checks.}$$

EXAMPLE 3

Divide $\dfrac{2a^3 + 9a^2 + 5a - 6}{2a + 3}$ using long division.

Solution

$$\begin{array}{r} a^2 \\ 2a + 3 \overline{)\, 2a^3 + 9a^2 + 5a - 6\,} \end{array}$$
How many times does $2a$ divide $2a^3$? $2a^3/2a = a^2$. Place a^2 in the quotient.

$$\begin{array}{r} a^2 \\ 2a + 3 \overline{)\, 2a^3 + 9a^2 + 5a - 6\,} \\ \underline{2a^3 + 3a^2} \\ 6a^2 + 5a \end{array}$$
Multiply each term in the divisor by a^2 to get $2a^3 + 3a^2$, subtract $2a^3 + 3a^2$ from $2a^3 + 9a^2$, and bring down the $5a$.

$$\begin{array}{r} a^2 \; + 3a \\ 2a + 3 \overline{)\, 2a^3 + 9a^2 + 5a - 6\,} \\ \underline{2a^3 + 3a^2} \\ 6a^2 + 5a \end{array}$$
How many times does $2a$ divide $6a^2$? $6a^2/2a = 3a$. Place the $+3a$ in the quotient.

$$\begin{array}{r}
a^2 + 3a \\
2a + 3\overline{)2a^3 + 9a^2 + 5a - 6} \\
\underline{2a^3 + 3a^2} \\
6a^2 + 5a \\
\underline{6a^2 + 9a} \\
-4a - 6
\end{array}$$

Multiply each term in the divisor by $3a$ to get $6a^2 + 9a$, subtract $6a^2 + 9a$ from $6a^2 + 5a$, and bring down the -6.

$$\begin{array}{r}
a^2 + 3a - 2 \\
2a + 3\overline{)2a^3 + 9a^2 + 5a - 6} \\
\underline{2a^3 + 3a^2} \\
6a^2 + 5a \\
\underline{6a^2 + 9a} \\
-4a - 6
\end{array}$$

How many times does $2a$ divide $-4a$? $-4a/2a = -2$. Place the -2 in the quotient.

$$\begin{array}{r}
a^2 + 3a - 2 \\
2a + 3\overline{)2a^3 + 9a^2 + 5a - 6} \\
\underline{2a^3 + 3a^2} \\
6a^2 + 5a \\
\underline{6a^2 + 9a} \\
-4a - 6 \\
\underline{-4a - 6} \\
0
\end{array}$$

Multiply each term in the divisor by -2 to get $-4a - 6$; subtract $-4a - 6$ from $-4a - 6$ to get 0.

Since the remainder is 0, the quotient is $a^2 + 3a - 2$. We can check the quotient by verifying that

$$\underbrace{(2a + 3)}_{\text{Divisor} \cdot}\underbrace{(a^2 + 3a - 2)}_{\text{quotient}} = \underbrace{2a^3 + 9a^2 + 5a - 6}_{\text{dividend}}$$ ■

E X A M P L E 4

Using long division form. Divide $\dfrac{3x^3 + 2x^2 - 3x + 8}{x - 2}$.

Solution

$$\begin{array}{r}
3x^2 + 8x + 13 \\
x - 2\overline{)3x^3 + 2x^2 - 3x + 8} \\
\underline{3x^3 - 6x^2} \\
8x^2 - 3x \\
\underline{8x^2 - 16x} \\
13x + 8 \\
\underline{13x - 26} \\
34
\end{array}$$

This division gives a quotient of $3x^2 + 8x + 13$ and a remainder of 34. It is common to form a fraction with the remainder as the numerator and the divisor as the denominator and to write the result as

$$3x^2 + 8x + 13 + \frac{34}{x - 2}$$

To check, we verify that

$$(x - 2)\left(3x^2 + 8x + 13 + \frac{34}{x - 2}\right) = 3x^3 + 2x^2 - 3x + 8$$

SELF CHECK

Divide $\dfrac{2a^3 + 3a^2 - a + 2}{a - 3}$.

Answer: $2a^2 + 9a + 26 + \frac{80}{a-3}$ ■

E X A M P L E 5

Arranging terms in descending powers. Divide $(-9x + 8x^3 + 10x^2 - 9) \div (3 + 2x)$.

Solution The division algorithm works best when the polynomials in the dividend and the divisor are written in descending powers of x. We can use the commutative property of addition to rearrange the terms. Then the division is routine:

$$
\begin{array}{r}
4x^2 - x - 3 \\
2x + 3\overline{)\,8x^3 + 10x^2 - 9x - 9} \\
\underline{8x^3 + 12x^2\phantom{{}-9x-9}} \\
-2x^2 - 9x\phantom{{}-9} \\
\underline{-2x^2 - 3x\phantom{{}-9}} \\
-6x - 9 \\
\underline{-6x - 9} \\
0
\end{array}
$$

Thus,

$$
\frac{-9x + 8x^3 + 10x^2 - 9}{3 + 2x} = 4x^2 - x - 3
$$

SELF CHECK   Divide $2 + 3a\overline{)\,-4a + 15a^2 + 18a^3 - 4}$.   *Answer:* $6a^2 + a - 2$ ∎

Missing Terms

If a power of the variable is missing in the dividend, it is helpful to insert "placeholder" terms, because they aid in the subtraction step of the long division procedure.

EXAMPLE 6

Introducing missing terms. Divide $8x^3 + 1$ by $2x + 1$.

Solution When we write the terms in the dividend in descending powers of x, we see that the terms involving x^2 and x are missing. We can introduce the terms $0x^2$ and $0x$ in the dividend or leave spaces for them. Then the division is routine.

$$
\begin{array}{r}
4x^2 - 2x + 1 \\
2x + 1\overline{)\,8x^3 + 0x^2 + 0x + 1} \\
\underline{8x^3 + 4x^2\phantom{{}+0x+1}} \\
-4x^2 + 0x\phantom{{}+1} \\
\underline{-4x^2 - 2x\phantom{{}+1}} \\
2x + 1 \\
\underline{2x + 1} \\
0
\end{array}
$$

Thus,

$$
\frac{8x^3 + 1}{2x + 1} = 4x^2 - 2x + 1
$$

SELF CHECK   Divide $27a^3 - 1$ by $3a - 1$.   *Answer:* $9a^2 + 3a + 1$ ∎

EXAMPLE 7

Arranging the terms in descending powers. Divide $-17x^2 + 5x + x^4 + 2$ by $x^2 - 1 + 4x$.

Solution We write the problem with the divisor and the dividend in descending powers of x. After introducing $0x^3$ for the missing term in the dividend, we proceed as follows:

$$
\begin{array}{r}
x^2 - 4x \\
x^2 + 4x - 1\overline{)\,x^4 + 0x^3 - 17x^2 + 5x + 2} \\
\underline{x^4 + 4x^3 - x^2\phantom{{}+5x+2}} \\
-4x^3 - 16x^2 + 5x\phantom{{}+2} \\
\underline{-4x^3 - 16x^2 + 4x\phantom{{}+2}} \\
x + 2
\end{array}
$$

This division gives a quotient of $x^2 - 4x$ and a remainder of $x + 2$.

$$\frac{-17x^2 + 5x + x^4 + 2}{x^2 - 1 + 4x} = x^2 - 4x + \frac{x + 2}{x^2 + 4x - 1}$$

SELF CHECK   Divide $\dfrac{2a^2 + 3a^3 + a^4 - 7 + a}{a^2 - 2a + 1}$.   *Answer:* $a^2 + 5a + 11 + \dfrac{18a - 18}{a^2 - 2a + 1}$ ■

STUDY SET

Section 6.7

VOCABULARY

In Exercises 1–4, fill in the blanks to make the statements true.

1. In the division

$$x - 4 \overline{)\, x^2 - x - 12} \quad \text{(quotient } x + 3\text{)}$$

$x^2 - x - 12$ is the _____, $x - 4$ is the _____, and $x + 3$ is the _____.

2. For the division shown here, the _____ is 1.

$$x + 4 \overline{)\, x^2 + 7x + 13} \quad \text{(quotient } x + 3\text{)}$$
$$\underline{x^2 + 4x}$$
$$3x + 13$$
$$\underline{3x + 12}$$
$$1$$

3. The division _____ is a repeating series of steps used to do a long division.

4. In the polynomial $4x^4 + 2x^3 - x^2 + x + 7$, the powers of x are written in _____ order.

CONCEPTS

In Exercises 5–6, fill in the blanks to make the statements true.

5. $\dfrac{a}{b} = \boxed{} \cdot a \quad (b \neq 0)$

6. Divisor $\cdot$ _____ + remainder = dividend

7. Suppose that after dividing $2x^3 + 5x^2 - 11x + 4$ by $2x - 1$, you obtain $x^2 + 3x - 4$. Show how multiplication can be used to check the result.

8. Consider the first step of the division process for

$$2x^2 - 1 \overline{)\, 4x^4 + 0x^3 + 0x^2 + 0x - 1}$$

How many times does $2x^2$ divide $4x^4$?

NOTATION

In Exercises 9–10, complete each solution.

9.
$$x + 4 \overline{)\, 2x^2 + 9x + 4} \quad \text{(quotient } 2x + 1\text{)}$$
$$\underline{\boxed{} + 8x}$$
$$\boxed{} + 4$$
$$\underline{x + \boxed{}}$$
$$0$$

10.
$$3x + 4 \overline{)\, 6x^2 + 5x - 4} \quad \text{(quotient } 2x - 1\text{)}$$
$$\underline{6x^2 + \boxed{}}$$
$$\boxed{} - 4$$
$$\underline{-3x - \boxed{}}$$
$$0$$

11. If a polynomial is divided by $3a - 2$ and the quotient is $3a^2 + 5$ with a remainder of 6, how do we write the result?

12. If a polynomial is divided by $3a - 2$ and the quotient is $3a^2 + 5$ with a remainder of -6, how do we write the result?

13. List three ways we can use symbols to write $x^2 - x - 12$ divided by $x - 4$.

14. Tell whether the statement below is true or false. Justify your answer.
$$2x^3 - 9 = 2x^3 + 0x^2 + 0x - 9$$

PRACTICE

In Exercises 15–26, do each division. Write each answer without using negative exponents.

15. $\dfrac{4x^2 y^3}{8x^5 y^2}$

16. $\dfrac{25x^4 y^7}{5xy^9}$

17. $-\dfrac{33a^2 b^2}{44a^4 b^2}$

18. $\dfrac{-63a^4}{81a^6 b^3}$

19. $\dfrac{4x + 6}{2}$

20. $\dfrac{11a^3 - 11a^2}{11}$

21. $\dfrac{4x^2 - x^3}{-6x}$

22. $\dfrac{5y^4 + 45y^3}{-15y^2}$

23. $\dfrac{12x^2y^3 + x^3y^2}{6xy}$

24. $\dfrac{54a^3y^2 - 18a^4y^3}{27a^2y^2}$

25. $\dfrac{24x^6y^7 - 12x^5y^{12} + 36xy}{48x^2y^3}$

26. $\dfrac{9x^4y^3 + 18x^2y - 27xy^4}{-9x^3y^3}$

In Exercises 27–58, do each division.

27. $\dfrac{x^2 + 5x + 6}{x + 3}$

28. $\dfrac{x^2 - 5x + 6}{x - 3}$

29. $\dfrac{x^2 + 10x + 21}{x + 3}$

30. $\dfrac{x^2 + 10x + 21}{x + 7}$

31. $\dfrac{6x^2 - x - 12}{2x + 3}$

32. $\dfrac{6x^2 - x - 12}{2x - 3}$

33. $\dfrac{3x^3 - 2x^2 + x - 6}{x - 1}$

34. $\dfrac{4a^3 + a^2 - 3a + 7}{a + 1}$

35. $\dfrac{6x^3 + 11x^2 - x - 2}{3x - 2}$

36. $\dfrac{6x^3 + 11x^2 - 9x - 5}{2x + 3}$

37. $\dfrac{6x^3 - x^2 - 6x - 9}{2x - 3}$

38. $\dfrac{16x^3 + 16x^2 - 9x - 5}{4x + 5}$

39. $(2a + 1 + a^2) \div (a + 1)$

40. $(a - 15 + 6a^2) \div (2a - 3)$

41. $(6y - 4 + 10y^2) \div (5y - 2)$

42. $(-10x + x^2 + 16) \div (x - 2)$

43. $\dfrac{-18x + 12 + 6x^2}{x - 1}$

44. $\dfrac{27x + 23x^2 + 6x^3}{2x + 3}$

45. $\dfrac{13x + 16x^4 + 3x^2 + 3}{4x + 3}$

46. $\dfrac{3x^2 + 9x^3 + 4x + 4}{3x + 2}$

47. $a^3 + 1$ divided by $a - 1$

48. $27a^3 - 8$ divided by $3a - 2$

49. $\dfrac{15a^3 - 29a^2 + 16}{3a - 4}$

50. $\dfrac{4x^3 - 12x^2 + 17x - 12}{2x - 3}$

51. $y - 2 \overline{)-24y + 24 + 6y^2}$

52. $3 - a \overline{)21a - a^2 - 54}$

53. $x^2 - 2 \overline{)x^6 - x^4 + 2x^2 - 8}$

54. $x^2 + 3 \overline{)x^6 + 2x^4 - 6x^2 - 9}$

55. $\dfrac{x^4 + 2x^3 + 4x^2 + 3x + 2}{x^2 + x + 2}$

56. $\dfrac{2x^4 + 3x^3 + 3x^2 - 5x - 3}{2x^2 - x - 1}$

57. $\dfrac{x^3 + 3x + 5x^2 + 6 + x^4}{x^2 + 3}$

58. $\dfrac{x^5 + 3x + 2}{x^3 + 1 + 2x}$

In Exercises 59–60, use a calculator to help find each quotient.

59. $x - 2 \overline{)9.8x^2 - 3.2x - 69.3}$

60. $2.5x - 3.7 \overline{)-22.25x^2 - 38.9x - 16.65}$

APPLICATIONS

61. ADVERTISING Find the length of one of the longer sides of the billboard shown in Illustration 1 if its area is given by $x^3 - 4x^2 + x + 6$.

ILLUSTRATION 1

$x + 1$

NEXT EXIT 3/4 MILES **Mobil**

62. MASONRY The steel trowel shown in Illustration 2 is in the shape of an isosceles triangle. Find the height if the area is given by $6 + 18t + t^2 + 3t^3$.

ILLUSTRATION 2

Height $6 + t^2$

63. WINTER TRAVEL Complete the table in Illustration 3, which lists the rate (mph), time traveled (hr), and distance traveled (mi) by an Alaskan trail guide.

ILLUSTRATION 3

	r	$\cdot$	t	$=$	d
Dog sled			$4x + 7$		$12x^2 + 13x - 14$
Snowshoes	$3x + 4$				$3x^2 + 19x + 20$

64. PRICING Complete the table in Illustration 4 for two items sold at a produce store.

ILLUSTRATION 4

	Price per lb ·	Number of lb =	Value
Cashews	$x^2 + 2x + 4$		$x^4 + 4x^2 + 16$
Sunflower seeds		$x^2 + 6$	$x^4 - x^2 - 42$

WRITING

65. Explain how to divide a monomial by a monomial.

66. Explain how to check the result of a division problem if there is a nonzero remainder.

REVIEW

In Exercises 67–70, simplify each expression.

67. $2(x^2 + 4x - 1) + 3(2x^2 - 2x + 2)$

68. $3(2a^2 - 3a + 2) - 4(2a^2 + 4a - 7)$

69. $-2(3y^3 - 2y + 7) - (y^2 + 2y - 4) + 4(y^3 + 2y - 1)$

70. $3(4y^3 + 3y - 2) + 2(3y^2 - y + 3) - 5(2y^3 - y^2 - 2)$

Expressions and Equations

In this chapter, we have discussed procedures for working with **rational expressions** and procedures for solving **rational equations**.

Rational Expressions

The **fundamental property of fractions** is used when simplifying rational expressions and when multiplying and dividing rational expressions: *We can divide out factors that are common to the numerator and the denominator of a fraction.*

1. a. Simplify $\dfrac{6x^2 + x - 2}{8x^2 + 2x - 3}$.

    b. What common factor was divided out?

2. a. Multiply $\dfrac{3d^2 - d - 2}{6d^2 - 5d - 6} \cdot \dfrac{4d^2 - 9}{2d^2 + 5d + 3}$.

    b. What common factors were divided out?

The fundamental property of fractions also states that *multiplying the numerator and denominator of a fraction by the same nonzero number does not change the value of the fraction.* We use this concept to "build" fractions when adding or subtracting rational expressions with unlike denominators, and when simplifying complex fractions.

3. a. Add $\dfrac{3}{x + 2} + \dfrac{5}{x - 4}$.

    b. By what did you multiply the first fraction to rewrite it in terms of the LCD? By what did you multiply the second fraction?

4. a. Use Method 2 to simplify $\dfrac{n - 1 - \dfrac{2}{n}}{\dfrac{n}{3}}$.

    b. By what did you multiply the numerator and denominator to simplify the complex fraction?

Rational Equations

The multiplication property of equality states that *if equal quantities are multiplied by the same nonzero number, the results will be equal quantities.* We use this property when solving rational equations. If we multiply both sides of the equation by the LCD of the rational expressions in the equation, we can clear it of fractions.

5. a. Solve $\dfrac{t - 3}{t - 2} - \dfrac{t - 3}{t} = \dfrac{1}{t}$.

    b. By what did you multiply both sides to clear the equation of fractions?

6. a. Solve $\dfrac{5}{2x^2 + x - 3} - \dfrac{x + 1}{x - 1} = \dfrac{2}{2x + 3} - 1$.

    b. By what did you multiply both sides to clear the equation of fractions?

7. a. Solve $\dfrac{x + 1}{x + 2} = \dfrac{x - 3}{x - 4}$.

    b. By what did you multiply both sides to clear the equation of fractions?

8. a. Solve $\dfrac{x^2}{a^2} - \dfrac{y^2}{b^2} = 1$ for a^2.

    b. By what did you multiply both sides to clear the equation of fractions?

Accent on Teamwork

Section 6.1

Vertical asymptotes Consider the rational function $f(x) = \frac{1}{x}$. Use your calculator to complete the two tables of values.

Table 1: x approaches 0 from the right

x	3	2	1	0.5	0.25	0.1	0.01	0.001
$f(x)$								

Table 2: x approaches 0 from the left

x	-3	-2	-1	-0.5	-0.25	-0.1	-0.01	-0.001
$f(x)$								

Plot the points from each table on the same graph and draw a smooth curve through them. Explain what happens to the function values $f(x)$ as x approaches 0 from the right and approaches 0 from the left. Label the vertical asymptote and explain why the function is not defined for $x = 0$.

Section 6.2

Cooking Find a simple recipe for a treat that you can make for your class. Use a proportion to determine the amount of each ingredient needed to make enough of the recipe for the exact number of people in your class. For example, if a recipe makes 2 dozen cookies and there are 35 students in your class, use the ratio $\frac{24}{35}$ in the proportion. Write the old recipe and the new recipe on separate pieces of poster board. Did the recipe serve the correct number of people? Tell the class how you made the calculations and what difficulties you encountered.

Section 6.3

Multiplying rational expressions Explain what is **wrong** with the student's work shown here. Then write a correct solution.

$$\frac{3}{4x} + \frac{2}{3x} = 12x\left(\frac{3}{4x} + \frac{2}{3x}\right)$$
$$= 12x \cdot \frac{3}{4x} + 12x \cdot \frac{2}{3x}$$
$$= 9 + 8$$
$$= 17$$

Section 6.4

Adding rational expressions Add the fractions by expressing them in terms of a common denominator $24b^3$. (Note: This is not the LCD.)

$$\frac{r}{4b^2} + \frac{s}{6b}$$

An extra step had to be performed because the lowest common denominator was not used. What was it?

Section 6.5

Complex fractions Each complex fraction in the list

$$1 + \frac{1}{2},\, 1 + \cfrac{1}{1 + \frac{1}{2}},\, 1 + \cfrac{1}{1 + \cfrac{1}{1 + \frac{1}{2}}},$$

$$1 + \cfrac{1}{1 + \cfrac{1}{1 + \cfrac{1}{1 + \frac{1}{2}}}},\, \ldots$$

can be simplified by using the value of the expression preceding it. For example, to simplify the second expression in the list, replace $1 + \frac{1}{2}$ with $\frac{3}{2}$. Show that the expressions in the list simplify to the fractions $\frac{3}{2}, \frac{5}{3}, \frac{8}{5}, \frac{13}{8}, \frac{21}{13}, \frac{34}{21}, \ldots$. Do you see a pattern? Can you predict the next fraction?

Section 6.6

Shared work Use a watch to time (in seconds) how long it takes to fill a pail using a garden hose. From a different faucet, with the water running at a different rate, determine how long it takes a second hose to fill the same pail. Use the concepts studied in Section 6.6 to determine how long it would take to fill the pail with both hoses. Then fill the pail using both hoses, keeping track of the time. How close is the actual time to the predicted time? What are some reasons for any difference between them? Make a video of this project and show it to your class.

Section 6.7

Factors Since 6 is a factor of 24, 6 divides 24 exactly with no remainder. Use this same reasoning to determine whether
a. $2x - 3$ is a factor of $10x^2 - x - 21$.
b. $x - 1$ is a factor of $x^5 - 1$.
c. $2x + y$ is a factor of $32x^5 + y^5$.

Section 6.1

Rational Functions and Simplifying Rational Expressions

CONCEPTS

Rational expressions are fractions that indicate the quotient of two polynomials.

A *rational function* is defined by a rational expression in one variable where the polynomial in the denominator is never zero.

To *simplify a rational expression:*
1. Factor the numerator and the denominator.
2. Divide out the common factors of the numerator and denominator by applying the *fundamental property of fractions:*

$$\frac{ak}{bk} = \frac{a}{b} \quad (b \neq 0 \text{ and } k \neq 0)$$

The quotient of any nonzero quantity and its opposite is -1.

REVIEW EXERCISES

1. Complete the table of values for the rational function $f(x) = \frac{4}{x}$ where $x > 0$. Then graph it. Label the horizontal asymptote.

x	$f(x)$
$\frac{1}{2}$	
1	
2	
3	
4	
5	
6	
7	
8	

2. Use a graphing calculator to graph the rational function $f(x) = \frac{3x+2}{x}$. From the graph, determine the equations of the horizontal and vertical asymptotes and the domain and range.

3. Simplify each rational expression.

a. $\dfrac{248x^2y}{576xy^2}$

b. $\dfrac{x^2 - 49}{x^2 + 14x + 49}$

c. $\dfrac{x^2 - 2x + 4}{2x^3 + 16}$

d. $\dfrac{x^2 + 6x + 36}{x^3 - 216}$

e. $\dfrac{ac - ad + bc - bd}{d^2 - c^2}$

f. $\dfrac{m^3 + m^2n - 2mn^2}{mn^2 + m^2n - 2m^3}$

g. $\dfrac{x - y}{y - x}$

h. $\dfrac{2m - 2n}{n - m}$

Section 6.2

Proportion and Variation

In a *proportion,* the product of the *extremes* is equal to the product of the *means.*

If two angles of one triangle have the same measure as two angles of a second triangle, the triangles are *similar.* The lengths of corresponding sides of similar triangles are proportional.

Direct variation: As one variable gets larger, the other gets larger as described by the equation $y = kx$, where k is the *constant of proportionality.*

4. Solve each proportion.

a. $\dfrac{x + 1}{8} = \dfrac{4x - 2}{24}$

b. $\dfrac{1}{x + 6} = \dfrac{x + 10}{12}$

5. SIMILAR TRIANGLES Find the height of a tree if it casts a 44-foot shadow when a 4-foot shrub casts a $2\frac{1}{2}$-foot shadow.

6. PROPERTY TAX The property tax in a certain county varies directly as assessed valuation. If a tax of \$1,575 is levied on a single-family home assessed at \$90,000, determine the property tax on an apartment complex assessed at \$312,000.

Inverse variation: As one variable gets larger, the other gets smaller as described by the equation

$$y = \frac{k}{x} \quad (k \text{ is a constant})$$

Joint variation: One variable varies with the product of several variables. For example, $y = kxz$ (k is a constant).

7. ELECTRICITY For a fixed voltage, the current in an electrical circuit varies inversely as the resistance in the circuit. If a certain circuit has a current of $2\frac{1}{2}$ amps when the resistance is 150 ohms, find the current in the circuit when the resistance is doubled.

8. HURRICANE WINDS The wind force on a vertical surface varies jointly as the area of the surface and the square of the wind's velocity. If a 10-mph wind exerts a force of 1.98 pounds on the sign shown in Illustration 1, find the force on the sign if the wind is blowing at 80 mph.

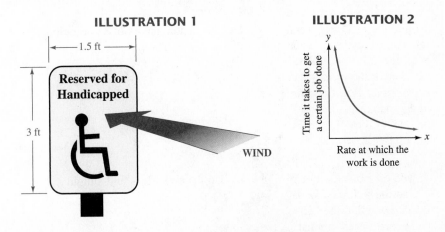

ILLUSTRATION 1

1.5 ft

Reserved for Handicapped

3 ft

WIND

ILLUSTRATION 2

Time it takes to get a certain job done

Rate at which the work is done

9. Does the graph in Illustration 2 show direct or inverse variation?

Combined variation: a combination of direct and inverse variation. For example,

$$y = \frac{kx}{z} \quad (k \text{ is a constant})$$

10. Assume that x_1 varies directly with the third power of t and inversely with x_2. Find the constant of variation if $x_1 = 1.6$ when $t = 8$ and $x_2 = 64$.

Section 6.3

To multiply two fractions, multiply the numerators and multiply the denominators:

$$\frac{a}{b} \cdot \frac{c}{d} = \frac{ac}{bd} \quad (b, d \neq 0)$$

To divide two fractions, invert the divisor and multiply:

$$\frac{a}{b} \div \frac{c}{d} = \frac{a}{b} \cdot \frac{d}{c}$$

$(b, d, c \neq 0)$

(Multiply the first fraction by the *reciprocal* of the second.)

Multiplying and Dividing Rational Expressions

11. Do the operations and simplify:

a. $\dfrac{x^2 + 4x + 4}{x^2 - x - 6} \cdot \dfrac{9 - x^2}{x^2 + 5x + 6}$

b. $\dfrac{2a^2 - 5a - 3}{a^2 - 9} \div \dfrac{2a^2 + 5a + 2}{2a^2 + 5a - 3}$

c. $\left(\dfrac{h - 2}{h^3 + 4} \right)^2$

d. $\dfrac{t^2 - 4}{t} \div (t + 2)$

e. $\dfrac{x^2 + 3x + 2}{x^2 - x - 6} \cdot \dfrac{3x^2 - 3x}{x^2 - 3x - 4} \div \dfrac{x^2 + 3x + 2}{x^2 - 2x - 8}$

Section 6.4

Adding and Subtracting Rational Expressions

To add (or subtract) two rational expressions with *like denominators,* add (or subtract) the numerators and keep the common denominator.

12. Do the operations and simplify.

   a. $\dfrac{5y}{x-y} - \dfrac{3}{x-y}$

   b. $\dfrac{3x-1}{x^2+2} + \dfrac{3(x-2)}{x^2+2}$

   c. $\dfrac{4}{t-3} + \dfrac{6}{3-t}$

   d. $\dfrac{p+3}{p^2+13p+12} - \dfrac{2p+4}{p^2+13p+12}$

To find the LCD of several rational expressions, factor each denominator and use each factor the greatest number of times that it appears in any one denominator. The product of these factors is the LCD.

13. The denominators of some rational expressions are given. Find the LCD.

   a. $15a^2h, \; 20ah^3$

   b. $ab^2 - ab, \; ab^2, \; b^2 - b$

   c. $x^2 - 4x - 5, \; x^2 - 25$

   d. $m^2 - 4m + 4, \; m^3 - 8$

To add or subtract rational expressions with *unlike denominators,* find the LCD and express each rational expression with a denominator that is the LCD. Add (or subtract) the resulting fractions and simplify the result if possible.

14. Do the operations and simplify.

   a. $9 - \dfrac{1}{a+1}$

   b. $\dfrac{x}{4z^2} + \dfrac{y}{6z}$

   c. $\dfrac{4x}{x-4} - \dfrac{3}{x+3}$

   d. $\dfrac{4}{3xy-6y} - \dfrac{4}{10-5x}$

   e. $\dfrac{-2(3+x)}{x^2+6x+9} + \dfrac{3(x+2)}{x^2-6x+9} - \dfrac{1}{x^2-9}$

Section 6.5

Complex Fractions

To *simplify a complex fraction:*

Method 1: Write the numerator and denominator as single fractions. Then divide the fractions and simplify.

Method 2: Multiply the numerator and denominator by the LCD of the fractions in the numerator and denominator. Then simplify the results.

15. Simplify each complex fraction.

   a. $\dfrac{\dfrac{p^2-9}{6pt}}{\dfrac{p^2+5p+6}{3pt}}$

   b. $\dfrac{\dfrac{1}{a}+\dfrac{2}{b}}{\dfrac{2}{a}-\dfrac{1}{b}}$

   c. $\dfrac{1-\dfrac{1}{x}-\dfrac{2}{x^2}}{1+\dfrac{4}{x}+\dfrac{3}{x^2}}$

   d. $\dfrac{(x-y)^{-2}}{x^{-2}-y^{-2}}$

   e. $\dfrac{3x-\dfrac{1}{3-\dfrac{x}{2}}}{\dfrac{3}{x-6}+x}$

Section 6.6

Equations Containing Rational Expressions

To solve a *rational equation*, multiply both sides by the LCD of the rational expressions in the equation to clear it of fractions.

Multiplying both sides of an equation by a quantity that contains a variable can lead to *extraneous* (false) solutions. All possible solutions of a rational equation must be checked.

16. Solve each equation, if possible.

a. $\dfrac{4}{x} - \dfrac{1}{10} = \dfrac{7}{2x}$

b. $\dfrac{5}{7+t} - 1 = \dfrac{-4}{t+7}$

c. $\dfrac{q-3}{6} = \dfrac{10}{q+4}$

d. $\dfrac{2}{x+5} - \dfrac{1}{6} = \dfrac{1}{x+4}$

e. $\dfrac{2(x-5)}{x-2} = \dfrac{6x+12}{4-x^2}$

f. $\dfrac{x+3}{x-5} + \dfrac{6+2x^2}{x^2-7x+10} = \dfrac{3x}{x-2}$

17. Solve each formula for the indicated variable.

a. $\dfrac{x^2}{a^2} - \dfrac{y^2}{b^2} = 1$ for y^2

b. $H = \dfrac{2ab}{a+b}$ for b

18. ADVERTISING See Illustration 3. A small plane pulling a banner can fly at a rate of 75 mph in calm air. Flying down the coast, with a tailwind, the plane flew 40 miles in the same time that it took to fly 35 miles up the coast, into a headwind. Find the rate of the wind.

ILLUSTRATION 3

WIND

19. TRIP LENGTH Traffic reduced a driver's usual speed by 10 mph, which lengthened her 200-mile trip by 1 hour. Find the driver's usual speed.

20. DRAINING A TANK If one outlet pipe can drain a tank in 24 hours and another pipe can drain the tank in 36 hours, how long will it take for both pipes to drain the tank?

21. INSTALLING SIDING Two men have estimated that they can side a house in 8 days. If one of them, who could have sided the house alone in 14 days, gets sick, how long will it take the other man to side the house alone?

22. METALLURGY The stiffness of the flagpole shown in Illustration 4 is given by the formula

$$k = \dfrac{1}{\dfrac{1}{k_1} + \dfrac{1}{k_2}}$$

where k_1 and k_2 are the individual stiffnesses of each section. If the design specifications require that the stiffness k of the entire pole be 1,900,000 in. lb/rad, what must the stiffness of Section 1 be?

ILLUSTRATION 4

Section 1
Stiffness k_1

Section 2
Stiffness
$k_2 = 4{,}200{,}000$ in. lb/rad

Section 6.7   Dividing Polynomials

To divide two *monomials:* Factor the numerator and denominator. Then divide out common factors. Or divide the coefficients and apply the rules for exponents.

To divide a *polynomial by a monomial,* multiply the polynomial by the reciprocal of the monomial. Remove parentheses and simplify each resulting rational expression.

The *long division* algorithm can be used to divide a *polynomial by a polynomial.* Write the powers of the variable in descending order. If a power of a variable is missing, insert a "place-holder" term with a coefficient of 0.

23. Do each division. Write each answer without negative exponents.

a. $\dfrac{25h^4k^7}{5hk^9}$

b. $(-5x^6y^3) \div (10x^3y^6)$

24. Find each quotient.

a. $\dfrac{36a + 32}{6}$

b. $\dfrac{30x^3y^2 - 15x^2y - 10xy^2}{-10xy}$

25. Find each quotient using long division.

a. $b + 5 \overline{)b^2 + 9b + 20}$

b. $\dfrac{-33v - 8v^2 + 3v^3 - 10}{1 + 3v}$

c. $x + 2 \overline{)x^3 + 8}$

d. $(2x^3 + 7x^2 + 3 + 4x) \div (2x + 3)$

CHAPTER 6

Test

In Problems 1–4, simplify each rational expression.

1. $\dfrac{-12x^2y^3z^2}{18x^3y^4z^2}$

2. $\dfrac{2x + 4}{x^2 - 4}$

3. $\dfrac{3y - 6z}{2z - y}$

4. $\dfrac{2x^2 + 7x + 3}{4x + 12}$

5. HOME RUNS In 1998, Mark McGwire of the St. Louis Cardinals baseball team hit 31 home runs in the first 68 games of the season. If he had continued at the same pace, how many home runs would he have hit at the end of a regular season of 162 games? (Round to the nearest home run.)

6. SOUND Sound intensity (loudness) varies inversely as the square of the distance from the source. If a rock band has a sound intensity of 100 decibels 30 feet away from the amplifier, find the sound intensity 60 feet away from the amplifier.

7. Graph the rational function $f(x) = \dfrac{2}{x}$ for $x > 0$. Label the horizontal asymptote.

8. Draw a possible graph showing that the weekly salary of a person *varies directly* with the number of hours worked during the week. Label the axes.

In Problems 9–18, do the operations and simplify, if necessary. Write all answers without negative exponents.

9. $\dfrac{x^2}{x^3z^2y^2} \cdot \dfrac{x^2z^4}{y^2z}$

10. $\dfrac{(x+1)(x+2)}{10} \cdot \dfrac{5}{x+2}$

11. $\dfrac{u^2+5u+6}{u^2-4} \cdot \dfrac{u^2-5u+6}{u^2-9}$

12. $\dfrac{x^3+y^3}{4} \div \dfrac{x^2-xy+y^2}{2x+2y}$

13. $\dfrac{xu+2u+3x+6}{u^2-9} \cdot \dfrac{2u-6}{x^2+3x+2}$

14. $\dfrac{a^2+7a+12}{a+3} \div \dfrac{16-a^2}{a-4}$

15. $\dfrac{-3t+4}{t^2+t-20} + \dfrac{6+5t}{t^2+t-20}$

16. $\dfrac{3w}{w-5} + \dfrac{w+10}{5-w}$

17. $8b-5+\dfrac{5b+4}{3b+1}$

18. $\dfrac{x+2}{x+1} - \dfrac{x+1}{x+2}$

In Problems 19–20, simplify each complex fraction.

19. $\dfrac{\dfrac{2u^2w^3}{v^2}}{\dfrac{4uw^4}{uv}}$

20. $\dfrac{\dfrac{4}{3k}+\dfrac{k}{k+1}}{\dfrac{k}{k+1}-\dfrac{3}{k}}$

In Problems 21–24, solve each equation.

21. $\dfrac{34}{x} + \dfrac{13}{20} = \dfrac{3}{2}$

22. $\dfrac{u-2}{u-3} + 3 = u + \dfrac{u-4}{3-u}$

23. $\dfrac{3}{x-2} = \dfrac{x+3}{2x}$

24. $\dfrac{4}{m^2-9} + \dfrac{5}{m^2-m-12} = \dfrac{7}{m^2-7m+12}$

In Problems 25–26, solve each formula for the indicated variable.

25. $\dfrac{x^2}{a^2} + \dfrac{y^2}{b^2} = 1$ for a^2

26. $\dfrac{1}{r} = \dfrac{1}{r_1} + \dfrac{1}{r_2}$ for r_2

27. ROOFING One roofing crew can finish a 2,800-square-foot roof in 12 hours, and another crew can do the job in 10 hours. If they work together, can they finish before a predicted rain in 5 hours? If not, how long will they have to work in the rain?

28. TOURING THE COUNTRYSIDE A man bicycles 5 mph faster than he can walk. He bicycles 24 miles and then hikes back along the same route in 11 hours. How fast does he walk?

29. Divide $\dfrac{18x^2y^3 - 12x^3y^2 + 9xy}{-3xy^4}$.

30. Divide $(y^3 - 48) \div (y + 2)$.

31. Explain why $x + 2$ can be divided out in the first expression and why it cannot be divided out in the second expression.

1st expression	2nd expression
$\dfrac{(x+2)(x-3)}{x+2}$	$\dfrac{(x+2)+(x-3)}{x+2}$

32. Explain what it means for two quantities to vary inversely. Give an example.

1. Solve $-3 = -\dfrac{9}{8}t$.

2. Solve $\dfrac{3x-4}{6} - \dfrac{x-2}{2} = \dfrac{-2x-3}{3}$.

3. AUTO SALES See the graph in Illustration 1. An automobile dealership is going to order 80 new Ford Escorts. According to the survey, exactly how many green Escorts should be purchased to meet the expected customer demand?

ILLUSTRATION 1

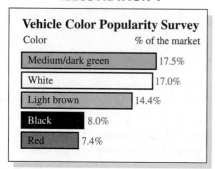

Vehicle Color Popularity Survey

Color	% of the market
Medium/dark green	17.5%
White	17.0%
Light brown	14.4%
Black	8.0%
Red	7.4%

Based on data from DuPont Automotive

4. LIFE EXPECTANCY Determine the predicted rate of change in the life expectancy of females during the years 2000–2050, as shown in the graph in Illustration 2.

ILLUSTRATION 2

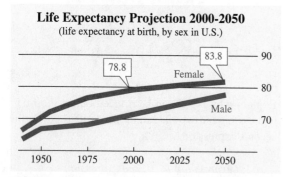

Life Expectancy Projection 2000-2050
(life expectancy at birth, by sex in U.S.)

78.8   83.8   90

Female

80

Male

70

1950   1975   2000   2025   2050

Based on data from the Social Security Administration, Office of Chief Actuary

In Exercises 5–6, write the equation of the line with the given properties. Express your answer in slope–intercept form.

5. Slope of -7, passing through $(7, 5)$

6. Passing through $(-4, 5)$ and $(2, -6)$

7. Draw the graph of the linear function $f(x) = \frac{2}{3}x - 2$. Then use interval notation to specify the domain and range.

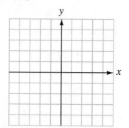

8. MECHANICAL ENGINEERING The tensions T_1 and T_2 (in pounds) in each of the ropes shown in Illustration 3 can be found by solving the system

$$\begin{cases} 0.6T_1 - 0.8T_2 = 0 \\ 0.8T_1 + 0.6T_2 = 100 \end{cases}$$

Find T_1 and T_2.

ILLUSTRATION 3

T_2 T_1

100 pounds

In Exercises 9–12, solve each inequality and show the solution set as a graph on the number line.

9. $-2x < -5$

10. $\left| \dfrac{3a}{5} - 2 \right| + 1 \geq \dfrac{6}{5}$

11. $5(x + 2) \leq 4(x + 1)$ and $11 + x < 0$

12. $x + 1 < -4$ or $x - 4 > 0$

In Exercises 13–16, simplify each expression.

13. $a^3b^2a^5b^2$

14. $\dfrac{a^3b^6}{a^7b^2}$

15. $\dfrac{1}{3^{-4}}$

16. $\left(\dfrac{2x^{-2}y^3}{x^2x^3y^4} \right)^{-3}$

In Exercises 17–18, write each number in standard notation.

17. 4.25×10^4

18. 7.12×10^{-4}

19. Express as a formula: *y varies directly with the product of x and z, and inversely with r.*

20. Evaluate the determinant $\begin{vmatrix} 3 & -2 \\ -2 & 4 \end{vmatrix}$.

21. Graph $y < 4 - x$.

22. Graph $f(x) = 2x^2 - 3$. Then use interval notation to specify the domain and range.

23. If $g(x) = -3x^3 + x - 4$, find $g(-2)$.

24. Find the degree of $3 + x^2y + 17x^3y^4$.

In Exercises 25–28, do the operations and simplify.

25. $(x^3 + 3x^2 - 2x + 7) + (x^3 - 2x^2 + 2x + 5)$

26. $(-5x^2 + 3x + 4) - (-2x^2 + 3x + 7)$

27. $(3x + 4)(2x - 5)$

28. $(2x^3 - 1)^2$

In Exercises 29–32, refer to Illustration 4. The graph shows the correction that must be made to a sundial reading to obtain accurate clock time. The difference is caused by the earth's orbit and tilted axis.

29. Is this the graph of a function?

30. During the year, what is the maximum number of minutes the sundial reading gets ahead of a clock?

31. During the year, what is the maximum number of minutes the sundial reading falls behind a clock?

ILLUSTRATION 4

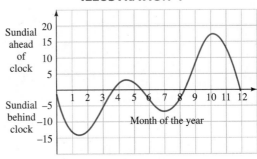

32. How many times during a year is the sundial reading exactly the same as a clock?

In Exercises 33–40, factor each expression.

33. $3r^2s^3 - 6rs^4$

34. $5(x - y) - a(x - y)$

35. $xu + yv + xv + yu$

36. $81x^4 - 16y^4$

37. $8x^3 - 27y^6$

38. $6x^2 + 5x - 6$

39. $9x^{2n} - 30x^n + 25$

40. $15x^2 - x - 6$

41. Solve $6x^2 + 7 = -23x$.

42. Solve $x^3 - 4x = 0$.

43. Solve $b^2x^2 + a^2y^2 = a^2b^2$ for b^2.

44. CAMPING The rectangular-shaped cooking surface of a small camping stove is 108 in.^{2}. If its length is 3 inches longer than its width, what are its dimensions?

In Exercises 45–48, simplify each expression.

45. $\dfrac{2x^2y + xy - 6y}{3x^2y + 5xy - 2y}$

46. $\dfrac{p^3 - q^3}{q^2 - p^2} \cdot \dfrac{q^2 + pq}{p^3 + p^2q + pq^2}$

47. $\dfrac{2}{x + y} + \dfrac{3}{x - y} - \dfrac{x - 3y}{x^2 - y^2}$

48. $\dfrac{\dfrac{a}{b} + b}{a - \dfrac{b}{a}}$

49. Solve $\dfrac{5x - 3}{x + 2} = \dfrac{5x + 3}{x - 2}$.

50. Solve $\dfrac{3}{x - 2} + \dfrac{x^2}{(x + 3)(x - 2)} = \dfrac{x + 4}{x + 3}$.

In Exercises 51–52, do the division.

51. $(x^2 + 9x + 20) \div (x + 5)$

52. $(2x^2 + 4x - x^3 + 3) \div (x - 1)$

53. ECONOMICS See Illustration 5. The controversial Phillips curve depicts the tradeoff between unemployment and inflation as seen by one school of economists. If unemployment drops to very low levels, what does the theoretical model predict will happen to the inflation rate?

ILLUSTRATION 5

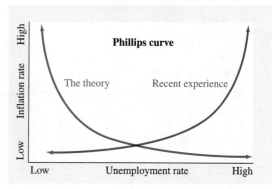

54. ECONOMICS See Exercise 53. The graph in Illustration 5 shows that economic factors have not followed the Phillips model in recent years. As unemployment has dropped to very low levels, what has happened to the inflation rate?

7

Radicals and Rational Exponents

CAMPUS CONNECTION

The Physics Department

In a physics class, students study relativity theory. This theory, primarily developed by Albert Einstein in the early 20th century, explains the relationship between matter and energy as well as the properties of light. One major conclusion of relativity theory is mathematically illustrated by an equation that contains a *rational exponent*. In this chapter, we discuss the meaning of rational (fractional) exponents and consider the relationship between rational exponents and radicals. Rational exponents and radicals occur in formulas and functions used in such diverse disciplines as carpentry, law enforcement, business, theater production, and physics.

Radical expressions have the form $\sqrt[n]{a}$. In this chapter, we will see how they are used to model many real-world situations.

▶ 7.1 Radical Expressions and Radical Functions

In this section, you will learn about

  Square roots ■ Square roots of expressions containing variables ■ The square root function ■ Cube roots ■ *n*th roots

Introduction   In this section, we will reverse the squaring process and learn how to find *square roots* of numbers. Then we will generalize the concept of root and consider cube roots, fourth roots, fifth roots, and so on. We will also discuss a new family of functions, called *radical functions,* and we will see how they have application in many disciplines.

Square Roots

When solving problems, we must often find what number must be squared to obtain a second number a. If such a number can be found, it is called a **square root of a.** For example,

■ 0 is a square root of 0, because $0^2 = 0$.
■ 4 is a square root of 16, because $4^2 = 16$.
■ -4 is a square root of 16, because $(-4)^2 = 16$.
■ $7xy$ is a square root of $49x^2y^2$, because $(7xy)^2 = 49x^2y^2$.
■ $-7xy$ is a square root of $49x^2y^2$, because $(-7xy)^2 = 49x^2y^2$.

The preceding examples illustrate the following definition.

Square root of a

> The number b is a **square root of a** if $b^2 = a$.

All positive numbers have two real-number square roots, one that is positive and one that is negative.

EXAMPLE 1   **Square roots.** Find the two square roots of 121.

Solution   The two square roots of 121 are 11 and -11, because

$$11^2 = 121 \qquad \text{and} \qquad (-11)^2 = 121$$

SELF CHECK   Find the square roots of 144.   *Answer:* 12, -12 ■

In the following definition, the symbol $\sqrt{}$ is called a **radical sign,** and the number x within the radical sign is called a **radicand.**

Principal square root

> If $x > 0$, the **principal square root of x** is the positive square root of x, denoted as $\sqrt{x}$.
>
> The principal square root of 0 is 0: $\sqrt{0} = 0$.

By definition, the principal square root of a positive number is always positive. Although 5 and -5 are both square roots of 25, only 5 is the principal square root. The radical expression $\sqrt{25}$ represents 5. The radical expression $-\sqrt{25}$ represents -5. When we write $\sqrt{25} = 5$ or $-\sqrt{25} = -5$, we say that we have *simplified the radical.*

EXAMPLE 2

Finding square roots. Simplify each radical.

a. $\sqrt{1} = 1$ **b.** $\sqrt{81} = 9$

c. $-\sqrt{81} = -9$ **d.** $-\sqrt{225} = -15$

e. $\sqrt{\dfrac{1}{4}} = \dfrac{1}{2}$ **f.** $-\sqrt{\dfrac{16}{121}} = -\dfrac{4}{11}$

g. $\sqrt{0.04} = 0.2$ **h.** $-\sqrt{0.0009} = -0.03$

SELF CHECK   Simplify   **a.** $-\sqrt{49}$ and   **b.** $\sqrt{\frac{25}{49}}$.   *Answers:*   **a.** -7,   **b.** $\frac{5}{7}$ ∎

Numbers such as 4, 9, 16, 49, and 1,600 are called **integer squares,** because each one is the square of an integer. The square root of every integer square is a rational number.

$$\sqrt{4} = 2, \qquad \sqrt{9} = 3, \qquad \sqrt{16} = 4, \qquad \sqrt{49} = 7, \qquad \sqrt{1,600} = 40$$

The square roots of many positive integers are not rational numbers. For example, $\sqrt{11}$ is an *irrational number.* To find an approximate value of $\sqrt{11}$, we enter 11 into a calculator and press the square root key . (On some calculators, the $\sqrt{}$ key must be pressed first.)

$$\sqrt{11} \approx 3.31662479$$

WARNING! Square roots of negative numbers are not real numbers. For example, $\sqrt{-9}$ is not a real number, because no real number squared equals -9. Square roots of negative numbers come from a set called the **imaginary numbers,** which we will discuss in the next chapter.

Square Roots of Expressions Containing Variables

If $x \neq 0$, the positive number x^2 has x and $-x$ for its two square roots. To denote the positive square root of $\sqrt{x^2}$, we must know whether x is positive or negative.

If $x > 0$, we can write

$$\sqrt{x^2} = x \qquad \sqrt{x^2} \text{ represents the positive square root of } x^2, \text{ which is } x.$$

If x is negative, then $-x > 0$, and we can write

$$\sqrt{x^2} = -x \qquad \sqrt{x^2} \text{ represents the positive square root of } x^2, \text{ which is } -x.$$

If we don't know whether x is positive or negative, we can use absolute value symbols to guarantee that $\sqrt{x^2}$ is positive.

Definition of $\sqrt{x^2}$

For any real number x,
$$\sqrt{x^2} = |x|$$

EXAMPLE 3

Simplifying radical expressions containing variables. Simplify
a. $\sqrt{16x^2}$, **b.** $\sqrt{x^2 + 2x + 1}$, and **c.** $\sqrt{m^4}$.

Solution

If x can be any real number, we have

a. $\sqrt{16x^2} = \sqrt{(4x)^2}$ Write the radicand $16x^2$ as $(4x)^2$.

$\qquad\qquad = |4x|$ Because $(|4x|)^2 = 16x^2$. Since x could be negative, absolute value symbols are needed.

$\qquad\qquad = 4|x|$ Since 4 is a positive constant in the product $4x$, we can write it outside the absolute value symbols.

b. $\sqrt{x^2 + 2x + 1}$

$\qquad\qquad = \sqrt{(x + 1)^2}$ Factor the radicand: $x^2 + 2x + 1 = (x + 1)^2$.

$\qquad\qquad = |x + 1|$ Since $x + 1$ can be negative (for example, when $x = -5$, $x + 1$ is -4), absolute value symbols are needed.

c. $\sqrt{m^4} = m^2$ Because $(m^2)^2 = m^4$. Since $m^2 \geq 0$, no absolute value symbols are needed.

SELF CHECK

Simplify **a.** $\sqrt{25a^2}$ and **b.** $\sqrt{16a^4}$.   *Answers:* **a.** $5|a|$, **b.** $4a^2$ ∎

If we are told that x represents a positive real number in parts a and b of Example 3, we do not need to use absolute value symbols to guarantee that the answers are positive.

$\sqrt{16x^2} = 4x$ If x is positive, $4x$ is positive.

$\sqrt{x^2 + 2x + 1} = x + 1$ If x is positive, $x + 1$ is positive.

The Square Root Function

Since there is one principal square root for every nonnegative real number x, the equation $f(x) = \sqrt{x}$ determines a function, called a **square root function.** Square root functions belong to a larger family of functions known as **radical functions.**

EXAMPLE 4

Graphing a square root function. Graph $f(x) = \sqrt{x}$ and find its domain and range.

Solution

To graph this square root function, we will evaluate it for several values of x. We begin with $x = 0$, since 0 is the smallest input for which $\sqrt{x}$ is defined.

$f(x) = \sqrt{x}$
$f(0) = \sqrt{0} = 0$

We enter the ordered pair $(0, 0)$ in the table of values in Figure 7-1. Then we continue the evaluation process for $x = 1, 4, 9$, and 16 and list the results in the table. After plotting all the ordered pairs, we draw a smooth curve through the points. This is the graph of function f (see Figure 7-1(a)). Since the equation defines a function, its graph passes the vertical line test.

We can use a graphing calculator with window settings of $[-1, 9]$ for x and $[-1, 9]$ for y to get the graph shown in Figure 7-1(b). From either graph, we can see that the domain and the range are the set of nonnegative real numbers. Expressed in interval notation, the domain is $[0, \infty)$, and the range is $[0, \infty)$.

FIGURE 7-1

$f(x) = \sqrt{x}$

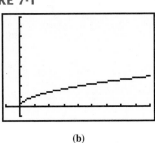

x	$f(x)$	$(x, f(x))$
0	0	$(0, 0)$
1	1	$(1, 1)$
4	2	$(4, 2)$
9	3	$(9, 3)$
16	4	$(16, 4)$

(b)

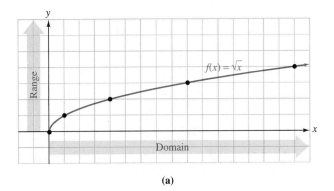

(a)

SELF CHECK
Graph $f(x) = \sqrt{x} + 2$. Then give its domain and range and compare the graph of $f(x) = \sqrt{x}$.

Answers:

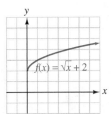

D: $[0, \infty)$, R: $[2, \infty)$; the graph is 2 units higher ∎

The graphs of many radical functions are translations or reflections of the square root function, $f(x) = \sqrt{x}$. For example, if $k > 0$,

■ The graph of $f(x) = \sqrt{x} + k$ is the graph of $f(x) = \sqrt{x}$ translated k units up.
■ The graph of $f(x) = \sqrt{x} - k$ is the graph of $f(x) = \sqrt{x}$ translated k units down.
■ The graph of $f(x) = \sqrt{x + k}$ is the graph of $f(x) = \sqrt{x}$ translated k units to the left.
■ The graph of $f(x) = \sqrt{x - k}$ is the graph of $f(x) = \sqrt{x}$ translated k units to the right.
■ The graph of $f(x) = -\sqrt{x}$ is the graph of $f(x) = \sqrt{x}$ reflected about the x-axis.

EXAMPLE 5

Graphing square root functions. Graph $f(x) = -\sqrt{x + 4} - 2$ and find its domain and range.

Solution
This graph will be the reflection of $f(x) = \sqrt{x}$ about the x-axis, translated 4 units to the left and 2 units down. See Figure 7-2(a). We can confirm this graph by using a graphing calculator with window settings of $[-5, 6]$ for x and $[-6, 2]$ for y to get the graph shown in Figure 7-2(b).

We can determine the domain of the function algebraically. Since the expression $\sqrt{x+4}$ is not a real number when $x + 4$ is negative, we must require that

$$x + 4 \geq 0$$

Solving for x, we have

$$x \geq -4 \qquad \text{The } x\text{-inputs must be real numbers greater than or equal to } -4.$$

Thus, the domain of $f(x) = -\sqrt{x+4} - 2$ is the interval $[-4, \infty)$.

From either graph, we can see that the domain is the interval $[-4, \infty)$ and that the range is the interval $(-\infty, -2]$.

FIGURE 7-2

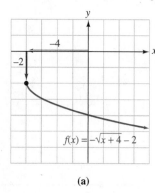

$$f(x) = -\sqrt{x+4} - 2$$

(a)

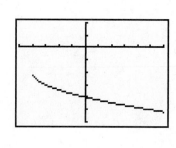

(b)

SELF CHECK   Graph $f(x) = \sqrt{x-2} - 4$. Then give the domain and the range.

Answers:

$$f(x) = \sqrt{x-2} - 4$$

D: $[2, \infty)$, R: $[-4, \infty)$ ∎

EXAMPLE 6

Square root functions. The **period of a pendulum** is the time required for the pendulum to swing back and forth to complete one cycle. The period (in seconds) is a function of the pendulum's length l (in feet) and is given by

$$f(l) = 2\pi \sqrt{\frac{l}{32}}$$

Find the period of the 5-foot-long pendulum of the clock shown in Figure 7-3.

FIGURE 7-3

Solution   To determine the period, we substitute 5 for l.

$$f(l) = 2\pi \sqrt{\frac{l}{32}}$$

$$f(5) = 2\pi \sqrt{\frac{5}{32}}$$

$$\approx 2.483647066 \qquad \text{Use a calculator to find an approximation.}$$

The period is approximately 2.5 seconds.

SELF CHECK   To the nearest hundredth, find the period of a pendulum that is 3 feet long.   *Answer:* 1.92 sec   ∎

ACCENT ON TECHNOLOGY *Evaluating a Square Root Function Using Its Graph*

To solve Example 6 with a graphing calculator with window settings of $[-2, 10]$ for x and $[-2, 10]$ for y, we graph the function $f(x) = 2\pi\sqrt{\frac{x}{32}}$, as in Figure 7-4(a). We then trace and move the cursor toward an x-value of 5 until we see the coordinates shown in Figure 7-4(b). The period is given by the y-value shown on the screen. By zooming in, we can get better results.

After entering $Y_1 = 2\pi\sqrt{\frac{x}{32}}$, we can also use the TABLE mode to evaluate the function. See Figure 7-4(c).

FIGURE 7-4

(a)

(b)

(c)

Cube Roots

The **cube root of x** is any number whose cube is x. For example,

- 4 is a cube root of 64, because $4^3 = 64$.
- $3x^2y$ is a cube root of $27x^6y^3$, because $(3x^2y)^3 = 27x^6y^3$.
- $-2y$ is a cube root of $-8y^3$, because $(-2y)^3 = -8y^3$.

Cube roots

The **cube root of x** is denoted as $\sqrt[3]{x}$ and is defined by

$$\sqrt[3]{x} = y \quad \text{if } y^3 = x$$

We note that 64 has two real-number square roots, 8 and -8. However, 64 has only one real-number cube root 4, because 4 is the only real number whose cube is 64. Since every real number has exactly one real cube root, it is unnecessary to use absolute value symbols when simplifying cube roots.

Definition of $\sqrt[3]{x^3}$

For any real number x,

$$\sqrt[3]{x^3} = x$$

E X A M P L E   7

Finding cube roots. Simplify each radical expression.

a. $\sqrt[3]{125} = 5$ Because $5^3 = 5 \cdot 5 \cdot 5 = 125$.

b. $\sqrt[3]{\dfrac{1}{8}} = \dfrac{1}{2}$ Because $\left(\frac{1}{2}\right)^3 = \frac{1}{2} \cdot \frac{1}{2} \cdot \frac{1}{2} = \frac{1}{8}$.

c. $\sqrt[3]{-27x^3} = -3x$ Because $(-3x)^3 = (-3x)(-3x)(-3x) = -27x^3$.

d. $\sqrt[3]{-\dfrac{8a^3}{27b^3}} = -\dfrac{2a}{3b}$ Because $\left(-\dfrac{2a}{3b}\right)^3 = \left(-\dfrac{2a}{3b}\right)\left(-\dfrac{2a}{3b}\right)\left(-\dfrac{2a}{3b}\right) = -\dfrac{8a^3}{27b^3}$.

e. $\sqrt[3]{0.216x^3y^6} = 0.6xy^2$ Because $(0.6xy^2)^3 = (0.6xy^2)(0.6xy^2)(0.6xy^2) = 0.216x^3y^6$.

SELF CHECK   Simplify   **a.** $\sqrt[3]{1,000}$,   **b.** $\sqrt[3]{\frac{1}{27}}$,   and   *Answers:*   **a.** 10,   **b.** $\frac{1}{3}$,
     c. $\sqrt[3]{125a^3}$.   **c.** 5a

The equation $f(x) = \sqrt[3]{x}$ defines a **cube root function.** From the graph shown in Figure 7-5(a), we can see that the domain and range of the function $f(x) = \sqrt[3]{x}$ are the set of real numbers. Note that the graph of $f(x) = \sqrt[3]{x}$ passes the vertical line test. Like square root functions, cube root functions are members of the family of radical functions. Figures 7-5(b) and 7-5(c) show several translations of the cube root function.

FIGURE 7-5

(a)

(b)

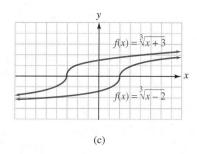

(c)

*n*th Roots

Just as there are square roots and cube roots, there are fourth roots, fifth roots, sixth roots, and so on.

When n is an odd natural number, the expression $\sqrt[n]{x}$ ($n > 1$) represents an **odd root.** Since every real number has just one real nth root when n is odd, we don't need to worry about absolute value symbols when finding odd roots. For example,

$\sqrt[5]{243} = \sqrt[5]{3^5} = 3$ Because $3^5 = 243$.

$\sqrt[7]{-128x^7} = \sqrt[7]{(-2x)^7} = -2x$ Because $(-2x)^7 = -128x^7$.

When n is an even natural number, the expression $\sqrt[n]{x}$ ($n > 1$, $x > 0$) represents an **even root.** In this case, there will be one positive and one negative real nth root. For example, the real sixth roots of 729 are 3 and -3, because $3^6 = 729$ and $(-3)^6 = 729$. When finding even roots, we can use absolute value symbols to guarantee that the nth root is positive.

$\sqrt[4]{(-3)^4} = |-3| = 3$ We could also simplify this as follows:
                              $\sqrt[4]{(-3)^4} = \sqrt[4]{81} = 3$.

$\sqrt[6]{729x^6} = \sqrt[6]{(3x)^6} = |3x| = 3|x|$ The absolute value symbols guarantee that the sixth root is positive.

In general, we have the following rules.

Rules for $\sqrt[n]{x^n}$

If x is a real number and $n > 1$, then

If n is an odd natural number, $\sqrt[n]{x^n} = x$.

If n is an even natural number, $\sqrt[n]{x^n} = |x|$.

In the radical expression $\sqrt[n]{x}$, n is called the **index** (or **order**) of the radical. When the index is 2, the radical is a square root, and we usually do not write the index.

$$\sqrt{x} = \sqrt[2]{x}$$

 WARNING! When n is even ($n > 1$) and $x < 0$, $\sqrt[n]{x}$ is not a real number. For example, $\sqrt[4]{-81}$ is not a real number, because no real number raised to the fourth power is -81.

EXAMPLE 8

Finding even and odd roots. Simplify each radical, if possible.

a. $\sqrt[4]{625} = 5$, because $5^4 = 625$ Read $\sqrt[4]{625}$ as "the fourth root of 625."

b. $\sqrt[4]{-1}$ is not a real number.   This is an even root of a negative number.

c. $\sqrt[5]{-32} = -2$, because $(-2)^5 = -32$ Read $\sqrt[5]{-32}$ as "the fifth root of -32."

d. $\sqrt[6]{\dfrac{1}{64}} = \dfrac{1}{2}$, because $\left(\dfrac{1}{2}\right)^6 = \dfrac{1}{64}$ Read $\sqrt[6]{\frac{1}{64}}$ as "the sixth root of $\frac{1}{64}$."

e. $\sqrt[7]{10^7} = 10$, because $10^7 = 10^7$ Read $\sqrt[7]{10^7}$ as "the seventh root of 10^7."

SELF CHECK Simplify **a.** $\sqrt[4]{\frac{1}{81}}$, **b.** $\sqrt[5]{10^5}$, and **c.** $\sqrt[6]{-64}$.   *Answers:* **a.** $\frac{1}{3}$, **b.** 10,
c. not a real number ∎

ACCENT ON TECHNOLOGY *Finding Roots*

The square root key $\boxed{\sqrt{\ \ }}$ on a scientific calculator can be used to evaluate square roots. To evaluate roots with an index greater than 2, we can use the root key $\boxed{\sqrt[x]{y}}$. For example, the function

$$r(V) = \sqrt[3]{\dfrac{3V}{4\pi}}$$

gives the radius of a sphere with volume V. To find the radius of the spherical propane tank shown in Figure 7-6, we substitute 113 for V to get

FIGURE 7-6

$$r(V) = \sqrt[3]{\dfrac{3V}{4\pi}}$$

$$r(113) = \sqrt[3]{\dfrac{3(113)}{4\pi}}$$

PROPANE
Capacity 113 ft³

To evaluate a root, we enter the radicand and press the root key $\boxed{\sqrt[x]{y}}$ followed by the index of the radical, which in this case is 3.

(continued)

Keystrokes 3 ☐× 113 ☐÷ ☐(4 ☐× ☐π ☐) ☐= 2nd $\sqrt[x]{y}$ 3 ☐=

```
2.999139118
```

To evaluate the cube root of $\frac{3(113)}{4\pi}$ using a graphing calculator, we enter these numbers and press these keys.

Keystrokes MATH 4 ☐(3 ☐× 113 ☐) ☐÷ ☐(4 ☐× 2nd ☐π ☐) ☐) ENTER

```
³√((3*113)/(4*π)
)
        2.999139118
```

The radius of the propane tank is about 3 feet.

EXAMPLE 9

Simplifying radical expressions containing variables. Simplify each radical expression. Assume that x can be any real number.

a. $\sqrt[5]{x^5} = x$ Since n is odd, absolute value symbols aren't needed.

b. $\sqrt[4]{16x^4} = |2x| = 2|x|$ Since n is even and x can be negative, absolute value symbols are needed to guarantee that the result is positive.

c. $\sqrt[6]{(x+4)^6} = |x+4|$ Absolute value symbols are needed to guarantee that the result is positive.

d. $\sqrt[3]{(x+1)^3} = x+1$ Since n is odd, absolute value symbols aren't needed.

SELF CHECK Simplify **a.** $\sqrt[4]{16a^8}$, **b.** $\sqrt[5]{(a+5)^5}$, and **c.** $\sqrt{(x^2+4x+4)^2}$.   *Answers:* **a.** $2a^2$, **b.** $a+5$, **c.** $(x+2)^2$ ∎

If we are told that x represents a positive real number in parts b and c of Example 9, we do not need to use absolute value symbols to guarantee that the results are positive.

$\sqrt[4]{16x^4} = 2x$ If x is positive, $2x$ is positive.

$\sqrt[6]{(x+4)^6} = x+4$ If x is positive, $x+4$ is positive.

We summarize the definitions concerning $\sqrt[n]{x}$ as follows.

Summary of the definitions of $\sqrt[n]{x}$

If n is a natural number greater than 1 and x is a real number, then

If $x > 0$, then $\sqrt[n]{x}$ is the positive number such that $\left(\sqrt[n]{x}\right)^n = x$.

If $x = 0$, then $\sqrt[n]{x} = 0$.

If $x < 0$ $\begin{cases} \text{and } n \text{ is odd, then } \sqrt[n]{x} \text{ is the real number such that } \left(\sqrt[n]{x}\right)^n = x. \\ \text{and } n \text{ is even, then } \sqrt[n]{x} \text{ is not a real number.} \end{cases}$

STUDY SET

Section 7.1

VOCABULARY

In Exercises 1–8, fill in the blanks to make the statements true.

1. $5x^2$ is the _____ of $25x^4$, because $(5x^2)^2 = 25x^4$.

2. $f(x) = \sqrt{x}$ and $g(t) = \sqrt[3]{t}$ are _____ functions.

3. The symbol $\sqrt{}$ is called a _____ sign.

4. In the expression $\sqrt[3]{27x^6}$, 3 is the _____ and $27x^6$ is the _____.

5. When n is an odd number, $\sqrt[n]{x}$ represents an _____ root.

6. When n is an _____ number, $\sqrt[n]{x}$ represents an even root.

7. When we write $\sqrt{b^2 + 6b + 9} = |b + 3|$, we say that we have _____ the radical.

8. 6 is the _____ of 216 because $6^3 = 216$.

CONCEPTS

In Exercises 9–16, fill in the blanks to make the statements true.

9. b is a square root of a if _____.

10. $\sqrt{0} =$ ___ and $\sqrt[3]{0} =$ ___

11. The number 25 has _____ square roots. The principal square root of 25 is the _____ square root of 25.

12. $\sqrt{-4}$ is not a real number, because no real number _____ equals -4.

13. $\sqrt[3]{x} = y$ if $y^3 =$ ___.

14. $\sqrt{x^2} =$ ___ and $\sqrt[3]{x^3} =$ ___

15. The graph of $f(x) = \sqrt{x} + 3$ is the graph of $f(x) = \sqrt{x}$ translated ___ units ___.

16. The graph of $f(x) = \sqrt{x + 5}$ is the graph of $f(x) = \sqrt{x}$ translated ___ units to the _____.

In Exercises 17–18, complete the table of values and then graph the radical function.

17. $f(x) = -\sqrt{x}$

x	y
0	
1	
4	
9	
16	

18. $f(x) = -\sqrt[3]{x}$

x	y
−8	
−1	
0	
1	
8	

NOTATION

In Exercises 19–22, translate each sentence into mathematical symbols.

19. The square root of x squared is the absolute value of x.

20. The cube root of x cubed is x.

21. f of x equals the square root of the quantity x minus five.

22. The fifth root of negative thirty-two is negative two.

PRACTICE

In Exercises 23–34, find each square root, if possible.

23. $\sqrt{121}$

24. $\sqrt{144}$

25. $-\sqrt{64}$

26. $-\sqrt{1}$

27. $\sqrt{\dfrac{1}{9}}$

28. $-\sqrt{\dfrac{4}{25}}$

29. $\sqrt{0.25}$

30. $\sqrt{0.16}$

31. $\sqrt{-25}$

32. $-\sqrt{-49}$

33. $\sqrt{(-4)^2}$

34. $\sqrt{(-9)^2}$

In Exercises 35–38, use a calculator to find each square root. Give the answer to four decimal places.

35. $\sqrt{12}$

36. $\sqrt{340}$

37. $\sqrt{679.25}$

38. $\sqrt{0.0063}$

In Exercises 39–46, find each square root. Assume that all variables are unrestricted, and use absolute value symbols when necessary.

39. $\sqrt{4x^2}$

40. $\sqrt{16y^4}$

41. $\sqrt{(t + 5)^2}$

42. $\sqrt{(a + 6)^2}$

43. $\sqrt{(-5b)^2}$

44. $\sqrt{(-8c)^2}$

45. $\sqrt{a^2 + 6a + 9}$

46. $\sqrt{x^2 + 10x + 25}$

In Exercises 47–54, find each value given that
$f(x) = \sqrt{x - 4}$ and $g(x) = \sqrt[3]{x - 4}$.

47. $f(4)$ **48.** $f(8)$

49. $f(20)$ **50.** $f(29)$

51. $g(12)$ **52.** $g(-4)$

53. $g(-996)$ **54.** $g(1,004)$

In Exercises 55–58, find each value given that
$f(x) = \sqrt{x^2 + 1}$ and $g(x) = \sqrt[3]{x^2 + 1}$. Give each
answer to four decimal places.

55. $f(4)$ **56.** $f(2.35)$

57. $g(6)$ **58.** $g(21.57)$

In Exercises 59–62, graph each radical function and find
its domain and range. Check your work with a graphing
calculator.

59. $f(x) = \sqrt{x + 4}$ **60.** $f(x) = -\sqrt{x - 1}$

61. $f(x) = -\sqrt[3]{x} - 3$ **62.** $f(x) = -\sqrt[3]{x} - 1$

In Exercises 63–78, simplify each cube root.

63. $\sqrt[3]{1}$ **64.** $\sqrt[3]{-8}$

65. $\sqrt[3]{-125}$ **66.** $\sqrt[3]{512}$

67. $\sqrt[3]{-\dfrac{8}{27}}$ **68.** $\sqrt[3]{\dfrac{125}{216}}$

69. $\sqrt[3]{0.064}$ **70.** $\sqrt[3]{0.001}$

71. $\sqrt[3]{8a^3}$ **72.** $\sqrt[3]{-27x^6}$

73. $\sqrt[3]{-1,000p^3q^3}$ **74.** $\sqrt[3]{343a^6b^3}$

75. $\sqrt[3]{-\dfrac{1}{8}m^6n^3}$ **76.** $\sqrt[3]{0.008z^9}$

77. $\sqrt[3]{-0.064s^9t^6}$ **78.** $\sqrt[3]{\dfrac{27}{1,000}a^6b^6}$

In Exercises 79–98, simplify each radical, if possible.
Assume that all variables represent positive real numbers.

79. $\sqrt[4]{81}$ **80.** $\sqrt[6]{64}$

81. $-\sqrt[5]{243}$ **82.** $-\sqrt[4]{625}$

83. $\sqrt[4]{-256}$ **84.** $\sqrt[6]{-729}$

85. $\sqrt[4]{\dfrac{16}{625}}$ **86.** $\sqrt[5]{-\dfrac{243}{32}}$

87. $-\sqrt[5]{-\dfrac{1}{32}}$ **88.** $-\sqrt[4]{\dfrac{81}{256}}$

89. $\sqrt[5]{32a^5}$ **90.** $\sqrt[5]{-32x^5}$

91. $\sqrt[4]{16a^4}$ **92.** $\sqrt[8]{x^{24}}$

93. $\sqrt[4]{k^{12}}$ **94.** $\sqrt[6]{64b^6}$

95. $\sqrt[4]{\dfrac{1}{16}m^4}$ **96.** $\sqrt[4]{\dfrac{1}{81}x^8}$

97. $\sqrt[25]{(x + 2)^{25}}$ **98.** $\sqrt[44]{(x + 4)^{44}}$

APPLICATIONS

Use a calculator to solve each problem. In each case,
round to the nearest tenth.

99. EMBROIDERY The radius r of a circle is given by
the formula

$$r = \sqrt{\dfrac{A}{\pi}}$$

where A is its area. Find the
diameter of the embroidery
hoop shown in Illustration 1
if there are 38.5 in.^{2} of
stretched fabric on which to
embroider.

ILLUSTRATION 1

100. SOFTBALL AND BASEBALL The length of a
diagonal of a square is given by the function
$d(s) = \sqrt{2s^2}$, where s is the length of a side of the
square. Find the distance from home plate to second
base on a softball diamond and on a baseball diamond. Illustration 2 gives the dimensions of each
type of infield.

ILLUSTRATION 2

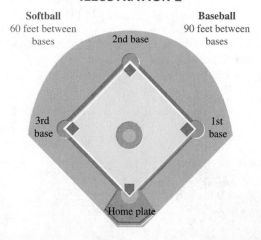

101. PULSE RATE The approximate pulse rate (in beats per minute) of an adult who is t inches tall is given by the function

$$p(t) = \frac{590}{\sqrt{t}}$$

The Guinness Book of World Records 1998 lists Ri Myong-hun of North Korea as the tallest living man, at 7 ft $8\frac{1}{2}$ in. Find his approximate pulse rate as predicted by the function.

102. THE GRAND CANYON The time t (in seconds) that it takes for an object to fall a distance of s feet is given by the formula

$$t = \frac{\sqrt{s}}{4}$$

In some places, the Grand Canyon is one mile (5,280 feet) deep. How long would it take a stone dropped over the edge of the canyon to hit bottom?

103. BIOLOGY Scientists will place five rats inside the controlled environment of a sealed hemisphere to study the rats' behavior. The function

$$d(V) = \sqrt[3]{12\left(\frac{V}{\pi}\right)}$$

gives the diameter of a hemisphere with volume V. Use the function to determine the diameter of the base of the hemisphere, if each rat requires 125 cubic feet of living space.

104. AQUARIUM The function

$$s(g) = \sqrt[3]{\frac{g}{7.5}}$$

determines how long (in feet) an edge of a cube-shaped tank must be if it is to hold g gallons of water. What dimensions should a cube-shaped aquarium have if it is to hold 1,250 gallons of water?

105. COLLECTIBLES The *effective rate of interest r* earned by an investment is given by the formula

$$r = \sqrt[n]{\frac{A}{P}} - 1$$

where P is the initial investment that grows to value A after n years. Determine the effective rate of interest earned by a collector on a Lladró porcelain figurine purchased for $800 and sold for $950 five years later.

106. LAW ENFORCEMENT The graphs of the two radical functions shown in Illustration 3 can be used to estimate the speed (in mph) of a car involved in an accident. Suppose a police accident report listed skid marks to be 220 feet long but failed to give the road conditions. Estimate the possible speeds the car was traveling prior to the brakes being applied.

ILLUSTRATION 3

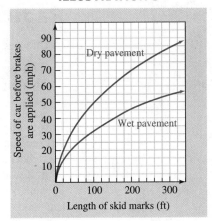

WRITING

107. Explain why 36 has two square roots, but $\sqrt{36}$ is just 6, not -6.

108. If x is any real number, then $\sqrt{x^2} = x$ is not correct. Explain.

109. Explain what is **wrong** with the graph in Illustration 4 if it is supposed to be the graph of $f(x) = \sqrt{x}$.

ILLUSTRATION 4

110. Explain how to estimate the domain and range of the radical function that is graphed in Illustration 5.

ILLUSTRATION 5

REVIEW

Do the operations.

111. $\dfrac{x^2 - x - 6}{x^2 - 2x - 3} \cdot \dfrac{x^2 - 1}{x^2 + x - 2}$

112. $\dfrac{x^2 - 3x - 4}{x^2 - 5x + 6} \div \dfrac{x^2 - 2x - 3}{x^2 - x - 2}$

113. $\dfrac{3}{m + 1} + \dfrac{3m}{m - 1}$

114. $\dfrac{2x + 3}{3x - 1} - \dfrac{x - 4}{2x + 1}$

▶ 7.2

Radical Equations

In this section, you will learn about

The power rule ■ Equations containing one radical ■ Equations containing two radicals ■ Solving formulas containing radicals

Introduction   Many situations can be modeled by equations that contain radicals. In this section, we will develop techniques to solve such equations. For example, to solve the radical equation $\sqrt{x} = 6$, we need to isolate x by undoing the operation performed on it. Recall that $\sqrt{x}$ represents the number that, when squared, gives x. Therefore, if we square $\sqrt{x}$, we will obtain x. From this observation, it is apparent that we can eliminate the radical on the left-hand side of the equation $\sqrt{x} = 6$ by squaring that side. Intuition tells us that we should also square the right-hand side. Squaring both sides of an equation is an application of the *power rule,* which we now consider in more detail.

The Power Rule

To solve equations containing radicals, we will use the **power rule.**

The power rule

If x, y, and n are real numbers and $x = y$, then

$$x^n = y^n$$

If we raise both sides of an equation to the same power, the resulting equation might not be equivalent to the original equation. For example, if we square both sides of the equation

1. $x = 3$ With a solution set of $\{3\}$

we obtain the equation

2. $x^2 = 9$ With a solution set of $\{3, -3\}$

Equations 1 and 2 are not equivalent, because they have different solution sets, and the solution -3 of Equation 2 does not satisfy Equation 1. Since raising both sides of an equation to the same power can produce an equation with roots that don't satisfy the original equation, we must always check each apparent solution in the original equation and discard any *extraneous solutions.*

Equations Containing One Radical

Radical equations contain a radical expression with a variable radicand. To solve radical equations, we apply the power rule.

E X A M P L E   1

Squaring both sides of an equation. Solve $\sqrt{x + 3} = 4$.

Solution   To eliminate the radical, we apply the power rule by squaring both sides of the equation and proceed as follows:

$$\sqrt{x + 3} = 4$$
$$\left(\sqrt{x + 3}\right)^2 = (4)^2 \quad \text{Square both sides.}$$
$$x + 3 = 16$$
$$x = 13 \quad \text{Subtract 3 from both sides.}$$

We must check the apparent solution of 13 to see whether it satisfies the original equation.

Check: $\sqrt{x + 3} = 4$

$\qquad \sqrt{13 + 3} \stackrel{?}{=} 4 \quad$ Substitute 13 for x.

$\qquad\qquad \sqrt{16} \stackrel{?}{=} 4$

$\qquad\qquad\quad 4 = 4$

Since 13 satisfies the original equation, it is a solution.

SELF CHECK   Solve $\sqrt{a - 2} = 3$.   *Answer:* 11   ▪

To solve an equation with radicals, we follow these steps.

Solving an equation containing radicals

1. Isolate one radical expression on one side of the equation.
2. Raise both sides of the equation to the power that is the same as the index of the radical.
3. Solve the resulting equation. If it still contains a radical, go back to step 1.
4. Check the results to eliminate extraneous solutions.

EXAMPLE 2

Amusement park ride. The distance d in feet that an object will fall in t seconds is given by the formula

FIGURE 7-7

$$t = \sqrt{\dfrac{d}{16}}$$

If the designers of the amusement park attraction shown in Figure 7-7 want the riders to experience 3 seconds of vertical "free fall," what length of vertical drop is needed?

Solution   We substitute 3 for t in the formula and solve for d.

$\qquad t = \sqrt{\dfrac{d}{16}}$

$\qquad 3 = \sqrt{\dfrac{d}{16}} \qquad$ Here the radical is isolated on the right-hand side.

$\qquad (3)^2 = \left(\sqrt{\dfrac{d}{16}} \right)^2 \qquad$ Raise both sides to the second power.

$\qquad 9 = \dfrac{d}{16} \qquad$ Simplify.

$\qquad 144 = d \qquad$ Solve the resulting equation by multiplying both sides by 16.

The amount of vertical drop needs to be 144 feet.

SELF CHECK   In Example 2, how long a vertical drop is needed if the riders are to "free fall" for 3.5 seconds?   *Answer:* 196 ft   ▪

| EXAMPLE 3 | **Isolating the radical.** Solve $\sqrt{3x + 1} + 1 = x$. |

Solution   We first subtract 1 from both sides to isolate the radical. Then, to eliminate the radical, we square both sides of the equation and proceed as follows:

$$\sqrt{3x + 1} + 1 = x$$

$$\sqrt{3x + 1} = x - 1 \qquad \text{Subtract 1 from both sides.}$$

$$\left(\sqrt{3x + 1}\right)^2 = (x - 1)^2 \qquad \text{Square both sides to eliminate the square root.}$$

$$3x + 1 = x^2 - 2x + 1 \qquad \begin{array}{l}\text{On the right-hand side, use the FOIL method:}\\ (x - 1)^2 = (x - 1)(x - 1) = x^2 - x - x + 1 =\\ x^2 - 2x + 1.\end{array}$$

$$0 = x^2 - 5x \qquad \begin{array}{l}\text{Subtract } 3x \text{ and 1 from both sides. This is a quadratic}\\ \text{equation. Use factoring to solve it.}\end{array}$$

$$0 = x(x - 5) \qquad \text{Factor } x^2 - 5x.$$

$$x = 0 \quad \text{or} \quad x - 5 = 0 \qquad \text{Set each factor each to 0.}$$

$$x = 0 \qquad\qquad x = 5$$

We must check each apparent solution to see whether it satisfies the original equation.

Check:
$$\sqrt{3x + 1} + 1 = x \qquad\qquad \sqrt{3x + 1} + 1 = x$$
$$\sqrt{3(0) + 1} + 1 \overset{?}{=} 0 \qquad\qquad \sqrt{3(5) + 1} + 1 \overset{?}{=} 5$$
$$\sqrt{1} + 1 \overset{?}{=} 0 \qquad\qquad \sqrt{16} + 1 \overset{?}{=} 5$$
$$2 \neq 0 \qquad\qquad\qquad 5 = 5$$

Since 0 does not check, it must be discarded. The only solution of the original equation is 5.

SELF CHECK   Solve $\sqrt{4x + 1} + 1 = x$. *Answer:* 6, 0 is extraneous ■

ACCENT ON TECHNOLOGY *Solving Equations Containing Radicals*

To find approximate solutions for $\sqrt{3x + 1} + 1 = x$ with a graphing calculator, we use window settings of $[-5, 10]$ for x and $[-2, 8]$ for y and graph the functions $f(x) = \sqrt{3x + 1} + 1$ and $g(x) = x$, as in Figure 7-8(a). We then trace to find the approximate x-coordinate of their intersection point, as in Figure 7-8(b). After repeated zooms, we will see that $x = 5$.

We can also use the INTERSECT feature to approximate the point of intersection of the graphs. See Figure 7-8(c). The intersection point of $(5, 5)$ implies that $x = 5$ is a solution of the radical equation. Check this result.

FIGURE 7-8

(a)   (b)   (c)

EXAMPLE 4

Cubing both sides. Solve $\sqrt[3]{x^3 + 7} = x + 1$.

Solution To eliminate the radical, we cube both sides of the equation and proceed as follows:

$$\sqrt[3]{x^3 + 7} = x + 1$$

$$\left(\sqrt[3]{x^3 + 7}\right)^3 = (x + 1)^3 \qquad \text{Cube both sides to eliminate the cube root.}$$

$$x^3 + 7 = x^3 + 3x^2 + 3x + 1 \qquad (x + 1)^3 = (x + 1)(x + 1)(x + 1).$$

$$0 = 3x^2 + 3x - 6 \qquad \text{Subtract } x^3 \text{ and 7 from both sides.}$$

$$0 = x^2 + x - 2 \qquad \text{Divide both sides by 3. To solve this quadratic equation, use factoring.}$$

$$0 = (x + 2)(x - 1) \qquad \text{Factor the trinomial.}$$

$$x + 2 = 0 \quad \text{or} \quad x - 1 = 0$$
$$x = -2 \qquad\qquad x = 1$$

We check each apparent solution to see whether it satisfies the original equation.

Check:
$$\sqrt[3]{x^3 + 7} = x + 1 \qquad\qquad \sqrt[3]{x^3 + 7} = x + 1$$
$$\sqrt[3]{(-2)^3 + 7} \stackrel{?}{=} -2 + 1 \qquad\qquad \sqrt[3]{1^3 + 7} \stackrel{?}{=} 1 + 1$$
$$\sqrt[3]{-8 + 7} \stackrel{?}{=} -1 \qquad\qquad \sqrt[3]{1 + 7} \stackrel{?}{=} 2$$
$$\sqrt[3]{-1} \stackrel{?}{=} -1 \qquad\qquad \sqrt[3]{8} \stackrel{?}{=} 2$$
$$-1 = -1 \qquad\qquad 2 = 2$$

Both solutions satisfy the original equation.

SELF CHECK Solve $\sqrt[3]{x^3 + 8} = x + 2$. *Answer:* 0, −2 ■

Equations Containing Two Radicals

EXAMPLE 5

Solving an equation containing two radicals. Solve $\sqrt{5x + 9} = 2\sqrt{3x + 4}$.

Solution Each radical is isolated on one side of the equation, so we square both sides to eliminate them.

$$\sqrt{5x + 9} = 2\sqrt{3x + 4}$$

$$\left(\sqrt{5x + 9}\right)^2 = \left(2\sqrt{3x + 4}\right)^2 \qquad \text{Square both sides.}$$

$$5x + 9 = 4(3x + 4) \qquad \begin{array}{l}\text{On the right-hand side:}\\ \left(2\sqrt{3x + 4}\right)^2 = 2^2\left(\sqrt{3x + 4}\right)^2 = 4(3x + 4).\end{array}$$

$$5x + 9 = 12x + 16 \qquad \text{Remove parentheses.}$$

$$-7 = 7x \qquad \text{Subtract } 5x \text{ and 16 from both sides.}$$

$$-1 = x \qquad \text{Divide both sides by 7.}$$

We check the solution by substituting −1 for x in the original equation.

$$\sqrt{5x + 9} = 2\sqrt{3x + 4}$$
$$\sqrt{5(-1) + 9} \stackrel{?}{=} 2\sqrt{3(-1) + 4} \qquad \text{Substitute } -1 \text{ for } x.$$
$$\sqrt{4} \stackrel{?}{=} 2\sqrt{1}$$
$$2 = 2$$

The solution checks.

SELF CHECK   Solve $\sqrt{x-4} = 2\sqrt{x-16}$.   *Answer:* 20   ∎

When more than one radical appears in an equation, it is often necessary to apply the power rule more than once.

EXAMPLE 6   **Solving equations containing two radicals.** Solve $\sqrt{x} + \sqrt{x+2} = 2$.

Solution   To remove the radicals, we must square both sides of the equation. This is easier to do if one radical is on each side of the equation. So we subtract $\sqrt{x}$ from both sides to isolate $\sqrt{x+2}$ on the left-hand side of the equation.

$$\sqrt{x} + \sqrt{x+2} = 2$$

$$\sqrt{x+2} = 2 - \sqrt{x} \qquad \text{Subtract } \sqrt{x} \text{ from both sides.}$$

$$\left(\sqrt{x+2}\right)^2 = \left(2 - \sqrt{x}\right)^2 \qquad \text{Square both sides to eliminate the square root.}$$

$$x + 2 = 4 - 4\sqrt{x} + x \qquad \text{Use FOIL: } \left(2 - \sqrt{x}\right)^2 = \left(2 - \sqrt{x}\right)\left(2 - \sqrt{x}\right) =$$
$$4 - 2\sqrt{x} - 2\sqrt{x} + x = 4 - 4\sqrt{x} + x.$$

$$2 = 4 - 4\sqrt{x} \qquad \text{Subtract } x \text{ from both sides.}$$

$$-2 = -4\sqrt{x} \qquad \text{Subtract 4 from both sides.}$$

$$\frac{1}{2} = \sqrt{x} \qquad \text{Divide both sides by } -4 \text{ and simplify.}$$

$$\frac{1}{4} = x \qquad \text{Square both sides.}$$

Check: $\sqrt{x} + \sqrt{x+2} = 2$

$$\sqrt{\frac{1}{4}} + \sqrt{\frac{1}{4} + 2} \overset{?}{=} 2$$

$$\frac{1}{2} + \sqrt{\frac{9}{4}} \overset{?}{=} 2$$

$$\frac{1}{2} + \frac{3}{2} \overset{?}{=} 2$$

$$2 = 2$$

The solution checks.

SELF CHECK   Solve $\sqrt{a} + \sqrt{a+3} = 3$.   *Answer:* 1   ∎

ACCENT ON TECHNOLOGY *Solving Equations Containing Radicals*

To find approximate solutions for $\sqrt{x} + \sqrt{x+2} = 5$ (an equation similar to that in Example 6) with a graphing calculator, we use window settings of $[-2, 10]$ for x and $[-2, 8]$ for y and graph the functions $f(x) = \sqrt{x} + \sqrt{x+2}$ and $g(x) = 5$. We then trace to find an approximation of the x-coordinate of their intersection point, as in Figure 7-9(a). From the figure, we can see that $x \approx 5.15$. We can zoom to get better results.

Figure 7-9(b) shows that the INTERSECT feature gives the approximate coordinates of the point of intersection of the two graphs as (5.29, 5). Therefore, an approximate solution of the radical equation is $x \approx 5.29$. Check its reasonableness.

FIGURE 7-9

(a)   (b)

Solving Formulas Containing Radicals

To *solve a formula for a variable* means to isolate that variable on one side of the equation, with all other quantities on the other side.

EXAMPLE 7

Depreciation rate. A piece of office equipment that is now worth V dollars originally cost C dollars 3 years ago. The rate r at which it has depreciated (lost value) is given by

$$r = 1 - \sqrt[3]{\frac{V}{C}}$$

Solve the formula for C.

Solution   We begin by isolating the cube root on the right-hand side of the equation.

$$r = 1 - \sqrt[3]{\frac{V}{C}}$$

$$r - 1 = -\sqrt[3]{\frac{V}{C}} \qquad \text{Subtract 1 from both sides.}$$

$$(r - 1)^3 = \left[-\sqrt[3]{\frac{V}{C}} \right]^3 \qquad \text{Cube both sides.}$$

$$(r - 1)^3 = -\frac{V}{C} \qquad \text{Simplify the right-hand side.}$$

$$C(r - 1)^3 = -V \qquad \text{Multiply both sides by } C.$$

$$C = -\frac{V}{(r - 1)^3} \qquad \text{Divide both sides by } (r - 1)^3.$$

SELF CHECK   A formula used in statistics to determine the necessary size of a sample to obtain the desired degree of accuracy is

$$e = z_0 \sqrt{\frac{pq}{n}}$$

Solve the formula for n.

Answer: $n = \dfrac{z_0^{2}pq}{e^2}$ ∎

STUDY SET

Section 7.2

VOCABULARY

In Exercises 1–4, fill in the blanks to make each statement true.

1. Equations such as $\sqrt{x+4} - 4 = 5$ and $\sqrt[3]{x+1} = 12$ are called _____ equations.

2. When solving equations containing radicals, try to _____ one radical expression on one side of the equation.

3. Squaring both sides of an equation can introduce _____ solutions.

4. To _____ an apparent solution means to substitute it into the original equation and see whether a true statement results.

CONCEPTS

5. What is the first step in solving each equation?
   a. $\sqrt{x+4} = 5$
   b. $\sqrt[3]{x+4} = 2$

6. Fill in the blank to make a true statement. $\sqrt{x}$ represents the number that, when squared, gives ___.

7. Simplify each expression.
   a. $\left(\sqrt{x}\right)^2$ b. $\left(\sqrt{x-5}\right)^2$
   c. $\left(4\sqrt{2x}\right)^2$ d. $\left(-\sqrt{x+3}\right)^2$

8. Simplify each expression.
   a. $\left(\sqrt[3]{x}\right)^3$ b. $\left(\sqrt[4]{x}\right)^4$
   c. $\left(-\sqrt[3]{2x}\right)^3$ d. $\left(2\sqrt[3]{x+3}\right)^3$

9. What is **wrong** with the student's work shown below?
   Solve $\sqrt{x+1} - 3 = 8$.
   $$\sqrt{x+1} = 11$$
   $$\left(\sqrt{x+1}\right)^2 = 11$$
   $$x + 1 = 11$$
   $$x = 10$$

10. Solve $\sqrt{x-2} + 2 = 4$ graphically, using the graphs in Illustration 1.

ILLUSTRATION 1

11. Use your own words to restate the power rule: If x, y, and n are real numbers and $x = y$, then $x^n = y^n$.

12. The first step of a student's solution is shown below. What is a better way to begin the solution?
    Solve $\sqrt{x} + \sqrt{x+22} = 12$.
    $$(\sqrt{x} + \sqrt{x+22})^2 = 12^2$$

NOTATION

In Exercises 13–14, complete each solution.

13. Solve $2\sqrt{x-2} = 4$.
    $$\left(\right)^2 = 4^2$$
    $$(x-2) = $$
    $$4x - = 16$$
    $$4x = $$
    $$x = 6$$

14. Solve $\sqrt{1-2x} = \sqrt{x+10}$.
    $$\left(\right)^2 = \left(\sqrt{x+10}\right)^2$$
    $$ = x + 10$$
    $$ = 9$$
    $$x = -3$$

PRACTICE

In Exercises 15–54, solve each equation. Write all apparent solutions. Cross out those that are extraneous.

15. $\sqrt{5x-6} = 2$ 16. $\sqrt{7x-10} = 12$
17. $\sqrt{6x+1} + 2 = 7$ 18. $\sqrt{6x+13} - 2 = 5$
19. $2\sqrt{4x+1} = \sqrt{x+4}$
20. $\sqrt{3(x+4)} = \sqrt{5x-12}$
21. $\sqrt[3]{7n-1} = 3$ 22. $\sqrt[3]{12m+4} = 4$
23. $\sqrt[4]{10p+1} = \sqrt[4]{11p-7}$
24. $\sqrt[4]{10y+6} = 2\sqrt[4]{y}$
25. $x = \dfrac{\sqrt{12x-5}}{2}$ 26. $x = \dfrac{\sqrt{16x-12}}{2}$
27. $\sqrt{x+2} - \sqrt{4-x} = 0$
28. $\sqrt{6-x} - \sqrt{2x+3} = 0$
29. $2\sqrt{x} = \sqrt{5x-16}$ 30. $3\sqrt{x} = \sqrt{3x+54}$
31. $r - 9 = \sqrt{2r-3}$ 32. $-s - 3 = 2\sqrt{5-s}$

33. $\sqrt{-5x + 24} = 6 - x$ **34.** $\sqrt{-x + 2} = x - 2$

35. $\sqrt{y + 2} = 4 - y$ **36.** $\sqrt{22y + 86} = y + 9$

37. $\sqrt[3]{x^3 - 7} = x - 1$ **38.** $\sqrt[3]{x^3 + 56} - 2 = x$

39. $\sqrt[4]{x^4 + 4x^2 - 4} = -x$ **40.** $\sqrt[4]{8x - 8} + 2 = 0$

41. $\sqrt[4]{12t + 4} + 2 = 0$ **42.** $u = \sqrt[4]{u^4 - 6u^2 + 24}$

43. $\sqrt{2y + 1} = 1 - 2\sqrt{y}$ **44.** $\sqrt{u} + 3 = \sqrt{u - 3}$

45. $\sqrt{y + 7} + 3 = \sqrt{y + 4}$
46. $1 + \sqrt{z} = \sqrt{z + 3}$
47. $2 + \sqrt{u} = \sqrt{2u + 7}$
48. $5r + 4 = \sqrt{5r + 20} + 4r$
49. $\sqrt{6t + 1} - 3\sqrt{t} = -1$
50. $\sqrt{4s + 1} - \sqrt{6s} = -1$
51. $\sqrt{2x + 5} + \sqrt{x + 2} = 5$
52. $\sqrt{2x + 5} + \sqrt{2x + 1} + 4 = 0$
53. $\sqrt{x - 5} - \sqrt{x + 3} = 4$
54. $\sqrt{x + 8} - \sqrt{x - 4} = -2$

In Exercises 55–62, solve each equation for the indicated variable.

55. $v = \sqrt{2gh}$ for h

56. $d = 1.4\sqrt{h}$ for h

57. $T = 2\pi\sqrt{\dfrac{l}{32}}$ for l

58. $d = \sqrt[3]{\dfrac{12V}{\pi}}$ for V

59. $r = \sqrt[3]{\dfrac{A}{P}} - 1$ for A

60. $r = \sqrt[3]{\dfrac{A}{P}} - 1$ for P

61. $L_A = L_B\sqrt{1 - \dfrac{v^2}{c^2}}$ for v^2

62. $R_1 = \sqrt{\dfrac{A}{\pi} - R_2^{\,2}}$ for A

APPLICATIONS

63. HIGHWAY DESIGN A curved concrete road will accommodate traffic traveling s mph if the radius of the curve is r feet, according to the formula $s = 3\sqrt{r}$. If engineers expect 40-mph traffic, what radius should they specify? Give the result to the nearest foot. (See Illustration 2.)

ILLUSTRATION 2

64. FORESTRY The higher a lookout tower is built, the farther an observer can see. See Illustration 3. That distance d (called the *horizon distance*, measured in miles) is related to the height h of the observer (measured in feet) by the formula $d = 1.4\sqrt{h}$. How tall must a lookout tower be to see the edge of the forest, 25 miles away? (Round to the nearest foot.)

ILLUSTRATION 3

65. WIND POWER The power generated by a certain windmill is related to the velocity of the wind by the formula

$$v = \sqrt[3]{\dfrac{P}{0.02}}$$

where P is the power (in watts) and v is the velocity of the wind (in mph). Find how much power the windmill is generating when the wind is 29 mph.

66. DIAMONDS The *effective rate of interest r* earned by an investment is given by the formula

$$r = \sqrt[n]{\dfrac{A}{P}} - 1$$

where P is the initial investment that grows to value A after n years. If a diamond buyer got \$4,000 for a 1.73-carat diamond that he had purchased 4 years earlier, and earned an annual rate of return of 6.5% on the investment, what did he originally pay for the diamond?

67. THEATER PRODUCTION The ropes, pulleys, and sandbags shown in Illustration 4 are part of a mechanical system used to raise and lower scenery for a stage play. For the scenery to be in the proper position, the following formula must apply:

$$w_2 = \sqrt{w_1{}^2 + w_3{}^2}$$

If $w_2 = 12.5$ lb and $w_3 = 7.5$ lb, find w_1.

ILLUSTRATION 4

68. CARPENTRY During construction, carpenters often brace walls as shown in Illustration 5, where the length of the brace is given by the formula

$$l = \sqrt{f^2 + h^2}$$

If a carpenter nails a 10-ft brace to the wall 6 feet above the floor, how far from the base of the wall should he nail the brace to the floor?

ILLUSTRATION 5

69. SUPPLY AND DEMAND The number of wrenches that will be produced at a given price can be predicted by the formula $s = \sqrt{5x}$, where s is the supply (in thousands) and x is the price (in dollars). The demand d for wrenches can be predicted by the formula $d = \sqrt{100 - 3x^2}$. Find the equilibrium price—that is, find the price at which supply will equal demand.

70. SUPPLY AND DEMAND The number of mirrors that will be produced at a given price can be predicted by the formula $s = \sqrt{23x}$, where s is the supply (in thousands) and x is the price (in dollars). The demand d for mirrors can be predicted by the formula $d = \sqrt{312 - 2x^2}$. Find the equilibrium price—that is, find the price at which supply will equal demand.

WRITING

71. If both sides of an equation are raised to the same power, the resulting equation might not be equivalent to the original equation. Explain.

72. Explain how the radical equation $\sqrt{2x - 1} = x$ can be solved graphically.

73. Explain how the table of values in Illustration 6 can be used to solve $\sqrt{4x - 3} - 2 = \sqrt{2x - 5}$ if $Y_1 = \sqrt{4x - 3} - 2$ and $Y_2 = \sqrt{2x - 5}$.

ILLUSTRATION 6

74. Explain how to use the graph of $f(x) = \sqrt[3]{x - 0.5} - 1$, shown in Illustration 7, to approximate the solution of $\sqrt[3]{x - 0.5} = 1$.

ILLUSTRATION 7

REVIEW

75. LIGHTING The intensity of the light reaching you from a light bulb varies inversely as the square of your distance from the bulb. If you are 5 feet away from a light bulb and the intensity is 40 foot-candles, what will the intensity be if you move 20 feet away from the bulb?

76. COMMITTEES What type of variation is shown in Illustration 8? As the number of people on this committee increased, what happened to its effectiveness?

ILLUSTRATION 8

77. TYPESETTING If 12-point type is 0.166044 inch tall, how tall is 30-point type?

78. GUITAR STRINGS The frequency of vibration of a string varies directly as the square root of the tension and inversely as the length of the string. Suppose a string 2.5 feet long, under a tension of 16 pounds, vibrates 25 times per second. Find k, the constant of proportionality.

▶ 7.3 Rational Exponents

In this section, you will learn about

> Rational exponents ■ Exponential expressions with variables in their bases ■ Rational exponents with numerators other than 1 ■ Negative rational exponents ■ Applying the rules for exponents ■ Simplifying radical expressions

Introduction We have worked with exponential expressions containing natural-number exponents, such as 5^3 and x^2. In Chapter 5, the definition of exponent was extended to include zero and negative integers, which gave meaning to expressions such as 8^{-3} and $(-9xy)^0$. In this section, we will again extend the definition of exponent—this time to include rational (fractional) exponents. We will see how expressions such as $9^{1/2}$, $\left(\frac{1}{16}\right)^{3/4}$, and $(-32x^5)^{-2/5}$ can be simplified by writing them in an equivalent radical form or by using the rules for exponents.

Rational Exponents

We have seen that positive-integer exponents indicate the number of times that a base is to be used as a factor in a product. For example, x^5 means that x is to be used as a factor five times.

$$x^5 = \overbrace{x \cdot x \cdot x \cdot x \cdot x}^{5 \text{ factors of } x}$$

Furthermore, we recall the following rules for exponents.

Rules for exponents

If there are no divisions by 0, then for all integers m and n,

1. $x^m x^n = x^{m+n}$ **2.** $(x^m)^n = x^{mn}$ **3.** $(xy)^n = x^n y^n$ **4.** $\left(\dfrac{x}{y}\right)^n = \dfrac{x^n}{y^n}$

5. $x^0 = 1 \; (x \neq 0)$ **6.** $x^{-n} = \dfrac{1}{x^n}$ **7.** $\dfrac{x^m}{x^n} = x^{m-n}$ **8.** $\left(\dfrac{x}{y}\right)^{-n} = \left(\dfrac{y}{x}\right)^n$

It is possible to raise many bases to fractional powers. Since we want fractional exponents to obey the same rules as integer exponents, the square of $10^{1/2}$ must be 10, because

$$(10^{1/2})^2 = 10^{(1/2)2} \quad \text{Keep the base and multiply the exponents.}$$
$$= 10^1 \qquad \tfrac{1}{2} \cdot 2 = 1.$$
$$= 10 \qquad 10^1 = 10.$$

However, we have seen that

$$\left(\sqrt{10}\right)^2 = 10$$

Since $(10^{1/2})^2$ and $\left(\sqrt{10}\right)^2$ both equal 10, we define $10^{1/2}$ to be $\sqrt{10}$. Likewise, we define

$$10^{1/3} \text{ to be } \sqrt[3]{10} \qquad \text{and} \qquad 10^{1/4} \text{ to be } \sqrt[4]{10}$$

Rational exponents

If n is a natural number greater than 1 and $\sqrt[n]{x}$ is a real number, then

$$x^{1/n} = \sqrt[n]{x}$$

In words, *a rational exponent of $\frac{1}{n}$ indicates that the nth root of the base should be found.*

Using this definition, we can simplify the exponential expression $8^{1/3}$. The first step is to write it as an equivalent expression in radical form and proceed as follows:

$8^{1/3} = \sqrt[3]{8}$ The base of the exponential expression, 8, is the radicand. The denominator of the fractional exponent, 3, is the index of the radical.

$\qquad = 2$

EXAMPLE 1

Simplifying expressions containing rational exponents. Write each expression in radical form and simplify, if possible.

a. $9^{1/2} = \sqrt{9}$
$\qquad = 3$

b. $-\left(\dfrac{16}{9}\right)^{1/2} = -\sqrt{\dfrac{16}{9}}$
$\qquad\qquad\qquad = -\dfrac{4}{3}$

c. $(-64)^{1/3} = \sqrt[3]{-64}$
$\qquad\qquad = -4$

d. $16^{1/4} = \sqrt[4]{16}$
$\qquad\qquad = 2$

e. $\left(\dfrac{1}{32}\right)^{1/5} = \sqrt[5]{\dfrac{1}{32}}$
$\qquad\qquad\quad = \dfrac{1}{2}$

f. $0^{1/8} = \sqrt[8]{0}$
$\qquad\quad = 0$

g. $y^{1/4} = \sqrt[4]{y}$

h. $-(2x^2)^{1/5} = -\sqrt[5]{2x^2}$

SELF CHECK Write each expression in radical form and simplify, if possible: **a.** $16^{1/2}$, **b.** $\left(-\frac{27}{8}\right)^{1/3}$, and **c.** $-(6x^3)^{1/4}$.   *Answers:* **a.** 4, **b.** $-\frac{3}{2}$, **c.** $-\sqrt[4]{6x^3}$ ∎

EXAMPLE 2

Writing roots using rational exponents. Write $\sqrt{5xyz}$ as an exponential expression with a rational exponent.

Solution The radicand is $5xyz$, so the base of the exponential expression is $5xyz$. The index of the radical is an understood 2, so the denominator of the fractional exponent is 2.

$$\sqrt{5xyz} = (5xyz)^{1/2}$$

SELF CHECK Write the radical with a fractional exponent: $\sqrt[6]{7ab}$.   *Answer:* $(7ab)^{1/6}$ ∎

Rational exponents appear in formulas used in many disciplines, such as science and engineering.

EXAMPLE 3

Satellites. See Figure 7-10. The formula

$$r = \left(\frac{GMP^2}{4\pi^2}\right)^{1/3}$$

gives the orbital radius (in meters) of a satellite circling the earth, where G and M are constants and P is the time in seconds for the satellite to make one complete revolution. Write the formula using a radical.

FIGURE 7-10

Earth

Solution   The fractional exponent $\frac{1}{3}$ has a denominator of 3, which indicates that we are to find the cube root of the base of the exponential expression. So we have

$$r = \sqrt[3]{\frac{GMP^2}{4\pi^2}}$$

■

Exponential Expressions with Variables in Their Bases

As with radicals, when n is an *odd natural number* in the expression $x^{1/n}$ ($n > 1$), there is exactly one real nth root, and we don't have to worry about absolute value symbols. When n is an *even natural number,* there are two nth roots. Since we want the expression $x^{1/n}$ to represent the positive nth root, we must often use absolute value symbols to guarantee that the simplified result is positive. Thus, if n is even,

$$(x^n)^{1/n} = |x|$$

When n is even and x is negative, the expression $x^{1/n}$ is not a real number.

EXAMPLE 4

A variable in the base. Simplify each exponential expression. Assume that the variables can be any real number.

a. $(-27x^3)^{1/3} = -3x$ Because $(-3x)^3 = -27x^3$. Since n is odd, no absolute value symbols are needed.

b. $(256a^8)^{1/8} = 2|a|$ Because $(2|a|)^8 = 256a^8$. Since n is even and a can be any real number, $2a$ can be negative. Thus, absolute value symbols are needed.

c. $[(y + 4)^2]^{1/2} = |y + 4|$ Because $|y + 4|^2 = (y + 4)^2$. Since n is even and y can be any real number, $y + 4$ can be negative. Thus, absolute value symbols are needed.

d. $(25b^4)^{1/2} = 5b^2$ Because $(5b^2)^2 = 25b^4$. Since $b^2 \geq 0$, no absolute value symbols are needed.

e. $(-256x^4)^{1/4}$ is not a real number.   Because no real number raised to the 4th power is $-256x^4$.

SELF CHECK   Simplify each expression: **a.** $(625a^4)^{1/4}$ and
b. $(b^4)^{1/2}$.   *Answers:* **a.** $5|a|$, **b.** b^2 ■

If we are told that the variables represent positive real numbers in parts b and c of Example 4, the absolute value symbols in the answers are not needed.

$(256a^8)^{1/8} = 2a$ If a represents a positive number, then $2a$ is positive.

$[(y + 4)^2]^{1/2} = y + 4$ If y represents a positive number, then $y + 4$ is positive.

We summarize the cases as follows.

Summary of the definitions of $x^{1/n}$

If n is a natural number greater than 1 and x is a real number,

If $x > 0$, then $x^{1/n}$ is the positive number such that $(x^{1/n})^n = x$.

If $x = 0$, then $x^{1/n} = 0$.

If $x < 0$ $\begin{cases} \text{and } n \text{ is odd, then } x^{1/n} \text{ is the real number such that } (x^{1/n})^n = x. \\ \text{and } n \text{ is even, then } x^{1/n} \text{ is not a real number.} \end{cases}$

Rational Exponents with Numerators Other Than 1

We can extend the definition of $x^{1/n}$ to include fractional exponents with numerators other than 1. For example, since $8^{2/3}$ can be written as $(8^{1/3})^2$, we have

$$8^{2/3} = (8^{1/3})^2$$
$$= \left(\sqrt[3]{8}\right)^2 \quad \text{Write } 8^{1/3} \text{ in radical form.}$$
$$= 2^2 \quad \text{Find the cube root first: } \sqrt[3]{8} = 2.$$
$$= 4 \quad \text{Then find the power.}$$

Thus, we can simplify $8^{2/3}$ by finding the second power of the cube root of 8.

The numerator of the rational
exponent is the power.

The base of the exponential expression is the radicand.

The denominator of the exponent
is the index of the radical.

We can also simplify $8^{2/3}$ by taking the cube root of 8 squared.

$$8^{2/3} = (8^2)^{1/3}$$
$$= 64^{1/3} \quad \text{Find the power first: } 8^2 = 64.$$
$$= \sqrt[3]{64} \quad \text{Write } 64^{1/3} \text{ in radical form.}$$
$$= 4 \quad \text{Now find the cube root.}$$

In general, we have the following rule.

**Changing from rational exponents
to radicals**

If m and n are positive integers ($n \neq 1$) and $\sqrt[n]{x}$ is a real number, then
$$x^{m/n} = \sqrt[n]{x^m} = \left(\sqrt[n]{x}\right)^m$$

Because of the previous definition, we can interpret $x^{m/n}$ in two ways:

1. $x^{m/n}$ means the nth root of the mth power of x.
2. $x^{m/n}$ means the mth power of the nth root of x.

EXAMPLE 5

Rational exponents with numerators other than 1. Simplify each expression in two ways.

a. $9^{3/2} = \left(\sqrt{9}\right)^3$ or $9^{3/2} = \sqrt{9^3}$
$\qquad = 3^3 \qquad\qquad = \sqrt{729}$
$\qquad = 27 \qquad\qquad = 27$

b. $\left(\dfrac{1}{16}\right)^{3/4} = \left(\sqrt[4]{\dfrac{1}{16}}\right)^3$ or $\left(\dfrac{1}{16}\right)^{3/4} = \left[\left(\dfrac{1}{16}\right)^3\right]^{1/4}$

$\qquad\qquad = \left(\dfrac{1}{2}\right)^3 \qquad\qquad\qquad = \left(\dfrac{1}{4,096}\right)^{1/4}$

$\qquad\qquad = \dfrac{1}{8} \qquad\qquad\qquad\qquad = \dfrac{1}{8}$

c. $(-8x^3)^{4/3} = \left(\sqrt[3]{-8x^3}\right)^4$ or $(-8x^3)^{4/3} = \sqrt[3]{(-8x^3)^4}$

$\qquad\qquad = (-2x)^4 \qquad\qquad\qquad = \sqrt[3]{4,096x^{12}}$

$\qquad\qquad = 16x^4 \qquad\qquad\qquad\quad = 16x^4$

SELF CHECK Simplify **a.** $16^{3/2}$ and **b.** $(-27x^6)^{2/3}$. *Answers:* **a.** 64, **b.** $9x^4$ ∎

To avoid large numbers, it is usually better to find the root of the base first, as shown with the first solution of each part in Example 5.

ACCENT ON TECHNOLOGY *Rational Exponents*

We can evaluate exponential expressions containing rational exponents using the exponential key $\boxed{y^x}$ or $\boxed{x^y}$ on a scientific calculator. For example, to evaluate $10^{2/3}$, we enter these numbers and press these keys:

Keystrokes $10 \boxed{y^x} \boxed{(} 2 \boxed{\div} 3 \boxed{)} \boxed{=}$ $\qquad\qquad$ $\boxed{4.641588834}$

Note that parentheses were used when entering the power. Without them, the calculator would interpret the entry as $10^2 \div 3$.

To evaluate the exponential expression using a graphing calculator, we use the $\boxed{\wedge}$ key, which raises a base to a power. Again, we use parentheses when entering the power.

Keystrokes $10 \boxed{\wedge} \boxed{(} 2 \boxed{\div} 3 \boxed{)} \boxed{\text{ENTER}}$ $\qquad$ $\boxed{\begin{array}{l} 10^{\wedge}(2/3) \\ \qquad 4.641588834 \end{array}}$

To the nearest hundredth, $10^{2/3} \approx 4.64$.

Negative Rational Exponents

To be consistent with the definition of negative-integer exponents, we define $x^{-m/n}$ as follows.

Definition of $x^{-m/n}$

If m and n are positive integers, $\frac{m}{n}$ is in simplified form, and $x^{1/n}$ is a real number, then

$$x^{-m/n} = \frac{1}{x^{m/n}} \quad \text{and} \quad \frac{1}{x^{-m/n}} = x^{m/n} \quad (x \neq 0)$$

EXAMPLE 6

Negative rational exponents. Write each expression without using negative exponents and simplify, if possible.

a. $64^{-1/2} = \dfrac{1}{64^{1/2}}$

$\qquad\qquad = \dfrac{1}{\sqrt{64}}$

$\qquad\qquad = \dfrac{1}{8}$

b. $(-16)^{-3/4}$ is not a real number, because $(-16)^{1/4}$ is not a real number.

c. $(-32x^5)^{-2/5} = \dfrac{1}{(-32x^5)^{2/5}}$

$\qquad\qquad\qquad = \dfrac{1}{[(-32x^5)^{1/5}]^2}$

$\qquad\qquad\qquad = \dfrac{1}{\left(\sqrt[5]{-32x^5}\right)^2}$

$\qquad\qquad\qquad = \dfrac{1}{(-2x)^2}$

$\qquad\qquad\qquad = \dfrac{1}{4x^2}$

d. $\dfrac{1}{16^{-3/2}} = 16^{3/2}$

$\qquad\qquad = (16^{1/2})^3$

$\qquad\qquad = \left(\sqrt{16}\right)^3$

$\qquad\qquad = 4^3$

$\qquad\qquad = 64$

SELF CHECK

Write without using negative exponents and simplify: **a.** $25^{-3/2}$ and   **b.** $(-27a^3)^{-2/3}$.

Answers: **a.** $\dfrac{1}{125}$,   **b.** $\dfrac{1}{9a^2}$ ∎

WARNING! By definition, 0^0 is undefined. A base of 0 raised to a negative power is also undefined. For example, 0^{-2} would equal $\frac{1}{0^2}$, which is undefined because we cannot divide by 0.

Applying the Rules for Exponents

We can use the rules for exponents to simplify many expressions with fractional exponents. If all variables represent positive numbers, no absolute value symbols are necessary.

EXAMPLE 7

Simplifying expressions by using the rules for exponents. Assume that all variables represent positive numbers. Write all answers without using negative exponents.

a. $5^{2/7}5^{3/7} = 5^{2/7+3/7}$ Use the rule $x^m x^n = x^{m+n}$.

$\qquad\quad = 5^{5/7}$ Add: $\frac{2}{7} + \frac{3}{7} = \frac{5}{7}$.

b. $(5^{2/7})^3 = 5^{(2/7)(3)}$ Use the rule $(x^m)^n = x^{mn}$.

         $= 5^{6/7}$ Multiply: $\frac{2}{7}(3) = \frac{6}{7}$.

c. $(a^{2/3}b^{1/2})^6 = (a^{2/3})^6(b^{1/2})^6$ Use the rule $(xy)^n = x^n y^n$.

         $= a^{12/3}b^{6/2}$ Use the rule $(x^m)^n = x^{mn}$ twice.

         $= a^4 b^3$ Simplify the exponents.

d. $\dfrac{a^{8/3}a^{1/3}}{a^2} = a^{8/3+1/3-2}$ Use the rules $x^m x^n = x^{m+n}$ and $\frac{x^m}{x^n} = x^{m-n}$.

         $= a^{8/3+1/3-6/3}$ $2 = \frac{6}{3}$.

         $= a^{3/3}$ $\frac{8}{3} + \frac{1}{3} - \frac{6}{3} = \frac{3}{3}$.

         $= a$ $\frac{3}{3} = 1$.

SELF CHECK   Simplify   **a.** $(x^{1/3}y^{3/2})^6$ and   **b.** $\dfrac{x^{5/3}x^{2/3}}{x^{1/3}}$.   *Answers:*   **a.** $x^2 y^9$,   **b.** x^2 ■

E X A M P L E   8   **Using the distributive property.** Assume all variables represent positive numbers and do the operations. Write all answers without negative exponents.

a. $a^{4/5}(a^{1/5} + a^{3/5}) = a^{4/5}a^{1/5} + a^{4/5}a^{3/5}$ Use the distributive property.

         $= a^{4/5+1/5} + a^{4/5+3/5}$ Use the rule $x^m x^n = x^{m+n}$.

         $= a^{5/5} + a^{7/5}$ Simplify the exponents.

         $= a + a^{7/5}$ We cannot add these terms because they are not like terms.

b. $x^{1/2}(x^{-1/2} + x^{1/2}) = x^{1/2}x^{-1/2} + x^{1/2}x^{1/2}$ Use the distributive property.

         $= x^{1/2+(-1/2)} + x^{1/2+1/2}$ Use the rule $x^m x^n = x^{m+n}$.

         $= x^0 + x^1$ Simplify each exponent.

         $= 1 + x$ $x^0 = 1$.   ■

Simplifying Radical Expressions

We can simplify many radical expressions by using the following steps.

Using rational exponents to simplify radicals

1. Change the radical expression into an exponential expression.
2. Simplify the rational exponents.
3. Change the exponential expression back into a radical.

E X A M P L E   9   **Simplifying radical expressions.** Simplify   **a.** $\sqrt[4]{3^2}$,   **b.** $\sqrt[8]{x^6}$,   and   **c.** $\sqrt[9]{27x^6 y^3}$.

Solution   **a.** $\sqrt[4]{3^2} = (3^2)^{1/4}$ Change the radical to an exponential expression.

         $= 3^{2/4}$ Use the rule $(x^m)^n = x^{mn}$.

         $= 3^{1/2}$ $\frac{2}{4} = \frac{1}{2}$.

         $= \sqrt{3}$ Change back to radical form.

b. $\sqrt[8]{x^6} = (x^6)^{1/8}$ — Change the radical to an exponential expression.

$= x^{6/8}$ — Use the rule $(x^m)^n = x^{mn}$.

$= x^{3/4}$ — $\frac{6}{8} = \frac{3}{4}$.

$= (x^3)^{1/4}$ — $\frac{3}{4} = 3\left(\frac{1}{4}\right)$.

$= \sqrt[4]{x^3}$ — Change back to radical form.

c. $\sqrt[9]{27x^6y^3} = (3^3x^6y^3)^{1/9}$ — Write 27 as 3^3 and change the radical to an exponential expression.

$= 3^{3/9}x^{6/9}y^{3/9}$ — Raise each factor to the $\frac{1}{9}$ power by multiplying the fractional exponents.

$= 3^{1/3}x^{2/3}y^{1/3}$ — Simplify each fractional exponent.

$= (3x^2y)^{1/3}$ — Use the rule $(xy)^n = x^ny^n$.

$= \sqrt[3]{3x^2y}$ — Change back to radical form.

SELF CHECK Simplify **a.** $\sqrt[6]{3^3}$ and **b.** $\sqrt[4]{64x^2y^2}$. *Answers:* **a.** $\sqrt{3}$, **b.** $\sqrt{8xy}$ ∎

STUDY SET

Section 7.3

VOCABULARY

In Exercises 1–4, fill in the blanks to make each statement true.

1. The expressions $4^{1/2}$ and $(-8)^{-2/3}$ have _____ exponents.

2. In the exponential expression $27^{4/3}$, 27 is the _____, and 4/3 is the _____.

3. In the radical expression $\sqrt[3]{4{,}096x^{12}}$, 3 is the _____, and $4{,}096x^{12}$ is the _____.

4. $32^{4/5}$ means the fourth _____ of the fifth _____ of 32.

CONCEPTS

5. Complete the table by writing the given expression in the alternate form.

Radical form	Exponential form
$\sqrt[5]{25}$	
	$(-27)^{2/3}$
$\left(\sqrt[4]{16}\right)^{-3}$	
	$81^{3/2}$
$-\sqrt{\frac{9}{64}}$	

6. In your own words, explain the two rules for rational exponents illustrated in the diagrams below.

a. $(-32)^{1/5} = \sqrt[5]{-32}$

b. $125^{4/3} = \left(\sqrt[3]{125}\right)^4$

7. Graph each number on the number line.

$$\left\{8^{2/3},\ (-125)^{1/3},\ -16^{-1/4},\ 4^{3/2},\ -\left(\frac{9}{100}\right)^{-1/2}\right\}$$

8. Evaluate $25^{3/2}$ in two ways. Which way is easier?

In Exercises 9–14, complete each rule for exponents.

9. $x^m x^n =$

10. $(x^m)^n =$

11. $\dfrac{x^m}{x^n} =$

12. $x^{-n} =$

13. $x^{1/n} =$

14. $x^{m/n} =$ $= \sqrt[n]{x^m}$

In Exercises 15–16, complete each table of values. Then graph the function.

15. $f(x) = x^{1/2}$

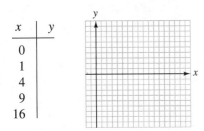

x	y
0	
1	
4	
9	
16	

16. $f(x) = x^{1/3}$

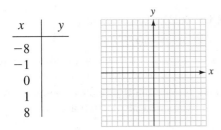

x	y
−8	
−1	
0	
1	
8	

NOTATION

In Exercises 17–18, complete each solution.

17. Simplify $(100a^4)^{3/2}$.

$$(100a^4)^{3/2} = \left(\boxed{}\right)^3$$
$$= \left(\boxed{}\right)^3$$
$$= 1{,}000a^6$$

18. Simplify $(m^{1/3}n^{1/2})^6$.

$$(m^{1/3}n^{1/2})^6 = \left(\boxed{}\right)^6(n^{1/2})^6$$
$$= m^{\boxed{}}n^{6/2}$$
$$= m^2n^3$$

PRACTICE

In Exercises 19–26, write each expression in radical form.

19. $x^{1/3}$ **20.** $b^{1/2}$

21. $(3x)^{1/4}$ **22.** $(4ab)^{1/6}$

23. $\left(\dfrac{1}{2}x^3y\right)^{1/4}$ **24.** $\left(\dfrac{3}{4}a^2b^2\right)^{1/5}$

25. $(x^2 + y^2)^{1/2}$ **26.** $(x^3 + y^3)^{1/3}$

In Exercises 27–34, change each radical to an exponential expression.

27. $\sqrt{m}$ **28.** $\sqrt[3]{r}$

29. $\sqrt[4]{3a}$ **30.** $3\sqrt[5]{a}$

31. $\sqrt[6]{\dfrac{1}{7}abc}$ **32.** $\sqrt[7]{\dfrac{3}{8}p^2q}$

33. $\sqrt[3]{a^2 - b^2}$ **34.** $\sqrt{x^2 + y^2}$

In Exercises 35–50, simplify each expression, if possible.

35. $4^{1/2}$ **36.** $25^{1/2}$

37. $8^{1/3}$ **38.** $125^{1/3}$

39. $16^{1/4}$ **40.** $625^{1/4}$

41. $32^{1/5}$ **42.** $0^{1/5}$

43. $\left(\dfrac{1}{4}\right)^{1/2}$ **44.** $\left(\dfrac{1}{16}\right)^{1/2}$

45. $-16^{1/4}$ **46.** $-125^{1/3}$

47. $(-64)^{1/2}$ **48.** $(-216)^{1/2}$

49. $(-27)^{1/3}$ **50.** $(-125)^{1/3}$

In Exercises 51–58, simplify each expression, if possible. Assume that all variables are unrestricted and use absolute value symbols when necessary.

51. $(25y^2)^{1/2}$ **52.** $(-27x^3)^{1/3}$

53. $(16x^4)^{1/4}$ **54.** $(-16x^4)^{1/2}$

55. $(243x^5)^{1/5}$ **56.** $[(x + 1)^4]^{1/4}$

57. $(-64x^8)^{1/4}$ **58.** $[(x + 5)^3]^{1/3}$

In Exercises 59–70, simplify each expression. Assume that all variables represent positive numbers.

59. $36^{3/2}$ **60.** $27^{2/3}$

61. $81^{3/4}$ **62.** $100^{3/2}$

63. $144^{3/2}$ **64.** $1{,}000^{2/3}$

65. $\left(\dfrac{1}{8}\right)^{2/3}$ **66.** $\left(\dfrac{4}{9}\right)^{3/2}$

67. $(25x^4)^{3/2}$ **68.** $(27a^3b^3)^{2/3}$

69. $\left(\dfrac{8x^3}{27}\right)^{2/3}$ **70.** $\left(\dfrac{27}{64y^6}\right)^{2/3}$

In Exercises 71–82, write each expression without using negative exponents. Assume that all variables represent positive numbers.

71. $4^{-1/2}$ **72.** $8^{-1/3}$

73. $4^{-3/2}$ **74.** $25^{-5/2}$

75. $(16x^2)^{-3/2}$ **76.** $(81c^4)^{-3/2}$

77. $(-27y^3)^{-2/3}$ **78.** $(-8z^9)^{-2/3}$

79. $\left(\dfrac{27}{8}\right)^{-4/3}$ **80.** $\left(\dfrac{25}{49}\right)^{-3/2}$

81. $\left(-\dfrac{8x^3}{27}\right)^{-1/3}$ **82.** $\left(\dfrac{16}{81y^4}\right)^{-3/4}$

In Exercises 83–86, use a calculator to evaluate each expression. Round to the nearest hundredth.

83. $\sqrt[3]{15}$

84. $\sqrt[4]{50.5}$

85. $\sqrt[5]{1.045}$

86. $\sqrt[5]{-1,000}$

In Exercises 87–106, do the operations. Write the answers without negative exponents. Assume that all variables represent positive numbers.

87. $5^{3/7}5^{2/7}$

88. $4^{2/5}4^{2/5}$

89. $(4^{1/5})^3$

90. $(3^{1/3})^5$

91. $\dfrac{9^{4/5}}{9^{3/5}}$

92. $\dfrac{7^{2/3}}{7^{1/2}}$

93. $6^{-2/3}6^{-4/3}$

94. $5^{1/3}5^{-5/3}$

95. $\dfrac{3^{4/3}3^{1/3}}{3^{2/3}}$

96. $\dfrac{2^{5/6}2^{1/3}}{2^{1/2}}$

97. $a^{2/3}a^{1/3}$

98. $b^{3/5}b^{1/5}$

99. $(a^{2/3})^{1/3}$

100. $(t^{4/5})^{10}$

101. $(a^{1/2}b^{1/3})^{3/2}$

102. $(mn^{-2/3})^{-3/5}$

103. $\dfrac{(4x^3y)^{1/2}}{(9xy)^{1/2}}$

104. $\dfrac{(27x^3y)^{1/3}}{(8xy^2)^{2/3}}$

105. $(27x^{-3})^{-1/3}$

106. $(16a^{-2})^{-1/2}$

In Exercises 107–110, do the multiplications. Assume that all variables are positive.

107. $y^{1/3}(y^{2/3} + y^{5/3})$

108. $y^{2/5}(y^{-2/5} + y^{3/5})$

109. $x^{3/5}(x^{7/5} - x^{2/5} + 1)$

110. $x^{4/3}(x^{2/3} + 3x^{5/3} - 4)$

In Exercises 111–114, use rational exponents to simplify each radical. Assume that all variables represent positive numbers.

111. $\sqrt[6]{p^3}$

112. $\sqrt[8]{q^2}$

113. $\sqrt[4]{25b^2}$

114. $\sqrt[9]{-8x^6}$

APPLICATIONS

115. A BALLISTIC PENDULUM See Illustration 1. The formula

$$v = \frac{m + M}{m}(2gh)^{1/2}$$

gives the velocity (in ft/sec) of a bullet with weight m fired into a block with weight M, that raises the height of the block h feet after the collision. The letter g represents the constant, 32. Find the velocity of the bullet to the nearest ft/sec.

ILLUSTRATION 1

$m = 0.0625$ lb $M = 6.0$ lb $h = 0.9$ ft

116. GEOGRAPHY The formula

$$A = [s(s - a)(s - b)(s - c)]^{1/2}$$

gives the area of a triangle with sides of length a, b, and c, where s is one-half of the perimeter. Estimate the area of Virginia (to the nearest square mile) using the data given in Illustration 2.

ILLUSTRATION 2

370 mi   220 mi   430 mi

117. RELATIVITY One of the concepts of relativity theory is that an object moving past an observer at a speed near the speed of light appears to have a larger mass because of its motion. If the mass of the object is m_0 when the object is at rest relative to the observer, its mass m will be given by the formula

$$m = m_0(1 - v^2/c^2)^{-1/2}$$

when it is moving with speed v (in miles per second) past the observer. The variable c is the speed of light, 186,000 mi/sec. If a proton with a rest mass of 1 unit is accelerated by a nuclear accelerator to a speed of 160,000 mi/sec, what mass will the technicians observe it to have? Round to the nearest hundredth.

118. LOGGING See Illustration 3. The width w and height h of the strongest rectangular beam that can be cut from a cylindrical log of radius a are given by

$$w = \frac{2a}{3}(3^{1/2}) \qquad h = a\left(\frac{8}{3}\right)^{1/2}$$

Find the width, height, and cross-sectional area of the strongest beam that can be cut from a log with *diameter* 4 feet. Round to the nearest hundredth.

ILLUSTRATION 3

119. CUBICLES The area of the base of a cube is given by the function $A(x) = V^{2/3}$, where V is the volume of the cube. In a preschool room, 18 children's cubicles like that shown in Illustration 4 are placed on the floor around the room. Estimate how much floor space is lost to the cubicles. Give your answer in square inches and in square feet.

ILLUSTRATION 4

Storage capacity 4,096 in.^{3}

120. CARPENTRY See Illustration 5. The length L of the longest board that can be carried horizontally around the right-angle corner of two intersecting hallways is given by the formula

$$L = (a^{2/3} + b^{2/3})^{3/2}$$

where a and b represent the widths of the hallways. Find the longest shelf that a carpenter can carry around the corner if $a = 40$ in. and $b = 64$ in. Give your result in inches and in feet. In each case, round to the nearest tenth.

ILLUSTRATION 5

WRITING

121. What is a rational exponent? Give some examples.

122. Explain how the root key $\sqrt[x]{y}$ on a scientific calculator can be used in combination with other keys to evaluate the expression $16^{3/4}$.

REVIEW

In Exercises 123–126, solve each inequality. Write each answer as an inequality.

123. $5x - 4 < 11$

124. $-2(3t - 5) \geq 8$

125. $\frac{4}{5}(r - 3) > \frac{2}{3}(r + 2)$

126. $-4 < 2x - 4 \leq 8$

▶ **7.4**

Simplifying and Combining Radical Expressions

FIGURE 7-11

Area = 12 cm^2

In this section, you will learn about

Properties of radicals ■ Simplifying radical expressions ■ Adding and subtracting radical expressions

Introduction The square shown in Figure 7-11 has an area of 12 square centimeters. We can determine the length of one side of the square in the following way:

$$A = s^2$$
$$12 = s^2 \qquad \text{Substitute 12 for } A.$$
$$\sqrt{12} = \sqrt{s^2} \qquad \text{Take the positive square root of both sides.}$$
$$\sqrt{12} = s \qquad \text{Each side of the square is } \sqrt{12} \text{ centimeters long.}$$

The form in which we express the length of a side of the square depends upon the situation. If an approximation is acceptable, we can use a calculator to find that $\sqrt{12} \approx 3.464101615$, and then we can round to a specified degree of accuracy. For example, to the nearest hundredth, each side is 3.46 centimeters long.

If the situation calls for the exact length, we must use a radical expression. As you will see in this section, it is common practice to write a radical expression such as $\sqrt{12}$ in *simplified* form. To simplify radicals, we will use the multiplication and division properties of radicals.

Properties of Radicals

Many properties of exponents have counterparts in radical notation. For example, because $a^{1/n}b^{1/n} = (ab)^{1/n}$, we have

1. $\sqrt[n]{a}\sqrt[n]{b} = \sqrt[n]{ab}$

For example, if x represents a positive real number,

$$\sqrt{5}\sqrt{5} = \sqrt{5 \cdot 5} = \sqrt{25} = 5$$

$$\sqrt[3]{7x}\sqrt[3]{49x^2} = \sqrt[3]{7x \cdot 49x^2} = \sqrt[3]{343x^3} = 7x$$

$$\sqrt[4]{2x^3}\sqrt[4]{8x} = \sqrt[4]{2x^3 \cdot 8x} = \sqrt[4]{16x^4} = 2x$$

These observations suggest the following rule.

Multiplication property of radicals

If $\sqrt[n]{a}$ and $\sqrt[n]{b}$ are real numbers, then

$$\sqrt[n]{ab} = \sqrt[n]{a}\sqrt[n]{b}$$

As long as all radical expressions represent real numbers, *the nth root of the product of two numbers is equal to the product of their nth roots.*

 WARNING! The multiplication property of radicals applies to the nth root of the product of two numbers. There is no such property for sums or differences. For example,

$$\sqrt{9 + 4} \neq \sqrt{9} + \sqrt{4} \qquad \sqrt{9 - 4} \neq \sqrt{9} - \sqrt{4}$$
$$\sqrt{13} \neq 3 + 2 \qquad \sqrt{5} \neq 3 - 2$$
$$\sqrt{13} \neq 5 \qquad \sqrt{5} \neq 1$$

Thus, $\sqrt{a + b} \neq \sqrt{a} + \sqrt{b}$ and $\sqrt{a - b} \neq \sqrt{a} - \sqrt{b}$.

A second property of radicals involves quotients. Because

$$\frac{a^{1/n}}{b^{1/n}} = \left(\frac{a}{b}\right)^{1/n}$$

it follows that

2. $\dfrac{\sqrt[n]{a}}{\sqrt[n]{b}} = \sqrt[n]{\dfrac{a}{b}} \quad (b \neq 0)$

For example, if x represents a positive real number,

$$\frac{\sqrt{8x^3}}{\sqrt{2x}} = \sqrt{\frac{8x^3}{2x}} = \sqrt{4x^2} = 2x$$

$$\frac{\sqrt[3]{54x^5}}{\sqrt[3]{2x^2}} = \sqrt[3]{\frac{54x^5}{2x^2}} = \sqrt[3]{27x^3} = 3x$$

Equation 2 suggests the following property.

Division property of radicals

If $\sqrt[n]{a}$ and $\sqrt[n]{b}$ are real numbers, then

$$\sqrt[n]{\frac{a}{b}} = \frac{\sqrt[n]{a}}{\sqrt[n]{b}} \quad (b \neq 0)$$

As long as all radical expressions represent real numbers, *the nth root of the quotient of two numbers is equal to the quotient of their nth roots.*

Simplifying Radical Expressions

A radical expression is said to be in simplest form when each of the following statements is true.

Simplified form of a radical expression

A radical expression is in simplest form when

1. No radicals appear in the denominator of a fraction.
2. The radicand contains no fractions or negative numbers.
3. Each factor in the radicand is to a power that is less than the index of the radical.

E X A M P L E   1

Simplifying radical expressions by using the multiplication property of radicals. Simplify   **a.** $\sqrt{12}$,   **b.** $\sqrt{98}$,   **c.** $\sqrt[3]{54}$,   and   **d.** $-\sqrt[4]{48}$.

Solution   **a.** Recall that numbers that are squares of integers, such as 1, 4, 9, 16, 25, and 36, are *perfect squares*. To simplify $\sqrt{12}$, we first factor 12 so that one factor is the largest perfect square that divides 12. Since 4 is the largest perfect square factor of 12, we write 12 as $4 \cdot 3$, use the multiplication property of radicals, and simplify.

$$\sqrt{12} = \sqrt{4 \cdot 3} \quad \text{Write 12 in factored form as } 4 \cdot 3.$$

$$= \sqrt{4}\sqrt{3} \quad \text{By the multiplication property of radicals, the square root of a product is equal to the product of the square roots: } \sqrt{4 \cdot 3} = \sqrt{4}\sqrt{3}.$$

$$= 2\sqrt{3} \quad \text{Simplify: } \sqrt{4} = 2.$$

We can show that $2\sqrt{3}$ is a square root of 12 as follows:

$$\left(2\sqrt{3}\right)^2 = (2)^2\left(\sqrt{3}\right)^2 \quad \text{Use a power rule of exponents: } (xy)^n = x^n y^n.$$
$$= 4(3) \qquad \sqrt{3} \text{ is the number that, when squared, gives 3.}$$
$$= 12$$

Thus, $\sqrt{12} = 2\sqrt{3}$.

b. The largest perfect square factor of 98 is 49. Thus,

$$\sqrt{98} = \sqrt{49 \cdot 2} \quad \text{Write 98 in factored form: } 98 = 49 \cdot 2.$$
$$= \sqrt{49}\sqrt{2} \quad \text{Apply the multiplication property of radicals:}$$
$$\sqrt{49 \cdot 2} = \sqrt{49}\sqrt{2}.$$
$$= 7\sqrt{2} \quad \text{Simplify: } \sqrt{49} = 7.$$

Check the result.

c. Numbers that are cubes of integers, such as 1, 8, 27, 64, 125, and 216, are called *perfect cubes*. Since the largest perfect cube factor of 54 is 27, we have

$$\sqrt[3]{54} = \sqrt[3]{27 \cdot 2} \quad \text{Write 54 as } 27 \cdot 2.$$
$$= \sqrt[3]{27}\sqrt[3]{2} \quad \text{By the multiplication property of radicals, the cube root of a}$$
$$\text{product is equal to the product of the cube roots:}$$
$$\sqrt[3]{27 \cdot 2} = \sqrt[3]{27}\sqrt[3]{2}.$$
$$= 3\sqrt[3]{2} \quad \text{Simplify: } \sqrt[3]{27} = 3.$$

We can show that $3\sqrt[3]{2}$ is a cube root of 54 as follows:

$$\left(3\sqrt[3]{2}\right)^3 = (3)^3 \left(\sqrt[3]{2}\right)^3 = 27(2) = 54$$

d. The largest perfect fourth-power factor of 48 is 16. Thus,

$$-\sqrt[4]{48} = -\sqrt[4]{16 \cdot 3} \quad \text{Write 48 as } 16 \cdot 3.$$
$$= -\sqrt[4]{16}\sqrt[4]{3} \quad \text{Apply the multiplication property of radicals.}$$
$$= -2\sqrt[4]{3} \quad \text{Simplify: } \sqrt[4]{16} = 2.$$

Check the result.

SELF CHECK   Simplify   **a.** $\sqrt{20}$,   **b.** $\sqrt[3]{24}$,   and   *Answers:*   **a.** $2\sqrt{5}$,   **b.** $2\sqrt[3]{3}$,   **c.** $\sqrt[5]{-128}$.   **c.** $-2\sqrt[5]{4}$

EXAMPLE 2   **Simplifying radical expressions.** Simplify **a.** $\sqrt{m^9}$, **b.** $\sqrt{128a^5}$, **c.** $\sqrt[3]{24x^5}$, and **d.** $\sqrt[5]{a^9b^5}$. Assume that all variables represent positive real numbers.

Solution   **a.** The largest perfect square factor of m^9 is m^8.

$$\sqrt{m^9} = \sqrt{m^8 \cdot m} \quad \text{Write } m^9 \text{ in factored form as } m^8 \cdot m.$$
$$= \sqrt{m^8}\sqrt{m} \quad \text{Apply the multiplication property of radicals.}$$
$$= m^4\sqrt{m} \quad \text{Simplify: } \sqrt{m^8} = m^4.$$

b. Since the largest perfect square factor of 128 is 64 and the largest perfect square factor of a^5 is a^4, the largest perfect square factor of $128a^5$ is $64a^4$. We write $128a^5$ as $64a^4 \cdot 2a$ and proceed as follows:

$$\sqrt{128a^5} = \sqrt{64a^4 \cdot 2a}$$
$$= \sqrt{64a^4}\sqrt{2a} \quad \text{Use the multiplication property of radicals.}$$
$$= 8a^2\sqrt{2a} \quad \text{Simplify: } \sqrt{64a^4} = 8a^2.$$

c. We write $24x^5$ as $8x^3 \cdot 3x^2$ and proceed as follows:

$$\sqrt[3]{24x^5} = \sqrt[3]{8x^3 \cdot 3x^2} \qquad 8x^3 \text{ is the largest perfect cube factor of } 24x^5.$$
$$= \sqrt[3]{8x^3}\sqrt[3]{3x^2} \qquad \text{Use the multiplication property of radicals.}$$
$$= 2x\sqrt[3]{3x^2} \qquad \text{Simplify: } \sqrt[3]{8x^3} = 2x.$$

d. The largest perfect fifth-power factor of a^9 is a^5, and b^5 is a fifth power.

$$\sqrt[5]{a^9b^5} = \sqrt[5]{a^5b^5 \cdot a^4} \qquad a^5b^5 \text{ is the largest perfect fifth-power factor of } a^9b^5.$$
$$= \sqrt[5]{a^5b^5}\sqrt[5]{a^4} \qquad \text{Use the multiplication property of radicals.}$$
$$= ab\sqrt[5]{a^4} \qquad \text{Simplify: } \sqrt[5]{a^5b^5} = ab.$$

SELF CHECK Simplify **a.** $\sqrt{98b^3}$, **b.** $\sqrt[3]{54y^5}$, and   *Answers:* **a.** $7b\sqrt{2b}$,
c. $\sqrt[4]{t^8u^{15}}$.   **b.** $3y\sqrt[3]{2y^2}$, **c.** $t^2u^3\sqrt[4]{u^3}$ ■

| **E X A M P L E 3** | **Simplifying radical expressions using the division property of radicals.** Simplify **a.** $\sqrt{\dfrac{7}{64}}$, **b.** $\sqrt{\dfrac{15}{49x^2}}$, and **c.** $\sqrt[3]{\dfrac{10x^2}{27y^6}}$. Assume that the variables represent positive real numbers. |

Solution **a.** We can write the square root of the quotient as the quotient of two square roots.

$$\sqrt{\frac{7}{64}} = \frac{\sqrt{7}}{\sqrt{64}} \qquad \text{Apply the division property of radicals.}$$

$$= \frac{\sqrt{7}}{8} \qquad \text{Simplify the denominator: } \sqrt{64} = 8.$$

b. $\sqrt{\dfrac{15}{49x^2}} = \dfrac{\sqrt{15}}{\sqrt{49x^2}} \qquad$ Apply the division property of radicals.

$$= \frac{\sqrt{15}}{7x} \qquad \text{Simplify the denominator: } \sqrt{49x^2} = 7x.$$

c. We can write the cube root of the quotient as the quotient of two cube roots. Since $y \neq 0$, we have

$$\sqrt[3]{\frac{10x^2}{27y^6}} = \frac{\sqrt[3]{10x^2}}{\sqrt[3]{27y^6}} \qquad \text{Use the division property of radicals.}$$

$$= \frac{\sqrt[3]{10x^2}}{3y^2} \qquad \text{Simplify the denominator.}$$

SELF CHECK Simplify **a.** $\sqrt{\dfrac{11}{36a^2}} \ (a > 0)$ and

b. $\sqrt[4]{\dfrac{a^3}{625y^{12}}} \ (a > 0, y > 0)$.   *Answers:* **a.** $\dfrac{\sqrt{11}}{6a}$, **b.** $\dfrac{\sqrt[4]{a^3}}{5y^3}$ ■

| **E X A M P L E 4** | **Simplifying radical expressions.** Simplify each expression. Assume that all variables represent positive numbers. **a.** $\dfrac{\sqrt{45xy^2}}{\sqrt{5x}}$ and **b.** $\dfrac{\sqrt[3]{-432x^5}}{\sqrt[3]{8x}}$. |

Solution   **a.** We can write the quotient of the square roots as the square root of a quotient.

$$\frac{\sqrt{45xy^2}}{\sqrt{5x}} = \sqrt{\frac{45xy^2}{5x}}$$ Use the division property of radicals.

$$= \sqrt{9y^2}$$ Simplify the fraction $\frac{45xy^2}{5x}$.

$$= 3y$$ Simplify the radical.

b. We can write the quotient of the cube roots as the cube root of a quotient.

$$\frac{\sqrt[3]{-432x^5}}{\sqrt[3]{8x}} = \sqrt[3]{\frac{-432x^5}{8x}}$$ Use the division property of radicals.

$$= \sqrt[3]{-54x^4}$$ Simplify the fraction $\frac{-432x^5}{8x}$.

$$= \sqrt[3]{-27x^3 \cdot 2x}$$ $-27x^3$ is the largest perfect cube that divides $-54x^4$.

$$= \sqrt[3]{-27x^3}\sqrt[3]{2x}$$ Use the multiplication property of radicals.

$$= -3x\sqrt[3]{2x}$$ Simplify: $\sqrt[3]{-27x^3} = -3x$.

SELF CHECK   Simplify each expression (assume that all variables represent positive numbers):

a. $\dfrac{\sqrt{50ab^2}}{\sqrt{2a}}$ and   **b.** $\dfrac{\sqrt[3]{-2{,}000x^5v^3}}{\sqrt[3]{2x}}$.

Answers: **a.** $5b$,   **b.** $-10xv\sqrt[3]{x}$ ∎

Adding and Subtracting Radical Expressions

Radical expressions with the same index and the same radicand are called **like** or **similar radicals**. For example, $3\sqrt{2}$ and $2\sqrt{2}$ are like radicals. However,

$3\sqrt{5}$ and $4\sqrt{2}$ are not like radicals, because the radicands are different.

$3\sqrt[4]{5}$ and $2\sqrt[3]{5}$ are not like radicals, because the indexes are different.

For a given expression containing two or more radical terms, we should attempt to combine like radicals, if possible. For example, to simplify the expression $3\sqrt{2} + 2\sqrt{2}$, we use the distributive property to factor out $\sqrt{2}$ and simplify.

$$3\sqrt{2} + 2\sqrt{2} = (3 + 2)\sqrt{2}$$
$$= 5\sqrt{2}$$

Radicals with the same index but different radicands can often be written as like radicals. For example, to simplify the expression $\sqrt{27} - \sqrt{12}$, we simplify both radicals first, and then we combine the like radicals.

$$\sqrt{27} - \sqrt{12} = \sqrt{9 \cdot 3} - \sqrt{4 \cdot 3}$$ Write 27 and 12 in factored form.

$$= \sqrt{9}\sqrt{3} - \sqrt{4}\sqrt{3}$$ Use the multiplication property of radicals.

$$= 3\sqrt{3} - 2\sqrt{3}$$ Simplify $\sqrt{9}$ and $\sqrt{4}$.

$$= (3 - 2)\sqrt{3}$$ Factor out $\sqrt{3}$.

$$= \sqrt{3}$$ $1\sqrt{3} = \sqrt{3}$.

As the previous examples suggest, we can use the following rule to add or subtract radicals.

Adding and subtracting radicals

To add or subtract radicals, simplify each radical expression and combine all like radicals. To add or subtract like radicals, combine the coefficients and keep the common radical.

| EXAMPLE 5 | **Combining like radicals.** Simplify $2\sqrt{12} - 3\sqrt{48} + 3\sqrt{3}$. |

Solution We simplify $2\sqrt{12}$ and $3\sqrt{48}$ separately and then combine like radicals.

$$\begin{aligned}
2\sqrt{12} - 3\sqrt{48} + 3\sqrt{3} &= 2\sqrt{4 \cdot 3} - 3\sqrt{16 \cdot 3} + 3\sqrt{3} \\
&= 2\sqrt{4}\sqrt{3} - 3\sqrt{16}\sqrt{3} + 3\sqrt{3} \\
&= 2(2)\sqrt{3} - 3(4)\sqrt{3} + 3\sqrt{3} \\
&= 4\sqrt{3} - 12\sqrt{3} + 3\sqrt{3} \\
\\
&= (4 - 12 + 3)\sqrt{3} \\
\\
&= -5\sqrt{3}
\end{aligned}$$

All three expressions have the same index and radicand.

Combine the coefficients of these like radicals and keep $\sqrt{3}$.

SELF CHECK Simplify $3\sqrt{75} - 2\sqrt{12} + 2\sqrt{48}$. *Answer:* $19\sqrt{3}$ ∎

| EXAMPLE 6 | **Combining like radicals.** Simplify $\sqrt[3]{16} - \sqrt[3]{54} + \sqrt[3]{24}$. |

Solution We begin by simplifying each radical expression separately:

$$\begin{aligned}
\sqrt[3]{16} - \sqrt[3]{54} + \sqrt[3]{24} &= \sqrt[3]{8 \cdot 2} - \sqrt[3]{27 \cdot 2} + \sqrt[3]{8 \cdot 3} \\
&= \sqrt[3]{8}\sqrt[3]{2} - \sqrt[3]{27}\sqrt[3]{2} + \sqrt[3]{8}\sqrt[3]{3} \\
&= 2\sqrt[3]{2} - 3\sqrt[3]{2} + 2\sqrt[3]{3}
\end{aligned}$$

Now we combine the two radical expressions that have the same index and radicand.

$$\sqrt[3]{16} - \sqrt[3]{54} + \sqrt[3]{24} = -\sqrt[3]{2} + 2\sqrt[3]{3}$$

SELF CHECK Simplify $\sqrt[3]{24} - \sqrt[3]{16} + \sqrt[3]{54}$. *Answer:* $2\sqrt[3]{3} + \sqrt[3]{2}$ ∎

WARNING! Even though the radical expressions $-\sqrt[3]{2}$ and $2\sqrt[3]{3}$ in the last line of Example 6 have the same index, we cannot combine them, because their radicands are different. Neither can we combine radical expressions having the same radicand but a different index. For example, the expression $\sqrt[3]{2} + \sqrt[4]{2}$ cannot be simplified.

| EXAMPLE 7 | **Combining like radicals.** Simplify $\sqrt[3]{16x^4} + \sqrt[3]{54x^4} - \sqrt[3]{-128x^4}$. |

Solution We simplify each radical expression separately, factor out $\sqrt[3]{2x}$, and simplify.

$$\begin{aligned}
&\sqrt[3]{16x^4} + \sqrt[3]{54x^4} - \sqrt[3]{-128x^4} \\
&= \sqrt[3]{8x^3 \cdot 2x} + \sqrt[3]{27x^3 \cdot 2x} - \sqrt[3]{-64x^3 \cdot 2x} \\
&= \sqrt[3]{8x^3}\sqrt[3]{2x} + \sqrt[3]{27x^3}\sqrt[3]{2x} - \sqrt[3]{-64x^3}\sqrt[3]{2x} \\
&= 2x\sqrt[3]{2x} + 3x\sqrt[3]{2x} + 4x\sqrt[3]{2x} \\
\\
&= (2x + 3x + 4x)\sqrt[3]{2x} \\
&= 9x\sqrt[3]{2x}
\end{aligned}$$

All three radicals have the same index and radicand.

In the parentheses, combine like terms.

SELF CHECK Simplify $\sqrt{32x^3} + \sqrt{50x^3} - \sqrt{18x^3}$. *Answer:* $6x\sqrt{2x}$ ∎

STUDY SET

Section 7.4

VOCABULARY

In Exercises 1–4, fill in the blanks to make each statement true.

1. Radical expressions such as $\sqrt[3]{4}$ and $6\sqrt[3]{4}$ with the same index and the same radicand are called _____ radicals.

2. Numbers such as 1, 4, 9, 16, 25, and 36 are called perfect _____. Numbers such as 1, 8, 27, 64, and 125 are called perfect _____.

3. The largest perfect square _____ of 27 is 9.

4. "To _____ $\sqrt{24}$" means to write it as $2\sqrt{6}$.

CONCEPTS

In Exercises 5–6, fill in the blanks to make each statement true.

5. $\sqrt[n]{ab} = $ ▨

   In words, the nth root of the _____ of two numbers is equal to the product of their nth _____.

6. $\sqrt[n]{\dfrac{a}{b}} = $ ▨

   In words, the nth root of the _____ of two numbers is equal to the quotient of their nth _____.

7. Consider the expressions

   $\sqrt{4 \cdot 5}$ and $\sqrt{4}\sqrt{5}$

   Which expression is
   a. the square root of a product?
   b. the product of square roots?
   c. How are these two expressions related?

8. Consider the expressions

   $\dfrac{\sqrt[3]{a}}{\sqrt[3]{x^2}}$ and $\sqrt[3]{\dfrac{a}{x^2}}$

   Which expression is

   a. the cube root of a quotient?

   b. the quotient of cube roots?

   c. How are these two expressions related?

9. **a.** Write two radical expressions that have the same radicand but a different index. Can the expressions be added?
   b. Write two radical expressions that have the same index but a different radicand. Can the expressions be added?

10. Explain the **mistake** in the student's solution shown below.

   Simplify $\sqrt[3]{54}$.
   $$\sqrt[3]{54} = \sqrt[3]{27 + 27}$$
   $$= \sqrt[3]{27} + \sqrt[3]{27}$$
   $$= 3 + 3$$
   $$= 6$$

NOTATION

In Exercises 11–12, complete each solution.

11. Simplify the radical expression.

   $$\sqrt[3]{32k^4} = \sqrt[3]{\boxed{} \cdot 4k}$$
   $$= \sqrt[3]{\boxed{}}\,\sqrt[3]{4k}$$
   $$= 2k\sqrt[3]{4k}$$

12. Simplify the radical expression.

   $$\frac{\sqrt{80s^2t^4}}{\sqrt{5s^2}} = \sqrt{\frac{80s^2t^4}{\boxed{}}}$$
   $$= \sqrt{\boxed{}}$$
   $$= 4t^2$$

PRACTICE

In Exercises 13–28, simplify each expression. Assume that all variables represent positive real numbers.

13. $\sqrt{6}\sqrt{6}$

14. $\sqrt{11}\sqrt{11}$

15. $\sqrt{t}\sqrt{t}$

16. $-\sqrt{z}\sqrt{z}$

17. $\sqrt[3]{5x^2}\sqrt[3]{25x}$

18. $\sqrt[4]{25a}\sqrt[4]{25a^3}$

19. $\dfrac{\sqrt{500}}{\sqrt{5}}$

20. $\dfrac{\sqrt{128}}{\sqrt{2}}$

21. $\dfrac{\sqrt{98x^3}}{\sqrt{2x}}$

22. $\dfrac{\sqrt{75y^5}}{\sqrt{3y}}$

23. $\dfrac{\sqrt{180ab^4}}{\sqrt{5ab^2}}$

24. $\dfrac{\sqrt{112ab^3}}{\sqrt{7ab}}$

25. $\dfrac{\sqrt[3]{48}}{\sqrt[3]{6}}$

26. $\dfrac{\sqrt[3]{64}}{\sqrt[3]{8}}$

27. $\dfrac{\sqrt[3]{189a^4}}{\sqrt[3]{7a}}$

28. $\dfrac{\sqrt[3]{243x^7}}{\sqrt[3]{9x}}$

In Exercises 29–48, simplify each radical expression.

29. $\sqrt{20}$ **30.** $\sqrt{8}$

31. $-\sqrt{200}$ **32.** $-\sqrt{250}$

33. $\sqrt[3]{80}$ **34.** $\sqrt[3]{270}$

35. $\sqrt[3]{-81}$ **36.** $\sqrt[3]{-72}$

37. $\sqrt[4]{32}$ **38.** $\sqrt[4]{48}$

39. $\sqrt[5]{96}$ **40.** $\sqrt[7]{256}$

41. $\sqrt{\dfrac{7}{9}}$ **42.** $\sqrt{\dfrac{3}{4}}$

43. $\sqrt[3]{\dfrac{7}{64}}$ **44.** $\sqrt[3]{\dfrac{4}{125}}$

45. $\sqrt[4]{\dfrac{3}{10,000}}$ **46.** $\sqrt[5]{\dfrac{4}{243}}$

47. $\sqrt[5]{\dfrac{3}{32}}$ **48.** $\sqrt[6]{\dfrac{5}{64}}$

In Exercises 49–68, simplify each radical expression. Assume that all variables represent positive numbers.

49. $\sqrt{50x^2}$ **50.** $\sqrt{75a^2}$

51. $\sqrt{32b}$ **52.** $\sqrt{80c}$

53. $-\sqrt{112a^3}$ **54.** $\sqrt{147a^5}$

55. $\sqrt{175a^2b^3}$ **56.** $\sqrt{128a^3b^5}$

57. $-\sqrt{300xy}$ **58.** $\sqrt{200x^2y}$

59. $\sqrt[3]{-54x^6}$ **60.** $-\sqrt[3]{-81a^3}$

61. $\sqrt[3]{16x^{12}y^3}$ **62.** $\sqrt[3]{40a^3b^6}$

63. $\sqrt[4]{32x^{12}y^4}$ **64.** $\sqrt[5]{64x^{10}y^5}$

65. $\sqrt{\dfrac{z^2}{16x^2}}$ **66.** $\sqrt{\dfrac{b^4}{64a^8}}$

67. $\sqrt[4]{\dfrac{5x}{16z^4}}$ **68.** $\sqrt[3]{\dfrac{11a^2}{125b^6}}$

In Exercises 69–104, simplify and combine like radicals. All variables represent positive numbers.

69. $4\sqrt{2x} + 6\sqrt{2x}$ **70.** $6\sqrt[3]{5y} + 3\sqrt[3]{5y}$

71. $8\sqrt[5]{7a^2} - 7\sqrt[5]{7a^2}$ **72.** $10\sqrt[6]{12xyz} - \sqrt[6]{12xyz}$

73. $\sqrt{2} - \sqrt{8}$ **74.** $\sqrt{20} - \sqrt{125}$

75. $\sqrt{98} - \sqrt{50}$ **76.** $\sqrt{72} - \sqrt{200}$

77. $3\sqrt{24} + \sqrt{54}$ **78.** $\sqrt{18} + 2\sqrt{50}$

79. $\sqrt[3]{24x} + \sqrt[3]{3x}$ **80.** $\sqrt[3]{16y} + \sqrt[3]{128y}$

81. $\sqrt[3]{32} - \sqrt[3]{108}$ **82.** $\sqrt[3]{80} - \sqrt[3]{10,000}$

83. $2\sqrt[3]{125} - 5\sqrt[3]{64}$ **84.** $3\sqrt[3]{27} + 12\sqrt[3]{216}$

85. $14\sqrt[4]{32} - 15\sqrt[4]{162}$ **86.** $23\sqrt[4]{768} + \sqrt[4]{48}$

87. $3\sqrt[4]{512} + 2\sqrt[4]{32}$ **88.** $4\sqrt[4]{243} - \sqrt[4]{48}$

89. $\sqrt{98} - \sqrt{50} - \sqrt{72}$

90. $\sqrt{20} + \sqrt{125} - \sqrt{80}$

91. $\sqrt{18t} + \sqrt{300t} - \sqrt{243t}$

92. $\sqrt{80m} - \sqrt{128m} + \sqrt{288m}$

93. $2\sqrt[3]{16} - \sqrt[3]{54} - 3\sqrt[3]{128}$

94. $\sqrt[4]{48} - \sqrt[4]{243} - \sqrt[4]{768}$

95. $\sqrt{25y^2z} - \sqrt{16y^2z}$ **96.** $\sqrt{25yz^2} + \sqrt{9yz^2}$

97. $\sqrt{36xy^2} + \sqrt{49xy^2}$ **98.** $3\sqrt{2x} - \sqrt{8x}$

99. $2\sqrt[3]{64a} + 2\sqrt[3]{8a}$ **100.** $3\sqrt[4]{x^4y} - 2\sqrt[4]{x^4y}$

101. $\sqrt{y^5} - \sqrt{9y^5} - \sqrt{25y^5}$

102. $\sqrt{8y^7} + \sqrt{32y^7} - \sqrt{2y^7}$

103. $\sqrt[5]{x^6y^2} + \sqrt[5]{32x^6y^2} + \sqrt[5]{x^6y^2}$

104. $\sqrt[3]{xy^4} + \sqrt[3]{8xy^4} - \sqrt[3]{27xy^4}$

APPLICATIONS

In Exercises 105–110, first give the exact answer, expressed as a simplified radical expression. Then give an approximation, rounded to the nearest tenth.

105. UMBRELLA The surface area of a cone is given by the formula $S = \pi r \sqrt{r^2 + h^2}$, where r is the radius of the base and h is its height. Use this formula to find the number of square feet of waterproof cloth used to make the umbrella shown in Illustration 1.

ILLUSTRATION 1

106. STRUCTURAL ENGINEERING Engineers have determined that two additional supports need to be added to strengthen a truss. See Illustration 2. Find the length L of each new support using the formula

$$L = \sqrt{\frac{b^2}{2} + \frac{c^2}{2} - \frac{a^2}{4}}$$

ILLUSTRATION 2

Joins at the midpoint of this segment

107. BLOW DRYER The current I (in amps), the power P (in watts), and the resistance R (in ohms) are related by the formula $I = \sqrt{\frac{P}{R}}$. What current is needed for a 1,200-watt hair dryer if the resistance is 16 ohms?

108. COMMUNICATIONS SATELLITE Engineers have determined that a spherical communications satellite needs to have a capacity of 565.2 cubic feet to house all of its operating systems. The volume V of a sphere is related to its radius r by the formula $r = \sqrt[3]{\frac{3V}{4\pi}}$. What radius must the satellite have to meet the engineer's specification? Use 3.14 for π.

109. DUCTWORK The pattern shown in Illustration 3 is laid out on a sheet of galvanized tin. Then it is cut out with snips and bent to make an air conditioning duct connection. Find the total length of the cut that must be made with the tin snips. (All measurements are in inches.)

ILLUSTRATION 3

110. OUTDOOR COOKING The diameter of a circle is given by the function $d(A) = 2\sqrt{\frac{A}{\pi}}$, where A is the area

of the circle. Find the difference between the diameters of the barbecue grills shown in Illustration 4.

ILLUSTRATION 4

Cooking area
147π in.^{3}

Cooking area
48π in.^{3}

WRITING

111. Explain why $\sqrt[3]{9x^4}$ is not in simplified form.

112. How are the procedures used to simplify $3x + 4x$ and $3\sqrt{x} + 4\sqrt{x}$ similar?

113. See Illustration 5. Explain how the graphs of $Y_1 = 3\sqrt{24x} + \sqrt{54x}$ (on the left) and $Y_2 = 9\sqrt{6x}$ (on the right) can be used to verify the simplification $3\sqrt{24x} + \sqrt{54x} = 9\sqrt{6x}$. In each graph, settings of $[-5, 20]$ for x and $[-5, 100]$ for y were used.

ILLUSTRATION 5

114. Explain how to verify algebraically that $\sqrt{200x^3y^5} = 10xy^2\sqrt{2xy}$

REVIEW

Do each operation.

115. $3x^2y^3(-5x^3y^{-4})$

116. $(2x^2 - 9x - 5) \cdot \dfrac{x}{2x^2 + x}$

117. $2p - 5\overline{)6p^2 - 7p - 25}$

118. $\dfrac{xy}{\dfrac{1}{x} - \dfrac{1}{y}}$

▶ 7.5 Multiplying and Dividing Radical Expressions

In this section, you will learn about

> Multiplying a monomial by a monomial ▪ Multiplying a polynomial by a monomial ▪ Multiplying a polynomial by a polynomial ▪ Rationalizing denominators ▪ Rationalizing binomial denominators ▪ Rationalizing numerators

Introduction In this section, we will discuss the methods used to multiply and divide radical expressions. Solving these problems often requires the use of procedures and properties studied earlier, such as simplifying radical expressions, combining like radicals, the FOIL method, and the distributive property.

Multiplying a Monomial by a Monomial

Radical expressions with the same index can be multiplied. For example, to find the product of $\sqrt{5}$ and $\sqrt{10}$, we proceed as follows:

$$\sqrt{5}\sqrt{10} = \sqrt{5 \cdot 10} \quad \text{By the multiplication property of radicals: } \sqrt[n]{a}\sqrt[n]{b} = \sqrt[n]{ab}.$$
$$= \sqrt{50} \quad \text{Multiply under the radical. Note that } \sqrt{50} \text{ can be simplified.}$$
$$= \sqrt{25 \cdot 2} \quad \text{Begin the process of simplifying } \sqrt{50} \text{ by factoring 50.}$$
$$= 5\sqrt{2} \quad \sqrt{25 \cdot 2} = \sqrt{25}\sqrt{2} = 5\sqrt{2}.$$

EXAMPLE 1 **Multiplying radical expressions.** Multiply $3\sqrt{6}$ by $2\sqrt{3}$.

Solution We use the commutative and associative properties of multiplication to multiply the coefficients and the radicals separately. Then we simplify any radicals in the product, if possible.

$$3\sqrt{6} \cdot 2\sqrt{3} = 3(2)\sqrt{6}\sqrt{3} \quad \text{Multiply the coefficients and multiply the radicals.}$$
$$= 6\sqrt{18} \quad 3(2) = 6 \text{ and } \sqrt{6}\sqrt{3} = \sqrt{18}.$$
$$= 6\sqrt{9}\sqrt{2} \quad \text{Simplify: } \sqrt{18} = \sqrt{9 \cdot 2} = \sqrt{9}\sqrt{2}.$$
$$= 6(3)\sqrt{2} \quad \sqrt{9} = 3.$$
$$= 18\sqrt{2} \quad \text{Multiply.}$$

SELF CHECK Multiply $-2\sqrt{7}$ by $5\sqrt{2}$. *Answer:* $-10\sqrt{14}$ ▪

Multiplying a Polynomial by a Monomial

To multiply a polynomial by a monomial, we use the distributive property to remove parentheses and then simplify each resulting term, if possible.

EXAMPLE 2 **Using the distributive property with radical expressions.**
Simplify $3\sqrt{3}(4\sqrt{8} - 5\sqrt{10})$.

Solution

$$3\sqrt{3}\left(4\sqrt{8} - 5\sqrt{10}\right)$$

$= 3\sqrt{3} \cdot 4\sqrt{8} - 3\sqrt{3} \cdot 5\sqrt{10}$ Use the distributive property.

$= 12\sqrt{24} - 15\sqrt{30}$ Multiply the coefficients and multiply the radicals.

$= 12\sqrt{4}\sqrt{6} - 15\sqrt{30}$ Simplify: $\sqrt{24} = \sqrt{4 \cdot 6} = \sqrt{4}\sqrt{6}$.

$= 12(2)\sqrt{6} - 15\sqrt{30}$ $\sqrt{4} = 2$.

$= 24\sqrt{6} - 15\sqrt{30}$

SELF CHECK   Simplify $4\sqrt{2}\left(3\sqrt{5} - 2\sqrt{8}\right)$.   *Answer:* $12\sqrt{10} - 32$ ■

Multiplying a Polynomial by a Polynomial

To multiply a binomial by a binomial, we use the FOIL method.

EXAMPLE 3   **Using the FOIL method with radical expressions.**
Multiply $\left(\sqrt{7} + \sqrt{2}\right)\left(\sqrt{7} - 3\sqrt{2}\right)$.

Solution

$$\left(\sqrt{7} + \sqrt{2}\right)\left(\sqrt{7} - 3\sqrt{2}\right)$$

$= \sqrt{7}\sqrt{7} - 3\sqrt{7}\sqrt{2} + \sqrt{2}\sqrt{7} - 3\sqrt{2}\sqrt{2}$ Use the FOIL method.

$= 7 - 3\sqrt{14} + \sqrt{14} - 3(2)$ Do each multiplication:
     $\sqrt{7}\sqrt{7} = \sqrt{49} = 7$ and
     $\sqrt{2}\sqrt{2} = \sqrt{4} = 2$.

$= 7 - 2\sqrt{14} - 6$ Combine like radicals:
     $-3\sqrt{14} + \sqrt{14} = -2\sqrt{14}$.

$= 1 - 2\sqrt{14}$ Combine like terms: $7 - 6 = 1$.

SELF CHECK   Multiply $\left(\sqrt{5} + 2\sqrt{3}\right)\left(\sqrt{5} - \sqrt{3}\right)$.   *Answer:* $-1 + \sqrt{15}$ ■

Technically, the expression $\sqrt{3x} - \sqrt{5}$ is not a polynomial, because the variable does not have a whole-number exponent $\left(\sqrt{3x} = 3^{1/2}x^{1/2}\right)$. However, we will multiply such expressions as if they were polynomials.

EXAMPLE 4   **Using the FOIL method with radical expressions.**
Multiply $\left(\sqrt{3x} - \sqrt{5}\right)\left(\sqrt{2x} + \sqrt{10}\right)$. Assume that $x > 0$.

Solution

$$\left(\sqrt{3x} - \sqrt{5}\right)\left(\sqrt{2x} + \sqrt{10}\right)$$

$= \sqrt{3x}\sqrt{2x} + \sqrt{3x}\sqrt{10} - \sqrt{5}\sqrt{2x} - \sqrt{5}\sqrt{10}$ Use FOIL.

$= \sqrt{6x^2} + \sqrt{30x} - \sqrt{10x} - \sqrt{50}$ Do each multiplication.

$= \sqrt{6}\sqrt{x^2} + \sqrt{30x} - \sqrt{10x} - \sqrt{25}\sqrt{2}$ Simplify $\sqrt{6x^2}$ and $\sqrt{50}$.

$= \sqrt{6}x + \sqrt{30x} - \sqrt{10x} - 5\sqrt{2}$

SELF CHECK   Multiply $\left(\sqrt{x} + 1\right)\left(\sqrt{x} - 3\right)$. Assume that $x > 0$.   *Answer:* $x - 2\sqrt{x} - 3$ ■

 WARNING! It is important to draw the radical sign carefully so that it completely covers the radicand, but no more than the radicand. To avoid confusion, we often write an expression such as $\sqrt{6}x$ in the form $x\sqrt{6}$.

Rationalizing Denominators

In Section 7.4, we saw that a radical expression is in simplified form when each of the following statements is true.

1. No radicals appear in the denominator of a fraction.

2. The radicand contains no fractions or negative numbers.

3. Each factor in the radicand appears to a power that is less than the index of the radical.

We now consider radical expressions that do not satisfy requirement 1 and radical expressions that do not satisfy requirement 2 of this list. We will introduce an algebraic technique, called *rationalizing the denominator,* that is used to write such expressions in an equivalent simplified form.

To divide radical expressions, we **rationalize the denominator** of a fraction to replace the denominator with a rational number. For example, to divide $\sqrt{70}$ by $\sqrt{3}$, we write the division as the fraction

$$\frac{\sqrt{70}}{\sqrt{3}}$$ The denominator is the irrational number $\sqrt{3}$. This radical expression is not in simplified form, because a radical appears in the denominator.

To eliminate the radical in the denominator, we multiply the numerator and the denominator by a number that will give a perfect square *under the radical in the denominator.* Because $3 \cdot 3 = 9$ and 9 is a perfect square, $\sqrt{3}$ is such a number.

$$\frac{\sqrt{70}}{\sqrt{3}} = \frac{\sqrt{70} \cdot \sqrt{3}}{\sqrt{3} \cdot \sqrt{3}}$$ Multiply numerator and denominator of the fraction by $\sqrt{3}$.

$$= \frac{\sqrt{210}}{\sqrt{9}}$$ Multiply the radicals in the numerator and in the denominator.

$$= \frac{\sqrt{210}}{3}$$ Simplify: $\sqrt{9} = 3$. The denominator is now the rational number 3.

Since there is no radical in the denominator and $\sqrt{210}$ cannot be simplified, the expression $\frac{\sqrt{210}}{3}$ is in simplest form, and the division is complete.

EXAMPLE 5

Division of radical expressions. Simplify by rationalizing the denominator:

a. $\sqrt{\dfrac{20}{7}}$ and **b.** $\dfrac{4}{\sqrt[3]{2}}$.

Solution **a.** This radical expression is not in simplified form, because the radicand contains a fraction. We begin by writing the square root of the quotient as the quotient of two square roots:

$$\sqrt{\frac{20}{7}} = \frac{\sqrt{20}}{\sqrt{7}}$$ Apply the division property of radicals: $\sqrt[n]{\dfrac{a}{b}} = \dfrac{\sqrt[n]{a}}{\sqrt[n]{b}}$.

To rationalize the denominator, we proceed as follows:

$$\frac{\sqrt{20}}{\sqrt{7}} = \frac{\sqrt{20} \cdot \sqrt{7}}{\sqrt{7} \cdot \sqrt{7}}$$ Multiply the numerator and denominator by $\sqrt{7}$.

$$= \frac{\sqrt{140}}{\sqrt{49}}$$ Multiply the radicals.

$$= \frac{2\sqrt{35}}{7}$$ Simplify: $\sqrt{140} = \sqrt{4 \cdot 35} = \sqrt{4}\sqrt{35} = 2\sqrt{35}$ and $\sqrt{49} = 7$.

b. Here, we must rationalize a denominator that is a cube root. We multiply the numerator and the denominator by a number that will give a perfect cube under the radical sign. Since $2 \cdot 4 = 8$ is a perfect cube, $\sqrt[3]{4}$ is such a number.

$$\frac{4}{\sqrt[3]{2}} = \frac{4 \cdot \sqrt[3]{4}}{\sqrt[3]{2} \cdot \sqrt[3]{4}} \qquad \text{Multiply numerator and denominator by } \sqrt[3]{4}.$$

$$= \frac{4\sqrt[3]{4}}{\sqrt[3]{8}} \qquad \text{Multiply the radicals in the denominator.}$$

$$= \frac{4\sqrt[3]{4}}{2} \qquad \text{Simplify: } \sqrt[3]{8} = 2.$$

$$= 2\sqrt[3]{4} \qquad \text{Simplify: } \frac{4\sqrt[3]{4}}{2} = \frac{\overset{1}{\cancel{2}} \cdot 2\sqrt[3]{4}}{\underset{1}{\cancel{2}}} = 2\sqrt[3]{4}.$$

SELF CHECK   Rationalize the denominator:

a. $\sqrt{\dfrac{24}{5}}$ and   **b.** $\dfrac{5}{\sqrt[4]{3}}$.   *Answers:*   **a.** $\dfrac{2\sqrt{30}}{5}$,   **b.** $\dfrac{5\sqrt[4]{27}}{3}$ ∎

E X A M P L E   6   **Photography** Many camera lenses (see Figure 7-12) have an adjustable opening called the **aperture,** which controls the amount of light passing through the lens. The *f*-number of a lens is its **focal length** divided by the diameter of its circular aperture:

$$f\text{-number} = \frac{f}{d} \qquad f \text{ is the focal length, and } d \text{ is the diameter of the aperture.}$$

A lens with a focal length of 12 centimeters and an aperture with a diameter of 6 centimeters has an *f*-number of $\frac{12}{6}$ and is an *f*/2 lens. If the area of the aperture is reduced to admit half as much light, the *f*-number of the lens will change. Find the new *f*-number.

Solution   We first find the area of the aperture when its diameter is 6 centimeters.

FIGURE 7-12

$A = \pi r^2$ The formula for the area of a circle.

$A = \pi(3)^2$ Since a radius is half the diameter, substitute 3 for r.

$A = 9\pi$

When the size of the aperture is reduced to admit half as much light, the area of the aperture will be $\frac{9\pi}{2}$ square centimeters. To find the diameter d of a circle with this area, we proceed as follows:

$A = \pi r^2$ The formula for the area of a circle.

$\dfrac{9\pi}{2} = \pi\left(\dfrac{d}{2}\right)^2$ Substitute $\dfrac{9\pi}{2}$ for A and $\dfrac{d}{2}$ for r.

$\dfrac{9\pi}{2} = \dfrac{\pi d^2}{4}$ $\left(\dfrac{d}{2}\right)^2 = \dfrac{d^2}{4}$.

$18 = d^2$ Multiply both sides by 4 and divide both sides by π.

$d = 3\sqrt{2}$ Take the positive square root of both sides. Then simplify: $\sqrt{18} = \sqrt{9}\sqrt{2} = 3\sqrt{2}$.

Since the focal length of the lens is still 12 centimeters and the diameter is now $3\sqrt{2}$ centimeters, the new f-number of the lens is

$$f\text{-number} = \frac{f}{d} = \frac{12}{3\sqrt{2}} \qquad \text{Substitute 12 for } f \text{ and } 3\sqrt{2} \text{ for } d.$$

$$= \frac{12\sqrt{2}}{3\sqrt{2}\sqrt{2}} \qquad \text{Rationalize the denominator.}$$

$$= \frac{12\sqrt{2}}{3(2)} \qquad \text{Simplify the denominator: } \sqrt{2}\sqrt{2} = 2.$$

$$= 2\sqrt{2} \qquad \text{Simplify: } \frac{12\sqrt{2}}{3(2)} = \frac{12\sqrt{2}}{6} = \frac{\overset{1}{\cancel{6}} \cdot 2\sqrt{2}}{\underset{1}{\cancel{6}}}.$$

$$\approx 2.828427125 \qquad \text{Use a calculator.}$$

The lens is now an $f/2.8$ lens. ∎

EXAMPLE 7

Rationalizing a variable denominator. Rationalize the denominator of $\dfrac{\sqrt{5xy^2}}{\sqrt{xy^3}}$ (x and y are positive numbers).

Solution In each case, we write the expression as a quotient of radicals and then simplify the radicand.

Method 1

$$\frac{\sqrt{5xy^2}}{\sqrt{xy^3}} = \sqrt{\frac{5xy^2}{xy^3}}$$

$$= \sqrt{\frac{5}{y}}$$

$$= \frac{\sqrt{5}}{\sqrt{y}}$$

$$= \frac{\sqrt{5}\sqrt{y}}{\sqrt{y}\sqrt{y}}$$

$$= \frac{\sqrt{5y}}{y}$$

Method 2

$$\frac{\sqrt{5xy^2}}{\sqrt{xy^3}} = \sqrt{\frac{5xy^2}{xy^3}}$$

$$= \sqrt{\frac{5}{y}}$$

$$= \sqrt{\frac{5 \cdot y}{y \cdot y}}$$

$$= \frac{\sqrt{5y}}{\sqrt{y^2}}$$

$$= \frac{\sqrt{5y}}{y}$$

SELF CHECK Rationalize the denominator of $\dfrac{\sqrt{4ab^3}}{\sqrt{2a^2b^2}}$. *Answer:* $\dfrac{\sqrt{2ab}}{a}$ ∎

EXAMPLE 8

Simplifying before rationalizing. Rationalize the denominator of $\sqrt{\dfrac{11}{20q^5}}$ ($q > 0$).

Solution   We write the expression as a quotient of two radicals. Then we simplify the radical in the denominator before rationalizing.

$$\sqrt{\frac{11}{20q^5}} = \frac{\sqrt{11}}{\sqrt{20q^5}}$$ The square root of a quotient is the quotient of the square roots.

$$= \frac{\sqrt{11}}{\sqrt{4q^4 \cdot 5q}}$$ To simplify $\sqrt{20q^5}$, write it as $\sqrt{4q^4 \cdot 5q}$.

$$= \frac{\sqrt{11}}{2q^2\sqrt{5q}}$$ $\sqrt{4q^4 \cdot 5q} = \sqrt{4q^4}\sqrt{5q} = 2q^2\sqrt{5q}$.

$$= \frac{\sqrt{11}\sqrt{5q}}{2q^2\sqrt{5q}\sqrt{5q}}$$ Rationalize the denominator.

$$= \frac{\sqrt{55q}}{2q^2(5q)}$$ Multiply the radicals: $\sqrt{5q}\sqrt{5q} = 5q$.

$$= \frac{\sqrt{55q}}{10q^3}$$ Multiply in the denominator.

SELF CHECK   Rationalize the denominator of $\sqrt[3]{\dfrac{1}{16h^4}}$.   *Answer:* $\dfrac{\sqrt[3]{4h^2}}{4h^2}$ ∎

EXAMPLE 9

Rationalizing a variable denominator. Rationalize the denominator of $\dfrac{\sqrt[3]{5}}{\sqrt[3]{9m}}$.

Solution   We multiply the numerator and the denominator by $\sqrt[3]{3m^2}$, which will produce a perfect cube, $27m^3$, under the radical sign in the denominator.

$$\frac{\sqrt[3]{5}}{\sqrt[3]{9m}} = \frac{\sqrt[3]{5}\sqrt[3]{3m^2}}{\sqrt[3]{9m}\sqrt[3]{3m^2}}$$ Multiply numerator and denominator by $\sqrt[3]{3m^2}$.

$$= \frac{\sqrt[3]{15m^2}}{\sqrt[3]{27m^3}}$$ Multiply the radicals.

$$= \frac{\sqrt[3]{15m^2}}{3m}$$ Simplify: $\sqrt[3]{27m^3} = 3m$.

SELF CHECK   Rationalize the denominator of $\dfrac{\sqrt{5}}{\sqrt{17b}}$.   *Answer:* $\dfrac{\sqrt{85b}}{17b}$ ∎

Rationalizing Binomial Denominators

To rationalize a denominator that contains a binomial expression involving square roots, we multiply its numerator and denominator by the *conjugate* of its denominator. **Conjugate binomials** are binomials with the same terms but with opposite signs between their terms.

Conjugate binomials

The **conjugate** of the binomial $a + b$ is $a - b$, and the conjugate of $a - b$ is $a + b$.

When we multiply a binomial expression involving square roots and its conjugate, the product does not contain a radical. For example, if we use the FOIL method to multiply $\sqrt{x} + 2$ and its conjugate, $\sqrt{x} - 2$, we have

$$\left(\sqrt{x} + 2\right)\left(\sqrt{x} - 2\right) = \sqrt{x}\sqrt{x} - 2\sqrt{x} + 2\sqrt{x} - 4$$
$$= x - 4 \qquad \text{Combine like radicals:} \\ -2\sqrt{x} + 2\sqrt{x} = 0.$$

EXAMPLE 10

Rationalizing binomial denominators. Rationalize the denominator of $\dfrac{1}{\sqrt{2} + 1}$.

Solution We multiply the numerator and denominator of the fraction by $\sqrt{2} - 1$, which is the conjugate of the denominator.

$$\frac{1}{\sqrt{2} + 1} = \frac{1\left(\sqrt{2} - 1\right)}{\left(\sqrt{2} + 1\right)\left(\sqrt{2} - 1\right)}$$

$$= \frac{\sqrt{2} - 1}{2 - 1} \qquad \begin{array}{l} \text{In the denominator, multiply the binomials:} \\ \left(\sqrt{2} + 1\right)\left(\sqrt{2} - 1\right) = \sqrt{2}\sqrt{2} - \sqrt{2} + \sqrt{2} - 1 = \\ 2 - 1. \end{array}$$

$$= \sqrt{2} - 1 \qquad \text{Simplify: } \dfrac{\sqrt{2} - 1}{2 - 1} = \dfrac{\sqrt{2} - 1}{1} = \sqrt{2} - 1.$$

SELF CHECK Rationalize the denominator of $\dfrac{2}{\sqrt{3} + 1}$.   *Answer:* $\sqrt{3} - 1$ ∎

EXAMPLE 11

Rationalizing binomial denominators. Rationalize the denominator of $\dfrac{\sqrt{x} + \sqrt{2}}{\sqrt{x} - \sqrt{2}}$ $(x > 0)$.

Solution We multiply the numerator and denominator by $\sqrt{x} + \sqrt{2}$, which is the conjugate of $\sqrt{x} - \sqrt{2}$, and simplify.

$$\frac{\sqrt{x} + \sqrt{2}}{\sqrt{x} - \sqrt{2}} = \frac{\left(\sqrt{x} + \sqrt{2}\right)\left(\sqrt{x} + \sqrt{2}\right)}{\left(\sqrt{x} - \sqrt{2}\right)\left(\sqrt{x} + \sqrt{2}\right)}$$

$$= \frac{x + \sqrt{2x} + \sqrt{2x} + 2}{x - 2} \qquad \begin{array}{l}\text{In the numerator and denominator, use the} \\ \text{FOIL method.}\end{array}$$

$$= \frac{x + 2\sqrt{2x} + 2}{x - 2} \qquad \text{In the numerator, combine like radicals.}$$

SELF CHECK Rationalize the denominator of $\dfrac{\sqrt{x} - \sqrt{2}}{\sqrt{x} + \sqrt{2}}$.   *Answer:* $\dfrac{x - 2\sqrt{2x} + 2}{x - 2}$ ∎

Rationalizing Numerators

In calculus, we sometimes have to rationalize a numerator by multiplying the numerator and denominator of the fraction by the conjugate of the numerator.

EXAMPLE 12 **Rationalizing numerators.** Rationalize the numerator of $\dfrac{\sqrt{x}-3}{\sqrt{x}}$ $(x > 0)$.

Solution We multiply the numerator and denominator by $\sqrt{x} + 3$, which is the conjugate of the numerator.

$$\frac{\sqrt{x}-3}{\sqrt{x}} = \frac{(\sqrt{x}-3)(\sqrt{x}+3)}{\sqrt{x}(\sqrt{x}+3)}$$

$$= \frac{x + 3\sqrt{x} - 3\sqrt{x} - 9}{x + 3\sqrt{x}}$$

$$= \frac{x - 9}{x + 3\sqrt{x}}$$

The final expression is not in simplified form. However, this nonsimplified form is often desirable in calculus.

SELF CHECK Rationalize the numerator of $\dfrac{\sqrt{x}+3}{\sqrt{x}}$. *Answer:* $\dfrac{x-9}{x-3\sqrt{x}}$ ∎

STUDY SET

Section 7.5

VOCABULARY

In Exercises 1–6, fill in the blanks to make each statement true.

1. To multiply $(\sqrt{3} + \sqrt{2})(\sqrt{3} - 2\sqrt{2})$, we can use the _____ method.

2. To multiply $2\sqrt{5}(3\sqrt{8} + \sqrt{3})$, use the _____ property to remove parentheses.

3. The denominator of the fraction $\dfrac{4}{\sqrt{5}}$ is an _____ number.

4. The _____ of $\sqrt{x} + 1$ is $\sqrt{x} - 1$.

5. To obtain a _____ cube under the radical in the denominator of $\dfrac{\sqrt[3]{7}}{\sqrt[3]{5n}}$, we multiply the numerator and denominator by $\sqrt[3]{25n^2}$.

6. To _____ the denominator of $\dfrac{4}{\sqrt{5}}$, we multiply the numerator and denominator by $\sqrt{5}$.

CONCEPTS

7. Do each operation, if possible.
   a. $4\sqrt{6} + 2\sqrt{6}$
   b. $4\sqrt{6}(2\sqrt{6})$
   c. $3\sqrt{2} - 2\sqrt{3}$
   d. $3\sqrt{2}(-2\sqrt{3})$

8. Do each operation, if possible.
   a. $5 + 6\sqrt[3]{6}$
   b. $5(6\sqrt[3]{6})$
   c. $\dfrac{30\sqrt[3]{15}}{5}$
   d. $\dfrac{\sqrt[3]{15}}{5}$

9. Consider $\dfrac{\sqrt{3}}{\sqrt{7}} = \dfrac{\sqrt{3}\sqrt{7}}{\sqrt{7}\sqrt{7}}$. Explain why the expressions on the left-hand side and the right-hand side of the equation are equal.

10. To rationalize the denominator of $\dfrac{\sqrt[4]{2}}{\sqrt[4]{3}}$, why wouldn't we multiply the numerator and denominator by $\dfrac{\sqrt[4]{3}}{\sqrt[4]{3}}$?

11. Explain why $\dfrac{\sqrt[3]{12}}{\sqrt[3]{5}}$ is not in simplified form.

12. Explain why $\sqrt{\dfrac{3a}{11k}}$ is not in simplified form.

NOTATION

In Exercises 13–14, fill in the blanks to make each statement true.

13. Multiply $5\sqrt{8} \cdot 7\sqrt{6}$.
$$5\sqrt{8} \cdot 7\sqrt{6} = 5(7)\sqrt{8\boxed{}}$$
$$= 35\sqrt{\boxed{}}$$
$$= 35\sqrt{\boxed{} \cdot 3}$$
$$= 35(\boxed{})\sqrt{3}$$
$$= 140\sqrt{3}$$

14. Rationalize the denominator of $\dfrac{9}{\sqrt[3]{4a^2}}$.
$$\frac{9}{\sqrt[3]{4a^2}} = \frac{9 \cdot \sqrt[3]{2a}}{\sqrt[3]{4a^2} \cdot \boxed{}}$$
$$= \frac{9\sqrt[3]{2a}}{\sqrt[3]{\boxed{}}}$$
$$= \frac{9\sqrt[3]{2a}}{2a}$$

PRACTICE

In Exercises 15–42, do each multiplication and simplify, if possible. All variables represent positive real numbers.

15. $\sqrt{11}\sqrt{11}$ **16.** $\sqrt{35}\sqrt{35}$

17. $(\sqrt{7})^2$ **18.** $(\sqrt{23})^2$

19. $\sqrt{2}\sqrt{8}$ **20.** $\sqrt{3}\sqrt{27}$

21. $\sqrt{5}\sqrt{10}$ **22.** $\sqrt{7}\sqrt{35}$

23. $2\sqrt{3}\sqrt{6}$ **24.** $-3\sqrt{11}\sqrt{33}$

25. $\sqrt[3]{5}\sqrt[3]{25}$ **26.** $-\sqrt[3]{7}\sqrt[3]{49}$

27. $(3\sqrt{2})^2$ **28.** $(2\sqrt{5})^2$

29. $(-2\sqrt{2})^2$ **30.** $(-3\sqrt{10})^2$

31. $(3\sqrt[3]{9})(2\sqrt[3]{3})$ **32.** $(2\sqrt[3]{16})(-\sqrt[3]{4})$

33. $\sqrt[3]{2}\sqrt[3]{12}$ **34.** $\sqrt[3]{3}\sqrt[3]{18}$

35. $\sqrt{ab^3}\sqrt{ab}$ **36.** $\sqrt{8x}\sqrt{2x^3y}$

37. $\sqrt{5ab}\sqrt{5a}$ **38.** $\sqrt{15rs^2}\sqrt{10r}$

39. $-4\sqrt[3]{5r^2s}(5\sqrt[3]{2r})$ **40.** $-\sqrt[3]{3xy^2}(-\sqrt[3]{9x^3})$

41. $\sqrt{x(x+3)}\sqrt{x^3(x+3)}$ **42.** $\sqrt{y^2(x+y)}\sqrt{(x+y)^3}$

In Exercises 43–60, do each multiplication and simplify. All variables represent positive real numbers.

43. $3\sqrt{5}(4 - \sqrt{5})$ **44.** $2\sqrt{7}(3\sqrt{7} - 1)$

45. $3\sqrt{2}(4\sqrt{6} + 2\sqrt{7})$ **46.** $-\sqrt{3}(\sqrt{7} - \sqrt{15})$

47. $-2\sqrt{5x}(4\sqrt{2x} - 3\sqrt{3})$

48. $3\sqrt{7t}(2\sqrt{7t} + 3\sqrt{3t^2})$

49. $(\sqrt{2} + 1)(\sqrt{2} - 3)$

50. $(2\sqrt{3} + 1)(\sqrt{3} - 1)$

51. $(\sqrt{5z} + \sqrt{3})(\sqrt{5z} + \sqrt{3})$

52. $(\sqrt{3p} - \sqrt{2})(\sqrt{3p} + \sqrt{2})$

53. $(\sqrt{3x} - \sqrt{2y})(\sqrt{3x} + \sqrt{2y})$

54. $(\sqrt{3m} + \sqrt{2n})(\sqrt{3m} + \sqrt{2n})$

55. $(2\sqrt{3a} - \sqrt{b})(\sqrt{3a} + 3\sqrt{b})$

56. $(5\sqrt{p} - \sqrt{3q})(\sqrt{p} + 2\sqrt{3q})$

57. $(3\sqrt{2r} - 2)^2$ **58.** $(2\sqrt{3t} + 5)^2$

59. $-2(\sqrt{3x} + \sqrt{3})^2$ **60.** $3(\sqrt{5x} - \sqrt{3})^2$

In Exercises 61–84, simplify each radical expression by rationalizing the denominator. All variables represent positive real numbers.

61. $\sqrt{\dfrac{1}{7}}$ **62.** $\sqrt{\dfrac{5}{3}}$

63. $\dfrac{6}{\sqrt{30}}$ **64.** $\dfrac{8}{\sqrt{10}}$

65. $\dfrac{\sqrt{5}}{\sqrt{8}}$ **66.** $\dfrac{\sqrt{3}}{\sqrt{50}}$

67. $\dfrac{\sqrt{8}}{\sqrt{2}}$ **68.** $\dfrac{\sqrt{27}}{\sqrt{3}}$

69. $\dfrac{1}{\sqrt[3]{2}}$ **70.** $\dfrac{2}{\sqrt[3]{6}}$

71. $\dfrac{3}{\sqrt[3]{9}}$ **72.** $\dfrac{2}{\sqrt[3]{a}}$

73. $\dfrac{\sqrt[3]{2}}{\sqrt[3]{9}}$ **74.** $\dfrac{\sqrt[3]{9}}{\sqrt[3]{54}}$

75. $\dfrac{\sqrt{8}}{\sqrt{xy}}$ **76.** $\dfrac{\sqrt{9xy}}{\sqrt{3x^2y}}$

77. $\dfrac{\sqrt{10xy^2}}{\sqrt{2xy^3}}$ **78.** $\dfrac{\sqrt{5ab^2c}}{\sqrt{10abc}}$

79. $\dfrac{\sqrt[3]{4a^2}}{\sqrt[3]{2ab}}$ **80.** $\dfrac{\sqrt[3]{9x}}{\sqrt[3]{3xy}}$

81. $\dfrac{1}{\sqrt[4]{4}}$ **82.** $\dfrac{1}{\sqrt[5]{2}}$

83. $\dfrac{1}{\sqrt[5]{16}}$ **84.** $\dfrac{4}{\sqrt[4]{32}}$

In Exercises 85–98, rationalize each denominator. All variables represent positive real numbers.

85. $\dfrac{1}{\sqrt{2}-1}$

86. $\dfrac{3}{\sqrt{3}-1}$

87. $\dfrac{\sqrt{2}}{\sqrt{5}+3}$

88. $\dfrac{\sqrt{3}}{\sqrt{3}-2}$

89. $\dfrac{\sqrt{3}+1}{\sqrt{3}-1}$

90. $\dfrac{\sqrt{2}-1}{\sqrt{2}+1}$

91. $\dfrac{\sqrt{7}-\sqrt{2}}{\sqrt{2}+\sqrt{7}}$

92. $\dfrac{\sqrt{3}+\sqrt{2}}{\sqrt{3}-\sqrt{2}}$

93. $\dfrac{2}{\sqrt{x}+1}$

94. $\dfrac{3}{\sqrt{x}-2}$

95. $\dfrac{2z-1}{\sqrt{2z}-1}$

96. $\dfrac{3t-1}{\sqrt{3t}+1}$

97. $\dfrac{\sqrt{x}-\sqrt{y}}{\sqrt{x}+\sqrt{y}}$

98. $\dfrac{\sqrt{x}+\sqrt{y}}{\sqrt{x}-\sqrt{y}}$

In Exercises 99–104, rationalize each numerator. All variables represent positive numbers.

99. $\dfrac{\sqrt{3}+1}{2}$

100. $\dfrac{\sqrt{5}-1}{2}$

101. $\dfrac{\sqrt{x}+3}{x}$

102. $\dfrac{2+\sqrt{x}}{5x}$

103. $\dfrac{\sqrt{x}+\sqrt{y}}{\sqrt{x}}$

104. $\dfrac{\sqrt{x}-\sqrt{y}}{\sqrt{x}+\sqrt{y}}$

APPLICATIONS

105. STATISTICS An example of a normal distribution curve, or *bell-shaped* curve, is shown in Illustration 1. A fraction that is part of the equation that models this curve is

$$\dfrac{1}{\sigma\sqrt{2\pi}}$$

where σ is a letter from the Greek alphabet. Rationalize the denominator of the fraction.

106. ANALYTIC GEOMETRY See Illustration 2. The length of the perpendicular segment drawn from $(-2, 2)$ to the line with equation $2x - 4y = 4$ is given by

$$L = \dfrac{|2(-2) + (-4)(2) + (-4)|}{\sqrt{(2)^2 + (-4)^2}}$$

Find L. Express the result in simplified radical form. Then give an approximation to the nearest tenth.

ILLUSTRATION 2

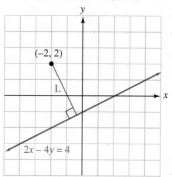

107. TRIGONOMETRY In trigonometry, we must often find the ratio of the lengths of two sides of right triangles. Use the information in Illustration 3 to find the ratio

$$\dfrac{\text{length of side } AC}{\text{length of side } AB}$$

Write the result in simplified radical form.

ILLUSTRATION 3

108. MECHANICAL ENGINEERING A measure of how fast the block shown in Illustration 4 will oscillate when the system is set in motion is given by the formula

$$\omega = \sqrt{\dfrac{k_1 + k_2}{m}}$$

where k_1 and k_2 indicate the stiffness of the springs and m is the mass of the block. Rationalize the right-hand side and restate the formula.

ILLUSTRATION 1

ILLUSTRATION 4

109. PHOTOGRAPHY In Example 6, we saw that a lens with a focal length of 12 centimeters and an aperture $3\sqrt{2}$ centimeters in diameter is an $f/2.8$ lens. Find the f-number if the area of the aperture is again cut in half.

110. ELECTRONICS A formula that is used when designing AC (alternating current) circuits is

$$f_0 = \frac{1}{2\pi} \sqrt{\frac{1}{LC}}$$

Rationalize the right-hand side and restate the formula.

WRITING

111. Explain why $\sqrt{m} \cdot \sqrt{m} = m$ but $\sqrt[3]{m} \cdot \sqrt[3]{m} \neq m$. (Assume that $m > 0$.)

112. Explain why the product of $\sqrt{m} + 3$ and $\sqrt{m} - 3$ does not contain a radical.

REVIEW

Solve the equation.

113. $\dfrac{2}{3 - a} = 1$

114. $5(s - 4) = -5(s - 4)$

115. $\dfrac{8}{b - 2} + \dfrac{3}{2 - b} = -\dfrac{1}{b}$

116. $\dfrac{2}{x - 2} + \dfrac{1}{x + 1} = \dfrac{1}{(x + 1)(x - 2)}$

▶ 7.6   Geometric Applications of Radicals

In this section, you will learn about

> The Pythagorean theorem ■ 45°–45°–90° triangles ■ 30°–60°–90° triangles ■ The distance formula

Introduction   In this section, we will consider several applications of square roots in geometry. Then we will find the distance between two points on a rectangular coordinate system, using a formula that contains a square root. We begin by considering an important theorem (mathematical statement) about right triangles.

The Pythagorean Theorem

If we know the lengths of two legs of a right triangle, we can always find the length of the **hypotenuse** (the side opposite the 90° angle) by using the **Pythagorean theorem.**

Pythagorean theorem

If a and b are the lengths of two legs of a right triangle and c is the length of the hypotenuse, then

$$a^2 + b^2 = c^2$$

In words, the Pythagorean theorem is expressed as follows:

In any right triangle, the square of the hypotenuse is equal to the sum of the squares of the two legs.

FIGURE 7-13

Suppose the right triangle shown in Figure 7-13 has legs of length 3 and 4 units. To find the length of the hypotenuse, we use the Pythagorean theorem.

$$a^2 + b^2 = c^2$$

$$3^2 + 4^2 = c^2 \quad \text{Substitute 3 for } a \text{ and 4 for } b.$$

$$9 + 16 = c^2$$

$$25 = c^2$$

To solve for c, we need to "undo" the operation performed on it. Since c is squared, we take the positive square root of both sides of the equation.

$$\sqrt{25} = \sqrt{c^2} \quad \text{Since } c \text{ represents the length of a side of a triangle, } c > 0. \text{ So } \sqrt{c^2} = c,$$
$$\text{because } c \cdot c = c^2.$$

$$5 = c$$

The length of the hypotenuse is 5 units.

EXAMPLE 1

Firefighting. To fight a forest fire, the forestry department plans to clear a rectangular fire break around the fire, as shown in Figure 7-14. Crews are equipped with mobile communications that have a 3,000-yard range. Can crews at points A and B remain in radio contact?

Solution　Points A, B, and C form a right triangle. To find the distance c from point A to point B, we can use the Pythagorean theorem, substituting 2,400 for a and 1,000 for b and solving for c.

FIGURE 7-14

A ... 1,000 yd ... *c* yd ... *C* ... 2,400 yd ... *B*

$$a^2 + b^2 = c^2$$

$$2{,}400^2 + 1{,}000^2 = c^2$$

$$5{,}760{,}000 + 1{,}000{,}000 = c^2$$

$$6{,}760{,}000 = c^2$$

$$\sqrt{6{,}760{,}000} = \sqrt{c^2} \quad \text{Take the positive square root of both sides.}$$

$$2{,}600 = c \quad \text{Use a calculator to find the square root.}$$

The two crews are 2,600 yards apart. Because this distance is less than the range of the radios, they can communicate by radio.

SELF CHECK　In Example 1, can the crews communicate if $b = 1{,}500$ yards?　　*Answer:* yes ∎

45°–45°–90° Triangles

An **isosceles right triangle** is a right triangle with two legs of equal length. Isosceles right triangles have angle measures of 45°, 45°, and 90°. If we know the length of one leg of an isosceles right triangle, we can use the Pythagorean theorem to find the length of the hypotenuse. Since the triangle shown in Figure 7-15 is a right triangle, we have

FIGURE 7-15

$$c^2 = a^2 + b^2$$

$$c^2 = a^2 + a^2 \quad \text{Both legs are } a \text{ units long, so replace } b \text{ with } a.$$

$$c^2 = 2a^2 \quad \text{Combine like terms.}$$

$$c = \sqrt{2a^2} \quad \text{Take the positive square root of both sides.}$$

$$c = a\sqrt{2} \quad \text{Simplify the radical: } \sqrt{2a^2} = \sqrt{2}\sqrt{a^2} = \sqrt{2}a = a\sqrt{2}.$$

Thus, *in an isosceles right triangle, the length of the hypotenuse is the length of one leg times* $\sqrt{2}$.

EXAMPLE 2

45°–45°–90° triangles. If one leg of the isosceles right triangle shown in Figure 7-15 is 10 feet long, find the length of the hypotenuse.

Solution   Since the length of the hypotenuse is the length of a leg times $\sqrt{2}$, we have

$$c = 10\sqrt{2}$$

The length of the hypotenuse is $10\sqrt{2}$ units. To two decimal places, the length is 14.14 units.

SELF CHECK   Find the length of the hypotenuse of an isosceles right triangle if one leg is 12 meters long.   *Answer:* $12\sqrt{2}$ m   ■

If the length of the hypotenuse of an isosceles right triangle is known, we can use the Pythagorean theorem to find the length of each leg.

EXAMPLE 3

Find the exact length of each leg of the isosceles right triangle shown in Figure 7-16.

Solution   We use the Pythagorean theorem.

FIGURE 7-16

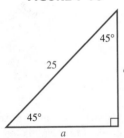

$$c^2 = a^2 + b^2$$

$$25^2 = a^2 + a^2$$ Since both legs are a units long, substitute a for b. The hypotenuse is 25 units long. Substitute 25 for c.

$$25^2 = 2a^2$$ Combine like terms.

$$\frac{625}{2} = a^2$$ Square 25 and divide both sides by 2.

$$\sqrt{\frac{625}{2}} = a$$ To solve for a, take the positive square root of both sides: $\sqrt{a^2} = a$.

$$\frac{\sqrt{625} \cdot \sqrt{2}}{\sqrt{2} \cdot \sqrt{2}} = a$$ Write $\sqrt{\frac{625}{2}}$ as $\frac{\sqrt{625}}{\sqrt{2}}$. Then rationalize the denominator.

$$\frac{25\sqrt{2}}{2} = a$$ In the numerator, simplify the radical: $\sqrt{625} = 25$. In the denominator, do the multiplication: $\sqrt{2} \cdot \sqrt{2} = 2$.

The exact length of each leg is $\dfrac{25\sqrt{2}}{2}$ units. To two decimal places, the length is 17.68 units.   ■

30°–60°–90° Triangles

From geometry, we know that an **equilateral triangle** is a triangle with three sides of equal length and three 60° angles. Each side of the equilateral triangle in Figure 7-17 is $2a$ units long. If an **altitude** is drawn to its base, the altitude bisects the base and divides the equilateral triangle into two 30°–60°–90° triangles. We can see that the shorter leg of each 30°–60°–90° triangle (the side *opposite* the 30° angle) is a units long. Thus,

The length of the shorter leg of a 30°–60°–90° right triangle is half as long as the hypotenuse.

FIGURE 7-17

We can discover another important relationship between the legs of a 30°–60°–90° triangle if we find the length of the altitude h in Figure 7-17. We begin by applying the Pythagorean theorem to one of the 30°–60°–90° triangles.

$$a^2 + b^2 = c^2$$

$$a^2 + h^2 = (2a)^2 \qquad \text{One leg is } h \text{ units long, so replace } b \text{ with } h. \text{ The hypotenuse is } 2a \text{ units long, so replace } c \text{ with } 2a.$$

$$a^2 + h^2 = 4a^2 \qquad (2a)^2 = (2a)(2a) = 4a^2.$$

$$h^2 = 3a^2 \qquad \text{Subtract } a^2 \text{ from both sides.}$$

$$h = \sqrt{3a^2} \qquad \text{Take the positive square root of both sides.}$$

$$h = a\sqrt{3} \qquad \text{Simplify the radical: } \sqrt{3a^2} = \sqrt{3}\sqrt{a^2} = a\sqrt{3}.$$

We see that the altitude—the longer side of the 30°–60°–90° triangle—is $\sqrt{3}$ times as long as the shorter leg. Thus,

The length of the longer leg of a 30°–60°–90° triangle is the length of the shorter leg times $\sqrt{3}$.

EXAMPLE 4

30°–60°–90° triangles. Find the length of the hypotenuse and the longer leg of the right triangle shown in Figure 7-18.

Solution   Since the shorter leg of a 30°–60°–90° triangle is half as long as the hypotenuse, the hypotenuse is 12 centimeters long.

Since the length of the longer leg is the length of the shorter leg times $\sqrt{3}$, the longer leg is $6\sqrt{3}$ (about 10.39) centimeters long.

FIGURE 7-18

SELF CHECK   Find the length of the hypotenuse and the longer leg of a 30°–60°–90° triangle if the shorter leg is 8 centimeters long.

Answers: 16 cm, $8\sqrt{3}$ cm   ∎

EXAMPLE 5

Stretching exercises. A doctor prescribed the back-strengthening exercise shown in Figure 7-19(a) for a patient. The patient was instructed to raise his leg to an angle of 60° and hold the position for 10 seconds. If the patient's leg is 36 inches long, how high off the floor will his foot be when his leg is held at the proper angle?

Solution   In Figure 7-19(b), we see that a 30°–60°–90° triangle, which we will call triangle *ABC*, models the situation. Since the side opposite the 30° angle of a 30°–60°–90° triangle is half as long as the hypotenuse, side *AC* is 18 inches long.

Since the length of the side opposite the 60° angle is the length of the side opposite the 30° angle times $\sqrt{3}$, side *BC* is $18\sqrt{3}$, or about 31 inches long. So the patient's foot will be about 31 inches from the floor when his leg is in the proper stretching position.

FIGURE 7-19

(a)

(b)

∎

The Distance Formula

With the *distance formula,* we can find the distance between any two points that are graphed on a rectangular coordinate system.

To find the distance d between points $P(x_1, y_1)$ and $Q(x_2, y_2)$ shown in Figure 7-20, we construct the right triangle PRQ. The distance between P and R is $|x_2 - x_1|$, and the distance between R and Q is $|y_2 - y_1|$. We apply the Pythagorean theorem to the right triangle PRQ to get

$$[d(PQ)]^2 = |x_2 - x_1|^2 + |y_2 - y_1|^2 \quad \text{Read } d(PQ) \text{ as "the distance between } P \text{ and } Q\text{."}$$
$$= (x_2 - x_1)^2 + (y_2 - y_1)^2 \quad \text{Because } |x_2 - x_1|^2 = (x_2 - x_1)^2 \text{ and } |y_2 - y_1|^2 = (y_2 - y_1)^2.$$

We take the positive square root of both sides:

1. $d(PQ) = \sqrt{(x_2 - x_1)^2 + (y_2 - y_1)^2}$

Equation 1 is the called the **distance formula.**

FIGURE 7-20

Distance formula

The distance between two points $P(x_1, y_1)$ and $Q(x_2, y_2)$ is given by the formula

$$d(PQ) = \sqrt{(x_2 - x_1)^2 + (y_2 - y_1)^2}$$

EXAMPLE 6

Using the distance formula. Find the distance between points $P(-2, 3)$ and $Q(4, -5)$.

Solution To find the distance, we can use the distance formula by substituting 4 for x_2, -2 for x_1, -5 for y_2, and 3 for y_1.

$$d(PQ) = \sqrt{(x_2 - x_1)^2 + (y_2 - y_1)^2}$$
$$= \sqrt{[4 - (-2)]^2 + (-5 - 3)^2}$$
$$= \sqrt{(4 + 2)^2 + (-5 - 3)^2}$$
$$= \sqrt{6^2 + (-8)^2}$$
$$= \sqrt{36 + 64}$$
$$= \sqrt{100}$$
$$= 10$$

The distance between P and Q is 10 units.

SELF CHECK Find the distance between $P(-2, -2)$ and $Q(3, 10)$. *Answer:* 13 ∎

EXAMPLE 7

Robotics. Computerized robots are used to weld the parts of an automobile chassis together on an automated production line. To do this, an imaginary coordinate system is superimposed on the side of the vehicle, and the robot is programmed to move to specific positions to make each weld. See Figure 7-21, which is scaled in inches. If the welder unit moves from point to point at an average rate of speed of 48 in./sec, how long will it take it to move from position 1 to position 2?

FIGURE 7-21

Solution   This is a uniform motion problem. We can use the formula $t = \frac{d}{r}$ to find the time it takes for the welder to move from position 1 (14, 57) to position 2 (154, 37).

We can use the distance formula to find the distance d that the welder unit moves.

$$d = \sqrt{(x_2 - x_1)^2 + (y_2 - y_1)^2}$$

$$d = \sqrt{(154 - 14)^2 + (37 - 57)^2} \quad \text{Substitute 154 for } x_2, \text{ 14 for } x_1, \text{ 37 for } y_2, \text{ and 57}$$
$$\text{for } y_1.$$

$$= \sqrt{140^2 + (-20)^2}$$

$$= \sqrt{20,000} \qquad\qquad 140^2 + (-20)^2 = 19,600 + 400 = 20,000.$$

$$= 100\sqrt{2} \qquad\qquad \text{Simplify: } \sqrt{20,000} = \sqrt{100 \cdot 100 \cdot 2} = 100\sqrt{2}.$$

The welder travels $100\sqrt{2}$ inches as it moves from position 1 to position 2. To find the time this will take, we divide the distance by the average rate of speed, 48 in./sec.

$$t = \frac{d}{r}$$

$$t = \frac{100\sqrt{2}}{48} \qquad \text{Substitute } 100\sqrt{2} \text{ for } d \text{ and 48 for } r.$$

$$t \approx 2.9 \qquad \text{Use a calculator to find an approximation to the nearest tenth.}$$

It will take the welder about 2.9 seconds to travel from position 1 to position 2.   ∎

STUDY SET

Section 7.6

VOCABULARY

In Exercises 1–4, fill in the blanks to make the statements true.

1. In a right triangle, the side opposite the 90° angle is called the _____.

2. An _____ triangle is a right triangle with two legs of equal length.

3. The _____ theorem states that in any right triangle, the square of the hypotenuse is equal to the sum of the squares of the lengths of the two legs.

4. An _____ triangle has three sides of equal length and three 60° angles.

CONCEPTS

In Exercises 5–12, fill in the blanks to make the statements true.

5. If a and b are the lengths of two legs of a right triangle and c is the length of the hypotenuse, then

_____.

6. In any right triangle, the square of the hypotenuse is equal to the _____ of the squares of the two _____.

7. In an isosceles right triangle, the length of the hypotenuse is the length of one leg times _____.

8. The shorter leg of a 30°–60°–90° triangle is _____ as long as the hypotenuse.

9. The length of the longer leg of a 30°–60°–90° triangle is the length of the shorter leg times _____.

10. The formula to find the distance between two points $P(x_1, y_1)$ and $Q(x_2, y_2)$ is _____.

11. In a right triangle, the shorter leg is opposite the ___ angle, and the longer leg is opposite the ___ angle.

12. An isosceles triangle has ___ sides of equal length.

13. **a.** To solve the equation $c^2 = 20$, where c represents the length of the hypotenuse of a right triangle, how do we "undo" the operation performed on c?

   b. What is the first step when solving the equation $25 + b^2 = 81$?

14. When the lengths of the sides of the triangle in Illustration 1 are substituted into the equation $a^2 + b^2 = c^2$, the result is a false statement. Explain why.

$$a^2 + b^2 = c^2$$
$$2^2 + 4^2 = 5^2$$
$$4 + 16 = 25$$
$$20 = 25$$

ILLUSTRATION 1

NOTATION

In Exercises 15–16, complete each solution.

15. Evaluate $\sqrt{(-1 - 3)^2 + [2 - (-4)]^2}$.

$$\sqrt{(-1 - 3)^2 + [2 - (-4)]^2} = \sqrt{(-4)^2 + [\;\;]^2}$$
$$= \sqrt{\;\;\;}$$
$$= \sqrt{\;\;\;} \cdot 13$$
$$= \;\;\sqrt{13}$$
$$\approx 7.21$$

16. Solve $8^2 + 4^2 = c^2$.

$$\;\;\; + 16 = c^2$$
$$\;\;\; = c^2$$
$$\sqrt{\;\;\;} = \sqrt{c^2}$$
$$\sqrt{\;\;\;} \cdot 5 = c$$
$$\;\;\sqrt{5} = c$$
$$c \approx 8.94$$

PRACTICE

In Exercises 17–20, the lengths of two sides of the right triangle ABC shown in Illustration 2 are given. Find the length of the missing side.

17. $a = 6$ ft and $b = 8$ ft

ILLUSTRATION 2

18. $a = 10$ cm and $c = 26$ cm

19. $b = 18$ m and $c = 82$ m

20. $a = 14$ in. and $c = 50$ in.

In Exercises 21–24, find the missing lengths in each triangle. Give the exact answer and then an approximation to two decimal places, when applicable.

21.

22.

23. 24.

In Exercises 25–28, find the missing lengths in each triangle. Give the answer to two decimal places.

25. 26.

27. 28.

29. **GEOMETRY** Find the exact length of the diagonal (in blue) of one of the *faces* of the cube shown in Illustration 3.

30. **GEOMETRY** Find the exact length of the diagonal (in green) of the cube shown in Illustration 3.

ILLUSTRATION 3

In Exercises 31–38, find the distance between P and Q.

31. $Q(0, 0)$, $P(3, -4)$

32. $Q(0, 0)$, $P(-12, 16)$

33. $P(-2, -8)$, $Q(3, 4)$

34. $P(-5, -2)$, $Q(7, 3)$

35. $P(6, 8)$, $Q(12, 16)$

36. $P(10, 4)$, $Q(2, -2)$

37. $Q(-3, 5)$, $P(-5, -5)$

38. $Q(2, -3)$, $P(4, -8)$

39. ISOSCELES TRIANGLE Use the distance formula to show that a triangle with vertices $(-2, 4)$, $(2, 8)$, and $(6, 4)$ is isosceles.

40. RIGHT TRIANGLE Use the distance formula and the Pythagorean theorem to show that a triangle with vertices $(2, 3)$, $(-3, 4)$, and $(1, -2)$ is a right triangle.

APPLICATIONS

In Exercises 41–46, give the exact answer. Then give an approximation to two decimal places.

41. WASHINGTON, DC The square in Illustration 4 shows the 100-square-mile site selected by George Washington in 1790 to serve as a permanent capital for the United States. In 1847, the part of the district lying on the west bank of the Potomac was returned to Virginia. Find the coordinates of each corner of the original square that outlined the District of Columbia.

ILLUSTRATION 4

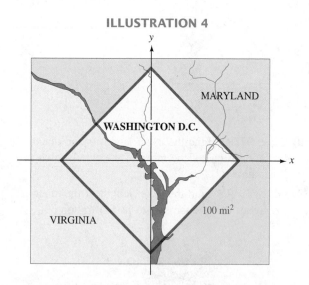

42. PAPER AIRPLANES Illustration 5 gives the directions for making a paper airplane from a square piece of paper with sides 8 inches long. Find the length l of the plane when it is completed.

ILLUSTRATION 5

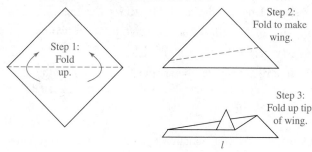

43. HARDWARE The sides of the regular hexagonal nut shown in Illustration 6 are 10 millimeters long. Find the height h of the nut.

ILLUSTRATION 6

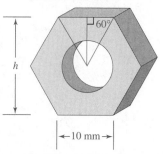

44. IRONING BOARD Find the height h of the ironing board shown in Illustration 7.

ILLUSTRATION 7

45. BASEBALL The baseball diamond shown in Illustration 8 is a square, 90 feet on a side. If the third baseman fields a ground ball 10 feet directly behind third base, how far must he throw the ball to throw a runner out at first base?

46. BASEBALL A shortstop fields a grounder at a point one-third of the way from second base to third base. (See Illustration 8.) How far will he have to throw the ball to make an out at first base?

ILLUSTRATION 8

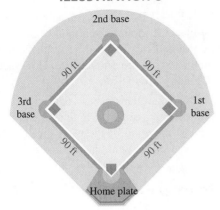

a. From the foot to the eye
b. From the belt to the hand holding the staff

c. From the shoulder to the symbol held in the hand

ILLUSTRATION 11

47. CLOTHESLINE A pair of damp jeans are hung on a clothesline to dry, as shown in Illustration 9. They pull the center down 1 foot. By how much is the line stretched?

ILLUSTRATION 9

48. FIREFIGHTING The base of the 37-foot ladder in Illustration 10 is 9 feet from the wall. Will the top reach a window ledge that is 35 feet above the ground? Verify your result.

ILLUSTRATION 10

49. ART HISTORY A figure displaying some of the basic characteristics of Egyptian art is shown in Illustration 11. Use the distance formula to find the following dimensions of the drawing. Round your answers to two decimal places.

50. PACKAGING The diagonal d of a rectangular box with dimensions $a \times b \times c$ is given by

$$d = \sqrt{a^2 + b^2 + c^2}$$

Will the umbrella fit in the shipping carton in Illustration 12? Verify your result.

ILLUSTRATION 12

51. PACKAGING An archaeologist wants to ship a 34-inch femur bone. Will it fit in a 4-inch-tall box that has a 24-inch-square base? (See Exercise 50.) Verify your result.

52. TELEPHONE SERVICE The telephone cable in Illustration 13 (on the next page) runs from A to B to C to D. How much cable is required to run from A to D directly?

ILLUSTRATION 13

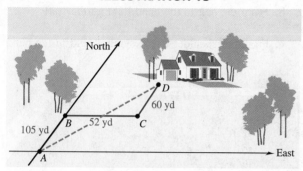

WRITING

53. State the Pythagorean theorem in words.

54. List the facts that you learned about special right triangles in this section.

REVIEW

55. DISCOUNT BUYING A repairman purchased some washing-machine motors for a total of $224. When the unit cost decreased by $4, he was able to buy one extra motor for the same total price. How many motors did he buy originally?

56. AVIATION An airplane can fly 650 miles with the wind in the same amount of time as it can fly 475 miles against the wind. If the wind speed is 40 mph, find the speed of the plane in still air.

57. Find the mean of 16, 6, 10, 4, 5, 13

58. Find the median of 16, 6, 10, 4, 5

Radicals

The expression $\sqrt[n]{a}$ is called a **radical expression.** In this chapter, we have discussed the properties and procedures used when simplifying radical expressions, solving radical equations, and writing radical expressions using rational exponents.

Expressions Containing Radicals

When working with expressions containing radicals, we must often apply the multiplication property and/or the division property of radicals to simplify the expression. Recall that

$$\sqrt[n]{ab} = \sqrt[n]{a}\sqrt[n]{b} \qquad \sqrt[n]{\frac{a}{b}} = \frac{\sqrt[n]{a}}{\sqrt[n]{b}} \quad (b \neq 0)$$

In Exercises 1–8, do each operation and simplify the expression.

1. Simplify: $\sqrt[3]{-54h^6}$.

2. Add: $2\sqrt[3]{64e} + 3\sqrt[3]{8e}$.

3. Subtract: $\sqrt{72} - \sqrt{200}$.

4. Multiply: $-4\sqrt[3]{5r^2s}\left(5\sqrt[3]{2r}\right)$.

5. Multiply: $\left(\sqrt{3s} - \sqrt{2t}\right)\left(\sqrt{3s} + \sqrt{2t}\right)$.

6. Multiply: $-\sqrt{3}\left(\sqrt{7} - \sqrt{5}\right)$.

7. Find the power: $\left(3\sqrt{2n} - 2\right)^2$.

8. Rationalize the denominator: $\dfrac{\sqrt[3]{9j}}{\sqrt[3]{3jk}}$.

Equations Containing Radicals

When solving radical equations, our objective is to rid the equation of the radical. This is achieved by using the *power rule:*

If x, y, and n are real numbers and $x = y$, then $x^n = y^n$.

If we raise both sides of an equation to the same power, the resulting equation might not be equivalent to the original equation. We must always check for extraneous solutions.

In Exercises 9–12, solve each radical equation, if possible.

9. $\sqrt{1 - 2g} = \sqrt{g + 10}$

10. $4 - \sqrt[3]{4 + 12x} = 0$

11. $\sqrt{y + 2} - 4 = -y$

12. $\sqrt[4]{12t + 4} + 2 = 0$

Radicals and Rational Exponents

Radicals can be written using rational (fractional) exponents, and exponential expressions having fractional exponents can be written in radical form. To do this, we use two rules for exponents introduced in this chapter.

$$x^{1/n} = \sqrt[n]{x} \qquad x^{m/n} = \sqrt[n]{x^m} = \left(\sqrt[n]{x}\right)^m$$

13. Express using a rational exponent: $\sqrt[3]{3}$.

14. Express in radical form: $5a^{2/5}$.

Accent on Teamwork

Section 7.1

A spiral of roots To do this project, you will need a piece of poster board, a protractor, a yardstick, and a pencil. Begin by drawing an isosceles right triangle near the right margin of the poster board. Label the length of each leg as 1 unit. (See Illustration 1.) Use the Pythagorean theorem to determine the length of the hypotenuse. Draw a second right triangle using the hypotenuse of the first triangle as one leg. Draw its second leg with a length of 1 unit. Find the length of the hypotenuse of triangle 2. Continue this process of creating right triangles, using the previous hypotenuse as one leg and drawing a new second leg of length 1 unit each time. Calculate the length of the resulting hypotenuse. What patterns, if any, do you see?

ILLUSTRATION 1

1 unit

3rd triangle

1 unit

2nd triangle

1st triangle

1 unit

1 unit

Section 7.2

Solving radical equations In this chapter, we solved equations that contained two radicals. The radicals in those equations always had the same index. That is not the case for the following two equations:

$$\sqrt[3]{2x} = \sqrt{x} \qquad \sqrt[4]{x} = \sqrt{\frac{x}{4}}$$

Brainstorm in your group to see if you can come up with a procedure that can be used to solve these equations. What are their solutions?

Section 7.3

Exponents Use the $\boxed{y^x}$ key on your calculator to approximate each of the following exponential expressions. Then write them in order from least to greatest. Which of the exponents are not rational exponents? Could you have written the expressions in increasing order without having to approximate them? Explain.

1. $3^{0.999}$
2. $3^{3.9}$
3. $3^{1\frac{3}{4}}$
4. $3^{\sqrt{2}}$
5. $3^{1.7}$
6. 3^{π}
7. $3^{\frac{2}{3}}$
8. $3^{\frac{4}{3}}$
9. $3^{2\sqrt{3}}$
10. $3^{0.1}$

Section 7.4

Common errors In each addition or subtraction problem below, tell what **mistake** was made. Compare each problem to a similar one involving variables to help clarify your explanation. For example, compare Problem 1 to $2a + 3a$ to help explain the correct procedure that should be used to simplify the expression.

1. $2\sqrt{5x} + 3\sqrt{5x} = 5\sqrt{10x}$
2. $30 + 30\sqrt[4]{2} = 60\sqrt[4]{2}$
3. $7\sqrt[3]{y^2} - 5\sqrt[3]{y^2} = 2$
4. $6\sqrt{11ab} - 3\sqrt{5ab} = 3\sqrt{6ab}$

Section 7.5

Multiplying radicals In this chapter, when we were asked to find the product of two radical expressions, the radicals always had the same index. Brainstorm in your group to see if you can come up with a procedure that can be used to find

$$\sqrt[3]{3} \cdot \sqrt{3}$$

Keep in mind two things: The indices must be the same to apply the multiplication property of radicals, and radical expressions can be written using rational exponents.

Section 7.6

Graphing in three dimensions A point is located in three-space by plotting ordered triples of numbers (x, y, z). The point $(2, 3, 4)$ is plotted in Illustration 2. In three-space, the distance formula is $d = \sqrt{(x_2 - x_1)^2 + (y_2 - y_1)^2 + (z_2 - z_1)^2}$. Use this formula to find the distance between **a.** $(3, 2, 1)$ and $(13, 12, 11)$ and **b.** $(-1, -2, -4)$ and $(6, -2, 5)$.

ILLUSTRATION 2

Section 7.1

Radical Expressions and Radical Functions

CONCEPTS

The number b is a *square root of a* if $b^2 = a$.

If $x > 0$, the *principal square root of x* is the positive square root of x, denoted $\sqrt{x}$. If x can be any real number, then $\sqrt{x^2} = |x|$.

The *cube root of x* is denoted as $\sqrt[3]{x}$ and is defined by

$$\sqrt[3]{x} = y \text{ if } y^3 = x$$

If n is an even natural number,

$$\sqrt[n]{a^n} = |a|$$

If n is an odd natural number,

$$\sqrt[n]{a^n} = a$$

If n is a natural number greater than 1 and x is a real number, then

- If $x > 0$, then $\sqrt[n]{x}$ is the positive number such that $\left(\sqrt[n]{x}\right)^n = x$.
- If $x = 0$, then $\sqrt[n]{x} = 0$.
- If $x < 0$, and n is odd, $\sqrt[n]{x}$ is the real number such that $\left(\sqrt[n]{x}\right)^n = x$.
- If $x < 0$, and n is even, $\sqrt[n]{x}$ is not a real number.

REVIEW EXERCISES

1. Simplify each radical expression, if possible. Assume that x can be any real number.

   a. $\sqrt{49}$

   b. $-\sqrt{121}$

   c. $\sqrt{\dfrac{225}{49}}$

   d. $\sqrt{-4}$

   e. $\sqrt{0.01}$

   f. $\sqrt{25x^2}$

   g. $\sqrt{x^8}$

   h. $\sqrt{x^2 + 4x + 4}$

2. Simplify each radical expression.

   a. $\sqrt[3]{-27}$

   b. $-\sqrt[3]{216}$

   c. $\sqrt[3]{64a^6b^3}$

   d. $\sqrt[3]{\dfrac{s^9}{125}}$

3. Simplify each radical expression, if possible. Assume that x and y can be any real number.

   a. $\sqrt[4]{625}$

   b. $\sqrt[5]{-32}$

   c. $\sqrt[4]{256x^8y^4}$

   d. $\sqrt{(-22y)^2}$

   e. $-\sqrt[4]{\dfrac{1}{16}}$

   f. $\sqrt[6]{-1}$

   g. $\sqrt{0}$

   h. $\sqrt[3]{0}$

4. GEOMETRY The side of a square with area A square feet is given by the function $s(A) = \sqrt{A}$. Find the *perimeter* of a square with an area of 144 square feet.

5. VOLUME OF A CUBE The total surface area of a cube is related to its volume V by the function $A(V) = 6\sqrt[3]{V^2}$. Find the surface area of a cube with a volume of 8 cm³.

6. Graph each radical function. Find the domain and range.

   a. $f(x) = \sqrt{x+2}$

   b. $f(x) = -\sqrt[3]{x} + 3$

Section 7.2

Radical Equations

The power rule:

   If $x = y$, then $x^n = y^n$.

Solving equations containing radicals:

1. Isolate one radical expression on one side of the equation.
2. Raise both sides of the equation to the power that is the same as the index.
3. Solve the resulting equation. If it still contains a radical, go back to step 1.
4. Check the solutions to eliminate *extraneous* solutions.

7. Solve each equation. Write all solutions. Cross out those that are extraneous.

   a. $\sqrt{7x - 10} - 1 = 11$

   b. $u = \sqrt{25u - 144}$

   c. $2\sqrt{y - 3} = \sqrt{2y + 1}$

   d. $\sqrt{z + 1} + \sqrt{z} = 2$

   e. $\sqrt[3]{x^3 + 56} - 2 = x$

   f. $\sqrt[4]{8x - 8} + 2 = 0$

8. Using the graphs of $f(x) = \sqrt{2x - 3}$ and $g(x) = -2x + 5$ in Illustration 1, estimate the solution of

   $$\sqrt{2x - 3} = -2x + 5$$

   Check the result.

ILLUSTRATION 1

9. Solve each equation for the indicated variable.

   a. $r = \sqrt{\dfrac{A}{P}} - 1$ for P

   b. $h = \sqrt[3]{\dfrac{12I}{b}}$ for I

10. **ELECTRONICS** The current I (measured in amperes) and the power P (measured in watts) are related by the formula

    $$I = \sqrt{\dfrac{P}{R}}$$

    Find the resistance R in a circuit if the current used by an electrical appliance that is rated at 980 watts is 7.4 amperes.

Section 7.3

Rational Exponents

If n ($n > 1$) is a natural number and $\sqrt[n]{x}$ is a real number, then

$$x^{1/n} = \sqrt[n]{x}$$

If n is a natural number greater than 1 and x is a real number,

- If $x > 0$, then $x^{1/n}$ is the positive number such that $(x^{1/n})^n = x$.
- If $x = 0$, then $x^{1/n} = 0$.
- If $x < 0$, and n is odd, then $x^{1/n}$ is the real number such that $(x^{1/n})^n = x$.
- If $x < 0$ and n is even, then $x^{1/n}$ is not a real number.

11. Write each expression in radical form.

    a. $t^{1/2}$

    b. $(5xy^3)^{1/4}$

12. Simplify each expression, if possible. Assume that all variables represent positive real numbers.

    a. $25^{1/2}$

    b. $-36^{1/2}$

    c. $(-36)^{1/2}$

    d. $1^{1/2}$

    e. $\left(\dfrac{9}{x^2}\right)^{1/2}$

    f. $\left(\dfrac{1}{27}\right)^{1/3}$

    g. $(-8)^{1/3}$

    h. $625^{1/4}$

    i. $(27a^3b)^{1/3}$

    j. $(81c^4d^4)^{1/4}$

If m and n are positive integers, $x > 0$, and $\frac{m}{n}$ is in simplest form,

$$x^{m/n} = \sqrt[n]{x^m} = \left(\sqrt[n]{x}\right)^m$$

$$x^{-m/n} = \frac{1}{x^{m/n}}$$

$$\frac{1}{x^{-m/n}} = x^{m/n} \quad (x \neq 0)$$

The *rules for exponents* can be used to simplify expressions with fractional exponents.

13. Simplify each expression, if possible. Assume that all variables represent positive real numbers.

a. $9^{3/2}$

b. $8^{-2/3}$

c. $-49^{5/2}$

d. $\dfrac{1}{100^{-1/2}}$

e. $\left(\dfrac{4}{9}\right)^{-3/2}$

f. $\dfrac{1}{25^{5/2}}$

g. $(25x^2y^4)^{3/2}$

h. $(8u^6v^3)^{-2/3}$

14. Do the operations. Write answers without negative exponents. Assume that all variables represent positive real numbers.

a. $5^{1/4}5^{1/2}$

b. $a^{3/7}a^{-2/7}$

c. $(k^{4/5})^{10}$

d. $\dfrac{(4g^3h)^{1/2}}{(9gh^{-1})^{1/2}}$

15. Do the multiplications. Assume all variables represent positive real numbers.

a. $u^{1/2}(u^{1/2} - u^{-1/2})$

b. $v^{2/3}(v^{1/3} + v^{4/3})$

16. Simplify $\sqrt[4]{\dfrac{a^2}{25b^2}}$. (The variables represent positive real numbers.)

17. Substitute the x- and y-coordinates of each point labeled in the graph in Illustration 2 into the equation

$$x^{2/3} + y^{2/3} = 32$$

Show that each one satisfies the equation.

ILLUSTRATION 2

$(-64, 64)$ $(64, 64)$

Section 7.4

Simplifying and Combining Radical Expressions

A radical is in *simplest form* when:
1. No radicals appear in a denominator.
2. The radicand contains no fractions or negative numbers.
3. Each factor in the radicand appears to a power less than the index.

Properties of radicals:
1. Multiplication:

$$\sqrt[n]{ab} = \sqrt[n]{a}\sqrt[n]{b}$$

2. Division:

$$\sqrt[n]{\frac{a}{b}} = \frac{\sqrt[n]{a}}{\sqrt[n]{b}} \quad (b \neq 0)$$

18. Simplify each expression. Assume that all variables represent positive real numbers.

a. $\sqrt{240}$

b. $\sqrt[3]{54}$

c. $\sqrt[4]{32}$

d. $-2\sqrt[5]{-96}$

e. $\sqrt{8x^5}$

f. $\sqrt[3]{r^{17}}$

g. $\sqrt[3]{16x^5y^4}$

h. $3\sqrt[3]{27j^7k}$

i. $\dfrac{\sqrt{32x^3}}{\sqrt{2x}}$

j. $\sqrt{\dfrac{17xy}{64a^4}}$

Like radicals can be combined by addition and subtraction.

Radicals that are not similar can often be converted to radicals that are similar and then combined.

19. Simplify and combine like radicals. Assume that all variables represent positive real numbers.

a. $\sqrt{2} + 2\sqrt{2}$

b. $6\sqrt{20} - \sqrt{5}$

c. $2\sqrt[3]{3} - \sqrt[3]{24}$

d. $-\sqrt[4]{32} - 2\sqrt[4]{162}$

e. $2x\sqrt{8} + 2\sqrt{200x^2} + \sqrt{50x^2}$

f. $\sqrt[3]{54} - 3\sqrt[3]{16} + 4\sqrt[3]{128}$

20. SEWING A corner of fabric is folded over to form a collar and stitched down as shown in Illustration 3. From the dimensions given in the figure, determine the exact number of inches of stitching that must be made. Then give an approximation to one decimal place. (All measurements are in inches.)

ILLUSTRATION 3

Stitch this flap down. $\sqrt{40}$ $\sqrt{32}$ $\sqrt{8}$

Section 7.5

Multiplying and Dividing Radical Expressions

If two radicals have the same index, they can be multiplied:

$$\sqrt[n]{a}\,\sqrt[n]{b} = \sqrt[n]{ab}$$

21. Simplify each expression. Assume that all variables represent positive real numbers.

a. $\sqrt{7}\sqrt{7}$

b. $\left(2\sqrt{5}\right)\left(3\sqrt{2}\right)$

c. $\left(-2\sqrt{8}\right)^2$

d. $2\sqrt{6}\sqrt{216}$

e. $\sqrt{9x}\sqrt{x}$

f. $\left(\sqrt{33}\right)^2$

g. $-\sqrt[3]{2x^2}\sqrt[3]{4x}$

h. $4\sqrt[3]{9}\sqrt[3]{9}$

i. $\sqrt{2}\left(\sqrt{8} - 3\right)$

j. $-\sqrt[4]{256x^5y^{11}}\sqrt[4]{625x^9y^3}$

k. $\left(\sqrt{3b} + \sqrt{3}\right)^2$

l. $\left(2\sqrt{u} + 3\right)\left(3\sqrt{u} - 4\right)$

If a radical appears in a denominator of a fraction, or if a radicand contains a fraction, we can write the radical in simplest form by *rationalizing the denominator.*

22. Rationalize each denominator.

a. $\dfrac{10}{\sqrt{3}}$

b. $\sqrt{\dfrac{3}{5}}$

c. $\dfrac{x}{\sqrt{xy}}$

d. $\dfrac{\sqrt[3]{uv}}{\sqrt[3]{u^5v^7}}$

To *rationalize the binomial denominator* of a fraction, multiply the numerator and the denominator by the conjugate of the binomial in the denominator.

e. $\dfrac{2}{\sqrt{2} - 1}$

f. $\dfrac{\sqrt{a} + 1}{\sqrt{a} - 1}$

23. Rationalize each numerator.

a. $\dfrac{3 - \sqrt{x}}{2}$

b. $\dfrac{\sqrt{a} - \sqrt{b}}{\sqrt{a}}$

24. VOLUME The formula relating the radius r of a sphere and its volume V is $r = \sqrt[3]{\dfrac{3V}{4\pi}}$. Write the radical in simplest form.

Section 7.6

Geometric Applications of Radicals

The Pythagorean theorem:

If *a* and *b* are the lengths of the *legs* of a right triangle and *c* is the length of the *hypotenuse,* then
$a^2 + b^2 = c^2$.

25. CARPENTRY The gable end of the roof shown in Illustration 4 is divided in half by a vertical brace, 8 feet in height. Find the length of the roof line.

26. SAILING A technique called *tacking* allows a sailboat to make progress into the wind. A sailboat follows the course in Illustration 5. Find *d*, the distance the boat advances into the wind after tacking.

ILLUSTRATION 4

ILLUSTRATION 5

In an *isosceles right triangle,* the length of the hypotenuse is the length of one leg times $\sqrt{2}$.

27. Find the length of the hypotenuse of an isosceles right triangle whose legs measure 7 meters.

The shorter leg of a *30°–60°–90° triangle* (the side opposite the 30° angle) is half as long as the hypotenuse. The longer leg (the side opposite the 60° angle) is the length of the shorter leg times $\sqrt{3}$.

28. The hypotenuse of a 30°–60°–90° triangle measures $12\sqrt{3}$ centimeters. Find the length of each leg.

29. Find *x* to two decimal places.

a.

b.

The distance formula:

$$d(PQ) = \sqrt{(x_2 - x_1)^2 + (y_2 - y_1)^2}$$

30. Find the distance between points *P* and *Q*.
a. $P(0, 0)$ and $Q(5, -12)$ **b.** $P(-4, 6)$ and $Q(-2, 8)$

CHAPTER 7

Test

1. Complete the table of values for $f(x) = \sqrt{x - 1}$. Then graph the function. Round to the nearest hundredth when necessary.

x	$f(x)$
1	
2	
3	
5	
10	
12	
17	

2. SUBMARINE The horizontal distance (measured in miles) that an observer can see is related to the height h (measured in feet) of the observer by the function $d(h) = 1.4\sqrt{h}$. If a submarine's periscope extends 4.7 feet above the surface of the ocean, how far is the horizon? Round to the nearest tenth.

In Problems 3–5, solve and check each solution.

3. $2\sqrt{x} = \sqrt{x + 1}$

4. $\sqrt[3]{6n + 4} - 4 = 0$

5. $1 - \sqrt{u} = \sqrt{u - 3}$

6. Solve $r = \sqrt[3]{\dfrac{GMt^2}{4\pi^2}}$ for G.

In Problems 7–12, simplify each expression. Assume that all variables represent positive real numbers, and write answers without using negative exponents.

7. $16^{1/4}$

8. $27^{2/3}$

9. $36^{-3/2}$

10. $\left(-\dfrac{8}{27}\right)^{-2/3}$

11. $\dfrac{2^{5/3}2^{1/6}}{2^{1/2}}$

12. $\dfrac{(8x^3y)^{1/2}(8xy^5)^{1/2}}{(x^3y^6)^{1/3}}$

In Problems 13–16, simplify each expression. Assume that the variables are unrestricted.

13. $\sqrt{x^2}$

14. $\sqrt{8x^2}$

15. $\sqrt[3]{54x^5}$

16. $\sqrt{18x^4y^8}$

In Problems 17–24, simplify each expression. Assume that all variables represent positive real numbers.

17. $\sqrt[3]{-64x^3y^6}$

18. $\sqrt{\dfrac{4a^2}{9}}$

19. $\sqrt[4]{-16}$

20. $\sqrt[5]{(t + 8)^5}$

21. $\sqrt{48}$

22. $\sqrt{250x^3y^5}$

23. $\dfrac{\sqrt[3]{24x^{15}y^4}}{\sqrt[3]{y}}$

24. $\sqrt{\dfrac{3a^5}{48a^7}}$

In Problems 25–28, simplify and combine like radicals. Assume that all variables represent positive real numbers.

25. $\sqrt{12} - \sqrt{27}$

26. $2\sqrt[3]{40} - \sqrt[3]{5,000} + 4\sqrt[3]{625}$

27. $2\sqrt{48y^5} - 3y\sqrt{12y^3}$

28. $\sqrt[4]{768z^5} + z\sqrt[4]{48z}$

In Problems 29–30, do each operation and simplify, if possible. All variables represent positive real numbers.

29. $-2\sqrt{xy}\left(3\sqrt{x} + \sqrt{xy^3}\right)$

30. $\left(3\sqrt{2} + \sqrt{3}\right)\left(2\sqrt{2} - 3\sqrt{3}\right)$

In Problems 31–32, rationalize each denominator.

31. $\dfrac{1}{\sqrt{5}}$

32. $\dfrac{3t - 1}{\sqrt{3t} - 1}$

In Problems 33–34, rationalize each numerator.

33. $\dfrac{\sqrt[3]{9}}{6}$

34. $\dfrac{\sqrt{5} + 3}{-4\sqrt{2}}$

In Problems 35–36, find x to two decimal places.

35.

36.

37. Find the distance between $(-2, 5)$ and $(22, 12)$.

38. PENDULUM The time t, in seconds, it takes for a pendulum to swing back and forth to complete one period is given by the formula $t = 2\pi\sqrt{\frac{l}{32}}$, where l is the length of the pendulum in feet. Find the exact period of a 4-foot-long pendulum. Then approximate the period to the nearest tenth.

39. SHIPPING CRATE The diagonal brace on the shipping crate in Illustration 1 is 53 inches. Find the height h of the crate.

40. Explain why, without having to perform any algebraic steps, it is obvious that the equation $\sqrt{x-8} = -10$ has no solutions.

ILLUSTRATION 1

h in. 53 in.

45 in.

8

Quadratic Equations, Functions, and Inequalities

CAMPUS CONNECTION

The Art Department

In a watercolor class, students learn that light, shadow, color, and perspective are fundamental components of an attractive painting. They also learn that the appropriate matting and frame can enhance their work. In this section, we will use mathematics to determine the dimensions of a uniform matting that is to have the same area as the picture it frames. To do this, we will write a quadratic equation and then solve it by *completing the square.* The technique of completing the square can be used to derive *the quadratic formula.* This formula is a valuable algebraic tool for solving any quadratic equation.

We have previously seen how to solve quadratic equations by factoring. In this chapter, we will discuss more general methods for solving quadratic equations, and we will consider the graphs of quadratic functions.

▶ 8.1   Completing the Square

In this section, you will learn about

A review of solving quadratic equations by factoring ■ The square root property ■ Completing the square ■ Solving equations by completing the square ■ Problem solving

Introduction We have seen that equations that involve first-degree polynomials, such as $12x - 4 = 0$, are called *linear equations*. We have also seen that equations that involve second-degree polynomials, such as $12x^2 - 4x = 0$, are called *quadratic equations*. In Chapter 5, we learned how to solve quadratic equations by factoring. However, as we shall see, the factoring method has its limitations. In this section, we will introduce a more general method that enables us to solve *any* quadratic equation.

A Review of Solving Quadratic Equations by Factoring

A *quadratic equation* is an equation of the form $ax^2 + bx + c = 0$ ($a \neq 0$), where a, b, and c are real numbers. We have discussed how to solve quadratic equations by factoring. For example, to solve $6x^2 - 7x - 3 = 0$, we proceed as follows:

$$6x^2 - 7x - 3 = 0$$
$$(2x - 3)(3x + 1) = 0 \qquad \text{Factor.}$$
$$2x - 3 = 0 \quad \text{or} \quad 3x + 1 = 0 \qquad \text{Set each factor equal to 0.}$$
$$x = \frac{3}{2} \qquad\qquad x = -\frac{1}{3} \qquad \text{Solve each linear equation.}$$

Many expressions do not factor as easily as $6x^2 - 7x - 3$. For example, it would be difficult to solve $2x^2 + 4x + 1 = 0$ by factoring, because $2x^2 + 4x + 1$ cannot be factored by using only integers. With this in mind, we will now develop another method of solving quadratic equations. It is based on the **square root property.**

The Square Root Property

To develop general methods for solving all quadratic equations, we first consider the equation $x^2 = c$. If $c > 0$, we can find the real solutions of $x^2 = c$ as follows:

$$x^2 = c$$
$$x^2 - c = 0 \qquad\qquad \text{Subtract } c \text{ from both sides.}$$
$$x^2 - \left(\sqrt{c}\right)^2 = 0 \qquad\qquad \text{Replace } c \text{ with } \left(\sqrt{c}\right)^2, \text{ since } c = \left(\sqrt{c}\right)^2.$$
$$\left(x + \sqrt{c}\right)\left(x - \sqrt{c}\right) = 0 \qquad\qquad \text{Factor the difference of two squares.}$$
$$x + \sqrt{c} = 0 \quad \text{or} \quad x - \sqrt{c} = 0 \qquad \text{Set each factor equal to 0.}$$
$$x = -\sqrt{c} \qquad\qquad x = \sqrt{c} \qquad \text{Solve each linear equation.}$$

The two solutions of $x^2 = c$ are $x = \sqrt{c}$ and $x = -\sqrt{c}$.

Square root property

If $c > 0$, the equation $x^2 = c$ has two real solutions. They are

$$x = \sqrt{c} \quad \text{and} \quad x = -\sqrt{c}$$

EXAMPLE 1

Solving a quadratic equation using the square root property.
Solve $x^2 - 12 = 0$.

Solution We can write the equation as $x^2 = 12$ and use the square root property to solve it.

$$x^2 - 12 = 0$$
$$x^2 = 12 \qquad \text{Add 12 to both sides.}$$
$$x = \sqrt{12} \quad \text{or} \quad x = -\sqrt{12} \quad \text{Use the square root property.}$$
$$x = 2\sqrt{3} \qquad x = -2\sqrt{3} \quad \text{Simplify: } \sqrt{12} = \sqrt{4}\sqrt{3} = 2\sqrt{3}.$$

Verify that each solution satisfies the original equation.

We can write the two solutions of the equation in a more concise form as $x = \pm 2\sqrt{3}$, where $\pm$ is read as "plus or minus."

SELF CHECK Solve $x^2 - 18 = 0$.   *Answer:* $\pm 3\sqrt{2}$ ∎

EXAMPLE 2

Phonograph records. Before compact disc (CD) technology, one way of recording music was by engraving grooves on thin vinyl discs called records. (See Figure 8-1.) The vinyl discs used for long-playing records had a surface area of about 111 square inches per side and were played at $33\frac{1}{3}$ revolutions per minute on a turntable. What is the radius of a long-playing record?

FIGURE 8-1

Solution The relationship between the area of a circle and its radius is given by the formula $A = \pi r^2$. We can find the radius of a record by substituting 111 for A and solving for r.

$$A = \pi r^2 \qquad \text{The formula for the area of a circle.}$$
$$\mathbf{111} = \pi r^2 \qquad \text{Substitute 111 for } A.$$
$$\frac{111}{\pi} = r^2 \qquad \text{To undo the multiplication by } \pi, \text{ divide both sides by } \pi.$$
$$r = \sqrt{\frac{111}{\pi}} \quad \text{or} \quad r = -\sqrt{\frac{111}{\pi}} \qquad \text{Use the square root property. Since the radius of the record cannot be negative, discard the second solution.}$$

The radius of a record is $\sqrt{\frac{111}{\pi}}$ inches—to the nearest tenth, 5.9 inches.   ∎

EXAMPLE 3

Using the square root property to solve an equation. Solve $(x - 3)^2 = 16$.

Solution
$$(x - 3)^2 = 16$$
$$x - 3 = \sqrt{16} \quad \text{or} \quad x - 3 = -\sqrt{16} \quad \text{Use the square root property.}$$
$$x - 3 = 4 \qquad\qquad x - 3 = -4 \qquad \text{Simplify: } \sqrt{16} = 4.$$
$$x = 3 + 4 \qquad\qquad x = 3 - 4 \qquad \text{Add 3 to both sides.}$$
$$x = 7 \qquad\qquad\quad x = -1 \qquad \text{Simplify.}$$

Verify that each solution satisfies the equation.

SELF CHECK Solve $(x + 2)^2 = 9$.   *Answer:* 1, -5 ∎

Completing the Square

All quadratic equations can be solved by **completing the square.** This method involves the special products

$$x^2 + 2ax + a^2 = (x + a)^2 \qquad \text{and} \qquad x^2 - 2ax + a^2 = (x - a)^2$$

The trinomials $x^2 + \mathbf{2}ax + a^2$ and $x^2 - \mathbf{2}ax + a^2$ are both perfect square trinomials, because both factor as the square of a binomial. In each case, the coefficient of the first term is 1, and if we take one-half of the coefficient of x in the middle term and square it, we obtain the third term.

$$\left[\frac{1}{2}(2a)\right]^2 = a^2 \qquad\qquad \left[\frac{1}{2}(-2a)\right]^2 = (-a)^2 = a^2$$

EXAMPLE 4

Completing the square. Add a number to make each binomial a perfect square trinomial: **a.** $x^2 + 10x$, **b.** $x^2 - 6x$, and **c.** $x^2 - 11x$.

Solution **a.** To make $x^2 + 10x$ a perfect square trinomial, we find one-half of 10, square it, and add that result to $x^2 + 10x$.

$$x^2 + 10x + \left[\frac{1}{2}(10)\right]^2 = x^2 + 10x + (5)^2 \quad \text{Simplify: } \tfrac{1}{2}(10) = 5.$$

$$= x^2 + 10x + 25 \quad \text{Note that } x^2 + 10x + 25 = (x + 5)^2.$$

b. To make $x^2 - 6x$ a perfect square trinomial, we find one-half of -6, square it, and add that result to $x^2 - 6x$.

$$x^2 - 6x + \left[\frac{1}{2}(-6)\right]^2 = x^2 - 6x + (-3)^2 \quad \text{Simplify: } \tfrac{1}{2}(-6) = -3.$$

$$= x^2 - 6x + 9 \quad \text{Note that } x^2 - 6x + 9 = (x - 3)^2.$$

c. To make $x^2 - 11x$ a perfect square trinomial, we find one-half of -11, square it, and add that result to $x^2 - 11x$.

$$x^2 - 11x + \left[\frac{1}{2}(-11)\right]^2$$

$$= x^2 - 11x + \left(-\frac{11}{2}\right)^2 \quad \text{Simplify: } \tfrac{1}{2}(-11) = -\tfrac{11}{2}.$$

$$= x^2 - 11x + \frac{121}{4} \quad \text{Note that } x^2 - 11x + \tfrac{121}{4} = \left(x - \tfrac{11}{2}\right)^2.$$

SELF CHECK Add a number to $a^2 - 5a$ to make it a perfect trinomial square.   *Answer:* $a^2 - 5a + \frac{25}{4}$ ∎

Solving Equations by Completing the Square

To solve an equation of the form $ax^2 + bx + c = 0$ by completing the square, we use the following steps.

> ### Completing the square
>
> 1. Make sure that the coefficient of x^2 (the **leading coefficient**) is 1. If it is not, make it 1 by dividing both sides of the equation by the coefficient of x^2.
> 2. If necessary, add a number to both sides of the equation so that the constant term is on the right-hand side of the equals sign.
> 3. Complete the square:
>   a. Find one-half of the coefficient of x and square it.
>   b. Add that square to both sides of the equation.
> 4. Factor the trinomial square.
> 5. Solve the resulting equation using the square root property.

EXAMPLE 5

Completing the square. Use completing the square to solve $x^2 + 8x + 7 = 0$.

Solution **Step 1:** In this example, the coefficient of x^2 is an understood 1.

Step 2: We add -7 to both sides so that the constant is on the right-hand side of the equals sign:

$$x^2 + 8x + 7 = 0$$
$$x^2 + 8x = -7$$

Step 3: The coefficient of x is 8, one-half of 8 is 4, and $4^2 = 16$. To complete the square, we add 16 to both sides.

$$x^2 + 8x + 16 = 16 - 7$$
1. $x^2 + 8x + 16 = 9$ Simplify: $16 - 7 = 9$.

Step 4: Since the left-hand side of Equation 1 is a perfect square trinomial, we can factor it to get $(x + 4)^2$.

$$x^2 + 8x + 16 = 9$$
2. $\quad\ (x + 4)^2 = 9$

Step 5: We then solve Equation 2 by using the square root property.

$$x + 4 = \pm\sqrt{9}$$
$$x + 4 = 3 \quad \text{or} \quad x + 4 = -3$$
$$x = -1 \qquad\qquad x = -7$$

Verify that both solutions satisfy the equation. ∎

EXAMPLE 6

The leading coefficient is not 1. Solve $6x^2 + 5x - 6 = 0$.

Solution **Step 1:** To make the coefficient of x^2 equal to 1, we divide both sides of the equation by 6.

$$6x^2 + 5x - 6 = 0$$
$$\frac{6x^2}{6} + \frac{5}{6}x - \frac{6}{6} = \frac{0}{6} \quad \text{Divide both sides by 6.}$$
$$x^2 + \frac{5}{6}x - 1 = 0 \quad \text{Simplify.}$$

Step 2: We add 1 to both sides so that the constant is on the right-hand side of the equals sign:

$$x^2 + \frac{5}{6}x = 1$$

Step 3: The coefficient of x is $\frac{5}{6}$, one-half of $\frac{5}{6}$ is $\frac{5}{12}$, and $\left(\frac{5}{12}\right)^2 = \frac{25}{144}$. To complete the square, we add $\frac{25}{144}$ to both sides.

$$x^2 + \frac{5}{6}x + \frac{25}{144} = 1 + \frac{25}{144}$$

3. $x^2 + \frac{5}{6}x + \frac{25}{144} = \frac{169}{144}$ Simplify: $1 + \frac{25}{144} = \frac{144}{144} + \frac{25}{144} = \frac{169}{144}$.

Step 4: Since the left-hand side of Equation 3 is a perfect square trinomial, we can factor it to get $\left(x + \frac{5}{12}\right)^2$.

4. $\left(x + \frac{5}{12}\right)^2 = \frac{169}{144}$

Step 5: We can solve Equation 4 by using the square root property.

$$x + \frac{5}{12} = \pm\sqrt{\frac{169}{144}}$$

$x + \frac{5}{12} = \frac{13}{12}$ or $x + \frac{5}{12} = -\frac{13}{12}$ Simplify: $\sqrt{\frac{169}{144}} = \frac{13}{12}$.

$x = -\frac{5}{12} + \frac{13}{12}$ $x = -\frac{5}{12} - \frac{13}{12}$ Subtract $\frac{5}{12}$ from both sides.

$x = \frac{8}{12}$ $x = -\frac{18}{12}$ Simplify.

$x = \frac{2}{3}$ $x = -\frac{3}{2}$ Simplify each fraction.

Verify that both solutions satisfy the original equation. ∎

EXAMPLE 7

Solving quadratic equations by completing the square. Solve $2x^2 + 4x + 1 = 0$.

Solution $2x^2 + 4x + 1 = 0$

$x^2 + 2x + \frac{1}{2} = 0$ Divide both sides by 2 to make the coefficient of x^2 equal to 1.

$x^2 + 2x = -\frac{1}{2}$ Subtract $\frac{1}{2}$ from both sides.

$x^2 + 2x + 1 = 1 - \frac{1}{2}$ Square one-half of the coefficient of x and add it to both sides.

$(x + 1)^2 = \frac{1}{2}$ Factor and combine like terms.

$x + 1 = \pm\sqrt{\frac{1}{2}}$ Apply the square root property.

To write $\sqrt{\frac{1}{2}}$ in simplified radical form, we write it as a quotient of square roots and then rationalize the denominator.

$$x + 1 = \frac{\sqrt{2}}{2} \qquad\qquad x + 1 = -\frac{\sqrt{2}}{2} \qquad \sqrt{\frac{1}{2}} = \frac{\sqrt{1}}{\sqrt{2}} = \frac{1 \cdot \sqrt{2}}{\sqrt{2}\sqrt{2}} = \frac{\sqrt{2}}{2}.$$

$$x = -1 + \frac{\sqrt{2}}{2} \qquad\qquad x = -1 - \frac{\sqrt{2}}{2} \quad \text{Subtract 1 from both sides.}$$

We can express each solution in an alternate form if we write -1 as a fraction with a denominator of 2.

$$x = -\frac{2}{2} + \frac{\sqrt{2}}{2} \qquad x = -\frac{2}{2} - \frac{\sqrt{2}}{2} \quad \text{Write } -1 \text{ as } -\frac{2}{2}.$$

$$x = \frac{-2 + \sqrt{2}}{2} \qquad x = \frac{-2 - \sqrt{2}}{2} \quad \begin{array}{l}\text{Add (subtract) the numerators and keep the}\\ \text{common denominator of 2.}\end{array}$$

The exact solutions are $x = \frac{-2 + \sqrt{2}}{2}$ or $x = \frac{-2 - \sqrt{2}}{2}$, or more concisely, $x = \frac{-2 \pm \sqrt{2}}{2}$. We can use a calculator to approximate them. To the nearest hundredth, $x \approx -0.29$ or $x \approx -1.71$.

SELF CHECK   Solve $3x^2 + 6x + 1 = 0$.   *Answer:* $\dfrac{-3 \pm \sqrt{6}}{3}$ ■

 WARNING! Recall that to simplify a fraction, we divide out common *factors* of the numerator and denominator. In Example 7, since -2 is a *term* of the numerator of $\frac{-2 + \sqrt{2}}{2}$, no further simplification of this expression can be made. In the Self Check, 2 is a common factor and can be divided out: $\dfrac{-6 \pm 2\sqrt{6}}{6} = \dfrac{\overset{1}{\cancel{2}}(-3 \pm \sqrt{6})}{\underset{1}{\cancel{2}(3)}} = \dfrac{-3 \pm \sqrt{6}}{3}$.

ACCENT ON TECHNOLOGY *Checking Solutions of Quadratic Equations*

We can use a graphing calculator to check the solutions of the quadratic equation $2x^2 + 4x + 1 = 0$ found in Example 7. After entering $Y_1 = 2x^2 + 4x + 1$, we call up the home screen by pressing $\boxed{\text{2nd}}$ QUIT. Then we press the $\boxed{\text{VARS}}$ key, arrow $\boxed{\blacktriangleright}$ to Y-VARS, and enter 1 to get the display shown in Figure 8-2(a). We evaluate $2x^2 + 4x + 1$ for $x = \frac{-2 + \sqrt{2}}{2}$ by inputting the solution using function notation, as shown in Figure 8-2(b). When $\boxed{\text{ENTER}}$ is pressed, the result of 0 is confirmation that $x = \frac{-2 + \sqrt{2}}{2}$ is a solution of the equation.

FIGURE 8-2

```
┌─────────────────┐      ┌─────────────────┐
│Y₁               │      │Y₁((-2+√(2))/2)  │
│                 │      │                0│
│                 │      │■                │
│                 │      │                 │
│                 │      │                 │
│                 │      │                 │
└─────────────────┘      └─────────────────┘
```

(a)   (b)

Problem Solving

EXAMPLE 8

Graduation announcement. In creating the announcement shown in Figure 8-3, the graphic artist wants to follow two design elements.

- A border of uniform width should surround the text.

- Equal areas should be devoted to the text and to the border.

To meet these requirements, how wide should the border be?

FIGURE 8-3

ANALYZE THE PROBLEM   The text occupies $4 \cdot 3 = 12$ in.^{2} of space. The border must also have an area of 12 in.^{2}.

FORM AN EQUATION   If we let x represent the width of the border, the length of the announcement is $(4 + 2x)$ inches and the width is $(3 + 2x)$ inches. We can now form the equation.

The area of the announcement	minus	the area of the text	equals	the area of the border.
$(4 + 2x)(3 + 2x)$	$-$	12	$=$	12

SOLVE THE EQUATION

$$(4 + 2x)(3 + 2x) - 12 = 12$$

$12 + 8x + 6x + 4x^2 - 12 = 12$ On the left-hand side, use the FOIL method.

$4x^2 + 14x = 12$ Combine like terms.

$2x^2 + 7x - 6 = 0$ Subtract 12 from both sides. Then divide both sides by 2.

We note that the trinomial on the left-hand side does not factor. We will solve the equation by completing the square.

$$x^2 + \frac{7}{2}x - 3 = 0$$ Divide both sides by 2 so that the coefficient of x^2 is 1.

$$x^2 + \frac{7}{2}x = 3$$ Add 3 to both sides.

$$x^2 + \frac{7}{2}x + \frac{49}{16} = 3 + \frac{49}{16}$$ One-half of $\frac{7}{2}$ is $\frac{7}{4}$. Square $\frac{7}{4}$, which is $\frac{49}{16}$, and add it to both sides.

$$\left(x + \frac{7}{4}\right)^2 = \frac{97}{16}$$ On the left-hand side, factor the trinomial. On the right-hand side, $3 = \frac{3 \cdot 16}{1 \cdot 16} = \frac{48}{16}$ and $\frac{48}{16} + \frac{49}{16} = \frac{97}{16}$.

$$x + \frac{7}{4} = \pm \frac{\sqrt{97}}{4}$$ Apply the square root property. On the right-hand side, $\sqrt{\frac{97}{16}} = \frac{\sqrt{97}}{\sqrt{16}} = \frac{\sqrt{97}}{4}$.

$$x = -\frac{7}{4} + \frac{\sqrt{97}}{4} \quad \text{or} \quad x = -\frac{7}{4} - \frac{\sqrt{97}}{4}$$ Subtract $\frac{7}{4}$ from both sides.

$$x = \frac{-7 + \sqrt{97}}{4} \qquad x = \frac{-7 - \sqrt{97}}{4}$$ Write each expression as a single fraction.

STATE THE CONCLUSION   The width of the border should be $\dfrac{-7 + \sqrt{97}}{4} \approx 0.71$ inch. (We discard the solution $\dfrac{-7 - \sqrt{97}}{4}$, since it is negative.)

CHECK THE RESULT   If the border is 0.71 inch wide, the announcement has an area of about $5.42 \cdot 4.42 \approx 23.96$ in.^{2}. If we subtract the area of the text from the area of the announcement, we get $23.96 - 12 = 11.96$ in.^{2}. This represents the area of the border, which was to be 12 in.^{2}. The answer seems reasonable.   ■

STUDY SET

Section 8.1

VOCABULARY

In Exercises 1–4, fill in the blanks to make the statements true.

1. An equation of the form $ax^2 + bx + c = 0$ $(a \neq 0)$ is called a _____ equation.

2. The symbol $\pm$ is read as "_____."

3. $x^2 + 6x + 9$ is called a _____ square trinomial because it factors as $(x + 3)^2$.

4. The _____ of x^2 in $x^2 - 12x + 36 = 0$ is 1, and the _____ term is 36.

CONCEPTS

In Exercises 5–8, fill in the blanks to make the statements true.

5. The solutions of $x^2 = c$, where $c > 0$, are _____ and _____.

6. To complete the square on x in $x^2 + 6x$, find one-half of ___, square it to get ___, and add ___ to get _____.

7. In the quadratic equation $3x^2 - 2x + 5 = 0$, $a =$ ___, $b =$ ___, and $c =$ ___.

8. In the quadratic equation $ax^2 + bx + c = 0$, a cannot be ___.

9. Check to see if $-2 + \sqrt{2}$ is a solution of $x^2 + 4x + 2 = 0$.

10. Check to see if $-3\sqrt{2}$ is a solution of $x^2 - 18 = 0$.

11. Find one-half of the coefficient of x and then square it.
    a. $x^2 + 12x$
    b. $x^2 - 5x$
    c. $x^2 - \dfrac{x}{2}$

12. Add a number to make each binomial a perfect square trinomial. Then factor the result.
    a. $x^2 + 8x$
    b. $x^2 - 8x$
    c. $x^2 - x$

13. What is the first step in solving the equation $x^2 + 12x = 35$
    a. by the factoring method?

    b. by completing the square?

14. Solve the equation $x^2 = 16$
    a. by the factoring method.
    b. by the square root method.

15. Explain the **error** in the work shown below.

$$\frac{4 \pm \sqrt{3}}{8} = \frac{\overset{1}{\cancel{4}} \pm \sqrt{3}}{\underset{1}{\cancel{4} \cdot 2}}$$

$$= \frac{1 \pm \sqrt{3}}{2}$$

16. Explain the **error** in the work shown below.

$$\frac{1 \pm \sqrt{5}}{5} = \frac{1 \pm \sqrt{\overset{1}{\cancel{5}}}}{\underset{1}{\cancel{5}}}$$

$$= \frac{1 \pm 1}{1}$$

In Exercises 17–18, note that a and b are solutions of the equation $(x - a)(x - b) = 0$.

17. Find a quadratic equation with solutions of 3 and 5.

18. Find a quadratic equation with solutions of -4 and 6.

NOTATION

19. In solving a quadratic equation, a student obtains $x = \pm 2\sqrt{5}$.
    a. How many solutions are represented by this notation? List them.
    b. Approximate the solutions to the nearest hundredth.

20. In solving a quadratic equation, a student obtains
$$x = \frac{-5 \pm \sqrt{7}}{3}.$$

a. How many solutions are represented by this notation? List them.

b. Approximate the solutions to the nearest hundredth.

PRACTICE

In Exercises 21–28, use factoring to solve each equation.

21. $6x^2 + 12x = 0$ **22.** $5x^2 + 11x = 0$
23. $2y^2 - 50 = 0$ **24.** $4y^2 - 64 = 0$
25. $r^2 + 6r + 8 = 0$ **26.** $x^2 + 9x + 20 = 0$

27. $2z^2 = -2 + 5z$ **28.** $3x^2 = 8 - 10x$

In Exercises 29–40, use the square root property to solve each equation.

29. $x^2 = 36$ **30.** $x^2 = 144$
31. $z^2 = 5$ **32.** $u^2 = 24$

33. $3x^2 - 16 = 0$ **34.** $5x^2 - 49 = 0$

35. $(x + 1)^2 = 1$ **36.** $(x - 1)^2 = 4$
37. $(s - 7)^2 - 9 = 0$ **38.** $(t + 4)^2 = 16$

39. $(x + 5)^2 - 3 = 0$ **40.** $(x + 3)^2 - 7 = 0$

In Exercises 41–44, use the square root property to solve for the indicated variable. Assume that all variables represent positive numbers. Express all radicals in simplified form.

41. $2d^2 = 3h$ for d **42.** $2x^2 = d^2$ for d

43. $E = mc^2$ for c **44.** $A = \pi r^2$ for r

In Exercises 45–62, use completing the square to solve each equation.

45. $x^2 + 2x - 8 = 0$ **46.** $x^2 + 6x + 5 = 0$

47. $x + 1 = 2x^2$ **48.** $-2 = 2x^2 - 5x$

49. $6x^2 + x - 2 = 0$ **50.** $9 - 6r = 8r^2$

51. $x^2 + 8x + 6 = 0$ **52.** $x^2 + 6x + 4 = 0$

53. $x^2 - 2x - 17 = 0$ **54.** $x^2 + 10x - 7 = 0$

55. $3x^2 - 6x = 1$ **56.** $2x^2 - 6x = -3$

57. $4x^2 - 4x - 7 = 0$ **58.** $2x^2 - 8x + 5 = 0$

59. $2x^2 + 5x - 2 = 0$ **60.** $4x^2 - 4x - 1 = 0$

61. $\dfrac{7x + 1}{5} = -x^2$ **62.** $\dfrac{3x^2}{8} = \dfrac{1}{8} - x$

APPLICATIONS

63. FLAG In 1912, an executive order by President Taft fixed the overall width and length of the U.S. flag in the ratio 1 to 1.9. (See Illustration 1.) If 100 square feet of cloth are to be used to make a U.S. flag, estimate its dimensions to the nearest $\frac{1}{4}$ foot.

ILLUSTRATION 1

1.9x

x

64. MOVIE STUNTS According to the *Guinness Book of World Records, 1998,* stuntman Dan Koko fell a distance of 312 feet into an airbag after jumping from the Vegas World Hotel and Casino. The distance d in feet traveled by a free-falling object in t seconds is given by the formula $d = 16t^2$. To the nearest tenth of a second, how long did the stuntman's free fall last?

65. ACCIDENTS The height h (in feet) of an object that is dropped from a height of s feet is given by the formula $h = s - 16t^2$, where t is the time the object has been falling. A 5-foot-tall woman on a sidewalk looks directly overhead and sees a window washer drop a bottle from 4 stories up. How long does she have to get out of the way? Round to the nearest tenth. (A story is 12 feet.)

66. GEOGRAPHY The surface area S of a sphere is given by the formula $S = 4\pi r^2$, where r is the radius of the sphere. An almanac lists the surface area of the earth as 196,938,800 square miles. Assuming the earth to be spherical, what is its radius to the nearest mile?

67. AUTOMOBILE ENGINE As the piston shown in Illustration 2 moves upward, it pushes a "cylinder" of a gasoline/air mixture that is ignited by the spark plug. The formula that gives the volume of a cylinder is $V = \pi r^2 h$, where r is the radius and h the height. Find the radius of the piston (to the nearest hundredth of an inch) if it displaces 47.75 cubic inches of gasoline/air mixture as it moves from its lowest to its highest point.

ILLUSTRATION 2

68. INVESTMENTS If P dollars are deposited in an account that pays an annual rate of interest r, then in n years, the amount of money A in the account is given by the formula $A = P(1 + r)^n$. A savings account was opened on January 3, 1996, with a deposit of $10,000 and closed on January 2, 1998, with an ending balance of $11,772.25. Find r, the rate of interest.

69. PICTURE FRAMING The matting around the picture in Illustration 3 has a uniform width. How wide is the matting if its area equals the area of the picture? Round to the nearest hundredth of an inch.

ILLUSTRATION 3

70. SWIMMING POOL See the advertisement in Illustration 4. How wide will the "free" concrete decking be if a uniform width is constructed around the perimeter of the pool? Round to the nearest hundredth of a yard. (*Hint:* Note the difference in units.)

ILLUSTRATION 4

71. DIMENSIONS OF A RECTANGLE A rectangle is 4 feet longer than it is wide, and its area is 20 square feet. Find its dimensions to the nearest tenth of a foot.

72. DIMENSIONS OF A TRIANGLE The height of a triangle is 4 meters longer than twice its base. Find the base and height if the area of the triangle is 10 square meters. Round to the nearest hundredth of a meter.

WRITING

73. Explain how to complete the square.

74. Tell why a cannot be 0 in the quadratic equation $ax^2 + bx + c = 0$.

REVIEW

Simplify each expression. All variables represent positive real numbers.

75. $\sqrt[3]{40a^3b^6}$

76. $\sqrt[3]{-27x^6}$

77. $\sqrt[8]{x^{24}}$

78. $\sqrt[4]{\dfrac{16}{625}}$

79. $\sqrt{175a^2b^3}$

80. $\sqrt{\dfrac{z^2}{16x^2}}$

▶ 8.2 The Quadratic Formula

In this section, you will learn about

> The quadratic formula ■ Solving quadratic equations using the quadratic formula ■ Problem solving

Introduction We can solve any quadratic equation by the method of completing the square, but the work is often tedious. In this section, we will develop a formula, called the *quadratic formula,* that lets us solve quadratic equations with much less effort.

The Quadratic Formula

To develop a formula we can use to solve quadratic equations, we solve the general quadratic equation $ax^2 + bx + c = 0$ for $a > 0$.

$$ax^2 + bx + c = 0$$

$$\frac{ax^2}{a} + \frac{bx}{a} + \frac{c}{a} = \frac{0}{a} \qquad \text{Since } a > 0, \text{ we can divide both sides by } a.$$

$$x^2 + \frac{bx}{a} = -\frac{c}{a} \qquad \frac{0}{a} = 0; \text{ subtract } \frac{c}{a} \text{ from both sides.}$$

$$x^2 + \frac{b}{a}x + \left(\frac{b}{2a}\right)^2 = \left(\frac{b}{2a}\right)^2 - \frac{c}{a} \qquad \text{Complete the square on } x. \text{ Half of } \frac{b}{a} \text{ is } \frac{b}{2a}. \text{ Add } \left(\frac{b}{2a}\right)^2 \text{ to both sides.}$$

$$x^2 + \frac{b}{a}x + \frac{b^2}{4a^2} = \frac{b^2}{4a^2} - \frac{4ac}{4aa} \qquad \text{Remove parentheses and get a common denominator of } 4a^2 \text{ on the right-hand side.}$$

1. $$\left(x + \frac{b}{2a}\right)^2 = \frac{b^2 - 4ac}{4a^2} \qquad \text{Factor the left-hand side and add the fractions on the right-hand side.}$$

We can solve Equation 1 using the square root property.

$$x + \frac{b}{2a} = \sqrt{\frac{b^2 - 4ac}{4a^2}} \qquad \text{or} \qquad x + \frac{b}{2a} = -\sqrt{\frac{b^2 - 4ac}{4a^2}}$$

$$x + \frac{b}{2a} = \frac{\sqrt{b^2 - 4ac}}{\sqrt{4a^2}} \qquad\qquad x + \frac{b}{2a} = -\frac{\sqrt{b^2 - 4ac}}{\sqrt{4a^2}}$$

$$x + \frac{b}{2a} = \frac{\sqrt{b^2 - 4ac}}{2a} \qquad\qquad x + \frac{b}{2a} = -\frac{\sqrt{b^2 - 4ac}}{2a}$$

$$x = -\frac{b}{2a} + \frac{\sqrt{b^2 - 4ac}}{2a} \qquad\qquad x = -\frac{b}{2a} - \frac{\sqrt{b^2 - 4ac}}{2a}$$

$$x = \frac{-b + \sqrt{b^2 - 4ac}}{2a} \qquad\qquad x = \frac{-b - \sqrt{b^2 - 4ac}}{2a}$$

If $a < 0$, it can be shown that the same two solutions are obtained. The result is called the **quadratic formula.**

The quadratic formula

The solutions of $ax^2 + bx + c = 0$ $(a \neq 0)$ are given by the formula

$$x = \frac{-b \pm \sqrt{b^2 - 4ac}}{2a} \qquad \text{Read the symbol } \pm \text{ as "plus or minus."}$$

WARNING! Be sure to draw the fraction bar under both parts of the numerator, and be sure to draw the radical sign exactly over $b^2 - 4ac$. Do not write the quadratic formula as

$$x = -b \pm \frac{\sqrt{b^2 - 4ac}}{2a} \quad \text{or as} \quad x = -b \pm \sqrt{\frac{b^2 - 4ac}{2a}}$$

Solving Quadratic Equations Using the Quadratic Formula

EXAMPLE 1

Solving quadratic equations. Solve $2x^2 - 3x - 5 = 0$ using the quadratic formula.

Solution   In this equation $a = 2$, $b = -3$, and $c = -5$.

$$x = \frac{-b \pm \sqrt{b^2 - 4ac}}{2a}$$

$$= \frac{-(-3) \pm \sqrt{(-3)^2 - 4(2)(-5)}}{2(2)} \qquad \text{Substitute 2 for } a, -3 \text{ for } b, \text{ and } -5 \text{ for } c.$$

$$= \frac{3 \pm \sqrt{9 + 40}}{4} \qquad \text{Simplify within the radical.}$$

$$= \frac{3 \pm \sqrt{49}}{4} \qquad \text{Simplify: } \sqrt{49} = 7.$$

$$= \frac{3 \pm 7}{4}$$

This result represents two solutions. To find the first solution, we add in the numerator. To find the second, we subtract in the numerator.

$$x = \frac{3 + 7}{4} \qquad \text{or} \qquad x = \frac{3 - 7}{4}$$

$$x = \frac{10}{4} \qquad\qquad x = \frac{-4}{4}$$

$$x = \frac{5}{2} \qquad\qquad x = -1$$

Verify that both solutions satisfy the original equation.

SELF CHECK   Solve $3x^2 - 5x - 2 = 0$.   *Answer:* $2, -\frac{1}{3}$ ■

When using the quadratic formula, we should write the equation in $ax^2 + bx + c = 0$ form (called **quadratic form**) so that a, b, and c can be determined.

EXAMPLE 2

Writing an equation in quadratic form first. Solve $\frac{1}{3}x^2 = -\frac{2}{3}x - \frac{1}{6}$.

Solution   We begin by writing the equation in quadratic form:

$$\frac{1}{3}x^2 = -\frac{2}{3}x - \frac{1}{6}$$

$$2x^2 = -4x - 1 \qquad \text{Multiply both sides by 6.}$$

$$2x^2 + 4x + 1 = 0 \qquad \text{Add } 4x \text{ and 1 to both sides.}$$

In this equation, $a = 2$, $b = 4$, and $c = 1$.

$$x = \frac{-b \pm \sqrt{b^2 - 4ac}}{2a}$$

$$= \frac{-4 \pm \sqrt{4^2 - 4(2)(1)}}{2(2)} \qquad \text{Substitute 2 for } a, \text{ 4 for } b, \text{ and 1 for } c.$$

$$= \frac{-4 \pm \sqrt{16 - 8}}{4} \qquad \text{Simplify within the radical.}$$

$$= \frac{-4 \pm \sqrt{8}}{4}$$

$$= \frac{-4 \pm 2\sqrt{2}}{4} \qquad \text{Simplify: } \sqrt{8} = \sqrt{4 \cdot 2} = 2\sqrt{2}.$$

$$= \frac{-2 \pm \sqrt{2}}{2} \qquad \frac{-4 \pm 2\sqrt{2}}{4} = \frac{2(-2 \pm \sqrt{2})}{4} = \frac{\overset{1}{\cancel{2}}(-2 \pm \sqrt{2})}{\underset{1}{\cancel{2} \cdot 2}} = \frac{-2 \pm \sqrt{2}}{2}.$$

The solutions are $x = \dfrac{-2 + \sqrt{2}}{2}$ or $x = \dfrac{-2 - \sqrt{2}}{2}$. We can approximate the solutions using a calculator. To two decimal places, $x \approx -0.29$ or $x \approx -1.71$.

SELF CHECK   Solve $\frac{1}{2}x^2 = \frac{1}{3}x + \frac{1}{2}$.   *Answer:* $\dfrac{1 \pm \sqrt{10}}{3}$ ∎

Problem Solving

EXAMPLE 3

Taking a shortcut. Instead of using the existing hallways, students are wearing a path through a planted quad area to walk 195 feet directly from the classrooms to the cafeteria. See Figure 8-4. If the length of the hallway from the office to the cafeteria is 105 feet longer than the hallway from the office to the classrooms, how much walking are the students saving by taking the shortcut?

FIGURE 8-4

ANALYZE THE PROBLEM   The two hallways and the shortcut form a right triangle with a hypotenuse 195 feet long. We will use the Pythagorean theorem to solve this problem.

FORM AN EQUATION   If we let x represent the length (in feet) of the hallway from the classrooms to the office, then the length of the hallway from the office to the cafeteria is $(x + 105)$ feet. Substituting these lengths into the Pythagorean theorem, we have

$$a^2 + b^2 = c^2 \qquad \text{The Pythagorean theorem.}$$

$$x^2 + (x + 105)^2 = 195^2 \qquad \text{Substitute } x \text{ for } a, (x + 105) \text{ for } b, \text{ and } 195 \text{ for } c.$$

$$x^2 + x^2 + 105x + 105x + 11,025 = 38,025 \qquad \text{Use the FOIL method to find } (x + 105)^2.$$

$$2x^2 + 210x + 11,025 = 38,025 \qquad \text{Combine like terms.}$$

$$2x^2 + 210x - 27,000 = 0 \qquad \text{Subtract 38,025 from both sides.}$$

$$x^2 + 105x - 13,500 = 0 \qquad \text{Divide both sides by 2.}$$

SOLVE THE EQUATION   To solve $x^2 + 105x - 13{,}500 = 0$, we will use the quadratic formula with $a = 1$, $b = 105$, and $c = -13{,}500$.

$$x = \frac{-b \pm \sqrt{b^2 - 4ac}}{2a}$$

$$x = \frac{-105 \pm \sqrt{(105)^2 - 4(1)(-\mathbf{13{,}500})}}{2(\mathbf{1})}$$

$$x = \frac{-105 \pm \sqrt{65{,}025}}{2} \qquad \text{Simplify: } (105)^2 - 4(1)(-13{,}500) = 11{,}025 + 54{,}000 = 65{,}025.$$

$$x = \frac{-105 \pm 255}{2} \qquad \text{Using a calculator, } \sqrt{65{,}025} = 255.$$

$$x = \frac{150}{2} \quad \text{or} \quad x = \frac{-360}{2}$$

$$x = 75 \qquad\qquad x = -180 \qquad \text{Since the length of the hallway can't be negative, discard the solution } x = -180.$$

STATE THE CONCLUSION   The length of the hallway from the classrooms to the office is 75 feet. The length of the hallway from the office to the cafeteria is $75 + 105 = 180$ feet. Instead of using the hallways, a distance of $75 + 180 = 255$ feet, the students are taking the 195-foot short-cut to the cafeteria, a savings of $(255 - 195)$, or 60 feet.

CHECK THE RESULT   The length of the 180-foot hallway is 105 feet longer than the length of the 75-foot hallway. The sum of the squares of the lengths of the hallways is $75^2 + 180^2 = 38{,}025$. This equals the square of the length of the 195-foot shortcut. The answer checks. ∎

EXAMPLE 4

Mass transit. A bus company has 4,000 passengers daily, each currently paying a 75¢ fare. For each 15¢ fare increase, the company estimates that it will lose 50 passengers. If the company needs to bring in $6,570 per day to stay in business, what fare must be charged to produce this amount of revenue?

ANALYZE THE PROBLEM   To understand how a fare increase affects the number of passengers, let's consider what happens if there are two fare increases. We organize the data in a table. The fares are expressed in terms of dollars.

Number of increases	New fare	Number of passengers
One $0.15 increase	$0.75 + $0.15(1) = $0.90	4,000 − 50(1) = 3,950
Two $0.15 increases	$0.75 + $0.15(2) = $1.05	4,000 − 50(2) = 3,900

In general, the new fare will be the old fare ($0.75) plus the number of fare increases times $0.15. The number of passengers who will pay the new fare is 4,000 minus 50 times the number of $0.15 fare increases.

FORM AN EQUATION   If we let x represent the number of $0.15 fare increases necessary to bring in $6,570 daily, then $(0.75 + 0.15x)$ is the fare that must be charged. The number of passengers who will pay this fare is $(4{,}000 - 50x)$. We can now form the equation.

The bus fare	times	the number of passengers who will pay that fare	equals	$6,570.
$(0.75 + 0.15x)$	·	$(4{,}000 - 50x)$	=	6,570

SOLVE THE EQUATION

$(0.75 + 0.15x)(4,000 - 50x) = 6,570$

$3,000 - 37.5x + 600x - 7.5x^2 = 6,570$ Use the FOIL method.

$-7.5x^2 + 562.5x + 3,000 = 6,570$ Combine like terms: $-37.5x + 600x = 562.5x$.

$-7.5x^2 + 562.5x - 3,570 = 0$ Subtract 6,570 from both sides.

$7.5x^2 - 562.5x + 3,570 = 0$ Multiply both sides by -1 so that a, 7.5, is positive.

To solve this equation, we will use the quadratic formula.

$$x = \frac{-b \pm \sqrt{b^2 - 4ac}}{2a}$$

$$x = \frac{-(-562.5) \pm \sqrt{(-562.5)^2 - 4(7.5)(3,570)}}{2(7.5)}$$ Substitute 7.5 for a, -562.5 for b, and 3,570 for c.

$$x = \frac{562.5 \pm \sqrt{209,306.25}}{15}$$ Simplify: $(-562.5)^2 - 4(7.5)(3,570) = 316,406.25 - 107,100 = 209,306.25$.

$$x = \frac{562.5 \pm 457.5}{15}$$ Using a calculator, $\sqrt{209,306.25} = 457.5$.

$$x = \frac{1,020}{15} \quad \text{or} \quad x = \frac{105}{15}$$

$$x = 68 \qquad\qquad x = 7$$

STATE THE CONCLUSION If there are 7 fifteen-cent increases in the fare, the new fare will be $0.75 + $0.15(7) = $1.80. If there are 68 fifteen-cent increases in the fare, the new fare will be $0.75 + $0.15(68) = $10.95. Although this fare would bring in the necessary revenue, a $10.95 bus fare is unreasonable, so we discard it.

CHECK THE RESULT A fare of $1.80 will be paid by $[4,000 - 50(7)] = 3,650$ bus riders. The amount of revenue brought in would be $1.80(3,650) = $6,570. The answer checks. ∎

EXAMPLE 5

Lawyers. The number of lawyers N in the United States each year from 1980 to 1996 is approximated by $N = 660x^2 + 11,500x + 580,000$, where $x = 0$ corresponds to the year 1980, $x = 1$ corresponds to 1981, $x = 2$ corresponds to 1982, and so on. (Thus, $0 \le x \le 16$.) In what year does this model indicate that the United States had three-quarters of a million (750,000) lawyers?

Solution We will substitute 750,000 for N in the equation. Then we can solve for x, which will give the number of years after 1980 that the United States had approximately 750,000 lawyers.

$N = 660x^2 + 11,500x + 580,000$

$750,000 = 660x^2 + 11,500x + 580,000$ Replace N with 750,000.

$0 = 660x^2 + 11,500x - 170,000$ Subtract 750,000 from both sides so that the equation is in quadratic form.

We can simplify the computations by dividing both sides of the equation by 20, the common factor of 660, 11,500, and 170,000.

$0 = 33x^2 + 575x - 8,500$ Divide both sides by 20.

We solve this equation using the quadratic formula.

$$x = \frac{-b \pm \sqrt{b^2 - 4ac}}{2a}$$

$$x = \frac{-575 \pm \sqrt{(575)^2 - 4(33)(-8,500)}}{2(33)}$$

Substitute 33 for a, 575 for b, and $-8,500$ for c.

$$x = \frac{-575 \pm \sqrt{330,625 + 1,122,000}}{66}$$

Simplify:
$(575)^2 - 4(33)(-8,500) = 330,625 + 1,122,000.$

$$x = \frac{-575 \pm \sqrt{1,452,625}}{66}$$

$$x \approx \frac{630}{66} \quad \text{or} \quad x \approx \frac{-1,780}{66}$$

Use a calculator.

$$x \approx 9.5 \qquad x \approx -27.0$$

Since the model is defined only for $0 \le x \le 16$, we discard the second solution.

In 9.5 years after 1980, or midway through 1989, the United States had approximately 750,000 lawyers. ■

STUDY SET

Section 8.2

VOCABULARY

In Exercises 1–2, fill in the blanks to make the statements true.

1. An equation of the form $ax^2 + bx + c = 0$ ($a \ne 0$) is a _____ equation.

2. $x = \dfrac{-b \pm \sqrt{b^2 - 4ac}}{2a}$ is called the _____ formula.

CONCEPTS

3. Consider the quadratic equation $4x^2 - 2x = 0$.
   a. Solve it by factoring.
   b. What are a, b, and c? Solve it using the quadratic formula.

4. Consider the quadratic equation $x^2 = 8$.
   a. Solve it by the square root method.
   b. What are a, b, and c? Solve it using the quadratic formula.

5. Tell whether each statement is true or false.
   a. Any quadratic equation can be solved by using the quadratic formula.
   b. Any quadratic equation can be solved by completing the square.

6. What is **wrong** with the beginning of the solution shown below?

   Solve $x^2 - 3x = 2$.
   $$a = 1 \qquad b = -3 \qquad c = 2$$

7. What form would the quadratic formula take if, in developing it, we began with the general quadratic equation expressed as $rc^2 + sc + t = 0$ ($r \ne 0$)?

8. A student used the quadratic formula to solve a quadratic equation and obtained $x = \dfrac{-2 \pm \sqrt{3}}{2}$.
   a. How many solutions does the equation have? What are they exactly?

   b. Graph the solutions on a number line.

9. Solve $x^2 - 4x + 1 = 0$ using the quadratic formula. Simplify the result.

10. Solve $x^2 + 4x + 2 = 0$ using the quadratic formula. Simplify the result.

NOTATION

11. On a quiz, students were asked to write the quadratic formula. What is **wrong** with each answer shown below?

a. $x = -b \pm \dfrac{\sqrt{b^2 - 4ac}}{2a}$

b. $x = \dfrac{-b\sqrt{b^2 - 4ac}}{2a}$

12. In reading $\dfrac{-b \pm \sqrt{b^2 - 4ac}}{2a}$, we say, "The

_____ of b, plus or _____ the

square _____ of b _____

minus ___ times a times c, all _____ $2a$."

PRACTICE

In Exercises 13–36, use the quadratic formula to solve each equation.

13. $x^2 + 3x + 2 = 0$ **14.** $x^2 - 3x + 2 = 0$

15. $x^2 + 12x = -36$ **16.** $y^2 - 18y = -81$

17. $2x^2 + 5x - 3 = 0$ **18.** $6x^2 - x - 1 = 0$

19. $5x^2 + 5x + 1 = 0$ **20.** $4w^2 + 6w + 1 = 0$

21. $8u = -4u^2 - 3$ **22.** $4t + 3 = 4t^2$

23. $16y^2 + 8y - 3 = 0$ **24.** $16x^2 + 16x + 3 = 0$

25. $x^2 - \dfrac{14}{15}x = \dfrac{8}{15}$ **26.** $x^2 = -\dfrac{5}{4}x + \dfrac{3}{2}$

27. $\dfrac{x^2}{2} + \dfrac{5}{2}x = -1$ **28.** $\dfrac{x^2}{8} - \dfrac{x}{4} = \dfrac{1}{2}$

29. $2x^2 - 1 = 3x$ **30.** $-9x = 2 - 3x^2$

31. $-x^2 + 10x = 18$ **32.** $-3x = \dfrac{x^2}{2} + 2$

33. $x^2 - 6x = 391$ **34.** $x^2 - 27x = 280$

35. $x^2 - \dfrac{5}{3} = -\dfrac{11}{6}x$ **36.** $x^2 - \dfrac{1}{2} = \dfrac{2}{3}x$

In Exercises 37–38, use the quadratic formula and a scientific calculator to solve each equation. Give all answers to the nearest hundredth.

37. $0.7x^2 - 3.5x - 25 = 0$
38. $-4.5x^2 + 0.2x + 3.75 = 0$

APPLICATIONS

39. MOVIE THEATER SCREEN The largest permanent movie screen is in the Panasonic Imax theater at Darling Harbor, Sydney, Australia. The rectangular screen has an area of 11,349 square feet. Find the dimensions of the screen if it is 20 feet longer than it is wide.

40. ROCK CONCERT During a 1997 tour, the rock group U2 used the world's largest LED (Light Emitting Diode) electronic screen as part of the stage backdrop. The screen, with an area of 9,520 square feet, had a length that was 2 feet more than three times its width. Find the dimensions of the LED screen.

41. CENTRAL PARK Central Park is one of New York's best-known landmarks. Rectangular in shape, its length is 5 times its width. When measured in miles, its perimeter numerically exceeds its area by 4.75. Find the dimensions of Central Park if we know that its width is less than 1 mile.

42. ANCIENT HISTORY One of the most important cities of the ancient world was Babylon. Greek historians wrote that the city was square-shaped. Measured in miles, its area numerically exceeded its perimeter by about 124. Find its dimensions. (Round to the nearest tenth.)

43. BADMINTON The person who wrote the instructions for setting up the badminton net shown in Illustration 1 forgot to give the specific dimensions for securing the pole. How long is the support string?

ILLUSTRATION 1

From the base of the pole, move out a distance of 1 inch less than half the length of the string, and place an anchor stake in the ground.

Move up the pole a distance that is 4 inches less than the length of the string. Secure the string to the pole.

String

44. RIGHT TRIANGLE The hypotenuse of a right triangle is 2.5 units long. The longer leg is 1.7 units longer than the shorter leg. Find the lengths of the sides of the triangle.

45. SCHOOL DANCE Tickets to a school dance cost $4, and the projected attendance is 300 persons. It is further projected that for every 10¢ increase in ticket price, the average attendance will decrease by 5. At what ticket price will the receipts from the dance be $1,248?

46. TICKET SALES A carnival at a county fair normally sells three thousand 25¢ ride tickets on a Saturday. For each 5¢ increase in price, management estimates that 80 fewer tickets will be sold. What increase in ticket price will produce $994 of revenue on Saturday?

47. MAGAZINE SALES The *Gazette's* profit is $20 per year for each of its 3,000 subscribers. Management estimates that the profit per subscriber will increase by 1¢ for each additional subscriber over the current 3,000. How many subscribers will bring a total profit of $120,000?

48. POLYGONS The five-sided polygon called a *pentagon,* shown in Illustration 2, has 5 diagonals. The number of diagonals d of a polygon of n sides is given by the formula

ILLUSTRATION 2

$$d = \frac{n(n-3)}{2}$$

Find the number of sides of a polygon if it has 275 diagonals.

49. INVESTMENT RATES A woman invests $1,000 in a mutual fund for which interest is compounded annually at a rate r. After one year, she deposits an additional $2,000. After two years, the balance in the account is

$$\$1,000(1 + r)^2 + \$2,000(1 + r)$$

If this amount is $3,368.10, find r.

50. METAL FABRICATION A box with no top is to be made by cutting a 2-inch square from each corner of the square sheet of metal shown in Illustration 3. After bending up the sides, the volume of the box is to be 220 cubic inches. How large should the piece of metal be? Round to the nearest hundredth.

51. RETIREMENT The labor force participation rate P (in percent) for men ages 55–64 from 1970 to 1996 is approximated by the quadratic equation $P = 0.027x^2 - 1.363x + 82.5$, where $x = 0$ corresponds to the year 1970, $x = 1$ corresponds to 1971, $x = 2$ corresponds to 1972, and so on. (Thus, $0 \le x \le 26$.)
   a. To the nearest percent, what percent of men ages 55–64 were in the workforce in 1970?
   b. When does the model indicate that 75% of the men ages 55–64 were part of the workforce?

52. SPACE PROGRAM The yearly budget B (in billions of dollars) for the National Aeronautics and Space Administration (NASA) is approximated by the quadratic equation $B = -0.1492x^2 + 1.8058x + 9$, where x is the number of years since 1988 and $0 \le x \le 8$.
   a. Use the model to find the approximate budget for NASA in 1996. Round to the nearest tenth.
   b. In what year does the model indicate that NASA's budget was about $12 billion?
   c. For the years 1988–1996, was NASA's budget ever $15 billion? Explain how you know.

WRITING

53. Explain why the quadratic formula, in most cases, is less tedious to use in solving a quadratic equation than is the method of completing the square.

54. On an exam, a student was asked to solve the equation $-4w^2 - 6w - 1 = 0$. Her first step was to multiply both sides of the equation by -1. She then used the quadratic formula to solve $4w^2 + 6w + 1 = 0$ instead. Is this a valid approach? Explain.

REVIEW

In Exercises 55–58, change each radical to an exponential expression.

55. $\sqrt{n}$

56. $\sqrt[7]{\dfrac{3}{8}r^2 s}$

57. $\sqrt[4]{3b}$

58. $3\sqrt[3]{c^2 - d^2}$

In Exercises 59–62, write each expression in radical form.

59. $t^{1/3}$

60. $\left(\dfrac{3}{4}m^2 n^2\right)^{1/5}$

61. $(3t)^{1/4}$

62. $(c^2 + d^2)^{1/2}$

ILLUSTRATION 3

▶ 8.3   Quadratic Functions and Their Graphs

In this section, you will learn about

> Quadratic functions ■ Graphs of $f(x) = ax^2$ ■ Graphs of $f(x) = ax^2 + k$ ■ Graphs of $f(x) = a(x - h)^2$ ■ Graphs of $f(x) = a(x - h)^2 + k$ ■ Graphs of $f(x) = ax^2 + bx + c$ ■ Determining minimum and maximum values ■ Solving quadratic equations graphically

Introduction   The graph in Figure 8-5 shows a trampolinist's distance from the ground (in relation to time) as she bounds into the air and then falls back down to the trampoline.

FIGURE 8-5

From the graph, we can see that the trampolinist is 14 feet above the ground 0.5 second after bounding upward and that her height above the ground after 1.75 seconds is 9 feet.

The parabola shown in Figure 8-5 is the graph of the *quadratic function* $s(t) = -16t^2 + 32t + 2$. In this section, we will discuss two forms in which quadratic functions are written and how to graph them.

Quadratic Functions

Quadratic functions

> A **quadratic function** is a second-degree polynomial function of the form
>
> $$f(x) = ax^2 + bx + c \qquad \text{or} \qquad y = ax^2 + bx + c \quad (a \neq 0)$$
>
> where a, b, and c are real numbers.

EXAMPLE 1

Quadratic functions as models. The quadratic function $s(t) = -16t^2 + 32t + 2$ gives the distance (in feet) that the trampolinist shown in Figure 8-5 is from the ground, t seconds after bounding upward. How far is she from the ground after being in the air for $\frac{3}{4}$ second?

Solution   To find her distance from the ground, we find the value of the function for $t = \frac{3}{4} = 0.75$.

$$s(t) = -16t^2 + 32t + 2$$
$$s(0.75) = -16(0.75)^2 + 32(0.75) + 2 \quad \text{Replace } t \text{ with } 0.75.$$
$$= -9 + 24 + 2$$
$$= 17$$

The trampolinist is 17 feet off the ground $\frac{3}{4}$ second after bounding upward.

SELF CHECK

Find the distance the trampolinist is from the ground after being in the air for $1\frac{1}{2}$ seconds.   *Answer:* 14 feet   ∎

We have seen that important information can be gained from the graph of a quadratic function. We now begin a discussion of how to graph such a function by considering the simplest case, quadratic functions of the form $f(x) = ax^2$.

Graphs of $f(x) = ax^2$

EXAMPLE 2

Graphing quadratic functions. Graph   **a.** $f(x) = x^2$,   **b.** $g(x) = 3x^2$, and   **c.** $h(x) = \frac{1}{3}x^2$.

Solution   We can make a table of ordered pairs that satisfy each equation, plot each point, and join them with a smooth curve, as in Figure 8-6. We note that the graph of $h(x) = \frac{1}{3}x^2$ is wider than the graph of $f(x) = x^2$, and that the graph of $g(x) = 3x^2$ is narrower than the graph of $f(x) = x^2$. In the function $f(x) = ax^2$, the smaller the value of $|a|$, the wider the graph.

FIGURE 8-6

$f(x) = x^2$

x	$f(x)$	$(x, f(x))$
-2	4	$(-2, 4)$
-1	1	$(-1, 1)$
0	0	$(0, 0)$
1	1	$(1, 1)$
2	4	$(2, 4)$

$g(x) = 3x^2$

x	$g(x)$	$(x, g(x))$
-2	12	$(-2, 12)$
-1	3	$(-1, 3)$
0	0	$(0, 0)$
1	3	$(1, 3)$
2	12	$(2, 12)$

$h(x) = \frac{1}{3}x^2$

x	$h(x)$	$(x, h(x))$
-2	$\frac{4}{3}$	$\left(-2, \frac{4}{3}\right)$
-1	$\frac{1}{3}$	$\left(-1, \frac{1}{3}\right)$
0	0	$(0, 0)$
1	$\frac{1}{3}$	$\left(1, \frac{1}{3}\right)$
2	$\frac{4}{3}$	$\left(2, \frac{4}{3}\right)$

∎

EXAMPLE 3

Graphing quadratic functions. Graph $f(x) = -3x^2$.

Solution We make a table of ordered pairs that satisfy the equation, plot each point, and join them with a smooth curve, as in Figure 8-7. We see that the parabola opens downward and has the same shape as the graph of $f(x) = 3x^2$.

FIGURE 8-7

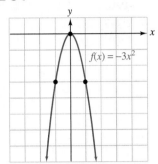

$f(x) = -3x^2$

x	$f(x)$	$(x, f(x))$
-2	-12	$(-2, -12)$
-1	-3	$(-1, -3)$
0	0	$(0, 0)$
1	-3	$(1, -3)$
2	-12	$(2, -12)$

SELF CHECK Graph $f(x) = -\dfrac{1}{3}x^2$.

Answer:

The graphs of quadratic functions of the form $f(x) = ax^2$ are parabolas. The lowest point of a parabola that opens upward, or the highest point of a parabola that opens downward, is called the **vertex** of the parabola. For example, the vertex of the parabola shown in Figure 8-7 is the point $(0, 0)$. The vertical line, called an **axis of symmetry,** that passes through the vertex divides the parabola into two congruent halves. The axis of symmetry of the parabola shown in Figure 8-7 is the line $x = 0$ (the y-axis).

The results of Examples 2 and 3 confirm the following facts.

The graph of $f(x) = ax^2$

The graph of $f(x) = ax^2$ is a parabola opening upward when $a > 0$ and downward when $a < 0$, with vertex at the point $(0, 0)$ and axis of symmetry the line $x = 0$.

Graphs of $f(x) = ax^2 + k$

EXAMPLE 4

Graphing quadratic functions. Graph **a.** $f(x) = 2x^2$, **b.** $g(x) = 2x^2 + 3$, and **c.** $h(x) = 2x^2 - 3$.

Solution We make a table of ordered pairs that satisfy each equation, plot each point, and join them with a smooth curve, as in Figure 8-8. We note that the graph of $g(x) = 2x^2 + 3$ is identical to the graph of $f(x) = 2x^2$, except that it has been translated 3 units upward. The graph of $h(x) = 2x^2 - 3$ is identical to the graph of $f(x) = 2x^2$, except that it has been translated 3 units downward. In each case, the axis of symmetry is the line $x = 0$.

FIGURE 8-8

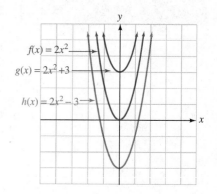

$f(x) = 2x^2$		
x	$f(x)$	$(x, f(x))$
-2	8	$(-2, 8)$
-1	2	$(-1, 2)$
0	0	$(0, 0)$
1	2	$(1, 2)$
2	8	$(2, 8)$

$g(x) = 2x^2 + 3$		
x	$g(x)$	$(x, g(x))$
-2	11	$(-2, 11)$
-1	5	$(-1, 5)$
0	3	$(0, 3)$
1	5	$(1, 5)$
2	11	$(2, 11)$

$h(x) = 2x^2 - 3$		
x	$h(x)$	$(x, h(x))$
-2	5	$(-2, 5)$
-1	-1	$(-1, -1)$
0	-3	$(0, -3)$
1	-1	$(1, -1)$
2	5	$(2, 5)$

The results of Example 4 confirm the following facts.

The graph of $f(x) = ax^2 + k$

The graph of $f(x) = ax^2 + k$ is a parabola having the same shape as $f(x) = ax^2$ but translated upward k units if k is positive and downward $|k|$ units if k is negative. The vertex is at the point $(0, k)$, and the axis of symmetry is the line $x = 0$.

Graphs of $f(x) = a(x - h)^2$

E X A M P L E   5

Graphing quadratic functions. Graph **a.** $f(x) = 2x^2$, **b.** $g(x) = 2(x - 3)^2$, and **c.** $h(x) = 2(x + 3)^2$.

Solution   We make a table of ordered pairs that satisfy each equation, plot each point, and join them with a smooth curve, as in Figure 8-9. We note that the graph of $g(x) = 2(x - 3)^2$ is identical to the graph of $f(x) = 2x^2$, except that it has been translated 3 units to the right. The graph of $h(x) = 2(x + 3)^2$ is identical to the graph of $f(x) = 2x^2$, except that it has been translated 3 units to the left.

FIGURE 8-9

$f(x) = 2x^2$		
x	$f(x)$	$(x, f(x))$
-2	8	$(-2, 8)$
-1	2	$(-1, 2)$
0	0	$(0, 0)$
1	2	$(1, 2)$
2	8	$(2, 8)$

$g(x) = 2(x - 3)^2$		
x	$g(x)$	$(x, g(x))$
1	8	$(1, 8)$
2	2	$(2, 2)$
3	0	$(3, 0)$
4	2	$(4, 2)$
5	8	$(5, 8)$

$h(x) = 2(x + 3)^2$		
x	$h(x)$	$(x, h(x))$
-5	8	$(-5, 8)$
-4	2	$(-4, 2)$
-3	0	$(-3, 0)$
-2	2	$(-2, 2)$
-1	8	$(-1, 8)$

The results of Example 5 confirm the following facts.

The graph of $f(x) = a(x - h)^2$

The graph of $f(x) = a(x - h)^2$ is a parabola having the same shape as $f(x) = ax^2$ but translated h units to the right if h is positive and $|h|$ units to the left if h is negative. The vertex is at the point $(0, h)$, and the axis of symmetry is the line $x = h$.

Graphs of $f(x) = a(x - h)^2 + k$

EXAMPLE 6

Graphing quadratic functions. Graph $f(x) = 2(x - 3)^2 - 4$.

Solution   The graph of $f(x) = 2(x - 3)^2 - 4$ is identical to the graph of $g(x) = 2(x - 3)^2$, except that it has been translated 4 units downward. The graph of $g(x) = 2(x - 3)^2$ is identical to the graph of $h(x) = 2x^2$, except that it has been translated 3 units to the right. Thus, to graph $f(x) = 2(x - 3)^2 - 4$, we can graph $h(x) = 2x^2$ and shift it 3 units to the right and then 4 units downward, as shown in Figure 8-10.

The vertex of the graph is the point $(3, -4)$, and the axis of symmetry is the line $x = 3$.

FIGURE 8-10

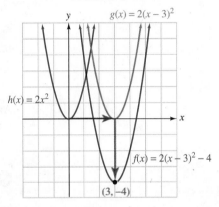

SELF CHECK   Graph $f(x) = 2(x + 3)^2 + 1$. Label the vertex and draw the axis of symmetry.

Answer:

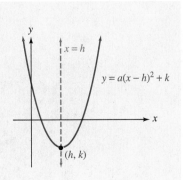

The results of Example 6 confirm the following facts.

The graph of $f(x) = a(x - h)^2 + k$

The graph of the quadratic function

$$f(x) = a(x - h)^2 + k$$

or

$$y = a(x - h)^2 + k$$

(where $a \neq 0$) is a parabola with vertex at (h, k). The parabola opens upward when $a > 0$ and downward when $a < 0$. The axis of symmetry is the line $x = h$.

EXAMPLE 7

Determining the vertex and the axis of symmetry. Consider the graph of $f(x) = -3(x + 1)^2 - 4$.

a. Does the graph open upward or downward?

b. What are the coordinates of the vertex?

c. What is the axis of symmetry?

Solution Rewriting the given function in $f(x) = a(x - h)^2 + k$ form, we have

Standard form requires a minus sign here.   Standard form requires a plus sign here.

$$f(x) = \underset{a}{-3} \, [x - \underset{h}{(-1)}\,]^2 + \underset{k}{(-4)}$$

a. Since $a = -3 < 0$, the parabola opens downward.

b. The vertex is $(h, k) = (-1, -4)$.

c. The axis of symmetry is the line $x = h$. In this case, $x = -1$.

SELF CHECK

Answer parts a, b, and c of Example 7 for the graph of $y = 6(x - 5)^2 + 1$.

Answers: **a.** upward, **b.** $(5, 1)$, **c.** $x = 5$ ∎

Graphs of $f(x) = ax^2 + bx + c$

To graph functions of the form $f(x) = ax^2 + bx + c$, we complete the square to write the function in the form $f(x) = a(x - h)^2 + k$.

EXAMPLE 8

Completing the square to determine the vertex. Graph $f(x) = 2x^2 - 4x - 1$.

Solution **Step 1:** We complete the square on x to write the given function in the form $f(x) = a(x - h)^2 + k$.

$$f(x) = 2x^2 - 4x - 1$$
$$f(x) = 2(x^2 - 2x) - 1 \quad \text{Factor 2 from } 2x^2 - 4x.$$

Now we complete the square on x by adding 1 inside the parentheses. Since this adds 2 to the right-hand side, we also subtract 2 from the right-hand side.

$$f(x) = 2(x^2 - 2x + 1) - 1 - 2 \quad \text{To complete the square, one-half of } -2 = -1 \text{ and } (-1)^2 = 1.$$

1. $$f(x) = 2(x - 1)^2 - 3 \quad \text{Factor } x^2 - 2x + 1 \text{ and combine like terms.}$$

Step 2: From Equation 1, we can see that $h = 1$ and $k = -3$, so the vertex will be at the point $(1, -3)$, and the axis of symmetry is $x = 1$. We plot the vertex and axis of symmetry on the coordinate system in Figure 8-11.

Step 3: Finally, we construct a table of values, plot the points, and draw the graph, which appears in the figure.

FIGURE 8-11

$$f(x) = 2x^2 - 4x - 1$$

x	$f(x)$	$(x, f(x))$
-1	5	$(-1, 5)$
0	-1	$(0, -1)$
2	-1	$(2, -1)$
3	5	$(3, 5)$

The x-coordinate of the vertex is 1. Choose values for x close to 1.

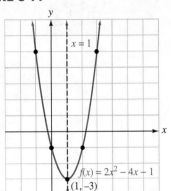

SELF CHECK   Graph $f(x) = 2x^2 - 4x + 1$.   *Answer:*

We can derive a formula for the vertex of the graph of $f(x) = ax^2 + bx + c$ by completing the square in the same manner as we did in Example 8. After using similar steps, the result is

$$f(x) = a\left[x - \left(-\frac{b}{2a} \right) \right]^2 + \frac{4ac - b^2}{4a}$$

$$\underset{h}{\uparrow} \qquad\qquad \underset{k}{\uparrow}$$

The x-coordinate of the vertex is $-\frac{b}{2a}$. The y-coordinate of the vertex is $\frac{4ac - b^2}{4a}$. However, we can also find the y-coordinate of the vertex by substituting the x-coordinate, $-\frac{b}{2a}$, for x in the quadratic function.

Formula for the vertex of a parabola

The vertex of the graph of the quadratic function $f(x) = ax^2 + bx + c$ is

$$\left(-\frac{b}{2a}, f\left(-\frac{b}{2a} \right) \right)$$

and the axis of symmetry of the parabola is the line $x = -\frac{b}{2a}$.

In Example 8, for the function $f(x) = 2x^2 - 4x - 1$, $a = 2$ and $b = -4$. To find the vertex of its graph, we compute

$$-\frac{b}{2a} = -\frac{-4}{2(2)} \qquad\qquad f\left(-\frac{b}{2a} \right) = f(1)$$

$$= -\frac{-4}{4} \qquad\qquad\qquad = 2(1)^2 - 4(1) - 1$$

$$= 1 \qquad\qquad\qquad\qquad = -3$$

The vertex is the point $(1, -3)$. This agrees with the result we obtained in Example 8 by completing the square.

ACCENT ON TECHNOLOGY *Finding the Vertex*

To graph the function $f(x) = 2x^2 + 6x - 3$ and find the coordinates of the vertex and the axis of symmetry of the parabola, we can use a graphing calculator with window settings of $[-10, 10]$ for x and $[-10, 10]$ for y. If we enter the function, we will obtain the graph shown in Figure 8-12(a).

We then trace to move the cursor to the lowest point on the graph, as shown in Figure 8-12(b). By zooming in, we can determine that the vertex is the point $\left(-\frac{3}{2}, -\frac{15}{2}\right)$ and that the line $x = -\frac{3}{2}$ is the axis of symmetry.

FIGURE 8-12

(a)   (b)

Determining Minimum and Maximum Values

It is often useful to know the smallest or largest possible value a quantity can assume. For example, companies try to minimize their costs and maximize their profits. If the quantity is expressed by a quadratic function, the vertex of the graph of the function gives its minimum or maximum value.

E X A M P L E   9

Minimizing costs. A glassworks that makes lead crystal vases has daily production costs given by the function $C(x) = 0.2x^2 - 10x + 650$, where x is the number of vases made each day. How many vases should be produced to minimize the per-day costs? What will the costs be?

Solution   The graph of $C(x) = 0.2x^2 - 10x + 650$ is a parabola opening upward. The vertex is the lowest point on the graph. To find the vertex, we compute

$$-\frac{b}{2a} = -\frac{-10}{2(0.2)} \quad b = -10 \text{ and } a = 0.2. \qquad f\left(-\frac{b}{2a}\right) = f(25)$$

$$= -\frac{-10}{0.4} \qquad\qquad\qquad = 0.2(25)^2 - 10(25) + 650$$

$$= 25 \qquad\qquad\qquad\qquad = 525$$

The vertex is the point $(25, 525)$, and it indicates that the costs are a minimum of \$525 when 25 vases are made daily.

To solve this problem with a graphing calculator with window settings of $[0, 50]$ for x and $[0, 1,000]$ for y, we graph the function $C(x) = 0.2x^2 - 10x + 650$. By using TRACE and ZOOM, we can locate the vertex of the graph. See Figure 8-13. The coordinates of the vertex indicate that the minimum cost is \$525 when the number of vases produced is 25.

FIGURE 8-13

EXAMPLE 10

Maximizing area. A kennel owner wants to build the rectangular pen shown in Figure 8-14(a) to house his dog. If he uses one side of his barn, find the maximum area that he can enclose with 80 feet of fencing.

Solution   We can let the width of the area be represented by w. Then the length is represented by $80 - 2w$. The function that gives the area the pen encloses is

$$A = (80 - 2w)w \quad A = lw.$$

We can find the maximum value of A by determining the vertex of the graph of the function. This time, we will find the vertex by completing the square.

$A = 80w - 2w^2$	Use the distributive property to remove parentheses.
$\quad = -2(w^2 - 40w)$	Factor out -2.
$\quad = -2(w^2 - 40w + \mathbf{400}) + \mathbf{800}$	Complete the square on w. Inside the parentheses, add $\left(\frac{-40}{2}\right)^2 = 400$. Since this subtracts 800 from the right-hand side, add 800 to the right-hand side.
$\quad = -2(w - 20)^2 + 800$	Factor $w^2 - 40w + 400$.

Thus, the coordinates of the vertex of the graph of the quadratic function are (20, 800), and the maximum area is 800 square feet.

To solve this problem with a graphing calculator with window settings of [0, 50] for x and [0, 1,000] for y, we graph the function $A = -2w^2 + 80w$ to get the graph in Figure 8-14(b). By using TRACE and ZOOM, we can determine the vertex of the graph, which indicates that the maximum area is 800 square feet when the width is 20 feet.

FIGURE 8-14

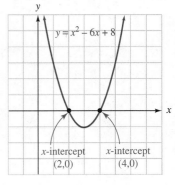

(a)

(b) ■

Solving Quadratic Equations Graphically

We can solve quadratic equations graphically. For example, the solutions of $x^2 - 6x + 8 = 0$ are the numbers x that will make y equal to 0 in the quadratic function $y = x^2 - 6x + 8$. To find these numbers, we inspect the graph of $y = x^2 - 6x + 8$ and locate the points on the graph with a y-coordinate of 0. In Figure 8-15, these points are (2, 0) and (4, 0), the x-intercepts of the graph. We can conclude that the x-coordinates of the x-intercepts, $x = 2$ and $x = 4$, are the solutions of $x^2 - 6x + 8 = 0$.

FIGURE 8-15

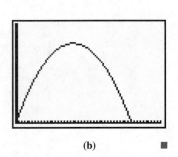

ACCENT ON TECHNOLOGY *Solving Quadratic Equations Graphically*

We can use a graphing calculator to find approximate solutions of quadratic equations. For example, the solutions of $0.7x^2 + 2x - 3.5 = \mathbf{0}$ are the numbers x that will make $y = 0$ in the quadratic function $y = 0.7x^2 + 2x - 3.5$. To approximate these numbers, we graph the quadratic function and read the x-intercepts from the graph.

We can use the standard window settings of $[-10, 10]$ for x and $[-10, 10]$ for y and graph the function. We then trace to move the cursor to each x-intercept, as in Figures 8-16(a) and 8-16(b). From the graph, we can read the approximate value of the x-coordinate of each x-intercept. For better results, we can zoom in.

FIGURE 8-16

(a)   (b)   (c)

We can also approximate the solutions of $0.7x^2 + 2x - 3.5 = 0$ using the ZERO feature. In Figure 8-16(c), we see that x-coordinate of the leftmost x-intercept of the graph is given as -4.082025. This means that an approximate solution of the equation is -4.08. To find the positive x-intercept, we use similar steps.

When solving quadratic equations graphically, we must consider three possibilities. If the graph of the associated quadratic function has two x-intercepts, the quadratic equation has two real-number solutions. Figure 8-17(a) shows an example of this. If the graph has one x-intercept, as shown in Figure 8-17(b), the equation has one real-number solution. Finally, if the graph does not have an x-intercept, as shown in Figure 8-17(c), the equation does not have any real-number solutions.

FIGURE 8-17

$x^2 + x - 2 = 0$
has two solutions,
$x = -2$ and $x = 1$.

$2x^2 + 12x + 18 = 0$
has one solution,
$x = -3$.

$-x^2 + 4x - 5 = 0$
has no real-number
solutions.

(a)   (b)   (c)

VOCABULARY

In Exercises 1–4, refer to Illustration 1. Fill in the blanks to make the statements true.

ILLUSTRATION 1

$f(x) = 2x^2 - 4x + 1$

$(1, -1)$

$x = 1$

1. The function $f(x) = 2x^2 - 4x + 1$ is graphed in Illustration 1. It is called a _____ function.

2. The graph in Illustration 1 is called a _____.

3. The lowest point on the graph is $(1, -1)$. This is called the _____.

4. The vertical line $x = 1$ divides the parabola into two halves. This line is called the _____.

CONCEPTS

In Exercises 5–6, make a table of values for each function. Then graph them on the same coordinate system.

5. $f(x) = x^2$, $g(x) = 2x^2$, $h(x) = \dfrac{1}{2}x^2$

6. $f(x) = -x^2$, $g(x) = -\dfrac{1}{4}x^2$, $h(x) = -4x^2$

In Exercises 7–8, make a table of values to graph function f. Then use a translation to graph the other two functions on the same coordinate system.

7. $f(x) = 4x^2$, $g(x) = 4x^2 + 3$, $h(x) = 4x^2 - 2$

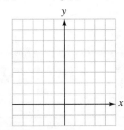

8. $f(x) = 3x^2$, $g(x) = 3(x + 2)^2$, $h(x) = 3(x - 3)^2$

In Exercises 9–10, make a table of values to graph function f. Then use a series of translations to graph function g on the same coordinate system.

9. $f(x) = -3x^2$, $g(x) = -3(x - 2)^2 - 1$

10. $f(x) = -\dfrac{1}{2}x^2$, $g(x) = -\dfrac{1}{2}(x + 1)^2 + 2$

11. Use the graph of $y = \frac{x^2}{10} - \frac{x}{5} - \frac{3}{2}$, shown below, to estimate the solutions of the quadratic equation $\frac{x^2}{10} - \frac{x}{5} - \frac{3}{2} = 0$. Check your answers.

12. Three quadratic equations are to be solved graphically. The graphs of their associated quadratic functions are shown below. Tell which graph indicates that the equation has
   a. two real solutions.
   b. one real solution.
   c. no real solutions.

i   ii   iii

NOTATION

13. The function $f(x) = 2(x + 1)^2 + 6$ is written in the form $f(x) = a(x - h)^2 + k$. Is $h = -1$ or is $h = 1$? Explain.

14. The vertex of a quadratic function $f(x) = ax^2 + bx + c$ is given by the formula $\left(-\frac{b}{2a}, f\left(-\frac{b}{2a}\right)\right)$. Explain what is meant by the notation $f\left(-\frac{b}{2a}\right)$.

PRACTICE

*In Exercises 15–24, find the coordinates of the vertex and the axis of symmetry of the graph of each function. If necessary, complete the square on x to write the equation in the form $y = a(x - h)^2 + k$. **Do not graph the equation,** but tell whether the graph will open upward or downward.*

15. $y = (x - 1)^2 + 2$

16. $y = 2(x - 2)^2 - 1$

17. $f(x) = 2(x + 3)^2 - 4$

18. $f(x) = -3(x + 1)^2 + 3$

19. $y = 2x^2 - 4x$

20. $y = 3x^2 - 3$

21. $y = -4x^2 + 16x + 5$

22. $y = 5x^2 + 20x + 25$

23. $y = 3x^2 + 4x + 2$

24. $y = -6x^2 + 5x - 7$

In Exercises 25–40, first determine the vertex and the axis of symmetry of the graph of the function. Then plot several points and complete the graph. (See Example 8.)

25. $f(x) = (x - 3)^2 + 2$ **26.** $f(x) = (x + 1)^2 - 2$

27. $f(x) = -(x - 2)^2$ **28.** $f(x) = -(x + 2)^2$

29. $f(x) = -2(x + 3)^2 + 4$ **30.** $f(x) = 2(x - 2)^2 - 4$

31. $f(x) = (x - 3)^2 + 2$ **32.** $f(x) = (x + 1)^2 - 2$

33. $f(x) = x^2 + x - 6$ **34.** $y = x^2 - x - 6$

35. $y = -x^2 + 2x + 3$

36. $f(x) = -x^2 + x + 2$

37. $f(x) = -3x^2 + 2x$

38. $f(x) = 5x + x^2$

39. $y = -12x^2 - 6x + 6$

40. $f(x) = -2x^2 + 4x + 3$

In Exercises 41–44, use a graphing calculator to find the coordinates of the vertex of the graph of each quadratic function. Round to the nearest hundredth.

41. $y = 2x^2 - x + 1$

42. $y = x^2 + 5x - 6$

43. $y = 7 + x - x^2$

44. $y = 2x^2 - 3x + 2$

In Exercises 45–48, use a graphing calculator to solve each equation. If an answer is not exact, round to the nearest hundredth.

45. $x^2 + x - 6 = 0$

46. $2x^2 - 5x - 3 = 0$

47. $0.5x^2 - 0.7x - 3 = 0$

48. $2x^2 - 0.5x - 2 = 0$

APPLICATIONS

49. FIREWORKS A fireworks shell is shot straight up with an initial velocity of 120 feet per second. Its height s after t seconds is given by the equation $s = 120t - 16t^2$. If the shell is designed to explode when it reaches its maximum height, how long after being fired, and at what height, will the fireworks appear in the sky?

50. BALLISTICS From the top of the building in Illustration 2, a ball is thrown straight up with an initial velocity of 32 feet per second. The equation

$$s = -16t^2 + 32t + 48$$

gives the height s of the ball t seconds after it is thrown. Find the maximum height reached by the ball and the time it takes for the ball to hit the ground.

ILLUSTRATION 2

51. FENCING A FIELD See Illustration 3. A farmer wants to fence in three sides of a rectangular field with 1,000 feet of fencing. The other side of the rectangle will be a river. If the enclosed area is to be maximum, find the dimensions of the field.

ILLUSTRATION 3

52. POLICE INVESTIGATION A police officer seals off the scene of a car collision using a roll of yellow police tape that is 300 feet long, as shown in Illustration 4. What dimensions should be used to seal off the maximum rectangular area around the collision? What is the maximum area?

ILLUSTRATION 4

53. OPERATING COSTS The cost C in dollars of operating a certain concrete-cutting machine is related to the number of minutes n the machine is run by the function

$$C(n) = 2.2n^2 - 66n + 655$$

For what number of minutes is the cost of running the machine a minimum? What is the minimum cost?

54. WATER USAGE The height (in feet) of the water level in a reservoir over a one-year period is modeled by the function

$$H(t) = 3.3(t - 9)^2 + 14$$

where $t = 1$ represents January, $t = 2$ represents February, and so on. How low did the water level get that year, and when did it reach the low mark?

55. U.S. ARMY The function

$$N(x) = -0.0534x^2 + 0.337x + 0.969$$

gives the number of active-duty military personnel in the United States Army (in millions) for the years 1965–1972, where $x = 0$ corresponds to 1965, $x = 1$ corresponds to 1966, and so on. For this period, when was the army's personnel strength level at its highest, and what was it? Historically, can you explain why?

56. SCHOOL ENROLLMENT The total annual enrollment (in millions) in U.S. elementary and secondary schools for the years 1975–1996 is given by the model

$$E = 0.058x^2 - 1.162x + 50.604$$

where $x = 0$ corresponds to 1975, $x = 1$ corresponds to 1976, and so on.
   a. For this period, when was enrollment the lowest? What was it?
   b. Use the model to complete the bar graph in Illustration 5.

ILLUSTRATION 5

57. MAXIMIZING REVENUE The revenue R received for selling x stereos is given by the formula

$$R = -\frac{x^2}{5} + 80x - 1,000$$

How many stereos must be sold to obtain the maximum revenue? Find the maximum revenue.

58. MAXIMIZING REVENUE When priced at \$30 each, a toy has annual sales of 4,000 units. The manufacturer estimates that each \$1 increase in cost will decrease sales by 100 units. Find the unit price that will maximize total revenue. (*Hint:* Total revenue = price · the number of units sold.)

WRITING

59. Use the example of a stream of water from a drinking fountain to explain the concepts of the vertex and the axis of symmetry of a parabola.

60. What are some quantities that are good to maximize? What are some quantities that are good to minimize?

61. A table of values for $f(x) = 2x^2 - 4x + 3$ is shown in Illustration 6. Explain why it appears that the vertex of the graph of f is the point $(1, 1)$

ILLUSTRATION 6

X	Y₁
.25	2.125
.5	1.5
.75	1.125
1	1
1.25	1.125
1.5	1.5
1.75	2.125

X=.25

62. Illustration 7 shows the graph of the quadratic function $y = -4x^2 + 12x$ with domain $[0, 3]$. Explain how the value of y changes as the value of x increases from 0 to 3.

ILLUSTRATION 7

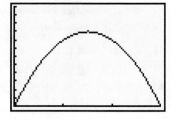

REVIEW

Simplify each expression.

63. $\sqrt{8a}\sqrt{2a^3b}$

64. $\left(\sqrt{23}\right)^2$

65. $\dfrac{\sqrt{3}}{\sqrt{50}}$

66. $\dfrac{3}{\sqrt[3]{9}}$

67. $3\left(\sqrt{5b} - \sqrt{3}\right)^2$

68. $-2\sqrt{5b}\left(4\sqrt{2b} - 3\sqrt{3}\right)$

▶ 8.4 Complex Numbers

In this section, you will learn about

> Imaginary numbers ■ Simplifying imaginary numbers ■ Complex numbers ■ Arithmetic of complex numbers ■ Complex conjugates ■ Rationalizing the denominator ■ Powers of i

Introduction We have seen that negative numbers do not have real-number square roots. For years, people believed that numbers like

$$\sqrt{-1}, \quad \sqrt{-3}, \quad \sqrt{-4}, \quad \text{and} \quad \sqrt{-9}$$

were nonsense. In the 17th century, René Descartes (1596–1650) called them *imaginary numbers*. Today, imaginary numbers have many important uses, such as describing the behavior of alternating current in electronics.

In this section, we will introduce imaginary numbers and discuss a broader set of numbers called *complex numbers*.

Imaginary Numbers

So far, all of our work with quadratic equations has involved only real numbers. However, the solutions of many quadratic equations are not real numbers.

EXAMPLE 1

A quadratic equation whose solution is not a real number. Solve $x^2 + x + 1 = 0$.

Solution   Because $x^2 + x + 1$ is prime and cannot be factored, we will use the quadratic formula, with $a = 1$, $b = 1$, and $c = 1$:

$$x = \frac{-b \pm \sqrt{b^2 - 4ac}}{2a}$$

$$= \frac{-1 \pm \sqrt{1^2 - 4(1)(1)}}{2(1)} \qquad \text{Substitute 1 for } a, \text{ 1 for } b, \text{ and 1 for } c.$$

$$= \frac{-1 \pm \sqrt{1 - 4}}{2} \qquad \text{Simplify within the radical.}$$

$$= \frac{-1 \pm \sqrt{-3}}{2}$$

$$x = \frac{-1 + \sqrt{-3}}{2} \quad \text{or} \quad x = \frac{-1 - \sqrt{-3}}{2}$$

Each solution contains the number $\sqrt{-3}$. Since no real number squared is -3, $\sqrt{-3}$ is not a real number, and these two solutions are not real numbers.

To solve equations such as $x^2 + x + 1 = 0$, we must define the square root of a negative number.

SELF CHECK   Solve $a^2 + 3a + 5 = 0$.   *Answer:* $\dfrac{-3 \pm \sqrt{-11}}{2}$ ■

Square roots of negative numbers are called **imaginary numbers.** The imaginary number $\sqrt{-1}$ is often denoted by the letter i:

$$i = \sqrt{-1}$$

Because i represents the square root of -1, it follows that

$$i^2 = -1$$

Simplifying Imaginary Numbers

We can use extensions of the multiplication and division properties of radicals discussed in Chapter 6 to simplify imaginary numbers. For example,

$$\sqrt{-25} = \sqrt{25(-1)} = \sqrt{25}\sqrt{-1} = 5i \qquad \text{Replace } \sqrt{-1} \text{ with } i.$$

$$\sqrt{\frac{-100}{49}} = \sqrt{\frac{100}{49}(-1)} = \frac{\sqrt{100}}{\sqrt{49}}\sqrt{-1} = \frac{10}{7}i$$

$$\sqrt{-3} = \sqrt{3(-1)} = \sqrt{3}\sqrt{-1} = \sqrt{3}i \qquad \begin{array}{l}\text{The expression } \sqrt{3}i \text{ can be} \\ \text{written as } i\sqrt{3} \text{ to make it clear} \\ \text{that } i \text{ is not part of the radicand.} \\ \text{Do not confuse } \sqrt{3}i \text{ with } \sqrt{3i}.\end{array}$$

These examples illustrate the following rules.

Properties of radicals

If at least one of a and b is a nonnegative real number, then

$$\sqrt{ab} = \sqrt{a}\sqrt{b} \qquad \text{and} \qquad \sqrt{\frac{a}{b}} = \frac{\sqrt{a}}{\sqrt{b}} \quad (b \neq 0)$$

EXAMPLE 2

Square roots of negative numbers. Write each expression in terms of i:
a. $\sqrt{-81}$, **b.** $-\sqrt{-8}$, and **c.** $3\sqrt{-\dfrac{98}{16}}$.

Solution
a. $\sqrt{-81} = \sqrt{81(-1)} = \sqrt{81}\sqrt{-1} = 9i$

b. $-\sqrt{-8} = -\sqrt{8(-1)} = -\sqrt{8}\sqrt{-1} = -\sqrt{8}i = -2\sqrt{2}i$

c. $3\sqrt{-\dfrac{98}{16}} = 3\sqrt{\dfrac{98}{16}(-1)} = \dfrac{3\sqrt{49}\sqrt{2}}{\sqrt{16}}\sqrt{-1} = \dfrac{21\sqrt{2}}{4}i$

SELF CHECK
Write each expression in terms of i: **a.** $-\sqrt{-4}$, *Answers:* **a.** $-2i$, **b.** $4\sqrt{7}i$,
b. $2\sqrt{-28}$, and **c.** $\sqrt{-\dfrac{27}{100}}$. **c.** $\dfrac{3\sqrt{3}}{10}i$ ∎

EXAMPLE 3

Solving quadratic equations. Solve $x^2 + 21 = 0$.

Solution
We can write the equation as $x^2 = -21$ and use the square root property to solve it.

$$x^2 + 21 = 0$$
$$x^2 = -21 \qquad \text{Subtract 21 from both sides.}$$
$$x = \pm\sqrt{-21} \qquad \begin{array}{l}\text{Apply the square root property. Note the square root of a} \\ \text{negative number.}\end{array}$$

This equation has imaginary solutions, which we can express in terms of i.

$$x = \pm\sqrt{21(-1)} \qquad \text{Use the multiplication property of radicals.}$$
$$x = \pm\sqrt{21}i \qquad \sqrt{21(-1)} = \sqrt{21}\sqrt{-1} = \sqrt{21}i.$$

Verify that $x = \sqrt{21}i$ and $x = -\sqrt{21}i$ satisfy the original equation.

SELF CHECK   Solve $x^2 + 75 = 0$.   *Answer:* $\pm 5\sqrt{3}i$ ∎

Complex Numbers

The imaginary numbers are a subset of a set of numbers called the **complex numbers.**

Complex numbers

> A **complex number** is any number that can be written in the form $a + bi$, where a and b are real numbers and $i = \sqrt{-1}$.
>
> In the complex number $a + bi$, a is called the **real part,** and b is called the **imaginary part.**

Some examples of complex numbers are

$3 + 4i$ $a = 3$ and $b = 4$.

$-6 - 5i$ Think of this as $-6 + (-5i)$, where $a = -6$ and $b = -5$. It is common to use $a - bi$ form as a substitute for $a + (-b)i$.

$\dfrac{1}{2} + 0i$ $a = \frac{1}{2}$ and $b = 0$.

$0 + \sqrt{2}i$ $a = 0$ and $b = \sqrt{2}$.

If $b = 0$, the complex number $a + bi$ is a real number. If $b \neq 0$ and $a = 0$, the complex number $0 + bi$ (or just bi) is an imaginary number. The relationship between the real numbers, the imaginary numbers, and the complex numbers is shown in Figure 8-18.

FIGURE 8-18
COMPLEX NUMBERS

Real numbers $a + 0i$	Imaginary numbers $0 + bi$ $(b \neq 0)$
$3, \dfrac{7}{3}, \pi, 125.345$	$4i, -12i, \sqrt{-4}$

$$4 + 7i, \quad 5 - 16i, \quad \frac{1}{32 - 12i}, \quad 15 + \sqrt{-25}$$

E X A M P L E   4

A quadratic equation with complex-number solutions. Solve $4t^2 - 6t + 3 = 0$. Express the solutions in $a + bi$ form.

Solution   We use the quadratic formula to solve the equation.

$$t = \frac{-b \pm \sqrt{b^2 - 4ac}}{2a}$$

$$= \frac{-(-6) \pm \sqrt{(-6)^2 - 4(4)(3)}}{2(4)} \qquad \text{Substitute 4 for } a, -6 \text{ for } b, \text{ and 3 for } c.$$

$$= \frac{6 \pm \sqrt{36 - 48}}{8} \qquad \text{Simplify within the radical.}$$

$$= \frac{6 \pm \sqrt{-12}}{8}$$

$$= \frac{6 \pm 2\sqrt{3}i}{8} \qquad \sqrt{-12} = \sqrt{12(-1)} = \sqrt{12}\sqrt{-1} = 2\sqrt{3}i.$$

$$= \frac{3 \pm \sqrt{3}i}{4} \qquad \frac{6 \pm 2\sqrt{3}i}{8} = \frac{\overset{1}{\cancel{2}}(3 \pm \sqrt{3}i)}{\cancel{2} \cdot 4} = \frac{3 \pm \sqrt{3}i}{4}.$$

Writing each solution as a complex number in $a + bi$ form,

$$t = \frac{3}{4} + \frac{\sqrt{3}}{4}i \quad \text{or} \quad t = \frac{3}{4} - \frac{\sqrt{3}}{4}i$$

SELF CHECK   Solve $a^2 + 2a + 3 = 0$.   *Answer:* $-1 \pm \sqrt{2}i$ ∎

Arithmetic of Complex Numbers

We now consider the rules used to add, subtract, multiply, and divide complex numbers.

Addition and subtraction of complex numbers

To add (or subtract) two complex numbers, add (or subtract) their real parts and add (or subtract) their imaginary parts.

$$(a + bi) + (c + di) = (a + c) + (b + d)i$$
$$(a + bi) - (c + di) = (a - c) + (b - d)i$$

EXAMPLE 5   **Adding and subtracting complex numbers.** Do the operations.

a. $(8 + 4i) + (12 + 8i) = (8 + 12) + (4 + 8)i$ Add the real parts. Add the imaginary parts.

$$= 20 + 12i$$

b. $\left(7 - \sqrt{-16}\right) + \left(9 + \sqrt{-4}\right)$

$= (7 - 4i) + (9 + 2i)$ Write $\sqrt{-16}$ and $\sqrt{-4}$ in terms of i.

$= (7 + 9) + (-4 + 2)i$ Add the real parts. Add the imaginary parts.

$= 16 - 2i$ Write $16 + (-2i)$ in the form $16 - 2i$.

c. $(-6 + i) - (3 + 2i) = (-6 - 3) + (1 - 2)i$:   Subtract the real parts. Subtract the imaginary parts.

$$= -9 - i$$

SELF CHECK   Do the operations:
a. $(3 - 5i) + (-2 + 7i)$ and   *Answers:*   **a.** $1 + 2i$,
b. $\left(3 - \sqrt{-25}\right) - \left(-2 + \sqrt{-49}\right)$.   **b.** $5 - 12i$ ∎

To multiply two imaginary numbers, they should first be expressed in terms of i. For example,

$$\sqrt{-2}\sqrt{-20} = \left(\sqrt{2}i\right)\left(2\sqrt{5}i\right) \quad \sqrt{-20} = \sqrt{20(-1)} = \sqrt{20}i = 2\sqrt{5}i.$$
$$= 2\sqrt{10}i^2$$
$$= -2\sqrt{10} \qquad i^2 = -1.$$

 WARNING! If a and b are both negative, then $\sqrt{ab} \neq \sqrt{a}\sqrt{b}$. For example, if $a = -16$ and $b = -4$,

$$\sqrt{(-16)(-4)} = \sqrt{64} = 8 \qquad \text{but} \qquad \sqrt{-16}\sqrt{-4} = (4i)(2i) = 8i^2 = 8(-1) = -8$$

To multiply a complex number by a real number, we use the distributive property to remove parentheses and then simplify. For example,

$$6(2 + 9i) = 6(2) + 6(9i) \quad \text{Use the distributive property.}$$
$$= 12 + 54i \quad \text{Simplify.}$$

To multiply a complex number by an imaginary number, we use the distributive property to remove parentheses and then simplify. For example,

$$-5i(4 - 8i) = -5i(4) - (-5i)8i \quad \text{Use the distributive property.}$$
$$= -20i + 40i^2 \quad \text{Simplify.}$$
$$= -40 - 20i \quad \text{Since } i^2 = -1, 40i^2 = 40(-1) = -40.$$

To multiply two complex numbers, we use the FOIL method to develop the following definition.

Multiplying complex numbers

Complex numbers are multiplied as if they were binomials, with $i^2 = -1$:

$$(a + bi)(c + di) = ac + adi + bci + bdi^2$$
$$= (ac - bd) + (ad + bc)i$$

EXAMPLE 6

Using the FOIL method to multiply complex numbers. Multiply the complex numbers.

a. $(2 + 3i)(3 - 2i) = 6 - 4i + 9i - 6i^2$ Use the FOIL method.

$= 6 + 5i - (-6)$ Combine the imaginary terms: $-4i + 9i = 5i$.
Simplify: $i^2 = -1$, so $6i^2 = -6$.

$= 6 + 5i + 6$

$= 12 + 5i$ Combine like terms.

b. $(-4 + 2i)(2 + i) = -8 - 4i + 4i + 2i^2$ Use the FOIL method.

$= -8 + 0i - 2$ $-4i + 4i = 0i$. Since $i^2 = -1$, $2i^2 = -2$.

$= -10 + 0i$

SELF CHECK   Multiply $(-2 + 3i)(3 - 2i)$.   *Answer:* $0 + 13i$ ∎

Complex Conjugates

Complex conjugates

The complex numbers $a + bi$ and $a - bi$ are called **complex conjugates.**

For example,

$3 + 4i$ and $3 - 4i$ are complex conjugates.

$5 - 7i$ and $5 + 7i$ are complex conjugates.

$8 + 17i$ and $8 - 17i$ are complex conjugates.

| EXAMPLE 7 | **Multiplying complex conjugates.** Find the product of $3 + i$ and its complex conjugate. |

Solution The complex conjugate of $3 + i$ is $3 - i$. We can find the product as follows:

$$(3 + i)(3 - i) = 9 - 3i + 3i - i^2 \quad \text{Use the FOIL method.}$$
$$= 9 - i^2 \quad \text{Combine like terms.}$$
$$= 9 - (-1) \quad i^2 = -1.$$
$$= 10$$

| SELF CHECK | Multiply $(2 + 3i)(2 - 3i)$. | *Answer:* 13 | ■ |

The product of the complex number $a + bi$ and its complex conjugate $a - bi$ is the real number $a^2 + b^2$, as the following work shows:

$$(a + bi)(a - bi) = a^2 - abi + abi - b^2i^2 \quad \text{Use the FOIL method.}$$
$$= a^2 - b^2(-1) \quad \text{Combine like terms. } i^2 = -1.$$
$$= a^2 + b^2$$

Rationalizing the Denominator

To divide complex numbers, we often have to rationalize a denominator.

| EXAMPLE 8 | **Rationalizing the denominator.** Divide and write the result in $a + bi$ form: $\dfrac{1}{3 + i}$. |

Solution The denominator of the fraction $\frac{1}{3+i}$ is not a rational number. We can rationalize the denominator by multiplying both the numerator and the denominator of the fraction by the complex conjugate of the denominator, which is $3 - i$.

$$\frac{1}{3 + i} = \frac{1}{3 + i} \cdot \frac{3 - i}{3 - i} \quad \frac{3-i}{3-i} = 1.$$

$$= \frac{3 - i}{9 - 3i + 3i - i^2} \quad \text{Multiply the numerators and multiply the denominators.}$$

$$= \frac{3 - i}{9 - (-1)} \quad i^2 = -1.$$

$$= \frac{3 - i}{10} \quad \text{Simplify in the denominator.}$$

$$= \frac{3}{10} - \frac{1}{10}i \quad \text{Write the complex number in } a + bi \text{ form.}$$

| SELF CHECK | Divide 1 by $5 - i$. | *Answer:* $\frac{5}{26} + \frac{1}{26}i$ | ■ |

EXAMPLE 9

Dividing complex numbers. Divide and write in $a + bi$ form: $\dfrac{3 - i}{2 + i}$.

Solution We multiply both the numerator and the denominator of the fraction by the complex conjugate of the denominator.

$$\frac{3 - i}{2 + i} = \frac{3 - i}{2 + i} \cdot \frac{2 - i}{2 - i} \qquad \tfrac{2-i}{2-i} = 1.$$

$$= \frac{6 - 3i - 2i + i^2}{4 - 2i + 2i - i^2} \qquad \text{Multiply the numerators and multiply the denominators.}$$

$$= \frac{5 - 5i}{4 - (-1)} \qquad i^2 = -1.$$

$$= \frac{5(1 - i)}{5} \qquad \text{Factor out 5 in the numerator and divide out the common factor of 5.}$$

$$= 1 - i \qquad \text{Simplify.}$$

SELF CHECK Divide $\dfrac{5 + 4i}{3 + 2i}$.

Answer: $\frac{23}{13} + \frac{2}{13}i$ ∎

EXAMPLE 10

Dividing complex numbers. Write $\dfrac{4 + \sqrt{-16}}{2 + \sqrt{-4}}$ in $a + bi$ form.

Solution $\dfrac{4 + \sqrt{-16}}{2 + \sqrt{-4}} = \dfrac{4 + 4i}{2 + 2i}$ Write the numerator and denominator in $a + bi$ form.

$$= \frac{2\overset{1}{\cancel{(2 + 2i)}}}{\underset{1}{\cancel{2 + 2i}}} \qquad \text{Factor out 2 in the numerator and divide out the common factor of } 2 + 2i.$$

$$= 2 + 0i$$

SELF CHECK Write $\dfrac{3 + \sqrt{-9}}{4 + \sqrt{-16}}$ in $a + bi$ form.

Answer: $\frac{3}{4} + 0i$ ∎

EXAMPLE 11

Division involving i. Find the quotient: $\dfrac{7}{2i}$. Express the result in $a + bi$ form.

Solution The denominator can be expressed as $0 + 2i$. Its conjugate is $0 - 2i$, or just $-2i$.

$$\frac{7}{2i} = \frac{7}{2i} \cdot \frac{-2i}{-2i} \qquad \tfrac{-2i}{-2i} = 1.$$

$$= \frac{-14i}{-4i^2}$$

$$= \frac{-14i}{4} \qquad i^2 = -1.$$

$$= \frac{-7i}{2} \qquad \text{Simplify.}$$

$$= 0 - \frac{7}{2}i \qquad \text{Write in } a + bi \text{ form.}$$

SELF CHECK Divide $\frac{5}{-i}$.

Answer: $0 + 5i$ ∎

Powers of i

The powers of i produce an interesting pattern:

$$i = \sqrt{-1} = i \qquad\qquad i^5 = i^4 i = 1i = i$$
$$i^2 = \left(\sqrt{-1}\right)^2 = -1 \qquad i^6 = i^4 i^2 = 1(-1) = -1$$
$$i^3 = i^2 i = -1i = -i \qquad\quad i^7 = i^4 i^3 = 1(-i) = -i$$
$$i^4 = i^2 i^2 = (-1)(-1) = 1 \qquad i^8 = i^4 i^4 = (1)(1) = 1$$

The pattern continues: $i, -1, -i, 1, \ldots$.

Larger powers of i can be simplified by using the fact that $i^4 = 1$. For example, to simplify i^{29}, we note that 29 divided by 4 gives a quotient of 7 and a remainder of 1. Thus, $29 = 4 \cdot 7 + 1$ and

$$i^{29} = i^{4 \cdot 7 + 1} \qquad 4 \cdot 7 = 28.$$
$$= (i^4)^7 \cdot i^1$$
$$= 1^7 \cdot i \qquad i^4 = 1.$$
$$= i \qquad\quad 1 \cdot i = i.$$

The result of this example illustrates the following fact.

Powers of i

If n is a natural number that has a remainder of r when divided by 4, then

$$i^n = i^r$$

EXAMPLE 12   **Simplifying powers of i.** Simplify i^{55}.

Solution   We divide 55 by 4 and get a remainder of 3. Therefore,

$$i^{55} = i^3 = -i$$

SELF CHECK   Simplify i^{62}.   *Answer:* -1 ∎

STUDY SET

Section 8.4

VOCABULARY

In Exercises 1–6, fill in the blanks to make the statements true.

1. $\sqrt{-1}$, $\sqrt{-3}$ and $\sqrt{-4}$ are examples of _____ numbers.

2. $3 + 5i$, $2 - 7i$, and $5 - \frac{1}{2}i$ are examples of _____ numbers.

3. The _____ part of $5 + 7i$ is 5. The _____ part is $7i$.

4. $6 + 3i$ and $6 - 3i$ are called complex _____.

5. _____ is a number whose square is -1.

6. i^{25} is called a _____ of i.

CONCEPTS

In Exercises 7–12, fill in the blanks to make the statements true.

7. $i = $

8. $i^2 = $

9. $i^3 = $

10. $i^4 = $

11. We multiply two complex numbers by using the _____ method.

12. To divide two complex numbers, we _____ the denominator.

13. Use a check to see if $x = -5\sqrt{2}i$ is a solution of $x^2 + 50 = 0$.

14. Give the complex conjugate of each number.
   a. $2 - 3i$ **b.** 2 **c.** $-3i$

15. Complete Illustration 1 to show the relationship between the real numbers, the imaginary numbers, the complex numbers, the rational numbers, and the irrational numbers.

ILLUSTRATION 1

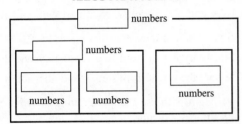

16. Tell whether each statement is true or false.
   a. Every complex number is a real number.
   b. Every real number is a complex number.
   c. Imaginary numbers can always be denoted using i.
   d. The square root of a negative number is an imaginary number.

17. Solve each equation.
   a. $x^2 - 1 = 0$ **b.** $x^2 + 1 = 0$

18. Is $\sqrt[3]{-64}$ an imaginary number? Explain.

NOTATION

In Exercises 19–20, complete each solution.

19. $(3 + 2i)(3 - i) = \boxed{} - 3i + \boxed{} - 2i^2$
   $= 9 + 3i + \boxed{}$
   $= 11 + 3i$

20. $\dfrac{3}{2 - i} = \dfrac{3}{2 - i} \cdot \dfrac{\boxed{}}{\boxed{}}$
   $= \dfrac{6 + \boxed{}}{4 + \boxed{} - 2i - i^2}$
   $= \dfrac{6 + 3i}{\boxed{}}$
   $= \dfrac{6}{5} + \dfrac{3}{5}i$

21. Tell whether each statement is true or false.
   a. $\sqrt{6}i = i\sqrt{6}$ **b.** $\sqrt{8}i = \sqrt{8i}$
   c. $\sqrt{-25} = -\sqrt{25}$ **d.** $-i = i$

22. Tell whether each statement is true or false.
   a. $\sqrt{-3}\sqrt{-2} = \sqrt{6}$
   b. $\sqrt{-36} + \sqrt{-25} = \sqrt{-61}$

PRACTICE

In Exercises 23–34, express each number in terms of i.

23. $\sqrt{-9}$ 24. $\sqrt{-4}$

25. $\sqrt{-7}$ 26. $\sqrt{-11}$

27. $\sqrt{-24}$ 28. $\sqrt{-28}$

29. $-\sqrt{-24}$ 30. $-\sqrt{-72}$

31. $5\sqrt{-81}$ 32. $6\sqrt{-49}$

33. $\sqrt{-\dfrac{25}{9}}$ 34. $-\sqrt{-\dfrac{121}{144}}$

In Exercises 35–42, simplify each expression.

35. $\sqrt{-1}\sqrt{-36}$ 36. $\sqrt{-9}\sqrt{-100}$

37. $\sqrt{-2}\sqrt{-6}$ 38. $\sqrt{-3}\sqrt{-6}$

39. $\dfrac{\sqrt{-25}}{\sqrt{-64}}$ 40. $\dfrac{\sqrt{-4}}{\sqrt{-1}}$

41. $-\dfrac{\sqrt{-400}}{\sqrt{-1}}$ 42. $-\dfrac{\sqrt{-225}}{\sqrt{-16}}$

In Exercises 43–54, solve each equation. Write all solutions in bi or a + bi form.

43. $x^2 + 9 = 0$ 44. $x^2 + 100 = 0$

45. $3x^2 = -16$ 46. $2x^2 = -25$

47. $x^2 + 2x + 2 = 0$ 48. $x^2 - 2x + 6 = 0$

49. $2x^2 + x + 1 = 0$ 50. $3x^2 + 2x + 1 = 0$

51. $3x^2 - 4x = -2$ 52. $2x^2 + 3x = -3$

53. $3x^2 - 2x = -3$ 54. $5x^2 = 2x - 1$

In Exercises 55–78, do the operations. Write all answers in a + bi form.

55. $(3 + 4i) + (5 - 6i)$ 56. $(5 + 3i) - (6 - 9i)$

57. $(7 - 3i) - (4 + 2i)$ 58. $(8 + 3i) + (-7 - 2i)$

59. $(6 - i) + (9 + 3i)$ **60.** $(5 - 4i) - (3 + 2i)$

61. $\left(8 + \sqrt{-25}\right) + \left(7 + \sqrt{-4}\right)$

62. $\left(-7 + \sqrt{-81}\right) - \left(-2 - \sqrt{-64}\right)$

63. $3(2 - i)$ **64.** $-4(3 + 4i)$

65. $-5i(5 - 5i)$ **66.** $2i(7 + 2i)$

67. $(2 + i)(3 - i)$ **68.** $(4 - i)(2 + i)$

69. $(3 - 2i)(2 + 3i)$ **70.** $(3 - i)(2 + 3i)$

71. $(4 + i)(3 - i)$ **72.** $(1 - 5i)(1 - 4i)$

73. $\left(2 - \sqrt{-16}\right)\left(3 + \sqrt{-4}\right)$

74. $\left(3 - \sqrt{-4}\right)\left(4 - \sqrt{-9}\right)$

75. $\left(2 + \sqrt{2}i\right)\left(3 - \sqrt{2}i\right)$ **76.** $\left(5 + \sqrt{3}i\right)\left(2 - \sqrt{3}i\right)$

77. $(2 + i)^2$ **78.** $(3 - 2i)^2$

In Exercises 79–102, write each expression in a + bi form.

79. $\dfrac{1}{i}$ **80.** $\dfrac{1}{i^3}$

81. $\dfrac{4}{5i^3}$ **82.** $\dfrac{3}{2i}$

83. $\dfrac{3i}{8\sqrt{-9}}$ **84.** $\dfrac{5i^3}{2\sqrt{-4}}$

85. $\dfrac{-3}{5i^5}$ **86.** $\dfrac{-4}{6i^7}$

87. $\dfrac{5}{2 - i}$ **88.** $\dfrac{26}{3 - 2i}$

89. $\dfrac{3}{5 + i}$ **90.** $\dfrac{-4}{7 - 2i}$

91. $\dfrac{-12}{7 - \sqrt{-1}}$ **92.** $\dfrac{4}{3 + \sqrt{-1}}$

93. $\dfrac{5i}{6 + 2i}$ **94.** $\dfrac{-4i}{2 - 6i}$

95. $\dfrac{-2i}{3 + 2i}$ **96.** $\dfrac{3i}{6 - i}$

97. $\dfrac{3 - 2i}{3 + 2i}$ **98.** $\dfrac{2 + 3i}{2 - 3i}$

99. $\dfrac{3 + 2i}{3 + i}$ **100.** $\dfrac{2 - 5i}{2 + 5i}$

101. $\dfrac{\sqrt{5} - \sqrt{3}i}{\sqrt{5} + \sqrt{3}i}$ **102.** $\dfrac{\sqrt{3} + \sqrt{2}i}{\sqrt{3} - \sqrt{2}i}$

In Exercises 103–110, simplify each expression.

103. i^{21} **104.** i^{19}

105. i^{27} **106.** i^{22}

107. i^{100} **108.** i^{42}

109. i^{97} **110.** i^{200}

APPLICATIONS

111. FRACTAL GEOMETRY Complex numbers are fundamental in the creation of the intricate geometric shape shown in Illustration 2, called a *fractal*. Fractal geometry is a rapidly expanding field with applications in science, medicine, and computer graphics. The process of creating this image is based on the following sequence of steps, which begins by picking any complex number, which we will call z.

1. Square z, and then add that result to z.

2. Square the result from step 1, and then add it to z.

3. Square the result from step 2, and then add it to z.

If we begin with the complex number i, what is the result after performing steps 1, 2, and 3?

ILLUSTRATION 2

112. ELECTRONICS The impedance Z in an AC (alternating current) circuit is a measure of how much the circuit impedes (hinders) the flow of current through it. The impedance is related to the voltage V and the current I by the formula

$$V = IZ$$

If a circuit has a current of $(0.5 + 2.0i)$ amps and an impedance of $(0.4 - 3.0i)$ ohms, find the voltage.

WRITING

113. What is an imaginary number?

114. In Example 1 of this section, what unusual situation illustrated the need to define the square root of a negative number?

115. What are the lengths of the two legs of a 30°–60°–90° triangle if the hypotenuse is 30 units long?

116. What are the lengths of the other two sides of a 45°–45°–90° triangle if one leg is 30 units long?

117. WIND SPEED A plane that can fly 200 mph in still air makes a 330-mile flight with a tail wind and re-turns, flying into the same wind. Find the speed of the wind if the total flying time is $3\frac{1}{3}$ hours.

118. FINDING RATES A student drove a distance of 135 miles at an average speed of 50 mph. How much faster would she have to drive on the return trip to save 30 minutes of driving time?

▶ 8.5   The Discriminant and Equations That Can Be Written in Quadratic Form

In this section, you will learn about

> The discriminant ■ Equations that can be written in quadratic form
> ■ Problem solving

Introduction   In this section, we will discuss how to predict what type of solutions a quadratic equation will have without actually solving the equation. We will then solve some special equations that can be written in quadratic form. Finally, we will use the equation-solving methods of this chapter to solve a shared-work problem.

The Discriminant

We can predict what type of solutions a particular quadratic equation will have without solving it. To see how, we suppose that the coefficients a, b, and c in the equation $ax^2 + bx + c = 0$ $(a \neq 0)$ are real numbers. Then the solutions of the equation are given by the quadratic formula

$$x = \frac{-b \pm \sqrt{b^2 - 4ac}}{2a} \quad (a \neq 0)$$

If $b^2 - 4ac \geq 0$, the solutions are real numbers. If $b^2 - 4ac < 0$, the solutions are nonreal complex numbers. Thus, the value of $b^2 - 4ac$, called the **discriminant,** determines the type of solutions for a particular quadratic equation.

The discriminant

If a, b, and c are real numbers and

if $b^2 - 4ac$ is . . .	*the solutions are . . .*
positive,	two different real numbers.
0,	two real numbers that are equal.
negative,	two different nonreal complex numbers that are complex conjugates.

If a, b, and c are rational numbers and

if $b^2 - 4ac$ is . . .	*the solutions are . . .*
a perfect square,	two different rational numbers.
positive and not a perfect square,	two different irrational numbers.

EXAMPLE 1	**Using the discriminant.** Determine the type of solutions for each equation: **a.** $x^2 + x + 1 = 0$ and **b.** $3x^2 + 5x + 2 = 0$.

Solution **a.** We calculate the discriminant for $x^2 + x + 1 = 0$:

$$b^2 - 4ac = 1^2 - 4(1)(1) \quad a = 1,\ b = 1,\ \text{and } c = 1.$$
$$= -3 \qquad \text{The result is a negative number.}$$

Since $b^2 - 4ac < 0$, the solutions of $x^2 + x + 1 = 0$ are two nonreal complex numbers that are complex conjugates.

b. For $3x^2 + 5x + 2 = 0$,

$$b^2 - 4ac = 5^2 - 4(3)(2) \quad a = 3,\ b = 5,\ \text{and } c = 2.$$
$$= 25 - 24$$
$$= 1 \qquad \text{The result is a positive number.}$$

Since $b^2 - 4ac > 0$ and $b^2 - 4ac$ is a perfect square, the two solutions of $3x^2 + 5x + 2 = 0$ are rational and unequal.

SELF CHECK Determine the type of solutions for **a.** $x^2 + x - 1 = 0$ and **b.** $4x^2 - 10x + 25 = 0$.

Answers: **a.** real numbers that are irrational and unequal, **b.** nonreal numbers that are complex conjugates ∎

Equations That Can Be Written in Quadratic Form

Many equations that are not quadratic can be written in quadratic form $(ax^2 + bx + c = 0)$ and then solved using the techniques discussed in previous sections. For example, a careful inspection of the equation $x^4 - 5x^2 + 4 = 0$ leads to the following observations:

In the leading term, this power of the variable is the square of . . .

$$x^4 - 5x^2 + 4 = 0$$

. . . the power of the variable in the middle term.

The last term is a constant.

Equations having these characteristics are said to be *quadratic in form*. One method used to solve them is to make a substitution.

$$x^4 - 5x^2 + 4 = 0 \quad \text{The given equation.}$$
$$(x^2)^2 - 5(x^2) + 4 = 0 \quad \text{Write } x^4 \text{ as } (x^2)^2.$$
$$y^2 - 5y + 4 = 0 \quad \text{Let } y = x^2. \text{ Replace each } x^2 \text{ with } y.$$

We can solve this quadratic equation by factoring.

$$(y - 4)(y - 1) = 0 \qquad \text{Factor } y^2 - 5y + 4.$$
$$y - 4 = 0 \quad \text{or} \quad y - 1 = 0 \quad \text{Set each factor equal to } 0.$$
$$y = 4 \qquad\qquad y = 1$$

These are *not* the solutions for x. To find x, we now "undo" the earlier substitutions by replacing each y with x^2. Then we solve for x.

$$x^2 = 4 \qquad \text{or} \quad x^2 = 1$$
$$x = \pm\sqrt{4} \qquad\quad x = \pm\sqrt{1} \quad \text{Use the square root property.}$$
$$x = \pm 2 \qquad\qquad x = \pm 1$$

This equation has four solutions: 1, −1, 2, and −2. Verify that each one satisfies the original equation.

EXAMPLE 2

Solving equations that are quadratic in form.   Solve $x - 7\sqrt{x} + 12 = 0$.

Solution   This equation is quadratic in form, because the power of the variable of the leading term is the square of the variable factor of the middle term: $x = (\sqrt{x})^2$. If we let $y = \sqrt{x}$, then $y^2 = x$. With this substitution, the equation

$$x - 7\sqrt{x} + 12 = 0$$

becomes a quadratic equation that can be solved by factoring.

$y^2 - 7y + 12 = 0$	Substitute y^2 for x and y for $\sqrt{x}$.
$(y - 3)(y - 4) = 0$	Factor $y^2 - 7y + 12 = 0$.
$y - 3 = 0$ or $y - 4 = 0$	Set each factor equal to 0.
$y = 3$ $y = 4$	

Replace each y with $\sqrt{x}$ and solve the radical equations by squaring both sides.

$$\sqrt{x} = 3 \quad \text{or} \quad \sqrt{x} = 4$$
$$x = 9 \qquad x = 16 \quad \text{Square both sides.}$$

Verify that both solutions satisfy the original equation.

SELF CHECK   Solve $x + x^{1/2} - 6 = 0$.   *Answer:* 4   ■

EXAMPLE 3

Solving equations that are quadratic in form.   Solve $2m^{2/3} - 2 = 3m^{1/3}$.

Solution   After writing the equation in descending powers of m, we see that

$$2m^{2/3} - 3m^{1/3} - 2 = 0$$

is quadratic in form, because $m^{2/3} = (m^{1/3})^2$. We will use the substitution $y = m^{1/3}$ to write this equation in quadratic form.

$2m^{2/3} - 3m^{1/3} - 2 = 0$	
$2(m^{1/3})^2 - 3m^{1/3} - 2 = 0$	
$2y^2 - 3y - 2 = 0$	Replace $m^{1/3}$ with y.
$(2y + 1)(y - 2) = 0$	Factor $2y^2 - 3y - 2 = 0$.
$2y + 1 = 0$ or $y - 2 = 0$	Set each factor equal to 0.
$y = -\dfrac{1}{2}$ $y = 2$	

Replace each y with $m^{1/3}$ and solve for m.

$$m^{1/3} = -\frac{1}{2} \qquad \text{or} \qquad m^{1/3} = 2$$

$$(m^{1/3})^3 = \left(-\frac{1}{2}\right)^3 \qquad (m^{1/3})^3 = (2)^3 \quad \begin{array}{l}\text{Recall that } m^{1/3} = \sqrt[3]{m}. \text{ To solve for } m, \text{ cube} \\ \text{both sides.}\end{array}$$

$$m = -\frac{1}{8} \qquad\qquad m = 8$$

Verify that both solutions satisfy the original equation.

SELF CHECK   Solve $a^{2/3} = -3a^{1/3} + 10$.   *Answer:* −125, 8   ■

EXAMPLE 4

Solving equations that are quadratic in form.
Solve $(4t + 2)^2 - 30(4t + 2) + 224 = 0$.

Solution If we make the substitution $y = 4t + 2$, the given equation becomes

$$y^2 - 30y + 224 = 0$$

which can be solved by using the quadratic formula.

$$y = \frac{-b \pm \sqrt{b^2 - 4ac}}{2a}$$

$$= \frac{-(-30) \pm \sqrt{(-30)^2 - 4(1)(224)}}{2(1)}$$ Substitute 1 for a, -30 for b, and 224 for c.

$$= \frac{30 \pm \sqrt{900 - 896}}{2}$$ Simplify within the radical.

$$= \frac{30 \pm 2}{2}$$ $\sqrt{900 - 896} = \sqrt{4} = 2$.

$$y = 16 \quad \text{or} \quad y = 14$$

To find t, we replace y with $4t + 2$ and solve for t.

$$4t + 2 = 16 \quad \text{or} \quad 4t + 2 = 14$$
$$4t = 14 \qquad\qquad 4t = 12$$
$$t = 3.5 \qquad\qquad t = 3$$

SELF CHECK Solve $(n + 3)^2 - 6(n + 3) = -8$. *Answer:* $-1, 1$ ■

Problem Solving

EXAMPLE 5

Household appliances. A water temperature control on a washing machine is shown in Figure 8-19. When the "warm" setting is selected, both the hot and cold water inlets open to fill the tub in 2 minutes 15 seconds. When the "cold" temperature setting is chosen, the cold water inlet fills the tub 45 seconds faster than when the "hot" setting is used. How long does it take to fill the washing machine with hot water?

ANALYZE THE PROBLEM The key to solving this problem is to determine how much of the tub is filled by each water temperature setting in 1 second. On the "warm" setting, when the hot and cold inlets are working together, the tub is filled in 2 minutes 15 seconds, or 135 seconds. So in 1 second, they fill $\frac{1}{135}$ of the tub.

FORM AN EQUATION Let x represent the number of seconds it takes to fill the tub when the "hot" temperature is chosen. In 1 second, the hot water inlet fills $\frac{1}{x}$ of the tub. Since the cold water inlet fills the tub in 45 seconds less time, the washing machine can be filled with cold water in $(x - 45)$ seconds. In 1 second, $\frac{1}{x - 45}$ of the tub is filled by the cold water inlet. We can now form an equation.

FIGURE 8-19
Electronic Temperature Control

Hot **Warm** Cold

Water Temp

What the hot water inlet pipe can do in 1 second	plus	what the cold water inlet pipe can do in 1 second	equals	what they can do together in 1 second.
$\dfrac{1}{x}$	$+$	$\dfrac{1}{x - 45}$	$=$	$\dfrac{1}{135}$

SOLVE THE EQUATION

$$\frac{1}{x} + \frac{1}{x-45} = \frac{1}{135}$$

$$135x(x-45)\left(\frac{1}{x} + \frac{1}{x-45}\right) = 135x(x-45)\left(\frac{1}{135}\right)$$ Multiply both sides by $135x(x-45)$ to clear the equation of fractions.

$$135(x-45) + 135x = x(x-45)$$ Simplify.

$$135x - 6{,}075 + 135x = x^2 - 45x$$ Use the distributive property to remove parentheses.

$$270x - 6{,}075 = x^2 - 45x$$ Combine like terms.

$$0 = x^2 - 315x + 6{,}075$$ Subtract $270x$ from both sides. Add $6{,}075$ to both sides.

To solve this equation, we will use the quadratic formula, with $a = 1$, $b = -315$, and $c = 6{,}075$.

$$x = \frac{-b \pm \sqrt{b^2 - 4ac}}{2a}$$

$$= \frac{-(-315) \pm \sqrt{(-315)^2 - 4(1)(6{,}075)}}{2(1)}$$ Substitute 1 for a, -315 for b, and $6{,}075$ for c.

$$= \frac{315 \pm \sqrt{99{,}225 - 24{,}300}}{2}$$ Simplify within the radical.

$$= \frac{315 \pm \sqrt{74{,}925}}{2}$$ $\sqrt{99{,}225 - 24{,}300} = \sqrt{74{,}925}$.

$$x \approx \frac{589}{2} \quad \text{or} \quad x \approx \frac{41}{2}$$

$$x \approx 294 \qquad\qquad x \approx 21$$

STATE THE CONCLUSION   We can disregard the solution of 21 seconds, because this would imply that the cold water inlet fills the tub in a negative number of seconds ($21 - 45 = -24$). Therefore, the hot water inlet fills the washing machine tub in about 294 seconds, which is 4 minutes 54 seconds.

CHECK THE RESULT   Use estimation to check the result.   ■

STUDY SET

Section 8.5

VOCABULARY

In Exercises 1–2, fill in the blanks to make the statements true.

1. For the quadratic equation $ax^2 + bx + c = 0$, the discriminant is _____.

2. When an equation is written in the form $ax^2 + bx + c = 0$, we say that it is written in _____ form.

CONCEPTS

In Exercises 3–6, consider the equation $ax^2 + bx + c = 0$, where a, b, and c are rational numbers, and fill in the blanks to make the statements true.

3. If $b^2 - 4ac < 0$, the solutions of the equation are nonreal complex _____.

4. If $b^2 - 4ac =$ ___, the solutions of the equation are equal real numbers.

5. If $b^2 - 4ac$ is a perfect square, the solutions are _____ numbers and _____.

6. If $b^2 - 4ac$ is positive and not a perfect square, the solutions are _____ numbers and _____.

7. Consider $x^4 - 3x^2 + 2 = 0$.
   a. What is the relationship between the powers of x in the first two terms on the left-hand side?
   b. Is this equation quadratic in form?

8. Consider $x^{2/3} + 4x^{1/3} - 5 = 0$.
   a. What is the relationship between the powers of x in the first two terms on the left-hand side?

   b. Is this equation quadratic in form?

NOTATION

In Exercises 9–10, complete each solution.

9. To find the type of solutions for the equation $x^2 + 5x + 6 = 0$, we compute the discriminant.
$$b^2 - 4ac = \boxed{}^2 - 4(1)\left(\boxed{}\right)$$
$$= 25 - \boxed{}$$
$$= 1$$

Since a, b, and c are rational numbers and the value of the discriminant is a perfect square, the solutions are _____ numbers and unequal.

10. Change $\dfrac{3}{4} + x = \dfrac{3x - 50}{4(x - 6)}$ to quadratic form.

$$\boxed{}\left(\frac{3}{4} + x\right) = \boxed{}\frac{3x - 50}{4(x - 6)}$$

$$3(x - 6) + 4x\left(\boxed{}\right) = 3x - 50$$

$$3x - \boxed{} + 4x^2 - \boxed{} = 3x - 50$$

$$4x^2 - 24x + \boxed{} = 0$$

$$\boxed{} - 6x + 8 = 0$$

PRACTICE

In Exercises 11–18, use the discriminant to determine what type of solutions exist for each quadratic equation. **Do not solve the equation.**

11. $4x^2 - 4x + 1 = 0$ 12. $6x^2 - 5x - 6 = 0$

13. $5x^2 + x + 2 = 0$ 14. $3x^2 + 10x - 2 = 0$

15. $2x^2 = 4x - 1$ 16. $9x^2 = 12x - 4$

17. $x(2x - 3) = 20$ 18. $x(x - 3) = -10$

19. Use the discriminant to determine whether the solutions of $1{,}492x^2 + 1{,}776x - 2{,}000 = 0$ are real numbers.

20. Use the discriminant to determine whether the solutions of $1{,}776x^2 - 1{,}492x + 2{,}000 = 0$ are real numbers.

In Exercises 21–50, solve each equation.

21. $x^4 - 17x^2 + 16 = 0$ 22. $x^4 - 10x^2 + 9 = 0$

23. $x^4 = 6x^2 - 5$ 24. $2x^4 + 24 = 26x^2$

25. $t^4 + 3t^2 = 28$ 26. $3h^4 + h^2 - 2 = 0$

27. $2x + \sqrt{x} - 3 = 0$ 28. $2x - \sqrt{x} - 1 = 0$

29. $3x + 5\sqrt{x} + 2 = 0$ 30. $3x - 4\sqrt{x} + 1 = 0$

31. $x - 6\sqrt{x} = -8$ 32. $x - 5x^{1/2} + 4 = 0$

33. $x^{2/3} + 5x^{1/3} + 6 = 0$ 34. $x^{2/3} - 7x^{1/3} + 12 = 0$

35. $a^{2/3} - 2a^{1/3} - 3 = 0$ 36. $r^{2/3} + 4r^{1/3} - 5 = 0$

37. $2(2x + 1)^2 - 7(2x + 1) + 6 = 0$

38. $3(2 - x)^2 + 10(2 - x) - 8 = 0$

39. $(c + 1)^2 - 4(c + 1) - 8 = 0$

40. $(k - 7)^2 + 6(k - 7) + 10 = 0$

41. $x + 5 + \dfrac{4}{x} = 0$ 42. $x - 4 + \dfrac{3}{x} = 0$

43. $\dfrac{1}{x + 2} + \dfrac{24}{x + 3} = 13$ 44. $\dfrac{3}{x} + \dfrac{4}{x + 1} = 2$

45. $\dfrac{2}{x - 1} + \dfrac{1}{x + 1} = 3$ 46. $\dfrac{3}{x - 2} - \dfrac{1}{x + 2} = 5$

47. $x^{-4} - 2x^{-2} + 1 = 0$ 48. $4x^{-4} + 1 = 5x^{-2}$

49. $x + \dfrac{2}{x - 2} = 0$ 50. $x + \dfrac{x + 5}{x - 3} = 0$

APPLICATIONS

51. FLOWER ARRANGING A florist needs to determine the height h of the flowers shown in Illustration 1. The radius r, the width w, and the height h of the circular-shaped arrangement are related by the formula

$$r = \frac{4h^2 + w^2}{8h}$$

If w is to be 34 inches and r is to be 18 inches, find h to the nearest tenth of an inch.

ILLUSTRATION 1

52. ARCHITECTURE A **golden rectangle** is said to be one of the most visually appealing of all geometric forms. The Parthenon, built by the Greeks in the 5th century B.C. and shown in Illustration 2, fits into a golden rectangle once its ruined triangular pediment is drawn in.

In a golden rectangle, the length l and width w must satisfy the equation

$$\frac{l}{w} = \frac{w}{l - w}$$

If a rectangular billboard is to have a width of 20 feet, what should its length be so that it is a golden rectangle? Round to the nearest tenth.

ILLUSTRATION 2

53. SNOWMOBILE A woman drives her snowmobile 150 miles at a rate of r mph. She could have gone the same distance in 2 hours less time if she had increased her speed by 20 mph. Find r.

54. BICYCLING Tina bicycles 160 miles at the rate of r mph. The same trip would have taken 2 hours longer if she had decreased her speed by 4 mph. Find r.

55. CROWD CONTROL After a sold-out performance at a county fair, security guards have found that the grandstand area can be emptied in 6 minutes if both the east and west exits are opened. If just the east exit is used, it takes 4 minutes longer to clear the grandstand than it does if just the west exit is opened. How long does it take to clear the grandstand if everyone must file through the east exit?

56. PAPER ROUTE When a father, in a car, and his son, on a bicycle, work together to distribute the morning edition, it takes them 35 minutes to complete a paper route. Working alone, it takes the son 25 minutes longer than the father. To the nearest minute, how long does it take the son to cover the paper route on his bicycle?

WRITING

57. Describe how to predict what type of solutions the equation $3x^2 - 4x + 5 = 0$ will have.

58. Explain how the method of substitution is used in this section to solve equations.

REVIEW

In Exercises 59–60, solve the equation.

59. $\dfrac{1}{4} + \dfrac{1}{t} = \dfrac{1}{2t}$

60. $\dfrac{p - 3}{3p} + \dfrac{1}{2p} = \dfrac{1}{4}$

61. Find the slope of the line passing through $(-2, -4)$ and $(3, 5)$.

62. Write the equation of the line passing through $(-2, -4)$ and $(3, 5)$ in general form.

▶ **8.6**

Quadratic and Other Nonlinear Inequalities

In this section, you will learn about

Solving quadratic inequalities ■ Solving rational inequalities
■ Graphs of nonlinear inequalities in two variables

Introduction If $a \neq 0$, inequalities of the form $ax^2 + bx + c < 0$ and $ax^2 + bx + c > 0$ are called *quadratic inequalities*. We will begin this section by showing how to solve these inequalities by making a sign chart. We will then show how to solve other nonlinear inequalities using the same technique. To conclude, we will show how to find graphical solutions of nonlinear inequalities containing two variables.

Solving Quadratic Inequalities

To solve the inequality $x^2 + x - 6 < 0$, we must find the values of x that make the inequality true. This can be done using a number line. We begin by factoring the trinomial to obtain

$$(x + 3)(x - 2) < 0$$

Since the product of $x + 3$ and $x - 2$ must be less than 0, the values of $x + 3$ and $x - 2$ must be opposite in sign. To find the intervals where this is true, we keep track of their signs by constructing the chart in Figure 8-20. The chart shows that

- $x - 2$ is 0 when $x = 2$, is positive when $x > 2$ (indicated with + signs in the figure), and is negative when $x < 2$ (indicated with − signs in the figure).
- $x + 3$ is 0 when $x = -3$, is positive when $x > -3$, and is negative when $x < -3$.

The only place where the values of the binomials are opposite in sign is in the interval $(-3, 2)$. Therefore, the solution set of the inequality can be denoted as

FIGURE 8-20

$$-3 < x < 2$$

The graph of the solution set is shown on the number line in Figure 8-20.

EXAMPLE 1

Constructing a sign chart. Solve $x^2 + 2x - 3 \geq 0$.

Solution We factor the trinomial to get $(x - 1)(x + 3)$ and construct a sign chart, as in Figure 8-21.

- $x - 1$ is 0 when $x = 1$, is positive when $x > 1$, and is negative when $x < 1$.
- $x + 3$ is 0 when $x = -3$, is positive when $x > -3$, and is negative when $x < -3$.

The product of $x - 1$ and $x + 3$ will be greater than 0 when the signs of the binomial factors are the same. This occurs in the intervals $(-\infty, -3)$ and $(1, \infty)$. The numbers -3 and 1 are also included, because they make the product equal to 0. Thus, the solution set is

$$(-\infty, -3] \cup [1, \infty) \qquad \text{or} \qquad x \leq -3 \text{ or } x \geq 1$$

The graph of the solution set is shown on the number line in Figure 8-21.

FIGURE 8-21

SELF CHECK Solve $x^2 + 2x - 15 > 0$ and graph the solution set.

Answer: $(-\infty, -5) \cup (3, \infty)$

Solving Rational Inequalities

Making a sign chart is useful for solving many inequalities that are neither linear nor quadratic. In the next three examples, we will use a sign chart to solve *rational inequalities*.

| EXAMPLE 2 | **Solving rational inequalities.** Solve $\frac{1}{x} < 6$. |

Solution   We subtract 6 from both sides to make the right-hand side equal to 0. We then find a common denominator and add:

$$\frac{1}{x} < 6$$

$$\frac{1}{x} - 6 < 0 \quad \text{Subtract 6 from both sides.}$$

$$\frac{1}{x} - \frac{6x}{x} < 0 \quad \text{Get a common denominator.}$$

$$\frac{1 - 6x}{x} < 0 \quad \text{Subtract the numerators and keep the common denominator.}$$

We now make a sign chart, as shown in Figure 8-22.

- The denominator x is 0 when $x = 0$, is positive when $x > 0$, and is negative when $x < 0$.
- The numerator $1 - 6x$ is 0 when $x = \frac{1}{6}$, is positive when $x < \frac{1}{6}$, and is negative when $x > \frac{1}{6}$.

The fraction $\frac{1-6x}{x}$ will be less than 0 when the numerator and denominator are opposite in sign. This occurs in the interval

$$(-\infty, 0) \cup \left(\frac{1}{6}, \infty\right) \quad \text{or} \quad x < 0 \text{ or } x > \frac{1}{6}$$

The graph of this interval is shown in Figure 8-22.

FIGURE 8-22

| SELF CHECK | Solve $\frac{3}{x} > 5$. | *Answer:* $\left(0, \frac{3}{5}\right)$ |

  WARNING! Since we don't know whether x is positive, 0, or negative, multiplying both sides of the inequality $\frac{1}{x} < 6$ by x is a three-case situation:

- If $x > 0$, then $1 < 6x$.
- If $x = 0$, then $\frac{1}{x}$ is undefined.
- If $x < 0$, then $1 > 6x$.

If we multiply both sides by x and solve the linear inequality $1 < 6x$, we are considering only one case and will get only part of the answer.

| EXAMPLE 3 | **Solving rational inequalities.** Solve $\frac{x^2 - 3x + 2}{x - 3} \geq 0$. |

Solution   We write the fraction with the numerator in factored form.

$$\frac{(x - 2)(x - 1)}{x - 3} \geq 0$$

FIGURE 8-23

To keep track of the signs of the three binomials, we construct the sign chart shown in Figure 8-23. The fraction will be positive in the intervals where all factors are positive, or where exactly two factors are negative. The numbers 1 and 2 are included, because they make the numerator (and thus the fraction) equal to 0. The number 3 is not included, because it gives a 0 in the denominator.

The solution is the interval $[1, 2] \cup (3, \infty)$. The graph appears in Figure 8-23.

SELF CHECK Solve $\dfrac{x + 2}{x^2 - 2x - 3} > 0$ and graph the solution set. *Answer:* $(-2, -1) \cup (3, \infty)$

E X A M P L E   4 **Solving rational inequalities.** Solve $\dfrac{3}{x - 1} < \dfrac{2}{x}$.

Solution We subtract $\frac{2}{x}$ from both sides to get 0 on the right-hand side and proceed as follows:

$$\frac{3}{x - 1} < \frac{2}{x}$$

$$\frac{3}{x - 1} - \frac{2}{x} < 0 \qquad \text{Subtract } \tfrac{2}{x} \text{ from both sides.}$$

$$\frac{3x}{(x - 1)x} - \frac{2(x - 1)}{x(x - 1)} < 0 \qquad \text{Get a common denominator.}$$

$$\frac{3x - 2x + 2}{x(x - 1)} < 0 \qquad \text{Keep the denominator and subtract the numerators.}$$

$$\frac{x + 2}{x(x - 1)} < 0 \qquad \text{Combine like terms.}$$

FIGURE 8-24

We can keep track of the signs of the three polynomials with the sign chart shown in Figure 8-24. The fraction will be negative in the intervals with either one or three negative factors. The numbers 0 and 1 are not included, because they give a 0 in the denominator, and the number -2 is not included, because it does not satisfy the inequality.

The solution is the interval $(-\infty, -2) \cup (0, 1)$, as shown in Figure 8-24.

SELF CHECK Solve $\frac{2}{x+1} > \frac{1}{x}$ and graph the solution set. *Answer:* $(-1, 0) \cup (1, \infty)$

ACCENT ON TECHNOLOGY *Solving Inequalities Graphically*

To approximate the solutions of the inequality $x^2 + 2x - 3 \geq 0$ (Example 1) by graphing, we can use the standard window settings of $[-10, 10]$ for x and $[-10, 10]$ for y and graph the quadratic function $y = x^2 + 2x - 3$, as in Figure 8-25. The solution of the inequality will be those numbers x for which the graph of $y = x^2 + 2x - 3$ lies above or on the x-axis. We can trace to find that this interval is $(-\infty, -3] \cup [1, \infty)$.

FIGURE 8-25

To approximate the solutions of $\frac{3}{x-1} < \frac{2}{x}$ (Example 4), we first write the inequality in the form

$$\frac{x+2}{x(x-1)} < 0$$

We use window settings of $[-5, 5]$ for x and $[-3, 3]$ for y and graph the function $y = \frac{x+2}{x(x-1)}$, as in Figure 8-26(a). The solution of the inequality will be those numbers x for which the graph lies below the x-axis.

We can trace to see that the graph is below the x-axis when x is less than -2. Since we cannot see the graph in the interval $0 < x < 1$, we redraw the graph using window settings of $[-1, 2]$ for x and $[-25, 10]$ for y. See Figure 8-26(b).

We can now see that the graph is below the x-axis in the interval $(0, 1)$. Thus, the solution of the inequality is the union of two intervals:

$$(-\infty, -2) \cup (0, 1)$$

FIGURE 8-26

(a)   (b)

Graphs of Nonlinear Inequalities in Two Variables

We now consider the graphs of nonlinear inequalities in two variables.

EXAMPLE 5

Graphing nonlinear inequalities in two variables. Graph $y < -x^2 + 4$.

Solution The graph of $y = -x^2 + 4$ is the parabolic boundary separating the region representing $y < -x^2 + 4$ and the region representing $y > -x^2 + 4$.

We graph the quadratic function $y = -x^2 + 4$ as a broken parabola, because equality is not permitted. Since the coordinates of the origin satisfy the inequality $y < -x^2 + 4$, the point $(0, 0)$ is in the graph. The complete graph is shown in Figure 8-27.

FIGURE 8-27

SELF CHECK Graph $y \geq -x^2 + 4$.

Answer:

EXAMPLE 6 | **Graphing nonlinear inequalities in two variables.** Graph $x \leq |y|$.

Solution We first graph $x = |y|$ as in Figure 8-28(a). We use a solid line, because equality is permitted. Because the origin is on the graph, we cannot use the origin as a test point. However, any other point, such as $(1, 0)$, will do. We substitute 1 for x and 0 for y into the inequality to get

$$x \leq |y|$$
$$1 \leq |0|$$
$$1 \leq 0$$

Since $1 \leq 0$ is a false statement, the point $(1, 0)$ does not satisfy the inequality and is not part of the graph. Thus, the graph of $x \leq |y|$ is to the left of the boundary. The complete graph is shown in Figure 8-28(b).

FIGURE 8-28

(a)

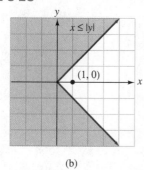

(b)

SELF CHECK | Graph $x \geq -|y|$.

Answer:

STUDY SET

Section 8.6

VOCABULARY

In Exercises 1–4, fill in the blanks to make the statements true.

1. Any inequality of the form $ax^2 + bx + c > 0$ $(a \neq 0)$ is called a _____ inequality.

2. The inequality $y < x^2 - 2x + 3$ is a nonlinear inequality in _____ variables.

3. The _____ $(3, 5)$ represents the real numbers between 3 and 5.

4. To decide which side of the boundary to shade when solving inequalities in two variables, we pick a _____ point.

CONCEPTS

In Exercises 5–8, fill in the blanks to make the statements true.

5. When $x > 3$, the binomial $x - 3$ is _____ than zero.

6. When $x < 3$, the binomial $x - 3$ is _____ than zero.

7. If $x = 0$, the fraction $\frac{1}{x}$ is _____.

8. To keep track of the signs of factors in a product or quotient, we can use a _____ chart.

9. Estimate the solution of the inequality $x^2 - x - 6 > 0$ using the graph of $y = x^2 - x - 6$, shown below.

10. Estimate the solution of the inequality $\frac{x-3}{x} \le 0$ using the graph of $y = \frac{x-3}{x}$, shown below.

NOTATION

In Exercises 11–12, consider the inequality
$(x + 2)(x - 3) > 0$.

11. Since the product of $x + 2$ and $x - 3$ is positive, the values of $x + 2$ and $x - 3$ are both positive or both negative.
   a. $x + 2 = 0$ when $x =$ ▢
   b. $x + 2 > 0$ when $x >$ ▢
   c. $x + 2 < 0$ when $x <$ ▢
   d. $x - 3 = 0$ when $x =$ ▢
   e. $x - 3 > 0$ when $x >$ ▢
   f. $x - 3 < 0$ when $x <$ ▢

12. Use the information in Exercise 11 to make a sign chart of the data and graph the solution set.

13. Match each inequality with the corresponding interval notation.
   a. $-4 \le x < 5$ **i.** $(-\infty, \infty)$
   b. $x \le -4$ or $x > 5$ **ii.** $[-4, \infty)$
   c. $x \ge -4$ **iii.** $[-4, 5)$
   d. $x < 5$ or $x \ge -4$ **iv.** $(-\infty, -4] \cup (5, \infty)$

14. What is the meaning of the symbol $\cup$?

PRACTICE

In Exercises 15–42, solve each inequality. Give each result in interval notation and graph the solution set.

15. $x^2 - 5x + 4 < 0$

16. $x^2 - 3x - 4 > 0$

17. $x^2 - 8x + 15 > 0$

18. $x^2 + 2x - 8 < 0$

19. $x^2 + x - 12 \le 0$

20. $x^2 - 8x \le -15$

21. $x^2 + 8x < -16$

22. $x^2 + 6x \ge -9$

23. $x^2 \ge 9$

24. $x^2 \ge 16$

25. $2x^2 - 50 < 0$

26. $3x^2 - 243 < 0$

27. $\dfrac{1}{x} < 2$

28. $\dfrac{1}{x} > 3$

29. $-\dfrac{5}{x} < 3$

30. $\dfrac{4}{x} \ge 8$

31. $\dfrac{x^2 - x - 12}{x - 1} < 0$

32. $\dfrac{x^2 + x - 6}{x - 4} \ge 0$

33. $\dfrac{6x^2 - 5x + 1}{2x + 1} > 0$

34. $\dfrac{6x^2 + 11x + 3}{3x - 1} < 0$

35. $\dfrac{3}{x - 2} < \dfrac{4}{x}$

36. $\dfrac{-6}{x + 1} \ge \dfrac{1}{x}$

37. $\dfrac{7}{x-3} \geq \dfrac{2}{x+4}$

38. $\dfrac{-5}{x-4} < \dfrac{3}{x+1}$

39. $\dfrac{x}{x+4} \leq \dfrac{1}{x+1}$

40. $\dfrac{x}{x+9} \geq \dfrac{1}{x+1}$

41. $(x+2)^2 > 0$

42. $(x-3)^2 < 0$

In Exercises 43–46, use a graphing calculator to solve each inequality. Give the answer in interval notation.

43. $x^2 - 2x - 3 < 0$ **44.** $x^2 + x - 6 > 0$

45. $\dfrac{x+3}{x-2} > 0$ **46.** $\dfrac{3}{x} < 2$

In Exercises 47–54, graph each inequality.

47. $y < x^2 + 1$ **48.** $y > x^2 - 3$

     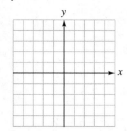

49. $y \leq x^2 + 5x + 6$ **50.** $y \geq x^2 + 5x + 4$

51. $y < |x+4|$ **52.** $y \geq |x-3|$

53. $y \leq -|x| + 2$ **54.** $y > |x| - 2$

APPLICATIONS

55. SUSPENSION BRIDGE If an x-axis is superimposed over the roadway of the Golden Gate Bridge, with the origin at the center of the bridge as shown in Illustration 1, the length L in feet of a vertical support cable can be approximated by the formula

$$L = \frac{1}{9{,}000}x^2 + 5$$

For the Golden Gate Bridge, $-2{,}100 < x < 2{,}100$. For what intervals along the x-axis are the vertical cables more than 95 feet long?

ILLUSTRATION 1

56. MALL The number of people n in a mall is modeled by the formula

$$n = -100x^2 + 1{,}200x$$

where x is the number of hours since the mall opened. If the mall opened at 9 A.M., when were there 2,000 or more people in it?

WRITING

57. Explain why $(x-4)(x+5)$ will be positive only when the signs of $x-4$ and $x+5$ are the same.

58. Explain how to find the graph of $y \geq x^2$.

59. The graph of $f(x) = x^2 - 3x + 4$ is shown in Illustration 2. Explain why the quadratic inequality $x^2 - 3x + 4 < 0$ has no solution.

60. What three important facts about the expression $x - 1$ are indicated by the chart in Illustration 3?

ILLUSTRATION 2

ILLUSTRATION 3

REVIEW

In Exercises 61–64, translate each statement into an equation.

61. *x* varies directly with *y*.

62. *y* varies inversely with *t*.

63. *t* varies jointly with *x* and *y*.

64. *d* varies directly with *t* and inversely with u^2.

Find the slope of the graph of each linear function.

65. $f(x) = 3x - 4$

66. $f(x) = -x$

Solving Quadratic Equations

We have discussed five methods for solving **quadratic equations.** Let's review each of them and list an advantage and a drawback of each method.

Factoring

■ It can be very quick and simple if the factoring pattern is evident.

■ Much of the time, $ax^2 + bx + c$ cannot be factored or is not easily factored.

In Exercises 1–3, solve the equation by factoring.

1. $4k^2 + 8k = 0$ 　　　　　**2.** $z^2 + 8z + 15 = 0$ 　　　　　**3.** $2r^2 + 5r = -3$

The Square Root Method

■ If the equation can be written in the form $x^2 = a$ or $(x + d)^2 = a$, where a is a constant, the square root method is a fast method, requiring few computations.

■ Most quadratic equations that we must solve are not written in either of these forms.

In Exercises 4–6, solve the equation by the square root method.

4. $u^2 = 24$ 　　　　　**5.** $(s - 7)^2 - 9 = 0$ 　　　　　**6.** $3x^2 - 16 = 0$

Completing the Square

■ It can be used to solve any quadratic equation.

■ Most often, it involves more steps than the other methods.

In Exercises 7–9, solve the equation by completing the square.

7. $x^2 + 10x - 7 = 0$ 　　　　　**8.** $4x^2 - 4x - 1 = 0$ 　　　　　**9.** $x^2 + 2x + 2 = 0$

The Quadratic Formula

■ It simply involves an evaluation of the expression $\dfrac{-b \pm \sqrt{b^2 - 4ac}}{2a}$.

■ If applicable, the factoring method and the square root method are usually faster.

In Exercises 10–12, solve the equation by using the quadratic formula.

10. $2x^2 - 1 = 3x$ 　　　　　**11.** $x^2 - 6x - 391 = 0$ 　　　　　**12.** $3x^2 + 2x + 1 = 0$

The Graphing Method

■ We can solve the equation using a graphing calculator. It doesn't require any computations.

■ It usually gives only approximations of the solutions.

13. Use the graph of $y = x^2 + x - 2$ to solve the equation $x^2 + x - 2 = 0$.

Accent on Teamwork

Section 8.1

Picture frames Each person in your group is to make an 8 in. × 10 in. collage that illustrates an important aspect of his or her life. Photographs, magazine pictures, drawings, and the like can be used to highlight family, hobbies, jobs, pets, etc. Assign each person the task of making a matting of uniform width to frame his or her collage. Some can make a matting whose area equals that of the picture. Others can make a matting that is half the area, or double the area. When completed, let each person briefly explain his or her collage to the other group members. See Example 8, Section 8.1, for some hints on how to do the mathematics.

Section 8.2

Team relay Have your group make a presentation to the class showing the derivation of the quadratic formula. (See page 561 of the text.) Begin with the general quadratic equation, $ax^2 + bx + c = 0$, and have each member of your group perform and explain several steps. Make a "baton" from a paper towel roll, like that shown in Illustration 1. As one student finishes with his or her segment of the derivation, the baton is passed to the next student to pick up where he or she left off.

ILLUSTRATION 1

Section 8.3

Determining the number of solutions graphically Experiment with various combinations of a's, b's, and c's to write quadratic functions of the form $f(x) = ax^2 + bx + c$ that, when graphed using a calculator, intersect the x-axis **a.** two times, **b.** one time, and **c.** no times. Then give the associated quadratic equation in each case and tell how many real-number solutions the equation has.

Section 8.4

Complex numbers Write a report about complex numbers that answers these questions: When were they first used and by whom? Why were they "invented"? What are some of their applications?

A book about the history of mathematics will be helpful. The school library or perhaps one of your mathematics instructors may have one you can borrow. Some colleges offer a course about complex numbers called Complex Variables. See if you can get a copy of the textbook. Germany once issued a postage stamp honoring Carl Gauss and featuring complex numbers. See if you can find a picture of it. Physics instructors or electronics instructors are two other possible resources for material.

Section 8.5

Solutions of a quadratic equation Consider the following property about the solutions of a quadratic equation: If

$$r_1 + r_2 = -\frac{b}{a} \quad \text{and} \quad r_1 r_2 = \frac{c}{a}$$

then r_1 and r_2 are the solutions of a quadratic equation $ax^2 + bx + c = 0$, with $a \neq 0$.

Use this property to show that
a. $\frac{3}{2}$ and $-\frac{1}{3}$ are solutions of $6x^2 - 7x - 3 = 0$.
b. $\sqrt{51}i$ and $-\sqrt{51}i$ are solutions of $x^2 + 51 = 0$.
c. $1 + 3\sqrt{2}$ and $1 - 3\sqrt{2}$ are solutions of $x^2 - 2x - 17 = 0$.

Section 8.6

Quadratic inequalities The fraction shown below has two factors in the numerator and two factors in the denominator.

$$\frac{(x - 1)(x + 4)}{(x + 2)(x + 1)}$$

a. With the four factors in mind, under what conditions will the fraction be positive?
b. With the four factors in mind, under what conditions will the fraction be negative?
c. Solve $\dfrac{(x - 1)(x + 4)}{(x + 2)(x + 1)} > 0$.

Section 8.1

CONCEPTS

The square root property:
If $c > 0$, the equation $x^2 = c$ has two real solutions:

$$x = \sqrt{c} \text{ and } x = -\sqrt{c}$$

To *complete the square:*
1. Make sure the coefficient of x^2 is 1.
2. Make sure the constant term is on the right-hand side of the equation.
3. Add the square of one-half of the coefficient of x to both sides.
4. Factor the trinomial.
5. Use the square root property.

Completing the Square

REVIEW EXERCISES

1. Solve each equation by factoring or using the square root property.
   a. $x^2 + 9x + 20 = 0$
   b. $6x^2 + 17x + 5 = 0$
   c. $x^2 = 28$
   d. $(t + 2)^2 = 36$
   e. $5a^2 + 11a = 0$
   f. $5x^2 - 49 = 0$

2. What number must be added to $x^2 - x$ to make it a perfect square?

3. Solve each equation by completing the square.
   a. $x^2 + 6x + 8 = 0$
   b. $2x^2 - 6x + 3 = 0$

4. HAPPY NEW YEAR As part of a New Year's Eve celebration, a huge ball is to be dropped from the top of a 605-foot-tall building at the proper moment so that it strikes the ground at exactly 12:00 midnight. The distance d in feet traveled by a free-falling object in t seconds is given by the formula $d = 16t^2$. To the nearest second, when should the ball be dropped from the building?

Section 8.2

The quadratic formula:
The solutions of $ax^2 + bx + c = 0$ are given by

$$x = \frac{-b \pm \sqrt{b^2 - 4ac}}{2a} \quad (a \neq 0)$$

The Quadratic Formula

5. Solve each equation using the quadratic formula.
   a. $-x^2 + 10x - 18 = 0$
   b. $x^2 - 10x = 0$
   c. $2x^2 + 13x = 7$
   d. $26y - 3y^2 = 2$

6. SPORTS POSTER The design specifications for a poster of tennis star Steffi Graf call for a 615-square-inch photograph to be surrounded by a blue border. (See Illustration 1.) The borders on the sides of the poster are to be half as wide as those at the top and bottom. Find the width of each border.

ILLUSTRATION 1	ILLUSTRATION 2

35 in.

23 in.

7. ACROBATS To begin his routine on a trapeze, an acrobat is catapulted upward as shown in Illustration 2. His distance d (in feet) from the arena floor during this maneuver is given by the formula $d = -16t^2 + 40t + 5$, where t is the time in seconds since being launched. If the trapeze bar is 25 feet in the air, at what two times will he be able to grab it? Round to the nearest tenth.

Section 8.3

A *quadratic function* is a second-degree polynomial function of the form $f(x) = ax^2 + bx + c$.

The graph of $f(x) = ax^2$ is a *parabola* opening upward when $a > 0$ and downward when $a < 0$, with *vertex* at the point $(0, 0)$ and *axis of symmetry* the line $x = 0$.

Each of the following functions has a graph that is the same shape as $f(x) = ax^2$ but involves a vertical or horizontal translation.
1. $f(x) = ax^2 + c$: translated upward if $c > 0$, downward if $c < 0$.
2. $f(x) = a(x - h)^2$: translated right if $h > 0$, and left if $h < 0$.

If $a \neq 0$, the graph of $f(x) = a(x - h)^2 + k$ is a parabola with vertex at (h, k). It opens upward when $a > 0$ and downward when $a < 0$.

The vertex of the graph of $f(x) = ax^2 + bx + c$ is

$$\left(-\frac{b}{2a}, f\left(-\frac{b}{2a}\right)\right)$$

and the axis of symmetry is the line $x = -\dfrac{b}{2a}$.

The vertex of the graph of a quadratic function gives the *minimum* or *maximum* value of the function.

Quadratic Functions and Their Graphs

8. AEROSPACE INDUSTRY See Illustration 3. The annual sales of the Boeing Company in billions of dollars for the years 1993–1997 can be modeled by the quadratic function

$$S(x) = 2.4375x^2 - 8.225x + 40.7$$

where x is the number of years since 1993. What were the annual sales for 1997?

ILLUSTRATION 3

9. Make a table of values to graph function f. Then use a series of translations to graph function g on the same coordinate system.
   a. $f(x) = 2x^2$
      $g(x) = 2x^2 - 3$
   b. $f(x) = -4x^2$
      $g(x) = -4(x - 2)^2 + 1$

10. First determine the vertex and the axis of symmetry of the graph of each function. Then plot several points and complete the graph.
    a. $y = -\left(x + \frac{3}{2}\right)^2 + \frac{5}{2}$
    b. $f(x) = 5x^2 + 10x - 1$

11. FARMING The number of farms in the United States for the years 1870–1970 is modeled by

$$N(x) = -1.46x^2 + 148.82x + 2,660$$

where $x = 0$ represents 1870, $x = 1$ represents 1871, and so on. For this period, when was the number of U.S. farms a maximum? How many farms were there?

12. Estimate the solutions of $-3x^2 - 5x + 2 = 0$ from the graph of $f(x) = -3x^2 - 5x + 2$, shown in Illustration 4.

ILLUSTRATION 4

Section 8.4   Complex Numbers

Square roots of negative numbers are called *imaginary numbers*.

$$i = \sqrt{-1} \quad \text{and} \quad i^2 = -1$$

13. Simplify each expression.

a. $\sqrt{-4}$

b. $\sqrt{-7}$

c. $-3\sqrt{-20}$

d. $\sqrt{-\dfrac{36}{49}}$

e. $\sqrt{-3}\sqrt{-3}$

f. $\dfrac{\sqrt{-64}}{\sqrt{-9}}$

A *complex number* is any number that can be written in the form $a + bi$, where a and b are real numbers and $i^2 = -1$.

14. Solve each equation.

a. $a^2 = -25$

b. $x^2 - 2x + 13 = 0$

Adding complex numbers:

$$(a + bi) + (c + di) = \\ (a + c) + (b + d)i$$

15. Do the operations and give all answers in $a + bi$ form.

a. $(5 + 4i) + (7 - 12i)$

b. $(-6 - 40i) - (-8 + 28i)$

c. $\left(-8 + \sqrt{-8}\right) + \left(6 - \sqrt{-32}\right)$

d. $2i(64 + 9i)$

Subtracting complex numbers:

$$(a + bi) - (c + di) = \\ (a - c) + (b - d)i$$

e. $(2 - 7i)(-3 + 4i)$

f. $\left(5 - \sqrt{-27}\right)\left(-6 + \sqrt{-12}\right)$

Multiplying complex numbers:

$$(a + bi)(c + di) = \\ (ac - bd) + (ad + bc)i$$

To divide complex numbers, we often have to *rationalize* a denominator by multiplying the numerator and the denominator by the *complex conjugate* of the denominator.

16. Write each expression in $a + bi$ form.

a. $\dfrac{6}{2 + i}$

b. $\dfrac{4 + i}{4 - i}$

c. $\dfrac{\sqrt{3} + \sqrt{-4}}{\sqrt{3} - \sqrt{-4}}$

d. $\dfrac{-2}{5i^3}$

The *powers of i* rotate through a cycle of four numbers: $i = i$, $i^2 = -1$, $i^3 = -i$, $i^4 = 1$

17. Simplify each power of i.

a. i^{65}

b. i^{48}

Section 8.5

The Discriminant and Equations That Can Be Written in Quadratic Form

The *discriminant* predicts the type of solutions of
$ax^2 + bx + c = 0$:

1. If $b^2 - 4ac > 0$, the solutions are unequal real numbers.
2. If $b^2 - 4ac = 0$, the solutions are equal real numbers.
3. If $b^2 - 4ac < 0$, the solutions are complex conjugates.

Many equations that are not quadratic can be written in quadratic form.

18. Use the discriminant to determine what types of solutions exist for each equation.

 a. $3x^2 + 4x - 3 = 0$

 b. $4x^2 - 5x + 7 = 0$

 c. $9x^2 - 12x + 4 = 0$

19. Solve each equation.

 a. $x - 13\sqrt{x} + 12 = 0$ **b.** $a^{2/3} + a^{1/3} - 6 = 0$

 c. $6x^4 - 19x^2 + 3 = 0$ **d.** $\dfrac{6}{x+2} + \dfrac{6}{x+1} = 5$

 e. $(x - 3)^2 - 8(x - 3) + 7 = 0$

20. WEEKLY CHORES Working together, two sisters can do the yard work at their house in 45 minutes. When the older girl does it all herself, she can complete the job in 20 minutes less time than it takes the younger girl working alone. How long does it take the older girl to do the yard work?

Section 8.6

Quadratic and Other Nonlinear Inequalities

To graph a *quadratic inequality in one variable,* get 0 on the right-hand side. Then factor the polynomial on the left-hand side. Use a *sign chart* to determine the solution set.

To solve rational inequalities, get 0 on the right-hand side and a single fraction on the left-hand side. Factor the numerator and denominator. Then use a sign chart to determine the solution set.

21. Solve each inequality. Give each result in interval notation and graph the solution set.

 a. $x^2 + 2x - 35 > 0$ **b.** $x^2 - 81 \leq 0$

 c. $\dfrac{3}{x} \leq 5$ **d.** $\dfrac{2x^2 - x - 28}{x - 1} > 0$

22. Use a graphing calculator to solve each inequality. Compare the results with Exercise 21.

 a. $x^2 + 2x - 35 > 0$ **b.** $\dfrac{2x^2 - x - 28}{x - 1} > 0$

To graph a *nonlinear inequality in two variables*, first graph the boundary. Then use a test point to determine which half-plane to shade.

23. Graph each inequality.

a. $y < \frac{1}{2}x^2 - 1$

b. $y \geq -|x|$

CHAPTER 8

Test

In Problems 1–2, solve each equation by factoring.

1. $3x^2 + 18x = 0$

2. $x(6x + 19) = -15$

3. Determine what number must be added to $x^2 + 24x$ to make it a perfect square.

4. Solve $x^2 - x - 1 = 0$ by completing the square.

5. Solve the equation $2x^2 - 8x = -5$ using the quadratic formula.

6. TABLECLOTH According to the *Guinness Book of World Records 1998*, the world's longest tablecloth was made in Illinois in 1990 and covered an area of 6,759 square feet. Its length was 3.5 feet more than 333 times its width. Find the dimensions of the tablecloth.

7. Simplify $\sqrt{-48}$.

8. Simplify i^{54}.

In Problems 9–14, do the operations. Give all answers in $a + bi$ form.

9. $(2 + 4i) + (-3 + 7i)$

10. $\left(3 - \sqrt{-9}\right) - \left(-1 + \sqrt{-16}\right)$

11. $2i(3 - 4i)$

12. $(3 + 2i)(-4 - i)$

13. $\frac{1}{i^3}$

14. $\frac{2 + i}{3 - i}$

15. Determine whether the solutions of $3x^2 + 5x + 17 = 0$ are real or nonreal.

16. Solve $x^2 = -12$.

17. Solve the equation $13 = 4t - t^2$.

18. Solve the equation $2y - 3\sqrt{y} + 1 = 0$.

In Problems 19–20, determine the vertex and the axis of symmetry of the graph of the function. Then graph it.

19. $f(x) = 2x^2 + x - 1$

20. $y = -3(x - 1)^2 - 2$

In Problems 21–22, solve the inequality and graph the solution set.

21. $x^2 - 2x - 8 > 0$

22. $\frac{x - 2}{x + 3} \leq 0$

23. Graph the inequality $y \le -x^2 + 3$.

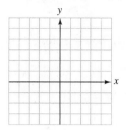

24. DRAWING An artist uses four equal-sized right triangles to block out a perspective drawing of an old hotel. See Illustration 1. For each triangle, one leg is 14 inches longer than the other, and the hypotenuse is 26 inches. On the centerline of the drawing, what is the length of the segment extending from the ground to the top of the building?

ILLUSTRATION 1

25. DISTRESS SIGNAL A flare is fired directly upward into the air from a boat that is experiencing engine problems. The height of the flare (in feet) above the water, t seconds after being fired, is given by the formula $h = -16t^2 + 112t + 15$. If the flare is designed to explode when it reaches its highest point, at what height will this occur?

26. What is an imaginary number? Give some examples.

27. The graph of a quadratic function of the form $f(x) = ax^2 + bx + c$ is shown in Illustration 2. Estimate the solutions of the corresponding quadratic equation $ax^2 + bx + c = 0$.

ILLUSTRATION 2

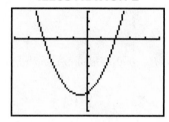

28. See Problem 27. Estimate the solution of the quadratic inequality $ax^2 + bx + c \le 0$.

CHAPTERS 1–8

Cumulative Review Exercises

In Exercises 1–2, write the equation of the line with the given properties.

1. $m = 3$, passing through $(-2, -4)$

2. Parallel to the graph of $2x + 3y = 6$ and passing through $(0, -2)$

3. AIRPORT TRAFFIC From the graph in Illustration 1, determine the projected average rate of change in the number of takeoffs and landings at Los Angeles International Airport for the years 2000–2015.

4. Solve the system by graphing.

$$\begin{cases} y = -\dfrac{5}{2}x + \dfrac{1}{2} \\ 2x - \dfrac{3}{2}y = 5 \end{cases}$$

ILLUSTRATION 1

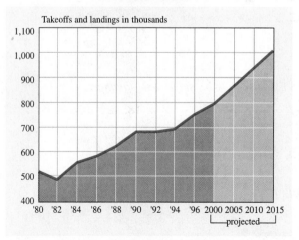

Based on data from *Los Angeles Times* (July 6, 1998) p. B3

5. Solve the system using Cramer's rule.

$$\begin{cases} x - y + z = 4 \\ x + 2y - z = -1 \\ x + y - 3z = -2 \end{cases}$$

6. Graph the solution set of the system
$$\begin{cases} 3x + 2y > 6 \\ x + 3y \le 2 \end{cases}$$

7. Solve the inequality $-9x + 6 > 16$. Give the result in interval notation and graph the solution set.

8. Solve $|2x - 5| \ge 25$. Give the result in interval notation and graph the solution set.

In Exercises 9–10, find the domain and range of each function.

9. $f(x) = 2x^2 - 3$
10. $y = -|x - 4|$

In Exercises 11–12, do each operation.

11. $(2a^2 + 4a - 7) - 2(3a^2 - 4a)$
12. $(3x + 2)(2x - 3)$

In Exercises 13–16, factor each expression.

13. $x^4 - 16y^4$
14. $15x^2 - 2x - 8$
15. $x^2 + 4y - xy - 4x$
16. $8x^6 + 125y^3$

In Exercises 17–20, solve each equation.

17. $x^2 - 5x - 6 = 0$
18. $6a^3 - 2a = a^2$

19. $\dfrac{x - 4}{x - 3} + \dfrac{x - 2}{x - 3} = x - 3$

20. $P + \dfrac{a}{V^2} = \dfrac{RT}{V - b}$ solve for b

In Exercises 21–22, simplify each expression.

21. $\dfrac{x^3 + y^3}{x^3 - y^3} \div \dfrac{x^2 - xy + y^2}{x^2 + xy + y^2}$

22. $\dfrac{1}{x + y} - \dfrac{1}{x - y} - \dfrac{2y}{y^2 - x^2}$

In Exercises 23–24, graph each function and give its domain and range.

23. $f(x) = x^3 + x^2 - 6x$
24. $f(x) = \dfrac{4}{x}$ for $x > 0$

25. LIGHT As light energy radiates away from its source, its intensity varies inversely as the square of the distance from the source. Illustration 2 shows that the light energy passing through an area 1 foot from the source spreads out over 4 units of area 2 feet from the source. That energy is therefore less intense 2 feet from the source than it was 1 foot from the source. Over how many units of area will the light energy spread out 3 feet from the source?

ILLUSTRATION 2

26. Graph the function $f(x) = \sqrt{x - 2}$ and give its domain and range.

In Exercises 27–30, simplify each expression.

27. $\sqrt[3]{-27x^3}$
28. $\sqrt{48t^3}$
29. $64^{-2/3}$
30. $\dfrac{x^{5/3} x^{1/2}}{x^{3/4}}$

In Exercises 31–34, simplify each expression.

31. $-3\sqrt[4]{32} - 2\sqrt[4]{162} + 5\sqrt[4]{48}$

32. $3\sqrt{2}\left(2\sqrt{3} - 4\sqrt{12}\right)$

33. $\dfrac{\sqrt{x} + 2}{\sqrt{x} - 1}$

34. $\dfrac{5}{\sqrt[3]{x}}$

In Exercises 35–36, solve each equation.

35. $5\sqrt{x + 2} = x + 8$

36. $\sqrt{x} + \sqrt{x + 2} = 2$

37. Find the length of the hypotenuse of the right triangle shown in Illustration 3.

ILLUSTRATION 3

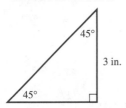

38. Find the length of the hypotenuse of the right triangle shown in Illustration 4.

ILLUSTRATION 4

39. Find the distance between $(-2, 6)$ and $(4, 14)$.

40. What number must be added to $x^2 + 6x$ to make a perfect square trinomial ?

41. Use the method of completing the square to solve the equation $2x^2 + x - 3 = 0$.

42. Use the quadratic formula to solve the equation $3x^2 + 4x - 1 = 0$.

43. Graph $y = \frac{1}{2}x^2 - x + 1$ and find the coordinates of its vertex.

44. Graph $f(x) = -x^2 - 4x$ and find the coordinates of its vertex.

In Exercises 45–46, write each expression in a + bi form.

45. $(3 + 5i) + (4 - 3i)$

46. $\dfrac{5}{3 - i}$

47. Simplify $\sqrt{-64}$.

48. Solve $a - 7a^{1/2} + 12 = 0$

49. The graph of $f(x) = 16x^2 + 24x + 9$ is shown in Illustration 5. Estimate the solution(s) of $16x^2 + 24x + 9 = 0$.

ILLUSTRATION 5

50. Use the graph in Illustration 5 to determine the solution of $16x^2 + 24x + 9 < 0$.

9

Exponential and Logarithmic Functions

CAMPUS CONNECTION

The Sociology Department

In a sociology course, students analyze surveys and statistical data to make observations and predictions about human social relations. One area of great interest to sociologists is how population growth affects social structures and institutions, such as the family, the workplace, and the schools. Population growth can often be mathematically modeled by an *exponential function*. In this chapter, we will examine exponential functions and their graphs. We will also introduce an important irrational number that is often used when writing an exponential function to describe growth or decay. Such functions have application in fields as diverse as banking, medicine, and nuclear energy.

In this chapter, we discuss the concept of function in more depth. We also introduce two new families of functions—exponential and logarithmic functions—which have applications in many areas.

▶ 9.1   Algebra and Composition of Functions

In this section, you will learn about

> Algebra of functions ■ Composition of functions ■ The identity function ■ Writing composite functions

Introduction   So far, we have defined functions with real-number domains and ranges and have graphed them on a rectangular coordinate system. Just as it is possible to perform operations on real numbers, it is also possible to perform operations on functions. We will begin this section by showing how to add, subtract, multiply, and divide functions. Then we will consider another method of combining functions, called *composition of functions.*

Algebra of Functions

We now consider how functions can be added, subtracted, multiplied, and divided.

Operations on functions

If the domains and ranges of functions f and g are subsets of the real numbers, then

The **sum** of f and g, denoted as $f + g,$ is defined by

$$(f + g)(x) = f(x) + g(x)$$

The **difference** of f and g, denoted as $f - g,$ is defined by

$$(f - g)(x) = f(x) - g(x)$$

The **product** of f and g, denoted as $f \cdot g,$ is defined by

$$(f \cdot g)(x) = f(x)g(x)$$

The **quotient** of f and g, denoted as $f/g,$ is defined by

$$(f/g)(x) = \frac{f(x)}{g(x)} \quad (g(x) \neq 0)$$

The domain of each of these functions is the set of real numbers x that are in the domain of both f and g. In the case of the quotient, there is the further restriction that $g(x) \neq 0$.

EXAMPLE 1   **Adding and subtracting functions.** Let $f(x) = 2x^2 + 1$ and $g(x) = 5x - 3$. Find each function and its domain: **a.** $f + g$ and   **b.** $f - g$.

Solution   **a.** $(f + g)(x) = f(x) + g(x)$

$$= (2x^2 + 1) + (5x - 3)$$

$$= 2x^2 + 5x - 2 \qquad \text{Combine like terms.}$$

The domain of $f + g$ is the set of real numbers that are in the domain of both f and g. Since the domain of both f and g is the interval $(-\infty, \infty)$, the domain of $f + g$ is also the interval $(-\infty, \infty)$.

b. $(f - g)(x) = f(x) - g(x)$

$$= (2x^2 + 1) - (5x - 3)$$

$$= 2x^2 + 1 - 5x + 3 \qquad \text{Remove parentheses.}$$

$$= 2x^2 - 5x + 4 \qquad \text{Combine like terms.}$$

Since the domain of both f and g is $(-\infty, \infty)$, the domain of $f - g$ is also the interval $(-\infty, \infty)$.

SELF CHECK   Let $f(x) = 3x - 2$ and $g(x) = 2x^2 + 3x$. Find   *Answers:* **a.** $2x^2 + 6x - 2$,
a. $f + g$ and   **b.** $f - g$.   **b.** $-2x^2 - 2$ ∎

EXAMPLE 2   **Multiplying and dividing functions.** Let $f(x) = 2x^2 + 1$ and $g(x) = 5x - 3$. Find each function and its domain: **a.** $f \cdot g$ and   **b.** f/g.

Solution   **a.** $(f \cdot g)(x) = f(x)g(x)$

$$= (2x^2 + 1)(5x - 3)$$

$$= 10x^3 - 6x^2 + 5x - 3 \quad \text{Use the FOIL method.}$$

The domain of $f \cdot g$ is the set of real numbers that are in the domain of both f and g. Since the domain of both f and g is the interval $(-\infty, \infty)$, the domain of $f \cdot g$ is also the interval $(-\infty, \infty)$.

b. $(f/g)(x) = \dfrac{f(x)}{g(x)}$

$$= \dfrac{2x^2 + 1}{5x - 3}$$

Since the denominator of the fraction cannot be 0, $x \neq \frac{3}{5}$. Thus, the domain of f/g is the interval $\left(-\infty, \frac{3}{5}\right) \cup \left(\frac{3}{5}, \infty\right)$.

SELF CHECK   Let $f(x) = 2x^2 - 3$ and $g(x) = x^2 + 1$. Find   *Answers:* **a.** $2x^4 - x^2 - 3$,
a. $f \cdot g$ and   **b.** f/g.   **b.** $\dfrac{2x^2 - 3}{x^2 + 1}$ ∎

Composition of Functions

We have seen that a function can be represented by a machine: We put in a number from the domain, and a number from the range comes out. For example, if we put the number 2 into the machine shown in Figure 9-1(a), the number $f(2) = 8$ comes out. In general, if we put x into the machine shown in Figure 9-1(b), the value $f(x)$ comes out.

Often one quantity is a function of a second quantity that depends, in turn, on a third quantity. For example, the cost of a car trip is a function of the gasoline consumed. The amount of gasoline consumed, in turn, is a function of the number of miles driven. Such chains of dependence can be analyzed mathematically as **compositions of functions.**

FIGURE 9-1

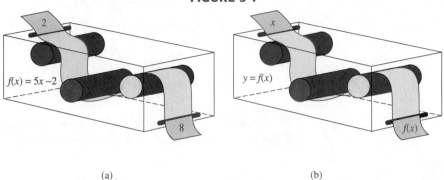

(a)   (b)

Suppose that $y = f(x)$ and $y = g(x)$ define two functions. Any number x in the domain of g will produce the corresponding value $g(x)$ in the range of g. If $g(x)$ is in the domain of function f, then $g(x)$ can be substituted into f, and a corresponding value $f(g(x))$ will be determined. This two-step process defines a new function, called a **composite function,** denoted by $f \circ g$. (This is read as "f composed with g.")

The function machines shown in Figure 9-2 illustrate the composition $f \circ g$. When we put a number x into the function g, $g(x)$ comes out. The value $g(x)$ goes into function f, which transforms $g(x)$ into $f(g(x))$. (This is read as "f of g of x.") If the function machines for g and f were connected to make a single machine, that machine would be named $f \circ g$.

FIGURE 9-2

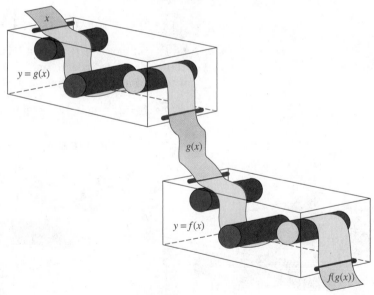

To be in the domain of the composite function $f \circ g$, a number x has to be in the domain of g. Also, the output of g must be in the domain of f. Thus, the domain of $f \circ g$ consists of those numbers x that are in the domain of g, and for which $g(x)$ is in the domain of f.

Composite functions

The **composite function $f \circ g$** is defined by

$$(f \circ g)(x) = f(g(x))$$

For example, if $f(x) = 4x$ and $g(x) = 3x + 2$, then

$$(f \circ g)(x) = f(g(x)) \qquad \text{or} \qquad (g \circ f)(x) = g(f(x))$$
$$= f(3x + 2) \qquad\qquad\qquad = g(4x)$$
$$= 4(3x + 2) \qquad\qquad\qquad = 3(4x) + 2$$
$$= 12x + 8 \qquad\qquad\qquad = 12x + 2$$

 WARNING! Note that in the previous example, $(f \circ g)(x) \neq (g \circ f)(x)$. This shows that the composition of functions is not commutative.

E X A M P L E   3

Evaluating composite functions. Let $f(x) = 2x + 1$ and $g(x) = x - 4$. Find **a.** $(f \circ g)(9)$, **b.** $(f \circ g)(x)$, and **c.** $(g \circ f)(-2)$.

Solution

a. $(f \circ g)(9)$ means $f(g(9))$. In Figure 9-3(a), function g receives the number 9, subtracts 4, and releases the number $g(9) = 5$. Then 5 goes into the f function, which doubles 5 and adds 1. The final result, 11, is the output of the composite function $f \circ g$:

$$(f \circ g)(9) = f(g(9)) = f(5) = 2(5) + 1 = 11$$

b. $(f \circ g)(x)$ means $f(g(x))$. In Figure 9-3(a), function g receives the number x, subtracts 4, and releases the number $x - 4$. Then $x - 4$ goes into the f function, which doubles $x - 4$ and adds 1. The final result, $2x - 7$, is the output of the composite function $f \circ g$.

$$(f \circ g)(x) = f(g(x)) = f(x - 4) = 2(x - 4) + 1 = 2x - 7$$

c. $(g \circ f)(-2)$ means $g(f(-2))$. In Figure 9-3(b), function f receives the number -2, doubles it and adds 1, and releases -3 into the g function. Function g subtracts 4 from -3 and outputs a final result of -7. Thus,

$$(g \circ f)(-2) = g(f(-2)) = g(-3) = -3 - 4 = -7$$

FIGURE 9-3

(a)

(b)

The Identity Function

The **identity function** is defined by the equation $I(x) = x$. Under this function, the value that is assigned to any real number x is x itself. For example $I(2) = 2$, $I(-3) = -3$, and $I(7.5) = 7.5$. If f is any function, the composition of f with the identity function is just the function f:

$$(f \circ I)(x) = (I \circ f)(x) = f(x)$$

EXAMPLE 4

The identity function. Let f be any function and let I be the identity function, $I(x) = x$. Show that **a.** $(f \circ I)(x) = f(x)$ and **b.** $(I \circ f)(x) = f(x)$.

Solution   **a.** $(f \circ I)(x)$ means $f(I(x))$. Because $I(x) = x$, we have

$$(f \circ I)(x) = f(I(x)) = f(x)$$

b. $(I \circ f)(x)$ means $I(f(x))$. Because I passes any number through unchanged, we have $I(f(x)) = f(x)$ and

$$(I \circ f)(x) = I(f(x)) = f(x)$$ ∎

Writing Composite Functions

EXAMPLE 5

Biological research. A laboratory specimen is stored in a refrigeration unit at a temperature of 15° Fahrenheit. Biologists remove the specimen and warm it at a controlled rate of 3° F per hour. Express the sample's Celsius temperature as a function of the time t since it was removed from refrigeration.

Solution   The temperature of the specimen is 15° F when the time $t = 0$. Because it warms at a rate of 3° F per hour, its initial temperature of 15° increases by $3t°$ F in t hours. The Fahrenheit temperature of the specimen is given by the function

$$F(t) = 3t + 15$$

The Celsius temperature C is a function of this Fahrenheit temperature F, given by the function

$$C(F) = \frac{5}{9}(F - 32)$$

To express the specimen's Celsius temperature as a function of *time,* we find the composite function

$$(C \circ F)(t) = C(F(t))$$
$$= C(3t + 15) \qquad \text{Substitute } 3t + 15 \text{ for } F(t).$$
$$= \frac{5}{9}[(3t + 15) - 32] \quad \text{Substitute } 3t + 15 \text{ for } F \text{ in } \tfrac{5}{9}(F - 32).$$
$$= \frac{5}{9}(3t - 17) \qquad \text{Simplify.}$$

The composite function, $C(t) = \frac{5}{9}(3t - 17)$, finds the temperature of the specimen in degrees Celsius t hours after it is removed from refrigeration. ∎

VOCABULARY

In Exercises 1–8, fill in the blanks to make the statements true.

1. The _____ of f and g, denoted as $f + g$, is defined by $(f + g)(x) = $ ▢.

2. The _____ of f and g, denoted as $f - g$, is defined by $(f - g)(x) = $ ▢.

3. The _____ of f and g, denoted as $f \cdot g$, is defined by $(f \cdot g)(x) = $ ▢.

4. The _____ of f and g, denoted as f/g, is defined by $(f/g)(x) = $ ▢.

5. In Exercises 1–3, the _____ of each function is the set of real numbers x that are in the domain of both f and g.

6. The _____ function $f \circ g$ is defined by $(f \circ g)(x) = f(g(x))$.

7. Under the _____ function, the value that is assigned to any real number x is x itself.

8. When reading the notation $f(g(x))$, we say "f ___ g ___ x."

CONCEPTS

9. Fill in the blanks to make the statements true.
   a. $(f \circ g)(x) = f(\ ▢\ )$
   b. To find $f(g(x))$, we first find _____ and then substitute that value for x in $f(x)$.

10. **a.** If $f(x) = 3x + 1$ and $g(x) = 1 - 2x$, find $f(g(3))$ and $g(f(3))$.
    b. Is the composition of functions commutative? Explain.

11. Complete the table of values for the identity function, $I(x) = x$. Then graph it.

x	$I(x)$
-3	
-2	
-1	
0	
1	
2	
3	

12. Fill in the blanks in the drawing of the function machines in Illustration 1 that show how to compute $g(f(-2))$.

ILLUSTRATION 1

$f(x) = 2 - 3x^2$

$g(x) = x + 10$

NOTATION

In Exercises 13–14, complete each solution.

13. Let $f(x) = 3x - 1$ and $g(x) = 2x + 3$. Find $f \cdot g$.

$$(f \cdot g)(x) = f(x) \cdot g(x)$$
$$= ▢ (2x + 3)$$
$$= 6x^2 + ▢ - ▢ - 3$$
$$= 6x^2 + 7x - 3$$

14. Let $f(x) = 3x - 1$ and $g(x) = 2x + 3$. Find $f \circ g$.

$$(f \circ g)(x) = f(g(x))$$
$$= f(\ ▢\ )$$
$$= 3(\ ▢\ ) - 1$$
$$= ▢ + ▢ - 1$$
$$= 6x + 8$$

PRACTICE

In Exercises 15–22, $f(x) = 3x$ and $g(x) = 4x$. Find each function and its domain.

15. $f + g$

16. $f - g$

17. $f \cdot g$

18. f/g

19. $g - f$

20. $g + f$

21. g/f

22. $g \cdot f$

In Exercises 23–30, f(x) = 2x + 1 and g(x) = x − 3.
Find each function and its domain.

23. $f + g$ **24.** $f - g$
25. $f \cdot g$ **26.** f/g

27. $g - f$ **28.** $g + f$

29. g/f

30. $g \cdot f$

In Exercises 31–34, f(x) = 3x − 2 and g(x) = 2x² + 1.
Find each function and its domain.

31. $f - g$ **32.** $f + g$

33. f/g **34.** $f \cdot g$

In Exercises 35–38, f(x) = x² − 1 and g(x) = x² − 4.
Find each function and its domain.

35. $f - g$ **36.** $f + g$

37. g/f

38. $g \cdot f$

In Exercises 39–50, f(x) = 2x + 1 and g(x) = x² − 1.
Find each value.

39. $(f \circ g)(2)$ **40.** $(g \circ f)(2)$
41. $(g \circ f)(-3)$ **42.** $(f \circ g)(-3)$
43. $(f \circ g)(0)$ **44.** $(g \circ f)(0)$
45. $(f \circ g)\left(\dfrac{1}{2}\right)$ **46.** $(g \circ f)\left(\dfrac{1}{3}\right)$
47. $(f \circ g)(x)$ **48.** $(g \circ f)(x)$
49. $(g \circ f)(2x)$ **50.** $(f \circ g)(2x)$

In Exercises 51–58, f(x) = 3x − 2 and g(x) = x² + x.
Find each value.

51. $(f \circ g)(4)$ **52.** $(g \circ f)(4)$
53. $(g \circ f)(-3)$ **54.** $(f \circ g)(-3)$
55. $(g \circ f)(0)$ **56.** $(f \circ g)(0)$
57. $(g \circ f)(x)$ **58.** $(f \circ g)(x)$

59. If $f(x) = x + 1$ and $g(x) = 2x - 5$, show that
$(f \circ g)(x) \neq (g \circ f)(x)$.

60. If $f(x) = x^2 + 1$ and $g(x) = 3x^2 - 2$, show that
$(f \circ g)(x) \neq (g \circ f)(x)$.

APPLICATIONS

61. METALLURGY A molten alloy must be cooled
slowly to control crystallization. When removed from
the furnace, its temperature is 2,700° F, and it will be
cooled at 200° per hour. Express the Celsius temperature as a function of the number of hours t since
cooling began.

62. WEATHER FORECASTING A high-pressure area
promises increasingly warmer weather for the next 48
hours. The temperature is now 34° Celsius and is expected to rise 1° every 6 hours. Express the Fahrenheit temperature as a function of the number of hours
from now. $\left(Hint:\ F = \frac{9}{5}C + 32.\right)$

63. VACATION MILEAGE COSTS
a. Use the graphs in Illustration 2 to determine the
cost of the gasoline consumed if a family drove
500 miles on a summer vacation.
b. Write a composition function that expresses the
cost of the gasoline consumed on the vacation as a
function of the miles driven.

ILLUSTRATION 2

64. HALLOWEEN COSTUMES The tables on the back
of the pattern package shown in Illustration 3 can be
used to determine the number of yards of material
needed to make a rabbit costume for a child.
a. How many yards of material are needed if the
child's chest measures 29 inches?
b. In this exercise, one quantity is a function of a
second quantity that depends, in turn, on a third
quantity. Explain this dependence.

ILLUSTRATION 3

PATTERN 9810   **Simplicity**

Costumes have front zipper, long raglan sleeves, elastic
sleeve and leg casings, hood. Fabrics: Fleece or suede

BODY MEASUREMENTS

Chest (in.)	21	22	23	25	26	27	$28\frac{1}{2}$	29	30
Pattern Size	2	3	4	6	7	8	10	11	12

YARDAGE NEEDED

Pattern Size	2-4	6-8	10-12
Yards	$2\frac{5}{8}$	$3\frac{3}{8}$	$3\frac{3}{4}$

WRITING

65. Exercise 63 illustrates a chain of dependence between the cost of the gasoline, the gasoline consumed, and the miles driven. Describe another chain of dependence that could be represented by a composition function.

66. Explain how to add, subtract, multiply, and divide two functions.

67. Write out in words how to say each of the following:

$(f \circ g)(2)$ $g(f(-8))$

68. If $Y_1 = f(x)$ and $Y_2 = g(x)$, explain how to use the tables of values shown in Illustration 4 to find $g(f(2))$.

ILLUSTRATION 4

X	Y₁	
-2	-2	
0	2	
2	6	
4	10	
6	14	
8	18	
10	22	

X= -2

X	Y₂	
-2	-7	
0	-6	
2	-5	
4	-4	
6	-3	
8	-2	
10	-1	

X= -2

REVIEW

Simplify each expression.

69. $\dfrac{3x^2 + x - 14}{4 - x^2}$

70. $\dfrac{2x^3 + 14x^2}{3 + 2x - x^2} \cdot \dfrac{x^2 - 3x}{x}$

71. $\dfrac{x^2 - 2x - 8}{3x^2 - x - 12} \div \dfrac{3x^2 + 5x - 2}{3x - 1}$

72. $\dfrac{x - 1}{1 + \dfrac{x}{x - 2}}$

▶ **9.2**

Inverses of Functions

In this section, you will learn about

One-to-one functions ■ The horizontal line test ■ Finding inverses of functions

Introduction   The function defined by $C = \frac{5}{9}(F - 32)$ is the formula that we use to convert degrees Fahrenheit to degrees Celsius. If we input a Fahrenheit reading into the formula, the output is a Celsius reading. For example, if we substitute 41° for F, we obtain a Celsius reading of 5°:

$$C = \frac{5}{9}(F - 32)$$

$$= \frac{5}{9}(41 - 32) \quad \text{Substitute 41 for } F.$$

$$= \frac{5}{9}(9)$$

$$= 5$$

If we want to find a Fahrenheit reading from a Celsius reading, we need a formula into which we can substitute a Celsius reading and have a Fahrenheit reading come out. Such a formula is $F = \frac{9}{5}C + 32$, which takes the Celsius reading of 5° and turns it back into a Fahrenheit reading of 41°.

$$F = \frac{9}{5}C + 32$$

$$= \frac{9}{5}(5) + 32 \quad \text{Substitute 5 for } C.$$

$$= 41$$

The functions defined by these two formulas do opposite things. The first turns 41° F into 5° C, and the second turns 5° C back into 41° F. For this reason, we say that the functions are *inverses* of each other.

In this section, we will show how to find inverses of one-to-one functions.

One-to-One Functions

Recall that for each input into a function, there is a single output. For some functions, different inputs have the same output, as shown in Figure 9-4(a). For other functions, different inputs have different outputs, as shown in Figure 9-4(b).

FIGURE 9-4

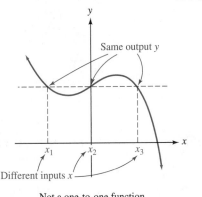

Not a one-to-one function

(a)

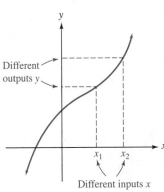

A one-to-one function

(b)

When every output of a function corresponds to exactly one input, we say that the function is *one-to-one*.

One-to-one functions

A function is called **one-to-one** if each input value of x in the domain determines a different output value of y in the range.

EXAMPLE 1

One-to-one functions. Determine whether the functions **a.** $f(x) = x^2$ and **b.** $f(x) = x^3$ are one-to-one.

Solution **a.** The function $f(x) = x^2$ is not one-to-one, because different input values x can determine the same output value y. For example, inputs of 3 and -3 produce the same output value of 9.

$$f(3) = 3^2 = 9 \quad \text{and} \quad f(-3) = (-3)^2 = 9$$

b. The function $f(x) = x^3$ is one-to-one, because different input values x determine different output values y. This is because different numbers have different cubes.

SELF CHECK Determine whether $f(x) = 2x + 3$ is one-to-one and explain why or why not.

Answer: yes, because different input values determine different output values ∎

The Horizontal Line Test

The **horizontal line test** can be used to decide whether the graph of a function represents a one-to-one function. If every horizontal line that intersects the graph of a function does so only once, the function is one-to-one. Otherwise, the function is not one-to-one. See Figure 9-5.

FIGURE 9-5

A one-to-one function

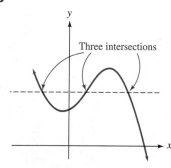

Not a one-to-one function

EXAMPLE 2

Determining one-to-one functions. Use the horizontal line test to decide whether the graphs in Figure 9-6 represent one-to-one functions.

Solution **a.** Because many horizontal lines intersect the graph shown in Figure 9-6(a) twice, the graph does not represent a one-to-one function.

b. Because each horizontal line that intersects the graph in Figure 9-6(b) does so exactly once, the graph represents a one-to-one function.

FIGURE 9-6

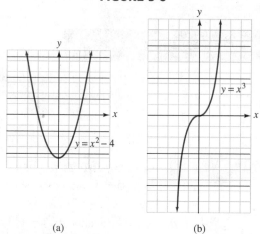

(a)   (b)

SELF CHECK Determine whether the following graphs represent one-to-one functions.

a.   **b.**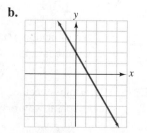

Answers: **a.** no, **b.** yes

■

Finding Inverses of Functions

If f is the function determined by the ordered pairs in the table shown in Figure 9-7(a), it turns the number 1 into 10, 2 into 20, and 3 into 30. Since the inverse of f must turn 10 back into 1, 20 back into 2, and 30 back into 3, it consists of the ordered pairs shown in Figure 9-7(b).

FIGURE 9-7

Function f			*Inverse of f*	
x	y		x	y
1	10		10	1
2	20		20	2
3	30		30	3

Domain Range   Domain Range

(a)   (b)

Note that the inverse of f is also a function.

We note that the domain of f and the range of its inverse is $\{1, 2, 3\}$. The range of f and the domain of its inverse is $\{10, 20, 30\}$.

This example suggests that to form the inverse of a function f, we simply interchange the coordinates of each ordered pair that determines f. When the inverse of a function is also a function, we call it **f inverse** and denote it with the symbol f^{-1}.

WARNING! The symbol $f^{-1}(x)$ is read as "the inverse of $f(x)$" or just "f inverse." The -1 in the notation $f^{-1}(x)$ is not an exponent. Remember that $f^{-1}(x) \neq \frac{1}{f(x)}$.

Finding the inverse of a one-to-one function

If a function is one-to-one, we find its inverse as follows:

1. If the function is written using function notation, replace $f(x)$ with y.
2. Interchange the variables x and y.
3. Solve the resulting equation for y.
4. We can substitute $f^{-1}(x)$ for y.

EXAMPLE 3

Finding the inverse of a function. If $f(x) = 4x + 2$, find the inverse of f and tell whether it is a function.

Solution   We proceed as follows:

$f(x) = 4x + 2$

$y = 4x + 2$ Step 1: Replace $f(x)$ with y.

$x = 4y + 2$ Step 2: Interchange the variables x and y.

To decide whether the inverse $x = 4y + 2$ is a function, we solve for y (Step 3).

$x = 4y + 2$

$x - 2 = 4y$ Subtract 2 from both sides.

1. $y = \dfrac{x - 2}{4}$ Divide both sides by 4 and write y on the left-hand side.

Because each input x that is substituted into Equation 1 gives one output y, the inverse of f is a function, so we can express it in the form

$$f^{-1}(x) = \frac{x - 2}{4} \quad \text{Step 4: Substitute } f^{-1}(x) \text{ for } y.$$

SELF CHECK

If $f(x) = -5x - 3$, find the inverse of f and tell whether it is a function.

Answers: $f^{-1}(x) = \frac{-x-3}{5}$, yes ∎

To emphasize an important relationship between a function and its inverse, we substitute some number x, such as $x = 3$, into the function $f(x) = 4x + 2$ of Example 3. The corresponding value of y produced is

$$f(3) = 4(3) + 2 = 14$$

If we substitute 14 into the inverse function, f^{-1}, the corresponding value of y that is produced is

$$f^{-1}(14) = \frac{14 - 2}{4} = 3$$

Thus, the function f turns 3 into 14, and the inverse function f^{-1} turns 14 back into 3. In general, the composition of a function and its inverse is the identity function.

To prove that $f(x) = 4x + 2$ and $f^{-1}(x) = \frac{x-2}{4}$ are inverse functions, we must show that their composition (in both directions) is the identity function:

$$(f \circ f^{-1})(x) = f(f^{-1}(x)) \qquad\qquad (f^{-1} \circ f)(x) = f^{-1}(f(x))$$

$$= f\left(\frac{x - 2}{4}\right) \qquad\qquad\qquad = f^{-1}(4x + 2)$$

$$= 4\left(\frac{x - 2}{4}\right) + 2 \qquad\qquad = \frac{4x + 2 - 2}{4}$$

$$= x - 2 + 2 \qquad\qquad\qquad = \frac{4x}{4}$$

$$= x \qquad\qquad\qquad\qquad\quad = x$$

Thus, $(f \circ f^{-1})(x) = (f^{-1} \circ f)(x) = x$, which is the identity function $I(x)$.

E X A M P L E   4

The graph of an inverse function. The set of all pairs (x, y) determined by $3x + 2y = 6$ is a function. Find its inverse function and graph the function and its inverse on one coordinate system.

Solution

To find the inverse function of $3x + 2y = 6$, we interchange x and y to obtain

$$3y + 2x = 6$$

and then solve the equation for y.

$$3y + 2x = 6$$

$$3y = -2x + 6$$

$$y = -\frac{2}{3}x + 2$$

The inverse function of $3x + 2y = 6$ is $y = -\frac{2}{3}x + 2$, which can also be written $f^{-1}(x) = -\frac{2}{3}x + 2$. The graphs of $3x + 2y = 6$ and its inverse, along with the graph of $y = x$, appear in Figure 9-8.

FIGURE 9-8

SELF CHECK

Find the inverse of the function defined by $2x - 3y = 6$. Graph the function and its inverse on one coordinate system. Also graph $y = x$.

Answer: $f^{-1}(x) = \frac{3}{2}x + 3$

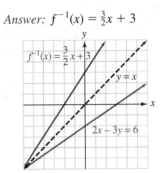

In Figure 9-8, we see that the graphs of $3x + 2y = 6$ and $y = -\frac{2}{3}x + 2$ (its inverse) are symmetric about the line $y = x$. That is, the graphs are mirror images (reflections) of each other with reference to $y = x$. This is always the case with a function and its inverse, because when the coordinates (a, b) satisfy an equation, the coordinates (b, a) will satisfy its inverse.

ACCENT ON TECHNOLOGY *Graphing the Inverse of a Function*

We can use a graphing calculator for an informal check of the result found in Example 4. First, we solve $3x + 2y = 6$ for y and enter it in the form $y = -\frac{3}{2}x + 3$. Then we enter what we believe to be the inverse function, $y = -\frac{2}{3}x + 2$, as well as the equation $y = x$. See Figure 9-9(a). Before graphing, we adjust the display so that the graphing grid will be composed of squares. (The standard window display is rectangular, having a grid with a unit width different from the unit height.) Check your owner's manual for directions on how to program the calculator to graph in the square mode.

In Figure 9-9(b), it appears that the two graphs are symmetric about the line $y = x$. Although it is not definitive, this visual check does help to validate that $y = -\frac{2}{3}x + 2$ is the inverse of the function determined by $3x + 2y = 6$.

FIGURE 9-9

(a) $Y_3 = x$ (b) $Y_1 = -\frac{3}{2}x + 3$ $Y_2 = -\frac{2}{3}x + 2$

In each example so far, the inverse of a function has been another function. This is not always true, as the following example will show.

EXAMPLE 5

Finding the inverse of a function. Find the inverse of the function determined by $f(x) = x^2$.

Solution

$y = x^2$ Replace $f(x)$ with y.

$x = y^2$ Interchange x and y.

$y = \pm\sqrt{x}$ Use the square root property and write y on the left-hand side.

When the inverse $y = \pm\sqrt{x}$ is graphed, as in Figure 9-10, we see that the graph does not pass the vertical line test. Thus, it is not a function.

The graph of $y = x^2$ is also shown in the figure. Note that it is not one-to-one. The graphs of $y = x^2$ and $y = \pm\sqrt{x}$ are symmetric about the line $y = x$.

FIGURE 9-10

SELF CHECK

Find the inverse of the function determined by $f(x) = 4x^2$.

Answer: $y = \pm\dfrac{\sqrt{x}}{2}$ ■

EXAMPLE 6

Finding the inverse of a function. Find the inverse of $f(x) = x^3$.

Solution

To find the inverse, we proceed as follows:

$y = x^3$ Replace $f(x)$ with y.

$x = y^3$ Interchange the variables x and y.

$\sqrt[3]{x} = y$ Take the cube root of both sides.

We note that to each number x there corresponds one real cube root. Thus, $y = \sqrt[3]{x}$ represents a function. Using $f^{-1}(x)$ notation, the inverse of $f(x) = x^3$ is

$$f^{-1}(x) = \sqrt[3]{x}$$

SELF CHECK

Find the inverse of $f(x) = x^5$.   *Answer:* $f^{-1}(x) = \sqrt[5]{x}$ ■

If a function is not one-to-one, it is often possible to make it one-to-one by restricting its domain.

EXAMPLE 7

Find the inverse of the function defined by $y = x^2$ with $x \geq 0$. Then tell whether it is a function. Graph the function and its inverse on one set of coordinate axes.

Solution

The inverse of the function $y = x^2$ with $x \geq 0$ is

$x = y^2$ with $y \geq 0$ Interchange the variables x and y.

This equation can be written in the form

$y = \pm\sqrt{x}$ with $y \geq 0$

Since $y \geq 0$, each number x gives only one value of y: $y = \sqrt{x}$. Thus, the inverse is a function.

The graphs of the two functions appear in Figure 9-11. The line $y = x$ is included so that we can see that the graphs are symmetric about the line $y = x$.

FIGURE 9-11

$y = x^2$ and $x \geq 0$

x	y	(x, y)
0	0	$(0, 0)$
1	1	$(1, 1)$
2	4	$(2, 4)$
3	9	$(3, 9)$

$x = y^2$ and $y \geq 0$

x	y	(x, y)
0	0	$(0, 0)$
1	1	$(1, 1)$
4	2	$(4, 2)$
9	3	$(9, 3)$

STUDY SET

Section 9.2

VOCABULARY

In Exercises 1–4, fill in the blanks to make the statements true.

1. A function is called _____ if each input determines a different output.

2. The _____ line test can be used to decide whether the graph of a function represents a one-to-one function.

3. The functions f and f^{-1} are _____.

4. An input value is an element of a function's _____. An output value is an element of a function's _____.

CONCEPTS

In Exercises 5–10, fill in the blanks to make the statements true.

5. If every horizontal line that intersects the graph of a function does so only _____, the function is one-to-one.

6. If any horizontal line that intersects the graph of a function does so more than once, the function is not _____.

7. If a function turns an input of 2 into an output of 5, the inverse function will turn an input of 5 into an output of ___.

8. The graphs of a function and its inverse are symmetrical about the line ▨.

9. $(f \circ f^{-1})(x) = $ ▨

10. To find the inverse of the function defined by $y = 2x - 3$, we begin by _____ x and y.

11. Use the table of values of the one-to-one function f to complete a table of values for f^{-1}.

x	$f(x)$		x	$f^{-1}(x)$
-6	-3		-3	
-4	-2		-2	
0	0		0	
2	1		1	
8	4		4	

12. How can we tell that function f is not one-to-one from the table of values?

x	$f(x)$
-2	4
-1	1
0	0
2	4
3	9

13. Is the inverse of a function always a function?

14. Name four points that the line $y = x$ passes through.

15. If f is a one-to-one function, and if $f(2) = 6$, then what is $f^{-1}(6)$?

16. If the point $(2, -4)$ is on the graph of the one-to-one function f, then what point is on the graph of f^{-1}?

17. Two functions are graphed on the square grid in Illustration 1 along with the line $y = x$. Explain why we know that the functions are not inverses of each other.

ILLUSTRATION 1

18. A table of values for a function f is shown in Illustration 2(a). A table of values for f^{-1} is shown Illustration 2(b). Explain how to use the tables to find $f^{-1}(f(4))$ and $f(f^{-1}(2))$.

ILLUSTRATION 2

(a)   (b)

NOTATION

In Exercises 19–20, complete each solution.

19. Find the inverse of $f(x) = 2x - 3$.

$$y = \boxed{} - 3 \quad \text{Replace } f(x) \text{ with } y.$$
$$x = \boxed{} - 3 \quad \text{Interchange the variables } x \text{ and } y.$$
$$x + \boxed{} = 2y \quad \text{Add 3 to both sides.}$$
$$\frac{x + 3}{2} = \boxed{} \quad \text{Divide both sides by 2.}$$

The inverse of $f(x) = 2x - 3$ is $\boxed{} = \dfrac{x + 3}{2}$.

20. Find the inverse of $f(x) = \sqrt[3]{x} + 2$.

$$\boxed{} = \sqrt[3]{x} + 2 \quad \text{Replace } f(x) \text{ with } y.$$
$$x = \sqrt[3]{\boxed{}} + 2 \quad \text{Interchange the variables } x \text{ and } y.$$
$$x - \boxed{} = \sqrt[3]{y} \quad \text{Subtract 2 from both sides.}$$
$$(x - 2)^3 = \boxed{} \quad \text{Cube both sides.}$$

The inverse of $f(x) = \sqrt[3]{x} + 2$ is $\boxed{} = (x - 2)^3$.

21. The symbol $f^{-1}(x)$ is read as "_____ f" or "f _____."

22. Explain the difference in the meaning of the -1 in the notation $f^{-1}(x)$ as compared to x^{-1}.

PRACTICE

In Exercises 23–26, determine whether the function is one-to-one.

23. $f(x) = 2x$ **24.** $f(x) = |x|$

25. $f(x) = x^4$ **26.** $f(x) = x^3 + 1$

In Exercises 27–30, each graph represents a function. Use the horizontal line test to decide whether the function is one-to-one.

27.

28.

29.

30.

In Exercises 31–34, tell whether the set of ordered pairs (x, y) is a function. Then find the inverse of the set of ordered pairs (x, y) and tell whether the inverse is a function.

31. $\{(3, 2), (2, 1), (1, 0)\}$

32. $\{(4, 1), (5, 1), (6, 1), (7, 1)\}$

33. $\{(1, 1), (2, 1), (3, 1), (4, 1)\}$

34. $\{(1, 2), (2, 3), (1, 3), (1, 5)\}$

In Exercises 35–40, find the inverse of the function and express it using $f^{-1}(x)$ notation.

35. $4x - 5y = 20$ **36.** $y + 1 = 5x$

37. $x + 4 = 5y$ **38.** $x = 3y + 1$

39. $f(x) = \dfrac{x - 4}{5}$ **40.** $f(x) = \dfrac{2x + 6}{3}$

In Exercises 41–44, find the inverse of each function and express it in terms of x and y. Then graph the function and its inverse on one coordinate system. Show the line of symmetry on the graph.

41. $y = 4x + 3$ **42.** $x = 3y - 1$

43. $2x + 3y = 9$

44. $3(x + y) = 2x + 4$

In Exercises 45–52, find the inverse of the function and tell whether it is a function. If it is a function, express it using $f^{-1}(x)$ notation.

45. $y = x^2 + 4$

46. $y = x^2 + 5$

47. $y = x^3$

48. $xy = 4$

49. $y = |x|$

50. $y = \sqrt[3]{x}$

51. $f(x) = 2x^3 - 3$

52. $f(x) = \dfrac{3}{x^3} - 1$

In Exercises 53–56, graph each equation and its inverse on one set of coordinate axes. Show the axis of symmetry.

53. $y = x^2 + 1$

54. $y = \dfrac{1}{4}x^2 - 3$

55. $y = \sqrt{x}$ $(x \geq 0)$

56. $y = |x|$

APPLICATIONS

57. INTERPERSONAL RELATIONSHIPS Feelings of anxiety in a relationship can increase or decrease, depending on what is going on in the relationship. The graph in Illustration 3 shows how a person's anxiety might vary as a relationship develops over time.
   a. Is this the graph of a function? Is its inverse a function?

b. Does each anxiety level correspond to exactly one point in time? Use the dashed lined labeled *Maximum threshold* to explain.

ILLUSTRATION 3

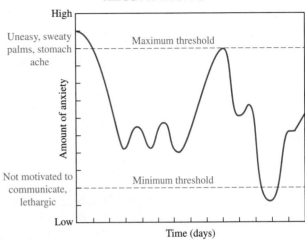

Based on information from Gudykunst, *Building Bridges: Interpersonal Skills for a Changing World* (Houghton Mifflin, 1994)

58. LIGHTING LEVELS The ability of the eye to see detail increases as the level of illumination increases. This relationship can be modeled by a function E, whose graph is shown in Illustration 4.
   a. From the graph, determine $E(240)$.
   b. Is function E one-to-one? Does E have an inverse?

   c. If the effectiveness of seeing in an office is 7, what is the illumination in the office? How can this question be asked using inverse function notation?

ILLUSTRATION 4

Based on information from *World Book Encyclopedia*

WRITING

59. What does it mean when we say that one graph is symmetric to the other about the line $y = x$?

60. Explain the purpose of the horizontal line test.

REVIEW

Simplify each expression and write the result in a + bi form.

61. $3 - \sqrt{-64}$

62. $(2 - 3i) + (4 + 5i)$

63. $(3 + 4i)(2 - 3i)$

64. $\dfrac{6 + 7i}{3 - 4i}$

65. $(6 - 8i)^2$

66. i^{100}

▶ 9.3   Exponential Functions

In this section, you will learn about

Irrational exponents ■ Exponential functions ■ Graphing exponential functions ■ Vertical and horizontal translations ■ Compound interest ■ Exponential functions as models

Introduction   The graph in Figure 9-12 shows the balance in a bank account in which $10,000 was invested in 1998 at 9%, compounded monthly. The graph shows that in the year 2008, the value of the account will be approximately $25,000, and in the year 2028, the value will be approximately $147,000. The curve shown in Figure 9-12 is the graph of a function called an *exponential function.*

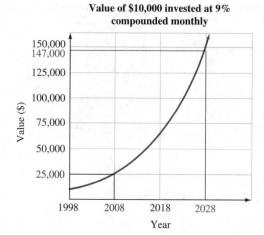

FIGURE 9-12

Value of $10,000 invested at 9% compounded monthly

Exponential functions are also suitable for modeling many other situations, such as population growth, the spread of an epidemic, the temperature of a heated object as it cools, and radioactive decay. Before we can discuss exponential functions in more detail, we must define irrational exponents.

Irrational Exponents

We have discussed expressions of the form b^x, where x is a rational number.

$8^{1/2}$ means "the square root of 8."

$5^{1/3}$ means "the cube root of 5."

$3^{-2/5} = \dfrac{1}{3^{2/5}}$ means "the reciprocal of the fifth root of 3^2."

To give meaning to b^x when x is an irrational number, we consider the expression $5^{\sqrt{2}}$ where $\sqrt{2}$ is the irrational number 1.414213562 . . .

Each number in the following list is defined, because each exponent is a rational number.

$$5^{1.4}, \quad 5^{1.41}, \quad 5^{1.414}, \quad 5^{1.4142}, \quad 5^{1.41421}, \quad . . .$$

Since the exponents are getting closer to $\sqrt{2}$, the numbers in this list are successively better approximations of $5^{\sqrt{2}}$. We can use a calculator to obtain a very good approximation.

ACCENT ON TECHNOLOGY *Evaluating Exponential Expressions*

To find the value of $5^{\sqrt{2}}$ with a scientific calculator, we enter these numbers and press these keys:

Keystrokes 5 $\boxed{y^x}$ 2 $\boxed{\sqrt{}}$ $\boxed{=}$ $\boxed{9.738517742}$

With a graphing calculator, we enter these numbers and press these keys:

Keystrokes 5 $\boxed{\wedge}$ $\boxed{\sqrt{}}$ 2 $\boxed{)}$ $\boxed{\text{Enter}}$ $\boxed{\begin{array}{l} 5 \wedge \sqrt{} (2) \\ \quad\quad 9.738517742 \end{array}}$

In general, if $b > 0$ and x is a real number, b^x represents a positive number. It can be shown that all of the familiar rules of exponents are also true for irrational exponents.

EXAMPLE 1

Exponential expressions having irrational exponents. Use the rules of exponents to simplify **a.** $\left(5^{\sqrt{2}}\right)^{\sqrt{2}}$ and **b.** $b^{\sqrt{3}} \cdot b^{\sqrt{12}}$.

Solution **a.** $\left(5^{\sqrt{2}}\right)^{\sqrt{2}} = 5^{\sqrt{2}\sqrt{2}}$ Keep the base and multiply the exponents.

$\qquad\qquad = 5^2$ $\sqrt{2}\sqrt{2} = \sqrt{4} = 2.$

$\qquad\qquad = 25$

b. $b^{\sqrt{3}} \cdot b^{\sqrt{12}} = b^{\sqrt{3}+\sqrt{12}}$ Keep the base and add the exponents.

$\qquad\qquad = b^{\sqrt{3}+2\sqrt{3}}$ $\sqrt{12} = \sqrt{4}\sqrt{3} = 2\sqrt{3}.$

$\qquad\qquad = b^{3\sqrt{3}}$ $\sqrt{3} + 2\sqrt{3} = 3\sqrt{3}.$

SELF CHECK Simplify: **a.** $\left(3^{\sqrt{2}}\right)^{\sqrt{8}}$ and **b.** $b^{\sqrt{2}} \cdot b^{\sqrt{18}}$.   *Answers:* **a.** 81, **b.** $b^{4\sqrt{2}}$ ∎

Exponential Functions

If $b > 0$ and $b \neq 1$, the function $f(x) = b^x$ is an **exponential function.** Since x can be any real number, its domain is the set of real numbers. This is the interval $(-\infty, \infty)$. Since b is positive, the value of $f(x)$ is positive, and the range is the set of positive numbers. This is the interval $(0, \infty)$.

Since $b \neq 1$, an exponential function cannot be the constant function $f(x) = 1^x$, in which $f(x) = 1$ for every real number x.

Exponential functions

> An **exponential function with base b** is defined by the equation
>
> $$f(x) = b^x \quad \text{or} \quad y = b^x \quad (b > 0, \ b \neq 1, \text{ and } x \text{ is a real number})$$
>
> The **domain of any exponential function** is the interval $(-\infty, \infty)$. The **range** is the interval $(0, \infty)$.

Graphing Exponential Functions

Since the domain and range of $f(x) = b^x$ are sets of real numbers, we can graph exponential functions on a rectangular coordinate system.

E X A M P L E 2 **Graphing exponential functions.** Graph $f(x) = 2^x$.

Solution To graph $f(x) = 2^x$, we find several points (x, y) whose coordinates satisfy the equation, plot the points, and join them with a smooth curve, as shown in Figure 9-13. For example, if $x = -1$, we have

$$f(x) = 2^x$$
$$f(-1) = 2^{-1}$$
$$= \frac{1}{2}$$

The point $\left(-1, \frac{1}{2}\right)$ is on the graph of $f(x) = 2^x$.

FIGURE 9-13

$f(x) = 2^x$

x	$f(x)$	$(x, f(x))$
-1	$\frac{1}{2}$	$\left(-1, \frac{1}{2}\right)$
0	1	$(0, 1)$
1	2	$(1, 2)$
2	4	$(2, 4)$
3	8	$(3, 8)$
4	16	$(4, 16)$

By looking at the graph, we can verify that the domain is the interval $(-\infty, \infty)$ and that the range is the interval $(0, \infty)$.

Note that as x decreases, the values of $f(x)$ decrease and approach 0. Thus, the x-axis is a horizontal asymptote of the graph.

Also note that the graph of $f(x) = 2^x$ passes through the points $(0, 1)$ and $(1, 2)$.

SELF CHECK   Graph $f(x) = 4^x$.

Answer:

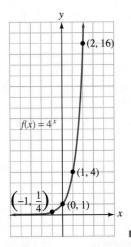

E X A M P L E 3   **Graphing exponential functions.** Graph $f(x) = \left(\dfrac{1}{2}\right)^x$.

Solution   We find and plot pairs (x, y) that satisfy the equation. The graph of $f(x) = \left(\dfrac{1}{2}\right)^x$ appears in Figure 9-14. For example, if $x = -4$, we have

$$f(x) = \left(\frac{1}{2}\right)^x$$

$$f(-4) = \left(\frac{1}{2}\right)^{-4}$$

$$= \left(\frac{2}{1}\right)^4$$

$$= 16$$

The point $(-4, 16)$ is on the graph of $f(x) = \left(\frac{1}{2}\right)^x$.

FIGURE 9-14

$f(x) = \left(\frac{1}{2}\right)^x$

x	$f(x)$	$(x, f(x))$
-4	16	$(-4, 16)$
-3	8	$(-3, 8)$
-2	4	$(-2, 4)$
-1	2	$(-1, 2)$
0	1	$(0, 1)$
1	$\frac{1}{2}$	$\left(1, \frac{1}{2}\right)$

By looking at the graph, we can verify that the domain is the interval $(-\infty, \infty)$ and that the range is the interval $(0, \infty)$.

In this case, as x increases, the values of $f(x)$ decrease and approach 0. The x-axis is a horizontal asymptote. Note that the graph of $f(x) = \left(\frac{1}{2}\right)^x$ passes through the points $(0, 1)$ and $\left(1, \frac{1}{2}\right)$.

SELF CHECK Graph $g(x) = \left(\frac{1}{3}\right)^x$.

Answer:

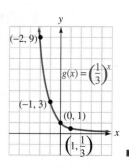

Examples 2 and 3 illustrate the following properties of exponential functions.

Properties of exponential functions

The domain of the exponential function $f(x) = b^x$ is the interval $(-\infty, \infty)$.

The range is the interval $(0, \infty)$.

The graph has a y-intercept of $(0, 1)$.

The x-axis is an asymptote of the graph.

The graph of $f(x) = b^x$ passes through the point $(1, b)$.

In Example 2 (where $b = 2$), the values of y increase as the values of x increase. Since the graph rises as we move to the right, we call the function an *increasing function*. When $b > 1$, the larger the value of b, the steeper the curve.

In Example 3 $\left(\text{where } b = \frac{1}{2}\right)$, the values of y decrease as the values of x increase. Since the graph drops as we move to the right, we call the function a *decreasing function*. When $0 < b < 1$, the smaller the value of b, the steeper the curve.

In general, the following is true.

Increasing and decreasing functions

If $b > 1$, then $f(x) = b^x$ is an **increasing function.**

If $0 < b < 1$, then $f(x) = b^x$ is a **decreasing function.**

An exponential function with base b is either increasing (for $b > 1$) or decreasing ($0 < b < 1$). Since different real numbers x determine different values of b^x, exponential functions are one-to-one.

ACCENT ON TECHNOLOGY *Graphing Exponential Functions*

To use a graphing calculator to graph $f(x) = \left(\frac{2}{3}\right)^x$ and $f(x) = \left(\frac{3}{2}\right)^x$, we enter the right-hand sides of the equations after the symbols $Y_1 =$ and $Y_2 =$. The screen will show the following equations.

$Y_1 = (2/3) \wedge X$

$Y_2 = (3/2) \wedge X$

FIGURE 9-15

If we use window settings of $[-10, 10]$ for x and $[-2, 10]$ for y and press the $\boxed{\text{GRAPH}}$ key, we will obtain the graph shown in Figure 9-15.

We note that the graph of $f(x) = \left(\frac{2}{3}\right)^x$ passes through the points $(0, 1)$ and $\left(1, \frac{2}{3}\right)$. Since $\frac{2}{3} < 1$, the function is decreasing.

The graph of $f(x) = \left(\frac{3}{2}\right)^x$ passes through the points $(0, 1)$ and $\left(1, \frac{3}{2}\right)$. Since $\frac{3}{2} > 1$, the function is increasing.

Since both graphs pass the horizontal line test, each function is one-to-one.

Vertical and Horizontal Translations

EXAMPLE 4

Translations. On one set of axes, graph $y = 2^x$ and $y = 2^x + 3$, and describe the translation.

Solution The graph of $y = 2^x + 3$ is identical to the graph of $y = 2^x$, except that it is translated 3 units upward. (See Figure 9-16.)

FIGURE 9-16

$y = 2^x$			$y = 2^x + 3$		
x	y	(x, y)	x	y	(x, y)
-4	$\frac{1}{16}$	$\left(-4, \frac{1}{16}\right)$	-4	$3\frac{1}{16}$	$\left(-4, 3\frac{1}{16}\right)$
0	1	$(0, 1)$	0	4	$(0, 4)$
2	4	$(2, 4)$	2	7	$(2, 7)$

SELF CHECK Graph $f(x) = 4^x$ and $g(x) = 4^x - 3$, and describe the translation.

Answer: The graph of $f(x) = 4^x$ is translated 3 units downward.

EXAMPLE 5

Translations. On one set of axes, graph $f(x) = 2^x$ and $g(x) = 2^{x+3}$, and describe the translation.

Solution   The graph of $g(x) = 2^{x+3}$ is identical to the graph of $f(x) = 2^x$, except that it is translated 3 units to the left. (See Figure 9-17.)

FIGURE 9-17

$f(x) = 2^x$

x	$f(x)$	$(x, f(x))$
-1	$\frac{1}{2}$	$\left(-1, \frac{1}{2}\right)$
0	1	$(0, 1)$
1	2	$(1, 2)$

$g(x) = 2^{x+3}$

x	$g(x)$	$(x, g(x))$
-1	4	$(-1, 4)$
0	8	$(0, 8)$
1	16	$(1, 16)$

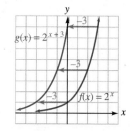

SELF CHECK   On one set of axes, graph $f(x) = 4^x$ and $g(x) = 4^{x-3}$, and describe the translation.

Answer: The graph of $f(x) = 4^x$ is translated 3 units to the right.

The graphs of $f(x) = kb^x$ and $f(x) = b^{kx}$ are vertical and horizontal stretchings of the graph of $f(x) = b^x$. To graph these functions, we can plot several points and join them with a smooth curve, or we can use a graphing calculator.

ACCENT ON TECHNOLOGY *Graphing Exponential Functions*

To use a graphing calculator to graph the exponential function $f(x) = 3(2^{x/3})$, we enter the right-hand side of the equation after the symbol $Y_1 =$. The display will show the equation

$$Y_1 = 3(2 \wedge (X/3))$$

If we use window settings of $[-10, 10]$ for x and $[-2, 18]$ for y and press the graph key, we will obtain the graph shown in Figure 9-18.

FIGURE 9-18

Compound Interest

If we deposit $\$P$ in an account paying an annual interest rate r, we can find the amount A in the account at the end of t years by using the formula $A = P + Prt$, or $A = P(1 + rt)$.

Suppose that we deposit $500 in such an account that pays interest every six months. Then $P = 500$, and after six months $\left(\frac{1}{2} \text{ year}\right)$, the amount in the account will be

$$A = 500(1 + rt)$$

$$= 500\left(1 + r \cdot \frac{1}{2}\right) \quad \text{Substitute } \tfrac{1}{2} \text{ for } t.$$

$$= 500\left(1 + \frac{r}{2}\right)$$

The account will begin the second six-month period with a value of $\$500\left(1 + \frac{r}{2}\right)$. After the second six-month period, the amount in the account will be

$$A = P(1 + rt)$$

$$A = \left[500\left(1 + \frac{r}{2}\right)\right]\left(1 + r \cdot \frac{1}{2}\right) \quad \text{Substitute } 500\left(1 + \tfrac{r}{2}\right) \text{ for } P \text{ and } \tfrac{1}{2} \text{ for } t.$$

$$= 500\left(1 + \frac{r}{2}\right)\left(1 + \frac{r}{2}\right)$$

$$= 500\left(1 + \frac{r}{2}\right)^2$$

At the end of a third six-month period, the amount in the account will be

$$A = 500\left(1 + \frac{r}{2}\right)^3$$

In this discussion, the earned interest is deposited back in the account and also earns interest. When this is the case, we say that the account is earning **compound interest.** The preceding discussion suggests the following formula for compound interest.

Formula for compound interest

If $\$P$ is deposited in an account and interest is paid k times a year at an annual rate r, the amount A in the account after t years is given by

$$A = P\left(1 + \frac{r}{k}\right)^{kt}$$

EXAMPLE 6

Saving for college. To save for college, parents invest $12,000 for their newborn child in a mutual fund that should average a 10% annual return. If the quarterly dividends are reinvested, how much will be available in 18 years?

Solution We substitute 12,000 for P, 0.10 for r, and 18 for t in the formula for compound interest and find A. Since interest is paid quarterly, $k = 4$.

$$A = P\left(1 + \frac{r}{k}\right)^{kt}$$

$$A = 12,000\left(1 + \frac{0.10}{4}\right)^{4(18)} \quad \text{Express } r = 10\% \text{ as a decimal.}$$

$$= 12,000(1 + 0.025)^{72}$$

$$= 12,000(1.025)^{72}$$

$$= 71,006.74 \qquad \text{Use a scientific calculator and press these keys:}$$
$$1.025 \boxed{y^x} \; 72 \boxed{=} \boxed{\times} \; 12,000 \boxed{=}.$$

In 18 years, the account will be worth $71,006.74.

SELF CHECK   In Example 6, how much would be available if
the parents invested $20,000?   *Answer:* $118,344.56   ■

In business applications, the initial amount of money deposited is often called the
present value (*PV*). The amount to which the money will grow is called the **future
value** (*FV*). The interest rate used for each compounding period is the **periodic interest rate** (*i*), and the number of times interest is compounded is the number of **compounding periods** (*n*). Using these definitions, we have an alternate formula for compound interest.

Formula for compound interest

$$FV = PV(1 + i)^n$$

This alternate formula appears on business calculators. To use this formula to solve
Example 6, we proceed as follows:

$FV = \boldsymbol{PV}(1 + \boldsymbol{i})^n$

$FV = \boldsymbol{12{,}000}(1 + \boldsymbol{0.025})^{\boldsymbol{72}}$ $i = \frac{0.10}{4} = 0.025$ and $n = 4(18) = 72$.

$\quad\; = 71{,}006.74$ Use a calculator to evaluate the expression.

ACCENT ON TECHNOLOGY *Solving Investment Problems*

Suppose $1 is deposited in an account earning 6% annual interest, compounded
monthly. To use a graphing calculator to estimate how much will be in the account
in 100 years, we can substitute 1 for *P*, 0.06 for *r*, and 12 for *k* in the formula

$$A = P\left(1 + \frac{r}{k}\right)^{kt}$$

$$A = 1\left(1 + \frac{0.06}{12}\right)^{12t}$$

and simplify to get

$$A = (1.005)^{12t}$$

FIGURE 9-19

We now graph $A = (1.005)^{12t}$ using window
settings of [0, 120] for *t* and [0, 400] for *A* to
obtain the graph shown in Figure 9-19. We can then trace and zoom to estimate that
$1 grows to be approximately $397 in 100 years. From the graph, we can see that
the money grows slowly in the early years and rapidly in the later years.

Exponential Functions as Models

EXAMPLE 7

Cellular telephones. For the years 1990–1996, the U.S. cellular telephone industry experienced "exponential growth." (See Figure 9-20.) The exponential function
$S(n) = 5.28(1.43)^n$ approximates the number of cellular telephone subscribers in millions, where *n* is the number of years since 1990 and $0 \le n \le 6$.

a. How many subscribers were there in 1990?

b. How many subscribers were there in 1996?

FIGURE 9-20

Based on information from *New York Times Almanac*, 1998

Solution   **a.** The year 1990 is 0 years after 1990. To find the number of subscribers in 1990, we substitute 0 for *n* in the function and find $S(0)$.

$$S(n) = 5.28(1.43)^n$$
$$S(0) = 5.28(1.43)^0$$
$$= 5.28 \cdot 1 \qquad (1.43)^0 = 1.$$
$$= 5.28$$

In 1990, there were approximately 5.28 million cellular telephone subscribers.

b. The year 1996 is 6 years after 1990. We need to find $S(6)$.

$$S(n) = 5.28(1.43)^n$$
$$S(6) = 5.28(1.43)^6 \qquad \text{Substitute 6 for } n.$$
$$\approx 45.14920914 \qquad \text{Use a calculator to find an approximation.}$$

In 1996, there were approximately 45.15 million cellular telephone subscribers.   ∎

STUDY SET

Section 9.3

VOCABULARY

In Exercises 1–8, refer to the graph of $f(x) = 3^x$ in Illustration 1.

ILLUSTRATION 1

$f(x) = 3^x$

1. What type of function is $f(x) = 3^x$?

2. What is the domain of the function?

3. What is the range of the function?

4. What is the *y*-intercept of the graph?

5. What is the *x*-intercept of the graph?

6. What is an asymptote of the graph?

7. Is *f* an increasing or a decreasing function?

8. The graph passes through the point $(1, y)$. What is *y*?

CONCEPTS

9. Graph the functions $f(x) = x^2$ and $g(x) = 2^x$ on the same set of coordinate axes.

10. Graph the functions $y = x^{\frac{1}{2}}$ and $y = \left(\frac{1}{2}\right)^x$ on the same set of coordinate axes.

11. What are the two formulas that are used to determine the amount of money in a savings account that is earning compound interest?

12. Explain the order in which the expression $20,000(1.036)^{72}$ should be evaluated.

13. Illustration 2 shows the graph of $f(x) = 2^x$, as well as two vertical translations of that graph. Using the notation $g(x)$ for one translation and $h(x)$ for the other, write the defining equation for each function.

ILLUSTRATION 2

14. Illustration 3 shows the graph of $f(x) = 2^x$, as well as a horizontal translation of that graph. Using the notation $g(x)$ for the translation, write its defining equation.

ILLUSTRATION 3

NOTATION

15. In $A = P\left(1 + \frac{r}{k}\right)^{kt}$, what is the base, and what is the exponent?

16. For an exponential function of the form $f(x) = b^x$, what are the restrictions on b?

PRACTICE

In Exercises 17–20, find each value to four decimal places.

17. $2^{\sqrt{2}}$ **18.** $7^{\sqrt{2}}$
19. $5^{\sqrt{5}}$ **20.** $6^{\sqrt{3}}$

In Exercises 21–24, simplify each expression.

21. $\left(2^{\sqrt{3}}\right)^{\sqrt{3}}$ **22.** $3^{\sqrt{2}}3^{\sqrt{18}}$
23. $7^{\sqrt{3}}7^{\sqrt{12}}$ **24.** $\left(3^{\sqrt{5}}\right)^{\sqrt{5}}$

In Exercises 25–32, graph each exponential function. Check your work with a graphing calculator.

25. $f(x) = 5^x$ **26.** $f(x) = 6^x$

27. $y = \left(\frac{1}{4}\right)^x$ **28.** $y = \left(\frac{1}{5}\right)^x$

29. $f(x) = 3^x - 2$ **30.** $y = 2^x + 1$

31. $f(x) = 3^{x-1}$ **32.** $f(x) = 2^{x+1}$

In Exercises 33–36, use a graphing calculator to graph each function. Tell whether the function is an increasing or a decreasing function.

33. $f(x) = \frac{1}{2}(3^{x/2})$ **34.** $f(x) = -3(2^{x/3})$

35. $y = 2(3^{-x/2})$

36. $y = -\dfrac{1}{4}(2^{-x/2})$

 APPLICATIONS

In Exercises 37–42, assume that there are no deposits or withdrawals.

37. COMPOUND INTEREST An initial deposit of $10,000 earns 8% interest, compounded quarterly. How much will be in the account after 10 years?

38. COMPOUND INTEREST An initial deposit of $10,000 earns 8% interest, compounded monthly. How much will be in the account after 10 years?

39. COMPARING INTEREST RATES How much more interest could $1,000 earn in 5 years, compounded quarterly, if the annual interest rate were $5\frac{1}{2}\%$ instead of 5%?

40. COMPARING SAVINGS PLANS Which institution in Illustration 4 provides the better investment?

ILLUSTRATION 4

> ### *Fidelity Savings & Loan*
> Earn 5.25%
> compounded monthly

> ### **Union Trust**
> Money Market Account
> paying 5.35%
> compounded annually

41. COMPOUND INTEREST If $1 had been invested on July 4, 1776, at 5% interest, compounded annually, what would it be worth on July 4, 2076?

42. FREQUENCY OF COMPOUNDING $10,000 is invested in each of two accounts, both paying 6% annual interest. In the first account, interest compounds quarterly, and in the second account, interest compounds daily. Find the difference between the accounts after 20 years.

43. WORLD POPULATION See the graph in Illustration 5.
   a. Estimate when the world's population reached $\frac{1}{2}$ billion and when it reached 1 billion.

   b. Estimate the world's population in the year 2000.

   c. What type of function does it appear could be used to model the population growth?

ILLUSTRATION 5

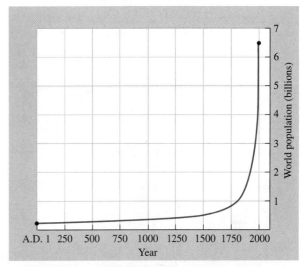

Based on data from *The Blue Planet* (Wiley, 1995)

44. THE STOCK MARKET The Dow Jones Industrial Average is a measure of how well the stock market is doing. Plot the following Dow milestones as ordered pairs of the form (year, average) on the graph in Illustration 6. What type of function does it appear could be used to model the growth of the stock market?

Year	Average	Year	Average
Jan. 1906	100	Feb. 1995	4,000
Mar. 1956	500	Nov. 1995	5,000
Nov. 1972	1,000	Oct. 1996	6,000
Jan. 1987	2,000	Feb. 1997	7,000
Apr. 1991	3,000	July 1997	8,000

ILLUSTRATION 6

45. MARKET VALUE OF A CAR The graph in Illustration 7 shows how the value of the average car depreciates as a percent of its original value over a 10-year period. It also shows the yearly maintenance costs as a percent of the car's value.

a. When is the car worth half of its purchase price?

b. When is the car worth a quarter of its purchase price?

c. When do the average yearly maintenance costs surpass the value of the car?

ILLUSTRATION 7

Based on data from the U.S. Department of Transportation

46. DIVING *Bottom time* is the time a scuba diver spends descending plus the actual time spent at a certain depth. On Illustration 8, graph the bottom time limits given in the table.

Bottom time limits			
Depth (ft)	Bottom time (min)	Depth (ft)	Bottom time (min)
30	no limit	80	40
35	310	90	30
40	200	100	25
50	100	110	20
60	60	120	15
70	50	130	10

47. BACTERIAL CULTURE A colony of 6 million bacteria is growing in a culture medium. See Illustration 9. The population P after t hours is given by the formula $P = (6 \times 10^6)(2.3)^t$. Find the population after 4 hours.

ILLUSTRATION 8

ILLUSTRATION 9

12:00 P.M.   4:00 P.M.

48. RADIOACTIVE DECAY A radioactive material decays according to the formula $A = A_0\left(\frac{2}{3}\right)^t$, where A_0 is the initial amount present and t is measured in years. Find the amount present in 5 years.

49. DISCHARGING A BATTERY The charge remaining in a battery decreases as the battery discharges. The charge C (in coulombs) after t days is given by the formula $C = (3 \times 10^{-4})(0.7)^t$. Find the charge after 5 days.

50. POPULATION GROWTH The population of North Rivers is decreasing exponentially according to the formula $P = 3,745(0.93)^t$, where t is measured in years from the present date. Find the population in 6 years, 9 months.

51. SALVAGE VALUE A small business purchased a computer for \$4,700. It is expected that its value each year will be 75% of its value in the preceding year. If the business disposes of the computer after 5 years, find its salvage value (the value after 5 years).

52. LOUISIANA PURCHASE In 1803, the United States negotiated the Louisiana Purchase with France. The country doubled its territory by adding 827,000 square miles of land for \$15 million. If the land appreciated at the rate of 6% each year, what would one square mile of land be worth in 1996?

WRITING

53. If world population is increasing exponentially, why is there cause for concern?

54. How do the graphs of $f(x) = 3^x$ and $g(x) = \left(\frac{1}{3}\right)^x$ differ? How are they similar?

REVIEW

In Exercises 55–58, refer to Illustration 10. Lines r and s are parallel.

55. Find x.

56. Find the measure of $\angle 1$.

57. Find the measure of $\angle 2$.

58. Find the measure of $\angle 3$.

ILLUSTRATION 10

▶ **9.4**

Base-*e* Exponential Functions

In this section, you will learn about

Continuous compound interest ■ The natural exponential function ■ Graphing the natural exponential function ■ Vertical and horizontal translations ■ Malthusian population growth ■ Base-*e* exponential function models

---

Introduction A special exponential function that appears in real-life applications is the base-*e* exponential function. In this section, we will show how to evaluate *e*, graph the base-*e* exponential function, and discuss one of its important applications in analyzing population growth.

Continuous Compound Interest

If a bank pays interest twice a year, we say that interest is compounded semiannually. If it pays interest four times a year, we say that interest is compounded quarterly. If it pays interest continuously (infinitely many times in a year), we say that interest is compounded continuously.

To develop the formula for continuous compound interest, we start with the formula

$$A = P\left(1 + \frac{r}{k}\right)^{kt}$$

The formula for compound interest: r is the annual rate and k is the number of times per year interest is paid.

and let $rn = k$. Since r and k are positive numbers, so is n.

$$A = P\left(1 + \frac{r}{rn}\right)^{rnt}$$

We can then simplify the fraction $\frac{r}{rn}$ and use the commutative property of multiplication to change the order of the exponents.

$$A = P\left(1 + \frac{1}{n}\right)^{nrt}$$

Finally, we can use a property of exponents to write the formula as

1. $A = P\left[\left(1 + \frac{1}{n}\right)^{n}\right]^{rt}$ Use the property $a^{mn} = (a^m)^n$.

To find the value of $\left(1 + \frac{1}{n}\right)^n$, we evaluate it for several values of n, as shown in Table 9-1.

TABLE 9-1

n	$\left(1 + \frac{1}{n}\right)^n$
1	2
2	2.25
4	2.44140625. . .
12	2.61303529. . .
365	2.71456748. . .
1,000	2.71692393. . .
100,000	2.71826830. . .
1,000,000	2.71828137. . .

The results suggest that as n gets larger, the value of $\left(1 + \frac{1}{n}\right)^n$ approaches the number 2.71828. . . . . This number is called e, which has the following value.

$$e = 2.718281828459. . .$$

In continuous compound interest, k (the number of compoundings) is infinitely large. Since k, r, and n are all positive, r is a fixed rate, and $k = rn$, as k gets very large (approaches infinity), so does n. Therefore, we can replace $\left(1 + \frac{1}{n}\right)^n$ in Equation 1 with e to get

$$A = Pe^{rt}$$

Formula for exponential growth

If a quantity P increases or decreases at an annual rate r, compounded continuously, the amount A after t years is given by

$$A = Pe^{rt}$$

If time is measured in years, then r is called the **annual growth rate.** If r is negative, the "growth" represents a decrease.

The Natural Exponential Function

The number e is often used as the base in applications of exponential functions.

The natural exponential function

The function defined by $f(x) = e^x$ is the **natural exponential function** (or the **base-e exponential function**) where $e = 2.71828. . . .$

The e^x key on a calculator is used to evaluate the natural exponential function.

ACCENT ON TECHNOLOGY *The Natural Exponential Function Key*

To compute the amount to which $12,000 will grow if invested for 18 years at 10% annual interest, compounded continuously, we substitute 12,000 for P, 0.10 for r, and 18 for t in the formula for exponential growth:

$$A = Pe^{rt}$$
$$A = 12{,}000e^{0.10(18)}$$
$$= 12{,}000e^{1.8}$$

To evaluate this expression, we enter these numbers and press these keys on a scientific calculator:

Keystrokes   1.8 $\boxed{e^x}$ $\boxed{\times}$ 12000 $\boxed{=}$ $\boxed{\text{72595.76957}}$

Using a graphing calculator, we enter these numbers and press these keys:

Keystrokes   12000 $\boxed{\times}$ $\boxed{\text{2nd}}$ $\boxed{e^x}$ 1.8 $\boxed{)}$ $\boxed{\text{ENTER}}$ $\boxed{\begin{array}{l}\text{12000*e^(1.8)}\\ \qquad\text{72595.76957}\end{array}}$

After 18 years, the account will contain $72,595.77. This is $1,589.03 more than the result in Example 6 in the previous section, where interest was compounded quarterly.

EXAMPLE 1

Continuous compound interest. If $25,000 accumulates interest at an annual rate of 8%, compounded continuously, find the balance in the account in 50 years.

Solution $A = Pe^{rt}$
$$A = 25{,}000e^{0.08(50)}$$
$$= 25{,}000e^4$$
$$\approx 1{,}364{,}953.75 \quad \text{Use a calculator.}$$

In 50 years, the balance will be $1,364,953.75—more than a million dollars.

SELF CHECK   In Example 1, find the balance in 60 years.   *Answer:* $3,037,760.44   ■

Graphing the Natural Exponential Function

To graph the natural exponential function, we plot several points and join them with a smooth curve, as shown in Figure 9-21. By looking at the graph, we can verify that $f(x) = e^x$ is an increasing function. The domain is the interval $(-\infty, \infty)$, and the range is the interval $(0, \infty)$. Note that as x decreases, the values of $f(x)$ decrease and approach 0. Thus, the x-axis is a horizontal asymptote of the graph.

FIGURE 9-21

$f(x) = e^x$

x	$f(x)$	$(x, f(x))$
-2	0.1	$(-2, 0.1)$
-1	0.4	$(-1, 0.4)$
0	1	$(0, 1)$
1	2.7	$(1, 2.7)$
2	7.4	$(2, 7.4)$

The outputs can be found using the $\boxed{e^x}$ key on a calculator. They are rounded to the nearest tenth.

Vertical and Horizontal Translations

We can illustrate the effects of vertical and horizontal translations of the natural exponential function by using a graphing calculator.

ACCENT ON TECHNOLOGY *Translations of the Natural Exponential Function*

Figure 9-22(a) shows the calculator graphs of $f(x) = e^x$, $g(x) = e^x + 5$, and $h(x) = e^x - 3$. To graph these functions with window settings of $[-3, 6]$ for x and $[-5, 15]$ for y, we enter the right-hand sides of the equations after the symbols $Y_1 =$, $Y_2 =$, and $Y_3 =$. The display will show

$$Y_1 = e \char94 (x)$$
$$Y_2 = e \char94 (x) + 5$$
$$Y_3 = e \char94 (x) - 3$$

After graphing these functions, we can see that the graph of $g(x) = e^x + 5$ is 5 units above the graph of $f(x) = e^x$, and that the graph of $h(x) = e^x - 3$ is 3 units below the graph of $f(x) = e^x$.

Figure 9-22(b) shows the calculator graphs of $f(x) = e^x$, $g(x) = e^{x+5}$, and $h(x) = e^{x-3}$. To graph these functions with window settings of $[-7, 10]$ for x and $[-5, 15]$ for y, we enter the right-hand sides of the equations after the symbols $Y_1 =$, $Y_2 =$, and $Y_3 =$. The display will show

$$Y_1 = e \char94 (x)$$
$$Y_2 = e \char94 (x + 5)$$
$$Y_3 = e \char94 (x - 3)$$

After graphing these functions, we can see that the graph of $g(x) = e^{x+5}$ is 5 units to the left of the graph of $f(x) = e^x$, and that the graph of $h(x) = e^{x-3}$ is 3 units to the right of the graph of $f(x) = e^x$.

FIGURE 9-22

(a)   (b)

ACCENT ON TECHNOLOGY *Graphing Exponential Functions*

Figure 9-23 shows the calculator graph of $f(x) = 3e^{-x/2}$. To graph this function with window settings of $[-7, 10]$ for x and $[-5, 15]$ for y, we enter the right-hand side of the equation after the symbol $Y_1 =$. The display will show the equation

$$Y_1 = 3(e \char94 (-x/2))$$

The calculator graph appears in Figure 9-23. Explain why the graph has a *y*-intercept of (0, 3).

FIGURE 9-23

Malthusian Population Growth

An equation based on the natural exponential function provides a model for **population growth.** In the **Malthusian model for population growth,** the future population of a colony is related to the present population by the formula $A = Pe^{rt}$.

EXAMPLE 2

City planning. The population of a city is currently 15,000, but changing economic conditions are causing the population to decrease 2% each year. If this trend continues, find the population in 30 years.

Solution

Since the population is decreasing 2% each year, the annual growth rate is -2%, or -0.02. We can substitute -0.02 for r, 30 for t, and 15,000 for P in the formula for exponential growth and find A.

$$A = Pe^{rt}$$
$$A = \mathbf{15,000}e^{-0.02(30)}$$
$$= 15,000e^{-0.6}$$
$$\approx 8,232.17$$

In 30 years, city planners expect a population of 8,232.

SELF CHECK

In Example 2, find the population in 50 years.   *Answer:* 5,518   ∎

The English economist Thomas Robert Malthus (1766–1834) pioneered in population study. He believed that poverty and starvation were unavoidable, because the human population tends to grow exponentially, but the food supply tends to grow linearly.

EXAMPLE 3

The Malthusian model. Suppose that a country with a population of 1,000 people is growing exponentially according to the formula

$$P = 1,000e^{0.02t}$$

where t is in years. Furthermore, assume that the food supply F, measured in adequate food per day per person, is growing linearly according to the formula

$$F = 30.625t + 2,000 \quad (t \text{ is time in years})$$

In how many years will the population outstrip the food supply?

Solution

We can use a graphing calculator with window settings of [0, 100] for x and [0, 10,000] for y. After graphing the functions, we obtain Figure 9-24(a). If we trace, as in Figure 9-24(b), we can find the point where the two graphs intersect. From the graph, we can see that the food supply will be adequate for about 71 years. At that time, the population of approximately 4,200 people will begin to have problems.

FIGURE 9-24

(a)   (b)

SELF CHECK

In Example 3, suppose that the population grows at a 3% rate. Use a graphing calculator to determine for how many years the food supply will be adequate.   *Answer:* about 38 years ∎

Base-*e* Exponential Function Models

E X A M P L E   4

Baking. A mother has just taken a cake out of the oven and set it on a rack to cool. The function $T(t) = 68 + 220e^{-0.2t}$ gives the cake's temperature in degrees Fahrenheit after it has cooled for t minutes. If her children will be home from school in 20 minutes, will the cake have cooled enough for the children to eat it?

Solution   When the children arrive home, the cake will have cooled for 20 minutes. To find the temperature of the cake at that time, we need to find $T(20)$.

$$T(t) = 68 + 220e^{-0.2t}$$
$$T(20) = 68 + 220e^{-0.2(20)} \quad \text{Substitute 20 for } t.$$
$$= 68 + 220e^{-4}$$
$$\approx 72.0 \qquad\qquad \text{Use a calculator.}$$

When the children return home, the temperature of the cake will be about 72°, and it can be eaten.   ∎

S T U D Y   S E T

Section 9.4

VOCABULARY

In Exercises 1–8, refer to the graph of $f(x) = e^x$ in Illustration 1.

ILLUSTRATION 1

$f(x) = e^x$

1. What is the name of the function $f(x) = e^x$?

2. What is the domain of the function?

3. What is the range of the function?

4. What is the y-intercept of the graph?

5. What is the x-intercept of the graph?

6. What is an asymptote of the graph?

7. Is f an increasing or a decreasing function?

8. The graph passes through the point $(1, y)$. What is y?

CONCEPTS

In Exercises 9–12, fill in the blanks to make the statements true.

9. In _____ compound interest, the number of compoundings is infinitely large.

10. The formula for continuous compound interest is $A = $ _____.

11. To two decimal places, the value of e is _____.

12. If n gets larger and larger, the value of $\left(1 + \frac{1}{n}\right)^n$ approaches the value of _____.

13. Graph each irrational number on the number line.
$$\left[\pi, e, \sqrt{2}, \frac{\sqrt{3}}{2}\right]$$

14. Complete the table of values. Round to the nearest hundredth.

x	-2	-1	0	1	2
e^x					

15. POPULATION OF THE UNITED STATES In Illustration 2, graph the U.S. census population figures shown in the table (in millions). What type of function does it appear could be used to model the population?

Year	Population	Year	Population
1790	3.9	1900	76.0
1800	5.3	1910	92.2
1810	7.2	1920	106.0
1820	9.6	1930	123.2
1830	12.9	1940	132.1
1840	17.0	1950	151.3
1850	23.1	1960	179.3
1860	31.4	1970	203.3
1870	38.5	1980	226.5
1880	50.1	1990	248.7
1890	62.9		

ILLUSTRATION 2

16. What is the Malthusian population growth formula?

17. The function $f(x) = e^x$ is graphed in Illustration 3, and the TRACE feature is used. What is the y-coordinate of the point on the graph having an x-coordinate of 1? What is the name given this number?

ILLUSTRATION 3

18. Illustration 4 shows a table of values for $f(x) = e^x$. As x decreases, what happens to the values of $f(x)$ listed in the Y_1 column? Will the value of $f(x)$ ever be 0 or negative?

ILLUSTRATION 4

X	Y1	
0	1	
-1	.36788	
-2	.13534	
-3	.04979	
-4	.01832	
-5	.00674	
-6	.00248	
X=0		

NOTATION

In Exercises 19–20, evaluate A in the formula $A = Pe^{rt}$ for the following values of r and t.

19. $P = 1,000$, $r = 0.09$ and $t = 10$

$$A = \boxed{}\, e^{(0.09)(\boxed{})}$$

$$= 1,000 e^{\boxed{}}$$

$$\approx \boxed{}\,(2.459603111)$$

$$\approx 2,459.603111$$

20. $P = 1,000$, $r = 0.12$ and $t = 50$

$$A = 1,000 e^{(\boxed{})(50)}$$

$$= \boxed{}\, e^{6}$$

$$\approx 1,000 \left(\boxed{}\right)$$

$$\approx 403,428.7935$$

PRACTICE

In Exercises 21–28, graph each function. Check your work with a graphing calculator. Compare each graph to the graph of $f(x) = e^x$.

21. $f(x) = e^x + 1$

22. $f(x) = e^x - 2$

23. $y = e^{x+3}$

24. $y = e^{x-5}$

25. $f(x) = -e^x$

26. $f(x) = -e^x + 1$

27. $f(x) = 2e^x$

28. $f(x) = \dfrac{1}{2}e^x$

APPLICATIONS

In Exercises 29–34, assume that there are no deposits or withdrawals.

29. CONTINUOUS COMPOUND INTEREST An initial investment of $5,000 earns 8.2% interest, compounded continuously. What will the investment be worth in 12 years?

30. CONTINUOUS COMPOUND INTEREST An initial investment of $2,000 earns 8% interest, compounded continuously. What will the investment be worth in 15 years?

31. COMPARISON OF COMPOUNDING METHODS An initial deposit of $5,000 grows at an annual rate of 8.5% for 5 years. Compare the final balances resulting from annual compounding and continuous compounding.

32. COMPARISON OF COMPOUNDING METHODS An initial deposit of $30,000 grows at an annual rate of 8% for 20 years. Compare the final balances resulting from annual compounding and continuous compounding.

33. DETERMINING THE INITIAL DEPOSIT An account now contains $11,180 and has been accumulating interest at 7% annual interest, compounded continuously, for 7 years. Find the initial deposit.

34. DETERMINING THE PREVIOUS BALANCE An account now contains $3,610 and has been accumulating interest at 8% annual interest, compounded continuously. How much was in the account 4 years ago?

35. WORLD POPULATION GROWTH The population of the earth is approximately 6 billion people and is growing at an annual rate of 1.9%. Assuming a Malthusian growth model, find the world population in 30 years.

36. WORLD POPULATION GROWTH The population of the earth is approximately 6 billion people and is growing at an annual rate of 1.9%. Assuming a Malthusian growth model, find the world population in 40 years.

37. POPULATION GROWTH The growth of a population is modeled by

$$P = 173e^{0.03t}$$

How large will the population be when $t = 20$?

38. POPULATION DECLINE The decline of a population is modeled by

$$P = (1.2 \times 10^6)e^{-0.008t}$$

How large will the population be when $t = 30$?

39. EPIDEMIC The spread of ungulate fever through a herd of cattle can be modeled by the formula

$$P = P_0 e^{0.27t} \quad (t \text{ is in days})$$

If a rancher does not act quickly to treat two cases, how many cattle will have the disease in 12 days?

40. OCEANOGRAPHY The width w (in millimeters) of successive growth spirals of the sea shell *Catapulus voluto*, shown in Illustration 5, is given by the exponential function

$$w = 1.54e^{0.503n}$$

where n is the spiral number. Find the width, to the nearest tenth of a millimeter, of the sixth spiral.

ILLUSTRATION 5

41. HALF-LIFE OF A DRUG The quantity of a prescription drug in the bloodstream of a patient t hours after it is administered can be modeled by an exponential function. (See the graph in Illustration 6.) Determine the time it takes to eliminate half of the initial dose from the body.

42. MEDICINE The concentration x of a certain prescription drug in an organ after t minutes is given by

$$x = 0.08(1 - e^{-0.1t})$$

Find the concentration of the drug at 30 minutes.

43. SKYDIVING Before the parachute opens, a skydiver's velocity v in meters per second is given by

$$v = 50(1 - e^{-0.2t})$$

Find the velocity after 20 seconds of free fall.

ILLUSTRATION 6

44. FREE FALL After t seconds a certain falling object has a velocity v given by

$$v = 50(1 - e^{-0.3t})$$

Which is falling faster after 2 seconds—the object or the skydiver in Exercise 43?

In Exercises 45–46, use a graphing calculator to solve each problem.

45. THE MALTHUSIAN MODEL In Example 3, suppose that better farming methods changed the formula for food growth to $y = 31x + 2,000$. How long would the food supply be adequate?

46. THE MALTHUSIAN MODEL In Example 3, suppose that a birth-control program changed the formula for population growth to $P = 1,000e^{0.01t}$. How long would the food supply be adequate?

WRITING

47. Explain why the graph of $y = e^x - 5$ is five units below the graph of $y = e^x$.

48. A feature article in a newspaper stated that the sport of snowboarding was growing *exponentially*. Explain what the author of the article meant by that.

REVIEW

Simplify each expression. Assume that all variables represent positive numbers.

49. $\sqrt{240x^5}$

50. $\sqrt[3]{-125x^5y^4}$

51. $4\sqrt{48y^3} - 3y\sqrt{12y}$

52. $\sqrt[4]{48z^5} + \sqrt[4]{768z^5}$

▶ **9.5**

Logarithmic Functions

In this section, you will learn about

Logarithms ■ Graphs of logarithmic functions ■ Vertical and horizontal translations ■ Base-10 logarithms ■ Applications of logarithms

Introduction A function that is closely related to the exponential function is the *logarithmic function*. It can be used to solve many application problems from fields such as electronics, seismology, business, and population growth.

Logarithms

Because an exponential function defined by $y = b^x$ (or $f(x) = b^x$) is one-to-one, it has an inverse function that is defined by the equation $x = b^y$. To express this inverse function in the form $y = f^{-1}(x)$, we must solve the equation $x = b^y$ for y. For this, we need the following definition.

Logarithmic functions

If $b > 0$ and $b \neq 1$, the **logarithmic function with base** b is defined by

$$y = \log_b x \quad \text{if and only if} \quad x = b^y$$

The **domain of the logarithmic function** is the interval $(0, \infty)$. The **range** is the interval $(-\infty, \infty)$.

Since the function $y = \log_b x$ is the inverse of the one-to-one exponential function $y = b^x$, the logarithmic function is also one-to-one.

WARNING! Since the domain of the logarithmic function is the set of positive numbers, it is impossible to find the logarithm of 0 or the logarithm of a negative number.

The previous definition guarantees that any pair (x, y) that satisfies the equation $y = \log_b x$ also satisfies the exponential equation $x = b^y$. Because of this relationship, a statement written in logarithmic form can be written in an equivalent exponential form.

Logarithmic form		*Exponential form*
$\log_7 1 = 0$	because	$1 = 7^0$
$\log_5 25 = 2$	because	$25 = 5^2$
$\log_5 \dfrac{1}{25} = -2$	because	$\dfrac{1}{25} = 5^{-2}$
$\log_{16} 4 = \dfrac{1}{2}$	because	$4 = 16^{1/2}$
$\log_2 \dfrac{1}{8} = -3$	because	$\dfrac{1}{8} = 2^{-3}$
$\log_b x = y$	because	$x = b^y$

In each of these examples, the logarithm of a number is an exponent. In fact,

$\log_b x$ is the exponent to which b is raised to get x.

Translating this statement into symbols, we have

$$b^{\log_b x} = x$$

EXAMPLE 1

Writing an equivalent exponential statement. Find y in each equation:
a. $\log_6 1 = y$, **b.** $\log_3 27 = y$, and **c.** $\log_5 \frac{1}{5} = y$.

Solution **a.** We can change the equation $\log_6 1 = y$ into the equivalent exponential equation $1 = 6^y$. Since $1 = 6^0$, it follows that $y = 0$. Thus,

$$\log_6 1 = 0$$

b. $\log_3 27 = y$ is equivalent to $27 = 3^y$. Since $27 = 3^3$, it follows that $3^y = 3^3$, and $y = 3$. Thus,

$$\log_3 27 = 3$$

c. $\log_5 \frac{1}{5} = y$ is equivalent to $\frac{1}{5} = 5^y$. Since $\frac{1}{5} = 5^{-1}$, it follows that $5^y = 5^{-1}$, and $y = -1$. Thus,

$$\log_5 \frac{1}{5} = -1$$

SELF CHECK Find y in each equation: **a.** $\log_3 9 = y$, *Answers:* **a.** 2, **b.** 4,
b. $\log_2 16 = y$, and **c.** $\log_5 \frac{1}{125} = y$. **c.** -3 ∎

EXAMPLE 2

Writing an equivalent exponential statement. Find the value of x in each equation: **a.** $\log_3 81 = x$, **b.** $\log_x 125 = 3$, and **c.** $\log_4 x = 3$.

Solution **a.** $\log_3 81 = x$ is equivalent to $3^x = 81$. Because $3^4 = 81$, it follows that $3^x = 3^4$. Thus, $x = 4$.

b. $\log_x 125 = 3$ is equivalent to $x^3 = 125$. Because $5^3 = 125$, it follows that $x^3 = 5^3$. Thus, $x = 5$.

c. $\log_4 x = 3$ is equivalent to $4^3 = x$. Because $4^3 = 64$, it follows that $x = 64$.

SELF CHECK Find x in each equation: **a.** $\log_2 32 = x$,
b. $\log_x 8 = 3$, and **c.** $\log_5 x = 2$. *Answers:* **a.** 5, **b.** 2, **c.** 25 ∎

EXAMPLE 3

Writing an equivalent exponential statement. Find the value of x in each equation: **a.** $\log_{1/3} x = 2$, **b.** $\log_{1/3} x = -2$, and **c.** $\log_{1/3} \frac{1}{27} = x$.

Solution **a.** $\log_{1/3} x = 2$ is equivalent to $\left(\frac{1}{3}\right)^2 = x$. Thus, $x = \frac{1}{9}$.

b. $\log_{1/3} x = -2$ is equivalent to $\left(\frac{1}{3}\right)^{-2} = x$. Thus,

$$x = \left(\frac{1}{3}\right)^{-2} = 3^2 = 9$$

c. $\log_{1/3} \frac{1}{27} = x$ is equivalent to $\left(\frac{1}{3}\right)^x = \frac{1}{27}$. Because $\left(\frac{1}{3}\right)^3 = \frac{1}{27}$, it follows that $x = 3$.

SELF CHECK Find x in each equation: **a.** $\log_{1/4} x = 3$ and
b. $\log_{1/4} x = -2$. *Answers:* **a.** $\frac{1}{64}$, **b.** 16 ∎

Graphs of Logarithmic Functions

To graph the logarithmic function $y = \log_2 x$, we calculate and plot several points with coordinates (x, y) that satisfy the equivalent equation $x = 2^y$. After joining these points with a smooth curve, we have the graph shown in Figure 9-25(a).

To graph $y = \log_{1/2} x$, we calculate and plot several points with coordinates (x, y) that satisfy the equation $x = \left(\frac{1}{2}\right)^y$. After joining these points with a smooth curve, we have the graph shown in Figure 9-25(b).

FIGURE 9-25

$y = \log_2 x$

x	y	(x, y)
$\frac{1}{4}$	-2	$\left(\frac{1}{4}, -2\right)$
$\frac{1}{2}$	-1	$\left(\frac{1}{2}, -1\right)$
1	0	$(1, 0)$
2	1	$(2, 1)$
4	2	$(4, 2)$
8	3	$(8, 3)$

Choose values for x that are integer powers of 2.

$y = \log_{1/2} x$

x	y	(x, y)
$\frac{1}{4}$	2	$\left(\frac{1}{4}, 2\right)$
$\frac{1}{2}$	1	$\left(\frac{1}{2}, 1\right)$
1	0	$(1, 0)$
2	-1	$(2, -1)$
4	-2	$(4, -2)$
8	-3	$(8, -3)$

Choose values for x that are integer powers of $\frac{1}{2}$.

(a)　　　　　　(b)

The graphs of all logarithmic functions are similar to those in Figure 9-25. If $b > 1$, the logarithmic function is increasing, as in Figure 9-26(a). If $0 < b < 1$, the logarithmic function is decreasing, as in Figure 9-26(b).

FIGURE 9-26

(a)　　　　　　(b)

The graph of $y = \log_b x$ (or $f(x) = \log_b x$) has the following properties.

1. It passes through the point $(1, 0)$.
2. It passes through the point $(b, 1)$.
3. The y-axis (the line $x = 0$) is an asymptote.
4. The domain is $(0, \infty)$ and the range is $(-\infty, \infty)$.

The exponential and logarithmic functions are inverses of each other, so their graphs have symmetry about the line $y = x$. The graphs of $f(x) = \log_b x$ and $g(x) = b^x$ are shown in Figure 9-27(a) when $b > 1$ and in Figure 9-27(b) when $0 < b < 1$.

FIGURE 9-27

(a)　　　　　　(b)

Vertical and Horizontal Translations

The graphs of many functions involving logarithms are translations of the basic logarithmic graphs.

EXAMPLE 4

Graphing logarithmic functions. Graph the function defined by $f(x) = 3 + \log_2 x$ and describe the translation.

Solution The graph of $f(x) = 3 + \log_2 x$ is identical to the graph of $y = \log_2 x$, except that it is translated 3 units upward. (See Figure 9-28.)

FIGURE 9-28

SELF CHECK Graph $y = \log_3 x - 2$ and describe the translation.

Answer: The graph of $y = \log_3 x$ is translated 2 units downward.

EXAMPLE 5

Graphing logarithmic functions. Graph $y = \log_{1/2} (x - 1)$ and describe the translation.

Solution The graph of $y = \log_{1/2} (x - 1)$ is identical to the graph of $y = \log_{1/2} x$, except that it is translated 1 unit to the right. (See Figure 9-29.)

FIGURE 9-29

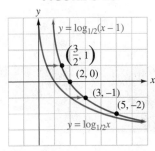

SELF CHECK Graph $f(x) = \log_{1/3} (x + 2)$ and describe the translation.

Answer: The graph of $g(x) = \log_{1/3} x$ is translated 2 units to the left.

Graphing calculators can draw graphs of logarithmic functions. To use a calculator to graph the logarithmic function $y = -2 + \log_{10} \left(\frac{1}{2}x\right)$, we enter the right-hand side of the equation after the symbol $Y_1 =$. The display will show the equation

$$Y_1 = -2 + \log (1/2x)$$

If we use window settings of $[-1, 5]$ for x and $[-4, 1]$ for y and press the $\boxed{\text{GRAPH}}$ key, we will obtain the graph shown in Figure 9-30.

FIGURE 9-30

Base-10 Logarithms

For computational purposes and in many applications, we will use base-10 logarithms (also called **common logarithms**). When the base b is not indicated in the notation $\log x$, we assume that $b = 10$:

$\log x$ means $\log_{10} x$

Because base-10 logarithms appear so often, it is a good idea to become familiar with the following base-10 logarithms:

Logarithmic form		*Exponential form*
$\log_{10} \dfrac{1}{100} = -2$	because	$10^{-2} = \dfrac{1}{100}$
$\log_{10} \dfrac{1}{10} = -1$	because	$10^{-1} = \dfrac{1}{10}$
$\log_{10} 1 = 0$	because	$10^0 = 1$
$\log_{10} 10 = 1$	because	$10^1 = 10$
$\log_{10} 100 = 2$	because	$10^2 = 100$
$\log_{10} 1,000 = 3$	because	$10^3 = 1,000$

In general, we have

$$\log_{10} 10^x = x$$

Before calculators, extensive tables provided logarithms of numbers. Today, logarithms are easy to find with a calculator. For example, to find $\log 2.34$ with a scientific calculator, we enter these numbers and press these keys:

Keystrokes   2.34 $\boxed{\text{LOG}}$ $\boxed{.369215857}$

To use a graphing calculator, we enter these numbers and press these keys:

Keystrokes $\boxed{\text{LOG}}$ 2.34 $\boxed{)}$ $\boxed{\text{ENTER}}$

$\boxed{\begin{array}{l} \text{log(2.34)} \\ \quad\quad .3692158574 \end{array}}$

To four decimal places, $\log 2.34 = 0.3692$.

EXAMPLE 6	**Using a calculator.** Find x in the equation $\log x = 0.3568$. Round to four decimal places.

Solution The equation $\log x = 0.3568$ is equivalent to $10^{0.3568} = x$. To find x with a scientific calculator, we enter these numbers and press these keys:

$$10 \quad \boxed{y^x} \quad .3568 \quad \boxed{=}$$

The display will read $\boxed{2.274049951}$. To four decimal places,

$$x = 2.2740$$

If your calculator has a $\boxed{10^x}$ key, enter .3568 and press it to get the same result.

SELF CHECK Solve $\log x = 1.87737$. Round to four decimal places. *Answer:* 75.3998 ∎

Applications of Logarithms

Common logarithms are used in electrical engineering to express the voltage gain (or loss) of an electronic device such as an amplifier. The unit of gain (or loss), called the **decibel,** is defined by a logarithmic relation.

Decibel voltage gain

If E_O is the output voltage of a device and E_I is the input voltage, the decibel voltage gain of the device (db gain) is given by

$$\text{db gain} = 20 \log \frac{E_O}{E_I}$$

EXAMPLE 7	**Finding db gain.** If the input to an amplifier is 0.5 volt and the output is 40 volts, find the decibel voltage gain of the amplifier.

Solution We can find the decibel voltage gain by substituting 0.5 for E_I and 40 for E_O into the formula for db gain:

$$\text{db gain} = 20 \log \frac{E_O}{E_I}$$

$$\text{db gain} = 20 \log \frac{40}{0.5}$$

$$= 20 \log 80$$

$$\approx 38 \qquad \text{Use a calculator.}$$

The amplifier provides a 38-decibel voltage gain. ∎

In seismology, common logarithms are used to measure the intensity of earthquakes on the **Richter scale.** The intensity of an earthquake is given by the following logarithmic function.

Richter scale

If R is the intensity of an earthquake, A is the amplitude (measured in micrometers), and P is the period (the time of one oscillation of the earth's surface measured in seconds), then

$$R = \log \frac{A}{P}$$

E X A M P L E   8

Measuring earthquakes. Find the measure on the Richter scale of an earthquake with an amplitude of 5,000 micrometers (0.5 centimeter) and a period of 0.1 second.

Solution   We substitute 5,000 for A and 0.1 for P in the Richter scale formula and simplify:

$$R = \log \frac{A}{P}$$

$$R = \log \frac{5,000}{0.1}$$

$$= \log 50,000$$

$$\approx 4.698970004 \quad \text{Use a calculator.}$$

The earthquake measures about 4.7 on the Richter scale.   ■

STUDY SET

Section 9.5

VOCABULARY

In Exercises 1–8, refer to the graph of $f(x) = \log_4 x$ in Illustration 1.

ILLUSTRATION 1

1. What type of function is $f(x) = \log_4 x$?
2. What is the domain of the function?
3. What is the range of the function?
4. What is the y-intercept of the graph?
5. What is the x-intercept of the graph?
6. What is an asymptote of the graph?
7. Is f an increasing or a decreasing function?
8. The graph passes through the point $(4, y)$. What is y?

CONCEPTS

In Exercises 9–12, fill in the blanks to make the statements true.

9. The equation $y = \log_b x$ is equivalent to the exponential equation _____.
10. $\log_b x$ is the _____ to which b is raised to get x.
11. The functions $f(x) = \log_b x$ and $f(x) = b^x$ are _____ functions.
12. The inverse of an exponential function is called a _____ function.

In Exercises 13–16, complete the table of values. If an evaluation is not possible, write "none."

13. $y = \log x$

x	y
$\frac{1}{100}$	
$\frac{1}{10}$	
1	
10	
100	

14. $f(x) = \log_5 x$

x	$f(x)$
$\frac{1}{25}$	
$\frac{1}{5}$	
1	
5	
25	

15. $f(x) = \log_6 x$

Input	Output
-6	
0	
$\frac{1}{216}$	
$\sqrt{6}$	
6^8	

16. $f(x) = \log_8 x$

x	$f(x)$
-8	
0	
$\frac{1}{8}$	
$\sqrt{8}$	
64	

17. Use a calculator to complete the table of values for $y = \log x$. Round to the nearest hundredth.

x	y
0.5	
1	
2	
3	
4	
5	
6	
7	
8	
9	
10	

18. Graph $y = \log x$ in Illustration 2. (See Exercise 17.) Note that the units on the x- and y-axes are different.

ILLUSTRATION 2

19. A table of values for $f(x) = \log x$ is shown in Illustration 3. As x decreases and gets close to 0, what happens to the values of $f(x)$?

ILLUSTRATION 3

X	Y1
1	0
.9	-.0458
.8	-.0969
.7	-.1549
.6	-.2218
.5	-.301
.4	-.3979

X=1

20. For each function, determine $f^{-1}(x)$.
  a. $f(x) = 10^x$ **b.** $f(x) = 3^x$

  c. $f(x) = \log x$ **d.** $f(x) = \log_2 x$

NOTATION

In Exercises 21–22, fill in the blanks to make the statements true.

21. The notation $\log x$ means $\log_{} x$.

22. $\log_{10} 10^x = $.

PRACTICE

In Exercises 23–30, write the equation in exponential form.

23. $\log_3 81 = 4$ **24.** $\log_7 7 = 1$

25. $\log_{12} 12 = 1$ **26.** $\log_6 36 = 2$

27. $\log_4 \dfrac{1}{64} = -3$ **28.** $\log_6 \dfrac{1}{36} = -2$

29. $\log 0.001 = -3$ **30.** $\log_3 243 = 5$

In Exercises 31–38, write the equation in logarithmic form.

31. $8^2 = 64$ **32.** $10^3 = 1,000$

33. $4^{-2} = \dfrac{1}{16}$ **34.** $3^{-4} = \dfrac{1}{81}$

35. $\left(\dfrac{1}{2}\right)^{-5} = 32$ **36.** $\left(\dfrac{1}{3}\right)^{-3} = 27$

37. $x^y = z$ **38.** $m^n = p$

In Exercises 39–66, find each value of x.

39. $\log_2 8 = x$ **40.** $\log_3 9 = x$

41. $\log_4 64 = x$ **42.** $\log_6 216 = x$

43. $\log_{1/2} \dfrac{1}{8} = x$ **44.** $\log_{1/3} \dfrac{1}{81} = x$

45. $\log_9 3 = x$ **46.** $\log_{125} 5 = x$

47. $\log_8 x = 2$ **48.** $\log_7 x = 0$

49. $\log_{25} x = \dfrac{1}{2}$ **50.** $\log_4 x = \dfrac{1}{2}$

51. $\log_5 x = -2$ **52.** $\log_3 x = -4$

53. $\log_{36} x = -\dfrac{1}{2}$ **54.** $\log_{27} x = -\dfrac{1}{3}$

55. $\log_{100} \dfrac{1}{1,000} = x$ **56.** $\log_{5/2} \dfrac{4}{25} = x$

57. $\log_{27} 9 = x$ **58.** $\log_{12} x = 0$

59. $\log_x 5^3 = 3$ **60.** $\log_x 5 = 1$

61. $\log_x \dfrac{9}{4} = 2$ **62.** $\log_x \dfrac{\sqrt{3}}{3} = \dfrac{1}{2}$

63. $\log_x \frac{1}{64} = -3$ **64.** $\log_x \frac{1}{100} = -2$

65. $\log_8 x = 0$ **66.** $\log_4 8 = x$

In Exercises 67–70, use a calculator to find each value. Give answers to four decimal places.

67. $\log 3.25$ **68.** $\log 0.57$

69. $\log 0.00467$ **70.** $\log 375.876$

In Exercises 71–74, use a calculator to find each value of y. Give answers to two decimal places.

71. $\log y = 1.4023$ **72.** $\log y = 0.926$

73. $\log y = -1.71$ **74.** $\log y = -0.5$

In Exercises 75–78, graph each function. Tell whether each function is an increasing or a decreasing function.

75. $f(x) = \log_3 x$ **76.** $f(x) = \log_{1/3} x$

77. $y = \log_{1/2} x$ **78.** $y = \log_4 x$

In Exercises 79–82, graph each function.

79. $f(x) = 3 + \log_3 x$ **80.** $f(x) = \log_{1/3} x - 1$

81. $y = \log_{1/2} (x - 2)$ **82.** $y = \log_4 (x + 2)$

In Exercises 83–86, graph each pair of inverse functions on a single coordinate system. Draw the axis of symmetry.

83. $y = 6^x$
$y = \log_6 x$

84. $y = 3^x$
$y = \log_3 x$

85. $f(x) = 5^x$
$g(x) = \log_5 x$

86. $f(x) = 8^x$
$g(x) = \log_8 x$

APPLICATIONS

87. FINDING INPUT VOLTAGE Find the db gain of an amplifier if the input voltage is 0.71 volts when the output voltage is 20 volts.

88. FINDING OUTPUT VOLTAGE Find the db gain of an amplifier if the output voltage is 2.8 volts when the input voltage is 0.05 volt.

89. db GAIN OF AN AMPLIFIER Find the db gain of the amplifier shown in Illustration 4.

ILLUSTRATION 4

90. db GAIN OF AN AMPLIFIER An amplifier produces an output of 80 volts when driven by an input of 0.12 volts. Find the amplifier's db gain.

91. THE RICHTER SCALE An earthquake has amplitude of 5,000 micrometers and a period of 0.2 second. Find its measure on the Richter scale.

92. EARTHQUAKE Find the period of an earthquake with amplitude of 80,000 micrometers that measures 6 on the Richter scale.

93. EARTHQUAKE An earthquake with a period of $\frac{1}{4}$ second measures 4 on the Richter scale. Find its amplitude.

94. MAJOR EARTHQUAKE In 1985, Mexico City experienced an earthquake of magnitude 8.1 on the Richter scale. In 1989, the San Francisco Bay area was rocked by an earthquake measuring 7.1. By what factor must the amplitude of an earthquake change to increase its severity by 1 point on the Richter scale? (Assume that the period remains constant.)

95. DEPRECIATION In business, equipment is often depreciated using the double declining-balance method. In this method, a piece of equipment with a life expectancy of N years, costing $\$C$, will depreciate to a value of $\$V$ in n years, where n is given by the formula

$$n = \frac{\log V - \log C}{\log\left(1 - \frac{2}{N}\right)}$$

A computer that cost $37,000 has a life expectancy of 5 years. If it has depreciated to a value of $8,000, how old is it?

96. DEPRECIATION A typewriter worth $470 when new had a life expectancy of 12 years. If it is now worth $189, how old is it? (See Exercise 95.)

97. TIME FOR MONEY TO GROW If $\$P$ is invested at the end of each year in an annuity earning annual interest at a rate r, the amount in the account will be $\$A$ after n years, where

$$n = \frac{\log\left[\dfrac{Ar}{P} + 1\right]}{\log(1 + r)}$$

If $1,000 is invested each year in an annuity earning 12% annual interest, how long will it take for the account to be worth $20,000?

98. TIME FOR MONEY TO GROW If $5,000 is invested each year in an annuity earning 8% annual interest, how long will it take for the account to be worth $50,000? (See Exercise 97.)

WRITING

99. Explain the mathematical relationship between $y = \log x$ and $y = 10^x$.

100. Explain why it is impossible to find the logarithm of a negative number.

REVIEW

Solve each equation.

101. $\sqrt[3]{6x + 4} = 4$
102. $\sqrt{3x - 4} = \sqrt{-7x + 2}$
103. $\sqrt{a + 1} - 1 = 3a$
104. $3 - \sqrt{t - 3} = \sqrt{t}$

▶ **9.6**

Base-*e* Logarithms

In this section, you will learn about

   Base-*e* logarithms ■ Graphing the natural logarithmic function
   ■ An application of base-*e* logarithms

Introduction   A special logarithmic function is the base-*e* logarithmic function. In this section, we will show how to evaluate base-*e* logarithms, graph the base-*e* logarithmic function, and solve some problems that involve the base-*e* logarithmic function.

Base-*e* Logarithms

We have seen the importance of the number e in mathematical models of events in nature. Base-*e* logarithms are just as important. They are called **natural logarithms** or **Napierian logarithms** after John Napier (1550–1617). They are usually written as ln x rather than $\log_e x$:

   ln x means $\log_e x$

Like all logarithmic functions, the domain of $f(x) = \ln x$ is the interval $(0, \infty)$, and the range is the interval $(-\infty, \infty)$.
   To find the base-*e* logarithms of numbers, we can use a calculator.

ACCENT ON TECHNOLOGY *Finding Base-e (Natural) Logarithms*

To use a scientific calculator to find the value of ln 2.34, we enter these numbers and press these keys:

Keystrokes   2.34 | LN | | .850150929 |

To use a graphing calculator, we enter these numbers and press these keys:

Keystrokes | LN | 2.34 |) | | ENTER |

```
ln(2.34)
        .8501509294
```

To four decimal places, ln 2.34 = 0.8502.

EXAMPLE 1

Evaluating base-*e* logarithms. Use a calculator to find each value:
a. ln 17.32   and   **b.** ln (−0.05).

Solution   **a.** Enter these numbers and press these keys:

Scientific calculator	*Graphing calculator*
17.32	LN		LN	17.32	)		ENTER	

Either way, the result is 2.851861903.

b. Enter these numbers and press these keys:

Scientific calculator	*Graphing calculator*
0.05	+/−		LN		LN		(−)	0.05	)		ENTER	

Either way, we obtain an error, because we cannot take the logarithm of a negative number.

SELF CHECK   Find each value to four decimal places: *Answers:* **a.** 1.1447, **b.** no
a. ln π and   **b.** ln 0.   value

EXAMPLE 2

Writing an equivalent exponential statement. Solve each equation:
a. ln $x = 1.335$ and   **b.** ln $x = -5.5$. Give each result to four decimal places.

Solution   **a.** Since the base of the natural logarithmic function is e, the equation ln $x = 1.335$ is equivalent to $e^{1.335} = x$. To use a scientific calculator to find x, press these keys:

    1.335 | e^x |

The display will read 3.799995946. To four decimal places,

    $x = 3.8000$

b. The equation ln $x = -5.5$ is equivalent to $e^{-5.5} = x$. To use a scientific calculator to find x, press these keys:

    5.5 | +/− | | e^x |

The display will read 0.004086771. To four decimal places,

    $x = 0.0041$

Solve **a.** $\ln x = 1.9344$ and **b.** $-3 = \ln x$. *Answers:* **a.** 6.9199,
Give each result to four decimal places. **b.** 0.0498 ∎

Graphing the Natural Logarithmic Function

The equation $y = \ln x$ is equivalent to the equation $x = e^y$. To get the graph of $\ln x$, we can plot points that satisfy the equation $x = e^y$ and join them with a smooth curve, as shown in Figure 9-31(a). Figure 9-31(b) shows the calculator graph of $y = \ln x$.

FIGURE 9-31

$y = \ln x$

x	y	(x, y)
$\frac{1}{e}$	-1	$\left(\frac{1}{e}, -1\right)$
1	0	$(1, 0)$
e	1	$(e, 1)$
e^2	2	$(e^2, 2)$

(a) (b)

FIGURE 9-32

The exponential function $f(x) = e^x$ and the natural logarithmic function $f(x) = \ln x$ are inverse functions. Figure 9-32 shows that their graphs are symmetric to the line $y = x$.

ACCENT ON TECHNOLOGY *Graphing Base-e Logarithmic Functions*

Many graphs of logarithmic functions involve translations of the graph of $y = \ln x$. For example, Figure 9-33 shows calculator graphs of the functions $y = \ln x$, $y = \ln x + 2$, and $y = \ln x - 3$.

FIGURE 9-33

The graph of $y = \ln x + 2$ is 2 units above the graph of $y = \ln x$.

The graph of $y = \ln x - 3$ is 3 units below the graph of $y = \ln x$.

Figure 9-34 shows the calculator graph of the functions $y = \ln x$, $y = \ln (x - 2)$, and $y = \ln (x + 3)$.

FIGURE 9-34

The graph of $y = \ln (x + 3)$ is 3 units to the left of the graph of $y = \ln x$.

The graph of $y = \ln (x - 2)$ is 2 units to the right of the graph of $y = \ln x$.

An Application of Base-*e* Logarithms

Base-*e* logarithms have many applications. If a population grows exponentially at a certain annual rate, the time required for the population to double is called the **doubling time.** It is given by the following formula.

Formula for doubling time

If r is the annual rate, compounded continuously, and t is time required for a population to double, then

$$t = \frac{\ln 2}{r}$$

EXAMPLE 3

Doubling time. The population of the earth is growing at the approximate rate of 2% per year. If this rate continues, how long will it take for the population to double?

Solution Because the population is growing at the rate of 2% per year, we substitute 0.02 for r in the formula for doubling time and simplify.

$$t = \frac{\ln 2}{r}$$

$$t = \frac{\ln 2}{0.02}$$

$$\approx 34.65735903 \quad \text{Use a calculator}$$

The population will double in about 35 years.

SELF CHECK See Example 3. If the population's annual growth rate could be reduced to 1.5% per year, what would be the doubling time? *Answer:* about 46 years ∎

EXAMPLE 4

Doubling time. How long will it take $1,000 to double at an annual rate of 8%, compounded continuously?

Solution We substitute 0.08 for r and simplify:

$$t = \frac{\ln 2}{r}$$

$$t = \frac{\ln 2}{0.08}$$

$$\approx 8.664339757 \quad \text{Use a calculator.}$$

It will take about $8\frac{2}{3}$ years for the money to double.

SELF CHECK In Example 4, how long will it take at 9%, compounded continuously? *Answer:* about 7.7 years ∎

STUDY SET
Section 9.6

VOCABULARY

In Exercises 1–2, fill in the blanks to make the statements true.

1. Base-*e* logarithms are often called _____ logarithms.

2. $f(x) = \ln x$ and $f(x) = e^x$ are _____ functions.

CONCEPTS

3. Use a calculator to complete the table of values for $f(x) = \ln x$. Round to the nearest hundredth.

x	$f(x)$
0.5	
1	
2	
3	
4	
5	
6	
7	
8	
9	
10	

4. Graph $f(x) = \ln x$ in Illustration 1. (See Exercise 3.) Note that the units on the *x*- and *y*-axes are different.

ILLUSTRATION 1

In Exercises 5–10, fill in the blanks to make the statements true.

5. The graph of $f(x) = \ln x$ has the _____ as an asymptote.

6. The domain of the function $f(x) = \ln x$ is the interval _____.

7. The range of the function $f(x) = \ln x$ is the interval _____.

8. The graph of $f(x) = \ln x$ passes through the point $(\ \ , 0)$.

9. The statement $y = \ln x$ is equivalent to the exponential statement _____.

10. The logarithm of a negative number is _____.

11. A table of values for $f(x) = \ln x$ is shown in Illustration 2. Explain why ERROR appears in the Y_1 column for the first three entries.

ILLUSTRATION 2

12. Illustration 3 shows the graph of $f(x) = \ln x$, as well as a vertical translation of that graph. Using the notation $g(x)$ for the translation, write the defining equation for the function.

ILLUSTRATION 3

13. In Illustration 4, $f(x) = \ln x$ was graphed, and the TRACE feature was used. What is the *x*-coordinate of the point on the graph having a *y*-coordinate of 1? What is the name given this number?

ILLUSTRATION 4

14. The graphs of $f(x) = \ln x$, $f(x) = e^x$, and $y = x$ are shown in Illustration 5. What phrase is used to describe the relationship between the graphs?

ILLUSTRATION 5

NOTATION

In Exercises 15–18, fill in the blanks to make the statements true.

15. $\ln 2$ means $\log\,\_\_\_ 2$.

16. $\log 2$ means $\log\,\_\_\_ 2$.

17. If a population grows exponentially at a rate r, the time it will take the population to double is given by the formula $t = $ _____.

18. To evaluate a base-10 logarithm with a calculator, use the _____ key. To evaluate a base-e logarithm, use the _____ key.

PRACTICE

In Exercises 19–26, use a calculator to find each value, if possible. Express all answers to four decimal places.

19. $\ln 35.15$ **20.** $\ln 0.675$

21. $\ln 0.00465$ **22.** $\ln 378.96$

23. $\ln 1.72$ **24.** $\ln 2.7$

25. $\ln (-0.1)$ **26.** $\ln (-10)$

In Exercises 27–34, use a calculator to find x. Express all answers to four decimal places.

27. $\ln x = 1.4023$ **28.** $\ln x = 2.6490$

29. $\ln x = 4.24$ **30.** $\ln x = 0.926$

31. $\ln x = -3.71$ **32.** $\ln x = -0.28$

33. $1.001 = \ln x$ **34.** $\ln x = -0.001$

In Exercises 35–38, use a graphing calculator to graph each function.

35. $y = \ln\left(\dfrac{1}{2}x\right)$ **36.** $y = \ln x^2$

37. $y = \ln (-x)$ **38.** $y = \ln (3x)$

APPLICATIONS

Use a calculator to solve each problem.

39. POPULATION GROWTH How long will it take the population of River City to double? See Illustration 6.

ILLUSTRATION 6

River City

A growing community

- 6 parks   • 12% annual growth
- 10 churches   • Low crime rate

40. DOUBLING MONEY How long will it take $1,000 to double if it is invested at an annual rate of 5% compounded continuously?

41. POPULATION GROWTH A population growing continuously at an annual rate r will triple in a time t given by the formula

$$t = \frac{\ln 3}{r}$$

How long will it take the population of a town to triple if it is growing at the rate of 12% per year?

42. TRIPLING MONEY Find the length of time for $25,000 to triple when it is invested at 6% annual interest, compounded continuously. See Exercise 41.

43. FORENSIC MEDICINE To estimate the number of hours t that a murder victim had been dead, a coroner used the formula

$$t = \frac{1}{0.25} \ln \frac{98.6 - T_s}{82 - T_s}$$

where T_s is the temperature of the surroundings where the body was found. If the crime took place in an apartment where the thermostat was set at 70° F, approximately how long ago did the murder occur?

44. MAKING JELL-O After the contents of a package of JELL-O are combined with boiling water, the mixture is placed in a refrigerator whose temperature remains a constant 42° F. Estimate the number of hours t that it will take for the JELL-O to cool to 50° F using the formula

$$t = -\frac{1}{0.9} \ln \frac{50 - T_r}{200 - T_r}$$

where T_r is the temperature of the refrigerator.

WRITING

45. Explain the difference between the functions $f(x) = \log x$ and $f(x) = \ln x$.

46. How are the functions $f(x) = \ln x$ and $f(x) = e^x$ related?

REVIEW

In Exercises 47–52, write the equation of the required line.

47. Parallel to $y = 5x - 8$ and passing through the origin

48. Having a slope of 7 and a y-intercept of 3

49. Passing through the point $(3, 2)$ and perpendicular to the line $y = \frac{2}{3}x - 12$

50. Parallel to the line $3x + 2y = 9$ and passing through the point $(-3, 5)$

51. Vertical line through the point $(2, 3)$

52. Horizontal line through the point $(2, 3)$

▶ 9.7 Properties of Logarithms

In this section, you will learn about

Properties of logarithms ■ The change-of-base formula
■ An application from chemistry

Introduction In this section, we will discuss eight properties of logarithms and use them to simplify logarithmic expressions. We will then show how to change a logarithm from one base to another. We conclude the section by solving some problems from the field of chemistry.

Properties of Logarithms

Since logarithms are exponents, the properties of exponents have counterparts in the theory of logarithms. We begin with four basic properties.

Properties of logarithms

If b is a positive number and $b \neq 1$, then

1. $\log_b 1 = 0$ **2.** $\log_b b = 1$
3. $\log_b b^x = x$ **4.** $b^{\log_b x} = x \ (x > 0)$

Properties 1 through 4 follow directly from the definition of a logarithm.

1. $\log_b 1 = 0$, because $b^0 = 1$.
2. $\log_b b = 1$, because $b^1 = b$.
3. $\log_b b^x = x$, because $b^x = b^x$.
4. $b^{\log_b x} = x$, because $\log_b x$ is the exponent to which b is raised to get x.

Properties 3 and 4 also indicate that the composition of the exponential and logarithmic functions (in both directions) is the identity function. This is expected, because the exponential and logarithmic functions are inverse functions.

EXAMPLE 1

Applying properties of logarithms. Simplify each expression: **a.** $\log_5 1$, **b.** $\log_3 3$, **c.** $\ln e^3$, and **d.** $b^{\log_b 7}$.

Solution   **a.** By property 1, $\log_5 1 = 0$, because $5^0 = 1$.

b. By property 2, $\log_3 3 = 1$, because $3^1 = 3$.

c. By property 3, $\ln e^3 = 3$, because $e^3 = e^3$.

d. By property 4, $b^{\log_b 7} = 7$, because $\log_b 7$ is the power to which b is raised to get 7.

SELF CHECK   Simplify   **a.** $\log_4 1$,   **b.** $\log_4 4$, **c.** $\log_2 2^4$,   and   **d.** $5^{\log_5 2}$.

Answers: **a.** 0,   **b.** 1,   **c.** 4, **d.** 2   ∎

The next two properties state that

The logarithm of a product is the sum of the logarithms.
The logarithm of a quotient is the difference of the logarithms.

Properties of logarithms

If M, N, and b are positive numbers and $b \neq 1$, then

5. $\log_b MN = \log_b M + \log_b N$ **6.** $\log_b \dfrac{M}{N} = \log_b M - \log_b N$

PROOF

To prove property 5, we let $x = \log_b M$ and $y = \log_b N$. We use the definition of logarithm to write each equation in exponential form.

$$M = b^x \quad \text{and} \quad N = b^y$$

Then $MN = b^x b^y$, and a property of exponents gives

$$MN = b^{x+y} \quad \text{Keep the base and add the exponents: } b^x b^y = b^{x+y}.$$

We write this exponential equation in logarithmic form as

$$\log_b MN = x + y$$

Substituting the values of x and y completes the proof.

$$\log_b MN = \log_b M + \log_b N \qquad \qquad \square$$

The proof of property 6 is similar.

In the following examples, we will use these properties to write a logarithm of a product as the sum of logarithms and a logarithm of a quotient as the difference of logarithms. Assume that all variables are positive and that $b \neq 1$.

| EXAMPLE 2 | **Applying the product rule.** Use the product rule for logarithms to rewrite each of the following: **a.** $\log_2 (2 \cdot 7)$ and **b.** $\log (100x)$. |

Solution **a.** $\log_2 (2 \cdot 7) = \log_2 2 + \log_2 7$ The log of a product is the sum of the logs.

$\qquad\qquad\qquad = 1 + \log_2 7$ $\log_2 2 = 1$.

b. $\log (100x) = \log 100 + \log x$ The log of a product is the sum of the logs.

$\qquad\qquad\quad = 2 + \log x$ $\log 100 = 2$.

SELF CHECK   Rewrite **a.** $\log_3 (4 \cdot 3)$ and **b.** $\log (1{,}000y)$.   *Answers:* **a.** $\log_3 4 + 1$, **b.** $3 + \log y$ ■

| EXAMPLE 3 | **Applying the quotient rule.** Use the quotient rule for logarithms to rewrite each of the following: **a.** $\ln \frac{10}{7}$ and **b.** $\log_4 \frac{x}{64}$. |

Solution **a.** $\ln \dfrac{10}{7} = \ln 10 - \ln 7$ The log of a quotient is the difference of the logs.

b. $\log_4 \dfrac{x}{64} = \log_4 x - \log_4 64$ The log of a quotient is the difference of the logs.

$\qquad\qquad = \log_4 x - 3$ $\log_4 64 = 3$.

SELF CHECK   Rewrite **a.** $\log_6 \frac{6}{5}$ and **b.** $\ln \frac{y}{100}$.   *Answers:* **a.** $1 - \log_6 5$, **b.** $\ln y - \ln 100$ ■

| EXAMPLE 4 | **Applying the product and quotient properties.** Use logarithm properties to rewrite each expression: **a.** $\log_b xyz$ and **b.** $\log \frac{xy}{z}$. |

Solution **a.** We observe that $\log_b xyz$ is the logarithm of a product.

$\qquad\log_b xyz = \log_b (xy)z$ Group the first two factors together.

$\qquad\qquad\quad = \log_b (xy) + \log_b z$ The log of a product is the sum of the logs.

$\qquad\qquad\quad = \log_b x + \log_b y + \log_b z$ The log of a product is the sum of the logs.

b. In the expression $\log \dfrac{xy}{z}$, we have the logarithm of a quotient.

$\qquad\log \dfrac{xy}{z} = \log (xy) - \log z$ The log of a quotient is the difference of the logs.

$\qquad\qquad\quad = (\log x + \log y) - \log z$ The log of a product is the sum of the logs.

$\qquad\qquad\quad = \log x + \log y - \log z$ Remove parentheses.

SELF CHECK   Rewrite $\log_b \dfrac{x}{yz}$.   *Answer:* $\log_b x - \log_b y - \log_b z$ ■

WARNING! By property 5 of logarithms, the logarithm of a *product* is equal to the *sum* of the logarithms. The logarithm of a sum or a difference usually does not simplify. In general,

$$\log_b (M + N) \neq \log_b M + \log_b N \quad \text{and} \quad \log_b (M - N) \neq \log_b M - \log_b N$$

By property 6, the logarithm of a *quotient* is equal to the *difference* of the logarithms. The logarithm of a quotient is not the quotient of the logarithms:

$$\log_b \frac{M}{N} \neq \frac{\log_b M}{\log_b N}$$

ACCENT ON TECHNOLOGY *Verifying Properties of Logarithms*

We can use a calculator to illustrate property 5 of logarithms by showing that

$$\ln [(3.7)(15.9)] = \ln 3.7 + \ln 15.9$$

We calculate the left- and right-hand sides of the equation separately and compare the results. To use a scientific calculator to find $\ln [(3.7)(15.9)]$, we enter these numbers and press these keys:

Keystrokes   3.7 $\boxed{\times}$ 15.9 $\boxed{=}$ $\boxed{\text{LN}}$ $\boxed{\text{4.074651929}}$

To find $\ln 3.7 + \ln 15.9$, we enter these numbers and press these keys:

Keystrokes   3.7 $\boxed{\text{LN}}$ $\boxed{+}$ 15.9 $\boxed{\text{LN}}$ $\boxed{=}$ $\boxed{\text{4.074651929}}$

Since the left- and right-hand sides are equal, the equation $\ln [(3.7)(15.9)] = \ln 3.7 + \ln 15.9$ is true.

Two more properties of logarithms state that

The logarithm of a power is the power times the logarithm.

If the logarithms of two numbers are equal, the numbers are equal.

Properties of logarithms

If M, p, and b are positive numbers and $b \neq 1$, then

7. $\log_b M^p = p \log_b M$ **8.** If $\log_b x = \log_b y$, then $x = y$.

PROOF

To prove property 7, we let $x = \log_b M$, write the expression in exponential form, and raise both sides to the pth power:

$$M = b^x$$
$$(M)^p = (b^x)^p \quad \text{Raise both sides to the } p\text{th power.}$$
$$M^p = b^{px} \quad \text{Keep the base and multiply the exponents.}$$

Using the definition of logarithms gives

$$\log_b M^p = px$$

Substituting the value for x completes the proof.

$$\log_b M^p = p \log_b M \qquad\qquad \square$$

Property 8 follows from the fact that the logarithmic function is a one-to-one function. Property 8 will be important in the next section, when we solve logarithmic equations.

EXAMPLE 5

Applying the power rule. Use the power rule for logarithms to rewrite each of the following: **a.** $\log_5 6^2$ and   **b.** $\log \sqrt{10}$.

Solution **a.** $\log_5 6^2 = 2 \log_5 6$ The log of a power is the power times the log.

b. $\log \sqrt{10} = \log (10)^{1/2}$ Write $\sqrt{10}$ using a fractional exponent: $\sqrt{10} = (10)^{1/2}$.

$\quad\quad\quad\quad = \dfrac{1}{2} \log 10$ The log of a power is the power times the log.

$\quad\quad\quad\quad = \dfrac{1}{2}$ Simplify: $\log 10 = 1$.

SELF CHECK   Rewrite **a.** $\ln x^4$ and **b.** $\log_2 \sqrt[3]{3}$.   *Answers:* **a.** $4 \ln x$, **b.** $\frac{1}{3} \log_2 3$ ■

EXAMPLE 6   **Applying properties of logarithms.** Use logarithm properties to rewrite each expression: **a.** $\log_b (x^2 y^3 z)$ and **b.** $\ln \dfrac{y^3 \sqrt{x}}{z}$.

Solution **a.** We begin by recognizing that $\log_b (x^2 y^3 z)$ is the logarithm of a product.

$$\log_b (x^2 y^3 z) = \log_b x^2 + \log_b y^3 + \log_b z$$ The log of a product is the sum of the logs.

$$= 2 \log_b x + 3 \log_b y + \log_b z$$ The log of a power is the power times the log.

b. The expression $\ln \dfrac{y^3 \sqrt{x}}{z}$ is the logarithm of a quotient.

$$\ln \frac{y^3 \sqrt{x}}{z} = \ln \left(y^3 \sqrt{x}\right) - \ln z$$ The log of a quotient is the difference of the logs.

$$= \ln y^3 + \ln \sqrt{x} - \ln z$$ The log of a product is the sum of the logs.

$$= \ln y^3 + \ln x^{1/2} - \ln z$$ Write $\sqrt{x}$ as $x^{1/2}$.

$$= 3 \ln y + \frac{1}{2} \ln x - \ln z$$ The log of a power is the power times the log.

SELF CHECK   Rewrite $\log \sqrt[4]{\dfrac{x^3 y}{z}}$.   *Answer:* $\frac{1}{4}(3 \log x + \log y - \log z)$ ■

We can use the properties of logarithms to combine several logarithms into one logarithm.

EXAMPLE 7   **Condensing logarithmic expressions.** Write each of the given expressions as one logarithm: **a.** $3 \log_a x + \frac{1}{2} \log_a y$ and **b.** $\frac{1}{2} \log_b (x - 2) - \log_b y + 3 \log_b z$.

Solution **a.** We begin by applying the power rule to each term of the expression.

$$3 \log_a x + \frac{1}{2} \log_a y$$

$$= \log_a x^3 + \log_a y^{1/2}$$ A power times a log is the log of the power.

$$= \log_a (x^3 y^{1/2})$$ The sum of two logs is the log of the product.

b. The first and third terms of this expression can be rewritten using the power rule of logarithms.

$$\frac{1}{2} \log_b (x - 2) - \log_b y + 3 \log_b z$$

$$= \log_b (x - 2)^{1/2} - \log_b y + \log_b z^3 \qquad \text{A power times a log is the log of the power.}$$

$$= \log_b \frac{(x - 2)^{1/2}}{y} + \log_b z^3 \qquad \text{The difference of two logs is the log of the quotient.}$$

$$= \log_b \frac{z^3 \sqrt{x - 2}}{y} \qquad \text{The sum of two logs is the log of the product. Write } (x - 2)^{1/2} \text{ as } \sqrt{x - 2}.$$

SELF CHECK   Write the expression as one logarithm: $2 \log_a x + \frac{1}{2} \log_a y - 2 \log_a (x - y)$.   *Answer:* $\log_a \dfrac{x^2 \sqrt{y}}{(x - y)^2}$ ∎

We summarize the properties of logarithms as follows.

Properties of logarithms

If b, M, and N are positive numbers and $b \neq 1$, then

1. $\log_b 1 = 0$
2. $\log_b b = 1$
3. $\log_b b^x = x$
4. $b^{\log_b x} = x$
5. $\log_b MN = \log_b M + \log_b N$
6. $\log_b \dfrac{M}{N} = \log_b M - \log_b N$
7. $\log_b M^p = p \log_b M$
8. If $\log_b x = \log_b y$, then $x = y$.

EXAMPLE 8   **Using properties of logarithms.** Given that $\log 2 \approx 0.3010$ and $\log 3 \approx 0.4771$, find approximations for   **a.** $\log 6$ and   **b.** $\log 18$.

Solution   **a.** $\log 6 = \log (2 \cdot 3)$ Write 6 using the factors 2 and 3.

$\quad\quad\quad\quad = \log 2 + \log 3$ The log of a product is the sum of the logs.

$\quad\quad\quad\quad \approx 0.3010 + 0.4771$ Substitute the value of each logarithm.

$\quad\quad\quad\quad \approx 0.7781$

b. $\log 18 = \log (2 \cdot 3^2)$ Write 18 using the factors 2 and 3.

$\quad\quad\quad\quad = \log 2 + \log 3^2$ The log of a product is the sum of the logs.

$\quad\quad\quad\quad = \log 2 + 2 \log 3$ The log of a power is the power times the log.

$\quad\quad\quad\quad \approx 0.3010 + 2(0.4771)$ Substitute the value of each logarithm.

$\quad\quad\quad\quad \approx 1.2552$

SELF CHECK   Find   **a.** $\log 1.5$ and   **b.** $\log 0.75$.   *Answers:*   **a.** 0.1761,   **b.** −0.1249 ∎

The Change-of-Base Formula

Most calculators can find common logarithms and natural logarithms. If we need to find a logarithm with some other base, we use a conversion formula.

If we know the base-a logarithm of a number, we can find its logarithm to some other base b by using a formula called the **change-of-base formula.**

Change-of-base formula

If a, b, and x are positive, $a \neq 1$, and $b \neq 1$, then

$$\log_b x = \frac{\log_a x}{\log_a b}$$

PROOF

To prove this formula, we begin with the equation $\log_b x = y$.

$$y = \log_b x$$
$$x = b^y \qquad \text{Change the equation from logarithmic to exponential form.}$$
$$\log_a x = \log_a b^y \qquad \text{Take the base-}a \text{ logarithm of both sides.}$$
$$\log_a x = y \log_a b \qquad \text{The log of a power is the power times the log.}$$
$$y = \frac{\log_a x}{\log_a b} \qquad \text{Divide both sides by } \log_a b.$$
$$\log_b x = \frac{\log_a x}{\log_a b} \qquad \text{Refer to the first equation and substitute } \log_b x \text{ for } y.$$

If we know logarithms to base a (for example, $a = 10$), we can find the logarithm of x to a new base b. We simply divide the base-a logarithm of x by the base-a logarithm of b.

 WARNING! $\dfrac{\log_a x}{\log_a b}$ means that one logarithm is to be divided by the other. They are not to be subtracted.

EXAMPLE 9

Using the change-of-base formula. Find $\log_3 5$.

Solution We can use base-10 logarithms to find a base-3 logarithm. To do this, we substitute 3 for b, 10 for a, and 5 for x in the change-of-base formula:

$$\log_b x = \frac{\log_a x}{\log_a b}$$

$$\log_3 5 = \frac{\log_{10} 5}{\log_{10} 3} \qquad b = 3, x = 5, \text{ and } a = 10.$$

$$\approx 1.464973521 \quad \text{Use a scientific calculator and press these keys:}$$
$$5 \;\boxed{\text{log}}\; \div \; 3 \;\boxed{\text{log}}\; \boxed{=} \;.$$

To four decimal places, $\log_3 5 = 1.4650$.

We can also use the natural logarithm function (base e) in the change-of-base formula to find a base-3 logarithm.

$$\log_b x = \frac{\log_a x}{\log_a b}$$

$$\log_3 5 = \frac{\ln 5}{\ln 3} \qquad \begin{array}{l} b = 3, x = 5, \text{ and } a = e. \\ \log_e 5 = \ln 5 \text{ and } \log_e 3 = \ln 3. \end{array}$$

$$\approx 1.464973521 \quad \text{Use a calculator.}$$

We obtain the same result.

SELF CHECK Find $\log_5 3$ to four decimal places. *Answer:* 0.6826 ∎

An Application from Chemistry

In chemistry, common logarithms are used to express the acidity of solutions. The more acidic a solution, the greater the concentration of hydrogen ions. This concentration is indicated indirectly by the *pH scale,* or *hydrogen ion index.* The pH of a solution is defined as follows.

pH of a solution

If $[H^+]$ is the hydrogen ion concentration in gram-ions per liter, then

$$pH = -\log [H^+]$$

EXAMPLE 10

Finding the pH of a solution. Find the pH of pure water, which has a hydrogen ion concentration of 10^{-7} gram-ions per liter.

Solution Since pure water has approximately 10^{-7} gram-ions per liter, its pH is

$pH = -\log [\mathbf{H^+}]$

$pH = -\log \mathbf{10^{-7}}$

$\quad = -(-7) \log 10$ The log of a power is the power times the log.

$\quad = -(-7) \cdot 1$ Simplify: $\log 10 = 1$.

$\quad = 7$ ■

EXAMPLE 11

Finding hydrogen ion concentration. Find the hydrogen ion concentration of seawater if its pH is 8.5.

Solution To find its hydrogen ion concentration, we solve the following equation for $[H^+]$.

$\mathbf{pH} = -\log [H^+]$

$\mathbf{8.5} = -\log [H^+]$

$-8.5 = \log [H^+]$ Multiply both sides by -1.

$[H^+] = 10^{-8.5}$ Change the equation to exponential form.

We can use a calculator to find that

$[H^+] \approx 3.2 \times 10^{-9}$ gram-ions per liter   ■

STUDY SET

Section 9.7

VOCABULARY

In Exercises 1–4, fill in the blanks to make the statements true.

1. The expression $\log_3 (4x)$ is the logarithm of a
_____.

2. The expression $\log_2 \frac{5}{x}$ is the logarithm of a
_____.

3. The expression $\log 4^x$ is the logarithm of a
_____.

4. In the expression $\log_5 4$, the number 5 is the
_____ of the logarithm.

CONCEPTS

In Exercises 5–16, fill in the blanks to make the statements true.

5. $\log_b 1 = $ ▢

6. $\log_b b = $ ▢

7. $\log_b MN = \log_b$ ▢ $+ \log_b$ ▢

8. $b^{\log_b x} = $ ▢

9. If $\log_b x = \log_b y$, then ▢ $= $ ▢.

10. $\log_b \dfrac{M}{N} = \log_b M$ ▢ $\log_b N$

11. $\log_b x^p = p \cdot \log_b$ ▢

12. $\log_b b^x = $ ▢

13. $\log_b (A + B)$ ▢ $\log_b A + \log_b B$

14. $\log_b A + \log_b B$ ▢ $\log_b AB$

15. $\log_b x = \dfrac{\log_a x}{\boxed{}}$

16. pH = ▢

17. Three logarithmic expressions have been evaluated, and the results are shown on the calculator display in Illustration 1. Show that each result is correct by writing the equivalent base-10 exponential statement.

ILLUSTRATION 1

```
log(1)
                    0
log(10)
                    1
log(10²)
                    2
■
```

18. Three logarithmic expressions have been evaluated, and the results are shown on the calculator display in Illustration 2. Show that each result is correct by writing the equivalent base-e exponential statement. (The notation $\ln(e^\wedge(2))$ means $\ln e^2$.)

ILLUSTRATION 2

```
ln(e^(2))
                    2
ln(e^(3))
                    3
ln(e^(4))
                    4
```

In Exercises 19–30, evaluate each expression.

19. $\log_4 1$

20. $\log_4 4$

21. $\log_4 4^7$

22. $\ln e^8$

23. $5^{\log_5 10}$

24. $8^{\log_8 10}$

25. $\log_5 5^2$

26. $\log_4 4^2$

27. $\ln e$

28. $\log_7 1$

29. $\log_3 3^7$

30. $5^{\log_5 8}$

NOTATION

In Exercises 31–32, complete each solution.

31. $\log_b rst = \log_b \left(\boxed{} \right) t$
$= \log_b (rs) + \log_b$ ▢
$= \log_b$ ▢ $+ \log_b$ ▢ $+ \log_b t$

32. $\log \dfrac{r}{st} = \log r - \log \left(\boxed{} \right)$
$= \log r - \left(\log \boxed{} + \log t \right)$
$= \log r - \log s \ \boxed{} \ \log t$

⊞ PRACTICE

In Exercises 33–38, use a calculator to verify each equation.

33. $\log [(2.5)(3.7)] = \log 2.5 + \log 3.7$

34. $\log 45.37 = \dfrac{\ln 45.37}{\ln 10}$

35. $\ln (2.25)^4 = 4 \ln 2.25$

36. $\ln \dfrac{11.3}{6.1} = \ln 11.3 - \ln 6.1$

37. $\log \sqrt{24.3} = \dfrac{1}{2} \log 24.3$

38. $\ln 8.75 = \dfrac{\log 8.75}{\log e}$

In Exercises 39–46, use the properties of logarithms to rewrite each expression. Assume that x, y, and z are positive numbers.

39. $\log_2 (4 \cdot 5)$

40. $\log_3 (27 \cdot 5)$

41. $\log_6 \dfrac{x}{36}$

42. $\log_8 \dfrac{y}{8}$

43. $\ln y^4$

44. $\ln z^9$

45. $\log \sqrt{5}$

46. $\log \sqrt[3]{7}$

In Exercises 47–58, assume that x, y, z, and b are positive numbers and $b \neq 1$. Use the properties of logarithms to write each expression in terms of the logarithms of x, y, and z.

47. $\log xyz$

48. $\log 4xz$

49. $\log_2 \dfrac{2x}{y}$

50. $\log_3 \dfrac{x}{yz}$

51. $\log x^3 y^2$

52. $\log xy^2 z^3$

53. $\log_b (xy)^{1/2}$

54. $\log_b x^3 y^{1/2}$

55. $\log_a \dfrac{\sqrt[3]{x}}{\sqrt[4]{yz}}$

56. $\log_b \sqrt[4]{\dfrac{x^3 y^2}{z^4}}$

57. $\ln x\sqrt{z}$ **58.** $\ln \sqrt{xy}$

In Exercises 59–66, assume that x, y, z, and b are positive numbers and $b \neq 1$. Use the properties of logarithms to write each expression as the logarithm of a single quantity.

59. $\log_2 (x + 1) - \log_2 x$

60. $\log_3 x + \log_3 (x + 2) - \log_3 8$

61. $2 \log x + \dfrac{1}{2} \log y$

62. $-2 \log x - 3 \log y + \log z$

63. $-3 \log_b x - 2 \log_b y + \dfrac{1}{2} \log_b z$

64. $3 \log_b (x + 1) - 2 \log_b (x + 2) + \log_b x$

65. $\ln \left(\dfrac{x}{z} + x \right) - \ln \left(\dfrac{y}{z} + y \right)$

66. $\ln (xy + y^2) - \ln (xz + yz) + \ln z$

In Exercises 67–72, tell whether the given statement is true. If a statement is false, explain why.

67. $\log xy = (\log x)(\log y)$

68. $\log ab = \log a + 1$

69. $\log_b (A - B) = \dfrac{\log_b A}{\log_b B}$

70. $\dfrac{\log_b A}{\log_b B} = \log_b A - \log_b B$

71. $\log_b \dfrac{A}{B} = \log_b A - \log_b B$

72. $\log_b 2 = \log_2 b$

In Exercises 73–80, assume that $\log_b 4 = 0.6021$, $\log_b 7 = 0.8451$, and $\log_b 9 = 0.9542$. Use these values and the properties of logarithms to find each value.

73. $\log_b 28$ **74.** $\log_b \dfrac{7}{4}$

75. $\log_b \dfrac{4}{63}$ **76.** $\log_b 36$

77. $\log_b \dfrac{63}{4}$ **78.** $\log_b 2.25$

79. $\log_b 64$ **80.** $\log_b 49$

In Exercises 81–88, use the change-of-base formula to find each logarithm to four decimal places.

81. $\log_3 7$ **82.** $\log_7 3$

83. $\log_{1/3} 3$ **84.** $\log_{1/2} 6$

85. $\log_3 8$ **86.** $\log_5 10$

87. $\log_{\sqrt{2}} \sqrt{5}$ **88.** $\log_\pi e$

APPLICATIONS

89. pH OF A SOLUTION Find the pH of a solution with a hydrogen ion concentration of 1.7×10^{-5} gram-ions per liter.

90. HYDROGEN ION CONCENTRATION Find the hydrogen ion concentration of a saturated solution of calcium hydroxide whose pH is 13.2.

91. AQUARIUM To test for safe pH levels in a freshwater aquarium, a test strip is compared with the scale shown in Illustration 3. Find the corresponding range in the hydrogen ion concentration.

ILLUSTRATION 3

AquaTest pH Kit

Safe range

6.4   6.8   7.2   7.6   8.0

92. pH OF SOUR PICKLES The hydrogen ion concentration of sour pickles is 6.31×10^{-4}. Find the pH.

WRITING

93. Explain the difference between a logarithm of a product and the product of logarithms.

94. How can the ⬛LOG⬛ key on a calculator be used to find $\log_2 7$?

REVIEW

Consider the line that passes through $P(-2, 3)$ and $Q(4, -4)$.

95. Find the slope of line PQ.

96. Find the distance PQ.

97. Find the midpoint of segment PQ.

98. Write the equation of line PQ.

▶ 9.8 Exponential and Logarithmic Equations

In this section, you will learn about

Solving exponential equations ■ Solving logarithmic equations
■ Radioactive decay ■ Population growth

Introduction An **exponential equation** is an equation that contains a variable in one of its exponents. Some examples of exponential equations are

$$3^x = 5, \qquad 6^{x-3} = 2^x, \qquad \text{and} \qquad 2^{x^2+2x} = \frac{1}{2}$$

A **logarithmic equation** is an equation with a logarithmic expression that contains a variable. Some examples of logarithmic equations are

$$\log 5x = 1, \qquad \log(3x + 2) - \log(2x - 3) = 0, \qquad \text{and} \qquad \frac{\log_2(5x - 6)}{\log_2 x} = 2$$

In this section, we will learn how to solve many of these equations.

Solving Exponential Equations

EXAMPLE 1

Taking the logarithm of each side of an equation. Solve $3^x = 5$.

Solution Since the logarithms of equal numbers are equal, we can take the common logarithm of each side of the equation. The power rule of logarithms then provides a way of moving the variable x from its position as an exponent to a position as a coefficient.

$$3^x = 5$$

$\log 3^x = \log 5$ Take the common logarithm of each side.

$x \log 3 = \log 5$ The log of a power is the power times the log: $\log 3^x = x \log 3$.

1. $x = \dfrac{\log 5}{\log 3}$ Divide both sides by log 3.

     $x \approx 1.464973521$ Use a calculator.

The solution is $\dfrac{\log 5}{\log 3}$. To four decimal places, $x = 1.4650$.

We can also take the natural logarithm of each side of the equation to solve for x.

$$3^x = 5$$

$\ln 3^x = \ln 5$ Take the natural logarithm of each side.

$x \ln 3 = \ln 5$ Use the power rule of logarithms: $\ln 3^x = x \ln 3$.

     $x = \dfrac{\ln 5}{\ln 3}$ Divide both sides by ln 3.

     $x \approx 1.464973521$ Use a calculator.

To check the solution, we substitute 1.4650 for x in 3^x and see if the result is close to 5. We can compute $3^{1.4650}$ by entering 3 $\boxed{y^x}$ 1.4650 $\boxed{=}$ on a scientific calculator. The result of 5.000145454 verifies that $x \approx 1.4650$ is an approximate solution of $3^x = 5$.

SELF CHECK   Solve $5^x = 4$. Give the answer to four decimal places.   *Answer:* 0.8614   ■

WARNING! A careless reading of Equation 1 leads to a common error. The right-hand side of Equation 1 calls for a division, not a subtraction.

$$\frac{\log 5}{\log 3} \quad \text{means} \quad (\log 5) \div (\log 3)$$

It is the expression $\log \frac{5}{3}$ that means $\log 5 - \log 3$.

EXAMPLE 2

Taking the logarithm of each side. Solve $6^{x-3} = 2^x$.

Solution

$$6^{x-3} = 2^x$$

$$\log 6^{x-3} = \log 2^x \qquad \text{Take the common logarithm of each side.}$$

$$(x - 3) \log 6 = x \log 2 \qquad \text{The log of a power is the power times the log.}$$

$$x \log 6 - 3 \log 6 = x \log 2 \qquad \text{Use the distributive property.}$$

$$x \log 6 - x \log 2 = 3 \log 6 \qquad \text{On both sides, add } 3 \log 6 \text{ and subtract } x \log 2.$$

$$x (\log 6 - \log 2) = 3 \log 6 \qquad \text{Factor out } x \text{ on the left-hand side.}$$

$$x = \frac{3 \log 6}{\log 6 - \log 2} \qquad \text{Divide both sides by } \log 6 - \log 2.$$

$$x \approx 4.892789261 \qquad \text{Use a calculator.}$$

SELF CHECK

Solve $5^{x-2} = 3^x$. *Answer:* $\frac{2 \log 5}{\log 5 - \log 3} \approx 6.3013$ ∎

EXAMPLE 3

Taking the natural logarithm of each side. Solve $e^{0.9t} = 8$.

Solution The exponential expression on the left-hand side has base e. In such cases, the computations are somewhat simpler if we take the natural logarithm of each side.

$$e^{0.9t} = 8$$

$$\ln e^{0.9t} = \ln 8 \qquad \text{Take the natural logarithm of each side.}$$

$$0.9t \ln e = \ln 8 \qquad \text{Use the power rule of logarithms: } e^{0.9t} = 0.9t \ln e.$$

$$0.9t \cdot 1 = \ln 8 \qquad \text{Simplify: } \ln e = 1.$$

$$0.9t = \ln 8$$

$$t = \frac{\ln 8}{0.9} \qquad \text{Divide both sides by } 0.9.$$

$$t \approx 2.310490602 \qquad \text{Use a calculator.}$$

To four decimal places, $t \approx 2.3105$.

SELF CHECK

Solve $e^{2.1t} = 35$. *Answer:* $\frac{\ln 35}{2.1} \approx 1.6930$ ∎

EXAMPLE 4

Exponential expressions with the same base. Solve $2^{x^2+2x} = \frac{1}{2}$.

Solution Since $\frac{1}{2} = 2^{-1}$, we can write the equation in the form

$$2^{x^2+2x} = 2^{-1} \qquad \text{Each side of the equation can be written as an exponential expression with base 2.}$$

Since equal quantities with equal bases have equal exponents, we have

$$x^2 + 2x = -1 \qquad \text{Equate the exponents.}$$

$$x^2 + 2x + 1 = 0 \qquad \text{Add 1 to both sides.}$$

$$(x + 1)(x + 1) = 0 \qquad \text{Factor the trinomial.}$$

$$x + 1 = 0 \quad \text{or} \quad x + 1 = 0 \qquad \text{Set each factor equal to 0.}$$

$$x = -1 \qquad\qquad x = -1$$

Verify that -1 satisfies the equation.

Solve $3^{x^2-2x} = \frac{1}{3}$. *Answer:* 1, 1 ∎

ACCENT ON TECHNOLOGY *Solving Exponential Equations Graphically*

To use a graphing calculator to approximate the solutions of $2^{x^2+2x} = \frac{1}{2}$ (see Example 4), we can subtract $\frac{1}{2}$ from both sides of the equation to get

$$2^{x^2+2x} - \frac{1}{2} = 0$$

and graph the corresponding function

$$y = 2^{x^2+2x} - \frac{1}{2}$$

If we use window settings of $[-4, 4]$ for x and $[-2, 6]$ for y, we obtain the graph shown in Figure 9-35(a).

Since the solutions of the equation are its x-intercepts, we can approximate the solutions by zooming in on the values of the x-intercepts, as in Figure 9-35(b). Since $x = -1$ is the only x-intercept, -1 is the only solution. In this case, we have found an exact solution.

FIGURE 9-35

(a)

(b)

(c)

We can also solve $2^{x^2 + 2x} = \frac{1}{2}$ using the INTERSECT feature found on most graphing calculators. After graphing $Y_1 = 2^{x^2 + 2x}$ and $Y_2 = \frac{1}{2}$, we select INTERSECT, which approximates the coordinates of the point of intersection of the two graphs. From the display shown in Figure 9-35(c), we can conclude that the solution is $x = -1$. Verify this by checking.

Solving Logarithmic Equations

EXAMPLE 5

Solving logarithmic equations. Solve $\log 5x = 3$.

Solution Recall that $\log 5x = \log_{10} 5x$. We can change the equation $\log 5x = 3$ into the equivalent base-10 exponential equation $10^3 = 5x$ and solve for x.

$$\log 5x = 3$$
$$10^3 = 5x$$
$$1{,}000 = 5x \quad \text{Simplify: } 10^3 = 1{,}000.$$
$$200 = x \quad \text{Divide both sides by 5.}$$

Check the result.

Solve $\log_2(x - 3) = -1$. *Answer:* $\frac{7}{2}$ ∎

EXAMPLE 6

Solving logarithmic equations. Solve $\log(3x + 2) - \log(2x - 3) = 0$.

Solution We isolate each logarithmic expression on one side of the equation.

$$\log(3x + 2) - \log(2x - 3) = 0$$
$$\log(3x + 2) = \log(2x - 3) \quad \text{Add } \log(2x - 3) \text{ to both sides.}$$

In the previous section, we saw that if the logarithms of two numbers are equal, the numbers are equal. So we have

$$(3x + 2) = (2x - 3) \quad \text{If } \log r = \log s, \text{ then } r = s.$$
$$x = -5 \quad \text{Subtract } 2x \text{ and } 2 \text{ from both sides.}$$

Check: $\quad \log(3x + 2) - \log(2x - 3) = 0$
$$\log[3(-5) + 2] - \log[2(-5) - 3] \overset{?}{=} 0$$
$$\log(-13) - \log(-13) \overset{?}{=} 0$$

Since the logarithm of a negative number does not exist, the apparent solution of -5 must be discarded. This equation has no solutions.

SELF CHECK

Solve $\log(5x + 2) - \log(7x - 2) = 0$. *Answer: 2* ∎

EXAMPLE 7

Using a property of logarithms to solve a logarithmic equation.
Solve $\log x + \log(x - 3) = 1$.

Solution
$$\log x + \log(x - 3) = 1$$
$$\log x(x - 3) = 1 \qquad \text{Use the product rule of logarithms.}$$
$$x(x - 3) = 10^1 \qquad \text{Use the definition of logarithms to change the equation to exponential form:}$$
$$\qquad\qquad\qquad \log x(x - 3) = \log_{10} x(x - 3).$$
$$x^2 - 3x - 10 = 0 \qquad \text{Remove parentheses and subtract 10 from both sides.}$$
$$(x + 2)(x - 5) = 0 \qquad \text{Factor the trinomial.}$$
$$x + 2 = 0 \quad \text{or} \quad x - 5 = 0 \qquad \text{Set each factor equal to 0.}$$
$$x = -2 \qquad\qquad x = 5$$

Check: The number -2 is not a solution, because it does not satisfy the equation (a negative number does not have a logarithm). We will check the remaining number, 5.

$$\log x + \log(x - 3) = 1$$
$$\log 5 + \log(5 - 3) \overset{?}{=} 1 \qquad \text{Substitute 5 for } x.$$
$$\log 5 + \log 2 \overset{?}{=} 1$$
$$\log 10 \overset{?}{=} 1 \qquad \text{Use the product rule of logarithms:}$$
$$\qquad\qquad\qquad \log 5 + \log 2 = \log(5 \cdot 2) = \log 10.$$
$$1 = 1 \qquad \text{Simplify: } \log 10 = 1.$$

Since 5 satisfies the equation, it is a solution.

SELF CHECK

Solve $\log x + \log(x + 3) = 1$. *Answer: 2* ∎

WARNING! Examples 6 and 7 illustrate that we must check the solutions of a logarithmic equation.

ACCENT ON TECHNOLOGY *Solving Logarithmic Equations Graphically*

To use a graphing calculator to approximate the solutions of $\log x + \log (x - 3) = 1$ (see Example 7), we can subtract 1 from both sides of the equation to get

$$\log x + \log (x - 3) - 1 = 0$$

and graph the corresponding function

$$y = \log x + \log (x - 3) - 1$$

If we use window settings of $[0, 20]$ for x and $[-2, 2]$ for y, we obtain the graph shown in Figure 9-36(a). Since the solution of the equation is the x-intercept, we can find the solution by zooming in on the value of the x-intercept. The solution is $x = 5$.

FIGURE 9-36

(a)

(b)

We can also solve $\log x + \log (x - 3) = 1$ using the INTERSECT feature. After graphing $Y_1 = \log x + \log (x - 3)$ and $Y_2 = 1$, we select INTERSECT, which approximates the coordinates of the point of intersection of the two graphs. From the display shown in Figure 9-36(b), we can conclude that the solution is $x = 5$. Verify this by checking.

EXAMPLE 8

Solving logarithmic equations. Solve $\dfrac{\log_2 (5x - 6)}{\log_2 x} = 2$.

Solution We can multiply both sides of the equation by $\log_2 x$ to get

$$\log_2 (5x - 6) = 2 \log_2 x$$

and apply the power rule of logarithms to get

$$\log_2 (5x - 6) = \log_2 x^2$$

By property 8 of logarithms, $5x - 6 = x^2$, because they have equal logarithms. Thus,

$$5x - 6 = x^2$$
$$0 = x^2 - 5x + 6$$
$$0 = (x - 3)(x - 2)$$
$$x - 3 = 0 \quad \text{or} \quad x - 2 = 0$$
$$x = 3 \qquad\qquad x = 2$$

Verify that both 2 and 3 satisfy the equation.

SELF CHECK Solve $\dfrac{\log_3 (5x + 6)}{\log_3 x} = 2$.

Answer: 6

Radioactive Decay

Experiments have determined the time it takes for half of a sample of a given radioactive material to decompose. This time is a constant, called the material's **half-life.**

When living organisms die, the oxygen/carbon dioxide cycle common to all living things ceases, and carbon-14, a radioactive isotope with a half-life of 5,700 years, is no longer absorbed. By measuring the amount of carbon-14 present in an ancient object, archaeologists can estimate the object's age by using the radioactive decay formula.

Radioactive decay formula

If A is the amount of radioactive material present at time t, A_0 was the amount present at $t = 0$, and h is the material's half-life, then

$$A = A_0 2^{-t/h}$$

EXAMPLE 9

Carbon-14 dating. How old is a wooden statue that retains only one-third of its original carbon-14 content?

Solution To find the time t when $A = \frac{1}{3}A_0$, we substitute $\frac{A_0}{3}$ for A and 5,700 for h in the radioactive decay formula and solve for t:

$$A = A_0 2^{-t/h}$$

$$\frac{A_0}{3} = A_0 2^{-t/5,700} \qquad \text{The half-life of carbon-14 is 5,700 years.}$$

$$1 = 3(2^{-t/5,700}) \qquad \text{Divide both sides by } A_0 \text{ and multiply both sides by 3.}$$

$$\log 1 = \log 3(2^{-t/5,700}) \qquad \text{Take the common logarithm of each side.}$$

$$0 = \log 3 + \log 2^{-t/5,700} \qquad \log 1 = 0, \text{ and use the product rule of logarithms.}$$

$$-\log 3 = -\frac{t}{5,700} \log 2 \qquad \text{Subtract log 3 from both sides and use the power rule of logarithms.}$$

$$5,700\left(\frac{\log 3}{\log 2}\right) = t \qquad \text{Multiply both sides by } -\frac{5,700}{\log 2}.$$

$$t \approx 9,034.286254 \qquad \text{Use a calculator.}$$

The statue is approximately 9,000 years old.

SELF CHECK

In Example 9, how old is a statue that retains 25% of its original carbon-14 content?   *Answer:* about 11,400 years   ∎

Population Growth

Recall that when there is sufficient food and space, populations of living organisms tend to increase exponentially according to the Malthusian growth model.

Malthusian growth model

If P is the population at some time t, P_0 is the initial population at $t = 0$, and k depends on the rate of growth, then

$$P = P_0 e^{kt}$$

| EXAMPLE 10 | **Population growth.** The bacteria in a laboratory culture increased from an initial population of 500 to 1,500 in 3 hours. How long will it take for the population to reach 10,000? |

Solution   We substitute 500 for P_0, 1,500 for P, and 3 for t and simplify to find k:

$$P = P_0 e^{kt}$$

$$1{,}500 = 500(e^{k3}) \qquad \text{Substitute 1,500 for } P, \text{ 500 for } P_0, \text{ and 3 for } t.$$

$$3 = e^{3k} \qquad \text{Divide both sides by 500.}$$

$$3k = \ln 3 \qquad \text{Change the equation from exponential to logarithmic form.}$$

$$k = \frac{\ln 3}{3} \qquad \text{Divide both sides by 3.}$$

To find when the population will reach 10,000, we substitute 10,000 for P, 500 for P_0, and $\frac{\ln 3}{3}$ for k in the equation $P = P_0 e^{kt}$ and solve for t:

$$P = P_0 e^{kt}$$

$$10{,}000 = 500 e^{[(\ln 3)/3]t}$$

$$20 = e^{[(\ln 3)/3]t} \qquad \text{Divide both sides by 500.}$$

$$\left(\frac{\ln 3}{3}\right) t = \ln 20 \qquad \text{Change the equation to logarithmic form.}$$

$$t = \frac{3 \ln 20}{\ln 3} \qquad \text{Multiply both sides by } \tfrac{3}{\ln 3}.$$

$$\approx 8.180499084 \qquad \text{Use a calculator.}$$

The culture will reach 10,000 bacteria in about 8 hours.

| SELF CHECK | In Example 10, how long will it take the population to reach 20,000? |

Answer: about 10 hours   ■

| EXAMPLE 11 | **Generation time.** If a medium is inoculated with a bacterial culture that contains 1,000 cells per milliliter, how many generations will pass by the time the culture has grown to a population of 1 million cells per milliliter? |

Solution   During bacterial reproduction, the time required for a population to double is called the *generation time*. If b bacteria are introduced into a medium, then after the generation time of the organism has elapsed, there are $2b$ cells. After another generation, there are $2(2b)$ or $4b$ cells, and so on. After n generations, the number of cells present will be

1. $B = b \cdot 2^n$

To find the number of generations that have passed while the population grows from b bacteria to B bacteria, we solve Equation 1 for n. To do this, we can take the common logarithm or the natural logarithm of each side.

$$\ln B = \ln (b \cdot 2^n) \qquad \text{Take the natural logarithm of each side.}$$

$$\ln B = \ln b + n \ln 2 \qquad \text{Apply the product and power rules of logarithms.}$$

$$\ln B - \ln b = n \ln 2 \qquad \text{Subtract } \ln b \text{ from both sides.}$$

$$n = \frac{1}{\ln 2} (\ln B - \ln b) \qquad \text{Multiply both sides by } \frac{1}{\ln 2}.$$

2. $$n = \frac{1}{\ln 2} \left(\ln \frac{B}{b} \right) \qquad \text{Use the quotient rule of logarithms.}$$

Equation 2 is a formula that gives the number of generations that will pass as the population grows from b bacteria to B bacteria.

To find the number of generations that have passed while a population of 1,000 cells per milliliter has grown to a population of 1 million cells per milliliter, we substitute 1,000 for b and 1,000,000 for B in Equation 2 and solve for n.

$$n = \frac{1}{\ln 2} \ln \frac{1,000,000}{1,000}$$

$$= \frac{1}{\ln 2} \ln 1,000 \qquad \text{Simplify.}$$

$$n \approx 9.965784285 \qquad$$

Use a calculator:
2 [ln] [1/x] [×] 1000 [ln] [=] .

Approximately 10 generations will have passed. ■

STUDY SET

Section 9.8

VOCABULARY

In Exercises 1–2, fill in the blanks to make the statements true.

1. An equation with a variable in its exponent, such as $3^{2x} = 8$, is called a(n) _____ equation.

2. An equation with a logarithmic expression that contains a variable, such as $\log_5 (2x - 3) = \log_5 (x + 4)$, is a(n) _____ equation.

CONCEPTS

In Exercises 3–4, fill in the blanks to make the statements true.

3. The formula for radioactive decay is $A =$ ▭ .

4. The formula for population growth is $P =$ ▭ .

5. Use the graphs in Illustration 1 to estimate the solution of $2^x = 3^{-x+3}$.

ILLUSTRATION 1

6. Use the graphs in Illustration 2 to estimate the solution of $3 \log (x - 1) = 2 \log x$.

ILLUSTRATION 2

7. Fill in the blanks to make the statements true. To solve $5^x = 21$, we can take the _____ of each side of the equation to get

$$\log 5^x = \log 21$$

The power rule for logarithms then provides a way of moving the variable x from its position as an _____ to a position as a coefficient.

8. [calculator] Use a calculator to determine whether $x \approx 2.5646$ is a solution of $2^{2x + 1} = 70$.

9. [calculator] Find $\dfrac{\log 8}{\log 5}$. Round to four decimal places.

10. [calculator] Find $\dfrac{2 \ln 12}{\ln 9}$. Round to four decimal places.

11. Simplify $\ln e$.

12. Does $\dfrac{\log 7}{\log 3} = \log 7 - \log 3$?

13. Write the corresponding base-10 exponential equation for $\log (x + 1) = 2$.

14. Write the corresponding base-e exponential equation for $\ln (x + 1) = 2$.

15. Solve each equation. Round to the nearest hundredth.
    a. $x^2 = 12$ **b.** $2^x = 12$

16. Solve each equation.
    a. $\log (x - 1) = 3$ **b.** $\log (x - 1) = \log 3$

17. Perform a check to see whether $x = -4$ is a solution of $\log_5 (x + 3) = \frac{1}{5}$.

18. Perform a check to see whether $x = -2$ is a solution of $5^{2x+3} = \frac{1}{5}$.

NOTATION

In Exercises 19–20, complete each solution.

19. Solve $2^x = 7$.

$$2^x = 7$$

$$\boxed{} 2^x = \log 7$$

$$x \boxed{} = \log 7$$

$$x = \frac{\log 7}{\log 2}$$

20. Solve $\log_2 (2x - 3) = \log_2 (x + 4)$.

$$\log_2 (2x - 3) = \log_2 (x + 4)$$

$$\boxed{} = x + 4$$

$$x = 7$$

PRACTICE

In Exercises 21–44, solve each exponential equation. Give answers to four decimal places when necessary.

21. $4^x = 5$
22. $7^x = 12$
23. $13^{x-1} = 2$
24. $5^{x+1} = 3$
25. $2^{x+1} = 3^x$
26. $5^{x-3} = 3^{2x}$
27. $2^x = 3^x$
28. $3^{2x} = 4^x$
29. $7^{x^2} = 10$
30. $8^{x^2} = 11$
31. $8^{x^2} = 9^x$
32. $5^{x^2} = 2^{5x}$
33. $e^{3x} = 9$
34. $e^{4x} = 60$
35. $e^{-0.2t} = 14.2$
36. $e^{0.3t} = 9.1$

37. $2^{x-2} = 64$
38. $3^{-3x+1} = 243$
39. $5^{4x} = \dfrac{1}{125}$
40. $8^{-x+1} = \dfrac{1}{64}$
41. $2^{x^2-2x} = 8$
42. $3^{x^2-3x} = 81$
43. $3^{x^2+4x} = \dfrac{1}{81}$
44. $7^{x^2+3x} = \dfrac{1}{49}$

In Exercises 45–48, use a graphing calculator to solve each equation. Give all answers to the nearest tenth.

45. $2^{x+1} = 7$
46. $3^{x-1} = 2^x$
47. $4(2^{x^2}) = 8^{3x}$
48. $3^x - 10 = 3^{-x}$

In Exercises 49–84, solve each logarithmic equation.

49. $\log (x + 2) = 4$
50. $\log 5x = 4$
51. $\log (7 - x) = 2$
52. $\log (2 - x) = 3$
53. $\ln x = 1$
54. $\ln x = 5$
55. $\ln (x + 1) = 3$
56. $\ln 2x = 5$

57. $\log 2x = \log 4$
58. $\log 3x = \log 9$
59. $\ln (3x + 1) = \ln (x + 7)$
60. $\ln (x^2 + 4x) = \ln (x^2 + 16)$

61. $\log (3 - 2x) - \log (x + 24) = 0$
62. $\log (3x + 5) - \log (2x + 6) = 0$
63. $\log \dfrac{4x + 1}{2x + 9} = 0$
64. $\log \dfrac{2 - 5x}{2(x + 8)} = 0$
65. $\log x^2 = 2$
66. $\log x^3 = 3$
67. $\log x + \log (x - 48) = 2$
68. $\log x + \log (x + 9) = 1$
69. $\log x + \log (x - 15) = 2$
70. $\log x + \log (x + 21) = 2$
71. $\log (x + 90) = 3 - \log x$
72. $\log (x - 90) = 3 - \log x$

73. $\log (x - 6) - \log (x - 2) = \log \dfrac{5}{x}$

74. $\log (3 - 2x) - \log (x + 9) = 0$

75. $\dfrac{\log (3x - 4)}{\log x} = 2$
76. $\dfrac{\log (8x - 7)}{\log x} = 2$

77. $\dfrac{\log (5x + 6)}{2} = \log x$
78. $\dfrac{1}{2} \log (4x + 5) = \log x$

79. $\log_3 x = \log_3 \left(\dfrac{1}{x}\right) + 4$

80. $\log_5 (7 + x) + \log_5 (8 - x) - \log_5 2 = 2$
81. $2 \log_2 x = 3 + \log_2 (x - 2)$
82. $2 \log_3 x - \log_3 (x - 4) = 2 + \log_3 2$
83. $\log (7y + 1) = 2 \log (y + 3) - \log 2$
84. $2 \log (y + 2) = \log (y + 2) - \log 12$

In Exercises 85–88, use a graphing calculator to solve each equation. If an answer is not exact, round to the nearest tenth.

85. $\log x + \log (x - 15) = 2$
86. $\log x + \log (x + 3) = 1$
87. $\ln (2x + 5) - \ln 3 = \ln (x - 1)$
88. $2 \log (x^2 + 4x) = 1$

APPLICATIONS

89. TRITIUM DECAY The half-life of tritium is 12.4 years. How long will it take for 25% of a sample of tritium to decompose?

90. RADIOACTIVE DECAY In two years, 20% of a radioactive element decays. Find its half-life.

91. THORIUM DECAY An isotope of thorium, ^{227}Th, has a half-life of 18.4 days. How long will it take for 80% of the sample to decompose?

92. LEAD DECAY An isotope of lead, ^{201}Pb, has a half-life of 8.4 hours. How many hours ago was there 30% more of the substance?

93. CARBON-14 DATING A bone fragment analyzed by archaeologists contains 60% of the carbon-14 that it is assumed to have had initially. How old is it?

94. CARBON-14 DATING Only 10% of the carbon-14 in a small wooden bowl remains. How old is the bowl?

95. COMPOUND INTEREST If $500 is deposited in an account paying 8.5% annual interest, compounded semiannually, how long will it take for the account to increase to $800?

96. CONTINUOUS COMPOUND INTEREST In Exercise 95, how long will it take if the interest is compounded continuously?

97. COMPOUND INTEREST If $1,300 is deposited in a savings account paying 9% interest, compounded quarterly, how long will it take the account to increase to $2,100?

98. COMPOUND INTEREST A sum of $5,000 deposited in an account grows to $7,000 in 5 years. Assuming annual compounding, what interest rate is being paid?

99. RULE OF SEVENTY A rule of thumb for finding how long it takes an investment to double is called the **rule of seventy.** To apply the rule, divide 70 by the interest rate written as a percent. At 5%, doubling requires $\frac{70}{5} = 14$ years to double an investment. At 7%, it takes $\frac{70}{7} = 10$ years. Explain why this formula works.

100. BACTERIAL GROWTH A bacterial culture grows according to the formula

$$P = P_0 a^t$$

If it takes 5 days for the culture to triple in size, how long will it take to double in size?

101. RODENT CONTROL The rodent population in a city is currently estimated at 30,000. If it is expected to double every 5 years, when will the population reach 1 million?

102. POPULATION GROWTH The population of a city is expected to triple every 15 years. When can the city planners expect the present population of 140 persons to double?

103. BACTERIAL CULTURE A bacterial culture doubles in size every 24 hours. By how much will it have increased in 36 hours?

104. OCEANOGRAPHY The intensity I of a light a distance x meters beneath the surface of a lake decreases exponentially. From Illustration 3, find the depth at which the intensity will be 20%.

105. MEDICINE If a medium is inoculated with a bacterial culture containing 500 cells per milliliter, how many generations will have passed by the time the culture contains 5×10^6 cells per milliliter?

ILLUSTRATION 3

106. MEDICINE If a medium is inoculated with a bacterial culture containing 800 cells per milliliter, how many generations will have passed by the time the culture contains 6×10^7 cells per milliliter?

107. NEWTON'S LAW OF COOLING Water initially at 100°C is left to cool in a room at temperature 60°C. After 3 minutes, the water temperature is 90°. The water temperature T is a function of time t given by

$$T = 60 + 40e^{kt}$$

Find k.

108. NEWTON'S LAW OF COOLING Refer to Exercise 107 and find the time for the water temperature to reach 70°C.

WRITING

109. Explain how to solve the equation $2^{x+1} = 31$.

110. Explain how to solve the equation $2^{x+1} = 32$.

REVIEW

Solve each equation.

111. $5x^2 - 25x = 0$ **112.** $4y^2 - 25 = 0$

113. $3p^2 + 10p = 8$ **114.** $4t^2 + 1 = -6t$

115. In Illustration 4, find the length of leg AC.

ILLUSTRATION 4

116. The amount of medicine a patient should take is often proportional to his or her weight. If a patient weighing 83 kilograms needs 150 milligrams of medicine, how much will be needed by a person weighing 99.6 kilograms?

Inverse Functions

One-to-One Functions

A function is *one-to-one* if each input value x in the domain determines a different output value y in the range. In Exercises 1–3, determine whether the function is one-to-one.

1. $f(x) = x^2$

2. $f(x) = |x|$

3.

$x = 4 - 2y$

The Inverse of a Function

If a function is one-to-one, its inverse is a function. To find the inverse of a function, interchange x and y and solve for y.

4. Find $f^{-1}(x)$ if $f(x) = -2x - 1$.

5. Given the table of values for a one-to-one function f, complete the table of values for f^{-1}.

x	-2	1	3
$f(x)$	4	-2	-6

x	4	-2	-6
$f^{-1}(x)$			

Exponential and Logarithmic Functions

The exponential function $y = b^x$ and $y = \log_b x$ (where $b > 0$ and $b \neq 1$) are inverse functions.

6. Write the exponential statement $10^3 = 1{,}000$ in logarithmic form.

7. Write the logarithmic statement $\log_2 \frac{1}{8} = -3$ in exponential form.

8. If $\log_4 y = \frac{1}{2}$, what is y?

9. If $\log_y \frac{9}{4} = 2$, what is y?

The Natural Exponential and Natural Logarithmic Functions

A special exponential function that is used in many real-life applications involving growth and decay is the base-e exponential function, $f(x) = e^x$. Its inverse is the natural logarithm function $f(x) = \ln x$.

10. What is an approximate value of e?

11. What is the base of the logarithmic function $f(x) = \ln x$?

12. Use a calculator to find x: $\ln x = -0.28$.

13. Graph $f(x) = e^x$ and $f(x) = \ln x$ on the coordinate system in Illustration 1. Label the axis of symmetry.

ILLUSTRATION 1

Accent on Teamwork

Section 9.1

Composition of functions Consider the functions $f(x) = x^2$, $g(x) = 2x + 1$, and $h(x) = 1 - x$. Determine whether the composition of functions f, g, and h is associative. That is, is the following true?

$$[f \circ (g \circ h)](x) \overset{?}{=} [(f \circ g) \circ h](x)$$

Section 9.2

One-to-one functions In newspapers, magazines, or books, find line graphs that are graphs of one-to-one functions and some that are not one-to-one. In each case, explain to the other members of your group what relationship the graph illustrates. Then tell whether the graph passes or fails the horizontal line test.

Section 9.3

Exponential functions On a piece of poster board, draw a rectangular coordinate system made up of a grid of 1-inch squares. Then graph each of the following exponential functions. How are the graphs similar? How are they different? For positive values of x, as the base increases, what happens to the steepness of the graph?

$$f(x) = 2^x \qquad f(x) = 3^x \qquad f(x) = 4^x$$
$$f(x) = 5^x \qquad f(x) = 6^x \qquad f(x) = 7^x$$

Section 9.4

Growth Make two copies of each of the graph shapes shown in Illustration 1.

ILLUSTRATION 1

Exponential growth
(a)

Linear growth
(b)

Think of two examples of quantities that, in general, have grown exponentially over the years. Label the horizontal and vertical axes of the graphs with the proper titles.

(You do not have to scale the axes in terms of units.) Do the same for two quantities that, in general, have grown linearly over the years. Share your observations with the other members of your group. See if they agree with your models.

Section 9.5

Logarithmic functions On a piece of poster board, draw a rectangular coordinate system made up of a grid of 1-inch squares. Then graph each of the following logarithmic functions. How are the graphs similar? How are they different? For positive values of x, as the base increases, what happens to the steepness of the graph?

$$f(x) = \log_2 x \qquad f(x) = \log_3 x \qquad f(x) = \log_4 x$$
$$f(x) = \log_5 x \qquad f(x) = \log_6 x \qquad f(x) = \log_7 x$$

Section 9.6

The number e The value of e can be calculated to any degree of accuracy by adding the terms of the following pattern:

$$1, 1, \frac{1}{2}, \frac{1}{2 \cdot 3}, \frac{1}{2 \cdot 3 \cdot 4}, \frac{1}{2 \cdot 3 \cdot 4 \cdot 5}, \cdots$$

The more terms that are added, the closer the sum will be to e. Write each of the first six terms in simplified form and then add them. To how many decimal places is the sum accurate?

Section 9.7

The properties of logarithms discussed in Section 9.7 hold for logarithms of any base. Use these properties to simplify each of the following natural logarithmic expressions.

a. $\ln x^y + \ln x^z$ **b.** $\ln x^y - \ln x^z$

c. $\ln e^{10}$ **d.** $\ln \dfrac{x}{y} + \ln \dfrac{y}{x}$

Section 9.8

To solve exponential equations involving e, it is convenient to take the natural logarithm (instead of the common logarithm) of both sides of the equation. Use this technique to solve each of the following equations. Give all answers to four decimal places.

a. $e^x = 10$ **b.** $e^{2x} = 15$

c. $e^{x+1} = 50.5$ **d.** $3e^{5x-2} = 3$

Section 9.1

Algebra and Composition of Functions

CONCEPTS

Functions can be added, subtracted, multiplied, and divided:

$$(f + g)(x) = f(x) + g(x)$$
$$(f - g)(x) = f(x) - g(x)$$
$$(f \cdot g)(x) = f(x)g(x)$$
$$(f/g)(x) = \frac{f(x)}{g(x)} \quad (g(x) \neq 0)$$

Composition of functions:

$$(f \circ g)(x) = f(g(x))$$

REVIEW EXERCISES

1. Let $f(x) = 2x$ and $g(x) = x + 1$. Find each function or value.
  a. $f + g$ **b.** $f - g$

  c. $f \cdot g$ **d.** f/g

  e. $(f \circ g)(1)$ **f.** $g(f(1))$

  g. $(f \circ g)(x)$ **h.** $(g \circ f)(x)$

Section 9.2

Inverses of Functions

A function is *one-to-one* if each input value x in the domain determines a different output value y in the range.

Horizontal line test: If every horizontal line that intersects the graph of a function does so only once, the function is one-to-one.

2. Determine whether the function is one-to-one.
  a. $f(x) = x^2 + 3$ **b.** $f(x) = x + 3$

3. Use the horizontal line test to decide whether the function is one-to-one.
  a.   **b.**

The graph of a function and its inverse are symmetric about the line $y = x$.

4. Given the graph of function f shown in Illustration 1, graph f^{-1} on the same coordinate axes. Label the axis of symmetry.

ILLUSTRATION 1

To *find the inverse of a function,* interchange the variables x and y and solve for y.

5. Find the inverse of each function.
  a. $f(x) = 6x - 3$ **b.** $f(x) = 4x + 5$

  c. $f(x) = x^3$ **d.** $y = 2x^2 - 1 \ (x \geq 0)$

Section 9.3

An *exponential function* with base b is defined by the equation

$$f(x) = b^x \quad (b > 0, \, b \neq 1)$$

If $b > 1$, then $f(x) = b^x$ is an *increasing function*.

If $0 < b < 1$, then $f(x) = b^x$ is a *decreasing function*.

Exponential functions are suitable models for describing many situations involving the *growth* or *decay* of a quantity.

Exponential Functions

6. Use properties of exponents to simplify each expression.

a. $5^{\sqrt{2}} \cdot 5^{\sqrt{2}}$

b. $\left(2^{\sqrt{5}}\right)^{\sqrt{2}}$

7. Graph each function. Then give the domain and the range.

a. $y = 3^x$

b. $f(x) = \left(\dfrac{1}{3}\right)^x$

8. Graph each function by using a translation.

a. $f(x) = \left(\dfrac{1}{2}\right)^x - 2$

b. $g(x) = \left(\dfrac{1}{2}\right)^{x+2}$

9. COAL PRODUCTION The table gives the number of tons of coal produced in the United States for the years 1800–1920. Graph the data in Illustration 2. What type of function does it appear could be used to model coal production over this period?

ILLUSTRATION 2

Year	Tons
1800	108,000
1810	178,000
1820	881,000
1830	1,334,000
1840	2,474,000
1850	8,356,000
1860	20,041,000

Year	Tons
1870	40,429,000
1880	79,407,000
1890	157,771,000
1900	269,684,000
1910	501,596,000
1920	658,265,000

Based on information from *World Book Encyclopedia*

(continued)

ILLUSTRATION 2 *(continued)*

Compound interest: If $P is the deposit, and interest is paid k times a year at an annual rate r, the amount A in the account after t years is given by

$$A = P\left(1 + \frac{r}{k}\right)^{kt}$$

10. COMPOUND INTEREST How much will $10,500 become if it earns 9% annual interest, compounded quarterly, for 60 years?

Section 9.4

Base-*e* Exponential Functions

The function defined by $f(x) = e^x$ is the *natural exponential function* where $e = 2.718281828459\ldots$.

11. MORTGAGE RATES The average annual interest rate on a 30-year fixed-rate home mortgage for the years 1980–1996 can be approximated by the function $r(t) = 13.9e^{-0.035t}$, where t is the number of years since 1980. To the nearest hundredth of a percent, what does this model predict was the 30-year fixed rate in 1995?

If a quantity increases or decreases at an annual rate r, *compounded continuously*, then the amount A after t years is given by

$$A = Pe^{rt}$$

12. INTEREST COMPOUNDED CONTINUOUSLY If $10,500 accumulates interest at an annual rate of 9%, compounded continuously, how much will be in the account in 60 years?

13. Graph each function. Give the domain and the range.
   a. $f(x) = e^x + 1$
   b. $f(x) = e^{x-3}$

Malthusian population growth is modeled by the formula

$$P = P_0 e^{kt}$$

14. POPULATION GROWTH The population of the United States is approximately 275,000,000 people. Find the population in 50 years if $k = 0.015$.

Section 9.5

Logarithmic Functions

If $b > 0$ and $b \neq 1$, then the *logarithmic function with base b* is defined by $y = \log_b x$.

$y = \log_b x$ means $x = b^y$.

$\log_b x$ is the exponent to which b is raised to get x.

$$b^{\log_b x} = x$$

15. Give the domain and range of the logarithmic function $f(x) = \log x$.

16. a. Write the statement $\log_4 64 = 3$ in exponential form.

   b. Write the statement $7^{-1} = \frac{1}{7}$ in logarithmic form.

17. Find each value, if possible.

   a. $\log_3 9$ **b.** $\log_9 \dfrac{1}{81}$ **c.** $\log_8 1$

   d. $\log_5 (-25)$ **e.** $\log_6 \sqrt{6}$ **f.** $\log 1{,}000$

18. Find the value of x in each equation.

   a. $\log_2 x = 5$ **b.** $\log_3 x = -4$ **c.** $\log_x 16 = 2$

   d. $\log_x \dfrac{1}{100} = -2$ **e.** $\log_9 3 = x$ **f.** $\log_{27} x = \dfrac{2}{3}$

19. Use a calculator to find the value of x to four decimal places.

   a. $\log 4.51 = x$ **b.** $\log x = 1.43$

If $b > 1$, then $f(x) = \log_b x$ is an *increasing function*. If $0 < b < 1$, then $f(x) = \log_b x$ is a *decreasing function*.

The exponential function $f(x) = b^x$ and the logarithmic function $f(x) = \log_b x$ are inverses of each other.

20. Graph each pair of equations on one set of coordinate axes.

   a. $y = 4^x$ and $y = \log_4 x$ **b.** $f(x) = \left(\dfrac{1}{3}\right)^x$ and $g(x) = \log_{1/3} x$

21. Graph each function.

   a. $f(x) = \log (x - 2)$ **b.** $f(x) = 3 + \log x$

Decibel voltage gain:

$$\text{db gain} = 20 \log \frac{E_o}{E_I}$$

22. ELECTRICAL ENGINEERING An amplifier has an output of 18 volts when the input is 0.04 volt. Find the db gain.

The Richter scale:

$$R = \log \frac{A}{P}$$

23. EARTHQUAKE An earthquake had a period of 0.3 second and an amplitude of 7,500 micrometers. Find its measure on the Richter scale.

Section 9.6

Natural logarithms:

$$\ln x \text{ means } \log_e x$$

Base-*e* Logarithms

24. Use a calculator to find each value to four decimal places.
a. $\ln 452$ **b.** $\ln 0.85$

25. Use a calculator to find the value of x to four decimal places.
a. $\ln x = 2.336$ **b.** $\ln x = -8.8$

26. What function is the inverse of $f(x) = \ln x$?

27. Graph each function.
a. $f(x) = 1 + \ln x$ **b.** $y = \ln (x + 1)$

Population doubling time:

$$t = \frac{\ln 2}{r}$$

28. POPULATION GROWTH How long will it take the population of the United States to double if the growth rate is 3% per year?

Section 9.7

Properties of logarithms: If b is a positive number and $b \neq 1$,

1. $\log_b 1 = 0$ **2.** $\log_b b = 1$

3. $\log_b b^x = x$ **4.** $b^{\log_b x} = x$

5. $\log_b MN = \log_b M + \log_b N$

6. $\log_b \dfrac{M}{N} = \log_b M - \log_b N$

7. $\log_b M^p = p \log_b M$

8. If $\log_b x = \log_b y$, then $x = y$.

Change-of-base formula:

$$\log_b x = \frac{\log_a x}{\log_a b}$$

pH scale:

$$pH = -\log [H^+]$$

Properties of Logarithms

29. Simplify each expression.
a. $\log_7 1$ **b.** $\log_7 7$
c. $\log_7 7^3$ **d.** $7^{\log_7 4}$
e. $\ln e^4$ **f.** $\ln e$

30. Use the properties of logarithms to rewrite each expression.
a. $\log_3 27x$ **b.** $\log \dfrac{100}{x}$

c. $\log_5 \sqrt{27}$ **d.** $\log 10ab$

31. Write each expression in terms of the logarithms of x, y, and z.
a. $\log_b \dfrac{x^2 y^3}{z}$ **b.** $\ln \sqrt{\dfrac{x}{yz^2}}$

32. Write each expression as the logarithm of one quantity.

a. $3 \log_2 x - 5 \log_2 y + 7 \log_2 z$

b. $-3 \log_b y - 7 \log_b z + \dfrac{1}{2} \log_b (x + 2)$

33. Assume that $\log_b 5 = 1.1609$ and $\log_b 8 = 1.5000$. Use these values and the properties of logarithms to find each value to four decimal places.
a. $\log_b 40$ **b.** $\log_b 64$

34. Find $\log_5 17$ to four decimal places.

35. pH OF GRAPEFRUIT The pH of grapefruit juice is about 3.1 Find its hydrogen ion concentration.

Section 9.8

Exponential and Logarithmic Equations

An *exponential equation* is an equation that contains a variable in one of its exponents.

A *logarithmic equation* is an equation with a logarithmic expression that contains a variable.

Carbon dating:

$$A = A_0 2^{-t/h}$$

36. Solve each equation for x. Give answers to four decimal places when necessary.

a. $3^x = 7$ **b.** $5^{x+2} = 625$

c. $2^x = 3^{x-1}$ **d.** $2^{x^2+4x} = \dfrac{1}{8}$

e. $e^x = 7$ **f.** $e^{-0.4t} = 25$

37. Solve each equation for x.

a. $\log (x - 4) = 2$ **b.** $\ln (2x - 3) = \ln 15$

c. $\log x + \log (29 - x) = 2$ **d.** $\log_2 x + \log_2 (x - 2) = 3$

e. $\dfrac{\log (7x - 12)}{\log x} = 2$ **f.** $\log_2 (x + 2) + \log_2 (x - 1) = 2$

g. $\log x + \log (x - 5) = \log 6$ **h.** $\log 3 - \log (x - 1) = -1$

38. CARBON-14 DATING A wooden statue found in Egypt has a carbon-14 content that is two-thirds of that found in living wood. If the half-life of carbon-14 is 5,700 years, how old is the statue?

39. The approximate coordinates of the points of intersection of the graphs of $f(x) = \log x$ and $g(x) = 1 - \log (7 - x)$ are shown in parts a and b of Illustration 3. Estimate the solutions of the logarithmic equation $\log x = 1 - \log (7 - x)$. Then check your answers.

ILLUSTRATION 3

(a)   (b)

CHAPTER 9

Test

In Problems 1–4, $f(x) = 4x$ and $g(x) = x - 1$. Find each function or value.

1. $g + f$ **2.** $g \cdot f$

3. Find $(g \circ f)(1)$.   **4.** Find $f(g(x))$.

In Problems 5–6, find the inverse of each function.

5. $3x + 2y = 12$ **6.** $f(x) = 3x^2 + 4$ $(x \geq 0)$

In Problems 7–8, graph each function.

7. $f(x) = 2^x + 1$ **8.** $f(x) = 2^{-x}$

9. RADIOACTIVE DECAY A radioactive material decays according to the formula $A = A_0(2)^{-t}$. How much of a 3-gram sample will be left in 6 years?

10. COMPOUND INTEREST An initial deposit of $1,000 earns 6% interest, compounded twice a year. How much will be in the account in one year?

11. Graph the function $f(x) = e^x$.

12. CONTINUOUS COMPOUNDING An account contains $2,000 and has been earning 8% interest, compounded continuously. How much will be in the account in 10 years?

In Problems 13–16, find x.

13. $\log_4 16 = x$

14. $\log_x 81 = 4$

15. $\log_3 x = -3$

16. $\ln x = 1$

17. Write the statement $\log_6 \frac{1}{36} = -2$ in exponential form.

18. Give the domain and range of the function $f(x) = \log x$.

In Problems 19–20, graph each function.

19. $f(x) = -\log_3 x$

20. $f(x) = \ln x$

21. Write the expression $\log a^2bc^3$ in terms of the logarithms of a, b, and c.

22. Write the expression $\frac{1}{2} \ln (a + 2) + \ln b - 3 \ln c$ as a logarithm of a single quantity.

23. Use the change-of-base formula to find $\log_7 3$ to four decimal places.

24. What function is the inverse of $y = 10^x$?

25. pH Find the pH of a solution with a hydrogen ion concentration of 3.7×10^{-7}. (*Hint:* pH = $-\log [H^+]$.)

26. ELECTRONICS Find the db gain of an amplifier when $E_O = 60$ volts and $E_I = 0.3$ volt. (*Hint:* db gain = $20 \log \left(\dfrac{E_O}{E_I}\right)$.)

In Problems 27–30, solve each equation. Round to four decimal places when necessary.

27. $5^x = 3$

28. $3^{x-1} = 27$

29. $\ln (5x + 2) = \ln (2x + 5)$

30. $\log x + \log (x - 9) = 1$

31. Illustration 1 shows the graphs of $y = \frac{1}{2} \ln (x - 1)$ and $y = \ln 2$ and the approximate coordinates of their point of intersection. Estimate the solution of the logarithmic equation $\frac{1}{2} \ln (x - 1) = \ln 2$.

ILLUSTRATION 1

Intersection
X=5 /   Y=.69314718

32. Show a check of your answer to Problem 31.

33. Give an example of a situation studied in this chapter that is modeled by a function with a graph that has the shape shown in Illustration 2. Label the axes. You do not have to scale the axes.

ILLUSTRATION 2

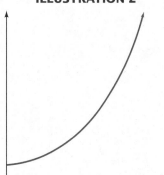

34. Consider the graph shown in Illustration 3.

    a. Is it the graph of a function?

    b. Is its inverse a function?

    c. What is $f^{-1}(260)$? What information does it give?

ILLUSTRATION 3

Relationship between car speed and tire temperature

10

Conic Sections;
More Graphing

CAMPUS CONNECTION

The Art Department

An art class wants to paint a mural on a wall at the local mall. After measuring the
wall, they decide to paint the mural on an elliptical background that is 10 feet wide
and 6 feet high. Fortunately, one of the students was studying conic sections in algebra class. Using her knowledge of conics, she was able to make this construction,
which is discussed on page 716.

In this chapter, we will discuss some graphs that do not represent functions. Since these curves are cross sections of a cone, they are called conic sections.

▶ 10.1   The Circle and the Parabola

In this section, you will learn about

Conic sections ■ The circle ■ Problem solving ■ The parabola ■ Problem solving

Introduction   We have seen that the graphs of linear functions are straight lines and that the graphs of quadratic functions are parabolas. In this section, we will discuss some special curves, called **conic sections**.

Conic Sections

The graphs of second-degree equations in x and y represent figures that were fully investigated in the 17th century by René Descartes (1596–1650) and Blaise Pascal (1623–1662). Descartes discovered that graphs of second-degree equations fall into one of several categories: a pair of lines, a point, a circle, a parabola, an ellipse, a hyperbola, or no graph at all. Because all of these graphs can be formed by the intersection of a plane and a right-circular cone, they are called **conic sections.** See Figure 10-1.

FIGURE 10-1

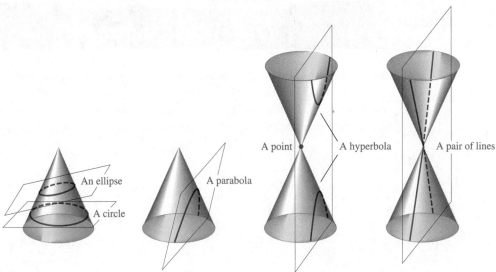

Conic sections have many applications. For example, everyone knows the importance of circular wheels and gears, pizza cutters, and Ferris wheels.

Parabolas can be rotated to generate dish-shaped surfaces called **paraboloids**. Any light or sound placed at the **focus** of a paraboloid is reflected outward in parallel paths, as shown in Figure 10-2(a). This property makes parabolic surfaces ideal for flashlight and headlight reflectors. It also makes parabolic surfaces good antennas, because signals captured by such antennas are concentrated at the focus. Parabolic mirrors are capable of concentrating the rays of the sun at a single point and thereby generating tremendous heat. This property is used in the design of certain solar furnaces.

Any object thrown upward and outward travels in a parabolic path, as shown in Figure 10-2(b). In architecture, many arches are parabolic in shape, because this gives strength. Cables that support suspension bridges hang in the form of a parabola. (See Figure 10-2(c).)

FIGURE 10-2

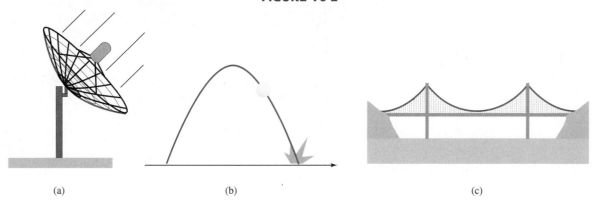

(a)   (b)   (c)

Ellipses have optical and acoustical properties that are useful in architecture and engineering. For example, many arches are portions of an ellipse, because the shape is pleasing to the eye. (See Figure 10-3(a).) The planets and many comets have elliptical orbits, as shown in Figure 10-3(b). Gears are often cut into elliptical shapes to provide nonuniform motion. (See Figure 10-3(c).)

FIGURE 10-3

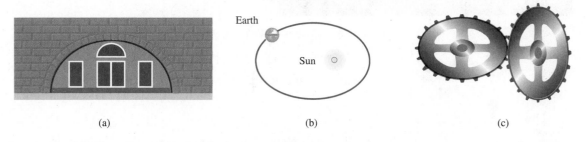

(a)   (b)   (c)

Hyperbolas serve as the basis of a navigational system known as LORAN (LOng RAnge Navigation). (See Figure 10-4.) They are also used to find the source of a distress signal, are the basis for the design of hypoid gears, and describe the orbits of some comets.

FIGURE 10-4

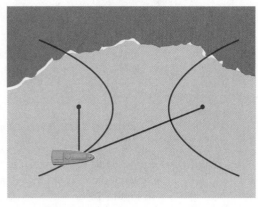

The Circle

The most common conic section is the circle.

The circle

A **circle** is the set of all points in a plane that are a fixed distance from a point called its **center.** The fixed distance is called the **radius** of the circle.

To develop the general equation of a circle, we will write the equation of a circle with a radius of r and with a center at some point $C(h, k)$, as in Figure 10-5. This task is equivalent to finding all points $P(x, y)$ such that the length of line segment CP is r. We can use the distance formula to find r.

$$r = \sqrt{(x - h)^2 + (y - k)^2}$$

We then square both sides to obtain

1. $r^2 = (x - h)^2 + (y - k)^2$

Equation 1 is called the **standard form of the equation of a circle** with a radius of r and center at the point with coordinates (h, k).

FIGURE 10-5

Standard equation of a circle with center at (h, k)

Any equation that can be written in the form

$$(x - h)^2 + (y - k)^2 = r^2$$

has a graph that is a circle with radius r and center at point (h, k).

If $r^2 = 0$, the graph reduces to a point called a **point circle.** If $r^2 < 0$, a circle does not exist. If both coordinates of the center are 0, the center of the circle is the origin.

Standard equation of a circle with center at (0, 0)

Any equation that can be written in the form

$$x^2 + y^2 = r^2$$

has a graph that is a circle with radius r and center at the origin.

E X A M P L E   1

Graphing circles. Graph $x^2 + y^2 = 25$.

Solution   Because this equation can be written in the form $x^2 + y^2 = r^2$, its graph is a circle with center at the origin. Since $r^2 = 25 = 5^2$, the circle has a radius of 5. The graph appears in Figure 10-6.

FIGURE 10-6

$x^2 + y^2 = 25$

x	y	(x, y)
-5	0	$(-5, 0)$
-4	± 3	$(-4, \pm 3)$
-3	± 4	$(-3, \pm 4)$
0	± 5	$(0, \pm 5)$
3	± 4	$(3, \pm 4)$
4	± 3	$(4, \pm 3)$
5	0	$(5, 0)$

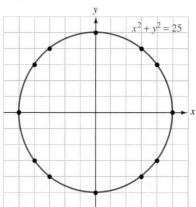

SELF CHECK   Graph $x^2 + y^2 = 4$.

Answer:

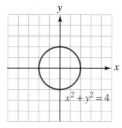

EXAMPLE 2

Finding equations of circles. Find the equation of the circle with radius 5 and center at $(3, 2)$.

Solution   We substitute 5 for r, 3 for h, and 2 for k in standard form and simplify.

$$(x - h)^2 + (y - k)^2 = r^2$$
$$(x - 3)^2 + (y - 2)^2 = 5^2$$
$$x^2 - 6x + 9 + y^2 - 4y + 4 = 25$$
$$x^2 + y^2 - 6x - 4y - 12 = 0$$

The equation is $x^2 + y^2 - 6x - 4y - 12 = 0$.

SELF CHECK   Find the equation of the circle with radius 6 and center at $(2, 3)$.

Answer: $x^2 + y^2 - 4x - 6y - 23 = 0$ ∎

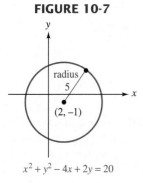

EXAMPLE 3

Graphing circles. Graph the equation $x^2 + y^2 - 4x + 2y = 20$.

Solution   To identify the curve, we complete the square on x and y and write the equation in standard form.

FIGURE 10-7

$$x^2 + y^2 - 4x + 2y = 20$$
$$x^2 - 4x + y^2 + 2y = 20$$

To complete the square on x and y, add 4 and 1 to both sides.

$$x^2 - 4x + \mathbf{4} + y^2 + 2y + \mathbf{1} = 20 + \mathbf{4} + \mathbf{1}$$
$$(x - 2)^2 + (y + 1)^2 = 25 \qquad \text{Factor } x^2 - 4x + 4 \text{ and } y^2 + 2y + 1.$$
$$(x - 2)^2 + [y - (-1)]^2 = 5^2$$

We can now see that this result is the equation of a circle with a radius of 5 and center at $h = 2$ and $k = -1$. If we plot the center and draw a circle with a radius of 5 units, we will obtain the circle shown in Figure 10-7.

$x^2 + y^2 - 4x + 2y = 20$

SELF CHECK   Write $x^2 + y^2 + 2x - 4y - 11 = 0$ in standard form and graph it.

Answer: $(x + 1)^2 + (y - 2)^2 = 16$

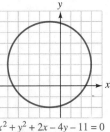

$x^2 + y^2 + 2x - 4y - 11 = 0$ ∎

ACCENT ON TECHNOLOGY *Graphing Circles*

Since the graphs of circles fail the vertical line test, their equations do not represent functions. It is somewhat more difficult to use a graphing calculator to graph equations that are not functions. For example, to graph the circle described by $(x - 1)^2 + (y - 2)^2 = 4$, we must split the equation into two functions and graph each one separately. We begin by solving the equation for y.

$$(x - 1)^2 + (y - 2)^2 = 4$$
$$(y - 2)^2 = 4 - (x - 1)^2 \qquad \text{Subtract } (x - 1)^2 \text{ from both sides.}$$
$$y - 2 = \pm \sqrt{4 - (x - 1)^2} \qquad \text{Take the square root of both sides.}$$
$$y = 2 \pm \sqrt{4 - (x - 1)^2} \qquad \text{Add 2 to both sides.}$$

This equation defines two functions. If we use window settings of $[-3, 5]$ for x and $[-3, 5]$ for y and graph the functions

$$y = 2 + \sqrt{4 - (x - 1)^2} \qquad \text{and} \qquad y = 2 - \sqrt{4 - (x - 1)^2}$$

we get the distorted circle shown in Figure 10-8(a). To get a better circle, we can use the graphing calculator's squaring feature, which gives an equal unit distance on both the x- and y-axes. Using this feature, we get the circle shown in Figure 10-8(b). Sometimes the two arcs will not join because of approximations made by the calculator at each endpoint.

FIGURE 10-8

(a)   (b)

Problem Solving

EXAMPLE 4

Radio translators. The broadcast area of a television station is bounded by the circle $x^2 + y^2 = 3,600$, where x and y are measured in miles. A translator station picks up the signal and retransmits it from the center of a circular area bounded by $(x + 30)^2 + (y - 40)^2 = 1,600$. Find the location of the translator and the greatest distance from the main transmitter that the signal can be received.

Solution   The coverage of the television station is bounded by $x^2 + y^2 = 60^2$, a circle centered at the origin with a radius of 60 miles, as shown in yellow in Figure 10-9. Because the translator is at the center of the circle $(x + 30)^2 + (y - 40)^2 = 1,600$, it is located at $(-30, 40)$, a point 30 miles west and 40 miles north of the television station. The radius of the translator's coverage is $\sqrt{1,600}$, or 40 miles.

As shown in Figure 10-9, the greatest distance of reception is the sum of A, the distance from the translator to the television station, and 40 miles, the radius of the translator's coverage.

To find A, we use the distance formula to find the distance between $(x_1, y_1) = (-30, 40)$ and the origin, $(x_2, y_2) = (0, 0)$.

FIGURE 10-9

$$A = \sqrt{(x_1 - x_2)^2 + (y_1 - y_2)^2}$$
$$A = \sqrt{(-30 - 0)^2 + (40 - 0)^2}$$
$$= \sqrt{(-30)^2 + 40^2}$$
$$= \sqrt{900 + 1,600}$$
$$= \sqrt{2,500}$$
$$= 50$$

The translator is located 50 miles from the television station, and it broadcasts the signal an additional 40 miles. The greatest reception distance is $50 + 40$, or 90 miles.

The Parabola

We have seen that equations of the form $y = a(x - h)^2 + k$ $(a \neq 0)$ represent parabolas with vertex at point (h, k). They open upward when $a > 0$ and downward when $a < 0$.

Equations of the form $x = a(y - k)^2 + h$ also represent parabolas with vertex at point (h, k). However, they open to the right when $a > 0$ and to the left when $a < 0$. Parabolas that open to the right or left do not represent functions, because their graphs fail the vertical line test.

Several types of parabolas are summarized in the following chart.

Equations of parabolas ($a > 0$)

Parabola opening	Vertex at origin	Vertex at (h, k)
Up	$y = ax^2$	$y = a(x - h)^2 + k$
Down	$y = -ax^2$	$y = -a(x - h)^2 + k$
Right	$x = ay^2$	$x = a(y - k)^2 + h$
Left	$x = -ay^2$	$x = -a(y - k)^2 + h$

E X A M P L E   5   **Graphing parabolas.** Graph   **a.** $x = \frac{1}{2}y^2$ and   **b.** $x = -2(y - 2)^2 + 3$.

Solution   **a.** We make a table of ordered pairs that satisfy the equation, plot each pair, and draw the parabola, as in Figure 10-10(a). Because the equation is of the form $x = ay^2$ with $a = \frac{1}{2} > 0$, the parabola opens to the right and has its vertex at the origin.

b. We make a table of ordered pairs that satisfy the equation, plot each pair, and draw the parabola, as in Figure 10-10(b). Because the equation is of the form $x = -a(y - k)^2 + h$, the parabola opens to the left and has its vertex at the point with coordinates $(3, 2)$.

FIGURE 10-10

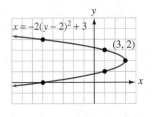

x	y	(x, y)
0	0	(0, 0)
2	2	(2, 2)
2	−2	(2, −2)
8	4	(8, 4)
8	−4	(8, −4)

x	y	(x, y)
−5	0	(−5, 0)
1	1	(1, 1)
3	2	(3, 2)
1	3	(1, 3)
−5	4	(−5, 4)

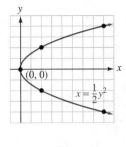

(a)

(b)

SELF CHECK   Graph $x = \frac{1}{2}(y - 1)^2 - 2$.   *Answer:*

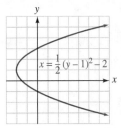

EXAMPLE 6

Graphing parabolas. Graph $y = -2x^2 + 12x - 15$.

Solution   Because the equation is not in standard form, the coordinates of the vertex are not obvious. To write the equation in standard form, we complete the square on x.

$$y = -2x^2 + 12x - 15$$
$$y = -2(x^2 - 6x) - 15 \qquad \text{Factor out } -2 \text{ from } -2x^2 + 12x.$$
$$y = -2(x^2 - 6x + 9) - 15 + 18 \quad \text{Subtract and add 18; } -2(9) = -18.$$
$$y = -2(x - 3)^2 + 3$$

Since the equation is written in the form $y = -a(x - h)^2 + k$, we can see that the parabola opens downward and has its vertex at (3, 3). The graph of the function is shown in Figure 10-11.

FIGURE 10-11

$$y = -2x^2 + 12x - 15$$

x	y	(x, y)
1	−5	(1, −5)
2	1	(2, 1)
3	3	(3, 3)
4	1	(4, 1)
5	−5	(5, −5)

$$y = -2x^2 + 12x - 15$$

SELF CHECK   Graph $y = 0.5x^2 - x - 1$.

Answer:

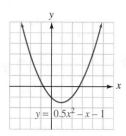

$y = 0.5x^2 - x - 1$

■

Problem Solving

EXAMPLE 7

Gateway Arch. The shape of the Gateway Arch in St. Louis is approximately a parabola, as shown in Figure 10-12(a). How high is the arch 100 feet from its foundation?

FIGURE 10-12

630 ft

630 ft

(a)

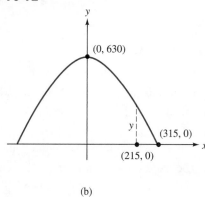

(0, 630)

(315, 0)

(215, 0)

(b)

Solution   We place the parabola on a coordinate system as in Figure 10-12(b), with ground level on the x-axis and the vertex of the parabola at the point $(h, k) = (0, 630)$. The equation of this downward-opening parabola has the form

$y = -a(x - h)^2 + k$

$y = -a(x - 0)^2 + \mathbf{630}$ Substitute 0 for h and 630 for k.

$y = -ax^2 + 630$ Simplify.

Because the Gateway Arch is 630 feet wide at its base, the parabola passes through the point $\left(\frac{630}{2}, 0\right)$, or (315, 0). To find a in the equation of the parabola, we proceed as follows:

$y = -ax^2 + 630$

$0 = -a \cdot \mathbf{315}^2 + \mathbf{630}$ Substitute 315 for x and 0 for y.

$\dfrac{-630}{315^2} = -a$ Subtract 630 from both sides and divide both sides by 315^2.

$\dfrac{2}{315} = a$ Multiply both sides by -1; $\dfrac{630}{315^2} = \dfrac{2}{315}$.

The equation of the parabola that approximates the shape of the Gateway Arch is

$$y = -\frac{2}{315}x^2 + 630$$

To find the height of the arch at a point 100 feet from its foundation, we substitute $315 - 100$, or 215, for x in the equation of the parabola and solve for y.

$$y = -\frac{2}{315}x^2 + 630$$

$$y = -\frac{2}{315}(\mathbf{215})^2 + 630$$

$$= 336.5079365 \qquad \text{Use a calculator.}$$

At a point 100 feet from the foundation, the height of the arch is about 337 feet.   ■

STUDY SET

Section 10.1

VOCABULARY

Fill in the blanks to make the statements true.

1. Intersections of a plane and a right-circular cone are called _____ sections.

2. A _____ is the set of all points in a plane that are a fixed distance from a point.

3. The fixed distance in Exercise 2 is called the _____ of the circle, and the point is called its _____.

4. Parabolas can be rotated to form dish-shaped surfaces called _____.

CONCEPTS

Fill in the blanks to make the statements true.

5. If _____ in the equation $x^2 + y^2 = r^2$, no circle exists.

6. The graph of $y = ax^2$ is a _____ with vertex at the _____ that opens _____.

7. The graph of $x = a(y - 2)^2 + 3$ is a _____ that has its vertex at _____ and opens to the _____.

8. The graph of $x = -a(y - 1)^2 - 3$ is a _____ that has its vertex at _____ and opens to the _____.

9. The standard equation of a circle with center at (h, k) is _____.

10. The standard equation of a circle with center at $(0, 0)$ is _____.

11. The standard equation of a parabola $(a > 0)$ that opens downward and has its vertex at the origin is _____.

12. The standard equation of a parabola $(a > 0)$ that opens to the left and has its vertex at (h, k) is _____.

NOTATION

Complete each solution.

13. Find the equation of a circle with radius 6 and center at $(2, 3)$.
$$(x - h)^2 + (y - k)^2 = r^2$$
$$\left(x - \boxed{}\right)^2 + (y - 3)^2 = \boxed{}^2$$
$$x^2 - \boxed{}x + 4 + y^2 - 6y + 9 = \boxed{}$$
$$x^2 + y^2 - 4x - 6y - 23 = 0$$

14. Write the equation of a parabola with vertex at $(2, 3)$, opening to the left, and with $a = 2$.
$$x = -a(y - k)^2 + h$$
$$x = -\boxed{}\left(y - \boxed{}\right) + 2$$
$$x = -2y + \boxed{} + 2$$
$$x = -2y + 8$$

PRACTICE

In Exercises 15–24, graph each equation.

15. $x^2 + y^2 = 9$

16. $x^2 + y^2 = 16$

17. $(x - 2)^2 + y^2 = 9$

18. $x^2 + (y - 3)^2 = 4$

32. Center at $(5, 3)$; radius 2
33. Center at $(-2, 6)$; radius 12
34. Center at $(5, -4)$; radius 6
35. Center at the origin; diameter $4\sqrt{2}$
36. Center at the origin; diameter $8\sqrt{3}$

19. $(x - 2)^2 + (y - 4)^2 = 4$

20. $(x - 3)^2 + (y - 2)^2 = 4$

In Exercises 37–44, graph each circle. Give the coordinates of the center.

37. $x^2 + y^2 + 2x - 8 = 0$ **38.** $x^2 + y^2 - 4y = 12$

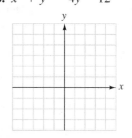

21. $(x + 3)^2 + (y - 1)^2 = 16$ **22.** $(x - 1)^2 + (y + 4)^2 = 9$

39. $9x^2 + 9y^2 - 12y = 5$ **40.** $4x^2 + 4y^2 + 4y = 15$

23. $x^2 + (y + 3)^2 = 1$

24. $(x + 4)^2 + y^2 = 1$

41. $x^2 + y^2 - 2x + 4y = -1$

In Exercises 25–28, use a graphing calculator to graph each equation.

25. $3x^2 + 3y^2 = 16$ **26.** $2x^2 + 2y^2 = 9$

42. $x^2 + y^2 + 4x + 2y = 4$

27. $(x + 1)^2 + y^2 = 16$ **28.** $x^2 + (y - 2)^2 = 4$

43. $x^2 + y^2 + 6x - 4y = -12$

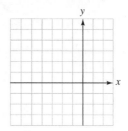

In Exercises 29–36, write the equation of the circle with the following properties.

29. Center at the origin; radius 1
30. Center at the origin; radius 4
31. Center at $(6, 8)$; radius 5

44. $x^2 + y^2 + 8x + 2y = -13$

In Exercises 45–56, find the vertex of each parabola and graph it.

45. $x = y^2$

46. $x = -y^2 + 1$

47. $x = -\dfrac{1}{4}y^2$

48. $x = 4y^2$

49. $y = x^2 + 4x + 5$

50. $y = -x^2 - 2x + 3$

51. $y = -x^2 - x + 1$

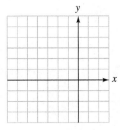

52. $x = \dfrac{1}{2}y^2 + 2y$

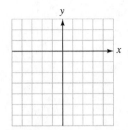

53. $y^2 + 4x - 6y = -1$

54. $x^2 - 2y - 2x = -7$

55. $y = 2(x - 1)^2 + 3$

56. $y = -2(x + 1)^2 + 2$

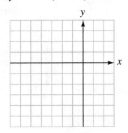

In Exercises 57–60, use a graphing calculator to graph each equation.

57. $x = 2y^2$

58. $x = y^2 - 4$

59. $x^2 - 2x + y = 6$

60. $x = -2(y - 1)^2 + 2$

APPLICATIONS

61. MESHING GEARS For design purposes, the large gear in Illustration 1 is the circle $x^2 + y^2 = 16$. The smaller gear is a circle centered at $(7, 0)$ and tangent to the larger circle. Find the equation of the smaller gear.

ILLUSTRATION 1

62. WIDTH OF A WALKWAY The walkway in Illustration 2 is bounded by the two circles $x^2 + y^2 = 2{,}500$ and $(x - 10)^2 + y^2 = 900$, measured in feet. Find the largest and the smallest width of the walkway.

ILLUSTRATION 2

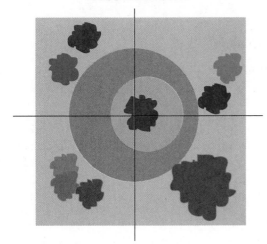

63. BROADCAST RANGES Radio stations applying for licensing may not use the same frequency if their broadcast areas overlap. One station's coverage is bounded by $x^2 + y^2 - 8x - 20y + 16 = 0$, and the other's by $x^2 + y^2 + 2x + 4y - 11 = 0$. May they be licensed for the same frequency?

64. HIGHWAY DESIGN Engineers want to join two sections of highway with a curve that is one-quarter of a circle, as in Illustration 3. The equation of the circle is $x^2 + y^2 - 16x - 20y + 155 = 0$, where distances are measured in kilometers. Find the locations (relative to the center of town) of the intersections of the highway with State and with Main.

ILLUSTRATION 3

65. FLIGHT OF A PROJECTILE The cannonball in Illustration 4 follows the parabolic trajectory $y = 30x - x^2$. Where does it land?

ILLUSTRATION 4

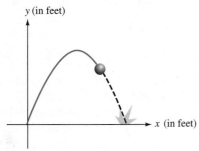

66. FLIGHT OF A PROJECTILE In Exercise 65, how high does the cannonball get?

67. ORBIT OF A COMET If the orbit of the comet shown in Illustration 5 is approximated by the equation $2y^2 - 9x = 18$, how far is it from the sun at the vertex of the orbit? Distances are measured in astronomical units (AU).

ILLUSTRATION 5

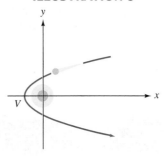

68. SATELLITE ANTENNA The cross section of the satellite antenna in Illustration 6 is a parabola given by the equation $y = \frac{1}{16}x^2$, with distances measured in feet. If the dish is 8 feet wide, how deep is it?

ILLUSTRATION 6

WRITING

69. Explain how to decide from its equation whether the graph of a parabola opens up, down, right, or left.

70. From the equation of a circle, explain how to determine the radius and the coordinates of the center.

71. On the day of an election, the following warning was posted in front of a school. Explain what it means.

> *No electioneering within a 1,000-foot radius of this polling place.*

72. What is meant by the *turning radius* of a truck?

REVIEW

Solve each equation.

73. $|3x - 4| = 11$ **74.** $\left| \dfrac{4 - 3x}{5} \right| = 12$

75. $|3x + 4| = |5x - 2|$ **76.** $|6 - 4x| = |x + 2|$

▶ 10.2

The Ellipse

In this section, you will learn about

The ellipse ■ Constructing an ellipse ■ Graphing ellipses
■ Problem solving

Introduction   A third conic section is an oval-shaped curve called an *ellipse*. Ellipses can be nearly "round" and look almost like a circle, or they can be long and narrow. In this section, we will learn how to construct ellipses and how to graph equations that represent ellipses.

The Ellipse

The ellipse

An **ellipse** is the set of all points P in a plane the sum of whose distances from two fixed points is a constant. See Figure 10-13, in which $d_1 + d_2$ is a constant.

Each of the two points is called a **focus**. Midway between the foci is the **center** of the ellipse.

FIGURE 10-13

Constructing an Ellipse

We can construct an ellipse by placing two thumbtacks fairly close together, as in Figure 10-14. We then tie each end of a piece of string to a thumbtack, catch the loop with the point of a pencil, and (keeping the string taut) draw the ellipse.

FIGURE 10-14

Using this method, we can construct an ellipse of any specific size. For example, to construct an ellipse that is 10 feet wide and 6 feet high, we must find the length of string to use and the distance between the thumbtacks.

To do this, we will let a represent the distance between the center and vertex V, as shown in Figure 10-15(a). We will also let c represent the distance between the center of the ellipse and either focus. When the pencil is at vertex V, the length of the string is $c + a + (a - c)$, or just $2a$. Because $2a$ is the 10-foot width of the ellipse, the string needs to be 10 feet long. The distance $2a$ is constant for any point on the ellipse, including point B shown in Figure 10-15(b).

FIGURE 10-15

(a)   (b)

From right triangle CBF' and the Pythagorean theorem, we can find c as follows:

$$a^2 = b^2 + c^2 \quad \text{or} \quad c = \sqrt{a^2 - b^2}$$

Since distance b is one-half of the height of the ellipse, $b = 3$. Since $2a = 10$, $a = 5$. We can now substitute $a = 5$ and $b = 3$ into the formula to find c:

$$
\begin{aligned}
c &= \sqrt{5^2 - 3^2} \\
&= \sqrt{25 - 9} \\
&= \sqrt{16} \\
&= 4
\end{aligned}
$$

Since $c = 4$, the distance between the thumbtacks must be 8 feet. We can construct the ellipse by tying a 10-foot string to thumbtacks that are 8 feet apart.

Graphing Ellipses

To graph ellipses, we can make a table of ordered pairs that satisfy the equation, plot them, and join the points with a smooth curve.

EXAMPLE 1

Graphing ellipses. Graph $\dfrac{x^2}{36} + \dfrac{y^2}{9} = 1$.

Solution First we note that the equation can be written in the form

$$\frac{x^2}{6^2} + \frac{y^2}{3^2} = 1 \quad 36 = 6^2 \text{ and } 9 = 3^2.$$

After making a table of ordered pairs that satisfy the equation, plotting each of them, and joining the points with a curve, we obtain the ellipse shown in Figure 10-16.

We note that the center of the ellipse is the origin, the ellipse intersects the x-axis at points $(6, 0)$ and $(-6, 0)$, and the ellipse intersects the y-axis at points $(0, 3)$ and $(0, -3)$.

FIGURE 10-16

$$\frac{x^2}{36} + \frac{y^2}{9} = 25$$

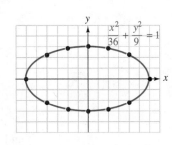

x	y	(x, y)
-6	0	$(-6, 0)$
-4	± 2.2	$(-4, \pm 2.2)$
-2	± 2.8	$(-2, \pm 2.8)$
0	± 3	$(0, \pm 3)$
2	± 2.8	$(2, \pm 2.8)$
4	± 2.2	$(4, \pm 2.2)$
6	0	$(6, 0)$

SELF CHECK Graph $\dfrac{x^2}{4} + \dfrac{y^2}{16} = 1$.

Answer:

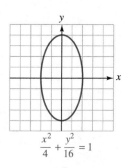

$\dfrac{x^2}{4} + \dfrac{y^2}{16} = 1$

Example 1 illustrates that the graph of

$$\frac{x^2}{a^2} + \frac{y^2}{b^2} = 1$$

is an ellipse centered at the origin. To find the x-intercepts of the graph, we can let $y = 0$ and solve for x.

$$\frac{x^2}{a^2} + \frac{0^2}{b^2} = 1 \qquad \text{Substitute 0 for } y.$$

$$\frac{x^2}{a^2} + 0 = 1 \qquad \frac{0^2}{b^2} = 0.$$

$$x^2 = a^2 \qquad \text{Simplify and multiply both sides by } a^2.$$

$$x = a \quad \text{or} \quad x = -a$$

The x-intercepts are $(a, 0)$ and $(-a, 0)$.

To find the y-intercepts of the graph, we can let $x = 0$ and solve for y.

$$\frac{0^2}{a^2} + \frac{y^2}{b^2} = 1$$

$$0 + \frac{y^2}{b^2} = 1$$

$$y^2 = b^2$$

$$y = b \quad \text{or} \quad y = -b$$

The y-intercepts are $(0, b)$ and $(0, -b)$.

In general, we have the following results.

Equations of an ellipse centered at the origin

The equation of an ellipse centered at the origin with x-intercepts at $V_1(a, 0)$ and $V_2(-a, 0)$ and with y-intercepts at $(0, b)$ and $(0, -b)$ is

$$\frac{x^2}{a^2} + \frac{y^2}{b^2} = 1 \quad (a > b > 0) \quad \text{See Figure 10-17(a).}$$

The equation of an ellipse centered at the origin with y-intercepts at $V_1(0, a)$ and $V_2(0, -a)$ and with x-intercepts at $(b, 0)$ and $(-b, 0)$ is

$$\frac{x^2}{b^2} + \frac{y^2}{a^2} = 1 \quad (a > b > 0) \quad \text{See Figure 10-17(b).}$$

In Figure 10-17, the points V_1 and V_2 are the **vertices** of the ellipse, the midpoint of V_1V_2 is the **center** of the ellipse, and the distance between the center of the ellipse and either vertex is a. The segment V_1V_2 is called the **major axis**, and the segment joining either $(0, b)$ and $(0, -b)$ or $(b, 0)$ and $(-b, 0)$ is called the **minor axis**.

FIGURE 10-17

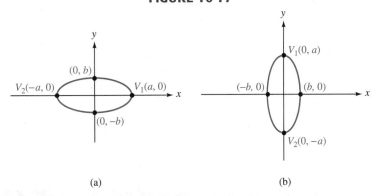

(a)   (b)

The equations for ellipses centered at (h, k) are as follows.

Equations of an ellipse centered at (h, k)

The equation of an ellipse centered at (h, k), with major axis parallel to the x-axis is

1. $\dfrac{(x - h)^2}{a^2} + \dfrac{(y - k)^2}{b^2} = 1$ $(a > b > 0)$

The equation of an ellipse centered at (h, k), with major axis parallel to the y-axis is

2. $\dfrac{(x - h)^2}{b^2} + \dfrac{(y - k)^2}{a^2} = 1$ $(a > b > 0)$

EXAMPLE 2

Graphing ellipses. Graph the ellipse $\dfrac{(x - 2)^2}{16} + \dfrac{(y + 3)^2}{25} = 1$.

Solution   We write the equation in the form

$$\frac{(x - 2)^2}{4^2} + \frac{[y - (-3)]^2}{5^2} = 1 \quad (5 > 4)$$

This is the equation of an ellipse centered at $(h, k) = (2, -3)$, with major axis parallel to the y-axis and with $b = 4$ and $a = 5$. We first plot the center, as shown in Figure 10-18. Since a is the distance from the center to the vertex, we can locate the vertices by counting 5 units above and 5 units below the center. The vertices are at points $(2, 2)$ and $(2, -8)$.

Since $b = 4$, we can locate two more points on the ellipse by counting 4 units to the left and 4 units to the right of the center. The points $(-2, -3)$ and $(6, -3)$ are also on the graph.

Using these four points as guides, we can draw the ellipse.

FIGURE 10-18

$$\frac{(x - 2)^2}{16} + \frac{(y + 3)^2}{25} = 1$$

x	y	(x, y)
2	2	$(2, 2)$
2	-8	$(2, -8)$
6	-3	$(6, -3)$
-2	-3	$(-2, -3)$

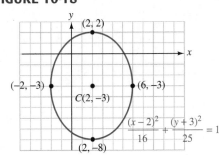

SELF CHECK   Graph $\dfrac{(x-1)^2}{9} + \dfrac{(y+2)^2}{16} = 1$.   *Answer:*

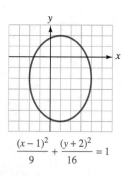

$$\dfrac{(x-1)^2}{9} + \dfrac{(y+2)^2}{16} = 1$$

■

ACCENT ON TECHNOLOGY *Graphing Ellipses*

To use a graphing calculator to graph

$$\dfrac{(x+2)^2}{4} + \dfrac{(y-1)^2}{25} = 1$$

we first clear the equation of fractions by multiplying both sides by 100 and solving for y.

$$25(x+2)^2 + 4(y-1)^2 = 100 \qquad \text{Multiply both sides by 100.}$$

$$4(y-1)^2 = 100 - 25(x+2)^2 \qquad \begin{array}{l}\text{Subtract } 25(x+2)^2 \text{ from}\\ \text{both sides.}\end{array}$$

$$(y-1)^2 = \dfrac{100 - 25(x+2)^2}{4} \qquad \text{Divide both sides by 4.}$$

$$y - 1 = \pm\dfrac{\sqrt{100 - 25(x+2)^2}}{2} \qquad \begin{array}{l}\text{Take the square root of}\\ \text{both sides.}\end{array}$$

$$y = 1 \pm\dfrac{\sqrt{100 - 25(x+2)^2}}{2} \qquad \text{Add 1 to both sides.}$$

If we use window settings $[-6, 6]$ for x and $[-6, 6]$ for y and graph the functions

$$y = 1 + \dfrac{\sqrt{100 - 25(x+2)^2}}{2} \qquad \text{and} \qquad y = 1 - \dfrac{\sqrt{100 - 25(x+2)^2}}{2}$$

we will obtain the ellipse shown in Figure 10-19.

FIGURE 10-19

We can complete the square to write the equations of many ellipses in standard form.

E X A M P L E   3

Writing an equation in standard form. Write $4x^2 + 9y^2 - 16x - 18y = 11$ in standard form to show that the equation represents an ellipse.

Solution   We write the equation in standard form by completing the square on x and y:

$$4x^2 + 9y^2 - 16x - 18y = 11$$

$$4x^2 - 16x + 9y^2 - 18y = 11 \qquad \text{Use the commutative property to rearrange terms.}$$

$$4(x^2 - 4x) + 9(y^2 - 2y) = 11 \qquad \text{Factor 4 from } 4x^2 - 16x \text{ and factor 9 from } 9y^2 - 18y \text{ to get coefficients of 1 for the squared terms.}$$

$$4(x^2 - 4x + 4) + 9(y^2 - 2y + 1) = 11 + \mathbf{16} + \mathbf{9} \qquad \text{Complete the square to make } x^2 - 4x \text{ and } y^2 - 2y \text{ perfect square trinomials. Since } 16 + 9 \text{ is added to the left-hand side, add } 16 + 9 \text{ to the right-hand side.}$$

$$4(x - 2)^2 + 9(y - 1)^2 = 36 \qquad \text{Factor } x^2 - 4x + 4 \text{ and } y^2 - 2y + 1.$$

$$\frac{(x - 2)^2}{9} + \frac{(y - 1)^2}{4} = 1 \qquad \text{Divide both sides by 36 and simplify.}$$

FIGURE 10-20

Since this equation matches Equation 1 on page 719, it represents an ellipse. Its graph is shown in Figure 10-20.

SELF CHECK   Graph $4x^2 - 8x + 9y^2 - 36y = -4$.   *Answer:*

Problem Solving

E X A M P L E   4

Landscape design. A landscape architect is designing an elliptical pool that will fit in the center of a 20-by-30-foot rectangular garden, leaving at least 5 feet of space on all sides. Find the equation of the ellipse.

Solution   We place the rectangular garden in a coordinate system, as in Figure 10-21 on the next page. To maintain 5 feet of clearance at the ends of the ellipse, the vertices must be the points $V_1(10, 0)$ and $V_2(-10, 0)$. Similarly, the y-intercepts are the points $(0, 5)$ and $(0, -5)$.

The equation of the ellipse has the form

$$\frac{x^2}{a^2} + \frac{y^2}{b^2} = 1$$

with $a = 10$ and $b = 5$. Thus, the equation of the boundary of the pool is

$$\frac{x^2}{100} + \frac{y^2}{25} = 1$$

FIGURE 10-21

STUDY SET

Section 10.2

VOCABULARY

Fill in the blanks to make the statements true.

1. An _____ is the set of all points in a plane the _____ of whose distances from two fixed points is a constant.

2. The fixed points in Exercise 1 are the _____ of the ellipse.

3. The midpoint of the line segment joining the foci of an ellipse is called the _____ of the ellipse.

4. The segment joining the vertices of an ellipse is called the _____ axis.

CONCEPTS

Fill in the blanks to make the statements true.

5. The elliptical graph of $\frac{x^2}{a^2} + \frac{y^2}{b^2} = 1$ has x-intercepts of _____.

6. The elliptical graph of $\frac{x^2}{a^2} + \frac{y^2}{b^2} = 1$ has y-intercepts of _____.

7. The center of the ellipse with an equation of

$$\frac{x^2}{a^2} + \frac{y^2}{b^2} = 1$$

is the point _____.

8. The center of the ellipse with an equation of

$$\frac{(x-h)^2}{a^2} + \frac{(y-k)^2}{b^2} = 1$$

is the point _____.

NOTATION

Complete each solution.

9. Write $16x^2 + 9y^2 - 144 = 0$ in standard form.

$$16x^2 + 9y^2 - 144 = 0$$
$$16x^2 + 9y^2 = \boxed{}$$
$$\frac{x^2}{9} + \frac{y^2}{\boxed{}} = 1$$

10. Write $9x^2 + 4y^2 - 18x + 8y = 23$ in standard form.

$$9x^2 + 4y^2 - 18x + 8y = 23$$
$$9x^2 - 18x + \boxed{}y^2 + 8y = 23$$
$$9\left(x^2 - \boxed{}x\right) + 4\left(y^2 + \boxed{}y\right) = 23$$
$$9\left(x^2 - 2x + \boxed{}\right) + 4(y^2 + 2y + 1) = 23 + 9 + 4$$
$$9(x-1)^2 + 4(y+1)^2 = \boxed{}$$
$$\frac{(x-1)^2}{4} + \frac{(y+1)^2}{9} = 1$$

PRACTICE

In Exercises 11–20, graph each equation.

11. $\dfrac{x^2}{4} + \dfrac{y^2}{9} = 1$

12. $x^2 + \dfrac{y^2}{9} = 1$

13. $x^2 + 9y^2 = 9$

14. $25x^2 + 9y^2 = 225$

15. $16x^2 + 4y^2 = 64$

16. $4x^2 + 9y^2 = 36$

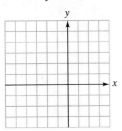

17. $\dfrac{(x - 2)^2}{9} + \dfrac{(y - 1)^2}{4} = 1$

18. $\dfrac{(x - 1)^2}{9} + \dfrac{(y - 3)^2}{4} = 1$

19. $(x + 1)^2 + 4(y + 2)^2 = 4$

20. $25(x + 1)^2 + 9y^2 = 225$

In Exercises 21–24, use a graphing calculator to graph each equation.

21. $\dfrac{x^2}{9} + \dfrac{y^2}{4} = 1$

22. $x^2 + 16y^2 = 16$

23. $\dfrac{x^2}{4} + \dfrac{(y - 1)^2}{9} = 1$

24. $\dfrac{(x + 1)^2}{9} + \dfrac{(y - 2)^2}{4} = 1$

In Exercises 25–28, write each equation in standard form and graph it.

25. $x^2 + 4y^2 - 4x + 8y + 4 = 0$

26. $x^2 + 4y^2 - 2x - 16y = -13$

27. $9x^2 + 4y^2 - 18x + 16y = 11$

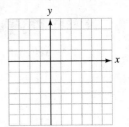

28. $16x^2 + 25y^2 - 160x - 200y + 400 = 0$

APPLICATIONS

29. DESIGNING AN UNDERPASS The arch of the underpass in Illustration 1 is a part of an ellipse. Find the equation of the arch.

ILLUSTRATION 1

10 ft

40 ft

30. CALCULATING CLEARANCE Find the height of the elliptical arch in Exercise 29 at a point 10 feet from the center of the roadway.

31. AREA OF AN ELLIPSE The area A of the ellipse

$$\frac{x^2}{a^2} + \frac{y^2}{b^2} = 1$$

is given by $A = \pi ab$. Find the area of the ellipse $9x^2 + 16y^2 = 144$.

32. AREA OF A TRACK The elliptical track in Illustration 2 is bounded by the ellipses $4x^2 + 9y^2 = 576$ and $9x^2 + 25y^2 = 900$. Find the area of the track. (See Exercise 31.)

ILLUSTRATION 2

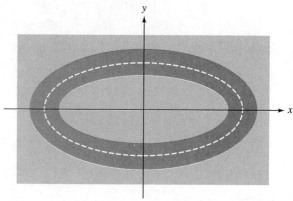

WRITING

33. Explain how to find the x- and y-intercepts of the graph of the ellipse $\frac{x^2}{a^2} + \frac{y^2}{b^2} = 1$.

34. Explain the relationship between the center, focus, and vertex of an ellipse.

REVIEW

Find each product.

35. $3x^{-2}y^2(4x^2 + 3y^{-2})$

36. $(2a^{-2} - b^{-2})(2a^{-2} + b^{-2})$

Write each expression without using negative exponents.

37. $\dfrac{x^{-2} + y^{-2}}{x^{-2} - y^{-2}}$

38. $\dfrac{2x^{-3} - 2y^{-3}}{4x^{-3} + 4y^{-3}}$

▶ 10.3   The Hyperbola

In this section, you will learn about

The hyperbola ■ Problem solving

Introduction   The final conic section, the *hyperbola*, is a curve with two branches. In this section, we will learn how to graph equations that represent hyperbolas.

The Hyperbola

The hyperbola

A **hyperbola** is the set of all points P in a plane for which the difference of the distances of each point from two fixed points is a constant. See Figure 10-22, in which $d_1 - d_2$ is a constant.

Each of the two points is called a **focus**. Midway between the foci is the **center** of the hyperbola.

FIGURE 10-22

The graph of the equation

$$\frac{x^2}{25} - \frac{y^2}{9} = 1$$

is a hyperbola. To graph the equation, we make a table of ordered pairs that satisfy the equation, plot each pair, and join the points with a smooth curve, as in Figure 10-23.

FIGURE 10-23

$$\frac{x^2}{25} - \frac{y^2}{9} = 1$$

x	y	(x, y)
-7	± 2.9	$(-7, \pm 2.9)$
-6	± 2.0	$(-6, \pm 2.0)$
-5	0	$(-5, 0)$
5	0	$(5, 0)$
6	± 2.0	$(6, \pm 2.0)$
7	± 2.9	$(7, \pm 2.9)$

This graph is centered at the origin and intersects the x-axis at $\left(\sqrt{25}, 0\right)$ and $\left(-\sqrt{25}, 0\right)$. After simplifying, these points are $(5, 0)$ and $(-5, 0)$. We also note that the graph does not intersect the y-axis.

It is possible to draw a hyperbola without plotting points. For example, if we want to graph the hyperbola with an equation of

$$\frac{x^2}{a^2} - \frac{y^2}{b^2} = 1$$

we first look at the x- and y-intercepts. To find the x-intercepts, we let $y = 0$ and solve for x:

$$\frac{x^2}{a^2} - \frac{0^2}{b^2} = 1$$

$$x^2 = a^2$$

$$x = \pm a$$

The hyperbola crosses the x-axis at the points $V_1(a, 0)$ and $V_2(-a, 0)$, called the **vertices** of the hyperbola. See Figure 10-24.

To attempt to find the y-intercepts, we let $x = 0$ and solve for y:

$$\frac{0^2}{a^2} - \frac{y^2}{b^2} = 1$$

$$y^2 = -b^2$$

$$y = \pm\sqrt{-b^2}$$

FIGURE 10-24

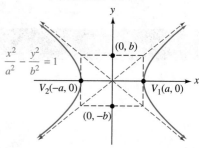

Since b^2 is always positive, $\sqrt{-b^2}$ is an imaginary number. This means that the hyperbola does not cross the y-axis.

If we construct a rectangle, called the **fundamental rectangle**, whose sides pass horizontally through $\pm b$ on the y-axis and vertically through $\pm a$ on the x-axis, the extended diagonals of the rectangle will be asymptotes of the hyperbola.

Equation of a hyperbola centered at the origin

Any equation that can be written in the form

$$\frac{x^2}{a^2} - \frac{y^2}{b^2} = 1$$

has a graph that is a hyperbola centered at the origin, as in Figure 10-25. The x-intercepts are the vertices $V_1(a, 0)$ and $V_2(-a, 0)$. There are no y-intercepts.

The asymptotes of the hyperbola are the extended diagonals of the rectangle in the figure.

FIGURE 10-25

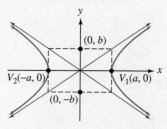

The branches of the hyperbola in previous discussions open to the left and to the right. It is possible for hyperbolas to have different orientations with respect to the x- and y-axes. For example, the branches of a hyperbola can open upward and downward. In that case, the following equation applies.

Equation of a hyperbola centered at the origin

Any equation that can be written in the form

$$\frac{y^2}{a^2} - \frac{x^2}{b^2} = 1$$

has a graph that is a hyperbola centered at the origin, as in Figure 10-26. The y-intercepts are the vertices $V_1(0, a)$ and $V_2(0, -a)$. There are no x-intercepts.

The asymptotes of the hyperbola are the extended diagonals of the rectangle shown in the figure.

FIGURE 10-26

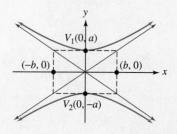

EXAMPLE 1

Graphing hyperbolas. Graph $9y^2 - 4x^2 = 36$.

Solution   To write the equation in standard form, we divide both sides by 36 to obtain

$$\frac{9y^2}{36} - \frac{4x^2}{36} = 1$$

$$\frac{y^2}{4} - \frac{x^2}{9} = 1 \quad \text{Simplify each fraction.}$$

FIGURE 10-27

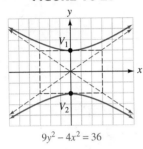

$9y^2 - 4x^2 = 36$

We then find the y-intercepts by letting $x = 0$ and solving for y:

$$\frac{y^2}{4} - \frac{0^2}{9} = 1$$

$$y^2 = 4$$

Thus, $y = \pm 2$, and the vertices of the hyperbola are $V_1(0, 2)$ and $V_2(0, -2)$. (See Figure 10-27.)

Since $\pm\sqrt{9} = \pm 3$, we use the points $(3, 0)$ and $(-3, 0)$ on the x-axis to help draw the fundamental rectangle. We then draw its extended diagonals and sketch the hyperbola.

SELF CHECK   Graph $9x^2 - 4y^2 = 36$.

Answer:

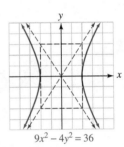

$9x^2 - 4y^2 = 36$ ∎

If a hyperbola is centered at a point with coordinates (h, k), the following equations apply.

Equations of hyperbolas centered at (h, k)

Any equation that can be written in the form

$$\frac{(x - h)^2}{a^2} - \frac{(y - k)^2}{b^2} = 1$$

is a hyperbola that has its center at (h, k) and opens left and right.

Any equation of the form

$$\frac{(y - k)^2}{a^2} - \frac{(x - h)^2}{b^2} = 1$$

is a hyperbola that has its center at (h, k) and opens up and down.

EXAMPLE 2

Graphing hyperbolas. Graph $\dfrac{(x - 3)^2}{16} - \dfrac{(y + 1)^2}{4} = 1$.

Solution   We write the equation in the form

$$\frac{(x-3)^2}{4^2} - \frac{[y-(-1)]^2}{2^2} = 1$$

to see that its graph will be a hyperbola centered at the point $(h, k) = (3, -1)$. Its vertices are located at $a = 4$ units to the right and left of the center, at $(7, -1)$ and $(-1, -1)$. Since $b = 2$, we can count 2 units above and below the center to locate points $(3, 1)$ and $(3, -3)$. With these points, we can draw the fundamental rectangle along with its extended diagonals. We can then sketch the hyperbola, as shown in Figure 10-28.

FIGURE 10-28

SELF CHECK   Graph $\dfrac{(x+2)^2}{9} - \dfrac{(y-1)^2}{4} = 1$.

Answer:

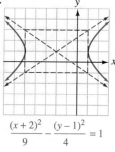

$$\frac{(x+2)^2}{9} - \frac{(y-1)^2}{4} = 1$$ ∎

EXAMPLE 3

Writing equations in standard form. Write the equation $x^2 - y^2 - 2x + 4y = 12$ in standard form to show that the equation represents a hyperbola. Then graph it.

Solution   We proceed as follows.

$$x^2 - y^2 - 2x + 4y = 12$$
$$x^2 - 2x - y^2 + 4y = 12 \quad \text{Use the commutative property to rearrange terms.}$$
$$x^2 - 2x - (y^2 - 4y) = 12 \quad \text{Factor } -1 \text{ from } -y^2 + 4y.$$

FIGURE 10-29

We then complete the square on x and y to make $x^2 - 2x$ and $y^2 - 4y$ perfect square trinomials.

$$x^2 - 2x + \mathbf{1} - (y^2 - 4y + \mathbf{4}) = 12 + \mathbf{1} - \mathbf{4}$$

We then factor $x^2 - 2x + 1$ and $y^2 - 4y + 4$ to get

$$(x-1)^2 - (y-2)^2 = 9$$
$$\frac{(x-1)^2}{9} - \frac{(y-2)^2}{9} = 1 \quad \text{Divide both sides by 9.}$$

This is the equation of a hyperbola with center at $(1, 2)$. Its graph is shown in Figure 10-29.

SELF CHECK   Graph $x^2 - 4y^2 + 2x - 8y = 7$.

Answer:

∎

There is a special type of hyperbola (also centered at the origin) that does not intersect either the x- or the y-axis. These hyperbolas have equations of the form $xy = k$, where $k \neq 0$.

EXAMPLE 4

Graphing hyperbolas. Graph $xy = -8$.

Solution We make a table of ordered pairs, plot each pair, and join the points with a smooth curve to obtain the hyperbola in Figure 10-30.

FIGURE 10-30

$xy = -8$

x	y	(x, y)
1	-8	$(1, -8)$
2	-4	$(2, -4)$
4	-2	$(4, -2)$
8	-1	$(8, -1)$
-1	8	$(-1, 8)$
-2	4	$(-2, 4)$
-4	2	$(-4, 2)$
-8	1	$(-8, 1)$

SELF CHECK Graph $xy = 6$.

Answer:

The result in Example 4 illustrates the following general equation.

Equations of hyperbolas of the form $xy = k$

Any equation of the form $xy = k$, where $k \neq 0$, has a graph that is a **hyperbola**, which does not intersect either the x- or the y-axis.

Problem Solving

EXAMPLE 5

Atomic structure. In an experiment that led to the discovery of the atomic structure of matter, Lord Rutherford (1871–1937) shot high–energy alpha particles toward a thin sheet of gold. Many of them were reflected, and Rutherford showed the existence of the nucleus of a gold atom. The alpha particle in Figure 10-31 is repelled by the nucleus at the origin; it travels along the hyperbolic path given by $4x^2 - y^2 = 16$. How close does the particle come to the nucleus?

FIGURE 10-31

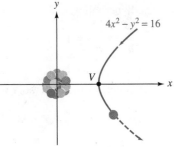

Solution   To find the distance from the nucleus at the origin, we must find the coordinates of the vertex V. To do so, we write the equation of the particle's path in standard form:

$$4x^2 - y^2 = 16$$

$$\frac{4x^2}{16} - \frac{y^2}{16} = \frac{16}{16} \quad \text{Divide both sides by 16.}$$

$$\frac{x^2}{4} - \frac{y^2}{16} = 1 \quad \text{Simplify.}$$

$$\frac{x^2}{2^2} - \frac{y^2}{4^2} = 1 \quad \text{Write 4 as } 2^2 \text{ and 16 as } 4^2.$$

This equation is in the form

$$\frac{x^2}{a^2} - \frac{y^2}{b^2} = 1$$

with $a = 2$. Thus, the vertex of the path is $(2, 0)$. The particle is never closer than 2 units from the nucleus.   ∎

STUDY SET

Section 10.3

VOCABULARY

Fill in each blank to make the statements true.

1. A _____ is the set of all points in a plane for which the _____ of the distances from two fixed points is a constant.

2. The fixed points in Exercise 1 are the _____ of the hyperbola.

3. The midpoint of the line segment joining the foci of a hyperbola is called the _____ of the hyperbola.

4. The rectangle whose sides pass horizontally through $\pm b$ on the y-axis and vertically through $\pm a$ on the x-axis is called the _____ rectangle.

CONCEPTS

Fill in the blanks to make the statements true.

5. The hyperbolic graph of

$$\frac{x^2}{a^2} - \frac{y^2}{b^2} = 1$$

has x-intercepts of _____.

6. The hyperbolic graph of

$$\frac{x^2}{a^2} - \frac{y^2}{b^2} = 1$$

has ___ y-intercepts.

7. The center of the hyperbola with an equation of

$$\frac{x^2}{a^2} - \frac{y^2}{b^2} = 1$$

is the point _____.

8. The center of the hyperbola with an equation of

$$\frac{(x - h)^2}{a^2} - \frac{(y - k)^2}{b^2} = 1$$

is the point _____.

NOTATION

Complete each solution.

9. Write the equation $4x^2 - 9y^2 - 36 = 0$ in standard form.

$$4x^2 - 9y^2 - 36 = 0$$

$$4x^2 - 9y^2 = \boxed{}$$

$$\frac{4x^2}{\boxed{}} - \frac{9y^2}{\boxed{}} = 1$$

$$\frac{x^2}{9} - \frac{y^2}{4} = 1$$

10. Write the equation $x^2 - 4y^2 + 2x - 63 = 0$ in standard form.

$$x^2 - 4y^2 + 2x - 63 = 0$$

$$x^2 + 2x - 4y^2 = \boxed{}$$

$$x^2 + 2x + \boxed{} - 4y^2 = 63 + \boxed{}$$

$$\left(\boxed{} \right)^2 - 4y^2 = 64$$

$$\frac{(x + 1)^2}{64} - \frac{y^2}{16} = 1$$

PRACTICE

In Exercises 11–22, graph each hyperbola.

11. $\dfrac{x^2}{9} - \dfrac{y^2}{4} = 1$

12. $\dfrac{x^2}{4} - \dfrac{y^2}{4} = 1$

13. $\dfrac{y^2}{4} - \dfrac{x^2}{9} = 1$

14. $\dfrac{y^2}{4} - \dfrac{x^2}{64} = 1$

15. $25x^2 - y^2 = 25$

16. $9x^2 - 4y^2 = 36$

17. $\dfrac{(x - 2)^2}{9} - \dfrac{y^2}{16} = 1$

18. $\dfrac{(x + 2)^2}{16} - \dfrac{(y - 3)^2}{25} = 1$

19. $4(x + 3)^2 - (y - 1)^2 = 4$

20. $(x + 5)^2 - 16y^2 = 16$

21. $xy = 8$

22. $xy = -10$

In Exercises 23–26, use a graphing calculator to graph each equation.

23. $\dfrac{x^2}{9} - \dfrac{y^2}{4} = 1$

24. $y^2 - 16x^2 = 16$

25. $\dfrac{x^2}{4} - \dfrac{(y-1)^2}{9} = 1$

26. $\dfrac{(y+1)^2}{9} - \dfrac{(x-2)^2}{4} = 1$

In Exercises 27–30, write each equation in standard form and graph it.

27. $4x^2 - y^2 + 8x - 4y = 4$

28. $x^2 - 9y^2 - 4x - 54y = 86$

29. $4y^2 - x^2 + 8y + 4x = 4$

30. $y^2 - 4x^2 - 4y - 8x = 4$

APPLICATIONS

31. ALPHA PARTICLES The particle in Illustration 1 approaches the nucleus at the origin along the path $9y^2 - x^2 = 81$. How close does the particle come to the nucleus?

ILLUSTRATION 1

32. LORAN By determining the difference of the distances between the ship in Illustration 2 and two radio transmitters, the LORAN system places the ship on the hyperbola $x^2 - 4y^2 = 576$. If the ship is also 5 miles out to sea, find its coordinates.

ILLUSTRATION 2

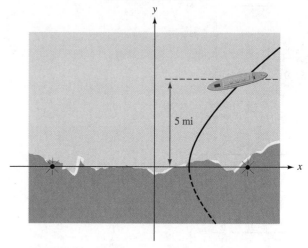

5 mi

33. SONIC BOOM The position of the sonic boom caused by the faster-than-sound aircraft in Illustration 3 is the hyperbola $y^2 - x^2 = 25$ in the coordinate system shown. How wide is the hyperbola 5 units from its vertex?

34. ELECTROSTATIC REPULSION Two similarly charged particles are shot together for an almost head-on collision, as in Illustration 4. They repel each other and travel the two branches of the hyperbola given by $x^2 - 4y^2 = 4$. How close do they get?

ILLUSTRATION 3

WRITING

35. Explain how to find the x- and y-intercepts of the graph of the ellipse

$$\frac{x^2}{a^2} + \frac{y^2}{b^2} = 1$$

36. Explain why the graph of the hyperbola

$$\frac{x^2}{y^2} - \frac{y^2}{b^2} = 1$$

has no y-intercept.

ILLUSTRATION 4

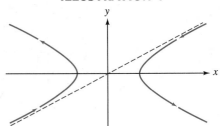

REVIEW

Factor each expression.

37. $-6x^4 + 9x^3 - 6x^2$

38. $4a^2 - b^2$

39. $15a^2 - 4ab - 4b^2$

40. $8p^3 - 27q^3$

▶ 10.4   Solving Simultaneous Second-Degree Equations

In this section, you will learn about

Solution by graphing ■ Solution by elimination

Introduction   In Chapter 3, we discussed how to solve systems of linear equations. To do this, we used the graphing, substitution, and addition methods. In this section, we will apply these methods to solve systems where at least one of the equations is nonlinear.

Solution by Graphing

We now discuss ways to solve systems of two equations in two variables where at least one of the equations is of second degree.

EXAMPLE 1

Solving systems of equations. Solve $\begin{cases} x^2 + y^2 = 25 \\ 2x + y = 10 \end{cases}$ by graphing.

Solution   The graph of $x^2 + y^2 = 25$ is a circle with center at the origin and radius of 5. The graph of $2x + y = 10$ is a line. Depending on whether the line is a secant (intersecting the circle at two points) or a tangent (intersecting the circle at one point) or does not intersect the circle at all, there are two, one, or no solutions to the system, respectively.

After graphing the circle and the line, as shown in Figure 10-32, we see that there are two intersection points, $P(3, 4)$ and $P'(5, 0)$, which is read as "P prime." The solutions of the system are

$$\begin{cases} x = 3 \\ y = 4 \end{cases} \text{ and } \begin{cases} x = 5 \\ y = 0 \end{cases}$$

Verify the results.

FIGURE 10-32

SELF CHECK Solve $\begin{cases} x^2 + y^2 = 13 \\ y = -\frac{1}{5}x + \frac{13}{5} \end{cases}$.

Answer: $(3, 2), (-2, 3)$

ACCENT ON TECHNOLOGY *Solving Systems of Equations*

To solve Example 1 with a graphing calculator, we graph the circle and the line on one set of coordinate axes. (See Figure 10-33(a).) We then trace to find the coordinates of the intersection points of the graphs. (See Figure 10-33(b) and Figure 10-33(c).)

We can zoom for better results.

FIGURE 10-33

(a)

(b)

(c)

Solution by Elimination

Algebraic methods can also be used to solve systems of equations.

EXAMPLE 2

Solving systems of equations. Solve $\begin{cases} x^2 + y^2 = 25 \\ 2x + y = 10 \end{cases}$.

Solution This system has one second-degree equation and one first-degree equation. We can solve this type of system by substitution. Solving the linear equation for y gives

$$2x + y = 10$$

1. $$y = -2x + 10$$

We can substitute $-2x + 10$ for y in the second-degree equation and solve the resulting quadratic equation for x:

$$x^2 + y^2 = 25$$
$$x^2 + (-2x + 10)^2 = 25$$
$$x^2 + 4x^2 - 40x + 100 = 25 \qquad (-2x + 10)(-2x + 10) = 4x^2 - 40x + 100.$$
$$5x^2 - 40x + 75 = 0 \qquad \text{Combine like terms and subtract 25 from both sides.}$$
$$x^2 - 8x + 15 = 0 \qquad \text{Divide both sides by 5.}$$
$$(x - 5)(x - 3) = 0 \qquad \text{Factor } x^2 - 8x + 15.$$
$$x - 5 = 0 \quad \text{or} \quad x - 3 = 0 \qquad \text{Set each factor equal to 0.}$$
$$x = 5 \qquad\qquad x = 3$$

If we substitute 5 for x in Equation 1, we get $y = 0$. If we substitute 3 for x in Equation 1, we get $y = 4$. The two solutions are

$$\begin{cases} x = 5 \\ y = 0 \end{cases} \quad \text{or} \quad \begin{cases} x = 3 \\ y = 4 \end{cases}$$

which can be written as $(5, 0)$ and $(3, 4)$.

SELF CHECK Solve: $\begin{cases} x^2 + y^2 = 13 \\ y = -\frac{1}{5}x + \frac{13}{5}. \end{cases}$ *Answer:* $(3, 2)$, $(-2, 3)$ ■

EXAMPLE 3   **Solving systems of equations.** Solve $\begin{cases} 4x^2 + 9y^2 = 5 \\ y = x^2 \end{cases}.$

Solution   We can solve this system by substitution.

$$4x^2 + 9y^2 = 5$$
$$4y + 9y^2 = 5 \qquad \text{Substitute } y \text{ for } x^2.$$
$$9y^2 + 4y - 5 = 0 \qquad \text{Subtract 5 from both sides.}$$
$$(9y - 5)(y + 1) = 0 \qquad \text{Factor } 9y^2 + 4y - 5.$$
$$9y - 5 = 0 \quad \text{or} \quad y + 1 = 0 \qquad \text{Set each factor equal to 0.}$$
$$y = \frac{5}{9} \qquad\qquad y = -1$$

Since $y = x^2$, the values of x are found by solving the equations

$$x^2 = \frac{5}{9} \quad \text{and} \quad x^2 = -1$$

Because $x^2 = -1$ has no real solutions, this possibility is discarded. The solutions of $x^2 = \frac{5}{9}$ are

$$x = \frac{\sqrt{5}}{3} \quad \text{or} \quad x = -\frac{\sqrt{5}}{3}$$

The solutions of the system are

$$\left(\frac{\sqrt{5}}{3}, \frac{5}{9} \right) \quad \text{and} \quad \left(-\frac{\sqrt{5}}{3}, \frac{5}{9} \right)$$

SELF CHECK Solve $\begin{cases} x^2 + y^2 = 20 \\ y = x^2 \end{cases}.$ *Answer:* $(2, 4)$, $(-2, 4)$ ■

| E X A M P L E   4 | **Solving systems of equations.** Solve $\begin{cases} 3x^2 + 2y^2 = 36 \\ 4x^2 - y^2 = 4 \end{cases}$. |

Solution   Since both equations are in the form $ax^2 + by^2 = c$, we can solve the system by addition.

We can copy the first equation and multiply the second equation by 2 to obtain the equivalent system

$$\begin{cases} 3x^2 + 2y^2 = 36 \\ 8x^2 - 2y^2 = 8 \end{cases}$$

We add the equations to eliminate y and solve the resulting equation for x:

$$11x^2 = 44$$
$$x^2 = 4$$
$$x = 2 \quad \text{or} \quad x = -2$$

To find y, we substitute 2 for x and then -2 for x in the first equation and proceed as follows:

For x = 2	**For x = -2**
$3x^2 + 2y^2 = 36$	$3x^2 + 2y^2 = 36$
$3(2)^2 + 2y^2 = 36$	$3(-2)^2 + 2y^2 = 36$
$12 + 2y^2 = 36$	$12 + 2y^2 = 36$
$2y^2 = 24$	$2y^2 = 24$
$y^2 = 12$	$y^2 = 12$
$y = \sqrt{12} \quad \text{or} \quad y = -\sqrt{12}$	$y = \sqrt{12} \quad \text{or} \quad y = -\sqrt{12}$
$y = 2\sqrt{3} \qquad y = -2\sqrt{3}$	$y = 2\sqrt{3} \qquad y = -2\sqrt{3}$

The four solutions of this system are

$$\left(2, 2\sqrt{3}\right), \quad \left(2, -2\sqrt{3}\right), \quad \left(-2, 2\sqrt{3}\right), \quad \text{and} \quad \left(-2, -2\sqrt{3}\right)$$

SELF CHECK   Solve $\begin{cases} x^2 + 4y^2 = 16 \\ x^2 - y^2 = 1 \end{cases}$.

Answer: $\left(2, \sqrt{3}\right), \left(2, -\sqrt{3}\right),$ $\left(-2, \sqrt{3}\right), \left(-2, -\sqrt{3}\right)$ ∎

STUDY SET

Section 10.4

VOCABULARY

Fill in the blanks to make the statements true.

1. We can solve systems of equations by _____, addition, or _____.

2. A _____ is a line that intersects a circle at two points.

3. A _____ is a line that intersects a circle at one point.

4. The equation $y = 3x + 4$ is a linear equation. The equation $y = 3x^2 + 4$ is a _____ equation.

CONCEPTS

Fill in the blanks to make the statements true.

5. At most, a line can intersect an ellipse at _____ points.

6. At most, an ellipse can intersect a parabola at _____ points.

7. At most, an ellipse can intersect a circle at _____ points.

8. At most, a hyperbola can intersect a circle at _____ points.

NOTATION

Complete each solution.

9. Solve $\begin{cases} x^2 + y^2 = 5 \\ y = 2x \end{cases}$.

$$x^2 + y^2 = 5$$
$$x^2 + \left(\rule{0.6cm}{0.3cm}\right)^2 = 5$$
$$x^2 + 4x^2 = \rule{0.6cm}{0.3cm}$$
$$\rule{0.6cm}{0.3cm}\,x^2 = 5$$
$$x^2 = \rule{0.6cm}{0.3cm}$$
$$x = 1 \quad \text{or} \quad x = -1$$

If $x = 1$, then
$$y = 2\left(\rule{0.5cm}{0.3cm}\right) = 2$$
If $x = -1$, then
$$y = 2\left(\rule{0.5cm}{0.3cm}\right) = -2$$
The solutions are $(1, 2)$ and $(-1, -2)$.

10. Solve $\begin{cases} x^2 + y^2 = 13 \\ x^2 - y^2 = 5 \end{cases}$

$$2x^2 = \rule{0.6cm}{0.3cm} \qquad \text{Add the equations.}$$
$$x^2 = \rule{0.6cm}{0.3cm}$$
$$x = \rule{0.6cm}{0.3cm} \quad \text{or} \quad x = \rule{0.6cm}{0.3cm}$$
$$2y^2 = \rule{0.6cm}{0.3cm} \qquad \text{Subtract the equations.}$$
$$y^2 = \rule{0.6cm}{0.3cm}$$
$$y = \rule{0.6cm}{0.3cm} \quad \text{or} \quad y = \rule{0.6cm}{0.3cm}$$
The solutions are
$$(3, 2), (3, -2), (-3, 2), (-3, -2)$$

PRACTICE

In Exercises 11–18, solve each system of equations by graphing.

11. $\begin{cases} 8x^2 + 32y^2 = 256 \\ x = 2y \end{cases}$

12. $\begin{cases} x^2 + y^2 = 2 \\ x + y = 2 \end{cases}$

13. $\begin{cases} x^2 + y^2 = 10 \\ y = 3x^2 \end{cases}$

14. $\begin{cases} x^2 + y^2 = 5 \\ x + y = 3 \end{cases}$

15. $\begin{cases} x^2 + y^2 = 25 \\ 12x^2 + 64y^2 = 768 \end{cases}$

16. $\begin{cases} x^2 + y^2 = 13 \\ y = x^2 - 1 \end{cases}$

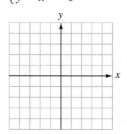

17. $\begin{cases} x^2 - 13 = -y^2 \\ y = 2x - 4 \end{cases}$

18. $\begin{cases} x^2 + y^2 = 20 \\ y = x^2 \end{cases}$

In Exercises 19–20, use a graphing calculator to solve each system.

19. $\begin{cases} x^2 - 6x - y = -5 \\ x^2 - 6x + y = -5 \end{cases}$

20. $\begin{cases} x^2 - y^2 = -5 \\ 3x^2 + 2y^2 = 30 \end{cases}$

In Exercises 21–46, solve each system of equations algebraically for real values of x and y.

21. $\begin{cases} 25x^2 + 9y^2 = 225 \\ 5x + 3y = 15 \end{cases}$

22. $\begin{cases} x^2 + y^2 = 20 \\ y = x^2 \end{cases}$

23. $\begin{cases} x^2 + y^2 = 2 \\ x + y = 2 \end{cases}$

24. $\begin{cases} x^2 + y^2 = 36 \\ 49x^2 + 36y^2 = 1,764 \end{cases}$

25. $\begin{cases} x^2 + y^2 = 5 \\ x + y = 3 \end{cases}$

26. $\begin{cases} x^2 - x - y = 2 \\ 4x - 3y = 0 \end{cases}$

27. $\begin{cases} x^2 + y^2 = 13 \\ y = x^2 - 1 \end{cases}$

28. $\begin{cases} x^2 + y^2 = 25 \\ 2x^2 - 3y^2 = 5 \end{cases}$

29. $\begin{cases} x^2 + y^2 = 30 \\ y = x^2 \end{cases}$

30. $\begin{cases} 9x^2 - 7y^2 = 81 \\ x^2 + y^2 = 9 \end{cases}$

31. $\begin{cases} x^2 + y^2 = 13 \\ x^2 - y^2 = 5 \end{cases}$ **32.** $\begin{cases} 2x^2 + y^2 = 6 \\ x^2 - y^2 = 3 \end{cases}$

33. $\begin{cases} x^2 + y^2 = 20 \\ x^2 - y^2 = -12 \end{cases}$ **34.** $\begin{cases} xy = -\dfrac{9}{2} \\ 3x + 2y = 6 \end{cases}$

35. $\begin{cases} y^2 = 40 - x^2 \\ y = x^2 - 10 \end{cases}$ **36.** $\begin{cases} x^2 - 6x - y = -5 \\ x^2 - 6x + y = -5 \end{cases}$

37. $\begin{cases} y = x^2 - 4 \\ x^2 - y^2 = -16 \end{cases}$ **38.** $\begin{cases} 6x^2 + 8y^2 = 182 \\ 8x^2 - 3y^2 = 24 \end{cases}$

39. $\begin{cases} x^2 - y^2 = -5 \\ 3x^2 + 2y^2 = 30 \end{cases}$ **40.** $\begin{cases} \dfrac{1}{x} + \dfrac{1}{y} = 5 \\ \dfrac{1}{x} - \dfrac{1}{y} = -3 \end{cases}$

41. $\begin{cases} \dfrac{1}{x} + \dfrac{2}{y} = 1 \\ \dfrac{2}{x} - \dfrac{1}{y} = \dfrac{1}{3} \end{cases}$ **42.** $\begin{cases} \dfrac{1}{x} + \dfrac{3}{y} = 4 \\ \dfrac{2}{x} - \dfrac{1}{y} = 7 \end{cases}$

43. $\begin{cases} 3y^2 = xy \\ 2x^2 + xy - 84 = 0 \end{cases}$ **44.** $\begin{cases} x^2 + y^2 = 10 \\ 2x^2 - 3y^2 = 5 \end{cases}$

45. $\begin{cases} xy = \dfrac{1}{6} \\ y + x = 5xy \end{cases}$ **46.** $\begin{cases} xy = \dfrac{1}{12} \\ y + x = 7xy \end{cases}$

47. INTEGER PROBLEM The product of two integers is 32, and their sum is 12. Find the integers.

48. NUMBER PROBLEM The sum of the squares of two numbers is 221, and the sum of the numbers is 9. Find the numbers.

APPLICATIONS

49. GEOMETRY The area of a rectangle is 63 square centimeters, and its perimeter is 32 centimeters. Find the dimensions of the rectangle.

50. INVESTING Grant receives $225 annual income from one investment. Jeff invested $500 more than Grant, but at an annual rate of 1% less. Jeff's annual income is $240. What is the amount and rate of Grant's investment?

51. INVESTING Carol receives $67.50 annual income from one investment. John invested $150 more than Carol at an annual rate of $1\frac{1}{2}\%$ more. John's annual income is $94.50. What is the amount and rate of Carol's investment? (*Hint:* There are two answers.)

52. ARTILLERY The shell fired from the base of the hill in Illustration 1 follows the parabolic path $y = -\frac{1}{6}x^2 + 2x$, with distances measured in miles. The hill has a slope of $\frac{1}{3}$. How far from the gun is the point of impact? (*Hint*: Find the coordinates of the point and then the distance.)

ILLUSTRATION 1

53. DRIVING RATES Jim drove 306 miles. Jim's brother made the same trip at a speed 17 mph slower than Jim did and required an extra $1\frac{1}{2}$ hours. What was Jim's rate and time?

54. FENCING PASTURES A rectangular pasture is to be fenced in along a riverbank, as shown in Illustration 2. If 260 feet of fencing is to enclose an area of 8,000 square feet, find the dimensions of the pasture.

ILLUSTRATION 2

WRITING

55. Describe the benefits of the graphical method for solving a system of equations.

56. Describe the drawbacks of the graphical method.

REVIEW

Simplify each radical expression. Assume that all variables represent positive numbers.

57. $\sqrt{200x^2} - 3\sqrt{98x^2}$

58. $a\sqrt{112a} - 5\sqrt{175a^3}$

59. $\dfrac{3t\sqrt{2t} - 2\sqrt{2t^3}}{\sqrt{18t} - \sqrt{2t}}$

60. $\sqrt[3]{\dfrac{x}{4}} + \sqrt[3]{\dfrac{x}{32}} - \sqrt[3]{\dfrac{x}{500}}$

Conic Sections

In this chapter, we have studied four conic sections: the circle, the parabola, the ellipse, and the hyperbola. They are called conic sections because they are formed by the intersection of a plane and a right-circular cone. These curves are graphs of second-degree equations in two variables.

In Exercises 1–9, classify the graph of each equation as a circle, a parabola, an ellipse, or a hyperbola.

1. $\dfrac{(x-2)^2}{9} + \dfrac{(y-1)^2}{4} = 1$

2. $(x-3)^2 + (y-2)^2 = 4$

3. $y = 4x^2 - 2x + 3$

4. $\dfrac{x^2}{4} - \dfrac{(y-1)^2}{9} = 1$

5. $4x^2 - y^2 + 8x - 4y = 4$

6. $\dfrac{x^2}{4} + \dfrac{y^2}{9} = 1$

7. $9x^2 + 4y^2 - 18x + 16y = 11$

8. $x^2 - 2y - 2x = -7$

9. $x^2 + y^2 + 8x + 2y = -13$

Equations of Conic Sections

When the equation of a conic section is written in standard form, important features of its graph are apparent.

10. Consider $(x+1)^2 + (y-2)^2 = 16$.
  a. What are the coordinates of the center of the circle?
  b. What is the radius of the circle?

11. Consider $x = \frac{1}{2}(y-1)^2 - 2$.
  a. What are the coordinates of the vertex of the parabola?
  b. In which direction does the parabola open?

12. Consider $\dfrac{x^2}{4} + \dfrac{y^2}{16} = 1$.
  a. What are the coordinates of the center of the ellipse?
  b. To which axis is the major axis of the ellipse parallel?
  c. How long is the major axis?
  d. How long is the minor axis?

13. Consider $\dfrac{(x+2)^2}{9} - \dfrac{(y-1)^2}{4} = 1$.
  a. What are the coordinates of the center of the hyperbola?
  b. In which direction do the branches of the hyperbola open?
  c. What are the dimensions of the fundamental rectangle?

Graphing Conic Sections

Use the results from Exercises 10–13 to graph each conic section.

14. $(x+1)^2 + (y-2)^2 = 16$

15. $x = \frac{1}{2}(y-1)^2 - 2$

16. $\dfrac{x^2}{4} + \dfrac{y^2}{16} = 1$

17. $\dfrac{(x+2)^2}{9} - \dfrac{(y-1)^2}{4} = 1$

Accent on Teamwork

Section 10.1

Constructing a parabola To construct a parabola, secure one end of a piece of string that is as long as a T-square to a large piece of paper. See Illustration 1. Attach the other end of the string to the upper end of the T-square. Then, hold the string taut against the T-square with a pencil and slide the T-square along the edge of the table. As the T-square moves, the pencil will trace a parabola.

Each point on a parabola is the same distance away from a given point as it is from a given line. With this method of construction, what is the given point? What is the given line?

ILLUSTRATION 1

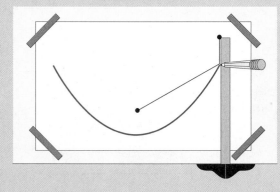

Section 10.2

Constructing ellipses We can construct an ellipse with two thumbtacks, a piece of string, and a pencil. Place the two thumbtacks fairly close together, as shown in Illustration 2. Catch the loop of the string with the point of a pencil and (keeping the string taut) draw the ellipse. Experiment by moving one of the thumbtacks farther away and then closer to the other thumbtack. How does the shape of the ellipse change?

For each point on an ellipse, the sum of the distances of the point from two given points is a constant. With this method of construction, what are the two given points? What is the constant?

ILLUSTRATION 2

Section 10.3

Conic section models Mold some clay into the shape of a right-circular cone, as shown in Illustration 3. With both hands, pull a thin wire through the clay to slice it in such a way that a circular shape results. Experiment with ways to slice the clay to obtain an elliptical shape, a parabolic shape, and one branch of a hyperbolic shape.

ILLUSTRATION 3

Section 10.4

Simultaneous systems Sketch the graphs of a system of second-degree equations that meet the following conditions, if possible.

The graph of one equation is a circle, the graph of the other equation is a parabola, and the system has
a. no solution.   **b.** one solution.
c. two solutions.   **d.** three solutions.
e. four solutions.   **f.** five solutions.

Simultaneous systems Sketch the graphs of a system of second-degree equations that meet the following conditions, if possible.

The graph of one equation is an ellipse, the graph of the other equation is a hyperbola, and the system has
a. no solution.   **b.** one solution.
c. two solutions.   **d.** three solutions.
e. four solutions.   **f.** five solutions.

Simultaneous systems Sketch the graphs of a system of second-degree equations that meet the following conditions, if possible.

The graph of one equation is a parabola, the graph of the other equation is a hyperbola, and the system has
a. no solution.   **b.** one solution.
c. two solutions.   **d.** three solutions.
e. four solutions.   **f.** five solutions.

Section 10.1

CONCEPTS

Equations of a circle:

$(x - h)^2 + (y - k)^2 = r^2$
   center (h, k), radius r

$x^2 + y^2 = r^2$
   center $(0, 0)$, radius r

The Circle and the Parabola

REVIEW EXERCISES

1. Graph each equation.

a. $(x - 1)^2 + (y + 2)^2 = 9$

b. $x^2 + y^2 = 16$

2. Write the equation in standard form and graph it.

$x^2 + y^2 + 4x - 2y = 4$

Equations of parabolas $(a > 0)$:

Parabola opening	Vertex at origin
Up	$y = ax^2$
Down	$y = -ax^2$
Right	$x = ay^2$
Left	$x = -ay^2$

Parabola opening	Vertex at (h, k)
Up	$y = a(x - h)^2 + k$
Down	$y = -a(x - h)^2 + k$
Right	$x = a(y - k)^2 + h$
Left	$x = -a(y - k)^2 + h$

3. Graph each equation.

a. $x = -3(y - 2)^2 + 5$

b. $x = 2(y + 1)^2 - 2$

Section 10.2

Equations of an ellipse:

Center at $(0, 0)$

$\dfrac{x^2}{a^2} + \dfrac{y^2}{b^2} = 1 \quad (a > b > 0)$

$\dfrac{x^2}{b^2} + \dfrac{y^2}{a^2} = 1 \quad (a > b > 0)$

The Ellipse

4. Graph each ellipse.

a. $9x^2 + 16y^2 = 144$

b. $\dfrac{(x - 2)^2}{4} + \dfrac{(y - 1)^2}{9} = 1$

Center at (h, k)

$$\frac{(x-h)^2}{a^2} + \frac{(y-k)^2}{b^2} = 1$$

$$\frac{(x-h)^2}{b^2} + \frac{(y-k)^2}{a^2} = 1$$

5. Write the equation in standard form and graph it.

$$4x^2 + 9y^2 + 8x - 18y = 23$$

Section 10.3 — The Hyperbola

Equations of a hyperbola:

Center at (0, 0)

$$\frac{x^2}{a^2} - \frac{y^2}{b^2} = 1$$

$$\frac{y^2}{b^2} - \frac{x^2}{a^2} = 1$$

Center at (h, k)

$$\frac{(x-h)^2}{a^2} - \frac{(y-k)^2}{b^2} = 1$$

$$\frac{(y-k)^2}{a^2} - \frac{(x-h)^2}{b^2} = 1$$

6. Graph each hyperbola.

a. $9x^2 - y^2 = -9$

b. $xy = 9$

7. Write the equation $4x^2 - 2y^2 + 8x - 8y = 8$ in standard form and tell whether its graph will be an ellipse or a hyperbola.

8. Write the equation in standard form and graph it.

$$9x^2 - 4y^2 - 18x - 8y = 31$$

Section 10.4 — Solving Simultaneous Second-Degree Equations

9. The graphs of $x^2 - y^2 = 9$ and $x^2 + y^2 = 9$ are shown in Illustration 1. Estimate the solutions of the system

$$\begin{cases} y^2 - x^2 = 9 \\ x^2 + y^2 = 9 \end{cases}$$

ILLUSTRATION 1

10. Solve each system.

a. $\begin{cases} 3x^2 + y^2 = 52 \\ x^2 - y^2 = 12 \end{cases}$

b. $\begin{cases} \dfrac{x^2}{16} + \dfrac{y^2}{12} = 1 \\ x^2 - \dfrac{y^2}{3} = 1 \end{cases}$

CHAPTER 10

Test

1. Find the center and the radius of the circle $(x - 2)^2 + (y + 3)^2 = 4$.

2. Find the center and the radius of the circle $x^2 + y^2 + 4x - 6y = 3$.

In Problems 3–6, graph each equation.

3. $(x + 1)^2 + (y - 2)^2 = 9$

4. $x = (y - 2)^2 - 1$

5. $9x^2 + 4y^2 = 36$

6. $\dfrac{(x - 2)^2}{9} - y^2 = 1$

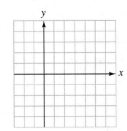

In Problems 7–8, write each equation in standard form and graph the equation.

7. $4x^2 + y^2 - 24x + 2y = -33$

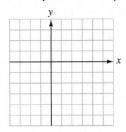

8. $x^2 - 9y^2 + 2x + 36y = 44$

In Problems 9–10, solve each system.

9. $\begin{cases} 2x - y = -2 \\ x^2 + y^2 = 16 + 4y \end{cases}$

10. $\begin{cases} x^2 + y^2 = 25 \\ 4x^2 - 9y = 0 \end{cases}$

11

Miscellaneous Topics

CAMPUS CONNECTION

The Political Science Department

In a political science class, students study the design of ballots used in city, county, state, and national elections. On a ballot, the names of the candidates running for a given office can be listed in many different orders. For example, they can be presented alphabetically, by party affiliation, or simply in random order. In this chapter, we will learn how to determine the number of ways in which a set of names can be arranged. This counting concept, known as a *permutation,* has many other important applications in areas such as designing license plates, assigning telephone numbers, and listing combinations for locks.

In this chapter, we introduce several topics that have applications in advanced mathematics and in certain occupations. The binomial theorem, permutations, and combinations are used in statistics. Arithmetic and geometric sequences are used in the mathematics of finance.

▶ 11.1 The Binomial Theorem

In this section, you will learn about

> Raising binomials to powers ■ Pascal's triangle ■ Factorial notation ■ The binomial theorem ■ Finding a particular term of a binomial expansion

Introduction In Chapter 5, we discussed how to raise binomials to positive-integer powers. For example, we learned that

$$(a + b)^2 = a^2 + 2ab + b^2$$

and that

$$
\begin{aligned}
(a + b)^3 &= (a + b)(a + b)^2 \\
&= (a + b)(a^2 + 2ab + b^2) \\
&= a^3 + 2a^2b + ab^2 + a^2b + 2ab^2 + b^3 \\
&= a^3 + 3a^2b + 3ab^2 + b^3
\end{aligned}
$$

In this section, we will learn how to raise binomials to positive-integer powers without doing the actual multiplications.

Raising Binomials to Powers

To see how to raise binomials to positive-integer powers, we consider the following binomial expansions:

$$(a + b)^0 = 1$$
$$(a + b)^1 = a + b$$
$$(a + b)^2 = a^2 + 2ab + b^2$$
$$(a + b)^3 = a^3 + 3a^2b + 3ab^2 + b^3$$
$$(a + b)^4 = a^4 + 4a^3b + 6a^2b^2 + 4ab^3 + b^4$$
$$(a + b)^5 = a^5 + 5a^4b + 10a^3b^2 + 10a^2b^3 + 5ab^4 + b^5$$
$$(a + b)^6 = a^6 + 6a^5b + 15a^4b^2 + 20a^3b^3 + 15a^2b^4 + 6ab^5 + b^6$$

Several patterns appear in these expansions:

1. Each expansion has one more term than the power of the binomial.
2. The degree of each term in each expansion is equal to the exponent of the binomial that is being expanded.
3. The first term in each expansion is a, raised to the power of the binomial.

4. The exponents on a decrease by one in each successive term. The exponents on b, beginning with $b^0 = 1$ in the first term, increase by one in each successive term. Thus, the variables have the pattern

$$a^n, a^{n-1}b, a^{n-2}b^2, \ldots, ab^{n-1}, b^n$$

Pascal's Triangle

To see another pattern, we write the coefficients of each expansion in the following triangular array:

$$
\begin{array}{ccccccccccccc}
&&&&&& 1 &&&&&& \\
&&&&& 1 && 1 &&&&& \\
&&&& 1 && 2 && 1 &&&& \\
&&& 1 && 3 && 3 && 1 &&& \\
&& 1 && 4 && 6 && 4 && 1 && \\
& 1 && 5 && 10 && 10 && 5 && 1 & \\
1 && 6 && 15 && 20 && 15 && 6 && 1
\end{array}
$$

In this array, called **Pascal's triangle,** each entry between the 1's is the sum of the closest pair of numbers in the line immediately above it. For example, the first 15 in the bottom row is the sum of the 5 and 10 immediately above it. Pascal's triangle continues with the same pattern forever. The next two lines are

$$
\begin{array}{ccccccccccccccc}
& 1 && 7 && 21 && 35 && 35 && 21 && 7 && 1 \\
1 && 8 && 28 && 56 && 70 && 56 && 28 && 8 && 1
\end{array}
$$

EXAMPLE 1

Raising a binomial to the 5th power. Expand $(x + y)^5$.

Solution The first term in the expansion is x^5, and the exponents on x decrease by one in each successive term. A y first appears in the second term, and the exponents on y increase by one in each successive term, concluding when the term y^5 is reached. Thus, the variables in the expansion are

$$x^5, x^4y, x^3y^2, x^2y^3, xy^4, y^5$$

The coefficients of these variables are given in Pascal's triangle in the row whose second entry is 5, the exponent of the binomial that is being expanded.

$$1 \quad 5 \quad 10 \quad 10 \quad 5 \quad 1$$

Combining this information gives the following expansion:

$$(x + y)^5 = x^5 + 5x^4y + 10x^3y^2 + 10x^2y^3 + 5xy^4 + y^5$$

SELF CHECK Expand $(x + y)^4$. *Answer:* $x^4 + 4x^3y + 6x^2y^2 + 4xy^3 + y^4$ ∎

EXAMPLE 2

Raising a binomial to the 4th power. Expand $(u - v)^4$.

Solution We note that $(u - v)^4$ can be written in the form $[u + (-v)]^4$. The variables in this expansion are

$$u^4, u^3(-v), u^2(-v)^2, u(-v)^3, (-v)^4$$

and the coefficients are given in Pascal's triangle in the row whose second entry is 4:

$$1 \quad 4 \quad 6 \quad 4 \quad 1$$

Thus, the required expansion is

$$(u - v)^4 = u^4 + 4u^3(-v) + 6u^2(-v)^2 + 4u(-v)^3 + (-v)^4$$
$$= u^4 - 4u^3v + 6u^2v^2 - 4uv^3 + v^4$$

SELF CHECK   Expand $(x - y)^5$.   *Answer:* $x^5 - 5x^4y + 10x^3y^2 - 10x^2y^3 + 5xy^4 - y^5$ ∎

Factorial Notation

Although Pascal's triangle gives the coefficients of the terms in a binomial expansion, it is not the best way to expand a binomial. To develop another way, we introduce **factorial notation.**

Factorial notation

If n is a natural number, the symbol $n!$ (read as n **factorial** or as **factorial** n) is defined as

$$n! = n(n - 1)(n - 2)(n - 3) \cdots (3)(2)(1)$$

E X A M P L E   3   **Evaluating factorials.** Find   **a.** 2!,   **b.** 5!,   **c.** $-9!$,   and   **d.** $(n - 2)!$.

Solution   **a.** $2! = 2 \cdot 1 = 2$

b. $5! = 5 \cdot 4 \cdot 3 \cdot 2 \cdot 1 = 120$

c. $-9! = -(9 \cdot 8 \cdot 7 \cdot 6 \cdot 5 \cdot 4 \cdot 3 \cdot 2 \cdot 1) = -362,880$

d. $(n - 2)! = (n - 2)(n - 3)(n - 4) \cdots \cdots 3 \cdot 2 \cdot 1$

 WARNING! According to the previous definition, part d is meaningful only if $n - 2$ is a natural number.

SELF CHECK   Find   **a.** 4!   and   **b.** $x!$.   *Answers:*   **a.** 24,
                                          b. $x(x - 1)(x - 2) \cdots \cdots 3 \cdot 2 \cdot 1$ ∎

We define zero factorial as follows.

Zero factorial

$$0! = 1$$

We note that

$$5 \cdot 4! = 5 \cdot 4 \cdot 3 \cdot 2 \cdot 1 = 5!$$
$$7 \cdot 6! = 7 \cdot 6 \cdot 5 \cdot 4 \cdot 3 \cdot 2 \cdot 1 = 7!$$
$$10 \cdot 9! = 10 \cdot 9 \cdot 8 \cdot 7 \cdot 6 \cdot 5 \cdot 4 \cdot 3 \cdot 2 \cdot 1 = 10!$$

These examples suggest the following property of factorials.

$n(n - 1)!$

If n is a positive integer, then $n(n - 1)! = n!$.

The Binomial Theorem

We now state the binomial theorem.

The binomial theorem

If n is any positive integer, then

$$(a + b)^n = a^n + \frac{n!}{1!(n-1)!} a^{n-1}b + \frac{n!}{2!(n-2)!} a^{n-2}b^2 + \frac{n!}{3!(n-3)!} a^{n-3}b^3$$

$$+ \cdots + \frac{n!}{r!(n-r)!} a^{n-r}b^r + \cdots + b^n$$

In the binomial theorem, the exponents on the variables follow the familiar pattern:

- The sum of the exponents on a and b in each term is n,
- the exponents on a decrease by 1 in each successive term, and
- the exponents on b increase by 1 in each successive term.

Only the method of finding the coefficients is different. Except for the first and last terms, the numerator of each coefficient is $n!$. If the exponent on b in a particular term is r, the denominator of the coefficient of that term is $r!(n-r)!$.

EXAMPLE 4

Expanding a binomial. Use the binomial theorem to expand $(a + b)^3$.

Solution   We can substitute directly into the binomial theorem and simplify:

$$(a + b)^3 = a^3 + \frac{3!}{1!(3-1)!} a^2b + \frac{3!}{2!(3-2)!} ab^2 + b^3$$

$$= a^3 + \frac{3!}{1! \cdot 2!} a^2b + \frac{3!}{2! \cdot 1!} ab^2 + b^3$$

$$= a^3 + \frac{3 \cdot 2 \cdot 1}{1 \cdot 2 \cdot 1} a^2b + \frac{3 \cdot 2 \cdot 1}{2 \cdot 1 \cdot 1} ab^2 + b^3$$

$$= a^3 + 3a^2b + 3ab^2 + b^3$$

SELF CHECK   Use the binomial theorem to expand
$(a + b)^4$.   *Answer:* $a^4 + 4a^3b + 6a^2b^2 + 4ab^3 + b^4$ ■

EXAMPLE 5

Expanding a binomial. Use the binomial theorem to expand $(x - y)^4$.

Solution   We can write $(x - y)^4$ in the form $[x + (-y)]^4$, substitute directly into the binomial theorem, and simplify:

$$(x - y)^4 = [x + (-y)]^4$$

$$= x^4 + \frac{4!}{1!(4-1)!} x^3(-y) + \frac{4!}{2!(4-2)!} x^2(-y)^2 + \frac{4!}{3!(4-3)!} x(-y)^3 + (-y)^4$$

$$= x^4 - \frac{4 \cdot 3!}{1! \cdot 3!} x^3y + \frac{4 \cdot 3 \cdot 2!}{2! \cdot 2!} x^2y^2 - \frac{4 \cdot 3!}{3! \cdot 1!} xy^3 + y^4$$

$$= x^4 - 4x^3y + 6x^2y^2 - 4xy^3 + y^4$$

SELF CHECK   Use the binomial theorem to expand $(x - y)^3$.   *Answer:* $x^3 - 3x^2y + 3xy^2 - y^3$ ■

EXAMPLE 6

Expanding a binomial. Use the binomial theorem to expand $(3u - 2v)^4$.

Solution

We write $(3u - 2v)^4$ in the form $[3u + (-2v)]^4$ and let $a = 3u$ and $b = -2v$. Then we can use the binomial theorem to expand $(a + b)^4$.

$$(a + b)^4 = a^4 + \frac{4!}{1!(4 - 1)!} a^3b + \frac{4!}{2!(4 - 2)!} a^2b^2 + \frac{4!}{3!(4 - 3)!} ab^3 + b^4$$
$$= a^4 + 4a^3b + 6a^2b^2 + 4ab^3 + b^4$$

Now we can substitute $3u$ for a and $-2v$ for b and simplify:

$$(3u - 2v)^4 = (3u)^4 + 4(3u)^3(-2v) + 6(3u)^2(-2v)^2 + 4(3u)(-2v)^3 + (-2v)^4$$
$$= 81u^4 - 216u^3v + 216u^2v^2 - 96uv^3 + 16v^4$$

SELF CHECK

Use the binomial theorem to expand $(2a - 3b)^3$.

Answer: $8a^3 - 36a^2b + 54ab^2 - 27b^3$ ∎

Finding a Particular Term of a Binomial Expansion

To find the fourth term of the expansion of $(a + b)^9$, we could raise the binomial $a + b$ to the 9th power and look at the fourth term. However, this task would be tedious. By using the binomial theorem, we can construct the fourth term without finding the complete expansion of $(a + b)^9$.

EXAMPLE 7

Finding the 4th term of a binomial expansion. Find the fourth term of the expansion of $(a + b)^9$.

Solution

Since b^1 appears in the second term, b^2 appears in the third term, and so on, the exponent on b in the fourth term is 3. Since the exponent on b added to the exponent on a must equal 9, the exponent on a must be 6. Thus, the variables of the fourth term are

a^6b^3 The sum of the exponents must be 9.

The coefficient of these variables is

$$\frac{n!}{r!(n - r)!} = \frac{9!}{3!(9 - 3)!}$$ The numerator of each coefficient is $n!$.
                           r is the exponent on b.

The complete fourth term is

$$\frac{9!}{3!(9 - 3)!} a^6b^3 = \frac{9 \cdot 8 \cdot 7 \cdot 6!}{3 \cdot 2 \cdot 1 \cdot 6!} a^6b^3$$
$$= 84a^6b^3$$

SELF CHECK

Find the third term of the expansion of $(a + b)^9$.   *Answer:* $36a^7b^2$ ∎

EXAMPLE 8

Finding the 6th term of a binomial expansion. Find the sixth term in the expansion of $(x - y)^7$.

Solution

We first find the sixth term of $[x + (-y)]^7$. In the sixth term, the exponent on $(-y)$ is 5. Thus, the variables in the sixth term are

$x^2(-y)^5$ The sum of the exponents must be 7.

The coefficient of these variables is

$$\frac{n!}{r!(n-r)!} = \frac{7!}{5!(7-5)!}$$

The numerator of each coefficient is $n!$.
r is the exponent on b.

The complete sixth term is

$$\frac{7!}{5!(7-5)!}x^2(-y)^5 = -\frac{7 \cdot 6 \cdot 5!}{5! \cdot 2 \cdot 1}x^2y^5$$
$$= -21x^2y^5$$

SELF CHECK   Find the fifth term of the expansion of $(a-b)^7$.   *Answer:* $35a^3b^4$ ∎

EXAMPLE 9   **Finding the 4th term of a binomial expansion.** Find the fourth term of the expansion of $(2x-3y)^6$.

Solution   We can let $a = 2x$ and $b = -3y$ and find the fourth term of the expansion of $(a+b)^6$:

$$\frac{6!}{3!(6-3)!}a^3b^3 = \frac{6 \cdot 5 \cdot 4 \cdot 3!}{3! \cdot 3 \cdot 2 \cdot 1}a^3b^3$$
$$= 20a^3b^3$$

We can now substitute $2x$ for a and $-3y$ for b and simplify:

$$20a^3b^3 = 20(2x)^3(-3y)^3$$
$$= -4{,}320x^3y^3$$

The fourth term is $-4{,}320x^3y^3$.

SELF CHECK   Find the third term of the expansion of $(2a-3b)^6$.   *Answer:* $2{,}160a^4b^2$ ∎

STUDY SET

Section 11.1

VOCABULARY

Fill in the blanks to make the statements true.

1. The expression $a+b$ is called a _____.

2. The triangular array that can be used to find the coefficients of a binomial expansion is called _____ triangle.

CONCEPTS

Fill in the blanks to make the statements true.

3. Every binomial expansion has ___ more term than the power of the binomial.

4. The first term in the expansion of $(a+b)^{20}$ is _____.

5. The degree of each term of a binomial expansion is equal to the _____ of the binomial that is being expanded.

6. In the expansion of $(a+b)^n$, the powers on a _____, and the powers on b _____.

7. In the expansion of $(a+b)^7$, the sum of the exponents on a and b is ___.

8. The coefficient of the fourth term of the expansion of $(a+b)^9$ is $9!$ divided by _____.

9. The symbol $5!$ is read as "_____."

10. $6 \cdot 5 \cdot 4 \cdot 3 \cdot 2 \cdot 1 =$ ___

11. $8! = 8 \cdot$ ___

12. $0! =$ ___

13. n_____ $= n!$

14. According to the binomial theorem, the third term of the expansion of $(a+b)^n$ is _____.

15. The exponent on b in the fourth term of the expansion of $(a+b)^6$ is ___.

16. The exponent on b in the fifth term of the expansion of $(a+b)^6$ is ___.

NOTATION

Complete the solution.

17. $(x + y)^3 = x^{\blacksquare} + \dfrac{\blacksquare}{1!(3-1)!}x^2y + \dfrac{\blacksquare}{2!(3-2)!}xy^2 + y^3$

$\qquad = x^3 + \dfrac{3\cdot 2\cdot 1}{1\cdot\blacksquare\cdot 1}x^2y + \dfrac{3\cdot 2\cdot 1}{2\cdot\blacksquare\cdot 1}xy^2 + y^3$

$\qquad = x^3 + 3x^2y + 3xy^2 + y^3$

18. Find the fifth term of the expansion of $(a - b)^7$.
   a. In the fifth term, the exponent on $(-b)$ is $\blacksquare$.
   b. Thus, the variables of the fifth term are $\blacksquare$.
   c. The coefficient of these variables is
   $$\dfrac{7!}{4!(7-4)!} = 35.$$
   d. The fifth term is $\blacksquare$.

PRACTICE

In Exercises 19–38, evaluate each expression.

19. $3!$

20. $7!$

21. $-5!$

22. $-6!$

23. $3! + 4!$

24. $2!(3!)$

25. $3!(4!)$

26. $4! + 4!$

27. $8(7!)$

28. $4!(5)$

29. $\dfrac{9!}{11!}$

30. $\dfrac{13!}{10!}$

31. $\dfrac{49!}{47!}$

32. $\dfrac{101!}{100!}$

33. $\dfrac{5!}{3!(5-3)!}$

34. $\dfrac{6!}{4!(6-4)!}$

35. $\dfrac{7!}{5!(7-5)!}$

36. $\dfrac{8!}{6!(8-6)!}$

37. $\dfrac{5!(8-5)!}{4!\cdot 7!}$

38. $\dfrac{6!\cdot 7!}{(8-3)!(7-4)!}$

In Exercises 39–54, use the binomial theorem to expand each expression.

39. $(x + y)^2$

40. $(x + y)^4$

41. $(x - y)^4$

42. $(x - y)^3$

43. $(2x + y)^3$

44. $(x + 2y)^3$

45. $(x - 2y)^3$

46. $(2x - y)^3$

47. $(2x + 3y)^3$

48. $(3x - 2y)^3$

49. $\left(\dfrac{x}{2} - \dfrac{y}{3}\right)^3$

50. $\left(\dfrac{x}{3} + \dfrac{y}{2}\right)^3$

51. $(3 + 2y)^4$

52. $(2x + 3)^4$

53. $\left(\dfrac{x}{3} - \dfrac{y}{2}\right)^4$

54. $\left(\dfrac{x}{2} + \dfrac{y}{3}\right)^4$

In Exercises 55–80, find the required term of each binomial expansion.

55. $(a + b)^3$; second term

56. $(a + b)^3$; third term

57. $(x - y)^4$; fourth term

58. $(x - y)^5$; second term

59. $(x + y)^6$; fifth term

60. $(x + y)^7$; fifth term

61. $(x - y)^8$; third term

62. $(x - y)^9$; seventh term

63. $(x + 3)^5$; third term

64. $(x - 2)^4$; second term

65. $(4x + y)^5$; third term

66. $(x + 4y)^5$; fourth term

67. $(x - 3y)^4$; second term

68. $(3x - y)^5$; third term

69. $(2x - 5)^7$; fourth term

70. $(2x + 3)^6$; sixth term

71. $(2x - 3y)^5$; fifth term

72. $(3x - 2y)^4$; second term

73. $\left(\dfrac{x}{2} - \dfrac{y}{3}\right)^4$; second term

74. $\left(\dfrac{x}{3} + \dfrac{y}{2}\right)^5$; fourth term

75. $(a + b)^n$; fourth term

76. $(a + b)^n$; third term

77. $(a - b)^n$; fifth term

78. $(a - b)^n$; sixth term

79. $(2a - 3b)^n$; fifth term

80. $(3a - 2b)^n$; fourth term

WRITING

81. Tell how to construct Pascal's triangle.

82. Tell how to find the variable part of each term in the expansion of $(r + s)^4$.

83. Tell how to find the coefficients of each term in the expansion of $(x + y)^5$.

84. Explain why the signs alternate in the expansion of $(x - y)^9$.

REVIEW

Find each value of x.

85. $\log_4 16 = x$

86. $\log_x 49 = 2$

87. $\log_{25} x = \dfrac{1}{2}$

88. $\log_{1/2} \dfrac{1}{8} = x$

Solve each system of equations.

89. $\begin{cases} 3x + 2y = 12 \\ 2x - y = 1 \end{cases}$

90. $\begin{cases} a + b + c = 6 \\ 2a + b + 3c = 11 \\ 3a - b - c = 6 \end{cases}$

Evaluate each determinant.

91. $\begin{vmatrix} 2 & -3 \\ 4 & -2 \end{vmatrix}$

92. $\begin{vmatrix} 1 & 2 & 3 \\ 4 & 5 & 0 \\ -1 & -2 & 1 \end{vmatrix}$

▶ 11.2   Arithmetic Sequences

In this section, you will learn about

  ■ Sequences and series ■ Arithmetic sequences ■ Arithmetic means
  ■ The sum of the first *n* terms of an arithmetic sequence
  ■ Summation notation

Introduction   In this section, we will discuss ordered lists called *sequences*. For example, a batting order in baseball is a sequence, because it determines who will bat first, bat second, and so on. When a sequence is a list of numbers, we can find their sum. When we put + signs between the numbers in a sequence, we form a sum that is called a *series*.

Sequences and Series

A mathematical **sequence** is a function whose domain is the set of natural numbers. For example, the function $f(n) = 3n + 2$ (where *n* is a natural number) is a sequence. If the natural numbers are substituted in order for *n*, the function $f(n) = 3n + 2$ generates the list

  5, 8, 11, 14, 17, . . .   The ellipsis . . . indicates that the list goes on forever.

Each number in the list is called a **term** of the sequence. Since this sequence has an unlimited number of terms, it is called an **infinite sequence**. A **finite sequence** has a specific number of terms.

  Examples of infinite sequences:

  $1^3, 2^3, 3^3, 4^3, 5^3, 6^3, . . .$ The ordered list of the cubes of the natural numbers.

  1, 1, 2, 3, 5, 8, 13, 21, . . .   The Fibonacci sequence.

The **Fibonacci sequence** is named after the 12th-century mathematician Leonardo of Pisa—also known as Fibonacci. Beginning with the 2, each term of the Fibonacci sequence is the sum of the two preceding terms.

  Examples of finite sequences:

  4, 8, 12, 16, 20, 24   The ordered list of the first six positive multiples of 4.

  2, 3, 5, 7, 11, 13, 17, 19, 23, 29   The ordered list of the first ten prime numbers.

When the commas between the terms of a sequence are replaced with + signs, we call the sum a **series.**

$4 + 8 + 12 + 16 + 20 + 24$ Since this series has a limited number of terms, it is a finite series.

$5 + 8 + 11 + 14 + 17 + \cdots$ Since this series has an unlimited number of terms, it is an infinite series.

Arithmetic Sequences

A sequence where each term is found by adding the same number to the preceding term is called an **arithmetic sequence.**

Arithmetic sequence

An **arithmetic sequence** is a sequence of the form

$$a, a + d, a + 2d, a + 3d, \ldots, a + (n - 1)d, \ldots$$

where a is the **first term,** $a + (n - 1)d$ is the **nth term,** and d is the **common difference.**

We note that the second term of an arithmetic sequence has an addend of $1d$, the third term has an addend of $2d$, the fourth term has an addend of $3d$, and the nth term has an addend of $(n - 1)d$. We also note that the difference between any two consecutive terms in an arithmetic sequence is d.

EXAMPLE 1

Finding terms of an arithmetic sequence. An arithmetic sequence has a first term of 5 and a common difference of 4.   **a.** Write the first six terms of the sequence.   **b.** Write the 25th term of the sequence.

Solution   **a.** Since the first term is $a = 5$ and the common difference is $d = 4$, the first six terms are

$$5, 5 + 4, 5 + 2(4), 5 + 3(4), 5 + 4(4), 5 + 5(4)$$

or

$$5, 9, 13, 17, 21, 25$$

b. The nth term is $a + (n - 1)d$. Since we want the 25th term, we let $n = 25$:

$n\text{th term} = a + (n - 1)d$

25th term $= 5 + (25 - 1)4$ Remember that $a = 5$ and $d = 4$.

$\phantom{25\text{th term}} = 5 + 24(4)$

$\phantom{25\text{th term}} = 5 + 96$

$\phantom{25\text{th term}} = 101$

SELF CHECK   **a.** Write the seventh term of the sequence of Example 1.   **b.** Write the 30th term.   *Answers:*   **a.** 29,   **b.** 121   ∎

EXAMPLE 2

Finding terms of an arithmetic sequence. The first three terms of an arithmetic sequence are 3, 8, and 13. Find   **a.** the 67th term and   **b.** the 100th term.

Solution   The common difference d is the difference between any two successive terms:

$$d = 8 - 3 = 13 - 8 = 5$$

a. We substitute 3 for a, 67 for n, and 5 for d in the formula for the nth term and simplify:

$$n\text{th term} = a + (n - 1)d$$
$$67\text{th term} = 3 + (67 - 1)5$$
$$= 3 + 66(5)$$
$$= 333$$

b. We substitute 3 for a, 100 for n, and 5 for d in the formula for the nth term and simplify:

$$n\text{th term} = a + (n - 1)d$$
$$100\text{th term} = 3 + (100 - 1)5$$
$$= 3 + 99(5)$$
$$= 498$$

SELF CHECK   Find the 50th term of the sequence in Example 2.   *Answer:* 248   ∎

EXAMPLE 3

Finding terms of an arithmetic sequence. The first term of an arithmetic sequence is 12, and the 50th term is 3,099. Write the first six terms of the sequence.

Solution   The key is to find the common difference. Because the 50th term of this sequence is 3,099, we can let $n = 50$ and solve the following equation for d:

$$50\text{th term} = a + (n - 1)d$$

$3,099 = 12 + (50 - 1)d$	
$3,099 = 12 + 49d$	Simplify.
$3,087 = 49d$	Subtract 12 from both sides.
$63 = d$	Divide both sides by 49.

The first term of the sequence is 12, and the common difference is 63. Its first six terms are

12, 75, 138, 201, 264, 327

SELF CHECK   The first term of an arithmetic sequence is 15, and the 12th term is 92. Write the first four terms of the sequence.   *Answer:* 15, 22, 29, 36   ∎

Arithmetic Means

If numbers are inserted between two numbers a and b to form an arithmetic sequence, the inserted numbers are called **arithmetic means** between a and b. If a single number is inserted between a and b, that number is called **the arithmetic mean** between a and b.

EXAMPLE 4

Inserting arithmetic means. Insert two arithmetic means between 6 and 27.

Solution   Here, the first term is $a = 6$, and the fourth term (or the last term) is $l = 27$. We must find the common difference so that the terms

$6, 6 + d, 6 + 2d, 27$

form an arithmetic sequence. To find d, we substitute 6 for a and 4 for n in the formula for the nth term:

$$n\text{th term} = a + (n - 1)d$$

$$4\text{th term} = 6 + (4 - 1)d$$

$$27 = 6 + 3d \qquad \text{Simplify.}$$

$$21 = 3d \qquad \text{Subtract 6 from both sides.}$$

$$7 = d \qquad \text{Divide both sides by 3.}$$

The two arithmetic means between 6 and 27 are

$$6 + d = 6 + 7 \qquad \text{or} \qquad 6 + 2d = 6 + 2(7)$$
$$= 13 \qquad\qquad\qquad = 6 + 14$$
$$\qquad\qquad\qquad\qquad = 20$$

The numbers 6, 13, 20, and 27 are the first four terms of an arithmetic sequence.

SELF CHECK   Insert two arithmetic means between 8 and 44.   *Answer:* 20, 32   ■

The Sum of the First n Terms of an Arithmetic Sequence

To develop a formula for evaluating the sum of the first n terms of an arithmetic sequence, we let S_n represent the sum of the first n terms of an arithmetic sequence:

$$S_n = \qquad a \qquad + \quad [a + d] \quad + \quad [a + 2d] \quad + \cdots + [a + (n - 1)d]$$

We write the same sum again, but in reverse order:

$$S_n = [a + (n - 1)d] + [a + (n - 2)d] + [a + (n - 3)d] + \cdots + a$$

Adding these equations together, term by term, we get

$$2S_n = [2a + (n - 1)d] + [2a + (n - 1)d] + [2a + (n - 1)d] + \cdots + [2a + (n - 1)d]$$

Because there are n equal terms on the right-hand side of the preceding equation, we can write

$$2S_n = n[2a + (n - 1)d]$$

$$2S_n = n[a + a + (n - 1)d] \qquad \text{Write } 2a \text{ as } a + a.$$

$$2S_n = n(a + l) \qquad\qquad \text{Substitute } l \text{ for } a + (n - 1)d, \text{ because } a + (n - 1)d \text{ is}$$
$$\qquad\qquad\qquad\qquad \text{the last term of the sequence.}$$

$$S_n = \frac{n(a + l)}{2}$$

This reasoning establishes the following formula.

Sum of the first n terms of an arithmetic sequence

The sum of the first n terms of an arithmetic sequence is given by the formula

$$S_n = \frac{n(a + l)}{2} \quad \text{with } l = a + (n - 1)d$$

where a is the first term, l is the last (or nth) term, and n is the number of terms in the sequence.

EXAMPLE 5

Find the sum of the first 40 terms of the arithmetic sequence 4, 10, 16, ...

Solution In this example, we let $a = 4$, $n = 40$, $d = 6$, and $l = 4 + (40 - 1)6 = 238$ and substitute these values into the formula for S_n:

$$S_n = \frac{n(a + l)}{2}$$

$$S_{40} = \frac{40(4 + 238)}{2}$$

$$= 20(242)$$

$$= 4,840$$

The sum of the first 40 terms is 4,840.

SELF CHECK Find the sum of the first 50 terms of the arithmetic sequence 3, 8, 13, ...   *Answer:* **6,275** ∎

Summation Notation

When the general term of a sequence is known, we can use a shorthand notation to write a series. This notation, called **summation notation,** involves the Greek letter Σ (sigma). The expression

$$\sum_{k=2}^{5} 3k \quad \text{Read as "the summation of } 3k \text{ as } k \text{ runs from 2 to 5."}$$

designates the sum of all terms obtained if we successively substitute the numbers 2, 3, 4, and 5 for k, called the **index of the summation.** Thus, we have

$$\sum_{k=2}^{5} 3k = 3(2) + 3(3) + 3(4) + 3(5)$$

$$= 6 + 9 + 12 + 15$$

$$= 42$$

EXAMPLE 6

Find each sum: **a.** $\displaystyle\sum_{k=3}^{5}(2k + 1)$, **b.** $\displaystyle\sum_{k=2}^{5} k^2$, and **c.** $\displaystyle\sum_{k=1}^{3}(3k^2 + 3)$.

Solution **a.** $\displaystyle\sum_{k=3}^{5}(2k + 1) = [2(3) + 1] + [2(4) + 1] + [2(5) + 1]$

$$= 7 + 9 + 11$$

$$= 27$$

b. $\displaystyle\sum_{k=2}^{5} k^2 = 2^2 + 3^2 + 4^2 + 5^2$

$$= 4 + 9 + 16 + 25$$

$$= 54$$

c. $\displaystyle\sum_{k=1}^{3}(3k^2 + 3) = [3(1^2) + 3] + [3(2^2) + 3] + [3(3^2) + 3]$

$$= 6 + 15 + 30$$

$$= 51$$

SELF CHECK Evaluate $\displaystyle\sum_{k=1}^{4}(2k^2 - 2)$.   *Answer:* **52** ∎

STUDY SET

Section 11.2

VOCABULARY

Fill in the blanks to make the statements true.

1. A _____ is a function whose domain is the set of natural numbers.

2. The sequence 1, 1, 2, 3, 5, 8, 13, 21, . . . is called the _____ sequence.

3. A sequence with an unlimited number of terms is called a(n) _____ sequence.

4. A sequence with a specific number of terms is called a(n) _____ sequence.

5. The sum of the terms of a sequence is called a _____.

6. If a number is inserted between a and b to form an arithmetic sequence, the number is called the arithmetic _____ between a and b.

CONCEPTS

Fill in the blanks to make the statements true.

7. The sequence 3, 9, 15, 21, . . . is an example of an infinite _____ sequence with a common _____ of 6.

8. The last term of an arithmetic sequence is given by the formula _____.

9. $3 + 7 + 11 + 14$ is an example of a _____ series.

10. The sum of the first n terms of an arithmetic sequence is given by the formula $S_n =$ _____.

NOTATION

In Exercises 11–12, fill in the blanks to make the statements true.

11. $\sum_{k=1}^{5} k$ means _____.

12. In the symbol $\sum_{k=1}^{5} (2k - 5)$, k is called the _____ of summation.

In Exercises 13–16, write the series associated with each summation.

13. $\sum_{k=1}^{3} k$

14. $\sum_{k=1}^{4} (3k)$

15. $\sum_{k=2}^{4} k^2$

16. $\sum_{k=3}^{5} (-2k)$

In Exercises 17–20, write the summation notation for each sum.

17. $3 + 4 + 5 + 6$

18. $1 + 4 + 9 + 16 + 25$

19. $2 + 4 + 6 + 8$

20. $-1 - 4 - 9 - 16 - 25 - 36$

PRACTICE

In Exercises 21–34, write the first five terms of each arithmetic sequence with the given properties.

21. $a = 3, d = 2$

22. $a = -2, d = 3$

23. $a = -5, d = -3$

24. $a = 8, d = -5$

25. $a = 5$, fifth term is 29

26. $a = 4$, sixth term is 39

27. $a = -4$, sixth term is -39

28. $a = -5$, fifth term is -37

29. $d = 7$, sixth term is -83

30. $d = 3$, seventh term is 12

31. $d = -3$, seventh term is 16

32. $d = -5$, seventh term is -12

33. The 19th term is 131 and the 20th term is 138.

34. The 16th term is 70 and the 18th term is 78.

35. Find the 30th term of the arithmetic sequence with $a = 7$ and $d = 12$.

36. Find the 55th term of the arithmetic sequence with $a = -5$ and $d = 4$.

37. Find the 37th term of the arithmetic sequence with a second term of -4 and a third term of -9.

38. Find the 40th term of the arithmetic sequence with a second term of 6 and a fourth term of 16.

39. Find the first term of the arithmetic sequence with a common difference of 11 if its 27th term is 263.

40. Find the common difference of the arithmetic sequence with a first term of -164 if its 36th term is -24.

41. Find the common difference of the arithmetic sequence with a first term of 40 if its 44th term is 556.

42. Find the first term of the arithmetic sequence with a common difference of -5 if its 23rd term is -625.

43. Insert three arithmetic means between 2 and 11.

44. Insert four arithmetic means between 5 and 25.

45. Insert four arithmetic means between 10 and 20.

46. Insert three arithmetic means between 20 and 30.

47. Find the arithmetic mean between 10 and 19.
48. Find the arithmetic mean between 5 and 23.
49. Find the arithmetic mean between -4.5 and 7.
50. Find the arithmetic mean between -6.3 and -5.2.

In Exercises 51–58, find the sum of the first n terms of each arithmetic sequence.

51. 1, 4, 7, . . .; $n = 30$
52. 2, 6, 10, . . .; $n = 28$
53. $-5, -1, 3, . . .$; $n = 17$
54. $-7, -1, 5, . . .$; $n = 15$
55. Second term is 7, third term is 12; $n = 12$
56. Second term is 5, fourth term is 9; $n = 16$
57. $f(n) = 2n + 1$, nth term is 31; n is a natural number

58. $f(n) = 4n + 3$, nth term is 23; n is a natural number

59. Find the sum of the first 50 natural numbers.
60. Find the sum of the first 100 natural numbers.
61. Find the sum of the first 50 odd natural numbers.

62. Find the sum of the first 50 even natural numbers.

In Exercises 63–70, find each sum.

63. $\displaystyle\sum_{k=1}^{4} (6k)$
64. $\displaystyle\sum_{k=2}^{5} (3k)$

65. $\displaystyle\sum_{k=3}^{4} k^3$
66. $\displaystyle\sum_{k=2}^{4} (-k^2)$

67. $\displaystyle\sum_{k=3}^{4} (k^2 + 3)$
68. $\displaystyle\sum_{k=2}^{6} (k^2 + 1)$

69. $\displaystyle\sum_{k=4}^{4} (2k + 4)$
70. $\displaystyle\sum_{k=3}^{5} (3k^2 - 7)$

APPLICATIONS

71. SAVING MONEY Yasmeen puts \$60 into a safety deposit box. After each succeeding month, she puts \$50 more in the box. Write the first six terms of an arithmetic sequence that gives the monthly amounts in her savings, and find her savings after 10 years.

72. INSTALLMENT LOANS Maria borrowed \$10,000, interest-free, from her mother. She agreed to pay back the loan in monthly installments of \$275. Write the first six terms of an arithmetic sequence that shows the balance due after each month, and find the balance due after 17 months.

73. DESIGNING PATIOS Each row of bricks in the triangular patio in Illustration 1 is to have one more brick than the previous row, ending with the longest row of 150 bricks. How many bricks will be needed?

ILLUSTRATION 1

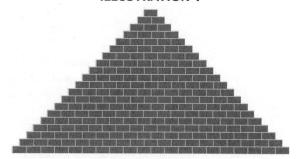

74. FALLING OBJECTS The equation $s = 16t^2$ represents the distance s in feet that an object will fall in t seconds. After 1 second, the object has fallen 16 feet. After 2 seconds, it has fallen 64 feet, and so on. Find the distance that the object will fall during the second and third seconds.

75. FALLING OBJECTS Refer to Exercise 74. How far will the object fall during the 12th second?

76. INTERIOR ANGLES The sums of the angles of several polygons are given in the table shown in Illustration 2. Assuming that the pattern continues, complete the table.

ILLUSTRATION 2

Figure	Number of sides	Sum of angles
Triangle	3	180°
Quadrilateral	4	360°
Pentagon	5	540°
Hexagon	6	720°
Octagon	8	
Dodecagon	12	

WRITING

77. Define an arithmetic sequence.

78. Develop the formula for finding the sum of the first n terms of an arithmetic sequence.

REVIEW

Do the operations and simplify, if possible.

79. $3(2x^2 - 4x + 7) + 4(3x^2 + 5x - 6)$,

80. $(2p + q)(3p^2 + 4pq - 3q^2)$

81. $\dfrac{3a + 4}{a - 2} + \dfrac{3a - 4}{a + 2}$

82. $2t - 3\overline{)8t^4 - 12t^3 + 8t^2 - 16t + 6}$

▶ 11.3

Geometric Sequences

In this section, you will learn about

> Geometric sequences ■ Geometric means ■ The sum of the first n terms of a geometric sequence ■ Population growth ■ Infinite geometric series

---

Introduction   In the previous section, we saw that the same number is added to each term of an arithmetic sequence to get the next term. In this section, we will consider another type of sequence where we multiply each term by the same number to get the next term. This type of sequence is called a *geometric sequence.* Two examples of geometric sequences are

$2, 6, 18, 54, \ldots$ This is an infinite geometric sequence where each term is found by multiplying the previous term by 3.

$27, 9, 3, 1, \dfrac{1}{3}, \dfrac{1}{9}$ This is a finite geometric sequence where each term is found by multiplying the previous term by $\frac{1}{3}$.

Geometric Sequences

Every term of a geometric sequence is found by multiplying the previous term by the same number.

Geometric sequence

> A **geometric sequence** is a sequence of the form
>
> $$a, ar, ar^2, ar^3, \ldots, ar^{n-1}, \ldots$$
>
> where a is the **first term,** ar^{n-1} is the **nth term,** and r is the **common ratio.**

We note that the second term of a geometric sequence has a factor of r^1, the third term has a factor of r^2, the fourth term has a factor of r^3, and the nth term has a factor of r^{n-1}. We also note that r is the quotient obtained when any term is divided by the previous term.

EXAMPLE 1

Finding terms of a geometric sequence. A geometric sequence has a first term of 5 and a common ration of 3.

a. Write the first five terms of the sequence.
b. Find the ninth term.

Solution **a.** Because the first term is $a = 5$ and the common ratio is $r = 3$, the first five terms are

$$5, 5(3), 5(3^2), 5(3^3), 5(3^4)$$

or

$$5, 15, 45, 135, 405$$

b. The nth term is ar^{n-1} with $a = 5$ and $r = 3$. Because we want the ninth term, we let $n = 9$:

$$n\text{th term} = ar^{n-1}$$
$$9\text{th term} = 5(3)^{9-1}$$
$$= 5(3)^8$$
$$= 5(6,561)$$
$$= 32,805$$

SELF CHECK

A geometric sequence has a first term of 3 and a common ratio of 4. **a.** Write the first four terms. **b.** Find the eighth term.

Answers: **a.** 3, 12, 48, 192, **b.** 49,152 ∎

EXAMPLE 2

Finding terms of a geometric sequence. The first three terms of a geometric sequence are 16, 4, and 1. Find the seventh term.

Solution To determine r, we divide the second term by the first term to obtain $\frac{4}{16} = \frac{1}{4}$. We substitute 16 for a, $\frac{1}{4}$ for r, and 7 for n in the formula for the nth term and simplify:

$$n\text{th term} = ar^{n-1}$$
$$7\text{th term} = 16\left(\frac{1}{4}\right)^{7-1}$$
$$= 16\left(\frac{1}{4}\right)^6$$
$$= 16\left(\frac{1}{4,096}\right)$$
$$= \frac{1}{256}$$

SELF CHECK

Find the tenth term of the geometric sequence of Example 2.

Answer: $\frac{1}{16,384}$ ∎

Geometric Means

If numbers are inserted between two numbers a and b to form a geometric sequence, the inserted numbers are called **geometric means** between a and b. If a single number is inserted between a and b, that number is called a **geometric mean** between a and b.

E X A M P L E   3

Inserting geometric means. Insert two geometric means between 7 and 1,512.

Solution   In this example, the first term is $a = 7$, and the fourth term (or last term) is $l = 1,512$. To find the common ratio r so that the terms

$$7, \quad 7r, \quad 7r^2, \quad 1,512$$

form a geometric sequence, we substitute 4 for n and 7 for a in the formula for the nth term of a geometric sequence and solve for r.

$$\textbf{nth term} = ar^{n-1}$$
$$\textbf{4th term} = 7r^{4-1}$$
$$1,512 = 7r^3$$
$$216 = r^3 \qquad \text{Divide both sides by 7.}$$
$$6 = r \qquad \text{Take the cube root of both sides.}$$

The two geometric means between 7 and 1,512 are

$$7r = 7(6) = \mathbf{42}$$

and

$$7r^2 = 7(6)^2 = 7(36) = \mathbf{252}$$

The numbers 7, 42, 252, and 1,512 are the first four terms of a geometric sequence.

SELF CHECK

Insert three geometric means between 1 and 16.   *Answer:* 2, 4, 8   ∎

E X A M P L E   4

Finding a geometric mean. Find a geometric mean between 2 and 20.

Solution   We want to find the middle term of the three-termed geometric sequence

$$2, \quad 2r, \quad 20$$

with $a = 2$, $l = 20$, and $n = 3$. To find r, we substitute these values into the formula for the nth term of a geometric sequence:

$$\textbf{nth term} = ar^{n-1}$$
$$\textbf{3rd term} = 2r^{3-1}$$
$$20 = 2r^2$$
$$10 = r^2 \qquad \text{Divide both sides by 2.}$$
$$\pm\sqrt{10} = r \qquad \text{Take the square root of both sides.}$$

Because r can be either $\sqrt{10}$ or $-\sqrt{10}$, there are two values for a geometric mean. They are

$$2r = 2\sqrt{10}$$

and

$$2r = -2\sqrt{10}$$

The numbers 2, $2\sqrt{10}$, 20 and 2, $-2\sqrt{10}$, 20 both form geometric sequences. The common ratio of the first sequence is $\sqrt{10}$, and the common ratio of the second sequence is $-\sqrt{10}$.

SELF CHECK

Find the positive geometric mean between 2 and 200.   *Answer:* 20   ∎

The Sum of the First n Terms of a Geometric Sequence

When we add the terms of a geometric sequence, we form a **geometric series.** There is a formula that gives the sum of the first n terms of a geometric sequence. To develop this formula, we let S_n represent the sum of the first n terms of a geometric sequence.

1. $S_n = a + ar + ar^2 + ar^3 + \cdots + ar^{n-1}$

We multiply both sides of Equation 1 by r to get

2. $S_n r = \qquad ar + ar^2 + ar^3 + \cdots + ar^{n-1} + ar^n$

We now subtract Equation 2 from Equation 1 and solve for S_n:

$$S_n - S_n r = a - ar^n$$
$$S_n(1 - r) = a - ar^n \quad \text{Factor out } S_n \text{ from the left-hand side.}$$
$$S_n = \frac{a - ar^n}{1 - r} \quad \text{Divide both sides by } 1 - r.$$

This reasoning establishes the following formula.

Sum of the first n terms of a geometric sequence

The sum of the first n terms of a geometric sequence is given by the formula

$$S_n = \frac{a - ar^n}{1 - r} \quad (r \neq 1)$$

where S_n is the sum, a is the first term, r is the common ratio, and n is the number of terms.

EXAMPLE 5

Evaluating a geometric series. Find the sum of the first six terms of the geometric sequence 250, 50, 10, ...

Solution In this geometric sequence, $a = 250$, $r = \frac{1}{5}$, and $n = 6$. We substitute these values into the formula for the sum of the first n terms of a geometric sequence and simplify:

$$S_n = \frac{a - ar^n}{1 - r}$$

$$S_6 = \frac{250 - 250\left(\dfrac{1}{5}\right)^6}{1 - \dfrac{1}{5}}$$

$$= \frac{250 - 250\left(\dfrac{1}{15{,}625}\right)}{\dfrac{4}{5}}$$

$$= \frac{5}{4}\left(250 - \frac{250}{15{,}625}\right)$$

$$= \frac{5}{4}\left(\frac{3{,}906{,}000}{15{,}625}\right)$$

$$= 312.48$$

The sum of the first six terms is 312.48.

SELF CHECK

Find the sum of the first five terms of the geometric sequence 100, 20, 4, ...

Answer: 124.96   ∎

Population Growth

EXAMPLE 6

The mayor of Leaf River (population 1,500) predicts a growth rate of 4% each year for
the next ten years. Find the expected population of Leaf River ten years from now.

Solution

Let P_0 be the initial population of Leaf River. After 1 year, there will be a different
population, P_1. The initial population (P_0) plus the growth (the product of P_0 and the
rate of growth, r) will equal this new population, P_1:

$$P_1 = P_0 + P_0 r = P_0(1 + r)$$

The population after 2 years will be P_2, and

$$P_2 = P_1 + P_1 r$$
$$= P_1(1 + r) \qquad \text{Factor out } P_1.$$
$$= P_0(1 + r)(1 + r) \quad \text{Remember that } P_1 = P_0(1 + r).$$
$$= P_0(1 + r)^2$$

The population after 3 years will be P_3, and

$$P_3 = P_2 + P_2 r$$
$$= P_2(1 + r) \qquad \text{Factor out } P_2.$$
$$= P_0(1 + r)^2(1 + r) \quad \text{Remember that } P_2 = P_0(1 + r)^2.$$
$$= P_0(1 + r)^3$$

The yearly population figures

$$P_0, \quad P_1, \quad P_2, \quad P_3, \ldots$$

or

$$P_0, \quad P_0(1 + r), \quad P_0(1 + r)^2, \quad P_0(1 + r)^3, \ldots$$

form a geometric sequence with a first term of P_0 and a common ratio of $1 + r$. The
population of Leaf River after 10 years is P_{10}, which is the 11th term of this sequence:

$$n\text{th term} = ar^{n-1}$$
$$P_{10} = 11\text{th term} = P_0(1 + r)^{10}$$
$$= 1,500(1 + 0.04)^{10}$$
$$= 1,500(1.04)^{10}$$
$$\approx 1,500(1.480244285) \quad \text{Use a calculator.}$$
$$\approx 2,220$$

The expected population ten years from now is 2,220 people.   ∎

Infinite Geometric Series

If we form the sum of the terms of an infinite geometric sequence, we get a series
called an **infinite geometric series**. For example, if the common ratio r is 3, we have

Infinite geometric sequence	*Infinite geometric series*
2, 6, 18, 54, 162, . . .	2 + 6 + 18 + 54 + 162 + · · ·

As the number of terms of this series gets larger, the value of the series gets larger. We can see that this is true by forming some **partial sums.**

The first partial sum, S_1, of the series is $S_1 = 2$.

The second partial sum, S_2, of the series is $S_2 = 2 + 6 = 8$.

The third partial sum, S_3, of the series is $S_3 = 2 + 6 + 18 = 26$.

The fourth partial sum, S_4, of the series is $S_4 = 2 + 6 + 18 + 54 = 80$.

We can now see that as the number of terms gets infinitely large, the value of the series will get infinitely large. The values of some infinite geometric series get closer and closer to a specific number as the number of terms approaches infinity. One such series is

$$\frac{3}{2} + \frac{3}{4} + \frac{3}{8} + \frac{3}{16} + \frac{3}{32} \ldots \quad \text{Here, } r = \tfrac{1}{2}.$$

To see that this is true, we form some partial sums.

The first partial sum is $S_1 = \frac{3}{2} = 1.5$.

The second partial sum is $S_2 = \frac{3}{2} + \frac{3}{4} = \frac{9}{4} = 2.25$.

The third partial sum is $S_3 = \frac{3}{2} + \frac{3}{4} + \frac{3}{8} = \frac{21}{8} = 2.625$.

The fourth partial sum is $S_4 = \frac{3}{2} + \frac{3}{4} + \frac{3}{8} + \frac{3}{16} = \frac{45}{16} = 2.8125$.

The fifth partial sum is $S_5 = \frac{3}{2} + \frac{3}{4} + \frac{3}{8} + \frac{3}{16} + \frac{3}{32} = \frac{93}{32} = 2.90625$.

As the number of terms in this series gets larger, the values of the partial sums will approach the number 3. We say that 3 is the **limit** of S_n as n approaches infinity, and we say that 3 is the **sum of the infinite geometric series.**

To develop a formula for finding the sum of an infinite geometric series, we consider the formula that gives the sum of the first n terms.

$$S_n = \frac{a - ar^n}{1 - r} \quad (r \neq 1)$$

If $|r| < 1$ and a is constant, the term ar^n in the above formula approaches 0 as n becomes very large. For example,

$$a\left(\frac{1}{2}\right)^1 = \frac{1}{2}a, \qquad a\left(\frac{1}{2}\right)^2 = \frac{1}{4}a, \qquad a\left(\frac{1}{2}\right)^3 = \frac{1}{8}a$$

and so on. Thus, when n is very large, the value of ar^n is negligible, and the term ar^n in the above formula can be ignored. This reasoning justifies the following formula.

Sum of an infinite geometric sequence

If a is the first term and r is the common ratio of an infinite geometric sequence, and if $|r| < 1$, the sum of the terms of the sequence is given by the formula

$$S = \frac{a}{1 - r}$$

EXAMPLE 7

Finding the sum of an infinite geometric sequence. Find the sum of the terms of the infinite geometric sequence 125, 25, 5, ...

Solution In this geometric sequence, $a = 125$ and $r = \frac{1}{5}$. Since $|r| = \left|\frac{1}{5}\right| = \frac{1}{5} < 1$, we can find the sum of all the terms of the sequence. We do this by substituting 125 for a and $\frac{1}{5}$ for r in the formula $S = \frac{a}{1-r}$ and simplifying:

$$S = \frac{a}{1-r} = \frac{125}{1 - \dfrac{1}{5}} = \frac{125}{\dfrac{4}{5}} = \frac{5}{4}(125) = \frac{625}{4}$$

The sum of the terms of the sequence 125, 25, 5, . . . is 156.25.

SELF CHECK Find the sum of the terms of the infinite geometric sequence 100, 20, 4, ...   *Answer:* 125   ∎

EXAMPLE 8

Finding the sum of an infinite geometric sequence. Find the sum of the infinite geometric sequence 64, −4, $\frac{1}{4}$, ...

Solution In this geometric sequence, $a = 64$ and $r = -\frac{1}{16}$. Since $|r| = \left|-\frac{1}{16}\right| = \frac{1}{16} < 1$, we can find the sum of all the terms of the sequence. We substitute 64 for a and $-\frac{1}{16}$ for r in the formula $S = \frac{a}{1-r}$ and simplify:

$$S = \frac{a}{1-r} = \frac{64}{1 - \left(-\dfrac{1}{16}\right)} = \frac{64}{\dfrac{17}{16}} = \frac{16}{17}(64) = \frac{1{,}024}{17}$$

The sum of the terms of the geometric sequence 64, −4, $\frac{1}{4}$, . . . is $\frac{1{,}024}{17}$.

SELF CHECK Find the sum of the infinite geometric sequence 81, 27, 9, ...   *Answer:* $\frac{243}{2}$ ∎

EXAMPLE 9

Changing a repeating decimal to a fraction. Change $0.\overline{8}$ to a common fraction.

Solution The decimal $0.\overline{8}$ can be written as the infinite series

$$0.\overline{8} = 0.888 \ldots = \frac{8}{10} + \frac{8}{100} + \frac{8}{1{,}000} + \cdots$$

where $a = \frac{8}{10}$ and $r = \frac{1}{10}$. Because $|r| = \left|\frac{1}{10}\right| = \frac{1}{10} < 1$, we can find the sum as follows:

$$S = \frac{a}{1-r} = \frac{\dfrac{8}{10}}{1 - \dfrac{1}{10}} = \frac{\dfrac{8}{10}}{\dfrac{9}{10}} = \frac{8}{9}$$

Thus, $0.\overline{8} = \frac{8}{9}$. Long division will verify that $\frac{8}{9} = 0.888 \ldots$.

SELF CHECK Change $0.\overline{6}$ to a common fraction.   *Answer:* $\frac{2}{3}$ ∎

EXAMPLE 10

Changing a repeating decimal to a fraction. Change $0.\overline{25}$ to a common fraction.

Solution The decimal $0.\overline{25}$ can be written as the infinite series

$$0.\overline{25} = 0.252525 \ldots = \frac{25}{100} + \frac{25}{10{,}000} + \frac{25}{1{,}000{,}000} + \cdots$$

where $a = \frac{25}{100}$ and $r = \frac{1}{100}$. Since $|r| = \left|\frac{1}{100}\right| = \frac{1}{100} < 1$, we can find the sum as follows:

$$S = \frac{a}{1-r} = \frac{\dfrac{25}{100}}{1 - \dfrac{1}{100}} = \frac{\dfrac{25}{100}}{\dfrac{99}{100}} = \frac{25}{99}$$

Thus, $0.\overline{25} = \frac{25}{99}$. Long division will verify that this is true.

SELF CHECK   Change $0.\overline{15}$ to a common fraction.   *Answer:* $\frac{5}{33}$ ■

STUDY SET

Section 11.3

VOCABULARY

Fill in the blanks to make the statements true.

1. A sequence of the form a, ar, ar^2, . . . is called a(n) _____ sequence.
2. If a geometric sequence has infinitely many terms, it is called a(n) _____ geometric sequence.
3. A number inserted between two numbers a and b is called a geometric _____ between a and b.
4. In a geometric sequence, r is called the _____.

CONCEPTS

Fill in the blanks to make the statements true.

5. The formula for the nth term of a geometric sequence is _____.
6. The sum of the first n terms of a geometric sequence

   is given by the formula _____.

7. The third partial sum of the sequence 2, 6, 18, 54, . . . is _____ = 26.
8. If we write the sum of the terms of an infinite geometric sequence, we form an infinite geometric _____.
9. The formula for the sum of the terms of an infinite geometric sequence is _____.
10. Write $0.\overline{75}$ as an infinite geometric series.

    _____.

NOTATION

Fill in the blanks to make the statements true.

11. In the formula for Exercise 6, a is the _____ term of the sequence, r is the common _____, and n is the number of _____.

12. The symbol S_3 represents the third _____ sum of an infinite geometric series.

PRACTICE

In Exercises 13–26, write the first five terms of each geometric sequence with the given properties.

13. $a = 3$, $r = 2$
14. $a = -2$, $r = 2$
15. $a = -5$, $r = \frac{1}{5}$
16. $a = 8$, $r = \frac{1}{2}$
17. $a = 2$, $r > 0$, third term is 32
18. $a = 3$, fourth term is 24
19. $a = -3$, fourth term is -192

20. $a = 2$, $r < 0$, third term is 50
21. $a = -64$, $r < 0$, fifth term is -4
22. $a = -64$, $r > 0$, fifth term is -4

23. $a = -64$, sixth term is -2
24. $a = -81$, sixth term is $\frac{1}{3}$
25. The second term is 10, and the third term is 50.

26. The third term is -27, and the fourth term is 81.

27. Find the tenth term of the geometric sequence with $a = 7$ and $r = 2$.
28. Find the 12th term of the geometric sequence with $a = 64$ and $r = \frac{1}{2}$.
29. Find the first term of the geometric sequence with a common ratio of -3 and an eighth term of -81.
30. Find the first term of the geometric sequence with a common ratio of 2 and a tenth term of 384.

31. Find the common ratio of the geometric sequence with a first term of -8 and a sixth term of $-1,944$.

32. Find the common ratio of the geometric sequence with a first term of 12 and a sixth term of $\frac{3}{8}$.

33. Insert three positive geometric means between 2 and 162.

34. Insert four geometric means between 3 and 96.

35. Insert four geometric means between -4 and $-12,500$.

36. Insert three geometric means (two positive and one negative) between -64 and $-1,024$.

37. Find the negative geometric mean between 2 and 128.

38. Find the positive geometric mean between 3 and 243.

39. Find the positive geometric mean between 10 and 20.

40. Find the negative geometric mean between 5 and 15.

41. Find a geometric mean, if possible, between -50 and 10.

42. Find a negative geometric mean, if possible, between -25 and -5.

In Exercises 43–54, find the sum of the first n terms of each geometric sequence.

43. $2, 6, 18, \ldots; n = 6$

44. $2, -6, 18, \ldots; n = 6$

45. $2, -6, 18, \ldots; n = 5$

46. $2, 6, 18, \ldots; n = 5$

47. $3, -6, 12, \ldots; n = 8$

48. $3, 6, 12, \ldots; n = 8$

49. $3, 6, 12, \ldots; n = 7$

50. $3, -6, 12, \ldots; n = 7$

51. The second term is 1 and the third term is $\frac{1}{5}$; $n = 4$.

52. The second term is 1 and the third term is 4; $n = 5$.

53. The third term is -2 and the fourth term is 1; $n = 6$.

54. The third term is -3 and the fourth term is 1; $n = 5$.

In Exercises 55–66, find the sum of each infinite geometric series, if possible.

55. $8 + 4 + 2 + \cdots$

56. $12 + 6 + 3 + \cdots$

57. $54 + 8 + 6 + \cdots$

58. $45 + 15 + 5 + \cdots$

59. $12 - 6 + 3 - \cdots$

60. $8 - 4 + 2 - \cdots$

61. $-45 + 15 - 5 + \cdots$

62. $-54 + 18 - 6 + \cdots$

63. $\frac{9}{2} + 6 + 8 + \cdots$

64. $-112 - 28 - 7 - \cdots$

65. $-\frac{27}{2} - 9 - 6 - \cdots$

66. $\frac{18}{25} + \frac{6}{5} + 2 + \cdots$

In Exercises 67–74, write each decimal in fraction form. Then check the answer by doing a long division.

67. $0.\overline{1}$

68. $0.\overline{2}$

69. $-0.\overline{3}$

70. $-0.\overline{4}$

71. $0.\overline{12}$

72. $0.\overline{21}$

73. $0.\overline{75}$

74. $0.\overline{57}$

APPLICATIONS

Use a calculator to solve each problem.

75. POPULATION GROWTH The population of Union is predicted to increase by 6% each year. What will the population of Union be 5 years from now if its current population is 500?

76. POPULATION DECLINE The population of Bensville is decreasing by 10% each year. If its current population is 98, what will the population be 8 years from now?

77. DECLINING SAVINGS John has $10,000 in a safety deposit box. Each year, he spends 12% of what is left in the box. How much will be in the box after 15 years?

78. SAVINGS GROWTH Sally has $5,000 in a savings account earning 12% annual interest. How much will be in her account 10 years from now? (Assume that Sally makes no deposits or withdrawals.)

79. HOUSE APPRECIATION A house appreciates by 6% each year. If the house is worth $70,000 today, how much will it be worth 12 years from now?

80. BOAT DEPRECIATION A motorboat that cost $5,000 when new depreciates at a rate of 9% per year. How much will the boat be worth in 5 years?

81. INSCRIBED SQUARES Each inscribed square in Illustration 1 joins the midpoints of the next larger square. The area of the first square, the largest, is 1. Find the area of the 12th square.

82. GENEALOGY The family tree in Illustration 2 spans 3 generations and lists 7 people. How many names would be listed in a family tree that spans 10 generations?

ILLUSTRATION 1

ILLUSTRATION 2

83. BOUNCING BALL On each bounce, the rubber ball in Illustration 3 rebounds to a height one-half of that from which it fell. Find the total vertical distance the ball travels.

ILLUSTRATION 3

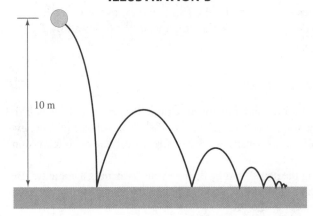

10 m

84. BOUNCING BALL A golf ball is dropped from a height of 12 feet. On each bounce, it returns to a height two-thirds of that from which it fell. Find the total vertical distance the ball travels.

85. PEST CONTROL To reduce the population of a destructive moth, biologists release 1,000 sterilized male moths each day into the environment. If 80% of these moths alive one day survive until the next, then after a long time the population of sterile males is the sum of the infinite geometric series

$$1,000 + 1,000(0.8) + 1,000(0.8)^2 + 1,000(0.8)^3 + \cdots$$

Find the long-term population.

86. PEST CONTROL If mild weather increases the day-to-day survival rate of the sterile male moths in Exercise 85 to 90%, find the long-term population.

WRITING

87. Define a geometric sequence.

88. Develop the formula for finding the sum of the first n terms of a geometric sequence.

89. Why must the common ratio be less than 1 before an infinite geometric sequence can have a sum?

90. If its common difference is not 0, can an infinite arithmetic sequence have a sum?

REVIEW

Solve each inequality.

91. $x^2 - 5x - 6 \le 0$

92. $a^2 - 7a + 12 \ge 0$

93. $\dfrac{x-4}{x+3} > 0$

94. $\dfrac{t^2 + t - 20}{t+2} < 0$

Determine whether each equation determines y to be a function of x.

95. $y = 3x^3 - 4$

96. $xy = 12$

97. $3x = y^2 + 4$

98. $x = |y|$

▶ 11.4   Permutations and Combinations

In this section, you will learn about

■ The multiplication principle for events ■ Permutations
■ Combinations ■ Alternative form of the binomial theorem

Introduction In this section, we will discuss methods of counting the different ways we can do something like lining up in a row or arranging books on a shelf. These kinds of problems are important in the fields of statistics, insurance, and telecommunications. Although one might think that counting problems are easy to solve, they can be deceptively difficult.

The Multiplication Principle for Events

Steven goes to the cafeteria for lunch. He has a choice of three different sandwiches (hamburger, hot dog, or ham and cheese) and four different beverages (cola, root beer, orange, or milk). His different options are shown in the *tree diagram* in Figure 11-1.

FIGURE 11-1

The tree diagram shows that there is a total of 12 different lunches to choose from. One possibility is a hamburger with a cola, and another is a hot dog with milk.

A situation that can have several different outcomes—such as choosing a sandwich—is called an **event.** Choosing a sandwich and choosing a beverage can be thought of as two events. The preceding example illustrates the **multiplication principle for events.**

Multiplication principle for events

Let E_1 and E_2 be two events. If E_1 can be done in a_1 ways, and if (after E_1 has occurred) E_2 can be done in a_2 ways, then the event "E_1 followed by E_2" can be done in $a_1 \cdot a_2$ ways.

EXAMPLE 1

Watching television. Before studying for an exam, Taylor plans to watch the evening news and then a situation comedy on television. If she has a choice of four news broadcasts and two comedies, in how many ways can she choose to watch television?

Solution Let E_1 be the event "watching the news" and E_2 be the event "watching a comedy." Because there are four ways to accomplish E_1 and two ways to accomplish E_2, the number of choices that Taylor has is $4 \cdot 2 = 8$.

SELF CHECK If Alex has 7 shirts and 5 pairs of pants, how many outfits could he wear? *Answer:* 35 ∎

The multiplication principle can be extended to any number of events. In Example 2, we use it to compute the number of ways in which we can arrange objects in a row.

EXAMPLE 2

Arranging books. In how many ways can we arrange five books on a shelf?

Solution We can fill the first space with any of the 5 books, the second space with any of the remaining 4 books, the third space with any of the remaining 3 books, the fourth space with any of the remaining 2 books, and the fifth space with the remaining 1 (or last) book. By the multiplication principle for events, the number of ways that the books can be arranged is

$$5 \cdot 4 \cdot 3 \cdot 2 \cdot 1 = 120$$

SELF CHECK In how many ways can 4 people line up in a row? *Answer:* 24 ∎

EXAMPLE 3 **Signal flags.** If a sailor has six flags (each of a different color) to hang on a flagpole, how many different signals can the sailor send by using four flags?

Solution The sailor must find the number of arrangements of 4 flags when there are 6 flags to choose from. The sailor can hang any one of the 6 flags in the top position, any one of the remaining 5 flags in the second position, any one of the remaining 4 flags in the third position, and any one of the remaining 3 flags in the lowest position. By the multiplication principle for events, the total number of signals that can be sent is

$$6 \cdot 5 \cdot 4 \cdot 3 = 360$$

SELF CHECK In Example 3, how many different signals can the sailor send if each signal uses three flags? *Answer:* 120 ∎

Permutations

When counting a number of possible arrangements such as books on a shelf or flags on a pole, we are finding the number of **permutations** of those objects. In Example 2, we found that the number of permutations of five books, using all five of them, is 120. In Example 3, we found that the number of permutations of six flags, using four of them, is 360.

The symbol $P(n, r)$, read as "the number of permutations of n things r at a time," is often used to express permutation problems. In Example 2, we found that $P(5, 5) = 120$. In Example 3, we found that $P(6, 4) = 360$.

EXAMPLE 4 **Signal flags.** If Sarah has seven flags (each of a different color) to hang on a flagpole, how many different signals can she send by using three flags?

Solution We must find $P(7, 3)$ (the number of permutations of 7 things 3 at a time). In the top position Sarah can hang any of the 7 flags, in the middle position any one of the remaining 6 flags, and in the bottom position any one of the remaining 5 flags. By the multiplication principle for events,

$$P(7, 3) = 7 \cdot 6 \cdot 5 = 210$$

Sarah can send 210 signals using only three of the seven flags.

SELF CHECK See Example 4. How many different signals can Sarah send using four flags? *Answer:* 840 ∎

Although it is correct to write $P(7, 3) = 7 \cdot 6 \cdot 5$, there is an advantage in changing the form of this answer to obtain a formula for computing $P(7, 3)$:

$$
\begin{aligned}
P(7, 3) &= 7 \cdot 6 \cdot 5 \\
&= \frac{7 \cdot 6 \cdot 5 \cdot 4 \cdot 3 \cdot 2 \cdot 1}{4 \cdot 3 \cdot 2 \cdot 1} \quad \text{Multiply both the numerator and denominator by } 4 \cdot 3 \cdot 2 \cdot 1. \\
&= \frac{7!}{4!} \\
&= \frac{7!}{(7 - 3)!}
\end{aligned}
$$

The generalization of this idea gives the following formula.

Finding $P(n, r)$

The number of permutations of n things r at a time is given by the formula

$$P(n, r) = \frac{n!}{(n - r)!}$$

EXAMPLE 5

Counting permutations. Compute **a.** $P(8, 2)$, **b.** $P(7, 5)$, **c.** $P(n, n)$, and **d.** $P(n, 0)$.

Solution
a. $P(8, 2) = \dfrac{8!}{(8 - 2)!}$

$\qquad = \dfrac{8 \cdot 7 \cdot 6!}{6!}$

$\qquad = 8 \cdot 7$

$\qquad = 56$

b. $P(7, 5) = \dfrac{7!}{(7 - 5)!}$

$\qquad = \dfrac{7 \cdot 6 \cdot 5 \cdot 4 \cdot 3 \cdot 2!}{2!}$

$\qquad = 7 \cdot 6 \cdot 5 \cdot 4 \cdot 3$

$\qquad = 2{,}520$

c. $P(n, n) = \dfrac{n!}{(n - n)!}$

$\qquad = \dfrac{n!}{0!}$

$\qquad = \dfrac{n!}{1}$

$\qquad = n!$

d. $P(n, 0) = \dfrac{n!}{(n - 0)!}$

$\qquad = \dfrac{n!}{n!}$

$\qquad = 1$

SELF CHECK Compute **a.** $P(10, 6)$ and **b.** $P(10, 10)$. *Answers:* **a.** 151,200, **b.** 1 ∎

Parts c and d of Example 5 establish the following formulas.

Finding $P(n, n)$ and $P(n, 0)$

The number of permutations of n things n at a time and n things 0 at a time are given by the formulas

$$P(n, n) = n! \quad \text{and} \quad P(n, 0) = 1$$

EXAMPLE 6

Television schedules. **a.** In how many ways can a television executive arrange the Saturday night lineup of 6 programs if there are 15 programs to choose from? **b.** If there are only 6 programs to choose from?

Solution
a. To find the number of permutations of 15 programs 6 at a time, we use the formula $P(n, r) = \frac{n!}{(n - r)!}$ with $n = 15$ and $r = 6$.

$$P(15, 6) = \frac{15!}{(15 - 6)!}$$

$$= \frac{15 \cdot 14 \cdot 13 \cdot 12 \cdot 11 \cdot 10 \cdot 9!}{9!}$$

$$= 15 \cdot 14 \cdot 13 \cdot 12 \cdot 11 \cdot 10$$

$$= 3{,}603{,}600$$

b. To find the number of permutations of 6 programs 6 at a time, we use the formula $P(n, n) = n!$ with $n = 6$.

$$P(6, 6) = 6! = 720$$

SELF CHECK

See Example 6. How many ways are there if the executive has 20 programs to choose from?   *Answer: 27,907,200*   ∎

Combinations

Suppose that Raul must read 4 books from a reading list of 10 books. The order in which he reads them is not important. For the moment, however, let's assume that order is important and find the number of permutations of 10 things 4 at a time:

$$P(10, 4) = \frac{10!}{(10 - 4)!}$$
$$= \frac{10 \cdot 9 \cdot 8 \cdot 7 \cdot 6!}{6!}$$
$$= 10 \cdot 9 \cdot 8 \cdot 7$$
$$= 5,040$$

If order is important, there are 5,040 ways of choosing 4 books when there are 10 books to choose from. However, because the order in which Raul reads the books does not matter, the previous result of 5,040 is too big. Since there are 24 (or 4!) ways of ordering the 4 books that are chosen, the result of 5,040 is exactly 24 (or 4!) times too big. Therefore, the number of choices that Raul has is the number of permutations of 10 things 4 at a time, divided by 24:

$$\frac{P(10, 4)}{24} = \frac{5,040}{24} = 210$$

Raul has 210 ways of choosing four books to read from the list of ten books.

In situations where order is *not* important, we are interested in **combinations,** not permutations. The symbols $C(n, r)$ and $\binom{n}{r}$ both mean the number of combinations of n things r at a time.

If a selection of r books is chosen from a total of n books, the number of possible selections is $C(n, r)$, and there are $r!$ arrangements of the r books in each selection. If we consider the selected books as an ordered grouping, the number of orderings is $P(n, r)$. Therefore, we have

1. $r! \cdot C(n, r) = P(n, r)$

We can divide both sides of Equation 1 by $r!$ to get the formula for finding $C(n, r)$:

$$C(n, r) = \binom{n}{r} = \frac{P(n, r)}{r!} = \frac{n!}{r!(n - r)!}$$

Finding $C(n, r)$

The number of combinations of n things r at a time is given by

$$C(n, r) = \frac{n!}{r!(n - r)!}$$

EXAMPLE 7

Counting combinations. Compute　**a.** $C(8, 5)$,　**b.** $\binom{7}{2}$,　**c.** $C(n, n)$,　and
d. $C(n, 0)$.

Solution　**a.** $C(8, 5) = \dfrac{8!}{5!(8 - 5)!}$

$$= \frac{8 \cdot 7 \cdot 6 \cdot 5!}{5! \cdot 3!}$$

$$= 8 \cdot 7$$

$$= 56$$

b. $\dbinom{7}{2} = \dfrac{7!}{2!(7 - 2)!}$

$$= \frac{7 \cdot 6 \cdot 5!}{2 \cdot 1 \cdot 5!}$$

$$= 21$$

c. $C(n, n) = \dfrac{n!}{n!(n - n)!}$

$$= \frac{n!}{n! \cdot 0!}$$

$$= \frac{1}{0!}$$

$$= \frac{1}{1}$$

$$= 1$$

d. $C(n, 0) = \dfrac{n!}{0!(n - 0)!}$

$$= \frac{n!}{0! \cdot n!}$$

$$= \frac{1}{0!}$$

$$= \frac{1}{1}$$

$$= 1$$

The symbol $C(n, 0)$ indicates that we choose 0 things from the available n things.

SELF CHECK　Compute　**a.** $C(9, 6)$　and　**b.** $C(10, 10)$.　　　*Answers:*　**a.** 84,　**b.** 1　■

Parts c and d of Example 7 establish the following formulas.

Finding $C(n, n)$ and $C(n, 0)$

The number of combinations of n things n at a time is 1. The number of combinations of n things 0 at a time is 1.

$$C(n, n) = 1 \quad \text{and} \quad C(n, 0) = 1$$

EXAMPLE 8

Choosing committees. If 15 students want to pick a committee of 4 students to plan a party, how many different committees are possible?

Solution　Since the ordering of people on each possible committee is not important, we find the number of combinations of 15 people 4 at a time:

$$C(15, 4) = \frac{15!}{4!(15 - 4)!}$$

$$= \frac{15 \cdot 14 \cdot 13 \cdot 12 \cdot 11!}{4 \cdot 3 \cdot 2 \cdot 1 \cdot 11!}$$

$$= \frac{15 \cdot 14 \cdot 13 \cdot 12}{4 \cdot 3 \cdot 2 \cdot 1}$$

$$= 1,365$$

There are 1,365 possible committees.

SELF CHECK　In how many ways can 20 students pick a committee of 5 students to plan a party?　　　*Answer:* 15,504　■

EXAMPLE 9

Choosing subcommittees. A committee in Congress consists of ten Democrats and eight Republicans. In how many ways can a subcommittee be chosen if it is to contain five Democrats and four Republicans?

Solution There are $C(10, 5)$ ways of choosing the 5 Democrats and $C(8, 4)$ ways of choosing the 4 Republicans. By the multiplication principle for events, there are $C(10, 5) \cdot C(8, 4)$ ways of choosing the subcommittee:

$$C(10, 5) \cdot C(8, 4) = \frac{10!}{5!(10 - 5)!} \cdot \frac{8!}{4!(8 - 4)!}$$

$$= \frac{10 \cdot 9 \cdot 8 \cdot 7 \cdot 6 \cdot 5!}{120 \cdot 5!} \cdot \frac{8 \cdot 7 \cdot 6 \cdot 5 \cdot 4!}{24 \cdot 4!}$$

$$= \frac{10 \cdot 9 \cdot 8 \cdot 7 \cdot 6}{120} \cdot \frac{8 \cdot 7 \cdot 6 \cdot 5}{24}$$

$$= 17{,}640$$

There are 17,640 possible subcommittees.

SELF CHECK See Example 9. In how many ways can a subcommittee be chosen if it is to contain four members from each party?   *Answer:* 14,700 ∎

Alternative Form of the Binomial Theorem

We have seen that the expansion of $(x + y)^3$ is

$$(x + y)^3 = 1x^3 + 3x^2y + 3xy^2 + 1y^3$$

and that

$$\binom{3}{0} = 1, \quad \binom{3}{1} = 3, \quad \binom{3}{2} = 3, \quad \text{and} \quad \binom{3}{3} = 1$$

Putting these facts together gives the following way of writing the expansion of $(x + y)^3$:

$$(x + y)^3 = \binom{3}{0}x^3 + \binom{3}{1}x^2y + \binom{3}{2}xy^2 + \binom{3}{3}y^3$$

Likewise, we have

$$(x + y)^4 = \binom{4}{0}x^4 + \binom{4}{1}x^3y + \binom{4}{2}x^2y^2 + \binom{4}{3}xy^3 + \binom{4}{4}y^4$$

The generalization of this idea allows us to state the binomial theorem in an alternative form using combinatorial notation.

The binomial theorem

If n is any positive integer, then

$$(a + b)^n = \binom{n}{0}a^n + \binom{n}{1}a^{n-1}b + \binom{n}{2}a^{n-2}b^2 + \cdots$$

$$+ \binom{n}{r}a^{n-r}b^r + \cdots + \binom{n}{n}b^n$$

EXAMPLE 10

Expanding binomials. Use the alternative form of the binomial theorem to expand $(x + y)^6$.

Solution
$$(x + y)^6 = \binom{6}{0}x^6 + \binom{6}{1}x^5y + \binom{6}{2}x^4y^2 + \binom{6}{3}x^3y^3 + \binom{6}{4}x^2y^4 + \binom{6}{5}xy^5 + \binom{6}{6}y^6$$
$$= x^6 + 6x^5y + 15x^4y^2 + 20x^3y^3 + 15x^2y^4 + 6xy^5 + y^6$$

SELF CHECK

Use the alternative form of the binomial theorem to expand $(a + b)^2$.

Answer: $a^2 + 2ab + b^2$ ∎

EXAMPLE 11

Expanding binomials. Use the alternative form of the binomial theorem to expand $(2x - y)^3$.

Solution
$$(2x - y)^3 = [2x + (-y)]^3$$
$$= \binom{3}{0}(2x)^3 + \binom{3}{1}(2x)^2(-y) + \binom{3}{2}(2x)(-y)^2 + \binom{3}{3}(-y)^3$$
$$= 1(2x)^3 + 3(4x^2)(-y) + 3(2x)(y^2) + (-y)^3$$
$$= 8x^3 - 12x^2y + 6xy^2 - y^3$$

SELF CHECK

Use the alternative form of the binomial theorem to expand $(3a + b)^3$.

Answer: $27a^3 + 27a^2b + 9ab^2 + b^3$ ∎

STUDY SET

Section 11.4

VOCABULARY

Fill in the blanks to make the statements true.

1. A _____ is an arrangement of objects.

2. When selecting objects when order is not important, we count _____.

CONCEPTS

Fill in the blanks to make the statements true.

3. If an event E_1 can be done in p ways and (after it occurs) a second event E_2 can be done in q ways, the event E_1 followed by E_2 can be done in _____ ways.

4. The symbol _____ means the number of permutations of n things taken r at a time.

5. The formula for the number of permutations of n things taken r at a time is _____.

6. $P(n, n) = $ ___

7. $P(n, 0) = $ ___

8. $8! = $ _____.

9. The symbol $C(n, r)$ or _____ means the number of _____ of n things taken r at a time.

10. The formula for the number of combinations of n things taken r at a time is _____.

11. $C(n, n) = $ ___

12. $C(n, 0) = $ ___

NOTATION

Complete each solution.

13. $P(6, 2) = \dfrac{\blacksquare}{(6 - 2)!}$

$= \dfrac{6 \cdot 5 \cdot 4!}{\blacksquare}$

$= 6 \cdot \blacksquare$

$= 30$

14. $C(6, 2) = \dfrac{\blacksquare}{\blacksquare(6 - 2)!}$

$= \dfrac{6 \cdot 5 \cdot 4!}{2 \cdot 1 \cdot \blacksquare}$

$= \blacksquare \cdot 5$

$= 15$

PRACTICE

In Exercises 15–36, evaluate each permutation or combination.

15. $P(3, 3)$ 16. $P(4, 4)$

17. $P(5, 3)$ 18. $P(3, 2)$

19. $P(2, 2) \cdot P(3, 3)$ 20. $P(3, 2) \cdot P(3, 3)$

21. $\dfrac{P(5, 3)}{P(4, 2)}$

22. $\dfrac{P(6, 2)}{P(5, 4)}$

23. $\dfrac{P(6, 2) \cdot P(7, 3)}{P(5, 1)}$

24. $\dfrac{P(8, 3)}{P(5, 3) \cdot P(4, 3)}$

25. $C(5, 3)$

26. $C(5, 4)$

27. $\dbinom{6}{3}$

28. $\dbinom{6}{4}$

29. $\dbinom{5}{4}\dbinom{5}{3}$

30. $\dbinom{6}{5}\dbinom{6}{4}$

31. $\dfrac{C(38, 37)}{C(19, 18)}$

32. $\dfrac{C(25, 23)}{C(40, 39)}$

33. $C(12, 0) \cdot C(12, 12)$

34. $\dfrac{C(8, 0)}{C(8, 1)}$

35. $C(n, 2)$

36. $C(n, 3)$

In Exercises 37–42, use the alternative form of the binomial theorem to expand each expression.

37. $(x + y)^4$

38. $(x - y)^2$

39. $(2x + y)^3$

40. $(2x + 1)^4$

41. $(3x - 2)^4$

42. $(3 - x^2)^3$

In Exercises 43–46, find the indicated term of the binomial expansion.

43. $(x - 5y)^5$; fourth term

44. $(2x - y)^5$; third term

45. $(x^2 - y^3)^4$; second term

46. $(x^3 - y^2)^4$; fourth term

APPLICATIONS

47. PLANNING AN EVENING Kristy plans to go to dinner and see a movie. In how many ways can she arrange her evening if she has a choice of five movies and seven restaurants?

48. TRAVEL CHOICES Paula has five ways to travel from New York to Chicago, three ways to travel from Chicago to Denver, and four ways to travel from Denver to Los Angeles. How many choices are available if she travels from New York to Los Angeles?

49. MAKING LICENSE PLATES How many six-digit license plates can be manufactured? Note that there are ten choices—0, 1, 2, 3, 4, 5, 6, 7, 8, 9—for each digit.

50. MAKING LICENSE PLATES How many six-digit license plates can be manufactured if no digit can be repeated?

51. MAKING LICENSE PLATES How many six-digit license plates can be manufactured if no license can begin with 0 and if no digit can be repeated?

52. MAKING LICENSE PLATES How many license plates can be manufactured with two letters followed by four digits?

53. PHONE NUMBERS How many seven-digit phone numbers are available in area code 815 if no phone number can begin with 0 or 1?

54. PHONE NUMBERS How many ten-digit phone numbers are available if area codes of 000 and 911 cannot be used and if no local number can begin with 0 or 1?

55. LINING UP In how many ways can six people be placed in a line?

56. ARRANGING BOOKS In how many ways can seven books be placed on a shelf?

57. ARRANGING BOOKS In how many ways can four novels and five biographies be arranged on a shelf if the novels are placed first?

58. MAKING A BALLOT In how many ways can six candidates for mayor and four candidates for the county board be arranged on a ballot if all of the candidates for mayor must be placed first?

59. COMBINATION LOCKS How many permutations does a combination lock have if each combination has three numbers, no two numbers of any combination are equal, and the lock has 25 numbers?

60. COMBINATION LOCKS How many permutations does a combination lock have if each combination has three numbers, no two numbers of any combination are equal, and the lock has 50 numbers?

61. ARRANGING APPOINTMENTS The receptionist at a dental office has only three appointment times available before next Tuesday, and ten patients have toothaches. In how many ways can the receptionist fill those appointments?

62. COMPUTERS In many computers, a *word* consists of 32 *bits*—a string of thirty-two 1's and 0's. How many different words are possible?

63. PALINDROMES A palindrome is any word, such as *madam* or *radar,* that reads the same backward and forward. How many five-digit numerical palindromes (like 13531) are there? (*Hint:* A leading 0 would be dropped.)

64. CALL LETTERS The call letters of a U.S. commercial radio station have 3 or 4 letters, and the first is either a W or a K. How many radio stations could this system support?

65. PLANNING A PICNIC A class of 14 students wants to pick a committee of 3 students to plan a picnic. How many committees are possible?

66. CHOOSING BOOKS Jeffrey must read 3 books from a reading list of 15 books. How many choices does he have?

67. FORMING COMMITTEES The number of three-person committees that can be formed from a group of persons is ten. How many persons are in the group?

68. FORMING COMMITTEES The number of three-person committees that can be formed from a group of persons is 20. How many persons are in the group?

69. WINNING A LOTTERY In one state lottery, anyone who picks the correct six numbers (in any order) wins. With the numbers 0 through 99 available, how many choices are possible?

70. TAKING TESTS The instructions on a test read, "Answer any ten of the following fifteen questions. Then choose one of the remaining questions for homework, and turn in its solution tomorrow." In how many ways can the questions be chosen?

71. FORMING COMMITTEES In how many ways can we select a committee of two men and two women from a group containing three men and four women?

72. FORMING COMMITTEES In how many ways can we select a committee of three men and two women from a group containing five men and three women?

73. CHOOSING CLOTHES In how many ways can we select 2 shirts and 3 neckties from a group of 12 shirts and 10 neckties?

74. CHOOSING CLOTHES In how many ways can we select five dresses and two coats from a wardrobe containing nine dresses and three coats?

WRITING

75. State the multiplication principle for events.

76. Explain why *permutation lock* would be a better name for a combination lock.

REVIEW

Find each value of x.

77. $|2x - 3| = 9$

78. $2x^2 - x = 15$

79. $\dfrac{3}{x - 5} = \dfrac{8}{x}$

80. $\dfrac{3}{x} = \dfrac{x - 2}{8}$

▶ 11.5   Probability

In this section, you will learn about

Probability

Introduction   In this section, we will discuss probabilities. The probability that an event will occur is a measure of the likelihood of that event. A tossed coin, for example, can land in two ways, either heads or tails. Because one of these two equally likely outcomes is heads, we expect that out of several tosses, about half will be heads. We say that the probability of obtaining heads in a single toss of the coin is $\frac{1}{2}$.

If records show that out of 100 days with weather conditions like today's, 30 have received rain, the weather service will report, "There is a $\frac{30}{100}$ or 30% probability of rain today."

Probability

Activities such as tossing a coin, rolling a die, drawing a card, and predicting rain are called **experiments**. For any experiment, a list of all possible outcomes is called a **sample space.** For example, the sample space S for the experiment of tossing two coins is the set

$S = \{(H, H), (H, T), (T, H), (T, T)\}$ There are four possible outcomes.

where the ordered pair (H, T) represents the outcome "heads on the first coin and tails on the second coin."

An **event** is a subset of the sample space of an experiment. For example, if E is the event "getting at least one heads" in the experiment of tossing two coins, then

$E = \{(H, H), (H, T), (T, H)\}$ There are 3 ways of getting at least one heads.

Because the outcome of getting at least one heads can occur in 3 out of 4 possible ways, we say that the **probability** of E is $\frac{3}{4}$, and we write

$$P(E) = P(\text{at least one heads}) = \frac{3}{4}$$

Probability of an event

If a sample space of an experiment has n distinct and equally likely outcomes and E is an event that occurs in s of those ways, the **probability of E** is

$$P(E) = \frac{s}{n}$$

Since $0 \leq s \leq n$, it follows that $0 \leq \frac{s}{n} \leq 1$. This implies that all probabilities have values from 0 to 1. If an event cannot happen, its probability is 0. If an event is certain to happen, its probability is 1.

EXAMPLE 1

Sample spaces. List the sample space of the experiment "rolling two dice a single time."

Solution   We can list ordered pairs and let the first number be the result on the first die and the second number the result on the second die. The sample space S is the following set of ordered pairs:

(1, 1) (1, 2) (1, 3) (1, 4) (1, 5) (1, 6)
(2, 1) (2, 2) (2, 3) (2, 4) (2, 5) (2, 6)
(3, 1) (3, 2) (3, 3) (3, 4) (3, 5) (3, 6)
(4, 1) (4, 2) (4, 3) (4, 4) (4, 5) (4, 6)
(5, 1) (5, 2) (5, 3) (5, 4) (5, 5) (5, 6)
(6, 1) (6, 2) (6, 3) (6, 4) (6, 5) (6, 6)

By counting, we see that the experiment has 36 equally likely possible outcomes.

SELF CHECK   How many pairs in the sample space have a sum of 4?   *Answer:* 3   ∎

EXAMPLE 2

Finding probabilities. Find the probability of the event "rolling a sum of 7 on one roll of two dice."

Solution   The sample space is listed in Example 1. In that sample space, the following six ordered pairs give a sum of 7:

(1, 6) (2, 5) (3, 4) (4, 3) (5, 2) (6, 1)

Since there are 6 ordered pairs whose numbers give a sum of 7 out of a total of 36 equally likely outcomes, we have

$$P(E) = P(\text{rolling a 7}) = \frac{s}{n} = \frac{6}{36} = \frac{1}{6}$$

SELF CHECK   See Example 2. Find the probability of rolling a sum of 4.   *Answer:* $\frac{1}{12}$ ∎

A standard playing deck of 52 cards has two red suits, hearts and diamonds, and two black suits, clubs and spades. Each suit has 13 cards, including the ace, king, queen, jack, and cards numbered from 2 to 10. We will refer to a standard deck of cards in many examples and exercises.

E X A M P L E   3

Finding probabilities. Find the probability of drawing an ace on one draw from a standard card deck.

Solution   Since there are four aces in the deck, the number of favorable outcomes is $s = 4$. Since there are 52 cards in the deck, the total number of possible outcomes is $n = 52$. The probability of drawing an ace is the ratio of the number of favorable outcomes to the number of possible outcomes.

$$P(\text{an ace}) = \frac{s}{n} = \frac{4}{52} = \frac{1}{13}$$

The probability of drawing and ace is $\frac{1}{13}$.

SELF CHECK   Find the probability of drawing a red ace on one draw from a standard card deck.   *Answer:* $\frac{1}{26}$ ■

E X A M P L E   4

Finding probabilities. Find the probability of drawing 5 cards, all hearts, from a standard card deck.

Solution   The number of ways we can draw 5 hearts from the 13 hearts is $C(13, 5)$, the number of combinations of 13 things taken 5 at a time. The number of ways to draw 5 cards from the deck is $C(52, 5)$, the number of combinations of 52 things taken 5 at a time. The probability of drawing 5 hearts is the ratio of the number of favorable outcomes to the number of possible outcomes.

$$P(5 \text{ hearts}) = \frac{s}{n} = \frac{C(13, 5)}{C(52, 5)}$$

$$P(5 \text{ hearts}) = \frac{\dfrac{13!}{5!8!}}{\dfrac{52!}{5!47!}}$$

$$= \frac{13!}{5!8!} \cdot \frac{5!47!}{52!}$$

$$= \frac{13 \cdot 12 \cdot 11 \cdot 10 \cdot 9 \cdot 8!}{8!} \cdot \frac{47!}{52 \cdot 51 \cdot 50 \cdot 49 \cdot 48 \cdot 47!}$$

$$= \frac{13 \cdot 12 \cdot 11 \cdot 10 \cdot 9}{52 \cdot 51 \cdot 50 \cdot 49 \cdot 48}$$

$$= \frac{33}{66,640}$$

The probability of drawing 5 hearts is $\frac{33}{66,640}$.

SELF CHECK   Find the probability of drawing 6 cards, all diamonds, from a standard card deck.   *Answer:* $\frac{33}{391,510}$ ■

STUDY SET

Section 11.5

VOCABULARY

Fill in the blanks to make the statements true.

1. An _____ is any activity for which the outcome is uncertain.

2. A list of all possible outcomes for an experiment is called a _____.

CONCEPTS

Fill in the blanks to make the statements true.

3. The probability of an event E is defined as $P(E) =$ _____.

4. If an event is certain to happen, its probability is ___.

5. If an event cannot happen, its probability is ___.

6. All probability values are between ___ and ___, inclusive.

NOTATION

Complete each solution.

7. Find the probability of drawing a black face card from a standard deck.
   a. The number of black face cards is ▢.
   b. The number of cards in the deck is ▢.
   c. The probability is ▢ or ▢.

8. Find the probability of drawing 4 aces from a standard card deck.
   a. The number of ways to draw 4 aces from 4 aces is $C(4, 4) =$ ___.
   b. The number of ways to draw 4 cards from 52 cards is $C(52, 4) =$ _____.
   c. The probability is _____.

PRACTICE

In Exercises 9–12, list the sample space of each experiment.

9. Rolling one die and tossing one coin

10. Tossing three coins

11. Selecting a letter of the alphabet

12. Picking a one-digit number

In Exercises 13–16, an ordinary die is rolled once. Find the probability of each event.

13. Rolling a 2

14. Rolling a number greater than 4

15. Rolling a number larger than 1 but less than 6

16. Rolling an odd number

In Exercises 17–20, balls numbered from 1 to 42 are placed in a container and stirred. If one is drawn at random, find the probability of each result.

17. The number is less than 20.

18. The number is less than 50.

19. The number is a prime number.

20. The number is less than 10 or greater than 40.

In Exercises 21–24, refer to the spinner in Illustration 1. If the spinner is spun, find the probability of each event. Assume that the spinner never stops on a line.

ILLUSTRATION 1

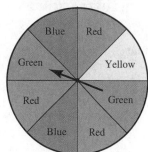

21. The spinner stops on red.

22. The spinner stops on green.

23. The spinner stops on orange.

24. The spinner stops on yellow.

In Exercises 25–32, find the probability of each event.

25. Rolling a sum of 4 on one roll of two dice

26. Drawing a diamond on one draw from a card deck

27. Drawing a red egg from a basket containing 5 red eggs and 7 blue eggs

28. Drawing a yellow egg from a basket containing 5 red eggs and 7 yellow eggs

29. Drawing 6 diamonds from a standard card deck without replacing the cards after each draw

30. Drawing 5 aces from a standard card deck without replacing the cards after each draw

31. Drawing 5 clubs from the black cards in a standard card deck

32. Drawing a face card from a standard card deck

In Exercises 33–38, assume that the probability that an airplane engine will fail during a torture test is $\frac{1}{2}$ and that the aircraft in question has 4 engines. In Exercises 34–38, find each probability.

33. Construct a sample space for the torture test.

34. All engines will survive the test.

35. Exactly 1 engine will survive.

36. Exactly 2 engines will survive.

37. Exactly 3 engines will survive.

38. No engines will survive.

39. Find the sum of the probabilities in Exercises 34 through 38.

In Exercises 40–42, assume that a survey of 282 people is taken to determine the opinions of doctors, teachers, and lawyers on a proposed piece of legislation, with the results shown in Illustration 2. A person is chosen at random from those surveyed. Refer to Illustration 2 to find each probability.

ILLUSTRATION 2

	Number that favor	Number that oppose	Number with no opinion	Total
Doctors	70	32	17	119
Teachers	83	24	10	117
Lawyers	23	15	8	46
Total	176	71	35	282

40. The person favors the legislation.

41. A doctor opposes the legislation.

42. A person who opposes the legislation is a lawyer.

APPLICATIONS

43. QUALITY CONTROL In a batch of 10 tires, 2 are known to be defective. If 4 tires are chosen at random, find the probability that all 4 tires are good.

44. MEDICINE Out of a group of 9 patients treated with a new drug, 4 suffered a relapse. Find the probability that 3 patients of this group, chosen at random, will remain disease-free.

WRITING

45. Explain why all probability values range from 0 to 1.

46. Explain the concept of probability.

REVIEW

Solve each equation.

47. $|x + 3| = 7$

48. $|3x - 2| = |2x - 3|$

In Exercises 49–50, solve each inequality. Give the answer in interval notation.

49. $|x - 3| < 7$

50. $|3x - 2| \geq 5$

The Language of Algebra

Algebra is a language in its own right. One of the keys to becoming a good algebra student is to know the vocabulary of algebra. In Exercises 1–26, match each instruction in column I with the most appropriate problem in column II. Each letter in column II is used only once.

Column I

1. Use the FOIL method.

2. Apply a rule for exponents to simplify.

3. Add the rational expressions.

4. Rationalize the denominator.

5. Factor completely.

6. Evaluate the expression for $a = -1$ and $b = -6$.

7. Express in lowest terms.

8. Solve for t.

9. Combine like terms.

10. Remove parentheses.

11. Solve the system by graphing.

12. Find $f(g(x))$.

13. Solve using the quadratic formula.

14. Identify the base and the exponent.

15. Write without a radical sign.

16. Write the equation of the line having the given slope and y-intercept.

17. Solve the inequality.

18. Complete the square to make a perfect square trinomial.

19. Find the slope of the line passing through the given points.

20. Use a property of logarithms to simplify.

21. Set each factor equal to zero and solve for x.

22. State the solution of the compound inequality using interval notation.

23. Find the inverse function, $h^{-1}(x)$.

24. Write using scientific notation.

25. Write the logarithmic statement in exponential form.

26. Find the sum of the first 6 terms of the sequence.

Column II

a. 2,300,000,000

b. e^3

c. $f(x) = x^2 + 1$ and $g(x) = 5 - 3x$

d. $-2x(3x^2 - 4x + 8)$

e. $4x - 7 > -3x - 7$

f. $\begin{cases} 2x = y - 5 \\ x + y = -1 \end{cases}$

g. $(x^2 - 5)(x^2 + 3)$

h. $(x + 2)(x - 10) = 0$

i. $\dfrac{x - 1}{2x^2} + \dfrac{x + 1}{8x}$

j. $\sqrt{4x^2}$

k. $\ln 6 + \ln x$

l. $\dfrac{10}{\sqrt{6} - \sqrt{2}}$

m. $2x - 8 + 6y - 14$

n. $(3, -2)$ and $(0, -5)$

o. $x^n \cdot x^{3n}$

p. $\dfrac{4x^2y}{16xy}$

q. $h(x) = 10^x$

r. $\log_2 8 = 3$

s. $m = \frac{2}{3}$ and passes through $(0, 2)$

t. 2, 6, 18, . . .

u. $3y^3 - 243b^6$

v. $x + 7 \geq 0$ and $-x < -1$

w. $x^2 - 3x - 4 = 0$

x. $Rt = cd + 2t$

y. $-2\pi a^2 b - 3b^3$

z. $x^2 + 4x$

Accent on Teamwork

Section 11.1

Pascal's triangle Find the sum of the numbers in each of the first ten rows of Pascal's triangle. What is the pattern? Then find the sum of the numbers in the designated diagonal rows of Pascal's triangle shown in Illustration 1. What is the pattern?

ILLUSTRATION 1

Factorial Tell whether each statement is true or false.
a. $8! - 6! = 2!$ **b.** $8! + 6! = 14!$

c. $\dfrac{8!}{6!} = 2!$ **d.** $8! \cdot 6! = 48!$

Sections 11.2 and 11.3

Sequences The terms of a sequence can be generated by substituting 1,000 for n, 2,000 for n, 3,000 for n, and so on, in the expression

$$\left(1 + \frac{1}{n}\right)^n$$

Use a calculator to make each computation for values of n from 1,000 to 20,000. The terms of the sequence get closer and closer to an important number that we studied in this course. What is that number?

Folding paper Suppose a piece of paper is 0.01 inch thick. If the paper is folded over itself 10 successive times, and if the thickness doubles after each fold, how thick will the result be?

Height of the rebound Drop a hard rubber ball from a height of 72 inches from the floor and measure how high it rebounds in inches. Plot the point $(1, 72)$ on graph paper to represent the first time the ball is dropped. Then plot $(2,$ rebound height in inches$)$ to represent the second time the ball is dropped. Continue this pattern, measuring how high the ball rebounds and plotting the data. Do the points on the graph form a pattern?

Section 11.4

Permutations and combinations Explain why $C(n, r) = C(n, n - r)$. Then determine if $P(n, r) = P(n, n - r)$.

Circular permutations In how many ways may four people be seated in a row?

Suppose the same four people are seated around a circular table. Some of the possible seating arrangements are really the same, because the relative position of each person to the others is not different. For an example of this, see Illustration 2. If we agree not to count any of the duplicate arrangements, in how many different ways can four people be seated around a circular table?

ILLUSTRATION 2

Steve   Sheryl   Spencer   Steve

Spencer   Sam   Sam   Sheryl

Section 11.5

Throwing dice Two standard dice are rolled, and the sum of the number of dots on the top faces of the dice is recorded.

a. Complete the table below, which gives the number of ways each sum can occur. For example, there are three ways to obtain a sum of 4: a ⊡ on die 1 and a ⋯ on die 2, a ⋯ on die 1 and a ⊡ on die 2, and a ⠢ on die 1 and a ⠢ on die 2.

Sum	2	3	4	5	6	7	8	9	10	11	12
Number of ways			3								

a. Use the table to determine the probability of obtaining a sum of 7 or 11 on one roll of the dice.

Odds For an experiment involving equally likely outcomes, the odds in favor of an event are defined to be the total number of favorable outcomes compared to the total number of unfavorable outcomes. Odds are usually expressed as a ratio of the form $1:3$ or $5:2$, for example. If a standard die is rolled, what are the odds in favor of the following events?
a. Getting a 1
b. Getting an even number
c. Getting a number less than 0
d. Getting a number greater than 6

Section 11.1 — The Binomial Theorem

CONCEPTS

The symbol $n!$ (*n factorial*) is defined as

$$n! = n(n - 1)(n - 2) \cdots 2 \cdot 1$$

where n is a natural number.

$$0! = 1$$

$n(n - 1)! = n!$ (n is a natural number)

The binomial theorem:

$$(a + b)^n =$$

$$a^n + \frac{n!}{1!(n - 1)!} a^{n-1}b$$

$$+ \frac{n!}{2!(n - 2)!} a^{n-2}b^2$$

$$+ \cdots + b^n$$

The binomial theorem can be used to find the nth term of a binomial expansion.

REVIEW EXERCISES

1. Evaluate each expression.

   a. $(4!)(3!)$ **b.** $\dfrac{5!}{3!}$

   c. $\dfrac{6!}{2!(6 - 2)!}$ **d.** $\dfrac{12!}{3!(12 - 3)!}$

   e. $(n - n)!$ **f.** $\dfrac{8!}{7!}$

2. Use the binomial theorem to find each expansion.
   a. $(x + y)^5$
   b. $(x - y)^4$
   c. $(4x - y)^3$
   d. $(x + 4y)^3$

3. Find the specified term in each expansion.
   a. $(x + y)^4$; third term
   b. $(x - y)^5$; fourth term
   c. $(3x - 4y)^3$; second term
   d. $(4x + 3y)^4$; third term

Section 11.2 — Arithmetic Sequences

An *arithmetic sequence* is a sequence of the form

$$a, a + d, a + 2d, \ldots, a + (n - 1)d, \ldots$$

where a is the first term, $a + (n - 1)d$ is the nth term, and d is the common difference.

If numbers are inserted between two given numbers a and b to form an arithmetic sequence, the inserted numbers are *arithmetic means* between a and b.

4. Find the eighth term of an arithmetic sequence whose first term is 7 and whose common difference is 5.

5. Write the first five terms of the arithmetic sequence whose ninth term is 242 and whose seventh term is 212.

6. Find two arithmetic means between 8 and 25.

The sum of the first n terms of an arithmetic sequence is given by

$$S_n = \frac{n(a + l)}{2} \quad \text{with}$$

$$l = a + (n - 1)d$$

where a is the first term, l is the last (or nth) term, and n is the number of terms in the sequence.

$$\sum_{k=1}^{n} f(k) = f(1) + f(2) + \cdots + f(n)$$

7. Find the sum of the first 20 terms of the sequence 11, 18, 25, ...

8. Find the sum of the first ten terms of the sequence 9, $6\frac{1}{2}$, 4, ...

9. Find each sum.

    a. $\displaystyle\sum_{k=4}^{6} \frac{1}{2}k$ **b.** $\displaystyle\sum_{k=2}^{5} 7k^2$

    c. $\displaystyle\sum_{k=1}^{4} (3k - 4)$ **d.** $\displaystyle\sum_{k=10}^{10} 36k$

Section 11.3   Geometric Sequences

A *geometric sequence* is a sequence of the form

$$a, ar, ar^2, ar^3, \ldots, ar^{n-1}, \ldots$$

where a is the first term, ar^{n-1} is the nth term, and r is the common ratio.

If numbers are inserted between a and b to form a geometric sequence, the inserted numbers are *geometric means* between a and b.

The sum of the first n terms of a geometric sequence is given by

$$S_n = \frac{a - ar^n}{1 - r} \quad (r \neq 1)$$

where S_n is the sum, a is the first term, r is the common ratio, and n is the number of terms in the sequence.

If r is the common ratio of an infinite geometric sequence, and if $|r| < 1$, the sum of the terms of the infinite geometric sequence is given by

$$S = \frac{a}{1 - r}$$

where a is the first term and r is the common ratio.

10. Write the first five terms of the geometric sequence whose fourth term is 3 and whose fifth term is $\frac{3}{2}$.

11. Find the sixth term of a geometric sequence with a first term of $\frac{1}{8}$ and a common ratio of 2.

12. Find two geometric means between -6 and 384.

13. Find the sum of the first seven terms of the sequence 162, 54, 18, ...

14. Find the sum of the first eight terms of the sequence $\dfrac{1}{8}$, $-\dfrac{1}{4}$, $\dfrac{1}{2}$, ...

15. Find the sum of the infinite geometric sequence 25, 20, 16, ...

16. Change the decimal $0.\overline{05}$ to a common fraction.

Section 11.4

The multiplication principle for events: If E_1 and E_2 are two events, and if E_1 can be done in a_1 ways and E_2 can be done in a_2 ways, then the event "E_1 followed by E_2" can be done in $a_1 \cdot a_2$ ways.

Formula for permutations:

$$P(n, r) = \frac{n!}{(n - r)!}$$

$P(n, n) = n!$ and $P(n, 0) = 1$

Formula for combinations:

$$C(n, r) = \binom{n}{r} = \frac{n!}{r!(n - r)!}$$

$$C(n, n) = \binom{n}{n} = 1$$

$$C(n, 0) = \binom{n}{0} = 1$$

Permutations and Combinations

17. If there are 17 flights from New York to Chicago and 8 flights from Chicago to San Francisco, in how many different ways could a passenger plan her trip?

18. Evaluate each expression.
    a. $P(7, 7)$
    b. $P(7, 0)$
    c. $P(8, 6)$
    d. $\dfrac{P(9, 6)}{P(10, 7)}$

19. Evaluate each expression.
    a. $C(7, 7)$
    b. $C(7, 0)$
    c. $\dbinom{8}{6}$
    d. $\dbinom{9}{6}$
    e. $C(6, 3) \cdot C(7, 3)$
    f. $\dfrac{C(7, 3)}{C(6, 3)}$

20. CAR DEPRECIATION A $5,000 car depreciates at the rate of 20% of the previous year's value. How much is the car worth after 5 years?

21. STOCK APPRECIATION The value of Mia's stock portfolio is expected to appreciate at the rate of 18% per year. How much will the portfolio be worth in 10 years if its current value is $25,700?

22. PLANTING CORN A farmer planted 300 acres in corn this year. He intends to plant an additional 75 acres in corn in each successive year until he has 1,200 acres in corn. In how many years will that be?

23. FALLING OBJECT If an object is in free fall, the sequence 16, 48, 80, . . . represents the distance in feet that object falls during the first second, during the second second, during the third second, and so on. How far will the object fall during the first ten seconds?

24. LINING UP In how many ways can five persons be arranged in a line?

25. LINING UP In how many ways can three men and five women be arranged in a line if the women are placed ahead of the men?

26. CHOOSING PEOPLE In how many ways can we pick three persons from a group of ten persons?

27. FORMING COMMITTEES In how many ways can we pick a committee of two Democrats and two Republicans from a group containing five Democrats and six Republicans?

Section 11.5 Probability

An event that cannot happen has a *probability* of 0. An event that is certain to happen has a probability of 1. All other events have probabilities between 0 and 1.

If S is the *sample space* of an experiment with n distinct and equally likely outcomes, and E is an event that occurs in s of those ways, then the probability of E is

$$P(E) = \frac{s}{n}$$

28. Find the probability of rolling an 11 on one roll of two dice.

29. Find the probability of living forever.

30. Find the probability of drawing a 10 from a standard deck of cards.

31. Find the probability of drawing a five-card poker hand that has 3 aces.

32. Find the probability of drawing 5 cards, all spades, from a standard card deck.

CHAPTER 11

Test

1. Evaluate $\dfrac{7!}{4!}$.

2. Evaluate $0!$.

3. Find the second term in the expansion of $(x - y)^5$.

4. Find the third term in the expansion of $(x + 2y)^4$.

5. Find the tenth term of an arithmetic sequence whose first three terms are 3, 10, and 17.

6. Find the sum of the first 12 terms of the sequence $-2, 3, 8, \dots$.

7. Find two arithmetic means between 2 and 98.

8. Evaluate $\displaystyle\sum_{k=1}^{3} (2k - 3)$.

9. Find the seventh term of the geometric sequence whose first three terms are $-\frac{1}{9}$, $-\frac{1}{3}$, and -1.

10. Find the sum of the first six terms of the sequence $\frac{1}{27}$, $\frac{1}{9}, \frac{1}{3}, \dots$.

11. Find two geometric means between 3 and 648.

12. Find the sum of all of the terms of the infinite geometric sequence $9, 3, 1, \dots$.

In Problems 13–20, find the value of each expression.

13. $P(5, 4)$

14. $P(8, 8)$

15. $C(6, 4)$

16. $C(8, 3)$

17. $C(6, 0) \cdot P(6, 5)$

18. $P(8, 7) \cdot C(8, 7)$

19. $\dfrac{P(6, 4)}{C(6, 4)}$

20. $\dfrac{C(9, 6)}{P(6, 4)}$

21. CHOOSING PEOPLE In how many ways can we pick 3 persons from a group of 7 persons?

22. CHOOSING COMMITTEES From a group of 5 men and 4 women, how many 3-person committees can be chosen that will include two women?

In Problems 23–26, find each probability.

23. Rolling a 5 on one roll of a die

24. Drawing a jack or a queen from a standard card deck

25. Receiving 5 hearts for a 5-card poker hand

26. Tossing 2 heads in 5 tosses of a fair coin

27. If the probability that a person is sick is 1, find the probability that the person is well.

28. What is a sample space of an experiment? Give an example.

Cumulative Review Exercises

In Exercises 1–4, consider the set $\left\{-\frac{4}{3}, \pi, 5.6, \sqrt{2}, 0, -23, e, 7i\right\}$. List the elements in the set that are

1. whole numbers

2. rational numbers

3. irrational numbers

4. real numbers

5. FINANCIAL PLANNING Ana has some money to invest. Her financial planner tells her that if she can come up with $3,000 more, she will qualify for an 11% annual interest rate. Otherwise, she will have to invest the money at 7.5% annual interest. The financial planner urges her to invest the larger amount, because the 11% investment would yield twice as much annual income as the 7.5% investment. How much does she originally have on hand to invest?

6. BOATING Use the graph in Illustration 1 to determine the average rate of change in the sound level of the engine of a boat in relation to rpm of the engine.

ILLUSTRATION 1

In Exercises 7–8, tell whether the graphs of the equations are parallel or perpendicular.

7. $3x - 4y = 12$, $y = \frac{3}{4}x - 5$

8. $y = 3x + 4$, $x = -3y + 4$

In Exercises 9–10, write the equation of the line with the given properties.

9. $m = -2$, passing through $(0, 5)$

10. Passing through $(8, -5)$ and $(-5, 4)$

11. Use substitution to solve $\begin{cases} 3x + y = 4 \\ 2x - 3y = -1 \end{cases}$.

12. Use addition to solve $\begin{cases} x + 2y = -2 \\ 2x - y = 6 \end{cases}$.

13. Solve using Cramer's rule: $\begin{cases} 4x - 3y = -1 \\ 3x + 4y = -7 \end{cases}$.

14. Solve $\begin{cases} x + y + z = 1 \\ 2x - y - z = -4 \\ x - 2y + z = 4 \end{cases}$.

15. The graphs of $y = 4(x - 5) - x - 2$ and $y = -(2x + 6) - 1$ are shown in Illustration 2. Use the information in the display to solve $4(x - 5) - x - 2 = -(2x + 6) - 1$ graphically.

ILLUSTRATION 2

16. Evaluate $\begin{vmatrix} 3 & -2 \\ 1 & -1 \end{vmatrix}$.

17. MARTIAL ARTS Find the measure of each angle of the triangle shown in Illustration 3.

ILLUSTRATION 3

This angle is 5° more than 5 times $\angle C$.

This angle is 5° larger than $\angle B$.

18. Solve $\begin{cases} 3x - 2y \leq 6 \\ y < -x + 2 \end{cases}$.

In Exercises 19–20, give the solution in interval notation and graph the solution set.

19. Solve $|5 - 3x| \leq 14$.

20. Solve $4.5x - 1 < -10$ or $6 - 2x \geq 12$.

In Exercises 21–24, do the operations.

21. $(4x - 3y)(3x + y)$

22. $(-2x^2y^3 + 6xy + 5y^2) - (-4x^2y^3 - 7xy + 2y^2)$

23. $(a - 2b)^2$

24. $(a + 2)(3a^2 + 4a - 2)$

In Exercises 25–26, factor the expression completely.

25. $3x^3y - 4x^2y^2 - 6x^2y + 8xy^2$

26. $256x^4y^4 - z^8$

27. Solve for λ: $\dfrac{A\lambda}{2} + 1 = 2d + 3\lambda$.

28. Complete the table of values for
$f(x) = -x^3 - x^2 + 6x$ and then graph the function.
What are the x- and y-intercepts of the graph?

x	$f(x)$
-4	
-3	
-2	
-1	
0	
1	
2	
3	

In Exercises 29–32, simplify.

29. $\left(\dfrac{4a^{-2}b}{3ab^{-3}} \right)^3$

30. $\dfrac{6x^2 + 13x + 6}{6 - 5x - 6x^2}$

31. $\dfrac{p^3 - q^3}{q^2 - p^2} \cdot \dfrac{q^2 + pq}{p^3 + p^2q + pq^2}$

32. $\dfrac{2}{a - 2} + \dfrac{3}{a + 2} - \dfrac{a - 1}{a^2 - 4}$

33. Solve $\dfrac{x - 4}{x - 3} + \dfrac{x - 2}{x - 3} = x - 3$.

34. Solve $\dfrac{1}{R} = \dfrac{1}{R_1} + \dfrac{1}{R_2} + \dfrac{1}{R_3}$ for R.

35. TIRE WEAR See Illustration 4.
   a. What type of function does it appear would model the relationship between the inflation of a tire and the percent of service it gives?
   b. At what percent(s) of inflation will a tire offer only 90% of its possible service?

ILLUSTRATION 4

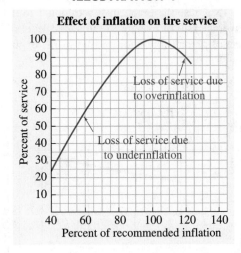

36. CHANGING DIAPERS Illustration 5 shows how to put a diaper on a baby. If the diaper is a square with sides 16 inches long, what is the largest waist size that this diaper can wrap around, assuming an overlap of 1 inch to pin the diaper?

37. Use the long division method to find
$$(2x^2 + 4x - x^3 + 3) \div (x - 1)$$

ILLUSTRATION 5

38. The graph of $f(x) = \sqrt{2x + 5} + \sqrt{x + 2} - 5$ is shown in Illustration 6. Use the information in the display to determine the solution of $\sqrt{2x + 5} = -\sqrt{x + 2} + 5$.

ILLUSTRATION 6

In Exercises 39–40, simplify each expression.

39. $\sqrt{98} + \sqrt{8} - \sqrt{32}$

40. $12\sqrt[3]{648x^4} + 3\sqrt[3]{81x^4}$

41. Evaluate $\left(\dfrac{25}{49}\right)^{-3/2}$.

42. Rationalize the denominator: $\dfrac{3t - 1}{\sqrt{3t} + 1}$.

In Exercises 43–44, write the expression in $a + bi$ form.

43. $\left(-7 + \sqrt{-81}\right) - \left(-2 - \sqrt{-64}\right)$

44. $\dfrac{2 - 5i}{2 + 5i}$

In Exercises 45–50, solve each equation.

45. $\sqrt{3a + 1} = a - 1$

46. $\sqrt{x + 3} - \sqrt{3} = \sqrt{x}$

47. $6a^2 + 5a - 6 = 0$

48. $4w^2 + 6w + 1 = 0$

49. $2(2x + 1)^2 - 7(2x + 1) + 6 = 0$

50. $3x^2 - 4x = -2$

51. If $f(x) = x^2 - 2$ and $g(x) = 2x + 1$, find $(f \circ g)(x)$.

52. Find the inverse function of $f(x) = 2x^3 - 1$.

53. Graph $f(x) = \left(\dfrac{1}{2}\right)^x$.

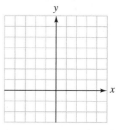

54. Graph $y = e^x$ and its inverse on the same coordinate system. Label the axis of symmetry.

55. Write $y = \log_2 x$ as an exponential equation.

56. Apply properties of logarithms to simplify $\log_6 \dfrac{x}{36}$.

In Exercises 57–60, find x.

57. $\log_x 25 = 2$ **58.** $\log_5 125 = x$

59. $\log_3 x = -3$ **60.** $\ln e = x$

61. Find the inverse of $y = \log_2 x$.

62. If $\log_{10} 10^x = y$, then y equals what quantity?

In Exercises 63–64, log 7 = 0.8451 and log 14 = 1.1461. Evaluate each expression without using a calculator or tables.

63. $\log 98$ **64.** $\log 2$

In Exercises 65–68, solve each equation. Round to four decimal places when necessary.

65. $2^{x+2} = 3^x$

66. $\log x + \log (x + 9) = 1$

67. $5^{4x} = \dfrac{1}{125}$

68. $\log_3 x = \log_3 \left(\dfrac{1}{x}\right) + 4$

69. Write the equation of the circle that has its center at $(1, 3)$ and passes through $(-2, -1)$.

70. Write the equation in standard form and then graph it.

$$x^2 - 9y^2 + 2x + 36y = 44$$

71. Graph $\dfrac{(x - 1)^2}{9} + \dfrac{(y - 3)^2}{4} = 1$.

72. Graph $y^2 + 4x - 6y = -1$.

In Exercises 73–74, use a calculator.

73. BOAT DEPRECIATION How much will a \$9,000 boat be worth after 9 years if it depreciates 12% per year?

74. Find $\log_6 8$.

75. Evaluate $\dfrac{6!7!}{5!}$.

76. Use the binomial theorem to expand $(3a - b)^4$.

77. Find the seventh term of the expansion of $(2x - y)^8$.

78. Find the 20th term of an arithmetic sequence with a first term of -11 and a common difference of 6.

79. Find the sum of the first 20 terms of an arithmetic sequence with a first term of 6 and a common difference of 3.

80. Insert two arithmetic means between -3 and 30.

81. Evaluate $\displaystyle\sum_{k=1}^{3} 3k^2$.

82. Evaluate $\displaystyle\sum_{k=3}^{5} (2k + 1)$.

83. Find the seventh term of a geometric sequence with a first term of $\frac{1}{27}$ and a common ratio of 3.

84. Find the sum of the first ten terms of the sequence $\frac{1}{64}$, $\frac{1}{32}$, $\frac{1}{16}$, ...

85. Insert two geometric means between -3 and 192.

86. Find the sum of all the terms of the sequence 9, 3, 1, ...

87. Evaluate $P(9, 3)$.

88. Evaluate $C(7, 4)$.

89. LINING UP In how many ways can seven people stand in a line?

90. FORMING A COMMITTEE In how many ways can a committee of three people be chosen from a group containing nine people?

Appendix I

Numerical Trigonometry

In this appendix, you will learn about

The sine ratio ■ The cosine ratio ■ The tangent ratio

Introduction The properties of similar triangles provide the basis for an area of mathematics called **trigonometry.** The word *trigonometry* means measuring triangles (trigons). Because trigonometry enables us to determine all sides and angles of a right triangle when only some are known, it is important in astronomy, surveying, carpentry, engineering, architecture, navigation, and physics.

Some special ratios involving the measures of the sides of a right triangle are used in trigonometry. However, before we discuss these trigonometric ratios, we must recall the following ideas.

■ Two triangles are **similar triangles** if they have the same shape.

■ If the measure of one acute angle in a right triangle is equal to the measure of an acute angle in a second right triangle, the triangles are similar.

■ The measures of corresponding sides of similar triangles are in proportion.

In Figure I-1, $\triangle ABC$ (read as "triangle ABC") and $\triangle DEF$ are right triangles, in which

$$\text{m}(\angle A) = \text{m}(\angle D) \quad \text{Read as "The measure of angle } A \text{ is equal to the measure of angle } D.\text{"}$$

Thus, the triangles are similar. Since the measures of corresponding sides of similar triangles are in proportion, we have

$$\frac{\text{m}(\overline{BC})}{\text{m}(\overline{EF})} = \frac{\text{m}(\overline{AB})}{\text{m}(\overline{DE})} \quad \text{Read m}(\overline{BC}) \text{ as "the measure of line segment } BC.\text{"}$$

By interchanging the means in this proportion, we obtain

1. $\dfrac{\text{m}(\overline{BC})}{\text{m}(\overline{AB})} = \dfrac{\text{m}(\overline{EF})}{\text{m}(\overline{DE})}$

FIGURE I-1

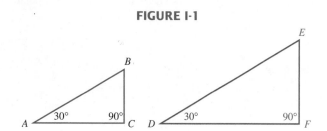

The fraction on the left-hand side of Equation 1 is the ratio of two sides of $\triangle ABC$. The fraction on the right-hand side is the ratio of the *corresponding two sides* of $\triangle DEF$. The values of these ratios depend on the size of the acute angle. They do not depend on the size of the triangles.

It can be shown that the measures of other pairs of sides in these triangles are also in proportion. In particular, the following proportions are true:

$$\frac{m(\overline{AC})}{m(\overline{AB})} = \frac{m(\overline{DF})}{m(\overline{DE})} \quad \text{and} \quad \frac{m(\overline{BC})}{m(\overline{AC})} = \frac{m(\overline{EF})}{m(\overline{DF})}$$

In right triangle ABC, shown in Figure I-2, the two acute angles are $\angle A$ and $\angle B$, and the right angle is $\angle C$. Side c, the side opposite the right angle, is called the **hypotenuse.** The other two sides are called **legs** of the triangle. The leg opposite $\angle A$ is a. Since leg b is one side of $\angle A$, we say that leg b is *adjacent* to $\angle A$.

FIGURE I-2

The symbol ⌐ indicates a right angle.

The Sine Ratio

The sine ratio

> The **sine of an acute angle of a right triangle** is the ratio of the length of the leg opposite the angle to the length of the hypotenuse. We can write this in formula form as
>
> $$\sin A = \frac{\text{length of leg opposite } \angle A}{\text{length of hypotenuse}}$$ The sine of $\angle A$ is abbreviated as $\sin A$.

EXAMPLE 1

Refer to the right triangle in Figure I-3, and find   **a.** $\sin A$ and   **b.** $\sin B$.

Solution   **a.** The length of the leg opposite $\angle A$ is 4 units, and the length of the hypotenuse is 5 units. Thus,

$$\sin A = \frac{\textbf{length of leg opposite } \angle A}{\textbf{length of hypotenuse}}$$

$$= \frac{4}{5}$$

Thus, $\sin A = \frac{4}{5}$.

FIGURE I-3

b. The length of the leg opposite $\angle B$ is 3 units, and the length of the hypotenuse is 5 units. Thus,

$$\sin B = \frac{\textbf{length of leg opposite } \angle B}{\textbf{length of hypotenuse}}$$

$$= \frac{3}{5}$$

Thus, $\sin B = \frac{3}{5}$.   ∎

E X A M P L E   2

Find sin 30°.

Solution

We draw a right triangle with an acute angle of 30°, as in Figure I-4. In this 30°–60° right triangle, the leg opposite the 30° angle is half as long as the hypotenuse.

Since the value of sin 30° depends on the size of the angle, not on the size of the triangle, we can assume that if the leg opposite the 30° angle is 1 unit long, the hypotenuse will be 2 units long.

FIGURE I-4

$$\sin A = \frac{\text{length of leg opposite } \angle A}{\text{length of hypotenuse}}$$

$$\sin 30° = \frac{1}{2}$$

Thus, sin 30° = $\frac{1}{2}$. ■

ACCENT ON TECHNOLOGY *Finding Sines of Angles*

A scientific calculator can be used to find the sine of any angle. For example, to find an approximate value of sin 57°, we set the calculator in degree mode, enter the number 57, and press the ⃞SIN key. The number that appears on the calculator's display is the sine of 57°.

Keystrokes   57 ⃞SIN | 0.83867057 |

On a graphing calculator, we enter these numbers and press these keys.

Keystrokes   ⃞SIN 57 ⃞) ⃞ENTER | sin (57) .8386705679 |

To four decimal places, sin 57° = 0.8387.

The sine ratio is useful for measuring certain distances indirectly.

E X A M P L E   3

A circus performer intends to walk up a taut cable to a platform 75 feet above the ground. If the cable makes an angle of 40° with the ground, how long is the cable?

Solution

The information is summarized in Figure I-5 on the next page. To find the value of *x*, we can use the sine ratio.

$$\sin 40° = \frac{\text{length of leg opposite } \angle A}{\text{length of hypotenuse}}$$

$$\sin 40° = \frac{75}{x}$$

We can now solve this equation for *x* as follows:

$$x(\sin 40°) = 75 \qquad \text{Multiply both sides by } x.$$

$$x = \frac{75}{\sin 40°} \qquad \text{Divide both sides by sin 40°.}$$

$$\approx \frac{75}{0.6427876097}$$

$$\approx 116.679287$$

The cable is approximately 117 feet long.

FIGURE I-5

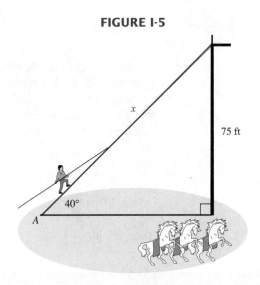

75 ft

x

40°

A

The Cosine Ratio

Another ratio of the measures of the sides of a right triangle is the **cosine ratio.**

The cosine ratio

The **cosine of an acute angle of a right triangle** is the ratio of the length of the leg adjacent the angle to length of the hypotenuse. We can write this in formula form as

$$\cos A = \frac{\text{length of leg adjacent } \angle A}{\text{length of hypotenuse}}$$ The cosine of $\angle A$ is abbreviated as $\cos A$.

EXAMPLE 4   Refer to the right triangle in Figure I-6 and find   **a.** $\cos A$ and   **b.** $\cos B$.

Solution   **a.** The length of the leg adjacent to $\angle A$ is 3 units, and the length of the hypotenuse is 5 units. Thus,

FIGURE I-6

$$\cos A = \frac{\text{length of leg adjacent } \angle A}{\text{length of hypotenuse}}$$

$$= \frac{3}{5}$$

Thus, $\cos A = \frac{3}{5}$.

b. The length of the leg adjacent to $\angle B$ is 4 units, and the length of the hypotenuse is 5 units. Thus,

$$\cos B = \frac{\text{length of leg adjacent } \angle B}{\text{length of hypotenuse}}$$

$$= \frac{4}{5}$$

Thus, $\cos B = \frac{4}{5}$.

EXAMPLE 5

Find cos 30°.

Solution   We draw a right triangle with an acute angle of 30°, as in Figure I-7. In the 30°–60° right triangle, the leg adjacent to the 30° angle is $\sqrt{3}$ times as long as the leg opposite the 30° angle. The hypotenuse is twice as long as the leg opposite the 30° angle.

Since cos 30° depends on the size of the angle, not the size of the triangle, we can assume that the length of the leg opposite the 30° angle is 1 unit long. Then the hypotenuse will be 2 units long and the adjacent leg $\sqrt{3}$ units long. The cosine of 30° is the cosine of $\angle A$. Thus,

FIGURE I-7

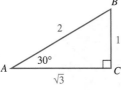

$$\cos A = \frac{\text{length of leg adjacent } \angle A}{\text{length of hypotenuse}}$$

$$= \frac{\sqrt{3}}{2}$$

Thus, $\cos 30° = \dfrac{\sqrt{3}}{2}$.

ACCENT ON TECHNOLOGY   *Finding Cosines of Angles*

A scientific calculator can be used to find the cosine of any angle. For example, to find an approximate value of cos 13°, we set the calculator in degree mode, enter the number 13, and press the $\boxed{\text{COS}}$ key. The number that appears on the calculator's display is the cosine of 13°.

Keystrokes   13 $\boxed{\text{COS}}$ $\boxed{\text{0.97437006}}$

On a graphing calculator, we enter these numbers and press these keys.

Keystrokes $\boxed{\text{COS}}$ 13 $\boxed{)}$ $\boxed{\text{ENTER}}$ $\boxed{\begin{array}{l} \cos(13) \\ \qquad\qquad .9743700648 \end{array}}$

To four decimal places, cos 13° = 0.9744.

Navigators and surveyors use the concept of bearing. See Figure I-8. The **bearing** of point *A* from point *O* (the observer) is the acute angle measured from the north–south line to the segment $\overline{OA}$. This bearing is denoted as N 30° E, read as "north 30° east" or "30° east of north."

The bearing of point *B* from *O* is N 75° W, the bearing of point *C* from *O* is S 20° W, and the bearing of point *D* from *O* is S 80° E.

FIGURE I-8

| EXAMPLE 6 | **Forest fires.** Two forest fire lookouts are on a north–south line, 17.2 miles apart. The bearing of a fire from the lookout at point A is S 27.0° W, and the bearing of the fire from the lookout at point B is N 63.0° W. How far is the fire from lookout B? |

Solution See Figure I-9. Since the sum of the three angles of any triangle is 180°, the angle at point F is 90°. It follows that $\triangle AFB$ is a right triangle with hypotenuse $\overline{AB}$. We can find how far B is from the fire by using the cosine ratio.

FIGURE I-9

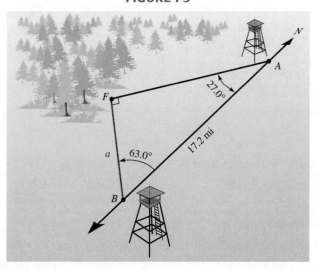

$$\cos 63.0° = \frac{\textbf{length of leg adjacent } \angle B}{\textbf{length of hypotenuse}}$$

$$\cos 63.0° = \frac{a}{17.2}$$

We can now solve this equation for a.

$$a = 17.2 \cos 63.0°$$
$$\approx 7.808636596$$

The fire is about 7.81 miles from lookout B. ∎

The Tangent Ratio

The last ratio we will discuss is the **tangent ratio.**

The tangent ratio

The **tangent of an acute angle of a right triangle** is the ratio of the length of the leg opposite the angle to the length of the leg adjacent to the angle. We can write this as the formula

$$\tan A = \frac{\text{length of leg opposite } \angle A}{\text{length of leg adjacent } \angle A}$$ The tangent of $\angle A$ is abbreviated as tan A.

E X A M P L E   7

Refer to the right triangle in Figure I-10. Find   **a.** tan A and   **b.** tan B.

Solution   **a.** The leg opposite $\angle A$ has a length of 4 centimeters. The leg adjacent to $\angle A$ has a length of 3 centimeters. Thus,

FIGURE I-10

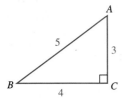

$$\tan A = \frac{\text{length of leg opposite } \angle A}{\text{length of leg adjacent to } \angle A}$$

$$= \frac{4}{3}$$

Thus, tan $A = \frac{4}{3}$.

b. The leg opposite $\angle B$ has a length of 3 centimeters. The leg adjacent to $\angle B$ has a length of 4 centimeters. Thus,

$$\tan B = \frac{\text{length of leg opposite } \angle B}{\text{length of leg adjacent to } \angle B}$$

$$= \frac{3}{4}$$

Thus, tan $B = \frac{3}{4}$.   ∎

E X A M P L E   8

Find tan 45°.

Solution   We draw a right triangle with an acute angle of 45°, as in Figure I-11. This triangle is an isosceles right triangle. Since the values of the ratios of the lengths of the legs of a right triangle depend on the angles of the triangle, not the lengths of the legs, we can assume that the two legs of the triangle are each 1 unit long. Then we have

FIGURE I-11

$$\tan A = \frac{\text{length of leg opposite } \angle A}{\text{length of leg adjacent to } \angle A}$$

$$\tan 45° = \frac{1}{1}$$

$$= 1$$

Thus, tan 45° = 1.   ∎

ACCENT ON TECHNOLOGY   *Finding Tangents of Angles*

A scientific calculator can be used to find the tangent of any angle. For example, to find an approximate value of tan 13°, we set the calculator in degree mode, enter the number 13, and press the $\boxed{\text{TAN}}$ key. The number that appears on the calculator's display is the tangent of 13°.

Keystrokes   13 $\boxed{\text{TAN}}$ $\boxed{\text{0.230868191}}$

On a graphing calculator, we enter these numbers and press these keys.

Keystrokes $\boxed{\text{TAN}}$ 13 $\boxed{)}$ $\boxed{\text{ENTER}}$ $\boxed{\begin{array}{l}\texttt{tan(13)} \\ \hphantom{xxxxx}\texttt{.2308681911}\end{array}}$

To four decimal places, tan 13° = 0.2309.

If an observer looks up at an object, the angle that the observer's line of sight makes with the horizontal is called the **angle of elevation**. If the observer looks down to see the object, the angle made with the horizontal is called the **angle of depression**. See Figure I-12.

FIGURE I-12

Horizontal line

Angle of → depression

Angle of ← elevation

Horizontal line

EXAMPLE 9

Height of a flagpole. From a location 23.0 meters from the base of a flagpole, the angle of elevation to its top is 37.5°. Find the height of the flagpole.

Solution We consider the right triangle shown in Figure I-13. The pole is the leg opposite the 37.5° angle, and the length of the adjacent leg is 23.0 meters. Since the tangent of the angle involves the lengths of the opposite and adjacent legs, we have

$$\tan 37.5° = \frac{\text{length of leg opposite } \angle A}{\text{length of leg adjacent } \angle A}$$

$$\tan 37.5° = \frac{h}{23.0}$$

FIGURE I-13

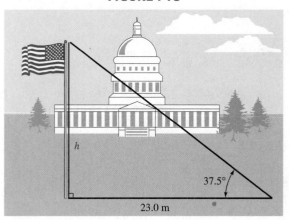

h

37.5°

23.0 m

We can solve this equation for h.

$$h = 23.0 \tan 37.5°$$
$$\approx 17.64852072$$

To the nearest tenth, the pole is 17.6 meters high. ■

STUDY SET

Appendix I

VOCABULARY

Fill in the blanks to make the statements true.

1. The word trigonometry means _____ triangles.

2. If two triangles have the same shape, they are called _____ triangles.

3. The _____ of an acute angle of a right triangle is the ratio of the length of the leg opposite the angle to the length of the hypotenuse.

4. The _____ of an acute angle of a right triangle is the ratio of the length of the leg opposite the angle to the length of the leg adjacent to the angle.

5. The _____ of an acute angle of a right triangle is the ratio of the length of the leg adjacent the angle to the length of the hypotenuse.

6. In a right triangle, the side opposite the right angle is called the _____.

7. The two sides of a right triangle that are opposite the acute angles are called _____.

8. The bearing of point A from point O is the acute angle measured from the _____ line to segment $\overline{OA}$.

9. When looking up, the angle that an observer's line of sight makes with the horizontal is called the angle of _____.

10. When looking down, the angle that an observer's line of sight makes with the horizontal is called the angle of _____.

CONCEPTS

Fill in the blanks to make the statements true.

11. If the measure of one acute angle in a right triangle is equal to the measure of an acute angle in the second triangle, the triangles are _____.

12. $\sin A =$ _____

13. $\cos A =$ _____

14. $\tan A =$ _____

15. In similar triangles, the lengths of corresponding sides are in _____.

16. Two sides of an _____ triangle are the same length.

NOTATION

Refer to Illustration 1 and complete each solution.

ILLUSTRATION 1

17. Find x.

$$\sin 50° = \frac{\boxed{}}{x}$$

$$\boxed{} \sin 50° = 40$$

$$x = \frac{40}{\boxed{}}$$

$$\approx \frac{\boxed{}}{0.7660444431}$$

$$\approx 52.21629157$$

18. Find y.

$$\boxed{}\, 50° = \frac{40}{y}$$

$$y \tan 50° = \boxed{}$$

$$y = \frac{40}{\tan \boxed{}}$$

$$\approx \frac{40}{\boxed{}}$$

$$\approx 33.56398524$$

PRACTICE

In Exercises 19–26, find each exact value without using a calculator.

19. $\sin 60°$

20. $\cos 60°$

21. $\tan 30°$

22. $\tan 60°$

23. $\sin 45°$

24. $\cos 45°$

25. $\sin 0°$

26. $\tan 0°$

In Exercises 27–34, use a calculator to find each value to four decimal places.

27. $\sin 73°$

28. $\cos 73°$

29. $\tan 73°$

30. $\sin 43.5°$

31. $\cos 43.5°$

32. $\tan 43.5°$

33. $\sin 90°$

34. $\cos 0°$

In Exercises 35–38, solve each triangle by finding all unknown sides and angles. Give the lengths of all sides to the nearest tenth.

35.

36.

37.

38.

APPLICATIONS

Remember that if two parallel lines are cut by a third line, alternate interior angles have the same measure.

39. HEIGHT OF A CLIFF A person on the edge of a cliff looks down at a boat on a lake. The angle of depression of the person's line of sight is 10°, and the line-of-sight distance from the person to the boat is 555 feet. To the nearest tenth, how high is the cliff? (Disregard the height of the person.)

40. HEIGHT OF A TREE The angle of elevation to the top of a tree from a point 40 feet from its base is 20.6°. To the nearest tenth, find the height of the tree.

41. DISTANCE TRAVELED A car drives up a long hill on a road that makes an angle of 7.5° with the horizontal. To the nearest tenth, how far will the car travel to reach the top if the horizontal distance traveled is 715 feet?

42. ALTITUDE LOST On an approach to an airport, a plane is descending at an angle of 3.5°. To the nearest tenth, how much altitude is lost as the plane travels a horizontal distance of 21.2 miles?

43. ANGLE OF DESCENT A plane loses 2,750 feet in altitude. If its angle of descent is 4°, how far did the plane fly horizontally? Give the result to the nearest hundred feet.

44. ANGLE OF ASCENT A train rises 230 feet as it travels up a steep 2.5° grade. To the nearest hundred feet, how far up the grade does the train travel? (*Hint:* 5,280 feet = 1 mile.)

45. METEOROLOGY See Illustration 2. The height h of the cloud can be measured by shining a light vertically and measuring the angle of elevation at a distance d from the light. If $d = 557$ feet and $m(\angle A) = 78.3°$, find h to the nearest ten.

ILLUSTRATION 2

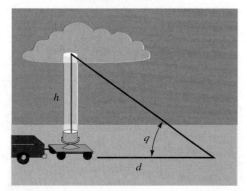

46. ASTRONOMY As viewed from the earth, the greatest angle between the sun and the planet Mercury is 28°. If the earth is 93 million miles from the sun, find the distance d between Mercury and the sun. Give the

result to the nearest million miles. (See Illustration 3.)

ILLUSTRATION 3

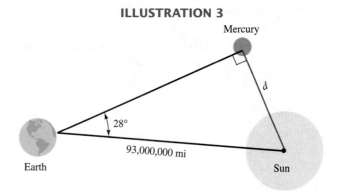

47. INDIRECT MEASUREMENT The sun visor in Latoya's car prevents her from seeing objects more than 17° above the horizon. When her car is 65 feet from an intersection, she can barely see the bottom of the stoplight hanging above the roadway. (See Illustration 4.) To the nearest foot, find the height of the stoplight.

ILLUSTRATION 4

48. NAVIGATION A ship leaves port on a bearing of S 12.7° E. To the nearest mile, how far south has the ship traveled during a trip of 327 miles?

49. BEARING The bearing of Madison, WI from Stevens Point, WI is S 4.1° E. If the distance between the cities is 108 miles, by how many miles is Madison east of Stevens Point? Give the answer to the nearest tenth.

50. BEARING A ship leaves port and sails due west. It then sails due south. At a certain moment, its bearing from port is S 62.9° W, and its distance from port is 275 miles. To the nearest mile, how far did the ship sail due south?

51. GEORGE WASHINGTON'S FACE Use the information in Illustration 5 to find the height of George Washington's face on Mount Rushmore. Give the answer to the nearest foot.

ILLUSTRATION 5

52. LENGTH OF AN ISLAND A plane is flying at an altitude of 5,120 feet. As it approaches an island, the navigator determines the angles of depression, as shown in Illustration 6. To the nearest tenth, find the length of the island in miles.

ILLUSTRATION 6

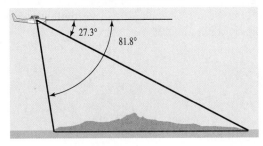

WRITING

53. Is $\sin A = \cos(90° - A)$? Explain.

54. Is $\tan A = \tan(90° - A)$? Explain.

55. In Illustration 7, $\sin(\angle A) = a$. Explain why the length of leg AB is $\frac{5}{a}$.

ILLUSTRATION 7

56. In Illustration 8, $\cos(\angle B) = b$. Explain why the length of leg AB is $\frac{7}{b}$.

ILLUSTRATION 8

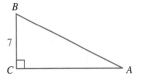

Appendix II

Synthetic Division

In this appendix, you will learn about

Synthetic division ■ The remainder theorem ■ The factor theorem

Introduction In Section 6.7, we discussed how to divide polynomials by polynomials. In this appendix, we will discuss a shortcut method, called **synthetic division,** that we can use to divide a polynomial by a binomial of the form $x - r$.

Synthetic Division

To see how synthetic division works, we consider the division of $4x^3 - 5x^2 - 11x + 20$ by $x - 2$.

$$
\begin{array}{r}
4x^2 + 3x - 5 \\
x - 2{\overline{\smash{\big)}\,4x^3 - 5x^2 - 11x + 20}} \\
\underline{4x^3 - 8x^2} \\
3x^2 - 11x \\
\underline{3x^2 - 6x} \\
-5x + 20 \\
\underline{-5x + 10} \\
10 \quad \text{(remainder)}
\end{array}
$$

$$
\begin{array}{r}
4 \quad 3 - 5 \\
1 - 2{\overline{\smash{\big)}\,4 - 5 - 11 \quad 20}} \\
\underline{4 - 8} \\
3 - 11 \\
\underline{3 - 6} \\
-5 \quad 20 \\
\underline{-5 \quad 10} \\
10 \quad \text{(remainder)}
\end{array}
$$

On the left is the long division, and on the right is the same division with the variables and their exponents removed. The various powers of x can be remembered without actually writing them, because the exponents of the terms in the divisor, dividend, and quotient were written in descending order.

We can further shorten the version on the right. The numbers printed in color need not be written, because they are duplicates of the numbers above them. Thus, we can write the division in the following form:

$$
\begin{array}{r}
4 \quad 3 - 5 \\
1 - 2{\overline{\smash{\big)}\,4 - 5 - 11 \quad 20}} \\
\underline{-8} \\
3 \\
\underline{-6} \\
-5 \\
\underline{10} \\
10
\end{array}
$$

We can shorten the process further by compressing the work vertically and eliminating the 1 (the coefficient of x in the divisor):

$$
\begin{array}{r}
4 \quad\ \ 3 \quad -5 \\
-2\overline{)4 \ \ -5 \ \ -11 \ \ \ 20} \\
-8 \ \ -6 \quad 10 \\
\hline
3 \quad -5 \quad 10
\end{array}
$$

If we write the 4 in the quotient on the bottom line, the bottom line gives the coefficients of the quotient and the remainder. If we eliminate the top line, the division appears as follows:

$$
\begin{array}{r}
\underline{-2}\ \big|\ \ \ 4 \quad -5 \quad -11 \quad 20 \\
-8 \quad -\ 6 \quad 10 \\
\hline
4 \quad\ \ 3 \quad -\ 5 \quad 10
\end{array}
$$

The bottom line was obtained by subtracting the middle line from the top line. If we replace the -2 in the divisor by $+2$, the division process will reverse the signs of every entry in the middle line, and then the bottom line can be obtained by addition. This gives the final form of the synthetic division.

$$
\begin{array}{r}
\underline{+2}\ \big|\ \ \ 4 \quad -5 \quad -11 \quad\ \ 20 \qquad \text{The coefficients of the dividend.}\\
8 \qquad 6 \quad -10 \\
\hline
4 \qquad 3 \quad -5\ \cdot \quad 10 \qquad \text{The coefficients of the quotient and the remainder.}
\end{array}
$$

Thus,

$$
\frac{4x^3 - 5x^2 - 11x + 20}{x - 2} = 4x^2 + 3x - 5 + \frac{10}{x - 2}
$$

EXAMPLE 1

Divide $6x^2 + 5x - 2$ by $x - 5$.

Solution   We write the coefficients in the dividend and the 5 in the divisor in the following form:

$$
\underline{5}\ \big|\ \ \ 6 \quad 5 \quad -2
$$

Then we follow these steps:

$$
\underline{5}\ \big|\ \ \ 6 \quad 5 \quad -2 \qquad \text{Begin by bringing down the 6.}
$$
$$
6
$$

$$
\underline{5}\ \big|\ \ \ 6 \quad 5 \quad -2 \qquad \text{Multiply 5 by 6 to get 30.}
$$
$$
\ \ \ 30
$$
$$
6
$$

$$
\underline{5}\ \big|\ \ \ 6 \quad 5 \quad -2 \qquad \text{Add 5 and 30 to get 35.}
$$
$$
\ \ \ 30
$$
$$
6 \quad 35
$$

$$
\underline{5}\ \big|\ \ \ 6 \quad 5 \quad -2 \qquad \text{Multiply 35 by 5 to get 175.}
$$
$$
\ \ \ 30 \quad 175
$$
$$
6 \quad 35
$$

$$
\underline{5}\ \big|\ \ \ 6 \quad 5 \quad -2 \qquad \text{Add } -2 \text{ and 175 to get 173.}
$$
$$
\ \ \ 30 \quad 175
$$
$$
6 \quad 35 \quad 173
$$

The numbers 6 and 35 represent the quotient $6x + 35$, and 173 is the remainder. Thus,

$$\frac{6x^2 + 5x - 2}{x - 5} = 6x + 35 + \frac{173}{x - 5}$$

∎

EXAMPLE 2

Divide $5x^3 + x^2 - 3$ by $x - 2$.

Solution We begin by writing

$2\rfloor\ \ 5\ \ \ 1\ \ \ \mathbf{0}\ \ \ -3$ Write 0 for the coefficient of x, the missing term.

and complete the division as follows:

$$
\begin{array}{c|cccc}
2 & 5 & 1 & 0 & -3 \\
  &   & 10 &   &   \\
\hline
  & 5 & 11 &   &
\end{array}
\qquad
\begin{array}{c|cccc}
2 & 5 & 1 & 0 & -3 \\
  &   & 10 & 22 &   \\
\hline
  & 5 & 11 & 22 &
\end{array}
\qquad
\begin{array}{c|cccc}
2 & 5 & 1 & 0 & -3 \\
  &   & 10 & 22 & 44 \\
\hline
  & 5 & 11 & 22 & 41
\end{array}
$$

Thus,

$$\frac{5x^3 + x^2 - 3}{x - 2} = 5x^2 + 11x + 22 + \frac{41}{x - 2}$$

∎

EXAMPLE 3

Divide $5x^2 + 6x^3 + 2 - 4x$ by $x + 2$.

Solution First, we write the dividend with the exponents in descending order.

$$6x^3 + 5x^2 - 4x + 2$$

Then we write the divisor in $x - r$ form: $x - (-2)$. Using synthetic division, we begin by writing

$$
\begin{array}{c|cccc}
-2 & 6 & 5 & -4 & 2 \\
\hline
  &   &   &   &
\end{array}
$$

and complete the division.

$$
\begin{array}{c|cccc}
-2 & 6 & 5 & -4 & 2 \\
  &   & -12 & 14 & -20 \\
\hline
  & 6 & -7 & 10 & -18
\end{array}
$$

Thus,

$$\frac{5x^2 + 6x^3 + 2 - 4x}{x + 2} = 6x^2 - 7x + 10 + \frac{-18}{x + 2}$$

∎

The Remainder Theorem

Synthetic division is important in mathematics because of the **remainder theorem.**

Remainder theorem

If a polynomial $P(x)$ is divided by $x - r$, the remainder is $P(r)$.

We illustrate the remainder theorem in the next example.

EXAMPLE 4

Let $P(x) = 2x^3 - 3x^2 - 2x + 1$. Find **a.** $P(3)$ and **b.** the remainder when $P(x)$ is divided by $x - 3$.

Solution **a.** $P(3) = 2(3)^3 - 3(3)^2 - 2(3) + 1$ Substitute 3 for x.
$$= 2(27) - 3(9) - 6 + 1$$
$$= 54 - 27 - 6 + 1$$
$$= 22$$

b. We use synthetic division to find the remainder when $P(x) = 2x^3 - 3x^2 - 2x + 1$ is divided by $x - 3$.

$$
\begin{array}{r|rrrr}
3 & 2 & -3 & -2 & 1 \\
  &   & 6 & 9 & 21 \\
\hline
  & 2 & 3 & 7 & 22
\end{array}
$$

The remainder is 22.
The results of parts a and b show that when $P(x)$ is divided by $x - 3$, the remainder is $P(3)$. ■

It is often easier to find $P(r)$ by using synthetic division than by substituting r for x in $P(x)$. This is especially true if r is a decimal.

The Factor Theorem

Recall that if two quantities are multiplied, each is called a **factor** of the product. Thus, $x - 2$ is one factor of $6x - 12$, because $6(x - 2) = 6x - 12$. A theorem, called the **factor theorem,** tells us how to find one factor of a polynomial if the remainder of a certain division is 0.

Factor theorem

If $P(x)$ is a polynomial in x, then
$$P(r) = 0 \quad \text{if and only if} \quad x - r \text{ is a factor of } P(x)$$

If $P(x)$ is a polynomial in x and if $P(r) = 0$, r is called a **zero of the polynomial.**

EXAMPLE 5

Let $P(x) = 3x^3 - 5x^2 + 3x - 10$. Show that **a.** $P(2) = 0$ and **b.** $x - 2$ is a factor of $P(x)$.

Solution   **a.** Use the remainder theorem to evaluate $P(2)$ by dividing $P(x) = 3x^3 - 5x^2 + 3x - 10$ by $x - 2$.

$$
\begin{array}{r|rrrr}
2 & 3 & -5 & 3 & -10 \\
  &   & 6 & 2 & 10 \\
\hline
  & 3 & 1 & 5 & 0 \\
\end{array}
$$

The remainder in this division is 0. By the remainder theorem, the remainder is $P(2)$. Thus, $P(2) = 0$, and 2 is a zero of the polynomial.

b. Because the remainder is 0, the numbers 3, 1, and 5 in the synthetic division in part a represent the quotient $3x^2 + x + 5$. Thus,

$$\underbrace{(x - 2)}_{\text{Divisor}\ \cdot} \cdot \underbrace{(3x^2 + x + 5)}_{\text{quotient}} + \underbrace{0}_{+\ \text{remainder}\ =} = \underbrace{3x^3 - 5x^2 + 3x - 10}_{\text{the dividend, } P(x)}$$

or

$$(x - 2)(3x^2 + x + 5) = 3x^3 - 5x^2 + 3x - 10$$

Thus, $x - 2$ is a factor of $P(x)$. ∎

The result in Example 5 is true, because the remainder, $P(2)$, is 0. If the remainder had not been 0, then $x - 2$ would not have been a factor of $P(x)$.

ACCENT ON TECHNOLOGY *Approximating Zeros of Polynomials*

We can use a graphing calculator to approximate the real zeros of a polynomial function $f(x)$. For example, to find the real zeros of $f(x) = 2x^3 - 6x^2 + 7x - 21$, we graph the function as in Figure II-1.

It is clear from the figure that the function f has a zero at $x = 3$.

$$
\begin{aligned}
f(3) &= 2(3)^3 - 6(3)^2 + 7(3) - 21 \quad \text{Substitute 3 for } x. \\
&= 2(27) - 6(9) + 21 - 21 \\
&= 0
\end{aligned}
$$

From the factor theorem, we know that $x - 3$ is a factor of the polynomial. To find the other factor, we can synthetically divide by 3.

$$
\begin{array}{r|rrrr}
3 & 2 & -6 & 7 & -21 \\
  &   & 6 & 0 & 21 \\
\hline
  & 2 & 0 & 7 & 0 \\
\end{array}
$$

Thus, $f(x) = (x - 3)(2x^2 + 7)$. Since $2x^2 + 7$ cannot be factored over the real numbers, we can conclude that 3 is the only real zero of the polynomial function.

FIGURE II-1

VOCABULARY

Fill in the blanks to make the statements true.

1. The expression $3x^3 - 2x^2 + 7x + 5$ is a _____ in x.

2. If $P(x)$ is a polynomial in x and if $P(x) = 0$, then r is called a _____ of the polynomial.

CONCEPTS

Fill in the blanks to make the statements true.

3. If a polynomial $P(x)$ is divided by $x - r$, the remainder is _____.

4. If $P(x)$ is a polynomial in x, then $P(r) = 0$ if and only if _____ is a factor of $P(x)$.

NOTATION

Complete each synthetic division.

5. $\underline{2|}$ 6 1 -23 2
$$ \boxed{}\ \ \boxed{}\ \ \ 6$$
$$ 6\ \ \ 13\ \ \ \ 3\ \ \ \boxed{}$$

6. $\underline{-3|}$ 2 -4 -25 15
$$ -6\ \ \ \ 30\ \ \ \boxed{}$$
$$ \boxed{}\ \ \boxed{}\ \ \ \boxed{}\ \ \ \ 0$$

PRACTICE

In Exercises 7–20, use synthetic division to do each division.

7. $(x^2 + x - 2) \div (x - 1)$

8. $(x^2 + x - 6) \div (x - 2)$

9. $(x^2 - 7x + 12) \div (x - 4)$

10. $(x^2 - 6x + 5) \div (x - 5)$

11. $(x^2 + 8 + 6x) \div (x + 4)$

12. $(x^2 - 15 - 2x) \div (x + 3)$

13. $(x^2 - 5x + 14) \div (x + 2)$

14. $(x^2 + 13x + 42) \div (x + 6)$

15. $(3x^3 - 10x^2 + 5x - 6) \div (x - 3)$

16. $(2x^3 - 9x^2 + 10x - 3) \div (x - 3)$

17. $(2x^3 - 5x - 6) \div (x - 2)$

18. $(4x^3 + 5x^2 - 1) \div (x + 2)$

19. $(5x^2 + 6x^3 + 4) \div (x + 1)$

20. $(4 - 3x^2 + x) \div (x - 4)$

In Exercises 21–26, use a calculator and synthetic division to do each division.

21. $(7.2x^2 - 2.1x + 0.5) \div (x - 0.2)$

22. $(8.1x^2 + 3.2x - 5.7) \div (x - 0.4)$

23. $(2.7x^2 + x - 5.2) \div (x + 1.7)$

24. $(1.3x^2 - 0.5x - 2.3) \div (x + 2.5)$

25. $(9x^3 - 25) \div (x + 57)$

26. $(0.5x^3 + x) \div (x - 2.3)$

In Exercises 27–34, let $P(x) = 2x^3 - 4x^2 + 2x - 1$. Evaluate $P(x)$ by substituting the given value of x into the polynomial and simplifying. Then evaluate the polynomial by using the remainder theorem and synthetic division.

27. $P(1)$ 28. $P(2)$

29. $P(-2)$ 30. $P(-1)$

31. $P(3)$ 32. $P(-4)$

33. $P(0)$ 34. $P(4)$

In Exercises 35–42, let $Q(x) = x^4 - 3x^3 + 2x^2 + x - 3$. Evaluate $Q(x)$ by substituting the given value of x into the polynomial and simplifying. Then evaluate the polynomial by using the remainder theorem and synthetic division.

35. $Q(-1)$ 36. $Q(1)$

37. $Q(2)$ 38. $Q(-2)$

39. $Q(3)$ 40. $Q(0)$

41. $Q(-3)$ 42. $Q(-4)$

In Exercises 43–50, use the remainder theorem and synthetic division to find $P(r)$.

43. $P(x) = x^3 - 4x^2 + x - 2; r = 2$

44. $P(x) = x^3 - 3x^2 + x + 1; r = 1$

45. $P(x) = 2x^3 + x + 2; r = 3$

46. $P(x) = x^3 + x^2 + 1; r = -2$

47. $P(x) = x^4 - 2x^3 + x^2 - 3x + 2; r = -2$

48. $P(x) = x^5 + 3x^4 - x^2 + 1; r = -1$

49. $P(x) = 3x^5 + 1; r = -\frac{1}{2}$

50. $P(x) = 5x^7 - 7x^4 + x^2 + 1; r = 2$

In Exercises 51–54, use the factor theorem and tell whether the first expression is a factor of $P(x)$.

51. $x - 3; P(x) = x^3 - 3x^2 + 5x - 15$

52. $x + 1; P(x) = x^3 + 2x^2 - 2x - 3$ (*Hint:* Write $x + 1$ as $x - (-1)$.)

53. $x + 2$; $P(x) = 3x^2 - 7x + 4$ (*Hint:* Write $x + 2$ as $x - (-2)$.)

54. x; $P(x) = 7x^3 - 5x^2 - 8x$ (*Hint:* $x = x - 0$.)

In Exercises 55–56, use a calculator to work each problem.

55. Find 2^6 by using synthetic division to evaluate the polynomial $P(x) = x^6$ at $x = 2$. Then check the answer by evaluating 2^6 with a calculator.

56. Find $(-3)^5$ by using synthetic division to evaluate the polynomial $P(x) = x^5$ at $x = -3$. Then check the answer by evaluating $(-3)^5$ with a calculator.

WRITING

57. If you are given $P(x)$, explain how to use synthetic division to calculate $P(a)$.

58. Explain the factor theorem.

Appendix III

Tables

TABLE A   Powers and Roots

n	n^2	$\sqrt{n}$	n^3	$\sqrt[3]{n}$	n	n^2	$\sqrt{n}$	n^3	$\sqrt[3]{n}$
1	1	1.000	1	1.000	51	2,601	7.141	132,651	3.708
2	4	1.414	8	1.260	52	2,704	7.211	140,608	3.733
3	9	1.732	27	1.442	53	2,809	7.280	148,877	3.756
4	16	2.000	64	1.587	54	2,916	7.348	157,464	3.780
5	25	2.236	125	1.710	55	3,025	7.416	166,375	3.803
6	36	2.449	216	1.817	56	3,136	7.483	175,616	3.826
7	49	2.646	343	1.913	57	3,249	7.550	185,193	3.849
8	64	2.828	512	2.000	58	3,364	7.616	195,112	3.871
9	81	3.000	729	2.080	59	3,481	7.681	205,379	3.893
10	100	3.162	1,000	2.154	60	3,600	7.746	216,000	3.915
11	121	3.317	1,331	2.224	61	3,721	7.810	226,981	3.936
12	144	3.464	1,728	2.289	62	3,844	7.874	238,328	3.958
13	169	3.606	2,197	2.351	63	3,969	7.937	250,047	3.979
14	196	3.742	2,744	2.410	64	4,096	8.000	262,144	4.000
15	225	3.873	3,375	2.466	65	4,225	8.062	274,625	4.021
16	256	4.000	4,096	2.520	66	4,356	8.124	287,496	4.041
17	289	4.123	4,913	2.571	67	4,489	8.185	300,763	4.062
18	324	4.243	5,832	2.621	68	4,624	8.246	314,432	4.082
19	361	4.359	6,859	2.668	69	4,761	8.307	328,509	4.102
20	400	4.472	8,000	2.714	70	4,900	8.367	343,000	4.121
21	441	4.583	9,261	2.759	71	5,041	8.426	357,911	4.141
22	484	4.690	10,648	2.802	72	5,184	8.485	373,248	4.160
23	529	4.796	12,167	2.844	73	5,329	8.544	389,017	4.179
24	576	4.899	13,824	2.884	74	5,476	8.602	405,224	4.198
25	625	5.000	15,625	2.924	75	5,625	8.660	421,875	4.217
26	676	5.099	17,576	2.962	76	5,776	8.718	438,976	4.236
27	729	5.196	19,683	3.000	77	5,929	8.775	456,533	4.254
28	784	5.292	21,952	3.037	78	6,084	8.832	474,552	4.273
29	841	5.385	24,389	3.072	79	6,241	8.888	493,039	4.291
30	900	5.477	27,000	3.107	80	6,400	8.944	512,000	4.309
31	961	5.568	29,791	3.141	81	6,561	9.000	531,441	4.327
32	1,024	5.657	32,768	3.175	82	6,724	9.055	551,368	4.344
33	1,089	5.745	35,937	3.208	83	6,889	9.110	571,787	4.362
34	1,156	5.831	39,304	3.240	84	7,056	9.165	592,704	4.380
35	1,225	5.916	42,875	3.271	85	7,225	9.220	614,125	4.397
36	1,296	6.000	46,656	3.302	86	7,396	9.274	636,056	4.414
37	1,369	6.083	50,653	3.332	87	7,569	9.327	658,503	4.431
38	1,444	6.164	54,872	3.362	88	7,744	9.381	681,472	4.448
39	1,521	6.245	59,319	3.391	89	7,921	9.434	704,969	4.465
40	1,600	6.325	64,000	3.420	90	8,100	9.487	729,000	4.481
41	1,681	6.403	68,921	3.448	91	8,281	9.539	753,571	4.498
42	1,764	6.481	74,088	3.476	92	8,464	9.592	778,688	4.514
43	1,849	6.557	79,507	3.503	93	8,649	9.644	804,357	4.531
44	1,936	6.633	85,184	3.530	94	8,836	9.695	830,584	4.547
45	2,025	6.708	91,125	3.557	95	9,025	9.747	857,375	4.563
46	2,116	6.782	97,336	3.583	96	9,216	9.798	884,736	4.579
47	2,209	6.856	103,823	3.609	97	9,409	9.849	912,673	4.595
48	2,304	6.928	110,592	3.634	98	9,604	9.899	941,192	4.610
49	2,401	7.000	117,649	3.659	99	9,801	9.950	970,299	4.626
50	2,500	7.071	125,000	3.684	100	10,000	10.000	1,000,000	4.642

TABLE B *(continued)*

N	0	1	2	3	4	5	6	7	8	9
5.5	.7404	.7412	.7419	.7427	.7435	.7443	.7451	.7459	.7466	.7474
5.6	.7482	.7490	.7497	.7505	.7513	.7520	.7528	.7536	.7543	.7551
5.7	.7559	.7566	.7574	.7582	.7589	.7597	.7604	.7612	.7619	.7627
5.8	.7634	.7642	.7649	.7657	.7664	.7672	.7679	.7686	.7694	.7701
5.9	.7709	.7716	.7723	.7731	.7738	.7745	.7752	.7760	.7767	.7774
6.0	.7782	.7789	.7796	.7803	.7810	.7818	.7825	.7832	.7839	.7846
6.1	.7853	.7860	.7868	.7875	.7882	.7889	.7896	.7903	.7910	.7917
6.2	.7924	.7931	.7938	.7945	.7952	.7959	.7966	.7973	.7980	.7987
6.3	.7993	.8000	.8007	.8014	.8021	.8028	.8035	.8041	.8048	.8055
6.4	.8062	.8069	.8075	.8082	.8089	.8096	.8102	.8109	.8116	.8122
6.5	.8129	.8136	.8142	.8149	.8156	.8162	.8169	.8176	.8182	.8189
6.6	.8195	.8202	.8209	.8215	.8222	.8228	.8235	.8241	.8248	.8254
6.7	.8261	.8267	.8274	.8280	.8287	.8293	.8299	.8306	.8312	.8319
6.8	.8325	.8331	.8338	.8344	.8351	.8357	.8363	.8370	.8376	.8382
6.9	.8388	.8395	.8401	.8407	.8414	.8420	.8426	.8432	.8439	.8445
7.0	.8451	.8457	.8463	.8470	.8476	.8482	.8488	.8494	.8500	.8506
7.1	.8513	.8519	.8525	.8531	.8537	.8543	.8549	.8555	.8561	.8567
7.2	.8573	.8579	.8585	.8591	.8597	.8603	.8609	.8615	.8621	.8627
7.3	.8633	.8639	.8645	.8651	.8657	.8663	.8669	.8675	.8681	.8686
7.4	.8692	.8698	.8704	.8710	.8716	.8722	.8727	.8733	.8739	.8745
7.5	.8751	.8756	.8762	.8768	.8774	.8779	.8785	.8791	.8797	.8802
7.6	.8808	.8814	.8820	.8825	.8831	.8837	.8842	.8848	.8854	.8859
7.7	.8865	.8871	.8876	.8882	.8887	.8893	.8899	.8904	.8910	.8915
7.8	.8921	.8927	.8932	.8938	.8943	.8949	.8954	.8960	.8965	.8971
7.9	.8976	.8982	.8987	.8993	.8998	.9004	.9009	.9015	.9020	.9025
8.0	.9031	.9036	.9042	.9047	.9053	.9058	.9063	.9069	.9074	.9079
8.1	.9085	.9090	.9096	.9101	.9106	.9112	.9117	.9122	.9128	.9133
8.2	.9138	.9143	.9149	.9154	.9159	.9165	.9170	.9175	.9180	.9186
8.3	.9191	.9196	.9201	.9206	.9212	.9217	.9222	.9227	.9232	.9238
8.4	.9243	.9248	.9253	.9258	.9263	.9269	.9274	.9279	.9284	.9289
8.5	.9294	.9299	.9304	.9309	.9315	.9320	.9325	.9330	.9335	.9340
8.6	.9345	.9350	.9355	.9360	.9365	.9370	.9375	.9380	.9385	.9390
8.7	.9395	.9400	.9405	.9410	.9415	.9420	.9425	.9430	.9435	.9440
8.8	.9445	.9450	.9455	.9460	.9465	.9469	.9474	.9479	.9484	.9489
8.9	.9494	.9499	.9504	.9509	.9513	.9518	.9523	.9528	.9533	.9538
9.0	.9542	.9547	.9552	.9557	.9562	.9566	.9571	.9576	.9581	.9586
9.1	.9590	.9595	.9600	.9605	.9609	.9614	.9619	.9624	.9628	.9633
9.2	.9638	.9643	.9647	.9652	.9657	.9661	.9666	.9671	.9675	.9680
9.3	.9685	.9689	.9694	.9699	.9703	.9708	.9713	.9717	.9722	.9727
9.4	.9731	.9736	.9741	.9745	.9750	.9754	.9759	.9763	.9768	.9773
9.5	.9777	.9782	.9786	.9791	.9795	.9800	.9805	.9809	.9814	.9818
9.6	.9823	.9827	.9832	.9836	.9841	.9845	.9850	.9854	.9859	.9863
9.7	.9868	.9872	.9877	.9881	.9886	.9890	.9894	.9899	.9903	.9908
9.8	.9912	.9917	.9921	.9926	.9930	.9934	.9939	.9943	.9948	.9952
9.9	.9956	.9961	.9965	.9969	.9974	.9978	.9983	.9987	.9991	.9996

TABLE B   Base-10 Logarithms

N	0	1	2	3	4	5	6	7	8	9
1.0	.0000	.0043	.0086	.0128	.0170	.0212	.0253	.0294	.0334	.0374
1.1	.0414	.0453	.0492	.0531	.0569	.0607	.0645	.0682	.0719	.0755
1.2	.0792	.0828	.0864	.0899	.0934	.0969	.1004	.1038	.1072	.1106
1.3	.1139	.1173	.1206	.1239	.1271	.1303	.1335	.1367	.1399	.1430
1.4	.1461	.1492	.1523	.1553	.1584	.1614	.1644	.1673	.1703	.1732
1.5	.1761	.1790	.1818	.1847	.1875	.1903	.1931	.1959	.1987	.2014
1.6	.2041	.2068	.2095	.2122	.2148	.2175	.2201	.2227	.2253	.2279
1.7	.2304	.2330	.2355	.2380	.2405	.2430	.2455	.2480	.2504	.2529
1.8	.2553	.2577	.2601	.2625	.2648	.2672	.2695	.2718	.2742	.2765
1.9	.2788	.2810	.2833	.2856	.2878	.2900	.2923	.2945	.2967	.2989
2.0	.3010	.3032	.3054	.3075	.3096	.3118	.3139	.3160	.3181	.3201
2.1	.3222	.3243	.3263	.3284	.3304	.3324	.3345	.3365	.3385	.3404
2.2	.3424	.3444	.3464	.3483	.3502	.3522	.3541	.3560	.3579	.3598
2.3	.3617	.3636	.3655	.3674	.3692	.3711	.3729	.3747	.3766	.3784
2.4	.3802	.3820	.3838	.3856	.3874	.3892	.3909	.3927	.3945	.3962
2.5	.3979	.3997	.4014	.4031	.4048	.4065	.4082	.4099	.4116	.4133
2.6	.4150	.4166	.4183	.4200	.4216	.4232	.4249	.4265	.4281	.4298
2.7	.4314	.4330	.4346	.4362	.4378	.4393	.4409	.4425	.4440	.4456
2.8	.4472	.4487	.4502	.4518	.4533	.4548	.4564	.4579	.4594	.4609
2.9	.4624	.4639	.4654	.4669	.4683	.4698	.4713	.4728	.4742	.4757
3.0	.4771	.4786	.4800	.4814	.4829	.4843	.4857	.4871	.4886	.4900
3.1	.4914	.4928	.4942	.4955	.4969	.4983	.4997	.5011	.5024	.5038
3.2	.5051	.5065	.5079	.5092	.5105	.5119	.5132	.5145	.5159	.5172
3.3	.5185	.5198	.5211	.5224	.5237	.5250	.5263	.5276	.5289	.5302
3.4	.5315	.5328	.5340	.5353	.5366	.5378	.5391	.5403	.5416	.5428
3.5	.5441	.5453	.5465	.5478	.5490	.5502	.5514	.5527	.5539	.5551
3.6	.5563	.5575	.5587	.5599	.5611	.5623	.5635	.5647	.5658	.5670
3.7	.5682	.5694	.5705	.5717	.5729	.5740	.5752	.5763	.5775	.5786
3.8	.5798	.5809	.5821	.5832	.5843	.5855	.5866	.5877	.5888	.5899
3.9	.5911	.5922	.5933	.5944	.5955	.5966	.5977	.5988	.5999	.6010
4.0	.6021	.6031	.6042	.6053	.6064	.6075	.6085	.6096	.6107	.6117
4.1	.6128	.6138	.6149	.6160	.6170	.6180	.6191	.6201	.6212	.6222
4.2	.6232	.6243	.6253	.6263	.6274	.6284	.6294	.6304	.6314	.6325
4.3	.6335	.6345	.6355	.6365	.6375	.6385	.6395	.6405	.6415	.6425
4.4	.6435	.6444	.6454	.6464	.6474	.6484	.6493	.6503	.6513	.6522
4.5	.6532	.6542	.6551	.6561	.6571	.6580	.6590	.6599	.6609	.6618
4.6	.6628	.6637	.6646	.6656	.6665	.6675	.6684	.6693	.6702	.6712
4.7	.6721	.6730	.6739	.6749	.6758	.6767	.6776	.6785	.6794	.6803
4.8	.6812	.6821	.6830	.6839	.6848	.6857	.6866	.6875	.6884	.6893
4.9	.6902	.6911	.6920	.6928	.6937	.6946	.6955	.6964	.6972	.6981
5.0	.6990	.6998	.7007	.7016	.7024	.7033	.7042	.7050	.7059	.7067
5.1	.7076	.7084	.7093	.7101	.7110	.7118	.7126	.7135	.7143	.7152
5.2	.7160	.7168	.7177	.7185	.7193	.7202	.7210	.7218	.7226	.7235
5.3	.7243	.7251	.7259	.7267	.7275	.7284	.7292	.7300	.7308	.7316
5.4	.7324	.7332	.7340	.7348	.7356	.7364	.7372	.7380	.7388	.7396

TABLE C Base-e Logarithms

N	0	1	2	3	4	5	6	7	8	9
1.0	.0000	.0100	.0198	.0296	.0392	.0488	.0583	.0677	.0770	.0862
1.1	.0953	.1044	.1133	.1222	.1310	.1398	.1484	.1570	.1655	.1740
1.2	.1823	.1906	.1989	.2070	.2151	.2231	.2311	.2390	.2469	.2546
1.3	.2624	.2700	.2776	.2852	.2927	.3001	.3075	.3148	.3221	.3293
1.4	.3365	.3436	.3507	.3577	.3646	.3716	.3784	.3853	.3920	.3988
1.5	.4055	.4121	.4187	.4253	.4318	.4383	.4447	.4511	.4574	.4637
1.6	.4700	.4762	.4824	.4886	.4947	.5008	.5068	.5128	.5188	.5247
1.7	.5306	.5365	.5423	.5481	.5539	.5596	.5653	.5710	.5766	.5822
1.8	.5878	.5933	.5988	.6043	.6098	.6152	.6206	.6259	.6313	.6366
1.9	.6419	.6471	.6523	.6575	.6627	.6678	.6729	.6780	.6831	.6881
2.0	.6931	.6981	.7031	.7080	.7129	.7178	.7227	.7275	.7324	.7372
2.1	.7419	.7467	.7514	.7561	.7608	.7655	.7701	.7747	.7793	.7839
2.2	.7885	.7930	.7975	.8020	.8065	.8109	.8154	.8198	.8242	.8286
2.3	.8329	.8372	.8416	.8459	.8502	.8544	.8587	.8629	.8671	.8713
2.4	.8755	.8796	.8838	.8879	.8920	.8961	.9002	.9042	.9083	.9123
2.5	.9163	.9203	.9243	.9282	.9322	.9361	.9400	.9439	.9478	.9517
2.6	.9555	.9594	.9632	.9670	.9708	.9746	.9783	.9821	.9858	.9895
2.7	.9933	.9969	1.0006	.0043	.0080	.0116	.0152	.0188	.0225	.0260
2.8	1.0296	.0332	.0367	.0403	.0438	.0473	.0508	.0543	.0578	.0613
2.9	.0647	.0682	.0716	.0750	.0784	.0818	.0852	.0886	.0919	.0953
3.0	1.0986	.1019	.1053	.1086	.1119	.1151	.1184	.1217	.1249	.1282
3.1	.1314	.1346	.1378	.1410	.1442	.1474	.1506	.1537	.1569	.1600
3.2	.1632	.1663	.1694	.1725	.1756	.1787	.1817	.1848	.1878	.1909
3.3	.1939	.1969	.2000	.2030	.2060	.2090	.2119	.2149	.2179	.2208
3.4	.2238	.2267	.2296	.2326	.2355	.2384	.2413	.2442	.2470	.2499
3.5	1.2528	.2556	.2585	.2613	.2641	.2669	.2698	.2726	.2754	.2782
3.6	.2809	.2837	.2865	.2892	.2920	.2947	.2975	.3002	.3029	.3056
3.7	.3083	.3110	.3137	.3164	.3191	.3218	.3244	.3271	.3297	.3324
3.8	.3350	.3376	.3403	.3429	.3455	.3481	.3507	.3533	.3558	.3584
3.9	.3610	.3635	.3661	.3686	.3712	.3737	.3762	.3788	.3813	.3838
4.0	1.3863	.3888	.3913	.3938	.3962	.3987	.4012	.4036	.4061	.4085
4.1	.4110	.4134	.4159	.4183	.4207	.4231	.4255	.4279	.4303	.4327
4.2	.4351	.4375	.4398	.4422	.4446	.4469	.4493	.4516	.4540	.4563
4.3	.4586	.4609	.4633	.4656	.4679	.4702	.4725	.4748	.4770	.4793
4.4	.4816	.4839	.4861	.4884	.4907	.4929	.4951	.4974	.4996	.5019
4.5	1.5041	.5063	.5085	.5107	.5129	.5151	.5173	.5195	.5217	.5239
4.6	.5261	.5282	.5304	.5326	.5347	.5369	.5390	.5412	.5433	.5454
4.7	.5476	.5497	.5518	.5539	.5560	.5581	.5602	.5623	.5644	.5665
4.8	.5686	.5707	.5728	.5748	.5769	.5790	.5810	.5831	.5851	.5872
4.9	.5892	.5913	.5933	.5953	.5974	.5994	.6014	.6034	.6054	.6074
5.0	1.6094	.6114	.6134	.6154	.6174	.6194	.6214	.6233	.6253	.6273
5.1	.6292	.6312	.6332	.6351	.6371	.6390	.6409	.6429	.6448	.6467
5.2	.6487	.6506	.6525	.6544	.6563	.6582	.6601	.6620	.6639	.6658
5.3	.6677	.6696	.6715	.6734	.6752	.6771	.6790	.6808	.6827	.6845
5.4	.6864	.6882	.6901	.6919	.6938	.6956	.6974	.6993	.7011	.7029

Use the properties of logarithms and ln $10 = 2.3026$ to find logarithms of numbers less than 1 or greater than 10.

TABLE C (continued)

N	0	1	2	3	4	5	6	7	8	9
5.5	1.7047	.7066	.7084	.7102	.7120	.7138	.7156	.7174	.7192	.7210
5.6	.7228	.7246	.7263	.7281	.7299	.7317	.7334	.7352	.7370	.7387
5.7	.7405	.7422	.7440	.7457	.7475	.7492	.7509	.7527	.7544	.7561
5.8	.7579	.7596	.7613	.7630	.7647	.7664	.7681	.7699	.7716	.7733
5.9	.7750	.7766	.7783	.7800	.7817	.7834	.7851	.7867	.7884	.7901
6.0	1.7918	.7934	.7951	.7967	.7984	.8001	.8017	.8034	.8050	.8066
6.1	.8083	.8099	.8116	.8132	.8148	.8165	.8181	.8197	.8213	.8229
6.2	.8245	.8262	.8278	.8294	.8310	.8326	.8342	.8358	.8374	.8390
6.3	.8405	.8421	.8437	.8453	.8469	.8485	.8500	.8516	.8532	.8547
6.4	.8563	.8579	.8594	.8610	.8625	.8641	.8656	.8672	.8687	.8703
6.5	1.8718	.8733	.8749	.8764	.8779	.8795	.8810	.8825	.8840	.8856
6.6	.8871	.8886	.8901	.8916	.8931	.8946	.8961	.8976	.8991	.9006
6.7	.9021	.9036	.9051	.9066	.9081	.9095	.9110	.9125	.9140	.9155
6.8	.9169	.9184	.9199	.9213	.9228	.9242	.9257	.9272	.9286	.9301
6.9	.9315	.9330	.9344	.9359	.9373	.9387	.9402	.9416	.9430	.9445
7.0	1.9459	.9473	.9488	.9502	.9516	.9530	.9544	.9559	.9573	.9587
7.1	.9601	.9615	.9629	.9643	.9657	.9671	.9685	.9699	.9713	.9727
7.2	.9741	.9755	.9769	.9782	.9796	.9810	.9824	.9838	.9851	.9865
7.3	.9879	.9892	.9906	.9920	.9933	.9947	.9961	.9974	.9988	2.0001
7.4	2.0015	.0028	.0042	.0055	.0069	.0082	.0096	.0109	.0122	.0136
7.5	2.0149	.0162	.0176	.0189	.0202	.0215	.0229	.0242	.0255	.0268
7.6	.0281	.0295	.0308	.0321	.0334	.0347	.0360	.0373	.0386	.0399
7.7	.0412	.0425	.0438	.0451	.0464	.0477	.0490	.0503	.0516	.0528
7.8	.0541	.0554	.0567	.0580	.0592	.0605	.0618	.0631	.0643	.0656
7.9	.0669	.0681	.0694	.0707	.0719	.0732	.0744	.0757	.0769	.0782
8.0	2.0794	.0807	.0819	.0832	.0844	.0857	.0869	.0882	.0894	.0906
8.1	.0919	.0931	.0943	.0956	.0968	.0980	.0992	.1005	.1017	.1029
8.2	.1041	.1054	.1066	.1078	.1090	.1102	.1114	.1126	.1138	.1150
8.3	.1163	.1175	.1187	.1199	.1211	.1223	.1235	.1247	.1258	.1270
8.4	.1282	.1294	.1306	.1318	.1330	.1342	.1353	.1365	.1377	.1389
8.5	2.1401	.1412	.1424	.1436	.1448	.1459	.1471	.1483	.1494	.1506
8.6	.1518	.1529	.1541	.1552	.1564	.1576	.1587	.1599	.1610	.1622
8.7	.1633	.1645	.1656	.1668	.1679	.1691	.1702	.1713	.1725	.1736
8.8	.1748	.1759	.1770	.1782	.1793	.1804	.1815	.1827	.1838	.1849
8.9	.1861	.1872	.1883	.1894	.1905	.1917	.1928	.1939	.1950	.1961
9.0	2.1972	.1983	.1994	.2006	.2017	.2028	.2039	.2050	.2061	.2072
9.1	.2083	.2094	.2105	.2116	.2127	.2138	.2148	.2159	.2170	.2181
9.2	.2192	.2203	.2214	.2225	.2235	.2246	.2257	.2268	.2279	.2289
9.3	.2300	.2311	.2322	.2332	.2343	.2354	.2364	.2375	.2386	.2396
9.4	.2407	.2418	.2428	.2439	.2450	.2460	.2471	.2481	.2492	.2502
9.5	2.2513	.2523	.2534	.2544	.2555	.2565	.2576	.2586	.2597	.2607
9.6	.2618	.2628	.2638	.2649	.2659	.2670	.2680	.2690	.2701	.2711
9.7	.2721	.2732	.2742	.2752	.2762	.2773	.2783	.2793	.2803	.2814
9.8	.2824	.2834	.2844	.2854	.2865	.2875	.2885	.2895	.2905	.2915
9.9	.2925	.2935	.2946	.2956	.2966	.2976	.2986	.2996	.3006	.3016

Appendix IV

Answers to Selected Exercises

STUDY SET   SECTION 1.1 (page 7)

1. equation   **3.** expressions   **5.** formula   **7.** perimeter   **9.** algebraic expression   **11.** equation   **13.** equation   **15.** algebraic expression
17. a. a line graph   **b.** 1 hour; 2 inches   **c.** 7 in.; 0 in.   **19.** $c = 13u + 24$ **21.** $w = \frac{c}{75}$ **23.** $A = t + 15$ **25.** $c = 12b$
27. $b = t - 10$; the height of the base is 10 ft less than the height of the tower   **29. a.** D **b.** 2, 4   **c.** multiplication, addition
31. 2, 6, 15   **33.** 22.44, 21.43, 0   **35.** 37 in.   **37.** 8 yd   **39. a.** The rental cost is the product of 10 and the number of hours it is
rented, increased by 20.   **b.** $C = 10h + 20$ **c.** 30, 40, 50, 60, 100

41. a. The measure of the angle on the scrap piece of molding is the difference of 180° and the measure of the angle on the finish
piece of molding.   **b.** $s = 180° - f$ **c.** 150, 135, 90, 45, 30

STUDY SET   SECTION 1.2 (page 18)

1. rational   **3.** absolute value   **5.** irrational   **7.** composite   **9.** 1, 2, 9   **11.** $-3, 0, 1, 2, 9$ **13.** $\sqrt{3}, \pi$ **15.** 2   **17.** 2   **19.** 9
21. nonrepeating, irrational   **23.** repeating, rational   **25.** $7 = \frac{7}{1}, -7\frac{3}{5} = \frac{-38}{5}, 0.007 = \frac{7}{1,000}, 700.1 = \frac{7,001}{10}$ **27.** 3.5 or -3.5

29.

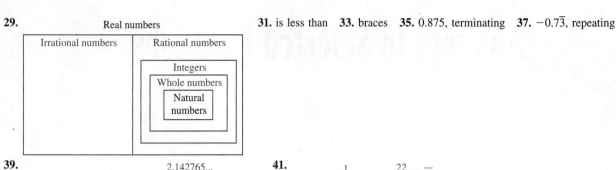

31. is less than   **33.** braces   **35.** 0.875, terminating   **37.** $-0.7\overline{3}$, repeating

39.

41.

43. **45.**

47.   **49.** $<$ **51.** $>$ **53.** $<$ **55.** $>$ **57.** $12 < 19$ **59.** $-5 \geq -6$

61. 20   **63.** -6 **65.** 5.9   **67.** $\frac{5}{4}$ **69.** $3\frac{1}{4} = 3.2500$, $\frac{51}{50} = 1.0200$, $\frac{15}{16} = 0.9375$, $2\frac{5}{8} = 2.6250$, $\frac{\pi}{4} = 0.7854$, $\sqrt{8} = 2.8284$ **75.** expression
77. 2.2, 8.5, 29.1

STUDY SET   SECTION 1.3 (page 33)

1. sum, difference   **3.** evaluate   **5.** squared, cubed   **7.** base, exponent   **9.** opposite   **11. a.** addition first and multiplication first
b. 12; multiplication is to be done before addition   **13. a.** area   **b.** volume   **15.** 8, 160   **17.** radical sign
19. 1, 8, 27, 64

21. -8 **23.** -4.3 **25.** -7 **27.** 0   **29.** -12 **31.** -2 **33.** $\frac{1}{6}$ **35.** $\frac{11}{10}$ **37.** $-\frac{6}{7}$

39. $\frac{24}{25}$ **41.** 144   **43.** -25 **45.** 32   **47.** 1.69   **49.** 8   **51.** $-\frac{3}{4}$ **53.** -17 **55.** 64   **57.** 13   **59.** -2 **61.** -8 **63.** 10,000   **65.** -39
67. -32 **69.** 5   **71.** $-\frac{1}{2}$ **73.** 2   **75.** 91,985   **77.** -24 **79.** -2 **81.** 61   **83.** 1   **85.** 10   **87.** 11.3 cm^2 **89.** 94.35 cm^3
91. 775.73 m^3 **93.** 25 ft^2 **95.** 1st term: area of bottom flap; 2nd term: area of left and right flaps; 3rd term: area of top flap; 4th term:
area of face; 42.5 in.^{2} **97.** $(967)   **99.** yes   **101.** 9   **105.** -7 and 3

6–year-old's dose   Adult dose

107. $\{\ldots, -2, -1, 0, 1, 2, \ldots\}$ **109.** true

STUDY SET   SECTION 1.4 (page 45)

1. like   **3.** term   **5.** simplify   **7.** undefined   **9.** $(x + y) + z = x + (y + z)$ **11.** $r(s + t) = rs + rt$ **13. a.** 5   **b.** -5 **15. a.** $\frac{16}{15}$
b. $-\frac{1}{20}$ **c.** 2   **d.** $\frac{1}{x}$ **17. a.** $2x^2$, $-x$, 6   **b.** 2, -1, 6   **19.** yes, $8x$ **21.** no   **23.** yes, $-2x^2$ **25.** no   **27.** multiplication by -1
29. $7 + 3$ **31.** $3 \cdot 2 + 3d$ **33.** c **35.** 1   **37.** $(8 + 7) + a$ **39.** $2(x + y)$ **41.** assoc. prop. of add.   **43.** distrib. prop.   **45.** $-4t + 12$
47. $-t + 3$ **49.** $2s - 6$ **51.** $0.7s + 1.4$ **53.** $4x - 5y + 1$ **55.** $72m$ **57.** $-45q$ **59.** $30bp$ **61.** $80ry$ **63.** $18x$ **65.** $13x^2$ **67.** 0
69. 0   **71.** $6x$ **73.** 0   **75.** $3.1h$ **77.** $\frac{9}{10}ab$ **79.** $\frac{14}{15}t$ **81.** $-4y + 36$ **83.** $7z - 20$ **85.** $14c + 62$ **87.** $-2x^2 + 15x$ **89.** $-2a + b - 2$
91. $-6p + 17$ **93. a.** $20(x + 6)$ m^2 **b.** $(20x + 120)$ m^2 **c.** $20(x + 6) = 20x + 120$; distrib. prop.   **99.** 0   **101.** $-\frac{7}{8}$ **103.** 1,000
105. -3

STUDY SET   SECTION 1.5 (page 57)

1. equation   **3.** satisfies   **5.** values   **7.** c, c **9. a.** $2y + 2$ **b.** 3   **c.** 18   **11.** No; a is not isolated on one side.   **13.** It clears the
equation of fractions.   **15.** $-2x$, 14, 14, 34, -2, -2 **17. a.** -1 **b.** $\frac{2}{3}$ **19.** yes   **21.** no   **23.** 28   **25.** -20 **27.** $\frac{1}{2}$ **29.** $-\frac{6}{5}$ **31.** 0
33. 0   **35.** 2   **37.** -9 **39.** -166 **41.** 2.52   **43.** $\frac{2}{3}$ **45.** -8 **47.** 1.395   **49.** 3   **51.** 13   **53.** -11 **55.** -8 **57.** -2 **59.** 24   **61.** 0

63. 6   **65.** $\frac{21}{19}$ **67.** 3   **69.** 24   **71.** identity   **73.** impossible   **75.** identity   **77.** $B = \dfrac{3V}{h}$ **79.** $t = \dfrac{I}{Pr}$ **81.** $w = \dfrac{P - 2l}{2}$

83. $B = \dfrac{2A}{h} - b$ or $B = \dfrac{2A - bh}{h}$ **85.** $x = \dfrac{y - b}{m}$ **87.** $v_0 = 2\bar{v} - v$ **89.** $l = \dfrac{a - S + Sr}{r}$ **91.** $l = \dfrac{2S - na}{n}$ or $l = \dfrac{2S}{n} - a$

93. $C = \frac{5(F - 32)}{9}$; 432, -179, 58, -89, 17, -66 **95.** $d = \dfrac{360A}{\pi(r_1{}^2 - r_2{}^2)}$; 140, 160   **97.** $n = \frac{PV}{R(T + 273)}$; 0.008, 0.090   **99.** $n = \frac{C - 6.50}{0.07}$;

621, 1,000, about 1,692.9 kwh   **101.** $h = \dfrac{A - 2\pi r^2}{2\pi r}$ **107.** $t - 4$ **109.** $-4b + 32$ **111.** $2.9b$ **113.** t

STUDY SET   SECTION 1.6 (page 67)

1. acute   **3.** complementary   **5.** right   **7.** angles   **9.** $d + 15$, $2d - 10$, $2d + 20$, $\frac{d}{2} - 10$, $2d$ **11. a.** $\frac{2}{3}x$ **b.** $2x$ **c.** $x + 2x + \frac{2}{3}x$
13. $26.5 - x$ **15.** Cheerios: \$294,000,000; Frosted Flakes: \$255,000,000   **17.** 20   **19.** 305 mi   **21.** 300 shares of BB,
200 shares of SS   **23.** 35 \$18 calculators, 50 \$87 calculators   **25.** 7 ft, 15 ft   **27.** 30°, 150°   **29.** 10°   **31.** 50°   **33.** 10   **35.** 60°
37. ii   **39.** 156 ft by 312 ft   **41.** 10 ft   **43.** 6 in.   **47.** repeating   **49.** $\{\ldots, -4, -3, -2, -1, 0, 1, 2, 3, 4, \ldots\}$ **51.** 0

STUDY SET   SECTION 1.7 (page 79)

1. principal   **3.** median   **5.** amount, base   **7. a.** It was in its first week of release.   **b.** Receipts increased 1,118%.   **c.** Receipts
decreased 31%.   **9. a.** \$25,000   **b.** \$(30,000 $- x$)   **11.** \$74.70, \$2.79p, \$2.59($p + 30$)   **13.** $x = 0.05 \cdot 10.56$ **15.** $32.5 = 0.74x$
17. 100, 100, 9x, 40,000, 16,000   **19.** $I = Prt$ **21.** $v = pn$ **23.** 375.6 quadrillion Btu   **25.** 20%   **27.** \$50   **29.** 9%   **31.** 11.6%, 8.5%
33. city: mean 31.7, median 29.5, mode 29; hwy: mean 39.6, median 40, mode 40   **35.** 84   **37.** \$2,000 at 8%, \$10,000 at 9%
39. \$45,000   **41.** \$100,000   **43.** $\frac{1}{4}$ hr = 15 min   **45.** $\frac{2}{3}$ hr   **47.** 3:30 P.M.   **49.** $1\frac{1}{2}$ hr   **51.** 20 lb of 95¢ candy; 10 lb of \$1.10 candy
53. 4,000 ft^3 of the premium mix, 2,000 ft^3 of sawdust   **55.** 10 oz   **57.** 2 gal   **63.** 0   **65.** 8

KEY CONCEPT (page 85)

1. given: 48 states, 4 more lie east of the Miss. River than west; find: how many states lie west   **2.** given: one angle is 5° more than
twice another, one angle is 90°, the sum of the angles' measures is 180°; find: the measure of the smallest angle   **3.** Let $x =$ the length
of the shortest piece.   **4.** Let $x =$ the number of miles he can drive.   **5. a.** subtraction   **b.** multiplication   **c.** addition   **d.** division
6. a. $d = rt$ **b.** $I = Prt$ **c.** $v = pn$ **d.** $P = 2l + 2w$ **7.** $0.15(15) + 0.50x = 0.40(15 + x)$; $0.15(15)$, $0.50x$, $0.40(15 + x)$
8. $450x + 500x = 2,850$

CHAPTER REVIEW (page 87)

1. a. $C = 2t + 15$ **b.** $l = \frac{25}{w}$ **c.** $P = u - 3$ **2.** 180, 195, 210, 225, 240   **3. a.**

b.

e. $-\sqrt{3}$, π, 0.13242368. . .   **f.** all   **g.** -5, $-\sqrt{3}$, $-\frac{2}{3}$, $-3.\overline{6}$ **h.** 2.4, 7, π, $\frac{15}{4}$, 0.13242368. . .   **i.** 7   **j.** none   **k.** 0   **l.** -5, 7
6. a. $>$ **b.** $<$ **7. a.** false   **b.** true   **8.**

20 21 22 23 24 25 26 27 28 29 30

4. 17.5 in.   **5. a.** 7   **b.** 0, 7   **c.** -5, 0, 7   **d.** -5, 0, 2.4, 7, $-\frac{2}{3}$, $-3.\overline{6}$, $\frac{15}{4}$

9.

$2.\overline{3}$ $\frac{3\pi}{4}$ $\sqrt{7}$ $\frac{8}{3}$ 2.75

2   2.5   3

10. a. 18   **b.** -6.26 **11. a.** -7 **b.** 10.1   **c.** $-\frac{3}{4}$ **d.** 2   **e.** 12.6   **f.** $-\frac{1}{32}$

g. 0.2   **h.** $-\frac{3}{56}$ **i.** -33 **j.** -5.7 **k.** -120 **l.** 1   **12. a.** -243 **b.** $\frac{4}{9}$ **c.** 0.027   **d.** -25 **13. a.** 2   **b.** -10 **c.** $\frac{3}{5}$ **d.** 0.8   **14. a.** 44
b. 1   **c.** -12 **d.** 58   **e.** 8   **f.** 3   **g.** -32 **h.** -64 **15. a.** 56   **b.** $-\frac{1}{2}$ **16. a.** 100 in.^{2} **b.** 251.3 in.^{3} **17. a.** $3x + 21$ **b.** $5t$ **c.** 0

d. $27 + (1 + 99)$ **e.** 1 **f.** m **g.** 1 **h.** 0 **i.** $-3(5 \cdot 2)$ **j.** $(z + t) \cdot t$ **18. a.** 1 **b.** 0 **c.** -25 **d.** undefined **19. a.** $8x + 48$
b. $-6x + 12$ **c.** $4 - 3y$ **d.** $3.6x - 2.4y$ **e.** $20c^2 - 10c + 10$ **f.** $2t + 6$ **20. a.** $48k$ **b.** $70xy$ **c.** $-189p$ **d.** $45a + 7$ **e.** 0
f. $3m - 48$ **g.** x **h.** $-24.54l$ **i.** $6t^3 + 4t^2$ **j.** $11h + 77$ **21. a.** yes **b.** no **22. a.** $\{-225\}$ **b.** $\{7.9\}$ **c.** $\{0.014\}$ **d.** $\{-4\}$
23. a. $-\frac{12}{5}$ **b.** -9 **c.** 8 **d.** $\frac{11}{7}$ **e.** 8 **f.** 12 **g.** 0.06 **h.** -8 **i.** 0 **j.** 3 **24. a.** impossible equation **b.** identity **25. a.** $h = \dfrac{V}{\pi r^2}$
b. $g = \dfrac{m - Y}{2}$ **c.** $x = \dfrac{T}{ab} - y$ or $x = \dfrac{T - aby}{ab}$ **d.** $r^3 = \dfrac{3V}{4\pi}$ **26.** O'Hare: 69 million; Atlanta: 63 million **27.** $245 - 5c$ **28.** $5,000;
$1,000x **29.** 42 ft, 45 ft, 48 ft, 51 ft **30.** 50°, 130° **31.** $18,000 at 10%, $7,000 at 9% **32.** children's: about 7.8% **33.** 503, 504.5,
505 **34.** 3 min **35.** 10 gal **36.** $1.95(x + 3)$

CHAPTER 1 TEST (page 92)

1. $s = T + 10$ **2.** $A = \frac{1}{2}bh$ **3.** $-2, 0, 5$ **4.** $-2, 0, -3\frac{3}{4}, 9.2, \frac{14}{5}, 5$ **5.** $\pi, -\sqrt{7}$ **6.** all
7.

8. **9.** -8

10. 5.5 **11.** 12.3 **12.** $\frac{4}{15}$ **13.** $\frac{11}{10}$ **14.** -64 **15.** $-\frac{8}{9}$ **16.** -35 **17.** 1 **18.** 100 mg **19.** commut. prop. of add. **20.** distrib. prop.
21. $11y$ **22.** $-45q$ **23.** -4 **24.** $-16 + 2a + 18b$ **25.** -12 **26.** 6 **27.** 6.1 **28.** 5.85 **29.** 5.6 **30.** $i = \dfrac{f(P - L)}{s}$ **31.** 4 cm by
9 cm **32.** 1,745% **33.** $4,000 **34.** 10 oz

STUDY SET   SECTION 2.1 (page 103)

1. ordered **3.** origin **5.** rectangular **7.** midpoint **9.** origin, right, down **11.** II **13.** $-3, -1, 1, 3$ **15.** $\dfrac{x_1 + x_2}{2}$ **17.** t: horizontal;
d: vertical **19.** x sub 1 **21–28.** **29.** $(2, 4)$ **31.** $(-2.5, -1.5)$ **33.** $(3, 0)$ **35.** $(0, 0)$ **37. a.** on the surface

b. diving **c.** 1,000 ft **d.** 500 ft **39. a.** 1990–1991, 1994–1995 **b.** 1990–1991 **c.** 1993–1996 **d.** Imports exceeded production by
about 1.7 million barrels per day. **41.** $(3, 4)$ **43.** $(9, 12)$ **45.** $\left(\frac{7}{2}, 6\right)$ **47.** $\left(\frac{1}{2}, -2\right)$ **49.** $(-4, 0)$ **51.** $(4, 1)$ **53.** $(-20, -3)$
55. Jonesville $(5, B)$, Easley $(1, B)$, Hodges $(2, E)$, Union $(6, C)$ **57. a.** $(2, -1)$ **b.** no **c.** yes **59. a.** $2 **b.** $4 **c.** $7 **d.** $9
61. tip of tail, front of engine, tip of wing **63. a.** Culture T **b.** Culture R **67.** 20 **69.** $\frac{1}{2}$ **71.** 0.7

STUDY SET   SECTION 2.2 (page 117)

1. satisfy **3.** y-intercept **5.** vertical **7. a.** 1, 1 **b.** 2, infinitely many **9.** $3x + 5y = -10$ **11. a.** $(-3, 0); (0, 4)$ **b.** false
13. a. $8x + 4y = -5$ was solved for y and entered as $Y_1 = -2x - \left(\frac{5}{4}\right)$. **b.** the y-intercept **15.** $-3, -2$ **17.** the y-axis **19.** 5, 4, 2
21. $0, -1, -2$ **23.** **25.** **27.**

29. **31.** **33.** **35.**

37.

39.

41.

43. 1.22   **45.** 4.67   **47.** $48

49.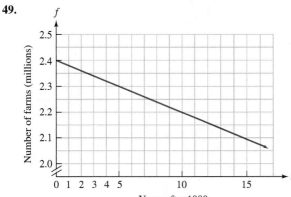

2.32 million, 2.14 million; the number of farms decreased   **51.** $162,500

53. 200   **55. a.** $c = 5t + 2$ **b.** 7, 12, 17, 22   **c.** $27

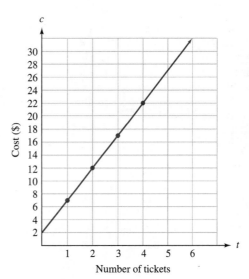

59. 11, 13, 17, 19, 23, 29
61. III   **63.** 80s **65.** $3x + 8$

STUDY SET   SECTION 2.3 (page 129)

1. slope   **3.** change   **5.** reciprocals   **7. a.** l_3; 0   **b.** l_2; undefined   **c.** l_1; 2   **d.** l_4; -3 **9. a.** college, 4 yr or more; about $24 a week/yr   **b.** less than 4 yr h.s.; about $6 a week/yr   **11.** $-\frac{4}{3}$ **13.** $m = \frac{y_2 - y_1}{x_2 - x_1}$ **15. a.** 6   **b.** 8   **c.** $\frac{3}{4}$ **17.** 3   **19.** -1 **21.** $-\frac{1}{3}$ **23.** 0

25. undefined   **27.** -1 **29.** $-\frac{3}{2}$ **31.** $\frac{3}{4}$ **33.** $\frac{1}{2}$ **35.** 0   **37.** perpendicular   **39.** neither   **41.** neither   **43.** parallel   **45.** perpendicular
47. neither   **49.** $\frac{3}{140}$, $\frac{1}{15}$, $\frac{1}{20}$; part 2   **51.** $\frac{1}{10}$; $\frac{1}{4}$ **53.** $\frac{3}{25}$; 4%   **55.** brace: $\frac{1}{2}$; support 1: -2; support 2: -1; yes, to support 1   **61.** 40 lb licorice; 20 lb gumdrops   **63.** 4 hr

STUDY SET   SECTION 2.4 (page 140)

1. $y - y_1 = m(x - x_1)$ **3.** perpendicular   **5.** no   **7.** $m = \frac{2}{3}$; $y + 3 = \frac{2}{3}(x + 2)$ **9.** $m = -\frac{2}{3}$; $(0, 1)$ **11.** yes   **13. a.** $(0, 0)$ **b.** none
15. No; the slopes are not negative reciprocals. Their product is not -1: $1(-0.9) = -0.9$.   **17.** $\frac{1}{3}x$, 2, 2, 1, $\frac{1}{3}$, -1 **19.** $y = 5x + 7$
21. $y = -3x + 6$ **23.** $y = x$ **25.** $y = \frac{7}{3}x - 3$ **27.** $y = \frac{2}{3}x + \frac{11}{3}$ **29.** $y = 3x + 17$ **31.** $y = -7x + 54$ **33.** $y = -4$
35. $y = -\frac{1}{2}x + 11$ **37.** $\frac{3}{2}$, $(0, -4)$ **39.** $-\frac{1}{3}$, $\left(0, -\frac{5}{6}\right)$ **41.** 1, $(0, -1)$

43. $\frac{2}{3}$, (0, 2)

45. $-\frac{3}{4}$, (0, −2)

47. parallel **49.** perpendicular **51.** neither

53. perpendicular **55.** $y = 4x$ **57.** $y = 4x − 3$ **59.** $y = \frac{4}{5}x − \frac{26}{5}$ **61.** $y = −\frac{1}{4}x$ **63.** $y = −\frac{1}{4}x + \frac{11}{2}$ **65.** $y = −\frac{5}{4}x + 3$
67. $y = −\frac{950}{3}x + 1{,}750$ **69.** $y = 1{,}811{,}250x + 36{,}225{,}000$ **71. a.** $B = \frac{1}{100}p − 195$ **b.** 905 **73. a.** $E = −\frac{1}{2}t + \frac{21}{2}$ **b.** The number of
errors is reduced by 1 for every 2 trials. **c.** On the 21st trial, the rat should make no errors. **75. a.** $y = 1.35x − 31.25$ **b.** When the
actual temperature is 0°F, the wind-chill temperature is about −31°F. **81.** $29,100 **83.** 9

STUDY SET   SECTION 2.5 (page 153)
1. function, one **3.** range **5.** x **7.** y **9.** $f(−1)$ **11. a.**

b. D = all real numbers greater than or equal to 0;
R = all real numbers greater than or equal to 2

13. −5, 25 **15.** of **17.** yes **19.** yes **21.** yes **23.** no **25.** 9, −3 **27.** 3, −5 **29.** 22, 2 **31.** 3, 11 **33.** 4, 9 **35.** 7, 26
37. 9, 16 **39.** 6, 15 **41.** 4, 4 **43.** 2, 2 **45.** $\frac{1}{5}$, 1 **47.** −2, $\frac{2}{5}$ **49.** 3.7, 1.1, 3.4 **51.** $−\frac{27}{64}$, $\frac{1}{216}$, $\frac{125}{8}$ **53.** $2w$, $2w + 2$
55. $3w − 5$, $3w − 2$ **57.** D = the set of all real numbers except 6 **59.** D = the set of all real numbers
61. D = the set of all real numbers **63.** D = the set of all real numbers **65.** D = {−2, 4, 6}, R = {3, 5, 7} **67.** D = the set of
all real numbers except 4, R = the set of all real numbers except 0 **69.** not a function **71.** a function **73.** a function
75. a function **77.**

D = the set of all real numbers, R = the set of all real numbers

79.

D = the set of all real numbers, R = the set of all real numbers
81. no **83.** yes **85.** between 20°C and 25°C

87. a. $I(b) = 1.75b − 50$ **b.** $142.50 **89. a.** (200, 25), (200, 90), (200, 105) **b.** It doesn't pass the vertical line test.
91. a. 6,595.50; the tax on an income of $35,000 is $6,595.50 **b.** D = greater than $24,650 and less than or equal to $59,750,
R = greater than or equal to $3,697.50 and less than or equal to $13,525.50 **c.** $T(a) = 13{,}525.50 + 0.31(a − 59{,}750)$ **93. a.** 624 ft
b. 0; the bullet strikes the ground 16 seconds after being shot **97.** $\frac{−15}{4}$ **99.** $\frac{1}{3}$

STUDY SET   SECTION 2.6 (page 166)
1. squaring **3.** absolute value **5.** vertical **7.** reflection **9.** 4, left **11.** 5, up **13. a.** 2 **b.** 0 **15.** −3 **17.** −3, 1, 2 **19. a.** 3
b. 0 **c.** 1.5 **21.**

D = the set of real numbers, R = the set of all real numbers greater than or equal to −3

23.   D = the set of real numbers,   **25.** 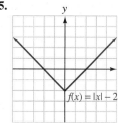   D = the set of real numbers,
$f(x) = (x-1)^3$ R = the set of real numbers $f(x) = |x| - 2$ R = the set of all real numbers greater than
or equal to −2

27.   D = the set of real numbers, R = the set of real numbers greater than or equal to 0
$f(x) = |x-1|$

29.   **31.**   **33.**   **35.**

37.   **39.**   **41.**   **43.**
$f(x) = x^2 - 5$ $f(x) = (x-1)^3$ $f(x) = |x-2| - 1$ $f(x) = (x+1)^3 - 2$

45.   **47.**   **49.** 4   **51.** −3   **53.**
$f(x) = -x^3$ $f(x) = -x^2$

Grand Ave.
210

55. a parabola   **61.** $W = T - ma$ **63.** $g = \dfrac{2(s - vt)}{t^2}$ **65.** $5.4 million

KEY CONCEPT (page 171)
1. a. correspondence, input, range, one, domain   **b.** dependent, independent   **3.** −17   **4.** $A = \pi r^2$ **5.** 60 ft   **6.** 2   **7.** $f(x) = 2x + 3$,
3   **8.** $A(r) = \pi r^2$ **9.** 0; the projectile will strike the ground 4 seconds after being shot into the air   **10.** −2, 4

CHAPTER REVIEW (page 173)
1.   **2.** −5, −5; 0, 0; −1, −1; 4, 4; 3, 3; 2, 2; 3, 3   **3. a.** 1 ft below its normal level   **b.** decreased by 3 ft

(0, 3)
(0, 0)   (2.5, 0)
(−2, −4) $\left(\dfrac{5}{2}, -1.75\right)$

c. from day 3 to the beginning of day 4 **4. a.** $10 increments **b.** $800 **5.** $(2, 2)$ **6.** $(2, -6)$ **7. a.**

b.

8. a.

b.

9. a.

b.

10. a. $9, 0, -9$ **b.** $-4, -\frac{5}{2}, -1$ **11.** slope of $l_1 = \frac{4}{5}$; slope of $l_2 = -\frac{8}{5}$ **12.** 1.375%/yr

13. a. 1 **b.** $-\frac{14}{9}$ **c.** 0 **d.** undefined **14. a.** $\frac{2}{3}$ **b.** -2 **c.** undefined **d.** 0 **15. a.** perpendicular **b.** parallel **16. a.** $y = 3x + 29$
b. $y = -\frac{13}{8}x + \frac{3}{4}$ **17. a.** $3x - 2y = 1$ **b.** $2x + 3y = -21$ **18.** $y = -\frac{3}{4}x - 3$; $m = -\frac{3}{4}$, $(0, -3)$

19. $y = -1,720x + 8,700$ **20. a.** yes **b.** yes **c.** no **d.** no **21. a.** -7 **b.** 18 **c.** 8 **d.** $3t + 2$ **22. a.** D = the set of real
numbers, R = the set of real numbers **b.** D = the set of real numbers, R = the set of all real numbers greater than or equal to 1
c. D = the set of all real numbers except 2, R = the set of all real numbers except 0 **d.** D = the set of real numbers, R = the set
of nonpositive real numbers **23. a.** function **b.** not a function **24.** $f(t) = 1.375t + 21.2$; 48.7% **25. a.** yes **b.** no

26. a.

b.

27.

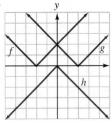

28. a. -4 **b.** 3 **29. a.** 4 **b.** -1 **c.** 2

CHAPTER 2 TEST (page 178)
1. 240 ft **2.** 1 sec and 7 sec **3.** about 256 ft **4.** 8 sec **5.** $\left(\frac{5}{2}, \frac{3}{2}\right)$ **6.** $(6, -9)$ **7.** $(5, 0), (0, -2)$

8.

9. 3   **10.** -1.5 degree/hr   **11.** $\frac{1}{2}$ **12.** $\frac{2}{3}$ **13.** undefined   **14.** 0   **15.** $y = \frac{2}{3}x - \frac{23}{3}$ **16.** $8x - y = -22$

17. $m = -\frac{1}{3}$, $\left(0, -\frac{3}{2}\right)$ **18.** neither   **19.** $y = \frac{3}{2}x$ **20.** no   **21.** D = the set of real numbers, R = the set of nonnegative real numbers
22. D = the set of real numbers, R = the set of real numbers   **23.** 10   **24.** -1 **25.** 3   **26.** $r^2 - 2r - 1$ **27.** function   **28.** not a
function   **29.**

30.

31. Find the x-coordinate of the x-intercept of the graph of
$y = 3(x - 2) - 2(-2 + x)$; 2

32. Find the x-coordinate of the point of intersection of the graph of $y = 3(x - 2) - 2(-2 + x)$ and $y = 1$; 3

CHAPTERS 1–2 CUMULATIVE REVIEW EXERCISES (page 180)

1. 1, 2, 6, 7   **2.** 0, 1, 2, 6, 7   **3.** -2, 0, 1, 2, $\frac{13}{12}$, 6, 7   **4.** $\sqrt{5}$, π **5.** -2 **6.** -2, 0, 1, 2, $\frac{13}{12}$, 6, 7, $\sqrt{5}$, π **7.** 2, 7   **8.** 6   **9.** -2, 0, 2, 6
10. 1, 7   **11.** -2 **12.** -2 **13.** 22   **14.** -2 **15.** $\frac{24}{25}$ **16.** 2   **17.** 4   **18.** -5 **19.** assoc. prop. of add.   **20.** distrib. prop.
21. commut. prop. of add.   **22.** assoc. prop. of mult.   **23.** $-5y$ **24.** $-28st$ **25.** 0   **26.** $z - 4$ **27.** 8   **28.** -27 **29.** -1 **30.** 6
31. $\frac{8}{3}$ **32.** 24   **33.** $a = \frac{2S}{n} - l$ or $a = \frac{2S - ln}{n}$ **34.** $h = \frac{2A}{b_1 + b_2}$ **35.** \$14,000   **36.** 39 mph going, 65 mph returning
37.

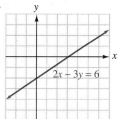

38. $-\frac{5}{6}$ **39.** $y = -\frac{7}{5}x + \frac{11}{5}$ **40.** $y = -3x - 3$ **41.** 5   **42.** -1 **43.** 3   **44.** $3r^2 + 2$

45. a function; D = the set of real numbers, R = the set of all real numbers less than or equal to 1

46. a function; D = the set of real numbers, R = the set of nonnegative real numbers

47. The points $(2, 20)$, $(3, 40)$ and $(6, 60)$ do not lie on a straight line.   **48. a.** Find the x-coordinate of the x-intercept of the graph of
$y = -7 - 5(x - 1) + 4x$; -2 **b.** Find the x-coordinate of the point of intersection of the graph of $y = -7 - 5(x - 1) + 4x$ and
$y = -4$; 2

STUDY SET   SECTION 3.1 (page 189)

1. system   **3.** inconsistent   **5.** dependent   **7. a.** true   **b.** false   **c.** true   **d.** true   **9. a.** -4, $(-4, 0)$; 2, $(0, 2)$; 3, $(2, 3)$ **b.** $(-4, 0)$;
$(0, 2)$ **11. a.** $\begin{cases} x + y = 5 \\ x - y = -1 \end{cases}$ (answers may vary)   **b.** $\begin{cases} x + y = 5 \\ 2x + 2y = 10 \end{cases}$ (answers may vary)   **c.** $\begin{cases} x + y = 5 \\ x + y = 4 \end{cases}$ (answers may vary)

13. brace   **15.** yes   **17.** no   **19.**   **21.**   **23.**

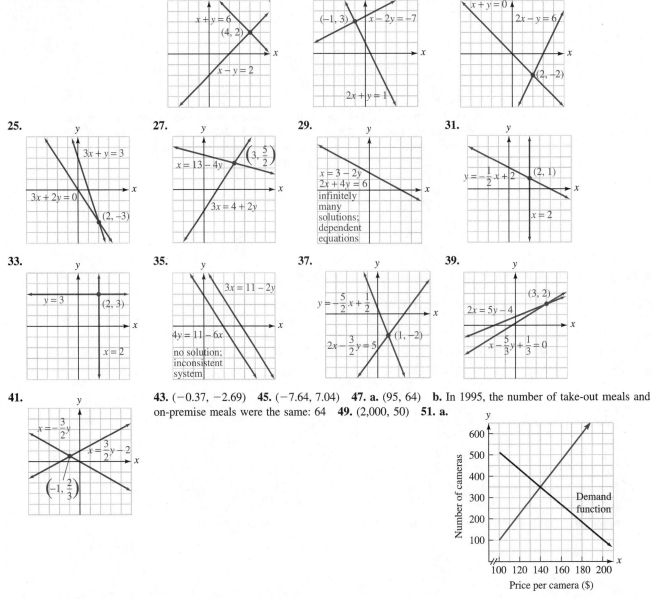

25.   **27.**   **29.**   **31.**

33.   **35.**   **37.**   **39.**

41.   **43.** $(-0.37, -2.69)$ **45.** $(-7.64, 7.04)$ **47. a.** $(95, 64)$ **b.** In 1995, the number of take-out meals and on-premise meals were the same: 64   **49.** $(2,000, 50)$ **51. a.**

b. $140   **c.** Supply increases, and demand decreases.   **53. a.** yes   **b.** $(3.75, -0.5)$ **c.** no   **57.** -3 **59.** 0   **61.** D = the set of real numbers, R = the set of all real numbers greater than or equal to -2 **63.** 40.5 cm^2

STUDY SET   SECTION 3.2 (page 202)

1. general   **3.** eliminated   **5. a.** 3; -4 (answers may vary)   **b.** 2; -3 (answers may vary)   **7.** addition method   **9. a.** ii   **b.** iii   **c.** i
11. $(2, 2)$ **13.** $(5, 3)$ **15.** $(-2, 4)$ **17.** no solution, inconsistent system   **19.** $(5, 2)$ **21.** $(-4, -2)$ **23.** $(1, 2)$ **25.** $\left(\frac{1}{2}, \frac{2}{3}\right)$ **27.** $\left(5, \frac{3}{2}\right)$
29. $\left(-2, \frac{3}{2}\right)$ **31.** infinitely many solutions, dependent equations   **33.** no solution, inconsistent system   **35.** $(4, 8)$ **37.** $(20, -12)$
39. $\left(\frac{2}{3}, \frac{3}{2}\right)$ **41.** $\left(\frac{1}{2}, -3\right)$ **43.** $(2, -3)$ **45.** $(9, -1)$ **47.** $(2, 3)$ **49.** $\left(-\frac{1}{3}, 1\right)$ **51.** $475, $800   **53.** dogs: 58 million; cats: 66 million
55. 16 m by 20 m   **57.** 75°, 25°   **59.** $3,000 at 10%, $5,000 at 12%   **61.** 45 mi   **63.** 85 racing bikes, 120 mountain bikes
65. 4,031   **67.** $4,666\frac{2}{3}$ books   **69.** 103   **71.** 148 g of the 0.2%, 37 g of the 0.7%   **73. a.** 590 units per month   **b.** 620 units per month
c. A (smaller loss)   **81.** $-\frac{5}{2}$ **83.** $-\frac{8}{5}$ **85.** $\frac{4}{3}$

STUDY SET   SECTION 3.3 (page 214)

1. system   **3.** three   **5.** dependent   **7. a.** no solution   **b.** no solution   **9.** $x + 2y - 3z = -6$ **11.** yes   **13.** $(1, 1, 2)$ **15.** $(0, 2, 2)$
17. $(3, 2, 1)$ **19.** no solution, inconsistent system   **21.** $(60, 30, 90)$ **23.** $(2, 4, 8)$ **25.** infinitely many solutions, dependent equations
27. $(2, 6, 9)$ **29.** 30 expensive, 50 middle-priced, 100 inexpensive   **31.** 2, 3, 1   **33.** 3 poles, 2 bears, 4 deer   **35.** 70%, 17%, 13%
37. a. infinitely many solutions, all lying on the line running down the binding   **b.** 3 parallel planes (shelves); no solution   **c.** each
pair of planes (cards) intersect; no solution   **d.** 3 planes (faces of die) intersect at a corner; 1 solution   **39.** $y = \frac{1}{2}x^2 - 2x - 1$
41. $x^2 + y^2 - 2x - 2y - 2 = 0$ **43.** $A = 40°, B = 60°, C = 80°$ **45.** 12, 15, 21

49.

51.

STUDY SET   SECTION 3.4 (page 224)

1. matrix   **3.** rows, columns   **5.** augmented   **7. a.** 2×3 **b.** 3×4 **9.** $\begin{cases} x - y = -10 \\ y = 6 \end{cases}$; $(-4, 6)$ **11.** It has no solution. The system is inconsistent.   **13. a.** multiply row 1 by $\frac{1}{3}$; $\begin{bmatrix} 1 & 2 & -3 & \vdots & 0 \\ 1 & 5 & -2 & \vdots & 1 \\ -2 & 2 & -2 & \vdots & 5 \end{bmatrix}$ **b.** to row 2, add -1 times row 1; $\begin{bmatrix} 1 & 2 & -3 & \vdots & 0 \\ 0 & 3 & 1 & \vdots & 1 \\ -2 & 2 & -2 & \vdots & 5 \end{bmatrix}$
15. $-1, 1, -5, 2, y, 4$ **17.** $(1, 1)$ **19.** $(2, -3)$ **21.** $(-1, -1)$ **23.** $(0, -3)$ **25.** $(1, 2, 3)$ **27.** $(4, 5, 4)$ **29.** $(2, 1, 0)$
31. $(-1, -1, 2)$ **33.** no solution, inconsistent system   **35.** infinitely many solutions, dependent equations   **37.** $(0, 1, 3)$
39. no solution, inconsistent system   **41.** $(-4, 8, 5)$ **43.** infinitely many solutions, dependent equations   **45.** $22°, 68°$ **47.** $40°, 65°, 75°$
49. $76°, 104°$ **51.** founder's circle: 100; box seats: 300; promenade: 400   **55.** $m = \dfrac{y_2 - y_1}{x_2 - x_1}$ $(x_2 \neq x_1)$ **57.** $y - y_1 = m(x - x_1)$

STUDY SET   SECTION 3.5 (page 234)

1. determinant   **3.** minor   **5.** rows, columns   **7.** dependent, inconsistent   **9.** $ad - bc$ **11.** $\begin{vmatrix} 3 & 4 \\ 2 & -3 \end{vmatrix}$ **13.** $\left(\frac{7}{11}, -\frac{5}{11}\right)$ **15.** 6, 30
17. 8   **19.** -2 **21.** 200   **23.** 6   **25.** 1   **27.** 26   **29.** 0   **31.** -79 **33.** $(4, 2)$ **35.** $\left(-\frac{1}{2}, \frac{1}{3}\right)$ **37.** no solution, inconsistent system
39. $(2, -1)$ **41.** $(1, 1, 2)$ **43.** $(3, 2, 1)$ **45.** $(3, -2, 1)$ **47.** $\left(-\frac{1}{2}, -1, -\frac{1}{2}\right)$ **49.** infinitely many solutions, dependent equations
51. no solution, inconsistent system   **53.** $(-2, 3, 1)$ **55.** 200 of the \$67 phones, 160 of the \$100 phones   **57.** \$5,000 in HiTech,
\$8,000 in SaveTel, \$7,000 in OilCo   **59.** -23 **61.** 26   **65.** no   **67.** yes   **69.** x **71.** y-intercept   **73.** $x; y$

KEY CONCEPT (page 238)

1. yes   **2.** no   **3.**

4. $\left(\frac{1}{3}, \frac{1}{4}\right)$ **5.** $(3, 2)$ **6.** $(-1, 0, 2)$ **7.** $(15, 2)$ **8.** $(3, 3, 2)$ **9.** The equations of the system are dependent. There are infinitely many solutions.
10. The system is inconsistent. There are no solutions.

CHAPTER REVIEW (page 240)

1. a. $(1, 3), (2, 1), (4, -3)$ (answers may vary)   **b.** $(0, -4), (2, -2), (4, 0)$ (answers may vary)   **c.** $(3, -1)$ **2.** President Clinton's job approval and disapproval ratings were the same: approximately 47% in 5/94 and approximately 48% in 5/95.
3. a.

b.

c.

d.

4. a. $(-1, 3)$ **b.** $(-3, -1)$ **c.** $(3, 4)$ **d.** infinitely many solutions, dependent equations   **5. a.** $(-3, 1)$ **b.** no solution, inconsistent system   **c.** $(9, -4)$ **d.** $\left(4, \frac{1}{2}\right)$ **6.** Using the addition method, the computations are easier.   **7.** $(-1, 0.7)$ (answers may vary); $\left(-1, \frac{2}{3}\right)$ **8.** A-H: 162 mi, A-SA: 83 mi   **9.** 8 mph, 2 mph   **10.** no   **11. a.** $(1, 2, 3)$ **b.** no solution, inconsistent system
c. $(-1, 1, 3)$ **d.** infinitely many solutions, dependent equations   **12.** 25 lb peanuts, 10 lb cashews, 15 lb Brazil nuts

13. a. $\begin{bmatrix} 5 & 4 & \vdots & 3 \\ 1 & -1 & \vdots & -3 \end{bmatrix}$ **b.** $\begin{bmatrix} 1 & 2 & 3 & \vdots & 6 \\ 1 & -3 & -1 & \vdots & 4 \\ 6 & 1 & -2 & \vdots & -1 \end{bmatrix}$ **14. a.** $(1, -3)$ **b.** $(5, -3, -2)$ **c.** infinitely many solutions, dependent equations

d. no solution, inconsistent system **15.** $4,000 at 6%, $6,000 at 12% **16. a.** 18 **b.** 38 **c.** -3 **d.** 28
17. a. (2, 1) **b.** no solution, inconsistent system **c.** (1, -2, 3) **d.** (-3, 2, 2) **18.** 2 cups mix A, 1 cup mix B, 1 cup mix C

CHAPTER 3 TEST (page 244)

1.

2. (7, 0) **3.** (2, -3) **4.** dependent **5.** no **6.** (3, 2, -1) **7.** 55, 70 **8.** 15 gal 40%, 5 gal 80%
9. (2, 2) **10.** (1, 0, -1) **11.** 22 **12.** 4 **13.** $\begin{vmatrix} -6 & -1 \\ -6 & 1 \end{vmatrix}$ **14.** $\begin{vmatrix} 1 & -1 \\ 3 & 1 \end{vmatrix}$ **15.** -3 **16.** 3
17. -1 **18.** C: 60, GA: 30, S: 10 **19.** Sales of Halloween and Christmas candy were the same in 1995 (about $840 million) and in 1997 (about $950 million). **21.** The system has no solution.

STUDY SET SECTION 4.1 (page 254)

1. inequality **3.** parenthesis **5.** linear **7.** is less than **9. a.** (4, ∞) **b.** ($-∞$, -4)
c. ($-∞$, 4] **11. a.** equation **b.** expression **c.** inequality **d.** expression **13.** i and ii **15. a.** yes **b.** yes **c.** yes
d. no **17. a.** the number of seriously injured ≤ 16 **b.** the number of references to carpools ≥ 10 **19.** -10, $\le$, $-∞$ **21. a.** It is an ordered pair. **b.** It is an interval. **23.** (-3, ∞) **25.** [20, ∞) **27.** [60, ∞)
29. $\left(-\frac{10}{3}, ∞\right)$ **31.** ($-∞$, 1) **33.** $\left[-\frac{2}{5}, ∞\right)$ **35.** [-2, ∞)
37. (6, ∞) **39.** ($-∞$, 1.5] **41.** ($-∞$, 20] **43.** ($-∞$, 10)
45. [-36, ∞) **47.** $\left(-∞, \frac{45}{7}\right]$ **49.** Midwest, South **51.** $6 + 45 \not> 52$ **53.** 8 hr **55.** 15 **57.** 13 hr
59. anything over $900 **61.** $x < 1$ **63.** $x \ge -4$ **69.** 4, 5, 3 **71.** 6, -6

STUDY SET SECTION 4.2 (page 264)

1. compound **3.** interval **5.** both **7.** reversed **9. a.** no **b.** yes **11. a.** no solution **b.** ($-∞$, ∞) **13. a.** ii **b.** iii **c.** i
15. a. **b.** **17.** $3 \not< -3$ **19.** (-2, 5] **21.** (-10, -9) **23.** (2, 3] **25.** no solution **27.** [5, ∞)
29. [1, 4] **31.** (8, 11) **33.** (-2, 5) **35.** [-4, 6)
37. [2, 2] **39.** (-0.7, 0.2] **41.** (-6, -3)
43. [-2, 4] **45.** ($-∞$, -2] $\cup$ (6, ∞) **47.** ($-∞$, -1) $\cup$ (2, ∞)
49. ($-∞$, 2) $\cup$ (7, ∞) **51.** ($-∞$, ∞) **53.** ($-∞$, 1) **55.** ($-∞$, ∞)
57. a. 128, 192 **b.** $32 \le s \le 48$ **59.** See doctor today. **61. a.** 1993, 1995, 1997 **b.** 1987, 1989, 1991, 1993, 1995, 1997
c. 1987, 1989 **d.** 1987, 1989, 1991 **67.** 85.7, 86, 86 **69.** 13.3 pts/game

STUDY SET SECTION 4.3 (page 276)

1. equation **3.** isolate **5.** 0 **7.** more than **9.** 5 **11. a.** yes **b.** no **c.** yes **d.** no **13. a.** ii **b.** iii **c.** i **15.** $|x| < 4$
17. $|x + 3| > 6$ **19.** 8 **21.** -0.02 **23.** $-\frac{31}{16}$ **25.** π **27.** **29.** **31.** 23, -23

$f(x) = |x| - 2$

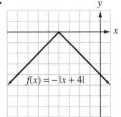

$f(x) = -|x + 4|$

33. 9.1, -2.9 **35.** $\frac{14}{3}$, -6 **37.** no solution **39.** 2, $-\frac{1}{2}$ **41.** -8 **43.** -4, -28 **45.** 0, -6 **47.** 40, -20 **49.** -2, $-\frac{4}{5}$ **51.** 0, -2
53. 0 **55.** $\frac{4}{3}$ **57.** (-4, 4) **59.** [-21, 3] **61.** $\left(-\frac{8}{3}, 4\right)$ **63.** no solution
65. ($-∞$, -3) $\cup$ (3, ∞) **67.** ($-∞$, -12) $\cup$ (36, ∞)
69. $\left(-∞, -\frac{16}{3}\right) \cup$ (4, ∞) **71.** ($-∞$, ∞) **73.** ($-∞$, -2] $\cup \left[\frac{10}{3}, ∞\right)$

75. $(-\infty, -2) \cup (5, \infty)$ **77.** $[-10, 14]$ **79.** $\left(-\frac{5}{3}, 1\right)$

81. $(-\infty, -24) \cup (-18, \infty)$ **83.** no solution   **85.** $70° \le t \le 86°$ **87. a.** $|c - 0.6°| \le 0.5°$

b. $[0.1°, 1.1°]$ **89. a.** 26.45%, 24.76%   **b.** It is less than or equal to 1%.   **97.** 50°, 130°

STUDY SET   SECTION 4.4 (page 285)

1. linear, two   **3.** edge   **5. a.** yes   **b.** no   **c.** yes   **d.** no   **7.** $m = 3$; $(0, -1)$ **9.** no   **11. a.** $x \ge 2$

b.   **13.**   **15.**   **17.**

19.   **21.**   **23.**   **25.**

27.   **29.**   **31.**   **33.** $3x + 2y > 6$ **35.** $x \le 3$

37.   **39.**   **41. a.** the Mississippi River   **b.** the area of the U.S. west of the Mississippi River

43. (5, 15), (15, 10), (20, 5)   **45.** (40, 80), (80, 80), (120, 40)   **49.** yes   **51.** (3, 1)

STUDY SET   SECTION 4.5 (page 293)

1. inequalities   **3.** intersect   **5. a.** yes   **b.** no   **c.** no   **d.** yes   **7. a.** false   **b.** true   **c.** true   **d.** false   **e.** true   **f.** true

9.   **11.**   **13.**   **15.**

17.   **19.**   **21.**   **23.**

25. **27.** **29.** **31.**

33. 1 $10 CD and 2 $15 CDs, 4 $10 CDs and 1 $15 CD

$$\begin{cases} 10x + 15y \ge 30 \\ 10x + 15y \le 60 \\ x \ge 0, y \ge 0 \end{cases}$$

35. 2 desk chairs and 4 side chairs, 1 desk chair and 5 side chairs   **39.** IV   **41.** II

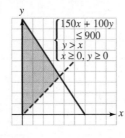

$$\begin{cases} 150x + 100y \\ \quad \le 900 \\ y > x \\ x \ge 0, y \ge 0 \end{cases}$$

KEY CONCEPT (page 297)

1. a. compound inequality   **b.** system of linear inequalities   **c.** absolute value inequality   **d.** linear inequality in two variables
e. linear inequality in one variable   **f.** double linear inequality   **g.** compound inequality   **h.** absolute value inequality   **i.** linear
inequality in two variables   **2. a.** yes   **b.** no   **c.** yes   **d.** no   **e.** yes   **f.** yes   **g.** no   **h.** no   **i.** yes   **3.**
3/2

4.

$2x + y \ge 4$

$2x + y = 4$

CHAPTER REVIEW (page 299)

1. a. $(-\infty, 3]$ ◄━┫━► **b.** $[4, \infty)$ ◄━┣━► **c.** $(-\infty, 20)$ ◄━━► **d.** $\left(-\infty, -\frac{51}{11}\right)$ ◄━━► **2.** $20,000 or more
      3   4   20   −51/11

3. a. yes   **b.** no   **4. a.** $[-10, -4)$ ◄━┣━━┫━► **b.** $(-\infty, -11)$ ◄━━► **5. a.** $\left(-\frac{1}{3}, 2\right)$ ◄━━━━━►
           −10 −4   −11   −1/3 2

b. $[1, 9]$ ◄━┣━━┫━► **6. a.** yes   **b.** no   **7. a.** $(-\infty, -5) \cup (4, \infty)$ ◄━━┃━━► **b.** $(-\infty, \infty)$ ◄━━┃━━►
   1 9   −5 0 4   0

8. $17 \le 4l \le 25$, 4.25 ft $\le l \le$ 6.25 ft   **9. a.** 7   **b.** $\frac{5}{16}$ **c.** −71.05   **d.** −12   **10. a.**   **b.**

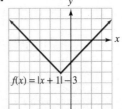

$f(x) = |x + 1| - 3$

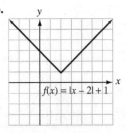

$f(x) = |x - 2| + 1$

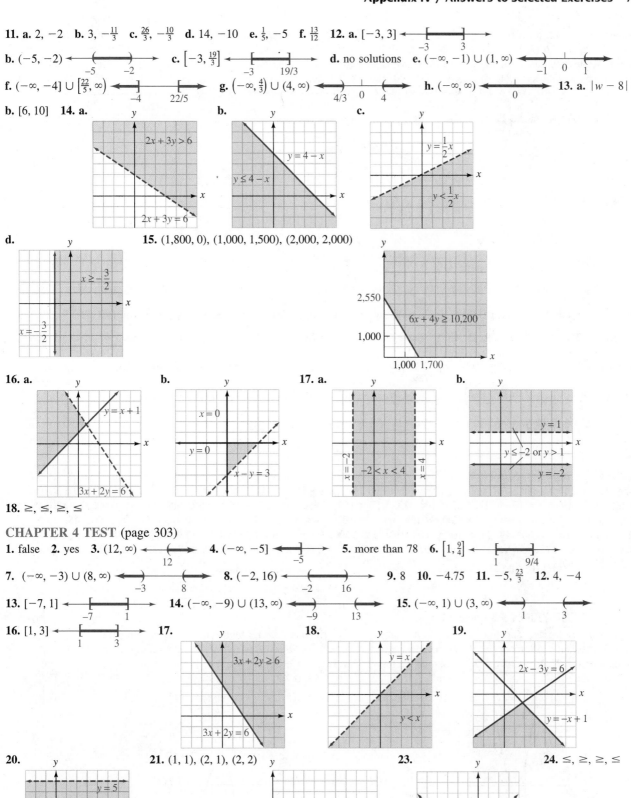

11. a. 2, −2   **b.** 3, $-\frac{11}{3}$ **c.** $\frac{26}{3}$, $-\frac{10}{3}$ **d.** 14, −10   **e.** $\frac{1}{5}$, −5   **f.** $\frac{13}{12}$ **12. a.** $[-3, 3]$

b. $(-5, -2)$ **c.** $\left[-3, \frac{19}{3}\right]$ **d.** no solutions   **e.** $(-\infty, -1) \cup (1, \infty)$

f. $(-\infty, -4] \cup \left[\frac{22}{5}, \infty\right)$ **g.** $\left(-\infty, \frac{4}{3}\right) \cup (4, \infty)$ **h.** $(-\infty, \infty)$ **13. a.** $|w - 8| \le 2$

b. $[6, 10]$ **14. a.**   **b.**   **c.**

d.   **15.** $(1{,}800, 0)$, $(1{,}000, 1{,}500)$, $(2{,}000, 2{,}000)$

16. a.   **b.**   **17. a.**   **b.**

18. $\ge$, $\le$, $\ge$, $\le$

CHAPTER 4 TEST (page 303)

1. false   **2.** yes   **3.** $(12, \infty)$ **4.** $(-\infty, -5]$ **5.** more than 78   **6.** $\left[1, \frac{9}{4}\right]$

7. $(-\infty, -3) \cup (8, \infty)$ **8.** $(-2, 16)$ **9.** 8   **10.** −4.75   **11.** −5, $\frac{23}{3}$ **12.** 4, −4

13. $[-7, 1]$ **14.** $(-\infty, -9) \cup (13, \infty)$ **15.** $(-\infty, 1) \cup (3, \infty)$

16. $[1, 3]$ **17.**   **18.**   **19.**

20.   **21.** $(1, 1)$, $(2, 1)$, $(2, 2)$ **23.**   **24.** $\le$, $\ge$, $\ge$, $\le$

CHAPTERS 1–4 CUMULATIVE REVIEW EXERCISES (page 304)

1. rational numbers: terminating and repeating decimals; irrational numbers: nonterminating, nonrepeating decimals   **2.** 0.125, 0.0625, 0.03125, 0.054125   **3.** 10   **4.** −6   **5.** $-26p^2 - 6p$ **6.** $-2a + b - 2$ **7.** 201 ft^2 **8.** $20,000   **9.** $\frac{26}{3}$ **10.** 3   **11.** 6   **12.** impossible

13. perpendicular   **14.** parallel   **15.** $y = \frac{1}{3}x + \frac{11}{3}$ **16.** $-\frac{8}{5}$ **17.** 9,000 prisoners/yr   **18.** 1990–1995, 72,200 prisoners/yr   **19.** 10
20. 14   **21.** (2, 1)   **22.** (3, 1)   **23.** (1, 1)   **24.** (−1, −1, 3)   **25.** (−1, −1)   **26.** (1, 2, −1)   **27.** (7, 23), in 1907 the percent of U.S.
workers in white-collar and farming jobs was the same (23%); (45, 42), in 1945 the percent of U.S. workers in white-collar and
blue-collar jobs was the same (42%)   **28.** $y = \frac{5}{9}x + 20$, about 81%   **29.** 750   **30.** 250 $5 tickets, 375 $3 tickets, 125 $2 tickets
31. $3, -\frac{3}{2}$ **32.** $-5, -\frac{3}{5}$ **33.** $(-\infty, 11]$ **34.** $(-3, 3)$ **35.** $\left[-\frac{2}{3}, 2\right]$

36. $(-\infty, -4) \cup (1, \infty)$ **37.**   **38.**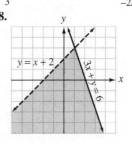

STUDY SET   SECTION 5.1 (page 315)

1. base, exponent   **3.** natural   **5.** x^{m+n} **7.** $x^n y^n$ **9.** 1   **11.** x^{m-n} **13.** $4^{-2} = \dfrac{1}{4^2}$ **15. a.** x^6 ft²   **b.** x^9 ft³   **17.** 9, (−2)

19. base is 5, exponent is 3   **21.** base is x, exponent is 5   **23.** base is b, exponent is 6   **25.** base is $\frac{n}{4}$, exponent is 3   **27.** 9   **29.** −9

31. 9   **33.** $\frac{1}{25}$ **35.** $-\frac{1}{25}$ **37.** $\frac{1}{25}$ **39.** 1   **41.** 1   **43.** $-32x^5$ **45.** x^5 **47.** x^{10} **49.** k^7 **51.** $2a^4 b^5$ **53.** p^{10} **55.** $x^5 y^4$ **57.** $\dfrac{1}{b^{72}}$ **59.** x^{28}

61. $\dfrac{s^3}{r^9}$ **63.** a^{20} **65.** $-\dfrac{1}{d^3}$ **67.** $27x^9 y^{12}$ **69.** $\dfrac{1}{729}m^6 n^{12}$ **71.** $\dfrac{a^{15}}{b^{10}}$ **73.** $\dfrac{a^6}{b^4}$ **75.** a^5 **77.** c^7 **79.** $\dfrac{9}{4}$ **81.** a^4 **83.** $\dfrac{1}{9x^3}$ **85.** $\dfrac{64b^{12}}{27a^9}$

87. 1   **89.** $-\dfrac{8a^{21}}{b^3}$ **91.** $-\dfrac{27}{8a^{18}}$ **93.** $\dfrac{1}{9x^3}$ **95.** 3.462825992   **97.** −244.140625   **105.** $10^{-2}, 10^{-3}, 10^{-4}, 10^{-5}, 10^{-6}, 10^{-7}, 10^{-8},$
10^{-9} **107.** $10^3 \cdot 26^3$; 17,576,000   **113.** $(-\infty, 1)$ **115.** $(-\infty, 20]$

STUDY SET   SECTION 5.2 (page 322)

1. scientific   **3.** 10^n **5.** left   **7.** 60.22 is not between 1 and 10.   **9.** 3.9×10^3 **11.** 7.8×10^{-3} **13.** 1.73×10^{14} **15.** 9.6×10^{-6}
17. 3.23×10^7 **19.** 6.0×10^{-4} **21.** 5.27×10^3 **23.** 3.17×10^{-4} **25.** 270   **27.** 0.00323   **29.** 796,000   **31.** 0.00037   **33.** 5.23
35. 23,650,000   **37.** 1.817×10^{12} **39.** 5.005×10^8 **41.** 5×10^{-8} **43.** 4.005×10^{20}; 400,500,000,000,000,000,000   **45.** 4.3×10^{-3};
0.0043   **47.** 6×10^3; 6,000   **49.** 3.6×10^{-5}; 0.000036   **51.** 2,600,000 to 1   **53.** 1.7×10^{12}, 3.9×10^9, 2.75×10^8, 3.12×10^8
55. 8.5×10^{-28} g   **57.** about 9.5×10^{15} m   **59. a.** 2.5×10^9 sec = 2,500,000,000 sec   **b.** about 79 years   **61.** 1.209×10^8 mi
63. 1.0×10^{21} **69.** $\left[1, \frac{9}{4}\right]$ **71.** $(-\infty, 2) \cup (7, \infty)$

STUDY SET   SECTION 5.3 (page 333)

1. polynomial   **3.** degree   **5.** cubic   **7.** like   **9.** monomial; 2   **11.** trinomial; 3   **13.** binomial; 2   **15.** monomial; 0
17. none of these; 10   **19.** binomial; 9   **21. a.** $-2x^4 - 5x^2 + 3x + 7$ **b.** $7a^3 x^5 - ax^3 - 5a^3 x^2 + a^2 x$ **23.** like terms, $10x$
25. unlike terms   **27.** like terms, $-5r^2 t^3$ **29.** unlike terms   **31.** $9x^2 + 3x - 2$ **33.** −1, −1, 1
35. 8, 2, 0, 2, 8   **37.** −42, 0, 12, 6, −6, −12, 0, 42   **39.**

$f(x) = 2x^2 - 4x + 2$

$f(x) = 2x^3 - 3x^2 - 11x + 6$

41. $x^2 - 5x + 6$ **43.** $-5a^2 + 4a + 4$ **45.** $5a^2 + 3a - 9$ **47.** $-y^3 + 4y^2 + 6$ **49.** $2pq - 2q$ **51.** $2x^2 y^3 + 13xy + 3y^2$
53. $4x^2 - 11$ **55.** $6x^3 - 6x^2 + 14x - 17$ **57.** $x^2 - 8x + 22$ **59.** $-3y^3 + 18y^2 - 28y + 35$ **61.** $2x$ **63.** 20 ft   **65.** 872 ft³
67. 2,160 in.³   **69. a.** $V(x) = 2,500x + 275,000$ **b.** $325,000   **77.** $[-5, 5]$ **79.** (−1, 9)

STUDY SET   SECTION 5.4 (page 342)

1. product   **3.** terms   **5.** factors   **7.** term   **9.** $x^2 + 2xy + y^2$ **11.** $x^2 - y^2$ **13.** $x^2 + 2x - 8$ **15.** $16a^2 + 24a + 9$ **17. a.** $2x, 4x$
b. $2x, -3$ **c.** $4, 4x$ **d.** $4, -3$ **19.** $-6a^3b$ **21.** $-15a^2b^2c^3$ **23.** $-120a^9b^3$ **25.** $-405x^7y^4$ **27.** $3x + 6$ **29.** $3x^3 + 9x^2$
31. $-6x^3 + 6x^2 - 4x$ **33.** $7r^3st + 7rs^3t - 7rst^3$ **35.** $-12m^4n^2 - 12m^3n^3$ **37.** $x^2 + 5x + 6$ **39.** $6t^2 + 5t - 6$ **41.** $6y^2 - 5yz + z^2$
43. $x^2 + 4x + 4$ **45.** $a^2 - 8a + 16$ **47.** $4a^2 + 4ab + b^2$ **49.** $x^2 - 4$ **51.** $x^4 + x^3 - x - 1$ **53.** $x^3 - y^3$ **55.** $6y^3 + 11y^2 + 9y + 2$
57. $8a^3 - b^3$ **59.** $a^3 - 3a^2b - ab^2 + 3b^3$ **61.** $2x^5 + x$ **63.** $\dfrac{3x}{y^4z} - \dfrac{x^6}{y^{10}z^2}$ **65.** $\dfrac{1}{x^2} - y^2$ **67.** $2x^3y^3 - \dfrac{2}{x^3y^3} + 3$ **69.** $x^{3n} - x^{2n}$
71. $x^{2n} - 1$ **73.** $x^{2n} - \dfrac{x^n}{y^n} - x^ny^n + 1$ **75.** $x^{4n} - y^{4n}$ **77.** $x^{2n} + x^n + x^ny^n + y^n$ **79.** $3x^2 + 12x$ **81.** $-p^2 + 4pq$ **83.** $3x^2 + 3x - 11$
85. $5x^2 - 36x + 7$ **87.** $9.2127x^2 - 7.7956x - 36.0315$ **89.** $299.29y^2 - 150.51y + 18.9225$ **91. a.** $(x + y)(x - y)$ **b.** $x(x - y)$;
$x^2 - xy$ **c.** $y(x - y)$; $xy - y^2$ **d.** They represent the same area. $(x + y)(x - y) = x^2 - y^2$
93. $x(12 - 2x)(12 - 2x) = 144x - 48x^2 + 4x^3$ **99.**   **101.**

STUDY SET   SECTION 5.5 (page 350)

1. factored   **3.** factor   **5.** greatest common factor   **7.** $6xy^2$ **9. a.** The terms inside the parentheses have a common factor of 2.
b. The terms inside the parentheses have a common factor of t.   **11.** 4   **13.** $x^2, 2$ **15.** $2 \cdot 3$ **17.** $3^3 \cdot 5$ **19.** 2^7 **21.** $5^2 \cdot 13$ **23.** 12
25. 2   **27.** $4a^2$ **29.** $6xy^2z^2$ **31.** $2(x + 4)$ **33.** $2x(x - 3)$ **35.** prime   **37.** $5x^2y(3 - 2y)$ **39.** prime   **41.** $3z(9z^2 + 4z + 1)$
43. $9x^7y^3(5x^3 - 7y^4 + 9x^3y^7)$ **45.** $-3(a + 2)$ **47.** $-x(3x + 1)$ **49.** $-3x(2x + y)$ **51.** $-6ab(3a + 2b)$
53. $-7u^2v^3z^2(9uv^3z^7 - 4v^4 + 3uz^2)$ **55.** $x^2(x^n + x^{n+1})$ **57.** $y^n(2y^2 - 3y^3)$ **59.** $x^{-2}(x^6 - 5x^8)$ **61.** $t^{-3}(t^8 + 4t^{-3})$
63. $4y^{-2n}(2y^{4n} + 3y^{2n} + 4)$ **65.** $(x + y)(4 + t)$ **67.** $(a - b)(r - s)$ **69.** $(m + n + p)(3 + x)$ **71.** $(u + v)(u + v - 1)$
73. $-(x + y)(a - b)$ **75.** $(x + y)(a + b)$ **77.** $(x + 2)(x + y)$ **79.** $(3 - c)(c + d)$ **81.** $(a + b)(a - 4)$ **83.** $(a + b)(x - 1)$
85. $(x + y)(x + y + z)$ **87.** $x(m + n)(p + q)$ **89.** $y(x + y)(x + y + 2z)$ **91.** $n(2n - p + 2m)(n^2p - 1)$ **93.** $r_1 = \dfrac{rr_2}{r_2 - r}$
95. $f = \dfrac{d_1d_2}{d_2 + d_1}$ **97.** $a^2 = \dfrac{b^2x^2}{b^2 - y^2}$ **99.** $r = \dfrac{S - a}{S - l}$ **101. a.** $\frac{1}{2}b_1h$ **b.** $\frac{1}{2}b_2h$ **c.** $\frac{1}{2}h(b_1 + b_2)$; the formula for the area of a trapezoid
103. $r^2(4 - \pi)$ **107.** a line   **109.** $(-\infty, -3)$ **111.** They are the same.

STUDY SET   SECTION 5.6 (page 358)

1. squares   **3.** 1, 4, 9, 16, 25, 36, 49, 64, 81, 100   **5.** $(x + 5)(x + 5) = x^2 + 10x + 25$ **7. a.** A common factor of 2 can be factored
out of each binomial.   **b.** $(1 - t^4)$ factors as a difference of two squares.   **9. a.** $5(p^2 + 4)$ **b.** $5(p + 2)(p - 2)$ **11.** $(p^2 - pq + q^2)$
13. $(p - q)$ **15.** $6y$, $+$, $7m$ **17.** $(x + 2)(x - 2)$ **19.** $(3y + 8)(3y - 8)$ **21.** prime   **23.** $(25a + 13b^2)(25a - 13b^2)$
25. $(9a^2 + 7b)(9a^2 - 7b)$ **27.** $(6x^2y + 7z^2)(6x^2y - 7z^2)$ **29.** $(x + y + z)(x + y - z)$ **31.** $(a - b + c)(a - b - c)$
33. $(x^2 + y^2)(x + y)(x - y)$ **35.** $(16x^2y^2 + z^4)(4xy + z^2)(4xy - z^2)$ **37.** $(a^x + b^x)(a^x - b^x)$ **39.** $(2x^n + 3y^n)(2x^n - 3y^n)$
41. $2(x + 12)(x - 12)$ **43.** $2x(x + 4)(x - 4)$ **45.** $5x(x + 5)(x - 5)$ **47.** $t^2(rs + x^2y)(rs - x^2y)$ **49.** $(a + b)(a - b + 1)$
51. $(a - b)(a + b + 2)$ **53.** $(2x + y)(1 + 2x - y)$ **55.** $(r + s)(r^2 - rs + s^2)$ **57.** $(x - 2y)(x^2 + 2xy + 4y^2)$
59. $(4a - 5b^2)(16a^2 + 20ab^2 + 25b^4)$ **61.** $(5xy^2 + 6z^3)(25x^2y^4 - 30xy^2z^3 + 36z^6)$ **63.** $(x^2 + y^2)(x^4 - x^2y^2 + y^4)$
65. $(a^b - c^b)(a^{2b} + a^bc^b + c^{2b})$ **67.** $(3x^n + y^n)(9x^{2n} - 3x^ny^n + y^{2n})$ **69.** $5(x + 5)(x^2 - 5x + 25)$ **71.** $4x^2(x - 4)(x^2 + 4x + 16)$
73. $2u^2(4v - t)(16v^2 + 4tv + t^2)$ **75.** $(a + b)(x + 3)(x^2 - 3x + 9)$ **77.** $\frac{4}{3}\pi(r_1 - r_2)(r_1^2 + r_1r_2 + r_2^2)$
81.   **83.**   **85.** $y = -4$

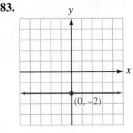

STUDY SET   SECTION 5.7 (page 368)

1. trinomial   **3.** lead   **5. a.** positive   **b.** negative   **c.** positive   **7.** integers   **9.** $2xy + y^2$ **11.** $x^2 - y^2$ **13.** 4, -4, 1   **15.** $(x + 2)$
17. $(x - 3)$ **19.** $(2a + 1)$ **21.** $(x + 1)^2$ **23.** $(a - 9)^2$ **25.** $(2y + 1)^2$ **27.** $(3b - 2)^2$ **29.** $(x - 3)(x - 2)$ **31.** $(x - 2)(x - 5)$
33. prime   **35.** $(x + 5)(x - 6)$ **37.** $(a + 10)(a - 5)$ **39.** $(x + 3y)(x - 7y)$ **41.** $3(x + 7)(x - 3)$ **43.** $x^2(b - 7)(b - 5)$

45. $-(a - 8)(a + 4)$ **47.** $-3(x - 3)(x - 2)$ **49.** $-2(p + 2q)(p - q)$ **51.** $(3y + 2)(2y + 1)$ **53.** $(4a - 3)(2a + 3)$
55. $(3x - 4y)(2x + y)$ **57.** prime **59.** $(4x - 3)(2x - 1)$ **61.** $(a + b)(a - 4b)$ **63.** $x(3x - 1)(x - 3)$ **65.** $-(3a + 2b)(a - b)$
67. $5(a - 3b)^2$ **69.** $x^2(7x - 8)(3x + 2)$ **71.** $(x^2 + 5)(x^2 + 3)$ **73.** $(y^2 - 10)(y^2 - 3)$ **75.** $(a + 3)(a - 3)(a + 2)(a - 2)$
77. $(x^n + 1)^2$ **79.** $(2a^{3n} + 1)(a^{3n} - 2)$ **81.** $(x^{2n} + y^{2n})^2$ **83.** $(3x^n - 1)(2x^n + 3)$ **85.** $(x + a + 1)^2$ **87.** $(a + b + 4)(a + b - 6)$
89. $(x + 2 + y)(x + 2 - y)$ **91.** $(x + 1 + 3z)(x + 1 - 3z)$ **93.** $(c + 2a - b)(c - 2a + b)$ **95.** $(a - 16)(a - 1)$ **97.** $(2u + 3)(u + 1)$
99. $(5r + 2s)(4r - 3s)$ **101.** $x + 3$ **105.** 5 **107.** $\frac{8}{3}$ **109.** $-26p^2 - 6p$

STUDY SET   SECTION 5.8 (page 373)

1. factoring **3.** cubes **5.** common **7.** trinomial **9.** Multiply the factors of $y^2z^3(x + 6)(x + 1)$ to see if the product is
$x^2y^2z^3 + 7xy^2z^3 + 6y^2z^3$. **11.** $3ab, 2b, 2a$ **13.** $(x + 4)^2$ **15.** $(2xy - 3)(4x^2y^2 + 6xy + 9)$ **17.** $(x - t)(y + s)$ **19.** $(5x + 4y)(5x - 4y)$
21. $(6x + 5)(2x + 7)$ **23.** $2(3x - 4)(x - 1)$ **25.** $y^2(2x + 1)(2x + 1)$ **27.** $(x + a^2y)(x^2 - a^2xy + a^4y^2)$ **29.** $2(x - 3)(x^2 + 3x + 9)$
31. $(a + b)(f + e)$ **33.** $(2x + 2y + 3)(x + y - 1)$ **35.** $(25x^2 + 16y^2)(5x + 4y)(5x - 4y)$ **37.** $36(x^2 + 1)(x + 1)(x - 1)$
39. $(a + 3)(a - 3)(a + 2)(a - 2)$ **41.** $(x + 3 + y)(x + 3 - y)$ **43.** $(2x + 1 + 2y)(2x + 1 - 2y)$ **45.** $(x + y + 1)(x - y - 1)$
47. $2(a^n - 3)(a^n + 4)$ **49.** $2(3a^n + 2b^n)(9a^{2n} - 6a^nb^n + 4b^{2n})$ **51.** $2(3m^{2x} + 1)(2m^{2x} + 1)$ **55.** perpendicular **57.** -225

STUDY SET   SECTION 5.9 (page 380)

1. quadratic **3.** At least one is 0. **5. a.** yes **b.** no **c.** yes **d.** no **7.** $3, -1$ **9.** $y + 6, y - 9, -6$ **11.** $0, -2$ **13.** $4, -4$
15. $0, -1$ **17.** $0, 5$ **19.** $-3, -5$ **21.** $-2, -4$ **23.** $-\frac{1}{3}, -3$ **25.** $\frac{1}{2}, 2$ **27.** $1, -\frac{1}{2}$ **29.** $3, 3$ **31.** $\frac{1}{4}, -\frac{3}{2}$ **33.** $2, -\frac{5}{6}$ **35.** $\frac{1}{3}, -1$
37. $2, \frac{1}{2}$ **39.** $\frac{1}{5}, -\frac{5}{3}$ **41.** $0, 0, -1$ **43.** $0, 7, -7$ **45.** $0, 7, -3$ **47.** $3, -3, 2, -2$ **49.** $0, -2, -3$ **51.** $0, \frac{5}{6}, -7$ **53.** 16, 18 or $-18,$
-16 **55.** 2.78, 0.72 **57.** 1 **59.** 10 in., 16 in. **61.** 3 ft **63.** 20 ft by 40 ft **65.** 11 sec and 19 sec **67.** 2 sec **69.** 6 m/sec **71.** 4
79. 25 ft^2

KEY CONCEPT (page 384)

1. $-5x^2 + 2x - 6$ **2.** $4s^3 - 3s^2 + 3s - 7$ **3.** $3m^2 + 5m - 12$ **4.** $3r^4st - 6r^2s^2t + 9r^2st^3$ **5.** $a^2 - 4ad + 4d^2$ **6.** $x^3 - 27y^3$
7. $6b^3 + 11b^2 + 9b + 2$ **8.** $-y^2 - 4y - 1$ **9.** 0; after falling for 6 seconds, the squeegee will strike the ground
10. $V(x) = 16x^3 + 12x^2 - 4x; 528$ **11.** $9, -9$ **12.** $0, 5$ **13.** $-3, -5$ **14.** $-\frac{3}{2}, -1$ **15.** $\frac{1}{3}, -1$ **16.** $1, 0, -9$

CHAPTER REVIEW (page 386)

1. a. 729 **b.** -32 **c.** -64 **d.** 15 **2. a.** x^6 **b.** a^5b^6 **c.** m^{18} **d.** t^{13} **e.** $9x^4y^6$ **f.** $\frac{x^{16}}{b^4}$ **g.** -3 **h.** $\frac{1}{x^{10}}$ **i.** $-\frac{1}{625}$ **j.** $70x^4$ **k.** $\frac{x^6}{9}$

l. $\frac{2}{x}$ **m.** $-c^{10}$ **n.** $\frac{25}{16}$ **o.** $\frac{1}{y^8}$ **p.** $\frac{-b^3}{8a^{21}}$ **3. a.** 1.93×10^{10} **b.** 2.735×10^{-8} **4. a.** 72,770,000 **b.** 0.0000000083 **5.** 7.6×10^2 sec
6. 1.67248×10^{-18} g **7.** 8.4×10^6 **8. a.** no **b.** yes **c.** yes **d.** no **9. a.** binomial, 2 **b.** monomial, 4 **c.** none of these, 4
d. trinomial, 8 **10.** 134 in.^{3} **11. a.**   **b.**   **12. a.** $5x^2 + 2x + 16$

$f(x) = x^2 - 2x$ $f(x) = x^3 - 3x^2 + 4$

b. $3x^2y^3 - 8x^2y + 8y$ **c.** $6x^3 + 4x^2 + x + 9$ **d.** $6k^4 - 6k^3 + 9k^2 - 2$ **13. a.** $-4a^3$ **b.** $6x^3y^2z^5$ **14. a.** $2x^4y^3 - 8x^2y^7$
b. $a^4b + 2a^3b^2 - a^2b^3$ **15. a.** $16x^2 + 14x - 15$ **b.** $6x^3 - 12x^2 + 4x - 8$ **c.** $25a^2 - 60a + 36$ **d.** $49c^4 - d^2$
e. $15x^4 - 22x^3 + 58x^2 - 40x$ **f.** $r^3 - 3r^2s - rs^2 + 3s^3$ **g.** $\frac{1}{a} - \frac{1}{a^2b^2} + ab - \frac{1}{b}$ **h.** $p^{2n} - p^nq^n + \frac{2p^n}{q^n} - 2$ **16.** $154x^2 + 27x + 1$
17. $2 \cdot 5^2 \cdot 7$ **18. a.** 6 **b.** $3xy^3$ **19. a.** $4(x + 2)$ **b.** $3x(x^2 - 2x + 3)$ **c.** prime **d.** $7a^3b(ab + 7)$ **e.** $5x^2(x + y)(1 - 3x)$
f. $9x^2y^3z^2(3xz + 9x^2y^2 - 10z^5)$ **20. a.** $-7(b - 2)$ **b.** $-7a^2b^2(a - b)^3(7a^2 - 7ab - 9b^2)$ **21.** $a^n(a^n - b^n)$ **22.** $x^{-1}(x^5 + x^3 - 1)$
23. a. $(x + 2)(y + 4)$ **b.** $(ry - a)(r - 1)$ **24.** $m_1 = \dfrac{mm_2}{m_2 - m}$ **25. a.** $(z + 4)(z - 4)$ **b.** $(y + 11)(y - 11)$ **c.** $(xy^2 + 8z^3)(xy^2 - 8z^3)$
d. prime **e.** $(c + a + b)(c - a - b)$ **f.** $3x^2(x^2 + 10)(x^2 - 10)$ **g.** $(2 + x^n)(2 - x^n)$ **h.** $(x^{2m} + y^m)(x^{2m} - y^m)$
26. a. $(t + 4)(t^2 - 4t + 16)$ **b.** $2y(x - 3z)(x^2 + 3xz + 9z^2)$ **c.** $(x^n - y^m)(x^{2m} + x^my^m + y^{2m})$ **d.** $(3 + a^{2n})(9 - 3a^{2n} + a^{4n})$
27. a. $(x + 5)^2$ **b.** $(a - 7)^2$ **28. a.** $(y + 20)(y + 1)$ **b.** $(z - 5)(z - 6)$ **c.** $-(x + 7)(x - 4)$ **d.** $(a - 8b)(a + 3b)$
e. $(4a - 1)(a - 1)$ **f.** prime **g.** $y(y + 2)(y - 1)$ **h.** $3(5x + y)(x - 4y)$ **i.** $2a^2(a + 3)(a - 1)$ **j.** $(v^2 - 7)(v^2 - 6)$
k. $(y^n - 3)(y^n + 2)$ **29.** $(s + t - 1)^2$ **30.** $(k + 1 + 3m)(k + 1 - 3m)$ **31. a.** $x(x - 1)(x + 6)$ **b.** $3y(x + 3)(x - 7)$
c. $(z - 2)(z + x + 2)$ **d.** $(x + 1 + p)(x + 1 - p)$ **e.** $(x + 2 + 2p^2)(x + 2 - 2p^2)$ **f.** $(y + 2)(y + 1 + x)$
g. $4(ab + 4)(a^2b^2 - 4ab + 16)$ **h.** $36(z^2 + 1)(z + 1)(z - 1)$ **i.** $(2x^n + 3)(3x^n - 1)$ **j.** $-3(x^{2n} + 4)(x^n + 2)(x^n - 2)$
32. $\frac{\pi}{2}h(r_1 + r_2)(r_1 - r_2)$ **33. a.** $0, \frac{3}{4}$ **b.** $6, -6$ **c.** $\frac{1}{2}, -\frac{5}{6}$ **d.** $\frac{2}{7}, 5$ **e.** $0, -\frac{2}{3}, \frac{4}{5}$ **f.** $-\frac{2}{3}, 7, 0$ **34.** 17 m by 20 m **35.** $-\frac{1}{2}, 1$

CHAPTER 5 TEST (page 391)

1. x^9 **2.** $-8x^6y^9$ **3.** $\dfrac{1}{m^5}$ **4.** $\dfrac{m^4}{9n^{10}}$ **5.** 4.706×10^{12} **6.** 0.000245 **7.** 1.45×10^{19} **8.** 1.116×10^7 mi/min   **9.** 5   **10.** 13

11. 110 ft   **12.** $3x^2 + 4x - 4$ **13.**

14. a.

b. $-2, 0$ **15.** $5y^2 + y - 1$

16. $-4u^2 + 2u - 14$ **17.** $-6x^4yz^4$ **18.** $-15a^3b^4 + 10a^3b^5$ **19.** $z^2 - 16$ **20.** $12x^{2n} - 17x^n + 6$ **21.** $16t^2 - 72t + 81$
22. $2s^3 - 2st^2$ **23.** $3x(1 + 2x)$ **24.** $3abc(4a^2b - abc + 2c^2)$ **25.** $(u - v)(r + s)$ **26.** $(a - y)(x + y)$ **27.** $(x + 7)(x - 7)$
28. $4(y^2 + 4)(y + 2)(y - 2)$ **29.** prime   **30.** $(b + 5)(b^2 - 5b + 25)$ **31.** $3(u - 2)(u^2 + 2u + 4)$ **32.** $(a - 6)(a + 1)$
33. $(3b + 2c)(2b - c)$ **34.** $(x + 3 + y)(x + 3 - y)$ **35.** $a^nb(a^nb - a^{2n})$ **36.** $(b^n + 2)(b^n - 1)$ **37.** $6, -1$ **38.** $\frac{3}{4}, -\frac{3}{2}$ **39.** $0, -4$

40. $v = \dfrac{v_1v_3}{v_3 + v_1}$ **41.** 11 ft by 22 ft

STUDY SET   SECTION 6.1 (page 402)
1. rational   **3.** asymptote   **5. a.** 1   **b.** 0.5   **c.** 0.25   **7. a.** 1   **b.** 1   **c.** 1   **d.** -1 **9.** 1: decreases then steadily increases; 2: increases,
decreases, then steadily increases; 3: steadily increases; 4: steadily decreases, approaching a cost of $2.00 per unit   **11. a.** true   **b.** true
13. 20 hr   **15.** 12 hr   **17.** 6, 3, 1.5, 1, 0.75, 0.6, 0.5

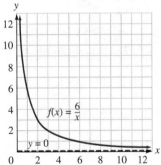

19. 3, 2, 1.5, 1.33, 1.25, 1.2, 1.17   **21.** $(-\infty, 0) \cup (0, \infty)$ **23.** $(-\infty, -2) \cup (-2, \infty)$

25. $(-\infty, 0) \cup (0, 1) \cup (1, \infty)$ **27.** $(-\infty, -7) \cup (-7, 8) \cup (8, \infty)$ **29.** $\frac{2}{3}$ **31.** $-\frac{28}{9}$ **33.** $4x^2$ **35.** $-\dfrac{4y}{3x}$ **37.** $-\dfrac{x}{2}$ **39.** -1 **41.** $\dfrac{1}{x - y}$

43. $\dfrac{5}{x - 2}$ **45.** $\dfrac{-3(x + 2)}{x + 1}$ **47.** in lowest terms   **49.** $x + 2$ **51.** $\dfrac{x + 1}{x + 3}$ **53.** $\dfrac{m - 2n}{n - 2m}$ **55.** $\dfrac{x + 4}{2(2x - 3)}$ **57.** $\dfrac{3(x - y)}{x + 2}$ **59.** $\dfrac{2x + 1}{2 - x}$

61. $\dfrac{a^2 - 3a + 9}{4(a - 3)}$ **63.** in lowest terms   **65.** $m + n$ **67.** $-\dfrac{m + n}{2m + n}$ or $\dfrac{-m - n}{2m + n}$ **69.** $\dfrac{x - y}{x + y}$ **71.** $\dfrac{2x - 3}{2y - 3}$ **73.** $\dfrac{3a + b}{y + b}$ **75.** 1
77. D: $(-\infty, 2) \cup (2, \infty)$; R: $(-\infty, 1) \cup (1, \infty)$ **79.** D: $(-\infty, -2) \cup (-2, 2) \cup (2, \infty)$; R: $(-\infty, \infty)$ **81. a.** $50,000   **b.** $200,000
83. a. $c(n) = 0.09n + 7.50$ **b.** $c(n) = \dfrac{0.09n + 7.50}{n}$ **c.** about 10¢   **85. a.** about 2.6 hr   **b.** about 4.6 hr   **89.** $3x(x - 3)$
91. $(3x^2 + 4y)(9x^4 - 12x^2y + 16y^2)$

STUDY SET SECTION 6.2 (page 415)

1. ratios **3.** extremes, means **5.** direct, inverse **7.** inverse **9.**

direct

Sales tax on the item

Price of an item

11.

inverse

Time needed to rewind a video cassette

Rewind speed of a VCR

13. $-7, 6, 18, -102$ **15.** 3 **17.** 5 **19.** 5 **21.** 39 **23.** $2, -2$ **25.** $4, -1$ **27.** $-\frac{5}{2}, -1$ **29.** no solution **31.** $A = kp^2$ **33.** $v = \dfrac{k}{r^2}$

35. $P = \dfrac{ka^2}{j^3}$ **37.** L varies jointly with m and n. **39.** R varies directly with L and inversely with d^2. **41.** 202, 172, 136 **43.** about 2

gal **45.** eye: 49.9 in.; seat: 17.6 in.; elbow: 27.8 in. **47.** 7.5 in. $\times$ 12.5 in. **49.** 555 ft **51.** $46\frac{7}{8}$ ft **53.** 880 ft **55.** 1,600 ft
57. 25 days **59.** 12 in.^{3} **61.** \$9,000 **63.** 3 ohms **65.** 0.275 in. **67.** 12.8 lb **71.** x^{10} **73.** -1

STUDY SET SECTION 6.3 (page 425)

1. divisor **3.** numerator **5.** multiply, $\frac{ac}{bd}$ **7.** 0 **9.** $(x - 5), (5x - 25), x(x + 3), 5(x - 5)$ **11.** yes, no, yes **13.** $\dfrac{5}{4}$ **15.** $-\dfrac{5}{6}$

17. $\dfrac{xy^2d}{c^3}$ **19.** $-\dfrac{x^{10}}{y^2}$ **21.** $x + 1$ **23.** 1 **25.** $\dfrac{x - 4}{x + 5}$ **27.** $-\dfrac{(a + 7)^2(a - 5)}{12x^2}$ **29.** $\dfrac{t - 1}{t + 1}$ **31.** $\dfrac{n + 2}{n + 1}$ **33.** $-\dfrac{x + 1}{x - 1}$ **35.** $\dfrac{1}{x + 1}$

37. $(x + 1)^2$ **39.** $\dfrac{(a + 7)^2(a - 5)}{12x^2}$ **41.** $\dfrac{x + y}{x - y}$ **43.** $\dfrac{a + b}{(x - 3)(c + d)}$ **45.** $-\dfrac{x + 1}{x + 3}$ or $\dfrac{-x - 1}{x + 3}$ **47.** $x^2(x + 3)$ **49.** $\dfrac{x + 2}{x - 2}$ **51.** $\dfrac{3x}{2}$

53. 1 **55.** $\dfrac{x^2 - 6x + 9}{x^6 + 8x^3 + 16}$ **57.** $\dfrac{4m^4 - 4m^3 - 11m^2 + 6m + 9}{x^4 - 2x^2 + 1}$ **59.** $k_1(k_1 + 2), k_2 + 6$ **65.** $-6a^5 + 2a^4$ **67.** $6g^2 - 5gn + n^2$

STUDY SET SECTION 6.4 (page 434)

1. divisible **3.** subtract, keep, $\dfrac{a - c}{b}$ **5.** same **7.** LCD **9. a.** ii **b.** adding or subtracting rational expressions **c.** simplifying a

rational expression **11.** $3x - 2, 3, 3x - 1$ **13.** $\frac{5}{2}$ **15.** $\frac{11}{4y}$ **17.** 2 **19.** 3 **21.** 3 **23.** $\dfrac{6x}{(x - 3)(x - 2)}$ **25.** $36x^2$

27. $x(x + 3)(x - 3)$ **29.** $(x + 3)^2(x^2 - 3x + 9)$ **31.** $(2x + 3)^2(x + 1)^2$ **33.** $\dfrac{5}{6}$ **35.** $\dfrac{21a - 8b}{14}$ **37.** $\dfrac{17}{12x}$ **39.** $\dfrac{9a^2 - 4b^2}{6ab}$

41. $\dfrac{3a - 5b}{a^2b^2}$ **43.** $\dfrac{3r + 2bs}{12b^2}$ **45.** $\dfrac{2(5a + 2b)}{21}$ **47.** $\dfrac{2(4x - 1)}{(x + 2)(x - 4)}$ **49.** $\dfrac{7x + 29}{(x + 5)(x + 7)}$ **51.** $\dfrac{4x + 1}{x}$ **53.** 2 **55.** $\dfrac{3(a - 1)}{3a - 2}$

57. $\dfrac{x(2x + 1)}{(x + 3)(x + 2)(x - 2)}$ **59.** $\dfrac{-x^2 + 11x + 8}{(3x + 2)(x + 1)(x - 3)}$ **61.** $\dfrac{2x^2 + 5x + 4}{x + 1}$ **63.** $\dfrac{x^2 - 5x - 5}{x - 5}$ **65.** $\dfrac{-2(2x^2 - 7x - 27)}{x(x + 3)(x - 3)}$

67. $\dfrac{11x^2 + 7x - 3}{(2x - 1)(3x + 2)}$ **69.** $\dfrac{2}{x + 1}$ **71.** $\dfrac{2b}{a + b}$ **73.** $\dfrac{7mn^2 - 7n^3 - 6m^2 + 3mn + n}{(m - n)^2}$ **75.** $\dfrac{2}{m - 1}$ **77.** $\dfrac{10r + 20}{r}; \dfrac{3t + 9}{t}$ **83.** 3, 3
85. $0, 0, -1$

STUDY SET SECTION 6.5 (page 444)

1. complex **3.** $\dfrac{t^2}{t^2}$ **5.** $y, y, \div, \frac{1}{6}$ **7.** $\div$ **9.** $\dfrac{2}{3}$ **11.** $-\dfrac{1}{7}$ **13.** $\dfrac{2y}{3z}$ **15.** $125b$ **17.** $-\dfrac{1}{y}$ **19.** $\dfrac{y - x}{x^2y^2}$ **21.** $\dfrac{b + a}{b}$ **23.** $\dfrac{y + x}{y - x}$ **25.** $y - x$

27. $-\dfrac{1}{a + b}$ **29.** $x^2 + x - 6$ **31.** $\dfrac{5x^2y^2}{xy + 1}$ **33.** -1 **35.** $\dfrac{x + 2}{x - 3}$ **37.** $\dfrac{a - 1}{a + 1}$ **39.** $\dfrac{2x^2 + 5x}{x^3 + x^2 + x + 1}$ **41.** $\dfrac{xy^2}{y - x}$ **43.** $\dfrac{x^2(xy^2 - 1)}{y^2(x^2y - 1)}$

45. $\dfrac{1}{x - y}$ **47.** $\dfrac{x - 9}{3}$ **49.** $-\dfrac{h}{4}$ **51.** $\dfrac{3a^2 + 2a}{2a + 1}$ **53.** $\dfrac{(-x^2 + 2x - 2)(3x + 2)}{(2 - x)(-3x^2 - 2x + 9)}$ **55.** $\dfrac{k_1k_2}{k_2 + k_1}$ **57.** $\dfrac{4d - 1}{3d - 1}$ **63.** 8 **65.** $2, -2, 3, -3$

STUDY SET SECTION 6.6 (page 456)

1. rational **3.** common denominator **5. a.** yes **b.** no **7.** $\frac{12}{x}, \frac{12}{x + 15}$ **9. a.** 7 **b.** 7 **11.** $30y, \frac{9}{2y}, \frac{10}{3y}, 7y, 35$ **13.** 12 **15.** 40 **17.** $\frac{1}{2}$
19. no solution **21.** $\frac{17}{25}$ **23.** 0 **25.** 1 **27.** 2 **29.** 2 **31.** $\frac{1}{3}$ **33.** 0 **35.** $2, -5$ **37.** $-4, 3$ **39.** $6, \frac{17}{3}$ **41.** $1, -11$

43. $r = \dfrac{E - IR_L}{I}$ **45.** $r = \dfrac{S - a}{S - l}$ **47.** $Q_1 = \dfrac{PQ_2}{1 + P}$ **49.** $R = \dfrac{R_1R_2R_3}{R_2R_3 + R_1R_3 + R_1R_2}$ **51.** $4\frac{8}{13}$ in. **53.** $L = \dfrac{SN - CN}{V - C}$, 8 years

55. a. $1\frac{7}{8}$ days **b.** Santos: \$412.50, Mays: \$375 **57.** about 110 sec **59.** $2\frac{4}{13}$ weeks **61.** 6 mph **63.** 35 mph and 45 mph **65.** 5 mph
67. 9 **73.** 9.0×10^9 **75.** 4.4×10^{-22}

STUDY SET   SECTION 6.7 (page 465)

1. dividend, divisor, quotient **3.** algorithm **5.** $\frac{1}{b}$ **7.** $(2x - 1)(x^2 + 3x - 4) = 2x^3 + 5x^2 - 11x + 4$ **9.** $2x^2$, x, 4 **11.** $3a^2 + 5 + \frac{6}{3a - 2}$

13. $\frac{x^2 - x - 12}{x - 4}$, $x - 4 \overline{)x^2 - x - 12}$, $(x^2 - x - 12) \div (x - 4)$ **15.** $\frac{y}{2x^3}$ **17.** $-\frac{3}{4a^2}$ **19.** $2x + 3$ **21.** $-\frac{2x}{3} + \frac{x^2}{6}$ **23.** $2xy^2 + \frac{x^2y}{6}$

25. $\frac{x^4y^4}{2} - \frac{x^3y^9}{4} + \frac{3}{4xy^2}$ **27.** $x + 2$ **29.** $x + 7$ **31.** $3x - 5 + \frac{3}{2x + 3}$ **33.** $3x^2 + x + 2 - \frac{4}{x - 1}$ **35.** $2x^2 + 5x + 3 + \frac{4}{3x - 2}$

37. $3x^2 + 4x + 3$ **39.** $a + 1$ **41.** $2y + 2$ **43.** $6x - 12$ **45.** $4x^3 - 3x^2 + 3x + 1$ **47.** $a^2 + a + 1 + \frac{2}{a - 1}$ **49.** $5a^2 - 3a - 4$

51. $6y - 12$ **53.** $x^4 + x^2 + 4$ **55.** $x^2 + x + 1$ **57.** $x^2 + x + 2$ **59.** $9.8x + 16.4 - \frac{36.5}{x - 2}$ **61.** $x^2 - 5x + 6$ **63.** $3x - 2$, $x + 5$

67. $8x^2 + 2x + 4$ **69.** $-2y^3 - y^2 + 10y - 14$

KEY CONCEPT (page 468)

1. a. $\frac{3x + 2}{4x + 3}$ **b.** $2x - 1$ **2. a.** $\frac{d - 1}{d + 1}$ **b.** $3d + 2$, $2d - 3$, $2d + 3$ **3. a.** $\frac{2(4x - 1)}{(x + 2)(x - 4)}$ **b.** $\frac{x - 4}{x - 4}$, $\frac{x + 2}{x + 2}$ **4. a.** $\frac{3(n^2 - n - 2)}{n^2}$

b. $3n$ **5. a.** 4 **b.** $t(t - 2)$ **6. a.** $\frac{1}{6}$ **b.** $(x - 1)(2x + 3)$ **7. a.** 1 **b.** $(x + 2)(x - 4)$ **8. a.** $a^2 = \frac{b^2x^2}{b^2 + y^2}$ **b.** a^2b^2

CHAPTER REVIEW (page 470)

1. 8, 4, 2, 1.33, 1, 0.8, 0.67, 0.57, $\frac{1}{2}$

2. $y = 3$, $x = 0$; D: $(-\infty, 0) \cup (0, \infty)$, R: $(-\infty, 3) \cup (3, \infty)$ **3. a.** $\frac{31x}{72y}$

b. $\frac{x - 7}{x + 7}$ **c.** $\frac{1}{2(x + 2)}$ **d.** $\frac{1}{x - 6}$ **e.** $\frac{-a - b}{c + d}$ **f.** $\frac{-m - 2n}{n + 2m}$ **g.** -1 **h.** -2 **4. a.** 5 **b.** -4, -12 **5.** 70.4 ft **6.** \$5,460

7. 1.25 amps **8.** 126.72 lb **9.** inverse variation **10.** 0.2 **11. a.** -1 **b.** $\frac{2a - 1}{a + 2}$ **c.** $\frac{h^2 - 4h + 4}{h^6 + 8h^3 + 16}$ **d.** $\frac{t - 2}{t}$ **e.** $\frac{3x(x - 1)}{(x - 3)(x + 1)}$

12. a. $\frac{5y - 3}{x - y}$ **b.** $\frac{6x - 7}{x^2 + 2}$ **c.** $-\frac{2}{t - 3}$ **d.** $-\frac{1}{p + 12}$ **13. a.** $60a^2h^3$ **b.** $ab^2(b - 1)$ **c.** $(x - 5)(x + 5)(x + 1)$

d. $(m^2 + 2m + 4)(m - 2)^2$ **14. a.** $\frac{9a + 8}{a + 1}$ **b.** $\frac{3x + 2yz}{12z^2}$ **c.** $\frac{4x^2 + 9x + 12}{(x - 4)(x + 3)}$ **d.** $\frac{4(3y + 5)}{15y(x - 2)}$ **e.** $\frac{x^2 + 26x + 3}{(x + 3)(x - 3)^2}$ **15. a.** $\frac{p - 3}{2(p + 2)}$

b. $\frac{b + 2a}{2b - a}$ **c.** $\frac{x - 2}{x + 3}$ **d.** $\frac{x^2y^2}{(x - y)^2(y^2 - x^2)}$ **e.** $\frac{3x^2 - 18x + 2}{x^2 - 6x + 3}$ **16. a.** 5 **b.** 2 **c.** 8, -9 **d.** -1, -2 **e.** 2 and -2 are extraneous.

f. 0 **17. a.** $y^2 = \frac{x^2b^2 - a^2b^2}{a^2}$ **b.** $b = \frac{Ha}{2a - H}$ **18.** 5 mph **19.** 50 mph **20.** $14\frac{2}{5}$ hr **21.** $18\frac{2}{3}$ days **22.** about $3,500,000$ in. lb/rad

23. a. $\frac{5h^3}{k^2}$ **b.** $-\frac{x^3}{2y^3}$ **24. a.** $6a + \frac{16}{3}$ **b.** $-3x^2y + \frac{3x}{2} + y$ **25. a.** $b + 4$ **b.** $v^2 - 3v - 10$ **c.** $x^2 - 2x + 4$

d. $x^2 + 2x - 1 + \frac{6}{2x + 3}$

CHAPTER 6 TEST (page 474)

1. $-\frac{2}{3xy}$ **2.** $\frac{2}{x - 2}$ **3.** -3 **4.** $\frac{2x + 1}{4}$ **5.** 74 **6.** 25 decibels **7.**

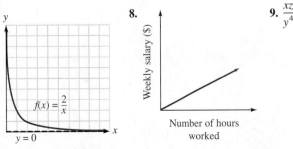

8.

9. $\frac{xz}{y^4}$

10. $\frac{x + 1}{2}$ **11.** 1 **12.** $\frac{(x + y)^2}{2}$ **13.** $\frac{2}{x + 1}$ **14.** -1 **15.** $\frac{2}{t - 4}$ **16.** 2 **17.** $\frac{24b^2 - 2b - 1}{3b + 1}$ **18.** $\frac{2x + 3}{(x + 1)(x + 2)}$ **19.** $\frac{u^2}{2vw}$

20. $\dfrac{3k^2 + 4k + 4}{3k^2 - 9k - 9}$ **21.** 40 **22.** 5; 3 is extraneous **23.** 6, −1 **24.** 26 **25.** $a^2 = \dfrac{x^2 b^2}{b^2 - y^2}$ **26.** $r_2 = \dfrac{rr_1}{r_1 - r}$ **27.** no, $\dfrac{5}{11}$ of an hour

28. 3 mph **29.** $\dfrac{-6x}{y} + \dfrac{4x^2}{y^2} - \dfrac{3}{y^3}$ **30.** $y^2 - 2y + 4 - \dfrac{56}{y + 2}$

CHAPTERS 1–6 CUMULATIVE REVIEW EXERCISES (page 476)

1. $\frac{8}{3}$ **2.** −2 **3.** 14 **4.** Life expectancy will increase 0.1 year each year during this period. **5.** $y = -7x + 54$ **6.** $y = -\frac{11}{6}x - \frac{7}{3}$
7. D: $(-\infty, \infty)$; R: $(-\infty, \infty)$ **8.** $T_1 = 80$, $T_2 = 60$ **9.** **10.**

11. **12.** **13.** $a^8 b^4$ **14.** $\dfrac{b^4}{a^4}$ **15.** 81 **16.** $\dfrac{x^{21} y^3}{8}$ **17.** 42,500 **18.** 0.000712 **19.** $y = \dfrac{kxz}{r}$ **20.** 8

21. **22.** D: $(-\infty, \infty)$; R: $[-3, \infty)$ **23.** 18 **24.** 7 **25.** $2x^3 + x^2 + 12$

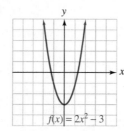

26. $-3x^2 - 3$ **27.** $6x^2 - 7x - 20$ **28.** $4x^6 - 4x^3 + 1$ **29.** yes **30.** about 17 **31.** about 14 **32.** 4 **33.** $3rs^3(r - 2s)$
34. $(x - y)(5 - a)$ **35.** $(x + y)(u + v)$ **36.** $(9x^2 + 4y^2)(3x + 2y)(3x - 2y)$ **37.** $(2x - 3y^2)(4x^2 + 6xy^2 + 9y^4)$ **38.** $(2x + 3)(3x - 2)$
39. $(3x^n - 5)^2$ **40.** $(5x + 3)(3x - 2)$ **41.** $-\dfrac{1}{3}, -\dfrac{7}{2}$ **42.** 0, 2, −2 **43.** $b^2 = \dfrac{a^2 y^2}{a^2 - x^2}$ **44.** 9 in. by 12 in. **45.** $\dfrac{2x - 3}{3x - 1}$ **46.** $-\dfrac{q}{p}$
47. $\dfrac{4}{x - y}$ **48.** $\dfrac{a^2 + ab^2}{a^2 b - b^2}$ **49.** 0 **50.** −17 **51.** $x + 4$ **52.** $-x^2 + x + 5 + \dfrac{8}{x - 1}$ **53.** It will rise sharply. **54.** It has dropped.

STUDY SET SECTION 7.1 (page 489)

1. square root **3.** radical **5.** odd **7.** simplified **9.** $b^2 = a$ **11.** two, positive **13.** x **15.** 3, up
17. 0, −1, −2, −3, −4 **19.** $\sqrt{x^2} = |x|$ **21.** $f(x) = \sqrt{x - 5}$ **23.** 11 **25.** −8 **27.** $\frac{1}{3}$ **29.** 0.5

31. not real **33.** 4 **35.** 3.4641 **37.** 26.0624 **39.** $2|x|$ **41.** $|t + 5|$ **43.** $5|b|$ **45.** $|a + 3|$ **47.** 0 **49.** 4 **51.** 2 **53.** −10
55. 4.1231 **57.** 3.3322 **59.** D: $[-4, \infty)$, R: $[0, \infty)$ **61.** D: $(-\infty, \infty)$, R: $(-\infty, \infty)$

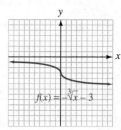

63. 1 **65.** −5 **67.** $-\frac{2}{3}$ **69.** 0.4 **71.** $2a$ **73.** $-10pq$ **75.** $-\frac{1}{2}m^2 n$ **77.** $-0.4s^3 t^2$ **79.** 3 **81.** −3 **83.** not real **85.** $\frac{2}{5}$ **87.** $\frac{1}{2}$
89. $2a$ **91.** $2a$ **93.** k^3 **95.** $\frac{1}{2}m$ **97.** $x + 2$ **99.** 7.0 in. **101.** about 61.3 beats/min **103.** 13.4 ft **105.** 3.5% **111.** 1
113. $\dfrac{3(m^2 + 2m - 1)}{(m + 1)(m - 1)}$

STUDY SET   SECTION 7.2 (page 498)

1. radical   **3.** extraneous   **5. a.** Square both sides.   **b.** Cube both sides.   **7. a.** x **b.** $x - 5$ **c.** $32x$ **d.** $x + 3$ **9.** Only one side of the equation was squared.   **11.** If we raise two equal quantities to the same power, the results are equal quantities.   **13.** $2\sqrt{x-2}$, 2, 4, 16, 8, 24   **15.** 2   **17.** 4   **19.** 0   **21.** 4   **23.** 8   **25.** $\frac{5}{2}, \frac{1}{2}$ **27.** 1   **29.** 16   **31.** 14, $\cancel{6}$ **33.** 4, 3   **35.** 2, $\cancel{7}$ **37.** 2, −1   **39.** −1, $\cancel{1}$

41. $\cancel{1}$, no solutions   **43.** 0, $\cancel{4}$ **45.** $-\cancel{3}$, no solutions   **47.** 1, 9   **49.** 4, $\cancel{0}$ **51.** 2, $\cancel{142}$ **53.** $\cancel{6}$, no solutions   **55.** $h = \dfrac{v^2}{2g}$ **57.** $l = \dfrac{8T^2}{\pi^2}$

59. $A = P(r + 1)^3$ **61.** $v^2 = c^2\left(1 - \dfrac{L_A{}^2}{L_B{}^2}\right)$ **63.** 178 ft   **65.** about 488 watts   **67.** 10 lb   **69.** $5   **75.** 2.5 foot-candles

77. 0.41511 in.

STUDY SET   SECTION 7.3 (page 508)

1. rational (or fractional)   **3.** index, radicand   **5.** $25^{1/5}$, $\left(\sqrt[3]{-27}\right)^2$, $16^{-3/4}$, $\left(\sqrt{81}\right)^3$, $-\left(\frac{9}{64}\right)^{1/2}$

7.   **9.** x^{m+n} **11.** x^{m-n} **13.** $\sqrt[n]{x}$

15. 0, 1, 2, 3, 4   **17.** $\sqrt{100a^4}$, $10a^2$ **19.** $\sqrt[3]{x}$ **21.** $\sqrt[4]{3x}$ **23.** $\sqrt[4]{\frac{1}{2}x^3 y}$ **25.** $\sqrt{x^2 + y^2}$ **27.** $m^{1/2}$

29. $(3a)^{1/4}$ **31.** $\left(\frac{1}{7}abc\right)^{1/6}$ **33.** $(a^2 - b^2)^{1/3}$ **35.** 2   **37.** 2   **39.** 2   **41.** 2   **43.** $\frac{1}{2}$ **45.** −2   **47.** not real   **49.** −3   **51.** $5|y|$ **53.** $2|x|$

55. $3x$ **57.** not real   **59.** 216   **61.** 27   **63.** 1,728   **65.** $\frac{1}{4}$ **67.** $125x^6$ **69.** $\dfrac{4x^2}{9}$ **71.** $\dfrac{1}{2}$ **73.** $\dfrac{1}{8}$ **75.** $\dfrac{1}{64x^3}$ **77.** $\dfrac{1}{9y^2}$ **79.** $\dfrac{16}{81}$

81. $-\dfrac{3}{2x}$ **83.** 2.47   **85.** 1.01   **87.** $5^{5/7}$ **89.** $4^{3/5}$ **91.** $9^{1/5}$ **93.** $\dfrac{1}{36}$ **95.** 3   **97.** a **99.** $a^{2/9}$ **101.** $a^{3/4}b^{1/2}$ **103.** $\dfrac{2x}{3}$ **105.** $\dfrac{1}{3}x$

107. $y + y^2$ **109.** $x^2 - x + x^{3/5}$ **111.** $\sqrt{p}$ **113.** $\sqrt{5b}$ **115.** 736 ft/sec   **117.** 1.96 units   **119.** 4,608 in.², 32 ft²   **123.** $x < 3$
125. $r > 28$

STUDY SET   SECTION 7.4 (page 518)

1. like   **3.** factor   **5.** $\sqrt[n]{a}\sqrt[n]{b}$, product, roots   **7. a.** $\sqrt{4 \cdot 5}$ **b.** $\sqrt{4}\sqrt{5}$ **c.** $\sqrt{4 \cdot 5} = \sqrt{4}\sqrt{5}$ **9. a.** $\sqrt{5}$, $\sqrt[3]{5}$ (answers may vary); no   **b.** $\sqrt{5}$, $\sqrt{6}$ (answers may vary); no   **11.** $8k^3$, $8k^3$ **13.** 6   **15.** t **17.** $5x$ **19.** 10   **21.** $7x$ **23.** $6b$ **25.** 2   **27.** $3a$ **29.** $2\sqrt{5}$

31. $-10\sqrt{2}$ **33.** $2\sqrt[3]{10}$ **35.** $-3\sqrt[3]{3}$ **37.** $2\sqrt[4]{2}$ **39.** $2\sqrt[5]{3}$ **41.** $\dfrac{\sqrt{7}}{3}$ **43.** $\dfrac{\sqrt[3]{7}}{4}$ **45.** $\dfrac{\sqrt[4]{3}}{10}$ **47.** $\dfrac{\sqrt[5]{3}}{2}$ **49.** $5x\sqrt{2}$ **51.** $4\sqrt{2b}$

53. $-4a\sqrt{7a}$ **55.** $5ab\sqrt{7b}$ **57.** $-10\sqrt{3xy}$ **59.** $-3x^2\sqrt[3]{2}$ **61.** $2x^4y\sqrt[3]{2}$ **63.** $2x^3y\sqrt[4]{2}$ **65.** $\dfrac{z}{4x}$ **67.** $\dfrac{\sqrt[4]{5x}}{2z}$ **69.** $10\sqrt{2x}$ **71.** $\sqrt[5]{7a^2}$

73. $-\sqrt{2}$ **75.** $2\sqrt{2}$ **77.** $9\sqrt{6}$ **79.** $3\sqrt[3]{3x}$ **81.** $-\sqrt[3]{4}$ **83.** −10   **85.** $-17\sqrt[4]{2}$ **87.** $16\sqrt[4]{2}$ **89.** $-4\sqrt{2}$ **91.** $3\sqrt{2t} + \sqrt{3t}$
93. $-11\sqrt[3]{2}$ **95.** $y\sqrt{z}$ **97.** $13y\sqrt{x}$ **99.** $12\sqrt[3]{a}$ **101.** $-7y^2\sqrt{y}$ **103.** $4x\sqrt[5]{xy^2}$ **105.** $8\pi\sqrt{5}$ ft²; 56.2 ft²   **107.** $5\sqrt{3}$ amps; 8.7 amps
109. $\left(26\sqrt{5} + 10\sqrt{3}\right)$ in.; 75.5 in.   **115.** $\dfrac{-15x^5}{y}$ **117.** $3p + 4 - \dfrac{5}{2p - 5}$

STUDY SET   SECTION 7.5 (page 528)

1. FOIL   **3.** irrational   **5.** perfect   **7. a.** $6\sqrt{6}$ **b.** 48   **c.** can't be simplified   **d.** $-6\sqrt{6}$ **9.** When the numerator and the denominator of a fraction are multiplied by the same nonzero number, the value of the fraction is not changed.   **11.** A radical appears in the denominator.   **13.** $\sqrt{6}$, 48, 16, 4   **15.** 11   **17.** 7   **19.** 4   **21.** $5\sqrt{2}$ **23.** $6\sqrt{2}$ **25.** 5   **27.** 18   **29.** 8   **31.** 18   **33.** $2\sqrt[3]{3}$
35. ab^2 **37.** $5a\sqrt{b}$ **39.** $-20r\sqrt[3]{10s}$ **41.** $x^2(x + 3)$ **43.** $12\sqrt{5} - 15$ **45.** $24\sqrt{3} + 6\sqrt{14}$ **47.** $-8x\sqrt{10} + 6\sqrt{15x}$
49. $-1 - 2\sqrt{2}$ **51.** $5z + 2\sqrt{15z} + 3$ **53.** $3x - 2y$ **55.** $6a + 5\sqrt{3ab} - 3b$ **57.** $18r - 12\sqrt{2r} + 4$ **59.** $-6x - 12\sqrt{x} - 6$

61. $\dfrac{\sqrt{7}}{7}$ **63.** $\dfrac{\sqrt{30}}{5}$ **65.** $\dfrac{\sqrt{10}}{4}$ **67.** 2   **69.** $\dfrac{\sqrt[3]{4}}{2}$ **71.** $\sqrt[3]{3}$ **73.** $\dfrac{\sqrt[3]{6}}{3}$ **75.** $\dfrac{2\sqrt{2xy}}{xy}$ **77.** $\dfrac{\sqrt{5y}}{y}$ **79.** $\dfrac{\sqrt[3]{2ab^2}}{b}$ **81.** $\dfrac{\sqrt[4]{4}}{2}$ **83.** $\dfrac{\sqrt[5]{2}}{2}$

85. $\sqrt{2} + 1$ **87.** $\dfrac{3\sqrt{2} - \sqrt{10}}{4}$ **89.** $2 + \sqrt{3}$ **91.** $\dfrac{9 - 2\sqrt{14}}{5}$ **93.** $\dfrac{2\left(\sqrt{x} - 1\right)}{x - 1}$ **95.** $\sqrt{2z} + 1$ **97.** $\dfrac{x - 2\sqrt{xy} + y}{x - y}$ **99.** $\dfrac{1}{\sqrt{3} - 1}$

101. $\dfrac{x - 9}{x\left(\sqrt{x} - 3\right)}$ **103.** $\dfrac{x - y}{\sqrt{x}\left(\sqrt{x} - \sqrt{y}\right)}$ **105.** $\dfrac{\sqrt{2\pi}}{2\pi\sigma}$ **107.** $\dfrac{\sqrt{2}}{2}$ **109.** $f/4$ **113.** 1   **115.** $\frac{1}{3}$

STUDY SET   SECTION 7.6 (page 536)

1. hypotenuse **3.** Pythagorean **5.** $a^2 + b^2 = c^2$ **7.** $\sqrt{2}$ **9.** $\sqrt{3}$ **11.** 30°, 60° **13. a.** Take the positive square root of both sides.
b. Subtract 25 from both sides. **15.** 6, 52, 4, 2 **17.** 10 ft **19.** 80 m **21.** $h = 2\sqrt{2} \approx 2.83$, $x = 2$ **23.** $x = 5\sqrt{3} \approx 8.66$, $h = 10$
25. $x = 4.69$, $y = 8.11$ **27.** $x = 12.11$, $y = 12.11$ **29.** $7\sqrt{2}$ cm **31.** 5 **33.** 13 **35.** 10 **37.** $2\sqrt{26}$ **41.** $\left(5\sqrt{2}, 0\right)$, $\left(0, 5\sqrt{2}\right)$,
$\left(-5\sqrt{2}, 0\right)$, $\left(0, -5\sqrt{2}\right)$; (7.07, 0), (0, 7.07), (−7.07, 0), (0, −7.07) **43.** $10\sqrt{3}$ mm, 17.32 mm **45.** $10\sqrt{181}$ ft, 134.54 ft **47.** about
0.13 ft **49. a.** 21.21 units **b.** 8.25 units **c.** 13.00 units **51.** yes **55.** 7 **57.** 9

KEY CONCEPT (page 541)

1. $-3h^2\sqrt[3]{2}$ **2.** $14\sqrt[3]{e}$ **3.** $-4\sqrt{2}$ **4.** $-20r\sqrt[3]{10s}$ **5.** $3s - 2t$ **6.** $-\sqrt{21} + \sqrt{15}$ **7.** $18n - 12\sqrt{2n} + 4$ **8.** $\dfrac{\sqrt[3]{3k^2}}{k}$ **9.** -3 **10.** 5
11. 2; 7 is extraneous **12.** no solutions **13.** $3^{1/3}$ **14.** $5\sqrt[5]{a^2}$

CHAPTER REVIEW (page 543)

1. a. 7 **b.** -11 **c.** $\frac{15}{7}$ **d.** not real **e.** 0.1 **f.** $5|x|$ **g.** x^4 **h.** $|x + 2|$ **2. a.** -3 **b.** -6 **c.** $4a^3b$ **d.** $\dfrac{s^3}{5}$ **3. a.** 5 **b.** -2
c. $4x^2|y|$ **d.** $22|y|$ **e.** $-\frac{1}{2}$ **f.** not real **g.** 0 **h.** 0 **4.** 48 ft **5.** 24 cm^2 **6. a.** D = $[-2, \infty)$, R = $[0, \infty)$

b. D = $(-\infty, \infty)$, R = $(-\infty, \infty)$

7. a. 22 **b.** 16, 9 **c.** $\frac{13}{2}$ **d.** $\frac{9}{16}$ **e.** 2, -4 **f.** $\mathscr{3}$, no solutions **8.** 2

9. a. $P = \dfrac{A}{(r + 1)^2}$ **b.** $I = \dfrac{h^3 b}{12}$ **10.** about 17.9 ohms **11. a.** $\sqrt{t}$ **b.** $\sqrt[4]{5xy^3}$ **12. a.** 5 **b.** -6 **c.** not real **d.** 1 **e.** $\dfrac{3}{x}$ **f.** $\dfrac{1}{3}$
g. -2 **h.** 5 **i.** $3ab^{1/3}$ **j.** $3cd$ **13. a.** 27 **b.** $\dfrac{1}{4}$ **c.** $-16{,}807$ **d.** 10 **e.** $\dfrac{27}{8}$ **f.** $\dfrac{1}{3{,}125}$ **g.** $125x^3y^6$ **h.** $\dfrac{1}{4u^4v^2}$ **14. a.** $5^{3/4}$ **b.** $a^{1/7}$
c. k^8 **d.** $\dfrac{2gh}{3}$ **15. a.** $u - 1$ **b.** $v + v^2$ **16.** $\sqrt{\dfrac{a}{5b}}$ **17.** Two true statements result: $16 = 16$. **18. a.** $4\sqrt{15}$ **b.** $3\sqrt[3]{2}$ **c.** $2\sqrt[4]{2}$
d. $4\sqrt[5]{3}$ **e.** $2x^2\sqrt{2x}$ **f.** $r^5\sqrt[3]{r^2}$ **g.** $2xy\sqrt[3]{2x^2y}$ **h.** $9j^2\sqrt[3]{jk}$ **i.** $4x$ **j.** $\dfrac{\sqrt{17xy}}{8a^2}$ **19. a.** $3\sqrt{2}$ **b.** $11\sqrt{5}$ **c.** 0 **d.** $-8\sqrt[4]{2}$ **e.** $29x\sqrt{2}$
f. $13\sqrt[3]{2}$ **20.** $\left(6\sqrt{2} + 2\sqrt{10}\right)$ in., 14.8 in. **21. a.** 7 **b.** $6\sqrt{10}$ **c.** 32 **d.** 72 **e.** $3x$ **f.** 33 **g.** $-2x$ **h.** $12\sqrt[3]{3}$ **i.** $4 - 3\sqrt{2}$
j. $-20x^3y^3\sqrt{xy}$ **k.** $3b + 6\sqrt{b} + 3$ **l.** $6u - 12 + \sqrt{u}$ **22. a.** $\dfrac{10\sqrt{3}}{3}$ **b.** $\dfrac{\sqrt{15}}{5}$ **c.** $\dfrac{\sqrt{xy}}{y}$ **d.** $\dfrac{\sqrt[3]{u^2}}{u^2v^2}$ **e.** $2\left(\sqrt{2} + 1\right)$
f. $\dfrac{a + 2\sqrt{a} + 1}{a - 1}$ **23. a.** $\dfrac{9 - x}{2\left(3 + \sqrt{x}\right)}$ **b.** $\dfrac{a - b}{a + \sqrt{ab}}$ **24.** $r = \dfrac{\sqrt[3]{6\pi^2 V}}{2\pi}$ **25.** 17 ft **26.** 88 yd **27.** $7\sqrt{2}$ m **28.** $6\sqrt{3}$ cm, 18 cm
29. a. 7.07 in. **b.** 8.66 cm **30. a.** 13 **b.** $2\sqrt{2}$

CHAPTER 7 TEST (page 548)

1. 0, 1, 1.41, 2, 3, 3.32, 4 **2.** 3.0 mi **3.** $\frac{1}{3}$ **4.** 10 **5.** 4 is extraneous. **6.** $G = \dfrac{4\pi^2 r^3}{Mt^2}$ **7.** 2 **8.** 9 **9.** $\dfrac{1}{216}$

10. $\dfrac{9}{4}$ **11.** $2^{4/3}$ **12.** $8xy$ **13.** $|x|$ **14.** $2|x|\sqrt{2}$ **15.** $3x\sqrt[3]{2x^2}$ **16.** $3x^2y^4\sqrt{2}$ **17.** $-4xy^2$ **18.** $\dfrac{2}{3}a$ **19.** not real **20.** $t + 8$

21. $4\sqrt{3}$ **22.** $5xy^2\sqrt{10xy}$ **23.** $2x^5y\sqrt[3]{3}$ **24.** $\dfrac{1}{4a}$ **25.** $-\sqrt{3}$ **26.** $14\sqrt[3]{5}$ **27.** $2y^2\sqrt{3y}$ **28.** $6z\sqrt[4]{3z}$ **29.** $-6x\sqrt{y} - 2xy^2$

30. $3 - 7\sqrt{6}$ **31.** $\dfrac{\sqrt{5}}{5}$ **32.** $\sqrt{3t} + 1$ **33.** $\dfrac{1}{2\sqrt[3]{3}}$ **34.** $\dfrac{1}{\sqrt{2}(\sqrt{5} - 3)}$ **35.** 9.24 cm   **36.** 8.67 cm   **37.** 25   **38.** $\dfrac{\pi\sqrt{2}}{2}$ sec, 2.22 sec

39. 28 in.

STUDY SET   SECTION 8.1 (page 558)

1. quadratic   **3.** perfect   **5.** $x = \sqrt{c}, x = -\sqrt{c}$ **7.** 3, -2, 5   **9.** It is a solution.   **11. a.** 36   **b.** $\dfrac{25}{4}$ **c.** $\dfrac{1}{16}$ **13. a.** Subtract 35 from both sides.   **b.** Add 36 to both sides.   **15.** 4 is not a factor of the numerator. It cannot be divided out.   **17.** $x^2 - 8x + 15 = 0$

19. a. 2; $2\sqrt{5}, -2\sqrt{5}$ **b.** ±4.47 **21.** 0, -2 **23.** 5, -5 **25.** -2, -4 **27.** 2, $\dfrac{1}{2}$ **29.** ±6 **31.** $\pm\sqrt{5}$ **33.** $\pm\dfrac{4\sqrt{3}}{3}$ **35.** 0, -2

37. 4, 10   **39.** $-5 \pm \sqrt{3}$ **41.** $d = \dfrac{\sqrt{6h}}{2}$ **43.** $c = \dfrac{\sqrt{Em}}{m}$ **45.** 2, -4 **47.** 1, $-\dfrac{1}{2}$ **49.** $\dfrac{1}{2}, -\dfrac{2}{3}$ **51.** $-4 \pm \sqrt{10}$ **53.** $1 \pm 3\sqrt{2}$

55. $\dfrac{3 \pm 2\sqrt{3}}{3}$ **57.** $\dfrac{1 \pm 2\sqrt{2}}{2}$ **59.** $\dfrac{-5 \pm \sqrt{41}}{4}$ **61.** $\dfrac{-7 \pm \sqrt{29}}{10}$ **63.** width: $7\frac{1}{4}$ ft; length: $13\frac{3}{4}$ ft   **65.** 1.6 sec   **67.** 1.70 in.

69. 0.92 in.   **71.** 2.9 ft, 6.9 ft   **75.** $2ab^2\sqrt[3]{5}$ **77.** x^3 **79.** $5ab\sqrt{7b}$

STUDY SET   SECTION 8.2 (page 566)

1. quadratic   **3. a.** 0, $\dfrac{1}{2}$ **b.** 4, -2, 0; 0, $\dfrac{1}{2}$ **5. a.** true   **b.** true   **7.** $c = \dfrac{-s \pm \sqrt{s^2 - 4rt}}{2r}$ **9.** $2 \pm \sqrt{3}$ **11. a.** The fraction bar wasn't drawn under both parts of the numerator.   **b.** A $\pm$ sign wasn't written between b and the radical.   **13.** -1, -2 **15.** -6, -6

17. $\dfrac{1}{2}$, -3 **19.** $\dfrac{-5 \pm \sqrt{5}}{10}$ **21.** $-\dfrac{3}{2}, -\dfrac{1}{2}$ **23.** $\dfrac{1}{4}, -\dfrac{3}{4}$ **25.** $\dfrac{4}{3}, -\dfrac{2}{5}$ **27.** $\dfrac{-5 \pm \sqrt{17}}{2}$ **29.** $\dfrac{3 \pm \sqrt{17}}{4}$ **31.** $5 \pm \sqrt{7}$ **33.** 23, -17

35. $\dfrac{2}{3}, -\dfrac{5}{2}$ **37.** 8.98, -3.98 **39.** 97 ft by 117 ft   **41.** 0.5 mi by 2.5 mi   **43.** 34 in.   **45.** \$4.80 or \$5.20   **47.** 4,000   **49.** 9%

51. a. 83%   **b.** early 1976   **55.** $n^{1/2}$ **57.** $(3b)^{1/4}$ **59.** $\sqrt[3]{t}$ **61.** $\sqrt[4]{3t}$

STUDY SET   SECTION 8.3 (page 579)

1. quadratic   **3.** vertex   **5.**

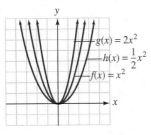

$g(x) = 2x^2$
$h(x) = \dfrac{1}{2}x^2$
$f(x) = x^2$

7.

$g(x) = 4x^2 + 3$
$f(x) = 4x^2$
$h(x) = 4x^2 - 2$

9.

$f(x) = -3x^2$
$g(x) = -3(x - 2)^2 - 1$

11. -3, 5   **13.** $h = -1$; $f(x) = 2[x - (-1)]^2 + 6$ **15.** $(1, 2)$, $x = 1$, upward   **17.** $(-3, -4)$, $x = -3$, upward

19. $(1, -2)$, $x = 1$, upward   **21.** $(2, 21)$, $x = 2$, downward   **23.** $\left(-\dfrac{2}{3}, \dfrac{2}{3}\right)$, $x = -\dfrac{2}{3}$, upward

25.

$(3, 2)$
$f(x) = (x - 3)^2 + 2$

27.

$(2, 0)$
$f(x) = -(x - 2)^2$

29.

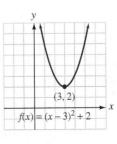

$(-3, 4)$
$f(x) = -2(x + 3)^2 + 4$

31.

$(3, 2)$
$f(x) = (x - 3)^2 + 2$

33.

$\left(-\dfrac{1}{2}, -\dfrac{25}{4}\right)$
$f(x) = x^2 + x - 6$

35.

$(1, 4)$
$y = -x^2 + 2x + 3$

37.

$\left(\dfrac{1}{3}, \dfrac{1}{3}\right)$
$f(x) = -3x^2 + 2x$

39.

$\left(-\dfrac{1}{4}, \dfrac{27}{4}\right)$
$y = -2x^2 - 6x + 6$

41. $(0.25, 0.88)$ **43.** $(0.50, 7.25)$ **45.** 2, -3 **47.** -1.85, 3.25   **49.** 3.75 sec, 225 ft   **51.** 250 ft by 500 ft   **53.** 15 min, \$160

55. 1968, 1.5 million; the U.S. involvement in the war in Vietnam was at its peak   **57.** 200, \$7,000   **63.** $4a^2\sqrt{b}$ **65.** $\dfrac{\sqrt{6}}{10}$

67. $15b - 6 + 9$

STUDY SET   SECTION 8.4 (page 590)

1. imaginary   **3.** real, imaginary   **5.** i or $\sqrt{-1}$ **7.** $\sqrt{-1}$ **9.** $-i$ **11.** FOIL   **13.** It is a solution.
15.

17. a. ± 1 **b.** $\pm i$ **19.** $9, 6i, 2$ **21. a.** true   **b.** false   **c.** false   **d.** false   **23.** $3i$
25. $\sqrt{7}i$ **27.** $2\sqrt{6}i$ **29.** $-2\sqrt{6}i$ **31.** $45i$ **33.** $\frac{5}{3}i$ **35.** -6 **37.** $-2\sqrt{3}$
39. $\frac{5}{8}$ **41.** -20 **43.** $\pm 3i$ **45.** $\pm \frac{4\sqrt{3}}{3}i$ **47.** $-1 \pm i$ **49.** $-\frac{1}{4} \pm \frac{\sqrt{7}}{4}i$
51. $\frac{2}{3} \pm \frac{\sqrt{2}}{3}i$ **53.** $\frac{1}{3} \pm \frac{2\sqrt{2}}{3}i$ **55.** $8 - 2i$ **57.** $3 - 5i$ **59.** $15 + 2i$
61. $15 + 7i$ **63.** $6 - 3i$ **65.** $-25 - 25i$ **67.** $7 + i$ **69.** $12 + 5i$ **71.** $13 - i$
73. $14 - 8i$ **75.** $8 + \sqrt{2}i$ **77.** $3 + 4i$ **79.** $0 - i$ **81.** $0 + \frac{4}{5}i$ **83.** $\frac{1}{8} - 0i$
85. $0 + \frac{3}{5}i$ **87.** $2 + i$ **89.** $\frac{15}{26} - \frac{3}{26}i$ **91.** $-\frac{42}{25} - \frac{6}{25}i$ **93.** $\frac{1}{4} + \frac{3}{4}i$ **95.** $-\frac{4}{13} - \frac{6}{13}i$ **97.** $\frac{5}{13} - \frac{12}{13}i$ **99.** $1110 + 310i$
101. $\frac{1}{4} - \frac{\sqrt{15}}{4}i$ **103.** i **105.** $-i$ **107.** 1 **109.** i **111.** $-1 + i$ **115.** 15 units, $15\sqrt{3}$ units   **117.** 20 mph

STUDY SET   SECTION 8.5 (page 597)

1. $b^2 - 4ac$ **3.** conjugates   **5.** rational, unequal   **7. a.** $x^4 = (x^2)^2$ **b.** yes   **9.** 5, 6, 24, rational   **11.** rational, equal   **13.** complex
conjugates   **15.** irrational, unequal   **17.** rational, unequal   **19.** yes   **21.** $1, -1, 4, -4$ **23.** $1, -1, \sqrt{5}, -\sqrt{5}$
25. $2, -2, \sqrt{7}i, -\sqrt{7}i$ **27.** 1   **29.** no solution   **31.** 16, 4   **33.** $-8, -27$ **35.** $-1, 27$ **37.** $\frac{1}{4}, \frac{1}{2}$ **39.** $1 - 2\sqrt{3}, 1 + 2\sqrt{3}$
41. $-1, -4$ **43.** $-1, -\frac{27}{13}$ **45.** $\frac{3 + \sqrt{57}}{6}, \frac{3 - \sqrt{57}}{6}$ **47.** $1, 1, -1, -1$ **49.** $1 \pm i$ **51.** 12.1 in.   **53.** 30 mph   **55.** 14.3 min
59. -2 **61.** $\frac{9}{5}$

STUDY SET   SECTION 8.6 (page 604)

1. quadratic   **3.** interval   **5.** greater   **7.** undefined   **9.** $(-\infty, -2) \cup (3, \infty)$ **11. a.** -2 **b.** -2 **c.** -2 **d.** 3   **e.** 3   **f.** 3   **13. a.** iii
b. iv   **c.** ii   **d.** i   **15.** $(1, 4)$,   **17.** $(-\infty, 3) \cup (5, \infty)$,   **19.** $[-4, 3]$,
21. no solutions   **23.** $(-\infty, -3] \cup [3, \infty)$,   **25.** $(-5, 5)$,
27. $(-\infty, 0) \cup (\frac{1}{2}, \infty)$,   **29.** $\left(-\infty, -\frac{5}{3}\right) \cup (0, \infty)$,   **31.** $(-\infty, -3) \cup (1, 4)$,
33. $\left(-\frac{1}{2}, \frac{1}{3}\right) \cup (\frac{1}{2}, \infty)$,   **35.** $(0, 2) \cup (8, \infty)$,   **37.** $\left[-\frac{34}{5}, -4\right) \cup (3, \infty)$,
39. $(-4, -2] \cup (-1, 2]$,   **41.** $(-\infty, -2) \cup (-2, \infty)$,   **43.** $(-1, 3)$ **45.** $(-\infty, -3) \cup (2, \infty)$

47.   **49.**   **51.**   **53.**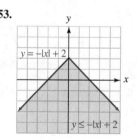

55. $(-2,100, -900) \cup (900, 2,100)$ **61.** $x = ky$ **63.** $t = kxy$ **65.** 3

KEY CONCEPT (page 608)

1. $0, -2$ **2.** $-3, -5$ **3.** $-\frac{3}{2}, -1$ **4.** $\pm 2\sqrt{6}$ **5.** $4, 10$ **6.** $\pm \frac{4\sqrt{3}}{3}$ **7.** $-5 \pm 4\sqrt{2}$ **8.** $\frac{1 \pm \sqrt{2}}{2}$ **9.** $-1 \pm i$ **10.** $\frac{3 \pm \sqrt{17}}{4}$
11. $23, -17$ **12.** $-\frac{1}{3} \pm \frac{\sqrt{2}}{3}i$ **13.** $1, -2$

CHAPTER REVIEW (page 610)

1. a. $-5, -4$ **b.** $-\frac{1}{3}, -\frac{5}{2}$ **c.** $\pm 2\sqrt{7}$ **d.** $4, -8$ **e.** $0, -\frac{11}{5}$ **f.** $\pm \frac{7\sqrt{5}}{5}$ **2.** $\frac{1}{4}$ **3. a.** $-4, -2$ **b.** $\frac{3 \pm \sqrt{3}}{2}$ **4.** 6 seconds before
midnight   **5. a.** $5 \pm \sqrt{7}$ **b.** $0, 10$ **c.** $\frac{1}{2}, -7$ **d.** $\frac{13 \pm \sqrt{163}}{3}$ **6.** sides: 1.25 in. wide; top/bottom: 2.5 in. wide   **7.** 0.7 sec, 1.8 sec

8. $46.8 billion   **9. a.**

b.

10. a.

b.

11. 1921, about 6,452   **12.** $-2, \frac{1}{3}$ **13. a.** $2i$ **b.** $\sqrt{7}i$ **c.** $-6\sqrt{5}i$ **d.** $\frac{6}{7}i$ **e.** -3 **f.** $\frac{8}{3}$ **14. a.** $\pm 5i$

b. $1 \pm 2\sqrt{3}i$ **15. a.** $12 - 8i$ **b.** $2 - 68i$ **c.** $-2 - 2\sqrt{2}i$ **d.** $-18 + 128i$ **e.** $22 + 29i$ **f.** $-12 + 28\sqrt{3}i$ **16. a.** $\dfrac{12}{5} - \dfrac{6}{5}i$

b. $\dfrac{15}{17} + \dfrac{8}{17}i$ **c.** $-\dfrac{1}{7} + \dfrac{4\sqrt{3}}{7}i$ **d.** $0 - \dfrac{2}{5}i$ **17. a.** i **b.** 1 **18. a.** irrational, unequal   **b.** complex conjugates   **c.** equal rational

numbers   **19. a.** $1, 144$ **b.** $8, -27$ **c.** $3, -\sqrt{3}, \dfrac{\sqrt{6}}{6}, -\dfrac{\sqrt{6}}{6}$ **d.** $1, -\frac{8}{5}$ **e.** $10, 4$ **20.** about 81 min

21. a. $(-\infty, -7) \cup (5, \infty)$,   **b.** $[-9, 9]$,   **c.** $(-\infty, 0) \cup [3/5, \infty)$,

d. $(-7/2, 1) \cup (4, \infty)$,   **22. a.**   **b.**

23. a.

b.

CHAPTER 8 TEST (page 614)

1. $0, -6$ **2.** $-\frac{3}{2}, -\frac{5}{3}$ **3.** 144 **4.** $\dfrac{1 \pm \sqrt{5}}{2}$ **5.** $\dfrac{4 \pm \sqrt{6}}{2}$ **6.** 4.5 ft by 1,502 ft   **7.** $4\sqrt{3}i$ **8.** -1 **9.** $-1 + 11i$ **10.** $4 - 7i$

11. $8 + 6i$ **12.** $-10 - 11i$ **13.** $0 + i$ **14.** $\frac{1}{2} + \frac{1}{2}i$ **15.** nonreal   **16.** $\pm 2\sqrt{3}i$ **17.** $2 \pm 3i$ **18.** $1, \frac{1}{4}$

19.

20.

21. $(-\infty, -2) \cup (4, \infty)$,

22. $(-3, 2]$,   **23.**

24. 20 in.   **25.** 211 ft   **27.** $-3, 2$ **28.** $[-3, 2]$

CHAPTERS 1-8 CUMULATIVE REVIEW EXERCISES (page 615)

1. $y = 3x + 2$ **2.** $y = -\frac{2}{3}x - 2$ **3.** an increase of about 13,333 a year **4.**

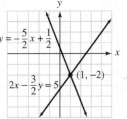

$$y = -\frac{5}{2}x + \frac{1}{2}$$

$$2x - \frac{3}{2}y = 5$$

$(1, -2)$

5. $(2, -1, 1)$

6.

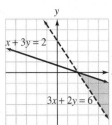

$x + 3y = 2$

$3x + 2y = 6$

7. $\left(-\infty, -\frac{10}{9}\right)$, $-10/9$

8. $(-\infty, -10] \cup [15, \infty)$, -10 15

9. $D = (-\infty, \infty)$, $R = [-3, \infty)$

10. $D = (-\infty, \infty)$, $R = (-\infty, 0]$ **11.** $-4a^2 + 12a - 7$ **12.** $6x^2 - 5x - 6$ **13.** $(x^2 + 4y^2)(x + 2y)(x - 2y)$

14. $(3x + 2)(5x - 4)$ **15.** $(x - y)(x - 4)$ **16.** $(2x^2 + 5y)(4x^4 - 10x^2y + 25y^2)$ **17.** $6, -1$ **18.** $0, \frac{2}{3}, -\frac{1}{2}$

19. 5; 3 is extraneous **20.** $b = \dfrac{-RTV^2 + aV + PV^3}{PV^2 + a}$ **21.** $\dfrac{x + y}{x - y}$ **22.** 0

23. $D = (-\infty, \infty)$, $R = (-\infty, \infty)$

$f(x) = x^3 + x^2 - 6x$

24. $D = (0, \infty)$, $R = (0, \infty)$

$f(x) = \dfrac{4}{x}$

25. 9 **26.** $D = [2, \infty)$, $R = [0, \infty)$

$f(x) = \sqrt{x - 2}$

27. $-3x$ **28.** $4t\sqrt{3t}$ **29.** $\frac{1}{16}$ **30.** $x^{17/12}$ **31.** $-12\sqrt[4]{2} + 10\sqrt[4]{3}$

32. $-18\sqrt{6}$ **33.** $\dfrac{x + 3\sqrt{x} + 2}{x - 1}$ **34.** $\dfrac{5\sqrt[3]{x^2}}{x}$ **35.** $2, 7$ **36.** $\frac{1}{4}$ **37.** $3\sqrt{2}$ in. **38.** $2\sqrt{3}$ in. **39.** 10 **40.** 9 **41.** $1, -\frac{3}{2}$ **42.** $\dfrac{-2 \pm \sqrt{7}}{3}$

43.

$\left(1, \frac{1}{2}\right)$

$y = \frac{1}{2}x^2 - x + 1$

44.

$f(x) = -x^2 - 4x$

$(-2, 4)$

45. $7 + 2i$ **46.** $\frac{3}{2} + \frac{1}{2}i$ **47.** $8i$ **48.** $9, 16$ **49.** $-\frac{3}{4}$ **50.** no solution

STUDY SET   SECTION 9.1 (page 624)

1. sum, $f(x) + g(x)$ **3.** product, $f(x)g(x)$ **5.** domain **7.** identity **9. a.** $g(x)$ **b.** $g(x)$

11. $-3, -2, -1, 0, 1, 2, 3$

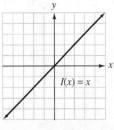

$I(x) = x$

13. $(3x - 1), 9x, 2x$ **15.** $7x, (-\infty, \infty)$ **17.** $12x^2, (-\infty, \infty)$ **19.** $x, (-\infty, \infty)$

21. $\frac{4}{3}, (-\infty, 0) \cup (0, \infty)$ **23.** $3x - 2, (-\infty, \infty)$

25. $2x^2 - 5x - 3$, $(-\infty, \infty)$ **27.** $-x - 4$, $(-\infty, \infty)$ **29.** $\dfrac{x - 3}{2x + 1}$, $(-\infty, -1/2) \cup (-1/2, \infty)$ **31.** $-2x^2 + 3x - 3$, $(-\infty, \infty)$

33. $\dfrac{3x - 2}{2x^2 + 1}$, $(-\infty, \infty)$ **35.** 3, $(-\infty, \infty)$ **37.** $\dfrac{x^2 - 4}{x^2 - 1}$, $(-\infty, -1) \cup (-1, 1) \cup (1, \infty)$ **39.** 7   **41.** 24   **43.** -1 **45.** $-\frac{1}{2}$ **47.** $2x^2 - 1$

49. $16x^2 + 8x$ **51.** 58   **53.** 110   **55.** 2   **57.** $9x^2 - 9x + 2$ **61.** $C(t) = \frac{5}{9}(2{,}668 - 200t)$

63. a. about \$37.50   **b.** $C(m) = \frac{1.50m}{20}$ **69.** $-\dfrac{3x + 7}{x + 2}$ **71.** $\dfrac{x - 4}{3x^2 - x - 12}$

STUDY SET   SECTION 9.2 (page 633)

1. one-to-one   **3.** inverses   **5.** once   **7.** 2   **9.** x **11.** $-6, -4, 0, 2, 8$ **13.** no   **15.** 2   **17.** The graphs are not symmetric about the line
$y = x$.   **19.** $2x, 2y, 3, y, f^{-1}(x)$ **21.** the inverse of, inverse   **23.** yes   **25.** no   **27.**

29.

31. yes, $\{(2, 3), (1, 2), (0, 1)\}$, yes   **33.** yes, $\{(1, 1), (1, 2), (1, 3), (1, 4)\}$, no   **35.** $f^{-1}(x) = \frac{5}{4}x + 5$ **37.** $f^{-1}(x) = 5x - 4$
39. $f^{-1}(x) = 5x + 4$ **41.**

43.

45. $y = \pm\sqrt{x - 4}$, no   **47.** $f^{-1}(x) = \sqrt[3]{x}$, yes

49. $x = |y|$, no   **51.** $f^{-1}(x) = \sqrt[3]{\dfrac{x + 3}{2}}$, yes   **53.**

55.

57. a. yes, no

b. No. Twice during this period, the person's anxiety level was at the maximum threshold value.   **61.** $3 - 8i$ **63.** $18 - i$ **65.** $-28 - 96i$

STUDY SET   SECTION 9.3 (page 645)

1. exponential   **3.** $(0, \infty)$ **5.** none   **7.** increasing   **9.**

11. $A = P\left(1 + \frac{r}{k}\right)^{kt}$, $FV = PV(1 + i)^n$

13. $g(x) = 2^x + 3$; $h(x) = 2^x - 2$ **15.** $\left(1 + \frac{r}{k}\right)$, kt **17.** 2.6651   **19.** 36.5548   **21.** 8   **23.** $7^{3\sqrt{3}}$
25.

27.

29.

31.

33. increasing   **35.** decreasing   **37.** \$22,080.40   **39.** \$32.03   **41.** \$2,273,996.13

43. a. about 1500, about 1825   **b.** 6.5 billion   **c.** exponential   **45. a.** at the end of the 2nd year   **b.** at the end of the 4th year
c. during the 7th year   **47.** 1.679046×10^8 **49.** 5.0421×10^{-5} coulombs   **51.** \$1,115.33   **55.** 40   **57.** 120°

STUDY SET   SECTION 9.4 (page 654)

1. the natural exponential function   **3.** $(0, \infty)$ **5.** none   **7.** increasing   **9.** continuous   **11.** 2.72
13.

15. an exponential function

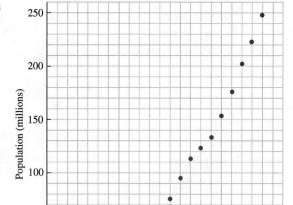

Year

17. 2.7182818. . .; e **19.** 1,000, 10, 0.9, 1,000

21.

23.

25.

27.

29. \$13,375.68   **31.** \$7,518.28 from annual compounding, \$7,647.95 from continuous compounding   **33.** \$6,849.16   **35.** 10.6 billion
37. 315   **39.** 51   **41.** 12 hr   **43.** 49 mps   **45.** about 72 yr   **49.** $4x^2\sqrt{15x}$ **51.** $10y\sqrt{3y}$

STUDY SET   SECTION 9.5 (page 664)

1. logarithmic   **3.** $(-\infty, \infty)$ **5.** $(1, 0)$ **7.** increasing   **9.** $x = b^y$ **11.** inverse   **13.** $-2, -1, 0, 1, 2$ **15.** none, none, $-3, \frac{1}{2}, 8$
17. $-0.30, 0, 0.30, 0.48, 0.60, 0.70, 0.78, 0.85, 0.90, 0.95, 1$ **19.** They decrease.   **21.** 10   **23.** $3^4 = 81$ **25.** $12^1 = 12$ **27.** $4^{-3} = \frac{1}{64}$
29. $10^{-3} = 0.001$ **31.** $\log_8 64 = 2$ **33.** $\log_4 \frac{1}{16} = -2$ **35.** $\log_{1/2} 32 = -5$ **37.** $\log_x z = y$ **39.** 3   **41.** 3   **43.** 3   **45.** $\frac{1}{2}$ **47.** 64
49. 5   **51.** $\frac{1}{25}$ **53.** $\frac{1}{6}$ **55.** $-\frac{3}{2}$ **57.** $\frac{2}{3}$ **59.** 5   **61.** $\frac{3}{2}$ **63.** 4   **65.** 1   **67.** 0.5119   **69.** -2.3307 **71.** 25.25   **73.** 0.02
75. increasing

77. decreasing

79.

81.

83.

85.

87. 29 db   **89.** 49.5 db   **91.** 4.4
93. 2,500 micrometers   **95.** 3 yr old
97. 10.8 yr   **101.** 10   **103.** 0; $-\frac{5}{9}$ does not
check

STUDY SET   SECTION 9.6 (page 671)

1. natural   **3.** $-0.69, 0, 0.69, 1.10, 1.39, 1.61, 1.79, 1.95, 2.08, 2.20, 2.30$ **5.** y-axis   **7.** $(-\infty, \infty)$ **9.** $e^y = x$ **11.** The logarithm of a negative number or zero is not defined.   **13.** $2.7182818. . .; e$ **15.** e **17.** $\frac{\ln 2}{r}$ **19.** 3.5596 **21.** -5.3709 **23.** 0.5423 **25.** undefined   **27.** 4.0645 **29.** 69.4079 **31.** 0.0245 **33.** 2.7210 **35.**   **37.**   **39.** 5.8 yr   **41.** 9.2 yr   **43.** about 3.5 hr

47. $y = 5x$ **49.** $y = -\frac{3}{2}x + \frac{13}{2}$ **51.** $x = 2$

STUDY SET   SECTION 9.7 (page 680)

1. product   **3.** power   **5.** 0 **7.** M, N **9.** x, y **11.** x **13.** $\neq$ **15.** $\log_a b$ **17.** $10^0 = 1, 10^1 = 10, 10^2 = 10^2$ **19.** 0 **21.** 7 **23.** 10 **25.** 2 **27.** 1 **29.** 7 **31.** rs, t, r, s **39.** $2 + \log_2 5$ **41.** $\log_6 x - 2$ **43.** $4 \ln y$ **45.** $\frac{1}{2}\log 5$ **47.** $\log x + \log y + \log z$ **49.** $1 + \log_2 x - \log_2 y$ **51.** $3 \log x + 2 \log y$ **53.** $\frac{1}{2}(\log_b x + \log_b y)$ **55.** $\frac{1}{3}\log_a x - \frac{1}{4}\log_a y - \frac{1}{4}\log_a z$ **57.** $\ln x + \frac{1}{2}\ln z$

59. $\log_2 \dfrac{x+1}{x}$ **61.** $\log x^2 y^{1/2}$ **63.** $\log_b \dfrac{z^{1/2}}{x^3 y^2}$ **65.** $\ln \dfrac{\frac{x}{z} + x}{\frac{y}{z} + y} = \ln \dfrac{x}{y}$ **67.** false   **69.** false   **71.** true   **73.** 1.4472 **75.** -1.1972

77. 1.1972 **79.** 1.8063 **81.** 1.7712 **83.** -1.0000 **85.** 1.8928 **87.** 2.3219 **89.** 4.8 **91.** from 2.5×10^{-8} to 1.6×10^{-7} **95.** $-\frac{7}{6}$ **97.** $\left(1, -\frac{1}{2}\right)$

STUDY SET   SECTION 9.8 (page 690)

1. exponential   **3.** $A_0 2^{-t/h}$ **5.** about 1.8 **7.** logarithm, exponent   **9.** 1.2920 **11.** 1 **13.** $10^2 = x + 1$ **15. a.** ± 3.46 **b.** 3.58 **17.** not a solution   **19.** $\log, \log 2$ **21.** 1.1610 **23.** 1.2702 **25.** 1.7095 **27.** 0 **29.** ± 1.0878 **31.** $0, 1.0566$ **33.** 0.7324 **35.** -13.2662 **37.** 8 **39.** $-\frac{3}{4}$ **41.** $3, -1$ **43.** $-2, -2$ **45.** 1.8 **47.** $8.8, 0.2$ **49.** $9,998$ **51.** -93 **53.** $e \approx 2.7183$ **55.** 19.0855 **57.** 2 **59.** 3 **61.** -7 **63.** 4 **65.** $10, -10$ **67.** 50 **69.** 20 **71.** 10 **73.** 10 **75.** no solution   **77.** 6 **79.** 9 **81.** 4 **83.** $1, 7$ **85.** 20 **87.** 8 **89.** 5.1 yr   **91.** 42.7 days   **93.** about $4,200$ yr   **95.** 5.6 yr   **97.** 5.4 yr   **99.** because $\ln 2 \approx 0.7$ **101.** 25.3 yr   **103.** 2.828 times larger   **105.** 13.3 **107.** $\frac{1}{3}\ln 0.75$ **111.** $0, 5$ **113.** $\frac{2}{3}, -4$ **115.** $\sqrt{137}$ in.

KEY CONCEPT (page 693)

1. no   **2.** no   **3.** yes   **4.** $f^{-1}(x) = \frac{-x-1}{2}$ **5.** $-2, 1, 3$ **6.** $\log 1,000 = 3$ **7.** $2^{-3} = \frac{1}{8}$ **8.** 2 **9.** $\frac{3}{2}$ **10.** 2.71828 **11.** e **12.** 0.7558
13.

CHAPTER REVIEW (page 695)

1. a. $(f + g)(x) = 3x + 1$ **b.** $(f - g)(x) = x - 1$ **c.** $(f \cdot g)(x) = 2x^2 + 2x$ **d.** $(f/g)(x) = \frac{2x}{x+1}$ **e.** 4 **f.** 3 **g.** $(f \circ g)(x) = 2(x + 1)$ **h.** $(g \circ f)(x) = 2x + 1$ **2. a.** no   **b.** yes   **3. a.** yes   **b.** no   **4.**   **5. a.** $f^{-1}(x) = \frac{x+3}{6}$ **b.** $f^{-1}(x) = \frac{x-5}{4}$

c. $f^{-1}(x) = \sqrt[3]{x}$ **d.** $y = \sqrt{\frac{x+1}{2}}$ **6. a.** $5^{2\sqrt{2}}$ **b.** $2^{\sqrt{10}}$ **7. a.** D: $(-\infty, \infty)$, R: $(0, \infty)$

b. D: $(-\infty, \infty)$, R: $(0, \infty)$

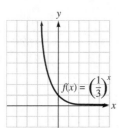

$f(x) = \left(\frac{1}{3}\right)^x$

8. a.

$y = \left(\frac{1}{2}\right)^x$

$f(x) = \left(\frac{1}{2}\right)^x - 2$

b.

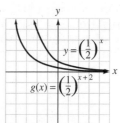

$y = \left(\frac{1}{2}\right)^x$

$g(x) = \left(\frac{1}{2}\right)^{x+2}$

9. an exponential function

Year

10. $2,189,703.45 **11.** 8.22% **12.** $2,324,767.37

13. a. D: $(-\infty, \infty)$, R: $(1, \infty)$

$f(x) = e^x + 1$

b. D: $(-\infty, \infty)$, R: $(0, \infty)$

$f(x) = e^{x-3}$

14. 582,175,004

15. D: $(0, \infty)$, R: $(-\infty, \infty)$ **16. a.** $4^3 = 64$ **b.** $\log_7 \frac{1}{7} = -1$ **17. a.** 2 **b.** -2 **c.** 0 **d.** not possible **e.** $\frac{1}{2}$ **f.** 3 **18. a.** 32 **b.** $\frac{1}{81}$
c. 4 **d.** 10 **e.** $\frac{1}{2}$ **f.** 9 **19. a.** 0.6542 **b.** 26.9153 **20. a.**

$y = 4^x$

$y = \log_4 x$

b.

$f(x) = \left(\frac{1}{3}\right)^x$

$g(x) = \log_{1/3} x$

22. about 53 **23.** about 4.4 **24. a.** 6.1137 **b.** -0.1625

21. a.

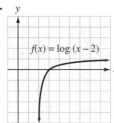

$f(x) = \log(x - 2)$

b.

$f(x) = 3 + \log x$

25. a. 10.3398 **b.** 0.0002 **26.** $f(x) = e^x$ **27. a.**

$f(x) = 1 + \ln x$

b.

$y = \ln(x + 1)$

28. about 23 yr

29. a. 0 **b.** 1 **c.** 3 **d.** 4 **e.** 4 **f.** 1 **30. a.** $3 + \log_3 x$ **b.** $2 - \log x$ **c.** $\frac{1}{2} \log_5 27$ **d.** $1 + \log a + \log b$

31. a. $2 \log_b x + 3 \log_b y - \log_b z$ **b.** $\frac{1}{2}(\ln x - \ln y - 2 \ln z)$ **32. a.** $\log_2 \frac{x^3 z^7}{y^5}$ **b.** $\log_b \frac{\sqrt{x + 2}}{y^3 z^7}$ **33. a.** 2.6609 **b.** 3.0000

34. 1.7604   **35.** about 7.9×10^{-4} gram-ions/liter   **36. a.** 1.7712   **b.** 2   **c.** 2.7095   **d.** $-3, -1$ **e.** 1.9459   **f.** -8.0472
37. a. 104   **b.** 9   **c.** 25, 4   **d.** 4   **e.** 4, 3   **f.** 2   **g.** 6   **h.** 31   **38.** about 3,300 yr   **39.** 2, 5

CHAPTER 9 TEST (page 700)

1. $(g + f)(x) = 5x - 1$ **2.** $(g \cdot f)(x) = 4x^2 - 4x$ **3.** 3   **4.** $4(x - 1)$ **5.** $y = \frac{12 - 2x}{3}$ **6.** $f^{-1}(x) = \sqrt{\dfrac{x - 4}{3}}$

7.

8.

9. $\frac{3}{64}$ g $= 0.046875$ g   **10.** \$1,060.90   **11.**

12. \$4,451.08   **13.** 2   **14.** 3   **15.** $\frac{1}{27}$ **16.** e **17.** $6^{-2} = \frac{1}{36}$ **18.** D: $(0, \infty)$, R: $(-\infty, \infty)$ **19.**

20.

21. $2 \log a + \log b + 3 \log c$ **22.** $\ln \dfrac{b\sqrt{a + 2}}{3}$ **23.** 0.5646   **24.** $y = \log x$ **25.** 6.4   **26.** 46
27. 0.6826   **28.** 4   **29.** 1   **30.** 10   **31.** 5^{c} **32.** $\frac{1}{2} \ln (5 - 1) = \frac{1}{2} \ln 4 = 0.69314718...$, which is $\ln 2$.
34. a. yes   **b.** yes   **c.** 80; when the temperature of the tire tread is 260°, the vehicle is traveling 80 mph

STUDY SET   SECTION 10.1 (page 712)

1. conic   **3.** radius, center   **5.** $r^2 < 0$ **7.** parabola, (3, 2), right   **9.** $(x - h)^2 + (y - k)^2 = r^2$ **11.** $y = -ax^2$ **13.** 2, 6, 4, 36

15.   **17.**   **19.**   **21.**

23.   **25.**   **27.**   **29.** $x^2 + y^2 = 1$ **31.** $(x - 6)^2 + (y - 8)^2 = 25$

33. $(x + 2)^2 + (y - 6)^2 = 144$ **35.** $x^2 + y^2 = 8$ **37.**   **39.**   **41.**

43.

$x^2 + y^2 + 6x - 4y = -12$

45.

47.

49.

51.

53.

55.

57. **59.**

61. $(x - 7)^2 + y^2 = 9$ **63.** no **65.** 30 ft away **67.** 2 AU **73.** 5, $-\frac{7}{3}$ **75.** 3, $-\frac{1}{4}$

STUDY SET SECTION 10.2 (page 722)

1. ellipse, sum **3.** center **5.** $(\pm a, 0)$ **7.** $(0, 0)$ **9.** 144, 16 **11.**

13.

15.

17.

19.

21. **23.**

25.

27.

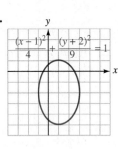

29. $y = \frac{1}{2}\sqrt{400 - x^2}$ **31.** 12π sq. units **35.** $12y^2 + \dfrac{9}{x^2}$ **37.** $\dfrac{y^2 + x^2}{y^2 - x^2}$

STUDY SET SECTION 10.3 (page 730)

1. hyperbola, difference **3.** center **5.** $(\pm a, 0)$ **7.** $(0, 0)$ **9.** 36, 36, 36 **11.**

13.

15.

$$25x^2 - y^2 = 25$$
or
$$x^2 - \frac{y^2}{25} = 1$$

17.
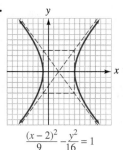
$$\frac{(x-2)^2}{9} - \frac{y^2}{16} = 1$$

19.

$$4(x+3)^2 - (y-1)^2 = 4$$
or $(x+3)^2 - \frac{(y-1)^2}{4} = 1$

21.

$xy = 8$

23. **25.** **27.**

$$\frac{(x+1)^2}{1} - \frac{(y+2)^2}{4} = 1$$

29.

$$\frac{(y+1)^2}{1} - \frac{(x-2)^2}{4} = 1$$

31. 3 units   **33.** $10\sqrt{3}$ units

37. $-3x^2(2x^2 - 3x + 2)$ **39.** $(5a + 2b)(3a - 2b)$

STUDY SET   SECTION 10.4 (page 736)

1. graphing, substitution   **3.** tangent   **5.** two   **7.** four   **9.** $2x$, 5, 5, 1, 1, -1 **11.**

$8x^2 + 32y^2 = 256$
$x = 2y$
$(4, 2)$
$(-4, -2)$

13.

$(-1, 3)$ $(1, 3)$
$y = 3x^2$
$x^2 + y^2 = 10$

15.
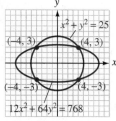
$x^2 + y^2 = 25$
$(-4, 3)$ $(4, 3)$
$(-4, -3)$ $(4, -3)$
$12x^2 + 64y^2 = 768$

17.
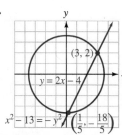
$(3, 2)$
$y = 2x - 4$
$x^2 - 13 = -y^2$ $\left(\frac{1}{5}, -\frac{18}{5}\right)$

19. $(1, 0), (5, 0)$

21. $(3, 0), (0, 5)$ **23.** $(1, 1)$

25. $(1, 2), (2, 1)$ **27.** $(-2, 3), (2, 3)$ **29.** $\left(\sqrt{5}, 5\right), \left(-\sqrt{5}, 5\right)$ **31.** $(3, 2), (3, -2), (-3, 2), (-3, -2)$ **33.** $(2, 4), (2, -4), (-2, 4)$, $(-2, -4)$ **35.** $\left(-\sqrt{15}, 5\right), \left(\sqrt{15}, 5\right), (-2, -6), (2, -6)$ **37.** $(0, -4), (-3, 5), (3, 5)$ **39.** $(-2, 3), (2, 3), (-2, -3), (2, -3)$

41. $(3, 3)$ **43.** $(6, 2), (-6, -2), \left(\sqrt{42}, 0\right), \left(-\sqrt{42}, 0\right)$ **45.** $\left(\frac{1}{2}, \frac{1}{3}\right), \left(\frac{1}{3}, \frac{1}{2}\right)$ **47.** 4, 8   **49.** 7 cm by 9 cm   **51.** either $750 at 9% or $900 at 7.5%   **53.** 68 mph, 4.5 hr   **57.** $-11x\sqrt{2}$ **59.** $\frac{1}{2}$

KEY CONCEPT (page 740)

1. ellipse   **2.** circle   **3.** parabola   **4.** hyperbola   **5.** hyperbola   **6.** ellipse   **7.** ellipse   **8.** parabola   **9.** circle   **10. a.** $(-1, 2)$ **b.** 4
11. a. $(-2, 1)$ **b.** right   **12. a.** $(0, 0)$ **b.** the y-axis   **c.** 8 units   **d.** 4 units   **13. a.** $(-2, 1)$ **b.** left and right   **c.** 6 units horizontally, 4 units vertically   **14.**
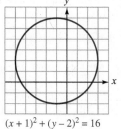
$(x+1)^2 + (y-2)^2 = 16$

15.
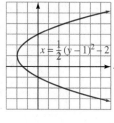
$x = \frac{1}{2}(y-1)^2 - 2$

16.
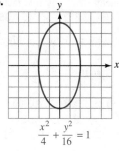
$\frac{x^2}{4} + \frac{y^2}{16} = 1$

17.
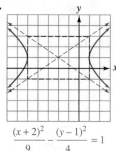
$\frac{(x+2)^2}{9} - \frac{(y-1)^2}{4} = 1$

CHAPTER REVIEW (page 742)

1. a.

b.

2.

3. a.

b.

4. a.

b.

5.

6. a.

b.

7. hyperbola **8.**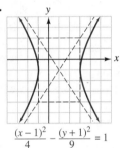

9. $(0, 3)$, $(0, -3)$

10. a. $(4, 2)$, $(4, -2)$, $(-4, 2)$, $(-4, -2)$ **b.** $(2, 3)$, $(2, -3)$, $(-2, 3)$, $(-2, -3)$

CHAPTER 10 TEST (page 744)

1. $(2, -3)$, 2 **2.** $(-2, 3)$, 4 **3.**

4.

5.

6.

7.

8.

9. $(2, 6)$, $(-2, -2)$ **10.** $(3, 4)$, $(-3, 4)$

STUDY SET   SECTION 11.1 (page 751)

1. binomial **3.** one **5.** exponent **7.** 7 **9.** five factorial **11.** 7! **13.** $(n - 1)!$ **15.** 3 **17.** 3, 3!, 3!, 2, 1 **19.** 6 **21.** -120 **23.** 30 **25.** 144 **27.** 40,320 **29.** $\frac{1}{110}$ **31.** 2,352 **33.** 10 **35.** 21 **37.** $\frac{1}{168}$ **39.** $x^2 + 2xy + y^2$ **41.** $x^4 - 4x^3y + 6x^2y^2 - 4xy^3 + y^4$

43. $8x^3 + 12x^2y + 6xy^2 + y^3$ **45.** $x^3 - 6x^2y + 12xy^2 - 8y^3$ **47.** $8x^3 + 36x^2y + 54xy^2 + 27y^3$ **49.** $\frac{x^3}{8} - \frac{x^2y}{4} + \frac{xy^2}{6} - \frac{y^3}{27}$

51. $81 + 216y + 216y^2 + 96y^3 + 16y^4$ **53.** $\frac{x^4}{81} - \frac{2x^3y}{27} + \frac{x^2y^2}{6} - \frac{xy^3}{6} + \frac{y^4}{16}$ **55.** $3a^2b$ **57.** $-4xy^3$ **59.** $15x^2y^4$ **61.** $28x^6y^2$

63. $90x^3$ **65.** $640x^3y^2$ **67.** $-12x^3y$ **69.** $-70,000x^4$ **71.** $810xy^4$ **73.** $-\frac{1}{6}x^3y$ **75.** $\frac{n!}{3!(n-3)!}a^{n-3}b^3$ **77.** $\frac{n!}{4!(n-4)!}a^{n-4}b^4$

79. $\frac{81(2^{n-4})n!}{4!(n-4)!}a^{n-4}b^4$ **85.** 2 **87.** 5 **89.** $(2, 3)$ **91.** 8

STUDY SET   SECTION 11.2 (page 758)

1. sequence   **3.** infinite   **5.** series   **7.** arithmetic, difference   **9.** finite   **11.** $1 + 2 + 3 + 4 + 5$ **13.** $1 + 2 + 3$ **15.** $4 + 9 + 16$
17. $\sum_{k=3}^{6} k$ **19.** $\sum_{k=1}^{4} 2k$ **21.** 3, 5, 7, 9, 11   **23.** $-5, -8, -11, -14, -17$ **25.** 5, 11, 17, 23, 29   **27.** $-4, -11, -18, -25, -32$
29. $-118, -111, -104, -97, -90$ **31.** 34, 31, 28, 25, 22   **33.** 5, 12, 19, 26, 33   **35.** 355   **37.** -179 **39.** -23 **41.** 12   **43.** $\frac{17}{4}, \frac{13}{2}, \frac{35}{4}$
45. 12, 14, 16, 18   **47.** $\frac{29}{2}$ **49.** $\frac{5}{4}$ **51.** 1,335   **53.** 459   **55.** 354   **57.** 255   **59.** 1,275   **61.** 2,500   **63.** 60   **65.** 91   **67.** 31   **69.** 12
71. 60, 110, 160, 210, 260, 310; \$6,060   **73.** 11,325   **75.** 368 ft   **79.** $18x^2 + 8x - 3$ **81.** $\dfrac{6a^2 + 16}{(a + 2)(a - 2)}$

STUDY SET   SECTION 11.3 (page 767)

1. geometric   **3.** mean   **5.** nth term $= ar^{n-1}$ **7.** $2 + 6 + 18$ **9.** $S = \frac{a}{1-r}$ **11.** first, ratio, terms   **13.** 3, 6, 12, 24, 48   **15.** $-5, -1,$
$-\frac{1}{5}, -\frac{1}{25}, -\frac{1}{125}$ **17.** 2, 8, 32, 128, 512   **19.** $-3, -12, -48, -192, -768$ **21.** $-64, 32, -16, 8, -4$ **23.** $-64, -32, -16, -8, -4$
25. 2, 10, 50, 250, 1,250   **27.** 3,584   **29.** $\frac{1}{27}$ **31.** 3   **33.** 6, 18, 54   **35.** $-20, -100, -500, -2,500$ **37.** -16 **39.** $10\sqrt{2}$
41. No geometric mean exists.   **43.** 728   **45.** 122   **47.** -255 **49.** 381   **51.** $\frac{156}{25}$ **53.** $-\frac{21}{4}$ **55.** 16   **57.** 81   **59.** 8   **61.** $-\frac{135}{4}$
63. no sum   **65.** $-\frac{81}{2}$ **67.** $\frac{1}{9}$ **69.** $-\frac{1}{3}$ **71.** $\frac{4}{33}$ **73.** $\frac{25}{33}$ **75.** about 669 people   **77.** \$1,469.74   **79.** \$140,853.75   **81.** $\left(\frac{1}{2}\right)^{11} \approx 0.0005$
83. 30 m   **85.** 5,000   **91.** $[-1, 6]$ **93.** $(-\infty, -3) \cup (4, \infty)$ **95.** yes   **97.** no

STUDY SET   SECTION 11.4 (page 776)

1. permutation   **3.** $p \cdot q$ **5.** $P(n, r) = \frac{n!}{(n-r)!}$ **7.** 1   **9.** $\binom{n}{r}$, combinations   **11.** 1   **13.** 6!, 4!, 5   **15.** 6   **17.** 60   **19.** 12   **21.** 5
23. 1,260   **25.** 10   **27.** 20   **29.** 50   **31.** 2   **33.** 1   **35.** $\frac{n!}{2!(n-2)!}$ **37.** $x^4 + 4x^3y + 6x^2y^2 + 4xy^3 + y^4$ **39.** $8x^3 + 12x^2y + 6xy^2 + y^3$
41. $81x^4 - 216x^3 + 216x^2 - 96x + 16$ **43.** $-1,250x^2y^3$ **45.** $-4x^6y^3$ **47.** 35   **49.** 1,000,000   **51.** 136,080   **53.** 8,000,000
55. 720   **57.** 2,880   **59.** 13,800   **61.** 720   **63.** 900   **65.** 364   **67.** 5   **69.** 1,192,052,400   **71.** 18   **73.** 7,920   **77.** 6, -3 **79.** 8

STUDY SET   SECTION 11.5 (page 781)

1. experiment   **3.** $\frac{s}{n}$ **5.** 0   **7.** 6, 52, $\frac{6}{52}$, $\frac{3}{26}$ **9.** {(1, H), (2, H), (3, H), (4, H), (5, H), (6, H), (1, T), (2, T), (4, T), (5, T), (6, T)}
11. (a, b, c, d, e, f, g, h, i, j, k, l, m, n, o, p, q, r, s, t, u, v, w, x, y, z)   **13.** $\frac{1}{6}$ **15.** $\frac{2}{3}$ **17.** $\frac{19}{42}$ **19.** $\frac{13}{42}$ **21.** $\frac{3}{8}$ **23.** 0   **25.** $\frac{1}{12}$ **27.** $\frac{5}{12}$
29. $\frac{33}{391,510}$ **31.** $\frac{9}{460}$ **35.** $\frac{1}{4}$ **37.** $\frac{1}{4}$ **39.** 1   **41.** $\frac{32}{119}$ **43.** $\frac{1}{3}$ **47.** $-10, 4$ **49.** $(-4, 10)$

KEY CONCEPT (page 783)

1. g   **2.** o   **3.** i   **4.** l   **5.** u   **6.** y   **7.** p   **8.** x   **9.** m   **10.** d   **11.** f   **12.** c   **13.** w   **14.** b   **15.** j   **16.** s   **17.** e   **18.** z   **19.** n   **20.** k
21. h   **22.** v   **23.** q   **24.** a   **25.** r   **26.** t

CHAPTER REVIEW (page 785)

1. a. 144   **b.** 20   **c.** 15   **d.** 220   **e.** 1   **f.** 8   **2. a.** $x^5 + 5x^4y + 10x^3y^2 + 10x^2y^3 + 5xy^4 + y^5$ **b.** $x^4 - 4x^3y + 6x^2y^2 - 4xy^3 + y^4$
c. $64x^3 - 48x^2y + 12xy^2 - y^3$ **d.** $x^3 + 12x^2y + 48xy^2 + 64y^3$ **3. a.** $6x^2y^2$ **b.** $-10x^2y^3$ **c.** $-108x^2y$ **d.** $864x^2y^2$ **4.** 42   **5.** 122,
137, 152, 167, 182   **6.** $\frac{41}{3}, \frac{58}{3}$ **7.** 1,550   **8.** $-\frac{45}{2}$ **9. a.** $\frac{15}{2}$ **b.** 378   **c.** 14   **d.** 360   **10.** 24, 12, 6, 3, $\frac{3}{2}$ **11.** 4   **12.** 24, -96 **13.** $\frac{2,186}{9}$
14. $-\frac{85}{8}$ **15.** 125   **16.** $\frac{5}{99}$ **17.** 136   **18. a.** 5,040   **b.** 1   **c.** 20,160   **d.** $\frac{1}{10}$ **19. a.** 1   **b.** 1   **c.** 28   **d.** 84   **e.** 700   **f.** $\frac{7}{4}$ **20.** \$1,638.40
21. \$134,509.57   **22.** 12 yr   **23.** 1,600 ft   **24.** 120   **25.** 720   **26.** 120   **27.** 150   **28.** $\frac{1}{18}$ **29.** 0   **30.** $\frac{1}{13}$ **31.** $\frac{1}{649,640}$ **32.** $\frac{63}{66,640}$

CHAPTER 11 TEST (page 788)

1. 210   **2.** 1   **3.** $-5x^4y$ **4.** $24x^2y^2$ **5.** 66   **6.** 306   **7.** 34, 66   **8.** 3   **9.** -81 **10.** $\frac{364}{27}$ **11.** 18, 108   **12.** $\frac{27}{2}$ **13.** 120   **14.** 40,320
15. 15   **16.** 56   **17.** 720   **18.** 322,560   **19.** 24   **20.** $\frac{7}{30}$ **21.** 35   **22.** 30   **23.** $\frac{1}{5}$ **24.** $\frac{2}{13}$ **25.** $\frac{33}{66,640}$ **26.** $\frac{5}{16}$ **27.** 0   **28.** a list of all
possible outcomes

CHAPTERS 1–11 CUMULATIVE REVIEW EXERCISES (page 789)

1. 0   **2.** $-\frac{4}{3}, 5.6, 0, -23$ **3.** $\pi, \sqrt{2}, e$ **4.** $-\frac{4}{3}, \pi, 5.6, \sqrt{2}, 0, -23, e$ **5.** \$8,250   **6.** $\frac{1}{120}$ db/rpm   **7.** parallel   **8.** perpendicular
9. $y = -2x + 5$ **10.** $y = -\frac{9}{13}x + \frac{7}{13}$ **11.** (1, 1)   **12.** (2, -2)   **13.** ($-1, -1$)   **14.** ($-1, -1, 3$)   **15.** 3   **16.** -1 **17.** 85°, 80°, 15°
18.   **19.** $\left[-3, \frac{19}{3}\right]$ **20.** $(-\infty, -2)$ **21.** $12x^2 - 5xy - 3y^2$

22. $2x^2y^3 + 13xy + 3y^2$ **23.** $a^2 - 4ab + 4b^2$ **24.** $3a^3 + 10a^2 + 6a - 4$ **25.** $xy(3x - 4y)(x - 2)$
26. $(16x^2y^2 + z^4)(4xy + z^2)(4xy - z^2)$ **27.** $\lambda = \frac{4d - 2}{A - 6}$

28. $(-3, 0)$, $(0, 0)$, $(2, 0)$; $(0, 0)$; 24, 0, -8, -6, 0, 4, 0, -18

29. $\dfrac{64b^{12}}{27a^9}$ **30.** $-\dfrac{3x + 2}{3x - 2}$ **31.** $-\dfrac{q}{p}$

32. $\dfrac{4a - 1}{(a + 2)(a - 2)}$ **33.** 5; 3 is extraneous

34. $R = \dfrac{R_1 R_2 R_3}{R_2 R_3 + R_1 R_3 + R_1 R_2}$ **35. a.** a quadratic function   **b.** at about 85% and 120% of the suggested inflation   **36.** about $21\frac{1}{2}$ in.

37. $-x^2 + x + 5 + \dfrac{8}{x - 1}$ **38.** 2   **39.** $5\sqrt{2}$ **40.** $81x\sqrt[3]{3x}$ **41.** $\frac{343}{125}$ **42.** $\sqrt{3t} - 1$ **43.** $-5 + 17i$ **44.** $-\frac{21}{29} - \frac{20}{29}i$

45. 5, 0 does not check   **46.** 0   **47.** $\frac{2}{3}$, $-\frac{3}{2}$ **48.** $\dfrac{-3 \pm \sqrt{5}}{4}$ **49.** $\frac{1}{4}$, $\frac{1}{2}$ **50.** $\dfrac{2 \pm \sqrt{2}i}{3}$ **51.** $4x^2 + 4x - 1$ **52.** $f^{-1}(x) = \sqrt[3]{\dfrac{x + 1}{2}}$

53.

54.

55. $2^y = x$ **56.** $\log_6 x - 2$ **57.** 5   **58.** 3   **59.** $\frac{1}{27}$ **60.** 1   **61.** $y = 2^x$

62. x **63.** 1.9912   **64.** 0.301   **65.** 3.4190   **66.** 1, -10 does not check   **67.** $-\frac{3}{4}$ **68.** 9   **69.** $x^2 + y^2 - 2x - 6y - 15 = 0$

70.

71.

72.

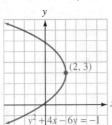

73. \$2,848.31   **74.** 1.16056   **75.** 30,240

76. $81a^4 - 108a^3 b + 54a^2 b^2 - 12ab^3 + b^4$ **77.** $112x^2 y^6$ **78.** 103   **79.** 690   **80.** 8, 19   **81.** 42   **82.** 27   **83.** 27   **84.** $\frac{1,023}{64}$
85. 12, -48 **86.** $\frac{27}{2}$ **87.** 504   **88.** 35   **89.** 5,040   **90.** 84

STUDY SET   APPENDIX I   (page A-8)

1. measuring   **3.** sine   **5.** cosine   **7.** legs   **9.** elevation   **11.** similar   **13.** $\dfrac{\text{length of leg adjacent to } \angle A}{\text{length of hypotenuse}}$ **15.** proportion   **17.** 40, x,

sin 50°, 40   **19.** $\dfrac{\sqrt{3}}{2}$ **21.** $\dfrac{\sqrt{3}}{3}$ **23.** $\dfrac{\sqrt{2}}{2}$ **25.** 0   **27.** 0.9563   **29.** 3.2709   **31.** 0.7254   **33.** 1   **35.** $B = 53°$, $a = 12.0$, $b = 16.0$
37. $B = 40°$, $b = 5.9$, $c = 9.1$ **39.** 96.4 ft   **41.** 721.2 ft   **43.** 39,300 ft   **45.** 2,690 ft   **47.** 22 ft   **49.** 7.7 mi   **51.** 61 ft

STUDY SET   APPENDIX II   (page A-17)

1. polynomial   **3.** $P(r)$ **5.** 12, 26, 8   **7.** $x + 2$ **9.** $x - 3$ **11.** $x + 2$ **13.** $x - 7 + \frac{28}{x + 2}$ **15.** $3x^2 - x + 2$ **17.** $2x^2 + 4x + 3$
19. $6x^2 - x + 1 + \frac{3}{x + 1}$ **21.** $7.2x - 0.66 + \frac{0.368}{x - 0.2}$ **23.** $2.7x - 3.59 + \frac{0.903}{x + 1.7}$ **25.** $9x^2 - 513x + 29,241 + \frac{-1,666,762}{x + 57}$ **27.** -1 **29.** -37
31. 23   **33.** -1 **35.** 2   **37.** -1 **39.** 18   **41.** 174   **43.** -8 **45.** 59   **47.** 44   **49.** $\frac{29}{32}$ **51.** yes   **53.** no   **55.** 64

Index